The derivative gives us information about how a function changes at a given instant and can be used to solve problems involving velocity and acceleration; marginal cost and profit; and the rate of change of a chemical reaction. Derivatives are the subjects of Chapters 2 through 4.

The Area Problem—The Basis of Integral Calculus

If we want to find the area of a rectangle or the area of a circle, formulas are available. (see Figure 2). But what if the figure is curvy, but not circular as in Figure 3? How do we find this area?

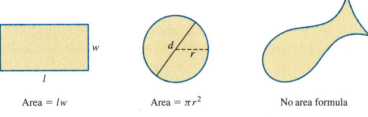

Area = lw Area = πr^2 No area formula

Figure 2   **Figure 3**

Calculus provides a way. Look at Figure 4(a). It shows the graph of $y = x^2$ from $x = 0$ to $x = 1$. Suppose we want to find the shaded area.

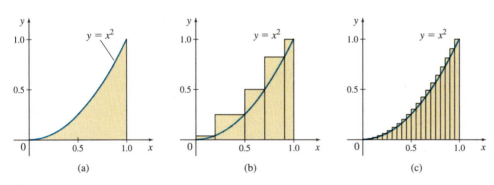

(a)   (b)   (c)

Figure 4

By subdividing the x-axis from 0 to 1 into small segments and drawing a rectangle of height x^2 above each segment, as in Figure 4(b), we can find the area of each rectangle and add them together. This sum approximates the shaded area in Figure 4(a). The smaller we make the segments of the x-axis and the more rectangles we draw, the better the approximation becomes. See Figure 4(c). This formulation leads to integral calculus, and the study of the integral of a function.

Two Problems—One Subject?

At first, differential calculus (the tangent problem) and integral calculus (the area problem) appear to be different, so why call both of them calculus? The Fundamental Theorem of Calculus establishes that the derivative and the integral are related. In fact, one of Newton's teachers, Isaac Barrow, recognized that the tangent problem and the area problem are closely related, and that derivatives and integrals are inverses of each other. Both Newton and Leibniz formalized this relationship between derivatives and integrals in the *Fundamental Theorem of Calculus*.

---

* Beckmann, P. (1976). *A History of* π (3rd. ed., p. 64). New York: St. Martin's Press.

SINGLE VARIABLE
CALCULUS
Early Transcendentals

MICHAEL SULLIVAN
Chicago State University

· · ·

KATHLEEN MIRANDA
State University of New York,
Old Westbury

W. H. Freeman and Company · New York

Senior Publisher: Ruth Baruth
Executive Acquisitions Editor: Terri Ward
Senior Development Editor: Andrew Sylvester
Development Editor: Tony Palermino
Associate Editor: Jorge Amaral
Editorial Assistant: Victoria Garvey
Market Development: Steven Rigolosi
Marketing Manager: Steve Thomas
Marketing Assistant: Samantha Zimbler
Senior Media Editor: Laura Judge
Associate Media Editor: Catriona Kaplan
Associate Media Editor: Courtney Elezovic
Assistant Media Editor: Liam Ferguson
Photo Editor: Christine Buese
Photo Researcher: Dena Betz
Art Director: Diana Blume
Project Editor: Robert Errera
Production Manager: Julia DeRosa
Illustration Coordinator: Janice Donnola
Illustrations: Network Graphics
Composition: Cenveo, Inc.
Printer: Quad Graphics

Library of Congress Control Number: 2013955445
ISBN-13: 978-1-4292-5433-5
ISBN-10: 1-4292-5433-5

W. H. Freeman and Company, 41 Madison Avenue, New York, NY 10010
Houndmills, Basingstoke RG21 6XS, England
www.whfreeman.com

Contents | *Calculus*

About the Authors

Michael Sullivan

Michael Sullivan, Emeritus Professor of Mathematics at Chicago State University, received a PhD in mathematics from the Illinois Institute of Technology. Before retiring, Mike taught at Chicago State for 35 years, where he honed an approach to teaching and writing that forms the foundation for his textbooks. Mike has been writing for over 35 years and currently has 15 books in print. His books have been awarded both Texty and McGuffey awards from TAA.

Mike is a member of the American Mathematical Association of Two Year Colleges, the American Mathematical Society, the Mathematical Association of America, and the Text and Academic Authors Association (TAA), and has received the TAA Lifetime Achievement Award in 2007. His influence in the field of mathematics extends to his four children: Kathleen, who teaches college mathematics; Michael III, who also teaches college mathematics, and who is his coauthor on two precalculus series; Dan, who is a sales director for a college textbook publishing company; and Colleen, who teaches middle-school and secondary school mathematics. Twelve grandchildren round out the family.

Mike would like to dedicate this text to his four children, 12 grandchildren, and future generations.

Kathleen Miranda

Kathleen Miranda, Ed.D. from St. John's University, is an Emeritus Associate Professor of the State University of New York (SUNY) where she taught for 25 years. Kathleen is a recipient of the prestigious New York State Chancellor's Award for Excellence in Teaching, and particularly enjoys teaching mathematics to underprepared and fearful students. In addition to her extensive classroom experience, Kathleen has worked as accuracy reviewer and solutions author on several mathematics textbooks, including Michael Sullivan's *Brief Calculus* and *Finite Mathematics*. Kathleen's goal is to help students unlock the complexities of calculus and appreciate its many applications.

Kathleen has four children: Edward, a plastic surgeon in San Francisco; James, an emergency medicine physician in Philadelphia; Kathleen, a chemical engineer working on vaccines; and Michael, a management consultant specializing in corporate strategy.

Kathleen would like to dedicate this text to her children and grandchildren.

Preface

The challenges facing instructors of calculus are daunting. Diversity among students, both in their mathematical preparedness for learning calculus and in their ultimate educational and career goals, is vast and growing. There is just not enough classroom time to teach every topic in the syllabus, to answer every student's questions, or to delve into the rich examples and applications that showcase the beauty and utility of calculus. As mathematics instructors we share these frustrations with you. As authors our goal is to create a student-oriented textbook that supports your teaching philosophy and promotes student success and confidence.

Promoting Student Success Is the Central Theme

Our goal is to write a mathematically precise calculus book that embraces proven pedagogical features to increase both student and instructor success. Many of these features are structural, but there are also many less obvious, intrinsic features embedded in the text.

- **The text is written to be read by students.** The language is simple, clear, and consistent. Definitions are simply stated and consistently used. Theorems are given names where appropriate. Numbering of definitions, equations, and theorems is kept to a minimum.

  The careful use of color throughout the text brings attention to important statements such as definitions and theorems. Important formulas are boxed, and procedures and summaries are called out so that a student can quickly look back and review the main points of the section.

- **The text is written to prepare students.** Whether students have educational goals that end in the social sciences, in the life and/or physical sciences, in engineering, or with a PhD in mathematics, the text provides ample practice, applications, and the mathematical precision and rigor required to prepare students to pursue their goals.

- **The text is written to be mastered by students.** Carefully used pedagogical features are found throughout the text. These features provide structure and form a carefully crafted learning system that helps students get the most out of their study. From our experience, students who use the features are more successful in calculus.

Pedagogical Features Promote Student Success

Just In Time Review Students forget; they often do not make connections. Instructors lament, "The students are not prepared!" So, throughout the text there are margin notes labeled **NEED TO REVIEW?** followed by a topic and page references. The **NEED TO REVIEW?** reference points to the discussion of a concept used in the current presentation.

NEED TO REVIEW? Summation notation is discussed in Appendix A.5, pp. A-38 to A-43.

The sum s_n of the areas of the n rectangles approximates the area A. That is,

$$A \approx s_n = f(c_1)\Delta x + f(c_2)\Delta x + \cdots + f(c_i)\Delta x + \cdots + f(c_n)\Delta x = \sum_{i=1}^{n} f(c_i)\Delta x$$

RECALL margin notes provide a quick refresher of key results that are being used in theorems, definitions, and examples.

RECALL $\sum_{i=1}^{n} i = \dfrac{n(n+1)}{2}$

Using summation properties, we get

$$S_n = \sum_{i=1}^{n} \frac{300}{n^2} i = \frac{300}{n^2} \sum_{i=1}^{n} i = \frac{300}{n^2} \frac{n(n+1)}{2} = 150\left(\frac{n+1}{n}\right) = 150\left(1 + \frac{1}{n}\right)$$

Additional margin notes are included throughout the text to help students understand and engage with the concepts. **IN WORDS** notes translate complex formulas, theorems, proofs, rules, and definitions using plain language that provide students with an alternate way to learn the concepts.

IN WORDS The average value $\bar{y} = \dfrac{1}{b-a}\displaystyle\int_a^b f(x)\,dx$ of a function f equals the value $f(u)$ in the Mean Value Theorem for Integrals.

DEFINITION Average Value of a Function over an Interval

Let f be a function that is continuous on the closed interval $[a, b]$. The **average value** $\bar{y}$ of f over $[a, b]$ is

$$\bar{y} = \frac{1}{b-a}\int_a^b f(x)\,dx \qquad (8)$$

CAUTION In writing an indefinite integral $\int f(x)\,dx$, remember to include the "dx."

CAUTIONS and **NOTES** provide supporting details or warnings about common pitfalls. **ORIGINS** give biographical information or historical details about key figures and discoveries in the development of calculus.

ORIGINS Willard F. Libby (1908–1980) grew up in California and went to college and graduate school at UC Berkeley. Libby was a physical chemist who taught at the University of Chicago and later at UCLA. While at Chicago, he developed the methods for using natural carbon-14 to date archaeological artifacts. Libby won the Nobel Prize in Chemistry in 1960 for this work.

Learning Objectives Students often need focus. Each section begins with a set of **Objectives** that serve as a broad outline of the section. The objectives help students study effectively by focusing attention on the concepts being covered. Each learning objective is supported by appropriate definitions, theorems, and proofs. One or more carefully chosen examples enhance the learning objective, and where appropriate, an application example is also included.

5.1 Area

OBJECTIVES *When you finish this section, you should be able to:*

1 Approximate the area under the graph of a function (p. 2)

2 Find the area under the graph of a function (p. 6)

Learning objectives help instructors prepare a syllabus that includes the important topics of calculus, and concentrate instruction on mastery of these topics. They are also helpful for answering the familiar question, "What is on the test?"

Effective Use of Color The text contains an abundance of graphs and illustrations that carefully utilize color to make concepts easier to visualize.

Dynamic Figures The text includes many pieces of art that students, can interact with through the online e-Book. These dynamic figures, indicated by the icon **DF** next to the figure label, illustrate select principles of calculus including limits, rates of change, solids of revolution, convergence, and divergence.

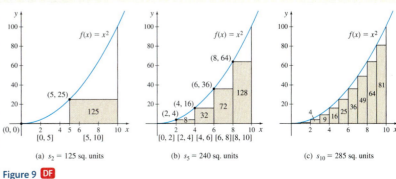

(a) $s_2 = 125$ sq. units   (b) $s_5 = 240$ sq. units   (c) $s_{10} = 285$ sq. units

Figure 9 **DF**

EXAMPLE 6   Using L'Hôpital's Rule to Find a One-Sided Limit

Find $\lim\limits_{x\to 0^+}\dfrac{\cot x}{\ln x}$.

$$\lim_{x\to 0^+}\frac{\cot x}{\ln x}=\lim_{x\to 0^+}\frac{\frac{d}{dx}\cot x}{\frac{d}{dx}\ln x}=\lim_{x\to 0^+}\frac{-\csc^2 x}{\frac{1}{x}}=-\lim_{x\to 0^+}\frac{x}{\sin^2 x}=-\lim_{x\to 0^+}\frac{\frac{d}{dx}x}{\frac{d}{dx}\sin^2 x}$$

L'Hôpital's Rule $\csc^2 x=\dfrac{1}{\sin^2 x}$ L'Hôpital's Rule

Examples with Detailed and Annotated Solutions: Examples are named according to their purpose. The text includes hundreds of examples with detailed step-by-step solutions. Additional annotations provide students with the formula used, or the reasoning needed, to perform a particular step of the solution. Where procedural steps for solving a type of problem are given in the text, these steps are followed in the solved examples. Often a figure or a graph is included to complement the solution.

CHAPTER 5 PROJECT   **Managing the Klamath River**

There is a gauge on the Klamath River, just downstream from the dam at Keno, Oregon. The U.S. Geological Survey has posted flow rates for this gauge every month since 1930. The averages of these monthly measurements since 1930 are given in Table 2. Notice that the data in Table 2 measure the rate of change of the volume V in cubic feet of water each second over one year; that is, the table gives $\dfrac{dV}{dt}=V'(t)$ in cubic feet per second, where t is in months.

Application Examples and Chapter Projects The inclusion of application examples motivates and underscores the conceptual theory, and powerfully conveys the message that calculus is a beneficial and a relevant subject to master. It is this combination of concepts with applied examples and exercises that gives students the tools not only to complete the exercises successfully, but also to comprehend the mathematics behind them.

This message is further reinforced by the case studies that open each chapter, demonstrating how major concepts apply to recognizable and often contemporary situations in biology, environmental studies, astronomy, engineering, technology, and other fields. Students see this case study again at the end of the chapter, where an extended project presents questions that guide students through a solution to the situation.

Immediate Reinforcement of Skills Following most examples there is the statement, **NOW WORK**. This callout directs students to a related exercise from the section's problem set. Math is best learned   **NOW WORK** Problem 35. actively. Doing a related problem immediately after working through a solved example not only keeps students engaged, but also enhances understanding and strengthens their ability to apply the objective. This practice also serves as a confidence-builder for students.

The Text Is Written with Ample Opportunity to Practice

Check Comprehension Before the Test Immediately after reading a section, we suggest that students assess their comprehension of the main points of the section by answering the **Concepts and Vocabulary** questions. These are a selection of quick fill in the blank, multiple-choice, and true/false questions. If a student has trouble answering these questions correctly, then a flag is raised immediately—go back and review the section or get help from the instructor.

Concepts and Vocabulary

1. *True or False* $\dfrac{f(x)}{g(x)}$ is an indeterminate form at c of the type $\dfrac{0}{0}$ if $\lim\limits_{x\to c}\dfrac{f(x)}{g(x)}$ does not exist.

Instructors may find Concept and Vocabulary problems useful for a class-opening, for a 5-minute quiz, or for iClicker responses to determine if students are prepared for class.

Skill Building

In Problems 5–18, find each derivative using Part 1 of the Fundamental Theorem of Calculus.

5. $\dfrac{d}{dx}\displaystyle\int_1^x \sqrt{t^2+1}\,dt$ 6. $\dfrac{d}{dx}\displaystyle\int_3^x \dfrac{t+1}{t}\,dt$

Practice, the Best Way to Learn Calculus The **Skill Building** exercises are grouped into subsets, usually corresponding to the objectives of the section, and are correlated to the solved examples. Working a variety of Skill Building problems increases students' computational skills and ability to choose the best approach to solve a problem. The result: Student success, which builds student confidence.

Applications and Extensions

51. **Uninhibited Growth** The population of a colony of mosquitoes obeys the uninhibited growth equation $\dfrac{dN}{dt}=kN$. If there are 1500 mosquitoes initially, and there are 2500 mosquitoes after 24 h, what is the mosquito population after 3 days?

Having mastered the computational skills, students are ready to tackle the **Application and Extension** problems. These are a diverse set of applied problems, some of which have been contributed by calculus students from various colleges and universities. The Application and Extension section also includes problems that use a slightly different approach to the section material, problems that require proof, and problems that extend the concepts of the section.

Challenge Problems

60. The floor function $f(x) = \lfloor x \rfloor$ is not continuous on $[0, 4]$. Show that $\int_0^4 f(x)\,dx$ exists.

Chapter Review

THINGS TO KNOW

5.1 Area

Definitions:

- Partition of an interval $[a, b]$ (p. 3)
- Area A under the graph of a function f from a to b (p. 6)

For students requiring more of a challenge, the **Challenge Problems** provide more difficult extensions of the section material. Often combined with concepts learned in previous chapters, Challenge Problems are intended to be thought-provoking problems that require some ingenuity to solve. They are designed to challenge the stronger students in the class.

Review, the Way to Reach Higher Levels of Learning Student success can be measured in many ways. Often in school student success is measured by a test. Recognizing this, each chapter, other than Chapter P, concludes with a **Chapter Review.** The review incorporates pedagogical features found in few advanced mathematics texts. Each Chapter Review consists of

- **Things to Know,** a detailed list of important definitions, formulas, and theorems contained in the chapter and page references to where they can be found.

- **Objectives Table,** which follows Things to Know. This table provides a section-by-section list of the learning objectives of the chapter (with page references), along with references to the worked examples and the review problems pertaining to the particular learning objective.

OBJECTIVES

Section	You should be able to …	Examples	Review Exercises
5.1	1 Approximate the area under the graph of a function (p. 344)	1, 2	7, 8
	2 Find the area under the graph of a function (p. 348)	3, 4	9, 10
5.2	1 Define a definite integral as the limit of Riemann sums (p. 353)	1, 2	11(a), (b)

- **Review Problems** that provide a comprehensive review of the key concepts of the chapter, matched to the learning objectives of each section.

By using the review exercises to study, a student can identify the objectives mastered and those that need additional work. By referring back to the objective(s) of problems missed, reviewing the material for that objective, reworking the **NOW WORK** **Problem(s),** and trying the review problem again, a student should have the skills and the confidence to take an exam or proceed to the next chapter.

Student success is ultimately measured by deep understanding and the ability to extend knowledge. When reviewing and incorporating previous knowledge, students begin to make connections. Once connections are formed, understanding and mastery follow.

In addition to these explicit pedagogical features, there are less obvious, but certainly no less significant, features that have been woven into the text. As educators and as students we realize that learning new material is often made more difficult when the style of presentation changes, so we have striven to keep the language and notation consistent throughout the book. Definitions are used in their entirety, both when presented and later when applied. This consistency is followed throughout the text and the exercises. Such attention to the consistency of language and notation is mirrored in our approach to writing precise mathematics while keeping the language clear and accessible to students.

The pedagogical features are tools that will aid students in their mastery of calculus. But it is the clarity of writing that allows students to master the understanding and to use the tools effectively. The accuracy of the mathematics and transparency in the writing will guarantee that any dedicated student can come to grips with the underlying theories, without having to decipher confusing dialogue.

It is our hope that you can encourage your students to read the textbook, use its features, practice the problems, and get help as soon as they need it. Our desire is that we, as a team, can build a cadre of confident and successful calculus students ready to pursue their dreams and goals!

Organizational Features That Set This Text Apart

Appendix A is a brief review of topics that students might have forgotten, but are used throughout calculus. The content of Appendix A consists of material studied prior to the introduction of functions in precalculus. Although definitions, theorems, and examples are provided in Appendix A, this appendix includes no exercises. The purpose of Appendix A is to refresh the student's memory of previously mastered concepts from prerequisite courses. You may wish to look at the content found in Appendix A. In particular, notice that summation notation, used in Chapter 5, is discussed in Section A.5

Chapter P deals with functions encountered in precalculus. It is designed either to be a quickly covered refresher at the beginning of the course or to be a just-in-time review used when needed. As such, the sections of Chapter P are lean and include only a limited number of practice problems that reinforce the reviewed topics.

Chapter 1 is dedicated to limits and continuity. Although most of the chapter addresses the idea of a limit and methods of finding limits, we state the formal ε-δ definition of limit in Section 1. We also include an example that investigates how close x must be to c to ensure that the difference between $f(x)$ and L is less than some prescribed number. But it is not until Section 6, after students are comfortable with the concept of a limit, the procedures of finding limits, and properties of limits, that the ε-δ definition of limit is repeated, discussed, and applied.

The Derivative has been split into two chapters (Chapters 2 and 3). This allows us to expand the coverage of the derivative and avoid a chapter of unwieldy length. Chapter 2 includes rates of change, the derivative as a function, and the Sum, Difference, Product, Quotient, and Simple Power rules for finding derivatives. It also contains derivatives of the exponential function and the trigonometric functions.

Chapter 3 covers the Chain Rule, implicit differentiation, the derivative of logarithmic and hyperbolic functions, differentials and linear approximations, and the approximation of zeros of functions using Newton's method.

Notice that there is also a section on Taylor polynomials (Section 3.5). We feel that the logical place for Taylor polynomials is immediately after differentials and linear approximations. It is natural to ask, "If a linear approximation to a function f near $x = x_0$ is good in a small interval surrounding x_0, then does a higher degree polynomial provide a better approximation of a function f over a wider interval?"

We begin Chapter 6, by finding the volume of a solid of revolution using the disk and washer method, followed by the shell method. We use both methods to solve several examples. This parallel approach enhances the student's appreciation of which method to choose when asked to find a volume. Only then do we introduce the slicing method, which is a generalization of the disk and washer method.

In Chapter 8, students are asked to recall the Taylor Polynomials studied earlier. The early exposure to the Taylor Polynomials serves two purposes here: First, it distinguishes a Taylor Polynomial from a Taylor Series; and second, it helps to make the idea of convergence of a Taylor Series easier to understand.

In addition to Chapter 16, Differential Equations, we provide examples of differential equations throughout the text. They are first introduced in Chapter 4: Applications of the Derivative, along with antiderivatives (Section 4.8). Again, the placement is a logical consequence of the relationship between derivatives and antiderivatives. At this point the idea of boundary values is introduced. Differential equations are revisited a second time in Chapter 5, The Integral, with the discussion of exponential growth (Section 5.5) and Newton's Law of Cooling (Section 5.6).

Acknowledgments

Ensuring the mathematical rigor, accuracy, and precision, as well as the complete coverage and clear language that we have strived for with this text, requires not only a great deal of time and effort on our part, but also on the part of an entire team of exceptional reviewers. First and foremost, Tony Palermino has provided exceptional editorial advice throughout the long development process. Mark Grinshpon has been invaluable for his mathematical insights, analytical expertise, and sheer dedication to the project; without his contributions, the text would be greatly diminished. The following people have also provided immeasurable support reading drafts of chapters and offering insight and advice critical to the success of this textbook:

James Brandt, *University of California, Davis*
William Cook, *Appalachian State University*
Mark Grinshpon, *Georgia State University*
Karl Havlak, *Angelo State University*
Steven L. Kent, *Youngstown State University*

Stephen Kokoska, *Bloomsburg University of Pennsylvania*
Larry Narici, *St. Johns University*
Frank Purcell, *Twin Prime Editorial*
William Rogge, *University of Nebraska, Lincoln*

In addition, we have had a host of individuals checking exercises at every stage of the text to ensure that they are accurate, interesting, and appropriate for a wide range of students:

Nick Belloit, *Florida State College at Jacksonville*
Jennifer Bowen, *The College of Wooster*
Brian Bradie, *Christopher Newport University*
James Bush, *Waynesburg University*
Deborah Carney, *Colorado School of Mines*
Julie Clark, *Hollins University*
Adam Coffman, *Indiana University-Purdue University, Fort Wayne*
Alain D'Amour, *Southern Connecticut State University*
John Davis, *Baylor University*
Jeffery Groah, *Lone Star College*
Tim Flaherty, *Carnegie Mellon University*
Melanie Fulton
Denise LeGrand, *University of Arkansas, Little Rock*
Serhiy Levkov and his team
Benjamin Levy, *Wentworth Institute of Technology*

Aihua Liu, *Montclair State University*
Joanne Lubben, *Dakota Weslyan University*
Betsy McCall, *Columbus State Community College*
Nicholas Ormes, *University of Denver*
Chihiro Oshima, *Santa Fe College*
Vadim Ponomarenko, *San Diego State University*
John Samons, *Florida State College at Jacksonville*
Ned Schillow, *Carbon Community College*
Mark Smith, *Miami University*
Timothy Trenary, *Regis University*
Kiryl Tsishchanka *and his team*
David Vinson, *Pellissippi Community College*
Bryan Wai-Kei, *University of South Carolina, Salkehatchie*
Lianwen Wang, *University of Central Missouri*
Kerry Wyckoff, *Brigham Young University*

Because the student is the ultimate beneficiary of this material, we would be remiss to neglect their input. We would like to thank students from the following schools who have provided us with feedback and suggestions for exercises throughout the text:

Barry University
Catholic University
Colorado State University
Murray State University

University of North Texas
University of South Florida
Southern Connecticut State University
Towson State University

Aside from commenting on exercises, a number of the applied exercises in the text were contributed by students. Although we were not able to use all of the submissions, we would like to thank all of the students who took the time and effort to participate in this project:

Boston University
Bronx Community College
Idaho State University
Lander University
Minnesota State University

Millikin University
University of Missouri—St. Louis
University of North Georgia
Trine University
University of Wisconsin—River Falls

Additional applied exercises, as well as many of the case study and chapter projects that open and close each chapter of this text were contributed by a team of creative individuals. Wayne Anderson, in particular, authored applied problems for every chapter in the text

and made himself available as a consultant at every request. We would like to thank all of
our exercise contributors for their work on this vital and exciting component of the text:

Wayne Anderson, *Sacramento City College* (Physics)
Allison Arnold, *University of Georgia* (Mathematics)
Kevin Cooper, *Washington State University* (Mathematics)
Adam Graham-Squire, *High Point University* (Mathematics)
Sergiy Klymchuk, *Auckland University of Technology* (Mathematics)
Eric Mandell, *Bowling Green State University* (Physics and Astronomy)

Eric Martell, *Millikin University* (Physics and Astronomy)
Barry McQuarrie, *University of Minnesota, Morris* (Mathematics)
Rachel Renee Miller, *Colorado School of Mines* (Physics)
Kanwal Singh, *Sarah Lawrence College* (Physics)
John Travis, *Mississippi College* (Mathematics)
Gordon Van Citters, *National Science Foundation*

Additional contributions to the writing of this text came from William Cook of Appalachian
State University, who contributed greatly to the parametric surfaces and surface integrals
section of Chapter 15, and to John Mitchell of Clark College, who provided recommendations
and material to help us better integrate technology into each chapter of the text.

We would like to thank the dozens of instructors who provided invaluable input as
reviewers of the manuscript and exercises and focus groups participants throughout the
development process:

Marwan A. Abu-Sawwa, *University of North Florida*
Jeongho Ahn, *Arkansas State University*
Martha Allen, *Georgia College & State University*
Roger C. Alperin, *San Jose State University*
Weam M. Al-Tameemi, *Texas A&M International University*
Robin Anderson, *Southwestern Illinois College*
Allison W. Arnold, *University of Georgia*
Mathai Augustine, *Cleveland State Community College*
Carroll Bandy, *Texas State University*
Scott E. Barnett, *Henry Ford Community College*
Emmanuel N. Barron, *Loyola University Chicago*
Abby Baumgardner, *Blinn College, Bryan*
Robert W. Bell, *Michigan State University*
Nicholas G. Belloit, *Florida State College at Jacksonville*
Daniel Birmajer, *Nazareth College of Rochester*
Laurie Boudreaux, *Nicholls State University*
Alain Bourget, *California State University at Fullerton*
David Boyd, *Valdosta State University*
Jim Brandt, *Southern Utah University*
Light R. Bryant, *Arizona Western College*
Kirby Bunas, *Santa Rosa Junior College*
Dietrich Burbulla, *University of Toronto*
Shawna M. Bynum, *Napa Valley College*
Luca Capogna, *University of Arkansas*
Jenna P. Carpenter, *Louisiana Tech University*
Nathan Carter, *Bentley University*
Stephen Chai, *Miles College*
E. William Chapin, Jr., *University of Maryland, Eastern Shore*
Sunil Kumar Chebolu, *Illinois State University*
William Cook, *Appalachian State University*
David A. Cox, *Amherst College*
Charles N. Curtis, *Missouri Southern State University*
Larry W. Cusick, *California State University, Fresno*
Rajendra Dahal, *Coastal Carolina University*
Ernesto Diaz, *Dominican University of California*
Robert Diaz, *Fullerton College*
Geoffrey D. Dietz, *Gannon University*
Della Duncan-Schnell, *California State University, Fresno*
Deborah A. Eckhardt, *St. Johns River State College*
Karen Ernst, *Hawkeye Community College*

Mark Farag, *Fairleigh Dickinson University*
Walden Freedman, *Humboldt State University*
Md Firozzaman, *Arizona State University*
Kseniya Fuhrman, *Milwaukee School of Engineering*
Douglas R. Furman, *SUNY Ulster Community College*
Tom Geil, *Milwaukee Area Technical College*
Jeff Gervasi, *Porterville College*
William T. Girton, *Florida Institute of Technology*
Darren Glass, *Gettysburg College*
Giséle Goldstein, *University of Memphis*
Jerome A. Goldstein, *University of Memphis*
Lourdes M. Gonzalez, *Miami Dade College*
Pavel Grinfeld, *Drexel University*
Mark Grinshpon, *Georgia State University*
Jeffrey M. Groah, *Lone Star College, Montgomery*
Gary Grohs, *Elgin Community College*
Paul Gunnells, *University of Massachusetts, Amherst*
Semion Gutman, *University of Oklahoma*
Christopher Hammond, *Connecticut College*
James Handley, *Montana Tech*
Alexander L. Hanhart, *New York University*
Gregory Hartman, *Virginia Military Institute*
Ladawn Haws, *California State University, Chico*
Mary Beth Headlee, *State College of Florida, Manatee-Sarasota*
Janice Hector, *DeAnza College*
Anders O.F. Hendrickson, *St. Norbert College*
Shahryar Heydari, *Piedmont College*
Max Hibbs, *Blinn College*
David Hobby, *SUNY at New Paltz*
Michael Holtfrerich, *Glendale Community College*
Keith E. Howard, *Mercer University*
Tracey Hoy, *College of Lake County*
Syed I. Hussain. *Orange Coast College*
Maiko Ishii, *Dawson College*
Nena Kabranski, *Tarrent County College*
William H. Kazez, *The University of Georgia*
Michael Keller, *University of Tulsa*
Eric S. Key, *University of Wisconsin, Milwaukee*
Raja Nicolas Khoury, *Collin College*
Michael Kirby, *Colorado State University*

Alex Kolesnik, *Ventura College*
Ashok Kumar, *Valdosta State University*
Geoffrey Laforte, *University of West Florida*
Tamara J. Lakins, *Allegheny College*
Justin Lambright, *Anderson University*
Carmen Latterell, *University of Minnesota, Duluth*
Glenn W. Ledder, *University of Nebraska, Lincoln*
Namyong Lee, *Minnesota State University, Mankato*
Aihua Li, *Montclair State University*
Rowan Lindley, *Westchester Community College*
Matthew Macauley, *Clemson University*
Filix Maisch, *Oregon State University*
Heath M. Martin, *University of Central Florida*
Vania Mascioni, *Ball State University*
Kate McGivney, *Shippensburg University*
Douglas B. Meade, *University of South Carolina*
Jie Miao, *Arkansas State University*
John Mitchell, *Clark College*
Val Mohanakumar, *Hillsborough Community College*
Catherine Moushon, *Elgin Community College*
Suzanne Mozdy, *Salt Lake Community College*
Gerald Mueller, *Columbus State Community College*
Kevin Nabb, *Moraine Valley Community College*
Bogdan G. Nita, *Montclair State University*
Charles Odion, *Houston Community College*
Giray Ökten, *Florida State University*
Kurt Overhiser, *Valencia College*
Edward W. Packel, *Lake Forest College*
Joshua Palmatier, *SUNY College at Oneonta*
Teresa Hales Peacock, *Nash Community College*
Chad Pierson, *University of Minnesota, Duluth*
Cynthia Piez, *University of Idaho*
Joni Burnette Pirnot, *State College of Florida, Manatee-Sarasota*
Jeffrey L. Poet, *Missouri Western State University*
Shirley Pomeranz, *The University of Tulsa*
Elise Price, *Tarrant County College, Southeast Campus*
Harald Proppe, *Concordia University*
Michael Radin, *Rochester Institute of Technology*
Jayakumar Ramanathan, *Eastern Michigan University*

Marc Renault, *Shippensburg University*
Suellen Robinson, *North Shore Community College*
William E. Rogge, *University of Nebraska, Lincoln*
Yongwu Rong, *George Washington University*
Amber Rosin, *California State Polytechnic University*
Richard J. Rossi, *Montana Tech of the University of Montana*
Bernard Rothman, *Ramapo College of New Jersey*
Dan Russow, *Arizona Western College*
Adnan H. Sabuwala, *California State University, Fresno*
Alan Saleski, *Loyola University Chicago*
Brandon Samples, *Georgia College & State University*
Jorge Sarmiento, *County College of Morris*
Ned Schillow, *Lehigh Carbon Community College*
Kristen R. Schreck, *Moraine Valley Community College*
Randy Scott, *Santiago Canyon College*
George F. Seelinger, *Illinois State University*
Rosa Garcia Seyfried, *Harrisburg Area Community College*
John Sumner, *University of Tampa*
Geraldine Taiani, *Pace University*
Barry A. Tesman, *Dickinson College*
Derrick Thacker, *Northeast State Community College*
Millicent P. Thomas, *Northwest University*
Tim Trenary, *Regis University*
Pamela Turner, *Hutchinson Community College*
Jen Tyne, *University of Maine*
David Unger, *Southern Illinois University - Edwardsville*
William Veczko, *St. Johns River State College*
James Vicknair, *California State University, San Bernardino*
Robert Vilardi, *Troy University, Montgomery*
Klaus Volpert, *Villanova University*
David Walnut, *George Mason University*
James Li-Ming Wang, *University of Alabama*
Jeffrey Xavier Watt, *Indiana University-Purdue University Indianapolis*
Qing Wang, *Shepherd University*
Rebecca Wong, *West Valley College*
Carolyn Yackel, *Mercer University*
Hong Zhang, *University of Wisconsin, Oshkosh*
Qiao Zhang, *Texas Christian University*
Qing Zhang, *University of Georgia*

Finally, we would like to thank the editorial, production, and marketing teams at W. H. Freeman, whose support, knowledge, and hard work were instrumental in the publication of this text: Ruth Baruth, Terri Ward, Andrew Sylvester, and Jorge Amaral on the Editorial side; Julia DeRosa, Diana Blume, Robert Errera, Christine Buese, and Janice Donnola on the Project Management, Design, and Production teams; and Steve Rigolosi, Steve Thomas, and Stephanie Ellis on the Marketing and Market Development teams. Thanks as well to the diligent group at Cenveo, lead by VK Owpee, for their expert composition, and to Ron Weickart at Network Graphics for his skill and creativity in executing the art program.

Michael Sullivan
Kathleen Miranda

SUPPLEMENTS

For Instructors

Instructor's Solutions Manual
Single Variable ISBN: 1-4641-4267-X
Multivariable ISBN: 1-4641-4266-1

Contains worked-out solutions to all exercises in the text.

Test Bank
Computerized (CD-ROM), ISBN: 1-4641-4272-6

Includes a comprehensive set of multiple-choice test items.

Instructor's Resource Manual
ISBN: 1-4641-8659-6

Provides sample course outlines, suggested class time, key points, lecture material, discussion topics, class activities, worksheets, projects, and questions to accompany the dynamic figures.

Instructor's Resource CD-ROM
ISBN: 1-4641-4274-2

Search and export all resources by key term or chapter. Includes text images, Instructor's Solutions Manual, Instructor's Resource Manual, and Test Bank.

For Students

Student Solutions Manual
Single Variable ISBN: 1-4641-4264-5
Multivariable ISBN: 1-4641-4265-3

Contains worked-out solutions to all odd-numbered exercises in the text.

Software Manuals
Maple™ and Mathematica® software manuals serve as basic introductions to popular mathematical software options.

ONLINE HOMEWORK OPTIONS

 W. H. Freeman's new online homework system, LaunchPad, offers our quality content curated and organized for easy assignability in a simple but powerful interface. We've taken what we've learned from thousands of instructors and the hundreds of thousands of students to create a new generation of W. H. Freeman/Macmillan technology.

Curated Units collect and organize videos, homework sets, dynamic figures, and e-book content to give instructors a course foundation to use as-is, or as a building block for their own learning units. Thousands of algorithmic exercises (with full algorithmic solutions) from the text can be assigned as online homework. Entire units worth of material and homework can be assigned in seconds drastically saving the amount of time it takes to get a course up and running.

Easily customizable. Instructors can customize LaunchPad units by adding quizzes and other activities from our vast wealth of resources. Instructors can also add a discussion board, a dropbox, and RSS feed, with a few clicks, allowing instructors to customize the student experience as much or as little as they'd like.

Useful Analytics. The gradebook quickly and easily allows instructors to look up performance metrics for classes, individual students, and individual assignments.

Intuitive interface and design. The student experience is simplified. Students navigation options and expectations are clearly laid out at all times ensuring they can never get lost in the system.

Assets integrated into LaunchPad include:

- **Interactive e-Book:** Every LaunchPad e-Book comes with powerful study tools for students, video and multimedia content, and easy customization for instructors. Students can search, highlight, and bookmark, making it easier to study and access key content. And instructors can make sure their class gets just the book they want to deliver: customize and rearrange chapters, add and share notes and discussions, and link to quizzes, activities, and other resources.

- **LEARNING**Curve LearningCurve provides students and instructors with powerful adaptive quizzing, a game-like format, direct links to the e-Book, and instant feedback. The quizzing system features questions tailored specifically to the text and adapts to students responses, providing material at different difficulty levels and topics based on student performance.

- **Dynamic Figures:** One hundred figures from the text have been recreated in a new interactive format for students and instructors to manipulate and explore, making the visual aspects and dimensions of calculus concepts easier to grasp. Brief tutorial videos accompany each figure and explain the concepts at work.

- **CalcClips:** These whiteboard tutorials provide animated and narrated step-by-step solutions to exercises that are based on key problems in the text.

- **SolutionMaster** The SolutionMaster tool offers an easy-to-use web-based version of the instructor's solutions, allowing instructors to generate a solution file for any set of homework exercises.

OTHER ONLINE HOMEWORK OPTIONS

WebAssign Premium www.webassign.net/freeman.com
WebAssign Premium integrates the book's exercises into the world's most popular and trusted online homework system, making it easy to assign algorithmically generated homework and quizzes. Algorithmic exercises offer the instructor optional algorithmic solutions. WebAssign Premium also offers access to resources, including Dynamic Figures, CalcClips whiteboard videos, tutorials, and "Show My Work" feature. In addition, WebAssign Premium is available with a fully customizable e-Book option that includes links to interactive applets and projects.

WeBWorK webwork.maa.org
W. H. Freeman offers thousands of algorithmically generated questions (with full solutions) though this free, open-source online homework system at the University of Rochester. Adopters also have access to a shared national library test bank with thousands of additional questions, including 1,500 problem sets matched to the book's table of contents.

Applications Index

Note: *Italics* indicates Example.

P

Preparing for Calculus

Until now, the mathematics you have encountered has centered mainly on algebra, geometry, and trigonometry. These subjects have a long history, well over 2000 years. But calculus is relatively new; it was developed less than 400 years ago.

Calculus deals with change and how the change in one quantity affects other quantities. Fundamental to these ideas are functions and their properties.

In Chapter P, we discuss many of the functions used in calculus. We also provide a review of techniques from precalculus used to obtain the graphs of functions and to transform known functions into new functions.

Your instructor may choose to cover all or part of the chapter. Regardless, throughout the text, you will see the NEED TO REVIEW? marginal notes. They reference specific topics, often discussed in Chapter P.

P.1 Functions and Their Graphs

OBJECTIVES *When you finish this section, you should be able to:*

1 Evaluate a function (p. 3)

2 Find the domain of a function (p. 4)

3 Identify the graph of a function (p. 5)

4 Analyze a piecewise-defined function (p. 7)

5 Obtain information from or about the graph of a function (p. 7)

6 Use properties of functions (p. 9)

7 Find the average rate of change of a function (p. 11)

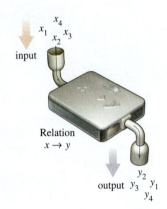

Relation
$x \rightarrow y$

Figure 1

Often there are situations where one variable is somehow linked to another variable. For example, the price of a gallon of gas is linked to the price of a barrel of oil. A person can be associated to her telephone number(s). The volume V of a sphere depends on its radius R. The force F exerted by an object corresponds to its acceleration a. These are examples of a **relation**, a correspondence between two sets called the **domain** and the **range**. If x is an element of the domain and y is an element of the range, and if a relation exists from x to y, then we say that y **corresponds** to x or that y **depends on** x, and we write $x \rightarrow y$. It is often helpful to think of x as the **input** and y as the **output** of the relation. See Figure 1.

Suppose an astronaut standing on the moon throws a rock 20 meters up and starts a stopwatch as the rock begins to fall back down. If x represents the number of seconds on the stopwatch and if y represents the altitude of the rock at that time, then there is a relation between time x and altitude y. If the altitude of the rock is measured at $x = 1$, 2, 2.5, 3, 4, and 5 seconds, then the altitude is approximately $y = 19.2$, 16.8, 15, 12.8, 7.2, and 0 meters, respectively.

The astronaut could express this relation numerically, graphically, or algebraically. The relation can be expressed by a table of numbers (see Table 1) or by the set of ordered pairs $\{(0, 20), (1, 19.2), (2, 16.8), (2.5, 15), (3, 12.8), (4, 7.2), (5, 0)\}$, where the first element of each pair denotes the time x and the second element denotes the altitude y. The relation also can be expressed visually, using either a graph, as in Figure 2, or a map, as in Figure 3. Finally, the relation can be expressed algebraically using the formula

$$y = 20 - 0.8x^2$$

TABLE 1

Time, x (in seconds)	Altitude, y (in meters)
0	20
1	19.2
2	16.8
2.5	15
3	12.8
4	7.2
5	0

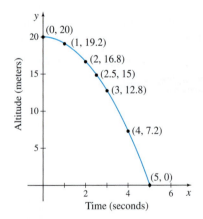

Figure 2

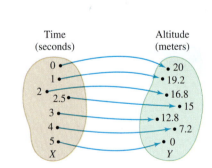

Figure 3

NOTE Not every relation is a function. If any element x in the set X corresponds to more than one element y in the set Y, then the relation is not a function.

In this example, notice that if X is the set of times from 0 to 5 seconds and Y is the set of altitudes from 0 to 20 meters, then each element of X corresponds to one and only one element of Y. Each given time value yields a **unique**, that is, exactly one, altitude value. Any relation with this property is called a *function from X into Y*.

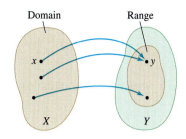

Figure 4

DEFINITION Function

Let X and Y be two nonempty sets.* A **function** f from X into Y is a relation that associates with each element of X exactly one element of Y.

The set X is called the **domain** of the function. For each element x in X, the corresponding element y in Y is called the **value** of the function at x, or the **image** of x. The set of all the images of the elements in the domain is called the **range** of the function. Since there may be elements in Y that are not images of any x in X, the range of a function is a subset of Y. See Figure 4.

1 Evaluate a Function

Functions are often denoted by letters such as f, F, g, and so on. If f is a function, then for each element x in the domain, the corresponding image in the range is denoted by the symbol $f(x)$, read "f of x." $f(x)$ is called the **value of f at x**. The variable x is called the **independent variable** or the **argument** because it can be assigned any element from the domain, while the variable y is called the **dependent variable**, because its value depends on x.

EXAMPLE 1   Evaluating a Function

For the function f defined by $f(x) = 2x^2 - 3x$, find:

(a) $f(5)$ **(b)** $f(x+h)$ **(c)** $f(x+h) - f(x)$ **(d)** $\dfrac{f(x+h) - f(x)}{h}, h \neq 0$

Solution (a) $f(5) = 2(5)^2 - 3(5) = 50 - 15 = 35$

(b) The function $f(x) = 2x^2 - 3x$ gives us a rule to follow. To find $f(x+h)$, expand $(x+h)^2$, multiply the result by 2, and then subtract the product of 3 and $(x+h)$.

$$f(x+h) = 2(x+h)^2 - 3(x+h) = 2(x^2 + 2hx + h^2) - 3x - 3h$$

In $f(x)$ replace x by $x + h$

$$= 2x^2 + 4hx + 2h^2 - 3x - 3h$$

(c) $f(x+h) - f(x) = [2x^2 + 4hx + 2h^2 - 3x - 3h] - [2x^2 - 3x] = 4hx + 2h^2 - 3h$

(d) $\dfrac{f(x+h) - f(x)}{h} = \dfrac{4hx + 2h^2 - 3h}{h} = \dfrac{h[4x + 2h - 3]}{h} = 4x + 2h - 3$

$h \neq 0$; divide out h

The expression in (d) is called the **difference quotient** of f. Difference quotients occur frequently in calculus, as we will see in Chapters 2 and 3.

NOW WORK   Problems **13** and **23**.

EXAMPLE 2   Finding the Amount of Gasoline in a Tank

A Shell station stores its gasoline in an underground tank that is a right circular cylinder lying on its side. The volume V of gasoline in the tank (in gallons) is given by the formula

$$V(h) = 40h^2 \sqrt{\frac{96}{h} - 0.608}$$

where h is the height (in inches) of the gasoline as measured on a depth stick. See Figure 5.

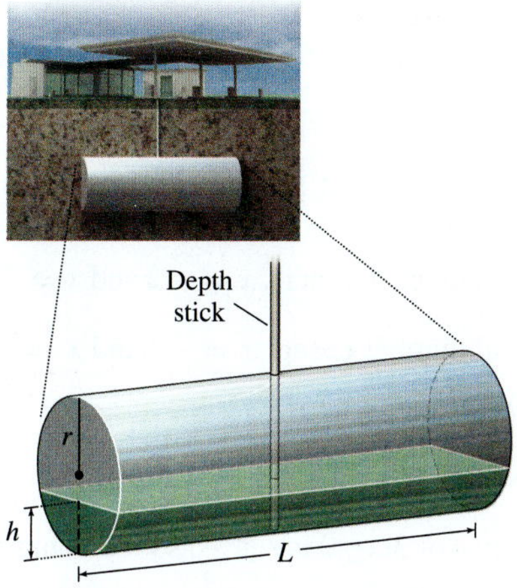

---

*The sets X and Y will usually be sets of real numbers, defining a **real function**. The two sets could also be sets of complex numbers, defining a **complex function**, or X could be a set of real numbers and Y a set of vectors, defining a **vector-valued function**. In the broad definition, X and Y can be any two sets.

Depth stick

Figure 5

(a) If $h = 12$ inches, how many gallons of gasoline are in the tank?

(b) If $h = 1$ inch, how many gallons of gasoline are in the tank?

Solution (a) We evaluate V when $h = 12$.

$$V(12) = 40(12)^2 \sqrt{\frac{96}{12} - 0.608} = 40 \cdot 144\sqrt{8 - 0.608} = 5760\sqrt{7.392} \approx 15{,}660$$

There are about 15,660 gallons of gasoline in the tank when the height of the gasoline in the tank is 12 inches.

(b) Evaluate V when $h = 1$.

$$V(1) = 40(1)^2 \sqrt{\frac{96}{1} - 0.608} = 40\sqrt{96 - 0.608} = 40\sqrt{95.392} \approx 391$$

There are about 391 gallons of gasoline in the tank when the height of the gasoline in the tank is 1 inch. ∎

Implicit Form of a Function

In general, a function f defined by an equation in x and y is said to be given **implicitly**. If it is possible to solve the equation for y in terms of x, then we write $y = f(x)$ and say the function is given **explicitly**. For example,

Implicit Form	Explicit Form
$x^2 - y = 6$	$y = f(x) = x^2 - 6$
$xy = 4$	$y = g(x) = \dfrac{4}{x}$

2 Find the Domain of a Function

In applications, the domain of a function is sometimes specified. For example, we might be interested in the population of a city from 1990 to 2012. The domain of the function is time, in years, and is restricted to the interval [1990, 2012]. Other times the domain is restricted by the context of the function itself. For example, the volume V of a sphere, given by the function $V = \frac{4}{3}\pi R^3$, makes sense only if the radius R is greater than 0. But often the domain of a function f is not specified; only the formula defining the function is given. In such cases, the **domain** of f is the largest set of real numbers for which the value $f(x)$ is defined and is a real number.

EXAMPLE 3 Finding the Domain of a Function

Find the domain of each of the following functions:

(a) $f(x) = x^2 + 5x$ (b) $g(x) = \dfrac{3x}{x^2 - 4}$

(c) $h(t) = \sqrt{4 - 3t}$ (d) $F(u) = \dfrac{5u}{\sqrt{u^2 - 1}}$

Solution (a) Since $f(x) = x^2 + 5x$ is defined for any real number x, the domain of f is the set of all real numbers.

(b) Since division by zero is not defined, $x^2 - 4$ cannot be 0, that is, $x \neq -2$ and $x \neq 2$. The function $g(x) = \dfrac{3x}{x^2 - 4}$ is defined for any real number except $x = -2$ and $x = 2$. So, the domain of g is the set of real numbers $\{x \mid x \neq -2, x \neq 2\}$.

(c) Since the square root of a negative number is not a real number, the value of $4 - 3t$ must be nonnegative. The solution of the inequality $4 - 3t \geq 0$ is $t \leq \dfrac{4}{3}$, so the domain of h is the set of real numbers $\left\{ t \mid t \leq \dfrac{4}{3} \right\}$ or the interval $\left(-\infty, \dfrac{4}{3} \right]$.

NEED TO REVIEW? Solving inequalities is discussed in Appendix A.1, pp. A-5 to A-8.

(d) Since the square root is in the denominator, the value of $u^2 - 1$ must be not only nonnegative; it also cannot equal zero. That is, $u^2 - 1 > 0$. The solution of the inequality $u^2 - 1 > 0$ is the set of real numbers $\{u | u < -1\} \cup \{u | u > 1\}$ or the set $(-\infty, -1) \cup (1, \infty)$. ∎

If x is in the domain of a function f, we say that f **is defined at** x, or $f(x)$ **exists**. If x is not in the domain of f, we say that f **is not defined at** x, or $f(x)$ **does not exist**. The domain of a function is expressed using inequalities, interval notation, set notation, or words, whichever is most convenient. Notice the various ways the domain was expressed in the solution to Example 3.

NEED TO REVIEW? Interval notation is discussed in Appendix A.1, p. A-5.

NOW WORK   Problem 17.

3 Identify the Graph of a Function

In applications, often a graph reveals the relationship between two variables more clearly than an equation. For example, Table 2 shows the average price of gasoline at a particular gas station in Texas (for the years 1980–2012 adjusted for inflation, based on 2008 dollars). If we plot these data and then connect the points, we obtain Figure 6.

TABLE 2

Year	Price	Year	Price	Year	Price
1980	3.41	1991	1.90	2002	1.86
1981	3.26	1992	1.82	2003	1.79
1982	3.15	1993	1.70	2004	2.13
1983	2.51	1994	1.85	2005	2.60
1984	2.51	1995	1.68	2006	2.62
1985	2.46	1996	1.87	2007	3.29
1986	1.63	1997	1.65	2008	2.10
1987	1.90	1998	1.50	2009	2.45
1988	1.77	1999	1.73	2010	2.97
1989	1.83	2000	1.85	2011	3.80
1990	2.25	2001	1.40	2012	3.91

Source: http://www.randomuseless.info/gasprice/gasprice.html

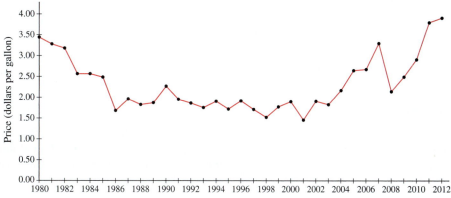

Average retail price of gasoline (2008 dollars)

Source: http://www.randomuseless.info/gasprice/gasprice.html

Figure 6

NEED TO REVIEW? The graph of an equation is discussed in Appendix A.3, pp. A-16 to A-20.

The graph shows that for each date on the horizontal axis there is only one price on the vertical axis. So, the graph represents a function, although the rule for determining the price from the year is not given.

When a function is defined by an equation in x and y, the **graph of the function** is the set of points (x, y) in the xy-plane that satisfy the equation.

But not every collection of points in the xy-plane represents the graph of a function. Recall that a relation is a function only if each element x in the domain corresponds to exactly one image y in the range. This means the graph of a function never contains two points with the same x-coordinate and different y-coordinates. Compare the graphs in Figures 7 and 8. In Figure 7 every number x is associated with exactly one number y, but in Figure 8 some numbers x are associated with three numbers y. Figure 7 shows the graph of a function; Figure 8 shows a graph that is not the graph of a function.

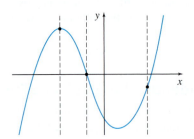

Figure 7 Function: Exactly one y for each x. Every vertical line intersects the graph in at most one point.

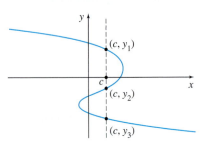

Figure 8 Not a function: $x = c$ has 3 y's associated with it. The vertical line $x = c$ intersects the graph in three points.

For a graph to be a *graph of a function*, it must satisfy the *Vertical-line Test*.

NOTE The phrase "if and only if" means the concepts on each side of the phrase are equivalent. That is, they have the same meaning.

THEOREM Vertical-line Test

A set of points in the xy-plane is the graph of a function if and only if every vertical line intersects the graph in at most one point.

EXAMPLE 4   Identifying the Graph of a Function

Which graphs in Figure 9 represent the graph of a function?

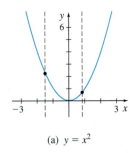

(a) $y = x^2$

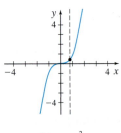

(b) $y = x^3$

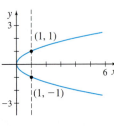

(c) $x = y^2$

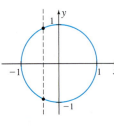

(d) $x^2 + y^2 = 1$

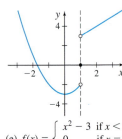

(e) $f(x) = \begin{cases} x^2 - 3 & \text{if } x < 1 \\ 0 & \text{if } x = 1 \\ x + 2 & \text{if } x > 1 \end{cases}$

Figure 9

Solution The graphs in Figure 9(a), 9(b), and 9(e) are graphs of functions because every vertical line intersects each graph in at most one point. The graphs in Figure 9(c) and 9(d) are not graphs of functions, because there is a vertical line that intersects each graph in more than one point. ■

NOW WORK Problems 31(a) and (b).

Notice that although the graph in Figure 9(e) represents a function, it looks different from the graphs in (a) and (b). The graph consists of two pieces plus a point and they are not connected. Also notice that different equations describe different pieces of the graph. Functions with graphs similar to the one in Figure 9(e) are called *piecewise-defined functions*.

4 Analyze a Piecewise-Defined Function

Sometimes a function is defined differently on different parts of its domain. For example, the *absolute value function* $f(x) = |x|$ is actually defined by two equations: $f(x) = x$ if $x \geq 0$ and $f(x) = -x$ if $x < 0$. These equations are usually combined into one expression as

$$f(x) = |x| = \begin{cases} x & \text{if } x \geq 0 \\ -x & \text{if } x < 0 \end{cases}$$

Figure 10 shows the graph of the absolute value function. Notice that the graph of f satisfies the Vertical-line Test.

When a function is defined by different equations on different parts of its domain, it is called a **piecewise-defined** function.

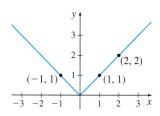

Figure 10 $f(x) = |x|$

RECALL x-intercepts are numbers on the x-axis at which a graph touches or crosses the x-axis.

EXAMPLE 5   Analyzing a Piecewise-Defined Function

The function f is defined as

$$f(x) = \begin{cases} 0 & \text{if } x < 0 \\ 10 & \text{if } 0 \leq x \leq 100 \\ 0.2x - 10 & \text{if } x > 100 \end{cases}$$

(a) Evaluate $f(-1)$, $f(100)$, and $f(200)$.

(b) Graph f.

(c) Find the domain, range, and the x- and y-intercepts of f.

Solution (a) $f(-1) = 0$; $f(100) = 10$; $f(200) = 0.2(200) - 10 = 30$

(b) The graph of f consists of three pieces corresponding to each equation in the definition. The graph is the horizontal line $y = 0$ on the interval $(-\infty, 0)$, the horizontal line $y = 10$ on the interval $[0, 100]$, and the line $y = 0.2x - 10$ on the interval $(100, \infty)$, as shown in Figure 11.

(c) f is a piecewise-defined function. Look at the values that x can take on: $x < 0$, $0 \leq x \leq 100$, $x > 100$. We conclude the domain of f is all real numbers. The range of f is the number 0 and all real numbers greater than or equal to 10. The x-intercepts are all the numbers in the interval $(-\infty, 0)$; the y-intercept is 10. ■

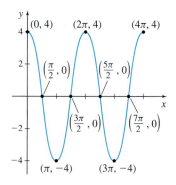

$$f(x) = \begin{cases} 0 & \text{if } x < 0 \\ 10 & \text{if } 0 \leq x \leq 100 \\ 0.2x - 10 & \text{if } x > 100 \end{cases}$$

Figure 11

NOW WORK Problem 33.

5 Obtain Information from or about the Graph of a Function

The graph of a function provides a great deal of information about the function. Reading and interpreting graphs is an essential skill for calculus.

EXAMPLE 6   Obtaining Information from the Graph of a Function

The graph of $y = f(x)$ is given in Figure 12. (x might represent time and y might represent the distance of the bob of a pendulum from its *at-rest* position. Negative values of y would indicate that the bob is to the left of its at-rest position; positive values of y would mean that the bob is to the right of its at-rest position.)

(a) What are $f(0)$, $f\left(\dfrac{3\pi}{2}\right)$, and $f(3\pi)$?

(b) What is the domain of f?

(c) What is the range of f?

(d) List the intercepts of the graph.

(e) How many times does the line $y = 2$ intersect the graph of f?

(f) For what values of x does $f(x) = -4$?

(g) For what values of x is $f(x) > 0$?

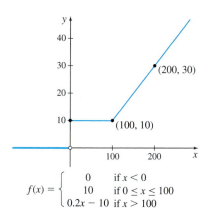

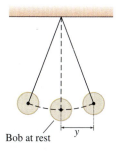

Figure 12

Solution (a) Since the point $(0, 4)$ is on the graph of f, the y-coordinate 4 is the value of f at 0; that is, $f(0) = 4$. Similarly, when $x = \dfrac{3\pi}{2}$, then $y = 0$, so $f\left(\dfrac{3\pi}{2}\right) = 0$, and when $x = 3\pi$, then $y = -4$, so $f(3\pi) = -4$.

(b) The points on the graph of f have x-coordinates between 0 and 4π inclusive. The domain of f is $\{x \mid 0 \le x \le 4\pi\}$ or the closed interval $[0, 4\pi]$.

(c) Every point on the graph of f has a y-coordinate between -4 and 4 inclusive. The range of f is $\{y \mid -4 \le y \le 4\}$ or the closed interval $[-4, 4]$.

RECALL Intercepts are points at which a graph crosses or touches a coordinate axis.

(d) The intercepts of the graph of f are $(0, 4)$, $\left(\dfrac{\pi}{2}, 0\right)$, $\left(\dfrac{3\pi}{2}, 0\right)$, $\left(\dfrac{5\pi}{2}, 0\right)$, and $\left(\dfrac{7\pi}{2}, 0\right)$.

(e) Draw the graph of the line $y = 2$ on the same set of coordinate axes as the graph of f. The line intersects the graph of f four times.

(f) Find points on the graph of f for which $y = f(x) = -4$; there are two such points: $(\pi, -4)$ and $(3\pi, -4)$. So $f(x) = -4$ when $x = \pi$ and when $x = 3\pi$.

(g) $f(x) > 0$ when the y-coordinate of a point (x, y) on the graph of f is positive. This occurs when x is in the set $\left[0, \dfrac{\pi}{2}\right) \cup \left(\dfrac{3\pi}{2}, \dfrac{5\pi}{2}\right) \cup \left(\dfrac{7\pi}{2}, 4\pi\right]$. ∎

NOW WORK Problems 37, 39, 41, 43, 45, 47, and 49.

EXAMPLE 7   Obtaining Information about the Graph of a Function

Consider the function $f(x) = \dfrac{x + 1}{x + 2}$.

(a) What is the domain of f?

(b) Is the point $\left(1, \dfrac{1}{2}\right)$ on the graph of f?

(c) If $x = 2$, what is $f(x)$? What is the corresponding point on the graph of f?

(d) If $f(x) = 2$, what is x? What is the corresponding point on the graph of f?

(e) What are the x-intercepts of the graph of f (if any)? What point(s) on the graph of f correspond(s) to the x-intercept(s)?

Solution (a) The domain of f consists of all real numbers except -2; that is, the set $\{x \mid x \ne -2\}$.

(b) When $x = 1$, then $f(1) = \dfrac{1 + 1}{1 + 2} = \dfrac{2}{3}$. The point $\left(1, \dfrac{2}{3}\right)$ is on the graph of f; the
$\underset{x = 1}{\uparrow}$
point $\left(1, \dfrac{1}{2}\right)$ is not on the graph of f.

(c) If $x = 2$, then $f(2) = \dfrac{2 + 1}{2 + 2} = \dfrac{3}{4}$. The point $\left(2, \dfrac{3}{4}\right)$ is on the graph of f.
$\underset{x = 2}{\uparrow}$

(d) If $f(x) = 2$, then $\dfrac{x + 1}{x + 2} = 2$. Solving for x, we find

$$x + 1 = 2(x + 2) = 2x + 4$$
$$x = -3$$

The point $(-3, 2)$ is on the graph of f.

(e) The x-intercepts of the graph of f occur when $y = 0$. That is, they are the solutions of the equation $f(x) = 0$. The x-intercepts are also called the real **zeros** or **roots** of the function f.

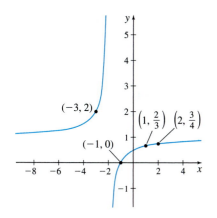

Figure 13 $f(x) = \dfrac{x+1}{x-2}$

The real zeros of the function $f(x) = \dfrac{x+1}{x+2}$ satisfy the equation $x + 1 = 0$ or $x = -1$. The only x-intercept is -1, so the point $(-1, 0)$ is on the graph of f. ∎

Figure 13 shows the graph of f.

NOW WORK Problems 55, 57, 59.

6 Use Properties of Functions

One of the goals of calculus is to develop techniques for graphing functions. Here we review some properties of functions that help obtain the graph of a function.

DEFINITION Even and Odd Functions

A function f is **even** if, for every number x in its domain, the number $-x$ is also in the domain and

$$\boxed{f(-x) = f(x)}$$

A function f is **odd** if, for every number x in its domain, the number $-x$ is also in the domain and

$$\boxed{f(-x) = -f(x)}$$

For example, $f(x) = x^2$ is an even function since

$$f(-x) = (-x)^2 = x^2 = f(x)$$

Also, $g(x) = x^3$ is an odd function since

$$g(-x) = (-x)^3 = -x^3 = -g(x)$$

NEED TO REVIEW? Symmetry of equations is discussed in Appendix A.3, pp. A-17 to A-18.

See Figure 14 for the graph of $f(x) = x^2$ and Figure 15 for the graph of $g(x) = x^3$. Notice that the graph of the even function $f(x) = x^2$ is symmetric with respect to the y-axis and the graph of the odd function $g(x) = x^3$ is symmetric with respect to the origin.

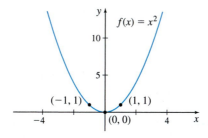

Figure 14 The function $f(x) = x^2$ is even. The graph of f is symmetric with respect to the y-axis.

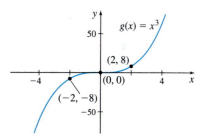

Figure 15 The function $g(x) = x^3$ is odd. The graph of g is symmetric with respect to the origin.

THEOREM Graphs of Even and Odd Functions

- A function is even if and only if its graph is symmetric with respect to the y-axis.
- A function is odd if and only if its graph is symmetric with respect to the origin.

EXAMPLE 8 Identifying Even and Odd Functions

Determine whether each of the following functions is even, odd, or neither. Then determine whether its graph is symmetric with respect to the y-axis, the origin, or neither.

(a) $f(x) = x^2 - 5$ **(b)** $g(x) = \dfrac{4x}{x^2 - 5}$ **(c)** $h(x) = \sqrt[3]{5x^3 - 1}$

(d) $F(x) = |x|$ **(e)** $H(x) = \dfrac{x^2 + 2x - 1}{(x - 5)^2}$

Solution **(a)** The domain of f is $(-\infty, \infty)$, so for every number x in its domain, $-x$ is also in the domain. Replace x by $-x$ and simplify.

$$f(-x) = (-x)^2 - 5 = x^2 - 5 = f(x)$$

Since $f(-x) = f(x)$, the function f is even. So the graph of f is symmetric with respect to the y-axis.

(b) The domain of g is $\{x \mid x \neq \pm\sqrt{5}\}$, so for every number x in its domain, $-x$ is also in the domain. Replace x by $-x$ and simplify.

$$g(-x) = \frac{4(-x)}{(-x)^2 - 5} = \frac{-4x}{x^2 - 5} = -g(x)$$

Since $g(-x) = -g(x)$, the function g is odd. So the graph of g is symmetric with respect to the origin.

(c) The domain of h is $(-\infty, \infty)$, so for every number x in its domain, $-x$ is also in the domain. Replace x by $-x$ and simplify.

$$h(-x) = \sqrt[3]{5(-x)^3 - 1} = \sqrt[3]{-5x^3 - 1} = \sqrt[3]{-(5x^3 + 1)} = -\sqrt[3]{5x^3 + 1}$$

Since $h(-x) \neq h(x)$ and $h(-x) \neq -h(x)$, the function h is neither even nor odd. The graph of h is not symmetric with respect to the y-axis and not symmetric with respect to the origin.

(d) The domain of F is $(-\infty, \infty)$, so for every number x in its domain, $-x$ is also in the domain. Replace x by $-x$ and simplify.

$$F(-x) = |-x| = |-1| \cdot |x| = |x| = F(x)$$

The function F is even. So the graph of F is symmetric with respect to the y-axis.

(e) The domain of H is $\{x \mid x \neq 5\}$. The number $x = -5$ is in the domain of H, but $x = 5$ is not in the domain. So the function H is neither even nor odd, and the graph of H is not symmetric with respect to the y-axis or the origin. ■

NOW WORK   **Problem 61.**

Another important property of a function is to know where it is increasing or decreasing.

DEFINITION

A function f is **increasing** on an interval I, if, for any choice of x_1 and x_2 in I, with $x_1 < x_2$, then $f(x_1) < f(x_2)$.

A function f is **decreasing** on an interval I, if, for any choice of x_1 and x_2 in I, with $x_1 < x_2$, then $f(x_1) > f(x_2)$.

A function f is **constant** on an interval I, if, for all choices of x in I, the values of $f(x)$ are equal.

Notice in the definition for an increasing (decreasing) function f, the value $f(x_1)$ is *strictly* less than (*strictly* greater than) the value $f(x_2)$. If a nonstrict inequality is used, we obtain the definitions for *nondecreasing* and *nonincreasing* functions.

DEFINITION

A function f is **nondecreasing** on an interval I, if, for any choice of x_1 and x_2 in I, with $x_1 < x_2$, then $f(x_1) \leq f(x_2)$.

A function f is **nonincreasing** on an interval I, if, for any choice of x_1 and x_2 in I, with $x_1 < x_2$, then $f(x_1) \geq f(x_2)$.

Figure 16 illustrates the definitions. In Chapter 4 we use calculus to find where a function is increasing or decreasing or is constant.

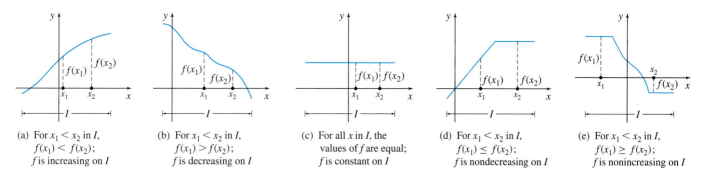

(a) For $x_1 < x_2$ in I,
$f(x_1) < f(x_2)$;
f is increasing on I

(b) For $x_1 < x_2$ in I,
$f(x_1) > f(x_2)$;
f is decreasing on I

(c) For all x in I, the
values of f are equal;
f is constant on I

(d) For $x_1 < x_2$ in I,
$f(x_1) \leq f(x_2)$;
f is nondecreasing on I

(e) For $x_1 < x_2$ in I,
$f(x_1) \geq f(x_2)$;
f is nonincreasing on I

Figure 16

NOTE From left to right, the graph of an increasing function goes up, the graph of a decreasing function goes down, and the graph of a constant function remains at a fixed height. From left to right, the graph of a nondecreasing function never goes down, and the graph of a nonincreasing function never goes up.

NOTE $\dfrac{\Delta y}{\Delta x}$ represents the change in y with respect to x.

NOW WORK Problem 51.

7 Find the Average Rate of Change of a Function

The average rate of change of a function plays an important role in calculus. It provides information about how a change in the independent variable x of a function $y = f(x)$ causes a change in the dependent variable y. We use the symbol Δx, read "delta x," to represent the change in x and Δy to represent the change in y. Then by forming the quotient $\dfrac{\Delta y}{\Delta x}$, we arrive at an *average rate of change*.

DEFINITION Average Rate of Change

If a and b, where $a \neq b$, are in the domain of a function $y = f(x)$, the **average rate of change of** f from a to b is defined as

$$\frac{\Delta y}{\Delta x} = \frac{f(b) - f(a)}{b - a} \qquad a \neq b$$

EXAMPLE 9 Finding the Average Rate of Change

Find the average rate of change of $f(x) = 3x^2$:

(a) From 1 to 3   **(b)** From 1 to x, $x \neq 1$

Solution **(a)** The average rate of change of $f(x) = 3x^2$ from 1 to 3 is

$$\frac{\Delta y}{\Delta x} = \frac{f(3) - f(1)}{3 - 1} = \frac{27 - 3}{3 - 1} = \frac{24}{2} = 12$$

See Figure 17.

Notice that the average rate of change of $f(x) = 3x^2$ from 1 to 3 is the slope of the line containing the points $(1, 3)$ and $(3, 27)$.

(b) The average rate of change of $f(x) = 3x^2$ from 1 to x is

$$\frac{\Delta y}{\Delta x} = \frac{f(x) - f(1)}{x - 1} = \frac{3x^2 - 3}{x - 1} = \frac{3(x^2 - 1)}{x - 1}$$

$$= \frac{3(x - 1)(x + 1)}{x - 1} = 3(x + 1) = 3x + 3$$

Figure 17 $f(x) = 3x^2$

provided $x \neq 1$. ∎

NOW WORK Problem 65.

EXAMPLE 10   Analyzing a Cost Function

The weekly cost C, in dollars, of manufacturing x lightbulbs is

$$C(x) = 7500 + \sqrt{125x}$$

(a) Find the average rate of change of the weekly cost C of manufacturing from 100 to 101 lightbulbs.

(b) Find the average rate of change of the weekly cost C of manufacturing from 1000 to 1001 lightbulbs.

(c) Interpret the results from parts (a) and (b).

Solution (a) The weekly cost of manufacturing 100 lightbulbs is

$$C(100) = 7500 + \sqrt{125 \cdot 100} = 7500 + \sqrt{12{,}500} \approx \$7611.80$$

The weekly cost of manufacturing 101 lightbulbs is

$$C(101) = 7500 + \sqrt{125 \cdot 101} = 7500 + \sqrt{12{,}625} \approx \$7612.36$$

The average rate of change of the weekly cost C from 100 to 101 is

$$\frac{\Delta C}{\Delta x} = \frac{C(101) - C(100)}{101 - 100} \approx \frac{7612.36 - 7611.80}{1} = \$0.56$$

(b) The weekly cost of manufacturing 1000 lightbulbs is

$$C(1000) = 7500 + \sqrt{125 \cdot 1000} = 7500 + \sqrt{125{,}000} \approx \$7853.55$$

The weekly cost of manufacturing 1001 lightbulbs is

$$C(1001) = 7500 + \sqrt{125 \cdot 1001} = 7500 + \sqrt{125{,}125} \approx \$7853.73$$

The average rate of change of the weekly cost C from 1000 to 1001 is

$$\frac{\Delta C}{\Delta x} = \frac{C(1001) - C(1000)}{1001 - 1000} \approx \frac{7853.73 - 7853.55}{1} = \$0.18$$

(c) Part (a) tells us that the cost of manufacturing the 101st lightbulb is $0.56. From (b) we learn that the cost of manufacturing the 1001st lightbulb is only $0.18. The unit cost per lightbulb decreases as the number of lightbulbs manufactured per week increases. ∎

NOW WORK   Problem 73.

P.1 Assess Your Understanding

Concepts and Vocabulary

1. If f is a function defined by $y = f(x)$, then x is called the _____ variable and y is the _____ variable.

2. *True or False*   The independent variable is sometimes referred to as the argument of the function.

3. *True or False*   If no domain is specified for a function f, then the domain of f is taken to be the set of all real numbers.

4. *True or False*   The domain of the function $f(x) = \dfrac{3(x^2 - 1)}{x - 1}$ is $\{x \mid x \neq \pm 1\}$.

5. *True or False*   A function can have more than one y-intercept.

6. A set of points in the xy-plane is the graph of a function if and only if every _____ line intersects the graph in at most one point.

7. If the point $(5, -3)$ is on the graph of f, then $f(\underline{\quad}) = \underline{\quad}$.

8. Find a so that the point $(-1, 2)$ is on the graph of $f(x) = ax^2 + 4$.

9. *Multiple Choice*   A function f is [(a) increasing (b) decreasing (c) nonincreasing (d) nondecreasing (e) constant] on an interval I if, for any choice of x_1 and x_2 in I, with $x_1 < x_2$, then $f(x_1) < f(x_2)$.

10. *Multiple Choice*   A function f is [(a) even (b) odd (c) neither even nor odd] if for every number x in its domain, the number $-x$ is also in the domain and $f(-x) = f(x)$. A function f is [(a) even

---

1. = NOW WORK problem   ⟨/∿⟩ = Graphing technology recommended   ⟨CAS⟩ = Computer Algebra System recommended

(b) odd **(c)** neither even nor odd] if for every number x in its domain, the number $-x$ is also in the domain and $f(-x) = -f(x)$.

11. *True or False*   Even functions have graphs that are symmetric with respect to the origin.

12. The average rate of change of $f(x) = 2x^3 - 3$ from 0 to 2 is _____.

Practice Problems

In Problems 13–16, for each function find:

(a) $f(0)$ **(b)** $f(-x)$ **(c)** $-f(x)$
(d) $f(x+1)$ **(e)** $f(x+h)$

13. $f(x) = 3x^2 + 2x - 4$ **14.** $f(x) = \dfrac{x}{x^2 + 1}$

15. $f(x) = |x| + 4$ **16.** $f(x) = \sqrt{3 - x}$

In Problems 17–22, find the domain of each function.

17. $f(x) = x^3 - 1$ **18.** $f(x) = \dfrac{x}{x^2 + 1}$

19. $v(t) = \sqrt{t^2 - 9}$ **20.** $g(x) = \sqrt{\dfrac{2}{x-1}}$

21. $h(x) = \dfrac{x+2}{x^3 - 4x}$ **22.** $s(t) = \dfrac{\sqrt{t+1}}{t-5}$

In Problems 23–28, find the difference quotient of f. That is, find
$$\dfrac{f(x+h) - f(x)}{h}, h \neq 0.$$

23. $f(x) = -3x + 1$ **24.** $f(x) = \dfrac{1}{x+3}$

25. $f(x) = \sqrt{x+7}$ **26.** $f(x) = \dfrac{2}{\sqrt{x+7}}$

27. $f(x) = x^2 + 2x$ **28.** $f(x) = (2x+3)^2$

In Problems 29–32, determine whether the graph is that of a function by using the Vertical-line Test. If it is, use the graph to find
(a) *the domain and range*
(b) *the intercepts, if any*
(c) *any symmetry with respect to the x-axis, y-axis, or the origin.*

29.

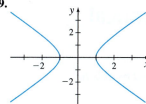

30.

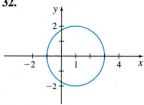

31.

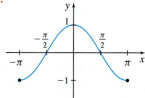

32.

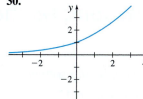

In Problems 33–36, for each piecewise-defined function:
(a) *Find $f(-1)$, $f(0)$, $f(1)$ and $f(8)$.*
(b) *Graph f.*
(c) *Find the domain, range, and intercepts of f.*

33. $f(x) = \begin{cases} x+3 & \text{if } -2 \leq x < 1 \\ 5 & \text{if } x = 1 \\ -x + 2 & \text{if } x > 1 \end{cases}$

34. $f(x) = \begin{cases} 2x+5 & \text{if } -3 \leq x < 0 \\ -3 & \text{if } x = 0 \\ -5x & \text{if } x > 0 \end{cases}$

35. $f(x) = \begin{cases} 1+x & \text{if } x < 0 \\ x^2 & \text{if } x \geq 0 \end{cases}$

36. $f(x) = \begin{cases} \dfrac{1}{x} & \text{if } x < 0 \\ \sqrt[3]{x} & \text{if } x \geq 0 \end{cases}$

In Problems 37–54, use the graph of the function f to answer the following questions.

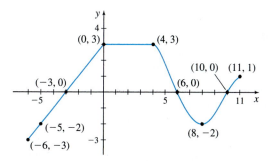

37. Find $f(0)$ and $f(-6)$.

38. Is $f(3)$ positive or negative?

39. Is $f(-4)$ positive or negative?

40. For what values of x is $f(x) = 0$?

41. For what values of x is $f(x) > 0$?

42. What is the domain of f?

43. What is the range of f?

44. What are the x-intercepts?

45. What is the y-intercept?

46. How often does the line $y = \dfrac{1}{2}$ intersect the graph?

47. How often does the line $x = 5$ intersect the graph?

48. For what values of x does $f(x) = 3$?

49. For what values of x does $f(x) = -2$?

50. On what interval(s) is the function f increasing?

51. On what interval(s) is the function f decreasing?

52. On what interval(s) is the function f constant?

53. On what interval(s) is the function f nonincreasing?

54. On what interval(s) is the function f nondecreasing?

In Problems 55–60, answer the questions about the function

$$g(x) = \frac{x+2}{x-6}.$$

55. What is the domain of g?

56. Is the point $(3, 14)$ on the graph of g?

57. If $x = 4$, what is $g(x)$? What is the corresponding point on the graph of g?

58. If $g(x) = 2$, what is x? What is(are) the corresponding point(s) on the graph of g?

59. List the x-intercepts, if any, of the graph of g.

60. What is the y-intercept, if there is one, of the graph of g?

In Problems 61–64, determine whether the function is even, odd, or neither. Then determine whether its graph is symmetric with respect to the y-axis, the origin, or neither.

61. $h(x) = \dfrac{x}{x^2 - 1}$

62. $f(x) = \sqrt[3]{3x^2 + 1}$

63. $G(x) = \sqrt{x}$

64. $F(x) = \dfrac{2x}{|x|}$

65. Find the average rate of change of $f(x) = -2x^2 + 4$:
  (a) From 1 to 2   **(b)** From 1 to 3
  (c) From 1 to 4   **(d)** From 1 to x, $x \neq 1$

66. Find the average rate of change of $s(t) = 20 - 0.8t^2$:
  (a) From 1 to 4   **(b)** From 1 to 3
  (c) From 1 to 2   **(d)** From 1 to t, $t \neq 1$

In Problems 67–72, the graph of a piecewise-defined function is given. Write a definition for each piecewise-defined function. Then state the domain and the range of the function.

67.

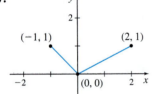

68.

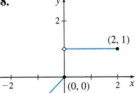

69.

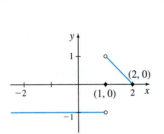

70.

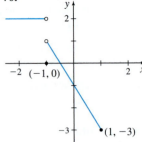

71.

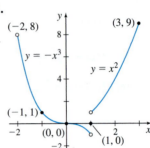

72.

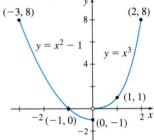

73. The monthly cost C, in dollars, of manufacturing x road bikes is given by the function
$$C(x) = 0.004x^3 - 0.6x^2 + 250x + 100{,}500$$

  (a) Find the average rate of change of the cost C of manufacturing from 100 to 101 road bikes.

  (b) Find the average rate of change of the cost C of manufacturing from 500 to 501 road bikes.

  (c) Interpret the results from parts (a) and (b).

74. The weekly cost in dollars to produce x tons of steel is given by the function
$$C(x) = \frac{1}{10}x^2 + 5x + 1500$$

  (a) Find the average rate of change of the cost C of producing from 500 to 501 tons of steel.

  (b) Find the average rate of change of the cost C of producing from 1000 to 1001 tons of steel.

  (c) Interpret the results from parts (a) and (b).

P.2 Library of Functions; Mathematical Modeling

OBJECTIVES *When you finish this section, you should be able to:*

1 Develop a library of functions (p. 15)

2 Analyze a polynomial function and its graph (p. 17)

3 Find the domain and the intercepts of a rational function (p. 19)

4 Construct a mathematical model (p. 20)

When a collection of functions have common properties, they can be "grouped together" as belonging to a **class of functions**. Polynomial functions, exponential functions, and trigonometric functions are examples of classes of functions. As we investigate principles of calculus, we will find that often a principle applies to all functions in a class in the same way.

1 Develop a Library of Functions

Most of the functions in this section will be familiar to you; several might be new. Although the list may seem familiar, pay special attention to the domain of each function and to its properties, particularly to the shape of each graph. Knowing these graphs lays the foundation for later graphing techniques.

Constant Function $\boxed{f(x) = A}$ *A* is a real number

The domain of a **constant function** is the set of all real numbers; its range is the single number *A*. The graph of a constant function is a horizontal line whose *y*-intercept is *A*; it has no *x*-intercept if $A \neq 0$. A constant function is an even function. See Figure 18.

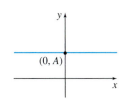

Figure 18 $f(x) = A$

Identity Function $\boxed{f(x) = x}$

The domain and the range of the **identity function** are the set of all real numbers. Its graph is the line through the origin whose slope is $m = 1$. Its only intercept is $(0, 0)$. The identity function is an odd function; its graph is symmetric with respect to the origin. It is increasing over its domain. See Figure 19.

Notice that the graph of $f(x) = x$ bisects quadrants I and III.

The graphs of both the constant function and the identity function are straight lines. These functions belong to the class of *linear functions*:

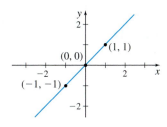

Figure 19 $f(x) = x$

Linear Functions $\boxed{f(x) = mx + b}$ *m* and *b* are real numbers

The domain of a **linear function** is the set of all real numbers. The graph of a linear function is a line with slope *m* and *y*-intercept *b*. If $m > 0$, f is an increasing function, and if $m < 0$, f is a decreasing function. If $m = 0$, then f is a constant function, and its graph is the horizontal line, $y = b$. See Figure 20.

NEED TO REVIEW? Equations of lines are discussed in Appendix A.3, pp. A-18 to A-21.

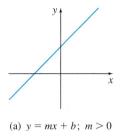   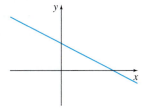   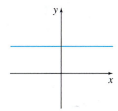

(a) $y = mx + b$; $m > 0$ (b) $y = mx + b$; $m < 0$ (c) $y = mx + b$; $m = 0$

Figure 20

In a *power function*, the independent variable *x* is raised to a power.

Power Functions $\boxed{f(x) = x^a}$ *a* is a real number

Below we examine power functions $f(x) = x^n$, where $n \geq 1$ is a positive integer. The domain of these power functions is the set of all real numbers. The only intercept of their graph is the point $(0, 0)$.

If f is a power function and *n* is a positive odd integer, then f is an odd function whose range is the set of all real numbers and the graph of f is symmetric with respect to the origin. The points $(-1, -1)$, $(0, 0)$, and $(1, 1)$ are on the graph of f. As *x* becomes unbounded in the negative direction, f also becomes unbounded in the negative direction. Similarly, as *x* becomes unbounded in the positive direction, f also becomes unbounded in the positive direction. The function f is increasing over its domain.

The graphs of several odd power functions are given in Figure 21.

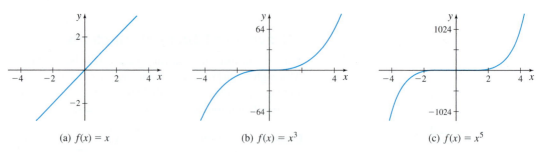

(a) $f(x) = x$ (b) $f(x) = x^3$ (c) $f(x) = x^5$

Figure 21

If f is a power function and n is a positive even integer, then f is an even function whose range is $\{y \mid y \geq 0\}$. The graph of f is symmetric with respect to the y-axis. The points $(-1, 1)$, $(0, 0)$, and $(1, 1)$ are on the graph of f. As x becomes unbounded in either the negative direction or the positive direction, f becomes unbounded in the positive direction. The function is decreasing on the interval $(-\infty, 0]$ and is increasing on the interval $[0, \infty)$.

The graphs of several even power functions are shown in Figure 22.

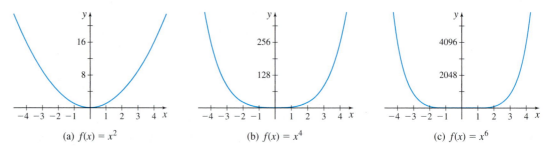

(a) $f(x) = x^2$ (b) $f(x) = x^4$ (c) $f(x) = x^6$

Figure 22

Look closely at Figures 21 and 22. As the integer exponent n increases, the graph of f is flatter (closer to the x-axis) when x is in the interval $(-1, 1)$ and steeper when x is in interval $(-\infty, -1)$ or $(1, \infty)$.

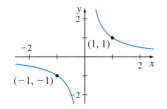

Figure 23 $f(x) = \dfrac{1}{x}$

The Reciprocal Function

$$f(x) = \frac{1}{x}$$

The domain and the range of the **reciprocal function** (the power function $f(x) = x^a$, $a = -1$) are the set of all nonzero real numbers. The graph has no intercepts. The reciprocal function is an odd function so the graph is symmetric with respect to the origin. The function is decreasing on $(-\infty, 0)$ and on $(0, \infty)$. See Figure 23.

Root Functions

$$f(x) = x^{1/n} = \sqrt[n]{x} \qquad n \geq 2 \text{ is a positive integer}$$

Root functions are also power functions. If $n = 2$, $f(x) = x^{1/2} = \sqrt{x}$ is the **square root function**. The domain and the range of the square root function are the set of nonnegative real numbers. The intercept is the point $(0, 0)$. The square root function is neither even nor odd; it is increasing on the interval $[0, \infty)$.

See Figure 24.

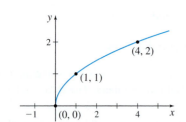

Figure 24 $f(x) = \sqrt{x}$

For root functions whose index n is a positive even integer, the domain and the range are the set of nonnegative real numbers. The intercept is the point $(0, 0)$. Such functions are neither even nor odd; they are increasing on their domain, the interval $[0, \infty)$.

If $n = 3$, $f(x) = x^{1/3} = \sqrt[3]{x}$ is the **cube root function**. The domain and range of the cube root function are all real numbers. The intercept of the graph of the cube root function is the point $(0, 0)$.

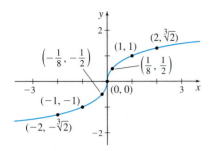

Figure 25 $f(x) = \sqrt[3]{x}$

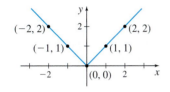

Figure 26 $f(x) = |x|$

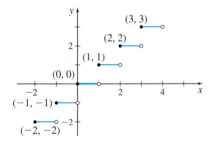

Figure 27 $f(x) = \lfloor x \rfloor$

IN WORDS The floor function can be thought of as the "rounding down" function. The ceiling function can be thought of as the "rounding up" function.

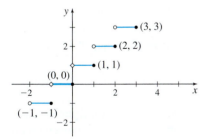

Figure 28 $f(x) = \lceil x \rceil$

NOTE When graphing a function with a graphing utility, the option of using a connected mode or dot mode exists. When graphing the floor function and other functions with discontinuities, use dot mode to prevent the grapher from connecting the dots when f changes from one integer to the next.

Because $f(-x) = \sqrt[3]{-x} = -\sqrt[3]{x} = -f(x)$, the cube root function is odd. Its graph is symmetric with respect to the origin. The cube root function is increasing on the interval $(-\infty, \infty)$. See Figure 25.

For root functions whose index n is a positive odd integer, the domain and the range are the set of all real numbers. The intercept is the point $(0, 0)$. Such functions are odd so their graphs are symmetric with respect to the origin. They are increasing on the interval $(-\infty, \infty)$.

Absolute Value Function $\boxed{f(x) = |x|}$

The **absolute value function** is defined as the piecewise-defined function

$$f(x) = |x| = \begin{cases} x & \text{if } x \geq 0 \\ -x & \text{if } x < 0 \end{cases}$$

or as the root function

$$\boxed{f(x) = |x| = \sqrt{x^2}}$$

The domain of the absolute value function is the set of all real numbers. The range is the set of nonnegative real numbers. The intercept of the graph of f is the point $(0, 0)$. Because $f(-x) = |-x| = |x| = f(x)$, the absolute value function is even. Its graph is symmetric with respect to the y-axis. See Figure 26.

NOW WORK Problems **11** and **15**.

The **floor function**, also known as the **greatest integer function**, is defined as the largest integer less than or equal to x:

$$f(x) = \lfloor x \rfloor = \text{largest integer less than or equal to } x$$

The domain of the floor function $\lfloor x \rfloor$ is the set of all real numbers; the range is the set of all integers. The y-intercept of $\lfloor x \rfloor$ is 0, and the x-intercepts are the numbers in the interval $[0, 1)$. The floor function is constant on every interval of the form $[k, k + 1)$, where k is an integer, and is nondecreasing on its domain. See Figure 27.

The **ceiling function** is defined as the smallest integer greater than or equal to x:

$$f(x) = \lceil x \rceil = \text{smallest integer greater than or equal to } x$$

The domain of the ceiling function $\lceil x \rceil$ is the set of all real numbers; the range is the set of integers. The y-intercept of $\lceil x \rceil$ is 0, and the x-intercepts are the numbers in the interval $(-1, 0]$. The ceiling function is constant on every interval of the form $(k, k+1]$, where k is an integer, and is nondecreasing on its domain. See Figure 28.

For example, for the floor function $\lfloor 5 \rfloor = 5$ and $\lfloor 4.9 \rfloor = 4$, but for the ceiling function $\lceil 5 \rceil = 5$ and $\lceil 4.9 \rceil = 5$.

The floor and ceiling functions are examples of **step functions**. At each integer the function has a *discontinuity*. That is, at integers the function jumps from one value to another without taking on any of the intermediate values.

2 Analyze a Polynomial Function and Its Graph

A **monomial** is a function of the form $y = ax^n$, where $a \neq 0$ is a real number and $n \geq 0$ is an integer. *Polynomial functions* are formed by adding a finite number of monomials.

DEFINITION Polynomial Function

A **polynomial function** is a function of the form

$$\boxed{f(x) = a_n x^n + a_{n-1} x^{n-1} + \cdots + a_1 x + a_0}$$

where $a_n, a_{n-1}, \ldots, a_1, a_0$ are real numbers and n is a nonnegative integer. The domain of a polynomial function is the set of all real numbers.

If $a_n \neq 0$, then a_n is called the **leading coefficient of** f, and the polynomial has **degree** n.

The constant function $f(x) = A$, where $A \neq 0$, is a polynomial function of degree 0. The constant function $f(x) = 0$ is the **zero polynomial function** and has no degree. Its graph is a horizontal line containing the point $(0, 0)$.

If the degree of a polynomial function is 1, then it is a linear function of the form $f(x) = mx + b$, where $m \neq 0$. The graph of a linear function is a straight line with slope m and y-intercept b.

Any polynomial function f of degree 2 can be written in the form

$$f(x) = ax^2 + bx + c$$

where a, b, and c are constants and $a \neq 0$. The square function $f(x) = x^2$ is a polynomial function of degree 2. Polynomial functions of degree 2 are also called **quadratic functions**. The graph of a quadratic function is known as a **parabola** and is symmetric about its **axis of symmetry**, the vertical line $x = -\dfrac{b}{2a}$. Figure 29 shows the graphs of typical parabolas and their axis of symmetry. The x-intercepts, if any, of a quadratic function satisfy the quadratic equation $ax^2 + bx + c = 0$.

NEED TO REVIEW? Quadratic equations and the discriminant are discussed in Appendix A.1, p. A-3.

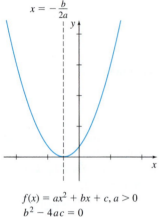

$f(x) = ax^2 + bx + c, a > 0$
$b^2 - 4ac = 0$
One x-intercept

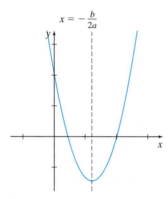

$f(x) = ax^2 + bx + c, a > 0$
$b^2 - 4ac > 0$
Two x-intercepts

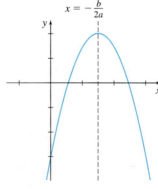

$f(x) = ax^2 + bx + c; a < 0$
$b^2 - 4ac > 0$
Two x-intercepts

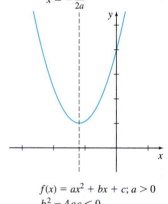

$f(x) = ax^2 + bx + c; a > 0$
$b^2 - 4ac < 0$
No x-intercepts

Figure 29

Graphs of polynomial functions have many properties in common.

The zeros of a polynomial function f give insight into the graph of f. If r is a real zero of a polynomial function f, then r is an x-intercept of the graph of f, and $(x - r)$ is a factor of f.

NOTE Finding the zeros of a polynomial function f is easy if f is linear, quadratic, or in factored form. Otherwise, finding the zeros can be difficult.

If $(x - r)$ occurs more than once in the factored form of f, then r is called a **repeated** or **multiple zero of** f. In particular, if $(x - r)^m$ is a factor of f, but $(x - r)^{m+1}$, where $m \geq 1$ is an integer, is not a factor of f, then r is called a **zero of multiplicity m of** f. When the multiplicity of a zero r is an even integer, the graph of f will touch (but not cross) the x-axis at r; when the multiplicity of r is an odd integer, then the graph of f will cross the x-axis at r. See Figure 30 on page 19.

EXAMPLE 1   **Analyzing the Graph of a Polynomial Function**

For the polynomial function $f(x) = x^2(x - 4)(x + 1)$:

(a) Find the x- and y-intercepts of the graph of f.

(b) Determine whether the graph crosses or touches the x-axis at each x-intercept.

(c) Plot at least one point to the left and right of each x-intercept and connect the points to obtain the graph.

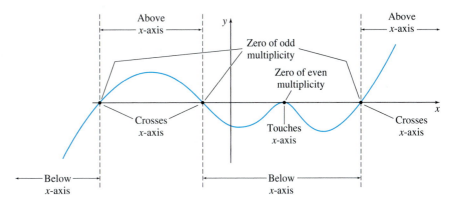

Figure 30

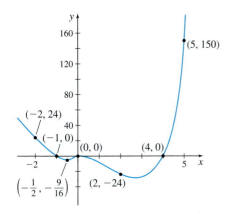

Figure 31

Solution (a) The y-intercept is $f(0) = 0$. The x-intercepts are the zeros of the function: $0, 4,$ and -1.

(b) 0 is a zero of multiplicity 2; the graph of f will touch the x-axis at 0. The numbers 4 and -1 are zeros of multiplicity 1; the graph of f will cross the x-axis at 4 and -1.

(c) Since $f(-2) = 24$, $f\left(-\dfrac{1}{2}\right) = -\dfrac{9}{16}$, $f(2) = -24$, and $f(5) = 150$, the points

$(-2, 24)$, $\left(-\dfrac{1}{2}, -\dfrac{9}{16}\right)$, $(2, -24)$, and $(5, 150)$ are on the graph. See Figure 31. ∎

NOW WORK Problem 21.

3 Find the Domain and the Intercepts of a Rational Function

The quotient of two polynomial functions p and q is called a *rational function*.

DEFINITION Rational Function

A **rational function** is a function of the form

$$R(x) = \frac{p(x)}{q(x)}$$

where p and q are polynomial functions and q is not the zero polynomial. The domain of R is the set of all real numbers, except those for which the denominator q is 0.

If $R(x) = \dfrac{p(x)}{q(x)}$ is a rational function, the real zeros, if any, of the numerator, which are also in the domain of R, are the x-intercepts of the graph of R.

EXAMPLE 2   **Finding the Domain and the Intercepts of a Rational Function**

Find the domain and the intercepts (if any) of each rational function:

(a) $R(x) = \dfrac{2x^2 - 4}{x^2 - 4}$ **(b)** $R(x) = \dfrac{x}{x^2 + 1}$ **(c)** $R(x) = \dfrac{x^2 - 1}{x - 1}$

Solution (a) The domain of $R(x) = \dfrac{2x^2 - 4}{x^2 - 4}$ is $\{x \mid x \neq -2; x \neq 2\}$. Since 0 is in the domain of R and $R(0) = 1$, the y-intercept is 1. The zeros of R are solutions of the equation $2x^2 - 4 = 0$ or $x^2 = 2$. Since $-\sqrt{2}$ and $\sqrt{2}$ are in the domain of R, the x-intercepts are $-\sqrt{2}$ and $\sqrt{2}$.

(b) The domain of $R(x) = \dfrac{x}{x^2 + 1}$ is the set of all real numbers. Since 0 is in the domain of R and $R(0) = 0$, the y-intercept is 0, and the x-intercept is also 0.

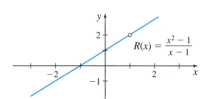

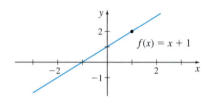

Figure 32

Figure 33

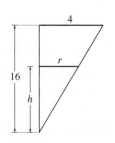

Figure 34

(c) The domain of $R(x) = \dfrac{x^2 - 1}{x - 1}$ is $\{x \mid x \neq 1\}$. Since 0 is in the domain of R and $R(0) = 1$, the y-intercept is 1. The x-intercept(s), if any, satisfy the equation

$$x^2 - 1 = 0$$
$$x^2 = 1$$
$$x = -1 \quad \text{or} \quad x = 1$$

Since 1 is not in the domain of R, the only x-intercept is -1. ∎

Figure 32 shows the graphs of $R(x) = \dfrac{x^2 - 1}{x - 1}$ and $f(x) = x + 1$. Notice the "hole" in the graph of R at the point $(1, 2)$. Also, $R(x) = \dfrac{x^2 - 1}{x - 1}$ and $f(x) = x + 1$ are not the same function: their domains are different. The domain of R is $\{x \mid x \neq 1\}$ and the domain of f is the set of all real numbers.

NOW WORK   **Problem 27.**

Algebraic Functions

Every function discussed so far belongs to a broad class of functions called **algebraic functions**. A function f is called algebraic if it can be expressed in terms of sums, differences, products, quotients, powers, or roots of polynomial functions. For example, the function f defined by

$$f(x) = \frac{3x^3 - x^2(x + 1)^{4/3}}{\sqrt{x^4 + 2}}$$

is an algebraic function. Functions that are not algebraic are called **transcendental functions**. Examples of transcendental functions include the trigonometric functions, exponential functions, and logarithmic functions, which are discussed in the later sections of this chapter.

4 Construct a Mathematical Model

Problems in engineering and the sciences often can by solved using mathematical models that involve functions. To build a model, verbal descriptions must be translated into the language of mathematics by assigning symbols to represent the independent and dependent variables and then finding a function that relates the variables.

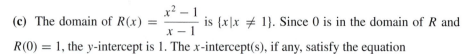

EXAMPLE 3   **Constructing a Model from a Verbal Description**

A liquid is poured into a container in the shape of a right circular cone with radius 4 meters and height 16 meters, as shown in Figure 33. Express the volume V of the liquid as a function of the height h of the liquid.

Solution The formula for the volume of a right circular cone of radius r and height h is

$$V = \frac{1}{3}\pi r^2 h$$

The volume depends on two variables, r and h. To express V as a function of h only, we use the fact that a cross section of the cone and the liquid form two similar triangles. See Figure 34.

Corresponding sides of similar triangles are in proportion. Since the cone's radius is 4 meters and its height is 16 meters, we have

$$\frac{r}{h} = \frac{4}{16} = \frac{1}{4}$$
$$r = \frac{1}{4}h$$

NEED TO REVIEW? Similar triangles
and geometry formulas are discussed in
Appendix A.2, pp. A-13 to A-14.

Then

$$V = \frac{1}{3}\pi r^2 h = \frac{1}{3}\pi \left(\frac{1}{4}h\right)^2 h = \frac{1}{48}\pi h^3$$

$$r = \frac{1}{4}h$$

So $V = V(h) = \dfrac{1}{48}\pi h^3$ expresses the volume V as a function of the height of the liquid.

Since h is measured in meters, V will be expressed in cubic meters. ∎

NOW WORK Problem 29.

In many applications, data are collected and are used to build the mathematical model. If the data involve two variables, the first step is to plot ordered pairs using rectangular coordinates. The resulting graph is called a **scatter plot**.

Scatter plots are used to help suggest the type of relation that exists between the two variables. Once the general shape of the relation is recognized, a function can be chosen whose graph closely resembles the shape in the scatter plot.

EXAMPLE 4   **Identifying the Shape of a Scatter Plot**

Determine whether you would model the relation between the two variables shown in each scatter plot in Figure 35 with a linear function or a quadratic function.

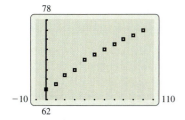
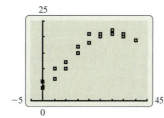
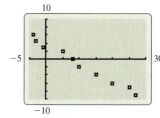

Figure 35

Solution For each scatter plot, we choose a function whose graph closely resembles the shape of the scatter plot. See Figure 36.

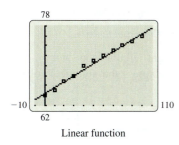
Linear function

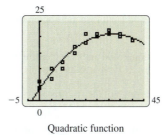
Quadratic function

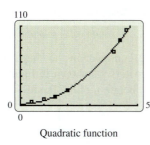
Quadratic function

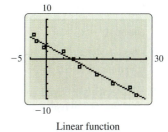
Linear function

Figure 36

∎

EXAMPLE 5   **Building a Function from Data**

The data shown in Table 3 on page 22 measure crop yield for various amounts of fertilizer:

(a) Draw a scatter plot of the data and determine a possible type of relation that may exist between the two variables.

(b) Use technology to find the function of best fit to these data.

Solution (a) Figure 37 on page 22 shows the scatter plot. The data suggest the graph of a quadratic function.

(b) The graphing calculator screen in Figure 38 shows that the quadratic function of best fit is

$$Y(x) = -0.017x^2 + 1.0765x + 3.8939$$

where x represents the amount of fertilizer used and Y represents crop yield. The graph of the quadratic model is illustrated in Figure 39.

TABLE 3

Plot	Fertilizer, x (Pounds/100 ft^2)	Yield (Bushels)
1	0	4
2	0	6
3	5	10
4	5	7
5	10	12
6	10	10
7	15	15
8	15	17
9	20	18
10	20	21
11	25	20
12	25	21
13	30	21
14	30	22
15	35	21
16	35	20
17	40	19
18	40	19

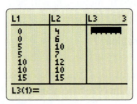 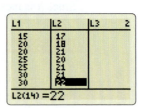 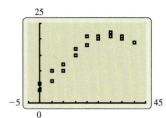

Figure 37

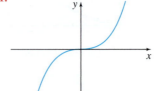 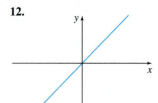

Figure 38   **Figure 39** ∎

NOW WORK Problem 31.

P.2 Assess Your Understanding

Concepts and Vocabulary

1. *Multiple Choice* The function $f(x) = x^2$ is [**(a)** increasing, **(b)** decreasing, **(c)** neither] on the interval $(0, \infty)$.

2. *True or False* The floor function $f(x) = \lfloor x \rfloor$ is an example of a step function.

3. *True or False* The cube function is odd and is increasing on the interval $(-\infty, \infty)$.

4. *True or False* The cube root function is odd and is decreasing on the interval $(-\infty, \infty)$.

5. *True or False* The domain and the range of the reciprocal function are all real numbers.

6. A number r for which $f(r) = 0$ is called a(n) _____ of the function f.

7. *Multiple Choice* If r is a zero of even multiplicity of a function f, the graph of f [**(a)** crosses, **(b)** touches, **(c)** doesn't intersect] the x-axis at r.

8. *True or False* The x-intercepts of the graph of a polynomial function are called zeros of the function.

9. *True or False* The function $f(x) = \left[x + \sqrt[5]{x^2 - \pi}\right]^{2/3}$ is an algebraic function.

10. *True or False* The domain of every rational function is the set of all real numbers.

Practice Problems

In Problems 11–18, match each graph to its function:

A. *Constant function*   B. *Identity function*

C. *Square function*   D. *Cube function*

E. *Square root function*   F. *Reciprocal function*

G. *Absolute value function*   H. *Cube root function*

11.

12.

13.

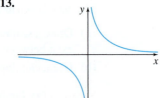

14.

1. = NOW WORK problem   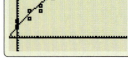 = Graphing technology recommended   CAS = Computer Algebra System recommended

15.
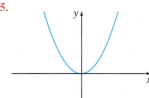

16.
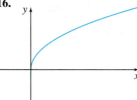

17.
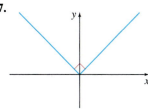

18.
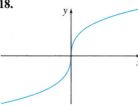

19. If $f(x) = \lfloor 2x \rfloor$, find
   (a) $f(1.2)$,   (b) $f(1.6)$,   (c) $f(-1.8)$.

20. If $f(x) = \left\lceil \dfrac{x}{2} \right\rceil$, find
   (a) $f(1.2)$,   (b) $f(1.6)$,   (c) $f(-1.8)$.

In Problems 21 and 22, for each polynomial function f:
(a) List each real zero and its multiplicity.
(b) Find the x- and y-intercepts of the graph of f.
(c) Determine whether the graph of f crosses or touches the x-axis at each x-intercept.

21. $f(x) = 3(x-7)(x+4)^3$ **22.** $f(x) = 4x(x^2+1)(x-2)^3$

In Problems 23 and 24, decide which of the polynomial functions in the list might have the given graph. (More than one answer is possible.)

23. (a) $f(x) = -4x(x-1)(x-2)$

   (b) $f(x) = x^2(x-1)^2(x-2)$

   (c) $f(x) = 3x(x-1)(x-2)$

   (d) $f(x) = x(x-1)^2(x-2)^2$

   (e) $f(x) = x^3(x-1)(x-2)$

   (f) $f(x) = -x(1-x)(x-2)$

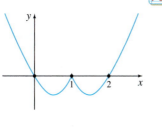

24. (a) $f(x) = 2x^3(x-1)(x-2)^2$

   (b) $f(x) = x^2(x-1)(x-2)$

   (c) $f(x) = x^3(x-1)^2(x-2)$

   (d) $f(x) = x^2(x-1)^2(x-2)^2$

   (e) $f(x) = 5x(x-1)^2(x-2)$

   (f) $f(x) = -2x(x-1)^2(2-x)$

In Problems 25–28, find the domain and the intercepts of each rational function.

25. $R(x) = \dfrac{5x^2}{x+3}$ **26.** $H(x) = \dfrac{-4x^2}{(x-2)(x+4)}$

27. $R(x) = \dfrac{3x^2 - x}{x^2 + 4}$ **28.** $R(x) = \dfrac{3(x^2 - x - 6)}{4(x^2 - 9)}$

29. Constructing a Model   The rectangle shown in the figure has one corner in quadrant I on the graph of $y = 16 - x^2$, another corner at the origin, and corners on both the positive y-axis and the positive x-axis. As the corner on $y = 16 - x^2$ changes, a variety of rectangles are obtained.

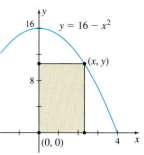

   (a) Express the area A of the rectangles as a function of x.

   (b) What is the domain of A?

30. Constructing a Model   The rectangle shown in the figure is inscribed in a semicircle of radius 2. Let $P = (x, y)$ be the point in quadrant I that is a vertex of the rectangle and is on the circle. As the point (x, y) on the circle changes, a variety of rectangles are obtained.

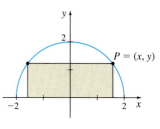

   (a) Express the area A of the rectangles as a function of x.

   (b) Express the perimeter p of the rectangles as a function of x.

31. Height of a Ball   A ballplayer throws a ball at an inclination of $45°$ to the horizontal. The following data represent the height h (in feet) of the ball at the instant that it has traveled x feet horizontally:

Distance, x	20	40	60	80	100	120	140	160	180	200
Height, h	25	40	55	65	71	77	77	75	71	64

   (a) Draw a scatter plot of the data. Comment on the type of relation that may exist between the two variables.

   (b) Use technology to verify that the quadratic function of best fit to these data is

$$h(x) = -0.0037x^2 + 1.03x + 5.7$$

   Use this function to determine the horizontal distance the ball travels before it hits the ground.

   (c) Approximate the height of the ball when it has traveled 10 feet.

32. Educational Attainment   The following data represent the percentage of the U.S. population whose age is x (in years) who did not have a high school diploma as of January 2011:

Age, x	30	40	50	60	70	80
Percentage without a High School Diploma, P	11.6	11.7	10.4	10.4	17.0	24.6

Source: U.S. Census Bureau.

   (a) Draw a scatter plot of the data, treating age as the independent variable. Comment on the type of relation that may exist between the two variables.

   (b) Use technology to verify that the cubic function of best fit to these data is

$$P(x) = 0.00026x^3 - 0.0303x^2 + 1.0877x - 0.5071$$

   (c) Use this model to predict the percentage of 35-year-olds who do not have a high school diploma.

P.3 Operations on Functions; Graphing Techniques

OBJECTIVES *When you finish this section, you should be able to:*

1 Form the sum, difference, product, and quotient of two functions (p. 24)
2 Form a composite function (p. 25)
3 Transform the graph of a function with vertical and horizontal shifts (p. 27)
4 Transform the graph of a function with compressions and stretches (p. 29)
5 Transform the graph of a function by reflecting it about the *x*-axis and the *y*-axis (p. 30)

1 Form the Sum, Difference, Product, and Quotient of Two Functions

Functions, like numbers, can be added, subtracted, multiplied, and divided. For example, the polynomial function $F(x) = x^2 + 4x$ is the sum of the two functions $f(x) = x^2$ and $g(x) = 4x$. The rational function $R(x) = \dfrac{x^2}{x^3 - 1}$, $x \neq 1$, is the quotient of the two functions $f(x) = x^2$ and $g(x) = x^3 - 1$.

DEFINITION Operations on Functions

If f and g are two functions, their sum, $f + g$; their difference, $f - g$; their product, $f \cdot g$; and their quotient, $\dfrac{f}{g}$, are defined by

- Sum: $(f + g)(x) = f(x) + g(x)$ • Difference: $(f - g)(x) = f(x) - g(x)$

- Product: $(f \cdot g)(x) = f(x) \cdot g(x)$ • Quotient: $\left(\dfrac{f}{g}\right)(x) = \dfrac{f(x)}{g(x)}$, $g(x) \neq 0$

For every operation except division, the domain of the resulting function consists of the intersection of the domains of f and g. The domain of a quotient $\dfrac{f}{g}$ consists of the numbers x that are common to the domains of both f and g, but excludes the numbers x for which $g(x) = 0$.

EXAMPLE 1 Forming the Sum, Difference, Product, and Quotient of Two Functions

Let f and g be two functions defined as

$$f(x) = \sqrt{x - 1} \qquad \text{and} \qquad g(x) = \sqrt{4 - x}$$

Find the following functions and determine their domain:

(a) $(f + g)(x)$ **(b)** $(f - g)(x)$ **(c)** $(f \cdot g)(x)$ **(d)** $\left(\dfrac{f}{g}\right)(x)$

Solution The domain of f is $\{x \mid x \geq 1\}$, and the domain of g is $\{x \mid x \leq 4\}$.

(a) $(f + g)(x) = f(x) + g(x) = \sqrt{x - 1} + \sqrt{4 - x}$. The domain of $(f + g)(x)$ is the closed interval $[1, 4]$.

(b) $(f - g)(x) = f(x) - g(x) = \sqrt{x - 1} - \sqrt{4 - x}$. The domain of $(f - g)(x)$ is the closed interval $[1, 4]$.

(c) $(f \cdot g)(x) = f(x) \cdot g(x) = (\sqrt{x-1})(\sqrt{4-x}) = \sqrt{-x^2 + 5x - 4}$. The domain of $(f \cdot g)(x)$ is the closed interval $[1, 4]$.

(d) $\left(\dfrac{f}{g}\right)(x) = \dfrac{f(x)}{g(x)} = \dfrac{\sqrt{x-1}}{\sqrt{4-x}} = \dfrac{\sqrt{-x^2 + 5x - 4}}{4 - x}$. The domain of $\left(\dfrac{f}{g}\right)(x)$ is the half-open interval $[1, 4)$.   ■

NOW WORK Problem 11.

2 Form a Composite Function

Suppose an oil tanker is leaking, and your job requires you to find the area of the circular oil spill surrounding the tanker. You determine that the radius of the spill is increasing at a rate of 3 meters per minute. That is, the radius r of the spill is a function of the time t in minutes since the leak began, and can be written as $r(t) = 3t$.

For example, after 20 minutes the radius of the spill is $r(20) = 3 \cdot 20 = 60$ meters. Recall that the area A of a circle is a function of its radius r; that is, $A(r) = \pi r^2$. So, the area of the oil spill after 20 minutes is $A(60) = \pi(60^2) = 3600\pi$ square meters. Notice that the argument r of the function A is itself a function, and that the area A of the oil spill is found at any time t by evaluating the function $A = A(r(t))$.

Functions such as $A = A(r(t))$ are called *composite functions*.

Another example of a composite function is $y = (2x + 3)^2$. If $y = f(u) = u^2$ and $u = g(x) = 2x + 3$, then by substituting $g(x) = 2x + 3$ for u, we obtain the original function:

$$y = f(u) = f(g(x)) = (2x + 3)^2$$
$$\uparrow \qquad\qquad\qquad \uparrow$$
$$u = g(x) \quad g(x) = 2x + 3$$

This substitution process is called *composition*.

In general, suppose that f and g are two functions and that x is a number in the domain of g. By evaluating g at x, we obtain $g(x)$. Now, if $g(x)$ is in the domain of the function f, we can evaluate f at $g(x)$, obtaining $f(g(x))$. The correspondence from x to $f(g(x))$ is called *composition*.

DEFINITION Composite Function

Given two functions f and g, the **composite function**, denoted by $f \circ g$ (read "f composed with g") is defined by

$$\boxed{(f \circ g)(x) = f(g(x))}$$

The domain of $f \circ g$ is the set of all numbers x in the domain of g for which $g(x)$ is in the domain of f.

Figure 40 illustrates the definition. Only those numbers x in the domain of g for which $g(x)$ is in the domain of f are in the domain of $f \circ g$. As a result, the domain of $f \circ g$ is a subset of the domain of g, and the range of $f \circ g$ is a subset of the range of f.

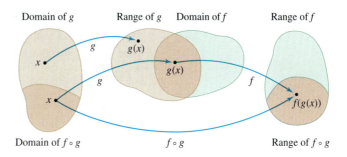

Figure 40

NOTE The "inside" function g, in $f(g(x))$, is evaluated first.

EXAMPLE 2   Evaluating a Composite Function

Suppose that $f(x) = \dfrac{1}{x+2}$ and $g(x) = \dfrac{4}{x-1}$.

(a) $(f \circ g)(0) = f(g(0)) = f(-4) = \dfrac{1}{-4+2} = -\dfrac{1}{2}$ $g(x) = \dfrac{4}{x-1}; g(0) = -4$

(b) $(g \circ f)(1) = g(f(1)) = g\left(\dfrac{1}{3}\right) = \dfrac{4}{\dfrac{1}{3}-1} = \dfrac{4}{-\dfrac{2}{3}} = -6$

$\quad\quad\quad\quad\quad\quad f(x) = \dfrac{1}{x+2}; f(1) = \dfrac{1}{3}$

(c) $(f \circ f)(1) = f(f(1)) = f\left(\dfrac{1}{3}\right) = \dfrac{1}{\dfrac{1}{3}+2} = \dfrac{1}{\dfrac{7}{3}} = \dfrac{3}{7}$

(d) $(g \circ g)(-3) = g(g(-3)) = g(-1) = \dfrac{4}{-1-1} = -2$

$\quad\quad\quad\quad\quad\quad g(x) = \dfrac{4}{x-1}; g(-3) = \dfrac{4}{-3-1} = -1$ ∎

NOW WORK Problem 15.

EXAMPLE 3   Finding the Domain of a Composite Function

Suppose that $f(x) = \dfrac{1}{x+2}$ and $g(x) = \dfrac{4}{x-1}$. Find $f \circ g$ and its domain.

Solution

$$(f \circ g)(x) = f(g(x)) = \frac{1}{g(x)+2} = \frac{1}{\dfrac{4}{x-1}+2} = \frac{x-1}{4+2(x-1)} = \frac{x-1}{2x+2}$$

To find the domain of $f \circ g$, first note that the domain of g is $\{x \,|\, x \neq 1\}$, so we exclude 1 from the domain of $f \circ g$. Next note that the domain of f is $\{x \,|\, x \neq -2\}$, which means $g(x)$ cannot equal -2. To determine what additional values of x to exclude, we solve the equation $g(x) = -2$:

$$\frac{4}{x-1} = -2 \qquad\qquad g(x) = -2$$
$$4 = -2(x-1)$$
$$4 = -2x + 2$$
$$2x = -2$$
$$x = -1$$

We also exclude -1 from the domain of $f \circ g$.

The domain of $f \circ g$ is $\{x \,|\, x \neq -1, x \neq 1\}$.

We could also find the domain of $f \circ g$ by first finding the domain of g: $\{x \,|\, x \neq 1\}$. So, exclude 1 from the domain of $f \circ g$. Then looking at $(f \circ g)(x) = \dfrac{x-1}{2x+2} = \dfrac{x-1}{2(x+1)}$, notice that $x \neq -1$, so we exclude -1 from the domain of $f \circ g$. Therefore, the domain of $f \circ g$ is $\{x \,|\, x \neq -1, x \neq 1\}$. ∎

NOW WORK Problem 23.

In general, the composition of two functions f and g is not commutative. That is, $f \circ g$ almost never equals $g \circ f$. For example, in Example 3,

$$(g \circ f)(x) = g(f(x)) = \frac{4}{f(x) - 1} = \frac{4}{\dfrac{1}{x+2} - 1} = \frac{4(x+2)}{1 - (x+2)} = -\frac{4(x+2)}{x+1}$$

Functions f and g for which $f \circ g = g \circ f$ will be discussed in the next section.

Some techniques in calculus require us to "decompose" a composite function. For example, the function $H(x) = \sqrt{x+1}$ is the composition $f \circ g$ of the functions $f(x) = \sqrt{x}$ and $g(x) = x + 1$.

EXAMPLE 4   Decomposing a Composite Function

Find functions f and g so that $f \circ g = F$ when:

(a) $F(x) = \dfrac{1}{x+1}$ **(b)** $F(x) = (x^3 - 4x - 1)^{100}$ **(c)** $F(t) = \sqrt{2-t}$

Solution (a) If we let $f(x) = \dfrac{1}{x}$ and $g(x) = x + 1$, then

$$(f \circ g)(x) = f(g(x)) = \frac{1}{g(x)} = \frac{1}{x+1} = F(x)$$

(b) If we let $f(x) = x^{100}$ and $g(x) = x^3 - 4x - 1$, then

$$(f \circ g)(x) = f(g(x)) = f(x^3 - 4x - 1) = (x^3 - 4x - 1)^{100} = F(x)$$

(c) If we let $f(t) = \sqrt{t}$ and $g(t) = 2 - t$, then

$$(f \circ g)(t) = f(g(t)) = f(2 - t) = \sqrt{2-t} = F(t)$$ ■

Although the functions f and g chosen in Example 4 are not unique, there is usually a "natural" selection for f and g that first comes to mind. When decomposing a composite function, the "natural" selection for g is often an expression inside parentheses, in a denominator, or under a radical.

NOW WORK   **Problem 27.**

3 Transform the Graph of a Function with Vertical and Horizontal Shifts

At times we need to graph a function that is very similar to a function with a known graph. Often techniques, called **transformations,** can be used to draw the new graph.

First we consider *translations*. **Translations** shift the graph from one position to another without changing its shape, size, or direction.

For example, let f be a function with a known graph, say, $f(x) = x^2$. If k is a positive number, then adding k to f adds k to each y-coordinate, causing the graph of f to **shift vertically up** k units. On the other hand, subtracting k from f subtracts k from each y-coordinate, causing the graph of f to **shift vertically down** k units. See Figure 41 on page 28.

So, adding (or subtracting) a positive constant to a function shifts the graph of the original function vertically up (or down). Now we investigate how to shift the graph of a function right or left.

Again, let f be a function with a known graph, say, $f(x) = x^2$, and let h be a positive number. To shift the graph to the right h units, subtract h from the argument of f. In other words, replace the argument x of a function f by $x - h$, $h > 0$. The graph of the new function $y = f(x - h)$ is the graph of f **shifted horizontally right** h units. On the other hand, if we replace the argument x of a function f by $x + h$, $h > 0$, the graph of the new function $y = f(x + h)$ is the graph of f **shifted horizontally left** h units. See Figure 42 on page 28.

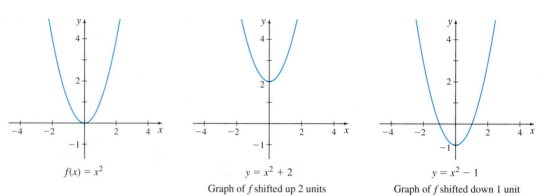

$f(x) = x^2$

$y = x^2 + 2$
Graph of f shifted up 2 units

$y = x^2 - 1$
Graph of f shifted down 1 unit

Figure 41

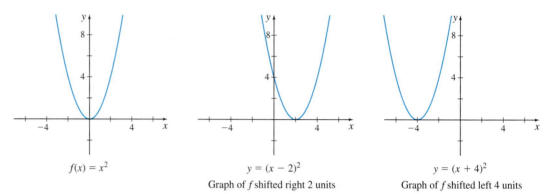

$f(x) = x^2$

$y = (x - 2)^2$
Graph of f shifted right 2 units

$y = (x + 4)^2$
Graph of f shifted left 4 units

Figure 42

The graph of a function f can be moved anywhere in the plane by combining vertical and horizontal shifts.

EXAMPLE 5   Combining Vertical and Horizontal Shifts

Use transformations to graph the function $f(x) = (x + 3)^2 - 5$.

Solution Graph f in steps:

- Observe that f is basically a square function, so begin by graphing $y = x^2$ in Figure 43(a).
- Replace the argument x with $x + 3$ to obtain $y = (x + 3)^2$. This shifts the graph of f horizontally to the left 3 units, as shown in Figure 43(b).
- Finally, subtract 5 from each y-coordinate, which shifts the graph in Figure 42(b) vertically down 5 units, and results in the graph of $f(x) = (x + 3)^2 - 5$ shown in Figure 43(c).

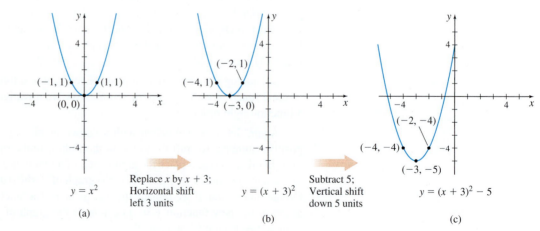

$y = x^2$

(a)

Replace x by $x + 3$;
Horizontal shift
left 3 units

$y = (x + 3)^2$

(b)

Subtract 5;
Vertical shift
down 5 units

$y = (x + 3)^2 - 5$

(c)

Figure 43

Notice the points plotted in Figure 43. Using key points can be helpful in keeping track of the transformation that has taken place.

NOW WORK   Problem 39.

4 Transform the Graph of a Function with Compressions and Stretches

When a function f is multiplied by a positive number a, the graph of the new function $y = af(x)$ is obtained by multiplying each y-coordinate on the graph of f by a. The new graph is a **vertically compressed** version of the graph of f if $0 < a < 1$, and is a **vertically stretched** version of the graph of f if $a > 1$. Compressions and stretches change the proportions of a graph.

For example, the graph of $f(x) = x^2$ is shown in Figure 44(a). Multiplying f by $a = \dfrac{1}{2}$ produces a new function $y = \dfrac{1}{2}f(x) = \dfrac{1}{2}x^2$, which vertically compresses the graph of f by a factor of $\dfrac{1}{2}$, as shown in Figure 44(b). On the other hand, if $a = 3$, then multiplying f by 3 produces a new function $y = 3f(x) = 3x^2$, and the graph of f is vertically stretched by a factor of 3, as shown in Figure 44(c).

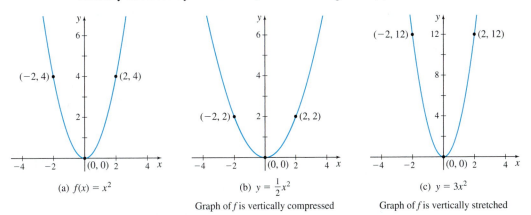

(a) $f(x) = x^2$

(b) $y = \dfrac{1}{2}x^2$

Graph of f is vertically compressed

(c) $y = 3x^2$

Graph of f is vertically stretched

Figure 44

If the argument x of a function f is multiplied by a positive number a, the graph of the new function $y = f(ax)$ is a **horizontal compression** of the graph of f when $a > 1$, and a **horizontal stretch** of the graph of f when $0 < a < 1$.

For example, the graph of $y = f(2x) = (2x)^2 = 4x^2$ is a horizontal compression of the graph of $f(x) = x^2$. See Figure 45(a) and 45(b). On the other hand, if $a = \dfrac{1}{3}$, then the graph of $y = \left(\dfrac{1}{3}x\right)^2 = \dfrac{1}{9}x^2$ is a horizontal stretch of the graph of $f(x) = x^2$. See Figure 45(c).

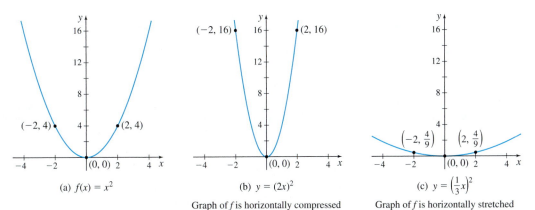

(a) $f(x) = x^2$

(b) $y = (2x)^2$

Graph of f is horizontally compressed

(c) $y = \left(\dfrac{1}{3}x\right)^2$

Graph of f is horizontally stretched

Figure 45

5 Transform the Graph of a Function by Reflecting It about the x-axis or the y-axis

The third type of transformation, reflection about the x- or y-axis, changes the orientation of the graph of the function f but keeps the shape and the size of the graph intact.

When a function f is multiplied by -1, the graph of the new function $y = -f(x)$ is the **reflection about the x-axis** of the graph of f. For example, if $f(x) = \sqrt{x}$, then the graph of the new function $y = -f(x) = -\sqrt{x}$ is the reflection of the graph of f about the x-axis. See Figures 46(a) and 46(b).

If the argument x of a function f is multiplied by -1, then the graph of the new function $y = f(-x)$ is the **reflection about the y-axis** of the graph of f. For example, if $f(x) = \sqrt{x}$, then the graph of the new function $y = f(-x) = \sqrt{-x}$ is the reflection of the graph of f about the y-axis. See Figures 46(a) and 46(c). Notice in this example that the domain of $y = \sqrt{-x}$ is all real numbers for which $-x \geq 0$, or equivalently, $x \leq 0$.

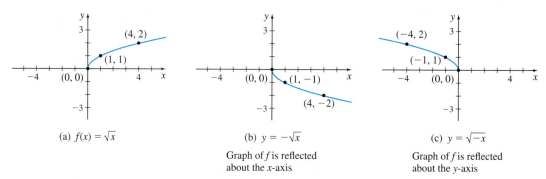

(a) $f(x) = \sqrt{x}$

(b) $y = -\sqrt{x}$
Graph of f is reflected about the x-axis

(c) $y = \sqrt{-x}$
Graph of f is reflected about the y-axis

Figure 46

EXAMPLE 6   Combining Transformations

Use transformations to graph the function $f(x) = \sqrt{1 - x} + 2$.

Solution We graph f in steps:

- Observe that f is basically a square root function, so we begin by graphing $y = \sqrt{x}$. See Figure 47(a).
- Now we replace the argument x with $x + 1$ to obtain $y = \sqrt{x + 1}$, which shifts the graph of $y = \sqrt{x}$ horizontally to the left 1 unit, as shown in Figure 47(b).
- Then we replace x with $-x$ to obtain $y = \sqrt{-x + 1} = \sqrt{1 - x}$, which reflects the graph about the y-axis. See Figure 47(c).
- Finally, we add 2 to each y-coordinate, which shifts the graph vertically up 2 units, and results in the graph of $f(x) = \sqrt{1 - x} + 2$ shown in Figure 47(d).

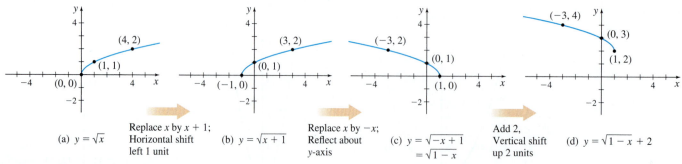

(a) $y = \sqrt{x}$

Replace x by $x + 1$;
Horizontal shift
left 1 unit

(b) $y = \sqrt{x + 1}$

Replace x by $-x$;
Reflect about
y-axis

(c) $y = \sqrt{-x + 1}$
  $= \sqrt{1 - x}$

Add 2,
Vertical shift
up 2 units

(d) $y = \sqrt{1 - x} + 2$

Figure 47

NOW WORK   Problem 43.

P.3 Assess Your Understanding

Concepts and Vocabulary

1. If the domain of a function f is $\{x \mid 0 \le x \le 7\}$ and the domain of a function g is $\{x \mid -2 \le x \le 5\}$, then the domain of the sum function $f + g$ is —————.

2. *True or False* If f and g are functions, then the domain of $\dfrac{f}{g}$ consists of all numbers x that are in the domains of both f and g.

3. *True or False* The domain of $f \cdot g$ consists of the numbers x that are in the domains of both f and g.

4. *True or False* The domain of the composite function $f \circ g$ is the same as the domain of $g(x)$.

5. *True or False* The graph of $y = -f(x)$ is the reflection of the graph of $y = f(x)$ about the x-axis.

6. *True or False* To obtain the graph of $y = f(x + 2) - 3$, shift the graph of $y = f(x)$ horizontally to the right 2 units and vertically down 3 units.

7. *True or False* Suppose the x-intercepts of the graph of the function f are -3 and 2. Then the x-intercepts of the graph of the function $y = 2f(x)$ are -3 and 2.

8. Suppose that the graph of a function f is known. Then the graph of the function $y = f(x - 2)$ can be obtained by a(n)————— shift of the graph of f to the ————— a distance of 2 units.

9. Suppose that the graph of a function f is known. Then the graph of the function $y = f(-x)$ can be obtained by a reflection about the —————-axis of the graph of f.

10. Suppose the x-intercepts of the graph of the function f are -2, 1, and 5. The x-intercepts of $y = f(x + 3)$ are —————, —————, and —————.

Practice Problems

In Problems 11–14, the functions f and g are given. Find each of the following functions and determine their domain:

(a) $(f + g)(x)$ **(b)** $(f - g)(x)$

(c) $(f \cdot g)(x)$ **(d)** $\left(\dfrac{f}{g}\right)(x)$

11. $f(x) = 3x + 4$ and $g(x) = 2x - 3$

12. $f(x) = 1 + \dfrac{1}{x}$ and $g(x) = \dfrac{1}{x}$

13. $f(x) = \sqrt{x + 1}$ and $g(x) = \dfrac{2}{x}$

14. $f(x) = |x|$ and $g(x) = x$

In Problems 15 and 16, for each of the functions f and g, find:

(a) $(f \circ g)(4)$ **(b)** $(g \circ f)(2)$
(c) $(f \circ f)(1)$ **(d)** $(g \circ g)(0)$

15. $f(x) = 2x$ and $g(x) = 3x^2 + 1$

16. $f(x) = \dfrac{3}{x + 1}$ and $g(x) = \sqrt{x}$

In Problems 17 and 18, evaluate each expression using the values given in the table.

17.

x	-3	-2	-1	0	1	2	3
$f(x)$	-7	-5	-3	-1	3	5	7
$g(x)$	8	3	0	-1	0	3	8

(a) $(f \circ g)(1)$ **(b)** $(f \circ g)(-1)$
(c) $(g \circ f)(-1)$ **(d)** $(g \circ f)(1)$
(e) $(g \circ g)(-2)$ **(f)** $(f \circ f)(-1)$

18.

x	-3	-2	-1	0	1	2	3
$f(x)$	11	9	7	5	3	1	-1
$g(x)$	-8	-3	0	1	0	-3	-8

(a) $(f \circ g)(1)$ **(b)** $(f \circ g)(2)$
(c) $(g \circ f)(2)$ **(d)** $(g \circ f)(3)$
(e) $(g \circ g)(1)$ **(f)** $(f \circ f)(3)$

In Problems 19 and 20, evaluate each composite function using the graphs of $y = f(x)$ and $y = g(x)$ shown in the figure below.

19. (a) $(g \circ f)(-1)$ **(b)** $(g \circ f)(6)$
   (c) $(f \circ g)(6)$ **(d)** $(f \circ g)(4)$

20. (a) $(g \circ f)(1)$ **(b)** $(g \circ f)(5)$
   (c) $(f \circ g)(7)$ **(d)** $(f \circ g)(2)$

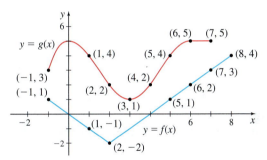

In Problems 21–26, for the given functions f and g, find:

(a) $f \circ g$ **(b)** $g \circ f$ **(c)** $f \circ f$ **(d)** $g \circ g$

State the domain of each composite function.

21. $f(x) = 3x + 1$ and $g(x) = 8x$

22. $f(x) = -x$ and $g(x) = 2x - 4$

23. $f(x) = x^2 + 1$ and $g(x) = \sqrt{x - 1}$

24. $f(x) = 2x + 3$ and $g(x) = \sqrt{x}$

25. $f(x) = \dfrac{x}{x - 1}$ and $g(x) = \dfrac{2}{x}$

26. $f(x) = \dfrac{1}{x + 3}$ and $g(x) = -\dfrac{2}{x}$

In Problems 27–32, find functions f and g so that $f \circ g = F$.

27. $F(x) = (2x + 3)^4$ **28.** $F(x) = (1 + x^2)^3$

29. $F(x) = \sqrt{x^2 + 1}$ **30.** $F(x) = \sqrt{1 - x^2}$

31. $F(x) = |2x + 1|$ **32.** $F(x) = |2x^2 + 3|$

---

1. = NOW WORK problem   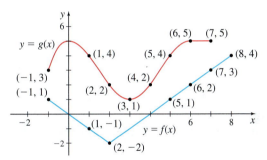 = Graphing technology recommended   CAS = Computer Algebra System recommended

In Problems 33–46, graph each function using the graphing techniques of shifting, compressing, stretching, and/or reflecting. Begin with the graph of a basic function and show all stages.

33. $f(x) = x^3 + 2$

34. $g(x) = x^3 - 1$

35. $h(x) = \sqrt{x} - 2$

36. $f(x) = \sqrt{x + 1}$

37. $g(x) = 4\sqrt{x}$

38. $f(x) = \dfrac{1}{2}\sqrt{x}$

39. $f(x) = (x - 1)^3 + 2$

40. $g(x) = 3(x - 2)^2 + 1$

41. $h(x) = \dfrac{1}{2x}$

42. $f(x) = \dfrac{4}{x} + 2$

43. $G(x) = 2|1 - x|$

44. $g(x) = -(x + 1)^3 - 1$

45. $g(x) = -4\sqrt{x - 1}$

46. $f(x) = 4\sqrt{2 - x}$

In Problems 47 and 48, the graph of a function f is illustrated. Use the graph of f as the first step in graphing each of the following functions:

(a) $F(x) = f(x) + 3$

(b) $G(x) = f(x + 2)$

(c) $P(x) = -f(x)$

(d) $H(x) = f(x + 1) - 2$

(e) $Q(x) = \dfrac{1}{2}f(x)$

(f) $g(x) = f(-x)$

(g) $h(x) = f(2x)$

47.

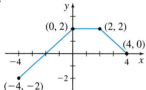

48.

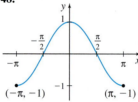

 49. Period of a Pendulum   The period T (in seconds) of a simple pendulum is a function of its length l (in meters) defined by the equation

$$T = 2\pi\sqrt{\frac{l}{g}}$$

where $g \approx 9.8$ meters/second^2 is the acceleration due to gravity.

(a) Use a graphing utility to graph the function $T = T(l)$.

(b) Now graph the functions $T = T(l + 1)$, $T = T(l + 2)$, and $T = T(l + 3)$.

(c) Discuss how adding to the length l changes the period T.

(d) Now graph the functions $T = T(2l)$, $T = T(3l)$, and $T = T(4l)$.

(e) Discuss how multiplying the length l by 2, 3, and 4 changes the period T.

50. Suppose $(1, 3)$ is a point on the graph of $y = g(x)$.

(a) What point is on the graph of $y = g(x + 3) - 5$?

(b) What point is on the graph of $y = -2g(x - 2) + 1$?

(c) What point is on the graph of $y = g(2x + 3)$?

P.4 Inverse Functions

OBJECTIVES *When you finish this section, you should be able to:*

1 Determine whether a function is one-to-one (p. 32)

2 Determine the inverse of a function defined by a set of ordered pairs (p. 34)

3 Obtain the graph of the inverse function from the graph of a one-to-one function (p. 35)

4 Find the inverse of a one-to-one function defined by an equation (p. 35)

1 Determine Whether a Function Is One-to-One

By definition, for a function $y = f(x)$, if x is in the domain of f, then x has one, and only one, image y in the range. If a function f also has the property that no y in the range of f is the image of more than one x in the domain, then the function is called a *one-to-one function*.

DEFINITION One-to-One Function

A function f is a **one-to-one function** if any two different inputs in the domain correspond to two different outputs in the range. That is, if $x_1 \neq x_2$, then $f(x_1) \neq f(x_2)$.

IN WORDS A function is not one-to-one if there are two different inputs in the domain corresponding to the same output.

Figure 48 illustrates the distinction among one-to-one functions, functions that are not one-to-one, and relations that are not functions.

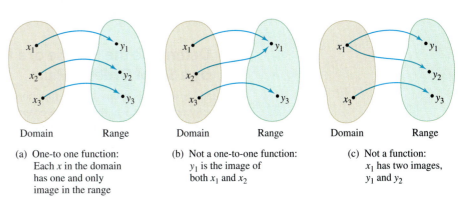

(a) One-to one function: Each x in the domain has one and only image in the range

(b) Not a one-to-one function: y_1 is the image of both x_1 and x_2

(c) Not a function: x_1 has two images, y_1 and y_2

Figure 48

If the graph of a function f is known, there is a simple test called the *Horizontal-line Test,* to determine whether f is a one-to-one function.

THEOREM Horizontal-line Test

The graph of a function in the xy-plane is the graph of a one-to-one function if and only if every horizontal line intersects the graph in at most one point.

EXAMPLE 1   Determining Whether a Function Is One-to-One

Determine whether each of these functions is one-to-one:

(a) $f(x) = x^2$ **(b)** $g(x) = x^3$

Solution (a) Figure 49 illustrates the Horizontal-line Test for the graph of $f(x) = x^2$. The horizontal line $y = 1$ intersects the graph of f twice, at $(1, 1)$ and at $(-1, 1)$, so f is not one-to-one.

(b) Figure 50 illustrates the horizontal-line test for the graph of $g(x) = x^3$. Because every horizontal line intersects the graph of g exactly once, it follows that g is one-to-one. ∎

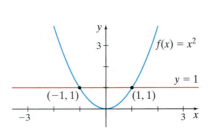

A horizontal line intersects the graph twice; f is not one-to-one

Figure 49

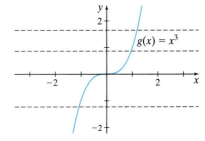

Every horizontal line intersects the graph exactly once; g is one-to-one

Figure 50

NOW WORK   Problem 9.

Notice that the one-to-one function $g(x) = x^3$ also is an increasing function on its domain. Because an increasing (or decreasing) function will always have different y-values for different x-values, a function that is increasing (or decreasing) on an interval is also a one-to-one function on that interval.

THEOREM One-to-One Function

- A function that is increasing on an interval I is a one-to-one function on I.
- A function that is decreasing on an interval I is a one-to-one function on I.

2 Determine the Inverse of a Function Defined by a Set of Ordered Pairs

Suppose that f is a one-to-one function. Then to each x in the domain of f, there is exactly one image y in the range (because f is a function); and to each y in the range of f, there is exactly one x in the domain (because f is one-to-one). The correspondence from the range of f back to the domain of f is also a function, called the *inverse function of f*. The symbol f^{-1} is used to denote the inverse of f.

DEFINITION Inverse Function

Let f be a one-to-one function. The **inverse of** f, denoted by f^{-1}, is the function defined on the range of f for which

$$\boxed{x = f^{-1}(y) \quad \text{if and only if} \quad y = f(x)}$$

NOTE f^{-1} is not the reciprocal function. That is, $f^{-1}(x) \neq \dfrac{1}{f(x)}$. The reciprocal function $\dfrac{1}{f(x)}$ is written $[f(x)]^{-1}$.

We will discuss how to find inverses for three representations of functions: (1) sets of ordered pairs, (2) graphs, and (3) equations. We begin with finding the inverse of a function represented by a set of ordered pairs.

If the function f is a set of ordered pairs (x, y), then the inverse of f, denoted f^{-1}, is the set of ordered pairs (y, x).

EXAMPLE 2 Finding the Inverse of a Function Defined by a Set of Ordered Pairs

Find the inverse of the one-to-one function:

$$\{(-3, -5), (-1, 1), (0, 2), (1, 3)\}$$

State the domain and the range of the function and its inverse.

Solution The inverse of the function is found by interchanging the entries in each ordered pair. The inverse is

$$\{(-5, -3), (1, -1), (2, 0), (3, 1)\}$$

The domain of the function is $\{-3, -1, 0, 1\}$; the range of the function is $\{-5, 1, 2, 3\}$. The domain of the inverse function is $\{-5, 1, 2, 3\}$; the range of the inverse function is $\{-3, -1, 0, 1\}$. ∎

Figure 51 shows the one-to-one function (in blue) and its inverse (in red).

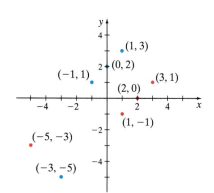

Figure 51

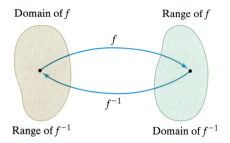

Figure 52

NOW WORK Problem 19.

Remember, if f is a one-to-one function, it has an inverse f^{-1}. See Figure 52. Based on the results of Example 2 and Figure 51, two properties of a one-to-one function f and its inverse function f^{-1} become apparent:

Domain of f = Range of f^{-1} Range of f = Domain of f^{-1}

The next theorem provides a means for verifying that two functions are inverses of one another.

THEOREM

Given a one-to-one function f and its inverse function f^{-1}, then

- $(f^{-1} \circ f)(x) = f^{-1}(f(x)) = x$ where x is in the domain of f
- $(f \circ f^{-1})(x) = f(f^{-1}(x)) = x$ where x is in the domain of f^{-1}

EXAMPLE 3 Verifying Inverse Functions

Verify that the inverse of $f(x) = \dfrac{1}{x-1}$ is $f^{-1}(x) = \dfrac{1}{x} + 1$. For what values of x is $f^{-1}(f(x)) = x$? For what values of x is $f(f^{-1}(x)) = x$?

Solution The domain of f is $\{x \mid x \neq 1\}$ and the domain of f^{-1} is $\{x \mid x \neq 0\}$. Now

$$f^{-1}(f(x)) = f^{-1}\left(\frac{1}{x-1}\right) = \frac{1}{\left(\dfrac{1}{x-1}\right)} + 1 = x - 1 + 1 = x, \quad \text{provided } x \neq 1$$

$$f(f^{-1}(x)) = f\left(\frac{1}{x} + 1\right) = \frac{1}{\left(\dfrac{1}{x} + 1\right) - 1} = \frac{1}{\dfrac{1}{x}} = x, \qquad\qquad \text{provided } x \neq 0 \quad\blacksquare$$

NOW WORK Problem 15.

3 Obtain the Graph of the Inverse Function from the Graph of a One-to-One Function

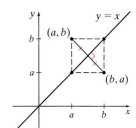

Figure 53

Suppose (a, b) is a point on the graph of a one-to-one function f defined by $y = f(x)$. Then $b = f(a)$. This means that $a = f^{-1}(b)$, so (b, a) is a point on the graph of the inverse function f^{-1}. Figure 53 shows the relationship between the point (a, b) on the graph of f and the point (b, a) on the graph of f^{-1}. The line segment containing (a, b) and (b, a) is perpendicular to the line $y = x$ and is bisected by the line $y = x$. (Do you see why?) The point (b, a) on the graph of f^{-1} is the reflection about the line $y = x$ of the point (a, b) on the graph of f.

THEOREM Symmetry of Inverse Functions

The graph of a one-to-one function f and the graph of its inverse function f^{-1} are symmetric with respect to the line $y = x$.

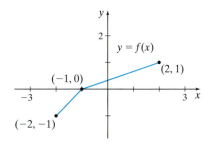

Figure 54

We can use this result to find the graph of f^{-1} given the graph of f. If we know the graph of f, then the graph of f^{-1} is obtained by reflecting the graph of f about the line $y = x$. See Figure 54.

EXAMPLE 4 Graphing the Inverse Function from the Graph of a Function

The graph in Figure 55 is that of a one-to-one function $y = f(x)$. Draw the graph of its inverse function.

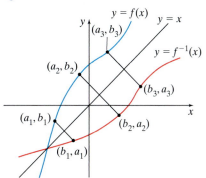

Figure 55

Solution Since the points $(-2, -1)$, $(-1, 0)$, and $(2, 1)$ are on the graph of f, the points $(-1, -2)$, $(0, -1)$, and $(1, 2)$ are on the graph of f^{-1}. Using the points and the fact that the graph of f^{-1} is the reflection about the line $y = x$ of the graph of f, draw the graph of f^{-1}, as shown in Figure 56. $\blacksquare$

NOW WORK Problem 27.

4 Find the Inverse of a One-to-One Function Defined by an Equation

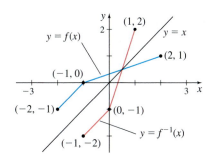

Figure 56

Since the graphs of a one-to-one function f and its inverse function f^{-1} are symmetric with respect to the line $y = x$, the inverse function f^{-1} can be obtained by interchanging the roles of x and y in f. If f is defined by the equation

$$y = f(x)$$

then f^{-1} is defined by the equation

$$x = f(y) \qquad \text{Interchange } x \text{ and } y$$

The equation $x = f(y)$ defines f^{-1} *implicitly*. If the implicit equation can be solved for y, we will have the *explicit* form of f^{-1}, that is,

$$y = f^{-1}(x)$$

Steps for Finding the Inverse of a One-to-One Function

Step 1 Write f in the form $y = f(x)$.

Step 2 Interchange the variables x and y to obtain $x = f(y)$ This equation defines the inverse function f^{-1} implicitly.

Step 3 If possible, solve the implicit equation for y in terms of x to obtain the explicit form of f^{-1}: $y = f^{-1}(x)$.

Step 4 Check the result by showing that $f^{-1}(f(x)) = x$ and $f(f^{-1}(x)) = x$.

EXAMPLE 5   **Finding the Inverse Function**

The function $f(x) = 2x^3 - 1$ is one-to-one. Find its inverse.

Solution We follow the steps given above.

Step 1 Write f as $y = 2x^3 - 1$.

Step 2 Interchange the variables x and y.

$$x = 2y^3 - 1$$

This equation defines f^{-1} implicitly.

Step 3 Solve the implicit form of the inverse function for y.

$$x + 1 = 2y^3$$

$$y^3 = \frac{x+1}{2}$$

$$y = \sqrt[3]{\frac{x+1}{2}} = f^{-1}(x)$$

Step 4 Check the result.

$$f^{-1}(f(x)) = f^{-1}(2x^3 - 1) = \sqrt[3]{\frac{(2x^3 - 1) + 1}{2}} = \sqrt[3]{\frac{2x^3}{2}} = \sqrt[3]{x^3} = x$$

$$f(f^{-1}(x)) = f\left(\sqrt[3]{\frac{x+1}{2}}\right) = 2\left(\sqrt[3]{\frac{x+1}{2}}\right)^3 - 1 = 2\left(\frac{x+1}{2}\right) - 1$$

$$= x + 1 - 1 = x$$

See Figure 57 for the graphs of f and f^{-1}.

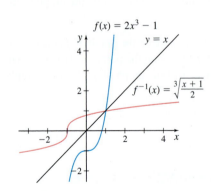

$f(x) = 2x^3 - 1$
$y = x$
$f^{-1}(x) = \sqrt[3]{\frac{x+1}{2}}$

Figure 57

NOW WORK   **Problem 31.**

If a function f is not one-to-one, it has no inverse function. But sometimes we can restrict the domain of such a function so that it is a one-to-one function. Then on the restricted domain the new function has an inverse function.

EXAMPLE 6   **Finding the Inverse of a Domain-Restricted Function**

Find the inverse of $f(x) = x^2$ if $x \geq 0$.

Solution The function $f(x) = x^2$ is not one-to-one (see Example 1(a)). However, by restricting the domain of f to $x \geq 0$, the new function f is one-to-one, so f^{-1} exists. To find f^{-1}, follow the steps.

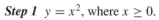

Step 1 $y = x^2$, where $x \geq 0$.

Step 2 Interchange the variables x and y: $x = y^2$, where $y \geq 0$. This is the inverse function written implicitly.

Step 3 Solve for y: $y = \sqrt{x} = f^{-1}(x)$. (Since $y \geq 0$, only the principal square root is obtained.)

Step 4 Check that $f^{-1}(x) = \sqrt{x}$ is the inverse function of f.

$$f^{-1}(f(x)) = \sqrt{f(x)} = \sqrt{x^2} = |x| = x, \qquad \text{where } x \geq 0$$
$$f(f^{-1}(x)) = [f^{-1}(x)]^2 = [\sqrt{x}]^2 = x, \qquad \text{where } x \geq 0 \qquad ■$$

The graphs of $f(x) = x^2$, $x \geq 0$, and $f^{-1}(x) = \sqrt{x}$ are shown in Figure 58.

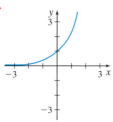

Figure 58

NEED TO REVIEW? Principal roots are discussed in Appendix A.1, p. A-9.

NOW WORK   **Problem 37.**

P.4 Assess Your Understanding

Concepts and Vocabulary

1. *True or False* If every vertical line intersects the graph of a function f at no more than one point, f is a one-to-one function.

2. If the domain of a one-to-one function f is $[4, \infty)$, the range of its inverse function f^{-1} is _____.

3. *True or False* If f and g are inverse functions, the domain of f is the same as the domain of g.

4. *True or False* If f and g are inverse functions, their graphs are symmetric with respect to the line $y = x$.

5. *True or False* If f and g are inverse functions, then $(f \circ g)(x) = f(x) \cdot g(x)$.

6. *True or False* If a function f is one-to-one, then $f(f^{-1}(x)) = x$, where x is in the domain of f.

7. Given a collection of points (x, y), explain how you would determine if it represents a one-to-one function $y = f(x)$.

8. Given the graph of a one-to-one function $y = f(x)$, explain how you would graph the inverse function f^{-1}.

Practice Problems

In Problems 9–14, the graph of a function f is given. Use the Horizontal-line Test to determine whether f is one-to one.

9.

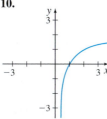

10.

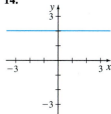

11.

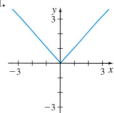

12.

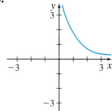

13.

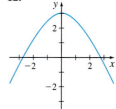

14.

In Problems 15–18, verify that the functions f and g are inverses of each other by showing that $(f \circ g)(x) = x$ and $(g \circ f)(x) = x$.

15. $f(x) = 3x + 4$; $g(x) = \dfrac{1}{3}(x - 4)$

16. $f(x) = x^3 - 8$; $g(x) = \sqrt[3]{x + 8}$

17. $f(x) = \dfrac{1}{x}$; $g(x) = \dfrac{1}{x}$

18. $f(x) = \dfrac{2x + 3}{x + 4}$; $g(x) = \dfrac{4x - 3}{2 - x}$

1. = NOW WORK problem   = Graphing technology recommended   CAS = Computer Algebra System recommended

*In Problems 19–22, **(a)** determine whether the function is one-to-one. If it is one-to-one, **(b)** find the inverse of each one-to-one function. **(c)** State the domain and the range of the function and its inverse.*

19. $\{(-3, 5), (-2, 9), (-1, 2), (0, 11), (1, -5)\}$

20. $\{(-2, 2), (-1, 6), (0, 8), (1, -3), (2, 8)\}$

21. $\{(-2, 1), (-3, 2), (-10, 0), (1, 9), (2, 1)\}$

22. $\{(-2, -8), (-1, -1), (0, 0), (1, 1), (2, 8)\}$

In Problems 23–28, the graph of a one-to-one function f is given. Draw the graph of the inverse function. For convenience, the graph of $y = x$ is also given.

23.

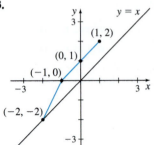

24.

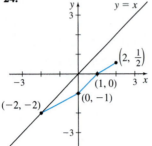

25.

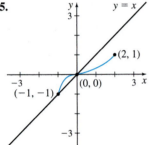

26.

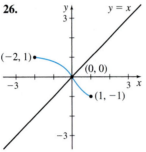

27.

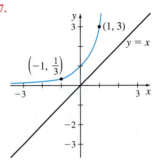

28.

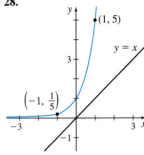

In Problems 29–38, the function f is one-to-one.

(a) Find its inverse and check the result.

(b) Find the domain and the range of f and the domain and the range of f^{-1}.

29. $f(x) = 4x + 2$

30. $f(x) = 1 - 3x$

31. $f(x) = \sqrt[3]{x + 10}$

32. $f(x) = 2x^3 + 4$

33. $f(x) = \dfrac{1}{x - 2}$

34. $f(x) = \dfrac{2x}{3x - 1}$

35. $f(x) = \dfrac{2x + 3}{x + 2}$

36. $f(x) = \dfrac{-3x - 4}{x - 2}$

37. $f(x) = x^2 + 4, x \geq 0$

38. $f(x) = (x - 2)^2 + 4, x \leq 2$

P.5 Exponential and Logarithmic Functions

OBJECTIVES *When you finish this section, you should be able to:*

1 Analyze an exponential function (p. 38)

2 Define the number e (p. 41)

3 Analyze a logarithmic function (p. 42)

4 Solve exponential equations and logarithmic equations (p. 45)

Here we begin our study of transcendental functions with the exponential and logarithmic functions. In Sections P.6 and P.7, we investigate the trigonometric functions and their inverse functions.

1 Analyze an Exponential Function

The expression a^r, where $a > 0$ is a fixed real number and $r = \dfrac{m}{n}$ is a rational number, in lowest terms with $n \geq 2$, is defined as

$$a^r = a^{m/n} = (a^{1/n})^m = (\sqrt[n]{a})^m$$

So, a function $f(x) = a^x$ can be defined so that its domain is the set of rational numbers. Our aim is to expand the domain of f to include both rational and irrational numbers, that is, to include all real numbers.

CAUTION Be careful to distinguish an exponential function $f(x) = a^x$, where $a > 0$ and $a \neq 1$, from a power function $g(x) = x^a$, where a is a real number. In $f(x) = a^x$ the independent variable x is the *exponent*; in $g(x) = x^a$ the independent variable x is the *base*.

Every irrational number x can be approximated by a rational number r formed by truncating (removing) all but a finite number of digits from x. For example, for $x = \pi$, we could use the rational numbers $r = 3.14$ or $r = 3.14159$, and so on. The closer r is to π, the better approximation a^r is to a^π. In general, we can make a^r as close as we please to a^x by choosing r sufficiently close to x. Using this argument, we can define an *exponential function* $f(x) = a^x$, where x includes all the rational numbers and all the irrational numbers.

DEFINITION Exponential Function

An **exponential function** is a function that can be expressed in the form

$$f(x) = a^x$$

where a is a positive real number and $a \neq 1$. The domain of f is the set of all real numbers.

NOTE The base $a = 1$ is excluded from the definition of an exponential function because $f(x) = 1^x = 1$ (a constant function). Bases that are negative are excluded because $a^{1/n}$, where $a < 0$ and n is an even integer, is not defined.

Examples of exponential functions are $f(x) = 2^x$, $g(x) = \left(\dfrac{2}{3}\right)^x$, and $h(x) = \pi^x$.

Consider the exponential function $f(x) = 2^x$. The domain of f is all real numbers; the range of f is the interval $(0, \infty)$. Some points on the graph of f are listed in Table 4. Since $2^x > 0$ for all x, the graph of f lies above the x-axis and has no x-intercept. The y-intercept is 1. Using this information, plot some points from Table 4 and connect them with a smooth curve, as shown in Figure 59.

TABLE 4

x	$f(x) = 2^x$	(x, y)
-10	$2^{-10} \approx 0.00098$	$(-10, 0.00098)$
-3	$2^{-3} = \dfrac{1}{8}$	$\left(-3, \dfrac{1}{8}\right)$
-2	$2^{-2} = \dfrac{1}{4}$	$\left(-2, \dfrac{1}{4}\right)$
-1	$2^{-1} = \dfrac{1}{2}$	$\left(-1, \dfrac{1}{2}\right)$
0	$2^0 = 1$	$(0, 1)$
1	$2^1 = 2$	$(1, 2)$
2	$2^2 = 4$	$(2, 4)$
3	$2^3 = 8$	$(3, 8)$
10	$2^{10} = 1024$	$(10, 1024)$

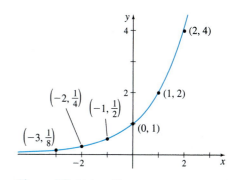

Figure 59 $f(x) = 2^x$

The graph of $f(x) = 2^x$ is typical of all exponential functions of the form $f(x) = a^x$ with $a > 1$, a few of which are graphed in Figure 60.

Notice that every graph in Figure 60 lies above the x-axis, passes through the point $(0, 1)$, and is increasing. Also notice that the graphs with larger bases are steeper when $x > 0$, but when $x < 0$, the graphs with larger bases are closer to the x-axis.

$y = 4^x$
$y = 3^x$
$y = 2^x$
$y = 1.5^x$
$(1, 4)$
$(1, 3)$
$(1, 2)$
$(1, 1.5)$
$\left(-1, \dfrac{2}{3}\right)$
$(0, 1)$
$\left(-1, \dfrac{1}{4}\right)$

Figure 60 $f(x) = a^x$; $a > 1$

All functions of the type $f(x) = a^x$, $a > 1$, have the following properties:

> **Properties of an Exponential Function** $f(x) = a^x$, $a > 1$
>
> - The domain is the set of all real numbers; the range is the set of positive real numbers.
> - There are no x-intercepts; the y-intercept is 1.
> - The exponential function f is increasing on the interval $(-\infty, \infty)$.
> - The graph of f contains the points $\left(-1, \dfrac{1}{a}\right)$, $(0, 1)$, and $(1, a)$.
> - $\dfrac{f(x+1)}{f(x)} = a$
> - Because $f(x) = a^x$ is a function, if $u = v$, then $a^u = a^v$.
> - Because $f(x) = a^x$ is a one-to-one function, if $a^u = a^v$, then $u = v$.

THEOREM Laws of Exponents

If u, v, a, and b are real numbers with $a > 0$ and $b > 0$, then

$$a^u \cdot a^v = a^{u+v} \quad \frac{a^u}{a^v} = a^{u-v} \quad (a^u)^v = a^{uv} \quad (ab)^u = a^u \cdot b^u \quad \left(\frac{a}{b}\right)^u = \frac{a^u}{b^u}$$

NEED TO REVIEW? The laws of exponents are discussed in Appendix A.1, pp. A-8 to A-9.

For example, we can use the Laws of Exponents to show the following property of an exponential function:

$$\frac{f(x+1)}{f(x)} = \frac{a^{x+1}}{a^x} = a^{(x+1)-x} = a^1 = a$$

EXAMPLE 1 Graphing an Exponential Function

Graph the exponential function $g(x) = \left(\dfrac{1}{2}\right)^x$.

Solution We begin by writing $\dfrac{1}{2}$ as 2^{-1}. Then

$$g(x) = \left(\frac{1}{2}\right)^x = (2^{-1})^x = 2^{-x}$$

Now we use the graph of $f(x) = 2^x$ shown in Figure 59 and reflect it about the y-axis to obtain the graph $g(x) = 2^{-x}$. See Figure 61.

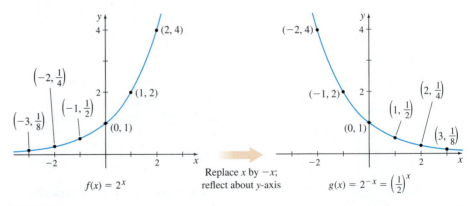

Replace x by $-x$;
reflect about y-axis

$f(x) = 2^x$

$g(x) = 2^{-x} = \left(\dfrac{1}{2}\right)^x$

Figure 61   ∎

NOW WORK Problem 19.

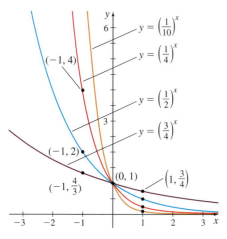

Figure 62 $f(x) = a^x, 0 < a < 1$

The graph of $g(x) = \left(\dfrac{1}{2}\right)^x$ in Figure 61 is typical of all exponential functions that have a base between 0 and 1. Figure 62 illustrates the graphs of several more exponential functions whose bases are between 0 and 1. Notice that the graphs with smaller bases are steeper when $x < 0$, but when $x > 0$, these graphs are closer to the x-axis.

Properties of an Exponential Function $f(x) = a^x, 0 < a < 1$

- The domain is the set of all real numbers; the range is the set of positive real numbers.
- There are no x-intercepts; the y-intercept is 1.
- The exponential function f is decreasing on the interval $(-\infty, \infty)$.
- The graph of f contains the points $\left(-1, \dfrac{1}{a}\right)$, $(0, 1)$, and $(1, a)$.
- $\dfrac{f(x+1)}{f(x)} = a$
- If $u = v$, then $a^u = a^v$.
- If $a^u = a^v$, then $u = v$.

EXAMPLE 2 Graphing an Exponential Function Using Transformations

Graph $f(x) = 3^{-x} - 2$ and determine the domain and range of f.

Solution We begin with the graph of $y = 3^x$. Figure 63 shows the steps.

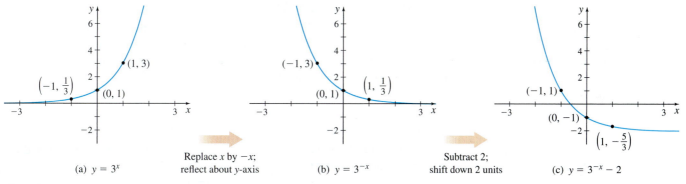

Figure 63

The domain of f is all real numbers, the range of f is the interval $(-2, \infty)$. ∎

NOW WORK Problem 27.

2 Define the Number e

In earlier courses you learned about an irrational number, called e. The number e is important because it appears in many applications and because it has properties that simplify computations in calculus.

To define e, consider the graphs of the functions $y = 2^x$ and $y = 3^x$ in Figure 64(a), where we have carefully drawn lines that just touch each graph at the point $(0, 1)$. (These lines are *tangent lines*, which we discuss in Chapter 1.) Notice that the slope of the tangent line to $y = 3^x$ is greater than 1 (approximately 1.10) and that the slope of the tangent line to the graph of $y = 2^x$ is less than 1 (approximately 0.69). Between these graphs there is an exponential function $y = a^x$, whose base is between 2 and 3, and whose tangent line to the graph at the point (0, 1) has a slope of exactly 1, as shown in Figure 64(b). The base of this exponential function is the number e.

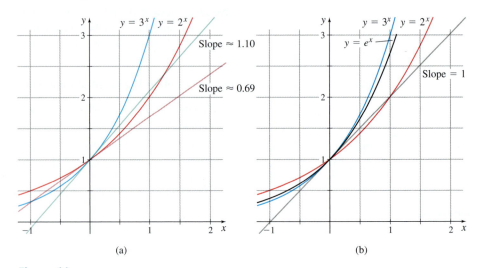

(a)   (b)

Figure 64

ORIGINS The irrational number we call e was named to honor the Swiss mathematician Leonhard Euler (1707–1783), who is considered one of the greatest mathematicians of all time. Euler was encouraged to study mathematics by his father. Considered the most prolific mathematician who ever lived, he made major contributions to many areas of mathematics. One of Euler's greatest gifts was his ability to explain difficult mathematical concepts in simple language.

DEFINITION The number e

The **number** e is defined as the base of the exponential function whose tangent line to the graph at the point (0, 1) has slope 1. The function $f(x) = e^x$ occurs with such frequency that it is usually referred to as *the* **exponential function**.

The number e is an irrational number, and in Chapter 3, we will show $e \approx 2.71828$.

3 Analyze a Logarithmic Function

Recall that a one-to-one function $y = f(x)$ has an inverse function that is defined implicitly by the equation $x = f(y)$. Since an exponential function $y = f(x) = a^x$, where $a > 0$ and $a \neq 1$, is a one-to-one function, it has an inverse function, called a *logarithmic function,* that is defined implicitly by the equation

$$x = a^y \qquad \text{where } a > 0 \quad \text{and} \quad a \neq 1$$

NEED TO REVIEW? Logarithms and their properties are discussed in Appendix A.1, pp. A-10 to A-11.

DEFINITION Logarithmic Function

The **logarithmic function with base** a, where $a > 0$ and $a \neq 1$, is denoted by $y = \log_a x$ and is defined by

$$y = \log_a x \quad \text{if and only if} \quad x = a^y$$

The domain of the logarithmic function $y = \log_a x$ is $x > 0$.

IN WORDS A logarithm is an exponent. That is, if $y = \log_a x$, then y is the exponent in $x = a^y$.

In other words, if $f(x) = a^x$, where $a > 0$ and $a \neq 1$, its inverse function is $f^{-1}(x) = \log_a x$.

If the base of a logarithmic function is the number e, then it is called the **natural logarithmic function**, and it is given a special symbol, **ln** (from the Latin, *logarithmus naturalis*). That is,

$$y = \ln x \quad \text{if and only if} \quad x = e^y$$

Since the exponential function $y = a^x$, $a > 0, a \neq 1$ and the logarithmic function $y = \log_a x$ are inverse functions, the following properties hold:

- $\log_a(a^x) = x$ for all real numbers x
- $a^{\log_a x} = x$ for all $x > 0$

Because exponential functions and logarithmic functions are inverses of each other, the graph of the logarithmic function $y = \log_a x$, $a > 0$ and $a \neq 1$, is the reflection of the graph of the exponential function $y = a^x$, about the line $y = x$, as shown in Figure 65.

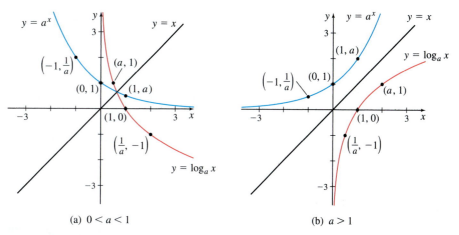

(a) $0 < a < 1$ (b) $a > 1$

Figure 65

Based on the graphs in Figure 65, we see that

$$\log_a 1 = 0 \qquad \log_a a = 1 \qquad a > 0 \quad \text{and} \quad a \neq 1$$

EXAMPLE 3   Graphing a Logarithmic Function

Graph:

(a) $f(x) = \log_2 x$ **(b)** $g(x) = \log_{1/3} x$ **(c)** $F(x) = \ln x$

Solution **(a)** To graph $f(x) = \log_2 x$, graph $y = 2^x$ and reflect it about the line $y = x$. See Figure 66(a).

(b) To graph $f(x) = \log_{1/3} x$, graph $y = \left(\dfrac{1}{3}\right)^x$ and reflect it about the line $y = x$. See Figure 66(b).

(c) To graph $F(x) = \ln x$, graph $y = e^x$ and reflect it about the line $y = x$. See Figure 66(c).

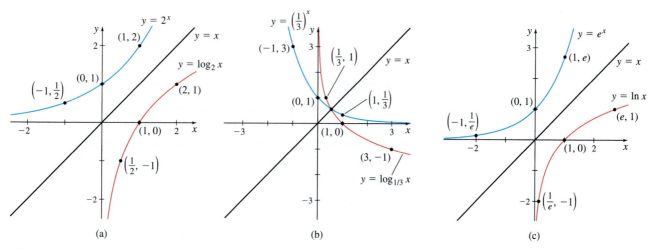

(a)   (b)   (c)

Figure 66

NOW WORK   **Problem 39.**

Since a logarithmic function is the inverse of an exponential function, it follows that:

- Domain of the logarithmic function = Range of the exponential function
$$= (0, \infty)$$
- Range of the logarithmic function = Domain of the exponential function
$$= (-\infty, \infty)$$

The domain of a logarithmic function is the set of *positive* real numbers, so the argument of a logarithmic function must be greater than zero.

EXAMPLE 4   Finding the Domain of a Logarithmic Function

Find the domain of each function:

(a) $F(x) = \log_2(x + 3)$ (b) $g(x) = \ln\left(\dfrac{1+x}{1-x}\right)$ (c) $h(x) = \log_{1/2}|x|$

Solution (a) The argument of a logarithm must be positive. So to find the domain of $F(x) = \log_2(x + 3)$, we solve the inequality $x + 3 > 0$. The domain of F is $\{x \mid x > -3\}$.

NEED TO REVIEW? Solving inequalities is discussed in Appendix A.1, pp. A-5 to A-8.

(b) Since $\ln\left(\dfrac{1+x}{1-x}\right)$ requires $\dfrac{1+x}{1-x} > 0$, we find the domain of g by solving the inequality $\dfrac{1+x}{1-x} > 0$. Since $\dfrac{1+x}{1-x}$ is not defined for $x = 1$, and the solution to the equation $\dfrac{1+x}{1-x} = 0$ is $x = -1$, we use -1 and 1 to separate the real number line into three intervals $(-\infty, -1)$, $(-1, 1)$, and $(1, \infty)$. Then we choose a test number in each interval, and evaluate the rational expression $\dfrac{1+x}{1-x}$ at these numbers to determine if the expression is positive or negative. For example, we chose the numbers -2, 0, and 2 and found that $\dfrac{1+x}{1-x} > 0$ on the interval $(-1, 1)$. See the table on the left. So the domain of $g(x) = \ln\left(\dfrac{1+x}{1-x}\right)$ is $\{x \mid -1 < x < 1\}$.

Interval	Test Number	Sign of $\dfrac{1+x}{1-x}$
$(-\infty, -1)$	-2	Negative
$(-1, 1)$	0	Positive
$(1, \infty)$	2	Negative

(c) $\log_{1/2}|x|$ requires $|x| > 0$. So the domain of $h(x) = \log_{1/2}|x|$ is $\{x \mid x \neq 0\}$. ∎

NOW WORK Problem 31.

Properties of a Logarithmic Function $f(x) = \log_a x$, $a > 0$ and $a \neq 1$

- The domain of f is the set of all positive real numbers; the range is the set of all real numbers.
- The x-intercept of the graph of f is 1. There is no y-intercept.
- A logarithmic function is decreasing on the interval $(0, \infty)$ if $0 < a < 1$ and increasing on the interval $(0, \infty)$ if $a > 1$.
- The graph of f contains the points $(1, 0)$, $(a, 1)$, and $\left(\dfrac{1}{a}, -1\right)$.
- Because $f(x) = \log_a x$ is a function, if $u = v$, then $\log_a u = \log_a v$.
- Because $f(x) = \log_a x$ is a one-to-one function, if $\log_a u = \log_a v$, then $u = v$.
- See Figure 67 for typical graphs with base a, $a > 1$; see Figure 68 for graphs with base a, $0 < a < 1$.

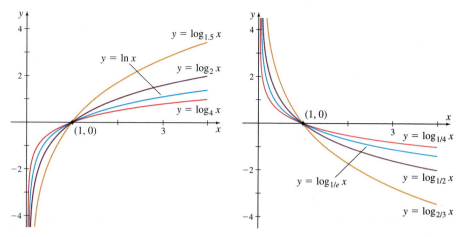

Figure 67 $y = \log_a x, a > 1$ **Figure 68** $y = \log_a x, 0 < a < 1$

4 Solve Exponential Equations and Logarithmic Equations

Equations that involve terms of the form a^x, $a > 0$, $a \neq 1$, are referred to as **exponential equations**. For example, the equation of the form $5^{2x+3} = 5^x$ is an exponential equation.

The one-to-one property of exponential functions

$$\text{if } a^u = a^v \quad \text{then } u = v \tag{1}$$

can be used to solve certain kinds of exponential equations.

> **IN WORDS** In an exponential equation, if the bases are equal, then the exponents are equal.

EXAMPLE 5   Solving Exponential Equations

Solve each exponential equation:

(a) $4^{2x-1} = 8^{x+3}$ **(b)** $e^{-x^2} = (e^x)^2 \cdot \dfrac{1}{e^3}$

Solution **(a)** We begin by expressing both sides of the equation with the same base so we can use the one-to-one property (1).

$$4^{2x-1} = 8^{x+3}$$
$$(2^2)^{2x-1} = (2^3)^{x+3} \qquad 4 = 2^2, 8 = 2^3$$
$$2^{2(2x-1)} = 2^{3(x+3)} \qquad (a^r)^s = a^{rs}$$
$$2(2x-1) = 3(x+3) \qquad \text{If } a^u = a^v, \text{ then } u = v.$$
$$4x - 2 = 3x + 9 \qquad \text{Simplify.}$$
$$x = 11 \qquad \text{Solve.}$$

The solution is 11.

(b) We use the Laws of Exponents to obtain the base e on the right side.

$$(e^x)^2 \cdot \frac{1}{e^3} = e^{2x} \cdot e^{-3} = e^{2x-3}$$

As a result,

$$e^{-x^2} = e^{2x-3}$$
$$-x^2 = 2x - 3 \qquad \text{If } a^u = a^v, \text{ then } u = v.$$
$$x^2 + 2x - 3 = 0$$
$$(x + 3)(x - 1) = 0$$
$$x = -3 \quad \text{or} \quad x = 1$$

The solution set is $\{-3, 1\}$. ∎

NOW WORK Problem 49.

To use the one-to-one property of exponential functions, each side of the equation must be written with the same base. Since for many exponential equations it is not possible to write each side with the same base, we need a different strategy to solve such equations.

EXAMPLE 6   Solving Exponential Equations

Solve the exponential equations:

(a) $10^{2x} = 50$ (b) $8 \cdot 3^x = 5$

Solution (a) Since 10 and 50 cannot be written with the same base, we write the exponential equation as a logarithm.

$$10^{2x} = 50 \quad \text{if and only if} \quad \log 50 = 2x$$

RECALL that logarithms to the base 10 are called **common logarithms** and are written without a subscript. That is, $x = \log_{10} y$ is written $x = \log y$.

Then $x = \dfrac{\log 50}{2}$ is an exact solution of the equation. Using a calculator, an approximate solution is $x = \dfrac{\log 50}{2} \approx 0.849$.

(b) It is impossible to write 8 and 5 as a power of 3, so we write the exponential equation as a logarithm.

$$8 \cdot 3^x = 5$$

$$3^x = \frac{5}{8}$$

$$\log_3 \frac{5}{8} = x$$

NEED TO REVIEW? The change-of-base formula, $\log_a u = \dfrac{\log_b u}{\log_b a}$, $a \neq 1$, $b \neq 1$, and u positive real numbers, is discussed in Appendix A.1, p. A-11.

Now we use the change-of-base formula to obtain the exact solution of the equation. An approximate solution can then be obtained using a calculator.

$$x = \log_3 \frac{5}{8} = \frac{\ln \frac{5}{8}}{\ln 3} \approx -0.428 \qquad \blacksquare$$

Alternatively, we could have solved each of the equations in Example 6 by taking the natural logarithm (or the common logarithm) of each side. For example,

$$8 \cdot 3^x = 5$$

$$3^x = \frac{5}{8}$$

$$\ln 3^x = \ln \frac{5}{8} \qquad \text{If } u = v, \text{ then } \log_a u = \log_a v.$$

$$x \ln 3 = \ln \frac{5}{8} \qquad \log_a u^r = r \log_a u$$

$$x = \frac{\ln \frac{5}{8}}{\ln 3}$$

NOW WORK Problem 51.

Equations that contain logarithms are called **logarithmic equations**. Care must be taken when solving logarithmic equations algebraically. In the expression $\log_a y$, remember that a and y are positive and $a \neq 1$. Be sure to check each apparent solution in the original equation and to discard any solutions that are extraneous.

Some logarithmic equations can be solved by changing the logarithmic equation to an exponential equation using the fact that $x = \log_a y$ if and only if $y = a^x$.

EXAMPLE 7 Solving Logarithmic Equations

Solve each equation:

(a) $\log_3(4x - 7) = 2$ **(b)** $\log_x 64 = 2$

Solution (a) We change the logarithmic equation to an exponential equation.

$$\log_3(4x - 7) = 2$$
$$4x - 7 = 3^2 \qquad \text{Change to an exponential equation.}$$
$$4x - 7 = 9$$
$$4x = 16$$
$$x = 4$$

Check: For $x = 4$, $\log_3(4x - 7) = \log_3(4 \cdot 4 - 7) = \log_3 9 = 2$, since $3^2 = 9$.
The solution is 4.

(b) We change the logarithmic equation to an exponential equation.

$$\log_x 64 = 2$$
$$x^2 = 64 \qquad \text{Change to an exponential equation.}$$
$$x = 8 \quad \text{or} \quad x = -8 \qquad \text{Solve.}$$

The base of a logarithm is always positive. As a result, we discard -8 and check the solution 8.

Check: For $x = 8$, $\log_8 64 = 2$, since $8^2 = 64$.
The solution is 8. ■

NOW WORK Problem 57.

The properties of logarithms that result from the fact that a logarithmic function is one-to-one can be used to solve some equations that contain two logarithms with the same base.

EXAMPLE 8 Solving a Logarithmic Equation

Solve the logarithmic equation $2 \ln x = \ln 9$.

Solution Each logarithm has the same base, so

$$2 \ln x = \ln 9$$
$$\ln x^2 = \ln 9 \qquad r \log_a u = \log_a u^r$$
$$x^2 = 9 \qquad \text{If } \log_a u = \log_a v, \text{ then } u = v.$$
$$x = 3 \quad \text{or} \quad x = -3$$

We discard the solution $x = -3$ since -3 is not in the domain of $f(x) = \ln x$. The solution is 3. ■

NOW WORK Problem 61.

Although, each of these equations was relatively easy to solve, this is not generally the case. Many solutions to exponential and logarithmic equations need to be approximated using technology.

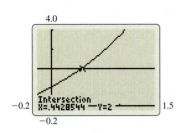

Figure 69

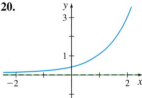 **EXAMPLE 9   Approximating the Solution to an Exponential Equation**

Solve $x + e^x = 2$. Express the solution rounded to three decimal places.

Solution We can approximate the solution to the equation by graphing the two functions $Y_1 = x + e^x$ and $Y_2 = 2$. Then we use graphing technology to approximate the intersection of the graphs. Since the function Y_1 is increasing (do you know why?) and the function Y_2 is constant, there will be only one point of intersection. Figure 69 shows the graphs of the two functions and their intersection. They intersect when $x \approx 0.4428544$, so the solution of the equation is 0.443 rounded to three decimal places. ∎

NOW WORK   Problem 65.

P.5 Assess Your Understanding

Concepts and Vocabulary

1. The graph of every exponential function $f(x) = a^x$, $a > 0$ and $a \neq 1$, passes through three points: _____, _____, and _____.

2. *True or False*   The graph of the exponential function
$f(x) = \left(\dfrac{3}{2}\right)^x$ is decreasing.

3. If $3^x = 3^4$, then $x = $ _____.

4. If $4^x = 8^2$ then $x = $ _____.

5. *True or False*   The graphs of $y = 3^x$ and $y = \left(\dfrac{1}{3}\right)^x$ are symmetric with respect to the line $y = x$.

6. *True or False*   The range of the exponential function $f(x) = a^x$, $a > 0$ and $a \neq 1$, is the set of all real numbers.

7. The number e is defined as the base of the exponential function f whose tangent line to the graph of f at the point $(0, 1)$ has slope _____.

8. The domain of the logarithmic function $f(x) = \log_a x$ is _____.

9. The graph of every logarithmic function $f(x) = \log_a x$, $a > 0$ and $a \neq 1$, passes through three points: _____, _____, and _____.

10. *Multiple Choice*   The graph of $f(x) = \log_2 x$ is [(a) increasing, (b) decreasing, (c) neither].

11. *True or False*   If $y = \log_a x$, then $y = a^x$.

12. *True or False*   The graph of $f(x) = \log_a x$, $a > 0$ and $a \neq 1$, has an x-intercept equal to 1 and no y-intercept.

13. *True or False* $\ln e^x = x$ for all real numbers.

14. $\ln e = $ _____.

15. Explain what the number e is.

16. What is the x-intercept of the function $h(x) = \ln(x + 1)$?

Practice Problems

17. Suppose that $g(x) = 4^x + 2$.

   (a)   What is $g(-1)$? What is the corresponding point on the graph of g?

(b)   If $g(x) = 66$, what is x? What is the corresponding point on the graph of g?

18. Suppose that $g(x) = 5^x - 3$.

   (a)   What is $g(-1)$? What is the corresponding point on the graph of g?

   (b)   If $g(x) = 122$, what is x? What is the corresponding point on the graph of g?

In Problems 19–24, the graph of an exponential function is given. Match each graph to one of the following functions:

(a) $y = 3^{-x}$ (b) $y = -3^x$ (c) $y = -3^{-x}$

(d) $y = 3^x - 1$ (e) $y = 3^{x-1}$ (f) $y = 1 - 3^x$

19.

20.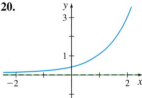

21.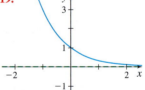

22.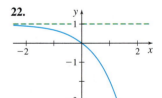

23.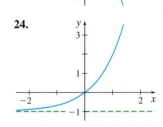

24.

In Problems 25–30, use transformations to graph each function. Find the domain and range.

25. $f(x) = 2^{x+2}$

26. $f(x) = 1 - 2^{-x/3}$

27. $f(x) = 4\left(\dfrac{1}{3}\right)^x$

28. $f(x) = \left(\dfrac{1}{2}\right)^{-x} + 1$

29. $f(x) = e^{-x}$

30. $f(x) = 5 - e^x$

---

1. = NOW WORK problem   = Graphing technology recommended   CAS = Computer Algebra System recommended

In Problems 31–34, find the domain of each function.

31. $F(x) = \log_2 x^2$

32. $g(x) = 8 + 5\ln(2x + 3)$

33. $f(x) = \ln(x - 1)$

34. $g(x) = \sqrt{\ln x}$

In Problems 35–40, the graph of a logarithmic function is given. Match each graph to one of the following functions:

(a) $y = \log_3 x$ **(b)** $y = \log_3(-x)$ **(c)** $y = -\log_3 x$

(d) $y = \log_3 x - 1$ **(e)** $y = \log_3(x - 1)$ **(f)** $y = 1 - \log_3 x$

35.
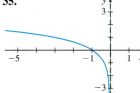

36.
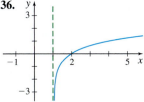

37.
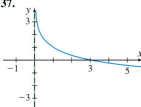

38.
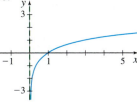

39.
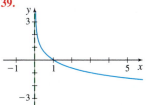

40.
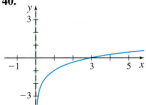

In Problems 41–44, use the given function f to:
(a) *Find the domain of f.*
(b) *Graph f.*
(c) *From the graph of f, determine the range of f.*
(d) *Find f^{-1}, the inverse of f.*
(e) *Use f^{-1} to find the range of f.*
(f) *Graph f^{-1}.*

41. $f(x) = \ln(x + 4)$

42. $f(x) = \dfrac{1}{2}\log(2x)$

43. $f(x) = 3e^x + 2$

44. $f(x) = 2^{x/3} + 4$

45. How does the transformation $y = \ln(x + c)$, $c > 0$, affect the x-intercept of the graph of the function $f(x) = \ln x$?

46. How does the transformation $y = e^{cx}$, $c > 0$, affect the y-intercept of the graph of the function $f(x) = e^x$?

In Problems 47–62, solve each equation.

47. $3^{x^2} = 9^x$

48. $5^{x^2+8} = 125^{2x}$

49. $e^{3x} = \dfrac{e^2}{e^x}$

50. $e^{4x} \cdot e^{x^2} = e^{12}$

51. $e^{1-2x} = 4$

52. $e^{1-x} = 5$

53. $5(2^{3x}) = 9$

54. $0.3(4^{0.2x}) = 0.2$

55. $3^{1-2x} = 4^x$

56. $2^{x+1} = 5^{1-2x}$

57. $\log_2(2x + 1) = 3$

58. $\log_3(3x - 2) = 2$

59. $\log_x\left(\dfrac{1}{8}\right) = 3$

60. $\log_x 64 = -3$

61. $\ln(2x + 3) = 2\ln 3$

62. $\dfrac{1}{2}\log_3 x = 2\log_3 2$

 In Problems 63–66, use graphing technology to solve each equation. Express your answer rounded to three decimal places.

63. $\log_5(x + 1) - \log_4(x - 2) = 1$ **64.** $\ln x = x$

65. $e^x + \ln x = 4$ **66.** $e^x = x^2$

 67. (a) If $f(x) = \ln(x + 4)$ and $g(x) = \ln(3x + 1)$, graph f and g on the same set of axes.

(b) Find the point(s) of intersection of the graphs of f and g by solving $f(x) = g(x)$. Label any intersection points on the graph drawn in (a).

(c) Based on the graph, solve $f(x) > g(x)$.

 68. (a) If $f(x) = 3^{x+1}$ and $g(x) = 2^{x+2}$, graph f and g on the same set of axes.

(b) Find the point(s) of intersection of the graphs of f and g by solving $f(x) = g(x)$. Round answers to three decimal places. Label any intersection points on the graph drawn in (a).

(c) Based on the graph, solve $f(x) > g(x)$.

P.6 Trigonometric Functions

OBJECTIVES *When you finish this section, you should be able to:*

1 Work with properties of trigonometric functions (p. 49)

2 Graph the trigonometric functions (p. 50)

NEED TO REVIEW? Trigonometric functions are discussed in Appendix A.4, pp. A-27 to A-35.

1 Work with Properties of Trigonometric Functions

Table 5 on page 50 lists the six trigonometric functions and the domain and range of each function.

TABLE 5

Function	Symbol	Domain	Range
sine	$y = \sin x$	All real numbers	$\{y \mid -1 \le y \le 1\}$
cosine	$y = \cos x$	All real numbers	$\{y \mid -1 \le y \le 1\}$
tangent	$y = \tan x$	$\left\{x \mid x \ne \text{ odd integer multiples of } \dfrac{\pi}{2}\right\}$	All real numbers
cosecant	$y = \csc x$	$\{x \mid x \ne \text{ integer multiples of } \pi\}$	$\{y \mid y \le -1 \text{ or } y \ge 1\}$
secant	$y = \sec x$	$\left\{x \mid x \ne \text{ odd integer multiples of } \dfrac{\pi}{2}\right\}$	$\{y \mid y \le -1 \text{ or } y \ge 1\}$
cotangent	$y = \cot x$	$\{x \mid x \ne \text{ integer multiples of } \pi\}$	All real numbers

An important property common to all trigonometric functions is that they are *periodic*.

DEFINITION

A function f is called **periodic** if there is a positive number p with the property that whenever x is in the domain of f, so is $x + p$, and

$$\boxed{f(x + p) = f(x)}$$

If there is a smallest number p with this property, it is called the (**fundamental**) **period** of f.

The sine, cosine, cosecant, and secant functions are periodic with period 2π; the tangent and cotangent functions are periodic with period π.

THEOREM Period of Trigonometric Functions

$$\begin{array}{lll} \sin(x + 2\pi) = \sin x & \cos(x + 2\pi) = \cos x & \tan(x + \pi) = \tan x \\ \csc(x + 2\pi) = \csc x & \sec(x + 2\pi) = \sec x & \cot(x + \pi) = \cot x \end{array}$$

Because the trigonometric functions are periodic, once the values of the function over one period are known, the values over the entire domain are known. This property is useful for graphing trigonometric functions.

The next result, also useful for graphing the trigonometric functions, is a consequence of the even-odd identities, namely, $\sin(-x) = -\sin x$ and $\cos(-x) = \cos x$. From these, we have

$$\tan(-x) = \frac{\sin(-x)}{\cos(-x)} = \frac{-\sin x}{\cos x} = -\tan x \qquad \sec(-x) = \frac{1}{\cos(-x)} = \frac{1}{\cos x} = \sec x$$

$$\cot(-x) = \frac{1}{\tan(-x)} = \frac{1}{-\tan x} = -\cot x \qquad \csc(-x) = \frac{1}{\sin(-x)} = \frac{1}{-\sin x} = -\csc x$$

THEOREM Even-Odd Properties of the Trigonometric Functions

The sine, tangent, cosecant, and cotangent functions are odd, so their graphs are symmetric with respect to the origin.

The cosine and secant functions are even, so their graphs are symmetric with respect to the y-axis.

NEED TO REVIEW? The values of the trigonometric functions for select numbers are discussed in Appendix A.4, pp. A-29 and A-31.

2 Graph the Trigonometric Functions

To graph $y = \sin x$, we use Table 6 to obtain points on the graph. Then we plot some of these points and connect them with a smooth curve. Since the sine function has a period of 2π, continue the graph to the left of 0 and to the right of 2π. See Figure 70.

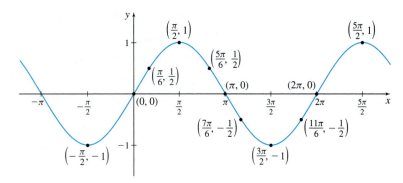

Figure 70 $f(x) = \sin x$

TABLE 6

x	$y = \sin x$	(x, y)
0	0	$(0, 0)$
$\dfrac{\pi}{6}$	$\dfrac{1}{2}$	$\left(\dfrac{\pi}{6}, \dfrac{1}{2}\right)$
$\dfrac{\pi}{2}$	1	$\left(\dfrac{\pi}{2}, 1\right)$
$\dfrac{5\pi}{6}$	$\dfrac{1}{2}$	$\left(\dfrac{5\pi}{6}, \dfrac{1}{2}\right)$
π	0	$(\pi, 0)$
$\dfrac{7\pi}{6}$	$-\dfrac{1}{2}$	$\left(\dfrac{7\pi}{6}, -\dfrac{1}{2}\right)$
$\dfrac{3\pi}{2}$	-1	$\left(\dfrac{3\pi}{2}, -1\right)$
$\dfrac{11\pi}{6}$	$-\dfrac{1}{2}$	$\left(\dfrac{11\pi}{6}, -\dfrac{1}{2}\right)$
2π	0	$(2\pi, 0)$

Notice the symmetry of the graph with respect to the origin. This is a consequence of f being an odd function.

The graph of $y = \sin x$ illustrates some facts about the sine function.

Properties of the Sine Function $f(x) = \sin x$

- The domain of f is the set of all real numbers.
- The range of f consists of all real numbers in the closed interval $[-1, 1]$.
- The sine function is an odd function, so its graph is symmetric with respect to the origin.
- The sine function has a period of 2π.
- The x-intercepts of f are $\ldots, -2\pi, -\pi, 0, \pi, 2\pi, 3\pi, \ldots$; the y-intercept is 0.
- The maximum value of f is 1 and occurs at $x = \ldots, -\dfrac{3\pi}{2}, \dfrac{\pi}{2}, \dfrac{5\pi}{2}, \dfrac{9\pi}{2}, \ldots$;

  the minimum value of f is -1 and occurs at $x = \ldots, -\dfrac{\pi}{2}, \dfrac{3\pi}{2}, \dfrac{7\pi}{2}, \dfrac{11\pi}{2}, \ldots$.

The graph of the cosine function is obtained in a similar way. Locate points on the graph of the cosine function $f(x) = \cos x$ for $0 \le x \le 2\pi$. Then connect the points with a smooth curve, and continue the graph to the left of 0 and to the right of 2π to obtain the graph of $y = \cos x$. See Figure 71.

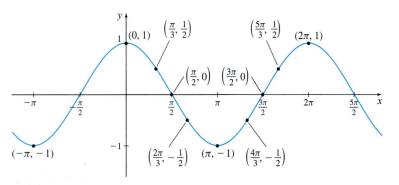

Figure 71 $f(x) = \cos x$

The graph of $y = \cos x$ illustrates some facts about the cosine function.

Properties of the Cosine Function $f(x) = \cos x$

- The domain of f is the set of all real numbers.
- The range of f consists of all real numbers in the closed interval $[-1, 1]$.
- The cosine function is an even function, so its graph is symmetric with respect to the y-axis.
- The cosine function has a period of 2π.
- The x-intercepts of f are $\ldots, -\dfrac{3\pi}{2}, -\dfrac{\pi}{2}, \dfrac{\pi}{2}, \dfrac{3\pi}{2}, \dfrac{5\pi}{2}, \ldots$; the y-intercept is 1.
- The maximum value of f is 1 and occurs at $x = \ldots, -2\pi, 0, 2\pi, 4\pi, 6\pi, \ldots$; the minimum value of f is -1 and occurs at $x = \ldots, -\pi, \pi, 3\pi, 5\pi, \ldots$.

Many variations of the sine and cosine functions can be graphed using transformations.

EXAMPLE 1   Graphing Variations of $f(x) = \sin x$ Using Transformations

Use the graph of $f(x) = \sin x$ to graph $g(x) = 2 \sin x$.

Solution Notice that $g(x) = 2f(x)$, so the graph of g is a vertical stretch of the graph of $f(x) = \sin x$. Figure 72 illustrates the transformation.

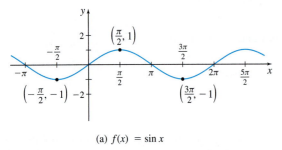

(a) $f(x) = \sin x$

Multiply by 2

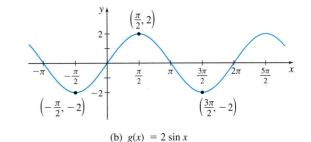

(b) $g(x) = 2 \sin x$

Figure 72

Notice that the values of $g(x) = 2 \sin x$ lie between -2 and 2, inclusive.

In general, the values of the functions $f(x) = A \sin x$ and $g(x) = A \cos x$, where $A \neq 0$, will satisfy the inequalities

$$-|A| \le A \sin x \le |A| \qquad \text{and} \qquad -|A| \le A \cos x \le |A|$$

respectively. The number $|A|$ is called the **amplitude** of $f(x) = A \sin x$ and of $g(x) = A \cos x$.

EXAMPLE 2   Graphing Variations of $f(x) = \cos x$ Using Transformations

Use the graph of $f(x) = \cos x$ to graph $g(x) = \cos(3x)$.

Solution The graph of $g(x) = \cos(3x)$ is a horizontal compression of the graph of $f(x) = \cos x$. Figure 73 shows the transformation.

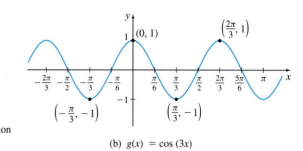

(a) $f(x) = \cos x$

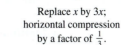

Replace x by $3x$;
horizontal compression
by a factor of $\frac{1}{3}$.

(b) $g(x) = \cos(3x)$

Figure 73

From the graph, we notice that the period of $g(x) = \cos(3x)$ is $\dfrac{2\pi}{3}$.

NOW WORK Problems **27** and **29**.

In general, if $\omega > 0$, the functions $f(x) = \sin(\omega x)$ and $g(x) = \cos(\omega x)$ have period $T = \dfrac{2\pi}{\omega}$. If $\omega > 1$, the graphs of $f(x) = \sin(\omega x)$ and $g(x) = \cos(\omega x)$ are horizontally compressed and the period of the functions is less than 2π. If $0 < \omega < 1$, the graphs of $f(x) = \sin(\omega x)$ and $g(x) = \cos(\omega x)$ are horizontally stretched, and the period of the functions is greater than 2π.

One period of the graph of $f(x) = \sin(\omega x)$ or $g(x) = \cos(\omega x)$ is called a **cycle**.

Sinusoidal Graphs

If we shift the graph of the function $y = \cos x$ to the right $\dfrac{\pi}{2}$ units, we obtain the graph of $y = \cos\left(x - \dfrac{\pi}{2}\right)$, as shown in Figure 74(a). Now look at the graph of $y = \sin x$ in Figure 74(b). Notice that the graph of $y = \sin x$ is the same as the graph of $y = \cos\left(x - \dfrac{\pi}{2}\right)$.

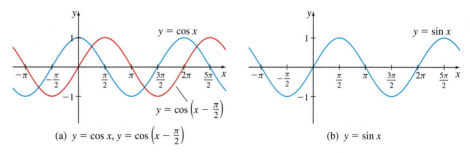

(a) $y = \cos x,\; y = \cos\left(x - \dfrac{\pi}{2}\right)$

(b) $y = \sin x$

Figure 74

Figure 74 suggests that

$$\sin x = \cos\left(x - \frac{\pi}{2}\right) \qquad (1)$$

NEED TO REVIEW? Trigonometric identities are discussed in Appendix A.4, pp. A-32 to A-35.

Proof To prove this identity, we use the difference formula for $\cos(\alpha - \beta)$ with $\alpha = x$ and $\beta = \dfrac{\pi}{2}$.

$$\cos\left(x - \frac{\pi}{2}\right) = \cos x \cos \frac{\pi}{2} + \sin x \sin \frac{\pi}{2} = \cos x \cdot 0 + \sin x \cdot 1 = \sin x \qquad \blacksquare$$

Because of this relationship, the graphs of $y = A\sin(\omega x)$ or $y = A\cos(\omega x)$ are referred to as **sinusoidal graphs,** and the functions and their variations are called **sinusoidal functions.**

THEOREM

For the graphs of $y = A\sin(\omega x)$ and $y = A\cos(\omega x)$, where $A \neq 0$ and $\omega > 0$,

$$\text{Amplitude} = |A| \qquad \text{Period} = T = \frac{2\pi}{\omega}$$

EXAMPLE 3   Finding the Amplitude and Period of a Sinusoidal Function

Determine the amplitude and period of $y = -3\sin(4\pi x)$.

Solution Comparing $y = -3\sin(4\pi x)$ to $y = A\sin(\omega x)$, we find that $A = -3$ and $\omega = 4\pi$. Then,

$$\text{Amplitude} = |A| = |-3| = 3 \qquad \text{Period} = T = \frac{2\pi}{\omega} = \frac{2\pi}{4\pi} = \frac{1}{2} \qquad \blacksquare$$

NOW WORK Problem 33.

TABLE 7

x	$y = \tan x$	(x, y)
0	0	$(0, 0)$
$\dfrac{\pi}{6}$	$\dfrac{\sqrt{3}}{3}$	$\left(\dfrac{\pi}{6}, \dfrac{\sqrt{3}}{3}\right)$
$\dfrac{\pi}{4}$	1	$\left(\dfrac{\pi}{4}, 1\right)$
$\dfrac{\pi}{3}$	$\sqrt{3}$	$\left(\dfrac{\pi}{3}, \sqrt{3}\right)$

The function $y = \tan x$ is an odd function with period π. It is not defined at the odd multiples of $\dfrac{\pi}{2}$. Do you see why? So, we construct Table 7 for $0 \leq x < \dfrac{\pi}{2}$. Then we plot the points from Table 7, connect them with a smooth curve, and reflect the graph about the origin, as shown in Figure 75. To investigate the behavior of $\tan x$ near $\dfrac{\pi}{2}$, we use the identity $\tan x = \dfrac{\sin x}{\cos x}$. When x is close to, but less than, $\dfrac{\pi}{2}$, $\sin x$ is close to 1 and $\cos x$ is a positive number close to 0, so the ratio $\dfrac{\sin x}{\cos x}$ is a large, positive number. The closer x gets to $\dfrac{\pi}{2}$, the larger $\tan x$ becomes. Figure 76 shows the graph of $y = \tan x$, $-\dfrac{\pi}{2} < x < \dfrac{\pi}{2}$. The complete graph of $y = \tan x$ is obtained by repeating the cycle drawn in Figure 76. See Figure 77.

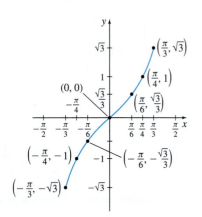

Figure 75 $y = \tan x$, $-\dfrac{\pi}{3} \leq x \leq \dfrac{\pi}{3}$

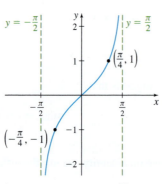

Figure 76 $y = \tan x$, $-\dfrac{\pi}{2} < x < \dfrac{\pi}{2}$

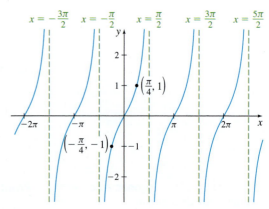

Figure 77 $y = \tan x$, $-\infty < x < \infty$, $x \neq$ odd integer multiples of $\dfrac{\pi}{2}$.

The graph of $y = \tan x$ illustrates the following properties of the tangent function:

> **Properties of the Tangent Function** $f(x) = \tan x$
>
> - The domain of f is the set of all real numbers, except odd multiples of $\dfrac{\pi}{2}$.
> - The range of f consists of all real numbers.
> - The tangent function is an odd function, so its graph is symmetric with respect to the origin.
> - The tangent function is periodic with period π.
> - The x-intercepts of f are $\ldots, -2\pi, -\pi, 0, \pi, 2\pi, 3\pi, \ldots$; the y-intercept is 0.

EXAMPLE 4 **Graphing Variations of $f(x) = \tan x$ Using Transformations**

Use the graph of $f(x) = \tan x$ to graph $g(x) = -\tan\left(x + \dfrac{\pi}{4}\right)$.

Solution Figure 78 illustrates the steps used in graphing $g(x) = -\tan\left(x + \dfrac{\pi}{4}\right)$.

- Begin by graphing $f(x) = \tan x$. See Figure 78(a).
- Replace the argument x by $x + \dfrac{\pi}{4}$ to obtain $y = \tan\left(x + \dfrac{\pi}{4}\right)$, which shifts the graph horizontally to the left $\dfrac{\pi}{4}$ unit, as shown in Figure 78(b).
- Multiply $\tan\left(x + \dfrac{\pi}{4}\right)$ by -1, which reflects the graph about the x-axis, and results in the graph of $y = -\tan\left(x + \dfrac{\pi}{4}\right)$, as shown in Figure 78(c).

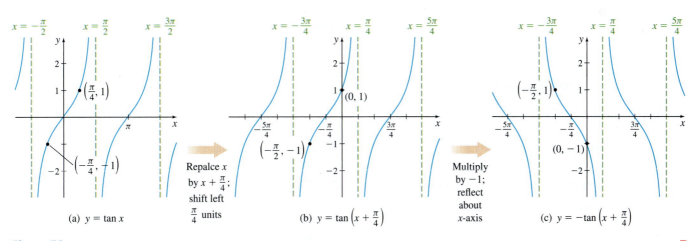

(a) $y = \tan x$ Repalce x by $x + \frac{\pi}{4}$; shift left $\frac{\pi}{4}$ units   (b) $y = \tan\left(x + \frac{\pi}{4}\right)$ Multiply by -1; reflect about x-axis   (c) $y = -\tan\left(x + \frac{\pi}{4}\right)$

Figure 78

NOW WORK Problem 31.

The graph of the cotangent function is obtained similarly. $y = \cot x$ is an odd function, with period π. Because $\cot x$ is not defined at integral multiples of π, graph y on the interval $(0, \pi)$ and then repeat the graph, as shown in Figure 79.

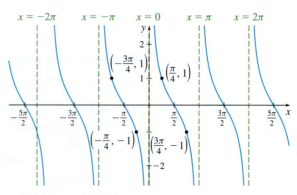

Figure 79 $f(x) = \cot x$, $-\infty < x < \infty$, $x \neq$ integer multiples of π.

The cosecant and secant functions, sometimes referred to as **reciprocal functions**, are graphed by using the reciprocal identities

$$\csc x = \frac{1}{\sin x} \quad \text{and} \quad \sec x = \frac{1}{\cos x}$$

The graphs of $y = \csc x$ and $y = \sec x$ are shown in Figures 80 and 81, respectively.

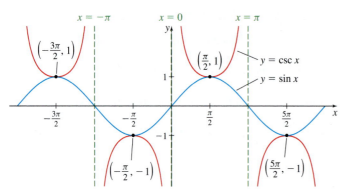

Figure 80

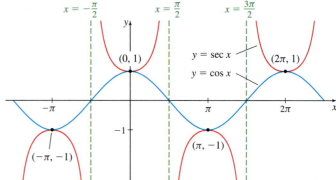

Figure 81

P.6 Assess Your Understanding

Concepts and Vocabulary

1. The sine, cosine, cosecant, and secant functions have period _____; the tangent and cotangent functions have period _____.

2. The domain of the tangent function $f(x) = \tan x$ is _____.

3. The range of the sine function $f(x) = \sin x$ is _____.

4. Explain why $\tan\left(\frac{\pi}{4} + 2\pi\right) = \tan\frac{\pi}{4}$.

5. *True or False* The range of the secant function is the set of all positive real numbers.

6. The function $f(x) = 3\cos(6x)$ has amplitude _____ and period _____.

7. *True or False* The graphs of $y = \sin x$ and $y = \cos x$ are identical except for a horizontal shift.

8. *True or False* The amplitude of the function $f(x) = 2\sin(\pi x)$ is 2 and its period is $\frac{\pi}{2}$.

9. *True or False* The graph of the sine function has infinitely many x-intercepts.

10. The graph of $y = \tan x$ is symmetric with respect to the _____.

11. The graph of $y = \sec x$ is symmetric with respect to the _____.

12. Explain, in your own words, what it means for a function to be periodic.

Practice Problems

In Problems 13–16, use the even-odd properties to find the exact value of each expression.

13. $\tan\left(-\frac{\pi}{4}\right)$

14. $\sin\left(-\frac{3\pi}{2}\right)$

15. $\csc\left(-\frac{\pi}{3}\right)$

16. $\cos\left(-\frac{\pi}{6}\right)$

---

1. = NOW WORK problem   = Graphing technology recommended   CAS = Computer Algebra System recommended

In Problems 17–20, if necessary, refer to a graph to answer each question.

17. What is the y-intercept of $f(x) = \tan x$?

18. Find the x-intercepts of $f(x) = \sin x$ on the interval $[-2\pi, 2\pi]$.

19. What is the smallest value of $f(x) = \cos x$?

20. For what numbers x, $-2\pi \le x \le 2\pi$, does $\sin x = 1$? Where in the interval $[-2\pi, 2\pi]$ does $\sin x = -1$?

In Problems 21–26, the graphs of six trigonometric functions are given. Match each graph to one of the following functions:

(a) $y = 2\sin\left(\dfrac{\pi}{2}x\right)$ **(b)** $y = 2\cos\left(\dfrac{\pi}{2}x\right)$

(c) $y = 3\cos(2x)$ **(d)** $y = -3\sin(2x)$

(e) $y = -2\cos\left(\dfrac{\pi}{2}x\right)$ **(f)** $y = -2\sin\left(\dfrac{1}{2}x\right)$

21.
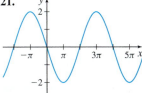

22.
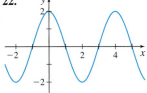

23.
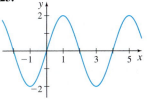

24.
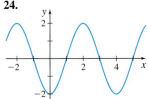

25.
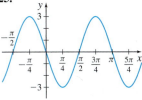

26.
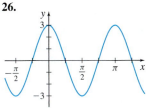

In Problems 27–32, graph each function using transformations. Be sure to label key points and show at least two periods.

27. $f(x) = 4\sin(\pi x)$ **28.** $f(x) = -3\cos x$

29. $f(x) = 3\cos(2x) - 4$ **30.** $f(x) = 4\sin(2x) + 2$

31. $f(x) = \tan\left(\dfrac{\pi}{2}x\right)$ **32.** $f(x) = 4\sec\left(\dfrac{1}{2}x\right)$

In Problems 33–36, determine the amplitude and period of each function.

33. $g(x) = \dfrac{1}{2}\cos(\pi x)$ **34.** $f(x) = \sin(2x)$

35. $g(x) = 3\sin x$ **36.** $f(x) = -2\cos\left(\dfrac{3}{2}x\right)$

In Problems 37 and 38, write the sine function that has the given properties.

37. Amplitude: 2, Period: π **38.** Amplitude: $\dfrac{1}{3}$, Period: 2

In Problems 39 and 40, write the cosine function that has the given properties.

39. Amplitude: $\dfrac{1}{2}$, Period: π **40.** Amplitude: 3, Period: 4π

In Problems 41–48, for each graph, find an equation involving the indicated trigonometric function.

41.
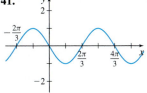
Sine function

42.
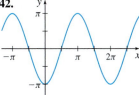
Cosine function

43.
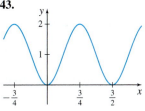
Cosine function

44.
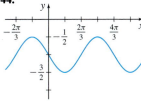
Sine function

45.
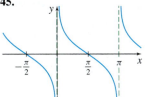
Cotangent function

46.
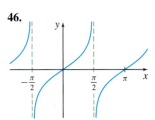
Tangent function

47.
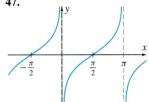
Tangent function

48.
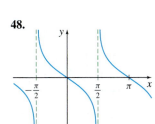
Cotangent function

P.7 Inverse Trigonometric Functions

OBJECTIVES *When you finish this section, you should be able to:*

1 Define the inverse trigonometric functions (p. 58)
2 Use the inverse trigonometric functions (p. 61)
3 Solve trigonometric equations (p. 61)

We now investigate the inverse trigonometric functions. In calculus the inverses of the sine function, the tangent function, and the secant function are used most often. So we define these inverse functions first.

1 Define the Inverse Trigonometric Functions

Although the trigonometric functions are not one-to-one, we can restrict the domains of the trigonometric functions so that each new function will be one-to-one on its restricted domain. Then an inverse function can be defined on each restricted domain.

The Inverse Sine Function

If the domain of the function $f(x) = \sin x$ is restricted to the interval $\left[-\dfrac{\pi}{2}, \dfrac{\pi}{2}\right]$, the function $y = \sin x$, $-\dfrac{\pi}{2} \le x \le \dfrac{\pi}{2}$, is one-to-one and so has an inverse function. See Figure 82.

To obtain an equation for the inverse of $y = \sin x$ on $-\dfrac{\pi}{2} \le x \le \dfrac{\pi}{2}$, we interchange x and y. Then the implicit form of the inverse is $x = \sin y$, $-\dfrac{\pi}{2} \le y \le \dfrac{\pi}{2}$. The explicit form of the inverse function is called the *inverse sine* of x and is written $y = \sin^{-1} x$.

IN WORDS We read $y = \sin^{-1} x$ as " y is the number (or angle) whose sine equals x."

DEFINITION Inverse Sine Function
The **inverse sine function**, denoted by $y = \sin^{-1} x$, or $y = \arcsin x$, is defined as

$$y = \sin^{-1} x \quad \text{if and only if} \quad x = \sin y$$
$$\text{where } -1 \le x \le 1 \quad \text{and} \quad -\dfrac{\pi}{2} \le y \le \dfrac{\pi}{2}$$

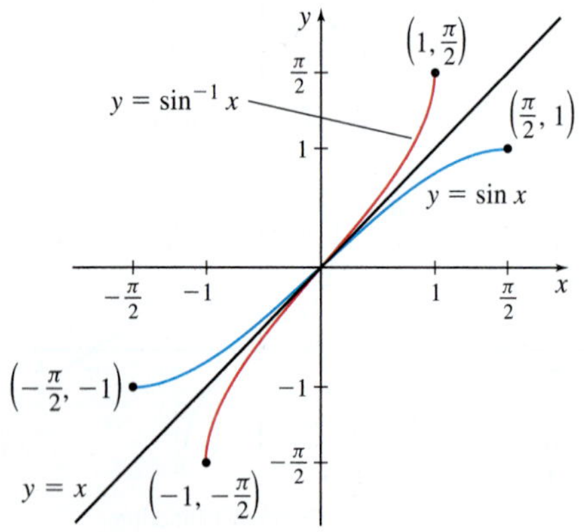

The domain of $y = \sin^{-1} x$ is the closed interval $[-1, 1]$, and the range is the closed interval $\left[-\dfrac{\pi}{2}, \dfrac{\pi}{2}\right]$. The graph of $y = \sin^{-1} x$ is the reflection of the restricted portion of the graph of $f(x) = \sin x$ about the line $y = x$. See Figure 83.

Because $y = \sin x$ defined on the closed interval $\left[-\dfrac{\pi}{2}, \dfrac{\pi}{2}\right]$ and $y = \sin^{-1} x$ are inverse functions,

$$\sin^{-1}(\sin x) = x \quad \text{if } x \text{ is in the closed interval } \left[-\dfrac{\pi}{2}, \dfrac{\pi}{2}\right] \tag{1}$$

$$\sin(\sin^{-1} x) = x \quad \text{if } x \text{ is in the closed interval } [-1, 1] \tag{2}$$

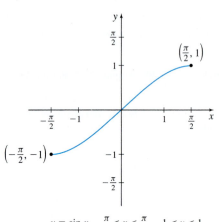

$y = \sin x$, $-\dfrac{\pi}{2} \le x \le \dfrac{\pi}{2}$, $-1 \le y \le 1$

Figure 82

$y = \sin^{-1} x$, $-1 \le x \le 1$, $-\dfrac{\pi}{2} \le y \le \dfrac{\pi}{2}$

Figure 83

CAUTION The superscript $^{-1}$ that appears in $y = \sin^{-1} x$ is not an exponent but the symbol used to denote the inverse function. (To avoid confusion, $y = \sin^{-1} x$ is sometimes written $y = \arcsin x$.)

EXAMPLE 1   Finding the Values of an Inverse Sine Function

Find the exact value of:

(a) $\sin^{-1} \dfrac{\sqrt{3}}{2}$ **(b)** $\sin^{-1}\left(-\dfrac{\sqrt{2}}{2}\right)$ **(c)** $\sin^{-1}\left(\sin \dfrac{\pi}{8}\right)$ **(d)** $\sin^{-1}\left(\sin \dfrac{5\pi}{8}\right)$

Solution **(a)** We seek an angle whose sine is $\dfrac{\sqrt{3}}{2}$. Since $-1 < \dfrac{\sqrt{3}}{2} < 1$, let $y = \sin^{-1} \dfrac{\sqrt{3}}{2}$.

Then by definition, $\sin y = \dfrac{\sqrt{3}}{2}$, where $-\dfrac{\pi}{2} \le y \le \dfrac{\pi}{2}$. Although $\sin y = \dfrac{\sqrt{3}}{2}$ has infinitely many solutions, the only number in the interval $\left[-\dfrac{\pi}{2}, \dfrac{\pi}{2}\right]$, for which $\sin y = \dfrac{\sqrt{3}}{2}$, is $\dfrac{\pi}{3}$. So $\sin^{-1} \dfrac{\sqrt{3}}{2} = \dfrac{\pi}{3}$.

(b) We seek an angle whose sine is $-\dfrac{\sqrt{2}}{2}$. Since $-1 < -\dfrac{\sqrt{2}}{2} < 1$, let $y = \sin^{-1}\left(-\dfrac{\sqrt{2}}{2}\right)$.

Then by definition, $\sin y = -\dfrac{\sqrt{2}}{2}$, where $-\dfrac{\pi}{2} \le y \le \dfrac{\pi}{2}$. The only number in the interval $\left[-\dfrac{\pi}{2}, \dfrac{\pi}{2}\right]$, whose sine is $-\dfrac{\sqrt{2}}{2}$, is $-\dfrac{\pi}{4}$. So $\sin^{-1}\left(-\dfrac{\sqrt{2}}{2}\right) = -\dfrac{\pi}{4}$.

(c) Since the number $\dfrac{\pi}{8}$ is in the interval $\left[-\dfrac{\pi}{2}, \dfrac{\pi}{2}\right]$, we use (1).

$$\sin^{-1}\left(\sin \dfrac{\pi}{8}\right) = \dfrac{\pi}{8}$$

(d) Since the number $\dfrac{5\pi}{8}$ is not in the closed interval $\left[-\dfrac{\pi}{2}, \dfrac{\pi}{2}\right]$, we cannot use (1). Instead, we find a number x in the interval $\left[-\dfrac{\pi}{2}, \dfrac{\pi}{2}\right]$ for which $\sin x = \sin \dfrac{5\pi}{8}$. Using Figure 84, we see that $\sin \dfrac{5\pi}{8} = y = \sin \dfrac{3\pi}{8}$. The number $\dfrac{3\pi}{8}$ is in the interval $\left[-\dfrac{\pi}{2}, \dfrac{\pi}{2}\right]$, so

$$\sin^{-1}\left(\sin \dfrac{5\pi}{8}\right) = \sin^{-1}\left(\sin \dfrac{3\pi}{8}\right) = \dfrac{3\pi}{8}$$ ∎

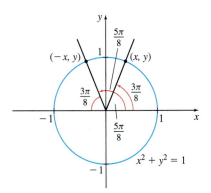

Figure 84

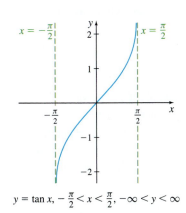

$y = \tan x,\ -\dfrac{\pi}{2} < x < \dfrac{\pi}{2},\ -\infty < y < \infty$

Figure 85

NOW WORK Problems 9 and 17.

The Inverse Tangent Function

The tangent function $y = \tan x$ is not one-to-one. To define the *inverse tangent function*, we restrict the domain of $y = \tan x$ to the open interval $\left(-\dfrac{\pi}{2}, \dfrac{\pi}{2}\right)$. On this interval, $y = \tan x$ is one-to-one. See Figure 85.

Now use the steps for finding the inverse of a one-to-one function:

Step 1 $y = \tan x,\ -\dfrac{\pi}{2} < x < \dfrac{\pi}{2}$.

Step 2 Interchange x and y, to obtain $x = \tan y,\ -\dfrac{\pi}{2} < y < \dfrac{\pi}{2}$, the implicit form of the inverse tangent function.

Step 3 The explicit form is called the *inverse tangent* of x, and is denoted by $y = \tan^{-1} x$.

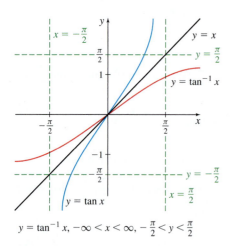

$y = \tan^{-1} x, -\infty < x < \infty, -\frac{\pi}{2} < y < \frac{\pi}{2}$

Figure 86

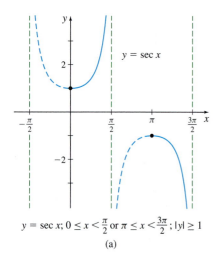

$y = \sec x; 0 \le x < \frac{\pi}{2}$ or $\pi \le x < \frac{3\pi}{2}; |y| \ge 1$

(a)

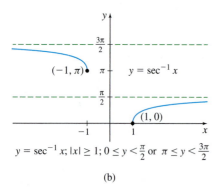

$y = \sec^{-1} x; |x| \ge 1; 0 \le y < \frac{\pi}{2}$ or $\pi \le y < \frac{3\pi}{2}$

(b)

Figure 87

DEFINITION Inverse Tangent Function

The **inverse tangent function**, symbolized by $y = \tan^{-1} x$, or $y = \arctan x$, is

$$y = \tan^{-1} x \quad \text{if and only if} \quad x = \tan y$$
$$\text{where} \; -\infty < x < \infty \quad \text{and} \quad -\frac{\pi}{2} < y < \frac{\pi}{2}$$

The domain of the function $y = \tan^{-1} x$ is the set of all real numbers and its range is $-\frac{\pi}{2} < y < \frac{\pi}{2}$. To graph $y = \tan^{-1} x$, reflect the graph of $y = \tan x$, $-\frac{\pi}{2} < x < \frac{\pi}{2}$, about the line $y = x$, as shown in Figure 86.

NOW WORK Problem 13.

The Inverse Secant Function

To define the *inverse secant function*, we restrict the domain of the function $y = \sec x$ to the set $\left[0, \frac{\pi}{2}\right) \cup \left[\pi, \frac{3\pi}{2}\right)$. See Figure 87(a). An equation for the inverse function is obtained by following the steps for finding the inverse of a one-to-one function. By interchanging x and y, we obtain the implicit form of the inverse function: $x = \sec y$, where $0 \le y < \frac{\pi}{2}$ or $\pi \le y < \frac{3\pi}{2}$. The explicit form is called the *inverse secant* of x and is written $y = \sec^{-1} x$. The graph of $y = \sec^{-1} x$ is given in Figure 87(b).

DEFINITION Inverse Secant Function

The **inverse secant function**, symbolized by $y = \sec^{-1} x$, or $y = \operatorname{arcsec} x$, is

$$y = \sec^{-1} x \quad \text{if and only if} \quad x = \sec y$$
$$\text{where} \; |x| \ge 1 \quad \text{and} \quad 0 \le y < \frac{\pi}{2} \quad \text{or} \quad \pi \le y < \frac{3\pi}{2}$$

You may wonder why we chose the restriction we did to define $y = \sec^{-1} x$. The reason is that this restriction not only makes $y = \sec x$ one-to-one, but it also makes $\tan x \ge 0$. With $\tan x \ge 0$, the derivative formula for the inverse secant function is simpler, as we will see in Chapter 3.

The remaining three inverse trigonometric functions are defined as follows.

DEFINITION

- **Inverse Cosine Function** $y = \cos^{-1} x$ if and only if $x = \cos y$,
  where $-1 \le x \le 1$ and $0 \le y \le \pi$
- **Inverse Cotangent Function** $y = \cot^{-1} x$ if and only if $x = \cot y$,
  where $-\infty < x < \infty$ and $0 < y < \pi$
- **Inverse Cosecant Function** $y = \csc^{-1} x$ if and only if $x = \csc y$,
  where $|x| \ge 1$ and
  $$-\pi < y \le -\frac{\pi}{2} \quad \text{or} \quad 0 < y \le \frac{\pi}{2}$$

The following three identities involve the inverse trigonometric functions. The proofs may be found in books on trigonometry.

THEOREM

- $\cos^{-1} x = \dfrac{\pi}{2} - \sin^{-1} x, \ -1 \leq x \leq 1$

- $\cot^{-1} x = \dfrac{\pi}{2} - \tan^{-1} x, \ -\infty < x < \infty$

- $\csc^{-1} x = \dfrac{\pi}{2} - \sec^{-1} x, \ \ |x| \geq 1$

The identities above are used whenever the inverse cosine, inverse cotangent, and inverse cosecant functions are needed. The graphs of the inverse cosine, inverse cotangent, and inverse cosecant functions can also be obtained by using these identities to transform the graphs of $y = \sin^{-1} x$, $y = \tan^{-1} x$, and $y = \sec^{-1} x$, respectively.

2 Use the Inverse Trigonometric Functions

EXAMPLE 2 Writing a Trigonometric Expression as an Algebraic Expression

Write $\sin(\tan^{-1} u)$ as an algebraic expression containing u.

Solution Let $\theta = \tan^{-1} u$ so that $\tan \theta = u$, where $-\dfrac{\pi}{2} < \theta < \dfrac{\pi}{2}$. We note that in the interval $\left(-\dfrac{\pi}{2}, \dfrac{\pi}{2}\right)$, $\sec \theta > 0$. Then

$$\sin(\tan^{-1} u) = \sin \theta = \sin \theta \cdot \dfrac{\cos \theta}{\cos \theta} = \tan \theta \cos \theta$$

$$= \dfrac{\tan \theta}{\sec \theta} \underset{\underset{\sec^2 \theta = 1 + \tan^2 \theta; \ \sec \theta > 0}{\uparrow}}{=} \dfrac{\tan \theta}{\sqrt{1 + \tan^2 \theta}} = \dfrac{u}{\sqrt{1 + u^2}}$$

An alternate method of obtaining the solution to Example 2 uses right triangles. Let $\theta = \tan^{-1} u$ so that $\tan \theta = u$, $-\dfrac{\pi}{2} < \theta < \dfrac{\pi}{2}$, and label the right triangles drawn in Figure 88. Using the Pythagorean Theorem, the hypotenuse of each triangle is $\sqrt{1 + u^2}$. Then $\sin(\tan^{-1} u) = \sin \theta = \dfrac{u}{\sqrt{1 + u^2}}$. ∎

Figure 88 $\tan \theta = u; \ -\dfrac{\pi}{2} < \theta < \dfrac{\pi}{2}$.

NOW WORK Problem 23.

3 Solve Trigonometric Equations

Inverse trigonometric functions can be used to solve trigonometric equations.

EXAMPLE 3 Solving Trigonometric Equations

Solve the equations:

(a) $\sin \theta = \dfrac{1}{2}$ **(b)** $\cos \theta = 0.4$

Give a general formula for all the solutions and list all the solutions in the interval $[-2\pi, 2\pi]$.

Solution (a) Use the inverse sine function $y = \sin^{-1} x$, $-\dfrac{\pi}{2} \le y \le \dfrac{\pi}{2}$.

$$\sin\theta = \frac{1}{2}$$

$$\theta = \sin^{-1}\frac{1}{2} \qquad -\frac{\pi}{2} \le \theta \le \frac{\pi}{2}$$

$$\theta = \frac{\pi}{6}$$

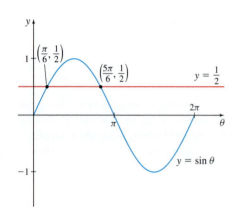

Figure 89

Over the interval $[0, 2\pi]$, there are two angles θ for which $\sin\theta = \dfrac{1}{2}$. See Figure 89.

All the solutions of $\sin\theta = \dfrac{1}{2}$ are given by the general formula

$$\theta = \frac{\pi}{6} + 2k\pi \quad \text{or} \quad \theta = \frac{5\pi}{6} + 2k\pi, \qquad \text{where } k \text{ is any integer}$$

The solutions in the interval $[-2\pi, 2\pi]$ are

$$\left\{ -\frac{11\pi}{6}, -\frac{7\pi}{6}, \frac{\pi}{6}, \frac{5\pi}{6} \right\}$$

(b) A calculator must be used to solve $\cos\theta = 0.4$. Then

$$\theta = \cos^{-1}(0.4) \approx 1.159279 \quad 0 \le \theta \le \pi$$

Rounded to three decimal places, $\theta = \cos^{-1} 0.4 = 1.159$ radians. But there is another angle θ in the interval $[0, 2\pi]$ for which $\cos\theta = 0.4$, namely, $\theta \approx 2\pi - 1.159 \approx 5.124$ radians.

Because the cosine function has period 2π, all the solutions of $\cos\theta = 0.4$ are given by the general formulas

$$\theta \approx 1.159 + 2k\pi \quad \text{or} \quad \theta \approx 5.124 + 2k\pi, \qquad \text{where } k \text{ is any integer}$$

The solutions in the interval $[-2\pi, 2\pi]$ are $\{-5.124, -1.159, 1.159, 5.124\}$. ∎

NOW WORK Problem 45.

EXAMPLE 4   Solving a Trigonometric Equation

Solve the equation $\sin(2\theta) = \dfrac{1}{2}$, where $0 \le \theta < 2\pi$.

Solution In the interval $[0, 2\pi)$, the sine function has the value $\dfrac{1}{2}$ at $\theta = \dfrac{\pi}{6}$ and at $\theta = \dfrac{5\pi}{6}$ as shown in Figure 90. Since the period of the sine function is 2π and the argument in the equation $\sin(2\theta) = \dfrac{1}{2}$ is 2θ, we write the general formula for all the solutions.

$$2\theta = \frac{\pi}{6} + 2k\pi \quad \text{or} \quad 2\theta = \frac{5\pi}{6} + 2k\pi, \qquad \text{where } k \text{ is any integer}$$

$$\theta = \frac{\pi}{12} + k\pi \qquad\qquad \theta = \frac{5\pi}{12} + k\pi$$

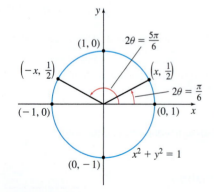

Figure 90

The solutions of $\sin(2\theta) = \dfrac{1}{2}$, $0 \le \theta < 2\pi$, are $\left\{ \dfrac{\pi}{12}, \dfrac{5\pi}{12}, \dfrac{13\pi}{12}, \dfrac{17\pi}{12} \right\}$. ∎

NOW WORK Problem 31.

NEED TO REVIEW? Trigonometric identities are discussed in Appendix Section A.4, pp. A-32 to A-35.

Many trigonometric equations can be solved by applying algebra techniques such as factoring or using the quadratic formula. It is often necessary to begin by using a trigonometric identity to express the given equation in a familiar form.

EXAMPLE 5   Solving a Trigonometric Equation

Solve the equation $3 \sin \theta - \cos^2 \theta = 3$, where $0 \le \theta < 2\pi$.

Solution The equation involves both sine and cosine functions. We use the Pythagorean identity $\sin^2 \theta + \cos^2 \theta = 1$ to rewrite the equation in terms of $\sin \theta$.

$$3 \sin \theta - \cos^2 \theta = 3$$
$$3 \sin \theta - (1 - \sin^2 \theta) = 3 \qquad \textcolor{blue}{\cos^2 \theta = 1 - \sin^2 \theta}$$
$$\sin^2 \theta + 3 \sin \theta - 4 = 0$$

This is a quadratic equation in $\sin \theta$. Factor the left side and solve for $\sin \theta$.

$$(\sin \theta + 4)(\sin \theta - 1) = 0$$
$$\sin \theta + 4 = 0 \qquad \text{or} \qquad \sin \theta - 1 = 0$$
$$\sin \theta = -4 \quad \text{or} \qquad \qquad \sin \theta = 1$$

The range of the sine function is $-1 \le y \le 1$, so $\sin \theta = -4$ has no solution. Solving $\sin \theta = 1$, we obtain

$$\theta = \sin^{-1} 1 = \frac{\pi}{2}$$

The only solution in the interval $[0, 2\pi)$ is $\frac{\pi}{2}$. ∎

NOW WORK   Problem 35.

P.7 Assess Your Understanding

Concepts and Vocabulary

1. $y = \sin^{-1} x$ if and only if $x =$ _____, where $-1 \le x \le 1$ and $-\dfrac{\pi}{2} \le y \le \dfrac{\pi}{2}$.

2. *True or False* $\sin^{-1}(\sin x) = x$, where $-1 \le x \le 1$.

3. *True or False*   The domain of $y = \sin^{-1} x$ is $-\dfrac{\pi}{2} \le x \le \dfrac{\pi}{2}$.

4. *True or False* $\sin(\sin^{-1} 0) = 0$.

5. *True or False* $y = \tan^{-1} x$ means $x = \tan y$, where $-\infty < x < \infty$ and $-\dfrac{\pi}{2} \le y \le \dfrac{\pi}{2}$.

6. *True or False*   The domain of the inverse tangent function is the set of all real numbers.

7. *True or False* $\sec^{-1} 0.5$ is not defined.

8. *True or False*   Trigonometric equations can have multiple solutions.

Practice Problems

In Problems 9–20, find the exact value of each expression.

9. $\sin^{-1}\left(\dfrac{\sqrt{2}}{2}\right)$

10. $\sin^{-1}\left(-\dfrac{1}{2}\right)$

11. $\sec^{-1} 2$

12. $\tan^{-1} \sqrt{3}$

13. $\tan^{-1}(-1)$

14. $\sec^{-1} \dfrac{2\sqrt{3}}{3}$

15. $\tan^{-1}\left(\tan \dfrac{\pi}{4}\right)$

16. $\tan^{-1}(\sin 0)$

17. $\sin\left(\sin^{-1} \dfrac{3}{5}\right)$

18. $\tan[\sec^{-1}(-3)]$

19. $\sin^{-1}\left(\sin \dfrac{4\pi}{5}\right)$

20. $\sec(\sec^{-1} 3)$

21. Write $\cos(\sin^{-1} u)$ as an algebraic expression containing u, where $|u| \le 1$.

22. Write $\tan(\sin^{-1} u)$ as an algebraic expression containing u, where $|u| \le 1$.

23. Write $\sec(\tan^{-1} u)$ as an algebraic expression containing u.

24. Show that $y = \sin^{-1} x$ is an odd function. That is, show $\sin^{-1}(-x) = -\sin^{-1} x$.

25. Show that $y = \tan^{-1} x$ is an odd function. That is, show $\tan^{-1}(-x) = -\tan^{-1} x$.

26. Given that $x = \sin^{-1} \dfrac{1}{2}$, find $\cos x$, $\tan x$, $\cot x$, $\sec x$, and $\csc x$.

In Problems 27–44, solve each equation on the interval $0 \le \theta < 2\pi$.

27. $\tan \theta = -\dfrac{\sqrt{3}}{3}$

28. $\sec \dfrac{3\theta}{2} = -2$

---

1. = NOW WORK problem   ⟁ = Graphing technology recommended   (CAS) = Computer Algebra System recommended

29. $2\sin\theta + 3 = 2$

30. $1 - \cos\theta = \dfrac{1}{2}$

31. $\sin(3\theta) = -1$

32. $\cos(2\theta) = \dfrac{1}{2}$

33. $4\cos^2\theta = 1$

34. $2\sin^2\theta - 1 = 0$

35. $2\sin^2\theta - 5\sin\theta + 3 = 0$

36. $2\cos^2\theta - 7\cos\theta - 4 = 0$

37. $1 + \sin\theta = 2\cos^2\theta$

38. $\sec^2\theta + \tan\theta = 0$

39. $\sin\theta + \cos\theta = 0$

40. $\tan\theta = \cot\theta$

41. $\cos(2\theta) + 6\sin^2\theta = 4$

42. $\cos(2\theta) = \cos\theta$

43. $\sin(2\theta) + \sin(4\theta) = 0$

44. $\cos(4\theta) - \cos(6\theta) = 0$

In Problems 45–48, use a calculator to solve each equation on the interval $0 \le \theta < 2\pi$. Round answers to three decimal places.

45. $\tan\theta = 5$

46. $\cos\theta = 0.6$

47. $2 + 3\sin\theta = 0$

48. $4 + \sec\theta = 0$

49. (a) On the same set of axes, graph $f(x) = 3\sin(2x) + 2$ and $g(x) = \dfrac{7}{2}$ on the interval $[0, \pi]$.

(b) Solve $f(x) = g(x)$ on the interval $[0, \pi]$, and label the points of intersection on the graph drawn in (a).

(c) Shade the region bounded by $f(x) = 3\sin(2x) + 2$ and $g(x) = \dfrac{7}{2}$ between the points found in (b) on the graph drawn in (a).

(d) Solve $f(x) > g(x)$ on the interval $[0, \pi]$.

50. (a) On the same set of axes, graph $f(x) = -4\cos x$ and $g(x) = 2\cos x + 3$ on the interval $[0, 2\pi]$.

(b) Solve $f(x) = g(x)$ on the interval $[0, 2\pi]$, and label the points of intersection on the graph drawn in (a).

(c) Shade the region bounded by $f(x) = -4\cos x$ and $g(x) = 2\cos x + 3$ between the points found in (b) on the graph drawn in (a).

(d) Solve $f(x) > g(x)$ on the interval $[0, 2\pi]$.

P.8 Technology Used in Calculus

Whether your instructor requires you to use a graphing calculator or a computer algebra system (CAS) or believes that calculus is learned best in the classical way, by hand, real applications of calculus—in engineering, science, economics, and statistics—will usually require the use of technology.

This text, as you see it, would not have been possible without technology. All the figures in the text were produced using technology—some on a graphing calculator, most with an interactive graphic system, others in Adobe Illustrator®. The equations and symbols were created and spaced by a computer using the MuPad CAS; the page numbering and printing were done electronically.

In this brief section, we outline some of the more popular technologies currently used in learning calculus.

Since the 1960s, portable computation devices have been available. Their introduction eliminated the need to perform long, tedious arithmetic calculations by hand. Many calculations that were previously impossible can now be done quickly and accurately.

Many calculators today, certainly those you are using, are not so much calculators but small, hand-held computers. They *numerically* manipulate data and mathematical expressions.

Most graphing calculators have the ability to:

- graph and compare functions of one variable.
- graph functions of one variable in several formats: rectangular, parametric, and polar.
- locate the intercepts, local maxima, and local minima of a graph.
- graph sequences and explore convergence.
- solve equations numerically.
- numerically solve a system of equations.
- find a function of best fit.
- find the derivative of a function at a particular number.
- numerically approximate a definite integral

As you read the text, you will see examples of graphs generated by a graphing calculator. You will also find problems marked with ⋀ , which alerts you that graphing technology is recommended.

In contrast to a calculator, a CAS *symbolically* manipulates mathematical expressions. This symbolic manipulation usually allows for exact mathematical solutions, as well as for numeric approximations. Most CAS systems are packaged with interactive graphing technology that can produce and manipulate two- and three-dimensional graphs.

There are many computer algebra systems available. They vary in versatility, ease of use, and price. Here, we outline the capabilities of several systems used in many colleges and universities.

Maple™ was developed in 1980 by the Symbolic Logic Group at the University of Waterloo in Ontario, Canada. It was the first CAS to use standard mathematical notation, and its source code is viewable. Maple™ is now owned and sold by **Maple**soft™. The latest version, Maple™17, was released in April 2012. Maple™ is marketed to mathematics educators, mathematicians, engineers, and scientists.

Maple™17's "Clickable Math" allows the user to enter mathematical expressions into the equation editor in standard mathematical notation using keystrokes, menus, and symbol palettes. Operations can be initiated using context-sensitive menus, and the output is annotated for future reference.

Maple™17 can be used to:

- generate two- and three-dimensional graphs using dropdown menus.
- resize, recenter, and rotate graphs using a mouse, and to overlay more than one graph on a set of axes.
- symbolically find limits, derivatives, and integrals with exact answers.
- numerically approximate limits, derivatives, and integrals.
- solve differential equations.
- perform dimensional analysis.

Maple™17 also includes step-by-step calculus tutorials and a programming language that allows the user to write programs and perform analysis.

Mathematica was developed in 1988 by Stephen Wolfram of Wolfram Research, Champaign, Illinois. It is probably the most complete computer algebra system available. Written in C, *Mathematica* is a computational software program that allows mathematicians, engineers, and scientists to compute symbolically, visually, and numerically to any precision. The current version of *Mathematica*, *Mathematica* 9, was released in 2013.

Mathematica 9 features a free-form linguistic input that needs no knowledge of syntax. *Mathematica* 9 can be used to:

- solve linear and nonlinear optimization problems.
- find limits, derivatives, and integrals, both definite and indefinite.
- solve differential equations.
- generate two- and three-dimensional graphs.
- rotate and resize three-dimensional graphs.
- analyze huge data sets.
- explore formulas and solve equations.
- analyze mathematical functions.

Wolfram|Alpha (http://www.wolframalpha.com/) is a Web-based derivative of *Mathematica,* developed by Wolfram Research in 2009. It is called a computational knowledge engine, and its aim is to organize data. Since it is written using *Mathematica*, it can generate graphs, solve equations, and perform calculus.

MATLAB® (short for matrix laboratory) is a technical computing language that supports vector and matrix operations and allows for object-oriented programming.

MATLAB® was developed at the University of New Mexico by Cleve Moler in the mid-1970s. Moler wanted to provide students access to matrix software without having to write a Fortran program. In 1984 Moler partnered with Jack Little to form the MathWorks™, Inc. After adding an interactive graphing system, sales of MATLAB® grew. MATLAB® is now used in such diverse products as cars, airplanes, cell phones, and financial derivatives. When packaged with the Symbolic Math Toolbox, MATLAB provides symbolic and numeric computing and extensive graphing capabilities. The current version of MATLAB® was released in March 2013. MATLAB® with the Symbolic Math Toolbox package can be used to:

- perform computations that require exact control over numeric accuracy to any number of digits.
- perform symbolic mathematical operations, including finding symbolic derivatives, integrals, and limits.
- interactively evaluate Riemann sums.
- find sums and products of series.
- perform Taylor series expansions.
- produce polar and parametric curves.
- produce contour and mesh surfaces.
- create animated two- and three-dimensional graphs.

The Symbolic Math Toolbox also provides users with access to the MuPad language for operating on symbolic mathematical expressions, extensive MuPad libraries in calculus and other areas, and the MuPad notebook interface with embedded text, graphics, and typeset mathematics.

MuPad (Multiprocessing Algebra Data Tool) is a CAS and high-precision decimal arithmetic program, as well as an interactive graphing system. It was developed in 1990 at the University of Paderborn, Germany. MuPad's syntax is modeled on the Pascal programming language and is similar to that used in Maple, but MuPad supports object-oriented programming. In 1997 MuPad was sold to SciFace Software GmbH & Co. KG. Although it is no longer sold as a stand-alone, MuPad is the CAS that drives the Symbolic Math Toolbox of MATLAB® and is the CAS used in Scientific WorkPlace®.

Sage (www.sagemath.org) is a free open-source CAS system that combines other open-source mathematics packages into a unified package written primarily in Python. Licensed under the General Public Licence, its stated mission is to "create a viable free open source alternative to Magma, Maple, Mathematica, and Matlab." The Sage project was begun in 2005 as a specialized system for number theory, and it continues to be developed. Sage can either be downloaded onto a computer or used through a Sage Network account. Although to use Sage requires entering code, there is a dropdown toggle that explains the code.

Until recently, only computers could support a CAS, but in the last decade or so, portable hand-held computer algebra systems have become available. These CAS look and work like calculators, but they compute symbolically and have enhanced graphic capabilities.

The TI-Nspire™ CAS, first released in 2007, is a hand-held system built with the *Derive*™ CAS. *Derive*™ was developed in 1988 by a software company now owned by Texas Instruments. *Derive*™, which is no longer sold independently, uses less memory than other CAS, so it works well in a hand-held device. The TI-Nspire™ CAS is used in schools and colleges because of its portability, affordability, and ease of use.

The Casio Prizm was released in 2010 and is another hand-held system used in many schools. Casio Corporation developed the first hand-held graphing calculator in 1985. In addition to the CAS, the Prizm includes applications to solve and graph differential equations.

The TI-Nspire™ CAS and the Casio Prizm can perform all the tasks of a graphing calculator. In addition, it can:

- obtain exact values in terms of variables x and y, and irrational numbers when performing algebra or calculus calculations.
- graph and rotate three-dimensional surfaces.
- find the derivative of a function.
- find an indefinite integral.

At appropriate places in the text, you will see problems marked CAS , which alerts you that a computer algebra system is recommended. If your instructor does not require a CAS, these problems can be omitted. However, they provide insight into calculus and will enrich your calculus experience.

1

Limits and Continuity

CHAPTER 1 PROJECT The Chapter Project on page 143 looks at a hypothetical situation of pollution in a lake and explores some legal arguments that might be made.

Oil Spills and Dispersant Chemicals

On April 20, 2010, the Deepwater Horizon drilling rig exploded and initiated the worst marine oil spill in recent history. Oil gushed from the well for three months and released millions of gallons of crude oil into the Gulf of Mexico. One technique used to help clean up during and after the spill was the use of the chemical dispersant Corexit. Oil dispersants allow the oil particles to spread more freely in the water, thus allowing the oil to biodegrade more quickly. Their use is debated, however, because some of their ingredients are carcinogens. Further, the use of oil dispersants can increase toxic hydrocarbon levels affecting sea life. Over time, the pollution caused by the oil spill and the dispersants will eventually diminish and sea life will return, more or less, to its previous condition. In the short term, however, pollution raises serious questions about the health of the local sea life and the safety of fish and shellfish for human consumption.

The concept of a limit is central to calculus. To understand calculus, it is essential to know what it means for a function to have a limit, and then how to find the limit of a function. Chapter 1 explains what a limit is, shows how to find the limit of a function, and demonstrates how to prove that limits exist using the definition of limit.

We begin the chapter using numerical and graphical approaches to explore the idea of a limit. Although these methods seem to work well, there are instances in which they fail to identify the correct limit. We conclude Section 1.1 with a precise definition of limit, the so-called ϵ-δ (epsilon-delta) definition.

In Section 1.2, we provide analytic techniques for finding limits. Some of the proofs of these techniques are found in Section 1.6, others in Appendix B. A limit found by correctly applying these analytic techniques is precise; there is no doubt that it is correct.

In Sections 1.3–1.5, we continue to study limits and some ways they are used. For example, we use limits to define *continuity*, an important property of a function.

Section 1.6 is dedicated to the ϵ-δ definition, which we use to show when a limit does, and does not, exist.

1.1 Limits of Functions Using Numerical and Graphical Techniques

OBJECTIVES *When you finish this section, you should be able to:*

1 Discuss the slope of a tangent line to a graph (p. 69)

2 Investigate a limit using a table of numbers (p. 71)

3 Investigate a limit using a graph (p. 73)

Calculus can be used to solve certain fundamental questions in geometry. Two of these questions are:

- Given a function f and a point P on its graph, what is the slope of the line tangent to the graph of f at P? See Figure 1.
- Given a nonnegative function f whose domain is the closed interval $[a, b]$, what is the area enclosed by the graph of f, the x-axis, and the vertical lines $x = a$ and $x = b$? See Figure 2.

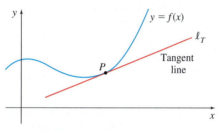

DF Figure 1

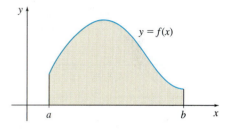

Figure 2

These questions, traditionally called the **tangent problem** and the **area problem**, were solved by Gottfried Wilhelm von Leibniz and Sir Isaac Newton during the late seventeenth and early eighteenth centuries. The solutions to the two seemingly different problems are both based on the idea of a limit. Their solutions not only are related to each other, but are also applicable to many other problems in science and geometry. Here, we begin to discuss the tangent problem. The discussion of the area problem begins in Chapter 5.

1 Discuss the Slope of a Tangent Line to a Graph

Notice that the line ℓ_T in Figure 1 just touches the graph of f at the point P. This unique line is the *tangent line* to the graph of f at P. But how is the tangent line defined?

In plane geometry, a tangent line to a circle is defined as a line having exactly one point in common with the circle, as shown in Figure 3. However, this definition does not work for graphs in general. For example, in Figure 4, on page 70, three lines ℓ_1, ℓ_2, and ℓ_3 contain the point P and have exactly one point in common with the graph of f, but they do not meet the requirement of just touching the graph at P. On the other hand, the line ℓ_T just touches the graph of f at P, but it intersects the graph at other points. It is the slope of the tangent line ℓ_T that distinguishes it from all other lines containing P.

So before defining a tangent line, we investigate its slope, which we denote by $m_{\tan}$. We begin with the graph of a function f, a point P on its graph, and the tangent line ℓ_T to f at P, as shown in Figure 5.

The tangent line ℓ_T to the graph of f at P must contain the point P. We denote the coordinates of P by $(c, f(c))$. Since finding slope requires two points, and we have only one point on the tangent line ℓ_T, we need another way to find the slope of ℓ_T.

Suppose we choose any point $Q = (x, f(x))$, other than P, on the graph of f, as shown in Figure 6. (Q can be to the left or to the right of P; we chose Q to be to the

Figure 3 Tangent line to a circle at the point P.

NEED TO REVIEW? The slope of a line is discussed in Appendix A.3, p. A-18.

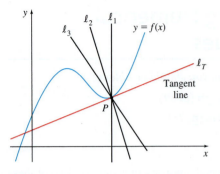

Figure 4

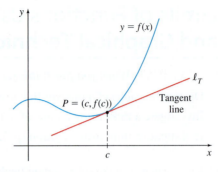

Figure 5

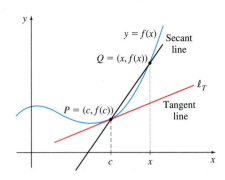

Figure 6

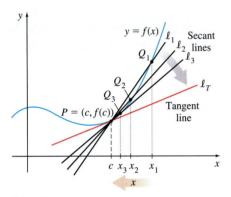

DF Figure 7

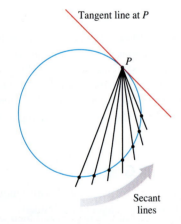

Figure 8

NOTE We discuss what it means for a limit to exist shortly.

NEED TO REVIEW? If $a < b$, the open interval (a, b) consists of all numbers x for which $a < x < b$. Interval notation is discussed in Appendix A.1, p. A-5.

right of P.) The line containing the points $P = (c, f(c))$ and $Q = (x, f(x))$ is called a **secant line** of the graph of f. The slope m_{sec} of this secant line is

$$m_{\text{sec}} = \frac{f(x) - f(c)}{x - c} \tag{1}$$

Figure 7 shows three different points Q_1, Q_2, and Q_3 on the graph of f that are successively closer to point P, and three associated secant lines ℓ_1, ℓ_2, and ℓ_3. The closer the point Q is to the point P, the closer the secant line is to the tangent line ℓ_T. The line ℓ_T, the *limiting position* of these secant lines, is the *tangent line to the graph* of f at P.

As Figure 8 illustrates, this new idea of a tangent line is consistent with the traditional definition of a tangent line to a circle.

If the limiting position of the secant lines is the tangent line, then the limit of the slopes of the secant lines should equal the slope of the tangent line. Notice in Figure 7 that as the point Q moves closer to the point P, the numbers x get closer to c. So, equation (1) suggests that

$$m_{\text{tan}} = [\text{Slope of the tangent line to } f \text{ at } P]$$

$$= \left[\text{Limit of } \frac{f(x) - f(c)}{x - c} \text{ as } x \text{ gets closer to } c\right]$$

In symbols, we write

$$m_{\text{tan}} = \lim_{x \to c} \frac{f(x) - f(c)}{x - c}$$

The notation $\lim_{x \to c}$ is read, "the limit as x approaches c."

The **tangent line** to the graph of a function f at a point $P = (c, f(c))$ is the line containing the point P whose slope is

$$m_{\text{tan}} = \lim_{x \to c} \frac{f(x) - f(c)}{x - c}$$

provided the limit exists.

We have begun to answer the tangent problem by introducing the idea of a *limit*. Now we describe the idea of a limit in more detail.

The Idea of a Limit

We begin by asking a question: What does it mean for a function f to have a limit L as x approaches some fixed number c? To answer the question, we need to be more precise about f, L, and c. The function f must be defined everywhere in an open interval containing the number c, except possibly at c, and L is a number. Using these restrictions, we introduce the notation

$$\lim_{x \to c} f(x) = L$$

which is read, "the limit as x approaches c of $f(x)$ is equal to the number L." The notation $\lim\limits_{x \to c} f(x) = L$ can be described as

The value $f(x)$ can be made as close as we please to L,
for x sufficiently close to c, but not equal to c.

See Figure 9.

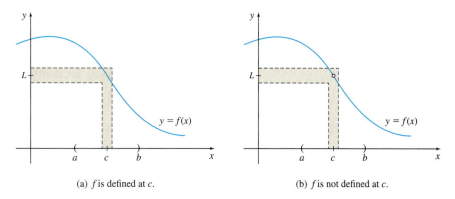

(a) f is defined at c.   (b) f is not defined at c.

Figure 9

2 Investigate a Limit Using a Table of Numbers

EXAMPLE 1   Investigating a Limit Using a Table of Numbers

Investigate $\lim\limits_{x \to 2}(2x + 5)$ using a table of numbers.

Solution We create Table 1 by evaluating $f(x) = 2x + 5$ at values of x near 2, choosing numbers x slightly less than 2 and numbers x slightly greater than 2.

TABLE 1

	numbers x slightly less than 2							numbers x slightly greater than 2			
x	1.99	1.999	1.9999	1.99999	$\rightarrow$	2	$\leftarrow$	2.00001	2.0001	2.001	2.01
$f(x) = 2x + 5$	8.98	8.998	8.9998	8.99998	$f(x)$ approaches 9			9.00002	9.0002	9.002	9.02

Table 1 suggests that the value of $f(x) = 2x + 5$ can be made "as close as we please" to 9 by choosing x "sufficiently close" to 2. This suggests that $\lim\limits_{x \to 2}(2x + 5) = 9$. ∎

NOW WORK   Problem 9.

In creating Table 1, first we used numbers x close to 2 but less than 2, and then we used numbers x close to 2 but greater than 2. When $x < 2$, we say, "x is approaching 2 from the left," and the number 9 is called the **left-hand limit**. When $x > 2$, we say, "x is approaching 2 from the right," and the number 9 is called the **right-hand limit**. Together, these are called the **one-sided limits** of f as x approaches 2.

One-sided limits are symbolized as follows. The left-hand limit, written

$$\lim_{x \to c^-} f(x) = L_{\text{left}}$$

is read, "The limit as x approaches c from the left of $f(x)$, equals L_{left}." It means that the value of f can be made as close as we please to the number L_{left} by choosing $x < c$ and sufficiently close to c.

Similarly, the right-hand limit, written

$$\lim_{x \to c^+} f(x) = L_{\text{right}}$$

and is read, "The limit as x approaches c from the right of $f(x)$, equals L_{right}." It means that the value of f can be made as close as we please to the number L_{right} by choosing $x > c$ and sufficiently close to c.

EXAMPLE 2   Investigating a Limit Using a Table of Numbers

Investigate $\displaystyle\lim_{x \to 0} \frac{e^x - 1}{x}$ using a table of numbers.

Solution The domain of $f(x) = \dfrac{e^x - 1}{x}$ is $\{x \mid x \neq 0\}$. So, f is defined everywhere in an open interval containing the number 0, except for 0.

We create Table 2, investigating the left-hand limit $\displaystyle\lim_{x \to 0^-} \frac{e^x - 1}{x}$ and the right-hand limit $\displaystyle\lim_{x \to 0^+} \frac{e^x - 1}{x}$. First, we evaluate f at numbers less than 0, but close to zero, and then at numbers greater than 0, but close to zero.

TABLE 2

	\multicolumn{4}{c}{x approaches 0 from the left $\rightarrow$}			\multicolumn{4}{c}{$\leftarrow$ x approaches 0 from the right}
x	-0.01	-0.001	-0.0001	-0.00001	$\rightarrow$ 0 $\leftarrow$		0.00001	0.0001	0.001	0.01
$f(x) = \dfrac{e^x - 1}{x}$	0.995	0.9995	0.99995	0.999995	$f(x)$ approaches 1		1.000005	1.00005	1.0005	1.005

Table 2 suggests that $\displaystyle\lim_{x \to 0^-} \frac{e^x - 1}{x} = 1$ and $\displaystyle\lim_{x \to 0^+} \frac{e^x - 1}{x} = 1$. This suggests

$$\lim_{x \to 0} \frac{e^x - 1}{x} = 1. \ \blacksquare$$

NOW WORK Problem 13.

EXAMPLE 3   Investigating a Limit Using a Table of Numbers

Investigate $\displaystyle\lim_{x \to 0} \frac{\sin x}{x}$ using a table of numbers.

Solution The domain of the function $f(x) = \dfrac{\sin x}{x}$ is $\{x \mid x \neq 0\}$. So, f is defined everywhere in an open interval containing 0, except for 0.

We create Table 3, by investigating one-sided limits of $\dfrac{\sin x}{x}$ as x approaches 0, choosing numbers (in radians) slightly less than 0 and numbers slightly greater than 0.

TABLE 3

	\multicolumn{3}{c}{x approaches 0 from the left $\rightarrow$}			\multicolumn{3}{c}{$\leftarrow$ x approaches 0 from the right}
x (in radians)	-0.02	-0.01	-0.005	$\rightarrow$ 0 $\leftarrow$		0.005	0.01	0.02
$f(x) = \dfrac{\sin x}{x}$	0.99993	0.99998	0.999996	$f(x)$ approaches 1		0.999996	0.99998	0.99993

NOTE $f(x) = \dfrac{\sin x}{x}$ is an even function, so the bottom row of Table 3 is symmetric about $x = 0$.

Table 3 suggests that $\displaystyle\lim_{x \to 0^-} f(x) = 1$ and $\displaystyle\lim_{x \to 0^+} f(x) = 1$. This suggests that

$$\lim_{x \to 0} \frac{\sin x}{x} = 1. \ \blacksquare$$

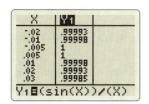

Figure 10

The table used to investigate the limit can be created using technology, as shown in Figure 10.

NOW WORK   Problem 15.

3 Investigate a Limit Using a Graph

The graph of a function can also help us investigate limits. Figure 11 shows the graphs of three different functions f, g, and h. Observe that in each function, *as x gets closer to c, whether from the left or from the right, the value of the function gets closer to the number L*. This is the key idea of a limit.

Notice in Figure 11(b) that the value of g at c does not affect the limit. Notice in Figure 11(c) that h is not defined at c, but the value of h gets closer to the number L for x sufficiently close to c. This suggests that the limit of each function as x approaches c is L even though the values of each function at c are different.

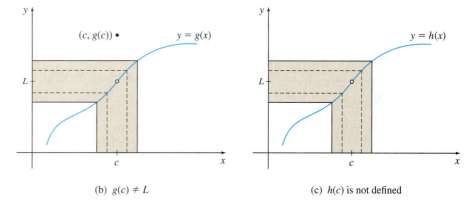

(a) $f(c) = L$ (b) $g(c) \neq L$ (c) $h(c)$ is not defined

Figure 11

EXAMPLE 4   Investigating a Limit Using a Graph

Use a graph to investigate $\lim\limits_{x \to 2} f(x)$ if $f(x) = \begin{cases} 3x + 1 & \text{if} \quad x \neq 2 \\ 10 & \text{if} \quad x = 2 \end{cases}$.

Solution The function f is a piecewise-defined function. Its graph is shown in Figure 12. Observe that as x approaches 2 from the left, the value of f is close to 7, and as x approaches 2 from the right, the value of f is close to 7. In fact, we can make the value of f as close as we please to 7 by choosing x sufficiently close to 2 but not equal to 2. This suggests $\lim\limits_{x \to 2} f(x) = 7$. ∎

If we use a table of numbers to investigate $\lim\limits_{x \to 2} f(x)$, the result is the same. See Table 4.

NEED TO REVIEW? Piecewise-defined functions are discussed in Section P.1, p. 7.

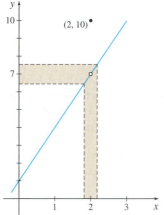

Figure 12 $f(x) = \begin{cases} 3x + 1 & \text{if } x \neq 2 \\ 10 & \text{if } x = 2 \end{cases}$

TABLE 4

	x approaches 2 from the left $\longrightarrow$					x approaches 2 from the right $\longleftarrow$			
x	1.99	1.999	1.9999	1.99999	$\to 2 \leftarrow$	2.00001	2.0001	2.001	2.01
$f(x)$	6.97	6.997	6.9997	6.99997	$f(x)$ approaches 7	7.00003	7.0003	7.003	7.03

Figure 12 shows that $f(2) = 10$, but that this value has no impact on the limit as x approaches 2.

We make the following observations:

- The limit L of a function $y = f(x)$ as x approaches a number c does not depend on the value of f at c.
- The limit L of a function $y = f(x)$ as x approaches a number c is unique; that is, a function cannot have more than one limit. (A proof of this property is given in Appendix B.)
- If there is *no single number* that the value of f approaches as x gets close to c, we say that *f has no limit as x approaches c*, or more simply, that the *limit of f does not exist at c.*

Examples 5 and 6 that follow illustrate situations in which a limit doesn't exist.

EXAMPLE 5   Investigating a Limit Using a Graph

Use a graph to investigate $\lim\limits_{x \to 0} f(x)$ if $f(x) = \begin{cases} x & \text{if} \quad x < 0 \\ 1 & \text{if} \quad x > 0 \end{cases}$.

Solution Figure 13 shows the graph of f. We first investigate the one-sided limits. The graph suggests that, as x approaches 0 from the left,

$$\lim_{x \to 0^-} f(x) = 0$$

and as x approaches 0 from the right,

$$\lim_{x \to 0^+} f(x) = 1$$

Since there is no single number that the values of f approach when x is close to 0, we conclude that $\lim\limits_{x \to 0} f(x)$ does not exist. ∎

Table 5 uses a numerical approach to support the conclusion that $\lim\limits_{x \to 0} f(x)$ does not exist.

Figure 13 $f(x) = \begin{cases} x & \text{if } x < 0 \\ 1 & \text{if } x > 0 \end{cases}$

TABLE 5

	x approaches 0 from the left					x approaches 0 from the right		
x	−0.01	−0.001	−0.0001	→ 0 ←		0.0001	0.001	0.01
$f(x)$	−0.01	−0.001	−0.0001	no single number		1	1	1

Examples 4 and 5 lead to the following result.

THEOREM

The limit L of a function $y = f(x)$ as x approaches a number c exists if and only if both one-sided limits exist at c and both one-sided limits are equal. That is,

$$\lim_{x \to c} f(x) = L \quad \text{if and only if} \quad \lim_{x \to c^-} f(x) = \lim_{x \to c^+} f(x) = L$$

NOW WORK Problems 25 and 31.

A one-sided limit is used to describe the behavior of functions such as $f(x) = \sqrt{x-1}$ near $x = 1$. Since the domain of f is $\{x \mid x \geq 1\}$, the left-hand limit, $\lim\limits_{x \to 1^-} \sqrt{x-1}$ makes no sense. But $\lim\limits_{x \to 1^+} \sqrt{x-1} = 0$ suggests how f behaves near and to the right of 1. See Figure 14 and Table 6. They suggest $\lim\limits_{x \to 1^+} f(x) = 0$.

Figure 14 $f(x) = \sqrt{x-1}$

TABLE 6

	x approaches 1 from the right					
x	1.009	1.0009	1.00009	1.000009	1.0000009	→ 1
$f(x) = \sqrt{x-1}$	0.0949	0.03	0.00949	0.003	0.000949	$f(x)$ approaches 0

Using numeric tables and/or graphs gives us an idea of what a limit might be. That is, these methods suggest a limit, but there are dangers in using these methods, as the following example illustrates.

EXAMPLE 6   **Investigating a Limit**

Investigate $\lim\limits_{x \to 0} \sin \dfrac{\pi}{x^2}$.

Solution   The domain of the function $f(x) = \sin \dfrac{\pi}{x^2}$ is $\{x \mid x \neq 0\}$. Suppose we let x approach zero in the following way:

	x approaches 0 from the left						x approaches 0 from the right			
x	$-\dfrac{1}{10}$	$-\dfrac{1}{100}$	$-\dfrac{1}{1000}$	$-\dfrac{1}{10{,}000}$	$\to \quad 0 \quad \leftarrow$		$\dfrac{1}{10}$	$\dfrac{1}{100}$	$\dfrac{1}{1000}$	$\dfrac{1}{10{,}000}$
$f(x) = \sin \dfrac{\pi}{x^2}$	0	0	0	0	$f(x)$ approaches 0		0	0	0	0

The table suggests that $\lim\limits_{x \to 0} \sin \dfrac{\pi}{x^2} = 0$. Now suppose we let x approach zero as follows:

	x approaches 0 from the left							x approaches 0 from the right				
x	$-\dfrac{2}{3}$	$-\dfrac{2}{5}$	$-\dfrac{2}{7}$	$-\dfrac{2}{9}$	$-\dfrac{2}{11}$	$\to \quad 0 \quad \leftarrow$		$\dfrac{2}{11}$	$\dfrac{2}{9}$	$\dfrac{2}{7}$	$\dfrac{2}{5}$	$\dfrac{2}{3}$
$f(x) = \sin \dfrac{\pi}{x^2}$	0.707	0.707	0.707	0.707	0.707	$f(x)$ approaches 0.707		0.707	0.707	0.707	0.707	0.707

This table suggests that $\lim\limits_{x \to 0} \sin \dfrac{\pi}{x^2} = \dfrac{\sqrt{2}}{2} \approx 0.707$.

In fact, by carefully selecting x, we can make f appear to approach any number in the interval $[-1, 1]$.

Now look at the graphs of $f(x) = \sin \dfrac{\pi}{x^2}$ shown in Figure 15. In Figure 15(a), the choice of $\lim\limits_{x \to 0} \sin \dfrac{\pi}{x^2} = 0$ seems reasonable. But in Figure 15(b), it appears that $\lim\limits_{x \to 0} \sin \dfrac{\pi}{x^2} = -\dfrac{1}{2}$. Figure 15(c) illustrates that the graph of f oscillates rapidly as x approaches 0. This suggests that the value of f does not approach a single number, and that $\lim\limits_{x \to 0} \sin \dfrac{\pi}{x^2}$ does not exist. ∎

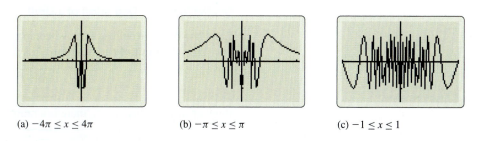

(a) $-4\pi \leq x \leq 4\pi$ (b) $-\pi \leq x \leq \pi$ (c) $-1 \leq x \leq 1$

Figure 15

NOW WORK   **Problem 55.**

So, how do we find a limit with certainty? The answer lies in giving a very precise definition of limit. The next example helps explain the definition.

EXAMPLE 7 Analyzing a Limit

In Example 1, we claimed that $\lim\limits_{x \to 2}(2x + 5) = 9$.

(a) How close must x be to 2, so that $f(x) = 2x + 5$ is within 0.1 of 9?

(b) How close must x be to 2, so that $f(x) = 2x + 5$ is within 0.05 of 9?

RECALL On the number line, the distance between two points with coordinates a and b is $|a - b|$.

Solution (a) The function $f(x) = 2x + 5$ is within 0.1 of 9, if the distance between $f(x)$ and 9 is less than 0.1 unit. That is, if $|f(x) - 9| \le 0.1$.

$$|(2x + 5) - 9| \le 0.1$$
$$|2x - 4| \le 0.1$$
$$|2(x - 2)| \le 0.1$$
$$|x - 2| \le \frac{0.1}{2} = 0.05$$
$$-0.05 \le x - 2 \le 0.05$$
$$1.95 \le x \le 2.05$$

NEED TO REVIEW? Inequalities involving absolute values are discussed in Appendix A.1, p. A-7.

So, if $1.95 \le x \le 2.05$, then $f(x)$ will be within 0.1 of 9.

(b) The function $f(x) = 2x + 5$ is within 0.05 of 9 if $|f(x) - 9| \le 0.05$. That is,

$$|(2x + 5) - 9| \le 0.05$$
$$|2x - 4| \le 0.05$$
$$|x - 2| \le \frac{0.05}{2} = 0.025$$

So, if $1.975 \le x \le 2.025$, then $f(x)$ will be within 0.05 of 9. ■

Notice that the closer we require f to be to the limit 9, the narrower the interval for x becomes. See Figure 16.

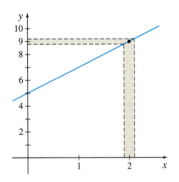

DF **Figure 16** $f(x) = 2x + 5$

NOW WORK Problem 57.

The discussion in Example 7 forms the basis of the definition of a limit. We state the definition here, but postpone the details until Section 1.6. It is customary to use the Greek letters ϵ (epsilon) and δ (delta) in the definition, so we call it the ϵ-δ *definition of a limit*.

DEFINITION ϵ-δ Definition of a Limit

Let f be a function defined everywhere in an open interval containing c, except possibly at c. Then the **limit of the function f as x approaches** c is the number L, written

$$\lim_{x \to c} f(x) = L$$

if, given any number $\epsilon > 0$, there is a number $\delta > 0$ so that

$$\text{whenever } 0 < |x - c| < \delta \quad \text{then } |f(x) - L| < \epsilon$$

Notice in the definition that f is defined everywhere in an open interval containing c except possibly at c. If f is defined at c and there is an open interval containing c that contains no other numbers in the domain of f, then $\lim\limits_{x \to c} f(x)$ does not exist.

1.1 Assess Your Understanding

Concepts and Vocabulary

1. *Multiple Choice* The limit as x approaches c of a function f is written symbolically as [(a) $\lim f(x)$, (b) $\lim\limits_{c \to x} f(x)$, (c) $\lim\limits_{x \to c} f(x)$]

2. *True or False* The tangent line to the graph of f at a point $P = (c, f(c))$ is the limiting position of the secant lines passing through P and a point $(x, f(x))$, $x \neq c$, as x moves closer to c.

3. *True or False* If f is not defined at $x = c$, then $\lim\limits_{x \to c} f(x)$ does not exist.

4. *True or False* The limit L of a function $y = f(x)$ as x approaches the number c depends on the value of f at c.

5. If $\lim\limits_{x \to c} \dfrac{f(x) - f(c)}{x - c}$ exists, it equals the _____ of the tangent line to the graph of f at the point $(c, f(c))$.

6. *True or False* The limit of a function $y = f(x)$ as x approaches a number c equals L if at least one of the one-sided limits as x approaches c equals L.

Skill Building

In Problems 7–12, complete each table and investigate the limit.

7. $\lim\limits_{x \to 1} 2x$

	x approaches 1 from the left		x approaches 1 from the right	
x	0.9 0.99 0.999 $\to$ 1		$\leftarrow$ 1.001 1.01 1.1	
$f(x) = 2x$				

8. $\lim\limits_{x \to 2} (x + 3)$

	x approaches 2 from the left		x approaches 2 from the right	
x	1.9 1.99 1.999 $\to$ 2		$\leftarrow$ 2.001 2.01 2.1	
$f(x) = x + 3$				

9. $\lim\limits_{x \to 0} (x^2 + 2)$

	x approaches 0 from the left		x approaches 0 from the right	
x	-0.1 -0.01 -0.001 $\to$ 0		$\leftarrow$ 0.001 0.01 0.1	
$f(x) = x^2 + 2$				

10. $\lim\limits_{x \to -1} (x^2 - 2)$

	x approaches -1 from the left		x approaches -1 from the right	
x	-1.1 -1.01 -1.001 $\to$ -1		$\leftarrow$ -0.999 -0.99 -0.9	
$f(x) = x^2 - 2$				

11. $\lim\limits_{x \to -3} \dfrac{x^2 - 9}{x + 3}$

	x approaches -3 from the left		x approaches -3 from the right	
x	-3.5 -3.1 -3.01 $\to$ -3		$\leftarrow$ -2.99 -2.9 -2.5	
$f(x) = \dfrac{x^2 - 9}{x + 3}$				

12. $\lim\limits_{x \to -1} \dfrac{x^3 + 1}{x + 1}$

	x approaches -1 from the left		x approaches -1 from the right	
x	-1.1 -1.01 -1.001 $\to$ -1		$\leftarrow$ -0.999 -0.99 -0.9	
$f(x) = \dfrac{x^3 + 1}{x + 1}$				

⃞ *In Problems 13–16, use technology to complete the table and investigate the limit.*

13. $\lim\limits_{x \to 0} \dfrac{2 - 2e^x}{x}$

	x approaches 0 from the left		x approaches 0 from the right	
x	-0.2 -0.1 -0.01 $\to$ 0		$\leftarrow$ 0.01 0.1 0.2	
$f(x) = \dfrac{2 - 2e^x}{x}$				

14. $\lim\limits_{x \to 1} \dfrac{\ln x}{x - 1}$

	x approaches 1 from the left		x approaches 1 from the right	
x	0.9 0.99 0.999 $\to$ 1		$\leftarrow$ 1.001 1.01 1.1	
$f(x) = \dfrac{\ln x}{x - 1}$				

15. $\lim\limits_{x \to 0} \dfrac{1 - \cos x}{x}$, where x is measured in radians

	x approaches 0 from the left		x approaches 0 from the right	
x (in radians)	-0.2 -0.1 -0.01 $\to$ 0		$\leftarrow$ 0.01 0.1 0.2	
$f(x) = \dfrac{1 - \cos x}{x}$				

16. $\lim\limits_{x \to 0} \dfrac{\sin x}{1 + \tan x}$, where x is measured in radians

	x approaches 0 from the left		x approaches 0 from the right	
x (in radians)	-0.2 -0.1 -0.01 $\to$ 0		$\leftarrow$ 0.01 0.1 0.2	
$f(x) = \dfrac{\sin x}{1 + \tan x}$				

---

1. = NOW WORK problem   ⃞ = Graphing technology recommended   CAS = Computer Algebra System recommended

In Problems 17–20, use the graph to investigate
(a) $\lim\limits_{x \to 2^-} f(x)$, *(b)* $\lim\limits_{x \to 2^+} f(x)$, *(c)* $\lim\limits_{x \to 2} f(x)$.

17.

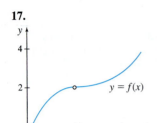

18.

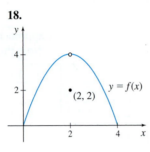

19.

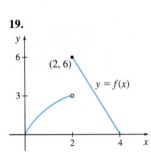

20.

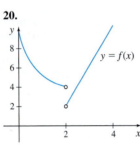

In Problems 21–28, use the graph to investigate $\lim\limits_{x \to c} f(x)$. *If the limit does not exist, explain why.*

21.

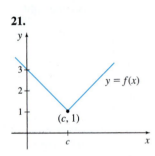

22.

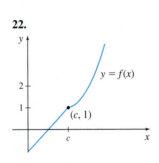

23.

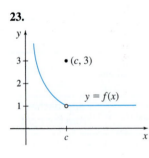

24.

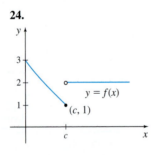

25.

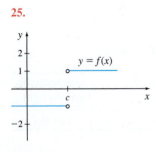

26.

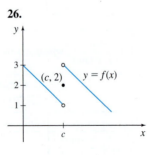

27.

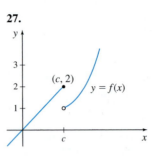

28.

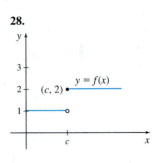

In Problems 29–36, use a graph to investigate $\lim\limits_{x \to c} f(x)$ *at the number c.*

29. $f(x) = \begin{cases} 2x + 5 & \text{if } x \le 2 \\ 4x + 1 & \text{if } x > 2 \end{cases}$ at $c = 2$

30. $f(x) = \begin{cases} 2x + 1 & \text{if } x \le 0 \\ 2x & \text{if } x > 0 \end{cases}$ at $c = 0$

31. $f(x) = \begin{cases} 3x - 1 & \text{if } x < 1 \\ 4 & \text{if } x = 1 \\ 4x & \text{if } x > 1 \end{cases}$ at $c = 1$

32. $f(x) = \begin{cases} x + 2 & \text{if } x < 2 \\ 4 & \text{if } x = 2 \\ x^2 & \text{if } x > 2 \end{cases}$ at $c = 2$

33. $f(x) = \begin{cases} 2x^2 & \text{if } x < 1 \\ 3x^2 - 1 & \text{if } x > 1 \end{cases}$ at $c = 1$

34. $f(x) = \begin{cases} x^3 & \text{if } x < -1 \\ x^2 - 1 & \text{if } x > -1 \end{cases}$ at $c = -1$

35. $f(x) = \begin{cases} x^2 & \text{if } x \le 0 \\ 2x + 1 & \text{if } x > 0 \end{cases}$ at $c = 0$

36. $f(x) = \begin{cases} x^2 & \text{if } x < 1 \\ 2 & \text{if } x = 1 \\ -3x + 2 & \text{if } x > 1 \end{cases}$ at $c = 1$

Applications and Extensions

In Problems 37–40, sketch a graph of a function with the given properties. Answers will vary.

37. $\lim\limits_{x \to 2} f(x) = 3$; $\quad \lim\limits_{x \to 3^-} f(x) = 3$; $\quad \lim\limits_{x \to 3^+} f(x) = 1$;
$f(2) = 3$; $\quad f(3) = 1$

38. $\lim\limits_{x \to -1} f(x) = 0$; $\quad \lim\limits_{x \to 2^-} f(x) = -2$; $\quad \lim\limits_{x \to 2^+} f(x) = -2$;
$f(-1)$ is not defined; $\quad f(2) = -2$

39. $\lim\limits_{x \to 1} f(x) = 4$; $\quad \lim\limits_{x \to 0^-} f(x) = -1$; $\quad \lim\limits_{x \to 0^+} f(x) = 0$;
$f(0) = -1$; $\quad f(1) = 2$

40. $\lim\limits_{x \to 2} f(x) = 2$; $\quad \lim\limits_{x \to -1} f(x) = 0$; $\quad \lim\limits_{x \to 1} f(x) = 1$;
$f(-1) = 1$; $\quad f(2) = 3$

In Problems 41–50, use either a graph or a table to investigate each limit.

41. $\lim\limits_{x \to 5^+} \dfrac{|x-5|}{x-5}$ **42.** $\lim\limits_{x \to 5^-} \dfrac{|x-5|}{x-5}$ **43.** $\lim\limits_{x \to \left(\frac{1}{2}\right)^-} \lfloor 2x \rfloor$

44. $\lim\limits_{x \to \left(\frac{1}{2}\right)^+} \lfloor 2x \rfloor$ **45.** $\lim\limits_{x \to \left(\frac{2}{3}\right)^-} \lfloor 2x \rfloor$ **46.** $\lim\limits_{x \to \left(\frac{2}{3}\right)^+} \lfloor 2x \rfloor$

47. $\lim\limits_{x \to 2^+} \sqrt{|x| - x}$ **48.** $\lim\limits_{x \to 2^-} \sqrt{|x| - x}$

49. $\lim\limits_{x \to 2^+} \sqrt[3]{\lfloor x \rfloor - x}$ **50.** $\lim\limits_{x \to 2^-} \sqrt[3]{\lfloor x \rfloor - x}$

51. Slope of a Tangent Line For $f(x) = 3x^2$:

(a) Find the slope of the secant line containing the points $(2, 12)$ and $(3, 27)$.

(b) Find the slope of the secant line containing the points $(2, 12)$ and $(x, f(x))$, $x \neq 2$.

(c) Create a table to investigate the slope of the tangent line to the graph of f at 2 using the result from (b).

(d) On the same set of axes, graph f, the tangent line to the graph of f at the point $(2, 12)$, and the secant line from (a).

52. Slope of a Tangent Line For $f(x) = x^3$:

(a) Find the slope of the secant line containing the points $(2, 8)$ and $(3, 27)$.

(b) Find the slope of the secant line containing the points $(2, 8)$ and $(x, f(x))$, $x \neq 2$.

(c) Create a table to investigate the slope of the tangent line to the graph of f at 2 using the result from (b).

(d) On the same set of axes, graph f, the tangent line to the graph of f at the point $(2, 8)$, and the secant line from (a).

53. Slope of a Tangent Line For $f(x) = \dfrac{1}{2}x^2 - 1$:

(a) Find the slope $m_{\sec}$ of the secant line containing the points $P = (2, f(2))$ and $Q = (2 + h, f(2 + h))$.

(b) Use the result from (a) to complete the following table:

h	-0.5	-0.1	-0.001	0.001	0.1	0.5
$m_{\sec}$						

(c) Investigate the limit of the slope of the secant line found in (a) as $h \to 0$.

(d) What is the slope of the tangent line to the graph of f at the point $P = (2, f(2))$?

(e) On the same set of axes, graph f and the tangent line to f at $P = (2, f(2))$.

54. Slope of a Tangent Line For $f(x) = x^2 - 1$:

(a) Find the slope $m_{\sec}$ of the secant line containing the points $P = (-1, f(-1))$ and $Q = (-1 + h, f(-1 + h))$.

(b) Use the result from (a) to complete the following table:

h	-0.1	-0.01	-0.001	-0.0001	0.0001	0.001	0.01	0.1
$m_{\sec}$								

(c) Investigate the limit of the slope of the secant line found in (a) as $h \to 0$.

(d) What is the slope of the tangent line to the graph of f at the point $P = (-1, f(-1))$?

(e) On the same set of axes, graph f and the tangent line to f at $P = (-1, f(-1))$.

55. (a) Investigate $\lim\limits_{x \to 0} \cos \dfrac{\pi}{x}$ by using a table and evaluating the function $f(x) = \cos \dfrac{\pi}{x}$ at

$$x = -\frac{1}{2}, -\frac{1}{4}, -\frac{1}{8}, -\frac{1}{10}, -\frac{1}{12}, \dots, \frac{1}{12}, \frac{1}{10}, \frac{1}{8}, \frac{1}{4}, \frac{1}{2}.$$

(b) Investigate $\lim\limits_{x \to 0} \cos \dfrac{\pi}{x}$ by using a table and evaluating the function $f(x) = \cos \dfrac{\pi}{x}$ at

$$x = -1, -\frac{1}{3}, -\frac{1}{5}, -\frac{1}{7}, -\frac{1}{9}, \dots, \frac{1}{9}, \frac{1}{7}, \frac{1}{5}, \frac{1}{3}, 1.$$

(c) Compare the results from (a) and (b). What do you conclude about the limit? Why do you think this happens? What is your view about using a table to draw a conclusion about limits?

(d) Use graphing technology to graph f. Begin with the x-window $[-2\pi, 2\pi]$ and the y-window $[-1, 1]$. If you were finding $\lim\limits_{x \to 0} f(x)$ using a graph, what would you conclude? Zoom in on the graph. Describe what you see. (*Hint*: Be sure your calculator is set to the radian mode.)

56. (a) Investigate $\lim\limits_{x \to 0} \cos \dfrac{\pi}{x^2}$ by using a table and evaluating the function $f(x) = \cos \dfrac{\pi}{x^2}$ at $x = -0.1, -0.01, -0.001,$ $-0.0001, 0.0001, 0.001, 0.01, 0.1.$

(b) Investigate $\lim\limits_{x \to 0} \cos \dfrac{\pi}{x^2}$ by using a table and evaluating the function $f(x) = \cos \dfrac{\pi}{x^2}$ at

$$x = -\frac{2}{3}, -\frac{2}{5}, -\frac{2}{7}, -\frac{2}{9}, \dots, \frac{2}{9}, \frac{2}{7}, \frac{2}{5}, \frac{2}{3}.$$

(c) Compare the results from (a) and (b). What do you conclude about the limit? Why do you think this happens? What is your view about using a table to draw a conclusion about limits?

(d) Use graphing technology to graph f. Begin with the x-window $[-2\pi, 2\pi]$ and the y-window $[-1, 1]$. If you were finding $\lim\limits_{x \to 0} f(x)$ using a graph, what would you conclude? Zoom in on the graph. Describe what you see. (*Hint*: Be sure your calculator is set to the radian mode.)

57. (a) Use a table to investigate $\lim\limits_{x \to 2} \dfrac{x - 8}{2}$.

(b) How close must x be to 2, so that $f(x)$ is within 0.1 of the limit?

(c) How close must x be to 2, so that $f(x)$ is within 0.01 of the limit?

58. (a) Use a table to investigate $\lim\limits_{x\to 2}(5-2x)$.

(b) How close must x be to 2, so that $f(x)$ is within 0.1 of the limit?

(c) How close must x be to 2, so that $f(x)$ is within 0.01 of the limit?

59. First-Class Mail As of January 2013, the U.S. Postal Service charged $0.46 postage for first-class letters weighing up to and including 1 ounce, plus a flat fee of $0.20 for each additional or partial ounce up to and including 3.5 ounces. First-class letter rates do not apply to letters weighing more than 3.5 ounces.

Source: U.S. Postal Service Notice 123.

(a) Find a function C that models the first-class postage charged, in dollars, for a letter weighing w ounces. Assume $w > 0$.

(b) What is the domain of C?

(c) Graph the function C.

(d) Use the graph to investigate $\lim\limits_{w\to 2^-} C(w)$ and $\lim\limits_{w\to 2^+} C(w)$. Do these suggest that $\lim\limits_{w\to 2} C(w)$ exists?

(e) Use the graph to investigate $\lim\limits_{w\to 0^+} C(w)$.

(f) Use the graph to investigate $\lim\limits_{w\to 3.5^-} C(w)$.

60. First-Class Mail As of January 2013, the U.S. Postal Service charged $0.92 postage for first-class retail flats (large envelopes) weighing up to and including 1 ounce, plus a flat fee of $0.20 for each additional or partial ounce up to and including 13 ounces. First-class rates do not apply to flats weighing more than 13 ounces.

(a) Find a function C that models the first-class postage charged, in dollars, for a large envelope weighing w ounces. Assume $w > 0$.

(b) What is the domain of C?

(c) Graph the function C.

(d) Use the graph to investigate $\lim\limits_{w\to 1^-} C(w)$ and $\lim\limits_{w\to 1^+} C(w)$. Do these suggest that $\lim\limits_{w\to 1} C(w)$ exists?

(e) Use the graph to investigate $\lim\limits_{w\to 12^-} C(w)$ and $\lim\limits_{w\to 12^+} C(w)$. Do these suggest that $\lim\limits_{w\to 12} C(w)$ exists?

(f) Use the graph to investigate $\lim\limits_{w\to 0^+} C(w)$.

(g) Use the graph to investigate $\lim\limits_{w\to 13^-} C(w)$.

Source: U.S. Postal Service Notice 123

61. Correlating Student Success to Study Time Professor Smith claims that a student's final exam score is a function of the time t (in hours) that the student studies. He claims that the closer to seven hours one studies, the closer to 100% the student scores on the final. He claims that studying significantly less than seven hours may cause one to be underprepared for the test, while studying significantly more than seven hours may cause "burnout."

(a) Write Professor Smith's claim symbolically as a limit.

(b) Write Professor Smith's claim using the ϵ-δ definition of limit.

Source: Submitted by the students of Millikin University.

62. The definition of the slope of the tangent line to the graph of $y = f(x)$ at the point $(c, f(c))$ is $m_{\tan} = \lim\limits_{x\to c}\dfrac{f(x)-f(c)}{x-c}$.

Another way to express this slope is to define a new variable $h = x - c$. Rewrite the slope of the tangent line $m_{\tan}$ using h and c.

63. If $f(2) = 6$, can you conclude anything about $\lim\limits_{x\to 2} f(x)$? Explain your reasoning.

64. If $\lim\limits_{x\to 2} f(x) = 6$, can you conclude anything about $f(2)$? Explain your reasoning.

65. The graph of $f(x) = \dfrac{x-3}{3-x}$ is a straight line with a point punched out.

(a) What straight line and what point?

(b) Use the graph of f to investigate the one-sided limits of f as x approaches 3.

(c) Does the graph suggest that $\lim\limits_{x\to 3} f(x)$ exists? If so, what is it?

66. (a) Use a table to investigate $\lim\limits_{x\to 0}(1+x)^{1/x}$.

(b) Use graphing technology to graph $g(x) = (1+x)^{1/x}$.

(c) What do (a) and (b) suggest about $\lim\limits_{x\to 0}(1+x)^{1/x}$?

(d) Find $\lim\limits_{x\to 0}(1+x)^{1/x}$. (CAS)

Challenge Problems

For Problems 67–70, investigate each of the following limits.

$$f(x) = \begin{cases} 1 & \text{if } x \text{ is an integer} \\ 0 & \text{if } x \text{ is not an integer} \end{cases}$$

67. $\lim\limits_{x\to 2} f(x)$ **68.** $\lim\limits_{x\to 1/2} f(x)$ **69.** $\lim\limits_{x\to 3} f(x)$ **70.** $\lim\limits_{x\to 0} f(x)$

1.2 Limits of Functions Using Properties of Limits

OBJECTIVES *When you finish this section, you should be able to:*

1 Find the limit of a sum, a difference, and a product (p. 82)
2 Find the limit of a power and the limit of a root (p. 84)
3 Find the limit of a polynomial (p. 86)
4 Find the limit of a quotient (p. 87)
5 Find the limit of an average rate of change (p. 89)
6 Find the limit of a difference quotient (p. 89)

In Section 1.1, we used a numerical approach (tables) and a graphical approach to investigate limits. We saw that these approaches are not always reliable. The only way to be sure a limit is correct is to use the ϵ-δ definition of a limit. In this section, we state without proof results based on the ϵ-δ definition. Some of the results are proved in Section 1.6 and others in Appendix B.

We begin with two basic limits.

THEOREM The Limit of a Constant

If $f(x) = A$, where A is a constant, then for any real number c,

$$\lim_{x \to c} f(x) = \lim_{x \to c} A = A \qquad (1)$$

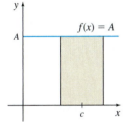

Figure 17 For x close to c, the value of f remains at A; $\lim_{x \to c} A = A$.

The theorem is proved in Section 1.6. See Figure 17 and Table 7 for graphical and numerical support of $\lim_{x \to c} A = A$.

TABLE 7

	x approaches c from the left				x approaches c from the right		
x	$c - 0.01$	$c - 0.001$	$c - 0.0001$	$\to c \leftarrow$	$c + 0.0001$	$c + 0.001$	$c + 0.01$
$f(x) = A$	A	A	A	$f(x)$ remains at A	A	A	A

For example,

$$\lim_{x \to 5} 2 = 2 \qquad \lim_{x \to \sqrt{2}} \frac{1}{3} = \frac{1}{3} \qquad \lim_{x \to 5}(-\pi) = -\pi$$

THEOREM The Limit of the Identity Function

If $f(x) = x$, then for any number c,

$$\lim_{x \to c} f(x) = \lim_{x \to c} x = c \qquad (2)$$

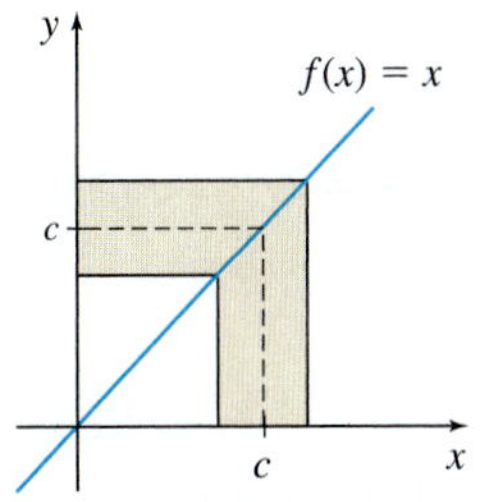

This theorem is proved in Section 1.6. See Figure 18 and Table 8 for graphical and numerical support of $\lim_{x \to c} x = c$.

Figure 18 For x close to c, the value of f is just as close to c; $\lim_{x \to c} x = c$.

TABLE 8

	x approaches c from the left				x approaches c from the right		
x	$c - 0.01$	$c - 0.001$	$c - 0.0001$	$\to c \leftarrow$	$c + 0.0001$	$c + 0.001$	$c + 0.01$
$f(x) = x$	$c - 0.01$	$c - 0.001$	$c - 0.0001$	$f(x)$ approaches c	$c + 0.0001$	$c + 0.001$	$c + 0.01$

For example,

$$\lim_{x \to -5} x = -5 \qquad \lim_{x \to \sqrt{3}} x = \sqrt{3} \qquad \lim_{x \to 0} x = 0$$

1 Find the Limit of a Sum, a Difference, and a Product

Many functions are combinations of sums, differences, and products of a constant function and the identity function. The following properties can be used to find the limit of such functions.

THEOREM Limit of a Sum

If f and g are functions for which $\lim_{x \to c} f(x)$ and $\lim_{x \to c} g(x)$ both exist, then $\lim_{x \to c}[f(x) + g(x)]$ exists and

$$\lim_{x \to c}[f(x) + g(x)] = \lim_{x \to c} f(x) + \lim_{x \to c} g(x)$$

IN WORDS The limit of the sum of two functions equals the sum of their limits.

A proof is given in Appendix B.

EXAMPLE 1 Finding the Limit of a Sum

Find $\lim_{x \to -3}(x + 4)$.

Solution $F(x) = x + 4$ is the sum of two functions $f(x) = x$ and $g(x) = 4$.

From the limits given in (1) and (2), we have

$$\lim_{x \to -3} f(x) = \lim_{x \to -3} x = -3 \qquad \text{and} \qquad \lim_{x \to -3} g(x) = \lim_{x \to -3} 4 = 4$$

Then, using the Limit of a Sum, we have

$$\lim_{x \to -3}(x + 4) = \lim_{x \to -3} x + \lim_{x \to -3} 4 = -3 + 4 = 1 \qquad \blacksquare$$

THEOREM Limit of a Difference

If f and g are functions for which $\lim_{x \to c} f(x)$ and $\lim_{x \to c} g(x)$ both exist, then $\lim_{x \to c}[f(x) - g(x)]$ exists and

$$\lim_{x \to c}[f(x) - g(x)] = \lim_{x \to c} f(x) - \lim_{x \to c} g(x)$$

IN WORDS The limit of the difference of two functions equals the difference of their limits.

EXAMPLE 2 Finding the Limit of a Difference

Find $\lim_{x \to 4}(6 - x)$.

Solution $F(x) = 6 - x$ is the difference of two functions $f(x) = 6$ and $g(x) = x$.

$$\lim_{x \to 4} f(x) = \lim_{x \to 4} 6 = 6 \qquad \text{and} \qquad \lim_{x \to 4} g(x) = \lim_{x \to 4} x = 4$$

Then, using the Limit of a Difference, we have

$$\lim_{x \to 4}(6 - x) = \lim_{x \to 4} 6 - \lim_{x \to 4} x = 6 - 4 = 2 \qquad \blacksquare$$

THEOREM Limit of a Product

If f and g are functions for which $\lim_{x \to c} f(x)$ and $\lim_{x \to c} g(x)$ both exist, then $\lim_{x \to c}[f(x) \cdot g(x)]$ exists and

$$\lim_{x \to c}[f(x) \cdot g(x)] = \lim_{x \to c} f(x) \cdot \lim_{x \to c} g(x)$$

IN WORDS The limit of the product of two functions equals the product of their limits.

A proof is given in Appendix B.

EXAMPLE 3   **Finding the Limit of a Product**

Find:

(a) $\lim\limits_{x \to 3} x^2$ **(b)** $\lim\limits_{x \to -5} (-4x)$

Solution (a) $F(x) = x^2$ is the product of two functions, $f(x) = x$ and $g(x) = x$. Then, using the Limit of a Product, we have

$$\lim_{x \to 3} x^2 = \lim_{x \to 3} x \cdot \lim_{x \to 3} x = (3)(3) = 9$$

(b) $F(x) = -4x$ is the product of two functions, $f(x) = -4$ and $g(x) = x$. Then, using the Limit of a Product, we have

$$\lim_{x \to -5} (-4x) = \lim_{x \to -5} (-4) \cdot \lim_{x \to -5} x = (-4)(-5) = 20 \qquad \blacksquare$$

A *corollary** of the Limit of a Product Theorem is the special case when $f(x) = k$ is a constant function.

COROLLARY Limit of a Constant Times a Function

If g is a function for which $\lim\limits_{x \to c} g(x)$ exists and if k is any real number, then $\lim\limits_{x \to c} [kg(x)]$ exists and

$$\boxed{\lim_{x \to c} [kg(x)] = k \lim_{x \to c} g(x)}$$

IN WORDS The limit of a constant times a function equals the constant times the limit of the function.

You are asked to prove this corollary in Problem 103.

Limit properties often are used in combination.

EXAMPLE 4   **Finding a Limit**

Find:

(a) $\lim\limits_{x \to 1} [2x(x + 4)]$ **(b)** $\lim\limits_{x \to 2^+} [4x(2 - x)]$

Solution (a)

$$\lim_{x \to 1} [(2x)(x + 4)] = \left[\lim_{x \to 1} (2x) \right] \left[\lim_{x \to 1} (x + 4) \right] \qquad \text{\color{blue}Limit of a Product}$$

$$= \left[2 \cdot \lim_{x \to 1} x \right] \cdot \left[\lim_{x \to 1} x + \lim_{x \to 1} 4 \right] \qquad \text{\color{blue}Limit of a Constant Times a Function, Limit of a Sum}$$

$$= (2 \cdot 1) \cdot (1 + 4) = 10 \qquad \text{\color{blue}Use (2) and (1), and simplify.}$$

NOTE The limit properties are also true for one-sided limits.

(b) We use properties of limits to find the one-sided limit.

$$\lim_{x \to 2^+} [4x(2 - x)] = 4 \lim_{x \to 2^+} [x(2 - x)] = 4 \left[\lim_{x \to 2^+} x \right] \left[\lim_{x \to 2^+} (2 - x) \right]$$

$$= 4 \cdot 2 \left[\lim_{x \to 2^+} 2 - \lim_{x \to 2^+} x \right] = 4 \cdot 2 \cdot (2 - 2) = 0 \qquad \blacksquare$$

NOW WORK **Problem 13.**

To find the limit of piecewise-defined functions at numbers where the defining equation changes requires the use of one-sided limits.

---

*A **corollary** is a theorem that follows directly from a previously proved theorem.

RECALL The limit L of a function $y = f(x)$ as x approaches a number c exists if and only if $\lim\limits_{x \to c^-} f(x) = \lim\limits_{x \to c^+} f(x) = L$.

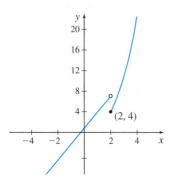

Figure 19 $f(x) = \begin{cases} 3x + 1 & \text{if } x < 2 \\ 2x(x-1) & \text{if } x \geq 2 \end{cases}$

ORIGINS Oliver Heaviside (1850–1925) was a self-taught mathematician and electrical engineer. He developed a branch of mathematics called **operational calculus** in which differential equations are solved by converting them to algebraic equations. Heaviside applied vector calculus to electrical engineering and, perhaps most significantly, he simplified *Maxwell's equations* to the form used by electrical engineers to this day. In 1902 Heaviside claimed there is a layer surrounding Earth from which radio signals bounce, allowing the signals to travel around the Earth. Heaviside's claim was proved true in 1923. The layer, contained in the ionosphere, is named the **Heaviside layer**. The function we discuss here is one of his minor contributions to mathematics and electrical engineering.

EXAMPLE 5 Finding a Limit

Find $\lim\limits_{x \to 2} f(x)$, if it exists, if

$$f(x) = \begin{cases} 3x + 1 & \text{if } x < 2 \\ 2x(x-1) & \text{if } x \geq 2 \end{cases}$$

Solution Since the rule for f changes at 2, we need to find the one-sided limits of f as x approaches 2.

For $x < 2$, we use the left-hand limit.

$$\lim_{x \to 2^-} f(x) = \lim_{x \to 2^-} (3x + 1) = \lim_{x \to 2^-} (3x) + \lim_{x \to 2^-} 1 = 3 \lim_{x \to 2^-} x + 1 = 3(2) + 1 = 7$$

For $x > 2$, we use the right-hand limit.

$$\lim_{x \to 2^+} f(x) = \lim_{x \to 2^+} [2x(x - 1)] = \lim_{x \to 2^+} (2x) \cdot \lim_{x \to 2^+} (x - 1)$$

$$= 2 \lim_{x \to 2^+} x \cdot \left[\lim_{x \to 2^+} x - \lim_{x \to 2^+} 1 \right] = 2 \cdot 2\,[2 - 1] = 4$$

Since $\lim\limits_{x \to 2^-} f(x) = 7 \neq \lim\limits_{x \to 2^+} f(x) = 4$, $\lim\limits_{x \to 2} f(x)$ does not exist. ∎

See Figure 19.

NOW WORK Problem 73.

The **Heaviside function**, $u_c(t) = \begin{cases} 0 & \text{if } t < c \\ 1 & \text{if } t \geq c \end{cases}$, is a step function that is used in electrical engineering to model a switch. The switch is off if $t < c$, and it is on if $t \geq c$.

EXAMPLE 6 Finding a Limit of the Heaviside Function

Find $\lim\limits_{t \to 0} u_0(t)$, where $u_0(t) = \begin{cases} 0 & \text{if } t < 0 \\ 1 & \text{if } t \geq 0 \end{cases}$

Solution Since this Heaviside function changes rules at $t = 0$, we find the one-sided limits as t approaches 0.

For $t < 0$, $\lim\limits_{t \to 0^-} u_0(t) = \lim\limits_{t \to 0^-} 0 = 0$ and for $t \geq 0$, $\lim\limits_{t \to 0^+} u_0(t) = \lim\limits_{t \to 0^+} 1 = 1$

Since the one-sided limits as t approaches 0 are not equal, $\lim\limits_{t \to 0} u_0(t)$ does not exist. ∎

NOW WORK Problem 81.

2 Find the Limit of a Power and the Limit of a Root

Using the Limit of a Product, if $\lim\limits_{x \to c} f(x)$ exists, then

$$\lim_{x \to c} [f(x)]^2 = \lim_{x \to c} [f(x) \cdot f(x)] = \lim_{x \to c} f(x) \cdot \lim_{x \to c} f(x) = \left[\lim_{x \to c} f(x) \right]^2$$

Repeated use of this property produces the next corollary.

COROLLARY Limit of a Power

If $\lim\limits_{x \to c} f(x)$ exists and if $n \geq 2$ is an integer, then

$$\boxed{\lim_{x \to c} [f(x)]^n = \left[\lim_{x \to c} f(x) \right]^n}$$

EXAMPLE 7 **Finding the Limit of a Power**

Find:

(a) $\lim_{x \to 2} x^5$ **(b)** $\lim_{x \to 1} (2x - 3)^3$ **(c)** $\lim_{x \to c} x^n$

Solution (a) $\lim_{x \to 2} x^5 = \left(\lim_{x \to 2} x \right)^5 = 2^5 = 32$

(b) $\lim_{x \to 1} (2x - 3)^3 = \left[\lim_{x \to 1} (2x - 3) \right]^3 = \left[\lim_{x \to 1} (2x) - \lim_{x \to 1} 3 \right]^3 = (2 - 3)^3 = -1$

(c) $\lim_{x \to c} x^n = \left[\lim_{x \to c} x \right]^n = c^n$ ∎

The result from Example 7(c) is worth remembering since it is used frequently:

$$\boxed{\lim_{x \to c} x^n = c^n}$$

where c is a number and n is a positive integer.

NOW WORK Problem 15.

THEOREM Limit of a Root

If $\lim_{x \to c} f(x)$ exists and if $n \ge 2$ is an integer, then

$$\boxed{\lim_{x \to c} \sqrt[n]{f(x)} = \sqrt[n]{\lim_{x \to c} f(x)}}$$

provided $f(x) \ge 0$ if n is even.

EXAMPLE 8 **Finding the Limit of** $f(x) = \sqrt[3]{x^2 + 11}$

Find $\lim_{x \to 4} \sqrt[3]{x^2 + 11}$.

Solution

$$\lim_{x \to 4} \sqrt[3]{x^2 + 11} \underset{\substack{\uparrow \\ \text{Limit of a Root}}}{=} \sqrt[3]{\lim_{x \to 4} (x^2 + 11)} \underset{\substack{\uparrow \\ \text{Limit of a Sum}}}{=} \sqrt[3]{\lim_{x \to 4} x^2 + \lim_{x \to 4} 11}$$

$$\underset{\substack{\uparrow \\ \lim_{x \to c} x^2 = c^2}}{=} \sqrt[3]{4^2 + 11} = \sqrt[3]{27} = 3$$

∎

NOW WORK Problem 19.

The Limit of a Power and the Limit of a Root are used together to find the limit of a function with a rational exponent.

THEOREM Limit of $[f(x)]^{m/n}$

If f is a function for which $\lim_{x \to c} f(x)$ exists and if $[f(x)]^{m/n}$ is defined for positive integers m and n, then

$$\boxed{\lim_{x \to c} [f(x)]^{m/n} = \left[\lim_{x \to c} f(x) \right]^{m/n}}$$

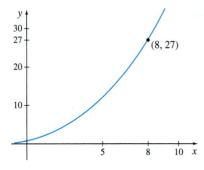

Figure 20 $f(x) = (x + 1)^{3/2}$

Finding the Limit of $f(x) = (x + 1)^{3/2}$

Find $\lim\limits_{x \to 8}(x + 1)^{3/2}$.

Solution Let $f(x) = x + 1$. Near 8, $x + 1 > 0$, so $(x + 1)^{3/2}$ is defined. Then

$$\lim_{x \to 8}[f(x)]^{3/2} = \lim_{x \to 8}(x + 1)^{3/2} = \left[\lim_{x \to 8}(x + 1)\right]^{3/2} = [8 + 1]^{3/2} = 9^{3/2} = 27$$

$$\lim_{x \to c}[f(x)]^{m/n} = \left[\lim_{x \to c} f(x)\right]^{m/n}$$

See Figure 20.

NOW WORK Problem 23.

3 Find the Limit of a Polynomial

Some limits can be found by substituting c for x. For example,

$$\lim_{x \to 2}(5x^2) = 5 \lim_{x \to 2} x^2 = 5 \cdot 2^2 = 20.$$

Since $\lim\limits_{x \to c} x^n = c^n$ if n is a positive integer, we can use the Limit of a Constant Times a Function to obtain a formula for the limit of a monomial $f(x) = ax^n$.

$$\lim_{x \to c}(ax^n) = ac^n$$

where a is any number.

Since a polynomial is the sum of monomials and the limit of a sum is the sum of the limits, we have the following result.

THEOREM Limit of a Polynomial Function

If P is a polynomial function, then

$$\lim_{x \to c} P(x) = P(c)$$

IN WORDS To find the limit of a polynomial as x approaches c, evaluate the polynomial at c.

for any number c.

Proof If P is a polynomial function, that is, if

$$P(x) = a_n x^n + a_{n-1} x^{n-1} + \cdots + a_1 x + a_0$$

where n is a nonnegative integer, then

$$\lim_{x \to c} P(x) = \lim_{x \to c}\left(a_n x^n + a_{n-1} x^{n-1} + \cdots + a_1 x + a_0\right)$$

$$= \lim_{x \to c}\left(a_n x^n\right) + \lim_{x \to c}\left(a_{n-1} x^{n-1}\right) + \cdots + \lim_{x \to c}(a_1 x) + \lim_{x \to c} a_0$$

$$= a_n c^n + a_{n-1} c^{n-1} + \cdots + a_1 c + a_0 \qquad \text{Limit of a Monomial}$$

$$= P(c)$$

Finding the Limit of a Polynomial

Find the limit of each polynomial:

(a) $\lim\limits_{x \to 3}(4x^2 - x + 2) = 4(3)^2 - 3 + 2 = 35$

(b) $\lim\limits_{x \to -1}(7x^5 + 4x^3 - 2x^2) = 7(-1)^5 + 4(-1)^3 - 2(-1)^2 = -13$

(c) $\lim\limits_{x \to 0}(10x^6 - 4x^5 - 8x + 5) = 10(0)^6 - 4(0)^5 - 8(0) + 5 = 5$

NOW WORK Problem 29.

4 Find the Limit of a Quotient

To find the limit of a rational function, which is the quotient of two polynomials, we use the following result.

IN WORDS The limit of the quotient of two functions equals the quotient of their limits, provided that the limit of the denominator is not zero.

THEOREM Limit of a Quotient

If f and g are functions for which $\lim\limits_{x \to c} f(x)$ and $\lim\limits_{x \to c} g(x)$ both exist, then $\lim\limits_{x \to c} \left[\dfrac{f(x)}{g(x)} \right]$ exists and

$$\lim_{x \to c} \left[\frac{f(x)}{g(x)} \right] = \frac{\lim\limits_{x \to c} f(x)}{\lim\limits_{x \to c} g(x)}$$

provided $\lim\limits_{x \to c} g(x) \neq 0$.

NEED TO REVIEW? Rational functions are discussed in Section P.2, pp. 19–20.

COROLLARY Limit of a Rational Function

If the number c is in the domain of a rational function $R(x) = \dfrac{p(x)}{q(x)}$, then

$$\lim_{x \to c} R(x) = R(c) \tag{3}$$

You are asked to prove this corollary in Problem 104.

EXAMPLE 11   Finding the Limit of a Rational Function

Find:

(a) $\lim\limits_{x \to 1} \dfrac{3x^3 - 2x + 1}{4x^2 + 5}$ **(b)** $\lim\limits_{x \to -2} \dfrac{2x + 4}{3x^2 - 1}$

Solution (a) Since 1 is in the domain of the rational function $R(x) = \dfrac{3x^3 - 2x + 1}{4x^2 + 5}$,

$$\lim_{x \to 1} R(x) = R(1) = \frac{3 - 2 + 1}{4 + 5} = \frac{2}{9}$$
$$\uparrow$$
$$\text{Use (3)}$$

(b) Since -2 is in the domain of the rational function $H(x) = \dfrac{2x + 4}{3x^2 - 1}$,

$$\lim_{x \to -2} H(x) = H(-2) = \frac{-4 + 4}{12 - 1} = \frac{0}{11} = 0$$
$$\uparrow$$
$$\text{Use (3)}$$ ∎

NOW WORK Problem 33.

EXAMPLE 12   Finding the Limit of a Quotient

Find $\lim\limits_{x \to 4} \dfrac{\sqrt{3x^2 + 1}}{x - 1}$.

Solution We seek the limit of the quotient of two functions. Since the limit of the denominator $\lim\limits_{x \to 4}(x - 1) \neq 0$, we use the Limit of a Quotient.

$$\lim_{x \to 4} \frac{\sqrt{3x^2 + 1}}{x - 1} = \frac{\lim\limits_{x \to 4} \sqrt{3x^2 + 1}}{\lim\limits_{x \to 4}(x - 1)} = \frac{\sqrt{\lim\limits_{x \to 4}(3x^2 + 1)}}{\lim\limits_{x \to 4}(x - 1)} = \frac{\sqrt{3 \cdot 4^2 + 1}}{4 - 1} = \frac{\sqrt{49}}{3} = \frac{7}{3}$$
$$\uparrow \uparrow$$
$$\text{Limit of a Quotient}\text{Limit of a Root}$$ ∎

NOW WORK Problem 31.

Based on these examples, you might be tempted to conclude that finding a limit as x approaches c is simply a matter of substituting the number c into the function. The next few examples show that substitution cannot always be used and other strategies need to be employed.

The limit of a rational function can be found using substitution, provided the number c being approached is in the domain of the rational function. The next example shows a strategy that can be tried when c is not in the domain.

EXAMPLE 13   Finding the Limit of a Rational Function

Find $\lim\limits_{x\to -2} \dfrac{x^2 + 5x + 6}{x^2 - 4}$.

Solution Since -2 is not in the domain of the rational function, (3) cannot be used. But this does not mean that the limit does not exist! Factoring the numerator and the denominator, we find

$$\frac{x^2 + 5x + 6}{x^2 - 4} = \frac{(x+2)(x+3)}{(x+2)(x-2)}$$

Since $x \neq -2$, and we are interested in the limit as x approaches -2, the factor $x + 2$ can be divided out. Then

$$\lim_{x\to -2} \frac{x^2 + 5x + 6}{x^2 - 4} = \lim_{x\to -2} \frac{(x+2)(x+3)}{(x+2)(x-2)} = \lim_{x\to -2} \frac{x+3}{x-2} = \frac{-2+3}{-2-2} = -\frac{1}{4}$$

$\uparrow$ Factor　　　　$\uparrow$ $x\neq -2$　　$\uparrow$ Use the Limit of a
Divide out $(x+2)$　Rational Function

NOW WORK Problem 35.

The Limit of a Quotient property can only be used when the limit of the denominator of the function is not zero. The next example illustrates a strategy to try if radicals are present.

EXAMPLE 14   Finding the Limit of a Quotient

Find $\lim\limits_{x\to 5} \dfrac{\sqrt{x} - \sqrt{5}}{x - 5}$.

Solution The domain of $h(x) = \dfrac{\sqrt{x} - \sqrt{5}}{x - 5}$ is $\{x \mid x \geq 0, x \neq 5\}$. Since the limit of the denominator is

$$\lim_{x\to 5} g(x) = \lim_{x\to 5}(x - 5) = 0$$

we cannot use the Limit of a Quotient property. A different strategy is necessary. We rationalize the numerator of the quotient.

$$\frac{\sqrt{x} - \sqrt{5}}{x - 5} = \frac{(\sqrt{x} - \sqrt{5})}{(x - 5)} \cdot \frac{(\sqrt{x} + \sqrt{5})}{(\sqrt{x} + \sqrt{5})} = \frac{x - 5}{(x-5)(\sqrt{x} + \sqrt{5})} = \frac{1}{\sqrt{x} + \sqrt{5}}$$
$\uparrow$ $x\neq 5$

Do you see why rationalizing the numerator works? It causes the term $x - 5$ to appear in the numerator, and since $x \neq 5$, the factor $x - 5$ can be divided out. Then

$$\lim_{x\to 5} \frac{\sqrt{x} - \sqrt{5}}{x - 5} = \lim_{x\to 5} \frac{1}{\sqrt{x} + \sqrt{5}} = \frac{1}{\sqrt{5} + \sqrt{5}} = \frac{1}{2\sqrt{5}} = \frac{\sqrt{5}}{10}$$
$\uparrow$ Use the Limit of a Quotient

NOTE When finding a limit, remember to include "$\lim\limits_{x\to c}$" at each step until you let $x \to c$.

NOW WORK Problem 41.

5 Find the Limit of an Average Rate of Change

The next two examples illustrate limits that we encounter in Chapter 2.

NEED TO REVIEW? Average rate of change is discussed in Section P.1, p. 11.

In Section P.1, we defined average rate of change: If a and b, where $a \neq b$, are in the domain of a function $y = f(x)$, the average rate of change of f from a to b is

$$\frac{\Delta y}{\Delta x} = \frac{f(b) - f(a)}{b - a} \qquad a \neq b$$

EXAMPLE 15   Finding the Limit of an Average Rate of Change

(a) Find the average rate of change of $f(x) = x^2 + 3x$ from 2 to x; $x \neq 2$.

(b) Find the limit as x approaches 2 of the average rate of change of $f(x) = x^2 + 3x$ from 2 to x.

Solution (a) The average rate of change of f from 2 to x is

$$\frac{\Delta y}{\Delta x} = \frac{f(x) - f(2)}{x - 2} = \frac{(x^2 + 3x) - [2^2 + 3(2)]}{x - 2} = \frac{x^2 + 3x - 10}{x - 2} = \frac{(x + 5)(x - 2)}{x - 2}$$

(b) The limit of the average rate of change is

$$\lim_{x \to 2} \frac{f(x) - f(2)}{x - 2} = \lim_{x \to 2} \frac{(x + 5)(x - 2)}{x - 2} = \lim_{x \to 2} (x + 5) = 7 \qquad ■$$

NOW WORK Problem 63.

6 Find the Limit of a Difference Quotient

In Section P.1, we defined the difference quotient for a function f as
$$\frac{f(x + h) - f(x)}{h}, h \neq 0.$$

EXAMPLE 16   Finding the Limit of a Difference Quotient

(a) For $f(x) = 2x^2 - 3x + 1$, find the difference quotient $\dfrac{f(x + h) - f(x)}{h}$, $h \neq 0$.

(b) Find the limit as h approaches 0 of the difference quotient of $f(x) = 2x^2 - 3x + 1$.

Solution (a) To find the difference quotient of f, we begin with $f(x + h)$.

$$f(x + h) = 2(x + h)^2 - 3(x + h) + 1 = 2(x^2 + 2xh + h^2) - 3x - 3h + 1$$
$$= 2x^2 + 4xh + 2h^2 - 3x - 3h + 1$$

Now

$$f(x+h) - f(x) = (2x^2 + 4xh + 2h^2 - 3x - 3h + 1) - (2x^2 - 3x + 1) = 4xh + 2h^2 - 3h$$

Then, the difference quotient is

$$\frac{f(x + h) - f(x)}{h} = \frac{4xh + 2h^2 - 3h}{h} = \frac{h(4x + 2h - 3)}{h} = 4x + 2h - 3, \quad h \neq 0$$

(b) $\displaystyle \lim_{h \to 0} \frac{f(x + h) - f(x)}{h} = \lim_{h \to 0} (4x + 2h - 3) = 4x + 0 - 3 = 4x - 3$ ■

NOW WORK Problem 71.

Summary

Two Basic Limits

- $\lim_{x \to c} A = A$, where A is a constant.
- $\lim_{x \to c} x = c$

Properties of Limits

If f and g are functions for which $\lim_{x \to c} f(x)$ and $\lim_{x \to c} g(x)$ both exist, and k is a constant, then

- **Limit of a Sum or a Difference:**
  $\lim_{x \to c}[f(x) \pm g(x)] = \lim_{x \to c} f(x) \pm \lim_{x \to c} g(x)$
- **Limit of a Product:** $\lim_{x \to c}[f(x) \cdot g(x)] = \lim_{x \to c} f(x) \cdot \lim_{x \to c} g(x)$
- **Limit of a Constant Times a Function:** $\lim_{x \to c}[kg(x)] = k \lim_{x \to c} g(x)$

- **Limit of a Power:** $\lim_{x \to c}[f(x)]^n = \left[\lim_{x \to c} f(x)\right]^n$
  where $n \geq 2$ is an integer
- **Limit of a Root:** $\lim_{x \to c} \sqrt[n]{f(x)} = \sqrt[n]{\lim_{x \to c} f(x)}$
  provided $f(x) \geq 0$ if $n \geq 2$ is even
- **Limit of $[f(x)]^{m/n}$:** $\lim_{x \to c}[f(x)]^{m/n} = \left[\lim_{x \to c} f(x)\right]^{m/n}$
  provided $[f(x)]^{m/n}$ is defined for positive integers m and n
- **Limit of a Quotient:** $\lim_{x \to c}\left[\dfrac{f(x)}{g(x)}\right] = \dfrac{\lim_{x \to c} f(x)}{\lim_{x \to c} g(x)}$
  provided $\lim_{x \to c} g(x) \neq 0$
- **Limit of a Polynomial Function:** $\lim_{x \to c} P(x) = P(c)$
- **Limit of a Rational Function:** $\lim_{x \to c} R(x) = R(c)$
  if c is in the domain of R

1.2 Assess Your Understanding

Concepts and Vocabulary

1. (a) $\lim_{x \to 4} (-3) =$ _____; (b) $\lim_{x \to 0} \pi =$ _____

2. If $\lim_{x \to c} f(x) = 3$, then $\lim_{x \to c}[f(x)]^5 =$ _____.

3. If $\lim_{x \to c} f(x) = 64$, then $\lim_{x \to c} \sqrt[3]{f(x)} =$ _____.

4. (a) $\lim_{x \to -1} x =$ _____; (b) $\lim_{x \to e} x =$ _____

5. (a) $\lim_{x \to 0} (x - 2) =$ ____; (b) $\lim_{x \to 1/2} (3 + x) =$ _____

6. (a) $\lim_{x \to 2} (-3x) =$ _____; (b) $\lim_{x \to 0} (3x) =$ _____

7. *True or False* If p is a polynomial function, then $\lim_{x \to 5} p(x) = p(5)$.

8. If the domain of a rational function R is $\{x \mid x \neq 0\}$, then $\lim_{x \to 2} R(x) = R(\_\_\_\_)$.

9. *True or False* Properties of limits cannot be used for one-sided limits.

10. *True or False* If $f(x) = \dfrac{(x+1)(x+2)}{x+1}$ and $g(x) = x + 2$, then $\lim_{x \to -1} f(x) = \lim_{x \to -1} g(x)$.

Skill Building

In Problems 11–44, find each limit using properties of limits.

11. $\lim_{x \to 3} [2(x + 4)]$

12. $\lim_{x \to -2} [3(x + 1)]$

13. $\lim_{x \to -2} [x(3x-1)(x + 2)]$

14. $\lim_{x \to -1} [x(x - 1)(x + 10)]$

15. $\lim_{t \to 1} (3t - 2)^3$

16. $\lim_{x \to 0} (-3x + 1)^2$

17. $\lim_{x \to 4} (3\sqrt{x})$

18. $\lim_{x \to 8} \left(\dfrac{1}{4}\sqrt[3]{x}\right)$

19. $\lim_{x \to 3} \sqrt{5x - 4}$

20. $\lim_{t \to 2} \sqrt{3t + 4}$

21. $\lim_{t \to 2} [t\sqrt{(5t + 3)(t + 4)}]$

22. $\lim_{t \to -1} [t\sqrt[3]{(t + 1)(2t - 1)}]$

23. $\lim_{x \to 3} (\sqrt{x} + x + 4)^{1/2}$

24. $\lim_{t \to 2} (t\sqrt{2t} + 4)^{1/3}$

25. $\lim_{t \to -1} [4t(t + 1)]^{2/3}$

26. $\lim_{x \to 0} (x^2 - 2x)^{3/5}$

27. $\lim_{t \to 1} (3t^2 - 2t + 4)$

28. $\lim_{x \to 0} (-3x^4 + 2x + 1)$

29. $\lim_{x \to \frac{1}{2}} (2x^4 - 8x^3 + 4x - 5)$

30. $\lim_{x \to -\frac{1}{3}} (27x^3 + 9x + 1)$

31. $\lim_{x \to 4} \dfrac{x^2 + 4}{\sqrt{x}}$

32. $\lim_{x \to 3} \dfrac{x^2 + 5}{\sqrt{3x}}$

33. $\lim_{x \to -2} \dfrac{2x^3 + 5x}{3x - 2}$

34. $\lim_{x \to 1} \dfrac{2x^4 - 1}{3x^3 + 2}$

35. $\lim_{x \to 2} \dfrac{x^2 - 4}{x - 2}$

36. $\lim_{x \to -2} \dfrac{x + 2}{x^2 - 4}$

37. $\lim_{x \to -1} \dfrac{x^3 - x}{x + 1}$

38. $\lim_{x \to -1} \dfrac{x^3 + x^2}{x^2 - 1}$

39. $\lim_{x \to -8} \left(\dfrac{2x}{x + 8} + \dfrac{16}{x + 8}\right)$

40. $\lim_{x \to 2} \left(\dfrac{3x}{x - 2} - \dfrac{6}{x - 2}\right)$

41. $\lim_{x \to 2} \dfrac{\sqrt{x} - \sqrt{2}}{x - 2}$

42. $\lim_{x \to 3} \dfrac{\sqrt{x} - \sqrt{3}}{x - 3}$

43. $\lim_{x \to 4} \dfrac{\sqrt{x + 5} - 3}{(x - 4)(x + 1)}$

44. $\lim_{x \to 3} \dfrac{\sqrt{x + 1} - 2}{x(x - 3)}$

In Problems 45–50, find each one-sided limit using properties of limits.

45. $\lim_{x \to 3^-} (x^2 - 4)$

46. $\lim_{x \to 2^+} (3x^2 + x)$

47. $\lim_{x \to 3^-} \dfrac{x^2 - 9}{x - 3}$

48. $\lim_{x \to 3^+} \dfrac{x^2 - 9}{x - 3}$

49. $\lim_{x \to 3^-} (\sqrt{9 - x^2} + x)^2$

50. $\lim_{x \to 2^+} (2\sqrt{x^2 - 4} + 3x)$

1. = NOW WORK problem   [/\/] = Graphing technology recommended   [CAS] = Computer Algebra System recommended

In Problems 51–58, use the information below to find each limit.

$$\lim_{x \to c} f(x) = 5 \qquad \lim_{x \to c} g(x) = 2 \qquad \lim_{x \to c} h(x) = 0$$

51. $\lim_{x \to c} [f(x) - 3g(x)]$

52. $\lim_{x \to c} [5f(x)]$

53. $\lim_{x \to c} [g(x)]^3$

54. $\lim_{x \to c} \dfrac{f(x)}{g(x) - h(x)}$

55. $\lim_{x \to c} \dfrac{h(x)}{g(x)}$

56. $\lim_{x \to c} [4f(x) \cdot g(x)]$

57. $\lim_{x \to c} \left[\dfrac{1}{g(x)}\right]^2$

58. $\lim_{x \to c} \sqrt[3]{5g(x) - 3}$

In Problems 59 and 60, use the graph of the functions and properties of limits to find each limit, if it exists. If the limit does not exist, say, "the limit does not exist," and explain why.

59. (a) $\lim_{x \to 4} [f(x) + g(x)]$

(b) $\lim_{x \to 4} \{f(x)[g(x) - h(x)]\}$

(c) $\lim_{x \to 4} [f(x) \cdot g(x)]$

(d) $\lim_{x \to 4} [2h(x)]$

(e) $\lim_{x \to 4} \dfrac{g(x)}{f(x)}$

(f) $\lim_{x \to 4} \dfrac{h(x)}{f(x)}$

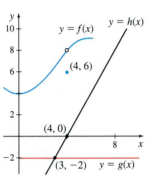

60. (a) $\lim_{x \to 3} \{2[f(x) + h(x)]\}$

(b) $\lim_{x \to 3^-} [g(x) + h(x)]$

(c) $\lim_{x \to 3} \sqrt[3]{h(x)}$

(d) $\lim_{x \to 3} \dfrac{f(x)}{h(x)}$

(e) $\lim_{x \to 3} [h(x)]^3$

(f) $\lim_{x \to 3} [f(x) - 2h(x)]^{3/2}$

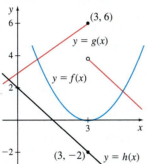

In Problems 61–66, for each function f, find the limit as x approaches c of the average rate of change of f from c to x. That is, find

$$\lim_{x \to c} \dfrac{f(x) - f(c)}{x - c}$$

61. $f(x) = 3x^2, \quad c = 1$

62. $f(x) = 8x^3, \quad c = 2$

63. $f(x) = -2x^2 + 4, \quad c = 1$

64. $f(x) = 20 - 0.8x^2, \quad c = 3$

65. $f(x) = \sqrt{x}, \quad c = 1$

66. $f(x) = \sqrt{2x}, \quad c = 5$

In Problems 67–72, find the limit of the difference quotient for each function f. That is, find $\lim_{h \to 0} \dfrac{f(x + h) - f(x)}{h}$.

67. $f(x) = 4x - 3$

68. $f(x) = 3x + 5$

69. $f(x) = 3x^2 + 4x + 1$

70. $f(x) = 2x^2 + x$

71. $f(x) = \dfrac{2}{x}$

72. $f(x) = \dfrac{3}{x^2}$

In Problems 73–80, find $\lim_{x \to c^-} f(x)$ *and* $\lim_{x \to c^+} f(x)$ *for the given number c. Based on the results, determine whether* $\lim_{x \to c} f(x)$ *exists.*

73. $f(x) = \begin{cases} 2x - 3 & \text{if } x \le 1 \\ 3 - x & \text{if } x > 1 \end{cases}$ at $c = 1$

74. $f(x) = \begin{cases} 5x + 2 & \text{if } x < 2 \\ 1 + 3x & \text{if } x \ge 2 \end{cases}$ at $c = 2$

75. $f(x) = \begin{cases} 3x - 1 & \text{if } x < 1 \\ 4 & \text{if } x = 1 \\ 2x & \text{if } x > 1 \end{cases}$ at $c = 1$

76. $f(x) = \begin{cases} 3x - 1 & \text{if } x < 1 \\ 2 & \text{if } x = 1 \\ 2x & \text{if } x > 1 \end{cases}$ at $c = 1$

77. $f(x) = \begin{cases} x - 1 & \text{if } x < 1 \\ \sqrt{x - 1} & \text{if } x > 1 \end{cases}$ at $c = 1$

78. $f(x) = \begin{cases} \sqrt{9 - x^2} & \text{if } 0 < x < 3 \\ \sqrt{x^2 - 9} & \text{if } x > 3 \end{cases}$ at $c = 3$

79. $f(x) = \begin{cases} \dfrac{x^2 - 9}{x - 3} & \text{if } x \ne 3 \\ 6 & \text{if } x = 3 \end{cases}$ at $c = 3$

80. $f(x) = \begin{cases} \dfrac{x - 2}{x^2 - 4} & \text{if } x \ne 2 \\ 1 & \text{if } x = 2 \end{cases}$ at $c = 2$

Applications and Extensions

Heaviside Functions *In Problems 81 and 82, find the limit of the given Heaviside function at c.*

81. $u_1(t) = \begin{cases} 0 & \text{if } t < 1 \\ 1 & \text{if } t \ge 1 \end{cases}$ at $c = 1$

82. $u_3(t) = \begin{cases} 0 & \text{if } t < 3 \\ 1 & \text{if } t \ge 3 \end{cases}$ at $c = 3$

In Problems 83–92, find each limit.

83. $\lim_{h \to 0} \dfrac{(x + h)^2 - x^2}{h}$

84. $\lim_{h \to 0} \dfrac{\sqrt{x + h} - \sqrt{x}}{h}$

85. $\lim_{h \to 0} \dfrac{\dfrac{1}{x + h} - \dfrac{1}{x}}{h}$

86. $\lim_{h \to 0} \dfrac{\dfrac{1}{(x + h)^3} - \dfrac{1}{x^3}}{h}$

87. $\lim_{x \to 0} \left[\dfrac{1}{x}\left(\dfrac{1}{4 + x} - \dfrac{1}{4}\right)\right]$

88. $\lim_{x \to -1} \left[\dfrac{2}{x + 1}\left(\dfrac{1}{3} - \dfrac{1}{x + 4}\right)\right]$

89. $\lim_{x \to 7} \dfrac{x - 7}{\sqrt{x + 2} - 3}$

90. $\lim_{x \to 2} \dfrac{x - 2}{\sqrt{x + 2} - 2}$

91. $\lim_{x \to 1} \dfrac{x^3 - 3x^2 + 3x - 1}{x^2 - 2x + 1}$

92. $\lim_{x \to -3} \dfrac{x^3 + 7x^2 + 15x + 9}{x^2 + 6x + 9}$

93. Cost of Water The Jericho Water District determines quarterly water costs, in dollars, using the following rate schedule:

Water used (in thousands of gallons)	Cost
$0 \le x \le 10$	$9.00
$10 < x \le 30$	$9.00 + 0.95$ for each thousand gallons in excess of 10,000 gallons
$30 < x \le 100$	$28.00 + 1.65$ for each thousand gallons in excess of 30,000 gallons
$x > 100$	$143.50 + 2.20$ for each thousand gallons in excess of 100,000 gallons

Source: Jericho Water District, Syosset, NY.

(a) Find a function C that models the quarterly cost, in dollars, of using x thousand gallons of water.

(b) What is the domain of the function C?

(c) Find each of the following limits. If the limit does not exist, explain why.

$$\lim_{x \to 5} C(x) \quad \lim_{x \to 10} C(x) \quad \lim_{x \to 30} C(x) \quad \lim_{x \to 100} C(x)$$

(d) What is $\lim_{x \to 0^+} C(x)$?

(e) Graph the function C.

94. Cost of Natural Gas In February 2012 Peoples Energy had the following monthly rate schedule for natural gas usage in single- and two-family residences with a single gas meter:

Monthly customer charge	$22.13
Per therm distribution charge	
≤ 50 therms	$0.25963 per therm
> 50 therms	$12.98 + 0.11806$ for each therm in excess of 50
Gas charge:	$0.3631 per therm

Source: Peoples Energy, Chicago, IL.

(a) Find a function C that models the monthly cost, in dollars, of using x therms of natural gas.

(b) What is the domain of the function C?

(c) Find $\lim_{x \to 50} C(x)$, if it exists. If the limit does not exist, explain why.

(d) What is $\lim_{x \to 0^+} C(x)$?

(e) Graph the function C.

95. Low-Temperature Physics In thermodynamics, the average molecular kinetic energy (energy of motion) of a gas having molecules of mass m is directly proportional to its temperature T on the absolute (or Kelvin) scale. This can be expressed as $\frac{1}{2}mv^2 = \frac{3}{2}kT$, where $v = v(T)$ is the speed of a typical molecule at time t and k is a constant, known as the **Boltzmann constant**.

(a) What limit does the molecular speed v approach as the gas temperature T approaches absolute zero (0 K or -273°C or -469°F)?

(b) What does this limit suggest about the behavior of a gas as its temperature approaches absolute zero?

96. For the function $f(x) = \begin{cases} 3x + 5 & \text{if } x \le 2 \\ 13 - x & \text{if } x > 2 \end{cases}$, find

(a) $\displaystyle\lim_{h \to 0^-} \frac{f(2+h) - f(2)}{h}$

(b) $\displaystyle\lim_{h \to 0^+} \frac{f(2+h) - f(2)}{h}$

(c) Does $\displaystyle\lim_{h \to 0} \frac{f(2+h) - f(2)}{h}$ exist?

97. Use the fact that $|x| = \begin{cases} x & \text{if } x \ge 0 \\ -x & \text{if } x < 0 \end{cases}$ to show that $\lim_{x \to 0} |x| = 0$.

98. Use the fact that $|x| = \sqrt{x^2}$ to show that $\lim_{x \to 0} |x| = 0$.

99. Find functions f and g for which $\lim_{x \to c} [f(x) + g(x)]$ may exist even though $\lim_{x \to c} f(x)$ and $\lim_{x \to c} g(x)$ do not exist.

100. Find functions f and g for which $\lim_{x \to c} [f(x)g(x)]$ may exist even though $\lim_{x \to c} f(x)$ and $\lim_{x \to c} g(x)$ do not exist.

101. Find functions f and g for which $\lim_{x \to c} \left[\dfrac{f(x)}{g(x)} \right]$ may exist even though $\lim_{x \to c} f(x)$ and $\lim_{x \to c} g(x)$ do not exist.

102. Find a function f for which $\lim_{x \to c} |f(x)|$ may exist even though $\lim_{x \to c} f(x)$ does not exist.

103. Prove that if g is a function for which $\lim_{x \to c} g(x)$ exists and if k is any real number, then $\lim_{x \to c} [kg(x)]$ exists and $\lim_{x \to c} [kg(x)] = k \lim_{x \to c} g(x)$.

104. Prove that if the number c is in the domain of a rational function $R(x) = \dfrac{p(x)}{q(x)}$, then $\lim_{x \to c} R(x) = R(c)$.

Challenge Problems

105. Find $\displaystyle\lim_{x \to a} \frac{x^n - a^n}{x - a}$, n a positive integer.

106. Find $\displaystyle\lim_{x \to -a} \frac{x^n + a^n}{x + a}$, n a positive integer.

107. Find $\displaystyle\lim_{x \to 1} \frac{x^m - 1}{x^n - 1}$, m, n positive integers.

108. Find $\displaystyle\lim_{x \to 0} \frac{\sqrt[3]{1+x} - 1}{x}$.

109. Find $\displaystyle\lim_{x \to 0} \frac{\sqrt{(1+ax)(1+bx)} - 1}{x}$.

110. Find $\displaystyle\lim_{x \to 0} \frac{\sqrt{(1+a_1x)(1+a_2x)\cdots(1+a_nx)} - 1}{x}$.

111. Find $\displaystyle\lim_{h \to 0} \frac{f(h) - f(0)}{h}$ if $f(x) = x|x|$.

1.3 Continuity

OBJECTIVES *When you finish this section, you should be able to:*

1 Determine whether a function is continuous at a number (p. 93)
2 Determine intervals on which a function is continuous (p. 96)
3 Use properties of continuity (p. 98)
4 Use the Intermediate Value Theorem (p. 100)

Sometimes $\lim\limits_{x \to c} f(x)$ equals $f(c)$ and sometimes it does not. In fact, $f(c)$ may not even be defined and yet $\lim\limits_{x \to c} f(x)$ may exist. In this section, we investigate the relationship between $\lim\limits_{x \to c} f(x)$ and $f(c)$. Figure 21 shows some possibilities.

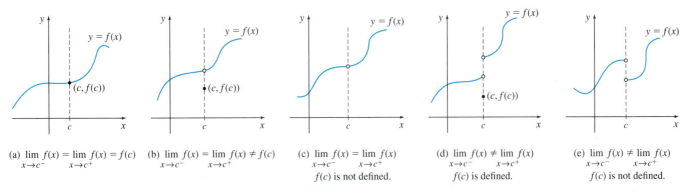

(a) $\lim\limits_{x \to c^-} f(x) = \lim\limits_{x \to c^+} f(x) = f(c)$

(b) $\lim\limits_{x \to c^-} f(x) = \lim\limits_{x \to c^+} f(x) \neq f(c)$

(c) $\lim\limits_{x \to c^-} f(x) = \lim\limits_{x \to c^+} f(x)$
$f(c)$ is not defined.

(d) $\lim\limits_{x \to c^-} f(x) \neq \lim\limits_{x \to c^+} f(x)$
$f(c)$ is defined.

(e) $\lim\limits_{x \to c^-} f(x) \neq \lim\limits_{x \to c^+} f(x)$
$f(c)$ is not defined.

Figure 21

Of these five graphs, the "nicest" one is Figure 21(a). There, $\lim\limits_{x \to c} f(x)$ exists and is equal to $f(c)$. Functions that have this property are said to be *continuous at the number c*. This agrees with the intuitive notion that a function is continuous if its graph can be drawn without lifting the pencil. The functions in Figures 21(b)–(e) are not continuous at c, since each has a break in the graph at c. This leads to the definition of *continuity at a number*.

DEFINITION Continuity at a Number

A function f is **continuous at a number** c if the following three conditions are met:

- $f(c)$ is defined (that is, c is in the domain of f)
- $\lim\limits_{x \to c} f(x)$ exists
- $\lim\limits_{x \to c} f(x) = f(c)$

If *any one* of these three conditions is not satisfied, then the function is **discontinuous at** c.

1 Determine Whether a Function Is Continuous at a Number

EXAMPLE 1 Determining Whether a Function Is Continuous at a Number

(a) Determine whether $f(x) = 3x^2 - 5x + 4$ is continuous at 1.

(b) Determine whether $g(x) = \dfrac{x^2 + 9}{x^2 - 4}$ is continuous at 2.

Solution (a) We begin by checking the conditions for continuity. First, 1 is in the domain of f and $f(1) = 2$. Second, $\lim\limits_{x \to 1} f(x) = \lim\limits_{x \to 1} (3x^2 - 5x + 4) = 2$, so $\lim\limits_{x \to 1} f(x)$ exists. Third, $\lim\limits_{x \to 1} f(x) = f(1)$. Since the three conditions are met, f is continuous at 1.

(b) Since 2 is not in the domain of g, the function g is discontinuous at 2. ∎

Figure 22 shows the graphs of f and g from Example 1. Notice that f is continuous at 1, and its graph is drawn without lifting the pencil. But the function g is discontinuous at 2, and to draw its graph, you must lift your pencil at $x = 2$.

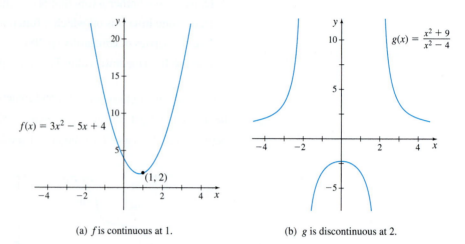

(a) f is continuous at 1.   (b) g is discontinuous at 2.

Figure 22

NOW WORK   **Problem** 19.

EXAMPLE 2   Determining Whether a Function Is Continuous at a Number

Determine if $f(x) = \dfrac{\sqrt{x^2 + 2}}{x^2 - 4}$ is continuous at the numbers -2, 0, and 2.

Solution   The domain of f is $\{x \mid x \neq -2, \ x \neq 2\}$. Since f is not defined at -2 and 2, the function f is not continuous at -2 and at 2. The number 0 is in the domain of f.

That is, f is defined at 0, and $f(0) = -\dfrac{\sqrt{2}}{4}$. Also,

$$\lim_{x \to 0} f(x) = \lim_{x \to 0} \frac{\sqrt{x^2 + 2}}{x^2 - 4} = \frac{\lim\limits_{x \to 0} \sqrt{x^2 + 2}}{\lim\limits_{x \to 0} (x^2 - 4)} = \frac{\sqrt{\lim\limits_{x \to 0} (x^2 + 2)}}{\lim\limits_{x \to 0} x^2 - \lim\limits_{x \to 0} 4}$$

$$= \frac{\sqrt{0 + 2}}{0 - 4} = -\frac{\sqrt{2}}{4} = f(0)$$

The three conditions of continuity at a number are met. So, the function f is continuous at 0. ∎

Figure 23 shows the graph of f.

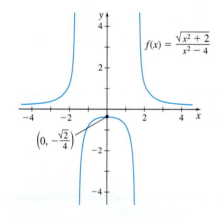

Figure 23 f is continuous at 0; f is discontinuous at -2 and 2.

NOW WORK   **Problem** 21.

EXAMPLE 3   Determining Whether a Piecewise-Defined Function Is Continuous

Determine if the function

$$f(x) = \begin{cases} \dfrac{x^2 - 9}{x - 3} & \text{if } x < 3 \\[2mm] 9 & \text{if } x = 3 \\[2mm] x^2 - 3 & \text{if } x > 3 \end{cases}$$

is continuous at 3.

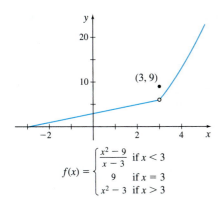

$$f(x) = \begin{cases} \dfrac{x^2 - 9}{x - 3} & \text{if } x < 3 \\ 9 & \text{if } x = 3 \\ x^2 - 3 & \text{if } x > 3 \end{cases}$$

Figure 24 f is discontinuous at 3.

Solution Since $f(3) = 9$, the function f is defined at 3. To check the second condition, we investigate the one-sided limits.

$$\lim_{x \to 3^-} f(x) = \lim_{x \to 3^-} \frac{x^2 - 9}{x - 3} = \lim_{x \to 3^-} \frac{(x - 3)(x + 3)}{x - 3} \underset{\substack{\uparrow \\ \text{Divide out } x - 3}}{=} \lim_{x \to 3^-} (x + 3) = 6$$

$$\lim_{x \to 3^+} f(x) = \lim_{x \to 3^+} (x^2 - 3) = 9 - 3 = 6$$

Since $\lim\limits_{x \to 3^-} f(x) = \lim\limits_{x \to 3^+} f(x)$, then $\lim\limits_{x \to 3} f(x)$ exists. But, $\lim\limits_{x \to 3} f(x) = 6$ and $f(3) = 9$, so the third condition of continuity is not satisfied. The function f is discontinuous at 3. ∎

Figure 24 shows the graph of f.

NOW WORK   **Problem 25.**

The discontinuity at $c = 3$ in Example 3 is called a *removable discontinuity* because we can redefine f at the number c to equal $\lim\limits_{x \to c} f(x)$ and make f continuous at c. So, in Example 3, if $f(3)$ is redefined to be 6, then f would be continuous at 3.

> **DEFINITION** Removable Discontinuity
>
> Let f be a function that is defined everywhere in an open interval containing c, except possibly at c. The number c is called a **removable discontinuity** of f if the function is discontinuous at c but $\lim\limits_{x \to c} f(x)$ exists. The discontinuity is removed by defining (or redefining) the value of f at c to be $\lim\limits_{x \to c} f(x)$.

NOW WORK   **Problems 13 and 35.**

EXAMPLE 4   Determining Whether a Function Is Continuous at a Number

Determine if the floor function $f(x) = \lfloor x \rfloor$ is continuous at 1.

NEED TO REVIEW? The floor function is discussed in Section P.2, p. 17.

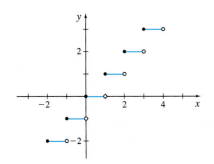

Figure 25 $f(x) = \lfloor x \rfloor$

Solution The floor function $f(x) = \lfloor x \rfloor$ = the greatest integer $\leq x$. The floor function f is defined at 1 and $f(1) = 1$. But

$$\lim_{x \to 1^-} f(x) = \lim_{x \to 1^-} \lfloor x \rfloor = 0 \qquad \text{and} \qquad \lim_{x \to 1^+} f(x) = \lim_{x \to 1^+} \lfloor x \rfloor = 1$$

So, $\lim\limits_{x \to 1} \lfloor x \rfloor$ does not exist. Since $\lim\limits_{x \to 1} \lfloor x \rfloor$ does not exist, f is discontinuous at 1. ∎

See Figure 25 for the graph of the floor function.

As Figure 25 illustrates, the floor function is discontinuous at each integer. Also, none of the discontinuities of the floor function is removable. Since at each integer the value of the floor function "jumps" to the next integer, without taking on any intermediate values, the discontinuity at integer values is called a **jump discontinuity**.

NOW WORK   **Problem 53.**

We have defined what it means for a function f to be *continuous* at a number. Now we define *one-sided continuity* at a number.

> **DEFINITION** One-Sided Continuity at a Number
>
> Let f be a function defined on the interval $(a, c]$. Then f is **continuous from the left at the number** c if
>
> $$\lim_{x \to c^-} f(x) = f(c)$$
>
> Let f be a function defined on the interval $[c, b)$. Then f is **continuous from the right at the number** c if
>
> $$\lim_{x \to c^+} f(x) = f(c)$$

In Example 4, we showed that the floor function $f(x) = \lfloor x \rfloor$ is discontinuous at $x = 1$. But since

$$f(1) = \lfloor 1 \rfloor = 1 \qquad \text{and} \qquad \lim_{x \to 1^+} f(x) = \lfloor x \rfloor = 1$$

the floor function is continuous from the right at 1. In fact, the floor function is discontinuous at each integer n, but it is continuous from the right at every integer n. (Do you see why?)

2 Determine Intervals on Which a Function Is Continuous

So far, we have considered only continuity at a number c. Now, we use one-sided continuity to define continuity on an interval.

> **DEFINITION Continuity on an Interval**
>
> - A function f is **continuous on an open interval** (a, b) if f is continuous at every number in (a, b).
> - A function f is **continuous on an interval** $[a, b)$ if f is continuous on the open interval (a, b) and continuous from the right at the number a.
> - A function f is **continuous on an interval** $(a, b]$ if f is continuous on the open interval (a, b) and continuous from the left at the number b.
> - A function f is **continuous on a closed interval** $[a, b]$ if f is continuous on the open interval (a, b), continuous from the right at a, and continuous from the left at b.

Figure 26 gives examples of graphs over different types of intervals.

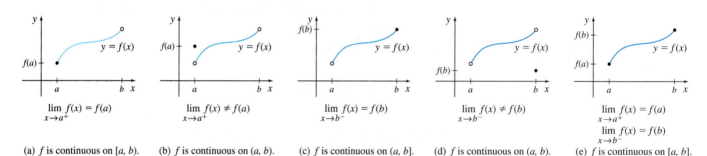

(a) f is continuous on $[a, b)$.   (b) f is continuous on (a, b).   (c) f is continuous on $(a, b]$.   (d) f is continuous on (a, b).   (e) f is continuous on $[a, b]$.

$$\lim_{x \to a^+} f(x) = f(a) \qquad \lim_{x \to a^+} f(x) \neq f(a) \qquad \lim_{x \to b^-} f(x) = f(b) \qquad \lim_{x \to b^-} f(x) \neq f(b) \qquad \lim_{x \to a^+} f(x) = f(a)$$
$$\lim_{x \to b^-} f(x) = f(b)$$

Figure 26

For example, the graph of the floor function $f(x) = \lfloor x \rfloor$ in Figure 25 illustrates that f is continuous on every interval $[n, n + 1)$, n an integer. In each interval, f is continuous from the right at the left endpoint n and is continuous at every number in the open interval $(n, n + 1)$.

EXAMPLE 5 Determining Whether a Function Is Continuous on a Closed Interval

Is the function $f(x) = \sqrt{4 - x^2}$ continuous on the closed interval $[-2, 2]$?

Solution The domain of f is $\{x \mid -2 \le x \le 2\}$. So, f is defined for every number in the closed interval $[-2, 2]$.

For any number c in the open interval $(-2, 2)$,

$$\lim_{x \to c} f(x) = \lim_{x \to c} \sqrt{4 - x^2} = \sqrt{\lim_{x \to c}(4 - x^2)} = \sqrt{4 - c^2} = f(c)$$

So, f is continuous on the open interval $(-2, 2)$.

To determine whether f is continuous on $[-2, 2]$, we investigate the limit from the right at -2 and the limit from the left at 2. Then,

$$\lim_{x \to -2^+} f(x) = \lim_{x \to -2^+} \sqrt{4 - x^2} = 0 = f(-2)$$

So, f is continuous from the right at -2. Similarly,

$$\lim_{x \to 2^-} f(x) = \lim_{x \to 2^-} \sqrt{4 - x^2} = 0 = f(2)$$

So, f is continuous from the left at 2. We conclude that f is continuous on the closed interval $[-2, 2]$. ∎

NOW WORK   Problem 37.

DEFINITION Continuity on a Domain

A function f is **continuous on its domain** if it is continuous at every number c in its domain.

EXAMPLE 6 Determining Whether $f(x) = \sqrt{x^2(x - 1)}$ Is Continuous on Its Domain

Determine if the function $f(x) = \sqrt{x^2(x - 1)}$ is continuous on its domain.

Solution The domain of $f(x) = \sqrt{x^2(x - 1)}$ is $\{x \mid x = 0\} \cup \{x \mid x \geq 1\}$. We need to determine whether f is continuous at the number 0 and whether f is continuous on the interval $[1, \infty)$.

At the number 0, there is an open interval containing 0 that contains no other number in the domain of f. [For example, use the interval $\left(-\dfrac{1}{2}, \dfrac{1}{2} \right)$.] This means $\lim\limits_{x \to 0} f(x)$ does not exist. So, f is discontinuous at 0.

For all numbers c in the open interval $(1, \infty)$ we have

$$f(c) = \sqrt{c^2(c - 1)}$$

and

$$\lim_{x \to c} \sqrt{x^2(x - 1)} = \sqrt{\lim_{x \to c}[x^2(x - 1)]} = \sqrt{c^2(c - 1)} = f(c)$$

So, f is continuous on the open interval $(1, \infty)$.

Now, at the number 1,

$$f(1) = 0 \qquad \text{and} \qquad \lim_{x \to 1^+} \sqrt{x^2(x - 1)} = 0$$

So, f is continuous from the right at 1.

The function $f(x) = \sqrt{x^2(x - 1)}$ is continuous on the interval $[1, \infty)$, but it is discontinuous at 0. So, f is not continuous on its domain. ∎

Figure 27 shows the graph of f. The discontinuity at 0 is subtle. It is neither a removable discontinuity nor a jump discontinuity.

When listing the properties of a function in Chapter P, we included the function's domain, its symmetry, and its zeros. Now we add continuity to the list by asking, "Where is the function continuous?" We answer this question here for two important classes of functions: polynomial functions and rational functions.

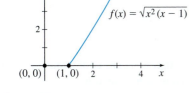

Figure 27 f is discontinuous at 0; f is continuous on $[1, \infty)$.

THEOREM

- A polynomial function is continuous for all real numbers.
- A rational function is continuous on its domain.

Proof If P is a polynomial function, its domain is the set of real numbers. For a polynomial function,

$$\lim_{x \to c} P(x) = P(c)$$

for any number c. That is, a polynomial function is continuous at every real number.

If $R(x) = \dfrac{p(x)}{q(x)}$ is a rational function, then $p(x)$ and $q(x)$ are polynomials and the domain of R is $\{x \mid q(x) \neq 0\}$. The Limit of a Rational Function (p. 87) states that for all c in the domain of a rational function,

$$\lim_{x \to c} R(x) = R(c)$$

So a rational function is continuous at every number in its domain. ■

To summarize:

- If a function is continuous *on an interval*, its graph has no holes or gaps on that interval.
- If a function is continuous *on its domain*, it will be continuous at every number in its domain; its graph may have holes or gaps at numbers that are not in the domain.

For example, the function $R(x) = \dfrac{x^2 - 2x + 1}{x - 1}$ is continuous on its domain $\{x \mid x \neq 1\}$ even though the graph has a hole at $(1, 0)$, as shown in Figure 28. The function $f(x) = \dfrac{1}{x}$ is continuous on its domain $\{x \mid x \neq 0\}$, as shown in Figure 29. Notice the behavior of the graph as x goes from negative numbers to positive numbers.

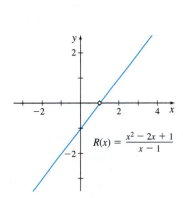

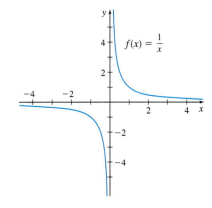

Figure 28 R has a hole at $(1, 0)$.

Figure 29 f is not defined at 0.

3 Use Properties of Continuity

So far we have shown that polynomial and rational functions are continuous on their domains. From these functions, we can build other continuous functions.

THEOREM Continuity of a Sum, Difference, Product, and Quotient

If the functions f and g are continuous at a number c, and if k is a real number, then the functions $f + g$, $f - g$, $f \cdot g$, and kf are also continuous at c. If $g(c) \neq 0$, the function $\dfrac{f}{g}$ is continuous at c.

The proofs of these properties are based on properties of limits. For example, the proof of the continuity of $f + g$ is based on the Limit of a Sum property. That is, if $\lim\limits_{x \to c} f(x)$ and $\lim\limits_{x \to c} g(x)$ exist, then $\lim\limits_{x \to c}[f(x) + g(x)] = \lim\limits_{x \to c} f(x) + \lim\limits_{x \to c} g(x)$.

EXAMPLE 7   Identifying Where Functions Are Continuous

Determine where each function is continuous:

(a) $F(x) = x^2 + 5 - \dfrac{x}{x^2 + 4}$ **(b)** $G(x) = x^3 + 2x + \dfrac{x^2}{x^2 - 1}$

Solution First we determine the domain of each function.

(a) F is the difference of the two functions $f(x) = x^2 + 5$ and $g(x) = \dfrac{x}{x^2 + 4}$, each of whose domain is the set of all real numbers. So, the domain of F is the set of all real numbers. Since f and g are continuous on their domains, the difference function F is continuous on its domain.

(b) G is the sum of the two functions $f(x) = x^3 + 2x$, whose domain is the set of all real numbers, and $g(x) = \dfrac{x^2}{x^2 - 1}$, whose domain is $\{x \mid x \neq -1,\ x \neq 1\}$. Since f and g are continuous on their domains, G is continuous on its domain, $\{x \mid x \neq -1,\ x \neq 1\}$. ■

NOW WORK Problem 45.

NEED TO REVIEW? Composite functions are discussed in Section P.3, pp. 25–27.

The continuity of a composite function depends on the continuity of its components.

THEOREM Continuity of a Composite Function

If a function g is continuous at c and a function f is continuous at $g(c)$, then the composite function $(f \circ g)(x) = f(g(x))$ is continuous at c.

EXAMPLE 8   Identifying Where Functions Are Continuous

Determine where each function is continuous:

(a) $F(x) = \sqrt{x^2 + 4}$ **(b)** $G(x) = \sqrt{x^2 - 1}$ **(c)** $H(x) = \dfrac{x^2 - 1}{x^2 - 4} + \sqrt{x - 1}$

Solution **(a)** $F = f \circ g$ is the composite of $f(x) = \sqrt{x}$ and $g(x) = x^2 + 4$. f is continuous for $x \geq 0$ and g is continuous for all real numbers. The domain of F is all real numbers and $F = (f \circ g)(x) = \sqrt{x^2 + 4}$ is continuous for all real numbers. That is, F is continuous on its domain.

(b) G is the composite of $f(x) = \sqrt{x}$ and $g(x) = x^2 - 1$. f is continuous for $x \geq 0$ and g is continuous for all real numbers. The domain of G is $\{x \mid x \geq 1\} \cup \{x \mid x \leq -1\}$ and $G = (f \circ g)(x) = \sqrt{x^2 - 1}$ is continuous on its domain.

(c) H is the sum of $f(x) = \dfrac{x^2 - 1}{x^2 - 4}$ and the function $g(x) = \sqrt{x - 1}$. The domain of f is $\{x \mid x \neq -2, x \neq 2\}$; f is continuous on its domain. The domain of g is $x \geq 1$. The domain of H is $\{x \mid 1 \leq x < 2\} \cup \{x \mid x > 2\}$; H is continuous on its domain. ■

NOW WORK Problem 47.

NEED TO REVIEW? Inverse functions are discussed in Section P.4, pp. 32–37.

Recall that for any function f that is one-to-one over its domain, its inverse f^{-1} is also a function, and the graphs of f and f^{-1} are symmetric with respect to the line $y = x$. It is intuitive that if f is continuous, then so is f^{-1}. See Figure 30 on page 100. The following theorem, whose proof is given in Appendix B, confirms this.

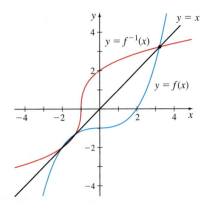

Figure 30

THEOREM Continuity of an Inverse Function

If f is a one-to-one function that is continuous on its domain, then its inverse function f^{-1} is also continuous on its domain.

---

4 Use the Intermediate Value Theorem

Functions that are continuous on a closed interval have many important properties. One of them is stated in the *Intermediate Value Theorem*. The proof of the Intermediate Value Theorem may be found in most books on advanced calculus.

THEOREM The Intermediate Value Theorem

Let f be a function that is continuous on a closed interval $[a, b]$ and $f(a) \neq f(b)$. If N is any number between $f(a)$ and $f(b)$, then there is at least one number c in the open interval (a, b) for which $f(c) = N$.

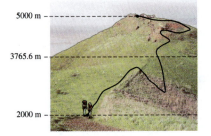

To get a better idea of this result, suppose you climb a mountain, starting at an elevation of 2000 meters and ending at an elevation of 5000 meters. No matter how many ups and downs you take as you climb, at some time your altitude must be 3765.6 meters, or any other number between 2000 and 5000.

In other words, a function f that is continuous on a closed interval $[a, b]$ must take on all values between $f(a)$ and $f(b)$. Figure 31 illustrates this and Figure 32 shows why the continuity of the function is crucial.

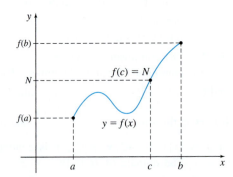

Figure 31 f takes on every value between $f(a)$ and $f(b)$.

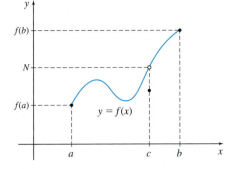

Figure 32 A discontinuity at c results in no number c in (a, b) for which $f(c) = N$.

An immediate application of the Intermediate Value Theorem involves locating the zeros of a function. Suppose a function f is continuous on the closed interval $[a, b]$ and $f(a)$ and $f(b)$ have opposite signs. Then by the Intermediate Value Theorem, there is at least one number c between a and b for which $f(c) = 0$. That is, f has at least one zero between a and b.

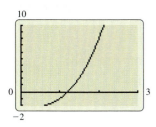

Figure 33 $f(x) = x^3 + x^2 - x - 2$

EXAMPLE 9   Using the Intermediate Value Theorem

Use the Intermediate Value Theorem to show that

$$f(x) = x^3 + x^2 - x - 2$$

has a zero between 1 and 2.

Solution Since f is a polynomial, it is continuous on the closed interval $[1, 2]$. Because $f(1) = -1$ and $f(2) = 8$ have opposite signs, the Intermediate Value Theorem states that $f(c) = 0$ for at least one number c in the interval $(1, 2)$. That is, f has at least one zero between 1 and 2. Figure 33 shows the graph of f on a graphing utility. ∎

NOW WORK **Problem 59.**

The Intermediate Value Theorem can be used to approximate a zero by dividing the interval $[a, b]$ into smaller subintervals. There are two popular methods of subdividing the interval $[a, b]$.

The **bisection method** bisects $[a, b]$, that is, divides $[a, b]$ into two equal subintervals, and compares the sign of $f\left(\dfrac{b-a}{2}\right)$ to the signs of the previously computed values $f(a)$ and $f(b)$. The subinterval whose endpoints have opposite signs is then bisected, and the process is repeated.

The second method divides $[a, b]$ into 10 subintervals of equal length and compares the signs of f evaluated at each of the eleven endpoints. The subinterval whose endpoints have opposite signs is then divided into 10 subintervals of equal length and the process is repeated.

We choose to use the second method because it lends itself well to the table feature of a graphing utility. You are asked to use the bisection method in Problems 103–110.

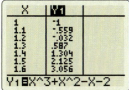

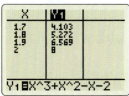

Figure 34

EXAMPLE 10   **Using the Intermediate Value Theorem to Approximate a Real Zero of a Function**

The function $f(x) = x^3 + x^2 - x - 2$ has a zero in the interval $(1, 2)$. Use the Intermediate Value Theorem to approximate the zero correct to three decimal places.

Solution Using the TABLE feature on a graphing utility, we subdivide the interval $[1, 2]$ into 10 subintervals, each of length 0.1. Then we find the subinterval whose endpoints have opposite signs, or the endpoint whose value equals 0 (in which case, the exact zero is found). From Figure 34, since $f(1.2) = -0.032$ and $f(1.3) = 0.587$, by the Intermediate Value Theorem, a zero lies in the interval $(1.2, 1.3)$. Correct to one decimal place, the zero is 1.2.

Repeat the process by subdividing the interval $[1.2, 1.3]$ into 10 subintervals, each of length 0.01. See Figure 35. We conclude that the zero is in the interval $(1.20, 1.21)$, so correct to two decimal places, the zero is 1.20.

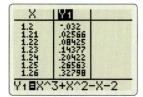

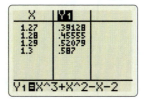

Figure 35

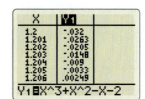

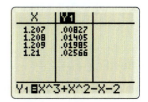

Figure 36

Now subdivide the interval $[1.20, 1.21]$ into 10 subintervals, each of length 0.001. See Figure 36.

We conclude that the zero of the function f is 1.205, correct to three decimal places. ∎

Notice that a benefit of the method used in Example 10 is that each additional iteration results in one additional decimal place of accuracy for the approximation.

NOW WORK Problem 65.

1.3 Assess Your Understanding

Concepts and Vocabulary

1. *True or False*   A polynomial function is continuous at every real number.

2. *True or False*   Piecewise-defined functions are never continuous at numbers where the function changes equations.

3. The three conditions necessary for a function f to be continuous at a number c are _____, _____, and _____.

4. *True or False*   If $f(x)$ is continuous at 0, then $g(x) = \dfrac{1}{4}f(x)$ is continuous at 0.

5. *True or False*   If f is a function defined everywhere in an open interval containing c, except possibly at c, then the number c is called a removable discontinuity of f if the function f is not continuous at c.

6. *True or False*   If a function f is discontinuous at a number c, then $\lim\limits_{x \to c} f(x)$ does not exist.

7. *True or False*   If a function f is continuous on an open interval (a, b), then it is continuous on the closed interval $[a, b]$.

8. *True or False*   If a function f is continuous on the closed interval $[a, b]$, then f is continuous on the open interval (a, b).

In Problems 9 and 10, explain whether each function is continuous or discontinuous on its domain.

9. The velocity of a ball thrown up into the air as a function of time, if the ball lands 5 seconds after it is thrown and stops.

10. The temperature of an oven used to bake a potato as a function of time.

11. *True or False*   If a function f is continuous on a closed interval $[a, b]$, then the Intermediate Value Theorem guarantees that the function takes on every value between $f(a)$ and $f(b)$.

12. *True or False*   If a function f is continuous on a closed interval $[a, b]$ and $f(a) \neq f(b)$, but both $f(a) > 0$ and $f(b) > 0$, then according to the Intermediate Value Theorem, f does not have a zero on the open interval (a, b).

Skill Building

In Problems 13–18, use the accompanying graph of $y = f(x)$.
(a) *Determine if f is continuous at c.*
(b) *If f is discontinuous at c, state which condition(s) of the definition of continuity is (are) not satisfied.*
(c) *If f is discontinuous at c, determine if the discontinuity is removable.*
(d) *If the discontinuity is removable, define (or redefine) f at c to make f continuous at c.*

13. $c = -3$ 14. $c = 0$

15. $c = 2$ 16. $c = 3$

17. $c = 4$ 18. $c = 5$

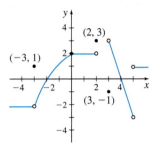

In Problems 19–32, determine whether the function f is continuous at c.

19. $f(x) = x^2 + 1$ at $c = -1$

20. $f(x) = x^3 - 5$ at $c = 5$

21. $f(x) = \dfrac{x}{x^2 + 4}$ at $c = -2$

22. $f(x) = \dfrac{x}{x - 2}$ at $c = 2$

23. $f(x) = \begin{cases} 2x + 5 & \text{if } x \leq 2 \\ 4x + 1 & \text{if } x > 2 \end{cases}$ at $c = 2$

24. $f(x) = \begin{cases} 2x + 1 & \text{if } x \leq 0 \\ 2x & \text{if } x > 0 \end{cases}$ at $c = 0$

25. $f(x) = \begin{cases} 3x - 1 & \text{if } x < 1 \\ 4 & \text{if } x = 1 \\ 2x & \text{if } x > 1 \end{cases}$ at $c = 1$

26. $f(x) = \begin{cases} 3x - 1 & \text{if } x < 1 \\ 2 & \text{if } x = 1 \\ 2x & \text{if } x > 1 \end{cases}$ at $c = 1$

27. $f(x) = \begin{cases} 3x - 1 & \text{if } x < 1 \\ 2x & \text{if } x > 1 \end{cases}$ at $c = 1$

28. $f(x) = \begin{cases} 3x - 1 & \text{if } x < 1 \\ 2 & \text{if } x = 1 \\ 3x & \text{if } x > 1 \end{cases}$ at $c = 1$

29. $f(x) = \begin{cases} x^2 & \text{if } x \leq 0 \\ 2x & \text{if } x > 0 \end{cases}$ at $c = 0$

30. $f(x) = \begin{cases} x^2 & \text{if } x < -1 \\ 2 & \text{if } x = -1 \\ -3x + 2 & \text{if } x > -1 \end{cases}$ at $c = -1$

31. $f(x) = \begin{cases} 4 - 3x^2 & \text{if } x < 0 \\ 4 & \text{if } x = 0 \\ \sqrt{\dfrac{16 - x^2}{4 - x}} & \text{if } 0 < x < 4 \end{cases}$ at $c = 0$

1. = NOW WORK problem   〔◺〕 = Graphing technology recommended   〔CAS〕 = Computer Algebra System recommended

32. $f(x) = \begin{cases} \sqrt{4+x} & \text{if } -4 \le x \le 4 \\ \sqrt{\dfrac{x^2 - 3x - 4}{x - 4}} & \text{if } x > 4 \end{cases}$ at $c = 4$

In Problems 33–36, each function f has a removable discontinuity at c. Define f(c) so that f is continuous at c.

33. $f(x) = \dfrac{x^2 - 4}{x - 2}$, $c = 2$

34. $f(x) = \dfrac{x^2 + x - 12}{x - 3}$, $c = 3$

35. $f(x) = \begin{cases} 1 + x & \text{if } x < 1 \\ 4 & \text{if } x = 1 \\ 2x & \text{if } x > 1 \end{cases}$ $c = 1$

36. $f(x) = \begin{cases} x^2 + 5x & \text{if } x < -1 \\ 0 & \text{if } x = -1 \\ x - 3 & \text{if } x > -1 \end{cases}$ $c = -1$

In Problems 37–40, determine if each function f is continuous on the given interval. If the answer is no, state the interval, if any, on which f is continuous.

37. $f(x) = \dfrac{x^2 - 9}{x - 3}$ on the interval $[-3, 3)$

38. $f(x) = 1 + \dfrac{1}{x}$ on the interval $[-1, 0)$

39. $f(x) = \dfrac{1}{\sqrt{x^2 - 9}}$ on the interval $[-3, 3]$

40. $f(x) = \sqrt{9 - x^2}$ on the interval $[-3, 3]$

In Problems 41–50, determine where each function f is continuous. First determine the domain of the function. Then support your decision using properties of continuity.

41. $f(x) = 2x^2 + 5x - \dfrac{1}{x}$

42. $f(x) = x + 1 + \dfrac{2x}{x^2 + 5}$

43. $f(x) = (x - 1)(x^2 + x + 1)$

44. $f(x) = \sqrt{x}(x^3 - 5)$

45. $f(x) = \dfrac{x - 9}{\sqrt{x} - 3}$

46. $f(x) = \dfrac{x - 4}{\sqrt{x} - 2}$

47. $f(x) = \sqrt{\dfrac{x^2 + 1}{2 - x}}$

48. $f(x) = \sqrt{\dfrac{4}{x^2 - 1}}$

49. $f(x) = (2x^2 + 5x - 3)^{2/3}$

50. $f(x) = (x + 2)^{1/2}$

In Problems 51–56, use the function

$$f(x) = \begin{cases} \sqrt{15 - 3x} & \text{if } x < 2 \\ \sqrt{5} & \text{if } x = 2 \\ 9 - x^2 & \text{if } 2 < x < 3 \\ \lfloor x - 2 \rfloor & \text{if } 3 \le x \end{cases}$$

51. Is f continuous at 0? Why or why not?

52. Is f continuous at 4? Why or why not?

53. Is f continuous at 3? Why or why not?

54. Is f continuous at 2? Why or why not?

55. Is f continuous at 1? Why or why not?

56. Is f continuous at 2.5? Why or why not?

In Problems 57 and 58:

(a) *Use graphing technology to graph f using a suitable scale on each axis.*

(b) *Based on the graph from (a), determine where f is continuous.*

(c) *Use the definition of continuity to determine where f is continuous.*

(d) *What advice would you give a fellow student about using graphing technology to determine where a function is continuous?*

57. $f(x) = \dfrac{x^3 - 8}{x - 2}$

58. $f(x) = \dfrac{x^2 - 3x + 2}{3x - 6}$

In Problems 59–64, use the Intermediate Value Theorem to determine which of the functions must have zeros in the given intervals. Indicate those for which the theorem gives no information. Do not attempt to locate the zeros.

59. $f(x) = x^3 - 3x$ on $[-2, 2]$

60. $f(x) = x^4 - 1$ on $[-2, 2]$

61. $f(x) = \dfrac{x}{(x + 1)^2} - 1$ on $[10, 20]$

62. $f(x) = x^3 - 2x^2 - x + 2$ on $[3, 4]$

63. $f(x) = \dfrac{x^3 - 1}{x - 1}$ on $[0, 2]$

64. $f(x) = \dfrac{x^2 + 3x + 2}{x^2 - 1}$ on $[-3, 0]$

In Problems 65–72, verify each function has a zero in the indicated interval. Then use the Intermediate Value Theorem to approximate the zero correct to three decimal places by repeatedly subdividing the interval containing the zero into 10 subintervals.

65. $f(x) = x^3 + 3x - 5$; interval: $(1, 2)$

66. $f(x) = x^3 - 4x + 2$; interval: $(1, 2)$

67. $f(x) = 2x^3 + 3x^2 + 4x - 1$; interval: $(0, 1)$

68. $f(x) = x^3 - x^2 - 2x + 1$; interval: $(0, 1)$

69. $f(x) = x^3 - 6x - 12$; interval: $(3, 4)$

70. $f(x) = 3x^3 + 5x - 40$; interval: $(2, 3)$

71. $f(x) = x^4 - 2x^3 + 21x - 23$; interval: $(1, 2)$

72. $f(x) = x^4 - x^3 + x - 2$; interval: $(1, 2)$

In Problems 73 and 74,

(a) *Use the Intermediate Value Theorem to show that f has a zero in the given interval.*

(b) *Use technology to find the zero rounded to three decimal places.*

73. $f(x) = \sqrt{x^2 + 4x} - 2$ in $(0, 1)$

74. $f(x) = x^3 - x + 2$ in $(-2, 0)$

Applications and Extensions

Heaviside Functions   *In Problems 75 and 76, determine whether the given Heaviside function is continuous at c.*

75. $u_1(t) = \begin{cases} 0 & \text{if } t < 1 \\ 1 & \text{if } t \ge 1 \end{cases}$ $c = 1$

76. $u_3(t) = \begin{cases} 0 & \text{if } t < 3 \\ 1 & \text{if } t \geq 3 \end{cases} \quad c = 3$

In Problems 77 and 78, determine where each function is continuous. Graph each function.

77. $f(x) = \begin{cases} 1 - x^2 & \text{if } |x| \leq 1 \\ x^2 - 1 & \text{if } |x| > 1 \end{cases}$

78. $f(x) = \begin{cases} \sqrt{4 - x^2} & \text{if } |x| \leq 2 \\ |x| - 2 & \text{if } |x| > 2 \end{cases}$

79. First-Class Mail As of January 2013, the U.S. Postal Service charged $0.46 postage for first-class letters weighing up to and including 1 ounce, plus a flat fee of $0.20 for each additional or partial ounce up to 3.5 ounces. First-class letter rates do not apply to letters weighing more than 3.5 ounces.

    (a) Find a function C that models the first-class postage charged for a letter weighing w ounces. Assume $w > 0$.

    (b) What is the domain of C?

    (c) Determine the intervals on which C is continuous.

    (d) At numbers where C is not continuous, what type of discontinuity does C have?

    (e) What are the practical implications of the answer to (d)?

    Source: U.S. Postal Service Notice 123.

80. First-Class Mail As of January 2013, the U.S. Postal Service charged $0.92 postage for first-class retail flats (large envelopes) weighing up to and including 1 ounce, plus a flat fee of $0.20 for each additional or partial ounce up to 13 ounces. First-class rates do not apply to flats weighing more than 13 ounces.

    (a) Find a function C that models the first-class postage charged for a large envelope weighing w ounces. Assume $w > 0$.

    (b) What is the domain of C?

    (c) Determine the intervals on which C is continuous.

    (d) At numbers where C is not continuous (if any), what type of discontinuity does C have?

    (e) What are the practical implications of the answer to (d)?

    Source: U.S. Postal Service Notice 123.

81. Cost of Natural Gas In February 2012 Peoples Energy had the following monthly rate schedule for natural gas usage in single- and two-family residences with a single gas meter:

Monthly customer charge	$22.13
Per therm distribution charge	
≤ 50 therms	$0.25963 per therm
> 50 therms	$12.98 + 0.11806 for each therm in excess of 50
Gas charge	$0.3631 per therm

Source: Peoples Energy, Chicago, IL.

    (a) Find a function C that models the monthly cost of using x therms of natural gas.

    (b) What is the domain of C?

    (c) Determine the intervals on which C is continuous.

    (d) At numbers where C is not continuous (if any), what type of discontinuity does C have?

    (e) What are the practical implications of the answer to (d)?

82. Cost of Water The Jericho Water District determines quarterly water costs, in dollars, using the following rate schedule:

Water used (in thousands of gallons)	Cost
$0 \leq x \leq 10$	$9.00
$10 < x \leq 30$	$9.00 + 0.95 for each thousand gallons in excess of 10,000 gallons
$30 < x \leq 100$	$28.00 + 1.65 for each thousand gallons in excess of 30,000 gallons
$x > 100$	$143.50 + 2.20 for each thousand gallons in excess of 100,000 gallons

Source: Jericho Water District, Syosset, NY.

    (a) Find a function C that models the quarterly cost of using x thousand gallons of water.

    (b) What is the domain of C?

    (c) Determine the intervals on which C is continuous.

    (d) At numbers where C is not continuous (if any), what type of discontinuity does C have?

    (e) What are the practical implications of the answer to (d)?

83. Gravity on Europa Europa, one of the larger satellites of Jupiter, has an icy surface and appears to have oceans beneath the ice. This makes it a candidate for possible extraterrestrial life. Because Europa is much smaller than most planets, its gravity is weaker. If we think of Europa as a sphere with uniform internal density, then inside the sphere, the gravitational field g is given by $g(r) = \dfrac{Gm}{R^3}r$, $0 \leq r < R$, where R is the radius of the sphere, r is the distance from the center of the sphere, and G is the universal gravitation constant. Outside a uniform sphere of mass m, the gravitational field g is given by

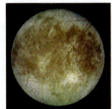

$$g(r) = \frac{Gm}{r^2}, \quad R < r.$$

    (a) For the gravitational field of Europa to be continuous at its surface, what must $g(r)$ equal? [*Hint:* Investigate $\lim\limits_{r \to R} g(r)$.]

    (b) Determine the gravitational field at Europa's surface. This will indicate the type of gravity environment organisms will experience. Use the following measured values: Europa's mass is 4.8×10^{22} kilograms, its radius is 1.569×10^6 meters, and $G = 6.67 \times 10^{-11}$.

    (c) Compare the result found in (b) to the gravitational field on Earth's surface, which is 9.8 meter/second^2. Is the gravity on Europa less than or greater than that on Earth?

84. Find constants A and B so that the function below is continuous for all x. Graph the resulting function.

$$f(x) = \begin{cases} (x-1)^2 & \text{if } -\infty < x < 0 \\ (A-x)^2 & \text{if } 0 \leq x < 1 \\ x+B & \text{if } 1 \leq x < \infty \end{cases}$$

85. Find constants A and B so that the function below is continuous for all x. Graph the resulting function.

$$f(x) = \begin{cases} x+A & \text{if } -\infty < x < 4 \\ (x-1)^2 & \text{if } 4 \leq x \leq 9 \\ Bx+1 & \text{if } 9 < x < \infty \end{cases}$$

86. For the function f below, find k so that f is continuous at 2.

$$f(x) = \begin{cases} \dfrac{\sqrt{2x+5} - \sqrt{x+7}}{x-2}, & x \geq -\dfrac{5}{2}, x \neq 2 \\ k & \text{if } x = 2 \end{cases}$$

87. Suppose $f(x) = \dfrac{x^2 - 6x - 16}{(x^2 - 7x - 8)\sqrt{x^2 - 4}}$.

    (a) For what numbers x is f defined?

    (b) For what numbers x is f discontinuous?

    (c) Which discontinuities found in (b) are removable?

88. Intermediate Value Theorem

    (a) Use the Intermediate Value Theorem to show that the function $f(x) = \sin x + x - 3$ has a zero in the interval $[0, \pi]$.

    (b) Approximate the zero rounded to three decimal places.

89. Intermediate Value Theorem

    (a) Use the Intermediate Value Theorem to show that the function $f(x) = e^x + x - 2$ has a zero in the interval $[0, 2]$.

    (b) Approximate the zero rounded to three decimal places.

90. Graph a function that is continuous on the closed interval $[5, 12]$, that is negative at both endpoints and has exactly three zeros in this interval. Does this contradict the Intermediate Value Theorem? Explain.

91. Graph a function that is continuous on the closed interval $[-1, 2]$, that is positive at both endpoints and has exactly two zeros in this interval. Does this contradict the Intermediate Value Theorem? Explain.

92. Graph a function that is continuous on the closed interval $[-2, 3]$, is positive at -2 and negative at 3 and has exactly two zeros in this interval. Is this possible? Does this contradict the Intermediate Value Theorem? Explain.

93. Graph a function that is continuous on the closed interval $[-5, 0]$, is negative at -5 and positive at 0 and has exactly three zeros in the interval. Is this possible? Does this contradict the Intermediate Value Theorem? Explain.

94. (a) Explain why the Intermediate Value Theorem gives no information about the zeros of the function $f(x) = x^4 - 1$ on the interval $[-2, 2]$.

    (b) Use technology to determine whether or not f has a zero on the interval $[-2, 2]$.

95. (a) Explain why the Intermediate Value Theorem gives no information about the zeros of the function $f(x) = \ln(x^2 + 2)$ on the interval $[-2, 2]$.

    (b) Use a graphing technology to determine whether or not f has a zero on the interval $[-2, 2]$.

96. Intermediate Value Theorem

    (a) Use the Intermediate Value Theorem to show that the functions $y = x^3$ and $y = 1 - x^2$ intersect somewhere between $x = 0$ and $x = 1$.

    (b) Use graphing technology to find the coordinates of the point of intersection rounded to three decimal places.

    (c) Use graphing technology to graph both functions on the same set of axes. Be sure the graph shows the point of intersection.

97. Intermediate Value Theorem An airplane is travelling at a speed of 620 miles per hour and then encounters a slight headwind that slows it to 608 miles per hour. After a few minutes, the headwind eases and the plane's speed increases to 614 miles per hour. Explain why the plane's speed is 610 miles per hour on at least two different occasions during the flight.

Source: Submitted by the students of Millikin University.

98. Suppose a function f is defined and continuous on the closed interval $[a, b]$. Is the function $h(x) = \dfrac{1}{f(x)}$ also continuous on the closed interval $[a, b]$? Discuss the continuity of h on $[a, b]$.

99. Given the two functions f and h:

$$f(x) = x^3 - 3x^2 - 4x + 12 \qquad h(x) = \begin{cases} \dfrac{f(x)}{x-3} & \text{if } x \neq 3 \\ p & \text{if } x = 3 \end{cases}$$

    (a) Find all the zeros of the function f.

    (b) Find the number p so that the function h is continuous at $x = 3$. Justify your answer.

    (c) Determine whether h, with the number found in (b), is even, odd, or neither. Justify your answer.

100. The function $f(x) = \dfrac{|x|}{x}$ is not defined at 0. Explain why it is impossible to define $f(0)$ so that f is continuous at 0.

101. Find two functions f and g that are each continuous at c, yet $\dfrac{f}{g}$ is not continuous at c.

102. Discuss the difference between a discontinuity that is removable and one that is nonremovable. Give an example of each.

Bisection Method for Approximating Zeros of a Function

Suppose the Intermediate Value Theorem indicates that a function f has a zero in the interval (a, b). The bisection method approximates the zero by evaluating f at the midpoint m_1 of the interval (a, b). If $f(m_1) = 0$, then m_1 is the zero we seek and the process ends. If $f(m_1) \neq 0$, then the sign of $f(m_1)$ is opposite that of either $f(a)$ or $f(b)$ (but not both), and the zero lies in that subinterval. Evaluate f at the midpoint m_2 of this subinterval. Continue bisecting the subinterval containing the zero until the desired degree of accuracy is obtained.

In Problems 103–110, use the bisection method three times to approximate the zero of each function in the given interval.

103. $f(x) = x^3 + 3x - 5$;   interval: $(1, 2)$

104. $f(x) = x^3 - 4x + 2$;   interval: $(1, 2)$

105. $f(x) = 2x^3 + 3x^2 + 4x - 1$;   interval: $(0, 1)$

106. $f(x) = x^3 - x^2 - 2x + 1$;   interval: $(0, 1)$

107. $f(x) = x^3 - 6x - 12$;   interval: $(3, 4)$

108. $f(x) = 3x^3 + 5x - 40$;   interval: $(2, 3)$

109. $f(x) = x^4 - 2x^3 + 21x - 23$;   interval $(1, 2)$

110. $f(x) = x^4 - x^3 + x - 2$;   interval: $(1, 2)$

111. Intermediate Value Theorem   Use the Intermediate Value Theorem to show that the function $f(x) = \sqrt{x^2 + 4x} - 2$ has a zero in the interval $[0, 1]$. Then approximate the zero correct to one decimal place.

112. Intermediate Value Theorem   Use the Intermediate Value Theorem to show that the function $f(x) = x^3 - x + 2$ has a zero in the interval $[-2, 0]$. Then approximate the zero correct to two decimal places.

113. Continuity of a Sum   If f and g are each continuous at c, prove that $f + g$ is continuous at c. (*Hint:* Use the Limit of a Sum Property.)

114. Intermediate Value Theorem   Suppose that the functions f and g are continuous on the interval $[a, b]$. If $f(a) < g(a)$ and $f(b) > g(b)$, prove that the graphs of $y = f(x)$ and $y = g(x)$ intersect somewhere between $x = a$ and $x = b$. [*Hint:* Define $h(x) = f(x) - g(x)$ and show $h(x) = 0$ for some x between a and b.]

Challenge Problems

115. Intermediate Value Theorem   Let $f(x) = \dfrac{1}{x-1} + \dfrac{1}{x-2}$.

Use the Intermediate Value Theorem to prove that there is a real number c between 1 and 2 for which $f(c) = 0$.

116. Intermediate Value Theorem   Prove that there is a real number c between 2.64 and 2.65 for which $c^2 = 7$.

117. Show that the existence of $\lim\limits_{h \to 0} \dfrac{f(a+h) - f(a)}{h}$ implies $f(x)$ is continuous at $x = a$.

118. Find constants A, B, C, and D so that the function below is continuous for all x. Sketch the graph of the resulting function.

$$f(x) = \begin{cases} \dfrac{x^2 + x - 2}{x - 1} & \text{if } -\infty < x < 1 \\ A & \text{if } x = 1 \\ B(x - C)^2 & \text{if } 1 < x < 4 \\ D & \text{if } x = 4 \\ 2x - 8 & \text{if } 4 < x < \infty \end{cases}$$

119. Let f be a function for which $0 \le f(x) \le 1$ for all x in $[0, 1]$. If f is continuous on $[0, 1]$, show that there exists at least one number c in $[0, 1]$ such that $f(c) = c$. [*Hint:* Let $g(x) = x - f(x)$.]

1.4 Limits and Continuity of Trigonometric, Exponential, and Logarithmic Functions

OBJECTIVES *When you finish this section, you should be able to:*

1 Use the Squeeze Theorem to find a limit (p. 106)
2 Find limits involving trigonometric functions (p. 108)
3 Determine where the trigonometric functions are continuous (p. 111)
4 Determine where an exponential or a logarithmic function is continuous (p. 113)

We have found limits using the basic limits $\lim\limits_{x \to c} A = A$ and $\lim\limits_{x \to c} x = c$ and properties of limits. But there are many limit problems that cannot be found by directly applying these techniques. To find such limits requires different results, such as the *Squeeze Theorem**, or basic limits involving trigonometric and exponential functions.

1 Use the Squeeze Theorem to Find a Limit

To use the Squeeze Theorem to find $\lim\limits_{x \to c} g(x)$, we need to know, or be able to find, two functions f and h that "sandwich" the function g between them for all x close to c. That is, in some interval containing c, the functions f, g, and h satisfy the inequality $f(x) \le g(x) \le h(x)$. Then if f and h have the same limit L as x approaches c, the function g is "squeezed" to the same limit L as x approaches c. See Figure 37.

We state the Squeeze Theorem here. The proof is given in Appendix B.

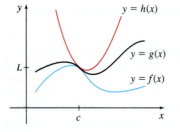

$\lim\limits_{x \to c} f(x) = L$, $\lim\limits_{x \to c} h(x) = L$, $\lim\limits_{x \to c} g(x) = L$

Figure 37

---

*The Squeeze Theorem is also known as the Sandwich Theorem and the Pinching Theorem.

THEOREM Squeeze Theorem

Suppose the functions f, g, and h have the property that for all x in an open interval containing c, except possibly at c,

$$f(x) \leq g(x) \leq h(x)$$

If

$$\lim_{x \to c} f(x) = \lim_{x \to c} h(x) = L$$

then

$$\lim_{x \to c} g(x) = L$$

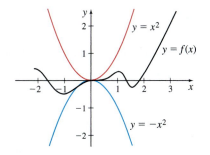

Figure 38

For example, suppose we wish to find $\lim\limits_{x \to 0} f(x)$, and we know that $-x^2 \leq f(x) \leq x^2$ for all $x \neq 0$. Since $\lim\limits_{x \to 0}(-x^2) = 0$ and $\lim\limits_{x \to 0} x^2 = 0$, the Squeeze Theorem tells us that $\lim\limits_{x \to 0} f(x) = 0$. Figure 38 illustrates how f is "squeezed" between $y = x^2$ and $y = -x^2$ near 0.

EXAMPLE 1   Using the Squeeze Theorem to Find a Limit

Use the Squeeze Theorem to find $\lim\limits_{x \to 0}\left(x \sin \dfrac{1}{x} \right)$.

Solution If $x \neq 0$, then $\sin \dfrac{1}{x}$ is defined. We seek two functions that "squeeze" $y = x \sin \dfrac{1}{x}$ near 0. Since $-1 \leq \sin x \leq 1$ for all x, we begin with the inequality

$$\left| \sin \frac{1}{x} \right| \leq 1 \qquad x \neq 0$$

Since $x \neq 0$ and we seek to squeeze $x \sin \dfrac{1}{x}$, we multiply both sides of the inequality by $|x|$, $x \neq 0$. Since $|x| > 0$, the direction of the inequality is preserved. Note that if we multiply $\left| \sin \dfrac{1}{x} \right| \leq 1$ by x, we would not know whether the inequality symbol would remain the same or be reversed since we do not know whether $x > 0$ or $x < 0$.

$$|x| \left| \sin \frac{1}{x} \right| \leq |x| \qquad \text{\textcolor{blue}{Multiply both sides by } } |x| > 0.$$

$$\left| x \sin \frac{1}{x} \right| \leq |x| \qquad \text{\textcolor{blue}{$|a| \cdot |b| = |a\, b|$.}}$$

$$-|x| \leq x \sin \frac{1}{x} \leq |x| \qquad \text{\textcolor{blue}{$|a| \leq b$ is equivalent to $-b \leq a \leq b$.}}$$

Now use the Squeeze Theorem with $f(x) = -|x|$, $g(x) = x \sin \dfrac{1}{x}$, and $h(x) = |x|$. Since $f(x) \leq g(x) \leq h(x)$ and

$$\lim_{x \to 0} f(x) = \lim_{x \to 0}(-|x|) = 0 \qquad \text{and} \qquad \lim_{x \to 0} h(x) = \lim_{x \to 0} |x| = 0$$

it follows that

$$\lim_{x \to 0} g(x) = \lim_{x \to 0}\left(x \cdot \sin \frac{1}{x} \right) = 0 \qquad \blacksquare$$

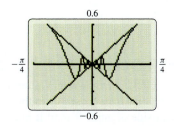

Figure 39 $g(x) = x \sin \dfrac{1}{x}$ is squeezed between $f(x) = -|x|$ and $h(x) = |x|$.

Figure 39 illustrates how $g(x) = x \sin \dfrac{1}{x}$ is squeezed between $y = -|x|$ and $y = |x|$.

NOW WORK Problem 5.

2 Find Limits Involving Trigonometric Functions

Knowing $\lim\limits_{x \to c} A = A$ and $\lim\limits_{x \to c} x = c$ helped us to find the limits of many algebraic functions. Knowing several basic trigonometric limits can help to find many limits involving trigonometric functions.

> **THEOREM** Two Basic Trigonometric Limits
>
> $$\lim\limits_{x \to 0} \sin x = 0 \qquad \lim\limits_{x \to 0} \cos x = 1$$

The graphs of $y = \sin x$ and $y = \cos x$ in Figure 40 illustrate that $\lim\limits_{x \to 0} \sin x = 0$ and $\lim\limits_{x \to 0} \cos x = 1$. The proofs of these limits both use the Squeeze Theorem. Problem 64 provides an outline of the proof that $\lim\limits_{x \to 0} \sin x = 0$.

A third basic trigonometric limit, $\lim\limits_{\theta \to 0} \dfrac{\sin \theta}{\theta} = 1$, is important in calculus. In Section 1.1, a table of numbers suggested that $\lim\limits_{\theta \to 0} \dfrac{\sin \theta}{\theta} = 1$. The function $f(\theta) = \dfrac{\sin \theta}{\theta}$, whose graph is given in Figure 41, is defined for all real numbers $\theta \neq 0$. The graph suggests $\lim\limits_{\theta \to 0} \dfrac{\sin \theta}{\theta} = 1$.

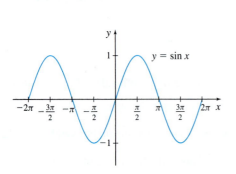

Figure 40

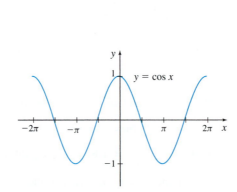

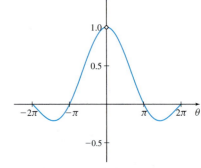

Figure 41 $f(\theta) = \dfrac{\sin \theta}{\theta}$

> **THEOREM**
>
> If θ is measured in radians, then
>
> $$\lim\limits_{\theta \to 0} \dfrac{\sin \theta}{\theta} = 1$$

Proof Although $\dfrac{\sin \theta}{\theta}$ is a quotient, we have no way to divide out θ. To find $\lim\limits_{\theta \to 0^+} \dfrac{\sin \theta}{\theta}$, we let θ be a positive acute central angle of a unit circle, as shown in Figure 42(a). Notice that COP is a sector of the circle. We add the point $B = (\cos \theta, 0)$ to the graph and form triangle BOP. Next we extend the terminal side of angle θ until it intersects the line $x = 1$ at the point D, forming a second triangle COD. The x-coordinate of D is 1. Since the length of the line segment $\overline{OC}$ is $OC = 1$, then the y-coordinate of D is

$$CD = \dfrac{CD}{OC} = \tan \theta. \text{ So, } D = (1, \tan \theta). \text{ See Figure 42(b).}$$

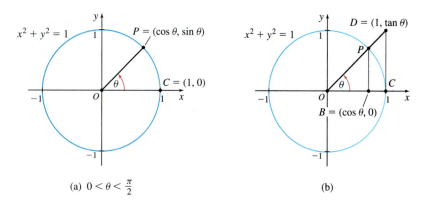

(a) $0 < \theta < \dfrac{\pi}{2}$ (b)

Figure 42

We see from Figure 42(b) that

$$\text{area of triangle } BOP \leq \text{area of sector } COP \leq \text{area of triangle } COD \qquad (1)$$

Each of these areas can be expressed in terms of θ as

$$\text{area of triangle } BOP = \frac{1}{2}\cos\theta \cdot \sin\theta \qquad \text{\textcolor{teal}{area of a triangle} } = \frac{1}{2}\text{ base} \times \text{height}$$

$$\text{area of sector } COP = \frac{\theta}{2} \qquad \text{\textcolor{teal}{area of sector } } A = \frac{1}{2}r^2\theta; r = 1$$

$$\text{area of triangle } COD = \frac{1}{2}\cdot 1 \cdot \tan\theta = \frac{\tan\theta}{2} \qquad \text{area of a triangle} = \frac{1}{2}\text{ base} \times \text{height}$$

So, for $0 < \theta < \dfrac{\pi}{2}$, we have

$$\frac{1}{2}\cos\theta \cdot \sin\theta \leq \frac{\theta}{2} \leq \frac{\tan\theta}{2} \qquad (1)$$

$$\cos\theta \cdot \sin\theta \leq \theta \leq \frac{\sin\theta}{\cos\theta} \qquad \text{Multiply all parts by 2; } \tan\theta = \frac{\sin\theta}{\cos\theta}.$$

$$\cos\theta \leq \frac{\theta}{\sin\theta} \leq \frac{1}{\cos\theta} \qquad \text{Divide all parts by } \sin\theta; \text{ since } 0 < \theta < \frac{\pi}{2}, \sin\theta > 0.$$

$$\frac{1}{\cos\theta} \geq \frac{\sin\theta}{\theta} \geq \cos\theta \qquad \text{Invert each term in the inequality; change the inequality signs.}$$

$$\cos\theta \leq \frac{\sin\theta}{\theta} \leq \frac{1}{\cos\theta} \qquad \text{Rewrite using } \leq.$$

Now we use the Squeeze Theorem with $f(\theta) = \cos\theta$, $g(\theta) = \dfrac{\sin\theta}{\theta}$, and $h(\theta) = \dfrac{1}{\cos\theta}$. Since $f(\theta) \leq g(\theta) \leq h(\theta)$, and since

$$\lim_{\theta \to 0^+} f(\theta) = \lim_{\theta \to 0^+} \cos\theta = 1 \quad \text{and} \quad \lim_{\theta \to 0^+} h(\theta) = \lim_{\theta \to 0^+}\frac{1}{\cos\theta} = \frac{\displaystyle\lim_{\theta \to 0^+} 1}{\displaystyle\lim_{\theta \to 0^+}\cos\theta} = \frac{1}{1} = 1$$

then by the Squeeze Theorem, $\displaystyle\lim_{\theta \to 0^+} g(\theta) = \lim_{\theta \to 0^+}\frac{\sin\theta}{\theta} = 1.$

Now we use the fact that $g(\theta) = \dfrac{\sin\theta}{\theta}$ is an even function, that is,

$$g(-\theta) = \frac{\sin(-\theta)}{-\theta} = \frac{-\sin\theta}{-\theta} = \frac{\sin\theta}{\theta} = g(\theta)$$

So,

$$\lim_{\theta \to 0^-}\frac{\sin\theta}{\theta} = \lim_{\theta \to 0^+}\frac{\sin(-\theta)}{-\theta} = \lim_{\theta \to 0^+}\frac{\sin\theta}{\theta} = 1$$

It follows that $\displaystyle\lim_{\theta \to 0}\frac{\sin\theta}{\theta} = 1.$ ∎

Since $\lim\limits_{\theta \to 0} \dfrac{\sin\theta}{\theta} = 1$, the ratio $\dfrac{\sin\theta}{\theta}$ is close to 1 for values of θ close to 0. That is, $\sin\theta \approx \theta$ for values of θ close to 0. Table 9 illustrates this property.

TABLE 9

	x near 0 on the left				x near 0 on the right			
θ (radians)	−0.5	−0.1	−0.01	−0.001	0.001	0.01	0.1	0.5
$\sin\theta$	−0.4794	−0.0998	−0.010	−0.001	0.001	0.010	0.0998	0.4794

The basic limit $\lim\limits_{\theta \to 0} \dfrac{\sin\theta}{\theta} = 1$ can be used to find the limits of similar expressions.

EXAMPLE 2   Finding the Limit of a Trigonometric Function

Find:

(a) $\lim\limits_{\theta \to 0} \dfrac{\sin(3\theta)}{\theta}$ **(b)** $\lim\limits_{\theta \to 0} \dfrac{\sin(5\theta)}{\sin(2\theta)}$

Solution (a) Since $\lim\limits_{\theta \to 0} \dfrac{\sin(3\theta)}{\theta}$ is not in the same form as $\lim\limits_{\theta \to 0} \dfrac{\sin\theta}{\theta}$, we multiply the numerator and the denominator by 3, and make the substitution $t = 3\theta$.

$$\lim_{\theta \to 0} \frac{\sin(3\theta)}{\theta} = \lim_{\theta \to 0} \frac{3\sin(3\theta)}{3\theta} \underset{\substack{\uparrow \\ t = 3\theta \\ t \to 0 \text{ as } \theta \to 0}}{=} \lim_{t \to 0} \left(3\frac{\sin t}{t}\right) = 3\lim_{t \to 0} \frac{\sin t}{t} \underset{\substack{\uparrow \\ \lim\limits_{t \to 0} \frac{\sin t}{t} = 1}}{=} (3)(1) = 3$$

(b) We begin by dividing the numerator and the denominator by θ. Then

$$\frac{\sin(5\theta)}{\sin(2\theta)} = \frac{\dfrac{\sin(5\theta)}{\theta}}{\dfrac{\sin(2\theta)}{\theta}}$$

Now we follow the approach in (a) on the numerator and on the denominator.

$$\lim_{\theta \to 0} \frac{\sin(5\theta)}{\theta} = \lim_{\theta \to 0} \frac{5\sin(5\theta)}{5\theta} \underset{\substack{\uparrow \\ t = 5\theta \\ t \to 0 \text{ as } \theta \to 0}}{=} \lim_{t \to 0} \frac{5\sin t}{t} = 5\lim_{t \to 0} \left(\frac{\sin t}{t}\right) = 5$$

$$\lim_{\theta \to 0} \frac{\sin(2\theta)}{\theta} = \lim_{\theta \to 0} \frac{2\sin(2\theta)}{2\theta} \underset{\substack{\uparrow \\ t = 2\theta \\ t \to 0 \text{ as } \theta \to 0}}{=} \lim_{t \to 0} \left(\frac{2\sin t}{t}\right) = 2\lim_{t \to 0} \frac{\sin t}{t} = 2$$

$$\lim_{\theta \to 0} \frac{\sin(5\theta)}{\sin(2\theta)} = \frac{\lim\limits_{\theta \to 0} \dfrac{\sin(5\theta)}{\theta}}{\lim\limits_{\theta \to 0} \dfrac{\sin(2\theta)}{\theta}} = \frac{5}{2}$$

NOW WORK Problems 23 and 25.

Example 3 establishes an important limit used in Chapter 2.

EXAMPLE 3   Finding a Basic Trigonometric Limit

Establish the formula

$$\lim_{\theta \to 0} \frac{\cos \theta - 1}{\theta} = 0$$

where θ is measured in radians.

Solution First we rewrite the expression $\dfrac{\cos \theta - 1}{\theta}$ as the product of two terms whose limits are known. For $\theta \neq 0$,

$$\frac{\cos \theta - 1}{\theta} = \left(\frac{\cos \theta - 1}{\theta} \right) \left(\frac{\cos \theta + 1}{\cos \theta + 1} \right)$$

$$= \frac{\cos^2 \theta - 1}{\theta(\cos \theta + 1)} \underset{\underset{\sin^2 \theta + \cos^2 \theta = 1}{\uparrow}}{=} \frac{-\sin^2 \theta}{\theta(\cos \theta + 1)} = \left(\frac{\sin \theta}{\theta} \right) \frac{(-\sin \theta)}{\cos \theta + 1}$$

Now we find the limit.

$$\lim_{\theta \to 0} \frac{\cos \theta - 1}{\theta} = \lim_{\theta \to 0} \left[\left(\frac{\sin \theta}{\theta} \right) \left(\frac{(-\sin \theta)}{\cos \theta + 1} \right) \right] = \left[\lim_{\theta \to 0} \frac{\sin \theta}{\theta} \right] \left[\lim_{\theta \to 0} \frac{-\sin \theta}{\cos \theta + 1} \right]$$

$$= 1 \cdot \frac{\displaystyle\lim_{\theta \to 0}(-\sin \theta)}{\displaystyle\lim_{\theta \to 0}(\cos \theta + 1)} = \frac{0}{2} = 0 \qquad \blacksquare$$

NOW WORK Problem 31.

3 Determine Where the Trigonometric Functions Are Continuous

The graphs of $f(x) = \sin x$ and $g(x) = \cos x$ shown on the left suggest that f and g are continuous on their domains, the set of all real numbers.

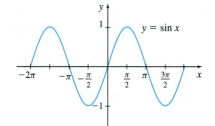

EXAMPLE 4   Showing $f(x) = \sin x$ Is Continuous at 0

- $f(0) = \sin 0 = 0$, so f is defined at 0.
- $\lim_{x \to 0} f(x) = \lim_{x \to 0} \sin x = 0$, so the limit at 0 exists.
- $\lim_{x \to 0} \sin x = \sin 0 = 0$.

Since all three conditions of continuity are satisfied, $f(x) = \sin x$ is continuous at 0. $\blacksquare$

In a similar way, we can show that $g(x) = \cos x$ is continuous at 0. That is, $\lim_{x \to 0} \cos x = \cos 0 = 1$.

You are asked to prove the following theorem in Problem 68.

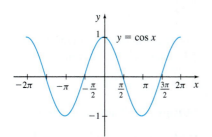

THEOREM

- The sine function $y = \sin x$ is continuous on its domain, all real numbers.
- The cosine function $y = \cos x$ is continuous on its domain, all real numbers.

Based on this theorem the following two limits can be added to the list of basic limits.

$$\lim_{x \to c} \sin x = \sin c \qquad \text{for all real numbers } c$$

$$\lim_{x \to c} \cos x = \cos c \qquad \text{for all real numbers } c$$

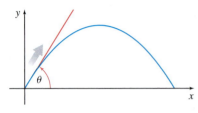

EXAMPLE 5   Application: Projectile Motion

An object is propelled from ground level at an angle θ, $0 < \theta < \dfrac{\pi}{2}$, to the horizontal with an initial velocity of 16 feet/second, as shown in Figure 43. The equations of the horizontal position $x = x(\theta)$ and the vertical position $y = y(\theta)$ of the object after t seconds are given by

$$x = x(\theta) = (16\cos\theta)t \qquad \text{and} \qquad y = y(\theta) = -16t^2 + (16\sin\theta)t$$

For a fixed time t:

(a) Find $\lim\limits_{\theta\to 0^+} x(\theta)$ and $\lim\limits_{\theta\to 0^+} y(\theta)$.

(b) Are the limits found in (a) consistent with what is expected physically?

(c) Find $\lim\limits_{\theta\to \frac{\pi}{2}^-} x(\theta)$ and $\lim\limits_{\theta\to \frac{\pi}{2}^-} y(\theta)$.

(d) Are the limits found in (c) consistent with what is expected physically?

Solution (a) $\quad\lim\limits_{\theta\to 0^+} x(\theta) = \left[\lim\limits_{\theta\to 0^+} (16\cos\theta)\right] t = 16t \qquad$ *t fixed*

$$\lim\limits_{\theta\to 0^+} y(\theta) = \lim\limits_{\theta\to 0^+} [-16t^2 + (16\sin\theta)t]$$

$$= \lim\limits_{\theta\to 0^+} (-16t^2) + \left[\lim\limits_{\theta\to 0^+} (16\sin\theta)\right]t = -16t^2$$

(b) The limits in (a) tell us that the horizontal position x of the object moves to the right ($x = 16t$) over time and that the vertical position y of the object immediately moves downward ($y = -16t^2$) into the ground. Neither of these conclusions is consistent with what we expect from the motion of an object propelled at an angle close to $\theta = 0$.

(c) $\lim\limits_{\theta\to \frac{\pi}{2}^-} x(\theta) = \left[\lim\limits_{\theta\to \frac{\pi}{2}^-} (16\cos\theta)\right] t = 0$

$$\lim\limits_{\theta\to \frac{\pi}{2}^-} y(\theta) = \lim\limits_{\theta\to \frac{\pi}{2}^-} \left[-16t^2 + (16\sin\theta)t\right] = \lim\limits_{\theta\to \frac{\pi}{2}^-} (-16t^2) + \left[\lim\limits_{\theta\to \frac{\pi}{2}^-} (16\sin\theta)\right]t$$

$$= -16t^2 + 16t$$

(d) The limits we found in (c) tell us that the horizontal position remains at 0. The vertical position $y = -16t^2 + 16t = -16t(t-1)$ tells us the object has a maximum height of 4 feet and returns to the ground after 1 second. (The parabola $y = -16t^2 + 16t$ opens down and has a maximum value at $t = \dfrac{-b}{2a} = \dfrac{-16}{-32} = \dfrac{1}{2}$ at which time $y = 4$.)

These conclusions are consistent with what we would expect from the motion of an object propelled close to the vertical. ∎

NOW WORK   Problem 55.

Using the facts that the sine and cosine functions are continuous for all real numbers, we can use basic trigonometric identities to determine where the remaining four trigonometric functions are continuous:

- $y = \tan x$: Since $\tan x = \dfrac{\sin x}{\cos x}$, from the Continuity of a Quotient property, $y = \tan x$ is continuous at all real numbers except those for which $\cos x = 0$.

  That is, it is continuous on its domain, all real numbers except odd multiples of $\dfrac{\pi}{2}$.

- $y = \sec x$: Since $\sec x = \dfrac{1}{\cos x}$, from the Continuity of a Quotient property, $y = \sec x$ is continuous at all real numbers except those for which $\cos x = 0$.

  That is, it is continuous on its domain, all real numbers except odd multiples of $\dfrac{\pi}{2}$.

- $y = \cot x$: Since $\cot x = \dfrac{\cos x}{\sin x}$, from the Continuity of a Quotient property, $y = \cot x$ is continuous at all real numbers except those for which $\sin x = 0$.

That is, it is continuous on its domain, all real numbers except integer multiples of π.

- $y = \csc x$: Since $\csc x = \dfrac{1}{\sin x}$, from the Continuity of a Quotient property, $y = \csc x$ is continuous at all real numbers except those for which $\sin x = 0$. That is, it is continuous on its domain, all real numbers except integer multiples of π.

Recall that a one-to-one function that is continuous on its domain has an inverse function that is continuous on its domain.

NEED TO REVIEW? Inverse trigonometric functions are discussed in Section P.7, pp. 58–63.

Since each of the six trigonometric functions is continuous on its domain, then each is continuous on the restricted domain used to define its inverse trigonometric function. This means the inverse trigonometric functions are continuous on their domains. These results are summarized in Table 10.

TABLE 10

Function	Domain	Properties
Sine	all real numbers	continuous on the interval $(-\infty, \infty)$
Cosine	all real numbers	continuous on the interval $(-\infty, \infty)$
Tangent	$\left\{ x \mid x \neq \text{odd integer multiples of } \dfrac{\pi}{2} \right\}$	continuous at all real numbers except odd multiples of $\dfrac{\pi}{2}$
Cosecant	$\{ x \mid x \neq \text{integer multiples of } \pi \}$	continuous at all real numbers except multiples of π
Secant	$\left\{ x \mid x \neq \text{odd integer multiples of } \dfrac{\pi}{2} \right\}$	continuous at all real numbers except odd multiples of $\dfrac{\pi}{2}$
Cotangent	$\{ x \mid x \neq \text{integer multiples of } \pi \}$	continuous at all real numbers except multiples of π
Inverse sine	$-1 \leq x \leq 1$	continuous on the closed interval $[-1, 1]$
Inverse cosine	$-1 \leq x \leq 1$	continuous on the closed interval $[-1, 1]$
Inverse tangent	all real numbers	continuous on the interval $(-\infty, \infty)$
Inverse cosecant	$|x| \geq 1$	continuous on the set $(-\infty, -1] \cup [1, \infty)$
Inverse secant	$|x| \geq 1$	continuous on the set $(-\infty, -1] \cup [1, \infty)$
Inverse cotangent	all real numbers	continuous on the interval $(-\infty, \infty)$

4 Determine Where an Exponential or a Logarithmic Function Is Continuous

The graphs of an exponential function $y = a^x$ and its inverse function $y = \log_a x$ are shown in Figure 44. The graphs suggest that an exponential function and a logarithmic function are continuous on their domains. We state the following theorem without proof.

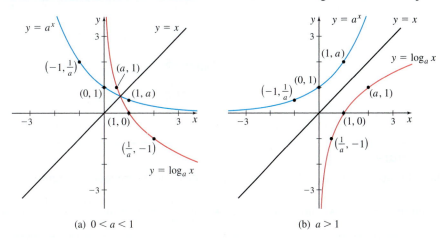

(a) $0 < a < 1$ (b) $a > 1$

Figure 44

THEOREM Continuity of Exponential and Logarithmic Functions

- An exponential function is continuous on its domain.
- A logarithmic function is continuous on its domain.

Based on this theorem, the following two limits can be added to the list of basic limits.

$$\lim_{x \to c} a^x = a^c \qquad \text{for all real numbers } c,\ a > 0,\ a \neq 1$$

and

$$\lim_{x \to c} \log_a x = \log_a c \qquad \text{for any real number } c > 0,\ a > 0,\ a \neq 1$$

EXAMPLE 6   Show that a Function Is Continuous

Show that:

(a) $f(x) = e^{2x}$ is continuous for all real numbers.

(b) $F(x) = \sqrt[3]{\ln x}$ is continuous for $x > 0$.

Solution (a) The domain of the exponential function is the set of all real numbers, so f is defined for any number c. That is, $f(c) = e^{2c}$. Also for any number c,

$$\lim_{x \to c} f(x) = \lim_{x \to c} e^{2x} = \lim_{x \to c}(e^x)^2 = \left[\lim_{x \to c} e^x\right]^2 = (e^c)^2 = e^{2c} = f(c)$$

Since $\lim_{x \to c} f(x) = f(c)$ for any number c, then f is continuous at all numbers c.

(b) The logarithmic function $f(x) = \ln x$ is continuous on its domain, the set of all positive real numbers. The function $g(x) = \sqrt[3]{x}$ is continuous on its domain, the set of all real numbers. Then for any real number $c > 0$, the composite function $F(x) = (g \circ f)(x) = \sqrt[3]{\ln x}$ is continuous at c. That is, F is continuous at all numbers $x > 0$. ■

Figures 45(a) and 45(b) illustrate the graphs of f and F.

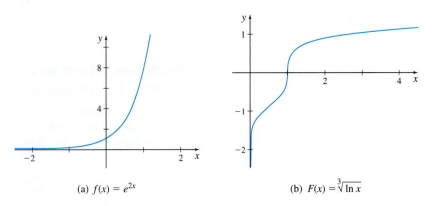

(a) $f(x) = e^{2x}$

(b) $F(x) = \sqrt[3]{\ln x}$

Figure 45

NOW WORK Problem 45.

Summary

Basic Limits   involving trigonometric, exponential, and logarithmic functions:

- $\displaystyle\lim_{x \to c} \sin x = \sin c$ • $\displaystyle\lim_{x \to c} \cos x = \cos c$ • $\displaystyle\lim_{\theta \to 0} \frac{\sin \theta}{\theta} = 1$ • $\displaystyle\lim_{\theta \to 0} \frac{\cos \theta - 1}{\theta} = 0$ • $\displaystyle\lim_{x \to c} a^x = a^c$ • $\displaystyle\lim_{x \to c} \log_a x = \log_a c,\ c > 0$

1.4 Assess Your Understanding

Concepts and Vocabulary

1. $\lim\limits_{x \to 0} \sin x = $ _____

2. *True or False* $\lim\limits_{x \to 0} \dfrac{\cos x - 1}{x} = 1$.

3. The Squeeze Theorem states that if the functions f, g, and h have the property $f(x) \le g(x) \le h(x)$ for all x in an open interval containing c, except possibly at c, and if $\lim\limits_{x \to c} f(x) = \lim\limits_{x \to c} h(x) = L$, then $\lim\limits_{x \to c} g(x) = $ _____.

4. *True or False* $f(x) = \csc x$ is continuous for all real numbers except $x = 0$.

Skill Building

In Problems 5–8, use the Squeeze Theorem to find each limit.

5. Suppose $-x^2 + 1 \le g(x) \le x^2 + 1$ for all x in an open interval containing 0. Find $\lim\limits_{x \to 0} g(x)$.

6. Suppose $-(x - 2)^2 - 3 \le g(x) \le (x - 2)^2 - 3$ for all x in an open interval containing 2. Find $\lim\limits_{x \to 2} g(x)$.

7. Suppose $\cos x \le g(x) \le 1$ for all x in an open interval containing 0. Find $\lim\limits_{x \to 0} g(x)$.

8. Suppose $-x^2 + 1 \le g(x) \le \sec x$ for all x in an open interval containing 0. Find $\lim\limits_{x \to 0} g(x)$.

In Problems 9–22, find each limit.

9. $\lim\limits_{x \to 0} (x^3 + \sin x)$

10. $\lim\limits_{x \to 0} (x^2 - \cos x)$

11. $\lim\limits_{x \to \pi/3} (\cos x + \sin x)$

12. $\lim\limits_{x \to \pi/3} (\sin x - \cos x)$

13. $\lim\limits_{x \to 0} \dfrac{\cos x}{1 + \sin x}$

14. $\lim\limits_{x \to 0} \dfrac{\sin x}{1 + \cos x}$

15. $\lim\limits_{x \to 0} \dfrac{3}{1 + e^x}$

16. $\lim\limits_{x \to 0} \dfrac{e^x - 1}{1 + e^x}$

17. $\lim\limits_{x \to 0} (e^x \sin x)$

18. $\lim\limits_{x \to 0} (e^{-x} \tan x)$

19. $\lim\limits_{x \to 1} \ln \left(\dfrac{e^x}{x} \right)$

20. $\lim\limits_{x \to 1} \ln \left(\dfrac{x}{e^x} \right)$

21. $\lim\limits_{x \to 0} \dfrac{e^{2x}}{1 + e^x}$

22. $\lim\limits_{x \to 0} \dfrac{1 - e^x}{1 - e^{2x}}$

In Problems 23–34, find each limit.

23. $\lim\limits_{x \to 0} \dfrac{\sin(7x)}{x}$

24. $\lim\limits_{x \to 0} \dfrac{\sin \dfrac{x}{3}}{x}$

25. $\lim\limits_{\theta \to 0} \dfrac{\theta + 3 \sin \theta}{2\theta}$

26. $\lim\limits_{x \to 0} \dfrac{2x - 5 \sin(3x)}{x}$

27. $\lim\limits_{\theta \to 0} \dfrac{\sin \theta}{\theta + \tan \theta}$

28. $\lim\limits_{\theta \to 0} \dfrac{\tan \theta}{\theta}$

29. $\lim\limits_{\theta \to 0} \dfrac{5}{\theta \cdot \csc \theta}$

30. $\lim\limits_{\theta \to 0} \dfrac{\sin(3\theta)}{\sin(2\theta)}$

31. $\lim\limits_{\theta \to 0} \dfrac{1 - \cos^2 \theta}{\theta}$

32. $\lim\limits_{\theta \to 0} \dfrac{\cos(4\theta) - 1}{2\theta}$

33. $\lim\limits_{\theta \to 0} (\theta \cdot \cot \theta)$

34. $\lim\limits_{\theta \to 0} \left[\sin \theta \left(\dfrac{\cot \theta - \csc \theta}{\theta} \right) \right]$

In Problems 35–38, determine whether f is continuous at the number c.

35. $f(x) = \begin{cases} 3 \cos x & \text{if } x < 0 \\ 3 & \text{if } x = 0 \\ x + 3 & \text{if } x > 0 \end{cases}$ at $c = 0$

36. $f(x) = \begin{cases} \cos x & \text{if } x < 0 \\ 0 & \text{if } x = 0 \\ e^x & \text{if } x > 0 \end{cases}$ at $c = 0$

37. $f(\theta) = \begin{cases} \sin \theta & \text{if } \theta \le \dfrac{\pi}{4} \\ \cos \theta & \text{if } \theta > \dfrac{\pi}{4} \end{cases}$ at $c = \dfrac{\pi}{4}$

38. $f(x) = \begin{cases} \tan^{-1} x & \text{if } x < 1 \\ \ln x & \text{if } x \ge 1 \end{cases}$ at $c = 1$

In Problems 39–46, determine where f is continuous.

39. $f(x) = \sin \left(\dfrac{x^2 - 4x}{x - 4} \right)$

40. $f(x) = \cos \left(\dfrac{x^2 - 5x + 1}{2x} \right)$

41. $f(\theta) = \dfrac{1}{1 + \sin \theta}$

42. $f(\theta) = \dfrac{1}{1 + \cos^2 \theta}$

43. $f(x) = \dfrac{\ln x}{x - 3}$

44. $f(x) = \ln(x^2 + 1)$

45. $f(x) = e^{-x} \sin x$

46. $f(x) = \dfrac{e^x}{1 + \sin^2 x}$

Applications and Extensions

In Problems 47–50, use the Squeeze Theorem to find each limit.

47. $\lim\limits_{x \to 0} \left(x^2 \sin \dfrac{1}{x} \right)$

48. $\lim\limits_{x \to 0} \left[x \left(1 - \cos \dfrac{1}{x} \right) \right]$

49. $\lim\limits_{x \to 0} \left[x^2 \left(1 - \cos \dfrac{1}{x} \right) \right]$

50. $\lim\limits_{x \to 0} \left[\sqrt{x^3 + 3x^2} \sin \left(\dfrac{1}{x} \right) \right]$

In Problems 51–54, show that each statement is true.

51. $\lim\limits_{x \to 0} \dfrac{\sin(ax)}{\sin(bx)} = \dfrac{a}{b}; b \ne 0$

52. $\lim\limits_{x \to 0} \dfrac{\cos(ax)}{\cos(bx)} = 1$

53. $\lim\limits_{x \to 0} \dfrac{\sin(ax)}{bx} = \dfrac{a}{b}; b \ne 0$

54. $\lim\limits_{x \to 0} \dfrac{1 - \cos(ax)}{bx} = 0; a \ne 0, b \ne 0$

55. **Projectile Motion** An object is propelled from ground level at an angle θ, $\dfrac{\pi}{4} < \theta < \dfrac{\pi}{2}$, up a ramp that is inclined to the horizontal at an angle of 45°. See the figure. If the object has an initial velocity of 10 feet/second, the equations of the horizontal position $x = x(\theta)$ and the vertical position $y = y(\theta)$ of the object after t seconds are given by

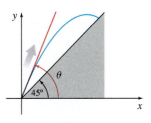

$x = x(\theta) = (10 \cos \theta)t$ and $y = y(\theta) = -16t^2 + (10 \sin \theta)t$

---

1. = NOW WORK problem   = Graphing technology recommended   CAS = Computer Algebra System recommended

For t fixed,

(a) Find $\lim\limits_{\theta \to \pi/4^+} x(\theta)$ and $\lim\limits_{\theta \to \pi/4^+} y(\theta)$.

(b) Are the limits found in (a) consistent with what is expected physically?

(c) Find $\lim\limits_{\theta \to \pi/2^-} x(\theta)$ and $\lim\limits_{\theta \to \pi/2^-} y(\theta)$.

(d) Are the limits found in (c) consistent with what is expected physically?

56. Show that $\lim\limits_{x \to 0} \dfrac{1 - \cos x}{x^2} = \dfrac{1}{2}$.

57. Squeeze Theorem If $0 \leq f(x) \leq 1$ for every number x, show that $\lim\limits_{x \to 0}[x^2 f(x)] = 0$.

58. Squeeze Theorem If $0 \leq f(x) \leq M$ for every x, show that $\lim\limits_{x \to 0}[x^2 f(x)] = 0$.

59. The function $f(x) = \dfrac{\sin(\pi x)}{x}$ is not defined at 0. Decide how to define $f(0)$ so that f is continuous at 0.

60. Define $f(0)$ and $f(1)$ so that the function $f(x) = \dfrac{\sin(\pi x)}{x(1 - x)}$ is continuous on the interval $[0, 1]$.

61. Is $f(x) = \begin{cases} \dfrac{\sin x}{x} & \text{if } x \neq 0 \\ 1 & \text{if } x = 0 \end{cases}$ continuous at 0?

62. Is $f(x) = \begin{cases} \dfrac{1 - \cos x}{x} & \text{if } x \neq 0 \\ 0 & \text{if } x = 0 \end{cases}$ continuous at 0?

63. Squeeze Theorem Show that $\lim\limits_{x \to 0}\left[x^n \sin\left(\dfrac{1}{x}\right)\right] = 0$, where n is a positive integer. (*Hint:* Look first at Problem 57.)

64. Prove $\lim\limits_{\theta \to 0} \sin \theta = 0$. (*Hint:*
Use a unit circle as shown
in the figure, first assuming
$0 < \theta < \dfrac{\pi}{2}$. Then use the
fact that $\sin \theta$ is less than
the length of the arc AP,
and the Squeeze Theorem,
to show that $\lim\limits_{\theta \to 0^+} \sin \theta = 0$.
Then use a similar argument
with $-\dfrac{\pi}{2} < \theta < 0$ to show that
$\lim\limits_{\theta \to 0^-} \sin \theta = 0$.)

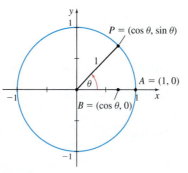

65. Prove $\lim\limits_{\theta \to 0} \cos \theta = 1$. Use either the proof outlined in Problem 64 or a proof using the result $\lim\limits_{\theta \to 0} \sin \theta = 0$ and a Pythagorean identity.

66. Without using limits, explain how you can decide whether $f(x) = \cos(5x^3 + 2x^2 - 8x + 1)$ is continuous.

67. Explain the Squeeze Theorem. Draw a graph to illustrate your explanation.

Challenge Problems

68. Use the Sum Formulas $\sin(a + b) = \sin a \cos b + \cos a \sin b$ and $\cos(a + b) = \cos a \cos b - \sin a \sin b$ to show that the sine function and cosine function are continuous on their domains.

69. Find $\lim\limits_{x \to 0} \dfrac{\sin x^2}{x}$.

70. Squeeze Theorem If $f(x) = \begin{cases} 1 & \text{if } x \text{ is rational} \\ 0 & \text{if } x \text{ is irrational} \end{cases}$ show that $\lim\limits_{x \to 0}[x f(x)] = 0$.

71. Suppose points A and B with coordinates $(0, 0)$ and $(1, 0)$, respectively, are given. Let n be a number greater than 0, and let θ be an angle with the property $0 < \theta < \dfrac{\pi}{1 + n}$. Construct a triangle ABC where $\overline{AC}$ and $\overline{AB}$ form the angle θ, and $\overline{CB}$ and $\overline{AB}$ form the angle $n\theta$ (see the figure below). Let D be the point of intersection of $\overline{AB}$ with the perpendicular from C to $\overline{AB}$. What is the limiting position of D as θ approaches 0?

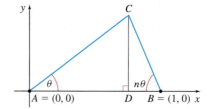

1.5 Infinite Limits; Limits at Infinity; Asymptotes

OBJECTIVES *When you finish this section, you should be able to:*

1 Investigate infinite limits (p. 117)

2 Find the vertical asymptotes of a function (p. 119)

3 Investigate limits at infinity (p. 120)

4 Find the horizontal asymptotes of a function (p. 125)

5 Find the asymptotes of a rational function using limits (p. 126)

RECALL The symbols ∞ (infinity) and $-\infty$ (negative infinity) are not numbers. The symbol ∞ expresses unboundedness in the positive direction, and $-\infty$ expresses unboundedness in the negative direction.

We have described $\lim_{x \to c} f(x) = L$ by saying if a function f is defined everywhere in an open interval containing c, except possibly at c, then the value $f(x)$ can be made as close as we please to L by choosing numbers x sufficiently close to c. In this section, we extend the language of limits to allow c to be ∞ or $-\infty$ (*limits at infinity*) and to allow L to be ∞ or $-\infty$ (*infinite limits*). These limits are useful for locating *asymptotes* that aid in graphing some functions.

Limits at infinity have practical use in many fields. They are used to determine what happens in *the long run*. For example, environmentalists are often interested in the effects of development on the long-term population of a certain species.

We begin with infinite limits.

1 Investigate Infinite Limits

Consider the function $f(x) = \dfrac{1}{x^2}$. Table 11 lists values of f for selected numbers x that are close to 0 and Figure 46 shows its graph.

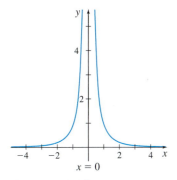

Figure 46 $f(x) = \dfrac{1}{x^2}$

TABLE 11

	x approaches 0 from the left				x approaches 0 from the right		
x	-0.01	-0.001	-0.0001	$\to 0 \leftarrow$	0.0001	0.001	0.01
$f(x) = \dfrac{1}{x^2}$	10,000	1,000,000	100,000,000	$f(x)$ becomes unbounded	100,000,000	1,000,000	10,000

As x approaches 0, the value of $\dfrac{1}{x^2}$ increases without bound. Since the value of $\dfrac{1}{x^2}$ is not approaching a single real number, the *limit* of $f(x)$ as x approaches 0 *does not exist*. However, since the numbers are increasing without bound, we describe the behavior of the function near zero by writing

$$\lim_{x \to 0} \frac{1}{x^2} = \infty$$

and say that $f(x) = \dfrac{1}{x^2}$ has an **infinite limit** as x approaches 0.

In other words, a function f has an infinite limit at c if f is defined everywhere in an open interval containing c, except possibly at c, and $f(x)$ becomes unbounded when x is sufficiently close to c.*

As a second example, consider the function $f(x) = \dfrac{1}{x}$. Table 12 shows values of f for selected numbers x that are close to 0 and Figure 47 shows its graph.

---

*A precise definition of an infinite limit is given in Section 1.6.

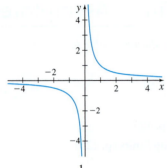

Figure 47 $f(x) = \dfrac{1}{x}$

TABLE 12

	x approaches 0 from the left						x approaches 0 from the right			
x	−0.01	−0.001	−0.0001	$\rightarrow$	0	$\leftarrow$	0.0001	0.001	0.01	0.1
$f(x) = \dfrac{1}{x}$	−100	−1000	−10,000	$f(x)$ becomes unbounded			10,000	1000	100	10

Here as x gets closer to 0 from the right, the value of $f(x) = \dfrac{1}{x}$ can be made as large as we please. That is, $\dfrac{1}{x}$ becomes unbounded in the positive direction. So,

$$\lim_{x \to 0^+} \frac{1}{x} = \infty$$

Similarly, the notation

$$\lim_{x \to 0^-} \frac{1}{x} = -\infty$$

is used to indicate that $\dfrac{1}{x}$ becomes unbounded in the negative direction as x approaches 0 from the left.

So, there are four possible one-sided infinite limits of a function f at c:

$$\lim_{x \to c^-} f(x) = \infty, \qquad \lim_{x \to c^-} f(x) = -\infty, \qquad \lim_{x \to c^+} f(x) = \infty, \qquad \lim_{x \to c^+} f(x) = -\infty$$

See Figure 48 for illustrations of these possibilities.

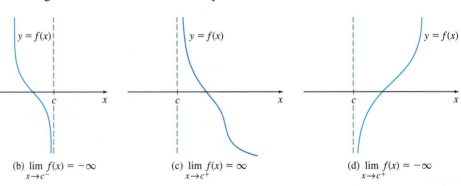

(a) $\displaystyle\lim_{x \to c^-} f(x) = \infty$ (b) $\displaystyle\lim_{x \to c^-} f(x) = -\infty$ (c) $\displaystyle\lim_{x \to c^+} f(x) = \infty$ (d) $\displaystyle\lim_{x \to c^+} f(x) = -\infty$

Figure 48

EXAMPLE 1 Investigating an Infinite Limit

Investigate $\displaystyle\lim_{x \to 0^+} \ln x$.

Solution The domain of $f(x) = \ln x$ is $\{x \mid x > 0\}$. Notice that the graph of $f(x) = \ln x$ in Figure 49 decreases without bound as x approaches 0 from the right. The graph suggests that

$$\lim_{x \to 0^+} \ln x = -\infty$$

∎

NOW WORK Problem 27.

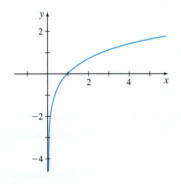

Figure 49 $f(x) = \ln x$

Based on the graphs of the trigonometric functions in Figure 50, we have the following infinite limits:

$$\lim_{x \to \pi/2^-} \tan x = \infty \qquad \lim_{x \to \pi/2^+} \tan x = -\infty \qquad \lim_{x \to \pi/2^-} \sec x = \infty \qquad \lim_{x \to \pi/2^+} \sec x = -\infty$$

$$\lim_{x \to 0^-} \csc x = -\infty \qquad \lim_{x \to 0^+} \csc x = \infty \qquad \lim_{x \to 0^-} \cot x = -\infty \qquad \lim_{x \to 0^+} \cot x = \infty$$

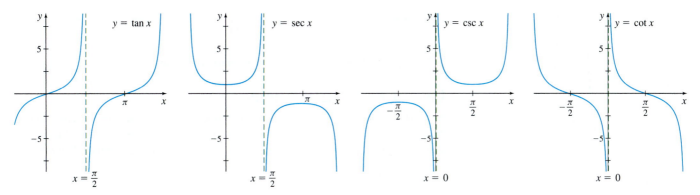

Figure 50

2 Find the Vertical Asymptotes of a Function

Figure 51 illustrates the possibilities that can occur when a function has an infinite limit at c. In each case, notice that the graph of f has a vertical asymptote at c.

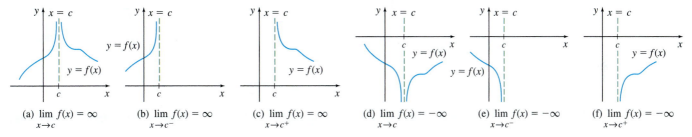

(a) $\lim\limits_{x \to c} f(x) = \infty$ (b) $\lim\limits_{x \to c^-} f(x) = \infty$ (c) $\lim\limits_{x \to c^+} f(x) = \infty$ (d) $\lim\limits_{x \to c} f(x) = -\infty$ (e) $\lim\limits_{x \to c^-} f(x) = -\infty$ (f) $\lim\limits_{x \to c^+} f(x) = -\infty$

Figure 51

DEFINITION Vertical Asymptote

The line $x = c$ is a **vertical asymptote** of the graph of the function f if any of the following is true:

$$\lim_{x \to c^-} f(x) = \infty \qquad \lim_{x \to c^+} f(x) = \infty \qquad \lim_{x \to c^-} f(x) = -\infty \qquad \lim_{x \to c^+} f(x) = -\infty$$

For rational functions, vertical asymptotes may occur where the denominator equals 0.

EXAMPLE 2 Finding a Vertical Asymptote

Find any vertical asymptote(s) of the graph of $f(x) = \dfrac{x}{(x - 3)^2}$.

Solution The domain of f is $\{x \,|\, x \neq 3\}$. Since 3 is the only number for which the denominator of f equals zero, we construct Table 13 and investigate the one-sided limits of f as x approaches 3. Table 13 suggests that

$$\lim_{x \to 3} \frac{x}{(x - 3)^2} = \infty$$

So, $x = 3$ is a vertical asymptote of the graph of f.

TABLE 13

	x approaches 3 from the left						x approaches 3 from the right		
x	2.9	2.99	2.999	$\to$	3	$\leftarrow$	3.001	3.01	3.1
$f(x) = \dfrac{x}{(x - 3)^2}$	290	29,900	2,999,000		$f(x)$ becomes unbounded		3,001,000	30,100	310

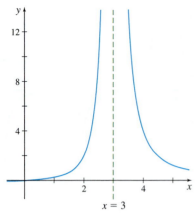

Figure 52 $f(x) = \dfrac{x}{(x-3)^2}$

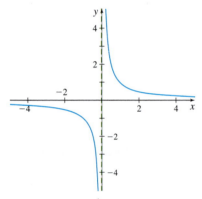

Figure 53 $f(x) = \dfrac{1}{x}$

Figure 52 shows the graph of $f(x) = \dfrac{x}{(x-3)^2}$ and its vertical asymptote. ■

NOW WORK Problems **15** and **63** (find any vertical asymptotes).

3 Investigate Limits at Infinity

Now we investigate what happens as x becomes unbounded in either the positive direction or the negative direction. Suppose as x becomes unbounded, the value of a function f approaches some real number L. Then the number L is called the *limit of f at infinity*.

For example, the graph of $f(x) = \dfrac{1}{x}$ in Figure 53 suggests that, as x becomes unbounded in either the positive direction or the negative direction, the values $f(x)$ get closer to 0. Table 14 illustrates this for a few numbers x.

TABLE 14

x	± 100	± 1000	$\pm 10{,}000$	$\pm 100{,}000$	x becomes unbounded
$f(x) = \dfrac{1}{x}$	± 0.01	± 0.001	± 0.0001	± 0.00001	$f(x)$ approaches 0

Since f can be made as close as we please to 0 by choosing x sufficiently large, we write

$$\lim_{x \to \infty} \frac{1}{x} = 0$$

and say that the limit as x approaches infinity of f is equal to 0. Similarly, as x becomes unbounded in the negative direction, the function $f(x) = \dfrac{1}{x}$ can be made as close as we please to 0, and we write

$$\lim_{x \to -\infty} \frac{1}{x} = 0$$

Although we do not prove $\displaystyle\lim_{x \to -\infty} \frac{1}{x} = 0$ here, it can be proved using the ϵ-δ Definition of a Limit at Infinity, and is left as an exercise in Section 1.6.

The limit properties discussed in Section 1.2, with "$x \to c$" replaced by "$x \to \infty$" or "$x \to -\infty$," hold for infinite limits. Although these properties are stated below for limits as $x \to \infty$, they are also valid for limits as $x \to -\infty$.

THEOREM Properties of Limits at Infinity

If k is a real number, $n \geq 2$ is an integer, and the functions f and g approach real numbers as $x \to \infty$, then the following properties are true:

- $\displaystyle\lim_{x \to \infty} A = A$ where A is a real number

- $\displaystyle\lim_{x \to \infty} [kf(x)] = k \lim_{x \to \infty} f(x)$

- $\displaystyle\lim_{x \to \infty} [f(x) \pm g(x)] = \lim_{x \to \infty} f(x) \pm \lim_{x \to \infty} g(x)$

- $\displaystyle\lim_{x \to \infty} [f(x)g(x)] = \left[\lim_{x \to \infty} f(x)\right]\left[\lim_{x \to \infty} g(x)\right]$

- $\displaystyle\lim_{x \to \infty} \frac{f(x)}{g(x)} = \frac{\displaystyle\lim_{x \to \infty} f(x)}{\displaystyle\lim_{x \to \infty} g(x)}$, provided $\displaystyle\lim_{x \to \infty} g(x) \neq 0$

- $\displaystyle\lim_{x \to \infty} [f(x)]^n = \left[\lim_{x \to \infty} f(x)\right]^n$

- $\displaystyle\lim_{x \to \infty} \sqrt[n]{f(x)} = \sqrt[n]{\lim_{x \to \infty} f(x)}$, where $f(x) \geq 0$ if n is even

EXAMPLE 3   Finding Limits at Infinity

Find:

(a) $\displaystyle\lim_{x\to-\infty}\frac{4}{x^2}$ (b) $\displaystyle\lim_{x\to\infty}\left(-\frac{10}{\sqrt{x}}\right)$ (c) $\displaystyle\lim_{x\to\infty}\left(2+\frac{3}{x}\right)$

Solution (a) $\displaystyle\lim_{x\to-\infty}\frac{4}{x^2}=4\left(\lim_{x\to-\infty}\frac{1}{x}\right)^2=4\cdot 0=0$

(b) $\displaystyle\lim_{x\to\infty}\left(-\frac{10}{\sqrt{x}}\right)=-10\lim_{x\to\infty}\frac{1}{\sqrt{x}}=-10\cdot\lim_{x\to\infty}\sqrt{\frac{1}{x}}=-10\cdot\sqrt{\lim_{x\to\infty}\frac{1}{x}}$

$\qquad\qquad = -10\cdot 0 = 0$

(c) $\displaystyle\lim_{x\to\infty}\left(2+\frac{3}{x}\right)=\lim_{x\to\infty}2+\lim_{x\to\infty}\frac{3}{x}=2+3\cdot\lim_{x\to\infty}\frac{1}{x}=2+3\cdot 0=2$ ∎

NOW WORK Problem 45.

EXAMPLE 4   Finding Limits at Infinity

Find:

(a) $\displaystyle\lim_{x\to\infty}\frac{3x^2-2x+8}{x^2+1}$ (b) $\displaystyle\lim_{x\to-\infty}\frac{4x^2-5x}{x^3+1}$

Solution (a) We find this limit by dividing each term of the numerator and the denominator by the term with the highest power of x that appears in the denominator, in this case, x^2. Then

$$\lim_{x\to\infty}\frac{3x^2-2x+8}{x^2+1}=\lim_{x\to\infty}\frac{\frac{3x^2-2x+8}{x^2}}{\frac{x^2+1}{x^2}}=\lim_{x\to\infty}\frac{3-\frac{2}{x}+\frac{8}{x^2}}{1+\frac{1}{x^2}}=\frac{\lim_{x\to\infty}\left[3-\frac{2}{x}+\frac{8}{x^2}\right]}{\lim_{x\to\infty}\left[1+\frac{1}{x^2}\right]}$$

Divide the numerator and denominator by x^2 / Limit of a Quotient

$$=\frac{\lim_{x\to\infty}3-\lim_{x\to\infty}\frac{2}{x}+\lim_{x\to\infty}\frac{8}{x^2}}{\lim_{x\to\infty}1+\lim_{x\to\infty}\frac{1}{x^2}}=\frac{3-2\lim_{x\to\infty}\frac{1}{x}+8\left(\lim_{x\to\infty}\frac{1}{x}\right)^2}{1+\left(\lim_{x\to\infty}\frac{1}{x}\right)^2}$$

$$=\frac{3-0+0}{1+0}=3$$

(b)

$$\lim_{x\to-\infty}\frac{4x^2-5x}{x^3+1}=\lim_{x\to-\infty}\frac{\frac{4x^2-5x}{x^3}}{\frac{x^3+1}{x^3}}=\lim_{x\to-\infty}\frac{\frac{4}{x}-\frac{5}{x^2}}{1+\frac{1}{x^3}}=\frac{\lim_{x\to-\infty}\left(\frac{4}{x}-\frac{5}{x^2}\right)}{\lim_{x\to-\infty}\left(1+\frac{3}{x^3}\right)}=\frac{0}{1}=0$$

Divide the numerator and denominator by x^3 ∎

The graphs of the functions $g(x)=\dfrac{3x^2-2x+8}{x^2+1}$ and $f(x)=\dfrac{4x^2-5x}{x^3+1}$ are shown in Figures 54(a) and 54(b), respectively. In each graph, notice how the graph of the function behaves as x becomes unbounded.

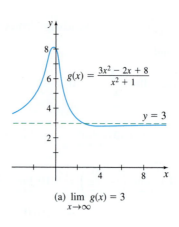

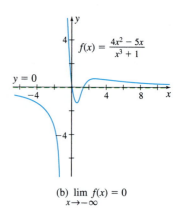

(a) $\lim_{x \to \infty} g(x) = 3$

(b) $\lim_{x \to -\infty} f(x) = 0$

Figure 54

NOW WORK **Problems 47 and 49.**

Limits at infinity have the following property.

If $p > 0$ is a rational number and k is any real number, then

$$\lim_{x \to \infty} \frac{k}{x^p} = 0 \quad \text{and} \quad \lim_{x \to -\infty} \frac{k}{x^p} = 0$$

provided x^p is defined when $x < 0$.

For example, $\lim_{x \to \infty} \dfrac{5}{x^3} = 0$, $\lim_{x \to \infty} \dfrac{-6}{x^{2/3}} = 0$, and $\lim_{x \to -\infty} \dfrac{4}{x^{8/3}} = 0$.

EXAMPLE 5 Finding the Limit at Infinity

Find $\lim_{x \to \infty} \dfrac{\sqrt{4x^2 + 10}}{x - 5}$.

Solution Divide each term of the numerator and the denominator by x, the term with the highest power of x that appears in the denominator. But remember, since $\sqrt{x^2} = |x| = x$ when $x \geq 0$, in the numerator the divisor in the square root will be x^2.

$$\lim_{x \to \infty} \frac{\sqrt{4x^2 + 10}}{x - 5} = \lim_{x \to \infty} \frac{\sqrt{\dfrac{4x^2 + 10}{x^2}}}{\dfrac{x - 5}{x}} = \lim_{x \to \infty} \frac{\sqrt{\dfrac{4x^2}{x^2} + \dfrac{10}{x^2}}}{1 - \dfrac{5}{x}} = \frac{\lim_{x \to \infty} \sqrt{4 + \dfrac{10}{x^2}}}{\lim_{x \to \infty} \left[1 - \dfrac{5}{x}\right]}$$

$$= \sqrt{\lim_{x \to \infty} \left[4 + \frac{10}{x^2}\right]} = \sqrt{4} = 2 \quad\blacksquare$$

The graph of $f(x) = \dfrac{\sqrt{4x^2 + 10}}{x - 5}$ and its behavior as $x \to \infty$ is shown in Figure 55.

$\lim_{x \to \infty} f(x) = 2$

Figure 55 $f(x) = \dfrac{\sqrt{4x^2 + 10}}{x - 5}$

NOW WORK **Problem 57.**

Infinite Limits at Infinity

In each of the examples above, $f(x)$ approached a real number as x became unbounded in either the positive or the negative direction. There are times, however, that this is not the case.

For example, consider the function $f(x) = x^2$. As x becomes unbounded in either the positive or negative direction, $f(x)$ increases without bound, as Figure 56 and Table 15 suggest. If a function grows without bound as x becomes unbounded, we say that f has an **infinite limit at infinity**.

TABLE 15

x	± 100	± 1000	$\pm 10,000$	x becomes unbounded
$f(x) = x^2$	10,000	1,000,000	100,000,000	$f(x)$ becomes unbounded

Compare the graph of $f(x) = x^2$ in Figure 56 to the graph of $g(x) = x^3$ shown in Figure 57. As x becomes unbounded in the positive direction, g increases without bound, but as x becomes unbounded in the negative direction g becomes unbounded in the negative direction.

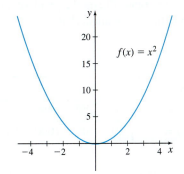

Figure 56 $\lim\limits_{x \to \infty} x^2 = \infty$; $\lim\limits_{x \to -\infty} x^2 = \infty$

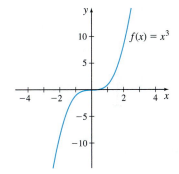

Figure 57 $\lim\limits_{x \to \infty} x^3 = \infty$; $\lim\limits_{x \to -\infty} x^3 = -\infty$

Infinite limits at infinity can take on any of the following four forms:

$$\lim_{x \to \infty} f(x) = \infty \qquad \lim_{x \to \infty} f(x) = -\infty \qquad \lim_{x \to -\infty} f(x) = \infty \qquad \lim_{x \to -\infty} f(x) = -\infty$$

EXAMPLE 6 **Finding a Limit at Infinity**

Find $\lim\limits_{x \to \infty} \dfrac{5x^4 - 3x^2}{2x^2 + 1}$.

Solution $R(x) = \dfrac{5x^4 - 3x^2}{2x^2 + 1}$ is a rational function defined for all real numbers. Find the limit by dividing each term of the numerator and the denominator by the term with the highest power of x that appears in the denominator, in this case, $2x^2$. Then

$$\lim_{x \to \infty} \frac{5x^4 - 3x^2}{2x^2 + 1} \underset{\substack{\uparrow \\ \text{Divide the numerator and} \\ \text{denominator by } 2x^2}}{=} \lim_{x \to \infty} \frac{\dfrac{5x^4 - 3x^2}{2x^2}}{\dfrac{2x^2 + 1}{2x^2}} = \lim_{x \to \infty} \frac{\dfrac{5x^2}{2} - \dfrac{3}{2}}{1 + \dfrac{1}{2x^2}} = \infty$$

because $\dfrac{5x^2}{2} - \dfrac{3}{2} \to \infty$ and $1 + \dfrac{1}{2x^2} \to 1$ as $x \to \infty$. ∎

The graph of R is shown in Figure 58.

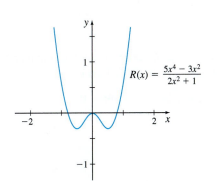

Figure 58 $\lim\limits_{x \to \infty} R(x) = \infty$

$R(x) = \dfrac{5x^4 - 3x^2}{2x^2 + 1}$

NOW WORK **Problem 59.**

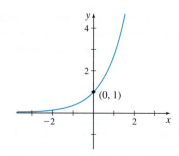

Figure 59 $f(x) = e^x$

The graph of the exponential function, shown in Figure 59, suggests that

$$\lim_{x \to -\infty} e^x = 0 \qquad \lim_{x \to \infty} e^x = \infty$$

These limits are supported by the information in Table 16.

TABLE 16

x	-1	-5	-10	-20	x approaches $-\infty$
$f(x) = e^x$	0.36788	0.00674	0.00005	-2×10^{-9}	$f(x)$ approaches 0
x	1	5	10	20	x approaches ∞
$f(x) = e^x$	$e \approx 2.71812$	148.41	22,026	4.85×10^8	$f(x)$ becomes unbounded

EXAMPLE 7   Finding the Limit at Infinity of $f(x) = \ln x$

Find $\displaystyle\lim_{x \to \infty} \ln x$.

Solution Table 17 and the graph of $f(x) = \ln x$ in Figure 60 suggest that $\ln x$ has an infinite limit at infinity. That is,

$$\lim_{x \to \infty} \ln x = \infty$$

TABLE 17

x	e^{10}	e^{100}	e^{1000}	$e^{10,000}$	$e^{100,000}$	$\to x$ becomes unbounded
$f(x) = \ln x$	10	100	1000	10,000	100,000	$\to f(x)$ becomes unbounded

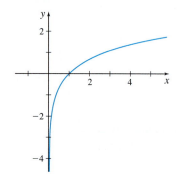

Figure 60 $f(x) = \ln x$

■

EXAMPLE 8   Application: Decomposition of Salt in Water

Salt (NaCl) decomposes in water into sodium (Na$^+$) ions and chloride (Cl$^-$) ions according to the law of uninhibited decay

$$A(t) = A_0 e^{kt}$$

where $A = A(t)$ is the amount (in kilograms) of salt present at time t (in hours), A_0 is the original amount of salt in the solution, and k is a negative number that represents the rate of decomposition.

(a) If initially there are 25 kilograms (kg) of salt and after 10 hours (h) there are 15 kg of salt remaining, how much salt is left after one day?

(b) How long will it take until $\dfrac{1}{2}$ kg of salt remains?

(c) Find $\displaystyle\lim_{t \to \infty} A(t)$.

(d) Interpret the answer found in (c).

NEED TO REVIEW? Solving exponential equations is discussed in Section P.5, pp. 45–46.

Solution **(a)** Initially, there are 25 kg of salt, so $A(0) = A_0 = 25$. To find the number k in $A(t) = A_0 e^{kt}$, we use the fact that at $t = 10$, then $A(10) = 15$. That is,

$$A(10) = 15 = 25e^{10k} \qquad A(t) = A_0 e^{kt},\ A_0 = 25;\ A(10) = 15$$

$$e^{10k} = \frac{3}{5}$$

$$10k = \ln \frac{3}{5}$$

$$k = \frac{1}{10} \ln 0.6$$

So, $A(t) = 25e^{\left(\frac{1}{10}\ln 0.6\right)t}$. The amount of salt that remains after one day (24 h) is

$$A(24) = 25e^{\left(\frac{1}{10}\ln 0.6\right)24} \approx 7.337 \text{ kilograms}$$

(b) We want to find t so that $A(t) = \dfrac{1}{2}$ kg. Then

$$\frac{1}{2} = 25e^{\left(\frac{1}{10}\ln 0.6\right)t}$$

$$e^{\left(\frac{1}{10}\ln 0.6\right)t} = \frac{1}{50}$$

$$\left(\frac{1}{10}\ln 0.6\right)t = \ln\frac{1}{50}$$

$$t \approx 76.582$$

After approximately 76.6 h, $\dfrac{1}{2}$ kg of salt will remain.

(c) Since $\dfrac{1}{10}\ln 0.6 \approx -0.051$, we have $\displaystyle\lim_{t\to\infty} A(t) = \lim_{t\to\infty}(25e^{-0.051t}) = \lim_{t\to\infty}\frac{25}{e^{0.051t}} = 0$

(d) As t becomes unbounded, the amount of salt in the water approaches 0 kg. Eventually, there will be no salt present in the water. ∎

NOW WORK **Problem 79.**

4 Find the Horizontal Asymptotes of a Function

Limits at infinity have an important geometric interpretation. When $\displaystyle\lim_{x\to\infty} f(x) = M$, it means that as x becomes unbounded in the positive direction, the value of $f(x)$ can be made as close as we please to a number M. That is, the graph of $y = f(x)$ is as close as we please to the horizontal line $y = M$ by choosing x sufficiently large. Similarly, $\displaystyle\lim_{x\to-\infty} f(x) = L$ means that the graph of $y = f(x)$ is as close as we please to the horizontal line $y = L$ for x unbounded in the negative direction. These lines are *horizontal asymptotes* of the graph of f.

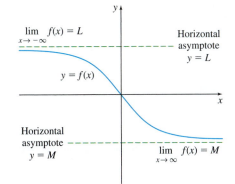

$\displaystyle\lim_{x\to-\infty} f(x) = L$

Horizontal asymptote $y = L$

$y = f(x)$

Horizontal asymptote $y = M$

$\displaystyle\lim_{x\to\infty} f(x) = M$

Figure 61

DEFINITION Horizontal Asymptote

The line $y = L$ is a **horizontal asymptote** of the graph of a function f for x unbounded in the negative direction if $\displaystyle\lim_{x\to-\infty} f(x) = L$.

The line $y = M$ is a **horizontal asymptote** of the graph of a function f for x unbounded in the positive direction if $\displaystyle\lim_{x\to\infty} f(x) = M$.

In Figure 61, $y = L$ is a horizontal asymptote as $x \to -\infty$ because $\displaystyle\lim_{x\to-\infty} f(x) = L$. The line $y = M$ is a horizontal asymptote as $x \to \infty$ because $\displaystyle\lim_{x\to\infty} f(x) = M$. To identify horizontal asymptotes, we find the limits of f at infinity.

EXAMPLE 9 Finding the Horizontal Asymptotes of a Function

Find the horizontal asymptotes, if any, of $f(x) = \dfrac{3x-2}{4x-1}$.

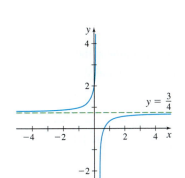

$y = \dfrac{3}{4}$

Figure 62 $f(x) = \dfrac{3x-2}{4x-1}$

Solution We examine the two limits at infinity: $\displaystyle\lim_{x\to-\infty}\frac{3x-2}{4x-1}$ and $\displaystyle\lim_{x\to\infty}\frac{3x-2}{4x-1}$.

Since $\displaystyle\lim_{x\to-\infty}\frac{3x-2}{4x-1} = \frac{3}{4}$, the line $y = \dfrac{3}{4}$ is a horizontal asymptote of the graph of f for x unbounded in the negative direction.

Since $\displaystyle\lim_{x\to\infty}\frac{3x-2}{4x-1} = \frac{3}{4}$, the line $y = \dfrac{3}{4}$ is a horizontal asymptote of the graph of f for x unbounded in the positive direction. ∎

Figure 62 shows the graph of $f(x) = \dfrac{3x-2}{4x-1}$ and the horizontal asymptote $y = \dfrac{3}{4}$.

NOW WORK **Problem 63 (find any horizontal asymptotes).**

5 Find the Asymptotes of a Rational Function Using Limits

In the next example we find the horizontal asymptotes and vertical asymptotes, if any, of a rational function.

EXAMPLE 10 **Finding the Asymptotes of a Rational Function Using Limits**

Find any asymptotes of the rational function $R(x) = \dfrac{3x^2 - 12}{2x^2 - 9x + 10}$.

Solution We begin by factoring R.

$$R(x) = \frac{3x^2 - 12}{2x^2 - 9x + 10} = \frac{3(x-2)(x+2)}{(2x-5)(x-2)}$$

The domain of R is $\left\{ x \mid x \neq \dfrac{5}{2} \text{ and } x \neq 2 \right\}$. Since R is a rational function, it is continuous on its domain, that is, all real numbers except $x = \dfrac{5}{2}$ and $x = 2$.

To check for vertical asymptotes, we find the limits as x approaches $\dfrac{5}{2}$ and 2. First we consider $\lim\limits_{x \to \frac{5}{2}^-} R(x)$.

$$\lim_{x \to \frac{5}{2}^-} R(x) = \lim_{x \to \frac{5}{2}^-} \left[\frac{3(x-2)(x+2)}{(2x-5)(x-2)} \right] = \lim_{x \to \frac{5}{2}^-} \left[\frac{3(x+2)}{(2x-5)} \right] = 3 \lim_{x \to \frac{5}{2}^-} \frac{x+2}{2x-5} = -\infty$$

That is, as x approaches $\dfrac{5}{2}$ from the left, R becomes unbounded in the negative direction. The graph of R has a vertical asymptote at $x = \dfrac{5}{2}$.

To determine the behavior to the right of $x = \dfrac{5}{2}$, we find the right-hand limit.

$$\lim_{x \to \frac{5}{2}^+} R(x) = \lim_{x \to \frac{5}{2}^+} \left[\frac{3(x+2)}{(2x-5)} \right] = 3 \lim_{x \to \frac{5}{2}^+} \frac{x+2}{2x-5} = \infty$$

As x approaches $\dfrac{5}{2}$ from the right, the graph of R becomes unbounded in the positive direction.

Next we consider $\lim\limits_{x \to 2} R(x)$.

$$\lim_{x \to 2} R(x) = \lim_{x \to 2} \frac{3(x-2)(x+2)}{(2x-5)(x-2)} = \lim_{x \to 2} \frac{3(x+2)}{2x-5} = \frac{3(2+2)}{2 \cdot 2 - 5} = \frac{12}{-1} = -12$$

The function R does not have a vertical asymptote at 2.

Since 2 is not in the domain of R, the graph of R has a **hole** at the point $(2, -12)$.

To check for horizontal asymptotes, we find the limits at infinity.

$$\lim_{x \to \infty} R(x) = \lim_{x \to \infty} \frac{3x^2 - 12}{2x^2 - 9x + 10} = \lim_{x \to \infty} \frac{\frac{3}{2} - \frac{6}{x^2}}{1 - \frac{9}{2x} + \frac{5}{x^2}} = \frac{\lim\limits_{x \to \infty} \left(\frac{3}{2} - \frac{6}{x^2} \right)}{\lim\limits_{x \to \infty} \left(1 - \frac{9}{2x} + \frac{5}{x^2} \right)}$$

↑ Divide the numerator and denominator by $2x^2$

$$= \frac{\frac{3}{2} - 0}{1 - 0 + 0} = \frac{3}{2}$$

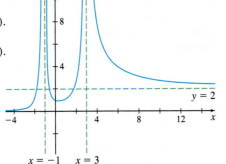

$$\lim_{x \to -\infty} R(x) = \lim_{x \to -\infty} \frac{3x^2 - 12}{2x^2 - 9x + 10} \underset{\substack{\uparrow \\ \text{Divide the numerator} \\ \text{and denominator by } 2x^2}}{=} \lim_{x \to -\infty} \frac{\dfrac{3}{2} - \dfrac{6}{x^2}}{1 - \dfrac{9}{2x} + \dfrac{5}{x^2}}$$

$$= \frac{\lim\limits_{x \to -\infty} \left(\dfrac{3}{2} - \dfrac{6}{x^2} \right)}{\lim\limits_{x \to -\infty} \left(1 - \dfrac{9}{2x} + \dfrac{5}{x^2} \right)} = \frac{3}{2}$$

The line $y = \dfrac{3}{2}$ is a horizontal asymptote of the graph of R for x unbounded in the negative direction and for x unbounded in the positive direction. ■

The graph of R and its asymptotes is shown in Figure 63. Notice the hole at the point $(2, -12)$.

Figure 63 $R(x) = \dfrac{3x^2 - 12}{2x^2 - 9x + 10}$

NOW WORK   **Problem 69.**

1.5 Assess Your Understanding

Concepts and Vocabulary

1. *True or False* ∞ is a number.

2. **(a)** $\lim\limits_{x \to 0^-} \dfrac{1}{x} = $ ____ ; **(b)** $\lim\limits_{x \to 0^+} \dfrac{1}{x} = $ ____ ; **(c)** $\lim\limits_{x \to 0^+} \ln x = $ ____

3. *True or False*   The graph of a rational function has a vertical asymptote at every number x at which the function is not defined.

4. If $\lim\limits_{x \to 4} f(x) = \infty$, then the line $x = 4$ is a(n) _____ asymptote of the graph of f.

5. **(a)** $\lim\limits_{x \to \infty} \dfrac{1}{x} = $ ____ ; **(b)** $\lim\limits_{x \to \infty} \dfrac{1}{x^2} = $ ____ ; **(c)** $\lim\limits_{x \to \infty} \ln x = $ ____

6. *True or False* $\lim\limits_{x \to -\infty} 5 = 0$.

7. **(a)** $\lim\limits_{x \to -\infty} e^x = $ ____ ; **(b)** $\lim\limits_{x \to \infty} e^x = $ ____ ; **(c)** $\lim\limits_{x \to \infty} e^{-x} = $ ____

8. *True or False*   The graph of a function can have at most two horizontal asymptotes.

Skill Building

In Problems 9–16, use the accompanying graph of $y = f(x)$.

9. Find $\lim\limits_{x \to \infty} f(x)$.

10. Find $\lim\limits_{x \to -\infty} f(x)$.

11. Find $\lim\limits_{x \to -1^-} f(x)$.

12. Find $\lim\limits_{x \to -1^+} f(x)$.

13. Find $\lim\limits_{x \to 3^-} f(x)$.

14. Find $\lim\limits_{x \to 3^+} f(x)$.

15. Identify all vertical asymptotes.

16. Identify all horizontal asymptotes.

In Problems 17–26, use the accompanying graph of $y = f(x)$.

17. Find $\lim\limits_{x \to \infty} f(x)$.

18. Find $\lim\limits_{x \to -\infty} f(x)$

19. Find $\lim\limits_{x \to -3^-} f(x)$.

20. Find $\lim\limits_{x \to -3^+} f(x)$.

21. Find $\lim\limits_{x \to 0^-} f(x)$.

22. Find $\lim\limits_{x \to 0^+} f(x)$.

23. Find $\lim\limits_{x \to 4^-} f(x)$.

24. Find $\lim\limits_{x \to 4^+} f(x)$.

25. Identify all vertical asymptotes.

26. Identify all horizontal asymptotes.

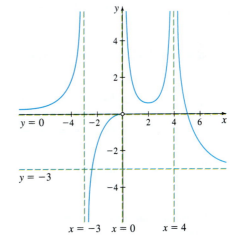

In Problems 27–42, find each limit.

27. $\lim\limits_{x \to 2^-} \dfrac{3x}{x - 2}$

28. $\lim\limits_{x \to -4^+} \dfrac{2x + 1}{x + 4}$

29. $\lim\limits_{x \to 2^+} \dfrac{5}{x^2 - 4}$

30. $\lim\limits_{x \to 1^-} \dfrac{2x}{x^3 - 1}$

31. $\lim\limits_{x \to -1^+} \dfrac{5x + 3}{x(x + 1)}$

32. $\lim\limits_{x \to 0^-} \dfrac{5x + 3}{5x(x - 1)}$

33. $\lim\limits_{x \to -3^-} \dfrac{1}{x^2 - 9}$

34. $\lim\limits_{x \to 2^+} \dfrac{x}{x^2 - 4}$

1. = NOW WORK problem   〰 = Graphing technology recommended   CAS = Computer Algebra System recommended

35. $\lim\limits_{x \to 3} \dfrac{1-x}{(3-x)^2}$

36. $\lim\limits_{x \to -1} \dfrac{x+2}{(x+1)^2}$

37. $\lim\limits_{x \to \pi^-} \cot x$

38. $\lim\limits_{x \to -\pi/2^-} \tan x$

39. $\lim\limits_{x \to \pi/2^+} \csc(2x)$

40. $\lim\limits_{x \to -\pi/2^-} \sec x$

41. $\lim\limits_{x \to -1^+} \ln(x+1)$

42. $\lim\limits_{x \to 1^+} \ln(x-1)$

In Problems 43–60, find each limit.

43. $\lim\limits_{x \to \infty} \dfrac{5}{x^2+4}$

44. $\lim\limits_{x \to -\infty} \dfrac{1}{x^2-9}$

45. $\lim\limits_{x \to \infty} \dfrac{2x+4}{5x}$

46. $\lim\limits_{x \to \infty} \dfrac{x+1}{x}$

47. $\lim\limits_{x \to \infty} \dfrac{x^3+x^2+2x-1}{x^3+x+1}$

48. $\lim\limits_{x \to \infty} \dfrac{2x^2-5x+2}{5x^2+7x-1}$

49. $\lim\limits_{x \to -\infty} \dfrac{x^2+1}{x^3-1}$

50. $\lim\limits_{x \to \infty} \dfrac{x^2-2x+1}{x^3+5x+4}$

51. $\lim\limits_{x \to \infty} \left[\dfrac{3x}{2x+5} - \dfrac{x^2+1}{4x^2+8} \right]$

52. $\lim\limits_{x \to \infty} \left[\dfrac{1}{x^2+x+4} - \dfrac{x+1}{3x-1} \right]$

53. $\lim\limits_{x \to -\infty} \left[2e^x \left(\dfrac{5x+1}{3x} \right) \right]$

54. $\lim\limits_{x \to -\infty} \left[e^x \left(\dfrac{x^2+x-3}{2x^3-x^2} \right) \right]$

55. $\lim\limits_{x \to \infty} \dfrac{\sqrt{x}+2}{3x-4}$

56. $\lim\limits_{x \to \infty} \dfrac{\sqrt{3x^3}+2}{x^2+6}$

57. $\lim\limits_{x \to \infty} \sqrt{\dfrac{3x^2-1}{x^2+4}}$

58. $\lim\limits_{x \to \infty} \left(\dfrac{16x^3+2x+1}{2x^3+3x} \right)^{2/3}$

59. $\lim\limits_{x \to -\infty} \dfrac{5x^3}{x^2+1}$

60. $\lim\limits_{x \to -\infty} \dfrac{x^4}{x-2}$

In Problems 61–66, find any horizontal or vertical asymptotes of the graph of f.

61. $f(x) = 3 + \dfrac{1}{x}$

62. $f(x) = 2 - \dfrac{1}{x^2}$

63. $f(x) = \dfrac{x^2}{x^2-1}$

64. $f(x) = \dfrac{2x^2-1}{x^2-1}$

65. $f(x) = \dfrac{\sqrt{2x^2-x+10}}{2x-3}$

66. $f(x) = \dfrac{\sqrt[3]{x^2+5x}}{x-6}$

In Problems 67–72, for each rational function R:
(a) Find the domain of R.
(b) Find any horizontal asymptotes of R.
(c) Find any vertical asymptotes of R
(d) Discuss the behavior of the graph at numbers where R is not defined.

67. $R(x) = \dfrac{-2x^2+1}{2x^3+4x^2}$

68. $R(x) = \dfrac{x^3}{x^4-1}$

69. $R(x) = \dfrac{x^2+3x-10}{2x^2-7x+6}$

70. $R(x) = \dfrac{x(x-1)^2}{(x+3)^3}$

71. $R(x) = \dfrac{x^3-1}{x-x^2}$

72. $R(x) = \dfrac{4x^5}{x^3-1}$

Applications and Extensions

In Problems 73 and 74:
(a) Sketch a graph of a function f that has the given properties.
(b) Define a function that describes the graph.

73. $f(3) = 0, \quad \lim\limits_{x \to \infty} f(x) = 1, \quad \lim\limits_{x \to -\infty} f(x) = 1,$
$\lim\limits_{x \to 1^-} f(x) = \infty, \quad \lim\limits_{x \to 1^+} f(x) = -\infty$

74. $f(2) = 0, \quad \lim\limits_{x \to \infty} f(x) = 0, \quad \lim\limits_{x \to -\infty} f(x) = 0,$
$\lim\limits_{x \to 0} f(x) = \infty, \quad \lim\limits_{x \to 5^-} f(x) = -\infty, \quad \lim\limits_{x \to 5^+} f(x) = \infty$

75. Newton's Law of Cooling Suppose an object is heated to a temperature u_0. Then at time $t = 0$, the object is put into a medium with a constant lower temperature T causing the object to cool. **Newton's Law of Cooling** states that the temperature u of the object at time t is given by $u = u(t) = (u_0 - T)e^{kt} + T$, where $k < 0$ is a constant.

(a) Find $\lim\limits_{t \to \infty} u(t)$. Is this the value you expected? Explain why or why not.

(b) Find $\lim\limits_{t \to 0^+} u(t)$. Is this the value you expected? Explain why or why not.

Source: Submitted by the students of Millikin University.

76. Environment A utility company burns coal to generate electricity. The cost C, in dollars, of removing $p\%$ of the pollutants emitted into the air is

$$C = \dfrac{70{,}000p}{100-p}, \qquad 0 \le p < 100$$

Find the cost of removing:

(a) 45% of the pollutants,

(b) 90% of the pollutants.

(c) Find $\lim\limits_{p \to 100^-} C$.

(d) Interpret the answer found in (c).

77. Pollution Control The cost C, in thousands of dollars, to remove a pollutant from a lake is

$$C(x) = \dfrac{5x}{100-x}, \qquad 0 \le x < 100$$

where x is the percent of pollutant removed. Find $\lim\limits_{x \to 100^-} C(x)$. Interpret your answer.

78. Population Model A rare species of insect was discovered in the Amazon Rain Forest. To protect the species, entomologists declared the insect endangered and transferred 25 insects to a protected area. The population P of the new colony t days after the transfer is

$$P(t) = \dfrac{50(1+0.5t)}{2+0.01t}$$

(a) What is the projected size of the colony after 1 year (365 days)?

(b) What is the largest population that the protected area can sustain? That is, find $\lim\limits_{t \to \infty} P(t)$.

(c) Graph the population P as a function of time t.

(d) Use the graph from (c) to describe the regeneration of the insect population. Does the graph support the answer to (b)?

79. Population of an Endangered Species Often environmentalists capture several members of an endangered species and transport them to a controlled environment where they can produce offspring and regenerate their population. Suppose six American bald eagles are captured, tagged, transported to Montana, and set free. Based on past experience, the environmentalists expect the population to grow according to the model

$$P(t) = \frac{500}{1 + 82.3e^{-0.162t}}$$

where t is measured in years.

(a) If the model is correct, how many bald eagles can the environment sustain? That is, find $\lim\limits_{t \to \infty} P(t)$.

(b) Graph the population P as a function of time t.

(c) Use the graph from (b) to describe the growth of the bald eagle population. Does the graph support the answer to (a)?

80. Hailstones Hailstones typically originate at an altitude of about 3000 meters (m). If a hailstone falls from 3000 m with no air resistance, its speed when it hits the ground would be about 240 meters/second (m/s), which is 540 miles/hour (mi/h)! That would be deadly! But air resistance slows the hailstone considerably. Using a simple model of air resistance, the speed $v = v(t)$ of a hailstone of mass m as a function of time t is given by $v(t) = \dfrac{mg}{k}(1 - e^{-kt/m})$ m/s, where $g = 9.8$ m/s^2 and k is a constant that depends on the size of the hailstone, its mass, and the conditions of the air. For a hailstone with a diameter $d = 1$ centimeter (cm) and mass m $= 4.8 \times 10^{-4}$ kg, k has been measured to be 3.4×10^{-4} kg/s.

(a) Determine the limiting speed of the hailstone by finding $\lim\limits_{t \to \infty} v(t)$. Express your answer in meters per second and miles per hour, using the fact that 1 mi/h ≈ 0.447 m/s. This speed is called the **terminal speed** of the hailstone.

(b) Graph $v = v(t)$. Does the graph support the answer to (a)?

81. Damped Harmonic Motion The motion of a spring is given by the function

$$x(t) = 1.2e^{-t/2}\cos t + 2.4e^{-t/2}\sin t$$

where x is the distance in meters from the the equilibrium position and t is the time in seconds.

(a) Graph $y = x(t)$. What is $\lim\limits_{t \to \infty} x(t)$, as suggested by the graph?

(b) Find $\lim\limits_{t \to \infty} x(t)$.

(c) Compare the results of (a) and (b). Is the answer to (b) supported by the graph in (a)?

82. Decomposition of Chlorine in a Pool Under certain water conditions, the free chlorine (hypochlorous acid, HOCl) in a swimming pool decomposes according to the law of uninhibited decay, $C = C(t) = C(0)e^{kt}$, where $C = C(t)$ is the amount (in parts per million, ppm) of free chlorine present at time t (in hours) and k is a negative number that represents the rate

of decomposition. After shocking his pool, Ben immediately tested the water and found the concentration of free chlorine to be $C_0 = C(0) = 2.5$ ppm. Twenty-four hours later, Ben tested the water again and found the amount of free chlorine to be 2.2 ppm.

(a) What amount of free chlorine will be left after 72 hours?

(b) When the free chlorine reaches 1.0 ppm, the pool should be shocked again. How long can Ben go before he must shock the pool again?

(c) Find $\lim\limits_{t \to \infty} C(t)$.

(d) Interpret the answer found in (c).

83. Decomposition of Sucrose Reacting with water in an acidic solution at 35°C, the amount A of sucrose ($C_{12}H_{22}O_{11}$) decomposes into glucose ($C_6H_{12}O_6$) and fructose ($C_6H_{12}O_6$) according to the law of uninhibited decay $A = A(t) = A(0)e^{kt}$, where $A = A(t)$ is the amount (in moles) of sucrose present at time t (in minutes) and k is a negative number that represents the rate of decomposition. An initial amount $A_0 = A(0) = 0.40$ mole of sucrose decomposes to 0.36 mole in 30 minutes.

(a) How much sucrose will remain after 2 hours?

(b) How long will it take until 0.10 mole of sucrose remains?

(c) Find $\lim\limits_{t \to \infty} A(t)$.

(d) Interpret the answer found in (c).

84. Macrophotography A camera lens can be approximated by a thin lens. A thin lens of focal length f obeys the thin-lens equation $\dfrac{1}{f} = \dfrac{1}{p} + \dfrac{1}{q}$, where $p > f$ is the distance from the lens to the object being photographed and q is the distance from the lens to the image formed by the lens. See the figure below. To photograph an object, the object's image must be formed on the photo sensors of the camera, which can only occur if q is positive.

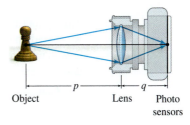

Object   Lens   Photo
                                   sensors

(a) Is the distance q of the image from the lens continuous as the distance of the object being photographed approaches the focal length f of the lens? (*Hint*: First solve the thin-lens equation for q and then find $\lim\limits_{p \to f^+} q$.)

(b) Use the result from (a) to explain why a camera (or any lens) cannot focus on an object placed close to its focal length.

In Problems 85 and 86, find conditions on a, b, c, and d so that the graph of f has no horizontal or vertical asymptotes.

85. $f(x) = \dfrac{ax^3 + b}{cx^4 + d}$ **86.** $f(x) = \dfrac{ax + b}{cx + d}$

87. Explain why the following properties are true. Give an example of each.

(a) If n is an even positive integer, then $\displaystyle\lim_{x \to c} \frac{1}{(x-c)^n} = \infty$.

(b) If n is an odd positive integer, then $\displaystyle\lim_{x \to c^-} \frac{1}{(x-c)^n} = -\infty$.

(c) If n is an odd positive integer, then $\displaystyle\lim_{x \to c^+} \frac{1}{(x-c)^n} = \infty$.

88. Explain why a rational function, whose numerator and denominator have no common zeros, will have vertical asymptotes at each point of discontinuity.

89. Explain why a polynomial function of degree 1 or higher cannot have any asymptotes.

90. If P and Q are polynomials of degree m and n, respectively, discuss $\displaystyle\lim_{x \to \infty} \frac{P(x)}{Q(x)}$ when:

(a) $m > n$ **(b)** $m = n$ **(c)** $m < n$

91. 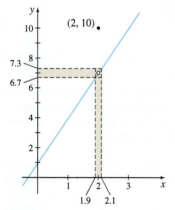 **(a)** Use a table to investigate $\displaystyle\lim_{x \to \infty} \left(1 + \frac{1}{x}\right)^x$.

[CAS] **(b)** Find $\displaystyle\lim_{x \to \infty} \left(1 + \frac{1}{x}\right)^x$.

(c) Compare the results from (a) and (b). Explain the possible causes of any discrepancy.

Challenge Problems

92. $\displaystyle\lim_{x \to \infty} \left(1 + \frac{1}{x}\right) = 1$, but $\displaystyle\lim_{x \to \infty} \left(1 + \frac{1}{x}\right)^x > 1$. Discuss why the property $\displaystyle\lim_{x \to \infty} [f(x)]^n = \left[\lim_{x \to \infty} f(x)\right]^n$ cannot be used to find the second limit.

93. Kinetic Energy   At low speeds the kinetic energy K, that is, the energy due to the motion of an object of mass m and speed v, is given by the formula $K = K(v) = \frac{1}{2}mv^2$. But this formula is only an approximation to the general formula, and works only for speeds much less than the speed of light, c. The general formula, which holds for all speeds, is

$$K_{\text{gen}}(v) = mc^2 \left[\frac{1}{\sqrt{1 - \dfrac{v^2}{c^2}}} - 1 \right]$$

(a) As an object is accelerated closer and closer to the speed of light, what does its kinetic energy K_{gen} approach?

(b) What does the result suggest about the possibility of reaching the speed of light?

1.6 The ϵ-δ Definition of a Limit

OBJECTIVES *When you finish this section, you should be able to:*

1 Use the ϵ-δ definition of a limit (p. 132)

Throughout the chapter, we stated that we could be sure a limit was correct only if it was based on the ϵ-δ definition of a limit. In this section, we examine this definition and how to use it to prove a limit exists, to verify the value of a limit, and to show that a limit does not exist.

Consider the function f defined by

$$f(x) = \begin{cases} 3x + 1 & \text{if } x \neq 2 \\ 10 & \text{if } x = 2 \end{cases}$$

whose graph is given in Figure 64.

As x gets closer to 2, the value $f(x)$ gets closer to 7. If in fact, by taking x close enough to 2, we can make $f(x)$ as close to 7 as we please, then $\displaystyle\lim_{x \to 2} f(x) = 7$.

Suppose we want $f(x)$ to differ from 7 by less than 0.3; that is,

$$-0.3 < f(x) - 7 < 0.3$$
$$6.7 < f(x) < 7.3$$

How close must x be to 2? First, we must require $x \neq 2$ because when $x = 2$, then $f(x) = f(2) = 10$, and we obtain $6.7 < 10 < 7.3$, which is impossible.

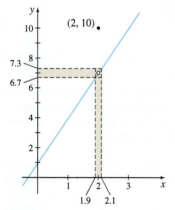

Figure 64 $f(x) = \begin{cases} 3x + 1 & \text{if } x \neq 2 \\ 10 & \text{if } x = 2 \end{cases}$

Then, when $x \neq 2$,

$$-0.3 < f(x) - 7 < 0.3$$
$$-0.3 < (3x + 1) - 7 < 0.3$$
$$-0.3 < 3x - 6 < 0.3$$
$$-0.3 < 3(x - 2) < 0.3$$
$$\frac{-0.3}{3} < x - 2 < \frac{0.3}{3}$$
$$-0.1 < x - 2 < 0.1$$
$$|x - 2| < 0.1$$

That is, whenever $x \neq 2$ and x differs from 2 by less than 0.1, then $f(x)$ differs from 7 by less than 0.3.

Now, generalizing the question, we ask, for $x \neq 2$, how close must x be to 2 to guarantee that $f(x)$ differs from 7 by less than any given positive number ϵ? (ϵ might be extremely small.) The statement "$f(x)$ differs from 7 by less than ϵ" means

$$-\epsilon < f(x) - 7 < \epsilon$$
$$7 - \epsilon < f(x) < 7 + \epsilon \qquad \text{Add 7 to each expression.}$$

When $x \neq 2$, then $f(x) = 3x + 1$, so

$$7 - \epsilon < 3x + 1 < 7 + \epsilon$$

Now we want the middle term to be $x - 2$. This is done as follows:

$$7 - \epsilon < 3x + 1 < 7 + \epsilon$$
$$6 - \epsilon < 3x < 6 + \epsilon \qquad \text{Subtract 1 from each expression.}$$
$$2 - \frac{\epsilon}{3} < x < 2 + \frac{\epsilon}{3} \qquad \text{Divide each expression by 3.}$$
$$-\frac{\epsilon}{3} < x - 2 < \frac{\epsilon}{3} \qquad \text{Subtract 2 from each expression.}$$
$$|x - 2| < \frac{\epsilon}{3}$$

The answer to our question is

"x must be within $\dfrac{\epsilon}{3}$ of 2, but not equal to 2, to guarantee that $f(x)$ is within ϵ of 7"

That is, whenever $x \neq 2$ and x differs from 2 by less than $\dfrac{\epsilon}{3}$, $x \neq 2$, then $f(x)$ differs from 7 by less than ϵ, which can be restated as

$$\text{whenever } \begin{bmatrix} x \neq 2 \text{ and } x \text{ differs from 2} \\ \text{by less than } \delta = \dfrac{\epsilon}{3} \end{bmatrix} \quad \text{then} \quad \begin{bmatrix} f(x) \text{ differs from 7} \\ \text{by less than } \epsilon \end{bmatrix}$$

Notation is used to shorten the statement. We shorten the phrase "ϵ is any given positive number" by writing $\epsilon > 0$. Then the statement on the right is written

$$|f(x) - 7| < \epsilon$$

Similarly, the statement "x differs from 2 by less than δ" is written $|x - 2| < \delta$. The statement "$x \neq 2$" is handled by writing

$$0 < |x - 2| < \delta$$

So for our example, we write

Given any $\epsilon > 0$, then there is a number $\delta > 0$ so that whenever $0 < |x - 2| < \delta$, then $|f(x) - 7| < \epsilon$.

Since the number $\delta = \dfrac{\epsilon}{3}$ satisfies the inequalities for any number $\epsilon > 0$, we conclude that $\lim\limits_{x \to 2} f(x) = 7$.

NOW WORK Problem 41.

This discussion explains the *definition of the limit of a function* given below.

DEFINITION Limit of a Function

Let f be a function defined everywhere in an open interval containing c, except possibly at c. Then the **limit as x approaches c of $f(x)$ is L**, written

$$\lim_{x \to c} f(x) = L$$

if given any number $\epsilon > 0$, there is a number $\delta > 0$ so that

$$\text{whenever} \quad 0 < |x - c| < \delta \qquad \text{then} \quad |f(x) - L| < \epsilon$$

This definition is commonly called the **ϵ-δ definition of a limit** of a function.

Figure 65 illustrates the definition for three choices of ϵ. Compare Figures 65(a) and 65(b). Notice that in Figure 65(b), the smaller ϵ requires a smaller δ. Figure 65(c) illustrates what happens if δ is too large for the choice of ϵ; here there are values of f, for example, at x_1 and x_2, for which $|f(x) - L| \not< \epsilon$.

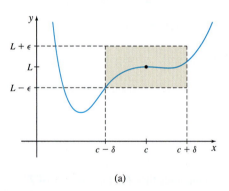

(a)

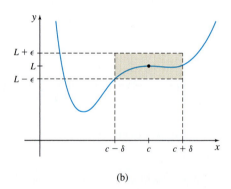

(b)

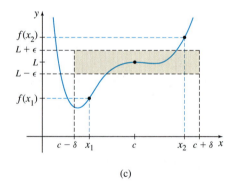

(c)

Figure 65

1 Use the ϵ-δ Definition of a Limit

EXAMPLE 1   Using the ϵ-δ Definition of a Limit

Use the ϵ-δ definition of a limit to prove $\lim\limits_{x \to -1} (1 - 2x) = 3$.

Solution Given any $\epsilon > 0$, we must show there is a number $\delta > 0$ so that

$$\text{whenever} \quad 0 < |x - (-1)| < \delta \qquad \text{then} \quad |(1 - 2x) - 3| < \epsilon$$

The idea is to find a connection between $|x - (-1)| = |x + 1|$ and $|(1 - 2x) - 3|$. Since

$$|(1 - 2x) - 3| = |-2x - 2| = |-2(x + 1)| = |-2| \cdot |x + 1| = 2|x + 1|$$

we see that for any $\epsilon > 0$,

$$\text{whenever} \quad |x - (-1)| = |x + 1| < \frac{\epsilon}{2} \qquad \text{then} \quad |(1 - 2x) - 3| < \epsilon$$

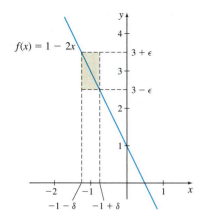

$f(x) = 1 - 2x$

Figure 66 $\lim\limits_{x \to -1} (1 - 2x) = 3$

That is, given any $\epsilon > 0$ there is a δ, $\delta = \dfrac{\epsilon}{2}$, so that whenever $0 < |x - (-1)| < \delta$, we have $|(1 - 2x) - 3| < \epsilon$. This proves that $\lim\limits_{x \to -1} (1 - 2x) = 3$. ∎

NOW WORK **Problem 21.**

A geometric interpretation of the ϵ-δ definition is shown in Figure 66. We see that whenever x on the horizontal axis is between $-1 - \delta$ and $-1 + \delta$, but not equal to -1, then $f(x)$ on the vertical axis is between the horizontal lines $y = 3 + \epsilon$ and $y = 3 - \epsilon$. So, $\lim\limits_{x \to -1} f(x) = 3$ describes the behavior of f near -1.

EXAMPLE 2 **Using the ϵ-δ Definition of a Limit**

Use the ϵ-δ definition of a limit to prove that:

(a) $\lim\limits_{x \to c} A = A$, where A and c are real numbers

(b) $\lim\limits_{x \to c} x = c$, where c is a real number

Solution **(a)** $f(x) = A$ is the constant function whose graph is a horizontal line. Given any $\epsilon > 0$, we must find $\delta > 0$ so that whenever $0 < |x - c| < \delta$, then $|f(x) - A| < \epsilon$.

Since $|A - A| = 0$, then $|f(x) - A| < \epsilon$ no matter what positive number δ is used. That is, any choice of δ guarantees that whenever $0 < |x - c| < \delta$, then $|f(x) - A| < \epsilon$.

(b) $f(x) = x$ is the identity function. Given any $\epsilon > 0$, we must find δ so that whenever $0 < |x - c| < \delta$, then $|f(x) - c| = |x - c| < \epsilon$. The easiest choice is to make $\delta = \epsilon$. That is, whenever $0 < |x - c| < \delta = \epsilon$, then $\underset{\underset{f(x)\,=\,x}{\uparrow}}{|f(x) - c|} = |x - c| < \epsilon$. ∎

Some observations about the ϵ-δ definition of a limit are given below:

- The limit of a function in no way depends on the value of the function at c.
- In general, the size of δ depends on the size of ϵ.
- For any ϵ, if a suitable δ has been found, any *smaller positive number* will also work for δ. That is, δ is not uniquely determined when ϵ is given.

EXAMPLE 3 **Using the ϵ-δ Definition of a Limit**

Prove: $\lim\limits_{x \to 2} x^2 = 4$.

Solution Given any $\epsilon > 0$, we must show there is a number $\delta > 0$ so that

$$\text{whenever} \quad 0 < |x - 2| < \delta \quad \text{then} \quad |x^2 - 4| < \epsilon$$

To establish a connection between $|x^2 - 4|$ and $|x - 2|$, we write $\left|x^2 - 4\right|$ as

$$\left|x^2 - 4\right| = |(x + 2)(x - 2)| = |x + 2| \cdot |x - 2|$$

Now, if we can find a number K for which $|x + 2| < K$, then we can choose $\delta = \dfrac{\epsilon}{K}$. To find K, we restrict x to some interval centered at 2. For example, suppose the distance between x and 2 is less than 1. Then

$$|x - 2| < 1$$
$$-1 < x - 2 < 1$$
$$1 < x < 3 \qquad \text{Simplify.}$$
$$1 + 2 < x + 2 < 3 + 2 \qquad \text{Add 2 to each part.}$$
$$3 < x + 2 < 5$$

In particular, we have $|x + 2| < 5$. It follows that whenever $|x - 2| < 1$,

$$|x^2 - 4| = |x + 2| \cdot |x - 2| < 5 |x - 2|$$

If $|x - 2| < \delta = \dfrac{\epsilon}{5}$, then $\left| x^2 - 4 \right| < 5 |x - 2| < 5 \cdot \dfrac{\epsilon}{5} = \epsilon$, as desired.

But before choosing $\delta = \dfrac{\epsilon}{5}$, we must remember that there are two constraints on $|x - 2|$. Namely,

$$|x - 2| < 1 \qquad \text{and} \qquad |x - 2| < \frac{\epsilon}{5}$$

To ensure that both inequalities are satisfied, we select δ to be the smaller of the numbers 1 and $\dfrac{\epsilon}{5}$, abbreviated as $\delta = \min \left\{ 1, \dfrac{\epsilon}{5} \right\}$. Now, whenever $|x - 2| < \delta = \min \left\{ 1, \dfrac{\epsilon}{5} \right\}$ then $|x^2 - 4| < \epsilon$ proving $\lim\limits_{x \to 2} x^2 = 4$. ∎

NOW WORK Problem 23.

In Example 3, the decision to restrict x so that $|x - 2| < 1$ was completely arbitrary. However, since we are looking for x close to 2, the interval chosen should be small. In Problem 40, you are asked to verify that if we had restricted x so that $|x - 2| < \dfrac{1}{3}$, then the choice for δ would be less than or equal to the smaller of $\dfrac{1}{3}$ and $\dfrac{3\epsilon}{13}$; that is,

$$\delta \le \min \left\{ \frac{1}{3}, \frac{3\epsilon}{13} \right\}.$$

EXAMPLE 4   **Using the ϵ-δ Definition of a Limit**

Prove $\lim\limits_{x \to c} \dfrac{1}{x} = \dfrac{1}{c}$, where $c > 0$.

Solution The domain of $f(x) = \dfrac{1}{x}$ is $\{x \mid x \neq 0\}$.

For any $\epsilon > 0$, we need to find a positive number δ so that whenever $0 < |x - c| < \delta$, then $\left| \dfrac{1}{x} - \dfrac{1}{c} \right| < \epsilon$. For $x \neq 0$, and $c > 0$, we have

$$\left| \frac{1}{x} - \frac{1}{c} \right| = \left| \frac{c - x}{xc} \right| = \frac{|c - x|}{|x| \cdot |c|} = \frac{|x - c|}{c|x|}$$

The idea is to find a connection between

$$|x - c| \qquad \text{and} \qquad \frac{|x - c|}{c|x|}$$

We proceed as in Example 3. Since we are interested in x near c, we restrict x to a small interval around c, say, $|x - c| < \dfrac{c}{2}$. Then,

$$-\frac{c}{2} < x - c < \frac{c}{2}$$

$$\frac{c}{2} < x < \frac{3c}{2} \qquad \text{\color{blue}{Add } } c \text{\color{blue}{ to each expression.}}$$

Since $c > 0$, then $x > \dfrac{c}{2} > 0$, and $\dfrac{1}{x} < \dfrac{2}{c}$. Now

$$\left| \frac{1}{x} - \frac{1}{c} \right| = \frac{|x - c|}{c|x|} < \frac{2}{c^2} \cdot |x - c| \qquad \text{\color{blue}{Substitute } } \frac{1}{|x|} < \frac{2}{c}.$$

We can make $\left| \dfrac{1}{x} - \dfrac{1}{c} \right| < \epsilon$ by choosing $\delta = \dfrac{c^2}{2}\epsilon$. Then

whenever $|x-c| < \delta = \dfrac{c^2}{2}\epsilon$,   we have $\left| \dfrac{1}{x} - \dfrac{1}{c} \right| < \dfrac{2}{c^2} \cdot |x-c| < \dfrac{2}{c^2} \cdot \left(\dfrac{c^2}{2} \cdot \epsilon \right) = \epsilon$

But remember, there are two restrictions on $|x - c|$.

$$|x - c| < \frac{c}{2} \qquad \text{and} \qquad |x - c| < \frac{c^2}{2} \cdot \epsilon$$

So, given any $\epsilon > 0$, we choose $\delta = \min\left(\dfrac{c}{2}, \dfrac{c^2}{2} \cdot \epsilon \right)$. Then whenever $0 < |x-c| < \delta$,

we have $\left| \dfrac{1}{x} - \dfrac{1}{c} \right| < \epsilon$. This proves $\lim\limits_{x \to c} \dfrac{1}{x} = \dfrac{1}{c}$, $c > 0$. ∎

NOW WORK **Problem 25.**

The ϵ-δ definition of a limit can be used to show that a limit does not exist, or that a mistake has been made in finding a limit. Example 5 illustrates how the ϵ-δ definition of a limit is used to discover a mistake in a computed limit.

EXAMPLE 5   Using the ϵ-δ Definition of a Limit

Use the ϵ-δ definition of a limit to prove the statement $\lim\limits_{x \to 3} (4x - 5) \neq 10$.

Solution We use a proof by contradiction.* Assume $\lim\limits_{x \to 3} (4x - 5) = 10$ and choose $\epsilon = 1$. (Any smaller positive number ϵ will also work.) Then there is a number $\delta > 0$, so that

whenever $0 < |x - 3| < \delta$ then $|(4x - 5) - 10| < 1$

We simplify the right inequality.

$$|(4x - 5) - 10| = |4x - 15| < 1$$
$$-1 < 4x - 15 < 1$$
$$14 < 4x < 16$$
$$3.5 < x < 4$$

NOTE For example, if $\delta = \dfrac{1}{4}$, then

$$3 - \frac{1}{4} < x < 3 + \frac{1}{4}$$
$$2.75 < x < 3.25$$

contradicting $3.5 < x < 4$.

According to our assumption, whenever $0 < |x - 3| < \delta$, then $3.5 < x < 4$. Regardless of the value of δ, the inequality $0 < |x - 3| < \delta$ is satisfied by a number x that is less than 3. This contradicts the fact that $3.5 < x < 4$. The contradiction means that $\lim\limits_{x \to 3} (4x - 5) \neq 10$. ∎

NOW WORK **Problem 33.**

The next example uses the ϵ-δ definition of a limit to show that a limit does not exist.

EXAMPLE 6   Using the ϵ-δ Definition of a Limit

The **Dirichlet function** is defined by $f(x) = \begin{cases} 1 & \text{if } x \text{ is rational} \\ 0 & \text{if } x \text{ is irrational} \end{cases}$.

Prove $\lim\limits_{x \to c} f(x)$ does not exist for any c.

Solution We use a proof by contradiction. That is, we assume that $\lim\limits_{x \to c} f(x)$ exists and show that this leads to a contradiction.

---

*In a proof by contradiction, we assume that the conclusion is not true and then show this leads to a contradiction.

Assume $\lim\limits_{x \to c} f(x) = L$ for some number c. Now if we are given $\epsilon = \dfrac{1}{2}$ (or any smaller positive number), then there is a positive number δ, so that

$$\text{whenever}\quad 0 < |x - c| < \delta \quad\text{then}\quad |f(x) - L| < \frac{1}{2}$$

Suppose x_1 is a rational number satisfying $0 < |x_1 - c| < \delta$, and x_2 is an irrational number satisfying $0 < |x_2 - c| < \delta$. Then from the definition of the function f,

$$f(x_1) = 1 \quad\text{and}\quad f(x_2) = 0$$

Using these values in the inequality $|f(x) - L| < \epsilon$, we get

$$|f(x_1) - L| = |1 - L| < \frac{1}{2} \quad\text{and}\quad |f(x_2) - L| = |0 - L| < \frac{1}{2}$$

$$-\frac{1}{2} < 1 - L < \frac{1}{2} \qquad\qquad -\frac{1}{2} < -L < \frac{1}{2}$$

$$-\frac{3}{2} < -L < -\frac{1}{2} \qquad\qquad \frac{1}{2} > L > -\frac{1}{2}$$

$$\frac{1}{2} < L < \frac{3}{2} \qquad\qquad -\frac{1}{2} < L < \frac{1}{2}$$

From the left inequality, we have $L > \dfrac{1}{2}$, and from the right inequality, we have $L < \dfrac{1}{2}$. Since it is impossible for both inequalities to be satisfied, we conclude that $\lim\limits_{x \to c} f(x)$ does not exist. ∎

EXAMPLE 7   Using the ϵ-δ Definition of a Limit

Prove that if $\lim\limits_{x \to c} f(x) > 0$, then there is an open interval around c, for which $f(x) > 0$ everywhere in the interval except possibly at c.

Solution Suppose $\lim\limits_{x \to c} f(x) = L > 0$. Then given any $\epsilon > 0$, there is a $\delta > 0$ so that

$$\text{whenever}\quad 0 < |x - c| < \delta \quad\text{then}\quad |f(x) - L| < \epsilon$$

If $\epsilon = \dfrac{L}{2}$, then from the definition of limit, there is a $\delta > 0$ so that

$$\text{whenever } 0 < |x - c| < \delta \quad \text{then } |f(x) - L| < \frac{L}{2} \quad\text{or equivalently,}\quad \frac{L}{2} < f(x) < \frac{3L}{2}$$

Since $\dfrac{L}{2} > 0$, the last statement proves our assertion that $f(x) > 0$ for all x in the interval $(c - \delta, c + \delta)$. ∎

In Problem 43, you are asked to prove the theorem stating that if $\lim\limits_{x \to c} f(x) < 0$, then there is an open interval around c, for which $f(x) < 0$ everywhere in the interval, except possibly at c.

We close this section with the ϵ-δ definitions of limits at infinity and infinite limits.

DEFINITION Limit at Infinity

Let f be a function defined on an open interval (b, ∞). Then f has a **limit at infinity**

$$\lim_{x \to \infty} f(x) = L$$

where L is a real number, if given any $\epsilon > 0$, there is a positive number M so that whenever $x > M$, then $|f(x) - L| < \epsilon$.

If f is a function defined on an open interval $(-\infty, a)$, then

$$\lim_{x \to -\infty} f(x) = L$$

if given any $\epsilon > 0$, there is a negative number N so that whenever $x < N$, then $|f(x) - L| < \epsilon$.

Figures 67 and 68 illustrate limits at infinity.

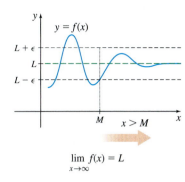

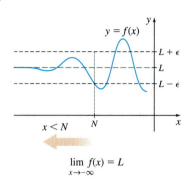

$$\lim_{x \to \infty} f(x) = L$$

$$\lim_{x \to -\infty} f(x) = L$$

Figure 67 For any $\epsilon > 0$, there is a positive number M so that whenever $x > M$, then $|f(x) - L| < \epsilon$.

Figure 68 For any $\epsilon > 0$, there is a negative number N so that whenever $x < N$, then $|f(x) - L| < \epsilon$.

DEFINITION Infinite Limit

Let f be a function defined everywhere on an open interval containing c, except possibly at c. Then $f(x)$ becomes unbounded in the positive direction (has an infinite limit) as x approaches c, written

$$\lim_{x \to c} f(x) = \infty$$

if, for every positive number M, a positive number δ exists so that

whenever $0 < |x - c| < \delta$ then $f(x) > M$.

Similarly, $f(x)$ becomes unbounded in the negative direction (has an infinite limit) as x approaches c, written

$$\lim_{x \to c} f(x) = -\infty$$

if, for every negative number N, a positive number δ exists so that

whenever $0 < |x - c| < \delta$ then $f(x) < N$.

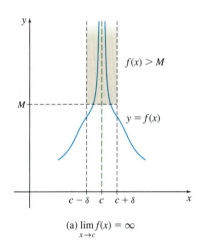

(a) $\lim\limits_{x \to c} f(x) = \infty$

Figure 69 illustrates infinite limits.

DEFINITION Infinite Limit at Infinity

Let f be a function defined on an open interval (b, ∞). Then f has an **infinite limit at infinity**

$$\lim_{x \to \infty} f(x) = \infty$$

if for any positive number M, there is a corresponding positive number N so that whenever $x > N$, then $f(x) > M$.

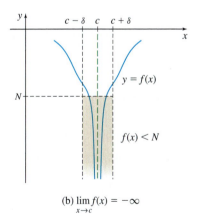

(b) $\lim\limits_{x \to c} f(x) = -\infty$

A similar definition applies for infinite limits at negative infinity.

Figure 69

1.6 Assess Your Understanding

Concepts and Vocabulary

1. *True or False* The limit of a function as x approaches c depends on the value of the function at c.

2. *True or False* In the ϵ-δ definition of a limit, we require $0 < |x - c|$ to ensure that $x \neq c$.

3. *True or False* In an ϵ-δ proof of a limit, the size of δ usually depends on the size of ϵ.

4. *True or False* When proving $\lim\limits_{x \to c} f(x) = L$ using the ϵ-δ definition of a limit, we try to find a connection between $|f(x) - c|$ and $|x - c|$.

1. = NOW WORK problem   〰 = Graphing technology recommended   CAS = Computer Algebra System recommended

5. *True or False*   Given any $\epsilon > 0$, suppose there is a $\delta > 0$ so that whenever $0 < |x - c| < \delta$, then $|f(x) - L| < \epsilon$. Then $\lim_{x \to c} f(x) = L$.

6. *True or False*   A function f has a limit L at infinity, if for any given $\epsilon > 0$, there is a positive number M so that whenever $x > M$, then $|f(x) - L| > \epsilon$.

Skill Building

In Problems 7–12, for each limit, find the largest δ that "works" for the given ϵ.

7. $\lim_{x \to 1} (2x) = 2$, $\epsilon = 0.01$

8. $\lim_{x \to 2} (-3x) = -6$, $\epsilon = 0.01$

9. $\lim_{x \to 2} (6x - 1) = 11$
$$\epsilon = \frac{1}{2}$$

10. $\lim_{x \to -3} (2 - 3x) = 11$
$$\epsilon = \frac{1}{3}$$

11. $\lim_{x \to 2} \left(-\frac{1}{2}x + 5 \right) = 4$
$$\epsilon = 0.01$$

12. $\lim_{x \to \frac{5}{6}} \left(3x + \frac{1}{2} \right) = 3$
$$\epsilon = 0.3$$

13. For the function $f(x) = 4x - 1$, we have $\lim_{x \to 3} f(x) = 11$. For each $\epsilon > 0$, find a $\delta > 0$ so that
$$\text{whenever} \quad 0 < |x - 3| < \delta \quad \text{then} \ |(4x - 1) - 11| < \epsilon$$

(a) $\epsilon = 0.1$ (b) $\epsilon = 0.01$

(c) $\epsilon = 0.001$ (d) $\epsilon > 0$ is arbitrary

14. For the function $f(x) = 2 - 5x$, we have $\lim_{x \to -2} f(x) = 12$. For each $\epsilon > 0$, find a $\delta > 0$ so that
$$\text{whenever} \ 0 < |x + 2| < \delta \quad \text{then} \ |(2 - 5x) - 12| < \epsilon$$

(a) $\epsilon = 0.2$ (b) $\epsilon = 0.02$

(c) $\epsilon = 0.002$ (d) $\epsilon > 0$ is arbitrary

15. For the function $f(x) = \dfrac{x^2 - 9}{x + 3}$, we have $\lim_{x \to -3} f(x) = -6$. For each $\epsilon > 0$, find a $\delta > 0$ so that
$$\text{whenever} \quad 0 < |x + 3| < \delta \quad \text{then} \ \left| \frac{x^2 - 9}{x + 3} - (-6) \right| < \epsilon$$

(a) $\epsilon = 0.1$ (b) $\epsilon = 0.01$ (c) $\epsilon > 0$ is arbitrary

16. For the function $f(x) = \dfrac{x^2 - 4}{x - 2}$, we have $\lim_{x \to 2} f(x) = 4$. For each $\epsilon > 0$, find a $\delta > 0$ so that
$$\text{whenever} \ 0 < |x - 2| < \delta \quad \text{then} \ \left| \frac{x^2 - 4}{x - 2} - 4 \right| < \epsilon$$

(a) $\epsilon = 0.1$ (b) $\epsilon = 0.01$ (c) $\epsilon > 0$ is arbitrary

In Problems 17–32, write a proof for each limit using the ϵ-δ definition of a limit.

17. $\lim_{x \to 2} (3x) = 6$

18. $\lim_{x \to 3} (4x) = 12$

19. $\lim_{x \to 0} (2x + 5) = 5$

20. $\lim_{x \to -1} (2 - 3x) = 5$

21. $\lim_{x \to -3} (-5x + 2) = 17$

22. $\lim_{x \to 2} (2x - 3) = 1$

23. $\lim_{x \to 2} (x^2 - 2x) = 0$

24. $\lim_{x \to 0} (x^2 + 3x) = 0$

25. $\lim_{x \to 1} \dfrac{1 + 2x}{3 - x} = \dfrac{3}{2}$

26. $\lim_{x \to 2} \dfrac{2x}{4 + x} = \dfrac{2}{3}$

27. $\lim_{x \to 0} \sqrt[3]{x} = 0$

28. $\lim_{x \to 1} \sqrt{2 - x} = 1$

29. $\lim_{x \to -1} x^2 = 1$

30. $\lim_{x \to 2} x^3 = 8$

31. $\lim_{x \to 3} \dfrac{1}{x} = \dfrac{1}{3}$

32. $\lim_{x \to 2} \dfrac{1}{x^2} = \dfrac{1}{4}$

33. Use the ϵ-δ definition of a limit to show that the statement $\lim_{x \to 3} (3x - 1) = 12$ is false.

34. Use the ϵ-δ definition of a limit to show that the statement $\lim_{x \to -2} (4x) = -7$ is false.

Applications and Extensions

35. Show that $\left| \dfrac{1}{x^2 + 9} - \dfrac{1}{18} \right| < \dfrac{7}{234} |x - 3|$ if $2 < x < 4$. Use this to show that $\lim_{x \to 3} \dfrac{1}{x^2 + 9} = \dfrac{1}{18}$.

36. Show that $|(2 + x)^2 - 4| \le 5 |x|$ if $-1 < x < 1$. Use this to show that $\lim_{x \to 0} (2 + x)^2 = 4$.

37. Show that $\left| \dfrac{1}{x^2 + 9} - \dfrac{1}{13} \right| \le \dfrac{1}{26} |x - 2|$ if $1 < x < 3$. Use this to show that $\lim_{x \to 2} \dfrac{1}{x^2 + 9} = \dfrac{1}{13}$.

38. Use the ϵ-δ definition of a limit to show that $\lim_{x \to 1} x^2 \ne 1.31$. (*Hint:* Use $\epsilon = 0.1$.)

39. If m and b are any constants, prove that
$$\lim_{x \to c} (mx + b) = mc + b$$

40. Verify that if x is restricted so that $|x - 2| < \dfrac{1}{3}$ in the proof of $\lim_{x \to 2} x^2 = 4$, then the choice for δ would be less than or equal to the smaller of $\dfrac{1}{3}$ and $\dfrac{3\epsilon}{13}$; that is, $\delta \le \min \left\{ \dfrac{1}{3}, \dfrac{3\epsilon}{13} \right\}$.

41. For $x \ne 3$, how close to 3 must x be to guarantee that $2x - 1$ differs from 5 by less than 0.1?

42. For $x \ne 0$, how close to 0 must x be to guarantee that 3^x differs from 1 by less than 0.1?

43. Prove that if $\lim_{x \to c} f(x) < 0$, then there is an open interval around c for which $f(x) < 0$ everywhere in the interval, except possibly at c.

44. Use the ϵ-δ definition of a limit at infinity to prove that
$$\lim_{x \to -\infty} \frac{1}{x} = 0.$$

45. Use the ϵ-δ definition of a limit at infinity to prove that

$$\lim_{x\to\infty}\left(-\frac{1}{\sqrt{x}}\right)=0.$$

46. For $\lim\limits_{x\to-\infty}\dfrac{1}{x^2}=0$, find a value of N that satisfies the ϵ-δ definition of limits at infinity for $\epsilon=0.1$.

47. Use the ϵ-δ definition of limit to prove that no number L exists so that $\lim\limits_{x\to0}\dfrac{1}{x}=L$.

48. Explain why in the ϵ-δ definition of a limit, the inequality $0<|x-c|<\delta$ has two strict inequality symbols.

49. The ϵ-δ definition of a limit states, in part, that f is defined everywhere in an open interval containing c, except possibly at c. Discuss the purpose of including the phrase, *except possibly at c*, and why it is necessary.

50. In the ϵ-δ definition of a limit, what does ϵ measure? What does δ measure? Give an example to support your explanation.

51. Discuss $\lim\limits_{x\to0}f(x)$ and $\lim\limits_{x\to1}f(x)$ if

$$f(x)=\begin{cases}x^2 & \text{if }x\text{ is rational}\\0 & \text{if }x\text{ is irrational}\end{cases}.$$

52. Discuss $\lim\limits_{x\to0}f(x)$ if $f(x)=\begin{cases}x^2 & \text{if }x\text{ is rational}\\\tan x & \text{if }x\text{ is irrational}\end{cases}$.

Challenge Problems

53. Use the ϵ-δ definition of limit to prove that $\lim\limits_{x\to1}(4x^3+3x^2-24x+22)=5$.

54. If $\lim\limits_{x\to c}f(x)=L$ and $\lim\limits_{x\to c}g(x)=M$, prove that $\lim\limits_{x\to c}[f(x)+g(x)]=L+M$. Use the ϵ-δ definition of limit.

55. For $\lim\limits_{x\to\infty}\dfrac{2-x}{\sqrt{5+4x^2}}=-\dfrac{1}{2}$, find a value of M that satisfies the ϵ-δ definition of limits at infinity for $\epsilon=0.01$.

56. Use the ϵ-δ definition of a limit to prove that the linear function $f(x)=ax+b$ is continuous everywhere.

57. Show that the function $f(x)=\begin{cases}0 & \text{if }x\text{ is rational}\\x & \text{if }x\text{ is irrational}\end{cases}$ is continuous only at $x=0$.

58. Suppose that f is defined on an interval (a,b) and there is a number K so that $|f(x)-f(c)|\le K|x-c|$ for all c in (a,b) and x in (a,b). Such a constant K is called a **Lipschitz constant**. Find a Lipschitz constant for $f(x)=x^3$ on $(0,2)$.

Chapter Review

THINGS TO KNOW

1.1 Limits of Functions Using Numerical and Graphical Techniques

- Slope of a secant line: $m_{\text{sec}}=\dfrac{f(x)-f(c)}{x-c}$ (p. 70)
- Slope of a tangent line $m_{\text{tan}}=\lim\limits_{x\to c}\dfrac{f(x)-f(c)}{x-c}$ (p. 70)
- The limit L of a function $y=f(x)$ as x approaches a number c does not depend on the value of f at c. (p. 74)
- The limit L of a function $y=f(x)$ as x approaches a number c is unique. A function cannot have more than one limit as x approaches c. (p. 74)
- The limit L of a function $y=f(x)$ as x approaches a number c exists if and only if both one-sided limits exist at c and both one-sided limits are equal. That is, $\lim\limits_{x\to c}f(x)=L$ if and only if $\lim\limits_{x\to c^-}f(x)=\lim\limits_{x\to c^+}f(x)=L$. (p. 74)

1.2 Limits of Functions Using Properties of Limits

Basic Limits

- $\lim\limits_{x\to c}A=A$, A a constant (p. 81)
- $\lim\limits_{x\to c}x=c$ (p. 81)

Properties of Limits If f and g are functions for which $\lim\limits_{x\to c}f(x)$ and $\lim\limits_{x\to c}g(x)$ both exist and if k is any real number, then:

- $\lim\limits_{x\to c}[f(x)\pm g(x)]=\lim\limits_{x\to c}f(x)\pm\lim\limits_{x\to c}g(x)$ (p. 82)
- $\lim\limits_{x\to c}[f(x)\cdot g(x)]=\lim\limits_{x\to c}f(x)\cdot\lim\limits_{x\to c}g(x)$ (p. 82)
- $\lim\limits_{x\to c}[kg(x)]=k\lim\limits_{x\to c}g(x)$ (p. 83)
- $\lim\limits_{x\to c}[f(x)]^n=\left[\lim\limits_{x\to c}f(x)\right]^n$, $n\ge2$ is an integer (p. 84)
- $\lim\limits_{x\to c}\sqrt[n]{f(x)}=\sqrt[n]{\lim\limits_{x\to c}f(x)}$, provided $f(x)\ge0$ if n is even (p. 85)
- $\lim\limits_{x\to c}[f(x)]^{m/n}=\left[\lim\limits_{x\to c}f(x)\right]^{m/n}$, provided $[f(x)]^{m/n}$ is defined for positive integers m and n (p. 85)
- $\lim\limits_{x\to c}\left[\dfrac{f(x)}{g(x)}\right]=\dfrac{\lim\limits_{x\to c}f(x)}{\lim\limits_{x\to c}g(x)}$, provided $\lim\limits_{x\to c}g(x)\ne0$ (p. 87)
- If P is a polynomial function, then $\lim\limits_{x\to c}P(x)=P(c)$. (p. 86)
- If R is a rational function and if c is in the domain of R, then $\lim\limits_{x\to c}R(x)=R(c)$. (p. 87)

1.3 Continuity

Definitions

- Continuity at a number (p. 93)
- Removable discontinuity (p. 95)
- One-sided continuity at a number (p. 95)
- Continuity on an interval (p. 96)
- Continuity on a domain (p. 97)

Properties of Continuity

- A polynomial function is continuous on its domain.   (p. 97)
- A rational function is continuous on its domain.   (p. 97)
- If the functions f and g are continuous at a number c, and if k is a real number, then the functions $f + g$, $f - g$, $f \cdot g$, and kf are also continuous at c. If $g(c) \neq 0$, the function $\dfrac{f}{g}$ is continuous at c.   (p. 98)
- If a function g is continuous at c and a function f is continuous at $g(c)$, then the composite function $(f \circ g)(x) = f(g(x))$ is continuous at c.   (p. 99)
- If f is a one-to-one function that is continuous on its domain, then its inverse function f^{-1} is also continuous on its domain. (p. 100)

The Intermediate Value Theorem   Let f be a function that is continuous on a closed interval $[a, b]$ with $f(a) \neq f(b)$. If N is any number between $f(a)$ and $f(b)$, then there is at least one number c in the open interval (a, b) for which $f(c) = N$.   (p. 100)

1.4 Limits and Continuity of Trigonometric, Exponential, and Logarithmic Functions

Basic Limits

- $\displaystyle\lim_{\theta \to 0} \frac{\sin \theta}{\theta} = 1$ (p. 108)
- $\displaystyle\lim_{x \to c} \sin x = \sin c$ (p. 111)
- $\displaystyle\lim_{x \to c} \cos x = \cos c$ (p. 111)
- $\displaystyle\lim_{\theta \to 0} \frac{\cos \theta - 1}{\theta} = 0$ (p. 111)
- $\displaystyle\lim_{x \to c} a^x = a^c$; $a > 0$, $a \neq 1$ (p. 114)
- $\displaystyle\lim_{x \to c} \log_a x = \log_a c$; $a > 0$, $a \neq 1$, and $c > 0$ (p. 114)

Squeeze Theorem   If the functions f, g, and h have the property that for all x in an open interval containing c, except possibly at c, $f(x) \leq g(x) \leq h(x)$, and if $\displaystyle\lim_{x \to c} f(x) = \lim_{x \to c} h(x) = L$, then $\displaystyle\lim_{x \to c} g(x) = L$.   (p. 107)

Properties of Continuity

- The six trigonometric functions are continuous on their domains.   (pp. 111–113)
- The six inverse trigonometric functions are continuous on their domains.   (p. 113)
- An exponential function is continuous on its domain.   (p. 114)
- A logarithmic function is continuous on its domain.   (p. 114)

1.5 Infinite Limits; Limits at Infinity; Asymptotes

Basic Limits

- $\displaystyle\lim_{x \to 0^-} \frac{1}{x} = -\infty$; $\displaystyle\lim_{x \to 0^+} \frac{1}{x} = \infty$ (p. 118)

- $\displaystyle\lim_{x \to \infty} \frac{1}{x} = 0$; $\displaystyle\lim_{x \to -\infty} \frac{1}{x} = 0$ (p. 120)
- $\displaystyle\lim_{x \to 0} \frac{1}{x^2} = \infty$ (p. 117)
- $\displaystyle\lim_{x \to 0^+} \ln x = -\infty$ (p. 118)
- $\displaystyle\lim_{x \to \infty} \ln x = \infty$ (p. 124)
- $\displaystyle\lim_{x \to -\infty} e^x = 0$; $\displaystyle\lim_{x \to \infty} e^x = \infty$ (p. 124)

Definitions

- Vertical asymptote   (p. 119)
- Horizontal asymptote   (p. 125)

Properties of Limits at Infinity   (p. 120): If k is a real number, $n \geq 2$ is an integer, and the functions f and g approach real numbers as $x \to \infty$, then:

- $\displaystyle\lim_{x \to \infty} A = A$, where A is a number
- $\displaystyle\lim_{x \to \infty} [kf(x)] = k \lim_{x \to \infty} f(x)$
- $\displaystyle\lim_{x \to \infty} [f(x) \pm g(x)] = \lim_{x \to \infty} f(x) \pm \lim_{x \to \infty} g(x)$
- $\displaystyle\lim_{x \to \infty} [f(x)g(x)] = \left[\lim_{x \to \infty} f(x)\right]\left[\lim_{x \to \infty} g(x)\right]$
- $\displaystyle\lim_{x \to \infty} \frac{f(x)}{g(x)} = \frac{\displaystyle\lim_{x \to \infty} f(x)}{\displaystyle\lim_{x \to \infty} g(x)}$ if $\displaystyle\lim_{x \to \infty} g(x) \neq 0$
- $\displaystyle\lim_{x \to \infty} [f(x)]^n = \left[\lim_{x \to \infty} f(x)\right]^n$
- $\displaystyle\lim_{x \to \infty} \sqrt[n]{f(x)} = \sqrt[n]{\lim_{x \to \infty} f(x)}$, where $f(x) \geq 0$ if n is even

1.6 The ϵ-δ Definition of a Limit

Definitions

- Limit of a Function   (p. 132)
- Limit at Infinity   (p. 136)
- Infinite Limit   (p. 137)
- Infinite Limit at Infinity   (p. 137)

Properties of limits

- If $\displaystyle\lim_{x \to c} f(x) > 0$, then there is an open interval around c, for which $f(x) > 0$ everywhere in the interval, except possibly at c.   (p. 136)
- If $\displaystyle\lim_{x \to c} f(x) < 0$, then there is an open interval around c, for which $f(x) < 0$ everywhere in the interval, except possibly at c.   (p. 136)

OBJECTIVES

Section	You should be able to …	Example	Review Exercises
1.1	1 Discuss the slope of a tangent line to a graph (p. 69)		4
	2 Investigate a limit using a table of numbers (p. 71)	1–3	1
	3 Investigate a limit using a graph (p. 73)	4–7	2, 3
1.2	1 Find the limit of a sum, a difference, and a product (p. 82)	1–6	8, 10, 12, 14, 22, 26, 29, 30, 47, 48
	2 Find the limit of a power and the limit of a root (p. 84)	7–9	11, 18, 28, 55
	3 Find the limit of a polynomial (p. 86)	10	10, 22
	4 Find the limit of a quotient (p. 87)	11–14	13–17, 19–21, 23–25, 27, 56
	5 Find the limit of an average rate of change (p. 89)	15	37
	6 Find the limit of a difference quotient (p. 89)	16	5, 6, 49
1.3	1 Determine whether a function is continuous at a number (p. 93)	1–4	31–36
	2 Determine intervals on which a function is continuous (p. 98)	5, 6	39–42
	3 Use properties of continuity (p. 98)	7, 8	39–42
	4 Use the Intermediate Value Theorem (p. 100)	9, 10	38, 44–46
1.4	1 Use the Squeeze Theorem to find a limit (p. 106)	1	7, 69
	2 Find limits involving trigonometric functions (p. 108)	2, 3	9, 51–55
	3 Determine where the trigonometric functions are continuous (p. 111)	4, 5	63–65
	4 Determine where an exponential or a logarithmic function is continuous (p. 113)	6	43
1.5	1 Investigate infinite limits (p. 117)	1	57, 58
	2 Find the vertical asymptotes of a function (p. 119)	2	61, 62
	3 Investigate limits at infinity (p. 120)	3–8	59, 60
	4 Find the horizontal asymptotes of a function (p. 125)	9	61, 62
	5 Find the asymptotes of a rational function using limits (p. 126)	10	67, 68
1.6	1 Use the ϵ-δ definition of a limit (p. 132)	1–7	50, 66

REVIEW EXERCISES

1. Use a table of numbers to investigate $\lim\limits_{x\to 0}\dfrac{1-\cos x}{1+\cos x}$.

In Problems 2 and 3, use a graph to investigate $\lim\limits_{x\to c} f(x)$.

2. $f(x) = \begin{cases} 2x-5 & \text{if } x < 1 \\ 6-9x & \text{if } x \geq 1 \end{cases}$ at $c=1$

3. $f(x) = \begin{cases} x^2+2 & \text{if } x < 2 \\ 2x+1 & \text{if } x \geq 2 \end{cases}$ at $c=2$

4. For $f(x) = x^2 - 3$:

(a) Find the slope of the secant line joining $(1, -2)$ and $(2, 1)$.

(b) Find the slope of the tangent line to the graph of f at $(1, -2)$.

In Problems 5 and 6, for each function find the limit of the difference quotient $\lim\limits_{h\to 0}\dfrac{f(x+h)-f(x)}{h}$.

5. $f(x) = \dfrac{3}{x}$

6. $f(x) = 3x^2 + 2x$

7. Find $\lim\limits_{x\to 0} f(x)$ if $1 + \sin x \leq f(x) \leq |x| + 1$

In Problems 8–22, find each limit.

8. $\lim\limits_{x\to 2}\left(2x - \dfrac{1}{x}\right)$

9. $\lim\limits_{x\to \pi}(x\cos x)$

10. $\lim\limits_{x\to -1}\left(x^3 + 3x^2 - x - 1\right)$

11. $\lim\limits_{x\to 0}\sqrt[3]{x(x+2)^3}$

12. $\lim\limits_{x\to 0}[(2x+3)(x^5+5x)]$

13. $\lim\limits_{x\to 3}\dfrac{x^3-27}{x-3}$

14. $\lim\limits_{x\to 3}\left(\dfrac{x^2}{x-3} - \dfrac{3x}{x-3}\right)$

15. $\lim\limits_{x\to 2}\dfrac{x^2-4}{x-2}$

16. $\lim\limits_{x\to -1}\dfrac{x^2+3x+2}{x^2+4x+3}$

17. $\lim\limits_{x\to -2}\dfrac{x^3+5x^2+6x}{x^2+x-2}$

18. $\lim\limits_{x\to 1}\left(x^2 - 3x + \dfrac{1}{x}\right)^{15}$

19. $\lim\limits_{x\to 2}\dfrac{3-\sqrt{x^2+5}}{x^2-4}$

20. $\lim\limits_{x\to 0}\left\{\dfrac{1}{x}\left[\dfrac{1}{(2+x)^2} - \dfrac{1}{4}\right]\right\}$

21. $\lim\limits_{x\to 0}\dfrac{(x+3)^2-9}{x}$

22. $\lim\limits_{x\to 1}[(x^3-3x^2+3x-1)(x+1)^2]$

In Problems 23–28, find each one-sided limit, if it exists.

23. $\lim\limits_{x\to -2^+}\dfrac{x^2+5x+6}{x+2}$

24. $\lim\limits_{x\to 5^+}\dfrac{|x-5|}{x-5}$

25. $\lim\limits_{x\to 1^-}\dfrac{|x-1|}{x-1}$

26. $\lim\limits_{x\to 3/2^+}\lfloor 2x \rfloor$

27. $\lim\limits_{x\to 4^-}\dfrac{x^2-16}{x-4}$

28. $\lim\limits_{x\to 1^+}\sqrt{x-1}$

In Problems 29 and 30, find $\lim\limits_{x\to c^-} f(x)$ and $\lim\limits_{x\to c^+} f(x)$ for the given c. Determine whether $\lim\limits_{x\to c} f(x)$ exists.

29. $f(x) = \begin{cases} 2x+3 & \text{if } x < 2 \\ 9-x & \text{if } x \geq 2 \end{cases}$ at $c=2$

30. $f(x) = \begin{cases} 3x + 1 & \text{if } x < 3 \\ 10 & \text{if } x = 3 \\ 4x - 2 & \text{if } x > 3 \end{cases}$ at $c = 3$

In Problems 31–36, determine whether f is continuous at c.

31. $f(x) = \begin{cases} 5x - 2 & \text{if } x < 1 \\ 5 & \text{if } x = 1 \\ 2x + 1 & \text{if } x > 1 \end{cases}$ at $c = 1$

32. $f(x) = \begin{cases} x^2 & \text{if } x < -1 \\ 2 & \text{if } x = -1 \\ -3x - 2 & \text{if } x > -1 \end{cases}$ at $c = -1$

33. $f(x) = \begin{cases} 4 - 3x^2 & \text{if } x < 0 \\ 4 & \text{if } x = 0 \\ \sqrt{16 - x^2} & \text{if } 0 < x \le 4 \end{cases}$ at $c = 0$

34. $f(x) = \begin{cases} \sqrt{4 + x} & \text{if } -4 \le x \le 4 \\ \sqrt{\dfrac{x^2 - 16}{x - 4}} & \text{if } x > 4 \end{cases}$ at $c = 4$

35. $f(x) = \lfloor 2x \rfloor$ at $c = \dfrac{1}{2}$ **36.** $f(x) = |x - 5|$ at $c = 5$

37. **(a)** Find the average rate of change of $f(x) = 2x^2 - 5x$ from 1 to x.

  (b) Find the limit as x approaches 1 of the average rate of change found in (a).

38. A function f is defined on the interval $[-1, 1]$ with the following properties: f is continuous on $[-1, 1]$ except at 0, negative at -1, positive at 1, but with no zeros. Does this contradict the Intermediate Value theorem?

In Problems 39–43 find all values x for which f(x) is continuous.

39. $f(x) = \dfrac{x}{x^3 - 27}$ **40.** $f(x) = \dfrac{x^2 - 3}{x^2 + 5x + 6}$

41. $f(x) = \dfrac{2x + 1}{x^3 + 4x^2 + 4x}$ **42.** $f(x) = \sqrt{x - 1}$

43. $f(x) = 2^{-x}$

44. Use the Intermediate Value Theorem to determine whether $2x^3 + 3x^2 - 23x - 42 = 0$ has a zero in the interval $[3, 4]$.

In Problems 45 and 46, use the Intermediate Value Theorem to approximate the zero correct to three decimal places.

45. $f(x) = 8x^4 - 2x^2 + 5x - 1$ on the interval $[0, 1]$.

46. $f(x) = 3x^3 - 10x + 9$; zero between -3 and -2.

47. Find $\lim\limits_{x \to 0^+} \dfrac{|x|}{x}(1 - x)$ and $\lim\limits_{x \to 0^-} \dfrac{|x|}{x}(1 - x)$. What can you say about $\lim\limits_{x \to 0} \dfrac{|x|}{x}(1 - x)$?

48. Find $\lim\limits_{x \to 2} \left(\dfrac{x^2}{x - 2} - \dfrac{2x}{x - 2} \right)$. Then comment on the statement that this limit is given by $\lim\limits_{x \to 2} \dfrac{x^2}{x - 2} - \lim\limits_{x \to 2} \dfrac{2x}{x - 2}$.

49. Find $\lim\limits_{h \to 0} \dfrac{f(x + h) - f(x)}{h}$ for $f(x) = \sqrt{x}$.

50. For $\lim\limits_{x \to 3}(2x + 1) = 7$, find the largest possible δ that "works" for $\epsilon = 0.01$.

In Problems 51–60, find each limit.

51. $\lim\limits_{x \to 0} \cos(\tan x)$ **52.** $\lim\limits_{x \to 0} \dfrac{\sin \dfrac{x}{4}}{x}$

53. $\lim\limits_{x \to 0} \dfrac{\tan(3x)}{\tan(4x)}$ **54.** $\lim\limits_{x \to 0} \dfrac{\cos \dfrac{x}{3} - 1}{x}$

55. $\lim\limits_{x \to 0} \left(\dfrac{\cos x - 1}{x} \right)^{10}$ **56.** $\lim\limits_{x \to 0} \dfrac{e^{4x} - 1}{e^x - 1}$

57. $\lim\limits_{x \to \pi/2^+} \tan x$ **58.** $\lim\limits_{x \to -3} \dfrac{2 + x}{(x + 3)^2}$

59. $\lim\limits_{x \to \infty} \dfrac{3x^3 - 2x + 1}{x^3 - 8}$ **60.** $\lim\limits_{x \to \infty} \dfrac{3x^4 + x}{2x^2}$

In Problems 61 and 62, find any vertical and horizontal asymptotes of f.

61. $f(x) = \dfrac{4x - 2}{x + 3}$ **62.** $f(x) = \dfrac{2x}{x^2 - 4}$

63. Let $f(x) = \begin{cases} \dfrac{\tan x}{2x} & \text{if } x \ne 0 \\ \dfrac{1}{2} & \text{if } x = 0 \end{cases}$. Is f continuous at 0?

64. Let $f(x) = \begin{cases} \dfrac{\sin(3x)}{x} & \text{if } x \ne 0 \\ 1 & \text{if } x = 0 \end{cases}$. Is f continuous at 0?

65. The function $f(x) = \dfrac{\cos\left(\pi x + \dfrac{\pi}{2}\right)}{x}$ is not defined at 0. Decide how to define $f(0)$ so that f is continuous at 0.

66. Use an ϵ-δ argument to show that the statement $\lim\limits_{x \to -3}(x^2 - 9) = -18$ is false.

67. **(a)** Sketch a graph of a function f that has the following properties:
$$f(-1) = 0, \quad \lim\limits_{x \to \infty} f(x) = 2, \quad \lim\limits_{x \to -\infty} f(x) = 2,$$
$$\lim\limits_{x \to 4^-} f(x) = -\infty, \quad \lim\limits_{x \to 4^+} f(x) = \infty$$

  (b) Define a function that describes your graph.

68. **(a)** Find the domain and the intercepts (if any) of
$$R(x) = \dfrac{2x^2 - 5x + 2}{5x^2 - x - 2}.$$

  (b) Discuss the behavior of the graph of R at numbers where R is not defined.

  (c) Find any vertical or horizontal asymptotes of the function R.

69. If $1 - x^2 \le f(x) \le \cos x$ for all x in the interval $-\dfrac{\pi}{2} < x < \dfrac{\pi}{2}$, show that $\lim\limits_{x \to 0} f(x) = 1$.

CHAPTER 1 PROJECT   Pollution in Clear Lake

The Toxic Waste Disposal Company (TWDC) specializes in the disposal of a particularly dangerous pollutant, Agent Yellow (AY). Unfortunately, instead of safely disposing of this pollutant, the company simply dumped AY in (formerly) Clear Lake.

Fortunately, they have been caught and are now defending themselves in court.

The facts below are not in dispute. As a result of TWDC's activity, the current concentration of AY in Clear Lake is now 10 ppm (parts per million). Clear Lake is part of a chain of rivers and lakes. Fresh water flows into Clear Lake and the contaminated water flows downstream from it. The Department of Environmental Protection estimates that the level of contamination in Clear Lake will fall by 20% each year. These facts can be modeled as

$$p(0) = 10 \qquad p(t+1) = 0.80p(t)$$

where $p = p(t)$, measured in ppm, is the concentration of pollutants in the lake at time t, in years.

1. Explain how the above equations model the facts.
2. Create a table showing the values of t for $t = 0, 1, 2, \ldots, 20$.
3. Show that $p(t) = 10(0.8)^t$.
4. Use graphing technology to graph $p = p(t)$.
5. What is $\lim_{t \to \infty} p(t)$?

Lawyers for TWDC looked at the results in 1–5 above and argued that their client has not done any real damage. They concluded that Clear Lake would eventually return to its former clear and unpolluted state. They even called in a mathematician, who wrote the following on a blackboard:

$$\lim_{t \to \infty} p(t) = 0$$

and explained that this bit of mathematics means, descriptively, that after many years the concentration of AY will, indeed, be close to zero.

Concerned citizens booed the mathematician's testimony. Fortunately, one of them has taken calculus and knows a little bit about limits. She noted that, although "after many years the concentration of AY will approach zero," the townspeople like to swim in Clear Lake and state regulations prohibit swimming unless the concentration of AY is below 2 ppm. She proposed a fine of $100,000 per year for each full year that the lake is unsafe for swimming. She also questioned the mathematician, saying, "Your testimony was correct as far as it went, but I remember from studying calculus that talking about the eventual concentration of AY after many, many years is only a small part of the story. The more precise meaning of your statement $\lim_{t \to \infty} p(t) = 0$ is that given some tolerance T for the concentration of AY, there is some time N (which may be very far in the future) so that for all $t > N$, $p(t) < T$."

6. Using the table or the graph for $p = p(t)$, find N so that if $t > N$, then $p(t) < 2$.
7. How much is the fine?

Her words were greeted by applause. The town manager sprang to his feet and noted that although a tolerance of 2 ppm was fine for swimming, the town used Clear Lake for its drinking water and until the concentration of AY dropped below 0.5 ppm, the water would be unsafe for drinking. He proposed a fine of $200,000 per year for each full year the water was unfit for drinking.

8. Using the table or the graph for $p = p(t)$, find N so that if $t > N$, then $p(t) < 0.5$.
9. How much is the fine?
10. How would you find if you were on the jury trying TWDC? If the jury found TWDC guilty, what fine would you recommend? Explain your answers.

2

The Derivative

CHAPTER 2 PROJECT In the Chapter Project on page 195 at the end of this chapter, we explore some of the physics at work that allowed engineers and pilots to successfully maneuver the Lunar Module to the Moon's surface.

The Apollo Lunar Module
"One Giant Leap for Mankind."

On May 25, 1961, in a special address to Congress, U.S. President John F. Kennedy proposed the goal ``before this decade is out, of landing a man on the Moon and returning him safely to the Earth.'' Roughly eight years later, on July 16, 1969, a Saturn V rocket launched from the Kennedy Space Center in Florida, carrying the *Apollo 11* spacecraft and three astronauts—Neil Armstrong, Buzz Aldrin, and Michael Collins—bound for the Moon.

The *Apollo* spacecraft had three parts: the Command Module with a cabin for the three astronauts; the Service Module that supported the Command Module with propulsion, electrical power, oxygen, and water; and the Lunar Module for landing on the Moon. After its launch, the spacecraft traveled for three days until it entered into lunar orbit. Armstrong and Aldrin then moved into the Lunar Module, which they landed in the flat expanse of the Sea of Tranquility. After more than 21 hours on the surface of the Moon, the first humans to touch the surface of the Moon crawled back into the Lunar Module and lifted off to rejoin the Command Module, which Collins had been piloting in lunar orbit. The three astronauts then headed back to Earth, where they splashed down in the Pacific Ocean on July 24.

Chapter 2 opens with an investigation of mathematical models involving change. First we consider velocity. Then we return to the tangent problem to find an equation of the tangent line to the graph of a function f at a point $P = (c, f(c))$. Remember in Section 1.1 we found that the slope of a tangent line was a limit,

$$m_{\tan} = \lim_{x \to c} \frac{f(x) - f(c)}{x - c}$$

This limit turns out to be one of the most significant ideas in calculus, the *derivative*.

In this chapter, we introduce interpretations of the derivative, consider the derivative as a function, and consider some properties of the derivative. By the end of the chapter, you will have a collection of basic derivative formulas and derivative rules that will be used throughout your study of calculus.

2.1 Rates of Change and the Derivative

OBJECTIVES *When you finish this section, you should be able to:*

1 Find instantaneous velocity (p. 145)

2 Find an equation of the tangent line to the graph of a function (p. 147)

3 Find the rate of change of a function (p. 149)

4 Find the derivative of a function at a number (p. 150)

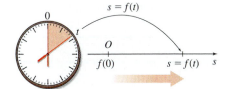

Figure 1 t is the travel time. s is the distance of the object from the origin at time t.

Everything in nature changes. Examples include climate change, change in the phases of the moon, and change in populations. To describe natural processes mathematically, the ideas of change and rate of change are often used. In this section, we use a limit to describe a rate of change and discover that seemingly unrelated interpretations of rates of change have a common basis we call a *derivative*.

Average velocity provides a physical example of an average rate of change. For example, consider an object moving along a horizontal line with the positive direction to the right, or moving along a vertical line with the positive direction upward. Motion of this sort is referred to as **rectilinear motion**. The object's location at time $t = 0$ is called its **initial position**. The initial position is usually marked as the origin O on the line. See Figure 1. We assume the distance s at time t of the object from the origin is given by a function $s = f(t)$. Here, s is the signed, or directed, distance (using some measure of distance such as centimeters, meters, feet, etc.) of the object from O at time t (in seconds or hours).

> **DEFINITION** Average Velocity
>
> The (signed) distance s from the origin at time t of an object in rectilinear motion is given by the function $s = f(t)$. If at time t_0 the object is at $s_0 = f(t_0)$ and at time t_1 the object is at $s_1 = f(t_1)$, then the change in time is $\Delta t = t_1 - t_0$ and the change in distance is $\Delta s = s_1 - s_0 = f(t_1) - f(t_0)$. The average rate of change of distance with respect to time is
>
> $$\frac{\Delta s}{\Delta t} = \frac{f(t_1) - f(t_0)}{t_1 - t_0} \qquad t_1 \neq t_0$$
>
> and is called the **average velocity** of the object over the interval $[t_0, t_1]$.

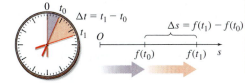

Figure 2 The average velocity is $\dfrac{\Delta s}{\Delta t}$.

See Figure 2.

EXAMPLE 1 Finding Average Velocity

The Mike O'Callaghan-Pat Tillman Memorial Bridge spanning the Colorado River opened on October 16, 2010. Having a span of 1900 feet, it is the longest arch bridge in the Western Hemisphere, and its roadway is 890 feet above the Colorado River.

If a rock falls from the roadway, the function $s = f(t) = 16t^2$ gives the distance s, in feet, that the rock falls after t seconds for $0 \leq t \leq 7.458$. Here, 7.458 seconds is the approximate time it takes the rock to fall 890 feet into the river. The average velocity of the rock during its fall is

$$\frac{\Delta s}{\Delta t} = \frac{f(7.458) - f(0)}{7.458 - 0} = \frac{890 - 0}{7.458} \approx 119.335 \text{ feet per second} \qquad \blacksquare$$

NOTE Here, the motion occurs along a vertical line with the positive direction downward.

1 Find Instantaneous Velocity

The average velocity of the rock in Example 1 approximates the average velocity over the interval $[0, 7.458]$. But the average velocity does not tell us about the velocity at any particular instant of time. That is, it gives no information about the rock's *instantaneous velocity*.

NOTE *Speed* and *velocity* are not the same thing. Speed measures how fast an object is moving and is defined as the absolute value of its velocity. Velocity measures both the speed and the direction of an object and may be a positive or a negative number or zero.

We can investigate the instantaneous velocity of the rock, say, at $t = 3$ seconds, by computing average velocities for short intervals of time beginning at $t = 3$. First we compute the average velocity for the interval beginning at $t = 3$ and ending at $t = 3.5$. The corresponding distances the rock has fallen are

$$f(3) = 16 \cdot 3^2 = 144 \text{ feet} \qquad \text{when } t = 3 \text{ seconds}$$

and

$$f(3.5) = 16 \cdot (3.5)^2 = 196 \text{ feet} \qquad \text{when } t = 3.5 \text{ seconds}$$

Then $\Delta t = 3.5 - 3.0 = 0.5$, and during this 0.5-second interval,

$$\text{Average velocity} = \frac{\Delta s}{\Delta t} = \frac{f(3.5) - f(3)}{3.5 - 3} = \frac{196 - 144}{0.5} = 104 \text{ feet per second}$$

Table 1 shows average velocities of the rock for smaller intervals of time.

TABLE 1

Time interval	Start $t_0 = 3$	End t	Δt	$\dfrac{\Delta s}{\Delta t} = \dfrac{f(t) - f(t_0)}{t - t_0} = \dfrac{16t^2 - 144}{t - 3}$
[3, 3.1]	3	3.1	0.1	$\dfrac{\Delta s}{\Delta t} = \dfrac{f(3.1) - f(3)}{3.1 - 3} = \dfrac{16 \cdot 3.1^2 - 144}{0.1} = 97.6$
[3, 3.01]	3	3.01	0.01	$\dfrac{\Delta s}{\Delta t} = \dfrac{f(3.01) - f(3)}{3.01 - 3} = \dfrac{16 \cdot 3.01^2 - 144}{0.01} = 96.16$
[3, 3.0001]	3	3.0001	0.0001	$\dfrac{\Delta s}{\Delta t} = \dfrac{f(3.0001) - f(3)}{3.0001 - 3} = \dfrac{16 \cdot 3.0001^2 - 144}{0.0001} = 96.0016$

The average velocity of 96.0016 over the time interval $\Delta t = 0.0001$ second should be very close to the instantaneous velocity of the rock at $t = 3$ seconds. As Δt gets closer to 0, the average velocity gets closer to the instantaneous velocity. So, to obtain the instantaneous velocity at $t = 3$ precisely, we use the limit of the average velocity as Δt approaches 0 or, equivalently, as t approaches 3.

$$\lim_{\Delta t \to 0} \frac{\Delta s}{\Delta t} = \lim_{t \to 3} \frac{f(t) - f(3)}{t - 3} = \lim_{t \to 3} \frac{16t^2 - 16 \cdot 3^2}{t - 3} = \lim_{t \to 3} \frac{16(t^2 - 9)}{t - 3}$$

$$= \lim_{t \to 3} \frac{16(t - 3)(t + 3)}{t - 3} = \lim_{t \to 3}[16(t + 3)] = 96$$

The rock's instantaneous velocity at $t = 3$ seconds is 96 ft/s.

We generalize this result to obtain a definition for instantaneous velocity.

DEFINITION Instantaneous Velocity

If $s = f(t)$ is a function that gives the distance s an object travels in time t, the **instantaneous velocity** v of the object at time t_0 is defined as the limit of the average velocity $\dfrac{\Delta s}{\Delta t}$ as Δt approaches 0. That is,

$$v = \lim_{\Delta t \to 0} \frac{\Delta s}{\Delta t} = \lim_{t \to t_0} \frac{f(t) - f(t_0)}{t - t_0} \tag{1}$$

provided the limit exists.

We usually shorten "instantaneous velocity" and just use the word "velocity."

NOW WORK Problem 7.

EXAMPLE 2   Finding Velocity

Find the velocity v of the falling rock from Example 1 at:

(a) $t_0 = 1$ second after it begins to fall.

(b) $t_0 = 7.4$ seconds, just before it hits the Colorado River.

(c) at any time t_0.

Solution (a) Use the definition of instantaneous velocity with $f(t) = 16t^2$ and $t_0 = 1$.

$$v = \lim_{\Delta t \to 0} \frac{\Delta s}{\Delta t} = \lim_{t \to 1} \frac{f(t) - f(1)}{t - 1} = \lim_{t \to 1} \frac{16t^2 - 16}{t - 1} = \lim_{t \to 1} \frac{16(t^2 - 1)}{t - 1}$$

$$= \lim_{t \to 1} \frac{16(t - 1)(t + 1)}{t - 1} = \lim_{t \to 1} [16(t + 1)] = 32$$

After 1 second, the velocity of the rock is 32 ft/s.

(b) For $t_0 = 7.4$ s,

$$v = \lim_{\Delta t \to 0} \frac{\Delta s}{\Delta t} = \lim_{t \to 7.4} \frac{f(t) - f(7.4)}{t - 7.4} = \lim_{t \to 7.4} \frac{16t^2 - 16 \cdot (7.4)^2}{t - 7.4}$$

$$= \lim_{t \to 7.4} \frac{16[t^2 - (7.4)^2]}{t - 7.4} = \lim_{t \to 7.4} \frac{16(t - 7.4)(t + 7.4)}{t - 7.4}$$

$$= \lim_{t \to 7.4} [16(t + 7.4)] = 16(14.8) = 236.8$$

After 7.4 seconds, the velocity of the rock is 236.8 ft/s.

NOTE Did you know? 236.8 ft/s is more than 161 mi/h!

(c)
$$v = \lim_{t \to t_0} \frac{f(t) - f(t_0)}{t - t_0} = \lim_{t \to t_0} \frac{16t^2 - 16t_0^2}{t - t_0} = \lim_{t \to t_0} \frac{16(t - t_0)(t + t_0)}{t - t_0}$$

$$= 16 \lim_{t \to t_0} (t + t_0) = 32t_0$$

At t_0 seconds, the velocity of the rock is $32t_0$ ft/s. ∎

NOW WORK **Problem 9.**

2 Find an Equation of the Tangent Line to the Graph of a Function

Consider the graph of the function $y = f(x)$ shown in Figure 3. The line connecting the two points $(c, f(c))$ and $(d, f(d))$ on the graph of f is a secant line and its slope is

$$m_{\text{sec}} = \frac{f(d) - f(c)}{d - c} \qquad d \ne c$$

Figure 4 shows there are many secant lines passing through the point $(c, f(c))$. If $(x, f(x))$ is any point on the graph of f with $x \ne c$, then as the number x moves closer to c, the point $(x, f(x))$ moves along the graph of f and approaches the point $(c, f(c))$. Suppose, as the points $(x, f(x))$ get closer to the point $(c, f(c))$, the associated secant lines approach a line. Then this line is the *tangent line* to the graph of f at $x = c$. Also, the slope m_{sec} of the secant lines approaches the slope m_{tan} of the tangent line.

Since the slope of the secant line connecting $(c, f(c))$ and $(x, f(x))$ is

$$\boxed{m_{\text{sec}} = \frac{f(x) - f(c)}{x - c} = \frac{\Delta y}{\Delta x} \qquad x \ne c}$$

the slope m_{tan} of the tangent line is

$$\boxed{m_{\text{tan}} = \lim_{x \to c} \frac{f(x) - f(c)}{x - c}}$$

provided the limit exists.

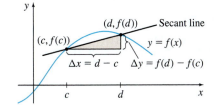

Figure 3 Slope of the secant line is $m_{\text{sec}} = \frac{f(d) - f(c)}{d - c} = \frac{\Delta y}{\Delta x}$.

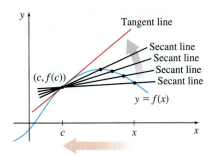

DF **Figure 4** The secant lines approach the tangent line.

DEFINITION Tangent Line

The **tangent line** to the graph of f at a point P is the line containing the point $P = (c, f(c))$ and having the slope

$$m_{\tan} = \lim_{x \to c} \frac{f(x) - f(c)}{x - c} \qquad (2)$$

provided the limit exists.*

The limit in equation (1) that defines instantaneous velocity and the limit in equation (2) that defines the slope of the tangent line occur so frequently that they are given a special notation $f'(c)$, read, "f prime of c," and called **prime notation**:

$$f'(c) = \lim_{x \to c} \frac{f(x) - f(c)}{x - c} \qquad (3)$$

THEOREM Equation of a Tangent Line

If $m_{\tan}$ exists, then an equation of the tangent line to the graph of f at the point $P = (c, f(c))$ is

$$y - f(c) = f'(c)(x - c)$$

EXAMPLE 3 Finding an Equation of the Tangent Line to the Graph of $f(x) = x^2$

(a) Find the slope of the tangent line to the graph of $f(x) = x^2$ at $c = 1$ and at $c = -2$.

(b) Use the results from (a) to find an equation of the tangent lines when $c = 1$ and $c = -2$.

(c) Graph f and the two tangent lines on the same set of axes.

Solution (a) At $c = 1$, the slope of the tangent line is

$$f'(1) = \lim_{x \to 1} \frac{f(x) - f(1)}{x - 1} \underset{\substack{\uparrow \\ f(x) = x^2}}{=} \lim_{x \to 1} \frac{x^2 - 1}{x - 1} = \lim_{x \to 1} \frac{(x - 1)(x + 1)}{x - 1} = \lim_{x \to 1} (x + 1) = 2$$

At $c = -2$, the slope of the tangent line is

$$f'(-2) = \lim_{x \to -2} \frac{f(x) - f(-2)}{x - (-2)} = \lim_{x \to -2} \frac{x^2 - (-2)^2}{x + 2} = \lim_{x \to -2} \frac{x^2 - 4}{x + 2}$$

$$= \lim_{x \to -2} (x - 2) = -4$$

NEED TO REVIEW? The point-slope form of a line is discussed in Appendix A.3, p. A-19.

(b) We use the results from (a) and the point-slope form of an equation of a line to obtain equations of the tangent lines. An equation of the tangent line containing the point $(1, f(1)) = (1, 1)$ is

$$y - f(1) = f'(1)(x - 1) \qquad \text{Point-slope form of an equation of the tangent line.}$$

$$y - 1 = 2(x - 1) \qquad f(1) = 1; \quad f'(1) = 2.$$

$$y = 2x - 1 \qquad \text{Simplify.}$$

---

* It is possible for the limit in (2) not to exist. The geometric significance of this is discussed in the next section.

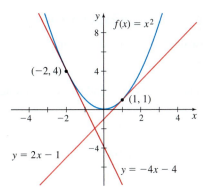

Figure 5

An equation of the tangent line containing the point $(-2, f(-2)) = (-2, 4)$ is

$$y - f(-2) = f'(-2)[x - (-2)]$$

$$y - 4 = -4 \cdot (x + 2) \qquad f(-2) = 4; \quad f'(-2) = -4$$

$$y = -4x - 4$$

(c) The graph of f and the two tangent lines are shown in Figure 5. ∎

NOW WORK Problem 17.

Velocity and the slope of a tangent line are each examples of the rate of change of a function.

3 Find the Rate of Change of a Function

Recall that the average rate of change of a function $y = f(x)$ over the interval from c to x is given by

$$\text{Average rate of change} = \frac{f(x) - f(c)}{x - c} \qquad x \neq c$$

DEFINITION Instantaneous Rate of Change

The **instantaneous rate of change** of f at c is the limit as x approaches c of the average rate of change. Symbolically, the instantaneous rate of change of f at c is

$$\lim_{x \to c} \frac{f(x) - f(c)}{x - c}$$

provided the limit exists.

The expression "instantaneous rate of change" is often shortened to *rate of change*.

Using prime notation, the **rate of change** of f at c is $f'(c) = \lim_{x \to c} \frac{f(x) - f(c)}{x - c}$.

EXAMPLE 4 Finding the Rate of Change of $f(x) = x^2 - 5x$

Find the rate of change of the function $f(x) = x^2 - 5x$ at:

(a) $c = 2$

(b) any real number c

Solution (a) For $c = 2$,

$$f(x) = x^2 - 5x \qquad \text{and} \qquad f(2) = 2^2 - 5 \cdot 2 = -6$$

The rate of change of f at $c = 2$ is

$$f'(2) = \lim_{x \to 2} \frac{f(x) - f(2)}{x - 2} = \lim_{x \to 2} \frac{(x^2 - 5x) - (-6)}{x - 2} = \lim_{x \to 2} \frac{x^2 - 5x + 6}{x - 2}$$

$$= \lim_{x \to 2} \frac{(x - 2)(x - 3)}{x - 2} = \lim_{x \to 2} (x - 3) = -1$$

(b) If c is any real number, then $f(c) = c^2 - 5c$, and the rate of change of f at c is

$$f'(c) = \lim_{x \to c} \frac{f(x) - f(c)}{x - c} = \lim_{x \to c} \frac{(x^2 - 5x) - (c^2 - 5c)}{x - c} = \lim_{x \to c} \frac{(x^2 - c^2) - 5(x - c)}{x - c}$$

$$= \lim_{x \to c} \frac{(x - c)(x + c) - 5(x - c)}{x - c} = \lim_{x \to c} \frac{(x - c)(x + c - 5)}{x - c} = \lim_{x \to c}(x + c - 5)$$

$$= 2c - 5 \qquad \blacksquare$$

NOW WORK Problem 25.

EXAMPLE 5   Finding the Rate of Change in a Biology Experiment

In a metabolic experiment, the mass M of glucose decreases according to the function

$$M(t) = 4.5 - 0.03t^2$$

where M is measured in grams (g) and t is the time in hours (h). Find the reaction rate $M'(t)$ at $t = 1$ h.

Solution The reaction rate at $t = 1$ is $M'(1)$.

$$M'(1) = \lim_{t \to 1} \frac{M(t) - M(1)}{t - 1} = \lim_{t \to 1} \frac{(4.5 - 0.03t^2) - (4.5 - 0.03)}{t - 1}$$

$$= \lim_{t \to 1} \frac{-0.03t^2 + 0.03}{t - 1} = \lim_{t \to 1} \frac{(-0.03)(t^2 - 1)}{t - 1} = \lim_{t \to 1} \frac{(-0.03)(t - 1)(t + 1)}{t - 1}$$

$$= (-0.03)(2) = -0.06$$

The reaction rate at $t = 1$ h is -0.06 g/h. That is, the mass M of glucose at $t = 1$ h is decreasing at the rate of 0.06 g/h. ■

NOW WORK Problem 43.

4 Find the Derivative of a Function at a Number

Velocity, slope of a tangent line, and rate of change of a function are all found using the same limit,

$$f'(c) = \lim_{x \to c} \frac{f(x) - f(c)}{x - c}$$

The common underlying idea is the mathematical concept of *derivative*.

DEFINITION Derivative of a Function at a Number

If $y = f(x)$ is a function and c is in the domain of f, then the **derivative** of f at c, denoted by $f'(c)$, is the number

$$f'(c) = \lim_{x \to c} \frac{f(x) - f(c)}{x - c}$$

provided this limit exists.

So far, we have given three interpretations of the derivative:

- *Physical interpretation*: When the distance s at time t of an object in rectilinear motion is given by the function $s = f(t)$, the derivative $f'(t_0)$ is the velocity of the object at time t_0.
- *Geometric interpretation*: If $y = f(x)$, the derivative $f'(c)$ is the slope of the tangent line to the graph of f at the point $(c, f(c))$.
- *Rate of change of a function interpretation*: If $y = f(x)$, the derivative $f'(c)$ is the rate of change of f at c.

EXAMPLE 6   Finding the Derivative of a Function at a Number

Find the derivative of $f(x) = 2x^3 - 5x$ at $x = 1$. That is, find $f'(1)$.

Solution Using the definition of a derivative, we have

$$f'(1) = \lim_{x \to 1} \frac{f(x) - f(1)}{x - 1} = \lim_{x \to 1} \frac{(2x^3 - 5x) - (-3)}{x - 1} = \lim_{x \to 1} \frac{2x^3 - 5x + 3}{x - 1} \qquad f(1) = 2 - 5 = -3$$

$$= \lim_{x \to 1} \frac{(x - 1)(2x^2 + 2x - 3)}{x - 1} = \lim_{x \to 1} (2x^2 + 2x - 3) = 2 + 2 - 3 = 1 \qquad ■$$

NOW WORK Problem 27.

EXAMPLE 7 **Finding an Equation of a Tangent Line**

(a) Find the derivative of $f(x) = \sqrt{2x}$ at $x = 8$.

(b) Use the derivative $f'(8)$ to find the equation of the tangent line to the graph of f at the point $(8, 4)$.

Solution **(a)** The derivative of f at 8 is

$$f'(8) = \lim_{x \to 8} \frac{f(x) - f(8)}{x - 8} = \lim_{\substack{x \to 8 \\ \uparrow}} \frac{\sqrt{2x} - 4}{x - 8} = \lim_{\substack{x \to 8 \\ \uparrow}} \frac{(\sqrt{2x} - 4)(\sqrt{2x} + 4)}{(x - 8)(\sqrt{2x} + 4)}$$

$$\underset{f(8) = \sqrt{2 \cdot 8} = 4}{} \qquad \underset{\substack{\text{Rationalize} \\ \text{the numerator.}}}{}$$

$$= \lim_{x \to 8} \frac{2x - 16}{(x - 8)(\sqrt{2x} + 4)} = \lim_{x \to 8} \frac{2(x - 8)}{(x - 8)(\sqrt{2x} + 4)} = \lim_{x \to 8} \frac{2}{\sqrt{2x} + 4} = \frac{1}{4}$$

(b) The slope of the tangent line to the graph of f at the point $(8, 4)$ is $f'(8) = \dfrac{1}{4}$. Using the point-slope form of a line, we get

$$y - f(8) = f'(8)(x - 8) \qquad\qquad y - y_1 = m(x - x_1)$$

$$y - 4 = \frac{1}{4}(x - 8) \qquad\qquad f(8) = 4; \quad f'(8) = \frac{1}{4}$$

$$y = \frac{1}{4}x - \frac{1}{4} \cdot 8 + 4$$

$$y = \frac{1}{4}x + 2 \qquad\qquad\qquad\qquad\qquad \blacksquare$$

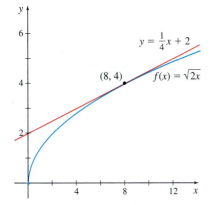

Figure 6

The graphs of f and the tangent line to the graph of f at $(8, 4)$ are shown in Figure 6.

2.1 Assess Your Understanding

Concepts and Vocabulary

1. *True or False* The derivative is used to find instantaneous velocity.

2. *True or False* The derivative can be used to find the rate of change of a function.

3. The notation $f'(c)$ is read f _____ of c; $f'(c)$ represents the _____ of the tangent line to the graph of f at the point _____ .

4. *True or False* If it exists, $\displaystyle\lim_{x \to 3} \frac{f(x) - f(3)}{x - 3}$ is the derivative of the function f at 3.

5. If $f(x) = x^2 - 3$, then $f'(3) = $ _____.

6. Velocity, the slope of a tangent line, and the rate of change of a function are three different interpretations of the mathematical concept called the _____.

Skill Building

7. Approximating Velocity An object in rectilinear motion moves according to the equation $s = 10t^2$ (s in centimeters). Approximate the velocity of the object at time $t_0 = 3$ seconds by letting Δt first equal 0.1 second, then 0.01 second, and finally 0.001 second. What limit does the velocity appear to be approaching? Organize the results in a table.

8. Approximating Velocity An object in rectilinear motion moves according to the equation $s = 5 - t^2$ (s in centimeters and t in seconds). Approximate the velocity of the object at time $t_0 = 1$ by letting Δt first equal 0.1, then 0.01, and finally 0.001. What limit does the velocity appear to be approaching? Organize the results in a table.

9. Rectilinear Motion As an object in rectilinear motion moves, the distance s (in meters) that it moves in t_0 seconds is given by $s = f(t) = 3t^2 + 4t$. Find the velocity v at $t_0 = 0$. At $t_0 = 2$. At any time t_0.

10. Rectilinear Motion As an object in rectilinear motion moves, the distance s (in meters) that it moves in t seconds is given by $s = f(t) = 2t^3 + 4$. Find the velocity v at $t_0 = 0$. At $t_0 = 3$. At any time t_0.

11. Rectilinear Motion As an object in rectilinear motion moves, its distance s from the origin at time t is given by the equation $s = s(t) = 3t^2 - \dfrac{1}{t}$, where s is in centimeters and t is in seconds. Find the velocity v of the object at $t_0 = 1$ and $t_0 = 4$.

12. Rectilinear Motion As an object in rectilinear motion moves, its distance s from the origin at time t is given by the equation $s = s(t) = \sqrt{4t}$, where s is in centimeters and t is in seconds. Find the velocity v of the object at $t_0 = 1$ and $t_0 = 4$.

1. = NOW WORK problem   ⟳ = Graphing technology recommended   CAS = Computer Algebra System recommended

In Problems 13–22, find an equation of the tangent line to the graph of each function at the indicated point. Graph each function and the tangent line.

13. $f(x) = 3x^2$ at $(-2, 12)$ **14.** $f(x) = x^2 + 2$ at $(-1, 3)$

15. $f(x) = x^3$ at $(-2, -8)$ **16.** $f(x) = x^3 + 1$ at $(1, 2)$

17. $f(x) = \dfrac{1}{x}$ at $(1, 1)$ **18.** $f(x) = \sqrt{x}$ at $(4, 2)$

19. $f(x) = \dfrac{1}{x+5}$ at $\left(1, \dfrac{1}{6}\right)$ **20.** $f(x) = \dfrac{2}{x+4}$ at $\left(1, \dfrac{2}{5}\right)$

21. $f(x) = \dfrac{1}{\sqrt{x}}$, at $(1, 1)$ **22.** $f(x) = \dfrac{1}{x^2}$ at $(1, 1)$

In Problems 23–26, find the rate of change of f at the indicated numbers.

23. $f(x) = 5x - 2$ at **(a)** $c = 0$, **(b)** $c = 2$, **(c)** c any real number

24. $f(x) = x^2 - 1$ at **(a)** $c = -1$, **(b)** $c = 1$, **(c)** c any real number

25. $f(x) = \dfrac{x^2}{x+3}$ at **(a)** $c = 0$, **(b)** $c = 1$,
   (c) c any real number, $c \neq -3$

26. $f(x) = \dfrac{x}{x^2 - 1}$ at **(a)** $c = 0$, **(b)** $c = 2$,
   (c) c any real number, $c \neq \pm 1$

In Problems 27–36, find the derivative of each function at the given number.

27. $f(x) = 2x + 3$ at 1 **28.** $f(x) = 3x - 5$ at 2

29. $f(x) = x^2 - 2$ at 0 **30.** $f(x) = 2x^2 + 4$ at 1

31. $f(x) = 3x^2 + x + 5$ at -1 **32.** $f(x) = 2x^2 - x - 7$ at -1

33. $f(x) = \sqrt{x}$ at 4 **34.** $f(x) = \dfrac{1}{x^2}$ at 2

35. $f(x) = \dfrac{2 - 5x}{1 + x}$ at 0 **36.** $f(x) = \dfrac{2 + 3x}{2 + x}$ at 1

37. The Princeton Dinky is the shortest rail line in the country. It runs for 2.7 miles, connecting Princeton University to the Princeton Junction railroad station. The Dinky starts from the university and moves north toward Princeton Junction. Its distance from Princeton is shown in the graph where the time t is in minutes and the distance s of the Dinky from Princeton University is in miles.

(a) When is the Dinky headed toward Princeton University?

(b) When is it headed toward Princeton Junction?

(c) When is the Dinky stopped?

(d) Find its average velocity on a trip from Princeton to Princeton Junction.

(e) Find its average velocity for the round trip shown in the graph, that is, from $t = 0$ to $t = 13$.

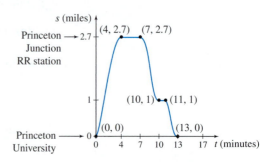

38. Barbara walks to the deli, which is six blocks east of her house. After walking two blocks, she realizes she left her phone on her desk, so she runs home. After getting the phone, closing and locking the door, Barbara starts on her way again. At the deli, she waits in line to buy a bottle of **vitaminwater**™, and then she jogs home. The graph below represents Barbara's journey. The time t is in minutes and s is Barbara's distance, in blocks, from home.

(a) At what times is she headed toward the deli?

(b) At what times is she headed home?

(c) When is the graph horizontal? What does this indicate?

(d) Find Barbara's average velocity from home until she starts back to get her phone.

(e) Find Barbara's average velocity from home to the deli after getting her phone.

(f) Find her average velocity from the deli to home.

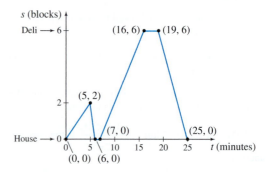

Applications and Extensions

39. Slope of a Tangent Line   The equation of the tangent line to the graph of a function f at $(2, 6)$ is $y = -3x + 12$. What is $f'(2)$?

40. Slope of a Tangent Line   The equation of the tangent line of a function f at $(3, 2)$ is $y = \dfrac{1}{3}x + 1$. What is $f'(3)$?

41. Tangent Line   Does the tangent line to the graph of $y = x^2$ at $(1, 1)$ pass through the point $(2, 5)$?

42. Tangent Line Does the tangent line to the graph of $y = x^3$ at (1, 1) pass through the point (2, 5)?

43. Respiration Rate A human being's respiration rate R (in breaths per minute) is given by $R = R(p) = 10.35 + 0.59p$, where p is the partial pressure of carbon dioxide in the lungs. Find the rate of change in respiration when $p = 50$.

44. Instantaneous Rate of Change The volume V of the right circular cylinder of height 5 m and radius r m shown in the figure is $V = V(r) = 5\pi r^2$. Find the instantaneous rate of change of the volume with respect to the radius when $r = 3$ m.

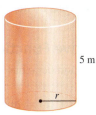

5 m

45. Market Share During a month-long advertising campaign, the total sales S of a magazine is modeled by the function $S(x) = 5x^2 + 100x + 10,000$, where x represents the number of days since the campaign began, $0 \le x \le 30$.

(a) What is the average rate of change of sales from $x = 10$ to $x = 20$ days?

(b) What is the instantaneous rate of change of sales when $x = 10$ days?

46. Demand Equation The demand equation for an item is $p = p(x) = 90 - 0.02x$, where p is the price in dollars and x is the number of units (in thousands) made.

(a) Assuming all units made can be sold, find the revenue function $R(x) = xp(x)$.

(b) **Marginal Revenue** Marginal revenue is defined as the additional revenue earned by selling an additional unit. If we use $R'(x)$ to measure the marginal revenue, find the marginal revenue when 1 million units are sold.

47. Gravity If a ball is dropped from the top of the Empire State Building, 1002 ft above the ground, the distance s (in feet) it falls after t seconds is $s(t) = 16t^2$.

(a) What is the average velocity of the ball for the first 2 s?

(b) How long does it take for the ball to hit the ground?

(c) What is the average velocity of the ball during the time it is falling?

(d) What is the velocity of the ball when it hits the ground?

48. Velocity A ball is thrown upward. Its height h in feet is given by $h(t) = 100t - 16t^2$, where t is the time elapsed in seconds.

(a) What is the velocity v of the ball at $t = 0$ s, $t = 1$ s, and $t = 4$ s?

(b) At what time t does the ball strike the ground?

(c) At what time t does the ball reach its highest point? (*Hint:* At the time the ball reaches its maximum height, it is stationary. So, its velocity $v = 0$.)

49. Gravity A rock is dropped from a height of 88.2 m and falls toward Earth in a straight line. In t seconds the rock falls $4.9t^2$ meters.

(a) What is the average velocity of the rock for the first 2 s?

(b) How long does it take for the rock to hit the ground?

(c) What is the average velocity of the rock during its fall?

(d) What is the velocity v of the rock when it hits the ground?

50. Velocity At a certain instant, the speedometer of an automobile reads V mi/h. During the next $\frac{1}{4}$ s the automobile travels 20 ft. Estimate V from this information.

51. Volume of a Cube A metal cube with each edge of length x centimeters is expanding uniformly as a consequence of being heated.

(a) Find the average rate of change of the volume of the cube with respect to an edge as x increases from 2.00 to 2.01 cm.

(b) Find the instantaneous rate of change of the volume of the cube with respect to an edge at the instant when $x = 2$ cm.

52. Rate of Change Show that the rate of change of a linear function $f(x) = mx + b$ is the slope m of the line $y = mx + b$.

53. Rate of Change Show that the rate of change of a quadratic function $f(x) = ax^2 + bx + c$ is a linear function of x.

54. Business The graph represents the demand d (in gallons) for olive oil as a function of the cost c in dollars per gallon of the oil.

(a) Interpret the derivative $d'(c)$.

(b) Which is larger, $d'(5)$ or $d'(30)$? Give an interpretation to $d'(5)$ and $d'(30)$.

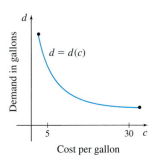

55. Agriculture The graph represents the diameter d (in centimeters) of a maturing peach as a function of the time t (in days) it is on the tree.

(a) Interpret the derivative $d'(t)$ as a rate of change.

(b) Which is larger, $d'(1)$ or $d'(20)$?

(c) Interpret both $d'(1)$ and $d'(20)$.

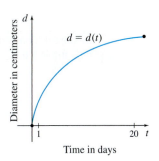

2.2 The Derivative as a Function

OBJECTIVES *When you finish this section, you should be able to:*

1 Define the derivative function (p. 154)
2 Graph the derivative function (p. 155)
3 Identify where a function has no derivative (p. 157)

1 Define the Derivative Function

The derivative of f at a real number c has been defined as the real number

$$f'(c) = \lim_{x \to c} \frac{f(x) - f(c)}{x - c} \qquad (1)$$

provided the limit exists. Next we show how to find the derivative of f at any real number. We begin by rewriting the expression $\dfrac{f(x) - f(c)}{x - c}$. Let $x = c + h, h \neq 0$. Then

$$\frac{f(x) - f(c)}{x - c} = \frac{f(c+h) - f(c)}{(c+h) - c} = \frac{f(c+h) - f(c)}{h}$$

Since $x = c + h$, as x approaches c, then h approaches 0. Equation (1) with these changes becomes

$$f'(c) = \lim_{x \to c} \frac{f(x) - f(c)}{x - c} = \lim_{h \to 0} \frac{f(c+h) - f(c)}{h}$$

So, we have the following alternate form for the derivative of f at a real number c.

$$f'(c) = \lim_{h \to 0} \frac{f(c+h) - f(c)}{h} \qquad (2)$$

See Figure 7.

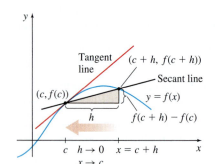

Figure 7 The slope of the tangent line at c is $f'(c) = \lim\limits_{h \to 0} \dfrac{f(c+h) - f(c)}{h}$.

NOTE Compare the solution and answer found in Example 1 to Example 4(b) on p. 149.

EXAMPLE 1 **Finding the Derivative of a Function at a Number c**

Find the derivative of the function $f(x) = x^2 - 5x$ at any real number c using form (2).

Solution Using form (2), we have

$$f'(c) = \lim_{h \to 0} \frac{f(c+h) - f(c)}{h} = \lim_{h \to 0} \frac{[(c+h)^2 - 5(c+h)] - (c^2 - 5c)}{h}$$

$$= \lim_{h \to 0} \frac{[(c^2 + 2ch + h^2) - 5c - 5h] - c^2 + 5c}{h} = \lim_{h \to 0} \frac{2ch + h^2 - 5h}{h}$$

$$= \lim_{h \to 0} \frac{h(2c + h - 5)}{h} = \lim_{h \to 0} (2c + h - 5) = 2c - 5 \qquad \blacksquare$$

NOW WORK Problem 9.

Based on Example 1, if $f(x) = x^2 - 5x$, then $f'(c) = 2c - 5$ for any choice of c. That is, the derivative f' is a function and, using x as the independent variable, we can write $f'(x) = 2x - 5$.

DEFINITION The Derivative Function f'

The **derivative function** f' of a function f is

$$f'(x) = \lim_{h \to 0} \frac{f(x+h) - f(x)}{h} \qquad (3)$$

IN WORDS In form (3) the derivative is the limit of a difference quotient.

provided the limit exists. If f has a derivative, then f is said to be **differentiable**.

The domain of the function f' is the set of real numbers in the domain of f for which the limit (3) exists. So the domain of f' is a subset of the domain of f.

We can use either form (1) or form (3) to find derivatives. However, if we want the derivative of f at a number c, we usually use form (1) to find $f'(c)$. If we want to find the derivative function of f, we usually use form (3) to find $f'(x)$. In this section, we use the definitions of the derivative, forms (1) and (3), to investigate derivatives. In the next section, we begin to develop formulas for finding the derivatives.

EXAMPLE 2   Finding the Derivative Function

NOTE The instruction "differentiate f" means to "find the derivative of f."

Differentiate $f(x) = \sqrt{x}$ and determine the domain of f'.

Solution The domain of f is $\{x \,|\, x \geq 0\}$. To find the derivative of f, we use form (3). Then

$$f'(x) = \lim_{h \to 0} \frac{f(x+h) - f(x)}{h} = \lim_{h \to 0} \frac{\sqrt{x+h} - \sqrt{x}}{h}$$

We rationalize the numerator to find the limit.

$$f'(x) = \lim_{h \to 0} \left[\frac{\sqrt{x+h} - \sqrt{x}}{h} \cdot \frac{\sqrt{x+h} + \sqrt{x}}{\sqrt{x+h} + \sqrt{x}} \right] = \lim_{h \to 0} \frac{(x+h) - x}{h(\sqrt{x+h} + \sqrt{x})}$$

$$= \lim_{h \to 0} \frac{h}{h(\sqrt{x+h} + \sqrt{x})} = \lim_{h \to 0} \frac{1}{\sqrt{x+h} + \sqrt{x}} = \frac{1}{2\sqrt{x}}$$

The limit does not exist when $x = 0$. But for all other x in the domain of f, the limit does exist. So, the domain of the derivative function $f'(x) = \dfrac{1}{2\sqrt{x}}$ is $\{x \,|\, x > 0\}$. ∎

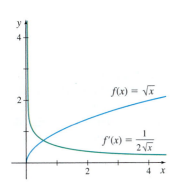

Figure 8

In Example 2, notice that the domain of the derivative function f' is a proper subset of the domain of the function f. The graphs of both f and f' are shown in Figure 8.

NOW WORK Problem 15.

EXAMPLE 3   Interpreting the Derivative as a Rate of Change

Show that the rate of change of the area of a circle with respect to its radius is equal to its circumference.

Solution The area $A = A(r)$ of a circle of radius r is $A(r) = \pi r^2$. The derivative function gives the rate of change of the area with respect to the radius.

$$A'(r) = \lim_{h \to 0} \frac{A(r+h) - A(r)}{h} = \lim_{h \to 0} \frac{\pi(r+h)^2 - \pi r^2}{h}$$

$$= \lim_{h \to 0} \frac{\pi(r^2 + 2rh + h^2) - \pi r^2}{h} = \lim_{h \to 0} \frac{\pi h(2r + h)}{h}$$

$$= \lim_{h \to 0} \pi(2r + h) = 2\pi r$$

The rate of change of the area of a circle with respect to its radius is the circumference of the circle, $2\pi r$. ∎

NOW WORK Problem 69.

2 Graph the Derivative Function

There is a relationship between the graph of a function and the graph of its derivative.

EXAMPLE 4   Graphing a Function and Its Derivative

Find f' if $f(x) = x^3 - 1$. Then graph $y = f(x)$ and $y = f'(x)$ on the same set of coordinate axes.

Solution $f(x) = x^3 - 1$ so

$$f(x + h) = (x + h)^3 - 1 = x^3 + 3hx^2 + 3h^2x + h^3 - 1$$

Using form (3), we find

$$f'(x) = \lim_{h \to 0} \frac{f(x + h) - f(x)}{h} = \lim_{h \to 0} \frac{(x^3 + 3hx^2 + 3h^2x + h^3 - 1) - (x^3 - 1)}{h}$$

$$= \lim_{h \to 0} \frac{3hx^2 + 3h^2x + h^3}{h} = \lim_{h \to 0} \frac{h(3x^2 + 3hx + h^2)}{h}$$

$$= \lim_{h \to 0} (3x^2 + 3hx + h^2) = 3x^2$$

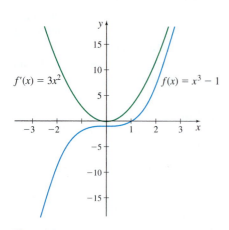

Figure 9

The graphs of f and f' are shown in Figure 9. ■

Figure 10 illustrates several tangent lines to the graph of $f(x) = x^3 - 1$. Observe that the tangent line to the graph of f at $(0, -1)$ is horizontal, so its slope is 0. Then $f'(0) = 0$, so the graph of f' contains the point $(0, 0)$. Also notice that every tangent line to the graph of f has a nonnegative slope, so $f'(x) \geq 0$. That is, the range of the function f' is $\{y \mid y \geq 0\}$. Finally, notice that the slope of each tangent line is the y-coordinate of the corresponding point on the graph of the derivative f'.

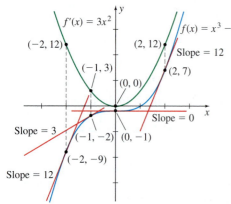

Figure 10

NOW WORK Problem 19.

With these ideas in mind, we can obtain a rough sketch of the derivative function f', even if we only know the graph of the function f.

EXAMPLE 5   Graphing the Derivative Function

Use the graph of the function $y = f(x)$, shown in Figure 11, to sketch the graph of the derivative function $y = f'(x)$.

Solution We begin by drawing tangent lines to the graph of f at the points shown in Figure 11. See the graph at the top of Figure 12. At the points $(-2, 3)$ and $\left(\frac{3}{2}, -2\right)$ the tangent lines are horizontal, so their slopes are 0. This means $f'(-2) = 0$ and $f'\left(\frac{3}{2}\right) = 0$, so the points $(-2, 0)$ and $\left(\frac{3}{2}, 0\right)$ are on the graph of the derivative function. Now we estimate the slope of the tangent lines at the other selected points. For example, at the point $(-4, -3)$. the slope of the tangent line is positive and the line is rather steep. We estimate the slope to be close to 6, and we plot the point $(-4, 6)$ on the bottom graph of Figure 12. Continue the process and then connect the points with a smooth curve. ■

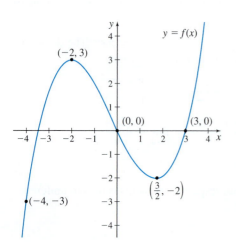

Figure 11

Notice in Figure 12 that at the points on the graph of f where the tangent lines are horizontal, the graph of the derivative f' intersects the x-axis. Also notice that wherever the graph of f is increasing, the slopes of the tangent lines are positive, that is, f' is

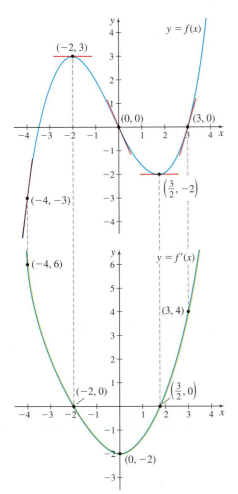

DF Figure 12

positive, so the graph of f' is above the x-axis. Similarly, wherever the graph of f is decreasing, the slopes of the tangent lines are negative, so the graph of f' is below the x-axis.

NOW WORK Problem 29.

3 Identify Where a Function Has No Derivative

A function f has no derivative at a number c if $\lim\limits_{x \to c} \dfrac{f(x) - f(c)}{x - c}$ does not exist. Two (of several) ways this can happen are:

- $\lim\limits_{x \to c^-} \dfrac{f(x) - f(c)}{x - c}$ exists and $\lim\limits_{x \to c^+} \dfrac{f(x) - f(c)}{x - c}$ exists, but they are not equal.
  One way this happens is if the graph of f has a *corner* at $(c, f(c))$. For example, the absolute value function $f(x) = |x|$ has a corner at $(0, 0)$ and so has no derivative at 0. See Figure 13. You are asked to prove this in Problem 71.

- The limit is infinite. One way this happens is if the graph of f has a *vertical tangent line* at $(c, f(c))$. For example, the cube root function $f(x) = \sqrt[3]{x}$ has a vertical tangent line at $(0, 0)$, and so it has no derivative at 0. See Figure 14. You are asked to prove this in Problem 72.

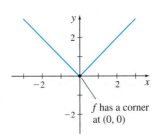

Figure 13 $f(x) = |x|$ **Figure 14** $f(x) = \sqrt[3]{x}$

EXAMPLE 6   **Identifying Where a Function Has No Derivative**

Given the piecewise defined function $f(x) = \begin{cases} -2x^2 + 4 & \text{if } x < 1 \\ x^2 + 1 & \text{if } x \geq 1 \end{cases}$, determine if $f'(1)$ exists.

Solution We investigate the limit

$$\lim_{x \to 1} \frac{f(x) - f(1)}{x - 1} = \lim_{x \to 1} \frac{f(x) - 2}{x - 1} \qquad \color{blue}{f(1) = 1^2 + 1 = 2}$$

If $x < 1$, then $f(x) = -2x^2 + 4$; if $x \geq 1$, then $f(x) = x^2 + 1$. So, it is necessary to investigate the one-sided limits at 1.

$$\lim_{x \to 1^-} \frac{f(x) - f(1)}{x - 1} = \lim_{x \to 1^-} \frac{(-2x^2 + 4) - 2}{x - 1} = \lim_{x \to 1^-} \frac{-2(x^2 - 1)}{x - 1}$$

$$= -2 \lim_{x \to 1^-} \frac{(x - 1)(x + 1)}{x - 1} = -2 \lim_{x \to 1^-} (x + 1) = -4$$

$$\lim_{x \to 1^+} \frac{f(x) - f(1)}{x - 1} = \lim_{x \to 1^+} \frac{(x^2 + 1) - 2}{x - 1} = \lim_{x \to 1^+} \frac{(x - 1)(x + 1)}{x - 1} = \lim_{x \to 1^+} (x + 1) = 2$$

Since the one-sided limits are not equal, $\lim\limits_{x \to 1} \dfrac{f(x) - f(1)}{x - 1}$ does not exist, and so $f'(1)$ does not exist. ∎

Figure 15 illustrates the graph of the function f from Example 6. At 1, where the

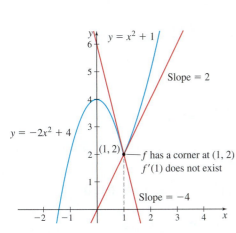

Figure 15 f has a corner at $(1, 2)$.

derivative does not exist (and so there is no tangent line), the graph of f has a corner. We usually say that the graph of f is not *smooth* at a corner.

NOW WORK   Problem 39.

Example 7 illustrates the behavior of the graph of a function f when the derivative at a number c does not exist because $\lim\limits_{x \to c} \dfrac{f(x) - f(c)}{x - c}$ is infinite.

EXAMPLE 7   Showing That a Function Has No Derivative

Show that $f(x) = \sqrt[5]{x + 2}$ has no derivative at $x = -2$.

Solution First we note that $f(-2) = \sqrt[5]{-2 + 2} = 0$. Then we investigate the limit.

$$\lim_{x \to -2} \frac{f(x) - f(-2)}{x - (-2)} = \lim_{x \to -2} \frac{\sqrt[5]{x + 2} - 0}{x + 2} = \lim_{x \to -2} \frac{\sqrt[5]{x + 2}}{x + 2} = \lim_{x \to -2} \frac{1}{\sqrt[5]{(x + 2)^4}} = \infty$$

Since the limit is infinite, the derivative of f does not exist at -2. ∎

Figure 16 shows that the tangent line to the graph of f at the point $(-2, 0)$ is vertical.

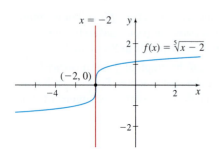

Figure 16 $f'(-2)$ does not exist; f has a vertical tangent line at the point $(-2, 0)$.

NEED TO REVIEW? Continuity is discussed in Section 1.3, pp. 93–99.

NOW WORK   Problem 35.

In Chapter 1, we investigated the continuity of a function. Here, we have been investigating the differentiability of a function. An important connection exists between continuity and differentiability.

THEOREM

If a function f has a derivative at a number c, then f is continuous at c.

Proof To show that f is continuous at c, we need to verify that $\lim\limits_{x \to c} f(x) = f(c)$. We begin by observing that if $x \neq c$, then

$$f(x) - f(c) = \left[\frac{f(x) - f(c)}{x - c} \right] (x - c)$$

We take the limit of both sides as $x \to c$, and use the fact that the limit of a product equals the product of the limits, (we show later that each limit exists).

$$\lim_{x \to c} [f(x) - f(c)] = \lim_{x \to c} \left\{ \left[\frac{f(x) - f(c)}{x - c} \right] (x - c) \right\}$$

$$= \left[\lim_{x \to c} \frac{f(x) - f(c)}{x - c} \right] \left[\lim_{x \to c} (x - c) \right]$$

Since f has a derivative at c, we know that

$$\lim_{x \to c} \frac{f(x) - f(c)}{x - c} = f'(c)$$

is a number. Also for any real number c, $\lim\limits_{x \to c} (x - c) = 0$. So

$$\lim_{x \to c} [f(x) - f(c)] = [f'(c)] \left[\lim_{x \to c} (x - c) \right] = f'(c) \cdot 0 = 0$$

That is, $\lim\limits_{x \to c} f(x) = f(c)$, so f is continuous at c. ∎

An equivalent statement of this theorem gives a condition under which a function has no derivative.

COROLLARY

If a function f is discontinuous at a number c, then f has no derivative at c.

Let's look at some of the possibilities. In Figure 17(a), the function f is continuous at the number 1 and it has a derivative at 1. The function g, graphed in Figure 17(b), is continuous at the number 0, but it has no derivative at 0. So continuity at a number c provides no prediction about differentiability. On the other hand, the function h graphed in Figure 17(c) illustrates the Corollary: If h is discontinuous at a number, it has no derivative at that number.

IN WORDS Differentiability implies continuity, but continuity does not imply differentiability.

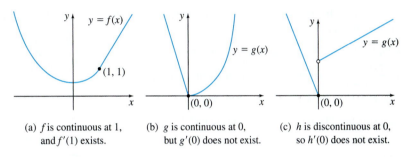

(a) f is continuous at 1, and $f'(1)$ exists.

(b) g is continuous at 0, but $g'(0)$ does not exist.

(c) h is discontinuous at 0, so $h'(0)$ does not exist.

Figure 17

The corollary is useful if we are seeking the derivative of a function f that we suspect is discontinuous at a number c. If we can show that f is discontinuous at c, then the corollary affirms that the function f has no derivative at c. For example, since the floor function $f(x) = \lfloor x \rfloor$ is discontinuous at every integer c, it has no derivative at an integer.

The **Heaviside function** $u_c(t) = \begin{cases} 0 & \text{if } t < c \\ 1 & \text{if } t \geq c \end{cases}$ is a step function that is used in electrical engineering to model a switch. The switch is off if $t < c$ and is on if $t \geq c$. See Figure 18.

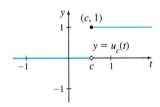

Figure 18 Heaviside function.

EXAMPLE 8   Determining If a Heaviside Function Has a Derivative at 0

Determine if the Heaviside function $u_0(t) = \begin{cases} 0 & \text{if } t < 0 \\ 1 & \text{if } t \geq 0 \end{cases}$ has a derivative at 0.

Solution Since $u_0(t)$ is discontinuous at 0, it has no derivative at 0. ∎

NOW WORK Problem 47.

Differentiability on a Closed Interval

Special attention needs to be paid to functions defined on a closed interval $[a, b]$. If a function f is defined on a closed interval $[a, b]$, we can investigate the derivative of f at every number c in the open interval (a, b). The endpoints must be handled separately. The right-hand derivative of f at a, for example, is defined as the right-hand limit

$$f'(a) = \lim_{x \to a^+} \frac{f(x) - f(a)}{x - a}$$

provided this limit exists. The left-hand derivative at b is handled similarly as the left-hand limit

$$f'(b) = \lim_{x \to b^-} \frac{f(x) - f(b)}{x - b}$$

provided this limit exists.

Using these definitions, we conclude that if f has a derivative at every number c in the closed interval $[a, b]$, then f is continuous on the closed interval $[a, b]$.

2.2 Assess Your Understanding

Concepts and Vocabulary

1. *True or False* The domain of a function f and the domain of its derivative function f' are always equal.

2. *True or False* If a function is continuous at a number c, then it is differentiable at c.

3. *Multiple Choice* If f is continuous at a number c and if $\lim\limits_{x \to c} \dfrac{f(x) - f(c)}{x - c}$ is infinite, then the graph of f has [(a) a horizontal, (b) a vertical, (c) no] tangent line at c.

4. The instruction "Differentiate f" means to find the _____ of f.

Skill Building

In Problems 5–10, find the rate of change of each function f at any real number c.

5. $f(x) = 10$
6. $f(x) = -4$
7. $f(x) = 2x + 3$
8. $f(x) = 3x - 5$
9. $f(x) = 2 - x^2$
10. $f(x) = 2x^2 + 4$

In Problems 11–16, differentiate each function f and determine the domain of f'. Use form (3) are page 154.

11. $f(x) = 5$
12. $f(x) = -2$
13. $f(x) = 3x^2 + x + 5$
14. $f(x) = 2x^2 - x - 7$
15. $f(x) = 5\sqrt{x - 1}$
16. $f(x) = 4\sqrt{x + 3}$

In Problems 17–22, differentiate each function f. Graph $y = f(x)$ and $y = f'(x)$ on the same set of coordinate axes.

17. $f(x) = \dfrac{1}{3}x + 1$
18. $f(x) = -4x - 5$
19. $f(x) = 2x^2 - 5x$
20. $f(x) = -3x^2 + 2$
21. $f(x) = x^3 - 8x$
22. $f(x) = -x^3 - 8$

In Problems 23–26, for each figure determine if the graphs represent a function f and its derivative f'. If they do, indicate which is the graph of f and which is the graph of f'.

23.

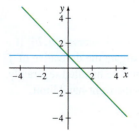

24.

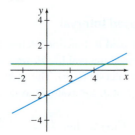

25.

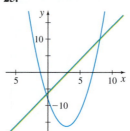

26.

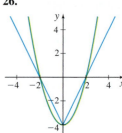

In Problems 27–30, use the graph of f to obtain the graph of f'.

27.

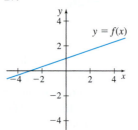

28.

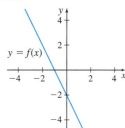

29.

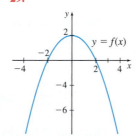

30.

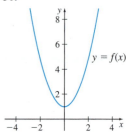

In Problems 31–34, the graph of a function f is given. Match each graph to the graph of its derivative f' in A–D.

31.

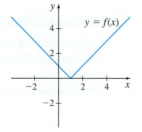

32.

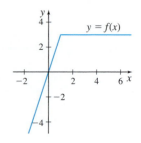

---

1. = NOW WORK problem   = Graphing technology recommended   **CAS** = Computer Algebra System recommended

33.

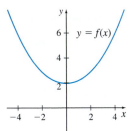

34.

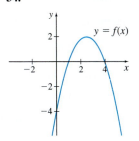

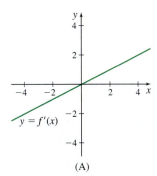

(A)

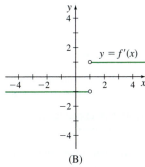

(B)

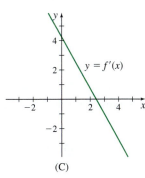

(C)

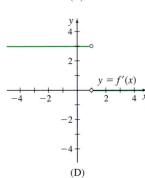

(D)

In Problems 35–44, determine whether each function f has a derivative at c. If it does, what is $f'(c)$? If it does not, give the reason why.

35. $f(x) = x^{2/3}$ at $c = -8$ **36.** $f(x) = 2x^{1/3}$ at $c = 0$

37. $f(x) = |x^2 - 4|$ at $c = 2$

38. $f(x) = |x^2 - 4|$ at $c = -2$

39. $f(x) = \begin{cases} 2x + 3 & \text{if } x < 1 \\ x^2 + 4 & \text{if } x \geq 1 \end{cases}$ at $c = 1$

40. $f(x) = \begin{cases} 3 - 4x & \text{if } x < -1 \\ 2x + 9 & \text{if } x \geq -1 \end{cases}$ at $c = -1$

41. $f(x) = \begin{cases} -4 + 2x & \text{if } x \leq \frac{1}{2} \\ 4x^2 - 4 & \text{if } x > \frac{1}{2} \end{cases}$ at $c = \frac{1}{2}$

42. $f(x) = \begin{cases} 2x^2 + 1 & \text{if } x < -1 \\ -1 - 4x & \text{if } x \geq -1 \end{cases}$ at $c = -1$

43. $f(x) = \begin{cases} 2x^2 + 1 & \text{if } x < -1 \\ 2 + 2x & \text{if } x \geq -1 \end{cases}$ at $c = -1$

44. $f(x) = \begin{cases} 5 - 2x & \text{if } x < 2 \\ x^2 & \text{if } x \geq 2 \end{cases}$ at $c = 2$

In Problems 45 and 46, use the given points $(c, f(c))$ on the graph of the function f.

(a) *For which numbers c does $\lim\limits_{x \to c} f(x)$ exist but f is not continuous at c?*

(b) *For which numbers c is f continuous at c but not differentiable at c?*

45.

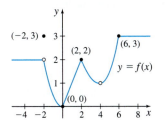

46.

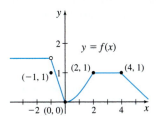

Heaviside Functions *In Problems 47 and 48:*

(a) *Determine if the given Heaviside function is differentiable at c.*

(b) *Give a physical interpretation of each function.*

47. $u_1(t) = \begin{cases} 0 & \text{if } t < 1 \\ 1 & \text{if } t \geq 1 \end{cases}$ at $c = 1$

48. $u_3(t) = \begin{cases} 0 & \text{if } t < 3 \\ 1 & \text{if } t \geq 3 \end{cases}$ at $c = 3$

In Problems 49–52, find the derivative of each function.

49. $f(x) = mx + b$ **50.** $f(x) = ax^2 + bx + c$

51. $f(x) = \dfrac{1}{x^2}$ **52.** $f(x) = \dfrac{1}{\sqrt{x}}$

Applications and Extensions

In Problems 53–60, each limit represents the derivative of a function f at some number c. Determine f and c in each case.

53. $\lim\limits_{h \to 0} \dfrac{(2+h)^2 - 4}{h}$ **54.** $\lim\limits_{h \to 0} \dfrac{(2+h)^3 - 8}{h}$

55. $\lim\limits_{x \to 1} \dfrac{x^2 - 1}{x - 1}$ **56.** $\lim\limits_{x \to 1} \dfrac{x^{14} - 1}{x - 1}$

57. $\lim\limits_{x \to \pi/6} \dfrac{\sin x - \frac{1}{2}}{x - \frac{\pi}{6}}$ **58.** $\lim\limits_{x \to \pi/4} \dfrac{\cos x - \frac{\sqrt{2}}{2}}{x - \frac{\pi}{4}}$

59. $\lim\limits_{x \to 0} \dfrac{2(x+2)^2 - (x+2) - 6}{x}$ **60.** $\lim\limits_{x \to 0} \dfrac{3x^3 - 2x}{x}$

61. For the function $f(x) = \begin{cases} x^3 & \text{if } x \leq 0 \\ x^2 & \text{if } x > 0 \end{cases}$, determine whether:

(a) f is continuous at 0.

(b) $f'(0)$ exists.

(c) Graph the function f and its derivative f'.

62. For the function $f(x) = \begin{cases} 2x & \text{if } x \leq 0 \\ x^2 & \text{if } x > 0 \end{cases}$, determine whether:

(a) f is continuous at 0.

(b) $f'(0)$ exists.

(c) Graph the function f and its derivative f'.

63. Velocity   The distance s (in feet) of an automobile from the origin at time t (in seconds) is given by

$$s = s(t) = \begin{cases} t^3 & \text{if } 0 \leq t < 5 \\ 125 & \text{if } t \geq 5 \end{cases}$$

(This could represent a crash test in which a vehicle is accelerated until it hits a brick wall at $t = 5$ s.)

(a) Find the velocity just before impact (at $t = 4.99$ s) and just after impact (at $t = 5.01$ s).

(b) Is the velocity function $v = s'(t)$ continuous at $t = 5$?

(c) How do you interpret the answer to (b)?

64. Population Growth   A simple model for population growth states that the rate of change of population size P with respect to time t is proportional to the population size. Express this statement as an equation involving a derivative.

65. Atmospheric Pressure   Atmospheric pressure p decreases as the distance x from the surface of Earth increases, and the rate of change of pressure with respect to altitude is proportional to the pressure. Express this law as an equation involving a derivative.

66. Electrical Current   Under certain conditions, an electric current I will die out at a rate (with respect to time t) that is proportional to the current remaining. Express this law as an equation involving a derivative.

67. Tangent Line   Let $f(x) = x^2 + 2$. Find all points on the graph of f for which the tangent line passes through the origin.

68. Tangent Line   Let $f(x) = x^2 - 2x + 1$. Find all points on the graph of f for which the tangent line passes through the point $(1, -1)$.

69. Area and Circumference of a Circle   A circle of radius r has area $A = \pi r^2$ and circumference $C = 2\pi r$. If the radius changes from r to $r + \Delta r$, find the:

(a) Change in area

(b) Change in circumference

(c) Average rate of change of area with respect to radius

(d) Average rate of change of circumference with respect to radius

(e) Rate of change of circumference with respect to radius

70. Volume of a Sphere   The volume V of a sphere of radius r is $V = \dfrac{4\pi r^3}{3}$. If the radius changes from r to $r + \Delta r$, find the:

(a) Change in volume

(b) Average rate of change of volume with respect to radius

(c) Rate of change of volume with respect to radius

71. Use the definition of the derivative to show that $f(x) = |x|$ has no derivative at 0.

72. Use the definition of the derivative to show that $f(x) = \sqrt[3]{x}$ has no derivative at 0.

73. If f is an even function that is differentiable at c, show that its derivative function is odd. That is, show $f'(-c) = -f'(c)$.

74. If f is an odd function that is differentiable at c, show that its derivative function is even. That is, show $f'(-c) = f'(c)$.

75. Tangent Lines and Derivatives   Let f and g be two functions, each with derivatives at c. State the relationship between their tangent lines at c if:

(a) $f'(c) = g'(c)$ **(b)** $f'(c) = -\dfrac{1}{g'(c)}$ $g'(c) \neq 0$

Challenge Problems

76. Let f be a function defined for all x. Suppose f has the following properties:

$$f(u + v) = f(u)f(v) \qquad f(0) = 1 \qquad f'(0) \text{ exists}$$

(a) Show that $f'(x)$ exists for all real numbers x.

(b) Show that $f'(x) = f'(0)f(x)$.

77. A function f is defined for all real numbers and has the following three properties:

$$f(1) = 5 \qquad f(3) = 21 \qquad f(a + b) - f(a) = kab + 2b^2$$

for all real numbers a and b where k is a fixed real number independent of a and b.

(a) Use $a = 1$ and $b = 2$ to find k.

(b) Find $f'(3)$.

(c) Find $f'(x)$ for all real x.

78. A function f is **periodic** if there is a positive number p so that $f(x + p) = f(x)$ for all x. Suppose f is differentiable. Show that if f is periodic with period p, then f' is also periodic with period p.

2.3 The Derivative of a Polynomial Function; The Derivative of $y = e^x$

OBJECTIVES *When you finish this section, you should be able to:*

1 Differentiate a constant function (p. 163)
2 Differentiate a power function (p. 164)
3 Differentiate the sum and the difference of two functions (p. 166)
4 Differentiate the exponential function $y = e^x$ (p. 168)

Finding the derivative of a function from the definition can become tedious, especially if the function f is complicated. Just as we did for limits, we derive some basic derivative formulas and some properties of derivatives that make finding a derivative simpler.

Before getting started, we introduce other notations commonly used for the derivative of a function $y = f(x)$. The most common ones are

$$y' \qquad \frac{dy}{dx} \qquad Df(x)$$

Leibniz notation $\dfrac{dy}{dx}$ may be written in several equivalent ways as

$$\frac{dy}{dx} = \frac{d}{dx}y = \frac{d}{dx}f(x)$$

where $\dfrac{d}{dx}$ is an instruction to find the derivative (with respect to the independent variable x) of the function $y = f(x)$.

In **operator notation** $Df(x)$, D is said to *operate* on the function, and the result is the derivative of f. To emphasize that the operation is performed with respect to the independent variable x, it is sometimes written $Df(x) = D_x f(x)$.

We use prime notation or Leibniz notation, or sometimes a mixture of the two, depending on which is more convenient. We do not use operator notation in this book.

1 Differentiate a Constant Function

See Figure 19. Since the graph of a constant function $f(x) = A$ is a horizontal line, the tangent line to f at any point is also a horizontal line, whose slope is 0. Since the derivative is the slope of the tangent line, the derivative of f is 0.

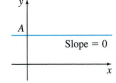

Figure 19 $f(x) = A$

IN WORDS The derivative of a constant is 0.

> **THEOREM** Derivative of a Constant Function
>
> If f is the constant function $f(x) = A$, then
>
> $$\boxed{f'(x) = 0}$$
>
> That is, if A is a constant, then
>
> $$\boxed{\frac{d}{dx}A = 0}$$

Proof If $f(x) = A$, then its derivative function is given by

$$f'(x) = \lim_{h \to 0} \frac{f(x+h) - f(x)}{h} = \lim_{h \to 0} \frac{A - A}{h} = 0 \qquad \blacksquare$$

The definition of a derivative, form (3)

$f(x) = A$
$f(x+h) = A$

Differentiating a Constant Function

(a) If $f(x) = \sqrt{3}$, then $f'(x) = 0$ **(b)** If $f(x) = -\dfrac{1}{2}$, then $f'(x) = 0$

(c) If $f(x) = \pi$, then $\dfrac{d}{dx}\pi = 0$ **(d)** If $f(x) = 0$, then $\dfrac{d}{dx}0 = 0$ ∎

2 Differentiate a Power Function

Next we take up the derivative of a power function $f(x) = x^n$, where $n \geq 1$ is an integer.

When $n = 1$, then $f(x) = x$ is the identity function and its graph is the line $y = x$, as shown in Figure 20.

The slope of the line $y = x$ is 1, so we would expect $f'(x) = 1$.

Proof $f'(x) = \dfrac{d}{dx}x = \lim_{h \to 0}\dfrac{f(x+h) - f(x)}{h} = \lim_{h \to 0}\dfrac{(x+h) - x}{h} = \lim_{h \to 0}\dfrac{h}{h} = 1$

$$\uparrow_{\substack{}} f(x) = x,\ f(x+h) = x+h$$

∎

Figure 20 $f(x) = x$

THEOREM Derivative of $f(x) = x$

If $f(x) = x$, then

$$f'(x) = \frac{d}{dx}x = 1$$

When $n = 2$, then $f(x) = x^2$ is the square function. The derivative of f is

$$f'(x) = \frac{d}{dx}x^2 = \lim_{h \to 0}\frac{(x+h)^2 - x^2}{h} = \lim_{h \to 0}\frac{x^2 + 2hx + h^2 - x^2}{h}$$

$$= \lim_{h \to 0}\frac{h(2x + h)}{h} = \lim_{h \to 0}(2x + h) = 2x$$

The slope of the tangent line to the graph of $f(x) = x^2$ is different for every number x. Figure 21 shows the graph of f and several of its tangent lines. Notice that the slope of each tangent line drawn is twice the value of x.

When $n = 3$, then $f(x) = x^3$ is the cube function. The derivative of f is

$$f'(x) = \lim_{h \to 0}\frac{(x+h)^3 - x^3}{h} = \lim_{h \to 0}\frac{x^3 + 3x^2h + 3xh^2 + h^3 - x^3}{h}$$

$$= \lim_{h \to 0}\frac{h(3x^2 + 3xh + h^2)}{h} = \lim_{h \to 0}(3x^2 + 3xh + h^2) = 3x^2$$

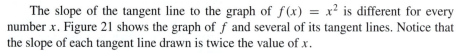

Notice that the derivative of each of these power functions is another power function, whose degree is 1 less than the degree of the original function and whose coefficient is the degree of the original function. This rule holds for all power functions as the following theorem, called the *Simple Power Rule*, indicates.

Figure 21 $f(x) = x^2$

THEOREM Simple Power Rule*

The derivative of the power function $y = x^n$, where $n \geq 1$ is an integer, is

$$y' = \frac{d}{dx}x^n = nx^{n-1}$$

IN WORDS The derivative of x raised to an integer power $n \geq 1$ is n times x raised to the power $n - 1$.

---

* $\dfrac{d}{dx}x^n = nx^{n-1}$ is true not only for positive integers n but also for any real number n. But the proof requires future results. As these are developed, we will expand the Power Rule to include an ever-widening set of numbers until we arrive at the fact it is true when n is a real number.

NEED TO REVIEW? The Binomial Theorem is discussed in Appendix A.5, pp. A-42 to A-43.

Proof If $f(x) = x^n$ and n is a positive integer, then $f(x + h) = (x + h)^n$. We use the Binomial Theorem to expand $(x + h)^n$. Then

$$f(x + h) = (x + h)^n = x^n + nx^{n-1}h + \frac{n(n-1)}{2}x^{n-2}h^2 + \frac{n(n-1)(n-2)}{6}x^{n-3}h^3 + \cdots + nxh^{n-1} + h^n$$

and

$$f'(x) = \lim_{h \to 0} \frac{f(x + h) - f(x)}{h}$$

$$= \lim_{h \to 0} \frac{\left[x^n + nx^{n-1}h + \frac{n(n-1)}{2}x^{n-2}h^2 + \frac{n(n-1)(n-2)}{6}x^{n-3}h^3 + \cdots + nxh^{n-1} + h^n\right] - x^n}{h}$$

$$= \lim_{h \to 0} \frac{nx^{n-1}h + \frac{n(n-1)}{2}x^{n-2}h^2 + \frac{n(n-1)(n-2)}{6}x^{n-3}h^3 + \cdots + nxh^{n-1} + h^n}{h} \qquad \text{Simplify.}$$

$$= \lim_{h \to 0} \frac{h\left[nx^{n-1} + \frac{n(n-1)}{2}x^{n-2}h + \frac{n(n-1)(n-2)}{6}x^{n-3}h^2 + \cdots + nxh^{n-2} + h^{n-1}\right]}{h} \qquad \text{Factor } h \text{ in the numerator.}$$

$$= \lim_{h \to 0} \left[nx^{n-1} + \frac{n(n-1)}{2}x^{n-2}h + \frac{n(n-1)(n-2)}{6}x^{n-3}h^2 + \cdots + nxh^{n-2} + h^{n-1}\right] \qquad \text{Divide out the common } h.$$

$$= nx^{n-1} \qquad \text{Take the limit. Only the first term remains.}$$

■

EXAMPLE 2   Differentiating a Power Function

(a) $\dfrac{d}{dx}x^5 = 5x^4$ **(b)** If $g(x) = x^{10}$, then $g'(x) = 10x^9$.   ■

NOW WORK   Problem 1.

But what if we want to find the derivative of the function $f(x) = ax^n$ when $a \neq 1$? The next theorem, called the *Constant Multiple Rule*, provides a way.

IN WORDS The derivative of a constant times a differentiable function f equals the constant times the derivative of f.

THEOREM   Constant Multiple Rule

If a function f is differentiable and k is a constant, then $F(x) = kf(x)$ is a function that is differentiable and

$$\boxed{F'(x) = kf'(x)}$$

Proof We use the definition of derivative.

$$F'(x) = \lim_{h \to 0} \frac{F(x + h) - F(x)}{h} = \lim_{h \to 0} \frac{kf(x + h) - kf(x)}{h}$$

$$= \lim_{h \to 0} \frac{k\left[f(x + h) - f(x)\right]}{h} = k \cdot \lim_{h \to 0} \frac{f(x + h) - f(x)}{h} = k \cdot f'(x) \qquad ■$$

Using Leibniz notation, the Constant Multiple Rule takes the form

$$\boxed{\frac{d}{dx}[kf(x)] = k\left[\frac{d}{dx}f(x)\right]}$$

A change in the symbol used for the independent variable does not affect the derivative formula. For example, $\dfrac{d}{dt}t^2 = 2t$ and $\dfrac{d}{du}u^5 = 5u^4$.

EXAMPLE 3   **Differentiating a Constant Times a Power Function**

Find the derivative of each power function:

(a) $f(x) = 5x^3$ **(b)** $g(u) = -\dfrac{1}{2}u^2$ **(c)** $u(x) = \pi^4 x^3$

Solution Notice that each of these functions involves the product of a constant and a power function. So, we use the Constant Multiple Rule followed by the Simple Power Rule.

(a) $f(x) = 5 \cdot x^3$, so $f'(x) = 5\left[\dfrac{d}{dx}x^3\right] = 5 \cdot 3x^2 = 15x^2$

(b) $g(u) = -\dfrac{1}{2} \cdot u^2$, so $g'(u) = -\dfrac{1}{2} \cdot \dfrac{d}{du}u^2 = -\dfrac{1}{2} \cdot 2u^1 = -u$

(c) $u(x) = \pi^4 x^3$, so $u'(x) = \pi^4 \cdot \underset{\substack{\uparrow \\ \pi \text{ is a constant}}}{\dfrac{d}{dx}x^3} = \pi^4 \cdot 3x^2 = 3\pi^4 x^2$ ■

NOW WORK   **Problem 31.**

3 Differentiate the Sum and the Difference of Two Functions

We can find the derivative of a function that is the sum of two functions whose derivatives are known by adding the derivatives of each function.

THEOREM Sum Rule

If two functions f and g are differentiable and if $F(x) = f(x) + g(x)$, then F is differentiable and

$$\boxed{F'(x) = f'(x) + g'(x)}$$

IN WORDS The derivative of the sum of two differentiable functions equals the sum of their derivatives. That is, $(f + g)' = f' + g'$.

Proof When $F(x) = f(x) + g(x)$, then

$$F(x + h) - F(x) = [f(x + h) + g(x + h)] - [f(x) + g(x)]$$
$$= [f(x + h) - f(x)] + [g(x + h) - g(x)]$$

So, the derivative of F is

$$F'(x) = \lim_{h \to 0} \frac{[f(x + h) - f(x)] + [g(x + h) - g(x)]}{h}$$

$$= \lim_{h \to 0} \frac{f(x + h) - f(x)}{h} + \lim_{h \to 0} \frac{g(x + h) - g(x)}{h} \qquad \text{The limit of a sum is the sum of the limits.}$$

$$= f'(x) + g'(x)$$ ■

In Leibniz notation, the Sum Rule takes the form

$$\boxed{\dfrac{d}{dx}[f(x) + g(x)] = \dfrac{d}{dx}f(x) + \dfrac{d}{dx}g(x)}$$

EXAMPLE 4   **Differentiating the Sum of Two Functions**

Find the derivative of $f(x) = 3x^2 + 8$.

Solution Here, f is the sum of $3x^2$ and 8. So, we begin by using the Sum Rule.

$$f'(x) = \underset{\substack{\uparrow \\ \text{Sum Rule}}}{\dfrac{d}{dx}(3x^2 + 8)} = \dfrac{d}{dx}(3x^2) + \dfrac{d}{dx}8 = \underset{\substack{\uparrow \\ \text{Constant Multiple Rule}}}{3\dfrac{d}{dx}x^2} + 0 = \underset{\substack{\uparrow \\ \text{Simple Power Rule}}}{3 \cdot 2x} = 6x$$ ■

NOW WORK   **Problem 7.**

THEOREM Difference Rule

If the functions f and g are differentiable and if $F(x) = f(x) - g(x)$, then F is differentiable, and $F'(x) = f'(x) - g'(x)$,

$$\frac{d}{dx}[f(x) - g(x)] = \frac{d}{dx}f(x) - \frac{d}{dx}g(x)$$

The proof of the Difference Rule is left as an exercise. (See Problem 78.)

The Sum and Difference Rules extend to sums (or differences) of more than two functions. That is, if the functions $f_1, f_2, \ldots, f_n$ are all differentiable, and $a_1, a_2, \ldots, a_n$ are constants, then

$$\frac{d}{dx}[a_1 f_1(x) + a_2 f_2(x) + \cdots + a_n f_n(x)] = a_1 \frac{d}{dx}f_1(x) + a_2\frac{d}{dx}f_2(x) + \cdots + a_n\frac{d}{dx}f_n(x)$$

Combining the rules for finding the derivative of a constant, a power function, and a sum or difference allows us to differentiate any polynomial function.

EXAMPLE 5   Differentiating a Polynomial Function

(a) Find the derivative of $f(x) = 2x^4 - 6x^2 + 2x - 3$.

(b) What is $f'(2)$?

(c) Find the slope of the tangent line to the graph of f at the point $(1, -5)$.

(d) Find an equation of the tangent line to the graph of f at the point $(1, -5)$.

(e) Use graphing technology to graph f and the tangent line to the graph of f at the point $(1, -5)$ on the same screen.

Solution (a)

$$f'(x) = \frac{d}{dx}(2x^4 - 6x^2 + 2x - 3) = \underset{\underset{\text{Sum \& Difference Rules}}{\uparrow}}{\frac{d}{dx}(2x^4)} - \frac{d}{dx}(6x^2) + \frac{d}{dx}(2x) - \frac{d}{dx}3$$

$$= \underset{\underset{\text{Constant Multiple Rule}}{\uparrow}}{2 \cdot \frac{d}{dx}x^4} - 6 \cdot \frac{d}{dx}x^2 + 2 \cdot \frac{d}{dx}x - 0$$

$$= \underset{\underset{\text{Simple Power Rule}}{\uparrow}}{2 \cdot 4x^3} - 6 \cdot 2x + 2 \cdot 1 = \underset{\underset{\text{Simplify}}{\uparrow}}{8x^3 - 12x + 2}$$

(b) $f'(2) = 8(2)^3 - 12(2) + 2 = 64 - 24 + 2 = 42$.

(c) The slope of the tangent line at the point $(1, -5)$ equals $f'(1)$.

$$f'(1) = 8(1)^3 - 12(1) + 2 = 8 - 12 + 2 = -2$$

(d) We use the point-slope form of an equation of a line to find an equation of the tangent line at $(1, -5)$.

$$y - (-5) = -2(x - 1)$$

$$y = -2(x - 1) - 5 = -2x + 2 - 5 = -2x - 3$$

The line $y = -2x - 3$ is tangent to the graph of $f(x) = 2x^4 - 6x^2 + 2x - 3$ at the point $(1, -5)$.

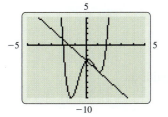

Figure 22 $f(x) = 2x^4 - 6x^2 + 2x - 3$

(e) The graphs of f and the tangent line to f at $(1, -5)$ are shown in Figure 22. ∎

NOW WORK Problem 33.

In some applications, we need to solve equations or inequalities involving the derivative of a function.

EXAMPLE 6   Solving Equations and Inequalities Involving Derivatives

(a) Find the points on the graph of $f(x) = 4x^3 - 12x^2 + 2$, where f has a horizontal tangent line.

(b) Where is $f'(x) > 0$? Where is $f'(x) < 0$?

Solution (a) The slope of a horizontal tangent line is 0. Since the derivative of f equals the slope of the tangent line, we need to find the numbers x for which $f'(x) = 0$.

$$f'(x) = 12x^2 - 24x = 12x(x - 2)$$

$$12x(x - 2) = 0 \qquad\qquad f'(x) = 0.$$

$$x = 0 \text{ or } x = 2 \qquad\qquad \text{Solve.}$$

At the points $(0, f(0)) = (0, 2)$ and $(2, f(2)) = (2, -14)$, the graph of the function $f(x) = 4x^3 - 12x^2 + 2$ has a horizontal tangent line.

(b) Since $f'(x) = 12x(x - 2)$ and we want to solve the inequalities $f'(x) > 0$ and $f'(x) < 0$, we use the zeros of f', 0 and 2, and form a table using the intervals $(-\infty, 0)$, $(0, 2)$, and $(2, \infty)$.

TABLE 2

Interval	$(-\infty, 0)$	$(0, 2)$	$(2, \infty)$
Sign of $f'(x) = 12x(x - 2)$	Positive	Negative	Positive

We conclude $f'(x) > 0$ on $(-\infty, 0) \cup (2, \infty)$ and $f'(x) < 0$ on $(0, 2)$. ∎

Figure 23 shows the graph of f.

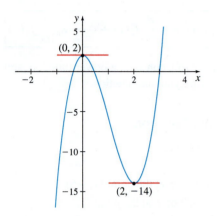

Figure 23 $f(x) = 4x^3 - 12x^2 + 2$

NOW WORK   Problem 37.

4 Differentiate the Exponential Function $y = e^x$

None of the differentiation rules developed so far allows us to find the derivative of an exponential function. To differentiate $f(x) = a^x$, we return to the definition of a derivative.

NEED TO REVIEW? Exponential functions are discussed in Section P.5, on pp. 38–42.

We begin by making some general observations about the derivative of $f(x) = a^x$, $a > 0$ and $a \neq 1$. We will then use these observations to find the derivative of the exponential function $y = e^x$.

Suppose $f(x) = a^x$, where $a > 0$ and $a \neq 1$. The derivative of f is

$$f'(x) = \lim_{h \to 0} \frac{f(x + h) - f(x)}{h} = \lim_{h \to 0} \frac{a^{x+h} - a^x}{h} = \lim_{\underset{a^{x+h} = a^x \cdot a^h}{h \to 0}} \frac{a^x \cdot a^h - a^x}{h}$$

$$= \lim_{\underset{\text{Factor out } a^x.}{h \to 0}} \left[a^x \cdot \frac{a^h - 1}{h} \right] = a^x \cdot \lim_{h \to 0} \frac{a^h - 1}{h}$$

provided $\lim_{h \to 0} \dfrac{a^h - 1}{h}$ exists.

Three observations about the derivative are significant:

- $f'(0) = a^0 \lim_{h \to 0} \dfrac{a^h - 1}{h} = \lim_{h \to 0} \dfrac{a^h - 1}{h}$.

- $f'(x)$ is a multiple of a^x. In fact, $\dfrac{d}{dx} a^x = f'(0) \cdot a^x$.

- If $f'(0)$ exists, then $f'(x)$ exists, and the domain of f' is the same as that of $f(x) = a^x$, all real numbers.

NEED TO REVIEW? The number e is discussed in Section P.5, pp. 41–42.

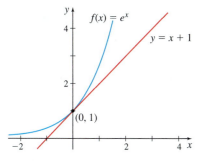

Figure 24

The slope of the tangent line to the graph of $f(x) = a^x$ at the point $(0, 1)$ is

$$f'(0) = \lim_{h \to 0} \frac{a^h - 1}{h},$$ and the value of this limit depends on the base a. In Section P.5, the number e was defined as that number for which the slope of the tangent line to the graph of $y = a^x$ at the point $(0, 1)$ equals 1. That is, the number e has the property that

$$\lim_{h \to 0} \frac{e^h - 1}{h} = 1$$

In other words, if $f(x) = e^x$, then $f'(0) = 1$. Figure 24 shows $f(x) = e^x$ and the tangent line $y = x + 1$ with slope 1 at the point $(0, 1)$.

Since $\dfrac{d}{dx} a^x = f'(0) \cdot a^x$, if $f(x) = e^x$, then $\dfrac{d}{dx} e^x = f'(0) \cdot e^x = 1 \cdot e^x = e^x$.

THEOREM Derivative of the Exponential Function $y = e^x$

The derivative of the exponential function $y = e^x$ is

$$y' = \frac{d}{dx} e^x = e^x \tag{1}$$

EXAMPLE 7   Differentiating an Expression Involving $y = e^x$

Find the derivative of $f(x) = 4e^x + x^3$.

Solution The function f is the sum of $4e^x$ and x^3. Then

$$f'(x) = \frac{d}{dx}(4e^x + x^3) = \underset{\text{Sum Rule}}{\frac{d}{dx}(4e^x)} + \frac{d}{dx}x^3 = \underset{\substack{\text{Constant Multiple Rule} \\ \text{Simple Power Rule}}}{4\frac{d}{dx}e^x + 3x^2} = \underset{\text{Use (1).}}{4e^x + 3x^2} = 4e^x + 3x^2 \quad \blacksquare$$

NOW WORK Problem 25.

Now we have the formula $\dfrac{d}{dx} e^x = e^x$. To find the derivative of $f(x) = a^x$, where $a > 0$ and $a \neq 1$, requires more information and is taken up in Chapter 3.

2.3 Assess Your Understanding

Concepts and Vocabulary

1. $\dfrac{d}{dx}\pi^2 = $ _____ ; $\dfrac{d}{dx}x^3 = $ _____ .

2. When n is a positive integer, the Simple Power Rule states that $\dfrac{d}{dx}x^n = $ _____ .

3. *True or False* The derivative of a power function of degree greater than 1 is also a power function.

4. If k is a constant and f is a differentiable function, then $\dfrac{d}{dx}[kf(x)] = $ _____ .

5. The derivative of $f(x) = e^x$ is _____ .

6. *True or False* The derivative of an exponential function $f(x) = a^x$, where $a > 0$ and $a \neq 1$, is always a constant multiple of a^x.

Skill Building

In Problems 7–26, find the derivative of each function using the formulas of this section. (a, b, c, and d, when they appear, are constants.)

7. $f(x) = 3x + \sqrt{2}$

8. $f(x) = 5x - \pi$

9. $f(x) = x^2 + 3x + 4$

10. $f(x) = 4x^4 + 2x^2 - 2$

11. $f(u) = 8u^5 - 5u + 1$

12. $f(u) = 9u^3 - 2u^2 + 4u + 4$

13. $f(s) = as^3 + \dfrac{3}{2}s^2$

14. $f(s) = 4 - \pi s^2$

15. $f(t) = \dfrac{1}{3}(t^5 - 8)$

16. $f(x) = \dfrac{1}{5}(x^7 - 3x^2 + 2)$

17. $f(t) = \dfrac{t^3 + 2}{5}$

18. $f(x) = \dfrac{x^7 - 5x}{9}$

19. $f(x) = \dfrac{x^3 + 2x + 1}{7}$

20. $f(x) = \dfrac{1}{a}(ax^2 + bx + c)$

---

1. = NOW WORK problem   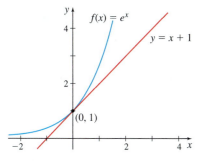 = Graphing technology recommended   CAS = Computer Algebra System recommended

21. $f(x) = ax^2 + bx + c$ **22.** $f(x) = ax^3 + bx^2 + cx + d$

23. $f(x) = 4e^x$ **24.** $f(x) = -\dfrac{1}{2}e^x$

25. $f(u) = 5u^2 - 2e^u$ **26.** $f(u) = 3e^u + 10$

In Problems 27–32, find each derivative.

27. $\dfrac{d}{dt}\left(\sqrt{3}t + \dfrac{1}{2}\right)$ **28.** $\dfrac{d}{dt}\left(\dfrac{2t^4 - 5}{8}\right)$

29. $\dfrac{dA}{dR}$ if $A(R) = \pi R^2$ **30.** $\dfrac{dC}{dR}$ if $C = 2\pi R$

31. $\dfrac{dV}{dr}$ if $V = \dfrac{4}{3}\pi r^3$ **32.** $\dfrac{dP}{dT}$ if $P = 0.2T$

In Problems 33–36:
(a) *Find the slope of the tangent line to the graph of each function f at the indicated point.*
(b) *Find an equation of the tangent line at the point.*
 (c) *Graph f and the tangent line found in (b) on the same set of axes.*

33. $f(x) = x^3 + 3x - 1$ at $(0, -1)$

34. $f(x) = x^4 + 2x - 1$ at $(1, 2)$

35. $f(x) = e^x + 5x$ at $(0, 1)$

36. $f(x) = 4 - e^x$ at $(0, 3)$

In Problems 37–42:
(a) *Find the points, if any, at which the graph of each function f has a horizontal tangent line.*
(b) *Find an equation for each horizontal tangent line.*
(c) *Solve the inequality $f'(x) > 0$.*
(d) *Solve the inequality $f'(x) < 0$.*
 (e) *Graph f and any horizontal lines found in (b) on the same set of axes.*
(f) *Describe the graph of f for the results obtained in parts (c) and (d).*

37. $f(x) = 3x^2 - 12x + 4$ **38.** $f(x) = x^2 + 4x - 3$

39. $f(x) = x + e^x$ **40.** $f(x) = 2e^x - 1$

41. $f(x) = x^3 - 3x + 2$ **42.** $f(x) = x^4 - 4x^3$

43. Rectilinear Motion   At t seconds, an object in rectilinear motion is s meters from the origin, where $s(t) = t^3 - t + 1$. Find the velocity of the object at $t = 0$ and at $t = 5$.

44. Rectilinear Motion   At t seconds, an object in rectilinear motion is s meters from the origin, where $s(t) = t^4 - t^3 + 1$. Find the velocity of the object at $t = 0$ and at $t = 1$.

Rectilinear Motion   *In Problems 45 and 46, each function describes the distance s from the origin at time t of an object in rectilinear motion:*
(a) *Find the velocity v of the object at any time t.*
(b) *When is the object at rest?*

45. $s(t) = 2 - 5t + t^2$ **46.** $s(t) = t^3 - \dfrac{9}{2}t^2 + 6t + 4$

In Problems 47 and 48, use the graphs to find each derivative.

47. Let $u(x) = f(x) + g(x)$ and $v(x) = f(x) - g(x)$.

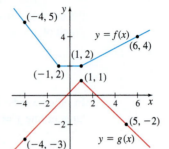

(a) $u'(0)$ **(b)** $u'(4)$
(c) $v'(-2)$ **(d)** $v'(6)$
(e) $3u'(5)$ **(f)** $-2v'(3)$

48. Let $F(t) = f(t) + g(t)$ and $G(t) = g(t) - f(t)$.

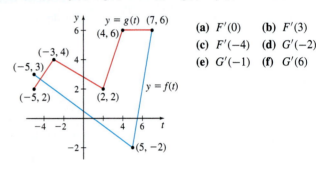

(a) $F'(0)$ **(b)** $F'(3)$
(c) $F'(-4)$ **(d)** $G'(-2)$
(e) $G'(-1)$ **(f)** $G'(6)$

In Problems 49 and 50, for each function f:
(a) *Find $f'(x)$ by expanding $f(x)$ and differentiating the polynomial.*
(b) *Find $f'(x)$ using a CAS.* **CAS**
(c) *Show that the results found in parts (a) and (b) are equivalent.*

49. $f(x) = (2x - 1)^3$ **50.** $f(x) = (x^2 + x)^4$

Applications and Extensions

In Problems 51–56, find each limit.

51. $\displaystyle\lim_{h\to 0} \dfrac{5\left(\dfrac{1}{2} + h\right)^8 - 5\left(\dfrac{1}{2}\right)^8}{h}$ **52.** $\displaystyle\lim_{h\to 0} \dfrac{6(2 + h)^5 - 6(2)^5}{h}$

53. $\displaystyle\lim_{h\to 0} \dfrac{\sqrt{3}(8 + h)^5 - \sqrt{3}(8)^5}{h}$ **54.** $\displaystyle\lim_{h\to 0} \dfrac{\pi(1 + h)^{10} - \pi}{h}$

55. $\displaystyle\lim_{h\to 0} \dfrac{a(x + h)^3 - ax^3}{h}$ **56.** $\displaystyle\lim_{h\to 0} \dfrac{b(x + h)^n - bx^n}{h}$

In Problems 57–62, find an equation of the tangent line(s) to the graph of the function f that is (are) parallel to the line L.

57. $f(x) = 3x^2 - x$; $L: y = 5x$

58. $f(x) = 2x^3 + 1$; $L: y = 6x - 1$

59. $f(x) = e^x$; $L: y - x - 5 = 0$

60. $f(x) = -2e^x$; $L: y + 2x - 8 = 0$

61. $f(x) = \dfrac{1}{3}x^3 - x^2$; $L: y = 3x - 2$

62. $f(x) = x^3 - x$; $L: x + y = 0$

63. Tangent Lines Let $f(x) = 4x^3 - 3x - 1$.

   (a) Find an equation of the tangent line to the graph of f at $x = 2$.

   (b) Find the coordinates of any points on the graph of f where the tangent line is parallel to $y = x + 12$.

   (c) Find an equation of the tangent line to the graph of f at any points found in (b).

   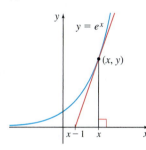 **(d)** Graph f, the tangent line found in (a), the line $y = x + 12$, and any tangent lines found in (c) on the same screen.

64. Tangent Lines Let $f(x) = x^3 + 2x^2 + x - 1$.

   (a) Find an equation of the tangent line to the graph of f at $x = 0$.

   (b) Find the coordinates of any points on the graph of f where the tangent line is parallel to $y = 3x - 2$.

   (c) Find an equation of the tangent line to the graph of f at any points found in (b).

   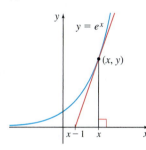 **(d)** Graph f, the tangent line found in (a), the line $y = 3x - 2$, and any tangent lines found in (c) on the same screen.

65. Tangent Line Show that the line perpendicular to the x-axis and containing the point (x, y) on the graph of $y = e^x$ and the tangent line to the graph of $y = e^x$ at the point (x, y) intersect the x-axis 1 unit apart. See the figure.

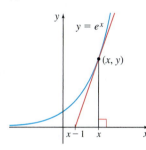

66. Tangent Line Show that the tangent line to the graph of $y = x^n$ at $(1, 1)$ has y-intercept $1 - n$.

67. Tangent Lines If n is an odd positive integer, show that the tangent lines to the graph of $y = x^n$ at $(1, 1)$ and at $(-1, -1)$ are parallel.

68. Tangent Line If the line $3x - 4y = 0$ is tangent to the graph of $y = x^3 + k$ in the first quadrant, find k.

69. Tangent Line Find the constants a, b, and c so that the graph of $y = ax^2 + bx + c$ contains the point $(-1, 1)$ and is tangent to the line $y = 2x$ at $(0, 0)$.

70. Tangent Line Let T be the line tangent to the graph of $y = x^3$ at the point $\left(\dfrac{1}{2}, \dfrac{1}{8}\right)$. At what other point Q on the graph of $y = x^3$ does the line T intersect the graph? What is the slope of the tangent line at Q?

71. Military Tactics A dive bomber is flying from right to left along the graph of $y = x^2$. When a rocket bomb is released, it follows a path that is approximately along the tangent line. Where should the pilot release the bomb if the target is at $(1, 0)$?

72. Military Tactics Answer the question in Problem 71 if the plane is flying from right to left along the graph of $y = x^3$.

73. Fluid Dynamics The velocity v of a liquid flowing through a cylindrical tube is given by the **Hagen–Poiseuille equation** $v = k(R^2 - r^2)$, where R is the radius of the tube, k is a constant that depends on the length of the tube and the velocity of the

liquid at its ends, and r is the variable distance of the liquid from the center of the tube.

   (a) Find the rate of change of v with respect to r at the center of the tube.

   (b) What is the rate of change halfway from the center to the wall of the tube?

   (c) What is the rate of change at the wall of the tube?

74. Rate of Change Water is leaking out of a swimming pool that measures 20 ft by 40 ft by 6 ft. The amount of water in the pool at a time t is $W(t) = 35,000 - 20t^2$ gallons, where t equals the number of hours since the pool was last filled. At what rate is the water leaking when $t = 2\,\mathrm{h}$?

75. Luminosity of the Sun The luminosity L of a star is the rate at which it radiates energy. This rate depends on the temperature T and surface area A of the star's photosphere (the gaseous surface that emits the light). Luminosity is modeled by the equation $L = \sigma A T^4$, where σ is a constant known as the **Stefan–Boltzmann constant**, and T is expressed in the absolute (Kelvin) scale for which $0\,\mathrm{K}$ is absolute zero. As with most stars, the Sun's temperature has gradually increased over the 6 billion years of its existence, causing its luminosity to slowly increase.

   (a) Find the rate at which the Sun's luminosity changes with respect to the temperature of its photosphere. Assume that the surface area A remains constant.

   (b) Find the rate of change at the present time. The temperature of the photosphere is presently $5800\,\mathrm{K}$ $(10,000\,°\mathrm{F})$, the radius of the photosphere is $r = 6.96 \times 10^8\,\mathrm{m}$, and $\sigma = 5.67 \times 10^{-8}\,\dfrac{\mathrm{W}}{\mathrm{m}^2\,\mathrm{K}^4}$.

   (c) Assuming that the rate found in (b) remains constant, how much would the luminosity change if its photosphere temperature increased by $1\,\mathrm{K}$ ($1\,°\mathrm{C}$ or $1.8\,°\mathrm{F}$)? Compare this change to the present luminosity of the Sun.

76. Medicine: Poiseuille's Equation The French physician Poiseuille discovered that the volume V of blood (in cubic centimeters per unit time) flowing through an artery with inner radius R (in centimeters) can be modeled by

$$V(R) = kR^4$$

where $k = \dfrac{\pi}{8vl}$ is constant (here, v represents the viscosity of blood and l is the length of the artery).

   (a) Find the rate of change of the volume V of blood flowing through the artery with respect to the radius R.

   (b) Find the rate of change when $R = 0.03$ and when $R = 0.04$.

   (c) If the radius of a partially clogged artery is increased from $0.03\,\mathrm{cm}$ to $0.04\,\mathrm{cm}$, estimate the effect on the rate of change of the volume V with respect to R of the blood flowing through the enlarged artery.

   (d) How do you interpret the results found in (b) and (c)?

77. Derivative of the Area   Let $f(x) = mx$, $m > 0$. Let $F(x)$, $x > 0$, be defined as the area of the shaded region in the figure. Find $F'(x)$.

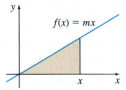

78. The Difference Rule   Prove that if f and g are differentiable functions and if $F(x) = f(x) - g(x)$, then $F'(x) = f'(x) - g'(x)$.

79. Simple Power Rule   Let $f(x) = x^n$, where n is a positive integer. Use a factoring principle to show that

$$f'(c) = \lim_{x \to c} \frac{f(x) - f(c)}{x - c} = nc^{n-1}.$$

Normal Lines   *Problems 80–87 involve the following discussion.*

*The **normal line** to the graph of a function f at a point $(c, f(c))$ is the line through $(c, f(c))$ perpendicular to the tangent line to the graph of f at $(c, f(c))$. See the figure. If f is a function whose derivative at c is $f'(c) \neq 0$, the slope of the normal line to the graph of f at $(c, f(c))$ is $-\dfrac{1}{f'(c)}$. Then an equation of the*

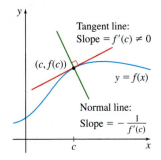

normal line to the graph of f at $(c, f(c))$ is

$$y - f(c) = -\frac{1}{f'(c)}(x - c).$$

In Problems 80–85, find the slope of the normal line to the graph of each function at the indicated point. Graph each function and show this normal line.

80. $f(x) = x^2 + 1$ at $(1, 2)$

81. $f(x) = x^2 - 1$ at $(-1, 0)$

82. $f(x) = x^2 - 2x$ at $(-1, 3)$

83. $f(x) = 2x^2 + x$ at $(1, 3)$

84. $f(x) = \dfrac{1}{x}$ at $(1, 1)$

85. $f(x) = \sqrt{x}$ at $(4, 2)$

86. Normal Lines   For what nonnegative number b is the line given by $y = -\dfrac{1}{3}x + b$ normal to the graph of $y = x^3$?

87. Normal Lines   Let N be the line normal to the graph of $y = x^2$ at the point $(-2, 4)$. At what other point Q does N meet the graph?

Challenge Problems

88. Tangent Line   Find a, b, c, d so that the tangent line to the graph of the cubic $y = ax^3 + bx^2 + cx + d$ at the point $(1, 0)$ is $y = 3x - 3$ and at the point $(2, 9)$ is $y = 18x - 27$.

89. Tangent Line   Find the fourth degree polynomial that contains the origin and to which the line $x + 2y = 14$ is tangent at both $x = 4$ and $x = -2$.

90. Tangent Lines   Find equations for all the lines containing the point $(1, 4)$ that are tangent to the graph of $y = x^3 - 10x^2 + 6x - 2$. At what points do each of the tangent lines touch the graph?

91. The line $x = c$, where $c > 0$, intersects the cubic $y = 2x^3 + 3x^2 - 9$ at the point P and intersects the parabola $y = 4x^2 + 4x + 5$ at the point Q, as shown in the figure below.

(a) If the line tangent to the cubic at the point P is parallel to the line tangent to the parabola at the point Q, find the number c.

(b) Write an equation for each of the two tangent lines described in (a).

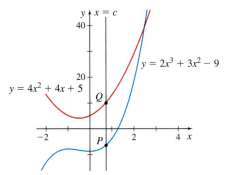

92. $f(x) = Ax^2 + B$, $A > 0$.

(a) Find c, $c > 0$, in terms of A so that the tangent lines to the graph of f at $(c, f(c))$ and $(-c, f(-c))$ are perpendicular.

(b) Find the slopes of the tangent lines in (a).

(c) Find the coordinates, in terms of A and B, of the point of intersection of the tangent lines in (a).

2.4 Differentiating the Product and the Quotient of Two Functions; Higher-Order Derivatives

OBJECTIVES *When you finish this section, you should be able to:*

1 Differentiate the product of two functions (p. 173)
2 Differentiate the quotient of two functions (p. 175)
3 Find higher-order derivatives (p. 177)
4 Work with acceleration (p. 179)

In this section, we obtain formulas for differentiating products and quotients of functions. As it turns out, the formulas are not what we might expect. The derivative of the product of two functions is *not* the product of their derivatives, and the derivative of the quotient of two functions is *not* the quotient of their derivatives.

1 Differentiate the Product of Two Functions

Consider the two functions $f(x) = 2x$ and $g(x) = x^3$. Both are differentiable, and their derivatives are $f'(x) = 2$ and $g'(x) = 3x^2$. Form the product

$$F(x) = f(x)g(x) = (2x)(x^3) = 2x^4$$

Now find F' using the Constant Multiple Rule and the Simple Power Rule.

$$F'(x) = 2(4x^3) = 8x^3$$

Notice that $f'(x)g'(x) = (2)(3x^2) = 6x^2$ is not equal to $F'(x) = \dfrac{d}{dx}[f(x)g(x)] = 8x^3$.

We conclude that the derivative of a product of two functions is *not* the product of their derivatives.

To find the derivative of the product of two differentiable functions f and g, we let $F(x) = f(x)g(x)$ and use the definition of a derivative, namely,

$$F'(x) = \lim_{h \to 0} \frac{[f(x+h)g(x+h)] - [f(x)g(x)]}{h}$$

We can express F' in an equivalent form that contains the difference quotients for f and g, by subtracting and adding $f(x+h)g(x)$ to the numerator.

$$F'(x) = \lim_{h \to 0} \frac{f(x+h)g(x+h) - f(x+h)g(x) + f(x+h)g(x) - f(x)g(x)}{h}$$

$$= \lim_{h \to 0} \frac{f(x+h)[g(x+h) - g(x)] + [f(x+h) - f(x)]g(x)}{h} \qquad \text{Group and factor.}$$

$$= \left[\lim_{h \to 0} f(x+h)\right]\left[\lim_{h \to 0} \frac{g(x+h) - g(x)}{h}\right] + \left[\lim_{h \to 0} \frac{f(x+h) - f(x)}{h}\right]\left[\lim_{h \to 0} g(x)\right] \qquad \text{Use properties of limits.}$$

$$= \left[\lim_{h \to 0} f(x+h)\right]g'(x) + f'(x)\left[\lim_{h \to 0} g(x)\right] \qquad \text{Definition of a derivative.}$$

$$= f(x)g'(x) + f'(x)g(x) \qquad \lim_{h \to 0} g(x) = g(x) \text{ as } h \text{ is not present; since } f \text{ is differentiable, it is continuous, so } \lim_{h \to 0} f(x+h) = f(x).$$

We have established the following formula.

THEOREM Product Rule

If f and g are differentiable functions and if $F(x) = f(x)g(x)$, then F is differentiable, and the derivative of the product F is

$$\boxed{F'(x) = [f(x)g(x)]' = f(x)g'(x) + f'(x)g(x)}$$

In Leibniz notation, the Product Rule takes the form

$$\boxed{\frac{d}{dx}F(x) = \frac{d}{dx}[f(x)g(x)] = f(x)\left[\frac{d}{dx}g(x)\right] + \left[\frac{d}{dx}f(x)\right]g(x)}$$

IN WORDS The derivative of the product of two differentiable functions equals the first function times the derivative of the second function plus the derivative of the first function times the second function. That is, $(fg)' = f(g') + (f')g$.

EXAMPLE 1   Differentiating the Product of Two Functions

Find y' if $y = (1 + x^2)e^x$.

Solution The function y is the product of two functions: a polynomial, $f(x) = 1 + x^2$, and the exponential function $g(x) = e^x$. By the Product Rule,

$$y' = \frac{d}{dx}[(1 + x^2)e^x] = (1 + x^2)\left[\frac{d}{dx}e^x\right] + \left[\frac{d}{dx}(1 + x^2)\right]e^x = (1 + x^2)e^x + 2xe^x$$
$$\uparrow$$
$$\text{Product Rule}$$

At this point, we have found the derivative, but it is customary to simplify the answer. Then

$$y' = (1 + x^2 + 2x)e^x = (x + 1)^2 e^x \qquad\blacksquare$$
$$\uparrow \qquad\qquad\qquad\qquad \uparrow$$
$$\text{Factor out } e^x. \qquad\qquad \text{Factor.}$$

NOW WORK Problem 9.

Do not use the Product Rule unnecessarily! When one of the factors is a constant, use the Constant Multiple Rule. For example, it is easier to work

$$\frac{d}{dx}[5(x^2 + 1)] = 5\frac{d}{dx}(x^2 + 1) = 5 \cdot 2x = 10x$$

than it is to work

$$\frac{d}{dx}[5(x^2 + 1)] = 5\frac{d}{dx}(x^2 + 1) + \left[\frac{d}{dx}5\right](x^2 + 1) = 5 \cdot 2x + 0 \cdot (x^2 + 1) = 10x$$

Also, it is easier to simplify $f(x) = x^2(4x - 3)$ before finding the derivative. That is, it is easier to work

$$\frac{d}{dx}[x^2(4x - 3)] = \frac{d}{dx}(4x^3 - 3x^2) = 12x^2 - 6x$$

than it is use the Product Rule

$$\frac{d}{dx}[x^2(4x - 3)] = x^2\frac{d}{dx}(4x - 3) + \left(\frac{d}{dx}x^2\right)(4x - 3) = (x^2)(4) + (2x)(4x - 3)$$

$$= 4x^2 + 8x^2 - 6x = 12x^2 - 6x$$

EXAMPLE 2   Differentiating a Product in Two Ways

Find the derivative of $F(v) = (5v^2 - v + 1)(v^3 - 1)$ in two ways:

(a) By using the Product Rule.

(b) By multiplying the factors of the function before finding its derivative.

Solution **(a)** F is the product of the two functions $f(v) = 5v^2 - v + 1$ and $g(v) = v^3 - 1$. Using the Product Rule, we get

$$F'(v) = (5v^2 - v + 1)\left[\frac{d}{dv}(v^3 - 1)\right] + \left[\frac{d}{dv}(5v^2 - v + 1)\right](v^3 - 1)$$

$$= (5v^2 - v + 1)(3v^2) + (10v - 1)(v^3 - 1)$$

$$= 15v^4 - 3v^3 + 3v^2 + 10v^4 - 10v - v^3 + 1$$

$$= 25v^4 - 4v^3 + 3v^2 - 10v + 1$$

(b) Here, we multiply the factors of F before differentiating.

$$F(v) = (5v^2 - v + 1)(v^3 - 1) = 5v^5 - v^4 + v^3 - 5v^2 + v - 1$$

Then

$$F'(v) = 25v^4 - 4v^3 + 3v^2 - 10v + 1 \qquad \blacksquare$$

Notice that the derivative is the same whether you differentiate and then simplify, or whether you multiply the factors and then differentiate. Use the approach that you find easier.

NOW WORK **Problem 13.**

2 Differentiate the Quotient of Two Functions

The derivative of the quotient of two functions is *not* equal to the quotient of their derivatives. Instead, the derivative of the quotient of two functions is found using the *Quotient Rule*.

THEOREM Quotient Rule

If two functions f and g are differentiable and if $F(x) = \dfrac{f(x)}{g(x)}$, $g(x) \neq 0$, then F is differentiable, and the derivative of the quotient F is

$$\boxed{F'(x) = \left[\frac{f(x)}{g(x)}\right]' = \frac{f'(x)g(x) - f(x)g'(x)}{[g(x)]^2}}$$

IN WORDS The derivative of a quotient of two functions is the derivative of the numerator times the denominator, minus the numerator times the derivative of the denominator, all divided by the denominator squared. That is,

$$\left(\frac{f}{g}\right)' = \frac{f'g - fg'}{g^2}.$$

In Leibniz notation, the Quotient Rule takes the form

$$\boxed{\frac{d}{dx}F(x) = \frac{d}{dx}\left[\frac{f(x)}{g(x)}\right] = \frac{\left[\frac{d}{dx}f(x)\right]g(x) - f(x)\left[\frac{d}{dx}g(x)\right]}{[g(x)]^2}}$$

Proof We use the definition of a derivative to find $F'(x)$.

$$F'(x) = \lim_{h \to 0} \frac{F(x+h) - F(x)}{h} = \lim_{h \to 0} \frac{\dfrac{f(x+h)}{g(x+h)} - \dfrac{f(x)}{g(x)}}{h}$$

$$F(x) = \frac{f(x)}{g(x)}$$

$$= \lim_{h \to 0} \frac{f(x+h)g(x) - f(x)g(x+h)}{h[g(x+h)g(x)]}$$

We write F' in an equivalent form that contains the difference quotients for f and g by subtracting and adding $f(x)g(x)$ to the numerator.

$$F'(x) = \lim_{h \to 0} \frac{f(x+h)g(x) - f(x)g(x) + f(x)g(x) - f(x)g(x+h)}{h[g(x+h)g(x)]}$$

Now we group and factor the numerator.

$$F'(x) = \lim_{h \to 0} \frac{[f(x+h) - f(x)]g(x) - f(x)[g(x+h) - g(x)]}{h[g(x+h)g(x)]}$$

$$= \lim_{h \to 0} \frac{\left[\dfrac{f(x+h) - f(x)}{h}\right] g(x) - f(x) \left[\dfrac{g(x+h) - g(x)}{h}\right]}{g(x+h)g(x)}$$

$$= \frac{\displaystyle\lim_{h \to 0}\left[\dfrac{f(x+h) - f(x)}{h}\right] \cdot \lim_{h \to 0} g(x) - \lim_{h \to 0} f(x) \cdot \lim_{h \to 0} \left[\dfrac{g(x+h) - g(x)}{h}\right]}{\displaystyle\lim_{h \to 0} g(x+h) \cdot \lim_{h \to 0} g(x)}$$

$$= \frac{f'(x)g(x) - f(x)g'(x)}{[g(x)]^2} \qquad\blacksquare$$

RECALL Since g is differentiable, it is continuous; so, $\lim\limits_{h \to 0} g(x+h) = g(x)$.

EXAMPLE 3   Differentiating the Quotient of Two Functions

Find y' if $y = \dfrac{x^2 + 1}{2x - 3}$.

Solution The function y is the quotient of $f(x) = x^2 + 1$ and $g(x) = 2x - 3$. Using the Quotient Rule, we have

$$y' = \frac{d}{dx}\left(\frac{x^2+1}{2x-3}\right) = \frac{\left[\dfrac{d}{dx}(x^2+1)\right](2x-3) - (x^2+1)\left[\dfrac{d}{dx}(2x-3)\right]}{(2x-3)^2}$$

$$= \frac{(2x)(2x-3) - (x^2+1)(2)}{(2x-3)^2} = \frac{4x^2 - 6x - 2x^2 - 2}{(2x-3)^2} = \frac{2x^2 - 6x - 2}{(2x-3)^2}$$

provided $x \neq \dfrac{3}{2}$. $\blacksquare$

NOW WORK Problem 23.

COROLLARY   Derivative of the Reciprocal of a Function

If a function g is differentiable, then

$$\frac{d}{dx}\left[\frac{1}{g(x)}\right] = -\frac{\dfrac{d}{dx}g(x)}{[g(x)]^2} = -\frac{g'(x)}{[g(x)]^2} \qquad (1)$$

provided $g(x) \neq 0$.

IN WORDS The derivative of the reciprocal of a function is the negative of the derivative of the denominator divided by the square of the denominator. That is, $\left(\dfrac{1}{g}\right)' = -\dfrac{g'}{g^2}$.

The proof of the corollary is left as an exercise. (See Problem 98.)

EXAMPLE 4   Differentiating the Reciprocal of a Function

(a) $\dfrac{d}{dx}\left(\dfrac{1}{x^2+x}\right) = -\dfrac{\dfrac{d}{dx}(x^2+x)}{(x^2+x)^2} = -\dfrac{2x+1}{(x^2+x)^2}$
      ↑ Use (1).

(b) $\dfrac{d}{dx}e^{-x} = \dfrac{d}{dx}\left(\dfrac{1}{e^x}\right) = -\dfrac{\dfrac{d}{dx}e^x}{(e^x)^2} = -\dfrac{e^x}{e^{2x}} = -\dfrac{1}{e^x} = -e^{-x}$
      ↑ Use (1).

$\blacksquare$

NOW WORK Problem 25.

Notice that the derivative of the reciprocal of a function f is *not* the reciprocal of the derivative. That is,

$$\frac{d}{dx}\left[\frac{1}{f(x)}\right] \neq \frac{1}{f'(x)}$$

The rule for the derivative of the reciprocal of a function allows us to extend the Simple Power Rule to all integers. Here is the proof.

Suppose n is a negative integer and $x \neq 0$. Then, $m = -n$ is a positive integer, and

$$y' = \frac{d}{dx}x^n = \frac{d}{dx}\frac{1}{x^m} = -\frac{\dfrac{d}{dx}(x^m)}{(x^m)^2} = -\frac{mx^{m-1}}{x^{2m}} = -mx^{m-1-2m} = -mx^{-m-1} = nx^{n-1}$$

$$\qquad\qquad\qquad\qquad\qquad \underset{\text{Use (1).}}{\uparrow} \quad \underset{\text{Simple Power Rule}}{\uparrow} \qquad\qquad\qquad\qquad\qquad\qquad \underset{\text{Substitute } n = -m.}{\uparrow}$$

THEOREM Power Rule

The derivative of $y = x^n$, where n any integer, is

$$\boxed{y' = \frac{d}{dx}x^n = nx^{n-1}}$$

EXAMPLE 5 Differentiating Using the Power Rule

(a) $\dfrac{d}{dx}x^{-1} = -x^{-2} = -\dfrac{1}{x^2}$

(b) $\dfrac{d}{du}\left(\dfrac{1}{u^2}\right) = \dfrac{d}{du}u^{-2} = -2u^{-3} = -\dfrac{2}{u^3}$

(c) $\dfrac{d}{ds}\left(\dfrac{4}{s^5}\right) = 4\dfrac{d}{ds}s^{-5} = 4\cdot(-5)\,s^{-6} = -20s^{-6} = -\dfrac{20}{s^6}$ ∎

NOW WORK Problem 31.

EXAMPLE 6 Using the Power Rule in Electrical Engineering

Ohm's Law states that the current I running through a wire is inversely proportional to the resistance R in the wire and can be written as $I = \dfrac{V}{R}$, where V is the voltage. Find the rate of change of I with respect to R when $V = 12$ volts.

Solution The rate of change of I with respect to R is the derivative $\dfrac{dI}{dR}$. We write Ohm's Law with $V = 12$ as $I = \dfrac{V}{R} = 12R^{-1}$ and use the Power Rule.

$$\frac{dI}{dR} = \frac{d}{dR}(12R^{-1}) = 12\cdot\frac{d}{dR}R^{-1} = 12(-1R^{-2}) = -\frac{12}{R^2}$$

The minus sign in $\dfrac{dI}{dR}$ indicates that the current I decreases as the resistance R in the wire increases. ∎

NOW WORK Problem 85.

3 Find Higher-Order Derivatives

Since the derivative f' is a function, it makes sense to ask about the derivative of f'. The derivative (if there is one) of f' is also a function called the **second derivative** of f and denoted by f'', read, "f double prime."

By continuing in this fashion, we can find the **third derivative** of f, the **fourth derivative** of f, and so on, provided that these derivatives exist. Collectively, these are called **higher-order derivatives**.

Leibniz notation can be used for higher-order derivatives as well. Table 3 summarizes the notation for higher-order derivatives.

TABLE 3

	Prime Notation		Leibniz Notation	
First Derivative	y'	$f'(x)$	$\dfrac{dy}{dx}$	$\dfrac{d}{dx}f(x)$
Second Derivative	y''	$f''(x)$	$\dfrac{d^2 y}{dx^2}$	$\dfrac{d^2}{dx^2}f(x)$
Third Derivative	y'''	$f'''(x)$	$\dfrac{d^3 y}{dx^3}$	$\dfrac{d^3}{dx^3}f(x)$
Fourth Derivative	$y^{(4)}$	$f^{(4)}(x)$	$\dfrac{d^4 y}{dx^4}$	$\dfrac{d^4}{dx^4}f(x)$
$\vdots$				
nth Derivative	$y^{(n)}$	$f^{(n)}(x)$	$\dfrac{d^n y}{dx^n}$	$\dfrac{d^n}{dx^n}f(x)$

EXAMPLE 7 **Finding Higher-Order Derivatives of a Power Function**

Find the second, third, and fourth derivatives of $y = 2x^3$.

Solution Since y is a power function, we use the Simple Power Rule and the Constant Multiple Rule to find each derivative. The first derivative is

$$y' = \frac{d}{dx}(2x^3) = 2 \cdot \frac{d}{dx}x^3 = 2 \cdot 3x^2 = 6x^2$$

The next three derivatives are

$$y'' = \frac{d^2}{dx^2}(2x^3) = \frac{d}{dx}(6x^2) = 6 \cdot \frac{d}{dx}x^2 = 6 \cdot 2x = 12x$$

$$y''' = \frac{d^3}{dx^3}(2x^3) = \frac{d}{dx}(12x) = 12$$

$$y^{(4)} = \frac{d^4}{dx^4}(2x^3) = \frac{d}{dx}12 = 0$$

■

All derivatives of this function f of order 4 or more equal 0. This result can be generalized. For a power function f of degree n, where n is a positive integer,

$$f(x) = x^n$$
$$f'(x) = nx^{n-1}$$
$$f''(x) = n(n-1)x^{n-2}$$
$$\vdots$$

NOTE If $n > 1$ is an integer, the product $n \cdot (n-1) \cdot (n-2) \cdot \cdots \cdot 3 \cdot 2 \cdot 1$ is often written $n!$ and is read, "n factorial." The **factorial symbol** $n!$ means $0! = 1$, $1! = 1$, and $n! = 1 \cdot 2 \cdot 3 \cdot \cdots \cdot (n-1) \cdot n$, where $n > 1$.

$$f^{(n)}(x) = n(n-1)(n-2)\ldots 3 \cdot 2 \cdot 1$$

The nth-order derivative of $f(x) = x^n$ is a constant function, so all derivatives of order greater than n equal 0.

It follows from this discussion that the nth derivative of a polynomial of degree n is a constant and that all derivatives of order $n + 1$ and higher equal 0.

NOW WORK Problem 41.

EXAMPLE 8   **Finding Higher-Order Derivatives**

Find the second and third derivatives of $y = (1 + x^2)e^x$.

Solution In Example 1, we found that $y' = (1 + x^2)e^x + 2xe^x = (x^2 + 2x + 1)e^x$. To find y'', we use the Product Rule with y'.

$$y'' = \frac{d}{dx}[(x^2 + 2x + 1)e^x] = (x^2 + 2x + 1)\left(\frac{d}{dx}e^x\right) + \left[\frac{d}{dx}(x^2 + 2x + 1)\right]e^x$$

$$\uparrow$$
$$\text{Product Rule}$$

$$= (x^2 + 2x + 1)e^x + (2x + 2)e^x = (x^2 + 4x + 3)e^x$$

$$y''' = \frac{d}{dx}[(x^2 + 4x + 3)e^x] = (x^2 + 4x + 3)\frac{d}{dx}e^x + \left[\frac{d}{dx}(x^2 + 4x + 3)\right]e^x$$

$$\uparrow$$
$$\text{Product Rule}$$

$$= (x^2 + 4x + 3)e^x + (2x + 4)e^x = (x^2 + 6x + 7)e^x$$ ■

NOW WORK Problem 45.

4 Work with Acceleration

For an object in rectilinear motion whose distance s from the origin at time t is $s = s(t)$, the derivative $s'(t)$ has a physical interpretation as the velocity of the object. The second derivative s'', which is the rate of change of the velocity s', is called *acceleration*.

DEFINITION Acceleration

When the distance s of an object from the origin at time t is given by a function $s = s(t)$, the first derivative $\dfrac{ds}{dt}$ is the velocity $v = v(t)$ of the object at time t.

The **acceleration** $a = a(t)$ of an object at time t is defined as the rate of change of velocity with respect to time. That is,

$$\boxed{a = a(t) = \frac{dv}{dt} = \frac{d}{dt}v = \frac{d}{dt}\left(\frac{ds}{dt}\right) = \frac{d^2s}{dt^2}}$$

IN WORDS Acceleration is the second derivative of distance with respect to time.

EXAMPLE 9   **Analyzing Vertical Motion**

A ball is propelled vertically upward from the ground with an initial velocity of 29.4 meters per second. The height s (in meters) of the ball above the ground is approximately $s = s(t) = -4.9t^2 + 29.4t$, where t is the number of seconds that elapse from the moment the ball is released.

(a) What is the velocity of the ball at time t? What is its velocity at $t = 1$ second?
(b) When will the ball reach its maximum height?
(c) What is the maximum height the ball reaches?
(d) What is the acceleration of the ball at any time t?
(e) How long is the ball in the air?
(f) What is the velocity of the ball upon impact with the ground?
(g) What is the total distance traveled by the ball?

Solution (a) Since $s = -4.9t^2 + 29.4t$, then

$$v = v(t) = \frac{ds}{dt} = -9.8t + 29.4$$

$$v(1) = -9.8 + 29.4 = 19.6$$

At $t = 1$ s, the velocity of the ball is 19.6 m/s.

(b) The ball reaches its maximum height when $v(t) = 0$.

$$v(t) = -9.8t + 29.4 = 0$$
$$9.8t = 29.4$$
$$t = 3$$

The ball reaches its maximum height after 3 seconds.

(c) The maximum height is

$$s = s(3) = -4.9(3^2) + 29.4(3) = 44.1$$

The maximum height of the ball is 44.1 m.

(d) The acceleration of the ball at any time t is

$$a = a(t) = \frac{d^2s}{dt^2} = \frac{dv}{dt} = \frac{d}{dt}(-9.8t + 29.4) = -9.8 \, \text{m/s}^2$$

(e) There are two ways to answer the question "How long is the ball in the air?" *First way:* Since it takes 3 s for the ball to reach its maximum altitude, it follows that it will take another 3 s to reach the ground, for a total time of 6 s in the air. *Second way:* When the ball reaches the ground, $s = s(t) = 0$. Solve for t:

$$s(t) = -4.9t^2 + 29.4t = 0$$
$$t(-4.9t + 29.4) = 0$$
$$t = 0 \quad \text{or} \quad t = \frac{29.4}{4.9} = 6$$

The ball is at ground level at $t = 0$ and at $t = 6$, so the ball is in the air for 6 seconds.

(f) Upon impact with the ground, $t = 6$ s. So the velocity is

$$v(6) = (-9.8)(6) + 29.4 = -29.4$$

Upon impact the direction of the ball is downward, and its speed is 29.4 m/s.

(g) The total distance traveled by the ball is

$$s(3) + s(3) = 2\,s(3) = 2(44.1) = 88.2 \, \text{m}$$

See Figure 25 for an illustration. ∎

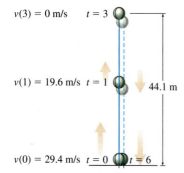

$v(3) = 0$ m/s $t = 3$

$v(1) = 19.6$ m/s $t = 1$ 44.1 m

$v(0) = 29.4$ m/s $t = 0$ $t = 6$

Figure 25

NOW WORK **Problem 83.**

In Example 9, the acceleration of the ball is constant. In fact, acceleration is the same for all falling objects at the same location, provided air resistance is not taken into account. In the sixteenth century, Galileo (1564–1642) discovered this by experimentation.* He also found that all falling bodies obey the law stating that the distance s they fall when dropped is proportional to the square of the time t it takes to fall that distance, and that the constant of proportionality c is the same for all objects. That is,

$$s = -ct^2$$

The velocity v of the falling object is

$$v = \frac{ds}{dt} = \frac{d}{dt}(-ct^2) = -2ct$$

NOTE The Earth is not perfectly round; it bulges slightly at the equator, and its mass is not distributed uniformly. As a result, the acceleration of a freely falling body varies slightly.

---

*In a famous legend, Galileo dropped a feather and a rock from the top of the Leaning Tower of Pisa, to show that the acceleration due to gravity is constant. He expected them to fall together, but he failed to account for air resistance that slowed the feather. In July 1971 Apollo 15 Astronaut David Scott repeated the experiment on the moon, where there is no air resistance. He dropped a hammer and a falcon feather from his shoulder height. Both hit the moon's surface at the same time. A video of this experiment may be found at the NASA website.

and its acceleration a is

$$a = \frac{dv}{dt} = \frac{d^2s}{dt^2} = -2c$$

which is a constant. Usually, we denote the constant $2c$ by g so $c = \frac{1}{2}g$. Then

$$\boxed{\quad a = -g \qquad v = -gt \qquad s = -\frac{1}{2}gt^2 \quad}$$

The number g is called the **acceleration due to gravity**. For our planet, g is approximately $32\,\text{ft/s}^2$, or $9.8\,\text{m/s}^2$. On the planet Jupiter, $g \approx 26.0\,\text{m/s}^2$, and on our moon, $g \approx 1.60\,\text{m/s}^2$.

2.4 Assess Your Understanding

Concepts and Vocabulary

1. *True or False* The derivative of a product is the product of the derivatives.

2. If $F(x) = f(x)g(x)$, then $F'(x) = $ _____ .

3. *True or False* $\dfrac{d}{dx}x^n = nx^{n+1}$, for any integer n.

4. If f and $g \neq 0$ are two differentiable functions, then
$$\frac{d}{dx}\left(\frac{f(x)}{g(x)}\right) = \underline{\quad\quad}.$$

5. *True or False* $f(x) = \dfrac{e^x}{x^2}$ can be differentiated using the
Quotient Rule or by writing $f(x) = \dfrac{e^x}{x^2} = x^{-2}e^x$ and using the
Product Rule.

6. If $g \neq 0$ is a differentiable function, then $\dfrac{d}{dx}\left[\dfrac{1}{g(x)}\right] = $ _____ .

7. If $f(x) = x$, then $f''(x) = $ _____ .

8. When an object in rectilinear motion is modeled by the function $s = s(t)$, then the acceleration a of the object at time t is given by $a = a(t) = $ _____ .

Skill Building

In Problems 9–40, find the derivative of each function.

9. $f(x) = xe^x$

10. $f(x) = x^2e^x$

11. $f(x) = x^2(x^3 - 1)$

12. $f(x) = x^4(x + 5)$

13. $f(x) = (3x^2 - 5)(2x + 1)$

14. $f(x) = (3x - 2)(4x + 5)$

15. $s(t) = (2t^5 - t)(t^3 - 2t + 1)$

16. $F(u) = (u^4 - 3u^2 + 1)(u^2 - u + 2)$

17. $f(x) = (x^3 + 1)(e^x + 1)$

18. $f(x) = (x^2 + 1)(e^x + x)$

19. $g(s) = \dfrac{2s}{s + 1}$

20. $F(z) = \dfrac{z + 1}{2z}$

21. $G(u) = \dfrac{1 - 2u}{1 + 2u}$

22. $f(w) = \dfrac{1 - w^2}{1 + w^2}$

23. $f(x) = \dfrac{4x^2 - 2}{3x + 4}$

24. $f(x) = \dfrac{-3x^3 - 1}{2x^2 + 1}$

25. $f(w) = \dfrac{1}{w^3 - 1}$

26. $g(v) = \dfrac{1}{v^2 + 5v - 1}$

27. $s(t) = t^{-3}$

28. $G(u) = u^{-4}$

29. $f(x) = -\dfrac{4}{e^x}$

30. $f(x) = \dfrac{3}{4e^x}$

31. $f(x) = \dfrac{10}{x^4} + \dfrac{3}{x^2}$

32. $f(x) = \dfrac{2}{x^5} - \dfrac{3}{x^3}$

33. $f(x) = 3x^3 - \dfrac{1}{3x^2}$

34. $f(x) = x^5 - \dfrac{5}{x^5}$

35. $s(t) = \dfrac{1}{t} - \dfrac{1}{t^2} + \dfrac{1}{t^3}$

36. $s(t) = \dfrac{1}{t + 2} + \dfrac{1}{t^2} + \dfrac{1}{t^3}$

37. $f(x) = \dfrac{e^x}{x^2}$

38. $f(x) = \dfrac{x^2}{e^x}$

39. $f(x) = \dfrac{x^2 + 1}{xe^x}$

40. $f(x) = \dfrac{xe^x}{x^2 - x}$

In Problems 41–54, find f' and f'' for each function.

41. $f(x) = 3x^2 + x - 2$

42. $f(x) = -5x^2 - 3x$

43. $f(x) = e^x - 3$

44. $f(x) = x - e^x$

45. $f(x) = (x + 5)e^x$

46. $f(x) = 3x^4e^x$

47. $f(x) = (2x + 1)(x^3 + 5)$

48. $f(x) = (3x - 5)(x^2 - 2)$

49. $f(x) = x + \dfrac{1}{x}$

50. $f(x) = x - \dfrac{1}{x}$

51. $f(t) = \dfrac{t^2 - 1}{t}$

52. $f(u) = \dfrac{u + 1}{u}$

53. $f(x) = \dfrac{e^x + x}{x}$

54. $f(x) = \dfrac{e^x}{x}$

55. Find y' and y'' for **(a)** $y = \dfrac{1}{x}$ and **(b)** $y = \dfrac{2x - 5}{x}$.

56. Find $\dfrac{dy}{dx}$ and $\dfrac{d^2y}{dx^2}$ for **(a)** $y = \dfrac{5}{x^2}$ and **(b)** $y = \dfrac{2 - 3x}{x}$.

Rectilinear Motion *In Problems 57–60, find the velocity v and acceleration a of an object in rectilinear motion whose distance s from the origin at time t is modeled by $s = s(t)$.*

57. $s(t) = 16t^2 + 20t$

58. $s(t) = 16t^2 + 10t + 1$

59. $s(t) = 4.9t^2 + 4t + 4$

60. $s(t) = 4.9t^2 + 5t$

---

1. = NOW WORK problem   〽 = Graphing technology recommended   CAS = Computer Algebra System recommended

In Problems 61–68, find the indicated derivative.

61. $f^{(4)}(x)$ if $f(x) = x^3 - 3x^2 + 2x - 5$

62. $f^{(5)}(x)$ if $f(x) = 4x^3 + x^2 - 1$

63. $\dfrac{d^8}{dt^8}\left(\dfrac{1}{8}t^8 - \dfrac{1}{7}t^7 + t^5 - t^3\right)$ **64.** $\dfrac{d^6}{dt^6}(t^6 + 5t^5 - 2t + 4)$

65. $\dfrac{d^7}{du^7}(e^u + u^2)$ **66.** $\dfrac{d^{10}}{du^{10}}(2e^u)$

67. $\dfrac{d^5}{dx^5}(-e^x)$ **68.** $\dfrac{d^8}{dx^8}(12x - e^x)$

In Problems 69–72:

(a) Find the slope of the tangent line for each function f at the given point.

(b) Find an equation of the tangent line to the graph of each function f at the given point.

(c) Find the points, if any, where the graph of the function has a horizontal tangent line.

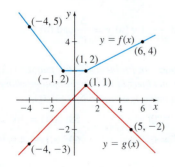 *(d) Graph each function, the tangent line found in (b), and any tangent lines found in (c) on the same set of axes.*

69. $f(x) = \dfrac{x^2}{x - 1}$ at $\left(-1, -\dfrac{1}{2}\right)$ **70.** $f(x) = \dfrac{x}{x + 1}$ at $(0, 0)$

71. $f(x) = \dfrac{x^3}{x + 1}$ at $\left(1, \dfrac{1}{2}\right)$ **72.** $f(x) = \dfrac{x^2 + 1}{x}$ at $\left(2, \dfrac{5}{2}\right)$

In Problems 73–80:

(a) Find the points, if any, at which the graph of each function f has a horizontal tangent line.

(b) Find an equation for each horizontal tangent line.

(c) Solve the inequality $f'(x) > 0$.

(d) Solve the inequality $f'(x) < 0$.

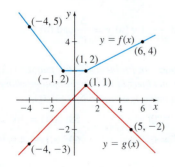 *(e) Graph f and any horizontal lines found in (b) on the same set of axes.*

(f) Describe the graph of f for the results obtained in (c) and (d).

73. $f(x) = (x + 1)(x^2 - x - 11)$ **74.** $f(x) = (3x^2 - 2)(2x + 1)$

75. $f(x) = \dfrac{x^2}{x + 1}$ **76.** $f(x) = \dfrac{x^2 + 1}{x}$

77. $f(x) = xe^x$ **78.** $f(x) = x^2 e^x$

79. $f(x) = \dfrac{x^2 - 3}{e^x}$ **80.** $f(x) = \dfrac{e^x}{x^2 + 1}$

In Problems 81 and 82, use the graphs to determine each derivative.

81. Let $u(x) = f(x) \cdot g(x)$ and $v(x) = \dfrac{g(x)}{f(x)}$.

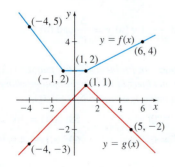

(a) $u'(0)$ **(b)** $u'(4)$

(c) $v'(-2)$ **(d)** $v'(6)$

(e) $\dfrac{d}{dx}\left(\dfrac{1}{f(x)}\right)$ at $x = -2$ **(f)** $\dfrac{d}{dx}\left(\dfrac{1}{g(x)}\right)$ at $x = 4$

82. Let $F(t) = f(t) \cdot g(t)$ and $G(t) = \dfrac{g(t)}{f(t)}$.

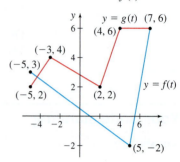

(a) $F'(0)$ **(b)** $F'(3)$

(c) $F'(-4)$ **(d)** $G'(-2)$

(e) $G'(-1)$ **(f)** $\dfrac{d}{dt}\left(\dfrac{1}{f(t)}\right)$ at $t = 3$

Applications and Extensions

83. Movement of an Object An object is propelled vertically upward from the ground with an initial velocity of 39.2 m/s. The distance s (in meters) of the object from the ground after t seconds is $s = s(t) = -4.9t^2 + 39.2t$.

(a) What is the velocity of the object at time t?

(b) When will the object reach its maximum height?

(c) What is the maximum height?

(d) What is the acceleration of the object at any time t?

(e) How long is the object in the air?

(f) What is the velocity of the object upon impact with the ground?

(g) What is the total distance traveled by the object?

84. Movement of a Ball A ball is thrown vertically upward from a height of 6 ft with an initial velocity of 80 ft/s. The distance s (in feet) of the ball from the ground after t seconds is $s = s(t) = 6 + 80t - 16t^2$.

(a) What is the velocity of the ball after 2 seconds?

(b) When will the ball reach its maximum height?

(c) What is the maximum height the ball reaches?

(d) What is the acceleration of the ball at any time t?

(e) How long is the ball in the air?

(f) What is the velocity of the ball upon impact with the ground?

(g) What is the total distance traveled by the ball?

85. Environmental Cost The cost C, in thousands of dollars, for the removal of a pollutant from a certain lake is given by the function

$$C(x) = \dfrac{5x}{110 - x}, \text{ where } x \text{ is the percent of pollutant removed.}$$

(a) What is the domain of C?

(b) Graph C.

(c) What is the cost to remove 80% of the pollutant?

(d) Find $C'(x)$, the rate of change of the cost C with respect to the amount of pollutant removed.

(e) Find the rate of change of the cost for removing 40%, 60%, 80%, and 90% of the pollutant.

(f) Interpret the answers found in (e).

86. **Investing in Fine Art**   The value V of a painting t years after it is purchased is modeled by the function

$$V(t) = \frac{100t^2 + 50}{t} + 400, \; 1 \le t \le 5$$

(a) Find the rate of change in the value V with respect to time.

(b) What is the rate of change in value after 1 year?

(c) What is the rate of change in value after 3 years?

(d) Interpret the answers in (b) and (c).

87. **Drug Concentration**   The concentration of a drug in a patient's blood t hours after injection is given by the

function $f(t) = \dfrac{0.4t}{2t^2 + 1}$ (in milligrams per liter).

(a) Find the rate of change of the concentration with respect to time.

(b) What is the rate of change of the concentration after 10 min? After 30 min? After 1 hour?

(c) Interpret the answers found in (b).

(d) Graph f for the first 5 hours after administering the drug.

(e) From the graph, approximate the time (in minutes) at which the concentration of the drug is highest. What is the highest concentration of the drug in the patient's blood?

88. **Population Growth**   A population of 1000 bacteria is introduced into a culture and grows in number according to the formula

$$P(t) = 1000\left(1 + \frac{4t}{100 + t^2}\right), \text{ where } t \text{ is measured in hours.}$$

(a) Find the rate of change in population with respect to time.

(b) What is the rate of change in population at $t = 1$, $t = 2$, $t = 3$, and $t = 4$?

(c) Interpret the answers found in (b).

(d) Graph $P = P(t)$, $0 \le t \le 20$.

(e) From the graph, approximate the time (in hours) when the population is the greatest. What is the maximum population of the bacteria in the culture?

89. **Economics**   The price-demand function for a popular e-book is given by $D(p) = \dfrac{100,000}{p^2 + 10p + 50}$, $5 \le p \le 20$, where $D = D(p)$ is the quantity demanded at the price p dollars.

(a) Find $D'(p)$, the rate of change of demand with respect to price.

(b) Find $D'(5)$, $D'(10)$, and $D'(15)$.

(c) Interpret the results found in (b).

90. **Intensity of Light**   The intensity of illumination I on a surface is inversely proportional to the square of the distance r from the surface to the source of light. If the intensity is 1000 units when the distance is 1 meter from the light, find the rate of change of the intensity with respect to the distance when the source is 10 meters from the surface.

91. **Ideal Gas Law**   The Ideal Gas Law, used in chemistry and thermodynamics, relates the pressure p, the volume V, and the absolute temperature T (in Kelvin) of a gas, using the equation $pV = nRT$, where n is the amount of gas (in moles) and $R = 8.31$ is the ideal gas constant. In an experiment, a spherical gas container of radius r meters is placed in a pressure chamber and is slowly compressed while keeping its temperature at 273 K.

(a) Find the rate of change of the pressure p with respect to the radius r of the chamber.

(*Hint:* The volume V of a sphere is $V = \dfrac{4}{3}\pi r^3$.)

(b) Interpret the sign of the answer found in (a).

(c) If the sphere contains 1.0 mol of gas, find the rate of change of the pressure when $r = \dfrac{1}{4}$ m. (*Note:* The metric unit of pressure is the pascal, Pa).

92. **Body Density**   The density ρ of an object is its mass m divided by its volume V; that is, $\rho = \dfrac{m}{V}$. If a person dives below the surface of the ocean, the water pressure on the diver will steadily increase, compressing the diver and therefore increasing body density. Suppose the diver is modeled as a sphere of radius r.

(a) Find the rate of change of the diver's body density with respect to the radius r of the sphere. (*Hint:* The volume V of a sphere is $V = \dfrac{4}{3}\pi r^3$.)

(b) Interpret the sign of the answer found in (a).

(c) Find the rate of change of the diver's body density when the radius is 45 cm and the mass is 80,000 g (80 kg).

Jerk and Snap   *Problems 93–96 use the following discussion:*
Suppose that an object is moving in rectilinear motion so that its distance s from the origin at time t is given by the function $s = s(t)$. The velocity $v = v(t)$ of the object at time t is the rate of change of the distance s with respect to time, namely,

$$v = v(t) = \frac{ds}{dt}.$$ The acceleration $a = a(t)$ of the object at time t is the rate of change of the velocity with respect to time, namely,

$$a = a(t) = \frac{dv}{dt} = \frac{d}{dt}\left(\frac{ds}{dt}\right) = \frac{d^2s}{dt^2}.$$

There are also physical interpretations to the third derivative and the fourth derivative of $s = s(t)$. The **jerk** $J = J(t)$ of the object at time t is the rate of change in the acceleration a with respect to time; that is,

$$J = J(t) = \frac{da}{dt} = \frac{d}{dt}\left(\frac{dv}{dt}\right) = \frac{d^2v}{dt^2} = \frac{d^3s}{dt^3}.$$

The **snap** $S = S(t)$ of the object at time t is the rate of change in the jerk J with respect to time; that is,

$$S = S(t) = \frac{dJ}{dt} = \frac{d^2a}{dt^2} = \frac{d^3v}{dt^3} = \frac{d^4s}{dt^4}.$$

Engineers take jerk into consideration when designing elevators, aircraft, and cars. In these cases, they try to minimize jerk, making for a smooth ride. But when designing thrill rides, such as roller coasters, the jerk is increased, making for an exciting experience.

93. Rectilinear Motion As an object in rectilinear motion moves, its distance s from the origin at time t is given by $s = s(t) = t^3 - t + 1$, where s is in meters and t is in seconds.

   (a) Find the velocity v, acceleration a, jerk J, and snap S of the object at time t.

   (b) When is the velocity of the object 0 m/s?

   (c) Find the acceleration of the object at $t = 2$ and at $t = 5$.

   (d) Does the jerk of the object ever equal 0 m/s³?

   (e) How would you interpret the snap for this object in rectilinear motion?

94. Rectilinear Motion As an object in rectilinear motion moves, its distance s from the origin at time t is given by $s = s(t) = \frac{1}{6}t^4 - t^2 + \frac{1}{2}t + 4$, where s is in meters and t is in seconds.

   (a) Find the velocity v, acceleration a, jerk J, and snap S of the object at any time t.

   (b) Find the velocity of the object at $t = 0$ and at $t = 3$.

   (c) Find the acceleration of the object at $t = 0$. Interpret your answer.

   (d) Is the jerk of the object constant? In your own words, explain what the jerk says about the acceleration of the object.

   (e) How would you interpret the snap for this object in rectilinear motion?

95. Elevator Ride Quality The ride quality of an elevator depends on several factors, two of which are acceleration and jerk. In a study of 367 persons riding in a 1600-kg elevator that moves at an average speed of 4 m/s, the majority of riders were comfortable in an elevator with vertical motion given by

$$s(t) = 4t + 0.8t^2 + 0.333t^3$$

   (a) Find the acceleration that the riders found acceptable.

   (b) Find the jerk that the riders found acceptable.

Source: *Elevator Ride Quality*, January 2007, http://www. lift-report.de/index.php/news/176/368/Elevator-Ride-Quality.

96. Elevator Ride Quality In a hospital, the effects of high acceleration or jerk may be harmful to patients, so the acceleration and jerk need to be lower than in standard elevators. It has been determined that a 1600-kg elevator that is installed in a hospital and that moves at an average speed of 4 m/s should have vertical motion

$$s(t) = 4t + 0.55t^2 + 0.1167t^3$$

   (a) Find the acceleration of a hospital elevator.

   (b) Find the jerk of a hospital elevator.

Source: *Elevator Ride Quality*, January 2007, http://www.lift-report.de/index.php/news/176/368/Elevator-Ride-Quality.

97. Current Density in a Wire The current density J in a wire is a measure of how much an electrical current is compressed as it flows through a wire and is modeled by the function $J(A) = \dfrac{I}{A}$, where I is the current (in amperes) and A is the cross-sectional area of the wire. In practice, current density, rather than merely current, is often important. For example, superconductors lose their superconductivity if the current density is too high.

   (a) As current flows through a wire, it heats the wire, causing it to expand in area A. If a constant current is maintained in a cylindrical wire, find the rate of change of the current density J with respect to the radius r of the wire.

   (b) Interpret the sign of the answer found in (a).

   (c) Find the rate of change of current density with respect to the radius r when a current of 2.5 amps flows through a wire of radius $r = 0.50$ millimeter.

98. Derivative of a Reciprocal, Function Prove that if a function g is differentiable, then $\dfrac{d}{dx}\left[\dfrac{1}{g(x)}\right] = -\dfrac{g'(x)}{[g(x)]^2}$, provided $g(x) \neq 0$.

99. Extended Product Rule Show that if f, g, and h are differentiable functions, then

$$\frac{d}{dx}[f(x)g(x)h(x)] = f(x)g(x)h'(x) + f(x)g'(x)h(x)$$
$$+ f'(x)g(x)h(x)$$

From this, deduce that

$$\frac{d}{dx}[f(x)]^3 = 3[f(x)]^2 f'(x)$$

In Problems 100–105, use the Extended Product Rule (Problem 99) to find y'.

100. $y = (x^2 + 1)(x - 1)(x + 5)$

101. $y = (x - 1)(x^2 + 5)(x^3 - 1)$

102. $y = (x^4 + 1)^3$ **103.** $y = (x^3 + 1)^3$

104. $y = (3x + 1)\left(1 + \dfrac{1}{x}\right)(x^{-5} + 1)$

105. $y = \left(1 - \dfrac{1}{x}\right)\left(1 - \dfrac{1}{x^2}\right)\left(1 - \dfrac{1}{x^3}\right)$

106. (Further) Extended Product Rule Write a formula for the derivative of the product of four differentiable functions. That is, find a formula for $\dfrac{d}{dx}[f_1(x)f_2(x)f_3(x)f_4(x)]$. Also find a formula for $\dfrac{d}{dx}[f(x)]^4$.

107. If f and g are differentiable functions, show that

if $F(x) = \dfrac{1}{f(x)g(x)}$, then

$$F'(x) = -F(x)\left[\frac{f'(x)}{f(x)} + \frac{g'(x)}{g(x)}\right]$$

provided $f(x) \neq 0$, $g(x) \neq 0$.

108. Higher-Order Derivatives   If $f(x) = \dfrac{1}{1-x}$, find a formula for the nth derivative of f. That is, find $f^{(n)}(x)$.

109. Let $f(x) = \dfrac{x^6 - x^4 + x^2}{x^4 + 1}$. Rewrite f in the form $(x^4 + 1)f(x) = x^6 - x^4 + x^2$. Now find $f'(x)$ without using the quotient rule.

110. If f and g are differentiable functions with $f \neq -g$, find the derivative of $\dfrac{fg}{f+g}$.

CAS 111. $f(x) = \dfrac{2x}{x+1}$.

(a) Use technology to find $f'(x)$.

(b) Simplify f' to a single fraction using either algebra or a CAS.

(c) Use technology to find $f^{(5)}(x)$. (*Hint:* Your CAS may have a method for finding higher-order derivatives without finding other derivatives first.)

Challenge Problems

112. Suppose f and g have derivatives up to the fourth order. Find the first four derivatives of the product fg and simplify the answers. In particular, show that the fourth derivative is

$$\frac{d^4}{dx^4}(fg) = f^{(4)}g + 4f^{(3)}g^{(1)} + 6f^{(2)}g^{(2)} + 4f^{(1)}g^{(3)} + fg^{(4)}$$

Identify a pattern for the higher-order derivatives of fg.

113. Suppose $f_1(x), \ldots, f_n(x)$ are differentiable functions.

(a) Find $\dfrac{d}{dx}[f_1(x) \cdot \cdots \cdot f_n(x)]$.

(b) Find $\dfrac{d}{dx}\left[\dfrac{1}{f_1(x) \cdot \cdots \cdot f_n(x)}\right]$.

114. Let a, b, c, and d be real numbers. Define

$$\begin{vmatrix} a & b \\ c & d \end{vmatrix} = ad - bc$$

This is called a 2×2 **determinant** and it arises in the study of linear equations. Let $f_1(x)$, $f_2(x)$, $f_3(x)$, and $f_4(x)$ be differentiable and let

$$D(x) = \begin{vmatrix} f_1(x) & f_2(x) \\ f_3(x) & f_4(x) \end{vmatrix}$$

Show that

$$D'(x) = \begin{vmatrix} f_1'(x) & f_2'(x) \\ f_3(x) & f_4(x) \end{vmatrix} + \begin{vmatrix} f_1(x) & f_2(x) \\ f_3'(x) & f_4'(x) \end{vmatrix}$$

115. Let $f_0(x) = x - 1$

$$f_1(x) = 1 + \frac{1}{x-1}$$

$$f_2(x) = 1 + \cfrac{1}{1 + \cfrac{1}{x-1}}$$

$$f_3(x) = 1 + \cfrac{1}{1 + \cfrac{1}{1 + \cfrac{1}{x-1}}}$$

(a) Write f_1, f_2, f_3, f_4 and f_5 in the form $\dfrac{ax+b}{cx+d}$.

(b) Using the results from (a), write the sequence of numbers representing the coefficients of x in the numerator, beginning with $f_0(x) = x - 1$.

(c) Write the sequence in (b) as a recursive sequence. (*Hint:* Look at the sums of consecutive terms.)

(d) Find f_0', f_1', f_2', f_3', f_4', and f_5'.

2.5 The Derivative of the Trigonometric Functions

OBJECTIVES *When you finish this section, you should be able to:*

1 Differentiate trigonometric functions (p. 185)

NEED TO REVIEW? The trigonometric functions are discussed in Section P.6, pp. 49–56.

1 Differentiate Trigonometric Functions

To find the derivatives of $y = \sin x$ and $y = \cos x$, we use the limits

$$\lim_{\theta \to 0} \frac{\sin \theta}{\theta} = 1 \quad \text{and} \quad \lim_{\theta \to 0} \frac{\cos \theta - 1}{\theta} = 0$$

that were established in Section 1.4.

THEOREM Derivative of $y = \sin x$

The derivative of $y = \sin x$ is $y' = \cos x$. That is,

$$y' = \frac{d}{dx}\sin x = \cos x$$

Proof

NEED TO REVIEW? Basic trigonometric identities are discussed in Appendix A.4, pp. A-32 to A-35.

$y' = \lim\limits_{h\to 0} \dfrac{\sin(x+h) - \sin x}{h}$ The definition of a derivative.

$= \lim\limits_{h\to 0} \dfrac{\sin x \cos h + \sin h \cos x - \sin x}{h}$ $\sin(A+B) = \sin A \cos B + \sin B \cos A$.

$= \lim\limits_{h\to 0}\left[\dfrac{\sin x \cos h - \sin x}{h} + \dfrac{\sin h \cos x}{h}\right]$ Rearrange terms.

$= \lim\limits_{h\to 0}\left[\sin x \cdot \dfrac{\cos h - 1}{h} + \dfrac{\sin h}{h}\cdot \cos x\right]$ Factor.

$= \left[\lim\limits_{h\to 0}\sin x\right]\left[\lim\limits_{h\to 0}\dfrac{\cos h - 1}{h}\right] + \left[\lim\limits_{h\to 0}\cos x\right]\left[\lim\limits_{h\to 0}\dfrac{\sin h}{h}\right]$ Use properties of limits.

$= \sin x \cdot 0 + \cos x \cdot 1 = \cos x$ $\lim\limits_{\theta\to 0}\dfrac{\cos\theta - 1}{\theta} = 0; \quad \lim\limits_{\theta\to 0}\dfrac{\sin\theta}{\theta} = 1.$ ∎

The geometry of the derivative $\frac{d}{dx}\sin x = \cos x$ is shown in Figure 26. On the graph of $f(x) = \sin x$, the horizontal tangents are marked as well as the tangent lines that have slopes of 1 and -1. The derivative function is plotted on the second graph and those points are connected with a smooth curve.

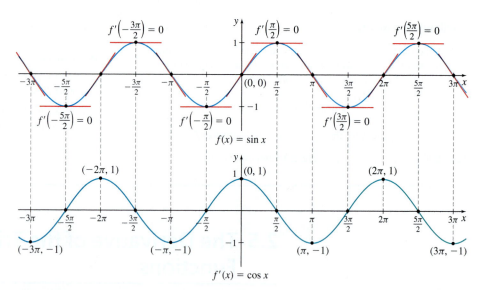

Figure 26

EXAMPLE 1   Differentiating the Sine Function

Find y' if:

(a) $y = x + 4\sin x$ **(b)** $y = x^2 \sin x$ **(c)** $y = \dfrac{\sin x}{x}$ **(d)** $y = e^x \sin x$

Solution (a) $y' = \dfrac{d}{dx}(x + 4\sin x) = \dfrac{d}{dx}x + \dfrac{d}{dx}(4\sin x) = 1 + 4\dfrac{d}{dx}\sin x = 1 + 4\cos x$

(b) $y' = \dfrac{d}{dx}(x^2 \sin x) = x^2\left[\dfrac{d}{dx}\sin x\right] + \left[\dfrac{d}{dx}x^2\right]\sin x = x^2\cos x + 2x\sin x$

(c) $y' = \dfrac{d}{dx}\left(\dfrac{\sin x}{x}\right) = \dfrac{\left[\dfrac{d}{dx}\sin x\right]\cdot x - \sin x \cdot \left[\dfrac{d}{dx}x\right]}{x^2} = \dfrac{x\cos x - \sin x}{x^2}$

(d) $y' = \dfrac{d}{dx}(e^x \sin x) = e^x \dfrac{d}{dx}\sin x + \left(\dfrac{d}{dx}e^x\right)\sin x = e^x \cos x + e^x \sin x$

$\quad\quad = e^x(\cos x + \sin x)$ ∎

NOW WORK Problems 5 and 29.

THEOREM Derivative of $y = \cos x$

The derivative of $y = \cos x$ is

$$y' = \dfrac{d}{dx}\cos x = -\sin x$$

You are asked to prove this in Problem 75.

EXAMPLE 2 Differentiating Trigonometric Functions

Find the derivative of each function:

(a) $f(x) = x^2 \cos x$ (b) $g(\theta) = \dfrac{\cos\theta}{1 - \sin\theta}$ (c) $F(t) = \dfrac{e^t}{\cos t}$

Solution (a) $f'(x) = \dfrac{d}{dx}(x^2 \cos x) = x^2 \dfrac{d}{dx}\cos x + \left(\dfrac{d}{dx}x^2\right)(\cos x)$

$\quad\quad\quad = x^2(-\sin x) + 2x\cos x = 2x\cos x - x^2 \sin x$

(b) $g'(\theta) = \dfrac{d}{d\theta}\left(\dfrac{\cos\theta}{1 - \sin\theta}\right) = \dfrac{\left(\dfrac{d}{d\theta}\cos\theta\right)(1 - \sin\theta) - (\cos\theta)\left[\dfrac{d}{d\theta}(1 - \sin\theta)\right]}{(1 - \sin\theta)^2}$

$\quad\quad = \dfrac{-\sin\theta\,(1 - \sin\theta) - \cos\theta\,(-\cos\theta)}{(1 - \sin\theta)^2} = \dfrac{-\sin\theta + \sin^2\theta + \cos^2\theta}{(1 - \sin\theta)^2}$

$\quad\quad = \dfrac{-\sin\theta + 1}{(1 - \sin\theta)^2} = \dfrac{1}{1 - \sin\theta}$

(c) $F'(t) = \dfrac{d}{dt}\left(\dfrac{e^t}{\cos t}\right) = \dfrac{\left(\dfrac{d}{dt}e^t\right)(\cos t) - e^t\left(\dfrac{d}{dt}\cos t\right)}{\cos^2 t} = \dfrac{e^t \cos t - e^t(-\sin t)}{\cos^2 t}$

$\quad\quad = \dfrac{e^t(\cos t + \sin t)}{\cos^2 t}$ ∎

NOW WORK Problem 13.

EXAMPLE 3 Identifying Horizontal Tangent Lines

Find all points on the graph of $f(x) = x + \sin x$ where the tangent line is horizontal.

Solution Since tangent lines are horizontal at points on the graph of f where $f'(x) = 0$, we begin by finding f':

$$f'(x) = 1 + \cos x$$

Now we solve the equation $f'(x) = 0$.

$$f'(x) = 1 + \cos x = 0$$
$$\cos x = -1$$
$$x = (2k + 1)\pi \quad\quad\quad \text{where } k \text{ is an integer}$$

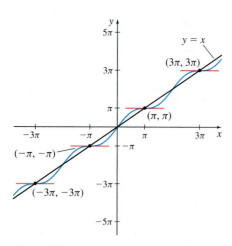

Figure 27 $f(x) = x + \sin x$

The graph of $f(x) = x + \sin x$ has a horizontal tangent line at each of the points $((2k+1)\pi, (2k+1)\pi)$, k an integer. See Figure 27 for the graph of f with the horizontal tangent lines marked. Notice that each of the points with a horizontal tangent line lies on the line $y = x$. ∎

NOW WORK Problem 57.

The derivatives of the remaining four trigonometric functions are obtained using trigonometric identities and basic derivative rules. We establish the formula for the derivative of $y = \tan x$ in Example 4. You are asked to prove formulas for the derivative of the secant function, the cosecant function, and the cotangent function in the exercises. (See Problems 76–78.)

EXAMPLE 4   Differentiating $y = \tan x$

Show that the derivative of $y = \tan x$ is

$$y' = \frac{d}{dx} \tan x = \sec^2 x$$

Solution

$$y' = \frac{d}{dx} \tan x \underset{\substack{\uparrow \\ \text{Identity}}}{=} \frac{d}{dx} \frac{\sin x}{\cos x} \underset{\substack{\uparrow \\ \text{Quotient Rule}}}{=} \frac{\left[\dfrac{d}{dx}\sin x\right]\cos x - \sin x\left[\dfrac{d}{dx}\cos x\right]}{\cos^2 x}$$

$$= \frac{\cos x \cdot \cos x - \sin x \cdot (-\sin x)}{\cos^2 x} = \frac{\cos^2 x + \sin^2 x}{\cos^2 x} = \frac{1}{\cos^2 x} = \sec^2 x \quad ∎$$

NOW WORK Problem 15.

Table 4 lists the derivatives of the six trigonometric functions along with the domain of each derivative.

TABLE 4

Derivative Function	Domain of the Derivative Function
$\dfrac{d}{dx}\sin x = \cos x$	$(-\infty, \infty)$
$\dfrac{d}{dx}\cos x = -\sin x$	$(-\infty, \infty)$
$\dfrac{d}{dx}\tan x = \sec^2 x$	$\left\{x \mid x \neq \dfrac{2k+1}{2}\pi, k \text{ an integer}\right\}$
$\dfrac{d}{dx}\cot x = -\csc^2 x$	$\{x \mid x \neq k\pi, k \text{ an integer}\}$
$\dfrac{d}{dx}\csc x = -\csc x \cot x$	$\{x \mid x \neq k\pi, k \text{ an integer}\}$
$\dfrac{d}{dx}\sec x = \sec x \tan x$	$\left\{x \mid x \neq \dfrac{2k+1}{2}\pi, k \text{ an integer}\right\}$

NOTE If the trigonometric function begins with the letter c, that is, cosine, cotangent, or cosecant, then its derivative has a minus sign.

Evaluating the Second Derivative of a Trigonometric
EXAMPLE 5 **Function**

Find $f''\left(\dfrac{\pi}{4}\right)$ if $f(x) = \sec x$.

Solution If $f(x) = \sec x$, then $f'(x) = \sec x \tan x$ and

$$f''(x) = \frac{d}{dx}(\sec x \tan x) = \sec x \left(\frac{d}{dx} \tan x \right) + \left(\frac{d}{dx} \sec x \right) \tan x$$

↑
Use the Product Rule.

$$= \sec x \cdot \sec^2 x + (\sec x \tan x) \tan x = \sec^3 x + \sec x \tan^2 x$$

Now,

$$f''\left(\frac{\pi}{4}\right) = \sec^3 \left(\frac{\pi}{4}\right) + \sec \left(\frac{\pi}{4}\right) \tan^2 \left(\frac{\pi}{4}\right) = (\sqrt{2})^3 + \sqrt{2} \cdot 1^2 = 2\sqrt{2} + \sqrt{2} = 3\sqrt{2}$$

↑
$\sec \dfrac{\pi}{4} = \sqrt{2}; \ \tan \dfrac{\pi}{4} = 1$

■

NOW WORK **Problem 45.**

Simple harmonic motion is a repetitive motion that can be modeled by a trigonometric function. A swinging pendulum and an oscillating spring are examples of simple harmonic motion.

EXAMPLE 6 **Analyzing Simple Harmonic Motion**

An object hangs on a spring, making it 2 m long in its equilibrium position. See Figure 28. If the object is pulled down 1 m and released, it will oscillate up and down. The length l of the spring after t seconds is modeled by the function $l(t) = 2 + \cos t$.

(a) How does the length of the spring vary?

(b) Find the velocity of the object.

(c) At what position is the speed of the object a maximum?

(d) Find the acceleration of the object.

(e) At what position is the acceleration equal to 0?

Solution (a) Since $l(t) = 2 + \cos t$ and $-1 \le \cos t \le 1$, the length of the spring oscillates between 1 m and 3 m.

(b) The velocity v of the object is

$$v = l'(t) = \frac{d}{dt}(2 + \cos t) = -\sin t$$

(c) Speed is the magnitude of velocity. Since $v = -\sin t$, the speed of the object is $|v| = |-\sin t| = |\sin t|$. Since $-1 \le \sin t \le 1$, the object moves the fastest when $|v| = |\sin t| = 1$. This occurs when $\sin t = \pm 1$ or, equivalently, when $\cos t = 0$. So, the speed is a maximum when $l(t) = 2$, that is, when the spring is at the equilibrium position.

(d) The acceleration a of the object is given by

$$a = l''(t) = \frac{d}{dt}l'(t) = \frac{d}{dt}(-\sin t) = -\cos t$$

(e) Since $a = -\cos t$, the acceleration is zero when $\cos t = 0$. This occurs when $l(t) = 2$, that is, when the spring is at the equilibrium position. At this time, the speed is maximum. ■

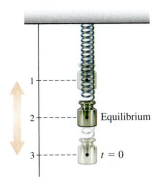
DF **Figure 28**

Figure 29 shows the graphs of the length of the spring $y = l(t)$, the velocity $y = v(t)$, and the acceleration $y = a(t)$.

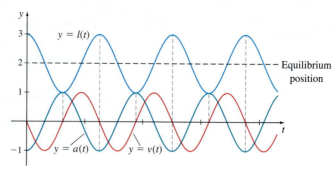

Figure 29

NOW WORK Problem 65.

2.5 Assess Your Understanding

Concepts and Vocabulary

1. *True or False* $\dfrac{d}{dx} \cos x = \sin x$

2. *True or False* $\dfrac{d}{dx} \tan x = \cot x$

3. *True or False* $\dfrac{d^2}{dx^2} \sin x = -\sin x$

4. *True or False* $\dfrac{d}{dx} \sin \dfrac{\pi}{3} = \cos \dfrac{\pi}{3}$

Skill Building

In Problems 5–8, find the derivative of f at c.

5. $f(x) = x - \sin x, \quad c = \pi$

6. $f(\theta) = 2 \sin \theta + \cos \theta, \quad c = \dfrac{\pi}{2}$

7. $f(\theta) = \dfrac{\cos \theta}{1 + \sin \theta}, \quad c = \dfrac{\pi}{3}$

8. $f(x) = \dfrac{\sin x}{1 + \cos x}, \quad c = \dfrac{5\pi}{6}$

In Problems 9–38, find y'.

9. $y = 3 \sin \theta - 2 \cos \theta$

10. $y = 4 \tan \theta + \sin \theta$

11. $y = \sin x \cos x$

12. $y = \cot x \tan x$

13. $y = t \cos t$

14. $y = t^2 \tan t$

15. $y = e^x \tan x$

16. $y = e^x \sec x$

17. $y = \pi \sec u \tan u$

18. $y = \pi u \tan u$

19. $y = \dfrac{\cot x}{x}$

20. $y = \dfrac{\csc x}{x}$

21. $y = x^2 \sin x$

22. $y = t^2 \tan t$

23. $y = t \tan t - \sqrt{3} \sec t$

24. $y = x \sec x + \sqrt{2} \cot x$

25. $y = \dfrac{\sin \theta}{1 - \cos \theta}$

26. $y = \dfrac{x}{\cos x}$

27. $y = \dfrac{\sin t}{1 + t}$

28. $y = \dfrac{\tan u}{1 + u}$

29. $y = \dfrac{\sin x}{e^x}$

30. $y = \dfrac{e^x}{\cos x}$

31. $y = \dfrac{\sin \theta + \cos \theta}{\sin \theta - \cos \theta}$

32. $y = \dfrac{\sin \theta - \cos \theta}{\sin \theta + \cos \theta}$

33. $y = \dfrac{\sec t}{1 + t \sin t}$

34. $y = \dfrac{\csc t}{1 + t \cos t}$

35. $y = \csc \theta \cot \theta$

36. $y = \tan \theta \cos \theta$

37. $y = \dfrac{1 + \tan x}{1 - \tan x}$

38. $y = \dfrac{\csc x - \cot x}{\csc x + \cot x}$

In Problems 39–50, find y''.

39. $y = \sin x$

40. $y = \cos x$

41. $y = \tan \theta$

42. $y = \sec \theta$

43. $y = t \sin t$

44. $y = t \cos t$

45. $y = e^x \sin x$

46. $y = e^x \cos x$

47. $y = 2 \sin u - 3 \cos u$

48. $y = 3 \sin u + 4 \cos u$

49. $y = a \sin x + b \cos x$

50. $y = a \sec \theta + b \tan \theta$

In Problems 51–56:

(a) *Find an equation of the tangent line to the graph of f at the indicated point.*

(b) *Graph the function and the tangent line on the same screen.*

51. $f(x) = \sin x$ at $(0, 0)$

52. $f(x) = \cos x$ at $\left(\dfrac{\pi}{3}, \dfrac{1}{2} \right)$

53. $f(x) = \tan x$ at $(0, 0)$

54. $f(x) = \tan x$ at $\left(\dfrac{\pi}{4}, 1 \right)$

55. $f(x) = \sin x + \cos x$ at $\left(\dfrac{\pi}{4}, \sqrt{2} \right)$

56. $f(x) = \sin x - \cos x$ at $\left(\dfrac{\pi}{4}, 0 \right)$

1. = NOW WORK problem   = Graphing technology recommended   CAS = Computer Algebra System recommended

In Problems 57–60:

(a) *Find all points on the graph of f where the tangent line is horizontal.*

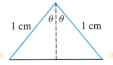 **(b)** *Graph the function and the horizontal tangent lines on the interval* $[-2\pi, 2\pi]$ *on the same screen.*

57. $f(x) = 2\sin x + \cos x$ **58.** $f(x) = \cos x - \sin x$

59. $f(x) = \sec x$ **60.** $f(x) = \csc x$

Applications and Extensions

In Problems 61 and 62, find the nth derivative of each function.

61. $f(x) = \sin x$ **62.** $f(\theta) = \cos \theta$

63. What is $\lim\limits_{h \to 0} \dfrac{\cos\left(\frac{\pi}{2} + h\right) - \cos\frac{\pi}{2}}{h}$?

64. What is $\lim\limits_{h \to 0} \dfrac{\sin(\pi + h) - \sin \pi}{h}$?

65. Simple Harmonic Motion The distance s (in meters) of an object from the origin at time t (in seconds) is modeled by the function $s(t) = \dfrac{1}{8}\cos t$.

(a) Find the velocity v of the object.

(b) When is the speed of the object at a maximum?

(c) Find the acceleration a of the object.

(d) When is the acceleration equal to 0?

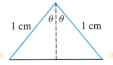 **(e)** Graph s, v, and a on the same screen.

66. Simple Harmonic Motion An object attached to a coiled spring is pulled down a distance $d = 5$ cm from its equilibrium position and then released as shown in the figure. The motion of the object at time t seconds is simple harmonic and is modeled by $d(t) = -5\cos t$.

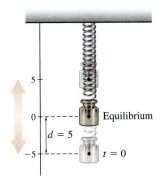

(a) As t varies from 0 to 2π, how does the length of the spring vary?

(b) Find the velocity v of the object.

(c) When is the speed of the object at a maximum?

(d) Find the acceleration a of the object.

(e) When is the acceleration equal to 0?

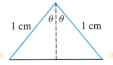 **(f)** Graph d, v, and a on the same screen.

67. Rate of Change A large, 8-ft high decorative mirror is placed on a wood floor and leaned against a wall. The weight of the mirror and the slickness of the floor cause the mirror to slip.

(a) If θ is the angle between the top of the mirror and the wall, and y is the distance from the floor to the top of the mirror, what is the rate of change of y with respect to θ?

(b) In feet/radian, how fast is the top of the mirror slipping down the wall when $\theta = \dfrac{\pi}{4}$?

68. Rate of Change The sides of an isosceles triangle are sliding outward. See the figure.

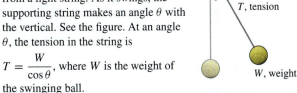

(a) Find the rate of change of the area of the triangle with respect to θ.

(b) How fast is the area changing when $\theta = \dfrac{\pi}{6}$?

69. Sea Waves Waves in deep water tend to have the symmetric form of the function $f(x) = \sin x$. As they approach shore, however, the sea floor creates drag which changes the shape of the wave. The trough of the wave widens and the height of the wave increases, so the top of the wave is no longer symmetric with the trough. This type of wave can be represented by a function such as

$$w(x) = \frac{4}{2 + \cos x}.$$

Source: http://www.crd.bc.ca/watersheds/protection/geology-processes/waves.htm

(a) Graph $w = w(x)$ for $0 \le x \le 4\pi$.

(b) What is the maximum and the minimum value of w?

(c) Find the values of x, $0 < x < 4\pi$, at which $w'(x) = 0$.

(d) Evaluate w' near the peak at π, using $x = \pi - 0.1$, and near the trough at 2π, using $x = 2\pi - 0.1$.

(e) Explain how these values confirm a non-symmetric wave shape.

70. Swinging Pendulum A simple pendulum is a small-sized ball swinging from a light string. As it swings, the supporting string makes an angle θ with the vertical. See the figure. At an angle θ, the tension in the string is

$T = \dfrac{W}{\cos \theta}$, where W is the weight of the swinging ball.

(a) Find the rate of change of the tension T with respect to θ when the pendulum is at its highest point ($\theta = \theta_{\max}$).

(b) Find the rate of change of the tension T with respect to θ when the pendulum is at its lowest point.

(c) What is the tension at the lowest point?

71. Restaurant Sales A restaurant in Naples, Florida is very busy during the winter months and extremely slow over the summer. But every year the restaurant grows its sales. Suppose over the next two years, the revenue R, in units of $\$10,000$, is projected to follow the model

$$R = R(t) = \sin t + 0.3t + 1 \qquad 0 \le t \le 12$$

where $t = 0$ corresponds to November 1, 2014; $t = 1$ corresponds to January 1, 2015; $t = 2$ corresponds to March 1, 2015; and so on.

(a) What is the projected revenue for November 1, 2014; March 1, 2015; September 1, 2015; and January 1, 2016?

(b) What is the rate of change of revenue with respect to time?

(c) What is the rate of change of revenue with respect to time for January 1, 2016?

 **(d)** Graph the revenue function and the derivative function $R' = R'(t)$.

(e) Does the graph of R support the facts that every year the restaurant grows its sales and that sales are higher during the winter and lower during the summer? Explain.

72. Polarizing Sunglasses
Polarizing sunglasses are filters that only transmit light for which the electric field oscillations are in a specific direction. Light gets polarized naturally by scattering off the molecules in the atmosphere and by reflecting off of many (but not all) types of surfaces. If light of intensity I_0 is already polarized in a certain direction, and the transmission direction of the polarizing filter makes an angle with that direction, then the intensity I of the light after passing through the filter is given by **Malus's Law**, $I(\theta) = I_0 \cos^2 \theta$.

(a) As you rotate a polarizing filter, θ changes. Find the rate of change of the light intensity I with respect to θ.

(b) Find both the intensity $I(\theta)$ and the rate of change of the intensity with respect to θ, for the angles $\theta = 0°, 45°$, and $90°$. (Remember to use radians for θ.)

73. If $y = \sin x$ and $y^{(n)}$ is the nth derivative of y with respect to x, find the smallest positive integer n for which $y^{(n)} = y$.

74. Use the identity $\sin A - \sin B = 2\cos \dfrac{A+B}{2} \sin \dfrac{A-B}{2}$, with $A = x + h$ and $B = x$, to prove that

$$\frac{d}{dx} \sin x = \lim_{h \to 0} \frac{\sin(x+h) - \sin x}{h} = \cos x$$

75. Use the definition of a derivative to prove $\dfrac{d}{dx} \cos x = -\sin x$.

76. Derivative of $y = \sec x$ Use a derivative rule to show that

$$\frac{d}{dx} \sec x = \sec x \tan x.$$

77. Derivative of $y = \csc x$ Use a derivative rule to show that

$$\frac{d}{dx} \csc x = -\csc x \cot x.$$

78. Derivative of $y = \cot x$ Use a derivative rule to show that

$$\frac{d}{dx} \cot x = -\csc^2 x.$$

79. Let $f(x) = \cos x$. Show that finding $f'(0)$ is the same as finding $\lim\limits_{x \to 0} \dfrac{\cos x - 1}{x}$.

80. Let $f(x) = \sin x$. Show that finding $f'(0)$ is the same as finding $\lim\limits_{x \to 0} \dfrac{\sin x}{x}$.

81. If $y = A \sin t + B \cos t$, where A and B are constants, show that $y'' + y = 0$.

Challenge Problem

82. For a differentiable function f, let f^* be the function defined by

$$f^*(x) = \lim_{h \to 0} \frac{f(x+h) - f(x-h)}{h}$$

(a) Find $f^*(x)$ for $f(x) = x^2 + x$.

(b) Find $f^*(x)$ for $f(x) = \cos x$.

(c) Write an equation that expresses the relationship between the functions f^* and f', where f' denotes the derivative of f. Justify your answer.

Chapter Review

THINGS TO KNOW

2.1 Rates of Change and the Derivative

- **Definition** Derivative of a function at a number (form 1)

$$f'(c) = \lim_{x \to c} \frac{f(x) - f(c)}{x - c}$$

  provided the limit exists. (p. 150)

Three Interpretations of the Derivative

- *Physical interpretation* When the distance s at time t of an object in rectilinear motion is given by the function $s = f(t)$, the derivative $f'(t_0)$ is the velocity of the object at time t_0. pp. 145–146
- *Geometric interpretation* If $y = f(x)$, the derivative $f'(c)$ is the slope of the tangent line to the graph of f at the point $(c, f(c))$. (pp. 147–148)
- *Rate of change of a function interpretation* If $y = f(x)$, the derivative $f'(c)$ is the rate of change of f at c. (pp. 149–150)

2.2 The Derivative as a Function

- **Definition of a derivative function** (form 3) (p. 154)

$$f'(x) = \lim_{h \to 0} \frac{f(x+h) - f(x)}{h}, \text{ provided the limit exists.}$$

- **Theorem** If a function f has a derivative at a number c, then f is continuous at c. (p. 158)
- **Corollary** If a function f is not continuous at a number c, then f has no derivative at c. (p. 159)

2.3 The Derivative of a Polynomial Function; The Derivative of $y = e^x$

- **Leibniz notation** $\dfrac{dy}{dx} = \dfrac{d}{dx} y = \dfrac{d}{dx} f(x)$ (p. 163)

• **Basic derivatives**

$$\frac{d}{dx}A = 0 \quad A \text{ is a constant (p. 163)} \qquad \frac{d}{dx}x = 1 \quad \text{(p. 164)}$$

$$\frac{d}{dx}e^x = e^x \quad \text{(p. 169)}$$

Properties of Derivatives

• **Simple Power Rule** $\frac{d}{dx}x^n = nx^{n-1}, \quad n$ an integer

   (pp. 164 and 177)

• **Sum Rule** $\frac{d}{dx}[f + g] = \frac{d}{dx}f + \frac{d}{dx}g$

   $(f + g)' = f' + g' \quad \text{(p. 166)}$

• **Difference Rule** $\frac{d}{dx}[f - g] = \frac{d}{dx}f - \frac{d}{dx}g$

   $(f - g)' = f' - g' \quad \text{(p. 167)}$

• **Constant Multiple Rule** $\frac{d}{dx}[kf] = k\frac{d}{dx}f$

   $(kf)' = k \cdot f'$

   k is a constant (p. 165)

2.4 Differentiating the Product and the Quotient of Two Functions; Higher-Order Derivatives

Properties of Derivatives

• **Product Rule** $\frac{d}{dx}(fg) = f\left(\frac{d}{dx}g\right) + \left(\frac{d}{dx}f\right)g$

   $(fg)' = fg' + f'g \quad \text{(p. 174)}$

• **Quotient Rule** $\frac{d}{dx}\left(\frac{f}{g}\right) = \dfrac{\left(\frac{d}{dx}f\right)g - f\left(\frac{d}{dx}g\right)}{g^2}$

   $$\left(\frac{f}{g}\right)' = \frac{f'g - fg'}{g^2},$$

   provided $g(x) \neq 0$ (p. 175)

• **Reciprocal Rule** $\frac{d}{dx}\left(\frac{1}{g}\right) = -\dfrac{\frac{d}{dx}g}{g^2}$

   $$\left(\frac{1}{g}\right)' = -\frac{g'}{g^2},$$

   provided $g(x) \neq 0$ (p. 176)

• *Higher-order derivatives*
   Prime notation: $f''(x), f'''(x), f^{(4)}(x), \ldots, f^{(n)}(x)$

   Liebniz notation: $\dfrac{d^2}{dx^2}[f(x)], \dfrac{d^3}{dx^3}[f(x)],$

   $\dfrac{d^4}{dx^4}[f(x)], \ldots, \dfrac{d^n}{dx^n}[f(x)]$ (p. 178)

2.5 The Derivative of the Trigonometric Functions

Basic Derivatives

$$\frac{d}{dx}\sin x = \cos x \text{ (p. 186)} \qquad \frac{d}{dx}\sec x = \sec x \tan x \text{ (p. 188)}$$

$$\frac{d}{dx}\cos x = -\sin x \text{ (p. 187)} \qquad \frac{d}{dx}\csc x = -\csc x \cot x \text{ (p. 188)}$$

$$\frac{d}{dx}\tan x = \sec^2 x \text{ (p. 188)} \qquad \frac{d}{dx}\cot x = -\csc^2 x \text{ (p. 188)}$$

OBJECTIVES

Section	You should be able to …	Example	Review Exercises
2.1	1 Find instantaneous velocity (p. 145)	1, 2	71(a), 72(a)
	2 Find an equation of the tangent line to the graph of a function (p. 147)	3	67–70
	3 Find the rate of change of a function (p. 149)	4, 5	1, 2, 73(a)
	4 Find the derivative of a function at a number (p. 150)	6, 7	3–8, 75
2.2	1 Define the derivative function (p. 154)	1–3	9–12, 77
	2 Graph the derivative function (p. 155)	4, 5	9–12, 15–18
	3 Identify where a function has no derivative (p. 157)	6–8	13, 14, 75
2.3	1 Differentiate a constant function (p. 163)	1	
	2 Differentiate a power function (p. 164)	2–3	19–22
	3 Differentiate the sum and the difference of two functions (p. 166)	4–6	23–26, 33, 34, 40, 51, 52, 67
	4 Differentiate the exponential function $y = e^x$ (p. 168)	7	44, 45, 53, 54, 56, 59, 69
2.4	1 Differentiate the product of two functions (p. 173)	1, 2	27, 28, 36, 46, 48–50, 53–56, 60
	2 Differentiate the quotient of two functions (p. 175)	3–6	29–35, 37–43, 47, 57–59, 68, 73, 74
	3 Find higher-order derivatives (p. 177)	7, 8	61–66, 71, 72, 76
	4 Work with acceleration (p. 179)	9	71, 72, 76
2.5	1 Differentiate trigonometric functions (p. 185)	1–6	49–60, 70

REVIEW EXERCISES

In Problems 1 and 2, use a definition of the derivative to find the rate of change of f at the indicated numbers.

1. $f(x) = \sqrt{x}$ at **(a)** $c = 1$,   **(b)** $c = 4$,   **(c)** c any positive real number

2. $f(x) = \dfrac{2}{x-1}$ at **(a)** $c = 0$,   **(b)** $c = 2$,   **(c)** c any real number, $c \neq 1$

In Problems 3–8, use a definition of the derivative to find the derivative of each function at the given number.

3. $F(x) = 2x + 5$ at 2

4. $f(x) = 4x^2 + 1$ at -1

5. $f(x) = 3x^2 + 5x$ at 0

6. $f(x) = \dfrac{3}{x}$ at 1

7. $f(x) = \sqrt{4x + 1}$ at 0

8. $f(x) = \dfrac{x+1}{2x-3}$ at 1

In Problems 9–12, use a definition of the derivative to find the derivative of each function. Graph f and f′ on the same set of axes.

9. $f(x) = x - 6$

10. $f(x) = 7 - 3x^2$

11. $f(x) = \dfrac{1}{2x^3}$

12. $f(x) = \pi$

In Problems 13 and 14, determine whether the function f has a derivative at c. If it does, find the derivative. If it does not, explain why. Graph each function.

13. $f(x) = |x^3 - 1|$ at $c = 1$

14. $f(x) = \begin{cases} 4 - 3x^2 & \text{if } x \leq -1 \\ -x^3 & \text{if } x > -1 \end{cases}$ at $c = -1$

In Problems 15 and 16, determine if the graphs represent a function f and its derivative f′. If they do, indicate which is the graph of f and which is the graph of f′.

15.   **16.**

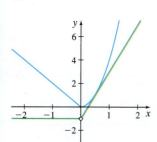

17. Use the information in the graph of $y = f(x)$ to sketch the graph of $y = f'(x)$.

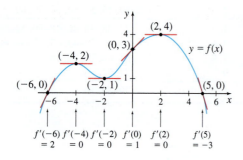

18. Match the graph of $y = f(x)$ with the graph of its derivative.

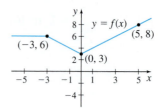

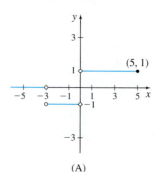 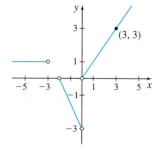

       (A)   (B)

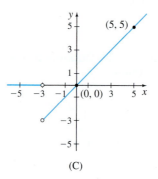 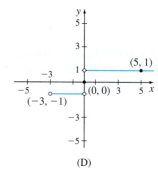

       (C)   (D)

In Problems 19–60, find the derivative of each function. Where a or b appears, it is a constant.

19. $f(x) = x^5$

20. $f(x) = ax^3$

21. $f(x) = \dfrac{x^4}{4}$

22. $f(x) = -6x^2$

23. $f(x) = 2x^2 - 3x$

24. $f(x) = 3x^3 + \dfrac{2}{3}x^2 - 5x + 7$

25. $F(x) = 7(x^2 - 4)$

26. $F(x) = \dfrac{5(x+6)}{7}$

27. $f(x) = 5(x^2 - 3x)(x - 6)$

28. $f(x) = (2x^3 + x)(x^2 - 5)$

29. $f(x) = \dfrac{6x^4 - 9x^2}{3x^3}$

30. $f(x) = \dfrac{2x + 2}{5x - 3}$

31. $f(x) = \dfrac{7x}{x - 5}$

32. $f(x) = 2x^{-12}$

33. $f(x) = 2x^2 - 5x^{-2}$

34. $f(x) = 2 + \dfrac{3}{x} + \dfrac{4}{x^2}$

35. $f(x) = \dfrac{a}{x} - \dfrac{b}{x^3}$

36. $f(x) = (x^3 - 1)^2$

37. $f(x) = \dfrac{3}{(x^2 - 3x)^2}$

38. $f(x) = \dfrac{x^2}{x + 1}$

39. $s(t) = \dfrac{t^3}{t - 2}$

40. $f(x) = 3x^{-2} + 2x^{-1} + 1$

41. $F(z) = \dfrac{1}{z^2 + 1}$

42. $f(v) = \dfrac{v - 1}{v^2 + 1}$

43. $g(z) = \dfrac{1}{1 - z + z^2}$

44. $f(x) = 3e^x + x^2$

45. $s(t) = 1 - e^t$

46. $f(x) = ae^x(2x^2 + 7x)$

47. $f(x) = \dfrac{1 + x}{e^x}$

48. $f(x) = (2xe^x)^2$

49. $f(x) = x \sin x$

50. $s(t) = \cos^2 t$

51. $G(u) = \tan u + \sec u$

52. $g(v) = \sin v - \dfrac{1}{3}\cos v$

53. $f(x) = e^x \sin x$

54. $f(x) = e^x \csc x$

55. $f(x) = 2 \sin x \cos x$

56. $f(x) = (e^x + b) \cos x$

57. $f(x) = \dfrac{\sin x}{\csc x}$

58. $f(x) = \dfrac{1 - \cot x}{1 + \cot x}$

59. $f(\theta) = \dfrac{\cos \theta}{2e^\theta}$

60. $f(\theta) = 4\theta \cot \theta \tan \theta$

In Problems 61–66, find the first derivative and the second derivative of each function.

61. $f(x) = (5x + 3)^2$

62. $f(x) = xe^x$

63. $g(u) = \dfrac{u}{2u + 1}$

64. $F(x) = e^x(\sin x + 2 \cos x)$

65. $f(u) = \dfrac{\cos u}{e^u}$

66. $F(x) = \dfrac{\sin x}{x}$

In Problems 67–70, for each function.
(a) Find an equation of the tangent line to the graph of the function at the indicated point.
 *(b) Graph the function and the tangent line on the same screen.*

67. $f(x) = 2x^2 - 3x + 7$ at $(-1, 12)$

68. $y = \dfrac{x^2 + 1}{2x - 1}$ at $\left(2, \dfrac{5}{3}\right)$

69. $f(x) = x^2 - e^x$ at $(0, -1)$

70. $s(t) = 1 + 2\sin t$ at $(\pi, 1)$

71. **Rectilinear Motion** The distance s (in meters) that an object in

rectilinear motion moves in time t (in seconds) is

$$s = f(t) = t^2 - 6t.$$

(a) Find the velocity at $t = 0$, at $t = 5$, and at any time t.

(b) Find the acceleration at any time t.

72. **Rectilinear Motion** As an object in rectilinear motion moves, its distance s from the origin at time t is $s(t) = t - t^2$, where s is in centimeters and t is in seconds.

(a) Find the velocity of the object at $t = 1$ second and $t = 3$ seconds.

(b) What is its acceleration at $t = 1$ and $t = 3$?

73. **Business** The price p in dollars per pound when x pounds of a commodity are demanded is modeled by the function

$$p(x) = \dfrac{10,000}{5x + 100} - 5, \text{ when between 0 and 90 pounds are}$$

demanded (purchased).

(a) Find the rate of change of price with respect to demand.

(b) What is the revenue function R? (Recall, revenue R equals price times amount purchased.)

(c) What is the marginal revenue R' at $x = 10$ and at $x = 40$ pounds?

74. If $f(x) = \dfrac{x - 1}{x + 1}$ for all $x \neq -1$, find $f'(1)$.

75. If $f(x) = 2 + |x - 3|$ for all x, determine whether the derivative f' exists at $x = 3$.

76. **Rectilinear Motion** An object in rectilinear motion moves according to the equation $s = 2t^3 - 15t^2 + 24t + 3$, where t is measured in minutes and s in meters. Determine:

(a) When is the object is at rest?

(b) Find the object's acceleration when $t = 3$.

77. Find the value of the limit below and specify the function f for which this is the derivative.

$$\lim_{\Delta x \to 0} \dfrac{[4 - 2(x + \Delta x)]^2 - (4 - 2x)^2}{\Delta x}$$

CHAPTER 2 PROJECT   The Lunar Module

The Lunar Module (LM) was a small spacecraft that detached from the Apollo Command Module and was designed to land on the Moon. Fast and accurate computations were needed to bring the LM from an orbiting speed of about 5500 ft/s
to a speed slow enough to land it within a few feet of a designated target on the Moon's surface. The LM carried a 70-lb computer to assist in guiding it successfully to its target. The approach to the target was split into three phases, each of which followed a

reference trajectory specified by NASA engineers.* The position and velocity of the LM were monitored by sensors that tracked its deviation from the preassigned path at each moment. Whenever the LM strayed from the reference trajectory, control thrusters were fired to reposition it. In other words, the LM's position and velocity were adjusted by changing its acceleration.

---

*A. R. Klumpp, "Apollo Lunar-Descent Guidance," MIT Charles Stark Draper Laboratory, R-695, June 1971,
http://www.hq.nasa.gov/alsj/ApolloDescentGuidnce.pdf.

The reference trajectory for each phase was specified by the engineers to have the form

$$r_{\text{ref}}(t) = R_T + V_T t + \frac{1}{2} A_T t^2 + \frac{1}{6} J_T t^3 + \frac{1}{24} S_T t^4 \qquad (1)$$

The variable r_{ref} represents the intended position of the LM at time t before the end of the landing phase. The engineers specified the end of the landing phase to take place at $t = 0$, so that during the phase, t was always negative. Note that the LM was landing in three dimensions, so there were actually three equations like (1). Since each of those equations had this same form, we will work in one dimension, assuming, for example, that r represents the distance of the LM above the surface of the Moon.

1. If the LM follows the reference trajectory, what is the reference velocity $v_{\text{ref}}(t)$?

2. What is the reference acceleration $a_{\text{ref}}(t)$?

3. The rate of change of acceleration is called **jerk**. Find the reference jerk $J_{\text{ref}}(t)$.

4. The rate of change of jerk is called **snap**. Find the reference snap $S_{\text{ref}}(t)$.

5. Evaluate $r_{\text{ref}}(t)$, $v_{\text{ref}}(t)$, $a_{\text{ref}}(t)$, $J_{\text{ref}}(t)$, and $S_{\text{ref}}(t)$ when $t = 0$.

The reference trajectory given in equation (1) is a fourth degree polynomial, the lowest degree polynomial that has enough free parameters to satisfy all the mission criteria. Now we see that the parameters $R_T = r_{\text{ref}}(0)$, $V_T = v_{\text{ref}}(0)$, $A_T = a_{\text{ref}}(0)$, $J_T = J_{\text{ref}}(0)$, and $S_T = S_{\text{ref}}(0)$. The five parameters in equation (1) are referred to as the **target parameters** since they provide the path the LM should follow.

But small variations in propulsion, mass, and countless other variables cause the LM to deviate from the predetermined path. To correct the LM's position and velocity, NASA engineers apply a force to the LM using rocket thrusters. That is, they changed the acceleration. (Remember Newton's second law, $F = ma$.) Engineers modeled the actual trajectory of the LM by

$$r(t) = R_T + V_T t + \frac{1}{2} A_T t^2 + \frac{1}{6} J_A t^3 + \frac{1}{24} S_A t^4 \qquad (2)$$

We know the target parameters for position, velocity, and acceleration. We need to find the actual parameters for jerk and snap to know the proper force (acceleration) to apply.

6. Find the actual velocity $v = v(t)$ of the LM.

7. Find the actual acceleration $a = a(t)$ of the LM.

8. Use equation (2) and the actual velocity found in Problem 6 to express J_A and S_A in terms of R_T, V_T, A_T, $r(t)$, and $v(t)$.

9. Use the results of Problems 7 and 8 to express the actual acceleration $a = a(t)$ in terms of R_T, V_T, A_T, $r(t)$, and $v(t)$.

The result found in Problem 9 provides the acceleration (force) required to keep the LM in its reference trajectory.

10. When riding in an elevator, the sensation one feels just before the elevator stops at a floor is jerk. Would you want jerk to be small or large in an elevator? Explain. Would you want jerk to be small or large on a roller coaster ride? Explain. How would you explain snap?

3 More About Derivatives

CHAPTER 3 PROJECT In the Chapter Project on page 253 we explore a basic model to predict world population and examine its accuracy.

World Population Growth

In the late 1700's Thomas Malthus predicted that population growth, if left unchecked, would outstrip food resources and lead to mass starvation. His prediction turned out to be incorrect, since in 1800 world population had not yet reached one billion and currently world population is in excess of seven billion. Malthus' most dire predictions did not come true largely because improvements in food production made his models inaccurate. On the other hand, his population growth model is still used today, and population growth remains an important issue in human progress.

In this chapter, we continue exploring properties of derivatives, beginning with the Chain Rule, which allows us to differentiate composite functions. The Chain Rule also provides the means to establish derivative formulas for exponential functions, logarithmic functions, and hyperbolic functions.

We also use the derivative to solve problems involving relative error in the measurements of two related variables, a way to approximate the real zeros of a function, and the approximation of functions by polynomials.

3.1 The Chain Rule

OBJECTIVES *When you finish this section, you should be able to:*

1 Differentiate a composite function (p. 198)

2 Differentiate $y = a^x$, $a > 0$, $a \neq 1$ (p. 202)

3 Use the Power Rule for functions to find a derivative (p. 202)

4 Use the Chain Rule for multiple composite functions (p. 204)

Using the differentiation rules developed so far, it would be difficult to differentiate the function

$$F(x) = (x^3 - 4x + 1)^{100}$$

NEED TO REVIEW? Composite functions and their properties are discussed in Section P.3, pp. 25–27.

But notice that F is the composite function: $y = f(u) = u^{100}$, $u = g(x) = x^3 - 4x + 1$, so $y = F(x) = (f \circ g)(x) = (x^3 - 4x + 1)^{100}$. In this section, we derive the *Chain Rule*, a result that enables us to find the derivative of a composite function. We use the Chain Rule to find the derivative in applications involving functions such as $A(t) = 102 - 90e^{-0.21t}$ (market penetration, Problem 101) and $v(t) = \dfrac{mg}{k}(1 - e^{-kt/m})$ (the terminal velocity of a falling object, Problem 103).

1 Differentiate a Composite Function

Suppose $y = (f \circ g)(x) = f(g(x))$ is a composite function, where $y = f(u)$ is a differentiable function of u, and $u = g(x)$ is a differentiable function of x. What then is the derivative of $(f \circ g)(x)$? It turns out that the derivative of the composite function $f \circ g$ is the product of the derivatives $f'(u) = f'(g(x))$ and $g'(x)$.

THEOREM Chain Rule

If a function g is differentiable at x_0 and a function f is differentiable at $g(x_0)$, then the composite function $f \circ g$ is differentiable at x_0 and

$$\boxed{(f \circ g)'(x_0) = f'(g(x_0)) \cdot g'(x_0)}$$

IN WORDS If you think of the function $y = f(u)$ as the outside function and the function $u = g(x)$ as the inside function, then the derivative of $f \circ g$ is the derivative of the outside function, evaluated at the inside function, times the derivative of the inside function.

That is, $\dfrac{d}{dx}(f \circ g)(x) = f'(g(x)) \cdot g'(x)$.

For differentiable functions $y = f(u)$ and $u = g(x)$, the Chain Rule, in Leibniz notation, takes the form

$$\boxed{\frac{dy}{dx} = \frac{dy}{du} \cdot \frac{du}{dx}}$$

where in $\dfrac{dy}{du}$ we substitute $u = g(x)$.

Partial Proof The Chain Rule is proved using the definition of a derivative. First we observe that if x changes by a small amount Δx, the corresponding change in $u = g(x)$ is Δu. That is, Δu depends on Δx. Also,

$$g'(x) = \frac{du}{dx} = \lim_{\Delta x \to 0} \frac{\Delta u}{\Delta x}$$

Since $y = f(u)$, the change Δu, which could equal 0, causes a change Δy. If Δu is never 0, then

$$f'(u) = \frac{dy}{du} = \lim_{\Delta u \to 0} \frac{\Delta y}{\Delta u}$$

To find $\dfrac{dy}{dx}$, we write

$$\frac{dy}{dx} = \lim_{\Delta x \to 0} \frac{\Delta y}{\Delta x} \underset{\substack{\uparrow \\ \Delta u \neq 0}}{=} \lim_{\Delta x \to 0} \left(\frac{\Delta y}{\Delta x} \cdot \frac{\Delta u}{\Delta u} \right) = \lim_{\Delta x \to 0} \left(\frac{\Delta y}{\Delta u} \cdot \frac{\Delta u}{\Delta x} \right)$$

$$= \left(\lim_{\Delta x \to 0} \frac{\Delta y}{\Delta u} \right) \left(\lim_{\Delta x \to 0} \frac{\Delta u}{\Delta x} \right)$$

Since the differentiable function $u = g(x)$ is continuous, $\Delta u \to 0$ as $\Delta x \to 0$, so in the first factor we can replace $\Delta x \to 0$ by $\Delta u \to 0$. Then

$$\frac{dy}{dx} = \left(\lim_{\Delta u \to 0} \frac{\Delta y}{\Delta u} \right) \left(\lim_{\Delta x \to 0} \frac{\Delta u}{\Delta x} \right) = \frac{dy}{du} \cdot \frac{du}{dx}$$

This proves the Chain Rule if Δu is never 0. To complete the proof, we need to consider the case when Δu may be 0. (This part of the proof is given in Appendix B.) ∎

EXAMPLE 1 Differentiating a Composite Function

Find the derivative of:

(a) $y = (x^3 - 4x + 1)^{100}$ **(b)** $y = \cos\left(3x - \dfrac{\pi}{4}\right)$

Solution **(a)** In the composite function $y = (x^3 - 4x + 1)^{100}$, let $u = x^3 - 4x + 1$. Then $y = u^{100}$. Now $\dfrac{dy}{du}$ and $\dfrac{du}{dx}$ are

$$\frac{dy}{du} = \frac{d}{du} u^{100} = 100u^{99} = \underset{\underset{\textstyle u = x^3 - 4x + 1}{\uparrow}}{100(x^3 - 4x + 1)^{99}}$$

and

$$\frac{du}{dx} = \frac{d}{dx}(x^3 - 4x + 1) = 3x^2 - 4$$

We use the Chain Rule to find $\dfrac{dy}{dx}$.

$$\frac{dy}{dx} \underset{\underset{\textstyle \text{Chain Rule}}{\uparrow}}{=} \frac{dy}{du} \cdot \frac{du}{dx} = 100(x^3 - 4x + 1)^{99}(3x^2 - 4)$$

(b) In the composite function $y = \cos\left(3x - \dfrac{\pi}{4}\right)$, let $u = 3x - \dfrac{\pi}{4}$. Then $y = \cos u$ and

$$\frac{dy}{du} = \frac{d}{du}\cos u = -\sin u = \underset{\underset{\textstyle u = 3x - \frac{\pi}{4}}{\uparrow}}{-\sin\left(3x - \frac{\pi}{4}\right)} \quad \text{and} \quad \frac{du}{dx} = \frac{d}{dx}\left(3x - \frac{\pi}{4}\right) = 3$$

Now we use the Chain Rule.

$$\frac{dy}{dx} \underset{\underset{\textstyle \text{Chain Rule}}{\uparrow}}{=} \frac{dy}{du} \cdot \frac{du}{dx} = -\sin\left(3x - \frac{\pi}{4}\right) \cdot 3 = -3\sin\left(3x - \frac{\pi}{4}\right)$$ ∎

NOW WORK Problems 9 and 37.

EXAMPLE 2 Differentiating a Composite Function

Find y' if:

(a) $y = e^{x^2 - 4}$ **(b)** $y = \sin(4e^x)$

Solution **(a)** For $y = e^{x^2 - 4}$, we let $u = x^2 - 4$. Then $y = e^u$ and

$$\frac{dy}{du} = \frac{d}{du} e^u = e^u = \underset{\underset{\textstyle u = x^2 - 4}{\uparrow}}{e^{x^2 - 4}} \quad \text{and} \quad \frac{du}{dx} = \frac{d}{dx}(x^2 - 4) = 2x$$

Using the Chain Rule, we get

$$y' = \frac{dy}{dx} = \frac{dy}{du} \cdot \frac{du}{dx} = e^{x^2-4} \cdot 2x = 2xe^{x^2-4}$$

(b) For $y = \sin(4e^x)$, we let $u = 4e^x$. Then $y = \sin u$ and

$$\frac{dy}{du} = \frac{d}{du}\sin u = \cos u = \cos(4e^x) \quad \text{and} \quad \frac{du}{dx} = \frac{d}{dx}(4e^x) = 4e^x$$

$$\uparrow$$
$$u = 4e^x$$

Using the Chain Rule, we get

$$y' = \frac{dy}{dx} = \frac{dy}{du} \cdot \frac{du}{dx} = \cos(4e^x) \cdot 4e^x = 4e^x \cos(4e^x)$$ ∎

NOW WORK Problem 41.

For composite functions $y = f(u(x))$, where f is an exponential or trigonometric function, the Chain Rule simplifies finding y'. For example,

- If $y = e^u$, where $u = u(x)$ is a differentiable function of x, then by the Chain Rule

$$y' = \frac{dy}{dx} = \frac{dy}{du} \cdot \frac{du}{dx} = e^u \frac{du}{dx}$$

That is,

$$\boxed{\frac{d}{dx}e^u = e^u \frac{du}{dx}}$$

- Similarly, if $u = u(x)$ is a differentiable function,

$$\boxed{\begin{array}{ll} \dfrac{d}{dx}\sin u = \cos u \dfrac{du}{dx} & \dfrac{d}{dx}\sec u = \sec u \tan u \dfrac{du}{dx} \\[2mm] \dfrac{d}{dx}\cos u = -\sin u \dfrac{du}{dx} & \dfrac{d}{dx}\csc u = -\csc u \cot u \dfrac{du}{dx} \\[2mm] \dfrac{d}{dx}\tan u = \sec^2 u \dfrac{du}{dx} & \dfrac{d}{dx}\cot u = -\csc^2 u \dfrac{du}{dx} \end{array}}$$

EXAMPLE 3   **Finding an Equation of a Tangent Line**

Find an equation of the tangent line to the graph of $y = 5e^{4x}$ at the point $(0, 5)$.

Solution   The slope of the tangent line to the graph of $y = f(x)$ at the point $(0, 5)$ is $f'(0)$.

$$f'(x) = \frac{d}{dx}(5e^{4x}) = 5\frac{d}{dx}e^{4x} = 5e^{4x} \cdot \frac{d}{dx}(4x) = 5e^{4x} \cdot 4 = 20e^{4x}$$

$$\uparrow \qquad\qquad \uparrow$$
$$\text{Constant} \qquad u = 4x;\ \frac{d}{dx}e^u = e^u \frac{du}{dx}$$
$$\text{Multiple Rule}$$

$m_{\tan} = f'(0) = 20e^0 = 20$. Using the point slope form of a line, we have

$$y - 5 = 20(x - 0) \qquad y - y_0 = m_{\tan}(x - x_0)$$
$$y = 20x + 5$$ ∎

The graph of $y = 5e^{4x}$ and the line $y = 20x + 5$ are shown in Figure 1.

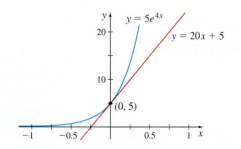

Figure 1

NOW WORK Problem 77.

EXAMPLE 4 **Application to Carbon-14 Dating**

All carbon on Earth contains some carbon-14, which is radioactive. When a living organism dies, the carbon-14 begins to decay at a fixed rate. The formula $P(t) = 100e^{-0.000121t}$ gives the percentage of carbon-14 present at time t years. Notice that when $t = 0$, the percentage of carbon-14 present is 100%. When the preserved bodies of 15-year-old La Doncella and her two children were found in Peru in 2005, 93.5% of the carbon-14 remained in their bodies, indicating that the three had died about 550 years earlier.

(a) What is the rate of change of the percentage of carbon-14 present in a 550-year-old fossil?

(b) What is the rate of change of the percentage of carbon-14 present in a 2000-year-old fossil?

Solution **(a)** The rate of change of P is given by its derivative

$$P'(t) = \frac{d}{dt}(100e^{-0.000121t}) = 100(-0.000121e^{-0.000121t}) = -0.0121e^{-0.000121t}$$

$$\frac{d}{dt}e^u = e^u\frac{du}{dt}$$

At $t = 550$ years,

$$P'(550) = -0.0121e^{-0.000121(550)} \approx -0.0113$$

The percentage of carbon-14 present in a 550-year-old fossil is decreasing at the rate of 1.13% per year.

(b) When $t = 2000$ years, the rate of change is

$$P'(2000) = -0.0121e^{-0.000121(2000)} \approx -0.0095$$

The percentage of carbon-14 present in a 2000-year-old fossil is decreasing at the rate of 0.95% per year. ∎

NOW WORK Problem 99.

When we first stated the Chain Rule, we expressed it two ways: using prime notation and using Leibniz notation. In each example so far, we have used Leibniz notation. But when solving numerical problems, using prime notation is often easier. In this form, the Chain Rule states that

$$\boxed{(f \circ g)'(x) = f'(g(x))g'(x)}$$

EXAMPLE 5 **Differentiating a Composite Function**

Suppose $h = f \circ g$. Find $h'(1)$ given that:

$$f(1) = 2 \quad f'(1) = 1 \quad f'(2) = -4 \quad g(1) = 2 \quad g'(1) = -3 \quad g'(2) = 5$$

Solution Based on the Chain Rule using prime notation, we have

$$h'(x_0) = (f \circ g)'(x_0) = f'(g(x_0))g'(x_0)$$

When $x_0 = 1$,

$$h'(1) = f'(g(1))g'(1) = f'(2) \cdot (-3) = (-4)(-3) = 12$$

$$g(1) = 2; g'(1) = -3 \qquad f'(2) = -4$$

∎

NOW WORK Problem 81.

NEED TO REVIEW? Properties of logarithms are discussed in Appendix A.1, pp. A-10 to A-11.

2 Differentiate $y = a^x, a > 0, a \neq 1$

The Chain Rule allows us to establish a formula for differentiating an exponential function $y = a^x$ for any base $a > 0$ and $a \neq 1$. We start with the following property of logarithms:

$$a^x = e^{\ln a^x} = e^{x \ln a} \qquad a > 0, a \neq 1$$

Then

$$\frac{d}{dx}a^x = \frac{d}{dx}e^{x \ln a} = e^{x \ln a}\frac{d}{dx}(x \ln a) = e^{x \ln a}\ln a = a^x \ln a$$

$$\underset{\frac{d}{dx}e^u = e^u\frac{du}{dx}}{\uparrow}$$

> **THEOREM Derivative of $y = a^x$**
>
> The derivative of an exponential function $y = a^x$, where $a > 0$ and $a \neq 1$, is
>
> $$y' = \frac{d}{dx}a^x = a^x \ln a$$

EXAMPLE 6 Differentiating Exponential Functions

Find the derivative of each function:

(a) $f(x) = 2^x$ **(b)** $F(x) = 3^{-x}$ **(c)** $g(x) = \left(\frac{1}{2}\right)^{x^2+1}$

Solution **(a)** f is an exponential function with base $a = 2$.

$$f'(x) = \frac{d}{dx}2^x = 2^x \ln 2 \qquad \underset{\frac{d}{dx}a^x = a^x \ln a}{}$$

(b) Since $F(x) = 3^{-x} = \dfrac{1}{3^x} = \left(\dfrac{1}{3}\right)^x$, F is an exponential function with base $\dfrac{1}{3}$. So,

$$F'(x) = \frac{d}{dx}\left(\frac{1}{3}\right)^x = \left(\frac{1}{3}\right)^x \ln\frac{1}{3} = \left(\frac{1}{3}\right)^x \ln 3^{-1} = -\left(\frac{1}{3}\right)^x \ln 3 = -\frac{1}{3^x}\ln 3$$

$$\underset{\frac{d}{dx}a^x = a^x \ln a}{\uparrow}$$

(c) $y = g(x) = \left(\dfrac{1}{2}\right)^{x^2+1}$ is a composite function. If $u = x^2 + 1$, then $y = \left(\dfrac{1}{2}\right)^u$ and

$$\frac{dy}{du} = \left(\frac{1}{2}\right)^u \ln\left(\frac{1}{2}\right) = -\left(\frac{1}{2}\right)^u \ln 2 = -\left(\frac{1}{2}\right)^{x^2+1}\ln 2 \quad \text{and} \quad \frac{du}{dx} = 2x$$

$$\underset{\frac{d}{du}a^u = a^u \ln a}{\uparrow} \qquad \underset{\ln\left(\frac{1}{2}\right) = -\ln 2}{\uparrow} \qquad \underset{u = x^2 + 1}{\uparrow}$$

So, by the Chain Rule,

$$g'(x) = \frac{dy}{dx} = \frac{dy}{du}\cdot\frac{du}{dx} = \left[-\left(\frac{1}{2}\right)^{x^2+1}\ln 2\right](2x) = (-\ln 2)\,x\left(\frac{1}{2}\right)^{x^2} \qquad \blacksquare$$

NOW WORK Problem 47.

3 Use the Power Rule for Functions to Find a Derivative

We use the Chain Rule to establish other derivative formulas, such as a formula for the derivative of a function raised to a power.

THEOREM Power Rule for Functions

If g is a differentiable function and n is an integer, then

$$\frac{d}{dx}[g(x)]^n = n[g(x)]^{n-1}g'(x)$$

Proof If $y = [g(x)]^n$, let $y = u^n$ and $u = g(x)$. Then

$$\frac{dy}{du} = nu^{n-1} = n[g(x)]^{n-1} \qquad \text{and} \qquad \frac{du}{dx} = g'(x)$$

By the Chain Rule,

$$y' = \frac{d}{dx}[g(x)]^n = \frac{dy}{du} \cdot \frac{du}{dx} = n[g(x)]^{n-1}g'(x) \qquad \blacksquare$$

EXAMPLE 7   Using the Power Rule for Functions to Find a Derivative

(a) If $f(x) = (3 - x^3)^{-5}$, then

$$f'(x) = \frac{d}{dx}(3 - x^3)^{-5} = -5(3 - x^3)^{-5-1} \cdot \frac{d}{dx}(3 - x^3)$$
$$\underset{\substack{\uparrow \\ \text{Power Rule} \\ \text{for Functions}}}{}$$

$$= -5(3 - x^3)^{-6} \cdot (-3x^2) = 15x^2(3 - x^3)^{-6} = \frac{15x^2}{(3 - x^3)^6}$$

(b) If $f(\theta) = \cos^3 \theta$, then $f(\theta) = (\cos \theta)^3$, and

$$f'(\theta) = \frac{d}{d\theta}(\cos \theta)^3 = 3(\cos \theta)^{3-1} \cdot \frac{d}{d\theta}\cos \theta = 3\cos^2 \theta \cdot (-\sin \theta)$$
$$\underset{\substack{\uparrow \\ \text{Power Rule} \\ \text{for Functions}}}{}$$

$$= -3\cos^2 \theta \sin \theta \qquad \blacksquare$$

NOW REWORK Example 7 using the Chain Rule and rework Example 1(a) using the Power Rule for Functions. Decide for yourself which method is easier.

NOW WORK Problem 21.

Often other derivative rules are used along with the Power Rule for Functions.

EXAMPLE 8   Using the Power Rule for Functions with Other Derivative Rules

Find the derivative of:

(a) $f(x) = e^x(x^2 + 1)^3$ **(b)** $g(x) = \left(\dfrac{3x + 2}{4x^2 - 5}\right)^5$

Solution (a) The function f is the product of e^x and $(x^2 + 1)^3$, so we first use the Product Rule.

$$f'(x) = e^x\left[\frac{d}{dx}(x^2 + 1)^3\right] + \left[\frac{d}{dx}e^x\right](x^2 + 1)^3$$
$$\underset{\substack{\uparrow \\ \text{Product Rule}}}{}$$

To complete the solution, we use the Power Rule for Functions to find $\dfrac{d}{dx}(x^2+1)^3$.

$$f'(x) = e^x\left[3(x^2+1)^2 \cdot \frac{d}{dx}(x^2+1)\right] + e^x(x^2+1)^3 \qquad \text{\color{blue}Power Rule for Functions}$$

$$= e^x[3(x^2+1)^2 \cdot 2x] + e^x(x^2+1)^3 = e^x[6x(x^2+1)^2 + (x^2+1)^3]$$

$$= e^x(x^2+1)^2[6x + x^2 + 1] = e^x(x^2+1)^2(x^2+6x+1)$$

(b) g is a function raised to a power, so we begin with the Power Rule for Functions.

$$g'(x) = \frac{d}{dx}\left(\frac{3x+2}{4x^2-5}\right)^5 = 5\left(\frac{3x+2}{4x^2-5}\right)^4\left[\frac{d}{dx}\left(\frac{3x+2}{4x^2-5}\right)\right] \qquad \text{\color{blue}Power Rule for Functions}$$

$$= 5\left(\frac{3x+2}{4x^2-5}\right)^4\left[\frac{(3)(4x^2-5) - (3x+2)(8x)}{(4x^2-5)^2}\right] \qquad \text{\color{blue}Quotient Rule}$$

$$= \frac{5(3x+2)^4[(12x^2-15) - (24x^2+16x)]}{(4x^2-5)^6}$$

$$= \frac{5(3x+2)^4[-12x^2 - 16x - 15]}{(4x^2-5)^6} \qquad \blacksquare$$

NOW WORK Problem 29.

4 Use the Chain Rule for Multiple Composite Functions

The Chain Rule can be extended to multiple composite functions. For example, if the functions

$$y = f(u) \qquad u = g(v) \qquad v = h(x)$$

are each differentiable functions of u, v, and x, respectively, then the composite function $y = (f \circ g \circ h)(x) = f(g(h(x)))$ is a differentiable function of x and

$$\boxed{y' = \frac{dy}{dx} = \frac{dy}{du} \cdot \frac{du}{dv} \cdot \frac{dv}{dx}}$$

where $u = g(v) = g(h(x))$ and $v = h(x)$. This "chain" of factors is the basis for the name Chain Rule.

EXAMPLE 9   Differentiating a Composite Function

Find y' if:

(a) $y = 5\cos^2(3x+2)$ **(b)** $y = \sin^3\left(\dfrac{\pi}{2}x\right)$

Solution (a) For $y = 5\cos^2(3x+2)$, we use the Chain Rule with $y = 5u^2$, $u = \cos v$, and $v = 3x+2$. Then $y = 5u^2 = 5\cos^2 v = 5\cos^2(3x+2)$ and

$$\frac{dy}{du} = \frac{d}{du}(5u^2) = 10u = 10\cos(3x+2)$$
$$\underset{\substack{\uparrow \\ u = \cos v \\ v = 3x+2}}{}$$

$$\frac{du}{dv} = \frac{d}{dv}\cos v = -\sin v = -\sin(3x+2)$$
$$\underset{\substack{\uparrow \\ v = 3x+2}}{}$$

$$\frac{dv}{dx} = \frac{d}{dx}(3x+2) = 3$$

Then

$$y' = \frac{dy}{dx} \underset{\text{Chain Rule}}{\uparrow} = \frac{dy}{du} \cdot \frac{du}{dv} \cdot \frac{dv}{dx} = [10\cos(3x+2)][-\sin(3x+2)][3]$$

$$= -30\cos(3x+2)\sin(3x+2)$$

(b) For $y = \sin^3\left(\frac{\pi}{2}x\right)$, we use the Chain Rule with $y = u^3$, $u = \sin v$, and $v = \frac{\pi}{2}x$.

Then $y = u^3 = (\sin v)^3 = \left[\sin\left(\frac{\pi}{2}x\right)\right]^3 = \sin^3\left(\frac{\pi}{2}x\right)$, and

$$\frac{dy}{du} = \frac{d}{du}u^3 = 3u^2 \underset{\substack{u = \sin v \\ v = \frac{\pi}{2}x}}{=} 3\left[\sin\left(\frac{\pi}{2}x\right)\right]^2 = 3\sin^2\left(\frac{\pi}{2}x\right)$$

$$\frac{du}{dv} = \frac{d}{dv}\sin v = \cos v \underset{v = \frac{\pi}{2}x}{=} \cos\left(\frac{\pi}{2}x\right)$$

$$\frac{dv}{dx} = \frac{d}{dx}\left(\frac{\pi}{2}x\right) = \frac{\pi}{2}$$

Then

$$y' = \frac{dy}{dx} \underset{\text{Chain Rule}}{\uparrow} = \frac{dy}{du} \cdot \frac{du}{dv} \cdot \frac{dv}{dx} = 3\sin^2\left(\frac{\pi}{2}x\right) \cdot \cos\left(\frac{\pi}{2}x\right) \cdot \left(\frac{\pi}{2}\right)$$

$$= \frac{3\pi}{2}\sin^2\left(\frac{\pi}{2}x\right)\cos\left(\frac{\pi}{2}x\right)$$

∎

NOW WORK Problem 59.

3.1 Assess Your Understanding

Concepts and Vocabulary

1. The derivative of a composite function $(f \circ g)(x)$ can be found using the _____ Rule.

2. *True or False* If $y = f(u)$ and $u = g(x)$ are differentiable functions, then $y = f(g(x))$ is differentiable.

3. *True or False* If $y = f(g(x))$ is a differentiable function, then $y' = f'(g'(x))$.

4. To find the derivative of $y = \tan(1 + \cos x)$, using the Chain Rule, begin with $y =$ _____ and $u =$ _____.

5. If $y = (x^3 + 4x + 1)^{100}$, then $y' =$ _____.

6. If $f(x) = e^{3x^2+5}$, then $f'(x) =$ _____.

7. *True or False* The Chain Rule can be applied to multiple composite functions.

8. $\dfrac{d}{dx}\sin x^2 =$ _____.

Skill Building

In Problems 9–14, write y as a function of x. Find $\dfrac{dy}{dx}$ using the Chain Rule.

9. $y = u^5$, $u = x^3 + 1$

10. $y = u^3$, $u = 2x + 5$

11. $y = \dfrac{u}{u+1}$, $u = x^2 + 1$

12. $y = \dfrac{u-1}{u}$, $u = x^2 - 1$

13. $y = (u+1)^2$, $u = \dfrac{1}{x}$

14. $y = (u^2 - 1)^3$, $u = \dfrac{1}{x+2}$

In Problems 15–32, find the derivative of each function using the Power Rule for Functions.

15. $f(x) = (3x + 5)^2$

16. $f(x) = (2x - 5)^3$

17. $f(x) = (6x - 5)^{-3}$

18. $f(t) = (4t + 1)^{-2}$

19. $g(x) = (x^2 + 5)^4$

20. $F(x) = (x^3 - 2)^5$

21. $f(u) = \left(u - \dfrac{1}{u}\right)^3$

22. $f(x) = \left(x + \dfrac{1}{x}\right)^3$

23. $g(x) = (4x + e^x)^3$

24. $F(x) = (e^x - x^2)^2$

25. $f(x) = \tan^2 x$

26. $f(x) = \sec^3 x$

27. $f(z) = (\tan z + \cos z)^2$

28. $f(z) = (e^z + 2\sin z)^3$

29. $y = (x^2 + 4)^2(2x^3 - 1)^3$

30. $y = (x^2 - 2)^3(3x^4 + 1)^2$

31. $y = \left(\dfrac{\sin x}{x}\right)^2$

32. $y = \left(\dfrac{x + \cos x}{x}\right)^5$

---

1. = NOW WORK problem   〰 = Graphing technology recommended   CAS = Computer Algebra System recommended

In Problems 33–54, find y'.

33. $y = \sin(4x)$

34. $y = \cos(5x)$

35. $y = 2\sin(x^2 + 2x - 1)$

36. $y = \dfrac{1}{2}\cos(x^3 - 2x + 5)$

37. $y = \sin\dfrac{1}{x}$

38. $y = \sin\dfrac{3}{x}$

39. $y = \sec(4x)$

40. $y = \cot(5x)$

41. $y = e^{1/x}$

42. $y = e^{1/x^2}$

43. $y = \dfrac{1}{x^4 - 2x + 1}$

44. $y = \dfrac{3}{x^5 + 2x^2 - 3}$

45. $y = \dfrac{100}{1 + 99e^{-x}}$

46. $y = \dfrac{1}{1 + 2e^{-x}}$

47. $y = 2^{\sin x}$

48. $y = (\sqrt{3})^{\cos x}$

49. $y = 6^{\sec x}$

50. $y = 3^{\tan x}$

51. $y = 5xe^{3x}$

52. $y = x^3 e^{2x}$

53. $y = x^2 \sin(4x)$

54. $y = x^2 \cos(4x)$

In Problems 55–58, find y' (a and b are constants).

55. $y = e^{-ax}\sin(bx)$

56. $y = e^{ax}\cos(-bx)$

57. $y = \dfrac{e^{ax} - 1}{e^{ax} + 1}$

58. $y = \dfrac{e^{-ax} + 1}{e^{bx} - 1}$

In Problems 59–62, write y as a function of x. Find $\dfrac{dy}{dx}$ using the Chain Rule.

59. $y = u^3, \ u = 3v^2 + 1, \ v = \dfrac{4}{x^2}$

60. $y = 3u, \ u = 3v^2 - 4, \ v = \dfrac{1}{x}$

61. $y = u^2 + 1, \ u = \dfrac{4}{v}, \ v = x^2$

62. $y = u^3 - 1, \ u = -\dfrac{2}{v}, \ v = x^3$

In Problems 63–70, find y'.

63. $y = e^{-2x}\cos(3x)$

64. $y = e^{\pi x}\tan(\pi x)$

65. $y = \cos(e^{x^2})$

66. $y = \tan(e^{x^2})$

67. $y = e^{\cos(4x)}$

68. $y = e^{\csc^2 x}$

69. $y = 4\sin^2(3x)$

70. $y = 2\cos^2(x^2)$

In Problems 71 and 72, find the derivative of each function by:
(a) Using the Chain Rule.
(b) Using the Power Rule for Functions.
(c) Expanding and then differentiating.
(d) Verify the answers from parts (a)–(c) are equal.

71. $y = (x^3 + 1)^2$

72. $y = (x^2 - 2)^3$

In Problems 73–78:
(a) Find an equation of the tangent line to the graph of f at the given point.
(b) Use graphing technology to graph f and the tangent line on the same screen.

73. $f(x) = (x^2 - 2x + 1)^5$ at $(1, 0)$

74. $f(x) = (x^3 - x^2 + x - 1)^{10}$ at $(0, 1)$

75. $f(x) = \dfrac{x}{(x^2 - 1)^3}$ at $\left(2, \dfrac{2}{27}\right)$

76. $f(x) = \dfrac{x^2}{(x^2 - 1)^2}$ at $\left(2, \dfrac{4}{9}\right)$

77. $f(x) = \sin(2x) + \cos\dfrac{x}{2}$ at $(0, 1)$

78. $f(x) = \sin^2 x + \cos^3 x$ at $\left(\dfrac{\pi}{2}, 1\right)$

In Problems 79 and 80, find the indicated derivative.

79. $\dfrac{d^2}{dx^2}\cos(x^5)$

80. $\dfrac{d^3}{dx^3}\sin^3 x$

81. Suppose $h = f \circ g$. Find $h'(1)$ if $f'(2) = 6$, $f(1) = 4$, $g(1) = 2$, and $g'(1) = -2$.

82. Suppose $h = f \circ g$. Find $h'(1)$ if $f'(3) = 4$, $f(1) = 1$, $g(1) = 3$, and $g'(1) = 3$.

83. Suppose $h = g \circ f$. Find $h'(0)$ if $f(0) = 3$, $f'(0) = -1$, $g(3) = 8$, and $g'(3) = 0$.

84. Suppose $h = g \circ f$. Find $h'(2)$ if $f(1) = 2$, $f'(1) = 4$, $f(2) = -3$, $f'(2) = 4$, $g(-3) = 1$, and $g'(-3) = 3$.

85. If $y = u^5 + u$ and $u = 4x^3 + x - 4$, find $\dfrac{dy}{dx}$ at $x = 1$.

86. If $y = e^u + 3u$ and $u = \cos x$, find $\dfrac{dy}{dx}$ at $x = 0$.

Applications and Extensions

In Problems 87–94, find the indicated derivative.

87. $\dfrac{d}{dx}f(x^2 + 1)$ (*Hint:* Let $u = x^2 + 1$.)

88. $\dfrac{d}{dx}f(1 - x^2)$

89. $\dfrac{d}{dx}f\left(\dfrac{x+1}{x-1}\right)$

90. $\dfrac{d}{dx}f\left(\dfrac{1-x}{1+x}\right)$

91. $\dfrac{d}{dx}f(\sin x)$

92. $\dfrac{d}{dx}f(\tan x)$

93. $\dfrac{d^2}{dx^2}f(\cos x)$

94. $\dfrac{d^2}{dx^2}f(e^x)$

95. Rectilinear Motion The distance s, in meters, of an object from the origin at time $t \geq 0$ seconds is given by $s = s(t) = A\cos(\omega t + \phi)$, where A, ω, and ϕ are constant.

(a) Find the velocity v of the object at time t.

(b) When is the velocity of the object 0?

(c) Find the acceleration a of the object at time t.

(d) When is the acceleration of the object 0?

96. Rectilinear Motion A bullet is fired horizontally into a bale of paper. The distance s (in meters) the bullet travels into the bale of paper in t seconds is given by

$$s = s(t) = 8 - (2 - t)^3, \quad 0 \leq t \leq 2.$$

(a) Find the velocity v of the bullet at any time t.

(b) Find the velocity of the bullet at $t = 1$ and at $t = 2$.

(c) Find the acceleration a of the bullet at any time t.

(d) Find the acceleration of the bullet at $t = 1$ and at $t = 2$.

(e) How far into the bale of paper did the bullet travel?

97. Rectilinear Motion   Find the acceleration a of a car if the distance s, in feet, it has traveled along a highway at time $t \geq 0$ seconds is given by

$$s(t) = \frac{80}{3}\left[t + \frac{3}{\pi}\sin\left(\frac{\pi}{6}t\right)\right]$$

98. Rectilinear Motion   An object moves in rectilinear motion so that at time $t \geq 0$ seconds, its distance from the origin is $s(t) = \sin e^t$, in feet.

(a) Find the velocity v and acceleration a of the object at any time t.

(b) At what time does the object first have zero velocity?

(c) What is the acceleration of the object at the time t found in (b)?

99. Resistance   The resistance R (measured in ohms) of an 80-meter-long electric wire of radius x (in centimeters) is given by the formula $R = R(x) = \dfrac{0.0048}{x^2}$. The radius x is given by $x = 0.1991 + 0.000003T$ where T is the temperature in Kelvin. How fast is R changing with respect to T when $T = 320 \,\mathrm{K}$?

100. Pendulum Motion in a Car   The motion of a pendulum swinging in the direction of motion of a car moving at a low, constant speed, can be modeled by

$$s = s(t) = 0.05\sin(2t) + 3t \qquad 0 \leq t \leq \pi$$

where s is the distance in meters and t is the time in seconds.

(a) Find the velocity v at $t = 0$, $t = \dfrac{\pi}{8}$, $t = \dfrac{\pi}{4}$, $t = \dfrac{\pi}{2}$, and $t = \pi$.

(b) Find the acceleration a at the times given in (a).

(c) Graph $s = s(t)$, $v = v(t)$, and $a = a(t)$ on the same screen.

Source: Mathematics students at Trine University.

101. Economics   The function $A(t) = 102 - 90\,e^{-0.21t}$ represents the relationship between A, the percentage of the market penetrated by second-generation smart phones, and t, the time in years, where $t = 0$ corresponds to the year 2010.

(a) Find $\lim\limits_{t \to \infty} A(t)$ and interpret the result.

(b) Graph the function $A = A(t)$, and explain how the graph supports the answer in (a).

(c) Find the rate of change of A with respect to time.

(d) Evaluate $A'(5)$ and $A'(10)$ and interpret these results.

(e) Graph the function $A' = A'(t)$, and explain how the graph supports the answers in (d).

102. Meteorology   The atmospheric pressure at a height of x meters above sea level is $P(x) = 10^4 e^{-0.00012x}$ kilograms per square meter. What is the rate of change of the pressure with respect to the height at $x = 500$ m? At $x = 750$ m?

103. Hailstones   Hailstones originate at an altitude of about 3000 m, although this varies. As they fall, air resistance slows down the hailstones considerably. In one model of air resistance, the speed of a hailstone of mass m as a function of time t is given by $v(t) = \dfrac{mg}{k}(1 - e^{-kt/m})$ m/s, where $g = 9.8 \,\mathrm{m/s}^2$ is the acceleration due to gravity and k is a constant that depends on the size of the hailstone and the conditions of the air.

(a) Find the acceleration $a(t)$ of a hailstone as a function of time t.

(b) Find $\lim\limits_{t \to \infty} v(t)$. What does this limit say about the speed of the hailstone?

(c) Find $\lim\limits_{t \to \infty} a(t)$. What does this limit say about the acceleration of the hailstone?

104. Mean Earnings   The mean earnings E, in dollars, of workers 18 years and over are given in the table below:

Year	1975	1980	1985	1990	1995	2000	2005	2010
Mean Earnings	8,552	12,665	17,181	21,793	26,792	32,604	41,231	49,733

Source: U.S. Bureau of the Census, Current Population Survey, 2012.

(a) Find the exponential function of best fit and show that it equals $E = E(t) = 9296(1.05)^t$, where t is the number of years since 1974.

(b) Find the rate of change of E with respect to t.

(c) Find the rate of change at $t = 26$ (year 2000).

(d) Find the rate of change at $t = 31$ (year 2005).

(e) Find the rate of change at $t = 36$ (year 2010).

(f) Compare the answers to (c), (d), and (e). Interpret each answer and explain the differences.

105. Rectilinear Motion   An object moves in rectilinear motion so that at time $t > 0$ its distance s from the origin is $s = s(t)$. The velocity v of the object is $v = \dfrac{ds}{dt}$, and its acceleration is $a = \dfrac{dv}{dt} = \dfrac{d^2 s}{dt^2}$. If the velocity $v = v(s)$ is expressed as a function of s, show that the acceleration a can be expressed as $a = v\dfrac{dv}{ds}$.

106. Student Approval   Professor Miller's student approval rating is modeled by the function $Q(t) = 21 + \dfrac{10\sin\left(\dfrac{2\pi t}{7}\right)}{\sqrt{t} - \sqrt{20}}$, where $0 \leq t \leq 16$ is the number of weeks since the semester began.

(a) Find $Q'(t)$.

(b) Evaluate $Q'(1)$, $Q'(5)$, and $Q'(10)$.

(c) Interpret the results obtained in (b).

(d) Use graphing technology to graph $Q(t)$ and $Q'(t)$.

(e) How would you explain the results in (d) to Professor Miller?

Source: Mathematics students at Millikin University, Decatur, Illinois.

107. Angular Velocity If the disk in the figure is rotated about the vertical through an angle θ, torsion in the wire attempts to turn the disk in the opposite direction. The motion θ at time t (assuming no friction or air resistance) obeys the equation

$$\theta(t) = \frac{\pi}{3}\cos\left(\frac{1}{2}\sqrt{\frac{2k}{5}}t\right)$$

where k is the coefficient of torsion of the wire.

(a) Find the angular velocity $\omega = \dfrac{d\theta}{dt}$ of the disk at any time t.

(b) What is the angular velocity at $t = 3$?

108. Harmonic Motion A weight hangs on a spring making it 2 m long when it is stretched out (see the figure). If the weight is pulled down and then released, the weight oscillates up and down, and the length l of the spring after t seconds is given by $l(t) = 2 + \cos(2\pi t)$.

$l = 2$ m

(a) Find the length l of the spring at the times $t = 0, \dfrac{1}{2}, 1, \dfrac{3}{2}$, and $\dfrac{5}{8}$.

(b) Find the velocity v of the weight at time $t = \dfrac{1}{4}$.

(c) Find the acceleration a of the weight at time $t = \dfrac{1}{4}$.

109. Find $F'(1)$ if $f(x) = \sin x$ and $F(t) = f(t^2 - 1)$.

110. Normal Line Find the point on the graph of $y = e^{-x}$ where the normal line to the graph passes through the origin.

111. Use the Chain Rule and the fact that $\cos x = \sin\left(\dfrac{\pi}{2} - x\right)$ to show that $\dfrac{d}{dx}\cos x = -\sin x$.

112. If $y = e^{2x}$, show that $y'' - 4y = 0$.

113. If $y = e^{-2x}$, show that $y'' - 4y = 0$.

114. If $y = Ae^{2x} + Be^{-2x}$, where A and B are constants, show that $y'' - 4y = 0$.

115. If $y = Ae^{ax} + Be^{-ax}$, where A, B, and a are constants, show that $y'' - a^2 y = 0$.

116. If $y = Ae^{2x} + Be^{3x}$, where A and B are constants, show that $y'' - 5y' + 6y = 0$.

117. If $y = Ae^{-2x} + Be^{-x}$, where A and B are constants, show that $y'' + 3y' + 2y = 0$.

118. If $y = A\sin(\omega t) + B\cos(\omega t)$, where A, B, and ω are constants, show that $y'' + \omega^2 y = 0$.

119. Show that $\dfrac{d}{dx}f(h(x)) = 2xg(x^2)$, if $\dfrac{d}{dx}f(x) = g(x)$ and $h(x) = x^2$.

120. Find the nth derivative of $f(x) = (2x + 3)^n$.

121. Find a general formula for the nth derivative of y.

(a) $y = e^{ax}$ (b) $y = e^{-ax}$

122. (a) What is $\dfrac{d^{10}}{dx^{10}}\sin(ax)$?

(b) What is $\dfrac{d^{25}}{dx^{25}}\sin(ax)$?

(c) Find the nth derivative of $f(x) = \sin(ax)$.

123. (a) What is $\dfrac{d^{11}}{dx^{11}}\cos(ax)$?

(b) What is $\dfrac{d^{12}}{dx^{12}}\cos(ax)$?

(c) Find the nth derivative of $f(x) = \cos(ax)$.

124. If $y = e^{-at}[A\sin(\omega t) + B\cos(\omega t)]$, where A, B, a, and ω are constants, find y' and y''.

125. Show that if a function f has the properties:

- $f(u + v) = f(u)f(v)$ for all choices of u and v
- $f(x) = 1 + xg(x)$, where $\lim\limits_{x\to 0} g(x) = 1$,

then $f' = f$.

Challenge Problems

126. Find the nth derivative of $f(x) = \dfrac{1}{3x - 4}$.

127. Let $f_1(x), \dots, f_n(x)$ be n differentiable functions. Find the derivative of $y = f_1(f_2(f_3(\dots(f_n(x)\dots))))$.

128. Let

$$f(x) = \begin{cases} x^2\sin\dfrac{1}{x} & \text{if } x \neq 0 \\ 0 & \text{if } x = 0 \end{cases}$$

Show that $f'(0)$ exists, but that $f'(x)$ is not continuous at 0.

129. Define f by

$$f(x) = \begin{cases} e^{-1/x^2} & \text{if } x \neq 0 \\ 0 & \text{if } x = 0 \end{cases}$$

Show that f is differentiable on $(-\infty, \infty)$ and find $f'(x)$ for each value of x. [*Hint:* To find $f'(0)$, use the definition of the derivative. Then show that $1 < x^2 e^{1/x^2}$ for $x \neq 0$.]

130. Suppose $f(x) = x^2$ and $g(x) = |x - 1|$. The functions f and g are continuous on their domains, the set of all real numbers.

(a) Is f differentiable at all real numbers? If not, where does f' not exist?

(b) Is g differentiable at all real numbers? If not, where does g' not exist?

(c) Can the Chain Rule be used to differentiate the composite function $(f \circ g)(x)$ for all x? Explain.

(d) Is the composite function $(f \circ g)(x)$ differentiable? If so, what is its derivative?

131. Suppose $f(x) = x^4$ and $g(x) = x^{1/3}$. The functions f and g are continuous on their domains, the set of all real numbers.

(a) Is f differentiable at all real numbers? If not, where does f' not exist?

(b) Is g differentiable at all real numbers? If not, where does g' not exist?

(c) Can the Chain Rule be used to differentiate the composite function $(f \circ g)(x)$ for all x? Explain.

(d) Is the composite function $(f \circ g)(x)$ differentiable? If so, what is its derivative?

132. The function $f(x) = e^x$ has the property $f'(x) = f(x)$. Give an example of another function $g(x)$ such that $g(x)$ is defined for all real x, $g'(x) = g(x)$, and $g(x) \neq f(x)$.

133. Harmonic Motion   The motion of the piston of an automobile engine is approximately simple harmonic. If the stroke of a piston (twice the amplitude) is $10\,\text{cm}$ and the angular velocity ω is 60 revolutions per second, then the motion of the piston is given by $s(t) = 5\sin(120\pi t)\,\text{cm}$.

(a) Find the acceleration a of the piston at the end of its stroke $\left(t = \frac{1}{240}\,\text{second}\right)$.

(b) If the piston weighs $1\,\text{kg}$, what resultant force must be exerted on it at this point? (*Hint:* Use Newton's Second Law, that is, $F = ma$.)

3.2 Implicit Differentiation; Derivatives of the Inverse Trigonometric Functions

OBJECTIVES   *When you finish this section, you should be able to:*

1 Find a derivative using implicit differentiation (p. 209)
2 Find higher-order derivatives using implicit differentiation (p. 212)
3 Differentiate functions with rational exponents (p. 213)
4 Find the derivative of an inverse function (p. 214)
5 Differentiate the inverse trigonometric functions (p. 216)

So far we have differentiated only functions $y = f(x)$ where the dependent variable y is expressed *explicitly* in terms of the independent variable x. There are functions that are not written in the form $y = f(x)$, but are written in the *implicit* form $F(x, y) = 0$. For example, x and y are related implicitly in the equations

$$xy - 4 = 0 \qquad y^2 + 3x^2 - 1 = 0 \qquad e^{x^2 - y^2} - \cos(xy) = 0$$

NEED TO REVIEW?   The implicit form of a function is discussed in Section P.1, p. 4.

The implicit form $xy - 4 = 0$ can easily be written explicitly as the function $y = \dfrac{4}{x}$.

Also, $y^2 + 3x^2 - 1 = 0$ can be written explicitly as the two functions $y_1 = \sqrt{1 - 3x^2}$ and $y_2 = -\sqrt{1 - 3x^2}$. In the equation $e^{x^2 - y^2} - \cos(xy) = 0$, it is impossible to express y as an explicit function of x. In this case, and in many others, we use the technique of *implicit differentiation* to find the derivative.

1 Find a Derivative Using Implicit Differentiation

The method used to differentiate an implicitly defined function is called **implicit differentiation**. It does not require rewriting the function explicitly, but it does require that the dependent variable y is a differentiable function of the independent variable x. So throughout the section, we make the assumption that there is a differentiable function $y = f(x)$ defined by the implicit equation. This assumption made, the method consists of differentiating both sides of the implicitly defined function with respect to x and then solving the resulting equation for $\dfrac{dy}{dx}$.

EXAMPLE 1 Finding a Derivative Using Implicit Differentiation

Find $\dfrac{dy}{dx}$ if $xy - 4 = 0$.

(a) Use implicit differentiation.

(b) Solve for y and then differentiate.

(c) Verify the results of (a) and (b) are the same.

Solution **(a)** To differentiate implicitly, we assume y is a differentiable function of x and differentiate both sides with respect to x.

$$\dfrac{d}{dx}(xy - 4) = \dfrac{d}{dx}0 \qquad \text{\color{blue}Differentiate both sides with respect to } x.$$

$$\dfrac{d}{dx}(xy) - \dfrac{d}{dx}4 = 0 \qquad \text{\color{blue}Use the Difference Rule.}$$

$$x \cdot \dfrac{d}{dx}y + \left(\dfrac{d}{dx}x\right)y - 0 = 0 \qquad \text{\color{blue}Use the Product Rule.}$$

$$x\dfrac{dy}{dx} + y = 0 \qquad \text{\color{blue}Simplify.}$$

$$\dfrac{dy}{dx} = -\dfrac{y}{x} \qquad \text{\color{blue}Solve for }\dfrac{dy}{dx}. \tag{1}$$

(b) Solve $xy - 4 = 0$ for y, obtaining $y = \dfrac{4}{x} = 4x^{-1}$. Then

$$\dfrac{dy}{dx} = \dfrac{d}{dx}(4x^{-1}) = -4x^{-2} = -\dfrac{4}{x^2} \tag{2}$$

(c) At first glance, the results in (1) and (2) appear to be different. However, since $xy - 4 = 0$, or equivalently, $y = \dfrac{4}{x}$, the result from (1) becomes

$$\dfrac{dy}{dx} \underset{\underset{(1)}{\uparrow}}{=} -\dfrac{y}{x} \underset{\underset{y = \frac{4}{x}}{\uparrow}}{=} -\dfrac{\frac{4}{x}}{x} = -\dfrac{4}{x^2}$$

which is the same as (2). ∎

In most instances, we will not know the explicit form of the function (as we did in Example 1) and so we will leave the derivative $\dfrac{dy}{dx}$ expressed in terms of x and y [as in (1)].

NOW WORK Problem 17.

The Power Rule for Functions is

$$\dfrac{d}{dx}[f(x)]^n = n[f(x)]^{n-1} f'(x)$$

where n is an integer. If $y = f(x)$, it takes the form

$$\boxed{\dfrac{d}{dx}y^n = ny^{n-1}\dfrac{dy}{dx}}$$

This is convenient notation to use with implicit differentiation when y^n appears.

$$\underset{\underset{n=1}{\uparrow}}{\dfrac{d}{dx}y = 1 \cdot \dfrac{dy}{dx} = \dfrac{dy}{dx}} \qquad \underset{\underset{n=2}{\uparrow}}{\dfrac{d}{dx}y^2 = 2y\dfrac{dy}{dx}} \qquad \underset{\underset{n=3}{\uparrow}}{\dfrac{d}{dx}y^3 = 3y^2\dfrac{dy}{dx}}$$

To differentiate an implicit function:

- Assume that y is a differentiable function of x.
- Differentiate both sides of the equation with respect to x.
- Solve the resulting equation for $y' = \dfrac{dy}{dx}$.

EXAMPLE 2   Finding a Derivative Using Implicit Differentiation

Find $\dfrac{dy}{dx}$ if $3x^2 + 4y^2 = 2x$.

Solution We assume that y is a differentiable function of x and differentiate both sides with respect to x.

$$\frac{d}{dx}(3x^2 + 4y^2) = \frac{d}{dx}(2x) \qquad \text{Differentiate both sides with respect to } x.$$

$$\frac{d}{dx}(3x^2) + \frac{d}{dx}(4y^2) = 2 \qquad \text{Sum Rule.}$$

$$3\frac{d}{dx}x^2 + 4\frac{d}{dx}y^2 = 2 \qquad \text{Constant Multiple Rule.}$$

$$6x + 4\left(2y\frac{dy}{dx}\right) = 2 \qquad \frac{d}{dx}y^2 = 2y\frac{dy}{dx}$$

$$6x + 8y\frac{dy}{dx} = 2 \qquad \text{Simplify.}$$

$$\frac{dy}{dx} = \frac{2 - 6x}{8y} = \frac{1 - 3x}{4y} \qquad \text{Solve for } \frac{dy}{dx}.$$

provided $y \neq 0$. ∎

Notice in Example 2 that $\dfrac{dy}{dx}$ is expressed in terms of x and y.

NOW WORK Problem 15.

EXAMPLE 3   Using Implicit Differentiation to Find an Equation of a Tangent Line

Find an equation of the tangent line to the graph of the ellipse $3x^2 + 4y^2 = 2x$ at the point $\left(\dfrac{1}{2}, -\dfrac{1}{4}\right)$.

Solution First we find the slope of the tangent line. We use the result from Example 2, and evaluate $\dfrac{dy}{dx} = \dfrac{1 - 3x}{4y}$ at $\left(\dfrac{1}{2}, -\dfrac{1}{4}\right)$.

$$\frac{dy}{dx} = \frac{1 - 3x}{4y} = \frac{1 - 3 \cdot \frac{1}{2}}{4 \cdot \left(-\frac{1}{4}\right)} = \frac{1}{2}$$

$$\uparrow$$
$$x = \frac{1}{2}, y = -\frac{1}{4}$$

The slope of the tangent line to the graph of $3x^2 + 4y^2 = 2x$ at the point $\left(\dfrac{1}{2}, -\dfrac{1}{4}\right)$ is $\dfrac{1}{2}$. An equation of the tangent line is

$$y + \frac{1}{4} = \frac{1}{2}\left(x - \frac{1}{2}\right)$$

$$y = \frac{1}{2}x - \frac{1}{2} \qquad\qquad ∎$$

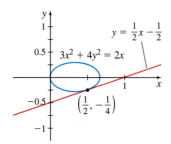

DF Figure 2

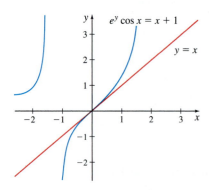

Figure 3

Figure 2 shows the graph of the ellipse $3x^2 + 4y^2 = 2x$ and the graph of the tangent line $y = \dfrac{1}{2}x - \dfrac{1}{2}$ at the point $\left(\dfrac{1}{2}, -\dfrac{1}{4}\right)$.

NOW WORK Problem 63.

EXAMPLE 4 Using Implicit Differentiation

(a) Find y' if $e^y \cos x = x + 1$.

(b) Find an equation of the tangent line to the graph at the point $(0, 0)$.

Solution (a) We use implicit differentiation:

$$\frac{d}{dx}(e^y \cos x) = \frac{d}{dx}(x + 1)$$

$$e^y \cdot \frac{d}{dx}(\cos x) + \left(\frac{d}{dx}e^y\right) \cdot \cos x = 1 \qquad \text{\color{blue}Use the Product Rule.}$$

$$e^y(-\sin x) + e^y y' \cdot \cos x = 1$$

$$(e^y \cos x)y' = 1 + e^y \sin x$$

$$y' = \frac{1 + e^y \sin x}{e^y \cos x}$$

(b) At the point $(0, 0)$, the derivative y' is $\dfrac{1 + e^0 \sin 0}{e^0 \cos 0} = \dfrac{1 + 1 \cdot 0}{1 \cdot 1} = 1$, so the slope of the tangent line to the graph at $(0, 0)$ is 1. An equation of the tangent line to the graph at the point $(0, 0)$ is $y = x$. ∎

The graph of $e^y \cos x = x + 1$ and the tangent line to the graph at $(0, 0)$ are shown in Figure 3.

2 Find Higher-Order Derivatives Using Implicit Differentiation

Implicit differentiation can be used to find higher-order derivatives.

EXAMPLE 5 Finding Higher-Order Derivatives

Use implicit differentiation to find y' and y'' if $y^2 - x^2 = 5$. Express y'' in terms of x and y.

Solution First, we assume there is a differentiable function $y = f(x)$ that satisfies $y^2 - x^2 = 5$. Now we find y'.

$$\frac{d}{dx}(y^2 - x^2) = \frac{d}{dx}5$$

$$\frac{d}{dx}y^2 - \frac{d}{dx}x^2 = 0$$

$$2yy' - 2x = 0 \qquad\qquad \frac{d}{dx}y^2 = 2y\frac{dy}{dx} = 2yy' \qquad (3)$$

$$y' = \frac{2x}{2y} = \frac{x}{y} \qquad\qquad \text{\color{blue}Solve for } y'. \qquad (4)$$

provided $y \neq 0$.

Equations (3) and (4) both involve y'. Either one can be used to find y''. We use (3) because it avoids differentiating a quotient.

$$\frac{d}{dx}(2yy' - 2x) = \frac{d}{dx}0$$

$$\frac{d}{dx}(yy') - \frac{d}{dx}x = 0$$

$$y \cdot \frac{d}{dx}y' + \left(\frac{d}{dx}y\right)y' - 1 = 0$$

$$yy'' + (y')^2 - 1 = 0$$

$$y'' = \frac{1 - (y')^2}{y} \tag{5}$$

provided $y \neq 0$. To express y'' in terms of x and y, we use (4) and substitute for y' in (5).

$$y'' = \frac{1 - \left(\dfrac{x}{y}\right)^2}{\underset{\underset{\displaystyle y' = \frac{x}{y}}{\uparrow}}{y}} = \frac{y^2 - x^2}{y^3} = \underset{\underset{\displaystyle y^2 - x^2 = 5}{\uparrow}}{\frac{5}{y^3}} \qquad \blacksquare$$

NOW WORK   **Problem 59.**

3 Differentiate Functions with Rational Exponents

In Section 2.4, we showed that the Simple Power Rule

$$\frac{d}{dx}x^n = nx^{n-1}$$

is true if the exponent n is any integer. We use implicit differentiation to extend this result for rational exponents.

THEOREM   **Power Rule for Rational Exponents**

If $y = x^{p/q}$, where $\dfrac{p}{q}$ is a rational number, then

$$\boxed{y' = \frac{d}{dx}x^{p/q} = \frac{p}{q} \cdot x^{(p/q)-1}} \tag{6}$$

provided that $x^{p/q}$ and $x^{(p/q)-1}$ are defined.

Proof   We begin with the function $y = x^{p/q}$, where p and $q > 0$ are integers. Now, we raise both sides of the equation to the power q to obtain

$$y^q = x^p$$

This is the implicit form of the function $y = x^{p/q}$. Assuming y is differentiable,* we can differentiate implicitly, obtaining

$$\frac{d}{dx}y^q = \frac{d}{dx}x^p$$

$$qy^{q-1}y' = px^{p-1}$$

---

*In Problem 113, you are asked to show that $y = x^{p/q}$ is differentiable.

Now solve for y'.

$$y' = \frac{px^{p-1}}{qy^{q-1}} \underset{\underset{y=x^{p/q}}{\uparrow}}{=} \frac{p}{q} \cdot \frac{x^{p-1}}{(x^{p/q})^{q-1}} = \frac{p}{q} \cdot \frac{x^{p-1}}{x^{p-(p/q)}} = \frac{p}{q} \cdot x^{p-1-[p-(p/q)]} = \frac{p}{q} \cdot x^{(p/q)-1} \quad \blacksquare$$

EXAMPLE 6   Differentiating Functions with Rational Exponents

(a) $\dfrac{d}{dx}\sqrt{x} = \dfrac{d}{dx}x^{1/2} = \dfrac{1}{2}x^{(1/2)-1} = \dfrac{1}{2}x^{-1/2} = \dfrac{1}{2x^{1/2}} = \dfrac{1}{2\sqrt{x}} = \dfrac{\sqrt{x}}{2x}$

(b) $\dfrac{d}{du}\sqrt[3]{u} = \dfrac{d}{du}u^{1/3} = \dfrac{1}{3}u^{-2/3} = \dfrac{1}{3\sqrt[3]{u^2}} = \dfrac{\sqrt[3]{u}}{3u}$

(c) $\dfrac{d}{dx}x^{5/2} = \dfrac{5}{2}x^{3/2}$

(d) $\dfrac{d}{ds}s^{-3/2} = -\dfrac{3}{2}s^{-5/2} = -\dfrac{3}{2s^{5/2}}$ ■

NOW WORK   Problem 31.

THEOREM   Power Rule for Functions

If u is a differentiable function of x and r is a rational number, then

$$\boxed{\dfrac{d}{dx}[u(x)]^r = r[u(x)]^{r-1}u'(x)}$$

provided u^r and u^{r-1} are defined.

EXAMPLE 7   Differentiating Functions Using the Power Rule

(a) $\dfrac{d}{ds}(s^3 - 2s + 1)^{5/3} = \dfrac{5}{3}(s^3 - 2s + 1)^{2/3}\dfrac{d}{ds}(s^3 - 2s + 1) = \dfrac{5}{3}(s^3 - 2s + 1)^{2/3}(3s^2 - 2)$

(b) $\dfrac{d}{dx}\sqrt[3]{x^4 - 3x + 5} = \dfrac{d}{dx}(x^4 - 3x + 5)^{1/3} = \dfrac{1}{3}(x^4 - 3x + 5)^{-2/3}\dfrac{d}{dx}(x^4 - 3x + 5)$

$$= \dfrac{4x^3 - 3}{3(x^4 - 3x + 5)^{2/3}}$$

(c) $\dfrac{d}{d\theta}[\tan(3\theta)]^{-3/4} = -\dfrac{3}{4}[\tan(3\theta)]^{-7/4}\dfrac{d}{d\theta}\tan(3\theta) = -\dfrac{3}{4}[\tan(3\theta)]^{-7/4} \cdot \sec^2(3\theta) \cdot 3$

$$= -\dfrac{9\sec^2(3\theta)}{4[\tan(3\theta)]^{7/4}} \qquad \blacksquare$$

NOW WORK   Problem 39.

4 Find the Derivative of an Inverse Function

NEED TO REVIEW? Inverse functions
are discussed in Section P.4, pp 32–37.

Suppose f is a function and g is its inverse function. Then

$$g(f(x)) = x$$

for all x in the domain of f.

If both f and g are differentiable, we can differentiate both sides with respect to x using the Chain Rule. Then

$$\frac{d}{dx}[g(f(x))] = g'(f(x)) \cdot f'(x) = 1$$

Use the Chain Rule on the left $\frac{d}{dx}x = 1$ on the right

Since the product of the derivatives is never 0, each function has a nonzero derivative on its domain.

Conversely, if a one-to-one function has a nonzero derivative, then its inverse function also has a nonzero derivative as stated in the following theorem.

THEOREM Derivative of an Inverse Function

Let $y = f(x)$ and $x = g(y)$ be inverse functions. Suppose f is differentiable on an open interval containing x_0 and $y_0 = f(x_0)$. If $f'(x_0) \neq 0$, then g is differentiable at $y_0 = f(x_0)$ and

$$\boxed{\frac{d}{dy}g(y_0) = g'(y_0) = \frac{1}{f'(x_0)}}$$

(7)

where the notation $\frac{d}{dy}g(y_0) = g'(y_0)$ means the value of $\frac{d}{dy}g(y)$ at y_0, and the notation $f'(x_0)$ means the value of $f'(x)$ at x_0.

In Leibniz notation, formula (7) has the simple form

$$\boxed{\frac{dx}{dy} = \frac{1}{\dfrac{dy}{dx}}}$$

We have two comments to make about this theorem, which is proved in Appendix B:

- If $y_0 = f(x_0)$ and $f'(x_0) \neq 0$ exists, then $\frac{d}{dy}g(y)$ exists at y_0.

- It gives formula (7) for finding $\frac{d}{dy}g(y)$ at y_0 without knowing a formula for $g = f^{-1}$, provided we can find x_0 and $f'(x_0)$.

EXAMPLE 8 Finding the Derivative of an Inverse Function

The function $f(x) = x^5 + x$ has an inverse function g. Find $g'(2)$.

Solution Using (7) with $y_0 = 2$, we get

$$g'(2) = \frac{1}{f'(x_0)} \qquad \text{where } 2 = f(x_0)$$

A solution of the equation

$$f(x_0) = x_0^5 + x_0 = 2$$

is $x_0 = 1$. Since $f'(x) = 5x^4 + 1$, then $f'(x_0) = f'(1) = 6$ and

$$g'(2) = \frac{1}{f'(1)} = \frac{1}{6} \qquad\blacksquare$$

Observe in Example 8 that the derivative of the inverse function g at 2 was evaluated without actually knowing a formula for g.

NOW WORK Problem 71.

5 Differentiate the Inverse Trigonometric Functions

Table 1 lists the inverse trigonometric functions and their domains.

NEED TO REVIEW? Inverse trigonometric functions are discussed in Section P.7, pp 58–61.

TABLE 1

f	Restricted Domain	f^{-1}	Domain
$f(x) = \sin x$	$\left[-\dfrac{\pi}{2}, \dfrac{\pi}{2}\right]$	$f^{-1}(x) = \sin^{-1} x$	$[-1, 1]$
$f(x) = \cos x$	$[0, \pi]$	$f^{-1}(x) = \cos^{-1} x$	$[-1, 1]$
$f(x) = \tan x$	$\left(-\dfrac{\pi}{2}, \dfrac{\pi}{2}\right)$	$f^{-1}(x) = \tan^{-1} x$	$(-\infty, \infty)$
$f(x) = \csc x$	$\left(-\pi, -\dfrac{\pi}{2}\right] \cup \left(0, \dfrac{\pi}{2}\right]$	$f^{-1}(x) = \csc^{-1} x$	$\lvert x \rvert \geq 1$
$f(x) = \sec x$	$\left[0, \dfrac{\pi}{2}\right) \cup \left[\pi, \dfrac{3\pi}{2}\right)$	$f^{-1}(x) = \sec^{-1} x$	$\lvert x \rvert \geq 1$
$f(x) = \cot x$	$(0, \pi)$	$f^{-1}(x) = \cot^{-1} x$	$(-\infty, \infty)$

To find the derivative of $y = \sin^{-1} x$, $-1 \leq x \leq 1$, $-\dfrac{\pi}{2} \leq y \leq \dfrac{\pi}{2}$, we write $\sin y = x$ and differentiate implicitly with respect to x.

$$\frac{d}{dx} \sin y = \frac{d}{dx} x$$

$$\cos y \cdot \frac{dy}{dx} = 1$$

$$\frac{dy}{dx} = \frac{1}{\cos y}$$

provided $\cos y \neq 0$. Since $\cos y = 0$ if $y = -\dfrac{\pi}{2}$ or $y = \dfrac{\pi}{2}$, we exclude these values.

Then $\dfrac{d}{dx} \sin^{-1} y = \dfrac{1}{\cos y}$, $-\dfrac{\pi}{2} < y < \dfrac{\pi}{2}$. Now, if $-\dfrac{\pi}{2} < y < \dfrac{\pi}{2}$, then $\cos y > 0$. Using a Pythagorean identity, we have

$$\cos^2 y = 1 - \sin^2 y$$

$$\cos y = \underset{\underset{\cos y > 0}{\uparrow}}{\sqrt{1 - \sin^2 y}} = \underset{\underset{\sin y = x}{\uparrow}}{\sqrt{1 - x^2}}$$

THEOREM Derivative of $y = \sin^{-1} x$

The derivative of the inverse sine function $y = \sin^{-1} x$ is

$$y' = \frac{d}{dx} \sin^{-1} x = \frac{1}{\sqrt{1 - x^2}} \qquad -1 < x < 1$$

EXAMPLE 9   Using the Chain Rule with the Inverse Sine Function

Find y' if:

(a) $y = \sin^{-1}(4x^2)$ **(b)** $y = e^{\sin^{-1} x}$

Solution **(a)** If $y = \sin^{-1} u$ and $u = 4x^2$, then $\dfrac{dy}{du} = \dfrac{1}{\sqrt{1 - u^2}}$ and $\dfrac{du}{dx} = 8x$. By the Chain Rule,

$$y' = \frac{dy}{dx} = \frac{dy}{du} \cdot \frac{du}{dx} = \left(\frac{1}{\sqrt{1 - u^2}}\right)(8x) = \frac{8x}{\sqrt{1 - 16x^4}} \qquad u = 4x^2$$

(b) If $y = e^u$ and $u = \sin^{-1} x$, then $\dfrac{dy}{du} = e^u$ and $\dfrac{du}{dx} = \dfrac{1}{\sqrt{1 - x^2}}$. By the Chain Rule,

$$y' = \frac{dy}{dx} = \frac{dy}{du} \cdot \frac{du}{dx} = e^u \cdot \frac{1}{\sqrt{1 - x^2}} \underset{\substack{\uparrow \\ u = \sin^{-1} x}}{=} \frac{e^{\sin^{-1} x}}{\sqrt{1 - x^2}}$$ ∎

NOW WORK Problem 51.

To derive a formula for the derivative of $y = \tan^{-1} x$, $-\infty < x < \infty$, $-\dfrac{\pi}{2} < y < \dfrac{\pi}{2}$, we write $\tan y = x$ and differentiate with respect to x. Then

$$\frac{d}{dx} \tan y = \frac{d}{dx} x$$

$$\sec^2 y \frac{dy}{dx} = 1$$

$$\frac{dy}{dx} = \frac{1}{\sec^2 y}$$

Since $-\dfrac{\pi}{2} < y < \dfrac{\pi}{2}$, then $\sec y \neq 0$.

Now we use the Pythagorean identity $\sec^2 y = 1 + \tan^2 y$, and substitute $x = \tan y$. Then

$$y' = \frac{d}{dx} \tan^{-1} x = \frac{1}{1 + x^2}$$

THEOREM Derivative of $y = \tan^{-1} x$

The derivative of the inverse tangent function $y = \tan^{-1} x$ is

$$\boxed{y' = \frac{d}{dx} \tan^{-1} x = \frac{1}{1 + x^2}}$$

The domain of y' is all real numbers.

EXAMPLE 10   Using the Chain Rule with the Inverse Tangent Function

Find y' if:

(a) $y = \tan^{-1}(4x)$ **(b)** $y = \sin(\tan^{-1} x)$

Solution **(a)** Let $y = \tan^{-1} u$ and $u = 4x$. Then $\dfrac{dy}{du} = \dfrac{1}{1 + u^2}$ and $\dfrac{du}{dx} = 4$. By the Chain Rule,

$$y' = \frac{dy}{dx} = \frac{dy}{du} \cdot \frac{du}{dx} = \frac{1}{1 + u^2} \cdot 4 = \frac{4}{1 + 16x^2}$$

(b) Let $y = \sin u$ and $u = \tan^{-1} x$. Then $\dfrac{dy}{du} = \cos u$ and $\dfrac{du}{dx} = \dfrac{1}{1 + x^2}$. By the Chain Rule,

$$y' = \frac{dy}{dx} = \frac{dy}{du} \cdot \frac{du}{dx} = \cos u \cdot \frac{1}{1 + x^2} = \frac{\cos(\tan^{-1} x)}{1 + x^2}$$ ∎

NOW WORK Problem 55.

Finally to find the derivative of $y = \sec^{-1} x$, $|x| \geq 1$, $0 \leq y < \dfrac{\pi}{2}$ or $\pi \leq y < \dfrac{3\pi}{2}$, we write $x = \sec y$ and use implicit differentiation.

$$\frac{d}{dx} x = \frac{d}{dx} \sec y$$

$$1 = \sec y \tan y \frac{dy}{dx}$$

We can solve for $\dfrac{dy}{dx}$ provided $\sec y \neq 0$ and $\tan y \neq 0$, or equivalently, $y \neq 0$ and $y \neq \pi$. That is,

$$\frac{dy}{dx} = \frac{1}{\sec y \tan y}, \qquad \text{provided } 0 < y < \frac{\pi}{2} \text{ or } \pi < y < \frac{3\pi}{2}.$$

With these restrictions on y, $\tan y > 0$.* Now we use the Pythagorean identity $1 + \tan^2 y = \sec^2 y$. Then $\tan y = \sqrt{\sec^2 y - 1}$ and

$$\frac{1}{\sec y \tan y} = \frac{1}{\sec y \sqrt{\sec^2 y - 1}} \underset{\underset{\sec y = x}{\uparrow}}{=} \frac{1}{x\sqrt{x^2 - 1}}$$

THEOREM Derivative of $y = \sec^{-1} x$

The derivative of the inverse secant function $y = \sec^{-1} x$ is

$$\boxed{\; y' = \frac{d}{dx}\sec^{-1} x = \frac{1}{x\sqrt{x^2 - 1}} \qquad |x| > 1 \;}$$

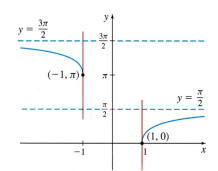

Figure 4 $y = \sec^{-1} x,\ |x| \ge 1,$
$0 \le y < \frac{\pi}{2} \text{ or } \pi \le y < \frac{3\pi}{2}$

Notice that $y = \sec^{-1} x$ is not differentiable when $x = \pm 1$. In fact, as Figure 4 shows, at the points $(-1, \pi)$ and $(1, 0)$, the graph of $y = \sec^{-1} x$ has vertical tangent lines.

The formulas for the derivatives of the three remaining inverse trigonometric functions can be obtained using the following identities:

- Since $\cos^{-1} x = \dfrac{\pi}{2} - \sin^{-1} x$, then $\dfrac{d}{dx}\cos^{-1} x = -\dfrac{1}{\sqrt{1 - x^2}}$, where $|x| < 1$.

- Since $\cot^{-1} x = \dfrac{\pi}{2} - \tan^{-1} x$, then $\dfrac{d}{dx}\cot^{-1} x = -\dfrac{1}{1 + x^2}$, where $-\infty < x < \infty$.

- Since $\csc^{-1} x = \dfrac{\pi}{2} - \sec^{-1} x$, then $\dfrac{d}{dx}\csc^{-1} x = -\dfrac{1}{x\sqrt{x^2 - 1}}$, where $|x| > 1$.

---

*The restricted domain of $y = \sec x$ was chosen as $\left\{ x \,\middle|\, 0 \le x < \dfrac{\pi}{2} \text{ or } \pi \le x < \dfrac{3\pi}{2} \right\}$ so that $\tan x > 0$.

3.2 Assess Your Understanding

Concepts and Vocabulary

1. *True or False* If f is a one-to-one differentiable function and if $f'(x) > 0$, then f has an inverse function whose derivative is positive.

2. *True or False* $\dfrac{d}{dx}\tan^{-1} x = \dfrac{1}{1 + x^2}$, where $-\infty < x < \infty$.

3. *True or False* Implicit differentiation is a technique for finding the derivative of an implicitly defined function.

4. $\dfrac{d}{dx}\sin^{-1} x = $ _____, $-1 < x < 1$.

5. *True or False* If $y^q = x^p$ for integers p and q, then $qy^{q-1} = px^{p-1}$.

6. $\dfrac{d}{dx}(3x^{1/3}) = $ _____.

Skill Building

In Problems 7–30, find $y' = \dfrac{dy}{dx}$ using implicit differentiation.

7. $x^2 + y^2 = 4$

8. $y^4 - 4x^2 = 4$

9. $e^y = \sin x$

10. $e^y = \tan x$

11. $e^{x+y} = y$

12. $e^{x+y} = x^2$

13. $x^2 y = 5$

14. $x^3 y = 8$

15. $x^2 - y^2 - xy = 2$

16. $x^2 - 4xy + y^2 = y$

17. $\dfrac{1}{x} + \dfrac{1}{y} = 1$

18. $\dfrac{1}{x} - \dfrac{1}{y} = 4$

19. $x^2 + y^2 = \dfrac{2y}{x}$

20. $x^2 + y^2 = \dfrac{2y^2}{x^2}$

21. $e^x \sin y + e^y \cos x = 4$

22. $e^y \cos x + e^{-x}\sin y = 10$

 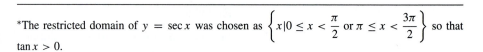

23. $(x^2 + y)^3 = y$

24. $(x + y^2)^3 = 3x$

25. $y = \tan(x - y)$

26. $y = \cos(x + y)$

27. $y = x \sin y$

28. $y = x \cos y$

29. $x^2 y = e^{xy}$

30. $ye^x = y - x$

In Problems 31–56, find y'.

31. $y = x^{2/3} + 4$

32. $y = x^{1/3} - 1$

33. $y = \sqrt[3]{x^2}$

34. $y = \sqrt[4]{x^5}$

35. $y = \sqrt[3]{x} - \dfrac{1}{\sqrt[3]{x}}$

36. $y = \sqrt{x} + \dfrac{1}{\sqrt{x}}$

37. $y = (x^3 - 1)^{1/2}$

38. $y = (x^2 - 1)^{1/3}$

39. $y = x\sqrt{x^2 - 1}$

40. $y = x\sqrt{x^3 + 1}$

41. $y = e^{\sqrt{x^2 - 9}}$

42. $y = \sqrt{e^x}$

43. $y = (x^2 \cos x)^{3/2}$

44. $y = (x^2 \sin x)^{3/2}$

45. $y = (x^2 - 3)^{3/2}(6x + 1)^{5/3}$

46. $y = \dfrac{(2x^3 - 1)^{4/3}}{(3x + 4)^{5/2}}$

47. $y = \sin^{-1}(4x)$

48. $y = \cos^{-1} x^2$

49. $y = \sec^{-1}(3x)$

50. $y = \tan^{-1}\left(\dfrac{1}{x}\right)$

51. $y = \sin^{-1} e^x$

52. $y = \sin^{-1}(1 - x^2)$

53. $y = x(\sin^{-1} x)$

54. $y = x \tan^{-1}(x + 1)$

55. $y = \tan^{-1}(\sin x)$

56. $y = \sin(\tan^{-1} x)$

In Problems 57–62, find y' and y''.

57. $x^2 + y^2 = 4$

58. $x^2 - y^2 = 1$

59. $xy^2 + yx^2 = 2$

60. $4xy = x^2 + y^2$

61. $y = \sqrt{x^2 + 1}$

62. $y = \sqrt{4 - x^2}$

In Problems 63–68 for each implicitly defined equation:

(a) Find the slope of the tangent line to the graph of the equation at the indicated point.

(b) Write an equation for this tangent line.

(c) Graph the tangent line on the same axes as the graph of the equation.

63. $x^2 + y^2 = 5$ at $(1, 2)$

64. $(x - 3)^2 + (y + 4)^2 = 25$ at $(0, 0)$

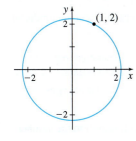

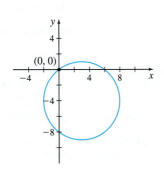

65. $x^2 - y^2 = 8$ at $(3, 1)$

66. $y^2 - 3x^2 = 6$ at $(1, -3)$

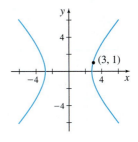

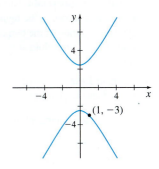

67. $\dfrac{x^2}{4} + \dfrac{y^2}{3} = 1$ at $\left(-1, \dfrac{3}{2}\right)$

68. $x^2 + \dfrac{y^2}{4} = 1$ at $\left(\dfrac{1}{2}, \sqrt{3}\right)$

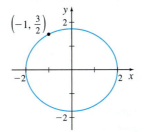

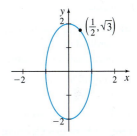

69. Find y' and y'' at the point $(-1, 1)$ on the graph of

$$3x^2 y + 2y^3 = 5x^2$$

70. Find y' and y'' at the point $(0, 0)$ on the graph of

$$4x^3 + 2y^3 = x + y.$$

In Problems 71–76, the functions f and g are inverse functions.

71. If $f(0) = 4$ and $f'(0) = -2$, find $g'(4)$.

72. If $f(1) = -2$ and $f'(1) = 4$, find $g'(-2)$.

73. If $g(3) = -2$ and $g'(3) = \dfrac{1}{2}$, find $f'(-2)$.

74. If $g(-1) = 0$ and $g'(-1) = -\dfrac{1}{3}$, find $f'(0)$.

75. The function $f(x) = x^3 + 2x$ has an inverse function g. Find $g'(0)$ and $g'(3)$.

76. The function $f(x) = 2x^3 + x - 3$ has an inverse function g. Find $g'(-3)$ and $g'(0)$.

Applications and Extensions

In Problems 77–84, find y' using the Power Rule.
(Hint: Use the fact that $|x| = \sqrt{x^2}$.)

77. $y = |3x|$

78. $y = |x^5|$

79. $y = |2x - 1|$

80. $y = |5 - x^2|$

81. $y = |\cos x|$

82. $y = |\sin x|$

83. $y = \sin|x|$

84. $|x| + |y| = 1$

85. Tangent Line to a Hypocycloid
The graph of $x^{2/3} + y^{2/3} = 5$ is called a **hypocycloid**. Part of its graph is shown in the figure. Find an equation of the tangent line to the hypocycloid at the point $(1, 8)$.

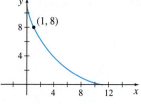

86. Tangent Line   At what point does the graph of
$$y = \frac{1}{\sqrt{x}}$$
have a tangent line parallel to the line $x + 16y = 5$? See the figure.

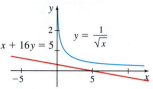

87. Tangent Line to a Cardioid
The graph of
$$(x^2 + y^2 + 2x)^2 = 4(x^2 + y^2),$$
called a **cardioid**, is shown in the figure.

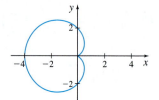

(a) Find all the points on the cardioid that have a horizontal tangent line.

(b) Find all the points on the cardioid that have a vertical tangent line.

(c) Find any point on the cardioid that has no tangent line.

88. Tangent Line to a Cardioid   The graph of
$$(x^2 + y^2 + y)^2 = (x^2 + y^2)$$
is a cardioid.

(a) Find all the points on the cardioid that have a horizontal tangent line.

(b) Find all the points on the cardioid that have a vertical tangent line.

(c) Find any point on the cardioid that has no tangent line.

[CAS] (d) Graph the cardioid and any horizontal or vertical tangent lines.

89. Tangent Line   For the equation $x + xy + 2y^2 = 6$:

(a) Find an expression for the slope of the tangent line at any point (x, y) on the graph.

(b) Write an equation for the line tangent to the graph at the point $(2, 1)$.

(c) Find the coordinates of any other point on this graph with slope equal to the slope at $(2, 1)$.

[CAS] (d) Graph the equation and the tangent lines found in parts (b) and (c) on the same screen.

90. Tangent Line to a Lemniscate
The graph of $(x^2 + y^2)^2 = x^2 - y^2$, called a **lemniscate**, is shown in the figure. There are exactly four points at which the tangent line to the lemniscate is horizontal. Find them.

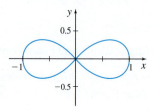

91. Tangent Line
(a) Find an equation for the tangent line to the graph of
$$y = \sin^{-1} \frac{x}{2} \text{ at the point } (0, 0).$$

(b) Graph $y = \sin^{-1} \frac{x}{2}$ and the tangent line at $(0, 0)$.

92. Physics   For ideal gases, **Boyle's law** states that pressure is inversely proportional to volume. A more realistic relationship between pressure P and volume V is given by the **van der Waals equation**
$$P + \frac{a}{V^2} = \frac{C}{V - b}$$
where C is the constant of proportionality, a is a constant that depends on molecular attraction, and b is a constant that depends on the size of the molecules. Find $\frac{dV}{dP}$, which measures the compressibility of the gas.

93. Rectilinear Motion   An object of mass m moves in rectilinear motion so that at time $t > 0$ its distance s from the origin and its velocity $v = \frac{ds}{dt}$ satisfy the equation
$$m\left(v^2 - v_0^2\right) = k\left(s_0^2 - s^2\right)$$
where k is a positive constant and v_0 and s_0 are the initial velocity and position, respectively, of the object. Show that if $v > 0$, then
$$ma = -ks$$
where $a = \frac{d^2 s}{dt^2}$ is the acceleration of the object.
[*Hint*: Differentiate the expression $m\left(v^2 - v_0^2\right) = k\left(s_0^2 - s^2\right)$ with respect to t.]

94. Price Function   It is estimated that t months from now the average price (in dollars) of a tablet will be given by
$$P(t) = \frac{300}{1 + \frac{1}{6}\sqrt{t}} + 100, \ 0 \le t \le 60.$$

(a) Find $P'(t)$

(b) Find $P'(0)$, $P'(16)$, and $P'(49)$ and interpret the results.

(c) Graph $P = P(t)$, and explain how the graph supports the answers in (b).

95. Production Function   The production of commodities sometimes requires several resources such as land, labor, and machinery. If there are two inputs that require amounts x and y, then the output z is given by the function of two variables: $z = f(x, y)$. Here, z is called a **production function**. For example, if x represents land, y represents capital, and z is the amount of a commodity produced, a possible production function is $z = x^{0.5}y^{0.4}$. Set z equal to a fixed amount produced and show that $\frac{dy}{dx} = -\frac{5y}{4x}$. This illustrates that the rate of change of capital with respect to land is always negative when the amount produced is fixed.

96. Learning Curve   The psychologist L. L. Thurstone suggested the following function for the time T it takes to memorize a list of n words: $T = f(n) = Cn\sqrt{n - b}$, where C and b are constants depending on the person and the task.

(a) Find the rate of change in time T with respect to the number n of words to be memorized.

(b) Suppose that for a certain person and a certain task, $C = 2$ and $b = 2$. Find $f'(10)$ and $f'(30)$.

(c) Interpret the results found in (c).

97. The Folium of Descartes   The graph of the equation $x^3 + y^3 = 2xy$ is called the **Folium of Descartes**.

(a) Find y'.

(b) Find an equation of the tangent line to the Folium of Descartes at the point $(1, 1)$.

(c) Find any points on the graph where the tangent line to the graph is horizontal.

CAS **(d)** Graph the equation $x^3 + y^3 = 2xy$. Explain how the graph supports the answers to (b) and (c).

98. If n is an even positive integer, show that the tangent line to the graph of $y = \sqrt[n]{x}$ at $(1, 1)$ is perpendicular to the tangent line to the graph of $y = x^n$ at $(-1, 1)$.

99. At what point(s), if any, is the line $y = x - 1$ parallel to the tangent line to the graph of $y = \sqrt{25 - x^2}$?

100. What is wrong with the following?
If $x + y = e^{x+y}$, then $1 + y' = e^{x+y}(1 + y')$.
Since $e^{x+y} > 0$, then $y' = -1$ for all x. Therefore, $x + y = e^{x+y}$ must be a line of slope -1.

101. Show that if a function y is differentiable, and x and y are related by the equation $x^n y^m + x^m y^n = k$, where k is a constant, then
$$\frac{dy}{dx} = -\frac{y(nx^r + my^r)}{x(mx^r + ny^r)} \quad \text{where} \quad r = n - m$$

102. If $g(x) = \cos^{-1}(\cos x)$, show that $g'(x) = \dfrac{\sin x}{|\sin x|}$.

103. Show that $\dfrac{d}{dx}\tan^{-1}(\cot x) = -1$.

104. Show that $\dfrac{d}{dx}\cot^{-1} x = \dfrac{d}{dx}\tan^{-1}\dfrac{1}{x}$ for all $x \neq 0$.

105. Establish the identity $\sin^{-1} x + \cos^{-1} x = \dfrac{\pi}{2}$ by showing that the derivative of $y = \sin^{-1} x + \cos^{-1} x$ is 0. Use the fact that when $x = 0$, then $y = \dfrac{\pi}{2}$.

106. Establish the identity $\tan^{-1} x + \cot^{-1} x = \dfrac{\pi}{2}$ by showing that the derivative of $y = \tan^{-1} x + \cot^{-1} x$ is 0. Use the fact that when $x = 1$, then $y = \dfrac{\pi}{2}$.

107. Tangent Line   Show that an equation for the tangent line at any point (x_0, y_0) on the ellipse $\dfrac{x^2}{a^2} + \dfrac{y^2}{b^2} = 1$ is $\dfrac{xx_0}{a^2} + \dfrac{yy_0}{b^2} = 1$.

108. Tangent Line   Show that the slope of the tangent line to a hypocycloid $x^{2/3} + y^{2/3} = a^{2/3}$, $a > 0$, at any point for which $x \neq 0$, is $-\dfrac{y^{1/3}}{x^{1/3}}$.

109. Tangent Line   Use implicit differentiation to show that the tangent line to a circle $x^2 + y^2 = R^2$ at any point P on the circle is perpendicular to OP, where O is the center of the circle.

Challenge Problems

110. Let $A = (2, 1)$ and $B = (5, 2)$ be points on the graph of $f(x) = \sqrt{x - 1}$. A line is moved upward on the graph so that it remains parallel to the secant line AB. Find the coordinates of the last point on the graph of f before the secant line loses contact with the graph.

Orthogonal Graphs   *Problems 111 and 112 require the following definition:*

*The graphs of two functions are said to be **orthogonal** if the tangent lines to the graphs are perpendicular at each point of intersection.*

111. (a) Show that the graphs of $xy = c_1$ and $-x^2 + y^2 = c_2$ are orthogonal, where c_1 and c_2 are positive constants.

CAS **(b)** Graph each function on one coordinate system for $c_1 = 1, 2, 3$ and $c_2 = 1, 9, 25$.

112. Find $a > 0$ so that the parabolas $y^2 = 2ax + a^2$ and $y^2 = a^2 - 2ax$ are orthogonal.

113. Show that if p and $q > 0$ are integers, then $y = x^{p/q}$ is a differentiable function of x.

114. We say that y is an **algebraic function** of x if it is a function that satisfies an equation of the form
$$P_0(x)y^n + P_1(x)y^{n-1} + \cdots + P_{n-1}(x)y + P_n(x) = 0$$
where $P_k(x)$, $k = 0, 1, 2, \ldots, n$, are polynomials. For example, $y = \sqrt{x}$ satisfies
$$y^2 - x = 0$$
Use implicit differentiation to obtain a formula for the derivative of an algebraic function.

115. Another way of finding the derivative of $y = \sqrt[n]{x}$ is to use inverse functions. The function $y = f(x) = x^n$, n a positive integer, has the derivative $f'(x) = nx^{n-1}$. So, if $x \neq 0$, then $f'(x) \neq 0$. The inverse function of f, namely, $x = g(y) = \sqrt[n]{y}$, is defined for all y, if n is odd, and for all $y \geq 0$, if n is even. Since this inverse function is differentiable for all $y \neq 0$, we have
$$g'(y) = \frac{d}{dy}\sqrt[n]{y} = \frac{1}{f'(x)} = \frac{1}{nx^{n-1}}$$
Since $nx^{n-1} = n\left(\sqrt[n]{y}\right)^{n-1} = ny^{(n-1)/n} = ny^{1-(1/n)}$, we have
$$\frac{d}{dy}\sqrt[n]{y} = \frac{d}{dy}y^{1/n} = \frac{1}{ny^{1-(1/n)}} = \frac{1}{n}y^{(1/n)-1}$$
Use the result from above and the Chain Rule to prove the formula
$$\frac{d}{dx}x^{p/q} = \frac{p}{q}x^{(p/q)-1}$$

116. (a) You might try to infer from Problem 104 that
$$\cot^{-1} x = \tan^{-1}\frac{1}{x} + C \text{ for all } x \neq 0, \text{ where } C \text{ is a constant.}$$
Show, however, that
$$\cot^{-1} x = \begin{cases} \tan^{-1}\dfrac{1}{x} & \text{if } x > 0 \\[2mm] \tan^{-1}\dfrac{1}{x} + \pi & \text{if } x < 0 \end{cases}$$

(b) What is an explanation of the incorrect inference?

3.3 Derivatives of Logarithmic Functions

OBJECTIVES *When you finish this section, you should be able to:*

1 Differentiate logarithmic functions (p. 222)
2 Use logarithmic differentiation (p. 225)
3 Express *e* as a limit (p. 227)

Recall that the function $f(x) = a^x$, where $a > 0$ and $a \neq 1$, is one-to-one and is defined for all real numbers. We know from Section 3.1 that $f'(x) = a^x \ln a$ and that the domain of $f'(x)$ is also the set of all real numbers. Moreover, since $f'(x) \neq 0$, the inverse of f, $y = \log_a x$, is differentiable on its domain $(0, \infty)$. See Figure 5.

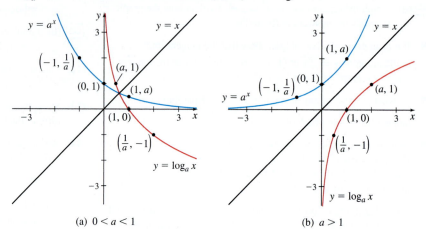

(a) $0 < a < 1$ (b) $a > 1$

Figure 5

NEED TO REVIEW? Logarithmic functions are discussed in Section P.5, pp. 42–45.

1 Differentiate Logarithmic Functions

To find $y' = \dfrac{d}{dx} \log_a x$, we use the fact that the following statements are equivalent:

$$y = \log_a x \quad \text{if and only} \quad \text{if } x = a^y$$

Now we differentiate the equation on the right using implicit differentiation.

$$\frac{d}{dx} x = \frac{d}{dx} a^y$$

$$1 = a^y \ln a \frac{dy}{dx} \qquad \text{Use the Chain Rule on the right.}$$

$$\frac{dy}{dx} = \frac{1}{a^y \ln a} \qquad \text{Solve for } \frac{dy}{dx}.$$

$$\frac{dy}{dx} = \frac{1}{x \ln a} \qquad x = a^y$$

THEOREM Derivative of a Logarithmic Function

The derivative of the logarithmic function $y = \log_a x$, $x > 0$, $a > 0$, and $a \neq 1$, is

$$\boxed{y' = \frac{d}{dx} \log_a x = \frac{1}{x \ln a} \qquad x > 0} \qquad (1)$$

When $a = e$, then $\ln a = \ln e = 1$, and (1) gives the formula for the derivative of the natural logarithm function $f(x) = \ln x$.

THEOREM Derivative of the Natural Logarithm Function

The derivative of the natural logarithm function $y = \ln x$, $x > 0$, is

$$\boxed{y' = \frac{d}{dx} \ln x = \frac{1}{x}}$$

EXAMPLE 1 Differentiating Logarithmic Functions

Find y' if:

(a) $y = (\ln x)^2$ **(b)** $y = \dfrac{1}{2}\log x$ **(c)** $\ln x + \ln y = 2x$

Solution (a) We use the Power Rule for Functions. Then

$$y' = \frac{d}{dx}(\ln x)^2 = 2\ln x \cdot \frac{d}{dx}\ln x = (2\ln x)\left(\frac{1}{x}\right) = \frac{2\ln x}{x}$$

$$\underset{\uparrow}{\frac{d}{dx}[u(x)]^n = n[u(x)]^{n-1}\frac{du}{dx}}$$

(b) Remember that $\log x = \log_{10} x$. Then $y = \dfrac{1}{2}\log x = \dfrac{1}{2}\log_{10} x$.

$$y' = \frac{1}{2}\left(\frac{d}{dx}\log_{10} x\right) = \frac{1}{2}\cdot\frac{1}{x\ln 10} = \frac{1}{2x\ln 10} \qquad \frac{d}{dx}\log_a x = \frac{1}{x\ln a}$$

(c) We assume y is a differentiable function of x and use implicit differentiation. Then

$$\frac{d}{dx}(\ln x + \ln y) = \frac{d}{dx}(2x)$$

$$\frac{d}{dx}\ln x + \frac{d}{dx}\ln y = 2$$

$$\frac{1}{x} + \frac{1}{y}\frac{dy}{dx} = 2 \qquad\qquad \frac{d}{dx}\ln y = \frac{1}{y}\frac{dy}{dx}$$

$$y' = \frac{dy}{dx} = y\left(2 - \frac{1}{x}\right) = \frac{y(2x-1)}{x} \qquad \text{Solve for } y' = \frac{dy}{dx}. \qquad\blacksquare$$

NOW WORK Problem 9.

If $y = \ln[u(x)]$, where $u = u(x) > 0$ is a differentiable function of x, then by the Chain Rule,

$$\frac{dy}{dx} = \frac{dy}{du}\cdot\frac{du}{dx} = \frac{1}{u}\frac{du}{dx} = \frac{u'(x)}{u(x)}$$

$$\boxed{\frac{d}{dx}\ln[u(x)] = \frac{u'(x)}{u(x)}}$$

EXAMPLE 2 Differentiating Logarithmic Functions

Find y' if:

(a) $y = \ln(5x)$ **(b)** $y = x\ln(x^2 + 1)$ **(c)** $y = \ln|x|$

Solution

(a)

$$y' = \frac{d}{dx}\ln(5x) = \frac{\frac{d}{dx}(5x)}{5x} = \frac{5}{5x} = \frac{1}{x} \qquad \frac{d}{dx}\ln u(x) = \frac{u'(x)}{u(x)}$$

(b) $y' = \dfrac{d}{dx}[x\ln(x^2+1)] = x\dfrac{d}{dx}\ln(x^2+1) + \left(\dfrac{d}{dx}x\right)\ln(x^2+1)$ Product Rule

$$= x\cdot\frac{\frac{d}{dx}(x^2+1)}{x^2+1} + 1\cdot\ln(x^2+1) = \frac{2x^2}{x^2+1} + \ln(x^2+1)$$

$$\underset{\uparrow}{\frac{d}{dx}\ln u(x) = \frac{u'(x)}{u(x)}}$$

(c)

$$y = \ln|x| = \begin{cases} \ln x & \text{if } x > 0 \\ \ln(-x) & \text{if } x < 0 \end{cases}$$

$$y' = \frac{d}{dx}\ln|x| = \begin{cases} \dfrac{1}{x} & \text{if } x > 0 \\ \dfrac{1}{-x}(-1) = \dfrac{1}{x} & \text{if } x < 0 \end{cases}$$

That is, $\dfrac{d}{dx}\ln|x| = \dfrac{1}{x}$ for all numbers $x \neq 0$. ■

Part (c) proves an important result that is used often:

$$\boxed{\frac{d}{dx}\ln|x| = \frac{1}{x} \qquad x \neq 0}$$

Look at the graph of $y = \ln|x|$ in Figure 6. As we move from left to right along the graph, starting where x is unbounded in the negative direction and ending near the y-axis, we see that the tangent lines to $y = \ln|x|$ go from being nearly horizontal with negative slope to being nearly vertical with a very large negative slope. This is reflected in the graph of the derivative $y' = \dfrac{1}{x}$, where $x < 0$: The graph starts close to zero and slightly negative and gets more negative as the y-axis is approached.

Similar remarks hold for $x > 0$. The graph of $y' = \dfrac{1}{x}$ just to the right of the y-axis is unbounded in the positive direction (the tangent lines to $y = \ln|x|$ are nearly vertical with positive slope). As x becomes unbounded in the positive direction, the graph of y' gets closer to zero but remains positive (the tangent lines to $y = \ln|x|$ are nearly horizontal with positive slope).

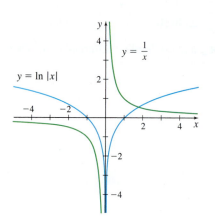

Figure 6

NOW WORK **Problem 17.**

NEED TO REVIEW? Properties of logarithms are discussed in Appendix A.1, pp. A-10 to A-11.

Properties of logarithms can sometimes be used to simplify the work needed to find a derivative.

EXAMPLE 3 Differentiating Logarithmic Functions

Find y' if $y = \ln\left[\dfrac{(2x-1)^3\sqrt{2x^4+1}}{x}\right]$.

NOTE In the remaining examples, we do not explicitly state the domain of a function containing a logarithm. Instead, we assume that the variable is restricted so all arguments for logarithmic functions are positive.

Solution Rather than attempting to use the Chain Rule, Quotient Rule, and Product Rule, we first simplify the right side by using properties of logarithms.

$$y = \ln\left[\frac{(2x-1)^3\sqrt{2x^4+1}}{x}\right] = \ln(2x-1)^3 + \ln\sqrt{2x^4+1} - \ln x$$

$$= 3\ln(2x-1) + \frac{1}{2}\ln(2x^4+1) - \ln x$$

Now we differentiate y.

$$y' = \frac{d}{dx}\left[3\ln(2x-1) + \frac{1}{2}\ln(2x^4+1) - \ln x\right]$$

$$= \frac{d}{dx}[3\ln(2x-1)] + \frac{d}{dx}\left[\frac{1}{2}\ln(2x^4+1)\right] - \frac{d}{dx}\ln x$$

$$= 3\cdot\frac{2}{2x-1} + \frac{1}{2}\cdot\frac{8x^3}{2x^4+1} - \frac{1}{x} = \frac{6}{2x-1} + \frac{4x^3}{2x^4+1} - \frac{1}{x}$$ ■

NOW WORK **Problem 27.**

2 Use Logarithmic Differentiation

Logarithms and their properties are very useful for finding derivatives of functions that involve products, quotients, or powers. This method, called **logarithmic differentiation**, uses the facts that the logarithm of a product is a sum, the logarithm of a quotient is a difference, and the logarithm of a power is a product.

ORIGINS Logarithmic differentiation was first used in 1697 by Johann Bernoulli (1667–1748) to find the derivative of $y = x^x$. Johann, a member of a famous family of mathematicians, was the younger brother of Jakob Bernoulli (1654–1705). He was also a contemporary of Newton, Leibniz, and the French mathematician Guillaume de L'Hôpital.

EXAMPLE 4   Finding Derivatives Using Logarithmic Differentiation

Find y' if $y = \dfrac{x^2\sqrt{5x+1}}{(3x-2)^3}$.

Solution It is easier to find y' if we take the natural logarithm of each side before differentiating. That is, we write

$$\ln y = \ln\left[\frac{x^2\sqrt{5x+1}}{(3x-2)^3}\right]$$

and simplify the equation using properties of logarithms.

$$\ln y = \ln[x^2\sqrt{5x+1}] - \ln(3x-2)^3 = \ln x^2 + \ln(5x+1)^{1/2} - \ln(3x-2)^3$$

$$= 2\ln x + \frac{1}{2}\ln(5x+1) - 3\ln(3x-2)$$

To find y', we use implicit differentiation.

$$\frac{d}{dx}\ln y = \frac{d}{dx}\left[2\ln x + \frac{1}{2}\ln(5x+1) - 3\ln(3x-2)\right]$$

$$\frac{y'}{y} = \frac{d}{dx}(2\ln x) + \frac{d}{dx}\left[\frac{1}{2}\ln(5x+1)\right] - \frac{d}{dx}[3\ln(3x-2)]$$

$$= \frac{2}{x} + \frac{5}{2(5x+1)} - \frac{9}{3x-2}$$

$$y' = y\left[\frac{2}{x} + \frac{5}{2(5x+1)} - \frac{9}{3x-2}\right] = \left[\frac{x^2\sqrt{5x+1}}{(3x-2)^3}\right]\left[\frac{2}{x} + \frac{5}{2(5x+1)} - \frac{9}{3x-2}\right]$$

Summarizing these steps, we arrive at the method of Logarithmic Differentiation.

Steps for Using Logarithmic Differentiation

Step 1 If the function $y = f(x)$ consists of products, quotients, and powers, take the natural logarithm of each side. Then simplify the equation using properties of logarithms.

Step 2 Differentiate implicitly, and use the fact that $\dfrac{d}{dx}\ln y = \dfrac{y'}{y}$.

Step 3 Solve for y', and replace y with $f(x)$.

NOW WORK Problem 51.

EXAMPLE 5   Using Logarithmic Differentiation

Find y' if $y = x^x$, $x > 0$.

Solution Notice that x^x is neither x raised to a fixed power a, nor a fixed base a raised to a variable power. We follow the steps for logarithmic differentiation:

Step 1 Take the natural logarithm of each side of $y = x^x$, and simplify:

$$\ln y = \ln x^x = x\ln x$$

Step 2 Differentiate implicitly.

$$\frac{d}{dx}\ln y = \frac{d}{dx}(x\ln x)$$

$$\frac{y'}{y} = x\cdot\frac{d}{dx}\ln x + \left(\frac{d}{dx}x\right)\cdot\ln x = x\left(\frac{1}{x}\right) + 1\cdot\ln x = 1 + \ln x$$

Step 3 Solve for y': $y' = y(1 + \ln x) = x^x(1 + \ln x)$.   ∎

A second approach to finding the derivative of $y = x^x$ is to use the fact that $y = x^x = e^{\ln x^x} = e^{x\ln x}$. See Problem 85.

NOW WORK Problem 57.

EXAMPLE 6   Finding an Equation of a Tangent Line

Find an equation of the tangent line to the graph of $f(x) = \dfrac{x\sqrt{x^2+3}}{1+x}$ at the point $(1, 1)$.

Solution The slope of the tangent line to the graph of $y = f(x)$ at the point $(1, 1)$ is $f'(1)$. Since the function consists of a product, a quotient, and a power, we follow the steps for logarithmic differentiation:

Step 1 Take the natural logarithm of each side, and simplify:

$$\ln y = \ln\frac{x\sqrt{x^2+3}}{1+x} = \ln x + \frac{1}{2}\ln(x^2+3) - \ln(1+x)$$

Step 2 Differentiate implicitly.

$$\frac{y'}{y} = \frac{1}{x} + \frac{1}{2}\cdot\frac{2x}{x^2+3} - \frac{1}{1+x} = \frac{1}{x} + \frac{x}{x^2+3} - \frac{1}{1+x}$$

Step 3 Solve for y' and simplify:

$$y' = f'(x) = y\left(\frac{1}{x} + \frac{x}{x^2+3} - \frac{1}{1+x}\right) = \frac{x\sqrt{x^2+3}}{1+x}\left(\frac{1}{x} + \frac{x}{x^2+3} - \frac{1}{1+x}\right)$$

Now we find the slope of the tangent line by evaluating $f'(1)$.

$$f'(1) = \frac{\sqrt{4}}{2}\left(1 + \frac{1}{4} - \frac{1}{2}\right) = \frac{3}{4}$$

Then an equation of the tangent line to the graph of f at the point $(1, 1)$ is

$$y - 1 = \frac{3}{4}(x - 1)$$

$$y = \frac{3}{4}x + \frac{1}{4}$$ ∎

See Figure 7 for the graphs of f and $y = \frac{3}{4}x + \frac{1}{4}$.

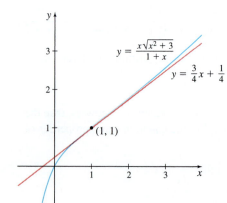

Figure 7

NOW WORK Problem 75.

We now prove the Power Rule for finding the derivative of $y = x^a$, where a is a real number.

THEOREM Power Rule

If a is a real number, then

$$\frac{d}{dx}x^a = ax^{a-1}$$

Proof Let $y = x^a$, a real. Then $\ln y = \ln x^a = a \ln x$.

$$\frac{y'}{y} = a\frac{1}{x}$$

$$y' = \frac{ay}{x} = \frac{ax^a}{x} = ax^{a-1}$$ ∎

EXAMPLE 7 Using the Power Rule

Find the derivative of:

(a) $y = x^{\sqrt{2}}$ **(b)** $y = (x^2 + 1)^\pi$

Solution **(a)** We differentiate y using the Power Rule.

$$y' = \frac{d}{dx}x^{\sqrt{2}} = \sqrt{2}x^{\sqrt{2}-1}$$

(b) The Power Rule for Functions also holds when the exponent is a real number. Then

$$y' = \frac{d}{dx}(x^2 + 1)^\pi = \pi(x^2 + 1)^{\pi-1}(2x) = 2\pi x(x^2 + 1)^{\pi-1}$$ ∎

3 Express e as a Limit

In Section P.5, we stated that the number e has this property: $\ln e = 1$. Here are two equivalent ways of writing e as a limit, both of which are used in later chapters.

THEOREM The Number e as a Limit

The number e can be expressed by either of the two limits:

$$\textbf{(a)} \ \lim_{h\to 0}(1 + h)^{1/h} = e \qquad \textbf{or} \qquad \textbf{(b)} \ \lim_{n\to\infty}\left(1 + \frac{1}{n}\right)^n = e \tag{2}$$

Proof **(a)** The derivative of $f(x) = \ln x$ is $f'(x) = \frac{1}{x}$ so $f'(1) = 1$. Then

$$f'(1) = \lim_{h\to 0}\frac{f(1 + h) - f(1)}{h} = \lim_{h\to 0}\frac{\ln(1 + h) - \ln 1}{h} = \lim_{h\to 0}\frac{\ln(1 + h) - 0}{h}$$

↑ Definition of a Derivative

$$= \lim_{h\to 0}\left[\frac{1}{h}\ln(1 + h)\right] = \lim_{h\to 0}\ln(1 + h)^{1/h}$$

Since $y = \ln x$ is continuous and $f'(1) = 1$,

$$\lim_{h\to 0}\ln(1 + h)^{1/h} = \ln\left[\lim_{h\to 0}(1 + h)^{1/h}\right] = 1$$

Since $\ln e = 1$,

$$\ln\left[\lim_{h\to 0}(1 + h)^{1/h}\right] = \ln e$$

Since $y = \ln x$ is a one-to-one function,

$$\lim_{h\to 0}(1 + h)^{1/h} = e$$

(b) The limit derived in (a) is valid when $h \to 0^+$. So if we let $n = \frac{1}{h}$, then as $h \to 0^+$, $n = \frac{1}{h} \to \infty$ and

$$e = \lim_{n\to\infty}\left(1 + \frac{1}{n}\right)^n$$ ∎

Table 2 shows the values of $\left(1 + \frac{1}{n}\right)^n$ for selected values of n.

TABLE 2

n	$\left(1 + \dfrac{1}{n}\right)^n$
10	2.593 742 46
1000	2.716 923 932
10,000	2.718 145 927
10^6	2.718 280 469

NOTE Correct to nine decimal places, $e = 2.718\,281\,828$.

EXAMPLE 8   **Expressing a Limit in Terms of e**

Express each limit in terms of the number e:

(a) $\displaystyle\lim_{h\to 0}(1+2h)^{1/h}$ **(b)** $\displaystyle\lim_{n\to\infty}\left(1+\frac{3}{n}\right)^{2n}$

Solution   **(a)** This limit resembles $\displaystyle\lim_{h\to 0}(1+h)^{1/h}$, and with some manipulation, it can be expressed in terms of $\displaystyle\lim_{h\to 0}(1+h)^{1/h}$.

$$(1+2h)^{1/h} = [(1+2h)^{1/(2h)}]^2$$

Now let $k=2h$, and note that $h\to 0$ is equivalent to $2h=k\to 0$. So,

$$\lim_{h\to 0}(1+2h)^{1/h} = \lim_{k\to 0}\left[(1+k)^{1/k}\right]^2 = \left[\lim_{k\to 0}(1+k)^{1/k}\right]^2 \underset{\underset{(2)}{\uparrow}}{=} e^2$$

(b) This limit resembles $\displaystyle\lim_{n\to\infty}\left(1+\frac{1}{n}\right)^n$. We rewrite it as follows:

$$\left(1+\frac{3}{n}\right)^{2n} = \left[\left(1+\frac{3}{n}\right)^{2n}\right]^{3/3} = \left[\left(1+\frac{3}{n}\right)^{n/3}\right]^6$$

Let $k=\dfrac{n}{3}$. Since $n\to\infty$ is equivalent to $\dfrac{n}{3}=k\to\infty$, we find that

$$\lim_{n\to\infty}\left(1+\frac{3}{n}\right)^{2n} = \lim_{k\to\infty}\left[\left(1+\frac{1}{k}\right)^k\right]^6 = \left[\lim_{k\to\infty}\left(1+\frac{1}{k}\right)^k\right]^6 \underset{\underset{(2)}{\uparrow}}{=} e^6 \quad■$$

NOW WORK   **Problem 77.**

The number e occurs in many applications. For example, in finance, the number e is used to find **continuously compounded interest**. See the discussion preceding Problem 89.

3.3 Assess Your Understanding

Concepts and Vocabulary

1. $\dfrac{d}{dx}\ln x =$ _____.

2. *True or False* $\dfrac{d}{dx}x^e = e\,x^{e-1}$.

3. *True or False* $\dfrac{d}{dx}\ln[x\sin^2 x] = \dfrac{d}{dx}\ln x \cdot \dfrac{d}{dx}\ln\sin^2 x$.

4. *True or False* $\dfrac{d}{dx}\ln\pi = \dfrac{1}{\pi}$.

5. $\dfrac{d}{dx}\ln|x| =$ _____ for all $x\neq 0$.

6. $\displaystyle\lim_{n\to\infty}\left(1+\frac{1}{n}\right)^n =$ _____.

Skill Building

In Problems 7–44, find y'.

7. $y=5\ln x$

8. $y=-3\ln x$

9. $y=\log_2 u$

10. $y=\log_3 u$

11. $y=(\cos x)(\ln x)$

12. $y=(\sin x)(\ln x)$

13. $y=\ln(3x)$

14. $y=\ln\dfrac{x}{2}$

15. $y=\ln(e^t-e^{-t})$

16. $y=\ln(e^{at}+e^{-at})$

17. $y=x\ln(x^2+4)$

18. $y=x\ln(x^2+5x+1)$

19. $y=v\ln\sqrt{v^2+1}$

20. $y=v\ln\sqrt[3]{3v+1}$

---

1. = NOW WORK problem   ◪ = Graphing technology recommended   **CAS** = Computer Algebra System recommended

21. $y = \dfrac{1}{2} \ln \dfrac{1+x}{1-x}$

22. $y = \dfrac{1}{2} \ln \dfrac{1+x^2}{1-x^2}$

23. $y = \ln(\ln x)$

24. $y = \ln\left(\ln \dfrac{1}{x}\right)$

25. $y = \ln \dfrac{x}{\sqrt{x^2+1}}$

26. $y = \ln \dfrac{4x^3}{\sqrt{x^2+4}}$

27. $y = \ln \dfrac{(x^2+1)^2}{x\sqrt{x^2-1}}$

28. $y = \ln \dfrac{x\sqrt{3x-1}}{(x^2+1)^3}$

29. $y = \ln(\sin\theta)$

30. $y = \ln(\cos\theta)$

31. $y = \ln(x + \sqrt{x^2+4})$

32. $y = \ln(\sqrt{x+1} + \sqrt{x})$

33. $y = \log_2(1 + x^2)$

34. $y = \log_2(x^2 - 1)$

35. $y = \tan^{-1}(\ln x)$

36. $y = \sin^{-1}(\ln x)$

37. $y = \ln(\tan^{-1} t)$

38. $y = \ln(\sin^{-1} t)$

39. $y = (\ln x)^{1/2}$

40. $y = (\ln x)^{-1/2}$

41. $y = \sin(\ln\theta)$

42. $y = \cos(\ln\theta)$

43. $y = x \ln \sqrt{\cos(2x)}$

44. $y = x^2 \ln \sqrt{\sin(2x)}$

In Problems 45–50, use implicit differentiation to find $y' = \dfrac{dy}{dx}$.

45. $x \ln y + y \ln x = 2$

46. $\dfrac{\ln y}{x} + \dfrac{\ln x}{y} = 2$

47. $\ln(x^2 + y^2) = x + y$

48. $\ln(x^2 - y^2) = x - y$

49. $\ln \dfrac{y}{x} = y$

50. $\ln \dfrac{y}{x} - \ln \dfrac{x}{y} = 1$

In Problems 51–72, use logarithmic differentiation to find y'. Assume that the variable is restricted so that all arguments of logarithm functions are positive.

51. $y = (x^2 + 1)^2 (2x^3 - 1)^4$

52. $y = (3x^2 + 4)^3 (x^2 + 1)^4$

53. $y = \dfrac{x^2(x^3 + 1)}{\sqrt{x^2+1}}$

54. $y = \dfrac{\sqrt{x}(x^3 + 2)^2}{\sqrt[3]{3x+4}}$

55. $y = \dfrac{x\cos x}{(x^2+1)^3 \sin x}$

56. $y = \dfrac{x\sin x}{(1 + e^x)^3 \cos x}$

57. $y = (3x)^x$

58. $y = (x - 1)^x$

59. $y = x^{\ln x}$

60. $y = (2x)^{\ln x}$

61. $y = x^{x^2}$

62. $y = (3x)^{\sqrt{x}}$

63. $y = x^{e^x}$

64. $y = (x^2 + 1)^{e^x}$

65. $y = x^{\sin x}$

66. $y = x^{\cos x}$

67. $y = (\sin x)^x$

68. $y = (\cos x)^x$

69. $y = (\sin x)^{\cos x}$

70. $y = (\sin x)^{\tan x}$

71. $x^y = 4$

72. $y^x = 10$

In Problems 73–76, find an equation of the tangent line to the graph of $y = f(x)$ at the given point.

73. $y = \ln(5x)$ at $\left(\dfrac{1}{5}, 0\right)$

74. $y = x \ln x$ at $(1, 0)$

75. $y = \dfrac{x^2\sqrt{3x-2}}{(x-1)^2}$ at $(2, 8)$

76. $y = \dfrac{x(\sqrt[3]{x} + 1)^2}{\sqrt{x+1}}$ at $(8, 24)$

In Problems 77–80, express each limit in terms of e.

77. $\displaystyle\lim_{n\to\infty} \left(1 + \dfrac{1}{n}\right)^{2n}$

78. $\displaystyle\lim_{n\to\infty} \left(1 + \dfrac{1}{n}\right)^{n/2}$

79. $\displaystyle\lim_{n\to\infty} \left(1 + \dfrac{1}{3n}\right)^{n}$

80. $\displaystyle\lim_{n\to\infty} \left(1 + \dfrac{4}{n}\right)^{n}$

Applications and Extensions

81. Find $\dfrac{d^{10}}{dx^{10}}(x^9 \ln x)$.

82. If $f(x) = \ln(x - 1)$, find $f^{(n)}(x)$.

83. If $y = \ln(x^2 + y^2)$, find the value of $\dfrac{dy}{dx}$ at the point $(1, 0)$.

84. If $f(x) = \tan\left(\ln x - \dfrac{1}{\ln x}\right)$, find $f'(e)$.

85. Find y' if $y = x^x$, $x > 0$, by using $y = x^x = e^{\ln x^x}$ and the Chain Rule.

86. If $y = \ln(kx)$, where $x > 0$ and $k > 0$ is a constant, show that $y' = \dfrac{1}{x}$.

In Problems 87 and 88, find y'. Assume that a is a constant.

87. $y = x \tan^{-1} \dfrac{x}{a} - \dfrac{1}{2} a \ln(x^2 + a^2)$, $a \neq 0$

88. $y = x \sin^{-1} \dfrac{x}{a} + a \ln \sqrt{a^2 - x^2}$, $|a| > |x|$, $a \neq 0$

Continuously Compounded Interest *In Problems 89 and 90, use the following discussion:*

*Suppose an initial investment, called the **principal** P, earns an annual rate of interest r, which is compounded n times per year. The interest earned on the principal P in the first compounding period is $P\left(\dfrac{r}{n}\right)$, and the resulting amount A of the investment after one compounding period is $A = P + P\left(\dfrac{r}{n}\right) = P\left(1 + \dfrac{r}{n}\right)$. After k compounding periods, the amount A of the investment is $A = P\left(1 + \dfrac{r}{n}\right)^k$. Since in t years there are nt compounding periods, the amount A after t years is*

$$A = P\left(1 + \dfrac{r}{n}\right)^{nt}$$

*When interest is compounded so that after t years the accumulated amount is $A = \displaystyle\lim_{n\to\infty} P\left(1 + \dfrac{r}{n}\right)^{nt}$, the interest is said to be **compounded continuously**.*

89. **(a)** Show that if the annual rate of interest r is compounded continuously, then the amount A after t years is $A = Pe^{rt}$, where P is the initial investment.

  (b) If an initial investment of $P = \$5000$ earns 2% interest compounded continuously, how much is the investment worth after 10 years?

  (c) How long does it take an investment of $\$10,000$ to double if it is invested at 2.4% compounded continuously?

(d) Show that the rate of change of A with respect to t when the interest rate r is compounded continuously is $\dfrac{dA}{dt} = rA$.

90. A bank offers a certificate of deposit (CD) that matures in 10 years with a rate of interest of 3% compounded continuously. (See Problem 89.) Suppose you purchase such a CD for $2000 in your IRA.

    (a) Write an equation that gives the amount A in the CD as a function of time t in years.

    (b) How much is the CD worth at maturity?

    (c) What is the rate of change of the amount A at $t = 3$? At $t = 5$? At $t = 8$?

    (d) Explain the results found in (c).

91. Sound Level of a Leaf Blower The loudness L, measured in decibels (dB), of a sound of intensity I is defined as

$$L(x) = 10 \log \frac{I(x)}{I_0},$$ where x is the distance in meters from

the source of the sound and $I_0 = 10^{-12}$ W/m^2 is the least intense sound that a human ear can detect. The intensity I is defined as the power P of the sound wave divided by the area A on which it falls. If the wave spreads out uniformly in all directions, that is, if it is spherical, the surface area is

$$A(x) = 4\pi x^2 \text{ m}^2, \text{ and } I(x) = \frac{P}{4\pi x^2} \text{ W/m}^2.$$

    (a) If you are 2.0 m from a noisy leaf blower and are walking away from it, at what rate is the loudness L changing with respect to distance x?

    (b) Interpret the sign of your answer.

92. Show that $\ln x + \ln y = 2x$ is equivalent to $xy = e^{2x}$. Use this equation to find y'. Compare this result to the solution found in Example 1(c).

93. If $\ln T = kt$, where k is a constant, show that $\dfrac{dT}{dt} = kT$.

94. Graph $y = \left(1 + \dfrac{1}{x}\right)^x$ and $y = e$ on the same set of axes. Explain how the graph supports the fact that $\lim\limits_{n \to \infty} \left(1 + \dfrac{1}{n}\right)^n = e$.

95. Power Rule for Functions Show that if u is a function of x that is differentiable and a is a real number, then

$$\frac{d}{dx}[u(x)]^a = a[u(x)]^{a-1} u'(x)$$

provided u^a and u^{a-1} are defined. [*Hint:* Let $|y| = |[u(x)]^a|$ and use logarithmic differentiation.]

96. Show that the tangent lines to the graphs of the family of parabolas $f(x) = -\dfrac{1}{2}x^2 + k$ are always perpendicular to the tangent lines to the graphs of the family of natural logarithms $g(x) = \ln(bx) + c$, where $b > 0$, k, and c are constants.

Source: Mathematics students at Millikin University, Decatur, Illinois.

Challenge Problems

97. Show that $2x - \ln(3 + 6e^x + 3e^{2x}) = C - 2\ln(1 + e^{-x})$ for some constant C.

98. If f and g are differentiable functions, and if $f(x) > 0$, show that

$$\frac{d}{dx} f(x)^{g(x)} = g(x)f(x)^{g(x)-1}f'(x) + f(x)^{g(x)}[\ln f(x)]g'(x)$$

3.4 Differentials; Linear Approximations; Newton's Method

OBJECTIVES *When you finish this section, you should be able to:*

1 Find the differential of a function and interpret it geometrically (p. 230)

2 Find the linear approximation to a function (p. 232)

3 Use differentials in applications (p. 233)

4 Use Newton's Method to approximate a real zero of a function (p. 234)

1 Find the Differential of a Function and Interpret It Geometrically

Recall that for a differentiable function $y = f(x)$, the derivative is defined as

$$\frac{dy}{dx} = f'(x) = \lim_{\Delta x \to 0} \frac{\Delta y}{\Delta x} = \lim_{\Delta x \to 0} \frac{f(x + \Delta x) - f(x)}{\Delta x}$$

That is, for Δx sufficiently close to 0, we can make $\dfrac{\Delta y}{\Delta x}$ as close as we please to $f'(x)$. This can be expressed as

$$\frac{\Delta y}{\Delta x} \approx f'(x) \qquad \text{when} \qquad \Delta x \approx 0, \quad \Delta x \neq 0$$

or, since $\Delta x \neq 0$, as

$$\boxed{\Delta y \approx f'(x)\Delta x \qquad \text{when} \qquad \Delta x \approx 0, \quad \Delta x \neq 0} \qquad (1)$$

DEFINITION

Let $y = f(x)$ be a differentiable function and let Δx denote a change in x.

The **differential of x**, denoted dx, is defined as $dx = \Delta x \neq 0$.

The **differential of y**, denoted dy, is defined as

$$\boxed{dy = f'(x)dx} \qquad (2)$$

From statement (1), if $\Delta x \,(= dx) \approx 0$, then $\Delta y \approx dy = f'(x)dx$. That is, when the change Δx in x is close to 0, then the differential dy is an approximation to the change Δy in y.

To see the geometric relationship between Δy and dy, we use Figure 8. There, $P = (x, y)$ is a point on the graph of $y = f(x)$ and $Q = (x + \Delta x, y + \Delta y)$ is a nearby point also on the graph of f. The change Δy in y

$$\Delta y = f(x + \Delta x) - f(x)$$

is the distance from M to Q. The slope of the tangent line to f at P is $f'(x) = \dfrac{dy}{dx}$. So numerically, the differential dy measures the distance from M to N. For Δx close to 0, $dy = f'(x)dx$ will be close to Δy.

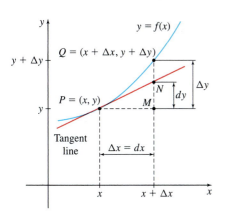

Figure 8

EXAMPLE 1   Finding and Interpreting Differentials Geometrically

For the function $f(x) = xe^x$:

(a) Find the differential dy.

(b) Compare dy to Δy when $x = 0$ and $\Delta x = 0.5$.

(c) Compare dy to Δy when $x = 0$ and $\Delta x = 0.1$.

(d) Compare dy to Δy when $x = 0$ and $\Delta x = 0.01$.

(e) Discuss the results.

Solution (a) $dy = f'(x)dx = (xe^x + e^x)dx = (x + 1)e^x dx$.

(b) See Figure 9(a). When $x = 0$ and $\Delta x = dx = 0.5$, then

$$dy = (x + 1)e^x dx = (0 + 1)e^0(0.5) = 0.5$$

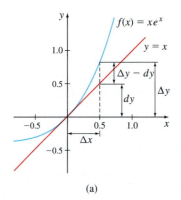

(a)

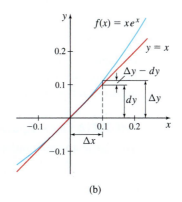

(b)

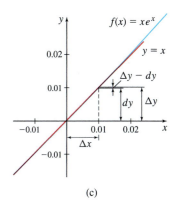

(c)

DF **Figure 9**

The tangent line rises by 0.5 as x changes from 0 to 0.5. The corresponding change in the height of the graph f is

$$\Delta y = f(x + \Delta x) - f(x) = f(0.5) - f(0) = 0.5e^{0.5} - 0 \approx 0.824$$
$$|\Delta y - dy| \approx |0.824 - 0.5| = 0.324$$

The graph of the tangent line is approximately 0.324 below the graph of f at $x = 0.5$.

(c) See Figure 9(b). When $x = 0$ and $\Delta x = dx = 0.1$, then

$$dy = (0 + 1)e^0(0.1) = 0.1$$

The tangent line rises by 0.1 as x changes from 0 to 0.1. The corresponding change in the height of the graph f is

$$\Delta y = f(x + \Delta x) - f(x) = f(0.1) - f(0) = 0.1e^{0.1} - 0 \approx 0.111$$
$$|\Delta y - dy| \approx |0.111 - 0.1| = 0.011$$

The graph of the tangent line is approximately 0.011 below the graph of f at $x = 0.1$.

(d) See Figure 9(c). When $x = 0$ and $\Delta x = dx = 0.01$, then

$$dy = (0 + 1)e^0(0.01) = 0.01$$

The tangent line rises by 0.01 as x changes from 0 to 0.01. The corresponding change in the height of the graph of f is

$$\Delta y = f(x + \Delta x) - f(x) = f(0.01) - f(0) = 0.01e^{0.01} - 0 \approx 0.0101$$
$$|\Delta y - dy| \approx |0.0101 - 0.01| = 0.0001$$

The graph of the tangent line is approximately 0.0001 below the graph of f at $x = 0.01$.

(e) The closer Δx is to 0, the closer dy is to Δy. So, we conclude that the closer Δx is to 0, the less the tangent line departs from the graph of the function. That is, we can use the tangent line to f at a point P as a *linear approximation* to f near P. ■

NOW WORK   Problems 9 and 17.

Example 1 shows that when dx is close to 0, the tangent line can be used as a linear approximation to the graph. We discuss next how to find this linear approximation.

2 Find the Linear Approximation to a Function

Suppose $y = f(x)$ is a differentiable function and suppose (x_0, y_0) is a point on the graph of f. Then

$$\Delta y = f(x) - f(x_0) \qquad \text{and} \qquad dy = f'(x_0)dx = f'(x_0)\Delta x = f'(x_0)(x - x_0)$$

If $dx = \Delta x$ is close to 0, then

$$\Delta y \approx dy$$
$$f(x) - f(x_0) \approx f'(x_0)(x - x_0)$$
$$f(x) \approx f(x_0) + f'(x_0)(x - x_0)$$

See Figure 10. Each figure shows the graph of $y = f(x)$ and the graph of the tangent line at (x_0, y_0). The first two figures show the difference between Δy and dy. The third figure illustrates that when Δx is close to 0, $\Delta y \approx dy$.

THEOREM Linear Approximation

The **linear approximation $L(x)$** to a differentiable function f at $x = x_0$ is given by

$$\boxed{L(x) = f(x_0) + f'(x_0)(x - x_0)} \qquad (3)$$

The closer x is to x_0, the better the approximation. The graph of the linear approximation $y = L(x)$ is the tangent line to the graph of f at x_0, and L is often

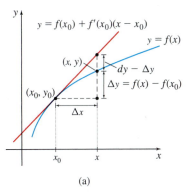

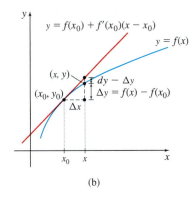

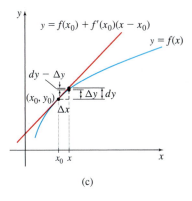

(a)   (b)   (c)

DF Figure 10

called the **linearization of f at x_0**. Although L provides a good approximation of f in an interval centered at x_0, the next example shows that as the interval widens, the accuracy of the approximation of L may decrease. In Section 3.5, we use higher-degree polynomials to extend the interval for which the approximation is efficient.

TABLE 3

$L(x) = x$	$f(x) = \sin x$	Error: $\lvert x - \sin x\rvert$
0.1	0.0998	0.0002
−0.3	−0.2955	0.0045
0.4	0.3894	0.0106
0.5	0.4794	0.0206
$\dfrac{\pi}{4} \approx 0.7854$	0.7071	0.0783

Figure 11

EXAMPLE 2 **Finding the Linear Approximation to a Function**

(a) Find the linear approximation $L(x)$ to $f(x) = \sin x$ near $x = 0$.

(b) Use $L(x)$ to approximate $\sin(-0.3)$, $\sin 0.1$, $\sin 0.4$, $\sin 0.5$, and $\sin \dfrac{\pi}{4}$.

(c) Graph f and L.

Solution **(a)** Since $f'(x) = \cos x$, then $f(0) = \sin 0 = 0$ and $f'(0) = \cos 0 = 1$. Using Equation (3), the linear approximation $L(x)$ to f at 0 is

$$L(x) = f(0) + f'(0)(x - 0) = x$$

So, for x close to 0, the function $f(x) = \sin x$ can be approximated by the line $L(x) = x$.

(b) The approximate values of $\sin x$ using $L(x) = x$, the true values of $\sin x$, and the absolute error in using the approximation are given in Table 3. From Table 3, we see that the further x is from 0, the worse the line $L(x) = x$ approximates $f(x) = \sin x$.

(c) See Figure 11 for the graphs of $f(x) = \sin x$ and $L(x) = x$. ∎

NOW WORK Problem **25**.

The next two examples show applications of differentials. In these examples, we use the differential to approximate the change.

3 Use Differentials in Applications

EXAMPLE 3 **Measuring Metal Loss in a Mechanical Process**

A spherical bearing has a radius of 3 cm when it is new. Use differentials to approximate the volume of the metal lost after the bearing wears down to a radius of 2.971 cm. Compare the approximation to the actual volume lost.

Solution The volume V of a sphere of radius R is $V = \dfrac{4}{3}\pi R^3$. As a machine operates, friction wears away part of the bearing. The exact volume of metal lost equals the change ΔV in the volume V of the sphere, when the change in the radius of the bearing is $\Delta R = 2.971 - 3 = -0.029$ cm. Since the change ΔR is small, we use the differential dV to approximate the change ΔV. Then

$$\Delta V \approx dV = 4\pi R^2 dR = (4\pi)(3^2)(-0.029) \approx -3.280$$
$$\underset{\uparrow}{dV = V'dR} \quad \underset{\uparrow}{dR = \Delta R}$$

The approximate loss in volume of the bearing is 3.28 cm^3.

The actual loss in volume ΔV is

$$\Delta V = V(R + \Delta R) - V(R) = \frac{4}{3}\pi \cdot 2.971^3 - \frac{4}{3}\pi \cdot 3^3 = \frac{4}{3}\pi(-0.7755) = -3.248\,\text{cm}^3$$

The approximate change in volume is correct to one decimal place. ∎

NOW WORK Problem 47.

The use of dy to approximate Δy when $\Delta x = dx$ is small is also helpful in approximating *error*. If Q is the quantity to be measured and if ΔQ is the change in Q, then the

$$\text{Relative error at } x_0 \text{ in } Q = \frac{|\Delta Q|}{Q(x_0)}$$

$$\text{Percentage error at } x_0 \text{ in } Q = \frac{|\Delta Q|}{Q(x_0)} \cdot 100\%$$

For example, if $Q = 50$ units and the change ΔQ in Q is measured to be 5 units, then

$$\text{Relative error at 50 in } Q = \frac{5}{50} = 0.10 \qquad \text{Percentage error at 50 in } Q = 10\%$$

When Δx is small, $dQ \approx \Delta Q$. The relative error and percentage error at x_0 in Q can be approximated by $\dfrac{|dQ|}{Q(x_0)}$ and $\dfrac{|dQ|}{Q(x_0)} \cdot 100\%$, respectively.

EXAMPLE 4   Measuring Error in a Manufacturing Process

A company manufactures spherical ball bearings of radius 3 cm. The customer accepts a tolerance of 1% in the radius. Use differentials to approximate the relative error for the surface area of the acceptable ball bearings.

Solution The tolerance of 1% in the radius R means that the relative error in the radius R must be within 0.01. That is, $\dfrac{|\Delta R|}{R} \le 0.01$. The surface area S of a sphere of radius R is given by the formula $S = 4\pi R^2$. We seek the relative error in S, $\dfrac{|\Delta S|}{S}$, which can be approximated by $\dfrac{|dS|}{S}$.

$$\frac{|\Delta S|}{S} \approx \underset{\substack{\uparrow \\ dS = S'dR}}{\frac{|dS|}{S}} = \frac{(8\pi R)|dR|}{4\pi R^2} = \frac{2|dR|}{R} = 2 \cdot \underset{\substack{\uparrow \\ \frac{|\Delta R|}{R} \le 0.01}}{\frac{|\Delta R|}{R}} \le 2(0.01) = 0.02$$

The relative error in the surface area will be less than or equal to 0.02. ∎

In Example 4, the tolerance of 1% in the radius of the ball bearing means the radius of the sphere must be somewhere between $3 - 0.01(3) = 2.97$ cm and $3 + 0.01(3) = 3.03$ cm. The corresponding 2% error in the surface area means the surface area lies within ± 0.02 of $S = 4\pi R^2 = 36\pi$ cm². That is, the surface area is between $35.28\pi \approx 110.84$ cm² and $36.72\pi \approx 115.36$ cm². Notice that a rather small error in the radius results in a more significant variation in the surface area!

NOW WORK Problem 59.

4 Use Newton's Method to Approximate a Real Zero of a Function

Newton's Method is used to approximate a real zero of a function. Suppose a function $y = f(x)$ is defined on a closed interval $[a, b]$, and its derivative f' is continuous on the interval (a, b). Also suppose that, from the Intermediate Value Theorem or from a

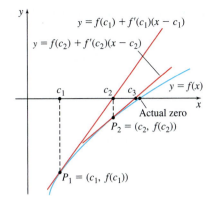

Figure 12

ORIGINS When Isaac Newton first introduced his method for finding real zeros in 1669, it was very different from what we did here. Newton's original method (also known as the **Newton-Raphson Method**) applied only to polynomials, and it did not use calculus! In 1690 Joseph Raphson simplified the method and made it recursive, again without calculus. It was not until 1740 that Thomas Simpson used calculus, giving the method the form we use here.

graph or by trial calculations, we know that the function f has a real zero in some open subinterval of (a, b) containing the number c_1.

Draw the tangent line to the graph of f at the point $P_1 = (c_1, f(c_1))$ and label as c_2 the x-intercept of the tangent line. See Figure 12. If c_1 is a *first approximation* to the required zero of f, then c_2 will be a better, or *second approximation* to the zero. Repeated use of this procedure often* generates increasingly accurate approximations to the zero we seek.

Suppose c_1 has been chosen. We seek a formula for finding the approximation c_2. The coordinates of P_1 are $(c_1, f(c_1))$, and the slope of the tangent line to the graph of f at point P_1 is $f'(c_1)$. So, the equation of the tangent line to graph of f at P_1 is

$$y - f(c_1) = f'(c_1)(x - c_1)$$

To find the x-intercept c_2, we let $y = 0$. Then c_2 satisfies the equation

$$-f(c_1) = f'(c_1)(c_2 - c_1)$$

$$c_2 = c_1 - \frac{f(c_1)}{f'(c_1)} \qquad \text{if } f'(c_1) \neq 0 \qquad \textcolor{teal}{\text{Solve for } c_2.}$$

Notice that we are using the linear approximation to f at the point $(c_1, f(c_1))$ to approximate the zero.

Now repeat the process by drawing the tangent line to the graph of f at the point $P_2 = (c_2, f(c_2))$ and label as c_3 the x-intercept of the tangent line. By continuing this process, we have Newton's method.

Newton's Method

Suppose a function f is defined on a closed interval $[a, b]$ and its first derivative f' is continuous and not equal to 0 on the open interval (a, b). If $x = c_1$ is a sufficiently close first approximation to a real zero of the function f in (a, b), then the formula

$$\boxed{c_2 = c_1 - \frac{f(c_1)}{f'(c_1)}}$$

gives a second approximation to the zero. The nth approximation to the zero is given by

$$\boxed{c_n = c_{n-1} - \frac{f(c_{n-1})}{f'(c_{n-1})}}$$

The formula in Newton's Method is a **recursive formula**, because the approximation is written in terms of the previous result. Since recursive processes are particularly useful in computer programs, variations of Newton's Method are used by many computer algebra systems and graphing utilities to find the zeros of a function.

Using Newton's Method to Approximate
EXAMPLE 5 **a Real Zero of a Function**

Use Newton's Method to find a fourth approximation to the positive real zero of the function $f(x) = x^3 + x^2 - x - 2$.

Solution Since $f(1) = -1$ and $f(2) = 8$, we know from the Intermediate Value Theorem that f has a zero in the interval $(1, 2)$. Also, since f is a polynomial function, both f and f' are differentiable functions. Now

$$f(x) = x^3 + x^2 - x - 2 \qquad \text{and} \qquad f'(x) = 3x^2 + 2x - 1$$

---

*See When Newton's Method Fails on page 236.

We choose $c_1 = 1.5$ as the first approximation and use Newton's Method. The second approximation to the zero is

$$c_2 = c_1 - \frac{f(c_1)}{f'(c_1)} = 1.5 - \frac{f(1.5)}{f'(1.5)} = 1.5 - \frac{2.125}{8.75} \approx 1.2571429$$

Use Newton's Method again, with $c_2 = 1.2571429$. Then $f(1.2571429) \approx 0.3100644$ and $f'(1.2571429) \approx 6.2555106$. The third approximation to the zero is

$$c_3 = c_2 - \frac{f(c_2)}{f'(c_2)} = 1.2571429 - \frac{0.3100641}{6.2555102} \approx 1.2075763$$

The fourth approximation to the zero is

$$c_4 = c_3 - \frac{f(c_3)}{f'(c_3)} = 1.2075763 - \frac{0.0116012}{5.7898745} \approx 1.2055726 \qquad ■$$

Using graphing technology, the zero of $f(x) = x^3 + x^2 - x - 2$ is given as $x = 1.2055694$. See Figure 13.

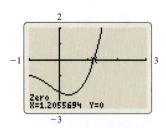

Figure 13 $f(x) = x^3 + x^2 - x - 2$

NOW WORK Problem **33**.

The table feature of a graphing utility can be used to take advantage of the recursive nature of Newton's Method and speed up the computation.

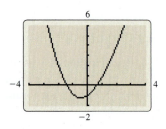 **EXAMPLE 6**   **Using Technology with Newton's Method**

The graph of the function $f(x) = \sin x + x^2 - 1$ is shown in Figure 14.

(a) Use the Intermediate Value Theorem to confirm that f has a zero in the interval $(0, 1)$.

(b) Use graphing technology with Newton's Method and a first approximation of $c_1 = 0.5$ to find a fourth approximation to the zero.

Solution   (a) The function $f(x) = \sin x + x^2 - 1$ is continuous on its domain, all real numbers, so it is continuous on the closed interval $[0, 1]$. Since $f(0) = -1$ and $f(1) = \sin 1 \approx 0.841$ have opposite signs, the Intermediate Value Theorem guarantees that f has a zero in the interval $(0, 1)$.

(b) We begin by finding $f'(x) = \cos x + 2x$. To use Newton's Method with a graphing utility, we enter

$$x - \frac{\sin x + x^2 - 1}{\cos x + 2x} \qquad x - \frac{f(x)}{f'(x)}$$

into the $Y =$ editor, as shown in Figure 15. We create a table by entering the initial value 0.5 in the X column. The graphing utility computes

$$0.5 - \frac{\sin 0.5 + 0.5^2 - 1}{\cos 0.5 + 2(0.5)} = 0.64410789$$

and displays 0.64411 in column Y_1 next to 0.5. The value Y_1 is the second approximation c_2 that we use in the next iteration. That is, we enter 0.64410789 in the X column of the next row, and the new entry in column Y_1 is the third approximation c_3. We repeat the process until we obtain the desired approximation. The fourth approximation to the zero of f is 0.63673, as shown in Figure 16. ■

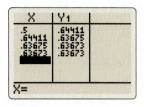

Figure 14 $f(x) = \sin x + x^2 - 1$

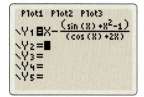

Figure 15

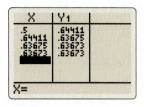

Figure 16

NOW WORK Problem **41**.

When Newton's Method Fails

You may wonder, does Newton's Method always work? The answer is no, as we mentioned earlier. The list below, while not exhaustive, gives some conditions under which Newton's Method fails.

- Newton's Method fails if the conditions of the theorem are not met

    (a) $f'(c_n) = 0$: Algebraically, division by 0 is not defined. Geometrically, the tangent line is parallel to the x-axis and so has no x-intercept.

    (b) $f'(c)$ is undefined: The process cannot be used.

- Newton's Method fails if the initial estimate c_1 may not be "good enough:"

    (a) Choosing an initial estimate too far from the required zero could result in approximating a different zero of the function.

    (b) The convergence could approach the zero so slowly that hundreds of iterations are necessary.

- Newton's Method fails if the terms oscillate between two values and so never get closer to the zero.

Problems 72–75 illustrate some of these possibilities.

3.4 Assess Your Understanding

Concepts and Vocabulary

1. *Multiple Choice*   If $y = f(x)$ is a differentiable function, the differential $dy = $ [(a) Δy, (b) Δx, (c) $f(x)dx$, (d) $f'(x)dx$].

2. A linear approximation to a differentiable function f near x_0 is given by the function $L(x) = $ _____.

3. *True or False*   The difference $|\Delta y - dy|$ measures the departure of the graph of $y = f(x)$ from the graph of the tangent line to f.

4. If Q is a quantity to be measured and ΔQ is the error made in measuring Q, then the relative error in the measurement is given by the ratio _____.

5. *True or False*   Newton's Method uses tangent lines to the graph of f to approximate the zeros of f.

6. *True or False*   Before using Newton's Method, we need a first approximation for the zero.

Skill Building

In Problems 7–16, find the differential dy of each function.

7. $y = x^3 - 2x + 1$

8. $y = e^x + 2x - 1$

9. $y = 4(x^2 + 1)^{3/2}$

10. $y = \sqrt{x^2 - 1}$

11. $y = 3\sin(2x) + x$

12. $y = \cos^2(3x) - x$

13. $y = e^{-x}$

14. $y = e^{\sin x}$

15. $y = xe^x$

16. $y = \dfrac{e^{-x}}{x}$

In Problems 17–22:

(a) *Find the differential dy for each function f.*

(b) *Evaluate dy and Δy at the given value of x when (i) $\Delta x = 0.5$, (ii) $\Delta x = 0.1$, and (iii) $\Delta x = 0.01$.*

(c) *Find the error $|\Delta y - dy|$ for each choice of $dx = \Delta x$.*

17. $f(x) = e^x$ at $x = 1$

18. $f(x) = e^{-x}$ at $x = 1$

19. $f(x) = x^{2/3}$ at $x = 2$

20. $f(x) = x^{-1/2}$ at $x = 1$

21. $f(x) = \cos x$ at $x = \pi$

22. $f(x) = \tan x$ at $x = 0$

In Problems 23–30:

(a) *Find the linear approximation $L(x)$ to f at x_0.*

(b) *Graph f and L on the same set of axes.*

23. $f(x) = (x + 1)^5$, $x_0 = 2$

24. $f(x) = x^3 - 1$, $x_0 = 0$

25. $f(x) = \sqrt{x}$, $x_0 = 4$

26. $f(x) = x^{2/3}$, $x_0 = 1$

27. $f(x) = \ln x$, $x_0 = 1$

28. $f(x) = e^x$, $x_0 = 1$

29. $f(x) = \cos x$, $x_0 = \dfrac{\pi}{3}$

30. $f(x) = \sin x$, $x_0 = \dfrac{\pi}{6}$

31. Approximate the change in:

    (a) $y = f(x) = x^2$ as x changes from 3 to 3.001.

    (b) $y = f(x) = \dfrac{1}{x + 2}$ as x changes from 2 to 1.98.

32. Approximate the change in:

    (a) $y = x^3$ as x changes from 3 to 3.01.

    (b) $y = \dfrac{1}{x - 1}$ as x changes from 2 to 1.98.

In Problems 33–40, for each function:

(a) *Use the Intermediate Value Theorem to confirm that a zero exists in the given interval.*

(b) *Use Newton's Method with the first approximation c_1 to find c_3, the third approximation to the real zero.*

33. $f(x) = x^3 + 3x - 5$, interval: $(1, 2)$. Let $c_1 = 1.5$.

34. $f(x) = x^3 - 4x + 2$, interval: $(1, 2)$. Let $c_1 = 1.5$.

35. $f(x) = 2x^3 + 3x^2 + 4x - 1$, interval: $(0, 1)$. Let $c_1 = 0.5$.

36. $f(x) = x^3 - x^2 - 2x + 1$, interval: $(0, 1)$. Let $c_1 = 0.5$

37. $f(x) = x^3 - 6x - 12$, interval: $(3, 4)$. Let $c_1 = 3.5$.

---

1. = NOW WORK problem   🗠 = Graphing technology recommended   CAS = Computer Algebra System recommended

38. $f(x) = 3x^3 + 5x - 40$, interval: $(2, 3)$. Let $c_1 = 2.5$.

39. $f(x) = x^4 - 2x^3 + 21x - 23$, interval: $(1, 2)$.
Use a first approximation c_1 of your choice.

40. $f(x) = x^4 - x^3 + x - 2$, interval: $(1, 2)$.
Use a first approximation c_1 of your choice.

In Problems 41–46, for each function:

(a) *Use the Intermediate Value Theorem to confirm that a zero exists in the given interval.*

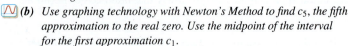 **(b)** *Use graphing technology with Newton's Method to find c_5, the fifth approximation to the real zero. Use the midpoint of the interval for the first approximation c_1.*

41. $f(x) = x + e^x$, interval: $(-1, 0)$

42. $f(x) = x - e^{-x}$, interval: $(0, 1)$

43. $f(x) = x^3 + \cos^2 x$, interval: $(-1, 0)$

44. $f(x) = x^2 + 2\sin x - 0.5$, interval: $(0, 1)$

45. $f(x) = 5 - \sqrt{x^2 + 2}$, interval: $(4, 5)$

46. $f(x) = 2x^2 + x^{2/3} - 4$, interval: $(1, 2)$

Applications and Extensions

47. Area of a Disk   A circular plate is heated and expands. If the radius of the plate increases from $R = 10$ cm to $R = 10.1$ cm, use differentials to approximate the increase in the area of the top surface.

48. Volume of a Cylinder   In a wooden block 3 cm thick, an existing circular hole with a radius of 2 cm is enlarged to a hole with a radius of 2.2 cm. Use differentials to approximate the volume of wood that is removed.

49. Volume of a Balloon   Use differentials to approximate the change in volume of a spherical balloon of radius 3 m as the balloon swells to a radius of 3.1 m.

50. Volume of a Paper Cup   A manufacturer produces paper cups in the shape of a right circular cone with a radius equal to one-fourth its height. Specifications call for the cups to have a top diameter of 4 cm. After production, it is discovered that the diameter measures only 3.8 cm. Use differentials to approximate the loss in capacity of the cup.

51. Volume of a Sphere

(a) Use differentials to approximate the volume of material needed to manufacture a hollow sphere if its inner radius is 2 m and its outer radius is 2.1 m.

(b) Is the approximation overestimating or underestimating the volume of material needed?

(c) Discuss the importance of knowing the answer to (b) if the manufacturer receives an order for 10,000 spheres.

52. Distance Traveled   A bee flies around a circle traced on an equator of a ball with a radius of 7 cm at a constant distance of 2 cm from the ball. An ant travels along the same circle but on the ball.

(a) Use differentials to approximate how many more centimeters the bee travels than the ant in one round trip.

(b) Does the linear approximation overestimate or underestimate the difference in the distances the bugs travel? Explain.

53. Estimating Height   To find the height of a building, the length of the shadow of a 3-m pole placed 9 m from the building is measured. See the figure. This measurement is found to be 1 m, with a percentage error of 1%. Use differentials to approximate the height of the building. What is the percentage error in the estimate?

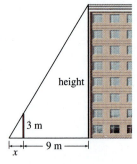

54. Pendulum Length   The period of the pendulum of a grandfather clock is

$$T = 2\pi\sqrt{\frac{l}{g}},$$ where l is the length (in

meters) of the pendulum, T is the period (in seconds), and g is the acceleration due to gravity (9.8 m/s^2). Suppose an increase in temperature increases the length l of the pendulum, a thin wire, by 1%. What is the corresponding percentage error in the period? How much time will the clock lose (or gain) each day?

l (in meters)

55. Pendulum Length   Refer to Problem 54. If the pendulum of a grandfather clock is normally 1 m long and the length is increased by 10 cm, use differentials to approximate the number of minutes the clock will lose (or gain) each day.

56. Luminosity of the Sun   The luminosity L of a star is the rate at which it radiates energy. This rate depends on the temperature T (in Kelvin, where 0 K is absolute zero) and the surface area A of the star's photosphere (the gaseous surface that emits the light). Luminosity at time t is given by the formula $L(t) = \sigma A T^4$, where σ is a constant, known as the **Stefan–Boltzmann constant**.

As with most stars, the Sun's temperature has gradually increased over the 5 billion years of its existence, causing its luminosity to slowly increase. For this problem, we assume that increased luminosity L is due only to an increase in temperature T. That is, we treat A as a constant.

(a) Find the rate of change of the temperature T of the Sun with respect to time t. Write the answer in terms of the rate of change of the Sun's luminosity L with respect to time t.

(b) 4.5 billion years ago, the Sun's luminosity was only 70% of what it is now. If the rate of change of luminosity L with respect to time t is constant, then $\dfrac{\Delta L}{\Delta t} = \dfrac{0.3 L_c}{\Delta t} = \dfrac{0.3 L_c}{4.5}$, where L_c is the current luminosity. Use differentials to approximate the current rate of change of the temperature T of the Sun in degrees per century.

57. Climbing a Mountain   Weight W is the force on an object due to the pull of gravity. On Earth, this force is given by Newton's Law of Universal Gravitation: $W = \dfrac{GmM}{r^2}$, where m is the mass of the object, $M = 5.974 \times 10^{24}$ kg is the mass of Earth, r is the distance of the object from the center of the Earth, and $G = 6.67 \times 10^{-11}$ m^3/(kg · s^2) is the universal gravitational constant. Suppose a person weighs 70 kg at sea level, that is, when $r = 6370$ km (the radius of Earth). Use differentials to approximate the person's weight at the top of Mount Everest, which is 8.8 km above sea level.

58. **Body Mass Index**   The **body mass index (BMI)** is given by the formula

$$\text{BMI} = 703\frac{m}{h^2}$$

where m is the person's weight in pounds and h is the person's height in inches. A BMI of 25 or less indicates that weight is normal, whereas a BMI greater than 25 indicates that a person is overweight.

(a) Suppose a man who is 5 ft 6 in. weighs 142 lb in the morning when he first wakes up, but he weighs 148 lb in the afternoon. Calculate his BMI in the morning and use differentials to approximate the change in his BMI in the afternoon. Round both answers to three decimal places.

(b) Did this linear approximation overestimate or underestimate the man's afternoon weight? Explain.

(c) A woman who weighs 165 lb estimates her height at 68 in. with a possible error of ±1.5 in. Calculate her BMI, assuming a height of 68 in. Then use differentials to approximate the possible error in her calculation of BMI. Round the answers to three decimal places.

(d) In the situations described in (a) and (c), how do you explain the classification of each person as normal or overweight?

59. **Percentage Error**   The radius of a spherical ball is found by measuring the volume of the sphere (by finding how much water it displaces). It is determined that the volume is 40 cubic centimeters (cm^3), with a tolerance of 1%. Find the percentage error in the radius of the sphere caused by the error in measuring the volume.

60. **Percentage Error**   The oil pan of a car has the shape of a hemisphere with a radius of 8 cm. The depth h of the oil is measured at 3 cm, with a percentage error of 10%. Approximate the percentage error in the volume. [*Hint:* The volume V for a spherical segment is $V = \frac{1}{3}\pi h^2(3R - h)$, where R is the radius of the sphere.]

61. **Percentage Error**   If the percentage error in measuring the edge of a cube is 2%, what is the percentage error in computing its volume?

62. **Focal Length**   To photograph an object, a camera's lens forms an image of the object on the camera's photo sensors. A camera lens can be approximated by a thin lens, which obeys the thin-lens equation $\frac{1}{f} = \frac{1}{p} + \frac{1}{q}$, where p is the distance from the lens to the object being photographed, q is the distance from the lens to the image of the object, and f is the focal length of the lens. A camera whose lens has a focal length of 50 mm is being used to photograph a dog. The dog is originally 15 m from the lens, but moves 0.33 m (about a foot) closer to the lens. Use differentials to approximate the distance the image of the dog moved.

Using Newton's Method to Solve Equations   *In Problems 63–66, use Newton's Method to solve each equation correct to three decimal places.*

63. $e^{-x} = \ln x$

64. $e^{-x} = x - 4$

65. $e^x = x^2$

66. $e^x = 2\cos x$

67. **Approximating e**   Use Newton's Method to approximate the value of e by finding the zero of the equation $\ln x - 1 = 0$. Use $c_1 = 3$ as the first approximation and find the fourth approximation to the zero. Compare the results from this approximation to the value of e obtained with a calculator.

68. Show that the linear approximation of a function $f(x) = (1 + x)^k$, where x is near 0 and k is any number, is given by $y = 1 + kx$.

69. Does it seem reasonable that if a first degree polynomial approximates a differentiable function in an interval near x_0, a higher-degree polynomial should approximate the function over a wider interval? Explain your reasoning.

70. Why does a function need to be differentiable at x_0 for a linear approximation to be used?

71. **Newton's Method**   Suppose you use Newton's Method to solve $f(x) = 0$ for a differentiable function f, and you obtain $x_{n+1} = x_n$. What can you conclude?

72. **When Newton's Method Fails**   Verify that the function $f(x) = -x^3 + 6x^2 - 9x + 6$ has a zero in the interval $(2, 5)$. Show that Newton's Method fails if an initial estimate of $c_1 = 2.9$ is chosen. Repeat Newton's Method with an initial estimate of $c_1 = 3.0$. Explain what occurs for each of these two choices. (The zero is near $x = 4.2$.)

73. **When Newton's Method Fails**   Show that Newton's Method fails if it is applied to $f(x) = x^3 - 2x + 2$ with an initial estimate of $c_1 = 0$.

74. **When Newton's Method Fails**   Show that Newton's Method fails if $f(x) = x^8 - 1$ if an initial estimate of $c_1 = 0.1$ is chosen. Explain what occurs.

75. **When Newton's Method Fails**   Show that Newton's Method fails if it is applied to $f(x) = (x - 1)^{1/3}$ with an initial estimate of $c_1 = 2$.

76. **Newton's Method**

(a) Use the Intermediate Value Theorem to show that $f(x) = x^4 + 2x^3 - 2x - 2$ has a zero in the interval $(-2, -1)$.

(b) Use Newton's Method to find c_3, a third approximation to the zero from (a).

(c) Explain why the initial approximation, $c_1 = -1$, cannot be used in (b).

Challenge Problems ———————

77. **Specific Gravity**   A solid wooden sphere of diameter d and specific gravity S sinks in water to a depth h, which is determined by the equation $2x^3 - 3x^2 - S = 0$, where $x = \frac{h}{d}$. Use Newton's Method to find a third approximation to h for a maple ball of diameter 6 in. for which $S = 0.786$.

78. **Kepler's Equation**   The equation $x - p\sin x = M$, called **Kepler's equation**, occurs in astronomy. Use Newton's Method to find a second approximation to x when $p = 0.2$ and $M = 0.85$. Use $c_1 = 1$ as your first approximation.

3.5 Taylor Polynomials

ORIGINS The Taylor polynomial is named after the English mathematician Brook Taylor (1685–1731). Taylor grew up in an affluent but strict home. He was home-schooled in the arts and the classics until he attended Cambridge University in 1703, where he studied mathematics. He was an accomplished musician and painter as well as a mathematician.

In the last section, we found that if a function f is differentiable in an interval (a, b), then a linear approximation $L(x)$ to f at a number x_0 in the interval is given by

$$L(x) = f(x_0) + f'(x_0)(x - x_0)$$

where $y = L(x)$ is the tangent line to the graph of f at the point $(x_0, f(x_0))$. This approximation for f is good if x is close enough to x_0, but as the interval widens, the accuracy of the estimate decreases.

The linear approximation $y = L(x)$ is a first degree polynomial function. If the function f has derivatives of orders 1 to n at x_0, then a polynomial function $P_n(x)$ of degree n can be used to approximate f near x_0.

To be sure that the polynomial approximation $P_n(x)$ fits f well, we require:

$$P_n(x_0) = f(x_0) \quad P_n'(x_0) = f'(x_0) \quad P_n''(x_0) = f''(x_0) \ldots P_n^{(n)}(x_0) = f^{(n)}(x_0)$$

That is, the value at x_0 of the function f and its first n derivatives equals the value at x_0 of the polynomial approximation and its first n derivatives.

NOTE The sum of the first two terms of a Taylor Polynomial is the linear approximation of f at x. That is, $P_1(x) = L(x)$.

DEFINITION Taylor Polynomial

If a function f and its first n derivatives are defined on an interval containing the number x_0, then

$$P_n(x) = f(x_0) + f'(x_0)(x - x_0) + \frac{f''(x_0)}{2!}(x - x_0)^2 + \cdots + \frac{f^{(n)}(x_0)}{n!}(x - x_0)^n$$

is called the nth **Taylor Polynomial** for f at x_0.

---

1 Find a Taylor Polynomial

EXAMPLE 1   Finding a Taylor Polynomial for $f(x) = \sqrt{x}$

Find the Taylor Polynomial $P_3(x)$ for $f(x) = \sqrt{x}$ at 1.

Solution   The first three derivatives of $f(x) = \sqrt{x}$ are

$$f'(x) = \frac{1}{2\sqrt{x}}, \qquad f''(x) = \frac{d}{dx}\left(\frac{x^{-1/2}}{2}\right) = -\frac{1}{4x^{3/2}}$$

$$f'''(x) = \frac{d}{dx}\left(-\frac{x^{-3/2}}{4}\right) = \frac{3}{8x^{5/2}}$$

Then

$$f(1) = 1 \qquad f'(1) = \frac{1}{2} \qquad f''(1) = -\frac{1}{4} \qquad f'''(1) = \frac{3}{8}$$

The Taylor Polynomial $P_3(x)$ for $f(x) = \sqrt{x}$ at 1 is

$$P_3(x) = f(1) + f'(1)(x - 1) + \frac{f''(1)}{2!}(x - 1)^2 + \frac{f'''(1)}{3!}(x - 1)^3$$

$$= 1 + \frac{x - 1}{2} - \frac{(x - 1)^2}{8} + \frac{(x - 1)^3}{16}$$

The graphs of the function f and the Taylor Polynomial P_3 are shown in Figure 17. Notice how $P_3(x) \approx f(x)$ for values of x near 1. ∎

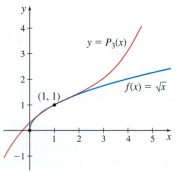

$P_3(x) = 1 + \frac{x - 1}{2} - \frac{(x - 1)^2}{8} + \frac{(x - 1)^3}{16}$

Figure 17

NOW WORK Problem 11.

EXAMPLE 2   Finding a Taylor Polynomial for $f(x) = \sin x$

Find the Taylor Polynomial $P_7(x)$ for $f(x) = \sin x$ at 0.

Solution   The derivatives of $f(x) = \sin x$ at 0 are

$$f(x) = \sin x \qquad f'(x) = \cos x \qquad f''(x) = -\sin x \qquad f'''(x) = -\cos x$$
$$f(0) = 0 \qquad\quad f'(0) = 1 \qquad\quad f''(0) = 0 \qquad\quad f'''(0) = -1$$

$$f^{(4)}(x) = \sin x \qquad f^{(5)}(x) = \cos x \qquad f^{(6)}(x) = -\sin x \qquad f^{(7)}(x) = -\cos x$$
$$f^{(4)}(0) = 0 \qquad\quad f^{(5)}(0) = 1 \qquad\quad f^{(6)}(0) = 0 \qquad\quad f^{(7)}(0) = -1$$

The Taylor Polynomial $P_7(x)$ for $\sin x$ at 0 is

$$P_7(x) = f(0) + f'(0)x + \frac{f''(0)}{2!}x^2 + \frac{f'''(0)}{3!}x^3 + \cdots + \frac{f^{(7)}(0)}{7!}x^7 = x - \frac{x^3}{3!} + \frac{x^5}{5!} - \frac{x^7}{7!}$$

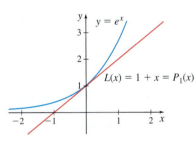

Figure 18 $P_7(x) = x - \dfrac{x^3}{3!} + \dfrac{x^5}{5!} - \dfrac{x^7}{7!}$

$y = P_7(x)$

$f(x) = \sin x$

$(0, 0)$

Figure 18 shows the graph of P_7 superimposed on the graph of $y = \sin x$. ∎

NOW WORK   Problem 13.

EXAMPLE 3   Finding a Taylor Polynomial for $f(x) = e^x$

Find the Taylor Polynomial $P_n(x)$ for $f(x) = e^x$ at 0.

Solution   The derivatives of $f(x) = e^x$ are

$$f(x) = e^x \qquad f'(x) = e^x \qquad f''(x) = e^x \qquad f'''(x) = e^x \dots f^{(n)}(x) = e^x$$
$$f(0) = 1 \qquad\quad f'(0) = 1 \qquad\quad f''(0) = 1 \qquad\quad f'''(0) = 1 \dots f^{(n)}(0) = 1$$

The Taylor Polynomial $P_n(x)$ at 0 is

$$P_n(x) = f(0) + f'(0)x + \frac{f''(0)}{2!}x^2 + \frac{f'''(0)}{3!}x^3 + \cdots + \frac{f^{(n)}(0)}{n!}x^n$$

$$= 1 + x + \frac{x^2}{2!} + \frac{x^3}{3!} + \cdots + \frac{x^n}{n!}$$ ∎

NOTE In Chapter 8 we investigate the relationship between a function f and its Taylor Polynomials.

Figure 19 illustrates how the graphs of the Taylor Polynomials $P_1(x)$, $P_2(x)$, and $P_3(x)$ compare to the graph of $f(x) = e^x$ near 0.

$y = e^x$

$L(x) = 1 + x = P_1(x)$

(a) Linear approximation $P_1(x)$

$y = e^x$

$P_2(x) = 1 + x + \frac{1}{2}x^2$

(b) Taylor polynomial $P_2(x)$

$y = e^x$

$P_3(x) = 1 + x + \frac{x^2}{2} + \frac{x^3}{6}$

(c) Taylor polynomial $P_3(x)$

DF Figure 19

3.5 Assess Your Understanding

Concepts and Vocabulary

1. *True or False*   If f is a function with both first and second derivatives defined on an interval containing x_0, then f can be approximated by the Taylor Polynomial $P_2(x) = f(x_0) + f'(x_0)(x - x_0) + f''(x_0)(x - x_0)^2$.

2. *True or False*   A Taylor Polynomial approximation P_n for a function f at x_0 has the following properties: $P_n(x_0) = f(x_0)$ and $P_n^k(x_0) = f^k(x_0)$ for derivatives of all orders from $k = 1$ to $k = n$.

---

1. = NOW WORK problem   = Graphing technology recommended   CAS = Computer Algebra System recommended

Skill Building

In Problems 3–28, for each function f find the Taylor Polynomial
$P_5(x)$ *for f at the given* x_0.

3. $f(x) = 3x^3 + 2x^2 - 6x + 5$ at $x_0 = 1$

4. $f(x) = 4x^3 - 2x^2 - 4$ at $x_0 = 1$

5. $f(x) = 2x^4 - 6x^3 + x$ at $x_0 = -1$

6. $f(x) = -3x^4 + 2x^2 - 5$ at $x_0 = -1$

7. $f(x) = x^5$ at $x_0 = 2$

8. $f(x) = x^6$ at $x_0 = 3$

9. $f(x) = \ln x$ at $x_0 = 1$

10. $f(x) = \ln(1 + x)$ at $x_0 = 0$

11. $f(x) = \dfrac{1}{x}$ at $x_0 = 1$

12. $f(x) = \dfrac{1}{x^2}$ at $x_0 = 1$

13. $f(x) = \cos x$ at $x_0 = 0$

14. $f(x) = \sin x$ at $x_0 = \dfrac{\pi}{4}$

15. $f(x) = e^{2x}$ at $x_0 = 0$

16. $f(x) = e^{-x}$ at $x_0 = 0$

17. $f(x) = \dfrac{1}{1 - x}$ at $x_0 = 0$

18. $f(x) = \dfrac{1}{1 + x}$ at $x_0 = 0$

19. $f(x) = \dfrac{1}{(1 + x)^2}$ at $x_0 = 0$

20. $f(x) = \dfrac{1}{1 + x^2}$ at $x_0 = 0$

21. $f(x) = x \ln x$ at $x_0 = 1$

22. $f(x) = xe^x$ at $x_0 = 1$

23. $f(x) = \sqrt{3 + x^2}$ at $x_0 = 1$

24. $f(x) = \sqrt{1 + x}$ at $x_0 = 0$

25. $f(x) = \tan x$ at $x_0 = \dfrac{\pi}{4}$

26. $f(x) = \sec x$ at $x_0 = 0$

27. $f(x) = \tan^{-1} x$ at $x_0 = 0$

28. $f(x) = \sin^{-1} x$ at $x_0 = 0$

Applications and Extensions

In Problems 29–32, express each polynomial as a polynomial in
$(x - 1)$ *by writing the Taylor Polynomial for f at 1.*

29. $f(x) = 3x^2 - 6x + 4$ **30.** $f(x) = 4x^2 - x + 1$

31. $f(x) = x^3 + x^2 - 8$ **32.** $f(x) = x^4 + 1$

In Problems 33 – 40:

(a) *find the indicated Taylor polynomial for each function f at* x_0.

CAS (b) *Graph f and the Taylor polynomial found in (a).*

33. $f(x) = \sin^{-1} x$ at $x_0 = \dfrac{1}{2}$, $P_4(x)$

34. $f(x) = \tan^{-1} x$ at $x_0 = 1$, $P_5(x)$

35. $f(x) = \dfrac{x}{\sqrt{x^2 + 3}}$ at $x_0 = 1$, $P_5(x)$

36. $f(x) = x\sqrt[3]{x^2 + 5}$ at $x_0 = 2$, $P_5(x)$

37. Uninhibited Decay $f(x) = 0.34e^{[(-\ln 2)/5600]x}$ at $x = 0$, $P_4(x)$

38. Uninhibited Growth $f(x) = 5000e^{0.04x}$ at $x = 0$, $P_4(x)$

39. Logistic Population Growth Model $f(x) = \dfrac{100}{1 + 30.2e^{-0.2x}}$ at $x_0 = 0$, $P_3(x)$

40. Gompertz Population Growth Model $f(x) = 100e^{-3e^{-0.2x}}$ at $x_0 = 0$, $P_3(x)$

41. The Taylor Polynomial $P_7(x)$ for $f(x) = \sin x$ at 0 has only terms with odd powers. See Example 2.

  (a) Discuss why this is true. Does this property hold for the Taylor Polynomial $P_n(x)$ for any number n?

  (b) Investigate the Taylor Polynomial $P_7(x)$ for f at $\dfrac{\pi}{2}$. Does this Taylor Polynomial have only odd terms? Explain why or why not.

42. The Taylor Polynomial $P_6(x)$ for $f(x) = \cos x$ at 0 has only terms with even powers. See Problem 13.

  (a) Discuss why this is true. Does this property hold for the Taylor Polynomial $P_n(x)$ for any n?

  (b) Investigate the Taylor Polynomial $P_6(x)$ for f at $\dfrac{\pi}{2}$. Does this Taylor Polynomial have only even terms? Explain why or why not.

Challenge Problems

43. The graphs of $y = \sin x$ and $y = \lambda x$ intersect near $x = \pi$ if λ is small. Let $f(x) = \sin x - \lambda x$. Find the Taylor Polynomial $P_2(x)$ for f at π, and use it to show that an approximate solution of the equation $\sin x = \lambda x$ is $x = \dfrac{\pi}{1 + \lambda}$.

44. The graphs of $y = \cot x$ and $y = \lambda x$ intersect near $x = \dfrac{\pi}{2}$ if λ is small. Let $f(x) = \cot x - \lambda x$. Find the Taylor Polynomial $P_2(x)$ for f at $\dfrac{\pi}{2}$, and use it to find an approximate solution of the equation $\cot x = \lambda x$.

3.6 Hyperbolic Functions

OBJECTIVES *When you finish this section, you should be able to:*

1 Define the hyperbolic functions (p. 243)
2 Establish identities for hyperbolic functions (p. 244)
3 Differentiate hyperbolic functions (p. 245)
4 Differentiate inverse hyperbolic functions (p. 246)

1 Define the Hyperbolic Functions

Functions involving certain combinations of e^x and e^{-x} occur so frequently in applied mathematics that they warrant special study. These functions, called *hyperbolic functions,* have properties similar to those of trigonometric functions. Because of this, they are named the *hyperbolic sine* (sinh), the *hyperbolic cosine* (cosh), the *hyperbolic tangent* (tanh), the *hyperbolic cotangent* (coth), the *hyperbolic cosecant* (csch), and the *hyperbolic secant* (sech).

DEFINITION

The **hyperbolic sine function** and **hyperbolic cosine function** are defined as

$$y = \sinh x = \frac{e^x - e^{-x}}{2} \qquad y = \cosh x = \frac{e^x + e^{-x}}{2}$$

Hyperbolic functions are related to a hyperbola in much the same way as the trigonometric functions (sometimes called *circular functions*) are related to the circle. Just as any point P on the unit circle $x^2 + y^2 = 1$ has coordinates $(\cos t, \sin t)$, as shown in Figure 20, a point P on the hyperbola $x^2 - y^2 = 1$ has coordinates $(\cosh t, \sinh t)$, as shown in Figure 21. Moreover, both the sector of the circle shown in Figure 20 and the shaded portion of Figure 21 have areas that each equal $\dfrac{t}{2}$.

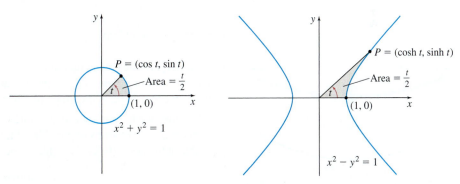

Figure 20   Figure 21

The functions $y = \sinh x$ and $y = \cosh x$ are defined for all real numbers. The hyperbolic sine function, $y = \sinh x$, is an odd function, so its graph is symmetric with respect to the origin; the hyperbolic cosine function, $y = \cosh x$, is an even function, so its graph is symmetric with respect to the y-axis. Their graphs may be found by combining the graphs of $y = \dfrac{e^x}{2}$ and $y = \dfrac{e^{-x}}{2}$, as illustrated in Figures 22 and 23, respectively. The range of $y = \sinh x$ is all real numbers, and the range of $y = \cosh x$ is the interval $[1, \infty)$.

The remaining four hyperbolic functions are combinations of the hyperbolic cosine and hyperbolic sine functions, and their relationships are similar to those of the trigonometric functions.

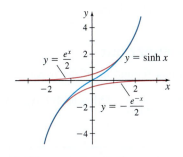

Figure 22

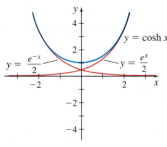

Figure 23

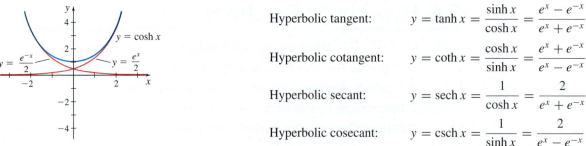

Hyperbolic tangent: $y = \tanh x = \dfrac{\sinh x}{\cosh x} = \dfrac{e^x - e^{-x}}{e^x + e^{-x}}$

Hyperbolic cotangent: $y = \coth x = \dfrac{\cosh x}{\sinh x} = \dfrac{e^x + e^{-x}}{e^x - e^{-x}}$

Hyperbolic secant: $y = \operatorname{sech} x = \dfrac{1}{\cosh x} = \dfrac{2}{e^x + e^{-x}}$

Hyperbolic cosecant: $y = \operatorname{csch} x = \dfrac{1}{\sinh x} = \dfrac{2}{e^x - e^{-x}}$

The graphs of these four hyperbolic functions are shown in Figures 24 through 27.

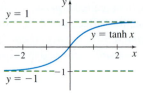

Figure 24

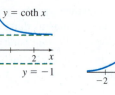

Figure 25

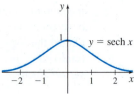

Figure 26

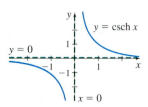

Figure 27

EXAMPLE 1   **Evaluating Hyperbolic Functions**

Find the exact value of:

(a) $\cosh 0$ **(b)** $\operatorname{sech} 0$ **(c)** $\tanh(\ln 2)$

Solution **(a)** $\cosh 0 = \dfrac{e^0 + e^0}{2} = \dfrac{2}{2} = 1$

(b) $\operatorname{sech} 0 = \dfrac{1}{\cosh 0} = \dfrac{1}{1} = 1$

(c) $\tanh(\ln 2) = \dfrac{e^{\ln 2} - e^{-\ln 2}}{e^{\ln 2} + e^{-\ln 2}} = \dfrac{2 - e^{\ln(1/2)}}{2 + e^{\ln(1/2)}} = \dfrac{2 - \dfrac{1}{2}}{2 + \dfrac{1}{2}} = \dfrac{3}{5}$ ∎

NOW WORK   Problem 9.

The hyperbolic cosine function has an interesting physical interpretation. If a cable or chain of uniform density is suspended at its ends, such as with high-voltage lines, it will assume the shape of the graph of a hyperbolic cosine.

Suppose we fix our coordinate system, as in Figure 28, so that the cable, which is suspended from endpoints of equal height, lies in the xy-plane, with the lowest point of the cable at the point $(0, b)$. Then the shape of the cable will be modeled by the equation

$$y = a \cosh \frac{x}{a} + b - a$$

where a is a constant that depends on the weight per unit length of the cable and on the tension or horizontal force holding the ends of the cable. The graph of this equation is called a **catenary**, from the Latin word *catena* that means "chain."

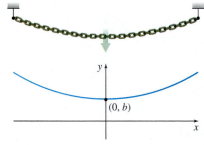

Figure 28 $y = a \cosh \dfrac{x}{a} + b - a$

2 Establish Identities for Hyperbolic Functions

There are identities for the hyperbolic functions that remind us of the trigonometric identities. For example, $\tanh x = \dfrac{\sinh x}{\cosh x}$ and $\coth x = \dfrac{1}{\tanh x}$. Other useful hyperbolic identities include

$$\cosh^2 x - \sinh^2 x = 1 \qquad \tanh^2 x + \operatorname{sech}^2 x = 1 \qquad \coth^2 x - \operatorname{csch}^2 x = 1$$

EXAMPLE 2   **Establishing Identities for Hyperbolic Functions**

Establish the identity $\cosh^2 x - \sinh^2 x = 1$.

Solution

$$\cosh^2 x - \sinh^2 x = \left(\frac{e^x + e^{-x}}{2}\right)^2 - \left(\frac{e^x - e^{-x}}{2}\right)^2$$

$$= \frac{e^{2x} + 2e^0 + e^{-2x}}{4} - \frac{e^{2x} - 2e^0 + e^{-2x}}{4} = \frac{2+2}{4} = 1 \quad \blacksquare$$

Numerous other identities involving hyperbolic functions can be established. We list some below.

Sum formulas:

$$\sinh(A + B) = \sinh A \cosh B + \cosh A \sinh B$$
$$\cosh(A + B) = \cosh A \cosh B + \sinh A \sinh B$$

Even/odd properties:

$$\sinh(-A) = -\sinh A \qquad \cosh(-A) = \cosh A$$

The derivations of these identities are left as exercises. (See Problems 17–20.)

NOW WORK   Problem 19.

3 Differentiate Hyperbolic Functions

Since the hyperbolic functions are algebraic combinations of e^x and e^{-x}, they are differentiable at all real numbers for which they are defined. For example,

$$\frac{d}{dx}\sinh x = \frac{d}{dx}\left(\frac{e^x - e^{-x}}{2}\right) = \frac{1}{2}\left[\frac{d}{dx}e^x - \frac{d}{dx}e^{-x}\right] = \frac{1}{2}(e^x + e^{-x}) = \cosh x$$

$$\frac{d}{dx}\cosh x = \frac{d}{dx}\left(\frac{e^x + e^{-x}}{2}\right) = \frac{1}{2}\left[\frac{d}{dx}e^x + \frac{d}{dx}e^{-x}\right] = \frac{1}{2}(e^x - e^{-x}) = \sinh x$$

$$\frac{d}{dx}\operatorname{csch} x = \frac{d}{dx}\left(\frac{1}{\sinh x}\right) = \frac{-\dfrac{d}{dx}\sinh x}{\sinh^2 x} = \frac{-\cosh x}{\sinh^2 x} = -\frac{1}{\sinh x}\cdot\frac{\cosh x}{\sinh x}$$

$$= -\operatorname{csch} x \coth x$$

The formulas for the derivatives of the hyperbolic functions are listed below.

$\dfrac{d}{dx}\sinh x = \cosh x$	$\dfrac{d}{dx}\tanh x = \operatorname{sech}^2 x$	$\dfrac{d}{dx}\operatorname{csch} x = -\operatorname{csch} x \coth x$
$\dfrac{d}{dx}\cosh x = \sinh x$	$\dfrac{d}{dx}\coth x = -\operatorname{csch}^2 x$	$\dfrac{d}{dx}\operatorname{sech} x = -\operatorname{sech} x \tanh x$

EXAMPLE 3   **Differentiating Hyperbolic Functions**

Find y'.

(a) $y = x^2 - 2\sinh x$ **(b)** $y = \cosh(x^2 + 1)$

Solution **(a)** $y' = \dfrac{d}{dx}(x^2 - 2\sinh x) = 2x - 2\dfrac{d}{dx}\sinh x = 2x - 2\cosh x$

(b) We use the Chain Rule with $y = \cosh u$ and $u = x^2 + 1$.

$$y' = \frac{dy}{dx} = \frac{dy}{du}\cdot\frac{du}{dx} = \sinh u \cdot 2x = 2x\sinh(x^2 + 1) \qquad \blacksquare$$

NOW WORK   Problem 31.

EXAMPLE 4   Finding the Angle Between a Catenary and Its Support

A cable is suspended between two poles of the same height that are 20 m apart, as shown in Figure 29(a). If the poles are placed at $(-10, 0)$ and $(10, 0)$, the equation that models the height of the cable is $y = 10 \cosh \dfrac{x}{10} + 15$. Find the angle θ at which the cable meets a pole.

Solution The slope of the tangent line to the catenary is given by

$$y' = \frac{d}{dx} \left(10 \cosh \frac{x}{10} + 15 \right) = 10 \cdot \frac{1}{10} \sinh \frac{x}{10} = \sinh \frac{x}{10}$$

At $x = 10$, the slope $m_{\tan}$ of the tangent line is $m_{\tan} = \sinh \dfrac{10}{10} = \sinh 1$.

The angle θ at which the cable meets the pole equals the angle between the tangent line and the pole. To find θ, we form a right triangle using the tangent line and the pole, as shown in Figure 29(b).

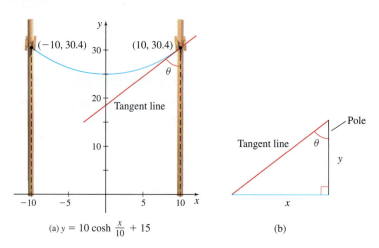

(a) $y = 10 \cosh \frac{x}{10} + 15$ (b)

Figure 29

From Figure 29(b), we find that the slope of the tangent line is $m_{\tan} = \dfrac{\Delta y}{\Delta x} = \sinh 1$.

Then $\tan \theta = \dfrac{\Delta x}{\Delta y} = \dfrac{1}{\sinh 1}$. So, $\theta = \tan^{-1} \left(\dfrac{1}{\sinh 1} \right) \approx 0.7050$ radians $\approx 40.4°$. ∎

NOW WORK Problem 51.

4 Differentiate Inverse Hyperbolic Functions

The graph of $y = \sinh x$, shown in Figure 22 on p. 243, suggests that every horizontal line intersects the graph at exactly one point. In fact, the function $y = \sinh x$ is one-to-one, and so it has an inverse function. We denote the inverse by $y = \sinh^{-1} x$ and define it as

NEED TO REVIEW? A discussion on one-to-one functions and inverse functions can be found in Section P.4, pp 32–37.

$$\boxed{y = \sinh^{-1} x \qquad \text{if and only if} \qquad x = \sinh y}$$

The domain of $y = \sinh^{-1} x$ is the set of real numbers, and the range is also the set of real numbers. See Figure 31(a).

The graph of $y = \cosh x$ (see Figure 23 on p. 244) shows that every horizontal line above $y = 1$ intersects the graph of $y = \cosh x$ at two points so $y = \cosh x$ is not one-to-one. However, if the domain of $y = \cosh x$ is restricted to the nonnegative values of x, we have a one-to-one function that has an inverse. We denote the inverse function by $y = \cosh^{-1} x$ and define it as

$$\boxed{y = \cosh^{-1} x \qquad \text{if and only if} \qquad x = \cosh y \qquad y \geq 0}$$

The domain of $y = \cosh^{-1} x$ is $x \geq 1$, and the range is $y \geq 0$. See Figure 31(b).

The other inverse hyperbolic functions are defined similarly. Their graphs are given in Figure 31(c)–(f).

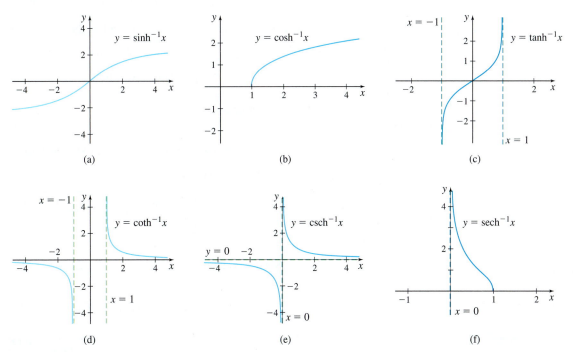

Figure 30

Since the hyperbolic functions are defined in terms of the exponential function, the inverse hyperbolic functions can be expressed in terms of natural logarithms.

$$y = \sinh^{-1} x = \ln\left(x + \sqrt{x^2 + 1}\right) \qquad \text{for all real } x$$

$$y = \cosh^{-1} x = \ln\left(x + \sqrt{x^2 - 1}\right) \qquad x \geq 1$$

$$y = \tanh^{-1} x = \frac{1}{2}\ln\left(\frac{1+x}{1-x}\right) \qquad |x| < 1$$

$$y = \coth^{-1} x = \frac{1}{2}\ln\left(\frac{x+1}{x-1}\right) \qquad |x| > 1$$

We show that $\sinh^{-1} x = \ln\left(x + \sqrt{x^2 + 1}\right)$ here, and leave the others as exercises. (See Problems 61–63.)

EXAMPLE 5 Expressing the Inverse Hyperbolic Sine Function as a Natural Logarithm

Express $y = \sinh^{-1} x$ as a natural logarithm.

Solution Since $y = \sinh^{-1} x$, where $x = \sinh y$, we have

$$x = \frac{e^y - e^{-y}}{2}$$

$$2xe^y = (e^y)^2 - 1 \qquad \text{Multiply both sides by } 2e^y.$$

$$(e^y)^2 - 2xe^y - 1 = 0$$

NEED TO REVIEW? The quadratic equation is discussed in Appendix A.1 (p. A3).

This is a quadratic equation in e^y. Use the quadratic formula and solve for e^y.

$$e^y = \frac{2x \pm \sqrt{4x^2 + 4}}{2}$$

$$e^y = x \pm \sqrt{x^2 + 1} \qquad \text{Simplify.}$$

Since $e^y > 0$ and $x < \sqrt{x^2 + 1}$ for all x, the minus sign on the right side is not possible. As a result, $e^y = x + \sqrt{x^2 + 1}$ so that

$$y = \sinh^{-1} x = \ln(x + \sqrt{x^2 + 1})$$

■

NOW WORK Problem 61.

EXAMPLE 6   Differentiating an Inverse Hyperbolic Sine Function

Show that if $y = \sinh^{-1} x$, then

$$y' = \frac{d}{dx} \sinh^{-1} x = \frac{1}{\sqrt{x^2 + 1}}.$$

Solution Since $y = \sinh^{-1} x = \ln\left(x + \sqrt{x^2 + 1}\right)$, we have

$$y' = \frac{d}{dx} \sinh^{-1} x = \frac{d}{dx}\left[\ln\left(x + \sqrt{x^2 + 1}\right)\right] = \frac{\dfrac{d}{dx}\left(x + \sqrt{x^2 + 1}\right)}{x + \sqrt{x^2 + 1}}$$

$$\uparrow$$
$$\frac{d}{dx} \ln(u) = \frac{u'(x)}{u(x)}$$

$$= \frac{1 + \dfrac{1}{2}(x^2 + 1)^{-1/2} \cdot 2x}{x + \sqrt{x^2 + 1}} = \frac{1 + \dfrac{x}{\sqrt{x^2 + 1}}}{x + \sqrt{x^2 + 1}} = \frac{\dfrac{\sqrt{x^2 + 1} + x}{\sqrt{x^2 + 1}}}{x + \sqrt{x^2 + 1}}$$

$$= \frac{x + \sqrt{x^2 + 1}}{(x + \sqrt{x^2 + 1})\sqrt{x^2 + 1}} = \frac{1}{\sqrt{x^2 + 1}}$$

■

NOW WORK Problem 43.

Similarly, we have the following derivative formulas. If $y = \cosh^{-1} x$, then

$$y' = \frac{d}{dx} \cosh^{-1} x = \frac{1}{\sqrt{x^2 - 1}} \qquad x > 1$$

If $y = \tanh^{-1} x$, then

$$y' = \frac{d}{dx} \tanh^{-1} x = \frac{1}{1 - x^2} \qquad |x| < 1$$

EXAMPLE 7   Differentiating an Inverse Hyperbolic Tangent Function

Find y' if $y = \tanh^{-1} \sqrt{x}$.

Solution We use the Chain Rule with $y = \tanh^{-1} u$ and $u = \sqrt{x}$. Then

$$y' = \frac{dy}{dx} = \frac{dy}{du} \cdot \frac{du}{dx} = \frac{d}{du} \tanh^{-1} u \cdot \frac{d}{dx} \sqrt{x}$$

$$= \frac{1}{1 - u^2} \cdot \frac{1}{2\sqrt{x}} = \frac{1}{1 - x} \cdot \frac{1}{2\sqrt{x}} = \frac{\sqrt{x}}{2x(1 - x)} \qquad u = \sqrt{x}$$

■

NOW WORK Problem 41.

3.6 Assess Your Understanding

Concepts and Vocabulary

1. *True or False* $\operatorname{csch} x = \dfrac{1}{\cosh x}$.

2. In terms of $\sinh x$ and $\cosh x$, $\tanh x =$ _____.

3. *True or False*   The domain of $y = \cosh x$ is $[1, \infty)$.

4. *Multiple Choice*   The function $y = \cosh x$ is [**(a)** even, **(b)** odd, **(c)** neither].

5. *True or False* $\cosh^2 x + \sinh^2 x = 1$.

6. *True or False* $\dfrac{d}{dx} \sinh x = \dfrac{1}{\sqrt{x^2 + 1}}$.

7. *True or False*   The function $y = \sinh^{-1} x$ is defined for all real numbers x.

8. *True or False*   The functions $f(x) = \tanh^{-1} x$ and $g(x) = \coth^{-1} x$ are identical except for their domains.

Skill Building

In Problems 9–14, find the exact value of each expression.

9. $\operatorname{csch}(\ln 3)$ **10.** $\operatorname{sech}(\ln 2)$ **11.** $\cosh^2(5) - \sinh^2(5)$

12. $\cosh(-\ln 2)$ **13.** $\tanh 0$ **14.** $\sinh\left(\ln \dfrac{1}{2}\right)$

In Problems 15–24, establish each identity.

15. $\tanh^2 x + \operatorname{sech}^2 x = 1$ **16.** $\coth^2 x - \operatorname{csch}^2 x = 1$

17. $\sinh(-A) = -\sinh A$ **18.** $\cosh(-A) = \cosh A$

19. $\sinh(A + B) = \sinh A \cosh B + \cosh A \sinh B$

20. $\cosh(A + B) = \cosh A \cosh B + \sinh A \sinh B$

21. $\sinh(2x) = 2 \sinh x \cosh x$

22. $\cosh(2x) = \cosh^2 x + \sinh^2 x$

23. $\cosh(3x) = 4 \cosh^3 x - 3 \cosh x$ **24.** $\tanh(2x) = \dfrac{2 \tanh x}{1 + \tanh^2 x}$

In Problems 25–48, find y'.

25. $y = \sinh(3x)$ **26.** $y = \cosh \dfrac{x}{2}$

27. $y = \cosh(x^2 + 1)$ **28.** $y = \cosh(2x^3 - 1)$

29. $y = \coth \dfrac{1}{x}$ **30.** $y = \tanh(x^2)$

31. $y = \sinh x \cosh(4x)$ **32.** $y = \sinh(2x) \cosh(-x)$

33. $y = \cosh^2 x$ **34.** $y = \tanh^2 x$

35. $y = e^x \cosh x$ **36.** $y = e^x (\cosh x + \sinh x)$

37. $y = x^2 \operatorname{sech} x$ **38.** $y = x^3 \tanh x$

39. $y = \cosh^{-1}(4x)$ **40.** $y = \sinh^{-1}(3x)$

41. $y = \tanh^{-1}(x^2 - 1)$ **42.** $y = \cosh^{-1}(2x + 1)$

43. $y = x \sinh^{-1} x$ **44.** $y = x^2 \cosh^{-1} x$

45. $y = \tanh^{-1}(\tan x)$ **46.** $y = \sinh^{-1}(\sin x)$

47. $y = \cosh^{-1}\left(\sqrt{x^2 - 1}\right), x > \sqrt{2}$ **48.** $y = \sinh^{-1}\left(\sqrt{x^2 + 1}\right)$

Applications and Extensions

49. **Taylor Polynomial**   Write the Taylor Polynomial $P_6(x)$ for $g(x) = \cosh x$ at 0.

50. **Taylor Polynomial**   Write the Taylor Polynomial $P_7(x)$ for $f(x) = \tanh x$ at 0.

51. **Catenary**   A cable is suspended between two supports of the same height that are 100 m apart. If the supports are placed at $(-50, 0)$ and $(50, 0)$, the equation that models the height of the cable is $y = 12 \cosh \dfrac{x}{12} + 20$. Find the angle θ at which the cable meets each support.

52. **Catenary**   A town hangs strings of holiday lights across the road between utility poles. Each set of poles is 12 m apart. The strings hang in catenaries modeled by $y = 15 \cosh \dfrac{x}{15} - 10$.

(a) Find the slope of the tangent line where the lights meet the pole.

(b) Find the angle at which the string of lights meets the pole.

53. **Catenary**   The famous Gateway Arch to the West in St. Louis, Missouri, is constructed in the shape of a modified inverted catenary. (Modified because the weight is not evenly dispersed throughout the arch.) If y is the height of the arch and $x = 0$ corresponds to the center of the arch (its highest point), an equation for the arch is given by

$$y = -68.767 \cosh\left(\dfrac{0.711x}{68.767}\right) + 693.859 \text{ ft},$$

(a) Find the maximum height of the arch.

(b) Find the width of the arch at ground level.

[CAS] (c) What is the slope of the arch 50 ft from its center?

[CAS] (d) What is the slope of the arch 200 ft from its center?

[CAS] (e) Find the angle (in degrees) that the arch makes with the ground. (Assume the ground is level.)

[N] (f) Graph the equation that models the Gateway Arch, and explain how the graph supports the answers found in (a)–(e).

54. Establish the identity $(\cosh x + \sinh x)^n = \cosh(nx) + \sinh(nx)$ for any real number n.

55. (a) Show that if $y = \cosh^{-1} x$, then $y' = \dfrac{1}{\sqrt{x^2 - 1}}$.

(b) What is the domain of y'?

56. (a) Show that if $y = \tanh^{-1} x$, then $y' = \dfrac{1}{1 - x^2}$.

(b) What is the domain of y'?

57. Show that $\dfrac{d}{dx} \tanh x = \operatorname{sech}^2 x$.

---

1. = NOW WORK problem   [N] = Graphing technology recommended   [CAS] = Computer Algebra System recommended

58. Show that $\dfrac{d}{dx} \coth x = -\operatorname{csch}^2 x$.

59. Show that $\dfrac{d}{dx} \operatorname{sech} x = -\operatorname{sech} x \tanh x$

60. Show that $\dfrac{d}{dx} \operatorname{csch} x = -\operatorname{csch} x \coth x$

61. Show that $\tanh^{-1} x = \dfrac{1}{2} \ln\left(\dfrac{1+x}{1-x}\right)$, $\quad -1 < x < 1$.

62. Show that $\cosh^{-1} x = \ln(x + \sqrt{x^2 - 1})$, $\quad x \geq 1$.

63. Show that $\coth^{-1} x = \dfrac{1}{2} \ln\left(\dfrac{x+1}{x-1}\right)$, $\quad |x| > 1$.

Challenge Problems

In Problems 64 and 65, find each limit.

64. $\displaystyle\lim_{x \to 0} \left(\dfrac{\sinh x}{x}\right)$ $\qquad$ **65.** $\displaystyle\lim_{x \to 0} \left(\dfrac{\cosh x - 1}{x}\right)$

66. (a) Sketch the graph of $y = f(x) = \dfrac{1}{2}(e^x + e^{-x})$.

(b) Let R be a point on the graph and let r, $r \neq 0$, be the x-coordinate of R. The tangent line to the graph of f at R crosses the x-axis at the point Q. Find the coordinates of Q in terms of r.

(c) If P is the point $(r, 0)$, find the length of the line segment PQ as a function of r and the limiting value of this length as r increases without bound.

67. What happens if you try to find the derivative of $f(x) = \sin^{-1}(\cosh x)$? Explain why this occurs.

68. Let $f(x) = x \sinh^{-1} x$.

(a) Show that f is an even function.

(b) Find $f'(x)$ and $f''(x)$.

Chapter Review

THINGS TO KNOW

3.1 The Chain Rule

Basic Derivative Formulas:

- $\dfrac{d}{dx} e^u = e^u \dfrac{du}{dx}$ (p. 200)

- $\dfrac{d}{dx} a^x = a^x \ln a \quad a > 0$ and $a \neq 1$ (p. 202)

Properties of Derivatives:

- Chain Rule: $(f \circ g)'(x) = f'(g(x)) \cdot g'(x)$ $\quad$ or
  $$\dfrac{dy}{dx} = \dfrac{dy}{du} \cdot \dfrac{du}{dx} \text{ (p. 198)}$$

- Power Rule for functions: $\dfrac{d}{dx}[g(x)]^n = n[g(x)]^{n-1} g'(x)$, where n is an integer (p. 203)

3.2 Implicit Differentiation; Derivatives of the Inverse Trigonometric Functions

Procedure: To differentiate an implicit function (p. 211):

- Assume y is a differentiable function of x.
- Differentiate both sides of the equation with respect to x.
- Solve the resulting equation for $y' = \dfrac{dy}{dx}$.

Basic Derivative Formulas:

- $\dfrac{d}{dx} \sin^{-1} x = \dfrac{1}{\sqrt{1-x^2}}$ $\quad -1 < x < 1$ (p. 216)

- $\dfrac{d}{dx} \tan^{-1} x = \dfrac{1}{1+x^2}$ (p. 217)

- $\dfrac{d}{dx} \sec^{-1} x = \dfrac{1}{x\sqrt{x^2-1}}$ $\quad |x| > 1$ (p. 218)

Properties of Derivatives:

- Power Rule for rational exponents: $\dfrac{d}{dx} x^{p/q} = \dfrac{p}{q} \cdot x^{(p/q)-1}$. provided $x^{p/q}$ and $x^{p/q-1}$ are defined. (p. 213)

- Power Rule for functions: $\dfrac{d}{dx}[u(x)]^r = r[u(x)]^{r-1} u'(x)$, r a rational number; provided u^r and u^{r-1} are defined. (p. 214)

Theorem: The derivative of an inverse function at a number (p. 215)

3.3 Derivatives of Logarithmic Functions

Basic Derivative Formulas:

- $\dfrac{d}{dx} \log_a x = \dfrac{1}{x \ln a}$, $a > 0, a \neq 1$ (p. 222)

- $\dfrac{d}{dx} \ln x = \dfrac{1}{x}$ (p. 222)

Steps for Using Logarithmic Differentiation (p. 225):

- **Step 1** If the function $y = f(x)$ consists of products, quotients, and powers, take the natural logarithm of each side. Then simplify using properties of logarithms.

- **Step 2** Differentiate implicitly, and use $\dfrac{d}{dx} \ln y = \dfrac{y'}{y}$.

- **Step 3** Solve for y', and replace y with $f(x)$.

Theorems:

- **Power Rule** If a is a real number, then $\dfrac{d}{dx} x^a = ax^{a-1}$. (p. 226)

- The number e can be expressed as
  $$\lim_{h \to 0}(1+h)^{1/h} = e \text{ or } \lim_{n \to \infty}\left(1 + \dfrac{1}{n}\right)^n = e. \text{ (p. 227)}$$

3.4 Differentials; Linear Approximations; Newton's Method

- The differential dx of x is defined as $dx = \Delta x \neq 0$, where Δx is the change in x. The differential dy of $y = f(x)$ is defined as $dy = f'(x)dx$. (p. 231)

- A **linear approximation** $L(x)$ to a differentiable function f near $x = x_0$ is given by $L(x) = f(x_0) + f'(x_0)(x - x_0)$. (p. 232)
- **Newton's Method** for finding the zero of a function. (p. 235)

3.5 Taylor Polynomials

- Taylor Polynomial $P_n(x)$ for f at x_0:

$$P_n(x) = f(x_0) + f'(x_0)(x - x_0) + \frac{f''(x_0)}{2!}(x - x_0)^2 + \cdots$$

$$+ \frac{f^{(n)}(x_0)}{n!}(x - x_0)^n \quad \text{(p. 240)}$$

3.6 Hyperbolic Functions

Definitions:

- Hyperbolic sine: $y = \sinh x = \dfrac{e^x - e^{-x}}{2}$ (p. 243)

- Hyperbolic cosine: $y = \cosh x = \dfrac{e^x + e^{-x}}{2}$ (p. 243)

Hyperbolic Identities (pp. 244–245):

- $\tanh x = \dfrac{\sinh x}{\cosh x}$
- $\coth x = \dfrac{\cosh x}{\sinh x}$
- $\operatorname{sech} x = \dfrac{1}{\cosh x}$
- $\operatorname{csch} x = \dfrac{1}{\sinh x}$
- $\cosh^2 x - \sinh^2 x = 1$
- $\tanh^2 x + \operatorname{sech}^2 x = 1$
- $\coth^2 x - \operatorname{csch}^2 x = 1$
- Sum Formulas:

$$\sinh(A + B) = \sinh A \cosh B + \cosh A \sinh B$$
$$\cosh(A + B) = \cosh A \cosh B + \sinh A \sinh B$$

- Even/odd Properties:

$$\sinh(-A) = -\sinh A \qquad \cosh(-A) = \cosh A$$

Inverse Hyperbolic Functions (p. 247):

- $y = \sinh^{-1} x = \ln\left(x + \sqrt{x^2 + 1}\right)$ for all real x
- $y = \cosh^{-1} x = \ln\left(x + \sqrt{x^2 - 1}\right)$ $x \geq 1$
- $y = \tanh^{-1} x = \dfrac{1}{2}\ln\left(\dfrac{1+x}{1-x}\right)$ $|x| < 1$
- $y = \coth^{-1} x = \dfrac{1}{2}\ln\left(\dfrac{x+1}{x-1}\right)$ $|x| > 1$

Basic Derivative Formulas (pp. 245, 248):

- $\dfrac{d}{dx}\sinh x = \cosh x$
- $\dfrac{d}{dx}\operatorname{sech} x = -\operatorname{sech} x \tanh x$
- $\dfrac{d}{dx}\cosh = \sinh x$
- $\dfrac{d}{dx}\operatorname{csch} x = -\operatorname{csch} x \coth x$
- $\dfrac{d}{dx}\tanh x = \operatorname{sech}^2 x$
- $\dfrac{d}{dx}\coth x = -\operatorname{csch}^2 x$
- $\dfrac{d}{dx}\sinh^{-1} x = \dfrac{1}{\sqrt{x^2 + 1}}$
- $\dfrac{d}{dx}\cosh^{-1} x = \dfrac{1}{\sqrt{x^2 - 1}}$ $x > 1$
- $\dfrac{d}{dx}\tanh^{-1} x = \dfrac{1}{1 - x^2}$ $|x| < 1$

OBJECTIVES

Section	You should be able to …	Example	Review Exercises
3.1	1 Differentiate a composite function (p. 198)	1–5	1, 13, 24
	2 Differentiate $y = a^x$, $a > 0$, $a \neq 1$ (p. 202)	6	19, 22
	3 Use the Power Rule for functions to find a derivative (p. 202)	7, 8	1, 11, 12, 14, 17
	4 Use the Chain Rule for multiple composite functions (p. 204)	9	15, 18, 61
3.2	1 Find a derivative using implicit differentiation (p. 209)	1–4	43–52, 73, 81
	2 Find higher-order derivatives using implicit differentiation (p. 212)	5	49–52
	3 Differentiate functions with rational exponents (p. 213)	6, 7	2–8, 15, 16, 61–64
	4 Find the derivative of an inverse function (p. 214)	8	53, 54
	5 Differentiate inverse trigonometric functions (p. 216)	9, 10	32–38
3.3	1 Differentiate logarithmic functions (p. 222)	1–3	20, 21, 23, 25–30, 52, 72
	2 Use logarithmic differentiation (p. 225)	4–7	9, 10, 31, 71
	3 Express e as a limit (p. 227)	8	55, 56
3.4	1 Find the differential of a function and interpret it geometrically (p. 230)	1	65, 69, 70
	2 Find the linear approximation to a function (p. 232)	2	67
	3 Use differentials in applications (p. 233)	3, 4	66, 68
	4 Use Newton's Method to approximate a real zero of a function (p. 234)	5, 6	78–80
3.5	1 Find a Taylor Polynomial (p. 240)	1–3	74–77
3.6	1 Define the hyperbolic functions (p. 243)	1	57, 58
	2 Establish identities for hyperbolic functions (p. 244)	2	59, 60
	3 Differentiate hyperbolic functions (p. 245)	3, 4	39–41
	4 Differentiate inverse hyperbolic functions (p. 246)	5–7	42

REVIEW EXERCISES

In Problems 1–42, find the derivative of each function. When a, b, or n appear, they are constants.

1. $y = (ax + b)^n$

2. $y = \sqrt{2ax}$

3. $y = x\sqrt{1 - x}$

4. $y = \dfrac{1}{\sqrt{x^2 + 1}}$

5. $y = (x^2 + 4)^{3/2}$

6. $F(x) = \dfrac{x^2}{\sqrt{x^2 - 1}}$

7. $z = \dfrac{\sqrt{2ax - x^2}}{x}$

8. $y = \sqrt{x} + \sqrt[3]{x}$

9. $y = (e^x - x)^{5x}$

10. $\phi(x) = \dfrac{(x^2 - a^2)^{3/2}}{\sqrt{x + a}}$

11. $f(x) = \dfrac{x^2}{(x - 1)^2}$

12. $u = (b^{1/2} - x^{1/2})^2$

13. $y = x\sec(2x)$

14. $u = \cos^3 x$

15. $y = \sqrt{a^2 \sin\left(\dfrac{x}{a}\right)}$

16. $\phi(z) = \sqrt{1 + \sin z}$

17. $u = \sin v - \dfrac{1}{3}\sin^3 v$

18. $y = \tan\sqrt{\dfrac{\pi}{x}}$

19. $y = (1.05)^x$

20. $v = \ln(y^2 + 1)$

21. $z = \ln(\sqrt{u^2 + 25} - u)$

22. $y = x^2 + 2^x$

23. $y = \ln[\sin(2x)]$

24. $f(x) = e^{-x}\sin(2x + \pi)$

25. $g(x) = \ln(x^2 - 2x)$

26. $y = \ln\dfrac{x^2 + 1}{x^2 - 1}$

27. $y = e^{-x}\ln x$

28. $w = \ln\left(\sqrt{x + 7} - \sqrt{x}\right)$

29. $y = \dfrac{1}{12}\ln\left(\dfrac{x}{\sqrt{144 - x^2}}\right)$

30. $y = \ln(\tan^2 x)$

31. $f(x) = \dfrac{e^x(x^2 + 4)}{(x - 2)}$

32. $y = \sin^{-1}(x - 1) + \sqrt{2x - x^2}$

33. $y = 2\sqrt{x} - 2\tan^{-1}\sqrt{x}$

34. $y = 4\tan^{-1}\dfrac{x}{2} + x$

35. $y = \sin^{-1}(2x - 1)$

36. $y = x^2\tan^{-1}\dfrac{1}{x}$

37. $y = x\tan^{-1} x - \ln\sqrt{1 + x^2}$

38. $y = \sqrt{1 - x^2}(\sin^{-1} x)$

39. $y = \tanh\dfrac{x}{2} + \dfrac{2x}{4 + x^2}$

40. $y = x\sinh x$

41. $y = \sqrt{\sinh x}$

42. $y = \sinh^{-1} e^x$

In Problems 43–48, find $y' = \dfrac{dy}{dx}$ using implicit differentiation.

43. $x = y^5 + y$

44. $x = \cos^5 y + \cos y$

45. $\ln x + \ln y = x\cos y$

46. $\tan(xy) = x$

47. $y = x + \sin(xy)$

48. $x = \ln(\csc y + \cot y)$

In Problems 49–52, find y' and y''.

49. $xy + 3y^2 = 10x$

50. $y^3 + y = x^2$

51. $xe^y = 4x^2$

52. $\ln(x + y) = 8x$

53. The function $f(x) = e^{2x}$ has an inverse function g. Find $g'(1)$.

54. The function $f(x) = \sin x$ defined on the restricted domain $\left[-\dfrac{\pi}{2}, \dfrac{\pi}{2}\right]$ has an inverse function g. Find $g'\left(\dfrac{1}{2}\right)$.

In Problems 55–56, express each limit in terms of the number e.

55. $\displaystyle\lim_{n \to \infty}\left(1 + \dfrac{2}{5n}\right)^n$

56. $\displaystyle\lim_{h \to 0}(1 + 3h)^{2/h}$

In Problems 57 and 58, find the exact value of each expression.

57. $\sinh 0$

58. $\cosh(\ln 3)$

In Problems 59 and 60, establish each identity.

59. $\sinh x + \cosh x = e^x$

60. $\tanh(x + y) = \dfrac{\tanh x + \tanh y}{1 + \tanh x \tanh y}$

61. If $f(x) = \sqrt{1 - \sin^2 x}$, find the domain of f'.

62. If $f(x) = x^{1/2}(x - 2)^{3/2}$ for all $x \geq 2$, find the domain of f'.

63. Let f be the function defined by $f(x) = \sqrt{1 + 6x}$.

   (a) What are the domain and the range of f?

   (b) Find the slope of the tangent line to the graph of f at $x = 4$.

   (c) Find the y-intercept of the tangent line to the graph of f at $x = 4$.

   (d) Give the coordinates of the point on the graph of f where the tangent line is parallel to the line $y = x + 12$.

64. **Tangent and Normal Lines** Find equations of the tangent and normal lines to the graph of $y = x\sqrt{x + (x - 1)^2}$ at the point $(2, 2\sqrt{3})$.

65. Find the differential dy if $x^3 + 2y^2 = x^2 y$.

66. **Measurement Error** If p is the period of a pendulum of length L, the acceleration due to gravity may be computed by the formula $g = \dfrac{(4\pi^2 L)}{p^2}$. If L is measured with negligible error, but a 2% error may occur in the measurement of p, what is the approximate percentage error in the computation of g?

67. **Linear Approximation** Find a linear approximation to
$$y = x + \ln x \text{ at } x = 1.$$

68. **Measurement Error** If the percentage error in measuring the edge of a cube is 5%, what is the percentage error in computing its volume?

69. For the function $f(x) = \tan x$:

   (a) Find the differential dy and Δy when $x = 0$.

   (b) Compare dy to Δy when $x = 0$ and (i) $\Delta x = 0.5$, (ii) $\Delta x = 0.1$, and (iii) $\Delta x = 0.01$.

70. For the function $f(x) = \ln x$:

   (a) Find the differential dy and Δy when $x = 1$.

   (b) Compare dy to Δy when $x = 1$ and (i) $\Delta x = 0.5$, (ii) $\Delta x = 0.1$, and (iii) $\Delta x = 0.01$.

71. If $f(x) = (x^2 + 1)^{(2 - 3x)}$, find $f'(1)$.

72. Find $\lim\limits_{x \to 2} \dfrac{\ln x - \ln 2}{x - 2}$.

73. Find y' at $x = \dfrac{\pi}{2}$ and $y = \pi$ if $x \sin y + y \cos x = 0$.

In Problems 74–77, find the Taylor Polynomial $P_n(x)$ for f at x_0 for the given n and x_0.

74. $f(x) = e^{2x}$; $n = 4$, $x_0 = 3$

75. $f(x) = \tan x$; $n = 4$, $x_0 = 0$

76. $f(x) = \dfrac{1}{1 + x}$; $n = 4$, $x_0 = 1$

77. $f(x) = \ln x$; $n = 6$, $x_0 = 2$

In Problems 78 and 79, for each function:

(a) Use the Intermediate Value Theorem to confirm that a zero exists in the given interval.

(b) Use Newton's Method to find c_3, the third approximation to the real zero.

78. $f(x) = 8x^4 - 2x^2 + 5x - 1$,   interval: $(0, 1)$. Let $c_1 = 0.5$.

79. $f(x) = 2 - x + \sin x$,   interval: $\left(\dfrac{\pi}{2}, \pi\right)$. Let $c_1 = \dfrac{\pi}{2}$.

80. **(a)** Use the Intermediate Value Theorem to confirm that the function $f(x) = 2\cos x - e^x$ has a zero in the interval $(0, 1)$.

   (b) Use graphing technology with Newton's Method to find c_5, the fifth approximation to the real zero. Use the midpoint of the interval for the first approximation c_1.

81. Tangent Line Find an equation of the tangent line to the graph of $4xy - y^2 = 3$ at the point $(1, 3)$.

CHAPTER 3 PROJECT   World Population

Law of Uninhibited Growth The Law of Uninhibited Growth states that, under certain conditions, the rate of change of a population is proportional to the size of the population at that time. One consequence of this law is that the time it takes for a population to double remains constant. For example, suppose a certain bacteria obeys the Law of Uninhibited Growth. Then if the bacteria take five hours to double from 100 organisms to 200 organisms, in the next five hours they will double again from 200 to 400. We can model this mathematically using the formula

$$P(t) = P_0 2^{t/D}$$

where $P(t)$ is the population at time t, P_0 is the population at time $t = 0$, and D is the doubling time.

If we use this formula to model population growth, a few observations are in order. For example, the model is continuous, but actual population growth is discrete. That is, an actual population would change from 100 to 101 individuals in an instant, as opposed to a model that has a continuous flow from 100 to 101. The model also produces fractional answers, whereas an actual population is counted in whole numbers. For large populations, however, the growth is continuous enough for the model to match real-world conditions, at least for a short time. In general, as growth continues, there are obstacles to growth at which point the model will fail to be accurate. Situations that follow the model of the Law of Uninhibited Growth vary from the introduction of invasive species into a new environment, to the spread of a deadly virus for which there is no immunization. Here, we investigate how accurately the model predicts world population.

1. The world population on January 1, 1959, was approximately 2.983435×10^9 persons and had a doubling time of $D = 40$ years. Use these data and the Law of Uninhibited Growth to write a formula for the world population $P = P(t)$. Use this model to solve Problems 2 through 4.

2. Find the rate of change of the world population $P = P(t)$ with respect to time t.

3. Find the rate of change on January 1, 2011 of the world population with respect to time. (Note that $t = 0$ is January 1, 1959.) Round the answer to the nearest whole number.

4. Approximate the world population at the beginning of 2011. Round the answer to the nearest person.

5. According to the United Nations, the world population on January 1, 2009, was 6.817727×10^9. Use $t_0 = 2009$, $P_0 = 6.817727 \times 10^9$, and $D = 40$ and find a new formula to model the world population $P = P(t)$.

6. Use the new model from Problem 5 to find the rate of change of the world population at the beginning of 2011.

7. Compare the results from Problems 3 and 6. Interpret and explain any discrepancy between the two rates of change.

8. Use the new model to approximate the world population at the beginning of 2011. Round the answer to the nearest person.

9. The State of World Population report produced by the United Nations Population Fund indicates that the world population was estimated to reach 7 billion on October 31, 2011. Use the model from Problem 5 to approximate the world population on November 1, 2011. Compare the predicted population to the actual world population of 7 billion on October 31, 2011.

10. Use the original model (1959 data) to approximate the world population on November 1, 2011.

11. Discuss possible reasons for the discrepancies in the approximations of the 2011 population and the official number released by the UN. Was it more accurate to use the 1959 data or the 2009 data? Why do you think one set of data gives better results than the other?

Source: *UN World Population Prospects*, 2010 Revision. © 2011, http://esa.un.org/wpp/unpp/panel_population.htm.

4 Applications of the Derivative

CHAPTER 4 PROJECT The Chapter Project on page 342 examines the two economic indicators, the Unemployment Rate and the GDP growth rate, and investigates relationships they might have to each other.

The U. S. Economy

There are many ways economists collect and report data about the U. S. economy. Two such reports are the Unemployment Rate and the Gross Domestic Product (GDP). The Unemployment Rate gives the percentage of people over the age of 16 who are unemployed and are actively looking for a job. The GDP growth rate measures the rate of increase (or decrease) in the market value of all goods and services produced over a period of time. The graphs on the right show the Unemployment Rate and the GDP growth rate in the United States from 2007–2012.

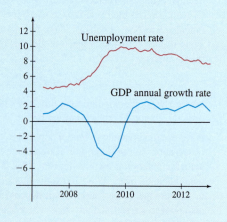

In Chapters 2 and 3, we developed a variety of formulas for finding derivatives. We also began to investigate how derivatives are applied, using derivatives to obtain polynomial approximations of functions and to approximate the zeros of a function.

In this chapter, we continue to explore applications of the derivative. We use the derivative to solve problems involving rates of change of variables that are related, to find optimal (minimum or maximum) values of cost functions or revenue functions, and to investigate properties of the graph of a function.

4.1 Related Rates

OBJECTIVE *When you finish this section, you should be able to:*

1 Solve related rate problems (p. 255)

In the natural sciences and in many of the social and behavioral sciences, there are quantities that are related to each other, but that vary with time. For example, the pressure of an ideal gas of fixed volume is proportional to the temperature, yet each of these variables may change over time. Problems involving the rates of change of related variables are called **related rate problems**. In a related rate problem, we seek the rate at which one of the variables is changing at a certain time, when the rates of change of the other variables are known.

1 Solve Related Rate Problems

We approach related rate problems by writing an equation involving the time-dependent variables. This equation is often obtained by investigating the geometric and/or physical conditions imposed by the problem. We then differentiate the equation with respect to time and obtain a new equation that involves the variables and their rates of change with respect to time.

For example, suppose an object falling into still water causes a circular ripple, as illustrated in Figure 1. Both the radius and the area of the circle created by the ripple increase with time and their rates of growth are related. We use the formula for the area of a circle

$$A = \pi r^2$$

to relate the radius and the area. Both A and r are functions of time t, and so the area of the circle can be expressed as

$$A(t) = \pi [r(t)]^2$$

Now we differentiate both sides with respect to t, obtaining

$$\frac{dA}{dt} = 2\pi r \frac{dr}{dt}$$

The derivatives (rates of change) are related by this equation, so we call them **related rates**. We can solve for one of these rates if the value of the other rate and the values of the variables are known.

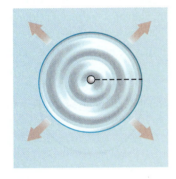

Figure 1

EXAMPLE 1 **Solving a Related Rate Problem**

A golfer hits a ball into a pond of still water, causing a circular ripple as shown in Figure 2. If the radius of the circle increases at the constant rate of 0.5 m/s, how fast is the area of the circle increasing when the radius of the ripple is 2 m?

Solution The quantities that are changing, that is, the variables of the problem, are

t = the time (in seconds) elapsed from the time when the ball hits the water

r = the radius (in meters) of the ripple after t seconds

A = the area (in square meters) of the circle formed by the ripple after t seconds

The rates of change with respect to time are

$\dfrac{dr}{dt}$ = the rate (in meters per second) at which the radius of the ripple is increasing

$\dfrac{dA}{dt}$ = the rate (in meters squared per second) at which the area of the circle is increasing

It is given that $\dfrac{dr}{dt} = 0.5$ m/s. We seek $\dfrac{dA}{dt}$ when $r = 2$ m.

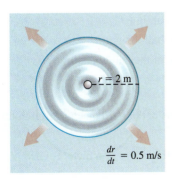

Figure 2

The relationship between A and r is given by the formula for the area of a circle:

$$A = \pi r^2$$

Since A and r are functions of t, we differentiate with respect to t to obtain

$$\frac{dA}{dt} = 2\pi r \frac{dr}{dt}$$

Since $\dfrac{dr}{dt} = 0.5 \, \text{m/s}$,

$$\frac{dA}{dt} = 2\pi r (0.5) = \pi r$$

When $r = 2$m,

$$\frac{dA}{dt} = \pi(2) = 2\pi$$

The area of the circle is increasing at a rate of about $6.283 \, \text{m}^2/\text{s}$. ∎

Steps for Solving a Related Rate Problem

Step 1 Read the problem carefully, perhaps several times. Pay particular attention to the rate you are asked to find. If possible, draw a picture to illustrate the problem.

Step 2 Identify the variables, assign symbols to them, and label the picture. Identify the rates of change as derivatives. Indicate what is given and what is asked for.

Step 3 Write an equation that relates the variables. It may be necessary to use more than one equation.

Step 4 Differentiate both sides of the equation(s).

Step 5 Substitute numerical values for the variables and the derivatives. Solve for the unknown rate.

CAUTION Numerical values cannot be substituted (Step 5) until after the equation has been differentiated (Step 4).

NOW WORK Problem 7.

EXAMPLE 2 Solving a Related Rate Problem

A spherical balloon is inflated at the rate of $10 \, \text{m}^3/\text{min}$. Find the rate at which the surface area of the balloon is increasing when the radius of the sphere is 3 m.

Solution We follow the steps for solving a related rate problem.

Step 1 Figure 3 shows a sketch of the balloon with its radius labeled.

Step 2 Identify the variables of the problem:

t = the time (in minutes) measured from the moment the balloon begins inflating

R = the radius (in meters) of the balloon at time t

V = the volume (in meters cubed) of the balloon at time t

S = the surface area (in meters squared) of the balloon at time t

Identify the rates of change:

$\dfrac{dR}{dt}$ = the rate of change of the radius of the balloon (in meters per minute)

$\dfrac{dV}{dt}$ = the rate of change of the volume of the balloon (in meters cubed per minute)

$\dfrac{dS}{dt}$ = the rate of change of the surface area of the balloon (in meters squared per minute)

We are given $\dfrac{dV}{dt} = 10 \, \text{m}^3/\text{min}$, and we seek $\dfrac{dS}{dt}$ when $R = 3$ m.

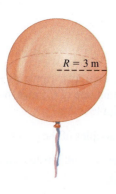

$R = 3$ m

Figure 3

Step 3 Since both the volume V of the balloon (a sphere) and its surface area S can be expressed in terms of the radius R, we use two equations to relate the variables.

NEED TO REVIEW? Geometry formulas are discussed in Appendix A.2, p. A-15.

$$V = \frac{4}{3}\pi R^3 \quad \text{and} \quad S = 4\pi R^2 \qquad \text{where } V, S, \text{ and } R \text{ are functions of } t$$

Step 4 Differentiate both sides of the equations with respect to time t.

$$\frac{dV}{dt} = 4\pi R^2 \frac{dR}{dt} \qquad \text{and} \qquad \frac{dS}{dt} = 8\pi R \frac{dR}{dt}$$

Combine the equations by solving for $\dfrac{dR}{dt}$ in the equation on the left and substituting the result into the equation for $\dfrac{dS}{dt}$ on the right. Then

$$\frac{dS}{dt} = 8\pi R \left(\frac{\frac{dV}{dt}}{4\pi R^2} \right) = \frac{2}{R}\frac{dV}{dt}$$

Step 5 Substitute $R = 3\,\text{m}$ and $\dfrac{dV}{dt} = 10\,\text{m}^3/\text{min}$.

$$\frac{dS}{dt} = \left(\frac{2}{3} \right)(10) \approx 6.667$$

When the radius of the balloon is 3 m, its surface area is increasing at the rate of about 6.667 m²/min. ∎

NOW WORK **Problems 11 and 17.**

EXAMPLE 3 **Solving a Related Rate Problem**

A rectangular swimming pool 10 m long and 5 m wide has a depth of 3 m at one end and 1 m at the other end. If water is pumped into the pool at the rate of 300 liters per minute (liter/min), at what rate is the water level rising when it is 1.5 m deep at the deep end?

Solution

Step 1 We draw a picture of the cross-sectional view of the pool, as shown in Figure 4.

Step 2 The width of the pool is 5 m, the water level (measured at the deep end) is h, the distance from the wall at the deep end to the edge of the water is L, and the volume of water in the pool is V. Each of the variables h, L, and V varies with respect to time t.

We are given $\dfrac{dV}{dt} = 300$ liter/min and are asked to find $\dfrac{dh}{dt}$ when $h = 1.5\,\text{m}$.

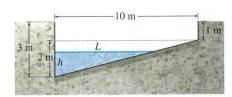

Figure 4

NEED TO REVIEW? Similar triangles are discussed in Appendix A.2, pp. A-13 to A-14.

Step 3 The volume V is related to L and h by the formula

$$V = (\text{Cross-sectional triangular area})(\text{width}) = \left(\frac{1}{2}Lh \right)(5) = \frac{5}{2}Lh$$

See Figure 4. Using similar triangles, L and h are related by the equation

$$\frac{L}{h} = \frac{10}{2}, \qquad \text{so} \quad L = 5h$$

Now we can write V as

$$V = \frac{5}{2}Lh = \frac{5}{2}(5h)h = \frac{25}{2}h^2 \tag{1}$$
$$\underset{L = 5h}{\uparrow}$$

Both V and h vary with time t.

Step 4 We differentiate both sides of equation (1) with respect to t.

$$\frac{dV}{dt} = 25h\frac{dh}{dt}$$

| **NOTE** 1000 liter $= 1\,\mathrm{m}^3$.

Step 5 Substitute $h = 1.5$ m and $\dfrac{dV}{dt} = 300$ liter/min $= \dfrac{300}{1000}\,\mathrm{m}^3/\mathrm{min} = 0.3\,\mathrm{m}^3/\mathrm{min}$.
Then

$$0.3 = 25(1.5)\frac{dh}{dt} \qquad \color{blue}{\frac{dV}{dt} = 25h\frac{dh}{dt}}$$

$$\frac{dh}{dt} = \frac{0.3}{25(1.5)} = 0.008$$

When the height of the water is 1.5 m, the water level is rising at a rate of 0.008 m/min. ∎

NOW WORK **Problem 23.**

EXAMPLE 4 **Solving a Related Rate Problem**

A person standing at the end of a pier is docking a boat by pulling a rope at the rate of 2 m/s. The end of the rope is 3 m above water level. See Figure 5(a). How fast is the boat approaching the base of the pier when 5 m of rope are left to pull in? (Assume the rope never sags, and that the rope is attached to the boat at water level.)

Solution

Step 1 Draw illustrations, like Figure 5, representing the problem. Label the sides of the triangle 3 and x, and label the hypotenuse of the triangle L.

Step 2 x is the distance (in meters) from the boat to the base of the pier, L is the length of the rope (in meters), and the distance between the water level and the person's hand is 3 m. Both x and L are changing with respect to time, so $\dfrac{dx}{dt}$ is the rate at which the boat approaches the pier, and $\dfrac{dL}{dt}$ is the rate at which the rope is pulled in.

We are given $\dfrac{dL}{dt} = 2$ m/s, and we seek $\dfrac{dx}{dt}$ when $L = 5$ m.

Step 3 The lengths 3, x, and L are the sides of a right triangle and, by the Pythagorean Theorem, are related by the equation

$$x^2 + 3^2 = L^2 \qquad (2)$$

Step 4 We differentiate both sides of equation (2) with respect to t.

$$2x\frac{dx}{dt} = 2L\frac{dL}{dt}$$

$$\frac{dx}{dt} = \frac{L}{x}\frac{dL}{dt} \qquad \color{blue}{\text{Solve for } \frac{dx}{dt}.}$$

Step 5 When $L = 5$, we use equation (2) to find $x = 4$. Since $\dfrac{dL}{dt} = 2$,

$$\frac{dx}{dt} = \frac{5}{4}(2) \qquad \color{blue}{L = 5, x = 4, \frac{dL}{dt} = 2}$$

$$\frac{dx}{dt} = 2.5$$

When 5 m of rope are left to be pulled in, the boat is approaching the pier at the rate of 2.5 m/s. ∎

NOW WORK **Problem 29.**

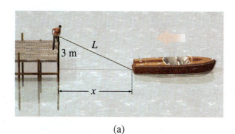

(a)

(b)

Figure 5

| **NEED TO REVIEW?** The Pythagorean Theorem is discussed in Appendix A.2, pp. A-12.

EXAMPLE 5   Solving a Related Rate Problem

A revolving light, located 5 km from a straight shoreline, turns at constant angular speed of 3 rad/min. With what speed is the spot of light moving along the shore when the beam makes an angle of 60° with the shoreline?

Solution Figure 6 illustrates the triangle that describes the problem.

x = the distance (in kilometers) of the beam of light from the point B

θ = the angle (in radians) the beam of light makes with AB

Both variables x and θ change with time t (in minutes). The rates of change are

$$\frac{dx}{dt} = \text{the speed of the spot of light along the shore (in kilometers per minute)}$$

$$\frac{d\theta}{dt} = \text{the angular speed of the beam of light (in radians per minute)}$$

We are given $\dfrac{d\theta}{dt} = 3\,\text{rad/min}$ and we seek $\dfrac{dx}{dt}$ when the angle $AOB = 60°$.

From Figure 6,

$$\tan \theta = \frac{x}{5} \qquad \text{so } x = 5 \tan \theta$$

Then

$$\frac{dx}{dt} = 5 \sec^2 \theta \, \frac{d\theta}{dt}$$

When $AOB = 60°$, angle $\theta = 30° = \dfrac{\pi}{6}$ rad, and

$$\frac{dx}{dt} = \frac{5}{\cos^2 \theta} \frac{d\theta}{dt} = \frac{5(3)}{\left(\cos \dfrac{\pi}{6}\right)^2} = \frac{15}{\dfrac{3}{4}} = 20$$

When $\theta = 30°$, the light is moving along the shore at a speed of $20\,\text{km/min}$. ∎

NOW WORK Problem 37.

RECALL For motion that is circular, angular speed ω is defined as the rate of change of a central angle θ of the circle with respect to time. That is, $\omega = \dfrac{d\theta}{dt}$, where θ is measured in radians.

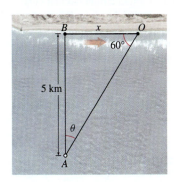

Figure 6

4.1 Assess Your Understanding

Concepts and Vocabulary

1. If a spherical balloon of volume V is inflated at a rate of $10\,\text{m}^3/\text{min}$, where t is the time (in minutes), what is the rate of change of V with respect to t?

2. For the balloon in Problem 1, if the radius r is increasing at the rate of $0.5\,\text{m/min}$, what is the rate of change of r with respect to t?

In Problems 3 and 4, x and y are differentiable functions of t.

Find $\dfrac{dx}{dt}$ *when* $x = 3$, $y = 4$, *and* $\dfrac{dy}{dt} = 2$.

3. $x^2 + y^2 = 25$

4. $x^3 y^2 = 432$

Skill Building

5. Suppose h is a differentiable function of t and suppose that $\dfrac{dh}{dt} = \dfrac{5}{16}\pi$ when $h = 8$. Find $\dfrac{dV}{dt}$ when $h = 8$ if $V = \dfrac{1}{12}\pi h^3$.

6. Suppose x and y are differentiable functions of t and suppose that when $t = 20$, $\dfrac{dx}{dt} = 5$, $\dfrac{dy}{dt} = 4$, $x = 150$, and $y = 80$. Find $\dfrac{ds}{dt}$ when $t = 20$ if $s^2 = x^2 + y^2$.

7. Suppose h is a differentiable function of t and suppose that when $h = 3$, $\dfrac{dh}{dt} = \dfrac{1}{12}$. Find $\dfrac{dV}{dt}$ when $h = 3$ if $V = 80h^2$.

8. Suppose x is a differentiable function of t and suppose that when $x = 15$, $\dfrac{dx}{dt} = 3$. Find $\dfrac{dy}{dt}$ when $x = 15$ if $y^2 = 625 - x^2$, $y \geq 0$.

9. **Volume of a Cube** If each edge of a cube is increasing at the constant rate of 3 cm/s, how fast is the volume of the cube increasing when the length x of an edge is 10 cm long?

10. **Volume of a Sphere** If the radius of a sphere is increasing at 1 cm/s, find the rate of change of its volume when the radius is 6 cm.

11. Radius of a Sphere   If the surface area of a sphere is shrinking at the constant rate of 0.1 cm²/h, find the rate of change of its radius when the radius is $\dfrac{20}{\pi}$ cm.

12. Surface Area of a Sphere   If the radius of a sphere is increasing at the constant rate of 2 cm/min, find the rate of change of its surface area when the radius is 100 cm.

13. Dimensions of a Triangle   Consider a right triangle with hypotenuse of (fixed) length 45 cm and variable legs of lengths x and y, respectively. If the leg of length x increases at the rate of 2 cm/min, at what rate is y changing when $x = 4$ cm?

14. Change in Area   The fixed sides of an isosceles triangle are of length 1 cm. (See the figure.) If the sides slide outward at a speed of 1 cm/min, at what rate is the area enclosed by the triangle changing when $\theta = 30°$?

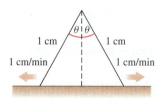

15. Area of a Triangle   An isosceles triangle has equal sides 4 cm long and the included angle is θ. If θ increases at the rate of 2°/min, at what rate is the area of the triangle changing when θ is 30°?

16. Area of a Rectangle   In a rectangle with a diagonal 15 cm long, one side is increasing at the rate of $2\sqrt{5}$ cm/s. Find the rate of change of the area when that side is 10 cm long.

17. Change in Surface Area   A spherical balloon filled with gas has a leak that causes the gas to escape at a rate of 1.5 m³/min. At what rate is the surface area of the balloon shrinking when the radius is 4 m?

18. Frozen Snow Ball   Suppose that the volume of a spherical ball of ice decreases (by melting) at a rate proportional to its surface area. Show that its radius decreases at a constant rate.

Applications and Extensions

19. Change in Inclination   A ladder 5 m long is leaning against a wall. If the lower end of the ladder slides away from the wall at the rate of 0.5 m/s, at what rate is the inclination θ of the ladder with respect to the ground changing when the lower end is 4 m from the wall?

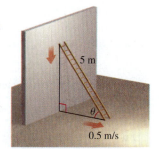

20. Angle of Elevation   A man 2 m tall walks horizontally at a constant rate of 1 m/s toward the base of a tower 25 m tall. When the man is 10 m from the tower, at what rate is the angle of elevation changing if that angle is measured from the horizontal to the line joining the top of the man's head to the top of the tower?

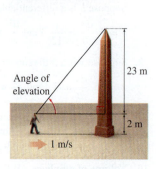

21. Filling a Pool   A public swimming pool is 30 m long and 5 m wide. Its depth is 3 m at the deep end and 1 m at the shallow end.

If water is pumped into the pool at the rate of 15 m³/min, how fast is the water level rising when it is 1 m deep at the deep end? Use Figure 4 as a guide.

22. Filling a Tank   Water is flowing into a vertical cylindrical tank of diameter 6 m at the rate of 5 m³/min. Find the rate at which the depth of the water is rising.

23. Fill Rate   A container in the form of a right circular cone (vertex down) has radius 4 m and height 16 m. See the figure. If water is poured into the container at the constant rate of 16 m³/min, how fast is the water level rising when the water is 8 m deep? (*Hint:* The volume V of a cone of radius r and height h is $V = \dfrac{1}{3}\pi r^2 h$.)

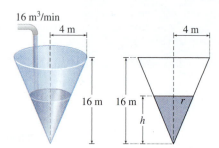

24. Building a Sand Pile   Sand is being poured onto the ground, forming a conical pile whose height equals one-fourth of the diameter of the base. If the sand is falling at the rate of 20 cm³/s, how fast is the height of the sand pile increasing when it is 3 cm high?

25. Is There a Leak?   A cistern in the shape of a cone 4 m deep and 2 m in diameter at the top is being filled with water at a constant rate of 3 m³/min.

(a)   At what rate is the water rising when the water in the tank is 3 m deep?

(b)   If, when the water is 3 m deep, it is observed that the water rises at a rate of 0.5 m/min, at what rate is water leaking from the tank?

26. Change in Area   The vertices of a rectangle are at $(0, 0)$, $(0, e^x)$, $(x, 0)$, and (x, e^x), $x > 0$. If x increases at the rate of 1 unit per second, at what rate is the area increasing when $x = 10$ units?

27. Distance from the Origin   An object is moving along the parabola $y^2 = 4(3 - x)$. When the object is at the point $(-1, 4)$, its y-coordinate is increasing at the rate of 3 units per second. How fast is the distance from the object to the origin changing at that instant?

28. Funneling Liquid   A conical funnel 15 cm in diameter and 15 cm deep is filled with a liquid that runs out at the rate of 5 cm³/min. At what rate is the depth of liquid changing when its depth is 8 cm?

29. Baseball   A baseball is hit along the third-base line with a speed of 100 ft/s. At what rate is the ball's distance from first base changing when it crosses third base? See the figure.

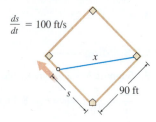

30. Flight of a Falcon   A peregrine falcon flies up from its trainer at an angle of $60°$ until it has flown 200 ft. It then levels off and continues to fly away. If the constant speed of the bird is 88 ft/s, how fast is the falcon moving away from the falconer after it is 6 seconds in flight. See the figure.

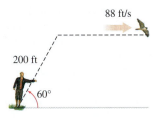

Source: http://www.rspb.org.uk.

31. Boyle's Law   A gas is said to be compressed adiabatically if there is no gain or loss of heat. When such a gas is diatomic (has two atoms per molecule), it satisfies the equation $PV^{1.4} = k$, where k is a constant, P is the pressure, and V is the volume. At a given instant, the pressure is $20 \, \text{kg/cm}^2$, the volume is $32 \, \text{cm}^3$, and the volume is decreasing at the rate of $2 \, \text{cm}^3/\text{min}$. At what rate is the pressure changing?

32. Heating a Plate   When a metal plate is heated, it expands. If the shape of the plate is circular and its radius is increasing at the rate of 0.02 cm/s, at what rate is the area of the top surface increasing when the radius is 3 cm?

33. Pollution   After a rupture, oil begins to escape from an underwater well. If, as the oil rises, it forms a circular slick whose radius increases at a rate of 0.42 ft/min, find the rate at which the area of the spill is increasing when the radius is 120 ft.

34. Flying a Kite   A girl flies a kite at a height 30 m above her hand. If the kite flies horizontally away from the girl at the rate of 2 m/s, at what rate is the string being let out when the length of the string released is 70 m? Assume that the string remains taut.

35. Falling Ladder   An 8-m ladder is leaning against a vertical wall. If a person pulls the base of the ladder away from the wall at the rate of 0.5 m/s, how fast is the top of the ladder moving down the wall when the base of the ladder is

(a) 3 m from the wall?   (b) 4 m from the wall?

(c) 6 m from the wall?

36. Beam from a Lighthouse   A light in a lighthouse 2000 m from a straight shoreline is rotating at 2 revolutions per minute. How fast is the beam moving along the shore when it passes a point 500 m from the point on the shore opposite the lighthouse? (*Hint:* One revolution $= 2\pi$ rad.)

37. Moving Radar Beam   A radar antenna, making one revolution every 5 seconds, is located on a ship that is 6 km from a straight shoreline. How fast is the radar beam moving along the shoreline when the beam makes an angle of $45°$ with the shore?

38. Tracking a Rocket   When a rocket is launched, it is tracked by a tracking dish on the ground located a distance D from the point of launch. The dish points toward the rocket and adjusts its angle of elevation θ to the horizontal (ground level) as the rocket rises. Suppose a rocket rises vertically at a constant speed of 2.0 m/s, with the tracking dish located 150 m from the launch point. Find the rate of change of the angle θ of elevation of the tracking dish with respect to time t (tracking rate) for each of the following:

(a) Just after launch.

(b) When the rocket is 100 m above the ground.

(c) When the rocket is 1.0 km above the ground.

(d) Use the results in (a)-(c) to describe the behavior of the tracking rate as the rocket climbs higher and higher. What limit does the tracking rate approach as the rocket gets extremely high?

39. Lengthening Shadow   A child, 1 m tall, is walking directly under a street lamp that is 6 m above the ground. If the child walks away from the light at the rate of 20 m/min, how fast is the child's shadow lengthening?

40. Approaching a Pole   A boy is walking toward the base of a pole 20 m high at the rate of 4 km/h. At what rate (in meters per second) is the distance from his feet to the top of the pole changing when he is 5 m from the pole?

41. Riding a Ferris Wheel   A Ferris wheel is 50 ft in diameter and its center is located 30 ft above the ground. See the image. If the wheel rotates once every 2 min, how fast is a passenger rising when he is 42.5 ft above the ground? How fast is he moving horizontally?

42. Approaching Cars   Two cars approach an intersection, one heading east at the rate of 30 km/h and the other heading south at the rate of 40 km/h. At what rate are the two cars approaching each other at the instant when the first car is 100 m from the intersection and the second car is 75 m from the intersection? Assume the cars maintain their respective speeds.

43. Parting Ways   An elevator in a building is located on the fifth floor, which is 25 m above the ground. A delivery truck is positioned directly beneath the elevator at street level. If, simultaneously, the elevator goes down at a speed of 5 m/s and the truck pulls away at a speed of 8 m/s, how fast will the elevator and the truck be separating 1 second later? Assume the speeds remain constant at all times.

44. Pulley   In order to lift a container 10 m, a rope is attached to the container and, with the help of a pulley, the container is hoisted. See the figure. If a person holds the end of the rope and walks away from beneath the pulley at the rate of 2 m/s, how fast is the container rising when the person is 5 m away? Assume the end of the rope in the person's hand was originally at the same height as the top of the container.

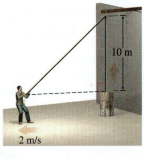

45. Business   A manufacturer of precision digital switches has daily cost C and revenue R functions of $C(x) = 10,000 + 3x$ and $R(x) = 5x - \dfrac{x^2}{2000}$, respectively, where x is the daily production of switches. Production is increasing at the rate of 50 switches per day when production is 1000 switches.

(a) Find the rate of increase in cost when production is 1000 switches per day.

(b) Find the rate of increase in revenue when production is 1000 switches per day.

(c) Find the rate of increase in profit when production is 1000 switches per day. (*Hint:* Profit = Revenue − Cost)

46. An Enormous Growing Black Hole In December 2011 astronomers announced the discovery of the two most massive black holes identified to date. The holes appear to be quasar* remnants, each having a mass equal to 10 billion Suns. Huge black holes typically grow by swallowing nearby matter, including whole stars. In this way, the size of the **event horizon** (the distance from the center of the black hole to the position at which no light can escape) increases. From general relativity, the radius R of the event horizon for a black hole of mass m is $R = \dfrac{2Gm}{c^2}$, where G is the gravitational constant and c is the speed of light in a vacuum.

(a) If one of these huge black holes swallows one Sun-like star per year, at what rate (in kilometers per year) will its event horizon grow? The mass of the Sun is 1.99×10^{30} kg, the speed of light in a vacuum is $c = 3.00 \times 10^8$ m/s, and $G = 6.67 \times 10^{-11}$ m³/(kg · s²).

(b) By what percent does the event horizon change per year?

47. Weight in Space An object that weighs K lb on the surface of Earth weighs approximately

$$W(R) = K\left(\frac{3960}{3960 + R}\right)^2$$

pounds when it is a distance of R mi from Earth's surface. Find the rate at which the weight of an object weighing 1000 lb on Earth's surface is changing when it is 50 mi above Earth's surface and is being lifted at the rate of 10 mi/s.

48. Pistons In a certain piston engine, the distance x in meters between the center of the crank shaft and the head of the piston is given by $x = \cos\theta + \sqrt{16 - \sin^2\theta}$, where θ is the angle between the crank and the path of the piston head. See the figure below.

(a) If θ increases at the constant rate of 45 radians/second, what is the speed of the piston head when $\theta = \dfrac{\pi}{6}$?

(b) Graph x as a function of θ on the interval $[0, \pi]$. Determine the maximum distance and the minimum distance of the piston head from the center of the crank shaft.

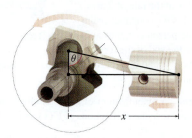

49. Tracking an Airplane A soldier at an anti-aircraft battery observes an airplane flying toward him at an altitude of 4500 ft. See the figure. When the angle of elevation of the battery is 30°, the soldier must increase the angle of elevation by 1°/second to keep the plane in sight. What is the ground speed of the plane?

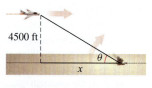

50. Change in the Angle of Elevation A hot air balloon is rising at a speed of 100 m/min. If an observer is standing 200 m from the lift-off point, what is the rate of change of the angle of elevation of the observer's line of sight when the balloon is 600 m high?

51. Rate of Rotation A searchlight is following a plane flying at an altitude of 3000 ft in a straight line over the light; the plane's velocity is 500 mi/h. At what rate is the searchlight turning when the distance between the light and the plane is 5000 ft? See the figure.

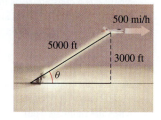

52. Police Chase A police car approaching an intersection at 80 ft/s spots a suspicious vehicle on the cross street. When the police car is 210 ft from the intersection, the policeman turns the spotlight on the vehicle, which is at that time just crossing the intersection at a constant rate of 60 ft/s. See the figure. How fast must the light beam be turning 2 s later in order to follow the suspicious vehicle?

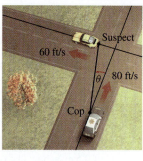

53. Change in Volume The height h and width x of an open box with a square base are related to its volume by the formula $V = hx^2$. Discuss how the volume changes

(a) if h decreases with time, but x remains constant.

(b) if both h and x change with time.

54. Rate of Change Let $y = 2e^{\cos x}$. If both x and y vary with time in such a way that y increases at a steady rate of 5 units per second, at what rate is x changing when $x = \dfrac{\pi}{2}$?

Challenge Problems

55. Moving Shadows The dome of an observatory is a hemisphere 60 ft in diameter. A boy is playing near the observatory at sunset. He throws a ball upward so that its shadow climbs to the highest point on the dome. How fast is the shadow moving along the dome $\dfrac{1}{2}$ second after the ball begins to fall? How did you use the fact that it was sunset in solving the problem? (*Note:* A ball falling from rest covers a distance $s = 16t^2$ ft in t seconds.)

---

*A **quasar** is an astronomical object that emits massive amounts of electromagnetic radiation.

56. Moving Shadows A railroad train is moving at a speed of 15 mi/h past a station 800 ft long. The track has the shape of the parabola $y^2 = 600x$ as shown in the figure. If the sun is just rising in the east, find how fast the shadow S of the locomotive L is moving along the wall at the instant it reaches the end of the wall.

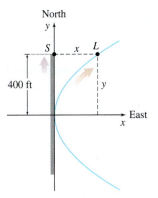

57. Change in Area The hands of a clock are 2 in and 3 in long. See the figure. As the hands move around the clock, they sweep out the triangle OAB. At what rate is the area of the triangle changing at 12:10 p.m.?

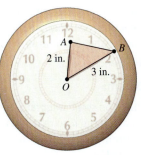

58. Distance A train is traveling northeast at a rate of 25 ft/s along a track that is elevated 20 ft above the ground. The track passes over the street below at an angle of 30°. See the figure. Five seconds after the train passes over the road, a car traveling east passes under the tracks going 40 ft/s. How fast are the train and the car separating after 3 seconds?

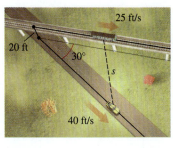

4.2 Maximum and Minimum Values; Critical Numbers

OBJECTIVES *When you finish this section, you should be able to:*

1 Identify absolute maximum and minimum values and local extreme values of a function (p. 263)

2 Find critical numbers (p. 267)

3 Find absolute maximum and absolute minimum values (p. 268)

Often problems in engineering and economics seek to find an optimal, or best, solution to a problem. For example, state and local governments try to set tax rates to optimize revenue. If a problem like this can be modeled by a function, then finding the maximum or the minimum values of the function solves the problem.

1 Identify Absolute Maximum and Minimum Values and Local Extreme Values of a Function

Figure 7 illustrates the graph of a function f defined on a closed interval $[a, b]$. Pay particular attention to the graph at the numbers x_1, x_2, and x_3. In small open intervals containing x_1 and x_3 the value of f is greatest at these numbers. We say that f has *local maxima* at x_1 and x_3, and that $f(x_1)$ and $f(x_3)$ are *local maximum values of f*. Similarly, in small open intervals containing x_2, the value of f is the least at x_2. We say f has a *local minimum* at x_2 and $f(x_2)$ is a *local minimum value of f*. ("Maxima" is the plural of "maximum"; "minima" is the plural of "minimum.")

On the closed interval $[a, b]$, the largest value of f is $f(x_3)$, while the smallest value of f is $f(b)$. These are called, respectively, the *absolute maximum value* and *absolute minimum value* of f on $[a, b]$.

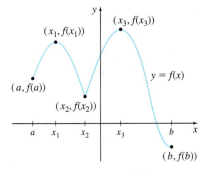

Figure 7

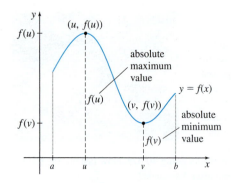

Figure 8 f is defined on $[a, b]$. For all x in $[a, b]$, $f(u) \geq f(x)$. For all x in $[a, b]$, $f(v) \leq f(x)$. $f(u)$ is the absolute maximum value of f. $f(v)$ is the absolute minimum value of f.

DEFINITION Absolute Extrema

Let f be a function defined on an interval I. If there is a number u in the interval for which $f(u) \geq f(x)$ for all x in I, then f has an **absolute maximum** at u and the number $f(u)$ is called the **absolute maximum value** of f on I.

If there is a number v in I for which $f(v) \leq f(x)$ for all x in I, then f has an **absolute minimum** at v and the number $f(v)$ is the **absolute minimum value** of f on I.

The values $f(u)$ and $f(v)$ are sometimes called **absolute extrema** or the **extreme values** of f on I. ("Extrema" is the plural of the Latin noun *extremum*.)

The *absolute* maximum value and the *absolute* minimum value, if they exist, are the largest and smallest values, respectively, of a function f on the interval I. See Figure 8. Contrast this idea with that of a *local* maximum value and a *local* minimum value. These are the largest and smallest values of f in *some open interval* in I. The next definition makes this distinction precise.

DEFINITION Local Extrema

Let f be a function defined on some interval I and let u and v be numbers in I. If there is an open interval in I containing u so that $f(u) \geq f(x)$ for all x in this open interval, then f has a **local maximum** (or **relative maximum**) at u, and the number $f(u)$ is called a **local maximum value**.

Similarly, if there is an open interval in I containing v so that $f(v) \leq f(x)$ for all x in this open interval, then f has a **local minimum** (or a **relative minimum**) at v, and the number $f(v)$ is called a **local minimum value.**

The term **local extreme value** describes either a local maximum value of f or a local minimum value of f.

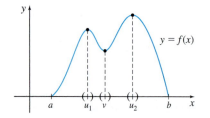

Figure 9 f has a local maximum at u_1 and at u_2. f has a local minimum at v.

Figure 9 illustrates the definition.

Notice in the definition of a local maximum that the interval that contains u is required to be *open*. Notice also in the definition of a local maximum that the value $f(u)$ must be greater than or equal to *all* other values of f in this open interval. The word "local" is used to emphasize that this condition holds on *some* open interval containing u. Similar remarks hold for a local minimum.

NEED TO REVIEW? Continuity is discussed in Section 1.3, pp. 93–102.

EXAMPLE 1 **Identifying Maximum and Minimum Values and Local Extreme Values from the Graph of a Function**

Figures 10–15 show the graphs of six different functions. For each function:

(a) Find the domain.

(b) Determine where the function is continuous.

(c) Identify the absolute maximum value and the absolute minimum value, if they exist.

(d) Identify any local extreme values.

Solution

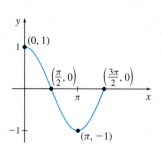

Figure 10

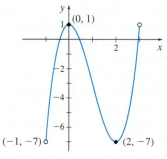

Figure 11

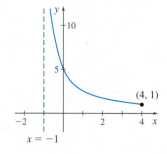

Figure 12

(a) Domain: $\left[0, \dfrac{3\pi}{2}\right]$

(b) Continuous on $\left[0, \dfrac{3\pi}{2}\right]$

(c) Absolute maximum value:
$f(0) = 1$
Absolute minimum value:
$f(\pi) = -1$

(d) No local maximum value;
local minimum value: $f(\pi) = -1$

(a) Domain: $(-1, 3)$

(b) Continuous on $(-1, 3)$

(c) Absolute maximum value:
$f(0) = 1$
Absolute minimum value:
$f(2) = -7$

(d) Local maximum value:
$f(0) = 1$
Local minimum value:
$f(2) = -7$

(a) Domain: $(-1, 4]$

(b) Continuous on $(-1, 4]$

(c) No absolute maximum value;
absolute minimum value:
$f(4) = 1$

(d) No local maximum value;
no local minimum value

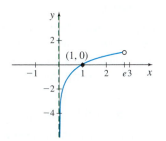

Figure 13

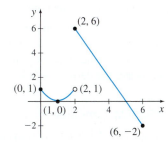

Figure 14

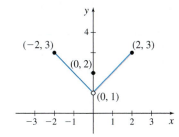

Figure 15

(a) Domain: $(0, e)$

(b) Continuous on $(0, e)$

(c) No absolute maximum value;
no absolute minimum value

(d) No local maximum value;
no local minimum value

(a) Domain: $[0, 6]$

(b) Continuous on $[0, 6]$ except
at $x = 2$

(c) Absolute maximum value:
$f(2) = 6$
Absolute minimum value:
$f(6) = -2$

(d) Local maximum value: $f(2) = 6$
Local minimum value: $f(1) = 0$

(a) Domain: $[-2, 2]$

(b) Continuous on $[-2, 2]$ except
at $x = 0$

(c) Absolute maximum value:
$f(-2) = 3, \quad f(2) = 3$
No absolute minimum value

(d) Local maximum value: $f(0) = 2$
No local minimum value

NOW WORK Problem 7.

Example 1 illustrates that a function f can have both an absolute maximum value and an absolute minimum value, can have one but not the other, or can have neither an absolute maximum value nor an absolute minimum value. The next theorem provides conditions for which a function f will always have absolute extrema. Although the theorem seems simple, the proof requires advanced topics and may be found in most advanced calculus books.

THEOREM Extreme Value Theorem

If a function f is continuous on a closed interval $[a, b]$, then f has an absolute maximum and an absolute minimum on $[a, b]$.

Look back at Example 1. Figure 10 illustrates a function that is continuous on a closed interval; it has an absolute maximum and an absolute minimum on the interval. But if a function is not continuous on a closed interval $[a, b]$, then the conclusion of the Extreme Value Theorem may or may not hold.

For example, the functions graphed in Figures 11–13 are all continuous, but not on a closed interval:

• In Figure 11, the function has both an absolute maximum and an absolute minimum.

• In Figure 12, the function has an absolute minimum but no absolute maximum.

• In Figure 13, the function has neither an absolute maximum nor an absolute minimum.

On the other hand, the functions graphed in Figures 14 and 15 are each defined on a closed interval, but neither is continuous on that interval:

- In Figure 14, the function has an absolute maximum and an absolute minimum.
- In Figure 15, the function has an absolute maximum but no absolute minimum.

NOW WORK Problem 9.

The Extreme Value Theorem is an example of an *existence theorem*. It states that, if a function is continuous on a closed interval, then extreme values exist. It does not tell us how to find these extreme values. Although we can locate both absolute and local extrema given the graph of a function, we need tools that allow us to locate the extreme values analytically, when only the function f is known.

Consider the function $y = f(x)$ whose graph is shown in Figure 16. There is a local maximum at x_1 and a local minimum at x_2. The derivative at these points is 0, since the tangent lines to the graph of f at x_1 and x_2 are horizontal. There is also a local maximum at x_3, where the derivative fails to exist. The next theorem provides the details.

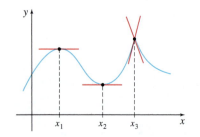

Figure 16

THEOREM Condition for a Local Maximum or a Local Minimum

If a function f has a local maximum or a local minimum at the number c, then either $f'(c) = 0$ or $f'(c)$ does not exist.

Proof Suppose f has a local maximum at c. Then, by definition,

$$f(c) \geq f(x)$$

for all x in some open interval containing c. Equivalently,

$$f(x) - f(c) \leq 0$$

The derivative of f at c may be written as

$$f'(c) = \lim_{x \to c} \frac{f(x) - f(c)}{x - c}$$

provided the limit exists. If this limit does not exist, then $f'(c)$ does not exist and there is nothing further to prove.

If this limit does exist, then

$$\lim_{x \to c^-} \frac{f(x) - f(c)}{x - c} = \lim_{x \to c^+} \frac{f(x) - f(c)}{x - c} \tag{1}$$

In the limit on the left, $x < c$ and $f(x) - f(c) \leq 0$, so $\dfrac{f(x) - f(c)}{x - c} \geq 0$ and

$$\lim_{x \to c^-} \frac{f(x) - f(c)}{x - c} \geq 0 \tag{2}$$

[Do you see why? If the limit were negative, then there would be an open interval about c, $x < c$, on which $f(x) - f(c) < 0$; refer to Section 1.6, Example 7, p. 136.]

Similarly, in the limit on the right side of (1), $x > c$ and $f(x) - f(c) \leq 0$, so $\dfrac{f(x) - f(c)}{x - c} \leq 0$ and

$$\lim_{x \to c^+} \frac{f(x) - f(c)}{x - c} \leq 0 \tag{3}$$

Since the limits (2) and (3) must be equal, we have

$$f'(c) = \lim_{x \to c} \frac{f(x) - f(c)}{x - c} = 0$$

The proof when f has a local minimum at c is similar and is left as an exercise (Problem 92). ∎

ORIGINS Pierre de Fermat (1601–1665), a lawyer whose contributions to mathematics were made in his spare time, ranks as one of the great "amateur" mathematicians. Although Fermat is often remembered for his famous "last theorem," he established many fundamental results in number theory and, with Pascal, cofounded the theory of probability. Since his work on calculus was the best done before Newton and Leibniz, he must be considered one of the principal founders of calculus.

For differentiable functions, the following theorem, often called Fermat's Theorem, is simpler.

Theorem If a differentiable function f has a local maximum or a local minimum at c, then $f'(c) = 0$.

In other words, for differentiable functions, local extreme values occur at points where the tangent line to the graph of f is horizontal.

As the theorems show, the numbers at which a function $f'(x) = 0$ or at which f' does not exist provide a clue for locating where f has local extrema. But unfortunately, knowing that $f'(c) = 0$ or that f' does not exist at c does not guarantee a local extremum occurs at c. For example, in Figure 17, $f'(x_3) = 0$, but f has neither a local maximum nor a local minimum at x_3. Similarly, $f'(x_4)$ does not exist, but f has neither a local maximum nor a local minimum at x_4.

Even though there may be no local extrema found at the numbers c for which $f'(c) = 0$ or $f'(c)$ do not exist, the collection of all such numbers provides *all* the *possibilities* where f *might* have local extreme values. For this reason, we call these numbers *critical numbers*.

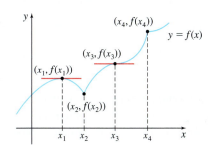

Figure 17

DEFINITION Critical Number

A **critical number** of a function f is a number c in the domain of f for which either $f'(c) = 0$ or $f'(c)$ does not exist.

2 Find Critical Numbers

EXAMPLE 2   Finding Critical Numbers

Find any critical numbers of the following functions:

(a) $f(x) = x^3 - 6x^2 + 9x + 2$ **(b)** $R(x) = \dfrac{1}{x - 2}$

(c) $g(x) = \dfrac{(x - 2)^{2/3}}{x}$ **(d)** $G(x) = \sin x$

Solution (a) Since f is a polynomial, it is differentiable at every real number. So, the critical numbers occur where $f'(x) = 0$.

$$f'(x) = 3x^2 - 12x + 9 = 3(x - 1)(x - 3)$$

$f'(x) = 0$ at $x = 1$ and $x = 3$; the numbers 1 and 3 are the critical numbers of f.

(b) The domain of $R(x) = \dfrac{1}{x - 2}$ is $\{x \mid x \neq 2\}$, and $R'(x) = -\dfrac{1}{(x - 2)^2}$. R' exists for all numbers x in the domain of R (remember, 2 is not in the domain of R). Since R' is never 0, R has no critical numbers.

(c) The domain of $g(x) = \dfrac{(x - 2)^{2/3}}{x}$ is $\{x \mid x \neq 0\}$, and the derivative of g is

$$g'(x) = \frac{x \cdot \left[\dfrac{2}{3}(x - 2)^{-1/3}\right] - 1 \cdot (x - 2)^{2/3}}{x^2} = \underset{\uparrow}{\frac{2x - 3(x - 2)}{3x^2(x - 2)^{1/3}}} = \frac{6 - x}{3x^2(x - 2)^{1/3}}$$

$$\text{Multiply by } \frac{3(x - 2)^{1/3}}{3(x - 2)^{1/3}}$$

268 Chapter 4 • Applications of the Derivative

Critical numbers occur where $g'(x) = 0$ or where $g'(x)$ does not exist. If $x = 6$, then $g'(6) = 0$. Next, $g'(x)$ does not exist where

$$3x^2(x-2)^{1/3} = 0$$
$$3x^2 = 0 \quad \text{or} \quad (x-2)^{1/3} = 0$$
$$x = 0 \quad \text{or} \quad x = 2$$

We ignore 0 since it is not in the domain of g. The critical numbers of g are 6 and 2.

(d) The domain of G is all real numbers. The function G is differentiable on its domain, so the critical numbers occur where $G'(x) = 0$. $G'(x) = \cos x$ and $\cos x = 0$ at $x = \pm\dfrac{\pi}{2}, \pm\dfrac{3\pi}{2}, \pm\dfrac{5\pi}{2}, \ldots$. This function has infinitely many critical numbers. ∎

NOW WORK Problem 25.

We are not yet ready to give a procedure for determining whether a function f has a local maximum, a local minimum, or neither at a critical number. This requires the *Mean Value Theorem*, which is the subject of the next section. However, the critical numbers do help us find the extreme values of a function f.

3 Find Absolute Maximum and Absolute Minimum Values

The following theorem provides a way to find the extreme values of a function f that is continuous on a closed interval $[a, b]$.

THEOREM Locating Extreme Values

Let f be a function that is continuous on a closed interval $[a, b]$. Then the absolute maximum value and the absolute minimum value of f are the largest and the smallest values, respectively, found among the following:

- The values of f at the critical numbers in the open interval (a, b)
- $f(a)$ and $f(b)$, the values of f at the endpoints a and b

For any function f satisfying the conditions of this theorem, the Extreme Value Theorem reveals that extreme values exist, and this theorem tells us how to find them.

Steps for Finding the Absolute Extreme Values of a Function f That Is Continuous on a Closed Interval $[a, b]$

Step 1 Locate all critical numbers in the open interval (a, b).

Step 2 Evaluate f at each critical number and at the endpoints a and b.

Step 3 The largest value is the absolute maximum value; the smallest value is the absolute minimum value.

EXAMPLE 3 Finding Absolute Maximum and Minimum Values

Find the absolute maximum value and the absolute minimum value of each function:

(a) $f(x) = x^3 - 6x^2 + 9x + 2$ on $[0, 2]$ **(b)** $g(x) = \dfrac{(x-2)^{2/3}}{x}$ on $[1, 10]$

Solution (a) The function f, a polynomial function, is continuous on the closed interval $[0, 2]$, so the Extreme Value Theorem guarantees that f has an absolute maximum value and an absolute minimum value on the interval. We follow the steps for finding the absolute extreme values to identify them.

Step 1 From Example 2(a), the critical numbers of f are 1 and 3. We exclude 3, since it is not in the interval $(0, 2)$.

Step 2 Find the value of f at the critical number 1 and at the endpoints 0 and 2:

$$f(1) = 6 \qquad f(0) = 2 \qquad f(2) = 4$$

Step 3 The largest value 6 is the absolute maximum value of f; the smallest value 2 is the absolute minimum value of f.

(b) The function g is continuous on the closed interval $[1, 10]$, so g has an absolute maximum and an absolute minimum on the interval.

Step 1 From Example 2(c), the critical numbers of g are 2 and 6. Both critical numbers are in the interval $(1, 10)$.

Step 2 We evaluate g at the critical numbers 2 and 6 and at the endpoints 1 and 10:

x	$\dfrac{(x-2)^{2/3}}{x}$	$g(x)$	
1	$\dfrac{(1-2)^{2/3}}{1} = (-1)^{2/3}$	1	← absolute maximum value
2	$\dfrac{(2-2)^{2/3}}{2}$	0	← absolute minimum value
6	$\dfrac{(6-2)^{2/3}}{6} = \dfrac{4^{2/3}}{6}$	≈ 0.42	
10	$\dfrac{(10-2)^{2/3}}{10} = \dfrac{8^{2/3}}{10} = \dfrac{4}{10}$	0.4	

Step 3 The largest value 1 is the absolute maximum value; the smallest value 0 is the absolute minimum value. ∎

NOW WORK Problem **49.**

For piecewise-defined functions f, we need to look carefully at the number(s) where the rules for the function change.

EXAMPLE 4   Finding Absolute Maximum and Minimum Values

Find the absolute maximum value and absolute minimum value of the function

$$f(x) = \begin{cases} 2x - 1 & \text{if } 0 \le x \le 2 \\ x^2 - 5x + 9 & \text{if } 2 < x \le 3 \end{cases}$$

Solution The function f is continuous on the closed interval $[0, 3]$. (You should verify this.) To find the absolute maximum value and absolute minimum value, we follow the three-step procedure.

Step 1 Find the critical numbers in the open interval $(0, 3)$:

- On the open interval $(0, 2)$: $f(x) = 2x - 1$ and $f'(x) = 2$. Since $f'(x) \neq 0$ on the interval $(0, 2)$, there are no critical numbers in $(0, 2)$.
- On the open interval $(2, 3)$: $f(x) = x^2 - 5x + 9$ and $f'(x) = 2x - 5$.

  Solving $f'(x) = 2x - 5 = 0$, we find $x = \dfrac{5}{2}$. Since $\dfrac{5}{2}$ is in the interval $(2, 3)$, $\dfrac{5}{2}$ is a critical number.

- At $x = 2$, the rule for f changes, so we investigate the one-sided limits of $\dfrac{f(x) - f(2)}{x - 2}$.

$$\lim_{x \to 2^-} \frac{f(x) - f(2)}{x - 2} = \lim_{x \to 2^-} \frac{(2x - 1) - 3}{x - 2} = \lim_{x \to 2^-} \frac{2x - 4}{x - 2}$$

$$= \lim_{x \to 2^-} \frac{2(x - 2)}{x - 2} = 2$$

$$\lim_{x \to 2^+} \frac{f(x) - f(2)}{x - 2} = \lim_{x \to 2^+} \frac{(x^2 - 5x + 9) - 3}{x - 2} = \lim_{x \to 2^+} \frac{x^2 - 5x + 6}{x - 2}$$

$$= \lim_{x \to 2^+} \frac{(x - 2)(x - 3)}{x - 2} = \lim_{x \to 2^+} (x - 3) = -1$$

The one-sided limits are not equal, so the derivative does not exist at 2; 2 is a critical number.

Step 2 Evaluate f at the critical numbers $\dfrac{5}{2}$ and 2 and at the endpoints 0 and 3.

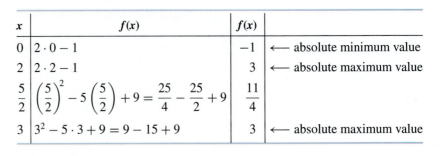

x	$f(x)$	$f(x)$	
0	$2 \cdot 0 - 1$	-1	← absolute minimum value
2	$2 \cdot 2 - 1$	3	← absolute maximum value
$\dfrac{5}{2}$	$\left(\dfrac{5}{2}\right)^2 - 5\left(\dfrac{5}{2}\right) + 9 = \dfrac{25}{4} - \dfrac{25}{2} + 9$	$\dfrac{11}{4}$	
3	$3^2 - 5 \cdot 3 + 9 = 9 - 15 + 9$	3	← absolute maximum value

Step 3 The largest value 3 is the absolute maximum value; the smallest value -1 is the absolute minimum value. ∎

The graph of f is shown in Figure 18. Notice that the absolute maximum occurs at 2 and at 3.

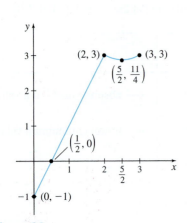

Figure 18

$$f(x) = \begin{cases} 2x - 1 & \text{if } 0 \le x \le 2 \\ x^2 - 5x + 9 & \text{if } 2 < x \le 3 \end{cases}$$

NOW WORK Problem 61.

EXAMPLE 5   Constructing a Rain Gutter

A rain gutter is to be constructed using a piece of aluminum 12 in wide. After marking a length of 4 in. from each edge, the piece of aluminum is bent up at an angle θ, as illustrated in Figure 19. The area A of a cross section of the opening, expressed as a function of θ, is

$$A(\theta) = 16 \sin \theta (\cos \theta + 1) \qquad 0 \le \theta \le \frac{\pi}{2}$$

Find the angle θ that maximizes the area A. (This bend will allow the most water to flow through the gutter.)

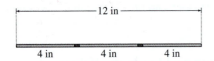

Figure 19

Solution The function $A = A(\theta)$ is continuous on the closed interval $\left[0, \dfrac{\pi}{2}\right]$. To find the angle θ that maximizes A, we follow the three-step procedure.

Step 1 We locate all critical numbers in the open interval $\left(0, \dfrac{\pi}{2}\right)$.

$$
\begin{aligned}
A'(\theta) &= 16 \sin \theta(-\sin \theta) + 16 \cos \theta(\cos \theta + 1) && \text{Product Rule} \\
&= 16[-\sin^2 \theta + \cos^2 \theta + \cos \theta] \\
&= 16[(\cos^2 \theta - 1) + \cos^2 \theta + \cos \theta] && -\sin^2 \theta = \cos^2 \theta - 1 \\
&= 16[2\cos^2 \theta + \cos \theta - 1] = 16(2\cos \theta - 1)(\cos \theta + 1)
\end{aligned}
$$

The critical numbers satisfy the equation $A'(\theta) = 0$, $0 < \theta < \dfrac{\pi}{2}$.

$$16(2\cos \theta - 1)(\cos \theta + 1) = 0$$

$$2\cos \theta - 1 = 0 \quad \text{or} \quad \cos \theta + 1 = 0$$

$$\cos \theta = \frac{1}{2} \qquad\qquad \cos \theta = -1$$

$$\theta = \frac{\pi}{3} \text{ or } \frac{5\pi}{3} \qquad\qquad \theta = \pi$$

Of these solutions, only $\dfrac{\pi}{3}$ is in the interval $\left(0, \dfrac{\pi}{2}\right)$. So, $\dfrac{\pi}{3}$ is the only critical number.

Step 2 We evaluate A at the critical number $\dfrac{\pi}{3}$ and at the endpoints 0 and $\dfrac{\pi}{2}$.

θ	$16\sin\theta(\cos\theta+1)$	$A(\theta)$	
0	$16\sin 0\,(\cos 0+1)=0$	0	
$\dfrac{\pi}{3}$	$16\sin\dfrac{\pi}{3}\left(\cos\dfrac{\pi}{3}+1\right)$		
	$=16\left(\dfrac{\sqrt{3}}{2}\right)\left(\dfrac{1}{2}+1\right)$		
	$=16\left(\dfrac{3\sqrt{3}}{4}\right)=12\sqrt{3}$	≈ 20.8	← absolute maximum value
$\dfrac{\pi}{2}$	$16\sin\dfrac{\pi}{2}\left(\cos\dfrac{\pi}{2}+1\right)$	16	
	$=16(1)(0+1)=16$		

Step 3 If the aluminum is bent at an angle of $\dfrac{\pi}{3}$, the area of the opening is maximum. The maximum area is about $20.8\ \text{in}^2$. ∎

NOW WORK Problem 71.

From physics, the volume V of fluid flowing through a pipe is related to the radius r of the pipe and the difference in pressure p at each end of the pipe. It is given by the equation

$$V = kpr^4$$

where k is a constant.

EXAMPLE 6   Analyzing a Cough

Figure 20

Coughing is caused by increased pressure in the lungs and is accompanied by a decrease in the diameter of the windpipe. See Figure 20. The radius r of the windpipe decreases with increased pressure p according to the formula $r_0 - r = cp$, where r_0 is the radius of the windpipe when there is no difference in pressure and c is a positive constant. The volume V of air flowing through the windpipe is

$$V = kpr^4$$

where k is a constant. Find the radius r that allows the most air to flow through the windpipe. Restrict r so that $0 < \dfrac{r_0}{2} \le r \le r_0$.

Solution Since $p = \dfrac{r_0 - r}{c}$, we can express V as a function of r:

$$V = V(r) = k\left(\frac{r_0 - r}{c}\right)r^4 = \frac{kr_0}{c}r^4 - \frac{k}{c}r^5 \qquad \frac{r_0}{2} \le r \le r_0$$

Now we find the absolute maximum of V on the interval $\left[\dfrac{r_0}{2}, r_0\right]$.

$$V'(r) = \frac{4k\,r_0}{c}r^3 - \frac{5k}{c}r^4 = \frac{k}{c}r^3(4r_0 - 5r)$$

The only critical number in the interval $\left(\dfrac{r_0}{2}, r_0\right)$ is $r = \dfrac{4r_0}{5}$.

We evaluate V at the critical number and at the endpoints, $\dfrac{r_0}{2}$ and r_0.

r	$V(r) = k\left(\dfrac{r_0 - r}{c}\right)r^4$
$\dfrac{r_0}{2}$	$k\left(\dfrac{r_0 - \dfrac{r_0}{2}}{c}\right)\left(\dfrac{r_0}{2}\right)^4 = \dfrac{k\,r_0^5}{32c}$
$\dfrac{4r_0}{5}$	$k\left(\dfrac{r_0 - \dfrac{4r_0}{5}}{c}\right)\left(\dfrac{4r_0}{5}\right)^4 = \dfrac{k}{c}\cdot\dfrac{4^4\,r_0^5}{5^5} = \dfrac{256\,k\,r_0^5}{3125c}$
r_0	0

The largest of these three values is $\dfrac{256\,kr_0^5}{3125c}$. So, the maximum air flow occurs when the radius of the windpipe is $\dfrac{4r_0}{5}$, that is, when the windpipe contracts by 20%. ∎

4.2 Assess Your Understanding

Concepts and Vocabulary

1. *True or False* Any function f that is defined on a closed interval $[a, b]$ will have both an absolute maximum value and an absolute minimum value.

2. *Multiple Choice* A number c in the domain of a function f is called a(n) [(**a**) extreme value, (**b**) critical number, (**c**) local number] of f if either $f'(c) = 0$ or $f'(c)$ does not exist.

3. *True or False* At a critical number, there is a local extreme value.

4. *True or False* If a function f is continuous on a closed interval $[a, b]$, then its absolute maximum value is found at a critical number.

5. *True or False* The Extreme Value Theorem tells us where the absolute maximum and absolute minimum can be found.

6. *True or False* If f is differentiable on the interval $(0, 4)$ and $f'(2) = 0$, then f has a local maximum or a local minimum at 2.

Skill Building

In Problems 7 and 8, use the graphs below to determine whether the function f has an absolute extremum and/or a local extremum or neither at x_1, x_2, x_3, x_4, x_5, x_6, x_7, and x_8.

7.

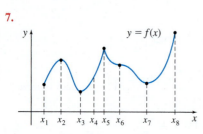

8.

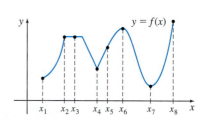

In Problems 9–12, provide a graph of a continuous function f that has the following properties:

9. domain $[0, 8]$, absolute maximum at 0, absolute minimum at 3, local minimum at 7.

10. domain $[-5, 5]$, absolute maximum at 3, absolute minimum at -3.

11. domain $[3, 10]$ and has no local extreme points.

12. no absolute extreme values, is differentiable at 4 and has a local minimum at 4, is not differentiable at 0, but has a local maximum at 0.

In Problems 13–36, find the critical numbers, if any, of each function.

13. $f(x) = x^2 - 8x$

14. $f(x) = 1 - 6x + x^2$

15. $f(x) = x^3 - 3x^2$

16. $f(x) = x^3 - 6x$

17. $f(x) = x^4 - 2x^2 + 1$

18. $f(x) = 3x^4 - 4x^3$

19. $f(x) = x^{2/3}$

20. $f(x) = x^{1/3}$

21. $f(x) = 2\sqrt{x}$

22. $f(x) = 4 - \sqrt{x}$

23. $f(x) = x + \sin x,\ 0 \le x \le \pi$

24. $f(x) = x - \cos x,\quad -\dfrac{\pi}{2} \le x \le \dfrac{\pi}{2}$

---

1. = NOW WORK problem   📈 = Graphing technology recommended   CAS = Computer Algebra System recommended

25. $f(x) = x\sqrt{1-x^2}$

26. $f(x) = x^2\sqrt{2-x}$

27. $f(x) = \dfrac{x^2}{x-1}$

28. $f(x) = \dfrac{x}{x^2-1}$

29. $f(x) = (x+3)^2(x-1)^{2/3}$

30. $f(x) = (x-1)^2(x+1)^{1/3}$

31. $f(x) = \dfrac{(x-3)^{1/3}}{x-1}$

32. $f(x) = \dfrac{(x+3)^{2/3}}{x+1}$

33. $f(x) = \dfrac{\sqrt[3]{x^2-9}}{x}$

34. $f(x) = \dfrac{\sqrt[3]{4-x^2}}{x}$

35. $f(x) = \begin{cases} 3x & \text{if } 0 \le x < 1 \\ 4-x & \text{if } 1 \le x \le 2 \end{cases}$

36. $f(x) = \begin{cases} x^2 & \text{if } 0 \le x < 1 \\ 1-x^2 & \text{if } 1 \le x \le 2 \end{cases}$

In Problems 37–64, find the absolute maximum value and absolute minimum value of each function on the indicated interval. Notice that the functions in Problems 37–58 are the same as those in Problems 13–34.

37. $f(x) = x^2 - 8x$ on $[-1, 10]$

38. $f(x) = 1 - 6x + x^2$ on $[0, 4]$

39. $f(x) = x^3 - 3x^2$ on $[1, 4]$

40. $f(x) = x^3 - 6x$ on $[-1, 1]$

41. $f(x) = x^4 - 2x^2 + 1$ on $[0, 2]$

42. $f(x) = 3x^4 - 4x^3$ on $[-2, 0]$

43. $f(x) = x^{2/3}$ on $[-1, 1]$

44. $f(x) = x^{1/3}$ on $[-1, 1]$

45. $f(x) = 2\sqrt{x}$ on $[1, 4]$

46. $f(x) = 4 - \sqrt{x}$ on $[0, 4]$

47. $f(x) = x + \sin x$ on $[0, \pi]$

48. $f(x) = x - \cos x$ on $\left[-\dfrac{\pi}{2}, \dfrac{\pi}{2}\right]$

49. $f(x) = x\sqrt{1-x^2}$ on $[-1, 1]$

50. $f(x) = x^2\sqrt{2-x}$ on $[0, 2]$

51. $f(x) = \dfrac{x^2}{x-1}$ on $\left[-1, \dfrac{1}{2}\right]$

52. $f(x) = \dfrac{x}{x^2-1}$ on $\left[-\dfrac{1}{2}, \dfrac{1}{2}\right]$

53. $f(x) = (x+3)^2(x-1)^{2/3}$ on $[-4, 5]$

54. $f(x) = (x-1)^2(x+1)^{1/3}$ on $[-2, 7]$

55. $f(x) = \dfrac{(x-3)^{1/3}}{x-1}$ on $[2, 11]$

56. $f(x) = \dfrac{(x+3)^{2/3}}{x+1}$ on $[-4, -2]$

57. $f(x) = \dfrac{\sqrt[3]{x^2-9}}{x}$ on $[3, 6]$

58. $f(x) = \dfrac{\sqrt[3]{4-x^2}}{x}$, on $[-4, -1]$

59. $f(x) = e^x - 3x$ on $[0, 1]$

60. $f(x) = e^{\cos x}$ on $[-\pi, 2\pi]$.

61. $f(x) = \begin{cases} 2x+1 & \text{if } 0 \le x < 1 \\ 3x & \text{if } 1 \le x \le 3 \end{cases}$

62. $f(x) = \begin{cases} x+3 & \text{if } -1 \le x \le 2 \\ 2x+1 & \text{if } 2 < x \le 4 \end{cases}$

63. $f(x) = \begin{cases} x^2 & \text{if } -2 \le x < 1 \\ x^3 & \text{if } 1 \le x \le 2 \end{cases}$

64. $f(x) = \begin{cases} x+2 & \text{if } -1 \le x < 0 \\ 2-x & \text{if } 0 \le x \le 1 \end{cases}$

Applications and Extensions

In Problems 65–68, for each function f:

(a) Find the derivative f'.

CAS (b) Use technology to find the critical numbers of f.

(c) Graph f and describe the behavior of f suggested by the graph at each critical number.

65. $f(x) = 3x^4 - 2x^3 - 21x^2 + 36x$

66. $f(x) = x^2 + 2x - \dfrac{2}{x}$

67. $f(x) = \dfrac{(x^2 - 5x + 2)\sqrt{x+5}}{\sqrt{x^2+2}}$

68. $f(x) = \dfrac{(x^2 - 9x + 16)\sqrt{x+3}}{\sqrt{x^2-4x+6}}$

In Problems 69 and 70, for each function f:

(a) Find the derivative f'.

(b) Use technology to find the absolute maximum value and the absolute minimum value of f on the closed interval $[0, 5]$.

(c) Graph f. Are the results from (b) supported by the graph?

69. $f(x) = x^4 - 12.4x^3 + 49.24x^2 - 68.64x$

70. $f(x) = e^{-x}\sin(2x) + e^{-x/2}\cos(2x)$

71. Cost of Fuel A truck has a top speed of 75 mi/h, and when traveling at the rate of x mi/h, it consumes fuel at the rate of

$$\frac{1}{200}\left(\frac{2500}{x} + x\right) \text{ gallon per mile. If the price of fuel is}$$

$3.60/$ gal, the cost C (in dollars) of driving 200 mi is given by

$$C(x) = (3.60)\left(\frac{2500}{x} + x\right)$$

(a) What is the most economical speed for the truck to travel? Use the interval $[10, 75]$.

(b) Graph the cost function C.

72. Trucking Costs   If the driver of the truck in Problem 71 is paid $28.00 per hour and wages are added to the cost of fuel, what is the most economical speed for the truck to travel?

73. Projectile Motion   An object is propelled upward at an angle θ, $45° < \theta < 90°$, to the horizontal with an initial velocity of v_0 ft/s from the base of an inclined plane that makes an angle of $45°$ to the horizontal. See the illustration below. If air resistance is ignored, the distance R that the object travels up the inclined plane is given by the function

$$R(\theta) = \frac{v_0^2 \sqrt{2}}{16} \cos\theta(\sin\theta - \cos\theta)$$

(a) Find the angle θ that maximizes R. What is the maximum value of R?

(b) Graph $R = R(\theta)$, $45° \le \theta \le 90°$, using $v_0 = 32$ ft/s.

(c) Does the graph support the result from (a)?

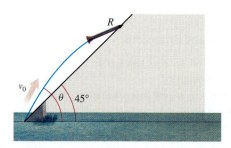

74. Height of a Cable   An electric cable is suspended between two poles of equal height that are 20 m apart, as illustrated in the figure. The shape of the cable is modeled by the equation

$y = 10\cosh\dfrac{x}{10} + 15$. If the

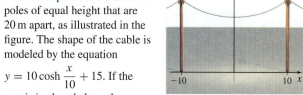

x-axis is placed along the ground and the two poles are at $(-10, 0)$ and $(10, 0)$, what is the height of the cable at its lowest point?

75. A Record Golf Stroke   The fastest golf ball speed ever recorded, 91.1 m/s (204 mi/h!), was a ball hit by Jason Zuback in 2007. When a ball is hit at an angle θ to the horizontal, $0° \le \theta \le 90°$, and lands at the same level from which it was hit, the horizontal range R of the ball is given

by $R = \dfrac{2v_0^2}{g}\sin\theta\cos\theta$, where v_0 is the initial speed of the ball

and $g = 9.8$ m/s^2 is the acceleration due to gravity.

(a) Show that the golf ball achieves its maximum range if the golfer hits it at an angle of $45°$.

(b) What is the maximum range that could be achieved by the record golf ball speed?

Source: http://www.guinnessworldrecords.com.

76. Optics   When light goes through a thin slit, it spreads out (diffracts). After passing through the slit, the intensity I of the

light on a distant screen is given by $I = I_0\left(\dfrac{\sin\alpha}{\alpha}\right)^2$, where

I_0 is the original intensity of the light and α depends on the angle away from the center.

(a) What is the intensity of the light as $\alpha \to 0$?

(b) Show that the bright spots, that is, the places where the intensity has a local maximum, occur when $\tan\alpha = \alpha$.

(c) Is the intensity of the light the same at each bright spot?

Economics   *In Problems 77 and 78, use the following discussion:*

In determining a tax rate on consumer goods, the government is always faced with the question, "What tax rate produces the largest tax revenue?" Imposing a tax may cause the price of the goods to increase and reduce the demand for the product. A very large tax may reduce the demand to zero, with the result that no tax is collected. On the other hand, if no tax is levied, there is no tax revenue at all. (Tax revenue R is the product of the tax rate t times the actual quantity q, in dollars, consumed.)

77. The government has determined that the relationship between the quantity q of a product consumed and the related tax rate t

is $t = \sqrt{27 - 3q^2}$. Find the tax rate that maximizes tax revenue. How much tax is generated by this rate?

78. On a particular product, government economists determine that the relationship between the tax rate t and the quantity q consumed is $t + 3q^2 = 18$. Find the tax rate that maximizes tax revenue and the revenue generated by the tax.

79. Catenary   A town hangs strings of holiday lights across the road between utility poles. Each set of poles is 12 m apart. The strings

hang in catenaries modeled by $y = 15\cosh\dfrac{x}{15} - 10$ with the

poles at $(\pm 6, 0)$. What is the height of the string of lights at its lowest point?

80. Harmonic Motion   An object of mass 1 kg moves in simple harmonic motion, with an amplitude $A = 0.24$ m and a period of 4 seconds. The position s of the object is given by $s(t) = A\cos(\omega t)$, where t is the time in seconds.

(a) Find the position of the object at time t and at time $t = 0.5$ seconds.

(b) Find the velocity $v = v(t)$ of the object.

(c) Find the velocity of the object when $t = 0.5$ seconds.

(d) Find the acceleration $a = a(t)$ of the object.

(e) Use Newton's Second Law of Motion, $F = ma$, to find the magnitude and direction of the force acting on the object when $t = 0.5$ second.

(f) Find the minimum time required for the object to move from its initial position to the point where $s = -0.12$ m.

(g) Find the velocity of the object when $s = -12$ m.

(CAS) 81. (a) Find the critical numbers of

$$f(x) = \frac{x^3}{3} - 0.055x^2 + 0.0028x - 4.$$

(b) Find the absolute extrema of f on the interval $[-1, 1]$.

(c) Graph f on the interval $[-1, 1]$.

CAS **82. Extreme Value** **(a)** Find the minimum value of $y = x - \cosh^{-1} x$.

(b) Graph $y = x - \cosh^{-1} x$.

83. Locating Extreme Values Find the absolute maximum value and the absolute minimum value of $f(x) = \sqrt{1 + x^2} + |x - 2|$ on $[0, 3]$, and determine where each occurs.

84. The function $f(x) = Ax^2 + Bx + C$ has a local minimum at 0, and its graph contains the points $(0, 2)$ and $(1, 8)$. Find A, B, and C.

85. (a) Determine the domain of the function
$f(x) = [(16 - x^2)(x^2 - 9)]^{1/2}$.

(b) Find the absolute maximum value of f on its domain.

86. Absolute Extreme Values Without finding them, explain why the function $f(x) = \sqrt{x(2 - x)}$ must have an absolute maximum value and an absolute minimum value. Then find the absolute extreme values in two ways (one with and one without calculus).

87. Put It Together If a function f is continuous on the closed interval $[a, b]$, which of the following is necessarily true?

(a) f is differentiable on the open interval (a, b).

(b) If $f(u)$ is an absolute maximum value of f, then $f'(u) = 0$.

(c) $\lim\limits_{x \to c} f(x) = f(\lim\limits_{x \to c} x)$ for $a < c < b$.

(d) $f'(x) = 0$, for some x, $a \le x \le b$.

(e) f has an absolute maximum value on $[a, b]$.

88. Write a paragraph that explains the similarities and differences between an absolute extreme value and a local extreme value.

89. Explain in your own words the method for finding the absolute extreme values of a continuous function that is defined on a closed interval.

90. A function f is defined and continuous on the closed interval $[a, b]$. Why can't $f(a)$ be a local extreme value on $[a, b]$?

91. Show that if f has a local minimum at c, then $g(x) = -f(x)$ has a local maximum at c.

92. Show that if f has a local minimum at c, then either $f'(c) = 0$ or $f'(c)$ does not exist.

Challenge Problem

93. (a) Prove that a rational function of the form $f(x) = \dfrac{ax^{2n} + b}{cx^n + d}$, $n \ge 1$ an integer, has at most five critical numbers.

(b) Give an example of such a rational function with exactly five critical numbers.

4.3 The Mean Value Theorem

OBJECTIVES *When you finish this section, you should be able to:*

1 Use Rolle's Theorem (p. 275)

2 Work with the Mean Value Theorem (p. 276)

3 Identify where a function is increasing and decreasing (p. 279)

ORIGINS Michel Rolle (1652–1719) was a French mathematician whose formal education was limited to elementary school. At age 24, he moved to Paris and married. To support his family, Rolle worked as an accountant, but he also began to study algebra and became interested in the theory of equations. Although Rolle is primarily remembered for the theorem proved here, he was also the first to use the notation $\sqrt[n]{\ }$ for the nth root.

In this section, we prove several theorems. The most significant of these, the *Mean Value Theorem*, is used to develop tests for locating local extreme values. To prove the Mean Value Theorem, we need *Rolle's Theorem*.

1 Use Rolle's Theorem

THEOREM Rolle's Theorem

Let f be a function defined on a closed interval $[a, b]$. If:

- f is continuous on $[a, b]$,
- f is differentiable on (a, b), and
- $f(a) = f(b)$,

then there is at least one number c in the open interval (a, b) for which $f'(c) = 0$.

Figure 21 $f'(c_1) = 0$; $f'(c_2) = 0$

Before we prove Rolle's Theorem, notice that the graph in Figure 21 meets the three conditions of Rolle's Theorem and has at least one number c at which $f'(c) = 0$.

In contrast, the graphs in Figure 22 show that the conclusion of Rolle's Theorem may not hold when one or more of the three conditions are not met.

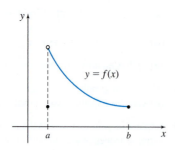

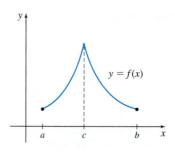

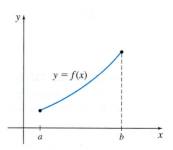

(a) f is defined on $[a, b]$.
  f is not continuous at a.
  f is differentiable on (a, b).
  $f(a) = f(b)$.
  No c in (a, b) at which $f'(c) = 0$.

(b) f is defined on $[a, b]$.
  f is continuous at $[a, b]$.
  f is not differentiable on (a, b),
    no derivative at c.
  $f(a) = f(b)$.
  No c in (a, b) at which $f'(c) = 0$.

(c) f is defined on $[a, b]$.
  f is continuous on $[a, b]$.
  f is differentiable on (a, b).
  $f(a) \neq f(b)$.
  No c in (a, b) at which $f'(c) = 0$.

Figure 22

Proof Because f is continuous on a closed interval $[a, b]$, the Extreme Value Theorem guarantees that f has an absolute maximum value and an absolute minimum value on $[a, b]$. There are two possibilities:

1. If f is a constant function on $[a, b]$, then $f'(x) = 0$ for all x in (a, b).
2. If f is not a constant function on $[a, b]$, then, because $f(a) = f(b)$, either the absolute maximum or the absolute minimum occurs at some number c in the open interval (a, b). Then $f(c)$ is a local maximum value (or a local minimum value), and, since f is differentiable on (a, b), $f'(c) = 0$. ∎

EXAMPLE 1   Using Rolle's Theorem

Find the x-intercepts of $f(x) = x^2 - 5x + 6$, and show that $f'(c) = 0$ for some number c belonging to the interval formed by the two x-intercepts. Find c.

Solution At the x-intercepts, $f(x) = 0$.

$$f(x) = x^2 - 5x + 6 = (x - 2)(x - 3) = 0$$

So, $x = 2$ and $x = 3$ are the x-intercepts of the graph of f, and $f(2) = f(3) = 0$.

Since f is a polynomial, it is continuous on the closed interval $[2, 3]$ formed by the x-intercepts and is differentiable on the open interval $(2, 3)$. The three conditions of Rolle's Theorem are satisfied, guaranteeing that there is a number c in the open interval $(2, 3)$ for which $f'(c) = 0$. Since $f'(x) = 2x - 5$, the number c for which $f'(x) = 0$ is $c = \dfrac{5}{2}$. ∎

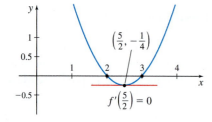

Figure 23 $f(x) = x^2 - 5x + 6$

See Figure 23 for the graph of f.

NOW WORK Problem 9.

2 Work with the Mean Value Theorem

The real importance of Rolle's Theorem is that it can be used to obtain other results, many of which have wide-ranging application. Perhaps the most important of these is the *Mean Value Theorem,* sometimes called the *Theorem of the Mean for Derivatives.*

We can provide a geometric example to motivate the Mean Value Theorem. Consider the graph of a function f that is continuous on a closed interval $[a, b]$ and differentiable on the open interval (a, b), as illustrated in Figure 24. Then there is at least one number c between a and b at which the slope $f'(c)$ of the tangent line to the graph of f equals the slope of the secant line joining the points $A = (a, f(a))$ and $B = (b, f(b))$. That is, there is at least one number c in the open interval (a, b) for which

$$f'(c) = \frac{f(b) - f(a)}{b - a}$$

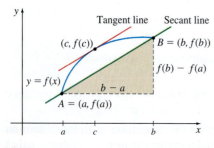

DF Figure 24

THEOREM Mean Value Theorem

Let f be a function defined on a closed interval $[a, b]$. If:

- f is continuous on $[a, b]$
- f is differentiable on (a, b),

then there is at least one number c in the open interval (a, b) for which

$$f'(c) = \frac{f(b) - f(a)}{b - a} \qquad (1)$$

Proof We begin by finding the slope $m_{\sec}$ of the secant line containing points $(a, f(a))$ and $(b, f(b))$.

$$m_{\sec} = \frac{f(b) - f(a)}{b - a}$$

Then using the point $(a, f(a))$ and the point slope form of an equation of a line, an equation of this secant line is

$$y - f(a) = \frac{f(b) - f(a)}{b - a}(x - a)$$

so

$$y = f(a) + \frac{f(b) - f(a)}{b - a}(x - a)$$

Now we construct a function g that satisfies the conditions of Rolle's Theorem, by defining g as

$$g(x) = f(x) - \underbrace{\left[f(a) + \frac{f(b) - f(a)}{b - a}(x - a) \right]}_{\substack{\text{Height of the secant line} \\ \text{containing } (a, f(a)) \text{ and } (b, f(b))}}$$

NOTE The function g has a geometric significance: Its value equals the vertical distance from the graph of $y = f(x)$ to the secant line joining $(a, f(a))$ to $(b, f(b))$. See Figure 25.

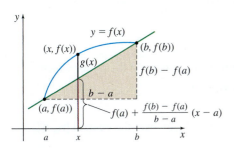

Figure 25

for all x on $[a, b]$.

Since f is continuous on $[a, b]$ and differentiable on (a, b), it follows that g is also continuous on $[a, b]$ and differentiable on (a, b). Also,

$$g(a) = f(a) - \left[f(a) + \frac{f(b) - f(a)}{b - a}(a - a) \right] = 0$$

$$g(b) = f(b) - \left[f(a) + \frac{f(b) - f(a)}{b - a}(b - a) \right] = 0$$

Now, since g satisfies the three conditions of Rolle's Theorem, there is a number c in (a, b) at which $g'(c) = 0$. Since $f(a)$ and $\dfrac{f(b) - f(a)}{b - a}$ are constants, $g'(x)$ is given by

$$g'(x) = f'(x) - \frac{f(b) - f(a)}{b - a}$$

Now we evaluate g' at c and use the fact that $g'(c) = 0$.

$$g'(c) = f'(c) - \left[\frac{f(b) - f(a)}{b - a} \right] = 0$$

$$f'(c) = \frac{f(b) - f(a)}{b - a} \qquad \blacksquare$$

The Mean Value Theorem is another example of an existence theorem. It does not tell us how to find the number c; it merely states that at least one number c exists. Often, the number c can be found.

EXAMPLE 2 Verifying the Mean Value Theorem

Verify that the function $f(x) = x^3 - 3x + 5$, $-1 \le x \le 1$ satisfies the conditions of the Mean Value Theorem. Find the number(s) c guaranteed by the Mean Value Theorem.

Solution Since f is a polynomial function, f is continuous on the closed interval $[-1, 1]$ and differentiable on the open interval $(-1, 1)$. The conditions of the Mean Value Theorem are met. Now,

$$f(-1) = 7 \qquad f(1) = 3 \qquad \text{and} \qquad f'(x) = 3x^2 - 3$$

The number(s) c in the open interval $(-1, 1)$ guaranteed by the Mean Value Theorem satisfy the equation

$$f'(c) = \frac{f(1) - f(-1)}{1 - (-1)}$$

$$3c^2 - 3 = \frac{3 - 7}{1 - (-1)} = \frac{-4}{2} = -2$$

$$3c^2 = 1$$

$$c = \sqrt{\frac{1}{3}} = \frac{\sqrt{3}}{3} \qquad \text{or} \qquad c = -\sqrt{\frac{1}{3}} = -\frac{\sqrt{3}}{3}$$

There are two numbers in the interval $(-1, 1)$ that satisfy the Mean Value Theorem. See Figure 26. ∎

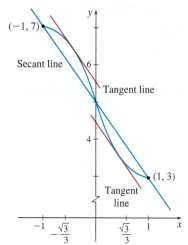

Figure 26 $f(x) = x^3 - 3x + 5$, $-1 \le x \le 1$

NOW WORK Problem 25.

The Mean Value Theorem can be applied to rectilinear motion. Suppose the function $s = f(t)$ models the distance s that an object has traveled from the origin at time t. If f is continuous on $[a, b]$ and differentiable on (a, b), the Mean Value Theorem tells us that there is at least one time t at which the velocity of the object equals its average velocity.

Let t_1 and t_2 be two distinct times in $[a, b]$. Then

$$\text{Average velocity over } [t_1, t_2] = \frac{f(t_2) - f(t_1)}{t_2 - t_1}$$

$$\text{Instantaneous velocity } v(t) = \frac{ds}{dt} = f'(t)$$

The Mean Value Theorem states there is a time t_0 in the interval (t_1, t_2) for which

$$f'(t_0) = \frac{f(t_2) - f(t_1)}{t_2 - t_1}$$

That is, at time t_0, the velocity equals the average velocity over the interval $[t_1, t_2]$.

EXAMPLE 3 Applying the Mean Value Theorem to Rectilinear Motion

Use the Mean Value Theorem to show that a car that travels 110 mi in 2 h must have had a speed of 55 miles per hour (mph) at least once during the 2 h.

Solution Let $s = f(t)$ represent the distance s the car has traveled after t hours. Its average velocity during the time period from 0 to 2 h is

$$\text{Average velocity} = \frac{f(2) - f(0)}{2 - 0} = \frac{110 - 0}{2 - 0} = 55 \text{ mph}$$

Using the Mean Value Theorem, there is a time t_0, $0 < t_0 < 2$, at which

$$f'(t_0) = \frac{f(2) - f(0)}{2 - 0} = 55$$

That is, the car had a velocity of 55 mph at least once during the 2 hour period. ∎

NOW WORK Problem 49.

The following corollaries are a consequence of the Mean Value Theorem.

IN WORDS If $f'(x) = 0$ for all numbers in (a, b), then f is a constant function.

COROLLARY

If a function f is continuous on the closed interval $[a, b]$ and is differentiable on the open interval (a, b), and if $f'(x) = 0$ for all numbers x in (a, b), then f is constant on (a, b).

Proof To show that f is a constant function, we must show that $f(x_1) = f(x_2)$ for any two numbers x_1 and x_2 in the interval (a, b).

Suppose $x_1 < x_2$. The conditions of the Mean Value Theorem are satisfied on the interval $[x_1, x_2]$, so there is a number c in this interval for which

$$f'(c) = \frac{f(x_2) - f(x_1)}{x_2 - x_1}$$

Since $f'(c) = 0$ for all numbers in (a, b), we have

$$0 = \frac{f(x_2) - f(x_1)}{x_2 - x_1}$$
$$0 = f(x_2) - f(x_1)$$
$$f(x_1) = f(x_2) \qquad \blacksquare$$

Applying the above corollary to rectilinear motion, if the velocity v of an object is zero over an interval (t_1, t_2), then the object does not move during this time interval.

IN WORDS If two functions f and g have the same derivative, then the functions differ by a constant, and the graph of f is a vertical shift of the graph of g.

COROLLARY

If the functions f and g are differentiable on an open interval (a, b) and if $f'(x) = g'(x)$ for all numbers x in (a, b), then there is a number C for which $f(x) = g(x) + C$ on (a, b).

Proof We define the function h so that $h(x) = f(x) - g(x)$. Then h is differentiable since it is the difference of two differentiable functions, and

$$h'(x) = f'(x) - g'(x) = 0$$

Since $h'(x) = 0$ for all numbers x in the interval (a, b), h is a constant function and we can write

$$h(x) = f(x) - g(x) = C \qquad \text{for some number } C$$

That is,

$$f(x) = g(x) + C \qquad \blacksquare$$

3 Identify Where a Function Is Increasing and Decreasing

NEED TO REVIEW? Increasing and decreasing functions are discussed in Section P.1, pp. 10–11.

A third corollary that follows from the Mean Value Theorem gives us a way to determine where a function is increasing and where a function is decreasing. Recall that functions are increasing or decreasing on *intervals*, either open or closed or half-open or half-closed.

COROLLARY Increasing/Decreasing Function Test

Let f be a function that is differentiable on the open interval (a, b):

- If $f'(x) > 0$ on (a, b), then f is increasing on (a, b).

- If $f'(x) < 0$ on (a, b), then f is decreasing on (a, b).

The proof of the first part of the corollary is given here. The proof of the second part is left for the exercises. See Problem 72.

Proof Since f is differentiable on (a, b), it is continuous on (a, b). To show that f is increasing on (a, b), we choose two numbers x_1 and x_2 in (a, b), with $x_1 < x_2$. We need to show that $f(x_1) < f(x_2)$.

Since x_1 and x_2 are in (a, b), f is continuous on the closed interval $[x_1, x_2]$ and differentiable on the open interval (x_1, x_2). By the Mean Value Theorem, there is a number c in the interval (x_1, x_2) for which

$$f'(c) = \frac{f(x_2) - f(x_1)}{x_2 - x_1}$$

Since $f'(x) > 0$ for all x in (a, b), it follows that $f'(c) > 0$. Since $x_1 < x_2$, then $x_2 - x_1 > 0$. As a result,

$$f(x_2) - f(x_1) > 0$$

$$f(x_1) < f(x_2)$$

So, f is increasing on (a, b). ∎

There are a few important observations to make about the Increasing/Decreasing Function Test:

- In using the Increasing/Decreasing Function Test, we determine open intervals on which the derivative is positive or negative.
- Suppose f is continuous on the closed interval $[a, b]$ and differentiable on the open interval (a, b).
  If $f'(x) > 0$ on (a, b), then f is increasing on the closed interval $[a, b]$.
  If $f'(x) < 0$ on (a, b), then f is decreasing on the closed interval $[a, b]$.
- The Increasing/Decreasing Function Test is valid if the interval (a, b) is $(-\infty, b)$ or (a, ∞) or $(-\infty, \infty)$.

EXAMPLE 4 | Identifying Where a Function Is Increasing and Decreasing

Determine where the function $f(x) = 2x^3 - 9x^2 + 12x - 5$ is increasing and where it is decreasing.

NEED TO REVIEW? Solving inequalities is discussed in Appendix A.1, pp. A-5 to A-8.

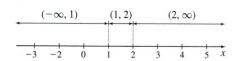

Figure 27

Solution The function f is a polynomial so f is continuous and differentiable at every real number. We find f'.

$$f'(x) = 6x^2 - 18x + 12 = 6(x - 2)(x - 1)$$

The Increasing/Decreasing Function Test states that f is increasing on intervals where $f'(x) > 0$ and that f is decreasing on intervals where $f'(x) < 0$. We solve these inequalities by using the numbers 1 and 2 to form three intervals, as shown in Figure 27. Then we determine the sign of $f'(x)$ on each interval, as shown in Table 1.

TABLE 1

Interval	Sign of $x - 1$	Sign of $x - 2$	Sign of $f'(x) =$ $6(x - 2)(x - 1)$	Conclusion
$(-\infty, 1)$	Negative $(-)$	Negative $(-)$	Positive $(+)$	f is increasing
$(1, 2)$	Positive $(+)$	Negative $(-)$	Negative $(-)$	f is decreasing
$(2, \infty)$	Positive $(+)$	Positive $(+)$	Positive $(+)$	f is increasing

We conclude that f is increasing on the intervals $(-\infty, 1)$ and $(2, \infty)$, and f is decreasing on the interval $(1, 2)$. ∎

The graph of f is shown in Figure 28. Notice that the graph of f has horizontal tangent lines at the points $(1, 0)$ and $(2, -1)$. Also, since f is continuous on its domain, we can

DF Figure 28 $f(x) = 2x^3 - 9x^2 + 12x - 5$

say that f is increasing on the intervals $(-\infty, 1]$ and $[2, \infty)$ and is decreasing on the interval $[1, 2]$.

EXAMPLE 5 Identifying Where a Function Is Increasing and Decreasing

Determine where the function $f(x) = (x^2 - 1)^{2/3}$ is increasing and where it is decreasing.

Solution f is continuous for all numbers x, and

$$f'(x) = \frac{2}{3}(x^2 - 1)^{-1/3}(2x) = \frac{4x}{3(x^2 - 1)^{1/3}}$$

The Increasing/Decreasing Function Test states that f is increasing on intervals where $f'(x) > 0$ and decreasing on intervals where $f'(x) < 0$. We solve these inequalities by using the numbers $-1, 0,$ and 1 to form four intervals. Then we determine the sign of f' in each interval, as shown in Table 2. We conclude that f is increasing on the intervals $(-1, 0)$ and $(1, \infty)$ and that f is decreasing on the intervals $(-\infty, -1)$ and $(0, 1)$.

TABLE 2

Interval	Sign of $4x$	Sign of $(x^2 - 1)^{1/3}$	Sign of $f'(x) = \dfrac{4x}{3(x^2 - 1)^{1/3}}$	Conclusion
$(-\infty, -1)$	Negative $(-)$	Positive $(+)$	Negative $(-)$	f is decreasing on $(-\infty, -1)$
$(-1, 0)$	Negative $(-)$	Negative $(-)$	Positive $(+)$	f is increasing on $(-1, 0)$
$(0, 1)$	Positive $(+)$	Negative $(-)$	Negative $(-)$	f is decreasing on $(0, 1)$
$(1, \infty)$	Positive $(+)$	Positive $(+)$	Positive $(+)$	f is increasing on $(1, \infty)$

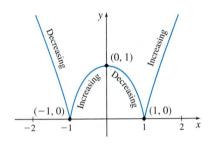

Figure 29 $f(x) = (x^2 - 1)^{2/3}$

Figure 29 shows the graph of f. Since $f'(0) = 0$, the graph of f has a horizontal tangent line at the point $(0, 1)$. Notice that f' does not exist at ± 1. Since f' becomes unbounded at -1 and 1, the graph of f has vertical tangent lines at the points $(-1, 0)$ and $(1, 0)$. Also since f is continuous on its domain, we can say that f is increasing on the intervals $[-1, 0]$ and $[1, \infty)$ and is decreasing on the intervals $(-\infty, 1]$ and $[0, 1]$.

NOW WORK Problem 33.

EXAMPLE 6 Determining Crop Yield*

A variation of the von Liebig model states that the yield $f(x)$ of a plant, measured in bushels, responds to the amount x of potassium in a fertilizer according to the following square root model:

$$f(x) = -0.057 - 0.417x + 0.852\sqrt{x}$$

For what amounts of potassium will the yield increase? For what amounts of potassium will the yield decrease?

Solution The yield is increasing when $f'(x) > 0$.

$$f'(x) = -0.417 + \frac{0.426}{\sqrt{x}} = \frac{-0.417\sqrt{x} + 0.426}{\sqrt{x}}$$

---

*__Source:__ Quirino Paris. (1992). The von Liebig Hypothesis. *American Journal of Agricultural Economics,* 74(4), 1019–1028.

Now, $f'(x) > 0$ when

$$-0.417\sqrt{x} + 0.426 > 0$$
$$0.417\sqrt{x} - 0.426 < 0$$
$$\sqrt{x} < 1.022$$
$$x < 1.044$$

The crop yield is increasing when the amount of potassium in the fertilizer is less than 1.044 and is decreasing when the amount of potassium in the fertilizer is greater than 1.044. ∎

4.3 Assess Your Understanding

Concepts and Vocabulary

1. *True or False* If a function f is defined and continuous on a closed interval $[a, b]$, differentiable on the open interval (a, b), and if $f(a) = f(b)$, then Rolle's Theorem guarantees that there is at least one number c in the interval (a, b) for which $f(c) = 0$.

2. In your own words, give a geometric interpretation of the Mean Value Theorem.

3. *True or False* If two functions f and g are differentiable on an open interval (a, b) and if $f'(x) = g'(x)$ for all numbers x in (a, b), then f and g differ by a constant.

4. *True or False* When the derivative f' is positive on an open interval I, then f is positive on I.

Skill Building

In Problems 5–16, verify that each function satisfies the three conditions of Rolle's Theorem on the given interval. Then find all numbers c in (a, b) guaranteed by Rolle's Theorem.

5. $f(x) = x^2 - 3x$ on $[0, 3]$

6. $f(x) = x^2 + 2x$ on $[-2, 0]$

7. $g(x) = x^2 - 2x - 2$ on $[0, 2]$

8. $g(x) = x^2 + 1$ on $[-1, 1]$

9. $f(x) = x^3 - x$ on $[-1, 0]$

10. $f(x) = x^3 - 4x$ on $[-2, 2]$

11. $f(t) = t^3 - t + 2$ on $[-1, 1]$

12. $f(t) = t^4 - 3$ on $[-2, 2]$

13. $s(t) = t^4 - 2t^2 + 1$ on $[-2, 2]$

14. $s(t) = t^4 + t^2$ on $[-2, 2]$

15. $f(x) = \sin(2x)$ on $[0, \pi]$

16. $f(x) = \sin x + \cos x$ on $[0, 2\pi]$

In Problems 17–20, state why Rolle's Theorem cannot be applied to the function f.

17. $f(x) = x^2 - 2x + 1$ on $[-2, 1]$

18. $f(x) = x^3 - 3x$ on $[2, 4]$

19. $f(x) = x^{1/3} - x$ on $[-1, 1]$

20. $f(x) = x^{2/5}$ on $[-1, 1]$

In Problems 21–30, verify that each function satisfies the conditions of the Mean Value Theorem on the closed interval. Then find all numbers c in (a, b) guaranteed by the Mean Value Theorem.

21. $f(x) = x^2 + 1$ on $[0, 2]$

22. $f(x) = x + 2 + \dfrac{3}{x - 1}$ on $[2, 7]$

23. $f(x) = \ln \sqrt{x}$ on $[1, e]$

24. $f(x) = xe^x$ on $[0, 1]$

25. $f(x) = x^3 - 5x^2 + 4x - 2$ on $[1, 3]$

26. $f(x) = x^3 - 7x^2 + 5x$ on $[-2, 2]$

27. $f(x) = \dfrac{x + 1}{x}$ on $[1, 3]$

28. $f(x) = \dfrac{x^2}{x + 1}$ on $[0, 1]$

29. $f(x) = \sqrt[3]{x^2}$ on $[1, 8]$

30. $f(x) = \sqrt{x - 2}$ on $[2, 4]$

In Problems 31–42, determine where each function is increasing and where each is decreasing.

31. $f(x) = x^3 + 6x^2 + 12 + 1$

32. $f(x) = -x^3 + 3x^2 + 4$

33. $f(x) = x^{2/3}(x^2 - 4)$

34. $f(x) = x^{1/3}(x^2 - 7)$

35. $f(x) = |x^3 + 3|$

36. $f(x) = |x^2 - 4|$

37. $f(x) = 3 \sin x$ on $[0, 2\pi]$

38. $f(x) = \cos(2x)$ on $[0, 2\pi]$

39. $f(x) = xe^x$

40. $g(x) = x + e^x$

41. $f(x) = e^x \sin x,\ \ 0 \le x \le 2\pi$

42. $f(x) = e^x \cos x,\ \ 0 \le x \le 2\pi$

Applications and Extensions

43. Show that the function $f(x) = 2x^3 - 6x^2 + 6x - 5$ is increasing for all x.

44. Show that the function $f(x) = x^3 - 3x^2 + 3x$ is increasing for all x.

45. Show that the function $f(x) = \dfrac{x}{x + 1}$ is increasing on any interval not containing $x = -1$.

46. Show that the function $f(x) = \dfrac{x + 1}{x}$ is decreasing on any interval not containing $x = 0$.

47. **Mean Value Theorem** Draw the graph of a function f that is continuous on $[a, b]$ but not differentiable on (a, b), and for which the conclusion of the Mean Value Theorem does not hold.

48. **Mean Value Theorem** Draw the graph of a function f that is differentiable on (a, b) but not continuous on $[a, b]$, and for which the conclusion of the Mean Value Theorem does not hold.

49. **Rectilinear Motion** An automobile travels 20 mi down a straight road at an average velocity of 40 mph. Show that the automobile must have a velocity of exactly 40 mph at some time during the trip. (Assume that the distance function is differentiable.)

1. = NOW WORK problem   ⟍⟋ = Graphing technology recommended   CAS = Computer Algebra System recommended

50. Rectilinear Motion Suppose a car is traveling on a highway. At 4:00 p.m., the car's speedometer reads 40 mph. At 4:12 p.m., it reads 60 mph. Show that at some time between 4:00 and 4:12 p.m., the acceleration was exactly $100 \, \text{mi/h}^2$.

51. Rectilinear Motion Two stock cars start a race at the same time and finish in a tie. If $f_1(t)$ is the position of one car at time t and $f_2(t)$ is the position of the second car at time t, show that at some time during the race they share the same velocity. (*Hint:* Set $f(t) = f_2(t) - f_1(t)$.)

52. Rectilinear Motion Suppose $s = f(t)$ is the distance s that an object has traveled from the origin at time t. If the object is at a specific location at $t = a$, and returns to that location at $t = b$, then $f(a) = f(b)$. Show that there is at least one time $t = c$, $a < c < b$ for which $f'(c) = 0$. That is, show that there is a time c when the velocity of the object is 0.

53. Loaded Beam The vertical deflection d (in feet), of a particular 5-foot-long loaded beam can be approximated by

$$d = d(x) = -\frac{1}{192}x^4 + \frac{25}{384}x^3 - \frac{25}{128}x^2$$

where x (in feet) is the distance from one end of the beam.

(a) Verify that the function $d = d(x)$ satisfies the conditions of Rolle's Theorem on the interval $[0, 5]$.

(b) What does the result in (a) say about the ends of the beam?

(c) Find all numbers c in $(0, 5)$ that satisfy the conclusion of Rolle's Theorem. Then find the deflection d at each number c.

(d) Graph the function d on the interval $[0, 5]$.

[CAS] **54.** For the function $f(x) = x^4 - 2x^3 - 4x^2 + 7x + 3$:

(a) Find the critical numbers of f rounded to three decimal places.

(b) Find the intervals where f is increasing and decreasing.

55. Rolle's Theorem Use Rolle's Theorem with the function $f(x) = (x - 1) \sin x$ on $[0, 1]$ to show that the equation $\tan x + x = 1$ has a solution in the interval $(0, 1)$.

56. Rolle's Theorem Use Rolle's Theorem to show that the function $f(x) = x^3 - 2$ has exactly one real zero.

57. Rolle's Theorem Use Rolle's Theorem to show that the function $f(x) = (x - 8)^3$ has exactly one real zero.

58. Rolle's Theorem Without finding the derivative, show that if $f(x) = (x^2 - 4x + 3)(x^2 + x + 1)$, then $f'(x) = 0$ for at least one number between 1 and 3. Check by finding the derivative and applying the Intermediate Value Theorem.

59. Rolle's Theorem Consider $f(x) = |x|$ on the interval $[-1, 1]$. Here, $f(1) = f(-1) = 1$ but there is no c in the interval $(-1, 1)$ at which $f'(c) = 0$. Explain why this does not contradict Rolle's Theorem.

60. Mean Value Theorem Consider $f(x) = x^{2/3}$ on the interval $[-1, 1]$. Verify that there is no c in $(-1, 1)$ for which

$$f'(c) = \frac{f(1) - f(-1)}{1 - (-1)}$$

Explain why this does not contradict the Mean Value Theorem.

61. Mean Value Theorem The Mean Value Theorem guarantees that there is a real number N in the interval $(0, 1)$ for which

$f'(N) = f(1) - f(0)$ if f is continuous on the interval $[0, 1]$ and differentiable on the interval $(0, 1)$. Find N if $f(x) = \sin^{-1} x$.

62. Mean Value Theorem Show that when the Mean Value Theorem is applied to the function $f(x) = Ax^2 + Bx + C$ in the interval $[a, b]$, the number c referred to in the theorem is the midpoint of the interval.

63. (a) Apply the Increasing/Decreasing Function Test to the function $f(x) = \sqrt{x}$. What do you conclude?

(b) Is f increasing on the interval $[0, \infty)$? Explain.

64. Explain why the function $f(x) = ax^4 + bx^3 + cx^2 + dx + e$ must have a zero between 0 and 1 if

$$\frac{a}{5} + \frac{b}{4} + \frac{c}{3} + \frac{d}{2} + e = 0$$

65. Put It Together If $f'(x)$ and $g'(x)$ exist and $f'(x) > g'(x)$ for all real x, then which of the following statements must be true about the graph of $y = f(x)$ and the graph of $y = g(x)$?

(a) They intersect exactly once.

(b) They intersect no more than once.

(c) They do not intersect.

(d) They could intersect more than once.

(e) They have a common tangent at each point of intersection.

66. Prove that there is no k for which the function

$$f(x) = x^3 - 3x + k$$

has two distinct zeros in the interval $[0, 1]$.

67. Show that $e^x > x^2$ for all $x > 0$.

68. Show that $e^x > 1 + x$ for all $x > 0$. (*Hint:* Show that $f(x) = e^x - 1 - x$ is an increasing function for $x > 0$.)

69. Show that $0 < \ln x < x$ for $x > 1$.

70. Show that $\tan \theta \geq \theta$ for all θ in the open interval $\left(0, \frac{\pi}{2}\right)$.

71. Let f be a function that is continuous on the closed interval $[a, b]$ and differentiable on the open interval (a, b). If $f(x) = 0$ for three different numbers x in (a, b), show that there must be at least two numbers in (a, b) at which $f'(x) = 0$.

72. Proof for the Increasing/Decreasing Function Test Let f be a function that is continuous on a closed interval $[a, b]$. Show that if $f'(x) < 0$ for all numbers in (a, b), then f is a decreasing function on (a, b). (See the Corollary on p. 279.)

73. Suppose that the domain of f is an open interval (a, b) and $f'(x) > 0$ for all x in the interval. Show that f cannot have an extreme value on (a, b).

Challenge Problems

74. Use Rolle's Theorem to show that between any two real zeros of a polynomial function f, there is a real zero of its derivative function f'.

75. Find where the general cubic $f(x) = ax^3 + bx^2 + cx + d$ is increasing and where it is decreasing by considering cases depending on the value of $b^2 - 3ac$. (*Hint:* $f'(x)$ is a quadratic function; examine its discriminant.)

76. Explain why the function $f(x) = x^n + ax + b$, where n is a positive even integer, has at most two distinct real zeros.

77. Explain why the function $f(x) = x^n + ax + b$, where n is a positive odd integer, has at most three distinct real zeros.

78. Explain why the function $f(x) = x^n + ax^2 + b$, where n is a positive odd integer, has at most three distinct real zeros.

79. Explain why the function $f(x) = x^n + ax^2 + b$, where n is a positive even integer, has at most four distinct real zeros.

80. **Mean Value Theorem**   Use the Mean Value Theorem to verify that

$$\frac{1}{9} < \sqrt{66} - 8 < \frac{1}{8}$$

(*Hint:* Consider $f(x) = \sqrt{x}$ on the interval $[64, 66]$.)

81. Given $f(x) = \dfrac{ax^n + b}{cx^n + d}$, where $n \geq 2$ is a positive integer and $ad - bc \neq 0$, find the critical numbers and the intervals on which f is increasing and decreasing.

82. Show that $x \leq \ln(1 + x)^{1+x}$ for all $x > -1$. (*Hint:* Consider $f(x) = -x + \ln(1+x)^{1+x}$.)

83. Show that $a \ln \dfrac{b}{a} \leq b - a < b \ln \dfrac{b}{a}$ if $0 < a < b$.

84. Show that for any positive integer n, $\dfrac{n}{n+1} < \ln\left(1 + \dfrac{1}{n}\right)^n < 1$.

4.4 Local Extrema and Concavity

OBJECTIVES   *When you finish this section, you should be able to:*

1 Use the First Derivative Test to find local extrema (p. 284)
2 Use the First Derivative Test with rectilinear motion (p. 286)
3 Determine the concavity of a function (p. 287)
4 Find inflection points (p. 290)
5 Use the Second Derivative Test to find local extrema (p. 291)

So far we know that if a function f defined on a closed interval has a local maximum or a local minimum at a number c in the open interval, then c is a critical number. We are now ready to see how the derivative is used to determine whether a function f has a local maximum, a local minimum, or neither at a critical number.

1 Use the First Derivative Test to Find Local Extrema

All local extreme values of a function f occur at critical numbers. While the value of f at each critical number is a candidate for being a local extreme value for f, not every critical number gives rise to a local extreme value.

How do we distinguish critical numbers that give rise to local extreme values from those that do not? And then how do we determine if a local extreme value is a local maximum value or a local minimum value? Figure 30 provides a clue. If you look from left to right along the graph of f, you see that the graph of f is increasing to the left of x_1, where a local maximum occurs, and is decreasing to its right. The function f is decreasing to the left of x_2, where a local minimum occurs, and is increasing to its right. So, knowing where a function f increases and decreases enables us to find local maximum values and local minimum values.

The next theorem is usually referred to as the *First Derivative Test*, since it relies on information obtained from the first derivative of a function.

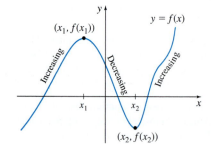

Figure 30

IN WORDS If c is a critical number of f and if f is increasing to the left of c and decreasing to the right of c, then $f(c)$ is a local maximum value. If f is decreasing to the left of c and increasing to the right of c, then $f(c)$ is a local minimum value.

THEOREM First Derivative Test

Let f be a function that is continuous on an interval I. Suppose that c is a critical number of f and (a, b) is an open interval in I containing c:

- If $f'(x) > 0$ for $a < x < c$ and $f'(x) < 0$ for $c < x < b$, then $f(c)$ is a local maximum value.
- If $f'(x) < 0$ for $a < x < c$ and $f'(x) > 0$ for $c < x < b$, then $f(c)$ is a local minimum value.
- If $f'(x)$ has the same sign on both sides of c, then $f(c)$ is neither a local maximum value nor a local minimum value.

Partial Proof If $f'(x) > 0$ on the interval (a, c), then f is increasing on (a, c). Also, if $f'(x) < 0$ on the interval (c, b), then f is decreasing on (c, b). So, for all x in (a, b), $f(x) \leq f(c)$. That is, $f(c)$ is a local maximum value. See Figure 31(a). ∎

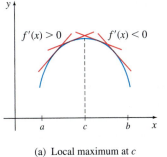
(a) Local maximum at c

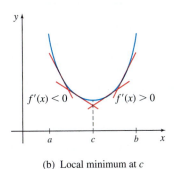
(b) Local minimum at c

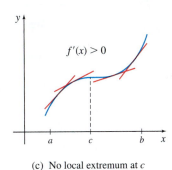
(c) No local extremum at c

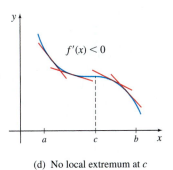
(d) No local extremum at c

Figure 31

The proofs of the second and third bullets are left as exercises (see Problems 130 and 131). See Figure 31(b) for an illustration of a local minimum value. In Figures 31(c) and 31(d), f' has the same sign on both sides of c, so f has neither a local maximum nor a local minimum at c.

EXAMPLE 1 Using the First Derivative Test to Find Local Extrema

Find the local extrema of $f(x) = x^4 - 4x^3$.

Solution Since f is a polynomial function, f is continuous and differentiable at every real number. We begin by finding the critical numbers of f.

$$f'(x) = 4x^3 - 12x^2 = 4x^2(x - 3)$$

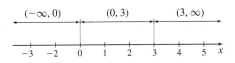

Figure 32

The critical numbers are 0 and 3. We use the critical numbers 0 and 3 to form three intervals, as shown in Figure 32. Then we determine where f is increasing and where it is decreasing by determining the sign of $f'(x)$ in each interval. See Table 3.

TABLE 3

Interval	Sign of x^2	Sign of $x - 3$	Sign of $f'(x) = 4x^2(x - 3)$	Conclusion
$(-\infty, 0)$	Positive (+)	Negative (−)	Negative (−)	f is decreasing
$(0, 3)$	Positive (+)	Negative (−)	Negative (−)	f is decreasing
$(3, \infty)$	Positive (+)	Positive (+)	Positive (+)	f is increasing

Using the First Derivative Test, f has neither a local maximum nor a local minimum at 0, and f has a local minimum at 3. The local minimum value is $f(3) = -27$. ∎

The graph of f is shown in Figure 33. Notice that the tangent lines to the graph of f are horizontal at the points $(0, 0)$ and $(3, -27)$.

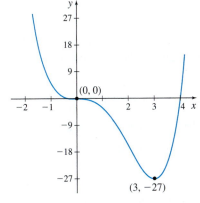

Figure 33 $f(x) = x^4 - 4x^3$

NOW WORK Problem 13.

EXAMPLE 2 Using the First Derivative Test to Find Local Extrema

Find the local extrema of $f(x) = x^{2/3}(x - 5)$.

Solution The domain of f is all real numbers and f is continuous on its domain.

$$f'(x) = \underset{\underset{\text{Use the Product Rule.}}{\uparrow}}{x^{2/3}} + \left(\frac{2}{3}\right) x^{-1/3}(x - 5) = \frac{3x + 2(x - 5)}{3x^{1/3}} = \frac{5}{3}\left(\frac{x - 2}{x^{1/3}}\right)$$

Since $f'(2) = 0$ and $f'(0)$ does not exist, the critical numbers are 0 and 2. The graph of f will have a horizontal tangent line at the point $(2, -3\sqrt[3]{4})$ and a vertical tangent line at the point $(0, 0)$.

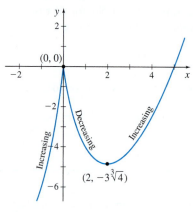

Figure 34 $f(x) = x^{2/3}(x - 5)$

Table 4 shows the intervals on which f is increasing and decreasing.

TABLE 4

Interval	Sign of $x - 2$	Sign of $x^{1/3}$	Sign of $f'(x) = \dfrac{5}{3}\left(\dfrac{x - 2}{x^{1/3}}\right)$	Conclusion
$(-\infty, 0)$	Negative $(-)$	Negative $(-)$	Positive $(+)$	f is increasing
$(0, 2)$	Negative $(-)$	Positive $(+)$	Negative $(-)$	f is decreasing
$(2, \infty)$	Positive $(+)$	Positive $(+)$	Positive $(+)$	f is increasing

By the First Derivative Test, f has a local maximum at 0 and a local minimum at 2; $f(0) = 0$ is a local maximum value and $f(2) = -3\sqrt[3]{4}$ is a local minimum value. ■

The graph of f is shown in Figure 34. Notice the vertical tangent line at the point $(0, 0)$ and the horizontal tangent line at the point $(2, -3\sqrt[3]{4})$.

NOW WORK   Problem 21.

2 Use the First Derivative Test with Rectilinear Motion

Increasing and decreasing functions can be used to investigate the motion of an object in rectilinear motion. Suppose the distance s of an object from the origin at time t is given by the function $s = f(t)$. We assume the motion is along a horizontal line with the positive direction to the right. The velocity v of the object is $v = \dfrac{ds}{dt}$.

- If $v = \dfrac{ds}{dt} > 0$, the distance s from the origin is increasing with time t, and the object moves to the right.

- If $v = \dfrac{ds}{dt} < 0$, the distance s from the origin is decreasing with time t, and the object moves to the left.

This information, along with the First Derivative Test, can be used to find the local extreme values of $s = f(t)$ and to determine at what times t the direction of the motion of the object changes.

Similarly, if the acceleration of the object, $a = \dfrac{dv}{dt} > 0$, then the velocity of the object is increasing, and if $a = \dfrac{dv}{dt} < 0$, then the velocity is decreasing. Again, this information, along with the First Derivative Test, is used to find the local extreme values of the velocity.

EXAMPLE 3   **Using the First Derivative Test with Rectilinear Motion**

Suppose the distance s of an object from the origin at time $t \geq 0$, in seconds, is given by

$$s = t^3 - 9t^2 + 15t + 3$$

(a) Determine the time intervals during which the object is moving to the right and to the left.
(b) When does the object reverse direction?
(c) When is the velocity of the object increasing and when is it decreasing?
(d) Draw a figure that illustrates the motion of the object.
(e) Draw a figure that illustrates the velocity of the object.

Solution (a) To investigate the motion, we find the velocity v.

$$v = \frac{ds}{dt} = 3t^2 - 18t + 15 = 3(t^2 - 6t + 5) = 3(t - 1)(t - 5)$$

The critical numbers are 1 and 5. We use Table 5 to describe the motion of the object.

TABLE 5

Time Interval	Sign of $t - 1$	Sign of $t - 5$	Velocity, v	Motion of the Object
$(0, 1)$	Negative $(-)$	Negative $(-)$	Positive $(+)$	To the right
$(1, 5)$	Positive $(+)$	Negative $(-)$	Negative $(-)$	To the left
$(5, \infty)$	Positive $(+)$	Positive $(+)$	Positive $(+)$	To the right

The object moves to the right for the first second and again after 5 seconds. The object moves to the left on the interval $(1, 5)$.

(b) The object reverses direction at $t = 1$ and $t = 5$.

(c) To determine when the velocity increases or decreases, we find the acceleration.

$$a = \frac{dv}{dt} = 6t - 18 = 6(t - 3)$$

Since $a < 0$ on the interval $(0, 3)$, the velocity v decreases for the first 3 seconds. On the interval $(3, \infty)$, $a > 0$, so the velocity v increases from 3 seconds onward.

(d) Figure 35 illustrates the motion of the object.

(e) Figure 36 illustrates the velocity of the object.

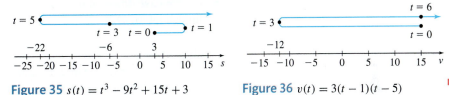

Figure 35 $s(t) = t^3 - 9t^2 + 15t + 3$ **Figure 36** $v(t) = 3(t - 1)(t - 5)$ ∎

An interesting phenomenon occurs during the first three seconds. The velocity of the object is decreasing on the interval $(0, 3)$, as shown Figure 36.

Now, we investigate the speed, namely, $|v(t)| = 3|t^2 - 6t + 5| = 3|(t - 1)(t - 5)|$ on the interval $(0, 3)$:

- If $0 < t < 1$, $|v(t)| = 3(t^2 - 6t + 5)$ and $\dfrac{d|v|}{dt} = 6(t - 3) < 0$, so the speed is decreasing.

- If $1 < t < 3$, $|v(t)| = -3(t^2 - 6t + 5)$ and $\dfrac{d|v|}{dt} = -6(t - 3) > 0$, so the speed is increasing.

That is, the velocity of the object is decreasing from $t = 0$ to $t = 3$, but its speed is decreasing from $t = 0$ to $t = 1$ and is increasing from $t = 1$ to $t = 3$.

NOW WORK Problem 35.

3 Determine the Concavity of a Function

Figure 37 shows the graphs of two familiar functions: $y = x^2, x \geq 0$, and $y = \sqrt{x}$. Each graph starts at the origin, passes through the point $(1, 1)$, and is increasing. But there is a noticeable difference in their shapes. The graph of $y = x^2$ bends upward, it is *concave up;* the graph of $y = \sqrt{x}$ bends downward, it is *concave down*.

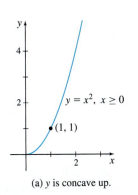

(a) y is concave up.

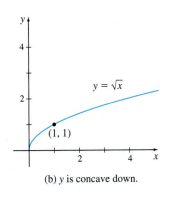

(b) y is concave down.

Figure 37

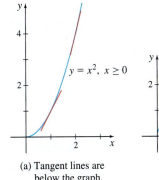

(a) Tangent lines are below the graph.

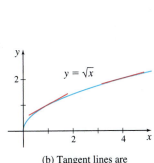

(b) Tangent lines are above the graph.

Figure 38

Suppose we draw tangent lines to the graphs of $y = x^2$ and $y = \sqrt{x}$, as shown in Figure 38. Notice that the graph of $y = x^2$ lies above all its tangent lines, and moving from left to right, the slopes of the tangent lines are increasing. That is, the derivative $y' = \dfrac{d}{dx}x^2$ is an increasing function.

On the other hand, the graph of $y = \sqrt{x}$ lies below all its tangent lines, and moving from left to right, the slopes of the tangent lines are decreasing. That is, the derivative $y' = \dfrac{d}{dx}\sqrt{x}$ is a decreasing function. This discussion leads to the following definition.

DEFINITION Concave Up; Concave Down.

Let f be a function that is continuous on a closed interval $[a, b]$ and differentiable on the open interval (a, b).

- f is **concave up** on (a, b) if the graph of f lies above each of its tangent lines throughout (a, b).
- f is **concave down** on (a, b) if the graph of f lies below each of its tangent lines throughout (a, b).

We can formulate a test to determine where a function f is concave up or concave down, provided f'' exists. Since f'' equals the rate of change of f', it follows that if $f''(x) > 0$ on an open interval, then f' is increasing on that interval, and if $f''(x) < 0$ on an open interval, then f' is decreasing on that interval. As we observed in Figure 38, when f' is increasing, the graph is concave up, and when f' is decreasing, the graph is concave down. These observations lead to the *test for concavity*.

THEOREM Test for Concavity

Let f be a function that is continuous on a closed interval $[a, b]$. Suppose f' and f'' exist on the open interval (a, b).

- If $f''(x) > 0$ on the interval (a, b), then f is concave up on (a, b).
- If $f''(x) < 0$ on the interval (a, b), then f is concave down on (a, b).

Proof Suppose $f''(x) > 0$ on the interval (a, b), and c is any fixed number in (a, b). An equation of the tangent line to f at the point $(c, f(c))$ is

$$y = f(c) + f'(c)(x - c)$$

We need to show that the graph of f lies above each of its tangent lines for all x in (a, b). That is, we need to show that

$$f(x) \geq f(c) + f'(c)(x - c) \qquad \text{for all } x \text{ in } (a, b)$$

If $x = c$, then $f(x) = f(c)$ and we are finished.

If $x \neq c$, then by applying the Mean Value Theorem to the function f, there is a number x_1 between c and x, for which

$$f'(x_1) = \frac{f(x) - f(c)}{x - c}$$

Now we solve for $f(x)$:

$$f(x) = f(c) + f'(x_1)(x - c) \tag{1}$$

There are two possibilities: Either $c < x_1 < x$ or $x < x_1 < c$.

Suppose $c < x_1 < x$. Since $f''(x) > 0$ on the interval (a, b), it follows that f' is increasing on (a, b). For $x_1 > c$, this means that $f'(x_1) > f'(c)$. As a result, from (1), we have

$$f(x) > f(c) + f'(c)(x - c)$$

That is, the graph of f lies above each of its tangent lines to the right of c in (a, b).

Similarly, if $x < x_1 < c$, then $f(x) > f(c) + f'(c)(x - c)$.

In all cases, $f(x) \geq f(c) + f'(c)(x - c)$ so f is concave up on (a, b).

The proof that if $f''(x) < 0$, then f is concave down is left as an exercise. See Problem 132. ■

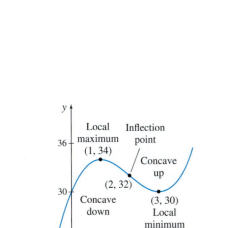

Figure 39 $f(x) = e^x$ is concave up on its domain.

EXAMPLE 4   Determining the Concavity of $f(x) = e^x$

Show that $f(x) = e^x$ is concave up on its domain.

Solution The domain of $f(x) = e^x$ is all real numbers. The first and second derivatives of f are

$$f'(x) = e^x \qquad f''(x) = e^x$$

Since $f''(x) > 0$ for all real numbers, by the Test for Concavity, f is concave up on its domain. ■

Figure 39 shows the graph of $f(x) = e^x$ and a selection of tangent lines to the graph. Notice that for any x, the graph of f lies above its tangent lines.

NOW WORK   Problem 45(a).

EXAMPLE 5   Finding Local Extrema and Determining Concavity

(a) Find any local extrema of the function $f(x) = x^3 - 6x^2 + 9x + 30$.
(b) Determine where $f(x) = x^3 - 6x^2 + 9x + 30$ is concave up and where it is concave down.

Solution (a) The first derivative of f is

$$f'(x) = 3x^2 - 12x + 9 = 3(x - 1)(x - 3)$$

So, 1 and 3 are critical numbers of f. Now,

- $f'(x) = 3(x - 1)(x - 3) > 0$ if $x < 1$ or $x > 3$.
- $f'(x) < 0$ if $1 < x < 3$.

So f is increasing on $(-\infty, 1)$ and on $(3, \infty)$; f is decreasing on $(1, 3)$.

At 1, f has a local maximum, and at 3, f has a local minimum. The local maximum value is $f(1) = 34$; the local minimum value is $f(3) = 30$.

(b) To determine concavity, we use the second derivative:

$$f''(x) = 6x - 12 = 6(x - 2)$$

Now, we solve the inequalities $f''(x) < 0$ and $f''(x) > 0$ and use the Test for Concavity.

- $f''(x) < 0$ if $x < 2$,   so f is concave down on $(-\infty, 2)$
- $f''(x) > 0$ if $x > 2$,   so f is concave up on $(2, \infty)$ ■

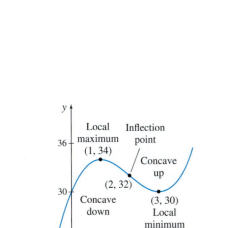

DF **Figure 40** $f(x) = x^3 - 6x^2 + 9x + 30$

Figure 40 shows the graph of f. The point $(2, 32)$ where the concavity of f changes is of special importance and is called an *inflection point*.

> **DEFINITION Inflection Point**
>
> Suppose f is a function that is differentiable on an open interval (a, b) containing c. If the concavity of f changes at the point $(c, f(c))$, then $(c, f(c))$ is an **inflection point** of f.

4 Find Inflection Points

If $(c, f(c))$ is an inflection point of f, then on one side of c the slopes of the tangent lines are increasing (or decreasing), and on the other side of c the slopes of the tangent lines are decreasing (or increasing). This means the derivative f' must have a local maximum or a local minimum at c. In either case, it follows that $f''(c) = 0$ or $f''(c)$ does not exist.

> **THEOREM A Condition for an Inflection Point**
>
> Let f denote a function that is differentiable on an open interval (a, b) containing c. If $(c, f(c))$ is an inflection point of f, then either $f''(c) = 0$ or f'' does not exist at c.

Notice the wording in the theorem. If you *know* that $(c, f(c))$ is an inflection point of f, then the second derivative of f at c is 0 or does not exist. The converse is not necessarily true. In other words, a number at which $f''(x) = 0$ or at which f'' does not exist will not always identify an inflection point.

Steps for Finding the Inflection Points of a Function f

Step 1 Find all numbers in the domain of f at which $f''(x) = 0$ or at which f'' does not exist.

Step 2 Use the Test for Concavity to determine the concavity of f on both sides of each of these numbers.

Step 3 If the concavity changes, there is an inflection point; otherwise, no inflection point exists.

EXAMPLE 6   Finding Inflection Points

Find the inflection points of $f(x) = x^{5/3}$.

Solution We follow the steps for finding an inflection point.

Step 1 The domain of f is all real numbers. The first and second derivatives of f are

$$f'(x) = \frac{5}{3}x^{2/3} \qquad f''(x) = \frac{10}{9}x^{-1/3} = \frac{10}{9x^{1/3}}$$

The second derivative of f does not exist when $x = 0$. So, $(0, 0)$ is a possible inflection point.

Step 2 Now use the Test for Concavity.

- If $x < 0$ then $f''(x) < 0$ so f is concave down on $(-\infty, 0)$.
- If $x > 0$ then $f''(x) > 0$ so f is concave up on $(0, \infty)$.

Step 3 Since the concavity of f changes at 0, we conclude that $(0, 0)$ is an inflection point of f.

Figure 41 shows the graph of f. ■

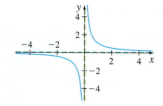

Figure 41 $f(x) = x^{5/3}$

Inflection point

$(0, 0)$

NOW WORK Problems 45(b) and 51.

A change in concavity does not, of itself, guarantee an inflection point. For example, Figure 42 shows the graph of $f(x) = \dfrac{1}{x}$. The first and second derivatives are $f'(x) = -\dfrac{1}{x^2}$ and $f''(x) = \dfrac{2}{x^3}$. Then

$$f''(x) = \frac{2}{x^3} < 0 \text{ if } x < 0 \text{ and } f''(x) = \frac{2}{x^3} > 0 \text{ if } x > 0.$$

So, f is concave down on $(-\infty, 0)$ and concave up on $(0, \infty)$, yet f has no inflection point at $x = 0$ because f is not defined at 0.

Figure 42 $f(x) = \dfrac{1}{x}$

5 Use the Second Derivative Test to Find Local Extrema

Suppose c is a critical number of f and $f'(c) = 0$. This means that the graph of f has a horizontal tangent line at the point $(c, f(c))$. If f'' exists on an open interval containing c and $f''(c) > 0$, then the graph of f is concave up on the interval, as shown in Figure 43(a). Intuitively, it would seem that $f(c)$ is a local minimum value of $f(c)$. If, on the other hand, $f''(c) < 0$, the graph of f is concave down on the interval, and $f(c)$ would appear to be a local maximum value of $f(c)$, as shown in Figure 43(b). The next theorem, known as the *Second Derivative Test,* confirms our intuition.

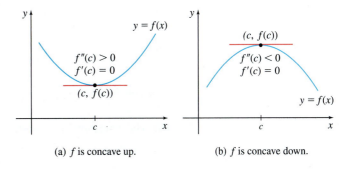

(a) f is concave up.   (b) f is concave down.

Figure 43

THEOREM Second Derivative Test

Let f be a function for which f' and f'' exist on an open interval (a, b). Suppose c lies in (a, b) and is a critical number of f:

- If $f''(c) < 0$, then $f(c)$ is a local maximum value.
- If $f''(c) > 0$, then $f(c)$ is a local minimum value.

Proof Suppose $f''(c) < 0$. Then

$$f''(c) = \lim_{x \to c} \frac{f'(x) - f'(c)}{x - c} < 0$$

Now since $\lim_{x \to c} \dfrac{f'(x) - f'(c)}{x - c} < 0$, there is an open interval about c for which

$$\frac{f'(x) - f'(c)}{x - c} < 0$$

everywhere in the interval, except possibly at c itself (refer to Example 7, p. 136, in Section 1.6). Since c is a critical number, $f'(c) = 0$, so

$$\frac{f'(x)}{x - c} < 0$$

For $x < c$ on this interval, $f'(x) > 0$, and for $x > c$ on this interval, $f'(x) < 0$. By the First Derivative Test, $f(c)$ is a local maximum value. ∎

In Problem 133, you are asked to prove that if $f''(c) > 0$, then $f(c)$ is a local minimum value.

EXAMPLE 7   **Using the Second Derivative Test to Identify Local Extrema**

(a) Determine where $f(x) = x - 2\cos x$, $0 \le x \le 2\pi$, is concave up and concave down.

(b) Find any inflection points.

(c) Use the Second Derivative Test to identify any local extreme values.

Solution **(a)** Since f is continuous on the closed interval $[0, 2\pi]$ and f' and f'' exist on the open interval $(0, 2\pi)$, we can use the Test for Concavity. The first and second derivatives of f are

$$f'(x) = \frac{d}{dx}(x - 2\cos x) = 1 + 2\sin x \qquad \text{and} \qquad f''(x) = \frac{d}{dx}(1 + 2\sin x) = 2\cos x$$

To determine concavity, we solve the inequalities $f''(x) < 0$ and $f''(x) > 0$. Since $f''(x) = 2\cos x$, we have

- $f''(x) > 0$ when $0 < x < \dfrac{\pi}{2}$ and $\dfrac{3\pi}{2} < x < 2\pi$

- $f''(x) < 0$ when $\dfrac{\pi}{2} < x < \dfrac{3\pi}{2}$

The function f is concave up on the intervals $\left(0, \dfrac{\pi}{2}\right)$ and $\left(\dfrac{3\pi}{2}, 2\pi\right)$, and f is concave down on the interval $\left(\dfrac{\pi}{2}, \dfrac{3\pi}{2}\right)$.

(b) The concavity of f changes at $\dfrac{\pi}{2}$ and $\dfrac{3\pi}{2}$, so the points $\left(\dfrac{\pi}{2}, \dfrac{\pi}{2}\right)$ and $\left(\dfrac{3\pi}{2}, \dfrac{3\pi}{2}\right)$ are inflection points of f.

(c) We find the critical numbers by solving the equation $f'(x) = 0$.

$$1 + 2\sin x = 0 \qquad 0 \le x \le 2\pi$$

$$\sin x = -\frac{1}{2}$$

$$x = \frac{7\pi}{6} \qquad \text{or} \qquad x = \frac{11\pi}{6}$$

Now using the Second Derivative Test, we get

$$f''\left(\frac{7\pi}{6}\right) = 2\cos\left(\frac{7\pi}{6}\right) = -\sqrt{3} < 0$$

and

$$f''\left(\frac{11\pi}{6}\right) = 2\cos\left(\frac{11\pi}{6}\right) = \sqrt{3} > 0$$

So,

$$f\left(\frac{7\pi}{6}\right) = \frac{7\pi}{6} - 2\cos\frac{7\pi}{6} = \frac{7\pi}{6} + \sqrt{3} \approx 5.4$$

is a local maximum value, and

$$f\left(\frac{11\pi}{6}\right) = \frac{11\pi}{6} - 2\cos\frac{11\pi}{6} = \frac{11\pi}{6} - \sqrt{3} \approx 4.03$$

is a local minimum value. ■

See Figure 44 for the graph of f.

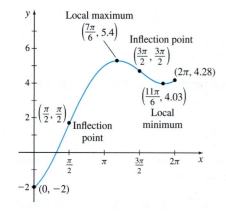

Figure 44 $y = x - 2\cos x,\ 0 \le x \le 2\pi$

NOW WORK **Problem 79.**

If the second derivative of the function does not exist, the Second Derivative Test cannot be used. If the second derivative exists at a critical number, but equals 0 there, the Second Derivative Test gives no information. In these cases, the First Derivative Test must be used to identify local extreme points. An example is the function $f(x) = x^4$. Both $f'(x) = 4x^3$ and $f''(x) = 12x^2$ exist for all real numbers. Since $f'(0) = 0$, 0 is a critical number of f. But $f''(0) = 0$, so the Second Derivative Test gives no information about the behavior of f at 0.

EXAMPLE 8   **Analyzing Monthly Sales**

Unit monthly sales R of a new product over a period of time are expected to follow the logistic function

$$R = R(t) = \frac{20{,}000}{1 + 50e^{-t}} - \frac{20{,}000}{51} \qquad t \geq 0$$

where t is measured in months.

(a) When are the monthly sales increasing? When are they decreasing?
(b) Find the rate of change of sales.
(c) When is the rate of change of sales R' increasing? When is it decreasing?
(d) When is the rate of change of sales a maximum?
(e) Find any inflection points of $R(t)$.
(f) Interpret the result found in (e).

Solution **(a)** We find $R'(t)$ and use the Increasing/Decreasing Function Test.

$$R'(t) = \frac{d}{dt}\left(\frac{20{,}000}{1 + 50e^{-t}} - \frac{20{,}000}{51}\right) = 20{,}000 \cdot \left[\frac{50e^{-t}}{(1 + 50e^{-t})^2}\right] = \frac{1{,}000{,}000e^{-t}}{(1 + 50e^{-t})^2}$$

Since $e^{-t} > 0$ for all $t \geq 0$, then $R'(t) > 0$ for $t \geq 0$. The sales function R is an increasing function. So, monthly sales are always increasing.

(b) The rate of change of sales is given by the derivative $R'(t) = \dfrac{1{,}000{,}000e^{-t}}{(1 + 50e^{-t})^2}, t \geq 0.$

(c) Using the Increasing/Decreasing Function Test with R', the rate of change of sales R' is increasing when its derivative $R''(t) > 0$; $R'(t)$ is decreasing when $R''(t) < 0$.

$$R''(t) = \frac{d}{dt}R'(t) = 1{,}000{,}000\left[\frac{-e^{-t}(1 + 50e^{-t})^2 + 100e^{-2t}(1 + 50e^{-t})}{(1 + 50e^{-t})^4}\right]$$

$$= 1{,}000{,}000e^{-t}\left[\frac{-1 - 50e^{-t} + 100e^{-t}}{(1 + 50e^{-t})^3}\right] = \frac{1{,}000{,}000e^{-t}}{(1 + 50e^{-t})^3}(50e^{-t} - 1)$$

Since $e^{-t} > 0$ for all t, the sign of R'' depends on the sign of $50e^{-t} - 1$.

$$
\begin{array}{cc}
50e^{-t} - 1 > 0 & 50e^{-t} - 1 < 0 \\
50e^{-t} > 1 & 50e^{-t} < 1 \\
50 > e^t & 50 < e^t \\
t < \ln 50 & t > \ln 50
\end{array}
$$

Since $R''(t) > 0$ for $t < \ln 50 \approx 3.9$ and $R''(t) < 0$ for $t > \ln 50 \approx 3.9$, the rate of change of sales is increasing for the first 3.9 months and is decreasing from 3.9 months on.

(d) The critical number of R' is $\ln 50 \approx 3.9$. Using the First Derivative Test, the rate of change of sales is a maximum about 3.9 months after the product is introduced.

(e) Since $R''(t) > 0$ for $t < \ln 50$ and $R''(t) < 0$ for $t > \ln 50$, the point $(\ln 50, 9608)$ is the inflection point of R.

(f) The sales function R is an increasing function, but at the inflection point $(\ln 50, 9608)$ the rate of change in sales begins to decrease.   ∎

See Figure 45 for the graphs of R and R'.

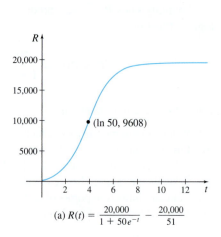

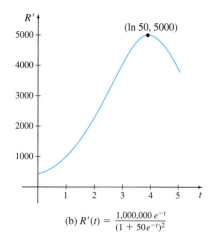

(a) $R(t) = \dfrac{20,000}{1 + 50e^{-t}} - \dfrac{20,000}{51}$

(b) $R'(t) = \dfrac{1,000,000\, e^{-t}}{(1 + 50e^{-t})^2}$

Figure 45

NOW WORK Problem 105.

4.4 Assess Your Understanding

Concepts and Vocabulary

1. *True or False*   If a function f is continuous on the interval $[a, b]$, differentiable on the interval (a, b), and changes from an increasing function to a decreasing function at the point $(c, f(c))$, then $(c, f(c))$ is an inflection point of f.

2. *True or False*   Suppose c is a critical number of f and (a, b) is an open interval containing c. If $f'(x)$ is positive on both sides of c, then $f(c)$ is a local maximum value.

3. *Multiple Choice*   Suppose a function f is continuous on a closed interval $[a, b]$ and differentiable on the open interval (a, b). If the graph of f lies above each of its tangent lines on the interval (a, b), then f is [(a) concave up, (b) concave down, (c) neither] on (a, b).

4. *Multiple Choice*   If the acceleration of an object in rectilinear motion is negative, then the velocity of the object is [(a) increasing, (b) decreasing, (c) neither].

5. *Multiple Choice*   Suppose f is a function that is differentiable on an open interval containing c and the concavity of f changes at the point $(c, f(c))$. Then $(c, f(c))$ is a(n) [(a) inflection point, (b) critical point, (c) both (d) neither] of f.

6. *Multiple Choice*   Suppose a function f is continuous on a closed interval $[a, b]$ and both f' and f'' exist on the open interval (a, b). If $f''(x) > 0$ on the interval (a, b), then f is [(a) increasing, (b) decreasing, (c) concave up, (d) concave down] on (a, b).

7. *True or False*   Suppose f is a function for which f' and f'' exist on an open interval (a, b) and suppose c, $a < c < b$, is a critical number of f. If $f''(c) = 0$, then the Second Derivative Test cannot be used to determine if there is a local extremum at c.

8. *True or False*   Suppose a function f is differentiable on the open interval (a, b). If either $f''(c) = 0$ or f'' does not exist at the number c in (a, b), then $(c, f(c))$ is an inflection point of f.

Skill Building

In Problems 9–12, the graph of a function f is given.

(a) *Identify the points where each function has a local maximum value, a local minimum value, or an inflection point.*

(b) *Identify the intervals on which each function is increasing, decreasing, concave up, or concave down.*

9.

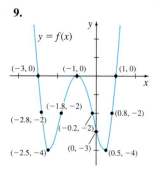

10.

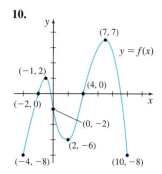

11.

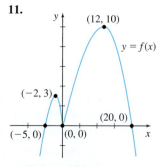

12.

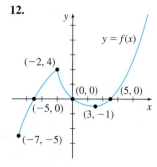

In Problems 13–30, for each function:

(a) *Find the critical numbers.*

(b) *Use the First Derivative Test to find any local extrema.*

13. $f(x) = x^3 - 6x^2 + 2$

14. $f(x) = x^3 + 6x^2 + 12 + 1$

---

1. = NOW WORK problem   📈 = Graphing technology recommended   CAS = Computer Algebra System recommended

15. $f(x) = 3x^4 - 4x^3$

16. $h(x) = x^4 + 2x^3 - 3$

17. $f(x) = (5 - 2x)e^x$

18. $f(x) = (x - 8)e^x$

19. $g(u) = u^{-1}e^u$

20. $f(x) = x^{-2}e^x$

21. $f(x) = x^{2/3} + x^{1/3}$

22. $f(x) = \dfrac{1}{2}x^{2/3} - x^{1/3}$

23. $g(x) = x^{2/3}(x^2 - 4)$

24. $f(x) = x^{1/3}(x^2 - 9)$

25. $f(x) = \dfrac{\ln x}{x^3}$

26. $h(x) = \dfrac{\ln x}{\sqrt{x^3}}$

27. $g(x) = |x^2 - 1|$

28. $f(x) = |x^2 - 4|$

29. $f(\theta) = \sin\theta - 2\cos\theta$

30. $f(x) = x + 2\sin x$

In Problems 31–38, the distance s of an object from the origin at time $t \geq 0$ (in seconds) is given. The motion is along a horizontal line with the positive direction to the right.

(a) *Determine the intervals during which the object moves to the right and the intervals during which it moves to the left.*

(b) *When does the object reverse direction?*

(c) *When is the velocity of the object increasing and when is it decreasing?*

(d) *Draw a figure to illustrate the motion of the object.*

(e) *Draw a figure to illustrate the velocity of the object.*

31. $s = t^2 - 2t + 3$

32. $s = 2t^2 + 8t - 7$

33. $s = 2t^3 + 6t^2 - 18t + 1$

34. $s = 3t^4 - 16t^3 + 24t^2$

35. $s = 2t - \dfrac{6}{t}, \quad t > 0$

36. $s = 3\sqrt{t} - \dfrac{1}{\sqrt{t}}, \quad t > 0$

37. $s = 2\sin(3t), \quad 0 \leq t \leq \dfrac{2\pi}{3}$

38. $s = 3\cos(\pi t), \quad 0 \leq t \leq 2$

In Problems 39–54, (a) determine the intervals on which each function is concave up and on which it is concave down;
(b) *find any points of inflection.*

39. $f(x) = x^2 - 2x + 5$

40. $f(x) = x^2 + 4x - 2$

41. $f(x) = x^3 - 9x^2 + 2$

42. $f(x) = x^3 - 6x^2 + 9x + 1$

43. $f(x) = x^4 - 4x^3 + 10$

44. $f(x) = 3x^4 - 8x^3 + 6x + 1$

45. $f(x) = x^{2/3}e^x$

46. $f(x) = x^{2/3}e^{-x}$

47. $f(x) = \dfrac{\ln x}{x^3}$

48. $f(x) = \dfrac{\ln x}{\sqrt{x^3}}$

49. $f(x) = x + \dfrac{1}{x}$

50. $f(x) = 2x^2 - \dfrac{1}{x}$

51. $f(x) = 3x^{1/3} + 2x$

52. $f(x) = x^{4/3} - 8x^{1/3}$

53. $f(x) = 3 - \dfrac{4}{x} + \dfrac{4}{x^2}$

54. $f(x) = (x - 1)^{3/2}$

In Problems 55–80:
(a) *Find the local extrema of f.*
(b) *Determine the intervals on which f is concave up and on which it is concave down.*
(c) *Find any points of inflection.*

55. $f(x) = 2x^3 - 6x^2 + 6x - 3$

56. $f(x) = 2x^3 + 9x^2 + 12x - 4$

57. $f(x) = x^4 - 4x$

58. $f(x) = x^4 + 4x$

59. $f(x) = 5x^4 - x^5$

60. $f(x) = 4x^6 + 6x^4$

61. $f(x) = 3x^5 - 20x^3$

62. $f(x) = 3x^5 + 5x^3$

63. $f(x) = x^2 e^x$

64. $f(x) = x^3 e^x$

65. $f(x) = \dfrac{e^x + e^{-x}}{2}$

66. $f(x) = \dfrac{e^x - e^{-x}}{2}$

67. $f(x) = 6x^{4/3} - 3x^{1/3}$

68. $f(x) = x^{2/3} - x^{1/3}$

69. $f(x) = x^{2/3}(x^2 - 8)$

70. $f(x) = x^{1/3}(x^2 - 2)$

71. $f(x) = x^2 - \ln x$

72. $f(x) = \ln x - x$

73. $f(x) = \dfrac{x}{(1 + x^2)^{5/2}}$

74. $f(x) = \dfrac{\sqrt{x}}{1 + x}$

75. $f(x) = x^2\sqrt{1 - x^2}$

76. $f(x) = x\sqrt{1 - x}$

77. $f(x) = \sin^2 x$

78. $f(x) = \cos^2 x$

79. $f(x) = x - 2\sin x, \quad 0 \leq x \leq 2\pi$

80. $f(x) = 2\cos^2 x - \sin^2 x, \quad 0 \leq x \leq 2\pi$

In Problems 81–84, find the local extrema of each function f by:
(a) *Using the First Derivative Test.*
(b) *Using the Second Derivative Test.*
(c) *Discuss which of the two tests you found easier.*

81. $f(x) = -2x^3 + 15x^2 - 36x + 7$

82. $f(x) = x^3 + 10x^2 + 25x - 25$

83. $f(x) = (x - 3)^2 e^x$

84. $f(x) = (x + 1)^2 e^{-x}$

Applications and Extensions

In Problems 85–96, sketch the graph of a continuous function f that has the given properties. Answers will vary.

85. *f is concave up on $(-\infty, \infty)$, increasing on $(-\infty, 0)$, decreasing on $(0, \infty)$, and $f(0) = 1$.*

86. *f is concave up on $(-\infty, 0)$, concave down on $(0, \infty)$, decreasing on $(-\infty, 0)$, increasing on $(0, \infty)$, and $f(0) = 1$.*

87. *f is concave down on $(-\infty, 1)$, concave up on $(1, \infty)$, decreasing on $(-\infty, 0)$, increasing on $(0, \infty)$, $f(0) = 1$, and $f(1) = 2$.*

88. *f is concave down on $(-\infty, 0)$, concave up on $(0, \infty)$, increasing on $(-\infty, \infty)$, and $f(0) = 1$ and $f(1) = 2$.*

89. $f'(x) > 0$ if $x < 0$; $f'(x) < 0$ if $x > 0$; $f''(x) > 0$ if $x < 0$; $f''(x) > 0$ if $x > 0$ and $f(0) = 1$.

90. $f'(x) > 0$ if $x < 0$; $f'(x) < 0$ if $x > 0$; $f''(x) > 0$ if $x < 0$; $f''(x) < 0$ if $x > 0$ and $f(0) = 1$.

91. $f''(0) = 0$; $f'(0) = 0$; $f''(x) > 0$ if $x < 0$; $f''(x) > 0$ if $x > 0$ and $f(0) = 1$.

92. $f''(0) = 0$; $f'(x) > 0$ if $x \neq 0$; $f''(x) < 0$ if $x < 0$; $f''(x) > 0$ if $x > 0$ and $f(0) = 1$.

93. $f'(0) = 0$; $f'(x) < 0$ if $x \neq 0$; $f''(x) > 0$ if $x < 0$; $f''(x) < 0$ if $x > 0$; $f(0) = 1$.

94. $f''(0) = 0$; $f'(0) = \frac{1}{2}$; $f''(x) > 0$ if $x < 0$;

$f''(x) < 0$ if $x > 0$ and $f(0) = 1$.

95. $f'(0)$ does not exist; $f''(x) > 0$ if $x < 0$; $f''(x) > 0$ if $x > 0$ and $f(0) = 1$.

96. $f'(0)$ does not exist; $f''(x) < 0$ if $x < 0$; $f''(x) > 0$ if $x > 0$ and $f(0) = 1$.

CAS *In Problems 97–100, for each function:*
(a) Determine the intervals on which f is concave up and on which it is concave down.
(b) Find any points of inflection.
(c) Graph f and describe the behavior of f at each inflection point.

97. $f(x) = e^{-(x-2)^2}$ **98.** $f(x) = x^2\sqrt{5-x}$

99. $f(x) = \dfrac{2-x}{2x^2 - 2x + 1}$ **100.** $f(x) = \dfrac{3x}{x^2 + 3x + 5}$

101. Critical Number Show that 0 is the only critical number of $f(x) = \sqrt[3]{x}$ and that f has no local extrema.

102. Critical Number Show that 0 is the only critical number of $f(x) = \sqrt[3]{x^2}$ and that f has a local minimum at 0.

103. Inflection Point For the function $f(x) = ax^3 + bx^2$, find a and b so that the point $(1, 6)$ is an inflection point of f.

104. Inflection Point For the cubic polynomial function $f(x) = ax^3 + bx^2 + cx + d$, find a, b, c, and d so that 0 is a critical number, $f(0) = 4$, and the point $(1, -2)$ is an inflection point of f.

105. Public Health In a town of 50,000 people, the number of people at time t who have influenza is $N(t) = \dfrac{10{,}000}{1 + 9999e^{-t}}$, where t is measured in days. Note that the flu is spread by the one person who has it at $t = 0$.

(a) Find the rate of change of the number of infected people.

(b) When is $N'(t)$ increasing? When is it decreasing?

(c) When is the rate of change of the number of infected people a maximum?

(d) Find any inflection points of $N(t)$.

(e) Interpret the result found in (d).

106. Business The profit P (in millions of dollars) generated by introducing a new technology is expected to follow the logistic function $P(t) = \dfrac{300}{1 + 50e^{-0.2t}}$, where t is the number of years after its release.

(a) When is annual profit increasing? When is it decreasing?

(b) Find the rate of change in profit.

(c) When is the rate of change in profit P' increasing? When is it decreasing?

(d) When is the rate of change in profit a maximum?

(e) Find any inflection points of $P(t)$.

(f) Interpret the result found in (e).

107. Population Model The following data represent the population of the United States:

Year	Population	Year	Population
1900	76,212,168	1960	179,323,175
1910	92,228,486	1970	203,302,031
1920	106,021,537	1980	226,542,203
1930	123,202,624	1990	248,709,873
1940	132,164,569	2000	281,421,906
1950	151,325,798	2010	308,745,538

An ecologist finds the data fit the logistic function

$$P(t) = \frac{679{,}021{,}144.9}{1 + 7.518e^{-0.0166t}}.$$

(a) Draw a scatterplot of the data using the years since 1900 as the independent variable and population as the dependent variable.

CAS (b) Verify that P is the logistic function of best fit.

(c) Find the rate of change in population.

(d) When is $P'(t)$ increasing? When is it decreasing?

(e) When is the rate of change in population a maximum?

(f) Find any inflection points of $P(t)$.

(g) Interpret the result found in (f).

Source: U.S. Census Bureau.

108. Biology The amount of yeast biomass in a culture after t hours is given in the table below.

Time (in hours)	Yeast Biomass	Time (in hours)	Yeast Biomass	Time (in hours)	Yeast Biomass
0	9.6	7	257.3	13	629.4
1	18.3	8	350.7	14	640.8
2	29.0	9	441.0	15	651.1
3	47.2	10	513.3	16	655.9
4	71.1	11	559.7	17	659.6
5	119.1	12	594.8	18	661.8
6	174.6				

Source: Tor Carlson, *Uber Geschwindigkeit und Grosse der Hefevermehrung in Wurrze, Biochemische Zeitschrift, Bd. 57*, 1913, pp. 313–334.

The logistic function $y = \dfrac{663.0}{1 + 71.6e^{-0.5470t}}$, where t is time, models the data.

(a) Draw a scatterplot of the data using time t as the independent variable.

CAS (b) Verify that y is the logistic function of best fit.

(c) Find the rate of change in biomass.

(d) When is $y'(t)$ increasing? When is it decreasing?

(e) When is the rate of change in the biomass a maximum?

(f) Find any inflection points of $y(t)$.

(g) Interpret the result found in (f).

109. U.S. Budget The United States budget documents the amount of money (revenue) the federal government takes in (through taxes, for example) and the amount (expenses) it pays out (for social programs, defense, etc.). When revenue exceeds expenses, we say there is a **budget surplus**, and when expenses exceed revenue, there is a **budget deficit**. The function

$$B = B(t) = -12.8t^3 + 163.4t^2 - 614.0t + 390.6$$

where $0 \le t \le 9$ approximates revenue minus expenses for the years 2000 to 2009, with $t = 0$ representing the year 2000 and B in billions of dollars.

(a) Find all the local extrema of B. (Round the answers to two decimal places.)

(b) Do the local extreme values represent a budget surplus or a budget deficit?

(c) Find the intervals on which B is concave up or concave down. Identify any inflection points of B.

(d) What does the concavity on either side of the inflection point(s) indicate about the rate of change of the budget? Is it increasing at an increasing rate? Increasing at a decreasing rate?

(CAS) (e) Graph the function B. Given that the total amount the government paid out in 2011 was 3.6 trillion dollars, does B seem to be an accurate predictor for the budget for the years 2010 and beyond?

110. If $f(x) = ax^3 + bx^2 + cx + d$, $a \ne 0$, how does the quantity $b^2 - 3ac$ determine the number of potential local extrema?

111. If $f(x) = ax^3 + bx^2 + cx + d$, $a \ne 0$, find a, b, c, and d so that f has a local minimum at 0, a local maximum at 4, and the graph contains the points $(0, 5)$ and $(4, 33)$.

112. Find the local minimum of the function

$$f(x) = \frac{2}{x} + \frac{8}{1-x}, \quad 0 < x < 1.$$

113. Find the local extrema and the inflection points of $y = \sqrt{3}\sin x + \cos x$, $0 \le x \le 2\pi$.

114. If $x > 0$ and $n > 1$, can the expression $x^n - n(x-1) - 1$ ever be negative?

115. Why must the First Derivative Test be used to find the local extreme values of the function $f(x) = x^{2/3}$?

116. Put It Together Which of the following is true of the function

$$f(x) = x^2 + e^{-2x}$$

(a) f is increasing (b) f is decreasing (c) f is discontinuous at 0

(d) f is concave up (e) f is concave down

117. Put It Together If a function f is continuous for all x and if f has a local maximum at $(-1, 4)$ and a local minimum at $(3, -2)$, which of the following statements must be true?

(a) The graph of f has an inflection point somewhere between $x = -1$ and $x = 3$.

(b) $f'(-1) = 0$.

(c) The graph of f has a horizontal asymptote.

(d) The graph of f has a horizontal tangent line at $x = 3$.

(e) The graph of f intersects both axes.

118. Vertex of a Parabola If $f(x) = ax^2 + bx + c$, $a \ne 0$, prove that f has a local maximum at $-\dfrac{b}{2a}$ if $a < 0$ and has a local minimum at $-\dfrac{b}{2a}$ if $a > 0$.

119. Show that $\sin x \le x$, $0 \le x \le 2\pi$. (*Hint:* Let $f(x) = x - \sin x$.)

120. Show that $1 - \dfrac{x^2}{2} \le \cos x$, $0 \le x \le 2\pi$. (*Hint:* Use the result of Problem 119.)

121. Show that $2\sqrt{x} > 3 - \dfrac{1}{x}$, for $x > 1$.

122. Use calculus to show that $x^2 - 8x + 21 > 0$ for all x.

123. Use calculus to show that $3x^4 - 4x^3 - 12x^2 + 40 > 0$ for all x.

124. Show that the function $f(x) = ax^2 + bx + c$, $a \ne 0$, has no inflection points. For what values of a is f concave up? For what values of a is f concave down?

125. Show that every polynomial function of degree 3 $f(x) = ax^3 + bx^2 + cx + d$ has exactly one inflection point.

126. Prove that a polynomial of degree $n \ge 3$ has at most $(n-1)$ critical numbers and at most $(n-2)$ inflection points.

127. Show that the function $f(x) = (x-a)^n$, a a constant, has exactly one inflection point if $n \ge 3$ is an odd integer.

128. Show that the function $f(x) = (x-a)^n$, a a constant, has no inflection point if $n \ge 2$ is an even integer.

129. Show that the function $f(x) = \dfrac{ax+b}{ax+d}$ has no critical points and no inflection points.

130. First Derivative Test Proof Let f be a function that is continuous on some interval I. Suppose c is a critical number of f and (a, b) is some open interval in I containing c. Prove that if $f'(x) < 0$ for $a < x < c$ and $f'(x) > 0$ for $c < x < b$, then $f(c)$ is a local minimum value.

131. First Derivative Test Proof Let f be a function that is continuous on some interval I. Suppose c is a critical number of f and (a, b) is some open interval in I containing c. Prove that if $f'(x)$ has the same sign on both sides of c, then $f(c)$ is neither a local maximum value nor a local minimum value.

132. Test of Concavity Proof Let f denote a function that is continuous on a closed interval $[a, b]$. Suppose f' and f'' exist on the open interval (a, b). Prove that if $f''(x) < 0$ on the interval (a, b), then f is concave down on (a, b).

133. Second Derivative Test Proof Let f be a function for which f' and f'' exist on an open interval (a, b). Suppose c lies in (a, b) and is a critical number of f. Prove that if $f''(c) > 0$, then $f(c)$ is a local minimum value.

Challenge Problems

134. Find the inflection point of $y = (x+1)\tan^{-1} x$.

135. Bernoulli's Inequality Prove **Bernoulli's inequality**: $(1+x)^n > 1 + nx$ for $x > -1$, $x \ne 0$, and $n > 1$. (*Hint:* Let $f(x) = (1+x)^n - (1+nx)$.)

4.5 Indeterminate Forms and L'Hôpital's Rule

OBJECTIVES *When you finish this section, you should be able to:*

1 Identify indeterminate forms of the type $\dfrac{0}{0}$ and $\dfrac{\infty}{\infty}$ (p. 298)

2 Use L'Hôpital's Rule to find a limit (p. 299)

3 Find the limit of an indeterminate form of the type $0 \cdot \infty$, $\infty - \infty$, 0^0, 1^∞, or ∞^0 (p. 303)

NEED TO REVIEW? Properties of limits are discussed in Section 1.2, pp. 87–88.

In this section, we reexamine the limit of a quotient and explore what we can do when previous limit theorems cannot be used.

1 Identify Indeterminate Forms of the Type $\dfrac{0}{0}$ and $\dfrac{\infty}{\infty}$

To find $\lim\limits_{x \to c} \dfrac{f(x)}{g(x)}$ we usually first try to use the Limit of a Quotient:

$$\lim_{x \to c} \frac{f(x)}{g(x)} = \frac{\lim\limits_{x \to c} f(x)}{\lim\limits_{x \to c} g(x)} \tag{1}$$

But this result cannot always be used. For example, to find $\lim\limits_{x \to 2} \dfrac{x^2 - 4}{x - 2}$, we cannot use equation (1), since the numerator and the denominator each approach 0 (resulting in the form $\dfrac{0}{0}$). Instead, we use algebra and obtain

$$\lim_{x \to 2} \frac{x^2 - 4}{x - 2} = \lim_{x \to 2} \frac{(x - 2)(x + 2)}{x - 2} = \lim_{x \to 2} (x + 2) = 4$$

NEED TO REVIEW? Limits at infinity and infinite limits are discussed in Section 1.5, pp. 117–125.

To find $\lim\limits_{x \to \infty} \dfrac{3x - 2}{x + 5}$, we cannot use equation (1), since the numerator and the denominator each become unbounded (resulting in the form $\dfrac{\infty}{\infty}$). Instead, we divide the numerator and denominator by x. Then

$$\lim_{x \to \infty} \frac{3x - 2}{x + 5} = \lim_{x \to \infty} \frac{3 - \dfrac{2}{x}}{1 + \dfrac{5}{x}} = \frac{\lim\limits_{x \to \infty} \left(3 - \dfrac{2}{x}\right)}{\lim\limits_{x \to \infty} \left(1 + \dfrac{5}{x}\right)} = 3$$

As a third example, to find $\lim\limits_{x \to 0} \dfrac{\sin x}{x}$, we cannot use equation (1) since it leads to the form $\dfrac{0}{0}$. Instead, we use a geometric argument (the Squeeze Theorem) to show that

$$\lim_{x \to 0} \frac{\sin x}{x} = 1$$

NOTE The word "indeterminate" conveys the idea that the limit cannot be found without additional work.

Whenever using $\lim\limits_{x \to c} \dfrac{f(x)}{g(x)} = \dfrac{\lim\limits_{x \to c} f(x)}{\lim\limits_{x \to c} g(x)}$ leads to the form $\dfrac{0}{0}$ or $\dfrac{\infty}{\infty}$, we say that $\dfrac{f}{g}$ is an *indeterminate form at c*. There are other indeterminate forms that we discuss in Objective 3.

DEFINITION Indeterminate Form at c of the Type $\dfrac{0}{0}$ or the Type $\dfrac{\infty}{\infty}$

If the functions f and g are each defined in an open interval containing the number c, except possibly at c, then the quotient $\dfrac{f(x)}{g(x)}$ is called an **indeterminate form at c of the type $\dfrac{0}{0}$** if

$$\lim_{x \to c} f(x) = 0 \qquad \text{and} \qquad \lim_{x \to c} g(x) = 0$$

and an **indeterminate form at c of the type $\dfrac{\infty}{\infty}$** if

$$\lim_{x \to c} f(x) = \pm\infty \qquad \text{and} \qquad \lim_{x \to c} g(x) = \pm\infty$$

NOTE $\dfrac{0}{0}$ and $\dfrac{\infty}{\infty}$ are symbols used to denote an indeterminate form.

These definitions also hold for limits at infinity.

EXAMPLE 1 Identifying Indeterminate Forms of the Types $\dfrac{0}{0}$ and $\dfrac{\infty}{\infty}$

(a) $\dfrac{\cos(3x) - 1}{2x}$ is an indeterminate form at 0 of the type $\dfrac{0}{0}$ since

$$\lim_{x \to 0}[\cos(3x) - 1] = 0 \text{ and } \lim_{x \to 0}(2x) = 0$$

(b) $\dfrac{x - 1}{x^2 + 2x - 3}$ is an indeterminate form at 1 of the type $\dfrac{0}{0}$ since

$$\lim_{x \to 1}(x - 1) = 0 \text{ and } \lim_{x \to 1}(x^2 + 2x - 3) = 0$$

(c) $\dfrac{x^2 - 2}{x - 3}$ is not an indeterminate form at 3 of the type $\dfrac{0}{0}$ since $\lim_{x \to 3}(x^2 - 2) \neq 0$.

(d) $\dfrac{x^2}{e^x}$ is an indeterminate form at ∞ of the type $\dfrac{\infty}{\infty}$ since

$$\lim_{x \to \infty} x^2 = \infty \text{ and } \lim_{x \to \infty} e^x = \infty$$ ∎

NOW WORK Problem 7.

2 Use L'Hôpital's Rule to Find a Limit

A theorem, named after the French mathematician Guillaume François de L'Hôpital (pronounced "low-p-tal"), provides a method for finding the limit of an indeterminate form.

THEOREM L'Hôpital's Rule

Suppose the functions f and g are differentiable on an open interval I containing the number c, except possibly at c, and $g'(x) \neq 0$ for all $x \neq c$ in I. Let L denote either a real number or $\pm\infty$, and suppose $\dfrac{f(x)}{g(x)}$ is an indeterminate form at c of the type $\dfrac{0}{0}$ or $\dfrac{\infty}{\infty}$. If $\lim\limits_{x \to c} \dfrac{f'(x)}{g'(x)} = L$, then $\lim\limits_{x \to c} \dfrac{f(x)}{g(x)} = L$.

L'Hôpital's Rule is also valid for limits at infinity and one-sided limits. A partial proof of L'Hôpital's rule is given in Appendix B. A limited proof is given here that assumes f' and g' are continuous at c and $g'(c) \neq 0$.

Proof Suppose $\lim\limits_{x \to c} f(x) = 0$ and $\lim\limits_{x \to c} g(x) = 0$. Then $f(c) = 0$ and $g(c) = 0$. Since both f' and g' are continuous at c and $g'(c) \neq 0$, we have

$$\lim_{x \to c} \frac{f'(x)}{g'(x)} \underset{\substack{\uparrow \\ \text{Quotient} \\ \text{Property}}}{=} \frac{\lim\limits_{x \to c} f'(x)}{\lim\limits_{x \to c} g'(x)} \underset{\substack{\uparrow \\ f', \ g' \\ \text{continuous}}}{=} \frac{f'(c)}{g'(c)} \underset{\substack{\uparrow \\ \text{Definition of} \\ \text{a derivative}}}{=} \frac{\lim\limits_{x \to c} \dfrac{f(x) - f(c)}{x - c}}{\lim\limits_{x \to c} \dfrac{g(x) - g(c)}{x - c}}$$

$$= \lim_{x \to c} \frac{\dfrac{f(x) - f(c)}{x - c}}{\dfrac{g(x) - g(c)}{x - c}} \underset{\substack{\uparrow \\ f(c) = 0 \\ g(c) = 0}}{=} \lim_{x \to c} \frac{f(x)}{g(x)} \qquad \blacksquare$$

ORIGINS Guillaume François de L'Hôpital (1661–1704) was a French nobleman. When he was 30, he hired Johann Bernoulli to tutor him in calculus. Several years later, he entered into a deal with Bernoulli. L'Hôpital paid Bernoulli an annual sum for Bernoulli to share his mathematical discoveries with him but no one else. In 1696 L'Hôpital published the first textbook on differential calculus. It was immensely popular, the last edition being published in 1781 (seventy-seven years after L'Hôpital's death). The book included the rule we study here. After L'Hôpital's death, Johann Bernoulli made his deal with L'Hôpital public and claimed that L'Hôpital's textbook was his own material. His position was dismissed because he often made such claims. However, in 1921, the manuscripts of Bernoulli's lectures to L'Hôpital were found, showing L'Hôpital's calculus book was indeed largely Johann Bernoulli's work.

Steps for Finding a Limit Using L'Hôpital's Rule

Step 1 Check that $\dfrac{f}{g}$ is an indeterminate form at c of the type $\dfrac{0}{0}$ or $\dfrac{\infty}{\infty}$. If it is not, do not use L'Hôpital's Rule.

Step 2 Differentiate f and g separately.

Step 3 Find $\lim\limits_{x \to c} \dfrac{f'(x)}{g'(x)}$. This limit is equal to $\lim\limits_{x \to c} \dfrac{f(x)}{g(x)}$, provided the limit is a number or ∞ or $-\infty$.

Step 4 If $\dfrac{f'}{g'}$ is an indeterminate form at c, repeat the process.

EXAMPLE 2   Using L'Hôpital's Rule to Find a Limit

Find $\lim\limits_{x \to 0} \dfrac{\tan x}{6x}$.

Solution We follow the steps for finding a limit using L'Hôpital's Rule.

Step 1 Since $\lim\limits_{x \to 0} \tan x = 0$ and $\lim\limits_{x \to 0} (6x) = 0$, the quotient $\dfrac{\tan x}{6x}$ is an indeterminate form at 0 of the type $\dfrac{0}{0}$.

Step 2 $\dfrac{d}{dx} \tan x = \sec^2 x$ and $\dfrac{d}{dx} (6x) = 6$.

Step 3 $\lim\limits_{x \to 0} \dfrac{\dfrac{d}{dx} \tan x}{\dfrac{d}{dx} (6x)} = \lim\limits_{x \to 0} \dfrac{\sec^2 x}{6} = \dfrac{1}{6}$.

It follows from L'Hôpital's Rule that $\lim\limits_{x \to 0} \dfrac{\tan x}{6x} = \dfrac{1}{6}$. $\blacksquare$

In the solution of Example 2, we were careful to determine that the limit of the ratio of the derivatives, that is, $\lim\limits_{x \to 0} \dfrac{\sec^2 x}{6}$, existed or became infinite before using L'Hôpital's Rule. However, the usual practice is to combine Step 2 and Step 3 as follows:

$$\lim_{x \to 0} \frac{\tan x}{6x} = \lim_{x \to 0} \frac{\dfrac{d}{dx} \tan x}{\dfrac{d}{dx} (6x)} = \lim_{x \to 0} \frac{\sec^2 x}{6} = \frac{1}{6}$$

At times, it is necessary to use L'Hôpital's Rule more than once.

EXAMPLE 3   Using L'Hôpital's Rule to Find a Limit

Find $\lim\limits_{x \to 0} \dfrac{\sin x - x}{x^2}$.

Solution We use the steps for finding a limit using L'Hôpital's Rule.

Step 1 Since $\lim\limits_{x \to 0} (\sin x - x) = 0$ and $\lim\limits_{x \to 0} x^2 = 0$, the expression $\dfrac{\sin x - x}{x^2}$ is an indeterminate form at 0 of the type $\dfrac{0}{0}$.

Steps 2 and 3 We use L'Hôpital's Rule.

$$\lim_{x \to 0} \frac{\sin x - x}{x^2} \underset{\substack{\uparrow \\ \text{L'Hôpital's} \\ \text{Rule}}}{=} \lim_{x \to 0} \frac{\dfrac{d}{dx}(\sin x - x)}{\dfrac{d}{dx}x^2} = \lim_{x \to 0} \frac{\cos x - 1}{2x} = \frac{1}{2} \lim_{x \to 0} \frac{\cos x - 1}{x}$$

Since $\lim\limits_{x \to 0}(\cos x - 1) = 0$ and $\lim\limits_{x \to 0} x = 0$, the expression $\dfrac{\cos x - 1}{x}$ is an indeterminate form at 0 of the type $\dfrac{0}{0}$. So, we use L'Hôpital's Rule again.

$$\lim_{x \to 0} \frac{\sin x - x}{x^2} = \frac{1}{2} \lim_{x \to 0} \frac{\cos x - 1}{x} \underset{\substack{\uparrow \\ \text{L'Hôpital's} \\ \text{Rule}}}{=} \frac{1}{2} \lim_{x \to 0} \frac{\dfrac{d}{dx}(\cos x - 1)}{\dfrac{d}{dx}x} = \frac{1}{2} \lim_{x \to 0} \frac{-\sin x}{1} = 0 \quad \blacksquare$$

NOW WORK Problem 29.

EXAMPLE 4   Using L'Hôpital's Rule to Find a Limit at Infinity

Find: **(a)** $\lim\limits_{x \to \infty} \dfrac{\ln x}{x}$ **(b)** $\lim\limits_{x \to \infty} \dfrac{x}{e^x}$ **(c)** $\lim\limits_{x \to \infty} \dfrac{e^x}{x}$

Solution (a) Since $\lim\limits_{x \to \infty} \ln x = \infty$ and $\lim\limits_{x \to \infty} x = \infty$, $\dfrac{\ln x}{x}$ is an indeterminate form at ∞ of the type $\dfrac{\infty}{\infty}$. Using L'Hôpital's Rule, we have

$$\lim_{x \to \infty} \frac{\ln x}{x} \underset{\substack{\uparrow \\ \text{L'Hôpital's} \\ \text{Rule}}}{=} \lim_{x \to \infty} \frac{\dfrac{d}{dx}\ln x}{\dfrac{d}{dx}x} = \lim_{x \to \infty} \frac{\dfrac{1}{x}}{1} = \lim_{x \to \infty} \frac{1}{x} = 0$$

(b) $\lim\limits_{x \to \infty} x = \infty$ and $\lim\limits_{x \to \infty} e^x = \infty$, so $\dfrac{x}{e^x}$ is an indeterminate form at ∞ of the type $\dfrac{\infty}{\infty}$. Using L'Hôpital's Rule, we have

$$\lim_{x \to \infty} \frac{x}{e^x} \underset{\substack{\uparrow \\ \text{L'Hôpital's} \\ \text{Rule}}}{=} \lim_{x \to \infty} \frac{\dfrac{d}{dx}x}{\dfrac{d}{dx}e^x} = \lim_{x \to \infty} \frac{1}{e^x} = 0$$

(c) From (b), we know that $\dfrac{e^x}{x}$ is an indeterminate form at ∞ of the type $\dfrac{\infty}{\infty}$. Using L'Hôpital's Rule, we have

$$\lim_{x \to \infty} \frac{e^x}{x} \underset{\substack{\uparrow \\ \text{L'Hôpital's} \\ \text{Rule}}}{=} \lim_{x \to \infty} \frac{\dfrac{d}{dx}e^x}{\dfrac{d}{dx}x} = \lim_{x \to \infty} \frac{e^x}{1} = \infty \quad \blacksquare$$

NOW WORK Problem 35.

The results from Example 4 tell us that $y = x$ grows faster than $y = \ln x$, and that $y = e^x$ grows faster than $y = x$. In fact, these inequalities are true for all positive powers of x. That is,

$$\lim_{x \to \infty} \frac{\ln x}{x^n} = 0 \qquad \lim_{x \to \infty} \frac{x^n}{e^x} = 0$$

for $n \geq 1$ an integer. You are asked to verify these results in Problems 95 and 96.

Sometimes simplifying first reduces the effort needed to find the limit.

EXAMPLE 5 Using L'Hôpital's Rule to Find a Limit

Find $\displaystyle\lim_{x \to 0} \frac{\tan x - \sin x}{x^2 \tan x}$.

Solution $\dfrac{\tan x - \sin x}{x^2 \tan x}$ is an indeterminate form at 0 of the type $\dfrac{0}{0}$. We simplify the expression before using L'Hôpital's Rule. Then it is easier to find the limit.

$$\frac{\tan x - \sin x}{x^2 \tan x} = \frac{\dfrac{\sin x}{\cos x} - \sin x}{x^2 \cdot \dfrac{\sin x}{\cos x}} = \frac{\dfrac{\sin x - \sin x \cos x}{\cos x}}{\dfrac{x^2 \sin x}{\cos x}} = \frac{\sin x\,(1 - \cos x)}{x^2 \sin x} = \frac{1 - \cos x}{x^2}$$

CAUTION Be sure to verify that a simplified form is an indeterminate form at c before using L'Hôpital's Rule.

Since $\dfrac{1 - \cos x}{x^2}$ is an indeterminate form at 0 of the type $\dfrac{0}{0}$, we use L'Hôpital's Rule.

$$\lim_{x \to 0} \frac{\tan x - \sin x}{x^2 \tan x} = \lim_{x \to 0} \frac{1 - \cos x}{x^2} \underset{\substack{\uparrow \\ \text{L'Hôpital's Rule}}}{=} \lim_{x \to 0} \frac{\dfrac{d}{dx}(1 - \cos x)}{\dfrac{d}{dx}x^2} = \lim_{x \to 0} \frac{\sin x}{2x}$$

$$= \frac{1}{2} \lim_{x \to 0} \frac{\sin x}{x} = \frac{1}{2} \qquad \lim_{x \to 0} \frac{\sin x}{x} = 1$$

Compare this solution to one that uses L'Hôpital's Rule at the start.

In Example 6, we use L'Hôpital's Rule to find a one-sided limit.

EXAMPLE 6 Using L'Hôpital's Rule to Find a One-Sided Limit

Find $\displaystyle\lim_{x \to 0^+} \frac{\cot x}{\ln x}$.

Solution Since $\displaystyle\lim_{x \to 0^+} \cot x = \infty$ and $\displaystyle\lim_{x \to 0^+} \ln x = -\infty$, $\dfrac{\cot x}{\ln x}$ is an indeterminate form at 0^+ of the type $\dfrac{\infty}{\infty}$. Using L'Hôpital's Rule, we find

$$\lim_{x \to 0^+} \frac{\cot x}{\ln x} \underset{\substack{\uparrow \\ \text{L'Hôpital's Rule}}}{=} \lim_{x \to 0^+} \frac{\dfrac{d}{dx}\cot x}{\dfrac{d}{dx}\ln x} = \lim_{x \to 0^+} \frac{-\csc^2 x}{\dfrac{1}{x}} \underset{\substack{\uparrow \\ \csc^2 x = \dfrac{1}{\sin^2 x}}}{=} -\lim_{x \to 0^+} \frac{x}{\sin^2 x} \underset{\substack{\uparrow \\ \text{L'Hôpital's Rule}}}{=} -\lim_{x \to 0^+} \frac{\dfrac{d}{dx}x}{\dfrac{d}{dx}\sin^2 x}$$

$$= -\lim_{x \to 0^+} \frac{1}{2 \sin x \cos x} = \left(-\frac{1}{2}\right) \left(\lim_{x \to 0^+} \frac{1}{\sin x}\right) \left(\lim_{x \to 0^+} \frac{1}{\cos x}\right)$$

$$= \left(-\frac{1}{2}\right) \left(\lim_{x \to 0^+} \frac{1}{\sin x}\right)(1) = -\infty$$

NOW WORK Problem 61.

3 Find the Limit of an Indeterminate Form of the Type $0 \cdot \infty$, $\infty - \infty$, 0^0, 1^∞, or ∞^0

L'Hôpital's Rule can only be used for indeterminate forms of the types $\dfrac{0}{0}$ and $\dfrac{\infty}{\infty}$. Indeterminate forms of the type $0 \cdot \infty$, $\infty - \infty$, 0^0, 1^∞, or ∞^0 need to be rewritten in one of the forms $\dfrac{0}{0}$ or $\dfrac{\infty}{\infty}$ before we can use L'Hôpital's Rule to find the limit.

Indeterminate Forms of the Type $0 \cdot \infty$

NOTE It might be tempting to argue that $0 \cdot \infty$ is 0, since "anything" times 0 is 0, but $0 \cdot \infty$ is not a product of numbers. That is, $0 \cdot \infty$ is not "zero times infinity"; it symbolizes "a quantity tending to zero" times "a quantity tending to infinity."

Suppose $\lim\limits_{x \to c} f(x) = 0$ and $\lim\limits_{x \to c} g(x) = \infty$. Then the product $f \cdot g$ is an indeterminate form at c of the type $0 \cdot \infty$. To find $\lim\limits_{x \to c} (f \cdot g)$, we rewrite the product $f \cdot g$ as one of the following quotients:

$$f \cdot g = \frac{f}{\dfrac{1}{g}} \qquad \text{or} \qquad f \cdot g = \frac{g}{\dfrac{1}{f}}$$

The right side of the equation on the left is an indeterminate form at c of the type $\dfrac{0}{0}$; the right side of the equation on the right is of the type $\dfrac{\infty}{\infty}$. Then we use L'Hôpital's Rule with one of these equations, choosing the one for which the derivatives are easier to find. If our first choice does not work, then we try the other one.

EXAMPLE 7   **Finding the Limit of an Indeterminate Form of the Type $0 \cdot \infty$**

Find:

(a) $\lim\limits_{x \to 0^+} (x \ln x)$ **(b)** $\lim\limits_{x \to \infty} \left(x \sin \dfrac{1}{x} \right)$

Solution (a) Since $\lim\limits_{x \to 0^+} x = 0$ and $\lim\limits_{x \to 0^+} \ln x = -\infty$, then $x \ln x$ is an indeterminate form at 0^+ of the type $0 \cdot \infty$. We change $x \ln x$ to an indeterminate form of the type $\dfrac{\infty}{\infty}$ by writing $x \ln x = \dfrac{\ln x}{\dfrac{1}{x}}$ and using L'Hôpital's Rule.

NOTE We choose to use $\dfrac{\ln x}{\dfrac{1}{x}}$ rather than $\dfrac{x}{\dfrac{1}{\ln x}}$ because it is easier to find the derivatives of $\ln x$ and $\dfrac{1}{x}$ than it is to find the derivatives of x and $\dfrac{1}{\ln x}$.

$$\lim\limits_{x \to 0^+} (x \ln x) = \lim\limits_{x \to 0^+} \frac{\ln x}{\dfrac{1}{x}} \underset{\substack{\uparrow \\ \text{L'Hôpital's Rule}}}{=} \lim\limits_{x \to 0^+} \frac{\dfrac{d}{dx} \ln x}{\dfrac{d}{dx} \dfrac{1}{x}} = \lim\limits_{x \to 0^+} \frac{\dfrac{1}{x}}{-\dfrac{1}{x^2}} \underset{\substack{\uparrow \\ \text{Simplify}}}{=} \lim\limits_{x \to 0^+} (-x) = 0$$

(b) Since $\lim\limits_{x \to \infty} x = \infty$ and $\lim\limits_{x \to \infty} \sin \dfrac{1}{x} = 0$, then $x \sin \dfrac{1}{x}$ is an indeterminate form at ∞ of the type $0 \cdot \infty$. We change $x \sin \dfrac{1}{x}$ to an indeterminate form of the type $\dfrac{0}{0}$ by writing

$$\lim\limits_{x \to \infty} x \sin \frac{1}{x} = \lim\limits_{x \to \infty} \frac{\sin \dfrac{1}{x}}{\dfrac{1}{x}} \underset{\substack{\uparrow \\ \text{Let } t = \frac{1}{x}}}{=} \lim\limits_{t \to 0^+} \frac{\sin t}{t} = 1$$

∎

NOTE Notice in the solution to (b) that we did not need to use L'Hôpital's Rule.

NOW WORK   Problem 45.

Indeterminate Forms of the Type $\infty - \infty$*

If the limit of a function results in the indeterminate form $\infty - \infty$, it is generally possible to rewrite the function as an indeterminate form of the type $\dfrac{0}{0}$ or $\dfrac{\infty}{\infty}$ by using algebra or trigonometry.

EXAMPLE 8 | **Finding the Limit of an Indeterminate Form of the Type $\infty - \infty$**

Find $\lim\limits_{x \to 0^+} \left(\dfrac{1}{x} - \dfrac{1}{\sin x} \right)$.

Solution Since $\lim\limits_{x \to 0^+} \dfrac{1}{x} = \infty$ and $\lim\limits_{x \to 0^+} \dfrac{1}{\sin x} = \infty$, then $\dfrac{1}{x} - \dfrac{1}{\sin x}$ is an indeterminate form at 0^+ of the type $\infty - \infty$. We rewrite the difference as a single fraction.

$$\lim_{x \to 0^+} \left(\frac{1}{x} - \frac{1}{\sin x} \right) = \lim_{x \to 0^+} \frac{\sin x - x}{x \sin x}$$

Then, $\dfrac{\sin x - x}{x \sin x}$ is an indeterminate form at 0^+ of the type $\dfrac{0}{0}$. Now we can use L'Hôpital's Rule.

$$\lim_{x \to 0^+} \left(\frac{1}{x} - \frac{1}{\sin x} \right) = \lim_{x \to 0^+} \frac{\sin x - x}{x \sin x} \underset{\uparrow}{=} \lim_{x \to 0^+} \frac{\dfrac{d}{dx}(\sin x - x)}{\dfrac{d}{dx}(x \sin x)} = \lim_{x \to 0^+} \frac{\cos x - 1}{x \cos x + \sin x}$$

L'Hôpital's Rule

$$\underset{\uparrow}{=} \lim_{x \to 0^+} \frac{\dfrac{d}{dx}(\cos x - 1)}{\dfrac{d}{dx}(x \cos x + \sin x)} = \lim_{x \to 0^+} \frac{-\sin x}{(-x \sin x + \cos x) + \cos x}$$

Type $\dfrac{0}{0}$; use L'Hôpital's Rule

$$= \lim_{x \to 0^+} \frac{\sin x}{x \sin x - 2 \cos x} = \frac{0}{-2} = 0$$ ∎

NOW WORK Problem 47.

Indeterminate Forms of the Type 1^∞, 0^0, or ∞^0

A function of the form $[f(x)]^{g(x)}$ may result in an indeterminate form of the type 1^∞, 0^0, or ∞^0. To find the limit of such a function, we let $y = [f(x)]^{g(x)}$ and take the natural logarithm of each side.

$$\ln y = \ln[f(x)]^{g(x)} = g(x) \ln f(x)$$

The expression on the right will then be an indeterminate form of the type $0 \cdot \infty$ and we find the limit using the method of Example 7.

NEED TO REVIEW? Properties of logarithms are discussed in Appendix A.1, pp. A-10 to A-11.

Steps for Finding $\lim\limits_{x \to c}[f(x)]^{g(x)}$ when $[f(x)]^{g(x)}$ is an Indeterminate Form at c of the Type 1^∞, 0^0, or ∞^0

Step 1 Let $y = [f(x)]^{g(x)}$ and take the natural logarithm of each side, obtaining $\ln y = g(x) \ln f(x)$.

Step 2 Find $\lim\limits_{x \to c} \ln y$.

Step 3 If $\lim\limits_{x \to c} \ln y = L$, then $\lim\limits_{x \to c} y = e^L$.

---

*The indeterminate form of the type $\infty - \infty$ is a convenient notation for any of the following: $\infty - \infty$, $-\infty - (-\infty)$, $\infty + (-\infty)$. Note that $\infty + \infty = \infty$ and $(-\infty) + (-\infty) = -\infty$ are not indeterminate forms.

These steps also can be used for limits at infinity and for one-sided limits.

EXAMPLE 9 **Finding the Limit of an Indeterminate Form of the Type 0^0**

Find $\lim\limits_{x \to 0^+} x^x$

Solution The expression x^x is an indeterminate form at 0^+ of the type 0^0. We follow the steps for finding $\lim\limits_{x \to c}[f(x)]^{g(x)}$.

Step 1 Let $y = x^x$. Then $\ln y = x \ln x$.

Step 2 $\lim\limits_{x \to 0^+} \ln y = \lim\limits_{x \to 0^+}(x \ln x) = 0$ [from Example 7(a)].

Step 3 Since $\lim\limits_{x \to 0^+} \ln y = 0$, $\lim\limits_{x \to 0^+} y = e^0 = 1$. ∎

> **CAUTION** Do not stop after finding $\lim\limits_{x \to c} \ln y (= L)$. Remember, we want to find $\lim\limits_{x \to c} y (= e^L)$.

NOW WORK Problem 51.

EXAMPLE 10 **Finding the Limit of an Indeterminate Form of the Type 1^∞**

Find $\lim\limits_{x \to 0^+}(1 + x)^{1/x}$.

Solution The expression $(1 + x)^{1/x}$ is an indeterminate form at 0^+ of the type 1^∞.

Step 1 Let $y = (1 + x)^{1/x}$. Then $\ln y = \dfrac{1}{x} \ln(1 + x)$.

Step 2 $\lim\limits_{x \to 0^+} \ln y = \lim\limits_{x \to 0^+} \dfrac{\ln(1 + x)}{x} = \lim\limits_{x \to 0^+} \dfrac{\dfrac{d}{dx} \ln(1 + x)}{\dfrac{d}{dx} x} = \lim\limits_{x \to 0^+} \dfrac{\dfrac{1}{1 + x}}{1} = 1$

Type $\dfrac{0}{0}$; use L'Hôpital's Rule

Step 3 Since $\lim\limits_{x \to 0^+} \ln y = 1$, $\lim\limits_{x \to 0^+} y = e^1 = e$. ∎

NOW WORK Problem 85.

4.5 Assess Your Understanding

Concepts and Vocabulary

1. *True or False* $\dfrac{f(x)}{g(x)}$ is an indeterminate form at c of the type $\dfrac{0}{0}$ if $\lim\limits_{x \to c} \dfrac{f(x)}{g(x)}$ does not exist.

2. *True or False* If $\dfrac{f(x)}{g(x)}$ is an indeterminate form at c of the type $\dfrac{0}{0}$, then L'Hôpital's Rule states that $\lim\limits_{x \to c} \dfrac{f(x)}{g(x)} = \lim\limits_{x \to c}\left[\dfrac{d}{dx}\left(\dfrac{f(x)}{g(x)}\right)\right]$.

3. *True or False* $\dfrac{1}{x}$ is an indeterminate form at 0.

4. *True or False* $x \ln x$ is not an indeterminate form at 0^+ because $\lim\limits_{x \to 0^+} x = 0$ and $\lim\limits_{x \to 0^+} \ln x = -\infty$, and $0 \cdot -\infty = 0$.

5. In your own words, explain why $\infty - \infty$ is an indeterminate form, but $\infty + \infty$ is not an indeterminate form.

6. In your own words, explain why $0 \cdot \infty \neq 0$.

Skill Building

In Problems 7–26:

(a) *Determine whether each expression is an indeterminate form at c.*

(b) *If it is, identify the type. If it is not an indeterminate form, say why.*

7. $\dfrac{1 - e^x}{x}$, $c = 0$

8. $\dfrac{1 - e^x}{x - 1}$, $c = 0$

9. $\dfrac{e^x}{x}$, $c = 0$

10. $\dfrac{e^x}{x}$, $c = \infty$

11. $\dfrac{\ln x}{x^2}$, $c = \infty$

12. $\dfrac{\ln(x + 1)}{e^x - 1}$, $c = 0$

13. $\dfrac{\sec x}{x}$, $c = 0$

14. $\dfrac{x}{\sec x - 1}$, $c = 0$

15. $\dfrac{\sin x(1 - \cos x)}{x^2}$, $c = 0$

16. $\dfrac{\sin x - 1}{\cos x}$, $c = \dfrac{\pi}{2}$

17. $\dfrac{\tan x - 1}{\sin(4x - \pi)}$, $c = \dfrac{\pi}{4}$

18. $\dfrac{e^x - e^{-x}}{1 - \cos x}$, $c = 0$

---

1. = NOW WORK problem   📈 = Graphing technology recommended   [CAS] = Computer Algebra System recommended

19. $x^2 e^{-x}$, $c = \infty$

20. $x \cot x$, $c = 0$

21. $\csc \dfrac{x}{2} - \cot \dfrac{x}{2}$, $c = 0$

22. $\dfrac{x}{x-1} + \dfrac{1}{\ln x}$, $c = 1$

23. $\left(\dfrac{1}{x^2}\right)^{\sin x}$, $c = 0$

24. $(e^x + x)^{1/x}$, $c = 0$

25. $(x^2 - 1)^x$, $c = 0$

26. $(\sin x)^x$, $c = 0$

In Problems 27–42, identify each quotient as an indeterminate form of the type $\dfrac{0}{0}$ or $\dfrac{\infty}{\infty}$. Then find the limit.

27. $\lim\limits_{x \to 2} \dfrac{x^2 + x - 6}{x^2 - 3x + 2}$

28. $\lim\limits_{x \to 1} \dfrac{2x^3 + 5x^2 - 4x - 3}{x^3 + x^2 - 10x + 8}$

29. $\lim\limits_{x \to 1} \dfrac{\ln x}{x^2 - 1}$

30. $\lim\limits_{x \to 0} \dfrac{\ln(1-x)}{e^x - 1}$

31. $\lim\limits_{x \to 0} \dfrac{e^x - e^{-x}}{\sin x}$

32. $\lim\limits_{x \to 0} \dfrac{\tan(2x)}{\ln(1+x)}$

33. $\lim\limits_{x \to 1} \dfrac{\sin(\pi x)}{x - 1}$

34. $\lim\limits_{x \to \pi} \dfrac{1 + \cos x}{\sin(2x)}$

35. $\lim\limits_{x \to \infty} \dfrac{x^2}{e^x}$

36. $\lim\limits_{x \to \infty} \dfrac{e^x}{x^4}$

37. $\lim\limits_{x \to \infty} \dfrac{\ln x}{e^x}$

38. $\lim\limits_{x \to \infty} \dfrac{x + \ln x}{x \ln x}$

39. $\lim\limits_{x \to 0} \dfrac{e^x - 1 - \sin x}{1 - \cos x}$

40. $\lim\limits_{x \to 0} \dfrac{e^x - e^{-x} - 2\sin x}{3x^3}$

41. $\lim\limits_{x \to 0} \dfrac{\sin x - x}{x^3}$

42. $\lim\limits_{x \to 0} \dfrac{x^3}{\cos x - 1}$

In Problems 43–58, identify each expression as an indeterminate form of the type $0 \cdot \infty$, $\infty - \infty$, 0^0, 1^∞, or ∞^0. Then find the limit.

43. $\lim\limits_{x \to 0^+} (x^2 \ln x)$

44. $\lim\limits_{x \to \infty} (x e^{-x})$

45. $\lim\limits_{x \to \infty} [x(e^{1/x} - 1)]$

46. $\lim\limits_{x \to \pi/2} [(1 - \sin x)\tan x]$

47. $\lim\limits_{x \to \pi/2} (\sec x - \tan x)$

48. $\lim\limits_{x \to 0} \left(\cot x - \dfrac{1}{x}\right)$

49. $\lim\limits_{x \to 1} \left(\dfrac{1}{\ln x} - \dfrac{x}{\ln x}\right)$

50. $\lim\limits_{x \to 0} \left(\dfrac{1}{x} - \dfrac{1}{e^x - 1}\right)$

51. $\lim\limits_{x \to 0^+} (2x)^{3x}$

52. $\lim\limits_{x \to 0} x^{x^2}$

53. $\lim\limits_{x \to \infty} (x+1)^{e^{-x}}$

54. $\lim\limits_{x \to \infty} (1 + x^2)^{1/x}$

55. $\lim\limits_{x \to 0^+} (\csc x)^{\sin x}$

56. $\lim\limits_{x \to \infty} x^{1/x}$

57. $\lim\limits_{x \to \pi/2^-} (\sin x)^{\tan x}$

58. $\lim\limits_{x \to 0} (\cos x)^{1/x}$

In Problems 59–90, find each limit.

59. $\lim\limits_{x \to 0^+} \dfrac{\cot x}{\cot(2x)}$

60. $\lim\limits_{x \to \infty} \dfrac{\ln(\ln x)}{\ln x}$

61. $\lim\limits_{x \to 1/2^-} \dfrac{\ln(1 - 2x)}{\tan(\pi x)}$

62. $\lim\limits_{x \to 1^-} \dfrac{\ln(1 - x)}{\cot(\pi x)}$

63. $\lim\limits_{x \to \infty} \dfrac{x^4 + x^3}{e^x + 1}$

64. $\lim\limits_{x \to \infty} \dfrac{x^2 + x - 1}{e^x + e^{-x}}$

65. $\lim\limits_{x \to 0} \dfrac{x e^{4x} - x}{1 - \cos(2x)}$

66. $\lim\limits_{x \to 0} \dfrac{x \tan x}{1 - \cos x}$

67. $\lim\limits_{x \to 0} \dfrac{\tan^{-1} x}{x}$

68. $\lim\limits_{x \to 0} \dfrac{\tan^{-1} x}{\sin^{-1} x}$

69. $\lim\limits_{x \to 0} \dfrac{\cos x - 1}{\cos(2x) - 1}$

70. $\lim\limits_{x \to 0} \dfrac{\tan x - \sin x}{x^3}$

71. $\lim\limits_{x \to 0^+} (x^{1/2} \ln x)$

72. $\lim\limits_{x \to \infty} [(x - 1)e^{-x^2}]$

73. $\lim\limits_{x \to \pi/2} [\tan x \ln(\sin x)]$

74. $\lim\limits_{x \to 0^+} [\sin x \ln(\sin x)]$

75. $\lim\limits_{x \to 0} [\csc x \ln(x + 1)]$

76. $\lim\limits_{x \to \pi/4} [(1 - \tan x)\sec(2x)]$

77. $\lim\limits_{x \to a} \left[(a^2 - x^2)\tan\left(\dfrac{\pi x}{2a}\right)\right]$

78. $\lim\limits_{x \to 1^+} \left[(1 - x)\tan\left(\dfrac{1}{2}\pi x\right)\right]$

79. $\lim\limits_{x \to 1} \left(\dfrac{1}{\ln x} - \dfrac{1}{x - 1}\right)$

80. $\lim\limits_{x \to 1} \left(\dfrac{x}{x - 1} - \dfrac{1}{\ln x}\right)$

81. $\lim\limits_{x \to \pi/2} \left(x \tan x - \dfrac{\pi}{2}\sec x\right)$

82. $\lim\limits_{x \to \pi} (\cot x - x \csc x)$

83. $\lim\limits_{x \to 1^-} (1 - x)^{\tan(\pi x)}$

84. $\lim\limits_{x \to 0^+} x^{\sqrt{x}}$

85. $\lim\limits_{x \to 0} \left(\dfrac{\sin x}{x}\right)^{1/x}$

86. $\lim\limits_{x \to \infty} \left(1 + \dfrac{5}{x} + \dfrac{3}{x^2}\right)^x$

87. $\lim\limits_{x \to (\pi/2)^-} (\tan x)^{\cos x}$

88. $\lim\limits_{x \to 0^+} (x^2 + x)^{-\ln x}$

89. $\lim\limits_{x \to 0} (\cosh x)^{e^x}$

90. $\lim\limits_{x \to 0} (\sinh x)^x$

Applications and Extensions

91. Wolf Population In 2002 there were 65 wolves in Wyoming outside of Yellowstone National Park, and in 2010 there were 247 wolves. Suppose the population w of wolves in the region at time t follows the logistic growth curve

$$w = w(t) = \dfrac{K e^{rt}}{\dfrac{K}{40} + e^{rt} - 1}$$

where $K = 366$, $r = 0.283$, and $t = 0$ represents the population in the year 2000.

(a) Find $\lim\limits_{t \to \infty} w(t)$.

(b) Interpret the answer found in (a).

(c) Use graphing technology to graph $w = w(t)$.

92. Skydiving The downward velocity v of a skydiver with nonlinear air resistance can be modeled by

$$v = v(t) = -A + RA \dfrac{e^{Bt+C} - 1}{e^{Bt+C} + 1}$$

where t is the time in seconds, and A, B, C, and R are positive constants with $R > 1$.

(a) Find $\lim\limits_{t \to \infty} v(t)$.

(b) Interpret the limit found in (a).

(c) If the velocity v is measured in feet per second, reasonable values of the constants are $A = 108.6$, $B = 0.554$,

$C = 0.804$, and $R = 2.62$. Graph the velocity of the skydiver with respect to time.

93. Electricity The equation governing the amount of current I (in amperes) in a simple RL circuit consisting of a resistance R (in ohms), an inductance L (in henrys), and an electromotive force E (in volts) is

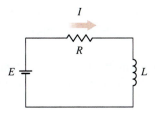

$$I = \frac{E}{R}(1 - e^{-Rt/L}).$$

(a) Find $\lim\limits_{t \to \infty} I(t)$ and $\lim\limits_{R \to 0^+} I(t)$.

(b) Interpret these limits.

94. Find $\lim\limits_{x \to 0} \dfrac{a^x - b^x}{x}$, where $a \neq 1$ and $b \neq 1$ are positive real numbers.

95. Show that $\lim\limits_{x \to \infty} \dfrac{\ln x}{x^n} = 0$, for $n \geq 1$ an integer.

96. Show that $\lim\limits_{x \to \infty} \dfrac{x^n}{e^x} = 0$ for $n \geq 1$ an integer.

97. Show that $\lim\limits_{x \to 0^+} (\cos x + 2 \sin x)^{\cot x} = e^2$.

98. Find $\lim\limits_{x \to \infty} \dfrac{P(x)}{e^x}$, where P is a polynomial function.

99. Find $\lim\limits_{x \to \infty} [\ln(x + 1) - \ln(x - 1)]$.

100. Show that $\lim\limits_{x \to 0^+} \dfrac{e^{-1/x^2}}{x} = 0$.   *Hint:* Write $\dfrac{e^{-1/x^2}}{x} = \dfrac{\frac{1}{x}}{e^{1/x^2}}$.

101. If n is an integer, show that $\lim\limits_{x \to 0^+} \dfrac{e^{-1/x^2}}{x^n} = 0$.

102. Show that $\lim\limits_{x \to \infty} \sqrt[x]{x} = 1$.

103. Show that $\lim\limits_{x \to \infty} \left(1 + \dfrac{a}{x}\right)^x = e^a$, a any real number.

104. Show that $\lim\limits_{x \to \infty} \left(\dfrac{x + a}{x - a}\right)^x = e^{2a}$, $a \neq 0$.

105. (a) Show that the function below has a derivative at 0. What is $f'(0)$?

$$f(x) = \begin{cases} e^{-1/x^2} & \text{if } x \neq 0 \\ 0 & \text{if } x = 0 \end{cases}$$

(b) Graph f using a graphing utility.

106. If $a, b \neq 0$ and $c > 0$ are real numbers, show that

$$\lim\limits_{x \to c} \dfrac{x^a - c^a}{x^b - c^b} = \dfrac{a}{b} c^{a-b}.$$

107. Prove L'Hôpital's rule when $\dfrac{f(x)}{g(x)}$ is an indeterminate form at $-\infty$ of the type $\dfrac{0}{0}$.

Challenge Problems

108. Explain why L'Hôpital's Rule does not apply to $\lim\limits_{x \to 0} \dfrac{x^2 \sin \frac{1}{x}}{\sin x}$.

109. Find each limit:

(a) $\lim\limits_{x \to \infty} \left(1 + \dfrac{1}{x}\right)^{-x^2}$ (b) $\lim\limits_{x \to \infty} \left(1 + \dfrac{\ln a}{x}\right)^x$, $a > 1$

(c) $\lim\limits_{x \to \infty} \left(1 + \dfrac{1}{x}\right)^{x^2}$ (d) $\lim\limits_{x \to \infty} \left(1 + \dfrac{\sin x}{x}\right)^x$

(e) $\lim\limits_{x \to \infty} (e^x)^{-1/\ln x}$ (f) $\lim\limits_{x \to \infty} \left[\left(\dfrac{1}{a}\right)^x\right]^{-1/x}$, $0 < a < 1$

(g) $\lim\limits_{x \to \infty} x^{1/x}$ (h) $\lim\limits_{x \to \infty} (a^x)^{1/x}$, $a > 1$

(i) $\lim\limits_{x \to \infty} [(2 + \sin x)^x]^{1/x}$ (j) $\lim\limits_{x \to 0^+} x^{-1/\ln x}$

110. Find constants A, B, C, and D so that

$$\lim\limits_{x \to 0} \dfrac{\sin(Ax) + Bx + Cx^2 + Dx^3}{x^5} = \dfrac{4}{15}.$$

111. A function f has derivatives of all orders.

(a) Find $\lim\limits_{h \to 0} \dfrac{f(x + 2h) - 2 f(x + h) + f(x)}{h^2}$.

(b) Find $\lim\limits_{h \to 0} \dfrac{f(x + 3h) - 3 f(x + 2h) + 3 f(x + h) - f(x)}{h^3}$.

(c) Generalize parts (a) and (b).

112. The formulas in Problem 111 can be used to approximate derivatives. Approximate $f'(2)$, $f''(2)$, and $f'''(2)$ from the table. The data are for $f(x) = \ln x$. Compare the exact values with your approximations.

x	2.0	2.1	2.2	2.3	2.4
$f(x)$	0.6931	0.7419	0.7885	0.8329	0.8755

113. Consider the function $f(t, x) = \dfrac{x^{t+1} - 1}{t + 1}$, where $x > 0$ and $t \neq -1$.

(a) For x fixed at x_0, show that $\lim\limits_{t \to -1} f(t, x) = \ln x_0$.

(b) For x fixed, define a function $F(t, x)$, where $x > 0$, that is continuous so $F(t, x) = f(t, x)$ for all t.

(c) For t fixed, show that $\dfrac{d}{dx} F(t, x) = x^t$, for $x > 0$ and all t.

Source: Michael W. Ecker (2012, September), Unifying Results via L'Hôpital's Rule. *Journal of the American Mathematical Association of Two Year Colleges*, 4(1) pp. 9–10.

4.6 Using Calculus to Graph Functions

OBJECTIVE *When you finish this section, you should be able to:*

1 Graph a function using calculus (p. 308)

In precalculus, algebra was used to graph a function. In this section, we combine precalculus methods with differential calculus to obtain a more detailed graph.

1 Graph a Function Using Calculus

The following steps are used to graph a function by hand or to analyze a computer-generated graph of a function.

Steps for Graphing a Function $y = f(x)$

Step 1 Find the domain of f and any intercepts.

Step 2 Identify any asymptotes, and examine the end behavior.

Step 3 Find $y' = f'(x)$ and use it to find any critical numbers of f. Determine the slope of the tangent line at the critical numbers. Also find f''.

Step 4 Use the Increasing/Decreasing Function Test to identify intervals on which f is increasing ($f' > 0$) and the intervals on which f is decreasing ($f' < 0$).

Step 5 Use either the First Derivative Test or the Second Derivative Test (if applicable) to identify local maximum values and local minimum values.

Step 6 Use the Test for Concavity to determine intervals where the function is concave up ($f'' > 0$) and concave down ($f'' < 0$). Identify any inflection points.

Step 7 Graph the function using the information gathered. Plot additional points as needed. It may be helpful to find the value of the derivative at such points.

EXAMPLE 1   Using Calculus to Graph a Polynomial Function

Graph $f(x) = 3x^4 - 8x^3$.

Solution We follow the steps for graphing a function.

Step 1 f is a fourth-degree polynomial; its domain is all real numbers. $f(0) = 0$, so the y-intercept is 0. We find the x-intercepts by solving $f(x) = 0$.

$$3x^4 - 8x^3 = 0$$
$$x^3(3x - 8) = 0$$
$$x = 0 \qquad \text{or} \qquad x = \frac{8}{3}$$

There are two x-intercepts: 0 and $\frac{8}{3}$. We plot the intercepts $(0, 0)$ and $\left(\frac{8}{3}, 0\right)$.

NOTE When graphing a function, piece together a sketch of the graph as you go, to ensure that the information is consistent and makes sense. If it appears to be contradictory, check your work. You may have made an error.

Step 2 Polynomials have no asymptotes, but the end behavior of f resembles the power function $y = 3x^4$.

Step 3 $f'(x) = 12x^3 - 24x^2 = 12x^2(x - 2)$ $\qquad$ $f''(x) = 36x^2 - 48x = 12x(3x - 4)$

For polynomials, the critical numbers occur when $f'(x) = 0$.

$$12x^2(x - 2) = 0$$
$$x = 0 \qquad \text{or} \qquad x = 2$$

The critical numbers are 0 and 2. At the points $(0, 0)$ and $(2, -16)$, the tangent lines are horizontal. We plot these points.

Step 4 To apply the Increasing/Decreasing Function Test, we use the critical numbers 0 and 2 to form three intervals.

Interval	Sign of f'	Conclusion
$(-\infty, 0)$	negative	f is decreasing on $(-\infty, 0)$
$(0, 2)$	negative	f is decreasing on $(0, 2)$
$(2, \infty)$	positive	f is increasing on $(2, \infty)$

Step 5 We use the First Derivative Test. Based on the table above, $f(0) = 0$ is neither a local maximum value nor a local minimum value, and $f(2) = -16$ is a local minimum value.

Step 6 $f''(x) = 36x^2 - 48x = 12x(3x - 4)$; the zeros of f'' are $x = 0$ and $x = \dfrac{4}{3}$.

To apply the Test for Concavity, we use the numbers 0 and $\dfrac{4}{3}$ to form three intervals.

Interval	Sign of f''	Conclusion
$(-\infty, 0)$	positive	f is concave up on $(-\infty, 0)$
$\left(0, \dfrac{4}{3}\right)$	negative	f is concave down on $\left(0, \dfrac{4}{3}\right)$
$\left(\dfrac{4}{3}, \infty\right)$	positive	f is concave up on $\left(\dfrac{4}{3}, \infty\right)$

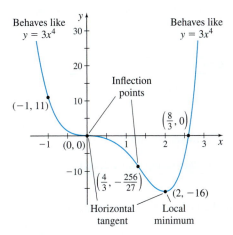

Behaves like $y = 3x^4$ (left)

Behaves like $y = 3x^4$ (right)

Inflection points

$(-1, 11)$

$\left(\dfrac{8}{3}, 0\right)$

$(0, 0)$

$\left(\dfrac{4}{3}, -\dfrac{256}{27}\right)$

$(2, -16)$

Horizontal tangent

Local minimum

DF Figure 46 $f(x) = 3x^4 - 8x^3$

The concavity of f changes at 0 and $\dfrac{4}{3}$, so the points $(0, 0)$ and $\left(\dfrac{4}{3}, -\dfrac{256}{27}\right)$ are inflection points. We plot the inflection points.

Step 7 Using the points plotted and the information about the shape of the graph, we graph the function. See Figure 46. ■

NOW WORK Problem 1.

EXAMPLE 2 Using Calculus to Graph a Rational Function

Graph $f(x) = \dfrac{6x^2 - 6}{x^3}$.

Solution We follow the steps for graphing a function.

Step 1 f is a rational function; the domain of f is $\{x \,|\, x \neq 0\}$. The x-intercepts are -1 and 1. There is no y-intercept. We plot the intercepts $(-1, 0)$ and $(1, 0)$.

Step 2 The degree of the numerator is less than the degree of the denominator, so f has a horizontal asymptote. We find the horizontal asymptotes by finding the limits at infinity. Using L'Hôpital's Rule, we get

$$\lim_{x \to \infty} \frac{6x^2 - 6}{x^3} = \lim_{x \to \infty} \frac{12x}{3x^2} = \lim_{x \to \infty} \frac{12}{6x} = 0$$

The line $y = 0$ is a horizontal asymptote as $x \to -\infty$ and as $x \to \infty$.

We identify vertical asymptotes by checking for infinite limits. $[\lim\limits_{x \to 0} f(x)$ is the only one to check—do you see why?]

$$\lim_{x \to 0^-} \frac{6x^2 - 6}{x^3} = \infty \qquad \lim_{x \to 0^+} \frac{6x^2 - 6}{x^3} = -\infty$$

The line $x = 0$ is a vertical asymptote. We draw the asymptotes on the graph.

Step 3 $f'(x) = \dfrac{d}{dx}\left(\dfrac{6x^2 - 6}{x^3}\right) = \dfrac{12x(x^3) - (6x^2 - 6)(3x^2)}{x^6}$

$$= \frac{6x^2(3 - x^2)}{x^6} = \frac{6(3 - x^2)}{x^4}$$

$$f''(x) = \frac{d}{dx}\left[\frac{6(3 - x^2)}{x^4}\right] = 6 \cdot \frac{(-2x)x^4 - (3 - x^2)(4x^3)}{x^8}$$

$$= \frac{12x^3(x^2 - 6)}{x^8} = \frac{12(x^2 - 6)}{x^5}$$

Critical numbers occur where $f'(x) = 0$ or where $f'(x)$ does not exist. Since $f'(x) = 0$ when $x = \pm\sqrt{3}$, there are two critical numbers, $\sqrt{3}$ and $-\sqrt{3}$. (The derivative does not exist at 0, but, since 0 is not in the domain of f, it is not a critical number.)

Step 4 To apply the Increasing/Decreasing Function Test, we use the numbers $-\sqrt{3}$, 0, and $\sqrt{3}$ to form four intervals.

Interval	Sign of f'	Conclusion
$(-\infty, -\sqrt{3})$	negative	f is decreasing on $(-\infty, -\sqrt{3})$
$(-\sqrt{3}, 0)$	positive	f is increasing on $(-\sqrt{3}, 0)$
$(0, \sqrt{3})$	positive	f is increasing on $(0, \sqrt{3})$
$(\sqrt{3}, \infty)$	negative	f is decreasing on $(\sqrt{3}, \infty)$

Step 5 We use the First Derivative Test. From the table in Step 4, we conclude

- $f(-\sqrt{3}) = -\dfrac{4}{\sqrt{3}} \approx -2.31$ is a local minimum value.

- $f(\sqrt{3}) = \dfrac{4}{\sqrt{3}} \approx 2.31$ is a local maximum value

Plot the local extreme points.

Step 6 We use the Test for Concavity. $f''(x) = \dfrac{12(x^2 - 6)}{x^5} = 0$ when $x = \pm\sqrt{6}$. Now use the numbers $-\sqrt{6}$, 0, and $\sqrt{6}$ to form four intervals.

Interval	Sign of f''	Conclusion
$(-\infty, -\sqrt{6})$	negative	f is concave down on $(-\infty, -\sqrt{6})$
$(-\sqrt{6}, 0)$	positive	f is concave up on $(-\sqrt{6}, 0)$
$(0, \sqrt{6})$	negative	f is concave down on $(0, \sqrt{6})$
$(\sqrt{6}, \infty)$	positive	f is concave up on $(\sqrt{6}, \infty)$

The concavity of the function f changes at $-\sqrt{6}$, 0, and $\sqrt{6}$. So, the points $\left(-\sqrt{6}, -\dfrac{5\sqrt{6}}{6}\right)$ and $\left(\sqrt{6}, \dfrac{5\sqrt{6}}{6}\right)$ are inflection points. Since 0 is not in the domain of f, there is no inflection point at 0. We plot the inflection points.

Step 7 We now use the information about where the function increases/decreases and the concavity to complete the graph. See Figure 47. Notice the apparent symmetry of the graph with respect to the origin. We can verify this symmetry by showing $f(-x) = -f(x)$ and concluding that f is an odd function.

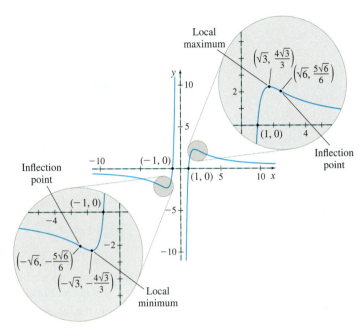

Figure 47 $f(x) = \dfrac{6x^2 - 6}{x^3}$ ∎

NOW WORK   **Problem 9.**

Some functions have a graph with an *oblique asymptote*. That is, the graph of the function approaches a line $y = mx + b$, $m \neq 0$, as x becomes unbounded in either the positive or negative direction.

In general, for a function $y = f(x)$, the line $y = mx + b$, $m \neq 0$, is an **oblique asymptote** of the graph of f, if

$$\lim_{x \to -\infty} [f(x) - (mx + b)] = 0 \qquad \text{or} \qquad \lim_{x \to \infty} [f(x) - (mx + b)] = 0$$

In Chapter 1, we saw that rational functions can have vertical or horizontal asymptotes. Rational functions can also have oblique asymptotes, but cannot have both a horizontal asymptote and an oblique asymptote. Rational functions for which the degree of the numerator is 1 more than the degree of the denominator have an oblique asymptote; rational functions for which the degree of the numerator is less than or equal to the degree of the denominator have a horizontal asymptote.

EXAMPLE 3   **Using Calculus to Graph a Rational Function**

Graph $f(x) = \dfrac{x^2 - 2x + 6}{x - 3}$.

Solution

Step 1 f is a rational function; the domain of f is $\{x \mid x \neq 3\}$. There are no x-intercepts, since $x^2 - 2x + 6 = 0$ has no real solutions. (Its discriminant is negative.) The y-intercept is $f(0) = -2$. Plot the intercept $(0, -2)$.

Step 2   We identify any vertical asymptotes by checking for infinite limits (3 is the only number to check). Since $\lim\limits_{x \to 3^-} \left(\dfrac{x^2 - 2x + 6}{x - 3} \right) = -\infty$ and $\lim\limits_{x \to 3^+} \left(\dfrac{x^2 - 2x + 6}{x - 3} \right) = \infty$, the line $x = 3$ is a vertical asymptote.

The degree of the numerator of f is 1 more than the degree of the denominator, so f will have an oblique asymptote. We divide $x^2 - 2x + 6$ by $x - 3$ to find the line $y = mx + b$.

$$f(x) = \frac{x^2 - 2x + 6}{x - 3} = x + 1 + \frac{9}{x - 3}$$

Since $\lim\limits_{x \to \infty} [f(x) - (x + 1)] = \lim\limits_{x \to \infty} \frac{9}{x - 3} = 0$, then $y = x + 1$ is an oblique asymptote of the graph of f. Draw the asymptotes on the graph.

Step 3 $f'(x) = \dfrac{d}{dx}\left(\dfrac{x^2 - 2x + 6}{x - 3}\right) = \dfrac{(2x - 2)(x - 3) - (x^2 - 2x + 6)(1)}{(x - 3)^2}$

$$= \frac{x^2 - 6x}{(x - 3)^2} = \frac{x(x - 6)}{(x - 3)^2}$$

$$f''(x) = \frac{d}{dx}\left[\frac{x^2 - 6x}{(x - 3)^2}\right] = \frac{(2x - 6)(x - 3)^2 - (x^2 - 6x)[2(x - 3)]}{(x - 3)^4}$$

$$= (x - 3)\frac{(2x^2 - 12x + 18) - (2x^2 - 12x)}{(x - 3)^4} = \frac{18}{(x - 3)^3}$$

$f'(x) = 0$ at $x = 0$ and at $x = 6$. So, 0 and 6 are critical numbers. The tangent lines are horizontal at 0 and at 6. Since 3 is not in the domain of f, 3 is not a critical number.

Step 4 To apply the Increasing/Decreasing Function Test, we use the numbers 0, 3, and 6 to form four intervals.

Interval	Sign of $f'(x)$	Conclusion
$(-\infty, 0)$	positive	f is increasing on $(-\infty, 0)$
$(0, 3)$	negative	f is decreasing on $(0, 3)$
$(3, 6)$	negative	f is decreasing on $(3, 6)$
$(6, \infty)$	positive	f is increasing on $(6, \infty)$

Step 5 Using the First Derivative Test, we find that $f(0) = -2$ is a local maximum value and $f(6) = 10$ is a local minimum value. Plot the points $(0, -2)$ and $(6, 10)$.

Step 6 $f''(x) = \dfrac{18}{(x - 3)^3}$. If $x < 3$, $f''(x) < 0$, and if $x > 3$, $f''(x) > 0$. From the Test for Concavity, we conclude f is concave down on the interval $(-\infty, 3)$ and f is concave up on the interval $(3, \infty)$. The concavity changes at 3, but 3 is not in the domain of f. So, the graph of f has no inflection point.

Step 7 We use the information to complete the graph of the function. See Figure 48. ∎

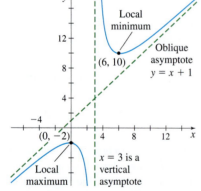

Figure 48 $f(x) = \dfrac{x^2 - 2x + 6}{x - 3}$

NOW WORK Problem 11.

EXAMPLE 4 Using Calculus to Graph a Function

Graph $f(x) = \dfrac{x}{\sqrt{x^2 + 4}}$.

Solution

Step 1 The domain of f is all real numbers. The only intercept is $(0, 0)$. So, we plot the point $(0, 0)$.

Step 2 We check for horizontal asymptotes. Since

$$\lim_{x \to \infty} f(x) = \lim_{x \to \infty} \frac{x}{\sqrt{x^2 + 4}} = \sqrt{\lim_{x \to \infty} \frac{x^2}{x^2 + 4}} = \sqrt{\lim_{x \to \infty} \frac{2x}{2x}} = 1$$

L'Hôpital's Rule

the line $y = 1$ is a horizontal asymptote as $x \to \infty$. Since $\lim\limits_{x \to -\infty} f(x) = -1$, the line $y = -1$ is also a horizontal asymptote as $x \to -\infty$. We draw the horizontal asymptotes on the graph.

Since f is defined for all real numbers, there are no vertical asymptotes.

Step 3 $f'(x) = \dfrac{d}{dx}\left(\dfrac{x}{\sqrt{x^2 + 4}}\right) = \dfrac{1 \cdot \sqrt{x^2 + 4} - x\left[\frac{1}{2}(x^2 + 4)^{-1/2} \cdot 2x\right]}{x^2 + 4}$

$$= \dfrac{\sqrt{x^2 + 4} - \dfrac{x^2}{\sqrt{x^2 + 4}}}{x^2 + 4} = \dfrac{x^2 + 4 - x^2}{(x^2 + 4)\sqrt{x^2 + 4}} = \dfrac{4}{(x^2 + 4)^{3/2}}$$

$$f''(x) = \dfrac{d}{dx}\left[\dfrac{4}{(x^2 + 4)^{3/2}}\right] = 4\left[-\dfrac{3}{2}(x^2 + 4)^{-5/2}\right] \cdot 2x = \dfrac{-12x}{(x^2 + 4)^{5/2}}$$

Since $f'(x)$ is never 0 and f' exists for all real numbers, there are no critical numbers.

Step 4 Since $f'(x) > 0$ for all x, f is increasing on $(-\infty, \infty)$.

Step 5 Because there are no critical numbers, there are no local extreme values.

Step 6 We test for concavity. Since $f''(x) > 0$ on the interval $(-\infty, 0)$, f is concave up on $(-\infty, 0)$; and since $f''(x) < 0$ on the interval $(0, \infty)$, f is concave down on $(0, \infty)$. The concavity changes at 0, and 0 is in the domain of f, so the point $(0, 0)$ is an inflection point of f. Plot the inflection point.

Step 7 Figure 49 shows the graph of f. ■

Notice the apparent symmetry of the graph with respect to the origin. We can verify this by showing $f(-x) = -f(x)$, that is, by showing f is an odd function.

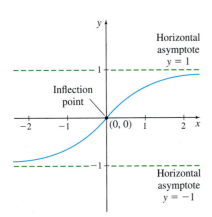

Figure 49 $f(x) = \dfrac{x}{\sqrt{x^2 + 4}}$

NOW WORK Problem 25.

EXAMPLE 5 **Using Calculus to Graph a Function**

Graph $f(x) = 4x^{1/3} - x^{4/3}$.

Solution

Step 1 The domain of f is all real numbers. Since $f(0) = 0$, the y-intercept is 0. Now $f(x) = 0$ when $4x^{1/3} - x^{4/3} = x^{1/3}(4 - x) = 0$ or when $x = 0$ or $x = 4$. So, the x-intercepts are 0 and 4. Plot the intercepts $(0, 0)$ and $(4, 0)$.

Step 2 Since $\lim\limits_{x \to \infty} f(x) = \lim\limits_{x \to \infty}[x^{1/3}(4 - x)] = -\infty$, there is no horizontal asymptote. Since the domain of f is all real numbers, there is no vertical asymptote.

Step 3 $f'(x) = \dfrac{d}{dx}(4x^{1/3} - x^{4/3}) = \dfrac{4}{3}x^{-2/3} - \dfrac{4}{3}x^{1/3} = \dfrac{4}{3}\left(\dfrac{1}{x^{2/3}} - x^{1/3}\right)$

$$= \dfrac{4}{3}\left(\dfrac{1 - x}{x^{2/3}}\right)$$

$$f''(x) = \dfrac{d}{dx}\left(\dfrac{4}{3}x^{-2/3} - \dfrac{4}{3}x^{1/3}\right) = -\dfrac{8}{9}x^{-5/3} - \dfrac{4}{9}x^{-2/3} = -\dfrac{4}{9}\left(\dfrac{2}{x^{5/3}} + \dfrac{1}{x^{2/3}}\right)$$

$$= -\dfrac{4}{9} \cdot \dfrac{2 + x}{x^{5/3}}$$

Since $f'(x) = \dfrac{4}{3}\left(\dfrac{1 - x}{x^{2/3}}\right) = 0$ when $x = 1$ and $f'(x)$ does not exist at $x = 0$, the critical numbers are 0 and 1. At the point $(1, 3)$, the tangent line to the graph is horizontal; at the point $(0, 0)$, the tangent line is vertical. Plot these points.

Step 4 To apply the Increasing/Decreasing Function Test, we use the critical numbers 0 and 1 to form three intervals.

Interval	Sign of f'	Conclusion
$(-\infty, 0)$	positive	f is increasing on $(-\infty, 0)$
$(0, 1)$	positive	f is increasing on $(0, 1)$
$(1, \infty)$	negative	f is decreasing on $(1, \infty)$

Step 5 By the First Derivative Test, $f(1) = 3$ is a local maximum value and $f(0) = 0$ is not a local extreme value.

Step 6 We now test for concavity by using the numbers -2 and 0 to form three intervals.

Interval	Sign of f''	Conclusion
$(-\infty, -2)$	negative	f is concave down on the interval $(-\infty, -2)$
$(-2, 0)$	positive	f is concave up on the interval $(-2, 0)$
$(0, \infty)$	negative	f is concave down on the interval $(0, \infty)$

The concavity changes at -2 and at 0. Since

$$f(-2) = 4(-2)^{1/3} - (-2)^{4/3} = 4\sqrt[3]{-2} - \sqrt[3]{16} \approx -7.56,$$

the inflection points are $(-2, -7.56)$ and $(0, 0)$. Plot the inflection points.

Step 7 The graph of f is given in Figure 50.   ∎

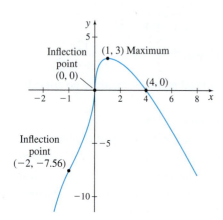

Figure 50 $f(x) = 4x^{1/3} - x^{4/3}$

NOW WORK   Problem 19.

EXAMPLE 6   Using Calculus to Graph a Trigonometric Function

Graph $f(x) = \sin x - \cos^2 x$, $0 \le x \le 2\pi$.

Solution

Step 1 The domain of f is $\{x \mid 0 \le x \le 2\pi\}$. Since $f(0) = \sin 0 - \cos^2 0 = -1$, the y-intercept is -1. The x-intercepts satisfy the equation

$$\sin x - \cos^2 x = \sin x - (1 - \sin^2 x) = \sin^2 x + \sin x - 1 = 0$$

This trigonometric equation is quadratic in form, so we use the quadratic formula.

$$\sin x = \frac{-1 \pm \sqrt{1 - 4(1)(-1)}}{2} = \frac{-1 \pm \sqrt{5}}{2}$$

Since $\dfrac{-1 - \sqrt{5}}{2} < -1$ and $-1 \le \sin x \le 1$, the x-intercepts occur at

$$x = \sin^{-1}\left(\frac{-1 + \sqrt{5}}{2}\right) \approx 0.67 \text{ and at } x = \pi - \sin^{-1}\left(\frac{-1 + \sqrt{5}}{2}\right) \approx 2.48.$$

Plot the intercepts.

NEED TO REVIEW? Trigonometric equations are discussed in Section P. 7, pp. 61–63.

Step 2 The function f has no asymptotes.

Step 3 $f'(x) = \dfrac{d}{dx}(\sin x - \cos^2 x) = \cos x + 2\cos x \sin x = \cos x(1 + 2\sin x)$

$f''(x) = \dfrac{d}{dx}[\cos x(1 + 2\sin x)] = 2\cos^2 x - \sin x(1 + 2\sin x)$

$= 2\cos^2 x - \sin x - 2\sin^2 x = -4\sin^2 x - \sin x + 2$

The critical numbers occur where $f'(x) = 0$. That is, where

$$\cos x = 0 \qquad \text{or} \qquad 1 + 2\sin x = 0$$

In the interval $[0, 2\pi]$: $\cos x = 0$ if $x = \dfrac{\pi}{2}$ or if $x = \dfrac{3\pi}{2}$;

and $1 + 2\sin x = 0$ when $\sin x = -\dfrac{1}{2}$, that is, when $x = \dfrac{7\pi}{6}$ or $x = \dfrac{11\pi}{6}$.

So, the critical numbers are $\dfrac{\pi}{2}$, $\dfrac{7\pi}{6}$, $\dfrac{3\pi}{2}$, and $\dfrac{11\pi}{6}$. At each of these numbers, the tangent lines are horizontal.

Step 4 To apply the Increasing/Decreasing Function Test, we use the numbers 0, $\dfrac{\pi}{2}$, $\dfrac{7\pi}{6}$, $\dfrac{3\pi}{2}$, $\dfrac{11\pi}{6}$, and 2π to form five intervals.

Interval	Sign of f'	Conclusion
$\left(0, \dfrac{\pi}{2}\right)$	positive	f is increasing on $\left(0, \dfrac{\pi}{2}\right)$
$\left(\dfrac{\pi}{2}, \dfrac{7\pi}{6}\right)$	negative	f is decreasing on $\left(\dfrac{\pi}{2}, \dfrac{7\pi}{6}\right)$
$\left(\dfrac{7\pi}{6}, \dfrac{3\pi}{2}\right)$	positive	f is increasing on $\left(\dfrac{7\pi}{6}, \dfrac{3\pi}{2}\right)$
$\left(\dfrac{3\pi}{2}, \dfrac{11\pi}{6}\right)$	negative	f is decreasing on $\left(\dfrac{3\pi}{2}, \dfrac{11\pi}{6}\right)$
$\left(\dfrac{11\pi}{6}, 2\pi\right)$	positive	f is increasing on $\left(\dfrac{11\pi}{6}, 2\pi\right)$

Step 5 By the First Derivative Test, $f\left(\dfrac{\pi}{2}\right) = 1$ and $f\left(\dfrac{3\pi}{2}\right) = -1$ are local maximum values, and $f\left(\dfrac{7\pi}{6}\right) = -\dfrac{5}{4}$ and $f\left(\dfrac{11\pi}{6}\right) = -\dfrac{5}{4}$ are local minimum values. Plot these points.

Step 6 We apply the Test for Concavity. To solve $f''(x) > 0$ and $f''(x) < 0$, we first solve the equation $f''(x) = -4\sin^2 x - \sin x + 2 = 0$, or equivalently,

$$4\sin^2 x + \sin x - 2 = 0 \qquad 0 \le x \le 2\pi$$

$$\sin x = \dfrac{-1 \pm \sqrt{1 + 32}}{8}$$

$$\sin x \approx 0.593 \qquad \text{or} \qquad \sin x \approx -0.843$$

$$x \approx 0.63 \qquad x \approx 2.51 \qquad x \approx 4.14 \qquad x \approx 5.28$$

We use these numbers to form five subintervals of $[0, 2\pi]$.

Interval	Sign of f''	Conclusion
$(0, 0.63)$	positive	f is concave up on the interval $(0, 0.63)$
$(0.63, 2.51)$	negative	f is concave down on the interval $(0.63, 2.51)$
$(2.51, 4.14)$	positive	f is concave up on the interval $(2.51, 4.14)$
$(4.14, 5.28)$	negative	f is concave down on the interval $(4.14, 5.28)$
$(5.28, 2\pi)$	positive	f is concave up on the interval $(5.28, 2\pi)$

The inflection points are $(0.63, -0.06)$, $(2.51, -0.06)$, $(4.14, -1.13)$, and $(5.28, -1.13)$. Plot the inflection points.

Step 7 The graph of f is given in Figure 51.

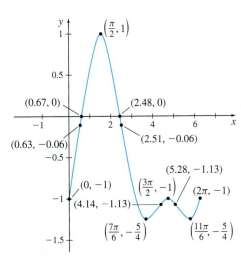

Figure 51 $f(x) = \sin x - \cos^2 x$, $0 \le x \le 2\pi$

NOTE The function f in Example 6 is periodic with period 2π. To graph this function over its unrestricted domain, the set of real numbers, repeat the graph in Figure 51 over intervals of length 2π.

NOW WORK Problem **33**.

EXAMPLE 7   Using Calculus to Graph a Function

Graph $f(x) = e^x(x^2 - 3)$.

Solution

Step 1 The domain of f is all real numbers. Since $f(0) = -3$, the y-intercept is -3. The x-intercepts occur where

$$f(x) = e^x(x^2 - 3) = 0$$
$$x = \sqrt{3} \quad \text{or} \quad x = -\sqrt{3}$$

Plot the intercepts.

Step 2 Since the domain of f is all real numbers, there are no vertical asymptotes. To determine if there is a horizontal asymptote, we find the limits at infinity.

$$\lim_{x \to -\infty} f(x) = \lim_{x \to -\infty} [e^x(x^2 - 3)]$$

Since $e^x(x^2 - 3)$ is an indeterminate form at $-\infty$ of the type $0 \cdot \infty$, we write $e^x(x^2 - 3) = \dfrac{x^2 - 3}{e^{-x}}$ and use L'Hôpital's Rule.

$$\lim_{x \to -\infty} f(x) = \lim_{x \to -\infty} [e^x(x^2 - 3)] = \lim_{x \to -\infty} \frac{x^2 - 3}{e^{-x}}$$

$$= \lim_{x \to -\infty} \frac{2x}{-e^{-x}} = \lim_{x \to -\infty} \frac{2}{e^{-x}} = 0$$

Type $\dfrac{\infty}{\infty}$; use L'Hôpital's Rule   Type $\dfrac{\infty}{\infty}$; use L'Hôpital's Rule

The line $y = 0$ is a horizontal asymptote as $x \to -\infty$.

Since $\lim_{x \to \infty} [e^x(x^2 - 3)] = \infty$, there is no horizontal asymptote as $x \to \infty$.

Draw the asymptote on the graph.

Step 3 $f'(x) = \dfrac{d}{dx}[e^x(x^2 - 3)] = e^x(2x) + e^x(x^2 - 3) = e^x(x^2 + 2x - 3)$

$$= e^x(x + 3)(x - 1)$$

$$f''(x) = \frac{d}{dx}[e^x(x^2 + 2x - 3)] = e^x(2x + 2) + e^x(x^2 + 2x - 3)$$

$$= e^x(x^2 + 4x - 1)$$

Solving $f'(x) = 0$, we find that the critical numbers are -3 and 1.

Step 4 To apply the Increasing/Decreasing Function Test, we use the critical numbers -3 and 1 to form three intervals.

Interval	Sign of f'	Conclusion
$(-\infty, -3)$	positive	f is increasing on the interval $(-\infty, -3)$
$(-3, 1)$	negative	f is decreasing on the interval $(-3, 1)$
$(1, \infty)$	positive	f is increasing on the interval $(1, \infty)$

Step 5 We use the First Derivative Test to identify the local extrema. From the table in Step 4, there is a local maximum at -3 and a local minimum at 1. Then $f(-3) = 6e^{-3} \approx 0.30$ is a local maximum value and $f(1) = -2e \approx -5.44$ is a local minimum value. Plot the local extrema.

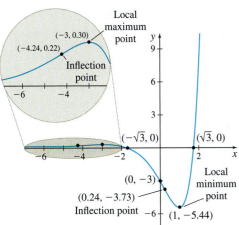

Figure 52 $f(x) = e^x(x^2 - 3)$

Step 6 To determine the concavity of f, first we solve $f''(x) = 0$. We find $x = -2 \pm \sqrt{5}$. Now we use the numbers $-2 - \sqrt{5} \approx -4.24$ and $-2 + \sqrt{5} \approx 0.24$ to form three intervals and apply the Test for Concavity.

Interval	Sign of f''	Conclusion
$(-\infty, -2 - \sqrt{5})$	positive	f is concave up on the interval $(-\infty, -2 - \sqrt{5})$
$(-2 - \sqrt{5}, -2 + \sqrt{5})$	negative	f is concave down on the interval $(-2 - \sqrt{5}, -2 + \sqrt{5})$
$(-2 + \sqrt{5}, \infty)$	positive	f is concave up on the interval $(-2 + \sqrt{5}, \infty)$

The inflection points are $(-4.24, 0.22)$ and $(0.24, -3.73)$. Plot the inflection points.

Step 7 The graph of f is given in Figure 52. ∎

NOW WORK Problem 39.

4.6 Assess Your Understanding

Skill Building

In Problems 1–42, use calculus to graph each function. Follow the steps for graphing a function.

1. $f(x) = x^4 - 6x^2 + 10$
2. $f(x) = x^5 - 3x^3 + 4$
3. $f(x) = \dfrac{1}{x - 2}$
4. $f(x) = \dfrac{2}{x + 2}$
5. $f(x) = \dfrac{2}{x^2 - 4}$
6. $f(x) = \dfrac{1}{x^2 - 1}$
7. $f(x) = \dfrac{2x - 1}{x + 1}$
8. $f(x) = \dfrac{x - 2}{x}$
9. $f(x) = \dfrac{x}{x^2 + 1}$
10. $f(x) = \dfrac{2x}{x^2 - 4}$
11. $f(x) = \dfrac{x^2 + 1}{2x}$
12. $f(x) = \dfrac{x^2 - 1}{2x}$
13. $f(x) = \dfrac{x^4 + 1}{x^2}$
14. $f(x) = \dfrac{x^3 + 1}{x + 1}$
15. $xy = x^2 + 2$
16. $xy = x^2 + x - 1$
17. $f(x) = \dfrac{x^2}{x + 3}$
18. $f(x) = \dfrac{3x^2 - 1}{x - 1}$
19. $f(x) = 1 + \dfrac{1}{x} + \dfrac{1}{x^2}$
20. $f(x) = \dfrac{2}{x} + \dfrac{1}{x^2}$
21. $f(x) = \sqrt{3 - x}$
22. $f(x) = x\sqrt{x + 2}$
23. $f(x) = x + \sqrt{x}$
24. $f(x) = \sqrt{x} - \sqrt{x + 1}$
25. $f(x) = \dfrac{x^2}{\sqrt{x + 1}}$
26. $f(x) = \dfrac{x}{\sqrt{x^2 + 2}}$
27. $f(x) = \dfrac{1}{(x + 1)(x - 2)}$
28. $f(x) = \dfrac{1}{(x - 1)(x + 3)}$

29. $f(x) = x^{2/3} + 3x^{1/3} + 2$
30. $f(x) = x^{5/3} - 5x^{2/3}$
31. $f(x) = \sin x - \cos x$
32. $f(x) = \sin x + \tan x$
33. $f(x) = \sin^2 x - \cos x$
34. $f(x) = \cos^2 x + \sin x$
35. $y = \ln x - x$
36. $y = x \ln x$
37. $f(x) = \ln(4 - x^2)$
38. $y = \ln(x^2 + 2)$
39. $f(x) = 3e^{3x}(5 - x)$
40. $f(x) = 3e^{-3x}(x - 4)$
41. $f(x) = e^{-x^2}$
42. $f(x) = e^{1/x}$

Applications and Extensions

CAS *In Problems 43–52, for each function:*

(a) *Use a CAS to graph the function.*
(b) *Identify any asymptotes.*
(c) *Use the graph to identify intervals on which the function increases or decreases and the intervals where the function is concave up or down.*
(d) *Approximate the local extreme values using the graph.*
(e) *Compare the approximate local maxima and local minima to the exact local extrema found by using calculus.*
(f) *Approximate any inflection points using the graph.*

43. $f(x) = \dfrac{x^{2/3}}{x - 1}$
44. $f(x) = \dfrac{5 - x}{x^2 + 3x + 4}$
45. $f(x) = x + \sin(2x)$
46. $f(x) = x - \cos x$
47. $f(x) = \ln(x\sqrt{x - 1})$
48. $f(x) = \ln(\tan^2 x)$
49. $f(x) = \sqrt[3]{\sin x}$
50. $f(x) = e^{-x} \cos x$
51. $y^2 = x^2(6 - x)$.
52. $y^2 = x^2(4 - x^2)$

1. = NOW WORK problem   ⟋⟍ = Graphing technology recommended   CAS = Computer Algebra System recommended

In Problems 53–56, graph a function f that is continuous on the interval [1, 6] and satisfies the given conditions.

53. $f'(2)$ does not exist

$f'(3) = -1$

$f''(3) = 0$

$f'(5) = 0$

$f''(x) < 0, \quad 2 < x < 3$

$f''(x) > 0, \quad x > 3$

54. $f'(2) = 0$

$f''(2) = 0$

$f'(3)$ does not exist

$f'(5) = 0$

$f''(x) > 0, \quad 2 < x < 3$

$f''(x) > 0, \quad x > 3$

55. $f'(2) = 0$

$\lim_{x \to 3^-} f'(x) = \infty$

$\lim_{x \to 3^+} f'(x) = \infty$

$f'(5) = 0$

$f''(x) > 0, \quad x < 3$

$f''(x) < 0, \quad x > 3$

56. $f'(2) = 0$

$\lim_{x \to 3^-} f'(x) = -\infty$

$\lim_{x \to 3^+} f'(x) = -\infty$

$f'(5) = 0$

$f''(x) < 0, \quad x < 3$

$f''(x) > 0, \quad x > 3$

57. Sketch the graph of a function f defined and continuous for $-1 \le x \le 2$ that satisfies the following conditions:

$f(-1) = 1 \quad f(1) = 2 \quad f(2) = 3 \quad f(0) = 0 \quad f\left(\frac{1}{2}\right) = 3$

$\lim_{x \to -1^+} f'(x) = -\infty \quad \lim_{x \to 1^-} f'(x) = -1 \quad \lim_{x \to 1^+} f'(x) = \infty$

f has a local minimum at 0 f has a local maximum at $\frac{1}{2}$

58. Graph of a Function   Which of the following is true about the graph of $y = \ln|x^2 - 1|$ in the interval $(-1, 1)$?

(a) The graph is increasing.

(b) The graph has a local minimum at $(0, 0)$.

(c) The graph has a range of all real numbers.

(d) The graph is concave down.

(e) The graph has an asymptote $x = 0$.

59. Properties of a Function   Suppose $f(x) = \dfrac{1}{x} + \ln x$ is defined only on the closed interval $\dfrac{1}{e} \le x \le e$.

(a) Determine the numbers x at which f has its absolute maximum and absolute minimum.

(b) For what numbers x is the graph concave up?

(c) Graph f.

60. Probability   The function $f(x) = \dfrac{1}{\sqrt{2\pi}} e^{-x^2/2}$, encountered in probability theory, is called the **standard normal density function**. Determine where this function is increasing and decreasing, find all local maxima and local minima, find all inflection points, and determine the intervals where f is concave up and concave down. Then graph the function.

In Problems 61–64, graph each function. Use L'Hôpital's Rule to find any asymptotes.

61. $f(x) = \dfrac{\sin 3x}{x\sqrt{4 - x^2}}$

62. $f(x) = x\sqrt{x}$

63. $f(x) = x^{1/x}$

64. $y = \dfrac{1}{x} \tan x \quad -\dfrac{\pi}{2} < x < \dfrac{\pi}{2}$

4.7 Optimization

OBJECTIVE   *When you finish this section, you should be able to:*

1 Solve optimization problems (p. 318)

Investigating maximum and/or minimum values of a function has been a recurring theme throughout most of this chapter. We continue this theme, using calculus to solve *optimization problems.*

Optimization is a process of determining the best solution to a problem. Often the best solution is one that maximizes or minimizes some quantity. For example, a business owner's goal usually involves minimizing cost while maximizing profit, or an engineer's goal may be to minimize the load on a beam. In this section, we model situations using a function and then find its maximum or minimum value.

1 Solve Optimization Problems

Even though each individual problem has unique features, it is possible to outline a general method for obtaining an optimal solution.

Steps for Solving Optimization Problems

Step 1 Read the problem until you understand it and can identify the quantity for which a maximum or minimum value is to be found. Assign a symbol to represent it.

Step 2 Assign symbols to represent the other variables in the problem. If possible, draw a picture and label it. Determine relationships among the variables.

Step 3 Express the quantity to be maximized or minimized as a function of one of the variables, and determine a meaningful domain for the function.

Step 4 Find the absolute maximum value or absolute minimum value of the function.

NOTE It is good practice to have in mind meaningful estimates of the answer.

Figure 53 Area $A = xy$

EXAMPLE 1   Maximizing an Area

A farmer with 4000 m of fencing wants to enclose a rectangular plot that borders a straight river, as shown in Figure 53. If the farmer does not fence the side along the river, what is the largest rectangular area that can be enclosed?

Solution

Step 1 The quantity to be maximized is the area; we denote it by A.

Step 2 We denote the dimensions of the rectangle by x and y (both in meters), with the length of the side y parallel to the river. The area is $A = xy$. Because there are 4000 m of fence available, the variables x and y are related by the equation

$$x + y + x = 4000$$
$$y = 4000 - 2x$$

Step 3 Now we express the area A as a function of x.

$$A = A(x) = x(4000 - 2x) = 4000x - 2x^2 \qquad A = xy, \quad y = 4000 - 2x$$

The domain of A is the closed interval $[0, 2000]$.

NOTE Since $A = A(x)$ is continuous on the closed interval $[0, 2000]$, it will have an absolute maximum.

Step 4 To find the number x that maximizes $A(x)$, we differentiate A with respect to x and find the critical numbers:

$$A'(x) = 4000 - 4x = 0$$

The critical number is $x = 1000$. The maximum value of A occurs either at the critical number or at an endpoint of the interval $[0, 2000]$.

$$A(1000) = 2,000,000 \qquad A(0) = 0 \qquad A(2000) = 0$$

The maximum value is $A(1000) = 2,000,000 \text{ m}^2$, which results from using a rectangular plot that measures 2000 m along the side parallel to the river and 1000 m along each of the other two sides. ∎

We could have reached the same conclusion by noting that $A''(x) = -4 < 0$ for all x, so A is always concave down, and the local maximum at $x = 1000$ must be the absolute maximum of the function.

NOW WORK Problem 3.

EXAMPLE 2   Maximizing a Volume

From each corner of a square piece of sheet metal 18 cm on a side, we remove a small square and turn up the edges to form an open box. What are the dimensions of the box with the largest volume?

Solution

Step 1 The quantity to be maximized is the volume of the box; we denote it by V. We denote the length of each side of the small squares by x and the length of each

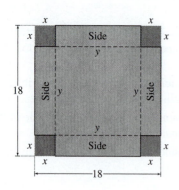

Figure 54 $y = 18 - 2x$

side after the small squares are removed by y, as shown in Figure 54. Both x and y are in centimeters.

Step 2 Then $y = (18 - 2x)$ cm. The height of the box is x cm, and the area of the base of the box is y^2 cm². So, the volume of the box is $V = xy^2$ cm³.

Step 3 To express V as a function of one variable, we substitute $y = 18 - 2x$ into the formula for V. Then the function to be maximized is

$$V = V(x) = x(18 - 2x)^2$$

Since both $x \geq 0$ and $18 - 2x \geq 0$, we find $x \leq 9$, meaning the domain of V is the closed interval $[0, 9]$. (All other numbers make no physical sense—do you see why?)

Step 4 To find the value of x that maximizes V, we differentiate V and find the critical numbers:

$$V'(x) = 2x(18 - 2x)(-2) + (18 - 2x)^2 = (18 - 2x)(18 - 6x)$$

Now, we solve $V'(x) = 0$ for x. The solutions are

$$x = 9 \qquad \text{or} \qquad x = 3$$

The only critical number in the open interval $(0, 9)$ is 3. We evaluate V at 3 and at the endpoints 0 and 9.

$$V(0) = 0 \qquad V(3) = 3(18 - 6)^2 = 432 \qquad V(9) = 0$$

The maximum volume is 432 cm³. The box with the maximum volume has a height of 3 cm. Since $y = 18 - 2(3) = 12$ cm, the base of the box measures 12 cm by 12 cm. ∎

NOW WORK Problem 9.

EXAMPLE 3 **Maximizing an Area**

A manufacturer makes a flexible square play yard (Figure 55a), that can be opened at one corner and attached at right angles to a wall or the side of a house, as shown in Figure 55(b). If each side is 3 m in length, the open configuration doubles the available area from 9 m² to 18 m². Is there a configuration that will more than double the play area?

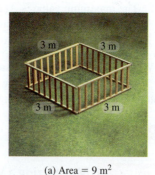

(a) Area = 9 m²

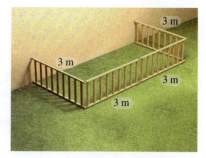

(b) Area = 18 m²

Figure 55

Solution

Step 1 We want to maximize the play area A.

Step 2 Since the play yard must be attached at right angles to the wall, the possible configurations depend on the amount of wall used as a fifth side, as shown in

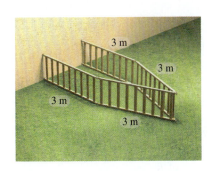

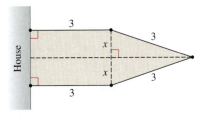

DF Figure 56

Figure 56. We use x to represent half the length (in meters) of the wall used as the fifth side.

Step 3 We partition the play area into two sections: a rectangle with area $3(2x) = 6x$ and a triangle with base $2x$ and altitude $\sqrt{3^2 - x^2} = \sqrt{9 - x^2}$. The play area A is the sum of the areas of the two sections. The area to be maximized is

$$A = A(x) = 6x + \frac{1}{2}(2x)\sqrt{9 - x^2} = 6x + x\sqrt{9 - x^2}$$

The domain of A is the closed interval $[0, 3]$.

Step 4 To find the maximum area of the play yard, we find the critical numbers of A.

$$A'(x) = 6 + x\left[\frac{1}{2}(-2x)(9 - x^2)^{-1/2}\right] + \sqrt{9 - x^2}$$

$$= 6 - \frac{x^2}{\sqrt{9 - x^2}} + \sqrt{9 - x^2} = \frac{6\sqrt{9 - x^2} - 2x^2 + 9}{\sqrt{9 - x^2}}$$

Critical numbers occur when $A'(x) = 0$ or where $A'(x)$ does not exist. A' does not exist at 3, and $A'(x) = 0$ at

$$6\sqrt{9 - x^2} - 2x^2 + 9 = 0$$

$$\sqrt{9 - x^2} = \frac{2x^2 - 9}{6}$$

$$9 - x^2 = \left(\frac{2x^2 - 9}{6}\right)^2 = \frac{4x^4 - 36x^2 + 81}{36}$$

$$324 - 36x^2 = 4x^4 - 36x^2 + 81$$

$$324 = 4x^4 + 81$$

$$x^4 = \frac{324 - 81}{4} = \frac{243}{4}$$

$$x = \sqrt[4]{\frac{243}{4}} \approx 2.792$$

The only critical number in the open interval $(0, 3)$ is $\sqrt[4]{\dfrac{243}{4}} \approx 2.792$.

Now we evaluate $A(x)$ at the endpoints 0 and 3 and at the critical number $x \approx 2.792$.

$$A(0) = 0 \qquad A(3) = 18 \qquad A(2.792) \approx 19.817$$

Using a wall of length $2x \approx 2(2.792) = 5.584$ m will maximize the area; the configuration shown in Figure 57 increases the play area by about 10% (from 18 to 19.817 m^2). ∎

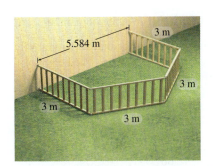

Figure 57 Area ≈ 19.817 m^2

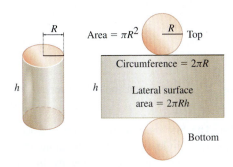

Figure 58

EXAMPLE 4   Minimizing Cost

A manufacturer needs to produce a cylindrical container with a capacity of 1000 cm^3. The top and bottom of the container are made of material that costs $0.05 per square centimeter, while the sides of the container are made of material costing $0.03 per square centimeter. Find the dimensions that will minimize the company's cost of producing the container.

Solution Figure 58 shows a cylindrical container and the area of its top, bottom, and lateral surfaces. As shown in the figure, we let h denote the height of the container and R denote the radius. The total area of the bottom and top is $2\pi R^2 \text{ cm}^2$. The area of the lateral surface of the can is $2\pi Rh \text{ cm}^2$.

NEED TO REVIEW? Geometry formulas are discussed in Appendix A.2, p. A-15.

The variables h and R are related. Since the volume of the cylinder is $1000\,\text{cm}^3$,

$$V = \pi R^2 h = 1000$$

$$h = \frac{1000}{\pi R^2}$$

The cost C, in dollars, of manufacturing the container is

$$C = (0.05)(2\pi R^2) + (0.03)(2\pi Rh) = 0.1\pi R^2 + 0.06\pi Rh$$

By substituting for h, we can express C as a function of R.

$$C = C(R) = 0.1\pi R^2 + (0.06\pi R)\left(\frac{1000}{\pi R^2}\right) = 0.1\pi R^2 + \frac{60}{R}$$

This is the function to be minimized. The domain of C is $\{R \mid R > 0\}$.

To find the minimum cost, we differentiate C with respect to R.

$$C'(R) = 0.2\pi R - \frac{60}{R^2} = \frac{0.2\pi R^3 - 60}{R^2}$$

Solve $C'(R) = 0$ to find the critical numbers.

$$0.2\pi R^3 - 60 = 0$$
$$R^3 = \frac{300}{\pi}$$
$$R = \sqrt[3]{\frac{300}{\pi}} \approx 4.571\ \text{cm}$$

Now we find $C''(x)$ and use the Second Derivative Test.

$$C''(R) = 0.2\pi + \frac{120}{R^3}$$
$$C''\left(\sqrt[3]{\frac{300}{\pi}}\right) = 0.2\pi + \frac{120\pi}{300} = 0.6\pi > 0$$

C has a local minimum at $\sqrt[3]{\dfrac{300}{\pi}}$. Since $C''(R) > 0$ for all R in the domain, the graph of C is always concave up, and the local minimum value is the absolute minimum value. The radius of the container that minimizes the cost is $R \approx 4.571$ cm. The height of the container that minimizes the cost of the material is

$$h = \frac{1000}{\pi R^2} \approx \frac{1000}{20.892\pi} \approx 15.236\ \text{cm}$$

The minimum cost of the container is

$$C\left(\sqrt[3]{\frac{300}{\pi}}\right) = 0.1\pi \left(\sqrt[3]{\frac{300}{\pi}}\right)^2 + \frac{60}{\sqrt[3]{\dfrac{300}{\pi}}} \approx \$19.69.$$ ∎

NOTE If the costs of the materials for the top, bottom, and lateral surfaces of a cylindrical container are all equal, then the minimum total cost occurs when the surface area is minimum. It can be shown (see Problem 39) that for any fixed volume, the minimum surface area of a cylindrical container is obtained when the height equals twice the radius.

NOW WORK Problem 11.

EXAMPLE 5 Maximizing Area

A rectangle is inscribed in a semicircle of radius 2. Find the dimensions of the rectangle that has the maximum area.

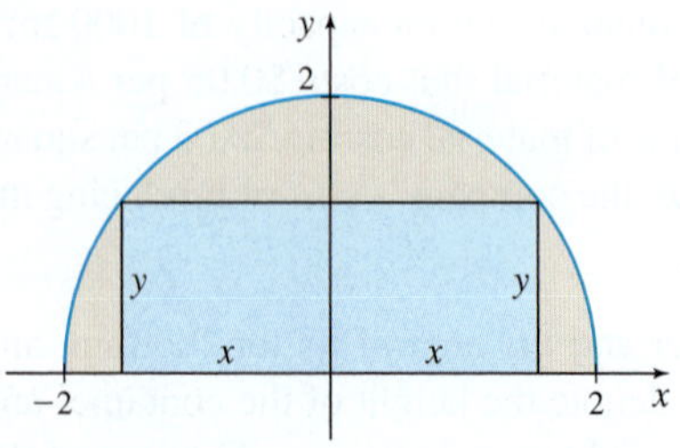

Solution We present two methods of solution: The first uses analytic geometry, the second uses trigonometry. To begin, we place the semicircle with its diameter along the x-axis and center at the origin. Then we inscribe a rectangle in the semicircle, as shown in Figure 59. The length of the inscribed rectangle is $2x$ and its height is y.

Figure 59

Analytic Geometry Method The area A of the inscribed rectangle is $A = 2xy$ and the equation of the semicircle is $x^2 + y^2 = 4$, $y \geq 0$. We solve for y in $x^2 + y^2 = 4$ and obtain $y = \sqrt{4 - x^2}$. Now we substitute this expression for y in the area formula for the rectangle to express A as a function of x alone.

$$A = A(x) = 2x\sqrt{4 - x^2} \qquad 0 \leq x \leq 2$$
$$\uparrow$$
$$A = 2xy; \; y = \sqrt{4 - x^2}$$

Then

$$A'(x) = 2\left[x\left(\frac{1}{2}\right)(4 - x^2)^{-1/2}(-2x) + \sqrt{4 - x^2}\right] = 2\left[\frac{-x^2}{\sqrt{4 - x^2}} + \sqrt{4 - x^2}\right]$$

$$= 2\left[\frac{-x^2 + (4 - x^2)}{\sqrt{4 - x^2}}\right] = 2\left[\frac{-2(x^2 - 2)}{\sqrt{4 - x^2}}\right] = -\frac{4(x^2 - 2)}{\sqrt{4 - x^2}}$$

The only critical number in the open interval $(0, 2)$ is $\sqrt{2}$, where $A'(\sqrt{2}) = 0$. $[-\sqrt{2}$ and -2 are not in the domain of A, and 2 is not in the open interval $(0, 2)$.] The values of A at the endpoints 0 and 2 and at the critical number $\sqrt{2}$ are

$$A(0) = 0 \qquad A(\sqrt{2}) = 4 \qquad A(2) = 0$$

The maximum area of the inscribed rectangle is 4, and it corresponds to the rectangle whose length is $2x = 2\sqrt{2}$ and whose height is $y = \sqrt{2}$.

Trigonometric Method Using Figure 59, we draw the radius $r = 2$ from O to the vertex of the rectangle and place the angle θ in the standard position, as shown in Figure 60. Then $x = 2\cos\theta$ and $y = 2\sin\theta$, $0 \leq \theta \leq \dfrac{\pi}{2}$. The area A of the rectangle is

$$A = 2xy = 2(2\cos\theta)(2\sin\theta) = 8\cos\theta\sin\theta = 4\sin(2\theta)$$
$$\uparrow$$
$$\sin(2\theta) = 2\sin\theta\cos\theta$$

Since the area A is a differentiable function of θ, we obtain the critical numbers by finding $A'(\theta)$ and solving the equation $A'(\theta) = 0$.

$$A'(\theta) = 8\cos(2\theta) = 0$$
$$\cos(2\theta) = 0$$
$$2\theta = \cos^{-1}0 = \frac{\pi}{2}$$
$$\theta = \frac{\pi}{4}$$

We now find $A''(\theta)$ and use the Second Derivative Test.

$$A''(\theta) = -16\sin(2\theta)$$
$$A''\left(\frac{\pi}{4}\right) = -16\sin\left(2\cdot\frac{\pi}{4}\right) = -16\sin\frac{\pi}{2} = -16 < 0$$

So at $\theta = \dfrac{\pi}{4}$, the area A is maximized and the maximum area of the inscribed rectangle is

$$A\left(\frac{\pi}{4}\right) = 4\sin\left(2\cdot\frac{\pi}{4}\right) = 4 \text{ square units}$$

The rectangle with the maximum area has

$$\text{length } 2x = 4\cos\frac{\pi}{4} = 2\sqrt{2} \text{ and height } y = 2\sin\frac{\pi}{4} = \sqrt{2} \qquad \blacksquare$$

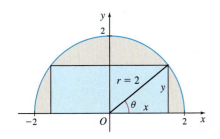

Figure 60

NOW WORK Problem 41.

ORIGINS In 1621 Willebrord Snell (c. 1580–1626) discovered the law of refraction, one of the basic principles of geometric optics. Pierre de Fermat (1601–1665) was able to prove the law mathematically using the principle that light follows the path that takes the least time.

EXAMPLE 6 Snell's Law of Refraction

Light travels at different speeds in different media (air, water, glass, etc.) Suppose that light travels from a point A in one medium, where its speed is c_1, to a point B in another medium, where its speed is c_2. See Figure 61. We use Fermat's principle that light always travels along the path that requires the least time to prove Snell's Law of Refraction.

$$\frac{\sin\theta_1}{c_1} = \frac{\sin\theta_2}{c_2}$$

Solution We position the coordinate system, as illustrated in Figure 62. The light passes from one medium to the other at the point P. Since the shortest distance between two points is a line, the path taken by the light is made up of two line segments—from $A = (0, a)$ to $P = (x, 0)$ and from $P = (x, 0)$ to $B = (k, -b)$, where a, b, and k are positive constants.

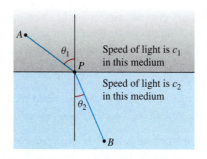

Figure 61

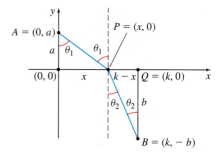

Figure 62

Since

$$\text{Time} = \frac{\text{Distance}}{\text{Speed}}$$

the travel time t_1 from $A = (0, a)$ to $P = (x, 0)$ is

$$t_1 = \frac{\sqrt{x^2 + a^2}}{c_1}$$

and the travel time t_2 from $P = (x, 0)$ to $B = (k, -b)$ is

$$t_2 = \frac{\sqrt{(k - x)^2 + b^2}}{c_2}$$

The total time $T = T(x)$ is given by

$$T(x) = t_1 + t_2 = \frac{\sqrt{x^2 + a^2}}{c_1} + \frac{\sqrt{(k - x)^2 + b^2}}{c_2}$$

To find the least time, we find the critical numbers of T.

$$T'(x) = \frac{x}{c_1\sqrt{x^2 + a^2}} - \frac{k - x}{c_2\sqrt{(k - x)^2 + b^2}} = 0 \qquad (1)$$

From Figure 62,

$$\frac{x}{\sqrt{x^2 + a^2}} = \sin\theta_1 \qquad \text{and} \qquad \frac{k - x}{\sqrt{(k - x)^2 + b^2}} = \sin\theta_2 \qquad (2)$$

Using the result from (2) in equation (1), we have

$$T'(x) = \frac{\sin\theta_1}{c_1} - \frac{\sin\theta_2}{c_2} = 0$$

$$\frac{\sin\theta_1}{c_1} = \frac{\sin\theta_2}{c_2}$$

Now to ensure that the minimum value of T occurs when $T'(x) = 0$, we show that $T''(x) > 0$. From (1),

$$T''(x) = \frac{d}{dx}\left[\frac{x}{c_1\sqrt{x^2 + a^2}}\right] - \frac{d}{dx}\left[\frac{k - x}{c_2\sqrt{(k - x)^2 + b^2}}\right]$$

$$= \frac{a^2}{c_1(x^2 + a^2)^{3/2}} + \frac{b^2}{c_2[(k - x)^2 + b^2]^{3/2}} > 0$$

Since $T''(x) > 0$ for all x, T is concave up for all x, and the minimum value of T occurs at the critical number. That is, T is a minimum when $\dfrac{\sin\theta_1}{c_1} = \dfrac{\sin\theta_2}{c_2}$. ∎

4.7 Assess Your Understanding

Applications and Extensions

1. **Maximizing Area**   The owner of a motel has 3000 m of fencing and wants to enclose a rectangular plot of land that borders a straight highway. If she does not fence the side along the highway, what is the largest area that can be enclosed?

2. **Maximizing Area**   If the motel owner in Problem 1 decides to also fence the side along the highway, except for 5 m to allow for access, what is the largest area that can be enclosed?

3. **Maximizing Area**   Find the dimensions of the rectangle with the largest area that can be enclosed on all sides by L meters of fencing.

4. **Maximizing the Area of a Triangle**   An isosceles triangle has a perimeter of fixed length L. What should the dimensions of the triangle be if its area is to be a maximum? See the figure.

5. **Maximizing Area**   A gardener with 200 m of available fencing wishes to enclose a rectangular field and then divide it into two plots with a fence parallel to one of the sides, as shown in the figure. What is the largest area that can be enclosed?

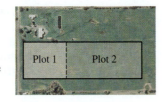

6. **Minimizing Fencing**   A realtor wishes to enclose 600 m² of land in a rectangular plot and then divide it into two plots with a fence parallel to one of the sides. What are the dimensions of the rectangular plot that require the least amount of fencing?

7. **Maximizing the Volume of a Box**   An open box with a square base is to be made from a square piece of cardboard that measures 12 cm on each side. A square will be cut out from each corner of the cardboard and the sides will be turned up to form the box. Find the dimensions that yield the maximum volume.

8. **Maximizing the Volume of a Box**   An open box with a rectangular base is to be made from a piece of tin measuring 30 cm by 45 cm by cutting out a square from each corner and turning up the sides. Find the dimensions that yield the maximum volume.

9. **Minimizing the Surface Area of a Box**   An open box with a square base is to have a volume of 2000 cm³. What should the dimensions of the box be if the amount of material used is to be a minimum?

10. **Minimizing the Surface Area of a Box**   If the box in Problem 9 is to be closed on top, what should the dimensions of the box be if the amount of material used is to be a minimum?

11. **Minimizing the Cost to Make a Can**   A cylindrical container that has a capacity of 10 m³ is to be produced. The top and bottom of the container are to be made of a material that costs $20 per square meter, while the side of the container is to be made of a material costing $15 per square meter. Find the dimensions that will minimize the cost of the material.

12. **Minimizing the Cost of Fencing**   A builder wishes to fence in 60,000 m² of land in a rectangular shape. For security reasons, the fence along the front part of the land will cost $20 per meter, while the fence for the other three sides will cost $10 per meter. How much of each type of fence should the builder buy to minimize the cost of the fence? What is the minimum cost?

13. **Maximizing Revenue**   A car rental agency has 24 cars (each an identical model). The owner of the agency finds that at a price of $18 per day, all the cars can be rented; however, for each $1 increase in rental cost, one of the cars is not rented. What should the agency charge to maximize income?

14. **Maximizing Revenue**   A charter flight club charges its members $200 per year. But for each new member in excess of 60, the charge for every member is reduced by $2. What number of members leads to a maximum revenue?

15. **Minimizing Distance**   Find the coordinates of the points on the graph of the parabola $y = x^2$ that are closest to the point $\left(2, \dfrac{1}{2}\right)$.

16. **Minimizing Distance**   Find the coordinates of the points on the graph of the parabola $y = 2x^2$ that are closest to the point $(1, 4)$.

17. **Minimizing Distance**   Find the coordinates of the points on the graph of the parabola $y = 4 - x^2$ that are closest to the point $(6, 2)$.

1. = NOW WORK problem   〰️ = Graphing technology recommended   [CAS] = Computer Algebra System recommended

18. Minimizing Distance Find the coordinates of the points on the graph of $y = \sqrt{x}$ that are closest to the point $(4, 0)$.

19. Minimizing Transportation Cost A truck has a top speed of 75 mi/h and, when traveling at a speed of x mi/h, consumes gasoline at the rate

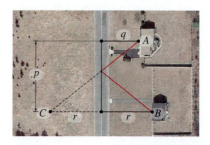

of $\dfrac{1}{200}\left[\dfrac{1600}{x} + x\right]$ gallons

per mile. The truck is to be taken on a 200-mi trip by a driver who is paid at the rate of \$$b$ per hour plus a commission of \$$c$. Since the time required for the trip at x mi/h is $\dfrac{200}{x}$, the cost C of the trip, when gasoline costs \$$a$ per gallon, is

$$C = C(x) = \left(\frac{1600}{x} + x\right)a + \frac{200}{x}b + c$$

Find the speed that minimizes the cost C under each of the following conditions:

(a) $a = \$3.50, b = 0, c = 0$

(b) $a = \$3.50, b = \$10.00, c = \$500$

(c) $a = \$4.00, b = \$20.00, c = 0$

20. Optimal Placement of a Cable Box A telephone company is asked to provide cable service to a customer whose house is located 2 km away from the road along which the cable lines run. The nearest cable box is located 5 km down the road. As shown in the figure below, let $5 - x$ denote the distance from the box to the connection so that x is the distance from this point to the point on the road closest to the house. If the cost to connect the cable line is \$500 per kilometer along the road and \$600 per kilometer away from the road, where along the road from the box should the company connect the cable line so as to minimize construction cost?

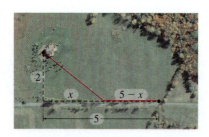

21. Minimizing a Path Two houses A and B on the same side of a road are a distance p apart, with distances q and r, respectively, from the center of the road, as shown in the figure in the next column. Find the length of the shortest path that goes from A to the road and then on to the other house B.

(a) Use calculus.

(b) Use only elementary geometry.

(*Hint:* Reflect B across the road to a point C that is also a distance r from the center of the road.)

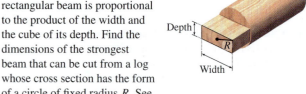

22. Minimizing Travel Time A small island is 3 km from the nearest point P on the straight shoreline of a large lake. A town is 12 km down the shore from P. See the figure. If a person on the island can row a boat 2.5 km/h and can walk 4 km/h, where should the boat be landed so that the person arrives in town in the shortest time?

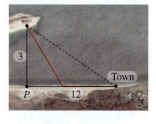

23. Supporting a Wall Find the length of the shortest beam that can be used to brace a wall if the beam is to pass over a second wall 2 m high and 5 m from the first wall. See the figure. What is the angle of elevation of the beam?

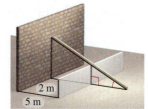

24. Maximizing the Strength of a Beam The strength of a rectangular beam is proportional to the product of the width and the cube of its depth. Find the dimensions of the strongest beam that can be cut from a log whose cross section has the form of a circle of fixed radius R. See the figure.

25. Maximizing the Strength of a Beam If the strength of a rectangular beam is proportional to the product of its width and the square of its depth, find the dimensions of the strongest beam that can be cut from a log whose cross section has the form of the ellipse $10x^2 + 9y^2 = 90$. (*Hint:* Choose width $= 2x$ and depth $= 2y$.)

26. Maximizing the Strength of a Beam The strength of a beam made from a certain wood is proportional to the product of its width and the cube of its depth. Find the dimensions of the rectangular cross section of the beam with maximum strength that can be cut from a log whose original cross section is in the form of the ellipse $b^2x^2 + a^2y^2 = a^2b^2$, $a \geq b$.

27. Pricing Wine A winemaker in Walla Walla, Washington, is producing the first vintage for her own label, and she needs to know how much to charge per case of wine. It costs her \$132 per case to make the wine. She understands from industry research and an assessment of her marketing list that she can sell

$$x = 1430 - \frac{11}{6}p \text{ cases of wine, where } p \text{ is the price of a case in}$$

dollars. She can make at most 1100 cases of wine in her production facility. How many cases of wine should she produce, and what price p should she charge per case to maximize her profit P? (*Hint:* Maximize the profit $P = xp - 132x$, where x equals the number of cases.)

28. Optimal Window Dimensions   A Norman window has the shape of a rectangle surmounted by a semicircle of diameter equal to the width of the rectangle, as shown in the figure. If the perimeter of the window is 10 m, what dimensions will admit the most light?

29. Maximizing Volume   The sides of a V-shaped trough are 28 cm wide. Find the angle between the sides of the trough that results in maximum capacity.

30. Maximizing Volume   A metal rain gutter is to have 10-cm sides and a 10-cm horizontal bottom, with the sides making equal angles with the bottom, as shown in the figure. How wide should the opening across the top be for maximum carrying capacity?

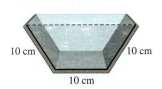

10 cm   10 cm

10 cm

31. Minimizing Construction Cost   A proposed tunnel with a fixed cross-sectional area is to have a horizontal floor, vertical walls of equal height, and a ceiling that is a semicircular cylinder. If the ceiling costs three times as much per square meter to build as the vertical walls and the floor, find the most economical ratio of the diameter of the semicircular cylinder to the height of the vertical walls.

32. Minimizing Construction Cost   An observatory is to be constructed in the form of a right circular cylinder surmounted by a hemispherical dome. If the hemispherical dome costs three times as much per square meter as the cylindrical wall, what are the most economical dimensions for a given volume? Neglect the floor.

33. Intensity of Light   The intensity of illumination at a point varies inversely as the square of the distance between the point and the light source. Two lights, one having an intensity eight times that of the other, are 6 m apart. At what point between the two lights is the total illumination least?

34. Drug Concentration   The concentration of a drug in the bloodstream t hours after injection into muscle tissue is given by $C(t) = \dfrac{2t}{16 + t^2}$. When is the concentration greatest?

35. Optimal Wire Length   A wire is to be cut into two pieces. One piece will be bent into a square, and the other piece will be bent into a circle. If the total area enclosed by the two pieces is to be 64 cm², what is the minimum length of wire that can be used? What is the maximum length of wire that can be used?

36. Optimal Wire Length   A wire is to be cut into two pieces. One piece will be bent into an equilateral triangle, and the other piece will be bent into a circle. If the total area enclosed by the two pieces is to be 64 cm², what is the minimum length of wire that can be used? What is the maximum length of wire that can be used?

37. Optimal Area   A wire 35 cm long is cut into two pieces. One piece is bent into the shape of a square, and the other piece is bent into the shape of a circle.

   (a) How should the wire be cut so that the area enclosed is a minimum?

   (b) How should the wire be cut so that the area enclosed is a maximum?

   (c) Graph the area enclosed as a function of the length of the piece of wire used to make the square. Show that the graph confirms the results of (a) and (b).

38. Optimal Area   A wire 35 cm long is cut into two pieces. One piece is bent into the shape of an equilateral triangle, and the other piece is bent into the shape of a circle.

   (a) How should the wire be cut so that the area enclosed is a minimum?

   (b) How should the wire be cut so that the area enclosed is a maximum?

   (c) Graph the area enclosed as a function of the length of the piece of wire used to make the triangle. Show that the graph confirms the results of (a) and (b).

39. Optimal Dimensions for a Can   Show that a cylindrical container of fixed volume V requires the least material (minimum surface area) when its height is twice its radius.

40. Maximizing Area   Find the triangle of largest area that has two sides along the positive coordinate axes if its hypotenuse is tangent to the graph of $y = 3e^{-x}$.

41. Maximizing Area   Find the largest area of a rectangle with one vertex on the parabola $y = 9 - x^2$, another at the origin, and the remaining two on the positive x-axis and positive y-axis, respectively. See the figure.

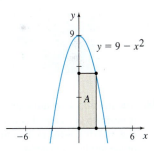

$y = 9 - x^2$

42. Maximizing Volume   Find the dimensions of the right circular cone of maximum volume having a slant height of 4 ft. See the figure.

43. Minimizing Distance   Let a and b be two positive real numbers. Find the line through the point (a, b) and connecting the points $(0, y_0)$ and $(x_0, 0)$ so that the distance from $(x_0, 0)$ to $(0, y_0)$ is a minimum. (In general, x_0 and y_0 will depend on the line.)

44. Maximizing Velocity   An object moves on the x-axis in such a way that its velocity at time t seconds, $t \geq 1$, is given by $v = \dfrac{\ln t}{t}$ cm/s. At what time t does the object attain its maximum velocity?

45. Physics   A heavy object of mass m is to be dragged along a horizontal surface by a rope making an angle θ to the horizontal. The force F required to move the object is given by the formula

$$F = \frac{cmg}{c \sin\theta + \cos\theta}$$

where g is the acceleration due to gravity and c is the **coefficient of friction** of the surface. Show that the force is least when $\tan\theta = c$.

46. Chemistry   A self-catalytic chemical reaction results in the formation of a product that causes its formation rate to increase. The reaction rate V of many self-catalytic chemicals is given by

$$V = kx(a - x)   0 \le x \le a$$

where k is a positive constant, a is the initial amount of the chemical, and x is the variable amount of the chemical. For what value of x is the reaction rate a maximum?

47. Optimal Viewing Angle   A picture 4 m in height is hung on a wall with the lower edge 3 m above the level of an observer's eye. How far from the wall should the observer stand in order to obtain the most favorable view? (That is, the picture should subtend the maximum angle.)

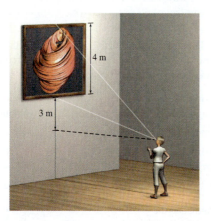

48. Maximizing Transmission Speed   Traditional telephone cable is made up of a core of copper wires covered by an insulating material. If x is the ratio of the radius of the core to the thickness of the insulating material, the speed v of signaling is

$v = kx^2 \ln\dfrac{1}{x}$, where k is a constant and $x \ge 1$. Determine the ratio x that results in maximum speed.

49. Absolute Minimum   If a, b, and c are positive constants, show that the minimum value of $f(x) = ae^{cx} + be^{-cx}$ is $2\sqrt{ab}$.

Challenge Problems

50. Maximizing Length   The figure shows two corridors meeting at a right angle. One has width 1 m, and the other, width 8 m. Find the length of the longest pipe that can be carried horizontally from one corridor, around the corner, and into the other corridor.

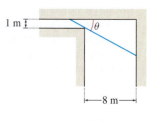

51. Optimal Height of a Lamp   In the figure, a circular area of radius 20 ft is surrounded by a walk. A light is placed above the center of the area. What height most strongly illuminates the walk? The intensity of illumination is given by $I = \dfrac{\sin\theta}{s}$, where s is the distance from the source and θ is the angle at which the light strikes the surface.

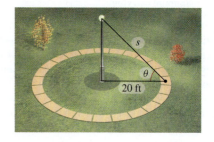

52. Maximizing Area   Show that the rectangle of largest area that can be inscribed under the graph of $y = e^{-x^2}$ has two of its vertices at the point of inflection of y. See the figure.

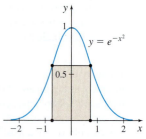

CAS 53. Minimizing Distance   Find the point (x, y) on the graph of $f(x) = e^{-x/2}$ that is closest to the point $(1, 8)$.

4.8 Antiderivatives; Differential Equations

OBJECTIVES   *When you finish this section, you should be able to:*

1 Find antiderivatives (p. 329)

2 Solve a differential equation (p. 331)

3 Solve applied problems modeled by differential equations (p. 333)

We have already learned that for each differentiable function f, there is a corresponding derivative function f'. We now consider the question: For a given function f, can we find a function F whose derivative is f? That is, is it possible to find a function F so that

$$F' = \frac{dF}{dx} = f?$$ If such a function F can be found, it is called an *antiderivative of f*.

DEFINITION Antiderivative

A function F is called an **antiderivative** of the function f if $F'(x) = f(x)$ for all x in the domain of f.

1 Find Antiderivatives

For example, an antiderivative of the function $f(x) = 2x$ is $F(x) = x^2$, since

$$F'(x) = \frac{d}{dx}x^2 = 2x$$

Another function F whose derivative is $2x$ is $F(x) = x^2 + 3$, since

$$F'(x) = \frac{d}{dx}(x^2 + 3) = 2x$$

This leads us to suspect that the function $f(x) = 2x$ has many antiderivatives. Indeed, any of the functions x^2 or $x^2 + \dfrac{1}{2}$ or $x^2 + 2$ or $x^2 + \sqrt{5}$ or $x^2 - 1$ has the property that its derivative is $2x$. Any function $F(x) = x^2 + C$, where C is a constant, is an antiderivative of $f(x) = 2x$.

Are there other antiderivatives of $2x$ that are not of the form $x^2 + C$? A corollary of the Mean Value Theorem (p. 279) tells us the answer is no.

COROLLARY

If f and g are differentiable functions and if $f'(x) = g'(x)$ for all numbers x in an interval (a, b), then there exists a number C for which $f(x) = g(x) + C$ on (a, b).

This result can be stated in the following way.

THEOREM

If a function F is an antiderivative of a function f defined on an interval I, then any other antiderivative of f has the form $F(x) + C$, where C is an (arbitrary) constant.

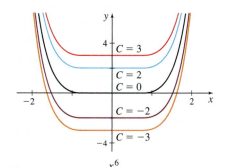

Figure 64 $F(x) = \dfrac{x^6}{6} + C$

All the antiderivatives of f can be obtained from the expression $F(x) + C$ by letting C range over all real numbers. For example, all the antiderivatives of $f(x) = x^5$ are of the form $F(x) = \dfrac{x^6}{6} + C$, where C is a constant. Figure 64 shows the graphs of

$$F(x) = \frac{x^6}{6} + C$$ for some numbers C. The antiderivatives of a function f are a family of functions, each one a vertical translation of the others.

EXAMPLE 1 **Finding the Antiderivatives of a Function**

Find all the antiderivatives of:

(a) $f(x) = 0$ **(b)** $g(\theta) = -\sin\theta$ **(c)** $h(x) = x^{1/2}$

Solution **(a)** Since the derivative of a constant function is 0, all the antiderivatives of f are of the form $F(x) = C$, where C is a constant.

(b) Since $\dfrac{d}{d\theta}\cos\theta = -\sin\theta$, all the antiderivatives of $g(\theta) = -\sin\theta$ are of the form $G(\theta) = \cos\theta + C$.

(c) The derivative of $\dfrac{2}{3}x^{3/2}$ is $\left(\dfrac{2}{3}\right)\left(\dfrac{3}{2}x^{\frac{3}{2}-1}\right) = x^{1/2}$. So, all the antiderivatives of

$h(x) = x^{1/2}$ are of the form $H(x) = \dfrac{2}{3}x^{3/2} + C$, where C is a constant. ∎

In Example 1(c), you may wonder how we knew to choose $\dfrac{2}{3}x^{3/2}$. For any real

number a, the Power Rule states $\dfrac{d}{dx}x^a = ax^{a-1}$. That is, differentiation reduces the

exponent by 1. Antidifferentiation is the inverse process, so it suggests we increase

the exponent by 1. This is how we obtain the factor $x^{3/2}$ of the antiderivative $\dfrac{2}{3}x^{3/2}$. The

factor $\dfrac{2}{3}$ is needed so that when we differentiate $\dfrac{2}{3}x^{3/2}$, the result is $\dfrac{2}{3} \cdot \dfrac{3}{2}x^{\frac{3}{2}-1} = x^{1/2}$.

NOW WORK   Problem 17.

Let $f(x) = x^a$, where a is a real number and $a \neq -1$. (The case for which $a = -1$ requires special attention.) Then the function $F(x)$ defined by

$$F(x) = \frac{x^{a+1}}{a+1} \quad a \neq -1$$

is an antiderivative of $f(x) = x^a$. That is, all the antiderivatives of $f(x) = x^a$, $a \neq -1$,

are of the form $\dfrac{x^{a+1}}{a+1} + C$, where C is a constant.

Now we consider the case when $a = -1$. We know that $\dfrac{d}{dx}\ln|x| = \dfrac{1}{x} = x^{-1}$, for

$x \neq 0$. So, all the antiderivatives of $f(x) = x^{-1}$ are of the form $\ln|x| + C$, provided $x \neq 0$.

Table 6 includes these results along with the antiderivatives of some other common functions.

TABLE 6

Function f	Antiderivatives F of f	Function f	Antiderivatives F of f
$f(x) = 0$	$F(x) = C$	$f(x) = \sec x \tan x$	$F(x) = \sec x + C$
$f(x) = 1$	$F(x) = x + C$	$f(x) = \csc x \cot x$	$F(x) = -\csc x + C$
$f(x) = x^a, \quad a \neq -1$	$F(x) = \dfrac{x^{a+1}}{a+1} + C$	$f(x) = \csc^2 x$	$F(x) = -\cot x + C$
$f(x) = x^{-1} = \dfrac{1}{x}$	$F(x) = \ln|x| + C$	$f(x) = \dfrac{1}{\sqrt{1-x^2}}, \quad |x| < 1$	$F(x) = \sin^{-1} x + C$
$f(x) = e^x$	$F(x) = e^x + C$	$f(x) = \dfrac{1}{1+x^2}$	$F(x) = \tan^{-1} x + C$
$f(x) = a^x$	$F(x) = \dfrac{a^x}{\ln a} + C, \quad a > 0, a \neq 1$	$f(x) = \dfrac{1}{x\sqrt{x^2-1}}, \quad |x| > 1$	$F(x) = \sec^{-1} x + C$
$f(x) = \sin x$	$F(x) = -\cos x + C$	$f(x) = \sinh x$	$F(x) = \cosh x + C$
$f(x) = \cos x$	$F(x) = \sin x + C$	$f(x) = \cosh x$	$F(x) = \sinh x + C$
$f(x) = \sec^2 x$	$F(x) = \tan x + C$		

The next two theorems are consequences of the properties of derivatives and the relationship between derivatives and antiderivatives.

IN WORDS An antiderivative of the sum of two functions equals the sum of the antiderivatives of the functions.

THEOREM Sum Rule

If the functions F_1 and F_2 are antiderivatives of the functions f_1 and f_2, respectively, then $F_1 + F_2$ is an antiderivative of $f_1 + f_2$.

The Sum Rule can be extended to any finite sum of functions. A similar result is also true for differences.

IN WORDS An antiderivative of a number times a function equals the number times an antiderivative of the function.

> **THEOREM Constant Multiple Rule**
>
> If k is a real number and if F is an antiderivative of f, then kF is an antiderivative of kf.

EXAMPLE 2   Finding the Antiderivatives of a Function

Find all the antiderivatives of $f(x) = e^x + \dfrac{6}{x^2} - \sin x$.

Solution Since f is the sum of three functions, we use the Sum Rule. That is, we find the antiderivatives of each function individually and then add.

$$f(x) = e^x + 6x^{-2} - \sin x$$

An antiderivative of e^x is e^x. An antiderivative of $6x^{-2}$ is

$$6 \cdot \frac{x^{-2+1}}{-2+1} = 6 \cdot \frac{x^{-1}}{-1} = -\frac{6}{x}$$

NOTE Since the antiderivatives of e^x are $e^x + C_1$, the antiderivatives of $\dfrac{6}{x^2}$ are $-\dfrac{6}{x} + C_2$, and the antiderivatives of $\sin x$ are $-\cos x + C_3$, the constant C in Example 2 is actually the sum of the constants C_1, C_2, and C_3.

Finally, an antiderivative of $\sin x$ is $-\cos x$. Then all the antiderivatives of the function f are given by

$$F(x) = e^x - \frac{6}{x} + \cos x + C$$

where C is a constant. ■

NOW WORK   Problem 27.

2 Solve a Differential Equation

In studies of physical, chemical, biological, and other phenomena, scientists attempt to find mathematical laws that describe and predict observed behavior. These laws often involve the derivatives of an unknown function F, which must be determined.

For example, suppose we seek all functions $y = F(x)$ for which

$$\frac{dy}{dx} = F'(x) = f(x)$$

An equation of the form $\dfrac{dy}{dx} = f(x)$ is an example of a *differential equation.*[*] Any function $y = F(x)$, for which $\dfrac{dy}{dx} = f(x)$, is a **solution** of the differential equation. The **general solution** of a differential equation $\dfrac{dy}{dx} = f(x)$ consists of all the antiderivatives of f.

For example, the general solution of the differential equation $\dfrac{dy}{dx} = 5x^2 + 2$ is

$$y = \frac{5}{3}x^3 + 2x + C$$

In the differential equation $\dfrac{dy}{dx} = 5x^2 + 2$, suppose the solution must satisfy the **boundary condition** when $x = 3$, then $y = 5$. We use the general solution as follows:

$$y = \frac{5}{3}x^3 + 2x + C$$

$$5 = \frac{5}{3}(3^3) + 2(3) + C = 45 + 6 + C \qquad x = 3, \quad y = 5$$

$$C = -46$$

***

[*]In Chapter 16, we give a more complete definition of a differential equation.

The **particular solution** of the differential equation $\dfrac{dy}{dx} = 5x^2+2$ satisfying the boundary condition when $x = 3$, then $y = 5$ is

$$y = \frac{5}{3}x^3 + 2x - 46$$

EXAMPLE 3 Solving a Differential Equation

Solve the differential equation $\dfrac{dy}{dx} = x^2 + 2x + 1$ with the boundary condition when $x = 3$, then $y = -1$.

Solution We begin by finding the general solution of the differential equation, namely

$$y = \frac{x^3}{3} + x^2 + x + C$$

To determine the number C, we use the boundary condition when $x = 3$, then $y = -1$.

$$-1 = \frac{3^3}{3} + 3^2 + 3 + C \qquad x = 3, \quad y = -1$$
$$C = -22$$

The particular solution of the differential equation with the boundary condition when $x = 3$, then $y = -1$, is

$$y = \frac{x^3}{3} + x^2 + x - 22$$ ∎

NOW WORK Problem 35.

Differential equations often involve higher-order derivatives. For example, the equation $\dfrac{d^2y}{dx^2} = 12x^2$ is an example of a *second-order differential equation*. The **order** of a differential equation is the order of the highest-order derivative of y appearing in the equation.

In solving higher-order differential equations, the number of arbitrary constants in the general solution equals the order of the differential equation. For particular solutions, a first-order differential equation requires one boundary condition; a second-order differential equation requires two boundary conditions; and so on.

EXAMPLE 4 Solving a Second-Order Differential Equation

Solve the differential equation $\dfrac{d^2y}{dx^2} = 12x^2$ with the boundary conditions when $x = 0$, then $y = 1$ and when $x = 3$, then $y = 8$.

Solution All the antiderivatives of $\dfrac{d^2y}{dx^2} = 12x^2$ are

$$\frac{dy}{dx} = 4x^3 + C_1$$

All the antiderivatives of $\dfrac{dy}{dx} = 4x^3 + C_1$ are

$$y = x^4 + C_1x + C_2$$

This is the general solution of the differential equation. To find C_1 and C_2 and the particular solution to the differential equation, we use the boundary conditions.

- When $x = 0$, $\quad 1 = 0^4 + C_1(0) + C_2 \quad$ so $\quad C_2 = 1$
- When $x = 3$, $\quad 8 = 3^4 + 3C_1 + 1 \quad$ so $\quad C_1 = -\dfrac{74}{3}$

The particular solution with the given boundary conditions is

$$y = x^4 - \frac{74}{3}x + 1$$ ■

NOW WORK Problem 39.

3 Solve Applied Problems Modeled by Differential Equations

Rectilinear Motion

Suppose the functions $s = s(t)$, $v = v(t)$, and $a = a(t)$ represent the distance s, velocity v, and acceleration a, respectively, of an object at time t. The three quantities s, v, and a are related by the differential equations

$$\frac{ds}{dt} = v(t) \qquad \text{and} \qquad \frac{dv}{dt} = a(t)$$

If the acceleration $a = a(t)$ is a known function of the time t, then the velocity can be found by solving the differential equation $\dfrac{dv}{dt} = a(t)$. Similarly, if the velocity $v = v(t)$ is a known function of t, then the distance s from the origin at time t is the solution of the differential equation $\dfrac{ds}{dt} = v(t)$.

In physical problems, boundary conditions are often the values of the velocity v and distance s at time $t = 0$. In such cases, $v(0) = v_0$ and $s(0) = s_0$ are referred to as **initial conditions**.

EXAMPLE 5   Solving a Rectilinear Motion Problem

Find the distance s of an object from the origin at time t if its acceleration a is

$$a(t) = 8t - 3$$

and the initial conditions are $v_0 = v(0) = 4$ and $s_0 = s(0) = 1$.

Solution First we solve the differential equation $\dfrac{dv}{dt} = a(t) = 8t - 3$ and use the initial condition $v_0 = v(0) = 4$.

$$v(t) = 4t^2 - 3t + C_1$$
$$v_0 = v(0) = 4(0)^2 - 3(0) + C_1 = 4$$
$$C_1 = 4$$

The velocity of the object at time t is $v(t) = 4t^2 - 3t + 4$.

The distance s of the object at time t satisfies the differential equation

$$\frac{ds}{dt} = v(t) = 4t^2 - 3t + 4$$

Then

$$s(t) = \frac{4}{3}t^3 - \frac{3}{2}t^2 + 4t + C_2$$

Using the initial condition, $s_0 = s(0) = 1$, we have

$$s_0 = s(0) = 0 - 0 + 0 + C_2 = 1$$
$$C_2 = 1$$

The distance s of the object from the origin at any time t is

$$s = s(t) = \frac{4}{3}t^3 - \frac{3}{2}t^2 + 4t + 1$$ ■

NOW WORK Problem 41.

EXAMPLE 6   Solving a Rectilinear Motion Problem

When the brakes of a car are applied, the car decelerates at a constant rate of $10 \, \text{m/s}^2$. If the car is to stop within $20 \, \text{m}$ after the brakes are applied, what is the maximum velocity the car could have been traveling? Express the answer in miles per hour.

Solution Let $s(t)$ represent the distance s in meters the car has traveled t seconds after the brakes are applied. Let v_0 be the velocity of the car at the time the brakes are applied ($t = 0$). Since the car decelerates at the rate of $10 \, \text{m/s}^2$, its acceleration a, in meters per second squared, is

$$a(t) = \frac{dv}{dt} = -10$$

We solve the differential equation for v.

$$v(t) = -10t + C_1$$

When $t = 0$, $v(0) = v_0$, the velocity of the car when the brakes are applied, so $C_1 = v_0$. Then

$$v(t) = \frac{ds}{dt} = -10t + v_0$$

Now we solve the differential equation $v(t) = \dfrac{ds}{dt}$ for s.

$$s(t) = -5t^2 + v_0 t + C_2$$

Since the distance s is measured from the point at which the brakes are applied, the second initial condition is $s(0) = 0$. Then $s(0) = -5 \cdot 0 + v_0 \cdot 0 + C_2 = 0$, so $C_2 = 0$. The distance s, in meters, the car travels t seconds after applying the brakes is

$$s(t) = -5t^2 + v_0 t$$

The car stops completely when its velocity equals 0. That is, when

$$v(t) = -10t + v_0 = 0$$

$$t = \frac{v_0}{10}$$

This is the time it takes the car to come to rest. Substituting $\dfrac{v_0}{10}$ for t in $s(t)$, the distance the car has traveled is

$$s\left(\frac{v_0}{10}\right) = -5\left(\frac{v_0}{10}\right)^2 + v_0\left(\frac{v_0}{10}\right) = \frac{v_0^2}{20}$$

If the car is to stop within $20 \, \text{m}$, then $s \leq 20$; that is, $\dfrac{v_0^2}{20} \leq 20$ or equivalently $v_0^2 \leq 400$. The maximum possible velocity v_0 for the car is $v_0 = 20 \, \text{m/s}$.

To express this in miles per hour, we proceed as follows:

$$v_0 = 20 \, \text{m/s} = \left(\frac{20 \, \text{m}}{\text{s}}\right)\left(\frac{1 \, \text{km}}{1000 \, \text{m}}\right)\left(\frac{3600 \, \text{s}}{1 \, \text{h}}\right)$$

$$= 72 \, \text{km/h} \approx \left(\frac{72 \, \text{km}}{\text{h}}\right)\left(\frac{1 \, \text{mi}}{1.6 \, \text{km}}\right) = 45 \, \text{mi/h}$$

The maximum possible velocity to stop within $20 \, \text{m}$ is $45 \, \text{mi/h}$. ∎

NOW WORK Problem 53.

Freely Falling Objects

An object falling toward Earth is a common example of motion with (nearly) constant acceleration. In the absence of air resistance, all objects, regardless of size, weight, or composition, fall with the same acceleration when released from the same point above

Earth's surface, and if the distance fallen is not too great, the acceleration remains constant throughout the fall. This ideal motion, in which air resistance and the small change in acceleration with altitude are neglected, is called **free fall**. The constant acceleration of a freely falling object is called the **acceleration due to gravity** and is denoted by the symbol g. Near Earth's surface, its magnitude is approximately $32\,\text{ft/s}^2$, or $9.8\,\text{m/s}^2$, and its direction is down toward the center of Earth.

EXAMPLE 7 Solving a Problem Involving Free Fall

A rock is thrown straight up with an initial velocity of 19.6 m/s from the roof of a building 24.5 m above ground level, as shown in Figure 64.

(a) How long does it take the rock to reach its maximum altitude?
(b) What is the maximum altitude of the rock?
(c) If the rock misses the edge of the building on the way down and eventually strikes the ground, what is the total time the rock is in the air?

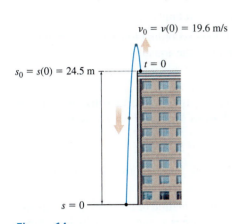

$v_0 = v(0) = 19.6$ m/s

$s_0 = s(0) = 24.5$ m

$t = 0$

$s = 0$

Figure 64

Solution To answer the questions, we need to find the velocity $v = v(t)$ and the distance $s = s(t)$ of the rock as functions of time. We begin measuring time when the rock is released. If s is the distance, in meters, of the rock from the ground, then since the rock is released at a height of 24.5 m, $s_0 = s(0) = 24.5$ m.

The initial velocity of the rock is given as $v_0 = v(0) = 19.6$ m/s. If air resistance is ignored, the only force acting on the rock is gravity. Since the acceleration due to gravity is $-9.8\,\text{m/s}^2$, the acceleration a of the rock is

$$a = \frac{dv}{dt} = -9.8$$

Solving the differential equation, we get

$$v(t) = -9.8t + v_0$$

Using the initial condition, $v_0 = v(0) = 19.6$ m/s, the velocity of the rock at any time t is

$$v(t) = -9.8t + 19.6$$

Now we solve the differential equation $\dfrac{ds}{dt} = v(t) = -9.8t + 19.6$. Then

$$s(t) = -4.9t^2 + 19.6t + s_0$$

Using the initial condition, $s(0) = 24.5$ m, the distance s of the rock from the ground at any time t is

$$s(t) = -4.9t^2 + 19.6t + 24.5$$

Now we can answer the questions.

(a) The rock reaches its maximum altitude when its velocity is 0.

$$v(t) = -9.8t + 19.6 = 0$$

$$t = 2$$

The rock reaches its maximum altitude at $t = 2$ seconds.

(b) To obtain the maximum altitude, we evaluate $s(2)$. The maximum altitude of the rock is

$$s(2) = -4.9(2^2) + 19.6(2) + 24.5 = 44.1 \text{ m}$$

(c) We find the total time the rock is in the air by solving $s(t) = 0$.

$$-4.9t^2 + 19.6t + 24.5 = 0$$
$$t^2 - 4t - 5 = 0$$
$$(t - 5)(t + 1) = 0$$
$$t = 5 \text{ or } t = -1$$

The only meaningful solution is $t = 5$. The rock is in the air for 5 seconds. ■

Now we examine the general problem of freely falling objects.

If F is the weight of an object of mass m, then according to Galileo, assuming air resistance is negligible, a freely falling object obeys the equation

$$F = -mg$$

where g is the acceleration due to gravity. The minus sign indicates that the object is falling. Also, according to **Newton's Second Law of Motion**, $F = ma$, so we have

$$ma = -mg \qquad \text{or} \qquad a = -g$$

where a is the acceleration of the object. We seek formulas for the velocity v and distance s from Earth of a freely falling object at time t.

Let $t = 0$ be the instant we begin to measure the motion of the object, and suppose at this instant the object's vertical distance above Earth is s_0 and its velocity is v_0. Since the acceleration $a = \dfrac{dv}{dt}$ and $a = -g$, we have

$$\boxed{a = \frac{dv}{dt} = -g} \qquad (1)$$

Solving the differential equation (1) for v and using the initial condition $v_0 = v(0)$, we obtain

$$\boxed{v(t) = -gt + v_0}$$

Now since $\dfrac{ds}{dt} = v(t)$, we have

$$\frac{ds}{dt} = -gt + v_0 \qquad (2)$$

Solving the differential equation (2) for s and using the initial condition $s_0 = s(0)$, we obtain

$$\boxed{s(t) = -\frac{1}{2}gt^2 + v_0 t + s_0}$$

NOW WORK   **Problem 55.**

Newton's First Law of Motion

We close this section with a special case of the *Law of Inertia*, which was originally stated by Galileo.

THEOREM Newton's First Law of Motion

If no force acts on a body, then a body at rest remains at rest and a body moving with constant velocity will continue to do so.

Proof The force F acting on a body of mass m is given by Newton's Second Law of Motion $F = ma$, where a is the acceleration of the body. If there is no force acting on the body, then $F = 0$. In this case, the acceleration a must be 0. But $a = \dfrac{dv}{dt}$, where v is the velocity of the body. So,

$$\frac{dv}{dt} = 0 \qquad \text{and} \qquad v = \text{Constant}$$

That is, the body is at rest ($v = 0$) or else in a state of uniform motion (v is a nonzero constant). ∎

4.8 Assess Your Understanding

Concepts and Vocabulary

1. A function F is called a(n) _____ of a function f if $F' = f$.

2. *True or False*   If F is an antiderivative of f, then $F(x) + C$, where C is a constant, is also an antiderivative of f.

3. All the antiderivatives of $y = x^{-1}$ are _____.

4. *True or False*   An antiderivative of $\sin x$ is $-\cos x + \pi$.

5. *True or False*   The general solution of a differential equation $\dfrac{dy}{dx} = f(x)$ consists of all the antiderivatives of $f'(x)$.

6. *True or False*   Free fall is an example of motion with constant acceleration.

7. *True or False*   To find a particular solution of a differential equation $\dfrac{dy}{dx} = f(x)$, we need a boundary condition.

8. *True or False*   If F_1 and F_2 are both antiderivatives of a function f on an interval I, then $F_1 - F_2 = C$, a constant.

Skill Building

In Problems 9–30, find all the antiderivatives of each function.

9. $f(x) = 2$

10. $f(x) = \dfrac{1}{2}$

11. $f(x) = 4x^5$

12. $f(x) = x$

13. $f(x) = 5x^{3/2}$

14. $f(x) = x^{5/2} + 2$

15. $f(x) = 2x^{-2}$

16. $f(x) = 3x^{-3}$

17. $f(x) = \sqrt{x}$

18. $f(x) = \dfrac{1}{\sqrt{x}}$

19. $f(x) = 4x^3 - 3x^2 + 1$

20. $f(x) = x^2 - x$

21. $f(x) = (2 - 3x)^2$

22. $f(x) = (3x - 1)^2$

23. $f(x) = \dfrac{3x - 2}{x}$

24. $f(x) = \dfrac{4x^{3/2} - 1}{x}$

25. $f(x) = 2x - 3\cos x$

26. $f(x) = 2\sin x - \cos x$

27. $f(x) = 4e^x + x$

28. $f(x) = e^{-x} + \sec^2 x$

29. $f(x) = \dfrac{7}{1 + x^2}$

30. $f(x) = x + \dfrac{10}{\sqrt{1 - x^2}}$

In Problems 31–40, find the particular solution of each differential equation having the given boundary condition(s).

31. $\dfrac{dy}{dx} = 3x^2 - 2x + 1$,   when $x = 2$, $y = 1$

32. $\dfrac{dv}{dt} = 3t^2 - 2t + 1$,   when $t = 1$, $v = 5$

33. $\dfrac{dy}{dx} = x^{1/3} + x\sqrt{x} - 2$,   when $x = 1$, $y = 2$

34. $s'(t) = t^4 + 4t^3 - 5$, $s(2) = 5$

35. $\dfrac{ds}{dt} = t^3 + \dfrac{1}{t^2}$,   when $t = 1$, $s = 2$

36. $\dfrac{dy}{dx} = \sqrt{x} - x\sqrt{x} + 1$,   when $x = 1$, $y = 0$

37. $f'(x) = x - 2\sin x$, $f(\pi) = 0$

38. $\dfrac{dy}{dx} = x^2 - 2\sin x$,   when $x = \pi$, $y = 0$

39. $\dfrac{d^2y}{dx^2} = e^x$,   when $x = 0$, $y = 2$,   when $x = 1$, $y = e$

40. $f''(\theta) = \sin\theta + \cos\theta$, $f'\left(\dfrac{\pi}{2}\right) = 2$ and $f(\pi) = 4$

In Problems 41–44, the acceleration of an object is given. Find the distance s of the object from the origin under the given initial conditions.

41. $a = -32\,\text{ft/s}^2$, $s(0) = 0\,\text{ft}$, $v(0) = 128\,\text{ft/s}$

42. $a = -980\,\text{cm/s}^2$, $s(0) = 5\,\text{cm}$, $v(0) = 1980\,\text{cm/s}$

43. $a = 3t\,\text{m/s}^2$, $s(0) = 2\,\text{m}$, $v(0) = 18\,\text{m/s}$

44. $a = 5t - 2\,\text{ft/s}^2$, $s(0) = 0\,\text{ft}$, $v(0) = 8\,\text{ft/s}$

Applications and Extensions

In Problems 45 and 46, find all the antiderivatives of each function. (Hint: Simplify first.)

45. $f(u) = \dfrac{u^2 + 10u + 21}{3u + 9}$

46. $f(t) = \dfrac{t^3 - 5t + 8}{t^5}$

In Problems 47 and 48, find the solution of each differential equation having the given boundary condition. (Hint: Simplify first.)

47. $f'(t) = \dfrac{t^4 + 3t - 1}{t}$ if $f(1) = \dfrac{1}{4}$

48. $g'(x) = \dfrac{x^2 - 1}{x^4 - 1}$ if $g(0) = 0$

49. Use the fact that
$$\frac{d}{dx}(x\cos x + \sin x) = -x\sin x + 2\cos x$$
to find F if
$$\frac{dF}{dx} = -x\sin x + 2\cos x \qquad \text{and} \qquad F(0) = 1$$

50. Use the fact that

$$\frac{d}{dx}\sin x^2 = 2x\cos x^2$$

to find h if

$$\frac{dh}{dx} = x\cos x^2 \quad \text{and} \quad h(0) = 2$$

51. Rectilinear Motion   A car decelerates at a constant rate of $10\,\text{m/s}^2$ when its brakes are applied. If the car must stop within 15 m after applying the brakes, what is the maximum allowable velocity for the car? Express the answer in m/s and in mi/h.

52. Rectilinear Motion   A car can accelerate from 0 to $60\,\text{km/h}$ in 10 seconds. If the acceleration is constant, how far does the car travel during this time?

53. Rectilinear Motion   A BMW 6 series can accelerate from 0 to 60 mph in 5 seconds. If the acceleration is constant, how far does the car travel during this time?

Source: BMW USA.

54. Free Fall   The 2-m high jump is common today. If this event were held on the Moon, where the acceleration due to gravity is $1.6\,\text{m/s}^2$, what height would be attained? Assume that an athlete can propel him- or herself with the same force on the Moon as on Earth.

55. Free Fall   The world's high jump record, set on July 27, 1993, by Cuban jumper Javier Sotomayor, is 2.45 m. If this event were held on the Moon, where the acceleration due to gravity is $1.6\,\text{m/s}^2$, what height would Sotomayor have attained? Assume that he propels himself with the same force on the Moon as on Earth.

56. Free Fall   A ball is thrown straight up from ground level, with an initial velocity of $19.6\,\text{m/s}$. How high is the ball thrown? How long will it take the ball to return to ground level?

57. Free Fall   A child throws a ball straight up. If the ball is to reach a height of 9.8 m, what is the minimum initial velocity that must be imparted to the ball? Assume the initial height of the ball is 1 m.

58. Free Fall   A ball thrown directly down from a roof 49 m high reaches the ground in 3 seconds. What is the initial velocity of the ball?

59. Inertia   A constant force is applied to an object that is initially at rest. If the mass of the object is 4 kg and if its velocity after 6 seconds is 12 m/s, determine the force applied to it.

60. Rectilinear Motion   Starting from rest, with what constant acceleration must a car move to travel 2 km in 2 min? (Give your answer in centimeters per second squared.)

61. Downhill Speed of a Skier   The down slope acceleration a of a skier is given by $a = a(t) = g\sin\theta$, where t is time, in seconds, $g = 9.8\,\text{m/s}^2$ is the acceleration due to gravity, and θ is the angle of the slope. If the skier starts from rest at the lift, points his skis straight down a 20° slope, and does not turn, how fast is he going after 5 seconds?

62. Free Fall   A child on top of a building 24 m high drops a rock and then 1 second later throws another rock straight down. What initial velocity must the second rock be given so that the dropped rock and the thrown rock hit the ground at the same time?

Challenge Problems

63. Radiation   Radiation, such as X-rays or the radiation from radioactivity, is absorbed as it passes through tissue or any other material. The rate of change in the intensity I of the radiation with respect to the depth x of tissue is directly proportional to the intensity I. This proportion can be expressed as an equation by introducing a positive constant of proportionality k, where k depends on the properties of the tissue and the type of radiation.

(a) Show that $\dfrac{dI}{dx} = -kI$, $k > 0$.

(b) Explain why the minus sign is necessary.

(c) Solve the differential equation in (a) to find the intensity I as a function of the depth x in the tissue. The intensity of the radiation when it enters the tissue is $I(0) = I_0$.

(d) Find the value of k if the intensity is reduced by 90% of its maximum value at a depth of $2.0\,\text{cm}$.

64. Moving Shadows   A lamp on a post 10 m high stands 25 m from a wall. A boy standing 5 m from the lamp and 20 m from the wall throws a ball straight up with an initial velocity of $19.6\,\text{m/s}$. The acceleration due to gravity is $a = -9.8\,\text{m/s}^2$. The ball is thrown up from an initial height of 1 m above ground.

(a) How fast is the shadow of the ball moving on the wall 3 seconds after the ball is released?

(b) Explain if the ball is moving up or down.

(c) How far is the ball above ground at $t = 3$ seconds?

Chapter Review

THINGS TO KNOW

4.1 Related Rates

Steps for solving a related rate problem (p. 256)

4.2 Maximum and Minimum Values; Critical Numbers

Definitions:

- Absolute maximum; absolute minimum (p. 264)
- Absolute maximum value; absolute minimum value (p. 264)
- Local maximum; local minimum (p. 264)
- Local maximum value; local minimum value (p. 264)
- Critical number (p. 267)

Theorems:

- **Extreme Value Theorem** If a function f is continuous on a closed interval $[a, b]$, then f has an absolute maximum and an absolute minimum on $[a, b]$. (p. 265)
- If a function f has a local maximum or a local minimum at the number c, then either $f'(c) = 0$ or $f'(c)$ does not exist. (p. 266)

Procedure:

- Steps for finding the absolute extreme values of a function f that is continuous on a closed interval $[a, b]$. (p. 268)

4.3 The Mean Value Theorem

- **Rolle's Theorem** (p. 275)
- **Mean Value Theorem** (p. 277)
- **Corollaries to the Mean Value Theorem**
  - If $f'(x) = 0$ for all numbers x in (a, b), then f is constant on (a, b). (p. 279)
  - If $f'(x) = g'(x)$ for all numbers x in (a, b), then there is a number C for which $f(x) = g(x) + C$ on (a, b). (p. 279)
- **Increasing/Decreasing Function Test** (p. 279)

4.4 Local Extrema and Concavity

Definitions:

- Concave up; concave down (p. 288)
- Inflection point (p. 289)

Theorems:

- First Derivative Test (p. 284)
- Test for concavity (p. 288)
- A condition for an inflection point (p. 290)
- Second Derivative Test (p. 291)

Procedure:

- Steps for finding the inflection points of a function (p. 290)

4.5 Indeterminate Forms and L'Hôpital's Rule

Definitions:

- Indeterminate form at c of the type $\dfrac{0}{0}$ or the type $\dfrac{\infty}{\infty}$ (p. 299)
- Indeterminate form at c of the type $0 \cdot \infty$ or $\infty - \infty$ (p. 304)
- Indeterminate form at c of the type 1^{∞}, 0^0, or ∞^0 (p. 304)

Theorem:

- L'Hôpital's Rule (p. 299)

Procedures:

- Steps for finding a limit using L'Hôpital's Rule (p. 300)

- Steps for finding $\lim\limits_{x \to c}[f(x)]^{g(x)}$, where $[f(x)]^{g(x)}$ is an indeterminate form at c of the type 1^{∞}, 0^0, or ∞^0 (p. 304)

4.6 Using Calculus to Graph Functions

Procedure:

- Steps for graphing a function $y = f(x)$. (p. 308)

4.7 Optimization

Procedure:

- Steps for solving optimization problems (p. 319)

4.8 Antiderivatives; Differential Equations

Definitions:

- Antiderivative (p. 329)
- General solution of a differential equation (p. 331)
- Particular solution of a differential equation (p. 332)
- Boundary condition (p. 331)
- Initial condition (p. 333)

Basic Antiderivatives   See Table 6 (p. 330)

Antidifferentiation Properties:

- **Sum Rule**: If functions F_1 and F_2 are antiderivatives of the functions f_1 and f_2, respectively, then $F_1 + F_2$ is an antiderivative of $f_1 + f_2$. (p. 330)
- **Constant Multiple Rule**: If k is a real number and if F is an antiderivative of f, then kF is an antiderivative of kf. (p. 331)

Theorems:

- If a function F is an antiderivative of a function f on an interval I, then any other antiderivative of f has the form $F(x) + C$, where C is an (arbitrary) constant. (p. 329)
- Newton's First Law of Motion (p. 336)
- Newton's Second Law of Motion (p. 336)

OBJECTIVES

Section	You should be able to ...	Examples	Review Exercises
4.1	1 Solve related rate problems (p. 255)	1–5	1–3
4.2	1 Identify absolute maximum and minimum values and local extreme values of a function (p. 263)	1	4, 5
	2 Find critical numbers (p. 267)	2	6(a), 7
	3 Find absolute maximum and absolute minimum values (p. 268)	3–6	8, 9
4.3	1 Use Rolle's Theorem (p. 275)	1	10
	2 Work with the Mean Value Theorem (p. 276)	2, 3	11, 12, 28
	3 Identify where a function is increasing and decreasing (p. 279)	4–6	23(a), 24(a)
4.4	1 Use the First Derivative Test to find local extrema (p. 284)	1, 2	6(b), 13(a)–15(a)
	2 Use the First Derivative Test with rectilinear motion (p. 286)	3	16
	3 Determine the concavity of a function (p. 287)	4, 5	23(b), 24(b)
	4 Find inflection points (p. 290)	6	23(c), 24(c)
	5 Use the Second Derivative Test to find local extrema (p. 291)	7–8	13(b)–15(b)
4.5	1 Identify indeterminate forms of the type $\dfrac{0}{0}$ and $\dfrac{\infty}{\infty}$ (p. 298)	1	41–44
	2 Use L'Hôpital's Rule to find a limit (p. 299)	2–6	45, 47, 49–52, 55
	3 Find the limit of an indeterminate form of the type $0 \cdot \infty$, $\infty - \infty$, 0^0, 1^{∞}, or ∞^0 (p. 303)	7–10	46, 48, 53, 54, 56

Section	You should be able to …	Examples	Review Exercises
4.6	1 Graph a function using calculus (p. 308)	1–7	17–22, 25, 26, 27
4.7	1 Solve optimization problems (p. 318)	1–6	29, 30, 61–63
4.8	1 Find antiderivatives (p. 329)	1, 2	31–38
	2 Solve a differential equation (p. 331)	3, 4	57–60
	3 Solve applied problems modeled by differential equations (p. 333)	5–7	39, 40, 64

REVIEW EXERCISES

1. **Related rates**   A spherical snowball is melting at the rate of $2\,\text{cm}^3/\text{min}$. How fast is the surface area changing when the radius is $5\,\text{cm}$?

2. **Related rates**   A lighthouse is $3\,\text{km}$ from a straight shoreline. Its light makes one revolution every 8 seconds. How fast is the light moving along the shore when it makes an angle of $30°$ with the shoreline?

3. **Related rates**   Two planes at the same altitude are approaching an airport, one from the north and one from the west. The plane from the north is flying at 250 mph and is 30 mi from the airport. The plane from the west is flying at $200\,\text{mi}/\text{h}$ and is 20 mi from the airport. How fast are the planes approaching each other at that instant?

In Problems 4 and 5, use the graphs below to determine whether each function has an absolute extremum and/or a local extremum or neither at the indicated points.

4.

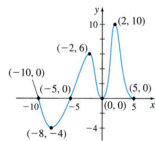

5.

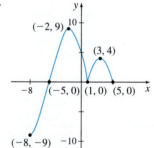

6. **Critical Numbers** $f(x) = \dfrac{x^2}{2x-1}$

   (a) Find all the critical numbers of f.

   (b) Find the local extrema of f.

7. **Critical Numbers**   Find all the critical numbers of $f(x) = \cos(2x)$ on the closed interval $[0, \pi]$.

In Problems 8 and 9, find the absolute maximum value and absolute minimum value of each function on the indicated interval.

8. $f(x) = x - \sin(2x)$ on $[0, 2\pi]$

9. $f(x) = \dfrac{3}{2}x^4 - 2x^3 - 6x^2 + 5$ on $[-2, 3]$

10. **Rolle's Theorem**   Verify that the hypotheses for Rolle's Theorem are satisfied for the function $f(x) = x^3 - 4x^2 + 4x$ on $[0, 2]$. Find the coordinates of the point(s) at which there is a horizontal tangent line to the graph of f.

11. **Mean Value Theorem**   Verify that the hypotheses for the Mean Value Theorem are satisfied for the function $f(x) = \dfrac{2x-1}{x}$ on the interval $[1, 4]$. Find a point on the graph of f that has a tangent with a slope equal to that of the secant line joining $(1, 1)$ to $\left(4, \dfrac{7}{4}\right)$.

12. **Mean Value Theorem**   Does the Mean Value Theorem apply to the function $f(x) = \sqrt{x}$ on the interval $[0, 9]$? If not, why not? If so, find the number c referred to in the theorem.

In Problems 13–15, find the local extrema of each function:
(a) *Using the First Derivative Test.*
(b) *Using the Second Derivative Test, if possible. If the Second Derivative Test cannot be used, explain why.*

13. $f(x) = x^3 - x^2 - 8x + 1$ 14. $f(x) = x^2 - 24x^{2/3}$

15. $f(x) = x^4 e^{-2x}$

16. **Rectilinear Motion**   The distance s of an object from the origin at time t is given by $s = s(t) = t^4 + 2t^3 - 36t^2$. Draw figures to illustrate the motion of the object and its velocity.

In Problems 17–22, graph each function. Follow the steps given in Section 4.6.

17. $f(x) = -x^3 - x^2 + 2x$ 18. $f(x) = x^{1/3}(x^2 - 9)$

19. $f(x) = xe^x$ 20. $f(x) = \dfrac{x-3}{x^2-4}$

21. $f(x) = x\sqrt{x-3}$ 22. $f(x) = x^3 - 3\ln x$

In Problems 23 and 24, for each function:
(a) *Determine the intervals where each function is increasing and decreasing.*
(b) *Determine the intervals on which each function f is concave up and concave down.*
(c) *Identify any inflection points.*

23. $f(x) = x^4 + 12x^2 + 36x - 11$

24. $f(x) = 3x^4 - 2x^3 - 24x^2 - 7x + 2$

25. If y is a function and $y' > 0$ for all x and $y'' < 0$ for all x, which of the following could be part of the graph of $y = f(x)$? See illustrations (A) through (D).

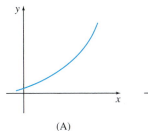

(A)

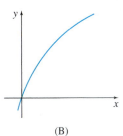

(B)

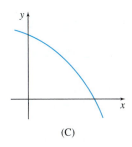

(C)

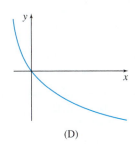

(D)

26. Sketch the graph of a function f that has the following properties:

$f(-3) = 2; \quad f(-1) = -5; \quad f(2) = -4;$

$f(6) = -1 \quad f'(-3) = f'(6) = 0$

$\lim_{x \to 0^-} f(x) = -\infty; \quad \lim_{x \to 0^+} f(x) = \infty;$

$f''(x) > 0$ if $x < -3$ or $0 < x < 4$

$f''(x) < 0$ if $-3 < x < 0$ or $4 < x$

27. Sketch the graph of a function f that has the following properties:

$f(-2) = 2; \quad f(5) = 1; \quad f(0) = 0$

$f'(x) > 0$ if $x < -2$ or $5 < x$

$f'(x) < 0$ if $-2 < x < 2$ or $2 < x < 5$

$f''(x) > 0$ if $x < 0$ or $2 < x$ and $f''(x) < 0$ if $0 < x < 2$

$\lim_{x \to 2^-} f(x) = -\infty \quad \lim_{x \to 2^+} f(x) = \infty$

28. Mean Value Theorem For the function $f(x) = x\sqrt{x+1}$, $0 \le x \le b$, the number c satisfying the Mean Value Theorem is $c = 3$. Find b.

29. Maximizing Volume An open box is to be made from a piece of cardboard by cutting squares out of each corner and folding up the sides. If the size of the cardboard is 2 ft by 3 ft, what size squares (in inches) should be cut out to maximize the volume of the box?

30. Minimizing Distance Find the point on the graph of $2y = x^2$ nearest to the point $(4, 1)$.

In Problems 31–38, find all the antiderivatives of each function.

31. $f(x) = 0$

32. $f(x) = x^{1/2}$

33. $f(x) = \cos x$

34. $f(x) = \sec x \tan x$

35. $f(x) = \dfrac{2}{x}$

36. $f(x) = -2x^{-3}$

37. $f(x) = 4x^3 - 9x^2 + 10x - 3$

38. $f(x) = e^x + \dfrac{4}{x}$

39. Velocity A box moves down an inclined plane with an acceleration $a(t) = t^2(t - 3)\,\text{cm/s}^2$. It covers a distance of 10 cm in 2 seconds. What was the original velocity of the box?

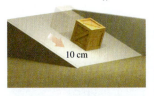

10 cm

40. Free Fall Two objects begin a free fall from rest at the same height 1 second apart. How long after the first object begins to fall will the two objects be 10 m apart?

In Problems 41–44, determine if the expression is an indeterminate form at 0. If it is, identify its type.

41. $\dfrac{xe^{3x} - x}{1 - \cos(2x)}$

42. $\left(\dfrac{1}{x}\right)^{\tan x}$

43. $\dfrac{1}{x^2} - \dfrac{1}{x^2 \sec x}$

44. $\dfrac{\tan x - x}{x - \sin x}$

In Problems 45–56, find each limit.

45. $\lim_{x \to \pi/2} \dfrac{\sec^2 x}{\sec^2(3x)}$

46. $\lim_{x \to 0} \left[\dfrac{2}{\sin^2 x} - \dfrac{1}{1 - \cos x}\right]$

47. $\lim_{x \to 0} \dfrac{e^x - e^{-x}}{\sin x}$

48. $\lim_{x \to 0^-} x \cot(\pi x)$

49. $\lim_{x \to 0} \dfrac{\tan x + \sec x - 1}{\tan x - \sec x + 1}$

50. $\lim_{x \to a} \dfrac{ax - x^2}{a^4 - 2a^3 x + 2ax^3 - x^4}$

51. $\lim_{x \to 0} \dfrac{x - \sin x}{x^3}$

52. $\lim_{x \to 0} \dfrac{\tan x - \sin x}{\sin^3 x}$

53. $\lim_{x \to \infty} (1 + 4x)^{2/x}$

54. $\lim_{x \to 1} \left[\dfrac{2}{x^2 - 1} - \dfrac{1}{x - 1}\right]$

55. $\lim_{x \to 4} \dfrac{x^2 - 16}{x^2 + x - 20}$

56. $\lim_{x \to 0^+} (\cot x)^x$

In Problems 57–60, find the solution of each differential equation having the given boundary conditions.

57. $\dfrac{dy}{dx} = e^x$, when $x = 0$, $y = 2$

58. $\dfrac{dy}{dx} = \dfrac{1}{2} \sec x \tan x$, when $x = 0$, $y = 7$

59. $\dfrac{dy}{dx} = \dfrac{2}{x}$, when $x = 1$, then $y = 4$

60. $\dfrac{d^2 y}{dx^2} = x^2 - 4$, when $x = 3$, $y = 2$, when $x = 2$, $y = 2$

61. Maximizing Profit   A manufacturer has determined that the cost C of producing x items is given by
$C(x) = 200 + 35x + 0.02x^2$ dollars.
Each item can be sold for \$78. How many items should she produce to maximize profit?

62. Optimization   The sales of a new stereo system over a period of time are expected to follow the logistic curve

$$f(x) = \frac{5000}{1 + 5e^{-x}} \qquad x \geq 0$$

where x is measured in years. In what year is the sales rate a maximum?

63. Maximum Area   Find the area of the rectangle of largest area in the fourth quadrant that has vertices at $(0, 0)$, $(x, 0)$, $x > 0$, and $(0, y)$, $y < 0$. The fourth vertex is on the graph of $y = \ln x$.

64. Differential Equation   A motorcycle accelerates at a constant rate from 0 to 72 km/h in 10 seconds. How far has it traveled in that time?

CHAPTER 4 PROJECT   The U.S. Economy

Economic data is quite variable and very complicated, making it difficult to model. The models we use here are rough approximations of the Unemployment Rate and Gross Domestic Product (GDP).

The unemployment rate can be modeled by

$$U = U(t) = \frac{5 \cos \dfrac{2t}{\pi}}{2 + \sin \dfrac{2t}{\pi}} + 6 \qquad 0 \leq t \leq 25$$

where t is in years and $t = 0$ represents January 1, 1985.

The GDP growth rate can be modeled by

$$G = G(t) = 5 \cos\left(\frac{2t}{\pi} + 5\right) + 2 \qquad 0 \leq t \leq 25$$

where t is in years and $t = 0$ represents January 1, 1985. If the GDP growth rate is increasing, the economy is **expanding** and if it is decreasing, the economy is **contracting**. A **recession** occurs when GDP growth rate is negative.

1. During what years was the U.S. economy in recession?
2. Find the critical numbers of $U = U(t)$.
3. Determine the intervals on which U is increasing and on which it is decreasing.
4. Find all the local extrema of U.
5. Find the critical numbers of $G = G(t)$.
6. Determine the intervals on which G is increasing and on which it is decreasing.
7. Find all the local extrema of G.

8. **(a)** Find the inflection points of G.

   (b) Describe in economic terms what the inflection points of G represent.

9. Graph $U = U(t)$ and $G = G(t)$ on the same set of coordinate axes.

10. Okun's Law states that an increase in the unemployment rate tends to coincide with a decrease in the GDP growth rate.

    (a) During years of recession, what is happening to the unemployment rate? Explain in economic terms why this makes sense.

    (b) Use your answer to part (a) to explain whether the functions U and G generally agree with Okun's Law.

    (c) If there are times when the functions do not satisfy Okun's Law, provide an explanation in economic terms for this.

11. What relationship, if any, exists between the inflection points of the GDP growth rate and the increase/ decrease of the unemployment rate? Explain in economic terms why such a relationship makes sense.

12. One school of economic thought holds the view that when the economy improves (increasing GDP), more jobs are created resulting in a lower unemployment rate. Others argue that once the unemployment rate improves (increases), more people are working and this increases GDP. Based on your analysis of the graphs of U and G, which position do you support?

For real data on the unemployment rate and GDP, see
http://data.bls.gov/pdq/SurveyOutputServlet and
http://www.tradingeconomics.com/unites-states/gdp-growth

To read Arthur Okun's paper, see
http://cowles.econ.yale.edu/p/cp/pϕ1b/pϕ19ϕ.pdf

5

The Integral

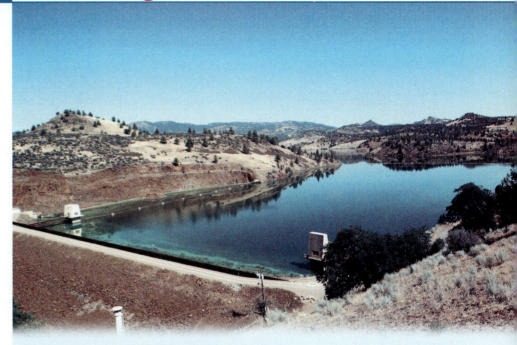

Managing the Klamath River

The Klamath River starts in the eastern lava plateaus of Oregon, passes through a farming region of that state, and then crosses northern California. Because it is the largest river in the region, runs through a variety of terrain, and has different land uses, it is important in a number of ways. Historically, the Klamath supported large salmon runs. It is used to irrigate agricultural lands, and to generate electric power for the region. Downstream, it runs through federal wild lands, and is used for fishing, rafting, and kayaking.

If a river is to be well-managed for such a variety of uses, its flow must be understood. For that reason, the U.S. Geological Survey (USGS) maintains a number of gauges along the Klamath. These typically measure the depth of the river, which can then be expressed as a rate of flow in cubic feet per second (ft^3/s). In order to understand the river flow fully, we must be able to find the total amount of water that flows down the river over any period of time.

CHAPTER 5 PROJECT The Chapter Project on page 403 examines ways to obtain the total flow over any period of time from data that provide flow rates.

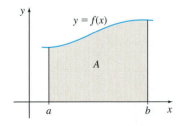

Figure 1 A is the area enclosed by the graph of f, the x-axis, and the lines $x = a$ and $x = b$.

We began our study of calculus by asking two questions from geometry. The first, the tangent problem, "What is the slope of the tangent line to the graph of a function?" led to the derivative of a function.

The second question was the area problem: Given a function f, defined and nonnegative on a closed interval $[a, b]$, what is the area enclosed by the graph of f, the x-axis, and the vertical lines $x = a$ and $x = b$? Figure 1 illustrates this area.

The first two sections of Chapter 5 show how the concept of the integral evolves from the area problem. At first glance, the area problem and the tangent problem look quite dissimilar. However, much of calculus is built on a surprising relationship between the two problems and their associated concepts. This relationship is the basis for the *Fundamental Theorem of Calculus*, discussed in Section 5.3.

343

5.1 Area

OBJECTIVES *When you finish this section, you should be able to:*

1 Approximate the area under the graph of a function (p. 344)

2 Find the area under the graph of a function (p. 348)

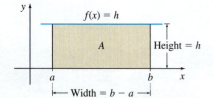

Figure 2 $A = h(b - a)$.

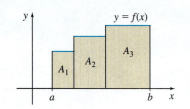

Figure 3 $A = A_1 + A_2 + A_3$

In this section, we present a method for finding the area enclosed by the graph of a function $y = f(x)$ that is nonnegative on a closed interval $[a, b]$, the x-axis, and the lines $x = a$ and $x = b$. The presentation uses summation notation ($\sum$), which is reviewed in Appendix A.5.

The area A of a rectangle with width w and height h is given by the geometry formula

$$A = hw$$

See Figure 2. The graph of a constant function $f(x) = h$, for some positive constant h, is a horizontal line that lies above the x-axis. The area enclosed by this line, the x-axis, and the lines $x = a$ and $x = b$ is the rectangle whose area A is the product of the width $(b - a)$ and the height h.

$$A = h(b - a)$$

If the graph of $y = f(x)$ consists of three horizontal lines, each of positive height as shown in Figure 3, the area A enclosed by the graph of f, the x-axis, and the lines $x = a$ and $x = b$ is the sum of the rectangular areas A_1, A_2, and A_3.

1 Approximate the Area Under the Graph of a Function

EXAMPLE 1 **Approximating the Area Under the Graph of a Function**

Approximate the area A enclosed by the graph of $f(x) = \dfrac{1}{2}x + 3$, the x-axis, and the lines $x = 2$ and $x = 4$.

Solution Figure 4 illustrates the area A to be approximated.

We begin by drawing a rectangle of width $4 - 2 = 2$ and height $f(2) = 4$. The area of the rectangle, $2 \cdot 4 = 8$, approximates the area A, but it underestimates A, as seen in Figure 5(a).

Alternatively, A can be approximated by a rectangle of width $4 - 2 = 2$ and height $f(4) = 5$. See Figure 5(b). This approximation of the area equals $2 \cdot 5 = 10$, but it overestimates A. We conclude that

$$8 < A < 10$$

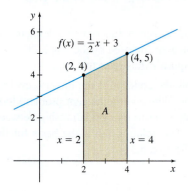

Figure 4

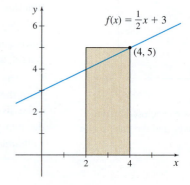

(a) The area A is underestimated.   (b) The area A is overestimated.

Figure 5

The approximation of the area A can be improved by dividing the closed interval $[2, 4]$ into two subintervals, $[2, 3]$ and $[3, 4]$. Now we draw two rectangles: one rectangle with width $3 - 2 = 1$ and height $f(2) = \dfrac{1}{2} \cdot 2 + 3 = 4$; the other rectangle with width $4 - 3 = 1$ and height $f(3) = \dfrac{1}{2} \cdot 3 + 3 = \dfrac{9}{2}$. As Figure 6(a) illustrates, the sum of the areas of the two rectangles

$$1 \cdot 4 + 1 \cdot \frac{9}{2} = \frac{17}{2} = 8.5$$

underestimates the area.

Now we repeat this process by drawing two rectangles, one of width 1 and height $f(3) = \dfrac{9}{2}$; the other of width 1 and height $f(4) = \dfrac{1}{2} \cdot 4 + 3 = 5$. As Figure 6(b) illustrates, the sum of the areas of these two rectangles,

$$1 \cdot \frac{9}{2} + 1 \cdot 5 = \frac{19}{2} = 9.5$$

overestimates the area. We conclude that

$$8.5 < A < 9.5$$

obtaining a better approximation to the area.

NOTE The actual area in Figure 4 is 9 square units, obtained by using the formula for the area A of a trapezoid with base b and parallel heights h_1 and h_2:

$$A = \frac{1}{2}b(h_1 + h_2) = \frac{1}{2}(2)(4 + 5) = 9.$$

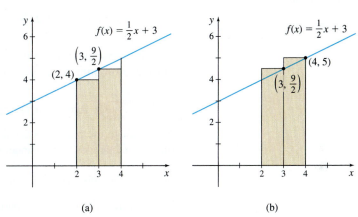

(a)   (b)

DF Figure 6

Observe that as the number n of subintervals of the interval increases, the approximation of the area A improves. For $n = 1$, the error in approximating A is 1 square unit, but for $n = 2$, the error is only 0.5 square unit.

NOW WORK **Problem 5.**

In general, the procedure for approximating the area A is based on the idea of summing the areas of rectangles. We shall refer to the area A enclosed by the graph of a function $y = f(x) \geq 0$, the x-axis, and the lines $x = a$ and $x = b$ as the **area under the graph of f from a to b.**

We make two assumptions about the function f:

- f is continuous on the closed interval $[a, b]$.
- f is nonnegative on the closed interval $[a, b]$.

We divide, or **partition**, the interval $[a, b]$ into n nonoverlapping subintervals:

$$[x_0, x_1], [x_1, x_2], \ldots, [x_{i-1}, x_i], \ldots, [x_{n-1}, x_n]$$

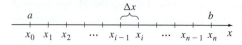

Figure 7

each of the same length. See Figure 7. Since there are n subintervals and the length of the interval $[a, b]$ is $b - a$, the common length Δx of each subinterval is

$$\Delta x = \frac{b - a}{n}$$

NEED TO REVIEW? The Extreme Value Theorem is discussed in Section 4.2, p. 265.

Since f is continuous on the closed interval $[a, b]$, it is continuous on every subinterval $[x_{i-1}, x_i]$ of $[a, b]$. By the Extreme Value Theorem, there is a number in each subinterval where f attains its absolute minimum. Label these numbers c_1, c_2, c_3, ..., c_n, so that $f(c_i)$ is the absolute minimum value of f in the subinterval $[x_{i-1}, x_i]$. Now construct n rectangles, each having Δx as its base and $f(c_i)$ as its height, as illustrated in Figure 8. This produces n narrow rectangles of uniform width $\Delta x = \dfrac{b - a}{n}$ and heights $f(c_1)$, $f(c_2)$, ..., $f(c_n)$, respectively. The areas of the n rectangles are

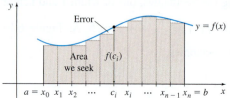

Figure 8 $f(c_i)$ is the absolute minimum value of f on $[x_{i-1}, x_i]$.

$$\text{Area of the first rectangle} = f(c_1)\Delta x$$

$$\text{Area of the second rectangle} = f(c_2)\Delta x$$

$$\vdots$$

$$\text{Area of the } n\text{th (and last) rectangle} = f(c_n)\Delta x$$

NEED TO REVIEW? Summation notation is discussed in Appendix A.5, pp. A-40 to A-42.

The sum s_n of the areas of the n rectangles approximates the area A. That is,

$$A \approx s_n = f(c_1)\Delta x + f(c_2)\Delta x + \cdots + f(c_i)\Delta x + \cdots + f(c_n)\Delta x = \sum_{i=1}^{n} f(c_i)\Delta x$$

Since the rectangles used to approximate the area A lie under the graph of f, the sum s_n, called a **lower sum**, *underestimates* A. That is, $s_n \leq A$.

EXAMPLE 2   Approximating Area Using Lower Sums

Approximate the area A under the graph of $f(x) = x^2$ from 0 to 10 by using lower sums s_n (rectangles that lie under the graph) for:

(a) $n = 2$ subintervals   **(b)** $n = 5$ subintervals   **(c)** $n = 10$ subintervals

Solution (a) For $n = 2$, we partition the closed interval $[0, 10]$ into two subintervals $[0, 5]$ and $[5, 10]$, each of length $\Delta x = \dfrac{10 - 0}{2} = 5$. See Figure 9(a). To compute s_2, we need to know where f attains its minimum value in each subinterval. Since f is an

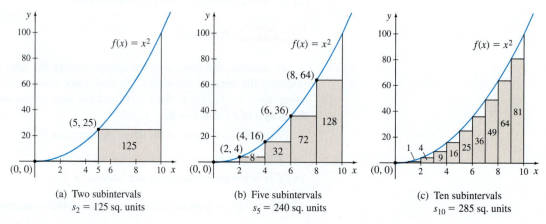

(a) Two subintervals
$s_2 = 125$ sq. units

(b) Five subintervals
$s_5 = 240$ sq. units

(c) Ten subintervals
$s_{10} = 285$ sq. units

DF Figure 9

increasing function, the absolute minimum is attained at the left endpoint of each subinterval. So, for $n = 2$, the minimum of f on $[0, 5]$ occurs at 0 and the minimum of f on $[5, 10]$ occurs at 5. The lower sum s_2 is

$$s_2 = \sum_{i=1}^{2} f(c_i)\Delta x = \Delta x \sum_{i=1}^{2} f(c_i) = 5[f(0) + f(5)] = 5(0 + 25) = 125$$

$$\uparrow \qquad\qquad \uparrow$$
$$\Delta x = 5 \qquad\qquad f(0) = 0$$
$$f(c_1) = f(0);\; f(c_2) = f(5) \quad f(5) = 25$$

(b) For $n = 5$, partition the interval $[0, 10]$ into five subintervals $[0, 2]$, $[2, 4]$, $[4, 6]$, $[6, 8]$, $[8, 10]$, each of length $\Delta x = \dfrac{10 - 0}{5} = 2$. See Figure 9(b). The lower sum s_5 is

$$s_5 = \sum_{i=1}^{5} f(c_i)\Delta x = \Delta x \sum_{i=1}^{5} f(c_i) = 2[f(0) + f(2) + f(4) + f(6) + f(8)]$$

$$= 2(0 + 4 + 16 + 36 + 64) = 240$$

(c) For $n = 10$, partition $[0, 10]$ into 10 subintervals, each of length $\Delta x = \dfrac{10 - 0}{10} = 1$. See Figure 9(c). The lower sum s_{10} is

$$s_{10} = \sum_{i=1}^{10} f(c_i)\Delta x = \Delta x \sum_{i=1}^{10} f(c_i) = 1[f(0) + f(1) + f(2) + \cdots + f(9)]$$

$$= 0 + 1 + 4 + 9 + 16 + 25 + 36 + 49 + 64 + 81 = 285 \qquad\blacksquare$$

NOW WORK Problem 13(a).

In general, as Figure 10(a) illustrates, the error due to using lower sums s_n (rectangles that lie below the graph of f) occurs because a portion of the area lies outside the rectangles. To improve the approximation of the area, we increase the number of subintervals. For example, in Figure 10(b), there are four subintervals and the error is reduced; in Figure 10(c), there are eight subintervals and the error is further reduced. So, by taking a finer and finer partition of the interval $[a, b]$, that is, by increasing n, the number of subintervals, without bound, we can make the sum of the areas of the rectangles as close as we please to the actual area. (A proof of this statement is usually found in books on advanced calculus.)

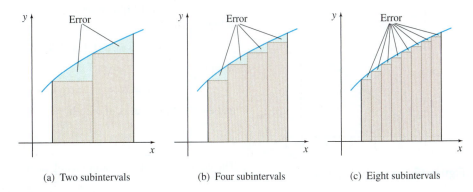

(a) Two subintervals   (b) Four subintervals   (c) Eight subintervals

Figure 10

In Section 5.2, we see that, for functions that are continuous on a closed interval, $\lim_{n \to \infty} s_n$ always exists. With this discussion in mind, we now define the area under the graph of a function f from a to b.

DEFINITION Area A Under the Graph of a Function from a to b

Suppose a function f is nonnegative and continuous on a closed interval $[a, b]$. Partition $[a, b]$ into n subintervals $[x_0, x_1], [x_1, x_2], \ldots, [x_{i-1}, x_i], \ldots, [x_{n-1}, x_n]$, each of length

$$\Delta x = \frac{b - a}{n}$$

In each subinterval $[x_{i-1}, x_i]$, let $f(c_i)$ equal the absolute minimum value of f on this subinterval. Form the lower sums

$$s_n = \sum_{i=1}^{n} f(c_i) \Delta x = f(c_1) \Delta x + \cdots + f(c_n) \Delta x$$

The **area A under the graph of f from a to b** is the number

$$\boxed{A = \lim_{n \to \infty} s_n}$$

The area A is defined using lower sums s_n (rectangles that lie below the graph of f). By a parallel argument, we can choose values $C_1, C_2, \ldots, C_n$ so that the height $f(C_i)$ of the ith rectangle is the absolute maximum value of f on the ith subinterval, as shown in Figure 11. The corresponding **upper sums S_n** (rectangles that lie above the graph of f) *overestimate* the area A. So, $S_n \geq A$. It can be shown that as n increases without bound, the limit of the upper sums S_n equals the limit of the lower sums s_n. That is,

$$\boxed{\lim_{n \to \infty} s_n = \lim_{n \to \infty} S_n = A}$$

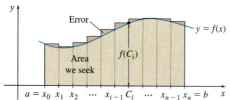

Figure 11 $f(C_i)$ is the absolute maximum value of f on $[x_{i-1}, x_i]$.

NOW WORK Problem 13(b).

2 Find the Area Under the Graph of a Function

In the next example, instead of using a specific number of rectangles to *approximate* area, we partition the interval $[a, b]$ into n subintervals, obtaining n rectangles. By letting $n \to \infty$, we find the *actual* area under the graph of f from a to b.

EXAMPLE 3   Finding Area Using Upper Sums

Find the area A under the graph of $f(x) = 3x$ from 0 to 10 using upper sums S_n (rectangles that lie above the graph of f). Then $A = \lim_{n \to \infty} S_n$.

Solution Figure 12 illustrates the area A. We partition the closed interval $[0, 10]$ into n subintervals

$$[x_0, x_1], [x_1, x_2], \ldots, [x_{i-1}, x_i], \ldots, [x_{n-1}, x_n]$$

where

$$0 = x_0 < x_1 < x_2 < \cdots < x_i < \cdots < x_{n-1} < x_n = 10$$

and each subinterval is of length

$$\Delta x = \frac{10 - 0}{n} = \frac{10}{n}$$

The coordinates of the endpoints of each subinterval, written in terms of n, are

$$x_0 = 0, \ x_1 = \frac{10}{n}, \ x_2 = 2\left(\frac{10}{n}\right), \ldots, x_{i-1} = (i - 1)\left(\frac{10}{n}\right),$$

$$x_i = i\left(\frac{10}{n}\right), \ldots, x_n = n\left(\frac{10}{n}\right) = 10$$

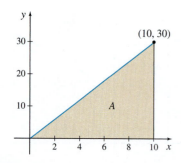

Figure 12 $f(x) = 3x, 0 \leq x \leq 10$

as illustrated in Figure 13.

$$\Delta x = \frac{10}{n}$$

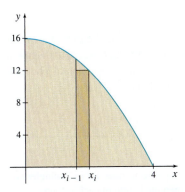

Figure 13

To find A using upper sums S_n (rectangles that lie above the graph of f), we need the absolute maximum value of f in each subinterval. Since $f(x) = 3x$ is an increasing function, the absolute maximum occurs at the right endpoint $x_i = i\left(\dfrac{10}{n}\right)$ of each subinterval. So,

$$S_n = \sum_{i=1}^{n} f(C_i)\Delta x = \sum_{i=1}^{n} 3x_i \cdot \frac{10}{n} = \sum_{i=1}^{n}\left(3 \cdot \frac{10i}{n}\right)\left(\frac{10}{n}\right) = \sum_{i=1}^{n}\frac{300}{n^2}i$$

$$\uparrow_{\Delta x = \frac{10}{n}} \qquad \uparrow_{x_i = \frac{10i}{n}}$$

NEED TO REVIEW? Summation properties are discussed in Appendix A.5, pp. A-40 to A-42.

Using summation properties, we get

$$S_n = \sum_{i=1}^{n}\frac{300}{n^2}i = \frac{300}{n^2}\sum_{i=1}^{n}i = \frac{300}{n^2}\frac{n(n+1)}{2} = 150\left(\frac{n+1}{n}\right) = 150\left(1+\frac{1}{n}\right)$$

RECALL $\displaystyle\sum_{i=1}^{n}i = \frac{n(n+1)}{2}$

Then

$$A = \lim_{n\to\infty} S_n = \lim_{n\to\infty}\left[150\left(1+\frac{1}{n}\right)\right] = 150\lim_{n\to\infty}\left(1+\frac{1}{n}\right) = 150$$

The area A under the graph of $f(x) = 3x$ from 0 to 10 is 150 square units. ∎

NOTE The area found in Example 3 is that of a triangle. So, we can verify that $A = 150$ by using the formula for the area A of a triangle with base b and height h:

$$A = \frac{1}{2}bh = \frac{1}{2}(10)(30) = 150$$

EXAMPLE 4   Finding Area Using Lower Sums

Find the area A under the graph of $f(x) = 16 - x^2$ from 0 to 4 by using lower sums s_n (rectangles that lie below the graph of f). Then $A = \lim\limits_{n\to\infty} s_n$.

Solution Figure 14 shows the area under the graph of f and a typical rectangle that lies below the graph. We partition the closed interval $[0, 4]$ into n subintervals

$$[x_0, x_1], [x_1, x_2], \ldots, [x_{i-1}, x_i], \ldots, [x_{n-1}, x_n]$$

where

$$0 = x_0 < x_1 < \cdots < x_i < \cdots < x_{n-1} < x_n = 4$$

and each interval is of length

$$\Delta x = \frac{4-0}{n} = \frac{4}{n}$$

As Figure 15 on page 350 illustrates, the endpoints of each subinterval, written in terms of n, are

$$x_0 = 0, \ x_1 = 1\left(\frac{4}{n}\right), \ x_2 = 2\left(\frac{4}{n}\right), \ldots, x_{i-1} = (i-1)\left(\frac{4}{n}\right),$$

$$x_i = i\left(\frac{4}{n}\right), \ldots, x_n = n\left(\frac{4}{n}\right) = 4$$

Figure 14 $f(x) = 16 - x^2, 0 \le x \le 4$

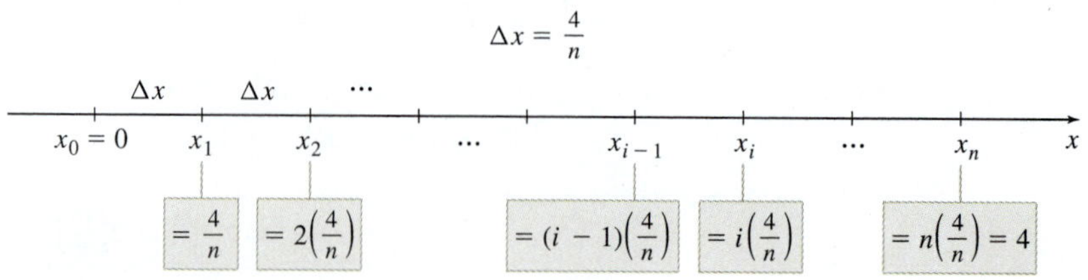

$$\Delta x = \frac{4}{n}$$

Figure 15

To find A using lower sums s_n (rectangles that lie below the graph of f), we must find the absolute minimum value of f on each subinterval. Since the function f is a decreasing function, the absolute minimum occurs at the right endpoint of each subinterval. So,

$$s_n = \sum_{i=1}^{n} f(c_i)\Delta x$$

Since $c_i = i\left(\dfrac{4}{n}\right) = \dfrac{4i}{n}$ and $\Delta x = \dfrac{4}{n}$, we have

$$s_n = \sum_{i=1}^{n} f(c_i)\Delta x = \sum_{i=1}^{n}\left[16 - \left(\frac{4i}{n}\right)^2\right]\left(\frac{4}{n}\right)$$

$\qquad c_i = \dfrac{4i}{n};\ f(c_i) = 16 - c_i^2$

$$= \sum_{i=1}^{n}\left[\frac{64}{n} - \frac{64i^2}{n^3}\right]$$

$$= \frac{64}{n}\sum_{i=1}^{n}1 - \frac{64}{n^3}\sum_{i=1}^{n}i^2$$

RECALL $\sum_{i=1}^{n}1 = n;$

$\sum_{i=1}^{n}i^2 = \dfrac{n(n+1)(2n+1)}{6}$

$$= \frac{64}{n}(n) - \frac{64}{n^3}\left[\frac{n(n+1)(2n+1)}{6}\right] = 64 - \frac{32}{3n^2}\left[2n^2 + 3n + 1\right]$$

$$= 64 - \frac{64}{3} - \frac{32}{n} - \frac{32}{3n^2} = \frac{128}{3} - \frac{32}{n} - \frac{32}{3n^2}$$

Then

$$A = \lim_{n\to\infty} s_n = \lim_{n\to\infty}\left(\frac{128}{3} - \frac{32}{n} - \frac{32}{3n^2}\right) = \frac{128}{3}$$

The area A under the graph of $f(x) = 16 - x^2$ from 0 to 4 is $\dfrac{128}{3}$ square units. ■

NOW WORK Problem 29.

The previous two examples illustrate just how complex it can be to find areas using lower sums and/or upper sums. In the next section, we define an *integral* and show how it can be used to find area. Then in Section 5.3, we present the Fundamental Theorem of Calculus, which provides a relatively simple way to find area.

5.1 Assess Your Understanding

Concepts and Vocabulary

1. Explain how rectangles can be used to approximate the area enclosed by the graph of a function $y = f(x) \geq 0$, the x-axis, and the lines $x = a$ and $x = b$.

2. *True or False* When a closed interval $[a, b]$ is partitioned into n subintervals each of the same length, the length of each subinterval is $\dfrac{a+b}{n}$.

3. If the closed interval $[-2, 4]$ is partitioned into 12 subintervals, each of the same length, then the length of each subinterval is _____.

4. *True or False* If the area A under the graph of a function f that is continuous and nonnegative on a closed interval $[a, b]$ is approximated using upper sums S_n, then $S_n \geq A$ and $A = \lim_{n\to\infty} S_n$.

1. = NOW WORK problem   = Graphing technology recommended   CAS = Computer Algebra System recommended

Skill Building

5. Approximate the area A enclosed by the graph of $f(x) = \frac{1}{2}x + 3$, the x-axis, and the lines $x = 2$ and $x = 4$ by partitioning the closed interval $[2, 4]$ into four subintervals:

$$\left[2, \frac{5}{2}\right], \left[\frac{5}{2}, 3\right], \left[3, \frac{7}{2}\right], \left[\frac{7}{2}, 4\right].$$

(a) Using the left endpoint of each subinterval, draw four small rectangles that lie below the graph of f and sum the areas of the four rectangles.

(b) Using the right endpoint of each subinterval, draw four small rectangles that lie above the graph of f and sum the areas of the four rectangles.

(c) Compare the answers from parts (a) and (b) to the exact area $A = 9$ and to the estimates obtained in Example 1.

6. Approximate the area A enclosed by the graph of $f(x) = 6 - 2x$, the x-axis, and the lines $x = 1$ and $x = 3$ by partitioning the closed interval $[1, 3]$ into four subintervals:

$$\left[1, \frac{3}{2}\right], \left[\frac{3}{2}, 2\right], \left[2, \frac{5}{2}\right], \left[\frac{5}{2}, 3\right].$$

(a) Using the right endpoint of each subinterval, draw four small rectangles that lie below the graph of f and sum the areas of the four rectangles.

(b) Using the left endpoint of each subinterval, draw four small rectangles that lie above the graph of f and sum the areas of the four rectangles.

(c) Compare the answers from parts (a) and (b) to the exact area $A = 4$.

In Problems 7 and 8, refer to the graphs below. Approximate the shaded area under the graph of f:
(a) By constructing rectangles using the left endpoint of each subinterval.
(b) By constructing rectangles using the right endpoint of each subinterval.

7.

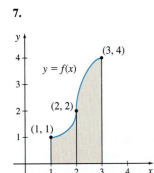

8.

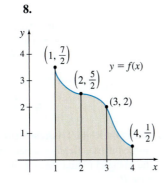

In Problems 9–12, partition each interval into n subintervals each of the same length.

9. $[1, 4]$ with $n = 3$

10. $[0, 9]$ with $n = 9$

11. $[-1, 4]$ with $n = 10$

12. $[-4, 4]$ with $n = 16$

In Problems 13 and 14, refer to the graphs. Using the indicated subintervals, approximate the shaded area:
(a) By using lower sums s_n (rectangles that lie below the graph of f).
(b) By using upper sums S_n (rectangles that lie above the graph of f).

13.

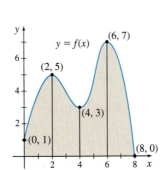

14.

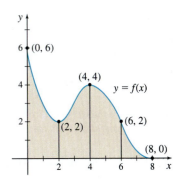

15. Area Under a Graph Consider the area under the graph of $y = x$ from 0 to 3.

(a) Sketch the graph and the area under the graph.

(b) Partition the interval $[0, 3]$ into n subintervals each of equal length.

(c) Show that $s_n = \sum_{i=1}^{n} (i - 1) \left(\frac{3}{n}\right)^2$.

(d) Show that $S_n = \sum_{i=1}^{n} i \left(\frac{3}{n}\right)^2$.

(e) Show that $\lim_{n \to \infty} s_n = \lim_{n \to \infty} S_n = \frac{9}{2}$.

16. Area Under a Graph Consider the area under the graph of $y = 4x$ from 0 to 5.

(a) Sketch the graph and the corresponding area.

(b) Partition the interval $[0, 5]$ into n subintervals each of equal length.

(c) Show that $s_n = \sum_{i=1}^{n} (i - 1) \frac{100}{n^2}$.

(d) Show that $S_n = \sum_{i=1}^{n} i \frac{100}{n^2}$.

(e) Show that $\lim_{n \to \infty} s_n = \lim_{n \to \infty} S_n = 50$.

In Problems 17–22, approximate the area A under the graph of each function f from a to b for $n = 4$ and $n = 8$ subintervals:
(a) By using lower sums s_n (rectangles that lie below the graph of f).
(b) By using upper sums S_n (rectangles that lie above the graph of f).

17. $f(x) = -x + 10$ on $[0, 8]$

18. $f(x) = 2x + 5$ on $[2, 6]$

19. $f(x) = 16 - x^2$ on $[0, 4]$

20. $f(x) = x^3$ on $[0, 8]$

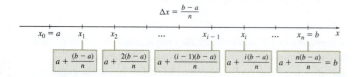

21. $f(x) = \cos x$ on $\left[-\dfrac{\pi}{2}, \dfrac{\pi}{2}\right]$ **22.** $f(x) = \sin x$ on $[0, \pi]$

23. Rework Example 3 by using lower sums s_n (rectangles that lie below the graph of f).

24. Rework Example 4 by using upper sums S_n (rectangles that lie above the graph of f).

In Problems 25–32, find the area A under the graph of f from a to b:

(a) *By using lower sums s_n (rectangles that lie below the graph of f).*

(b) *By using upper sums S_n (rectangles that lie above the graph of f).*

(c) *Compare the work required in (a) and (b). Which is easier? Could you have predicted this?*

25. $f(x) = 2x + 1$ from $a = 0$ to $b = 4$

26. $f(x) = 3x + 1$ from $a = 0$ to $b = 4$

27. $f(x) = 12 - 3x$ from $a = 0$ to $b = 4$

28. $f(x) = 5 - x$ from $a = 0$ to $b = 4$

29. $f(x) = 4x^2$ from $a = 0$ to $b = 2$

30. $f(x) = \dfrac{1}{2}x^2$ from $a = 0$ to $b = 3$

31. $f(x) = 4 - x^2$ from $a = 0$ to $b = 2$

32. $f(x) = 12 - x^2$ from $a = 0$ to $b = 3$

Applications and Extensions

In Problems 33–38, find the area under the graph of f from a to b. Partition the closed interval $[a, b]$ into n subintervals $[x_0, x_1], [x_1, x_2], \ldots, [x_{i-1}, x_i], \ldots, [x_{n-1}, x_n]$, where $a = x_0 < x_1 < \cdots < x_i < \cdots < x_{n-1} < x_n = b$, and each subinterval is of length $\Delta x = \dfrac{b - a}{n}$. As the figure below illustrates, the endpoints of each subinterval, written in terms of n, are

$$x_0 = a, \quad x_1 = a + \frac{b-a}{n}, \quad x_2 = a + 2\left(\frac{b-a}{n}\right), \ldots,$$

$$x_{i-1} = a + (i-1)\left(\frac{b-a}{n}\right), \quad x_i = a + i\left(\frac{b-a}{n}\right), \ldots,$$

$$x_n = a + n\left(\frac{b-a}{n}\right)$$

$$\Delta x = \frac{b-a}{n}$$

| $x_0 = a$ | x_1 | x_2 | $\cdots$ | x_{i-1} | x_i | $\cdots$ | $x_n = b$ | x |

| $a + \dfrac{(b-a)}{n}$ | $a + \dfrac{2(b-a)}{n}$ | $a + \dfrac{(i-1)(b-a)}{n}$ | $a + \dfrac{i(b-a)}{n}$ | $a + \dfrac{n(b-a)}{n} = b$ |

33. $f(x) = x + 3$ from $a = 1$ to $b = 3$

34. $f(x) = 3 - x$ from $a = 1$ to $b = 3$

35. $f(x) = 2x + 5$ from $a = -1$ to $b = 2$

36. $f(x) = 2 - 3x$ from $a = -2$ to $b = 0$

37. $f(x) = 2x^2 + 1$ from $a = 1$ to $b = 3$

38. $f(x) = 4 - x^2$ from $a = 1$ to $b = 2$

In Problems 39–42, approximate the area A under the graph of each function f by partitioning $[a, b]$ into 20 subintervals of equal length and using an upper sum.

39. $f(x) = xe^x$ on $[0, 8]$ **40.** $f(x) = \ln x$ on $[1, 3]$

41. $f(x) = \dfrac{1}{x}$ on $[1, 5]$ **42.** $f(x) = \dfrac{1}{x^2}$ on $[2, 6]$

43. (a) Graph $y = \dfrac{4}{x}$ from $x = 1$ to $x = 4$ and shade the area under its graph.

(b) Partition the interval $[1, 4]$ into n subintervals of equal length.

(c) Show that the lower sum s_n is $s_n = \displaystyle\sum_{i=1}^{n} \dfrac{4}{\left(1 + \dfrac{3i}{n}\right)}\left(\dfrac{3}{n}\right)$.

(d) Show that the upper sum S_n is

$$S_n = \sum_{i=1}^{n} \frac{4}{\left(1 + \dfrac{3(i-1)}{n}\right)}\left(\frac{3}{n}\right)$$

(e) Complete the following table:

n	5	10	50	100
s_n				
S_n				

(f) Use the table to give an upper and lower bound for the area.

Challenge Problems

44. Area Under a Graph Approximate the area under the graph of $f(x) = x$ from $a \geq 0$ to b by using lower sums s_n and upper sums S_n for a partition of $[a, b]$ into n subintervals, each of length $\dfrac{b-a}{n}$. Show that

$$s_n < \frac{b^2 - a^2}{2} < S_n$$

45. Area Under a Graph Approximate the area under the graph of $f(x) = x^2$ from $a \geq 0$ to b by using lower sums s_n and upper sums S_n for a partition of $[a, b]$ into n subintervals, each of length $\dfrac{b-a}{n}$. Show that

$$s_n < \frac{b^3 - a^3}{3} < S_n$$

46. Area of a Right Triangle Use lower sums s_n (rectangles that lie inside the triangle) and upper sums S_n (rectangles that lie outside the triangle) to find the area of a right triangle of height H and base B.

47. Area of a Trapezoid Use lower sums s_n (rectangles that lie inside the trapezoid) and upper sums S_n (rectangles that lie outside the trapezoid) to find the area of a trapezoid of heights H_1 and H_2 and base B.

5.2 The Definite Integral

OBJECTIVES *When you finish this section, you should be able to:*

1 Define a definite integral as the limit of Riemann sums (p. 353)
2 Find a definite integral using the limit of Riemann sums (p. 356)

The area A under the graph of $y = f(x)$ from a to b is obtained by finding

$$A = \lim_{n \to \infty} s_n = \lim_{n \to \infty} \sum_{i=1}^{n} f(c_i)\Delta x = \lim_{n \to \infty} S_n = \lim_{n \to \infty} \sum_{i=1}^{n} f(C_i)\Delta x \tag{1}$$

where the following assumptions are made:

- The function f is continuous on $[a, b]$.
- The function f is nonnegative on $[a, b]$.
- The closed interval $[a, b]$ is partitioned into n subintervals, each of length

$$\Delta x = \frac{b - a}{n}$$

- $f(c_i)$ is the absolute minimum value of f on the ith subinterval, $i = 1, 2, \ldots, n$.
- $f(C_i)$ is the absolute maximum value of f on the ith subinterval, $i = 1, 2, \ldots, n$.

In Section 5.1, we found the area A under the graph of f from a to b by choosing either the number c_i, where f has an absolute minimum on the ith subinterval, or the number C_i, where f has an absolute maximum on the ith subinterval. Suppose we arbitrarily choose a number u_i in each subinterval $[x_{i-1}, x_i]$, and draw rectangles of height $f(u_i)$ and width Δx. Then from the definitions of absolute minimum value and absolute maximum value

$$f(c_i) \le f(u_i) \le f(C_i)$$

and, since $\Delta x > 0$,

$$f(c_i)\Delta x \le f(u_i)\Delta x \le f(C_i)\Delta x$$

Then

$$\sum_{i=1}^{n} f(c_i)\Delta x \le \sum_{i=1}^{n} f(u_i)\Delta x \le \sum_{i=1}^{n} f(C_i)\Delta x$$

$$s_n \le \sum_{i=1}^{n} f(u_i)\Delta x \le S_n$$

Since $\lim\limits_{n \to \infty} s_n = \lim\limits_{n \to \infty} S_n = A$, by the Squeeze Theorem, we have

$$\lim_{n \to \infty} \sum_{i=1}^{n} f(u_i)\Delta x = A$$

In other words, we can use any number u_i in the ith subinterval to find the area A.

1 Define a Definite Integral as the Limit of Riemann Sums

We now investigate sums of the form

$$\sum_{i=1}^{n} f(u_i)\Delta x_i$$

using the following more general assumptions:

- The function f is not necessarily continuous on $[a, b]$.
- The function f is not necessarily nonnegative on $[a, b]$.

ORIGINS Riemann sums are named after the German mathematician Georg Friedrich Bernhard Riemann (1826–1866). Early in his life, Riemann was home schooled. At age 14, he was sent to a lyceum (high school) and then the University of Göttingen to study theology. Once at Göttingen, he asked for and received permission from his father to study mathematics. He completed his PhD under Karl Friedrich Gauss (1777–1855). In his thesis, Riemann used topology to analyze complex functions. Later he developed a theory of geometry to describe real space. His ideas were far ahead of their time and were not truly appreciated until they provided the mathematical framework for Einstein's Theory of Relativity.

NEED TO REVIEW? The Squeeze Theorem is discussed in Section 1.4, pp. 106–107.

- The lengths $\Delta x_i = x_i - x_{i-1}$ of the subintervals $[x_{i-1}, x_i]$, $i = 1, 2, \ldots, n$ of $[a, b]$ are not necessarily equal.
- The number u_i may be any number in the subinterval $[x_{i-1}, x_i]$, $i = 1, 2, \ldots, n$.

The sums $\sum_{i=1}^{n} f(u_i)\Delta x_i$, called **Riemann sums** for f on $[a, b]$, form the foundation of integral calculus.

EXAMPLE 1 Forming Riemann Sums

For the function $f(x) = x^2 - 3$, $0 \leq x \leq 6$, partition the interval $[0, 6]$ into 4 subintervals $[0, 1]$, $[1, 2]$, $[2, 4]$, $[4, 6]$ and form the Riemann sum for which

(a) u_i is the left endpoint of each subinterval.

(b) u_i is the midpoint of each subinterval.

Solution In forming Riemann sums $\sum_{i=1}^{n} f(u_i)\Delta x_i$, n is the number of subintervals in the partition, $f(u_i)$ is the value of f at the number u_i chosen in the ith subinterval, and Δx_i is the length of the ith subinterval.

(a) Figure 16 shows the graph of f, the partition of the interval $[0, 6]$ into the 4 subintervals, and values of $f(u_i)$ at the left endpoint of each subinterval, namely,

$$f(u_1) = f(0) = -3 \qquad f(u_2) = f(1) = -2$$
$$f(u_3) = f(2) = 1 \qquad f(u_4) = f(4) = 13$$

The 4 subintervals have length

$$\Delta x_1 = 1 - 0 = 1 \quad \Delta x_2 = 2 - 1 = 1 \quad \Delta x_3 = 4 - 2 = 2 \quad \Delta x_4 = 6 - 4 = 2$$

The Riemann sum is formed by adding the products $f(u_i)\Delta x_i$ for $i = 1, 2, 3, 4$.

$$\sum_{i=1}^{4} f(u_i)\Delta x_i = -3 \cdot 1 + (-2) \cdot 1 + 1 \cdot 2 + 13 \cdot 2 = 23$$

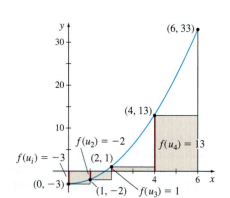

DF Figure 16 $f(x) = x^2 - 3$, $0 \leq x \leq 6$

(b) See Figure 17. If u_i is chosen as the midpoint of each subinterval, then the value of $f(u_i)$ at the midpoint of each subinterval is

$$f(u_1) = f\left(\frac{1}{2}\right) = -\frac{11}{4} \qquad f(u_2) = f\left(\frac{3}{2}\right) = -\frac{3}{4}$$
$$f(u_3) = f(3) = 6 \qquad f(u_4) = f(5) = 22$$

The Riemann sum formed by adding the products $f(u_i)\Delta x_i$ for $i = 1, 2, 3, 4$ is

$$\sum_{i=1}^{4} f(u_i)\Delta x_i = -\frac{11}{4} \cdot 1 + \left(-\frac{3}{4}\right) \cdot 1 + 6 \cdot 2 + 22 \cdot 2 = \frac{105}{2} = 52.5 \qquad \blacksquare$$

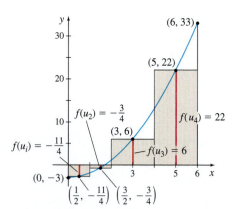

DF Figure 17 $f(x) = x^2 - 3$, $0 \leq x \leq 6$

NOW WORK Problem 9.

Suppose a function f is defined on a closed interval $[a, b]$, and we partition the interval $[a, b]$ into n subintervals

$$[x_0, x_1], [x_1, x_2], [x_2, x_3], \ldots, [x_{i-1}, x_i], \ldots, [x_{n-1}, x_n]$$

where

$$a = x_0 < x_1 < x_2 < \cdots < x_{i-1} < x_i < \cdots < x_{n-1} < x_n = b$$

These subintervals are not necessarily of the same length. Denote the length of the first subinterval by $\Delta x_1 = x_1 - x_0$, the length of the second subinterval by $\Delta x_2 = x_2 - x_1$, and so on. In general, the length of the ith subinterval is

$$\Delta x_i = x_i - x_{i-1}$$

for $i = 1, 2, \ldots, n$. This set of subintervals of the interval $[a, b]$ is called a **partition** of $[a, b]$. The length of the largest subinterval in a partition is called the **norm** of the partition and is denoted by max Δx_i.

DEFINITION Definite Integral

Let f be a function defined on the closed interval $[a, b]$. Partition $[a, b]$ into n subintervals of length $\Delta x_i = x_i - x_{i-1}$, $i = 1, 2, \ldots, n$. Choose a number u_i in each subinterval, evaluate $f(u_i)$, and form the Riemann sums $\sum_{i=1}^{n} f(u_i)\Delta x_i$.

If $\lim_{\max \Delta x_i \to 0} \sum_{i=1}^{n} f(u_i)\Delta x_i = I$ exists and does not depend on the choice of the partition or on the choice of u_i, then the number I is called the **Riemann integral** or **definite integral** of f from a to b and is denoted by the symbol $\int_a^b f(x)\,dx$. That is,

$$\int_a^b f(x)dx = \lim_{\max \Delta x_i \to 0} \sum_{i=1}^{n} f(u_i)\Delta x_i$$

When the above limit exists, then we say that f is **integrable over** $[a, b]$.

For the definite integral $\int_a^b f(x)\,dx$, the number a is called the **lower limit of integration**, the number b is called the **upper limit of integration**, the symbol $\int$ (an elongated S to remind you of summation) is called the **integral sign**, $f(x)$ is called the **integrand**, and dx is the differential of the independent variable x. The variable used in the definite integral is an *artificial* or a *dummy* variable because it may be replaced by any other symbol. For example,

$$\int_a^b f(x)\,dx \qquad \int_a^b f(t)\,dt \qquad \int_a^b f(s)\,ds \qquad \int_a^b f(\theta)\,d\theta$$

all denote the definite integral of f from a to b, and if any of them exist, they are all equal to the same number.

EXAMPLE 2   Expressing the Limit of Riemann Sums as a Definite Integral

(a) Find the Riemann sums for $f(x) = x^2 - 3$ on the closed interval $[0, 6]$ for a partition of $[0, 6]$ into n subintervals of length $\Delta x_i = x_i - x_{i-1}$, $i = 1, 2, \ldots, n$.

(b) Assuming that the limit of the Riemann sums exists as max $\Delta x_i \to 0$, express the limit as a definite integral.

Solution (a) The Riemann sums for $f(x) = x^2 - 3$ on the closed interval $[0, 6]$ are

$$\sum_{i=1}^{n} f(u_i)\Delta x_i = \sum_{i=1}^{n} (u_i^2 - 3)\Delta x_i$$

where $[0, 6]$ is partitioned into n subintervals $[x_{i-1}, x_i]$, and u_i is some number in the subinterval $[x_{i-1}, x_i]$, $i = 1, 2, \ldots, n$.

(b) Since $\lim_{\max \Delta x_i \to 0} \sum_{i=1}^{n} (u_i^2 - 3)\Delta x_i$ exists, then

$$\lim_{\max \Delta x_i \to 0} \sum_{i=1}^{n} (u_i^2 - 3)\Delta x_i = \int_0^6 (x^2 - 3)\,dx \qquad \blacksquare$$

NOW WORK Problem 15.

If f is integrable over $[a, b]$, then $\lim\limits_{\max \Delta x_i \to 0} \sum\limits_{i=1}^{n} f(u_i)\Delta x_i$ exists for any choice of u_i in the ith subinterval, so we are free to choose the u_i any way we please. The choices could be the left endpoint of each subinterval, or the right endpoint, or the midpoint, or any other number in each subinterval. Also, $\lim\limits_{\max \Delta x_i \to 0} \sum\limits_{i=1}^{n} f(u_i)\Delta x_i$ is independent of the partition of the closed interval $[a, b]$, provided $\max \Delta x_i$ can be made as close as we please to 0. It is these flexibilities that make the definite integral so important in engineering, physics, chemistry, geometry, and economics.

In defining the definite integral $\int_a^b f(x)\, dx$, we assumed that $a < b$. To remove this restriction, we give the following definitions.

DEFINITION

- If $f(a)$ is defined, then

$$\int_a^a f(x)\, dx = 0$$

- If $a > b$ and if $\int_b^a f(x)\, dx$ exists, then

$$\int_a^b f(x)\,dx = -\int_b^a f(x)\,dx$$

IN WORDS Interchanging the limits of integration reverses the sign of the definite integral.

For example,

$$\int_1^1 x^2\, dx = 0 \qquad \text{and} \qquad \int_3^2 x^2\, dx = -\int_2^3 x^2\, dx$$

Next we give a condition on the function f that guarantees f is integrable. The proof of this result may be found in advanced calculus texts.

THEOREM Existence of the Definite Integral

If a function f is continuous on a closed interval $[a, b]$, then the definite integral $\int_a^b f(x)\, dx$ exists.

The two conditions of the theorem deserve special attention. First, f is defined on a *closed* interval, and second, f is *continuous* on that interval. There are some functions that are continuous on an open interval (or even a half-open interval) for which the integral does not exist. For example, although $f(x) = \dfrac{1}{x^2}$ is continuous on $(0, 1)$ (and

IN WORDS If a function f is continuous on $[a, b]$, then it is integrable over $[a, b]$. But if f is not continuous on $[a, b]$, then f may or may not be integrable over $[a, b]$.

on $(0, 1]$), the definite integral $\int_0^1 \dfrac{1}{x^2}\, dx$ does not exist. Also, there are many examples of discontinuous functions for which the integral exists. (See Problems 64 and 65.)

2 Find a Definite Integral Using the Limit of Riemann Sums

Suppose f is integrable over the closed interval $[a, b]$. To find

$$\int_a^b f(x)\, dx = \lim\limits_{\max \Delta x_i \to 0} \sum\limits_{i=1}^{n} f(u_i)\Delta x_i$$

using Riemann sums, we usually partition $[a, b]$ into n subintervals, each of the same length $\Delta x = \dfrac{b - a}{n}$. Such a partition is called a **regular partition**. For a regular partition, the norm of the partition is

$$\max \Delta x_i = \frac{b - a}{n}$$

Since $\lim\limits_{n\to\infty} \dfrac{b-a}{n} = 0$, it follows that for a regular partition, the two statements

$$\max \Delta x_i \to 0 \qquad \text{and} \qquad n \to \infty$$

are interchangeable. As a result, for regular partitions $\Delta x = \dfrac{b-a}{n}$,

$$\int_a^b f(x)\,dx = \lim_{\max \Delta x_i \to 0} \sum_{i=1}^n f(u_i)\Delta x_i = \lim_{n\to\infty} \sum_{i=1}^n f(u_i)\Delta x$$

The next result uses Riemann sums to establish a formula to find the definite integral of a constant function.

THEOREM

If $f(x) = h$, where h is some constant, then

$$\int_a^b f(x)\,dx = \int_a^b h\,dx = h(b-a)$$

Proof The constant function $f(x) = h$ is continuous on the set of real numbers and so is integrable. We form the Riemann sums for f on the closed interval $[a, b]$ using a regular partition. Then $\Delta x_i = \Delta x = \dfrac{b-a}{n}$, $i = 1, 2, \ldots, n$. The Riemann sums of f on the interval $[a, b]$ are

$$\sum_{i=1}^n f(u_i)\Delta x_i = \sum_{i=1}^n f(u_i)\Delta x = \underset{\underset{f(x)=h}{\uparrow}}{\sum_{i=1}^n h\,\Delta x} = \sum_{i=1}^n h\left(\frac{b-a}{n}\right)$$

$$= h\left(\frac{b-a}{n}\right)\sum_{i=1}^n 1 = h\left(\frac{b-a}{n}\right)\cdot n = h(b-a)$$

Then

$$\lim_{n\to\infty} \sum_{i=1}^n f(u_i)\Delta x = \lim_{n\to\infty} [h(b-a)] = h(b-a)$$

So,

$$\int_a^b h\,dx = h(b-a) \qquad \blacksquare$$

For example,

$$\int_1^2 3\,dx = 3(2-1) = 3 \qquad \int_2^6 dx = 1(6-2) = 4 \qquad \int_{-3}^4 (-2)\,dx = (-2)[4-(-3)] = -14$$

NOW WORK **Problem 23.**

EXAMPLE 3 Finding a Definite Integral Using the Limit of Riemann Sums

Find $\displaystyle\int_0^3 (3x - 8)\,dx$.

Solution Since the integrand $f(x) = 3x - 8$ is continuous on the closed interval $[0, 3]$, the function f is integrable over $[0, 3]$. Although we can use any partition of $[0, 3]$ whose norm can be made as close to 0 as we please, and we can choose any u_i in each subinterval, we use a regular partition and choose u_i as the right endpoint of each subinterval. This will result in a simple expression for the Riemann sums.

We partition $[0, 3]$ into n subintervals, each of length $\Delta x = \dfrac{3 - 0}{n} = \dfrac{3}{n}$. The endpoints of each subinterval of the partition, written in terms of n, are

$$x_0 = 0, \quad x_1 = \frac{3}{n}, \quad x_2 = 2\left(\frac{3}{n}\right), \ldots, x_{i-1} = (i - 1)\left(\frac{3}{n}\right),$$

$$x_i = i\left(\frac{3}{n}\right), \ldots, x_n = n\left(\frac{3}{n}\right) = 3$$

The Riemann sums of $f(x) = 3x - 8$ from 0 to 3, using $u_i = x_i = \dfrac{3i}{n}$ (the right endpoint) and $\Delta x = \dfrac{3}{n}$, are

$$\sum_{i=1}^{n} f(u_i)\,\Delta x_i = \sum_{i=1}^{n} f(x_i)\,\Delta x = \underset{\underset{\Delta x = \frac{3}{n}}{\uparrow}}{\sum_{i=1}^{n}}(3x_i - 8)\frac{3}{n} = \underset{\underset{x_i = \frac{3i}{n}}{\uparrow}}{\sum_{i=1}^{n}}\left[3\left(\frac{3i}{n}\right) - 8\right]\frac{3}{n}$$

$$= \sum_{i=1}^{n}\left(\frac{27i}{n^2} - \frac{24}{n}\right) = \frac{27}{n^2}\sum_{i=1}^{n} i - \frac{24}{n}\sum_{i=1}^{n} 1$$

$$= \frac{27}{n^2} \cdot \frac{n(n+1)}{2} - \frac{24}{n} \cdot n = \frac{27}{2} + \frac{27}{2n} - 24 = -\frac{21}{2} + \frac{27}{2n}$$

Now

$$\int_0^3 (3x - 8)\,dx = \lim_{n\to\infty}\left(-\frac{21}{2} + \frac{27}{2n}\right) = -\frac{21}{2} \qquad \blacksquare$$

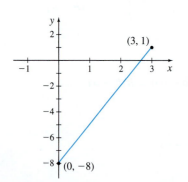

DF Figure 18 $f(x) = 3x - 8, 0 \le x \le 3$

NOW WORK Problem 45.

Figure 18 shows the graph of $f(x) = 3x - 8$ on $[0, 3]$. Since f is not nonnegative on $[0, 3]$, we cannot interpret $\int_0^3 (3x - 8)\,dx$ as an area. The fact that the answer is negative is further evidence that this is not an area problem. When finding a definite integral, do not presume it represents area. As you will see in Section 5.4 and in Chapter 6, the definite integral has many interpretations. Interestingly enough, definite integrals are used to find the volume of a solid of revolution, the length of a graph, the work done by a variable force, and other quantities.

EXAMPLE 4   Interpreting a Definite Integral

Determine if each definite integral can be interpreted as an area. If it can, describe the area; if it cannot, explain why.

(a) $\displaystyle\int_0^{3\pi/4} \cos x\,dx$ (b) $\displaystyle\int_2^{10} |x - 4|\,dx$

Solution (a) See Figure 19 on page 359. Since $\cos x < 0$ on the interval $\left(\dfrac{\pi}{2}, \dfrac{3\pi}{4}\right]$, the integral $\int_0^{3\pi/4} \cos x\,dx$ cannot be interpreted as area.

(b) See Figure 20. Since $|x - 4| \ge 0$ on the interval $[2, 10]$, the integral $\int_2^{10} |x - 4|\,dx$ can be interpreted as the area enclosed by the graph of $y = |x - 4|$, the x-axis, and the lines $x = 2$ and $x = 10$.

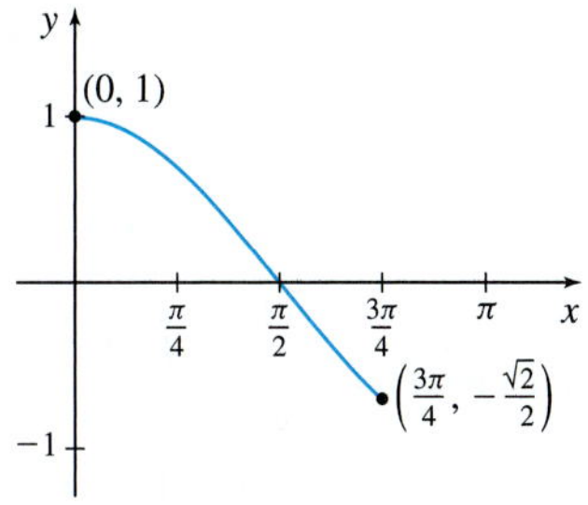

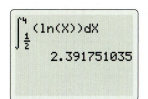

Figure 19 $f(x) = \cos x, 0 \le x \le \dfrac{3\pi}{4}$

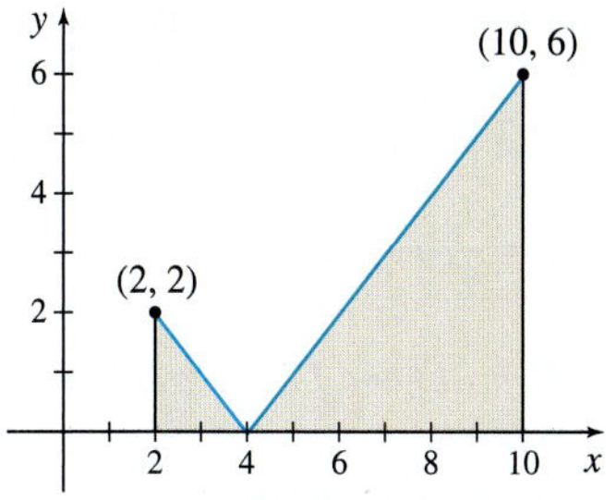

Figure 20 $f(x) = |x - 4|, 2 \le x \le 10$

NOW WORK   **Problem 33.**

EXAMPLE 5   **Finding a Definite Integral Using Technology**

(a) Use a graphing utility to find $\int_{1/2}^{4} \ln x \, dx$

(b) Use a computer algebra system to find $\int_{1/2}^{4} \ln x \, dx$.

Solution (a) A graphing utility, such as a TI-84, provides only an approximate numerical answer to the integral $\int_{1/2}^{4} \ln x \, dx$. As shown in Figure 21, $\int_{1/2}^{4} \ln x \, dx \approx 2.391751035$.

(b) Because a computer algebra system manipulates symbolically, it can find an exact value of the definite integral. Using MuPAD in Scientific WorkPlace, we get

$$\int_{1/2}^{4} \ln x \, dx = \frac{17}{2} \ln 2 - \frac{7}{2}$$

An approximate numerical value of the definite integral using MuPAD is 2.3918. ∎

Figure 21

NOW WORK   **Problem 43.**

5.2 Assess Your Understanding

Concepts and Vocabulary

1. If an interval $[a, b]$ is partitioned into n subintervals $[x_0, x_1]$, $[x_1, x_2], [x_2, x_3], \ldots, [x_{n-1}, x_n]$, where $a = x_0 < x_1 < x_2 < \cdots < x_{n-1} < x_n = b$, then the set of subintervals of the interval $[a, b]$ is called a(n) _____ of $[a, b]$.

2. *Multiple Choice* In a regular partition of $[0, 40]$ into 20 subintervals, $\Delta x = $ [(**a**) 20, (**b**) 40, (**c**) 2, (**d**) 4].

3. *True or False* A function f defined on the closed interval $[a, b]$ has an infinite number of Riemann sums.

4. In the notation for a definite integral $\int_a^b f(x) \, dx$, a is called the _____ _____; b is called the _____ _____; $\int$ is called the _____ _____; and $f(x)$ is called the _____.

5. If $f(a)$ is defined, $\int_a^a f(x) \, dx = $ _____.

6. *True or False* If a function f is integrable over a closed interval $[a, b]$, then $\int_a^b f(x) \, dx = \int_b^a f(x) \, dx$.

7. *True or False* If a function f is continuous on a closed interval $[a, b]$, then the definite integral $\int_a^b f(x) \, dx$ exists.

8. *Multiple Choice* Since $\int_0^2 (3x - 8) \, dx = -10$, then $\int_2^0 (3x - 8) \, dx = $ [(**a**) −10, (**b**) 10, (**c**) 5, (**d**) 0].

Skill Building

In Problems 9–12, find the Riemann sum for each function f for the partition and the numbers u_i listed.

9. $f(x) = x, 0 \le x \le 2$. Partition the interval $[0, 2]$ as follows:
$$x_0 = 0, x_1 = \frac{1}{4}, x_2 = \frac{1}{2}, x_3 = \frac{3}{4}, x_4 = 1, x_5 = 2;$$
$$\left[0, \frac{1}{4}\right], \left[\frac{1}{4}, \frac{1}{2}\right], \left[\frac{1}{2}, \frac{3}{4}\right], \left[\frac{3}{4}, 1\right], [1, 2]$$

and choose
$$u_1 = \frac{1}{8}, u_2 = \frac{3}{8}, u_3 = \frac{5}{8}, u_4 = \frac{7}{8}, u_5 = \frac{9}{8}.$$

10. $f(x) = x, 0 \le x \le 2$. Partition the interval $[0, 2]$ as follows:
$$\left[0, \frac{1}{2}\right], \left[\frac{1}{2}, 1\right], \left[1, \frac{3}{2}\right], \left[\frac{3}{2}, 2\right], \text{ and choose } u_1 = \frac{1}{2}, u_2 = 1,$$
$$u_3 = \frac{3}{2}, u_4 = 2.$$

11. $f(x) = x^2, -2 \le x \le 1$. Partition the interval $[-2, 1]$ as follows:
$$[-2, -1], [-1, 0], [0, 1] \text{ and choose } u_1 = -\frac{3}{2}, u_2 = -\frac{1}{2},$$
$$u_3 = \frac{1}{2}.$$

---

1. = NOW WORK problem   = Graphing technology recommended   CAS = Computer Algebra System recommended

12. $f(x) = x^2$, $1 \le x \le 2$. Partition the interval $[1, 2]$ as follows:

$\left[1, \frac{5}{4}\right]$, $\left[\frac{5}{4}, \frac{3}{2}\right]$, $\left[\frac{3}{2}, \frac{7}{4}\right]$, $\left[\frac{7}{4}, 2\right]$ and choose $u_1 = \frac{5}{4}$, $u_2 = \frac{3}{2}$,

$u_3 = \frac{7}{4}$, $u_4 = 2$.

In Problems 13 and 14, the graph of a function f defined on an interval $[a, b]$ is given.

(a) *Partition the interval $[a, b]$ into six subintervals (not necessarily of the same size using the points shown on each graph).*

(b) *Approximate $\int_a^b f(x)\, dx$ by choosing u_i as the left endpoint of each subinterval and using Riemann sums.*

(c) *Approximate $\int_a^b f(x)\, dx$ by choosing u_i as the right endpoint of each subinterval and using Riemann sums.*

13.

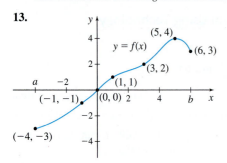

14.

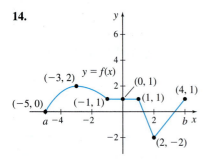

In Problems 15–22, write the limit of the Riemann sums as a definite integral. Here u_i is in the subinterval $[x_{i-1}, x_i]$, $i = 1, 2, \ldots n$ and $\Delta x_i = x_i - x_{i-1}$, $i = 1, 2, \ldots n$. Assume each limit exists.

15. $\displaystyle \lim_{\max \Delta x_i \to 0} \sum_{i=1}^{n} (e^{u_i} + 2)\Delta x_i$ on $[0, 2]$

16. $\displaystyle \lim_{\max \Delta x_i \to 0} \sum_{i=1}^{n} \ln u_i \, \Delta x_i$ on $[1, 8]$

17. $\displaystyle \lim_{\max \Delta x_i \to 0} \sum_{i=1}^{n} \cos u_i \, \Delta x_i$ on $[0, 2\pi]$

18. $\displaystyle \lim_{\max \Delta x_i \to 0} \sum_{i=1}^{n} (\cos u_i + \sin u_i)\, \Delta x_i$ on $[0, \pi]$

19. $\displaystyle \lim_{\max \Delta x_i \to 0} \sum_{i=1}^{n} \frac{2}{u_i^2} \Delta x_i$ on $[1, 4]$

20. $\displaystyle \lim_{\max \Delta x_i \to 0} \sum_{i=1}^{n} u_i^{1/3} \Delta x_i$ on $[0, 8]$

21. $\displaystyle \lim_{\max \Delta x_i \to 0} \sum_{i=1}^{n} u_i \ln u_i \, \Delta x_i$ on $[1, e]$

22. $\displaystyle \lim_{\max \Delta x_i \to 0} \sum_{i=1}^{n} \ln(u_i + 1)\Delta x_i$ on $[0, e]$

In Problems 23–28, find each definite integral.

23. $\displaystyle \int_{-3}^{4} e \, dx$

24. $\displaystyle \int_{0}^{3} (-\pi) \, dx$

25. $\displaystyle \int_{3}^{0} (-\pi) \, dt$

26. $\displaystyle \int_{7}^{2} 2 \, ds$

27. $\displaystyle \int_{4}^{4} 2\theta \, d\theta$

28. $\displaystyle \int_{-1}^{-1} 8 \, dr$

In Problems 29–32, the graph of a function is shown. Express the shaded area as a definite integral.

29.

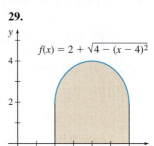

30.

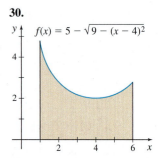

31.

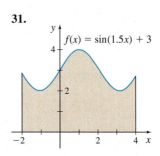

32.

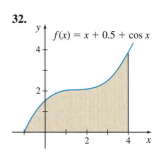

In Problems 33–38, determine which of the following definite integrals can be interpreted as area. For those that can, describe the area; for those that cannot, explain why.

33. $\displaystyle \int_{0}^{\pi} \sin x \, dx$

34. $\displaystyle \int_{-\pi/4}^{\pi/4} \tan x \, dx$

35. $\displaystyle \int_{1}^{4} (x - 2)^{1/3} dx$

36. $\displaystyle \int_{1}^{4} (x + 2)^{1/3} dx$

37. $\displaystyle \int_{1}^{4} (|x| - 2) \, dx$

38. $\displaystyle \int_{-2}^{4} |x| \, dx$

In Problems 39–44:

(a) *For each function defined on the given interval, use a regular partition to form Riemann sums $\displaystyle \sum_{i=1}^{n} f(u_i)\Delta x_i$.*

(b) *Express the limit as $n \to \infty$ of the Riemann sums as a definite integral.*

CAS **(c)** *Use a computer algebra system to find the value of the definite integral in (b).*

39. $f(x) = x^2 - 1$ on $[0, 2]$

40. $f(x) = x^3 - 2$ on $[0, 5]$

41. $f(x) = \sqrt{x + 1}$ on $[0, 3]$

42. $f(x) = \sin x$ on $[0, \pi]$

43. $f(x) = e^x$ on $[0, 2]$

44. $f(x) = e^{-x}$ on $[0, 1]$

In Problems 45 and 46, find each definite integral using Riemann sums.

45. $\displaystyle\int_0^1 (x-4)\,dx$ **46.** $\displaystyle\int_0^3 (3x-1)\,dx$

[CAS] *In Problems 47–50, for each function defined on the interval [a, b]:*

(a) *Complete the table of Riemann sums using a regular partition of [a, b].*

n	10	50	100
Using left endpoints			
Using right endpoints			
Using the midpoint			

(b) *Use a CAS to find the definite integral.*

47. $f(x) = 2 + \sqrt{x}$ on $[1, 5]$ **48.** $f(x) = e^x + e^{-x}$ on $[-1, 3]$

49. $f(x) = \dfrac{3}{1+x^2}$ on $[-1, 1]$ **50.** $f(x) = \dfrac{1}{\sqrt{x^2+4}}$ on $[0, 2]$

Applications and Extensions

51. Find an approximate value of $\displaystyle\int_1^2 \dfrac{1}{x}\,dx$ by finding Riemann sums corresponding to a partition of $[1, 2]$ into four subintervals, each of the same length, and evaluating the integrand at the midpoint of each subinterval. Compare your answer with the actual value, $\ln 2 = 0.6931 \ldots$.

52. (a) Find the approximate value of $\int_0^2 \sqrt{4-x^2}\,dx$ by finding Riemann sums corresponding to a partition of $[0, 2]$ into 16 subintervals, each of the same length, and evaluating the integrand at the left endpoint of each subinterval.

(b) Can $\int_0^2 \sqrt{4-x^2}\,dx$ be interpreted as area? If it can, describe the area; if it cannot, explain why.

(c) Find the actual value of $\int_0^2 \sqrt{4-x^2}\,dx$ by graphing $y = \sqrt{4-x^2}$ and using a familiar formula from geometry.

53. Units of an Integral In the definite integral $\int_0^5 F(x)\,dx$, F represents a force measured in newtons and x, $0 \le x \le 5$, is measured in meters. What are the units of $\int_0^5 F(x)\,dx$?

54. Units of an Integral In the definite integral $\int_0^{50} C(x)\,dx$, C represents the concentration of a drug in grams per liter and x, $0 \le x \le 50$, is measured in liters of alcohol. What are the units of $\int_0^{50} C(x)\,dx$?

55. Units of an Integral In the definite integral $\int_a^b v(t)\,dt$, v represents velocity measured in meters per second and time t is measured in seconds. What are the units of $\int_a^b v(t)\,dt$?

56. Units of an Integral In the definite integral $\int_a^b S(t)\,dt$, S represents the rate of sales of a corporation measured in millions of dollars per year and time t is measured in years. What are the units of $\int_a^b S(t)\,dt$?

[CAS] **57. Area**

(a) Graph the function $f(x) = 3 - \sqrt{6x - x^2}$.

(b) Find the area under the graph of f from 0 to 6.

(c) Confirm the answer to (b) using geometry.

[CAS] **58. Area**

(a) Graph the function $f(x) = \sqrt{4x - x^2} + 2$.

(b) Find the area under the graph of f from 0 to 4.

(c) Confirm the answer to (b) using geometry.

59. The interval $[1, 5]$ is partitioned into eight subintervals each of the same length.

(a) What is the largest Riemann sum of $f(x) = x^2$ that can be found using this partition?

(b) What is the smallest Riemann sum?

(c) Compute the average of these sums.

(d) What integral has been approximated, and what is the integral's exact value?

Challenge Problems

60. The floor function $f(x) = \lfloor x \rfloor$ is not continuous on $[0, 4]$. Show that $\int_0^4 f(x)\,dx$ exists.

61. Consider the **Dirichlet function** f, where
$$f(x) = \begin{cases} 1 & \text{if } x \text{ is rational} \\ 0 & \text{if } x \text{ is irrational} \end{cases}$$
Show that $\int_0^1 f(x)\,dx$ does not exist. (*Hint*: Evaluate the Riemann sums in two different ways: first by using rational numbers for u_i and then by using irrational numbers for u_i.)

62. It can be shown (with a certain amount of work) that if $f(x)$ is integrable on the interval $[a, b]$, then so is $|f(x)|$. Is the converse true?

63. If only regular partitions are allowed, then we could not always partition an interval $[a, b]$ in a way that automatically partitions subintervals $[a, c]$ and $[c, b]$ for $a < c < b$. Why not?

64. If f is a function that is continuous on a closed interval $[a, b]$, except at $x_1, x_2, \ldots, x_n$, $n \ge 1$ an integer, where it has a jump discontinuity, show that f is integrable on $[a, b]$.

65. If f is a function that is continuous on a closed interval $[a, b]$, except at $x_1, x_2, \ldots, x_n$, $n \ge 1$ an integer, where it has a removable discontinuity, show that f is integrable on $[a, b]$.

5.3 The Fundamental Theorem of Calculus

OBJECTIVES *When you finish this section, you should be able to:*

1 Use Part 1 of the Fundamental Theorem of Calculus (p. 363)

2 Use Part 2 of the Fundamental Theorem of Calculus (p. 365)

3 Interpret an integral using Part 2 of the Fundamental Theorem of Calculus (p. 365)

In this section, we discuss the Fundamental Theorem of Calculus, a method for finding integrals more easily, avoiding the need to find the limit of Riemann sums. The Fundamental Theorem is aptly named because it links the two branches of calculus: differential calculus and integral calculus. As it turns out, the Fundamental Theorem of Calculus has two parts, each of which relates an integral to an antiderivative.

Suppose f is a function that is continuous on a closed interval $[a, b]$. Then the definite integral $\int_a^b f(x)\,dx$ exists and is equal to a real number. Now if x denotes any number in $[a, b]$, the definite integral $\int_a^x f(t)\,dt$ exists and depends on x. That is, $\int_a^x f(t)\,dt$ is a function of x, which we name I, for "integral."

$$I(x) = \int_a^x f(t)\,dt$$

NEED TO REVIEW? Antiderivatives are discussed in Section 4.8, pp. 328–331.

The domain of I is the closed interval $[a, b]$. The integral that defines I has a *variable upper limit of integration* x. The t that appears in the integrand is a dummy variable. Surprisingly, when we differentiate I with respect to x, we get back the original function f. That is, $\int_a^x f(t)\,dt$ *is an antiderivative of* f.

THEOREM Fundamental Theorem of Calculus, Part 1

Let f be a function that is continuous on a closed interval $[a, b]$. The function I defined by

$$I(x) = \int_a^x f(t)\,dt$$

has the properties that it is continuous on $[a, b]$ and differentiable on (a, b). Moreover,

$$I'(x) = \frac{d}{dx}\left[\int_a^x f(t)\,dt\right] = f(x)$$

for all x in (a, b).

The proof of Part 1 of the Fundamental Theorem of Calculus is given in Appendix B. However, if the integral $\int_a^x f(t)\,dt$ represents area, we can interpret the theorem using geometry.

Figure 22 shows the graph of a function f that is nonnegative and continuous on a closed interval $[a, b]$. Then $I(x) = \int_a^x f(t)\,dt$ equals the area under the graph of f from a to x.

$$I(x) = \int_a^x f(t)\,dt = \text{the area under the graph of } f \text{ from } a \text{ to } x$$

$$I(x + h) = \int_a^{x+h} f(t)\,dt = \text{the area under the graph of } f \text{ from } a \text{ to } x + h$$

$$I(x + h) - I(x) = \text{the area under the graph of } f \text{ from } x \text{ to } x + h$$

$$\frac{I(x + h) - I(x)}{h} = \frac{\text{the area under the graph of } f \text{ from } x \text{ to } x + h}{h} \tag{1}$$

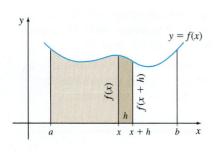

DF Figure 22

Based on the definition of a derivative,

$$\lim_{h \to 0} \left[\frac{I(x+h) - I(x)}{h} \right] = I'(x) \qquad (2)$$

Since f is continuous, $\lim_{h \to 0} f(x+h) = f(x)$. As $h \to 0$, the area under the graph of f from x to $x+h$ gets closer to the area of a rectangle with width h and height $f(x)$. That is,

$$\lim_{h \to 0} \frac{\text{the area under the graph of } f \text{ from } x \text{ to } x+h}{h} = \lim_{h \to 0} \frac{h[f(x)]}{h} = f(x) \qquad (3)$$

Combining (1), (2), and (3), it follows that $I'(x) = f(x)$.

1 Use Part 1 of the Fundamental Theorem of Calculus

EXAMPLE 1 **Using Part 1 of the Fundamental Theorem of Calculus**

(a) $\dfrac{d}{dx} \displaystyle\int_0^x \sqrt{t+1}\, dt = \sqrt{x+1}$ **(b)** $\dfrac{d}{dx} \displaystyle\int_2^x \dfrac{s^3 - 1}{2s^2 + s + 1}\, ds = \dfrac{x^3 - 1}{2x^2 + x + 1}$ ∎

NOW WORK Problem 5.

EXAMPLE 2 **Using Part 1 of the Fundamental Theorem of Calculus**

Find $\dfrac{d}{dx} \displaystyle\int_4^{3x^2+1} \sqrt{e^t + t}\, dt$.

NEED TO REVIEW? The Chain Rule is discussed in Section 3.1, pp. 198–200.

Solution The upper limit of integration is a function of x, so we use the Chain Rule along with Part 1 of the Fundamental Theorem of Calculus.

Let $y = \int_4^{3x^2+1} \sqrt{e^t + t}\, dt$ and $u(x) = 3x^2 + 1$. Then $y = \int_4^u \sqrt{e^t + t}\, dt$ and

$$\frac{d}{dx} \int_4^{3x^2+1} \sqrt{e^t + t}\, dt = \frac{dy}{dx} = \underset{\underset{\text{Chain Rule}}{\uparrow}}{\frac{dy}{du} \cdot \frac{du}{dx}} = \left[\frac{d}{du} \int_4^u \sqrt{e^t + t}\, dt \right] \cdot \frac{du}{dx}$$

$$= \underset{\underset{\substack{\text{Use the Fundamental}\\\text{Theorem}}}{\uparrow}}{\sqrt{e^u + u}} \cdot \frac{du}{dx} = \underset{\underset{u = 3x^2+1;\ \frac{du}{dx} = 6x}{\uparrow}}{\sqrt{e^{(3x^2+1)} + 3x^2 + 1} \cdot 6x}$$

∎

NOW WORK Problem 11.

EXAMPLE 3 **Using Part 1 of the Fundamental Theorem of Calculus**

Find $\dfrac{d}{dx} \displaystyle\int_{x^3}^5 (t^4 + 1)^{1/3}\, dt$.

Solution To use Part 1 of the Fundamental Theorem of Calculus, the variable must be part of the upper limit of integration. So, we use the fact that $\int_a^b f(x)\, dx = -\int_b^a f(x)\, dx$ to interchange the limits of integration.

$$\frac{d}{dx} \int_{x^3}^5 (t^4 + 1)^{1/3}\, dt = \frac{d}{dx} \left[-\int_5^{x^3} (t^4 + 1)^{1/3}\, dt \right] = -\frac{d}{dx} \int_5^{x^3} (t^4 + 1)^{1/3}\, dt$$

Now we use the Chain Rule. We let $y = \int_5^{x^3} (t^4 + 1)^{1/3} dt$ and $u(x) = x^3$.

$$\frac{d}{dx} \int_{x^3}^{5} (t^4 + 1)^{1/3}\, dt = -\frac{d}{dx} \int_5^{x^3} (t^4 + 1)^{1/3}\, dt = -\frac{dy}{dx} \underset{\underset{\text{\color{blue}Chain Rule}}{\uparrow}}{=} -\frac{dy}{du} \cdot \frac{du}{dx}$$

$$= -\frac{d}{du} \int_5^u (t^4 + 1)^{1/3} dt \cdot \frac{du}{dx}$$

$$= -(u^4 + 1)^{1/3} \cdot \frac{du}{dx} \qquad \text{\color{blue}Use the Fundamental Theorem}$$

$$= -(x^{12} + 1)^{1/3} \cdot 3x^2 \qquad \text{\color{blue}} u = x^3;\ \frac{du}{dx} = 3x^2$$

$$= -3x^2(x^{12} + 1)^{1/3} \qquad\qquad\qquad\qquad \blacksquare$$

NOTE In these examples, the differentiation is with respect to the variable that appears in the upper or lower limit of integration and the answer is a function of that variable.

NOW WORK Problem **15.**

Part 1 of the Fundamental Theorem of Calculus establishes a relationship between the derivative and the definite integral. Part 2 of the Fundamental Theorem of Calculus provides a method for finding a definite integral without using Riemann sums.

THEOREM Fundamental Theorem of Calculus, Part 2

Let f be a function that is continuous on a closed interval $[a, b]$. If F is any antiderivative of f on $[a, b]$, then

$$\int_a^b f(x)\, dx = F(b) - F(a)$$

Proof Let $I(x) = \int_a^x f(t)\, dt$. Then from Part 1 of the Fundamental Theorem of Calculus, $I = I(x)$ is continuous for $a \le x \le b$ and differentiable for $a < x < b$. So,

$$\frac{d}{dx} \int_a^x f(t)\, dt = f(x) \qquad a < x < b$$

RECALL Any two antiderivatives of a function differ by a constant.

That is, $\int_a^x f(t)\, dt$ is an antiderivative of f. So, if F is any antiderivative of f, then

$$F(x) = \int_a^x f(t)\, dt + C$$

where C is some constant. Since F is continuous on $[a, b]$, we have

$$F(a) = \int_a^a f(t)\, dt + C \qquad F(b) = \int_a^b f(t)\, dt + C$$

NOTE For any constant C, $[F(b)+C]-[F(a)+C] = F(b)-F(a)$. In other words, it does not matter which antiderivative of f is chosen when using Part 2 of the Fundamental Theorem of Calculus, since the same answer is obtained for every antiderivative.

Since, $\int_a^a f(t)\, dt = 0$, subtracting $F(a)$ from $F(b)$ gives

$$F(b) - F(a) = \int_a^b f(t)\, dt$$

Since t is a dummy variable, we can replace t by x and the result follows. $\blacksquare$

As an aid in computation, we introduce the notation

$$\int_a^b f(x)\, dx = \Big[F(x)\Big]_a^b = F(b) - F(a)$$

The notation $\Big[F(x)\Big]_a^b$ also suggests that to find $\int_a^b f(x)\, dx$, we first find an antiderivative $F(x)$ of $f(x)$. Then we write $\Big[F(x)\Big]_a^b$ to represent $F(b) - F(a)$.

2 Use Part 2 of the Fundamental Theorem of Calculus

EXAMPLE 4   **Using Part 2 of the Fundamental Theorem of Calculus**

Use Part 2 of the Fundamental Theorem of Calculus to find:

(a) $\displaystyle\int_{-2}^{1} x^2\,dx$ **(b)** $\displaystyle\int_{0}^{\pi/6} \cos x\,dx$ **(c)** $\displaystyle\int_{0}^{\sqrt{3}/2} \frac{1}{\sqrt{1-x^2}}\,dx$ **(d)** $\displaystyle\int_{1}^{2} \frac{1}{x}\,dx$

Solution   **(a)** An antiderivative of $f(x) = x^2$ is $F(x) = \dfrac{x^3}{3}$. By Part 2 of the Fundamental Theorem of Calculus,

$$\int_{-2}^{1} x^2\,dx = \left[\frac{x^3}{3}\right]_{-2}^{1} = \frac{1^3}{3} - \frac{(-2)^3}{3} = \frac{1}{3} + \frac{8}{3} = \frac{9}{3} = 3$$

(b) An antiderivative of $f(x) = \cos x$ is $F(x) = \sin x$. By Part 2 of the Fundamental Theorem of Calculus,

$$\int_{0}^{\pi/6} \cos x\,dx = \left[\sin x\right]_{0}^{\pi/6} = \sin\frac{\pi}{6} - \sin 0 = \frac{1}{2}$$

(c) An antiderivative of $f(x) = \dfrac{1}{\sqrt{1-x^2}}$ is $F(x) = \sin^{-1} x$, provided $|x| < 1$. By Part 2 of the Fundamental Theorem of Calculus,

$$\int_{0}^{\sqrt{3}/2} \frac{1}{\sqrt{1-x^2}}\,dx = \left[\sin^{-1} x\right]_{0}^{\sqrt{3}/2} = \sin^{-1}\frac{\sqrt{3}}{2} - \sin^{-1} 0 = \frac{\pi}{3} - 0 = \frac{\pi}{3}$$

(d) An antiderivative of $f(x) = \dfrac{1}{x}$ is $F(x) = \ln|x|$. By Part 2 of the Fundamental Theorem of Calculus,

$$\int_{1}^{2} \frac{1}{x}\,dx = \left[\ln|x|\right]_{1}^{2} = \ln 2 - \ln 1 = \ln 2 - 0 = \ln 2 \qquad\blacksquare$$

NOW WORK   Problems 25 and 31.

EXAMPLE 5   **Finding the Area Under a Graph**

Find the area under the graph of $f(x) = e^x$ from -1 to 1.

Solution   Figure 23 shows the graph of $f(x) = e^x$ on the closed interval $[-1, 1]$.

The area A under the graph of f from -1 to 1 is given by

$$A = \int_{-1}^{1} e^x\,dx = \left[e^x\right]_{-1}^{1} = e^1 - e^{-1} = e - \frac{1}{e} \approx 2.350 \qquad\blacksquare$$

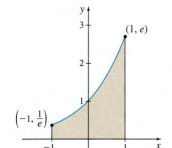

Figure 23 $f(x) = e^x,\ -1 \le x \le 1$

NOW WORK   Problem 45.

3 Interpret an Integral Using Part 2 of the Fundamental Theorem of Calculus

Part 2 of the Fundamental Theorem of Calculus states that, under certain conditions,

$$\int_{a}^{b} f(x)\,dx = F(b) - F(a) \qquad \text{where } F' = f$$

That is,

$$\int_{a}^{b} F'(x)\,dx = F(b) - F(a)$$

In other words,

> The integral from a to b of the rate of change of F equals the change in F from a to b.

EXAMPLE 6   Interpreting an Integral Whose Integrand Is a Rate of Change

(a) The function $P = P(t)$ relates the population P (in billions of persons) as a function of the time t (in years). Suppose $\int_0^{10} P'(t)\, dt = 3$. Since $P'(t)$ is the rate of change of the population with respect to time, then the change in population from $t = 0$ to $t = 10$ is 3 billion persons.

(b) The function $v = v(t)$ is the speed v (in meters per second) of an object at a time t (in seconds). If $\int_0^5 v(t)\, dt = 8$, then the object travels 8 m during the interval $0 \le t \le 5$. Do you see why? The speed $v = v(t)$ is the rate of change of distance s with respect to time t. That is, $v = \dfrac{ds}{dt}$. ■

This interpretation of an integral is important since it reveals how to go from a rate of change of a function F back to the change itself.

NOW WORK Problem 51.

5.3 Assess Your Understanding

Concepts and Vocabulary

1. According to Part 1 of the Fundamental Theorem of Calculus, if a function f is continuous on a closed interval $[a, b]$, then
$$\frac{d}{dx}\left[\int_a^x f(t)\, dt\right] = \underline{\hspace{1cm}} \text{ for all numbers } x \text{ in } (a, b).$$

2. *True or False* By Part 2 of the Fundamental Theorem of Calculus, $\int_a^b x\, dx = b - a$.

3. *True or False* By Part 2 of the Fundamental Theorem of Calculus, $\int_a^b f(x)\, dx = f(b) - f(a)$.

4. *True or False* $\int_a^b F'(x)\, dx$ can be interpreted as the rate of change in F from a to b.

Skill Building

In Problems 5–18, find each derivative using Part 1 of the Fundamental Theorem of Calculus.

5. $\dfrac{d}{dx}\int_1^x \sqrt{t^2 + 1}\, dt$

6. $\dfrac{d}{dx}\int_3^x \dfrac{t+1}{t}\, dt$

7. $\dfrac{d}{dt}\left[\int_0^t (3 + x^2)^{3/2}dx\right]$

8. $\dfrac{d}{dx}\left[\int_{-4}^x (t^3 + 8)^{1/3}\, dt\right]$

9. $\dfrac{d}{dx}\left[\int_1^x \ln u\, du\right]$

10. $\dfrac{d}{dt}\left[\int_4^t e^x dx\right]$

11. $\dfrac{d}{dx}\left[\int_1^{2x^3} \sqrt{t^2 + 1}\, dt\right]$

12. $\dfrac{d}{dx}\left[\int_1^{\sqrt{x}} \sqrt{t^4 + 5}\, dt\right]$

13. $\dfrac{d}{dx}\left[\int_2^{x^5} \sec t\, dt\right]$

14. $\dfrac{d}{dx}\left[\int_3^{1/x} \sin^5 t\, dt\right]$

15. $\dfrac{d}{dx}\left[\int_x^5 \sin(t^2)\, dt\right]$

16. $\dfrac{d}{dx}\left[\int_x^3 (t^2 - 5)^{10}\, dt\right]$

17. $\dfrac{d}{dx}\left[\int_{5x^2}^5 (6t)^{2/3}\, dt\right]$

18. $\dfrac{d}{dx}\left[\int_{x^2}^0 e^{10t}\, dt\right]$

In Problems 19–36, use Part 2 of the Fundamental Theorem of Calculus to find each definite integral.

19. $\displaystyle\int_{-2}^3 dx$

20. $\displaystyle\int_{-2}^3 2\, dx$

21. $\displaystyle\int_{-1}^2 x^3 dx$

22. $\displaystyle\int_1^3 \dfrac{1}{x^3}\, dx$

23. $\displaystyle\int_0^1 \sqrt{u}\, du$

24. $\displaystyle\int_1^8 \sqrt[3]{y}\, dy$

25. $\displaystyle\int_{\pi/6}^{\pi/2} \csc^2 x\, dx$

26. $\displaystyle\int_0^{\pi/2} \cos x\, dx$

27. $\displaystyle\int_0^{\pi/4} \sec x \tan x\, dx$

28. $\displaystyle\int_{\pi/6}^{\pi/2} \csc x \cot x\, dx$

29. $\displaystyle\int_{-1}^0 e^x dx$

30. $\displaystyle\int_{-1}^0 e^{-x} dx$

31. $\displaystyle\int_1^e \dfrac{1}{x}\, dx$

32. $\displaystyle\int_e^1 \dfrac{1}{x}\, dx$

33. $\displaystyle\int_0^1 \dfrac{1}{1 + x^2}\, dx$

34. $\displaystyle\int_0^{\sqrt{2}/2} \dfrac{1}{\sqrt{1 - x^2}}\, dx$

35. $\displaystyle\int_{-1}^8 x^{2/3}\, dx$

36. $\displaystyle\int_0^4 x^{3/2}\, dx$

---

1. = NOW WORK problem   ⌁ = Graphing technology recommended   CAS = Computer Algebra System recommended

In Problems 37–42, find $\int_a^b f(x)\,dx$ over the domain of f indicated in the graph.

37.

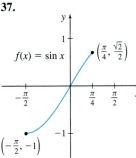

$f(x) = \sin x$

$\left(\frac{\pi}{4}, \frac{\sqrt{2}}{2}\right)$

$\left(-\frac{\pi}{2}, -1\right)$

38.

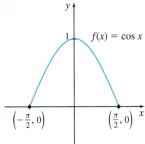

$f(x) = \cos x$

$\left(-\frac{\pi}{2}, 0\right)$ $\left(\frac{\pi}{2}, 0\right)$

39.

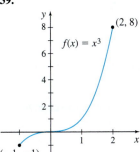

$f(x) = x^3$ (2, 8)

$(-1, -1)$

40.

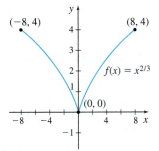

$(-8, 4)$ $(8, 4)$

$f(x) = x^{2/3}$

$(0, 0)$

41.

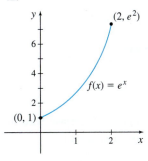

$(2, e^2)$

$f(x) = e^x$

$(0, 1)$

42.

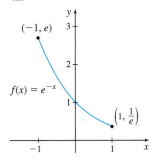

$(-1, e)$

$f(x) = e^{-x}$

$\left(1, \frac{1}{e}\right)$

43. Given that $f(x) = (2x^3 - 3)^2$ and $f'(x) = 12x^2(2x^3 - 3)$, find $\int_0^2 [12x^2(2x^3 - 3)]\,dx$.

44. Given that $f(x) = (x^2 + 5)^3$ and $f'(x) = 6x(x^2 + 5)^2$, find $\int_{-1}^2 6x(x^2 + 5)^2\,dx$.

Applications and Extensions

45. Area Find the area under the graph of $f(x) = \dfrac{1}{\sqrt{1 - x^2}}$ from 0 to $\dfrac{1}{2}$.

46. Area Find the area under the graph of $f(x) = \cosh x$ from -1 to 1.

47. Area Find the area under the graph of $f(x) = \dfrac{1}{x^2 + 1}$ from 0 to $\sqrt{3}$.

48. Area Find the area under the graph of $f(x) = \dfrac{1}{1 + x^2}$ from 0 to r, where $r > 0$. What happens as $r \to \infty$?

49. Area Find the area under the graph of $y = \dfrac{1}{\sqrt{x}}$ from $x = 1$ to $x = r$, where $r > 1$. Then examine the behavior of this area as $r \to \infty$.

50. Area Find the area under the graph of $y = \dfrac{1}{x^2}$ from $x = 1$ to $x = r$, where $r > 1$. Then examine the behavior of this area as $r \to \infty$.

51. Interpreting an Integral The function $R = R(t)$ models the rate of sales of a corporation measured in millions of dollars per year as a function of the time t in years. Interpret the integral $\int_0^2 R(t)\,dt = 23$.

52. Interpreting an Integral The function $v = v(t)$ models the speed v in meters per second of an object at a time t in seconds. Interpret the integral $\int_0^{10} v(t)\,dt = 4.8$.

53. Interpreting an Integral Helium is leaking from a large advertising balloon at a rate of $H(t)$ cubic centimeters per minute, where t is measured in minutes.

  (a) Write an integral that models the change in helium in the balloon over the interval $a \le t \le b$.

  (b) What are the units of the integral from (a)?

  (c) Interpret $\int_0^{300} H(t)\,dt = -100$.

54. Interpreting an Integral Water is being added to a reservoir at a rate of $w(t)$ kiloliters per hour, where t is measured in hours.

  (a) Write an integral that models the change in amount of water in the reservoir over the interval $a \le t \le b$.

  (b) What are the units of the integral from (a)?

  (c) Interpret $\int_0^{36} w(t)\,dt = 800$.

55. Free Fall The speed v of an object dropped from rest is given by $v(t) = 9.8t$, where v is in meters per second and time t is in seconds.

  (a) Express the distance traveled in the first 5.2 seconds as an integral.

  (b) Find the distance traveled in 5.2 seconds.

56. Area Find h so that the area under the graph of $y^2 = x^3$, $0 \le x \le 4$, $y \ge 0$, is equal to the area of a rectangle of base 4 and height h.

57. Area If P is a polynomial that is positive for $x > 0$, and for each $k > 0$ the area under the graph of P from $x = 0$ to $x = k$ is $k^3 + 3k^2 + 6k$, find P.

58. Put It Together If $f(x) = \int_0^x \dfrac{1}{\sqrt{t^3 + 2}}\,dt$, which of the following is *false*?

  (a) f is continuous at x for all $x \ge 0$

  (b) $f(1) > 0$

  (c) $f(0) = \dfrac{1}{\sqrt{2}}$

  (d) $f'(1) = \dfrac{1}{\sqrt{3}}$

In Problems 59–62:

(a) *Use Part of 2 the Fundamental Theorem of Calculus to find each definite integral.*

(b) *Determine whether the integrand is an even function, an odd function, or neither.*

(c) *Can you make a conjecture about the definite integrals in (a) based on the analysis from (b)? Look at Objective 3 in Section 5.6.*

59. $\int_0^4 x^2\,dx$ and $\int_{-4}^4 x^2\,dx$

60. $\int_0^4 x^3\,dx$ and $\int_{-4}^4 x^3\,dx$

61. $\int_0^{\pi/4} \sec^2 x\,dx$ and $\int_{-\pi/4}^{\pi/4} \sec^2 x\,dx$

62. $\int_0^{\pi/4} \sin x\,dx$ and $\int_{-\pi/4}^{\pi/4} \sin x\,dx$

63. Area Find c, $0 < c < 1$, so that the area under the graph of $y = x^2$ from 0 to c equals the area under the same graph from c to 1.

64. Area Let A be the area under the graph of $y = \dfrac{1}{x}$ from $x = m$ to $x = 2m$, $m > 0$. Which of the following is true about the area A?

(a) A is independent of m.

(b) A increases as m increases.

(c) A decreases as m increases.

(d) A decreases as m increases when $m < \dfrac{1}{2}$ and increases as m increases when $m > \dfrac{1}{2}$.

(e) A increases as m increases when $m < \dfrac{1}{2}$ and decreases as m increases when $m > \dfrac{1}{2}$.

65. Put It Together If F is a function whose derivative is continuous for all real x, find

$$\lim_{h \to 0} \frac{1}{h} \int_c^{c+h} F'(x)\,dx$$

66. Suppose the closed interval $\left[0, \dfrac{\pi}{2}\right]$ is partitioned into n subintervals, each of length Δx, and u_i is an arbitrary number in the subinterval $[x_{i-1}, x_i]$, $i = 1, 2, \ldots, n$. Explain why

$$\lim_{n \to \infty} \sum_{i=1}^n [(\cos u_i)\,\Delta x] = 1$$

67. The interval $[0, 4]$ is partitioned into n subintervals, each of length Δx, and a number u_i is chosen in the subinterval $[x_{i-1}, x_i]$, $i = 1, 2, \ldots, n$. Find $\displaystyle\lim_{n \to \infty} \sum_{i=1}^n (e^{u_i}\,\Delta x)$.

68. If u and v are differentiable functions and f is a continuous function, find a formula for

$$\frac{d}{dx}\left[\int_{u(x)}^{v(x)} f(t)\,dt\right]$$

69. Suppose that the graph of $y = f(x)$ contains the points $(0, 1)$ and $(2, 5)$. Find $\int_0^2 f'(x)\,dx$. (Assume that f' is continuous.)

70. If f' is continuous on the interval $[a, b]$, show that

$$\int_a^b f(x)f'(x)\,dx = \frac{1}{2}\left\{[f(b)]^2 - [f(a)]^2\right\}.$$

[*Hint:* Look at the derivative of $F(x) = \dfrac{[f(x)]^2}{2}$.]

71. If f'' is continuous on the interval $[a, b]$, show that

$$\int_a^b xf''(x)\,dx = bf'(b) - af'(a) - f(b) + f(a).$$

[*Hint:* Look at the derivative of $F(x) = xf'(x) - f(x)$.]

Challenge Problems

72. What conditions on f and f' guarantee that $f(x) = \int_0^x f'(t)\,dt$?

73. Suppose that F is an antiderivative of f on the interval $[a, b]$. Partition $[a, b]$ into n subintervals, each of length $\Delta x_i = x_i - x_{i-1}$, $i = 1, 2, \ldots, n$.

(a) Apply the Mean Value Theorem for derivatives to F in each subinterval $[x_{i-1}, x_i]$ to show that there is a point u_i in the subinterval for which $F(x_i) - F(x_{i-1}) = f(u_i)\Delta x_i$.

(b) Show that $\displaystyle\sum_{i=1}^n [F(x_i) - F(x_{i-1})] = F(b) - F(a)$.

(c) Use parts (a) and (b) to explain why

$$\int_a^b f(x)\,dx = F(b) - F(a).$$

(In this alternate proof of Part 2 of the Fundamental Theorem of Calculus, the continuity of f is not assumed.)

74. Given $y = \sqrt{x^2 - 1}(4 - x)$, $1 \le x \le a$, for what number a will $\int_1^a y\,dx$ have a maximum value?

75. Find $a > 0$, so that the area under the graph of $y = x + \dfrac{1}{x}$ from a to $(a + 1)$ is minimum.

76. If n is a known positive integer, for what number c is

$$\int_1^c x^{n-1}\,dx = \frac{1}{n}$$

77. Let $f(x) = \displaystyle\int_0^x \frac{dt}{\sqrt{1 - t^2}}$, $0 < x < 1$.

(a) Find $\dfrac{d}{dx} f(\sin x)$.

(b) Is f one-to-one?

(c) Does f have an inverse?

5.4 Properties of the Definite Integral

OBJECTIVES *When you finish this section, you should be able to:*

1 Use properties of the definite integral (p. 369)

2 Work with the Mean Value Theorem for Integrals (p. 372)

3 Find the average value of a function (p. 373)

We have seen that there are properties of limits that make it easier to find limits, properties of continuity that make it easier to determine continuity, and properties of derivatives that make it easier to find derivatives. Here, we investigate several properties of the definite integral that will make it easier to find integrals.

1 Use Properties of the Definite Integral

NOTE From now on, we will refer to Parts 1 and 2 of the Fundamental Theorem of Calculus simply as the **Fundamental Theorem of Calculus.**

Proofs of most of the properties in this section require the use of the definition of the definite integral. However, if the condition is added that the integrand is continuous, then the Fundamental Theorem of Calculus can be used to establish the properties. We will use this added condition in the proofs included here.

THEOREM The Integral of the Sum of Two Functions

If two functions f and g are continuous on the closed interval $[a, b]$, then

$$\int_a^b [f(x) + g(x)]\,dx = \int_a^b f(x)\,dx + \int_a^b g(x)\,dx \tag{1}$$

IN WORDS The integral of a sum equals the sum of the integrals.

Proof Since f and g are continuous on $[a, b]$, then the definite integral of each function exists. Let F and G be an antiderivative of f and g, respectively, on (a, b). Then $F' = f$ and $G' = g$ on (a, b).

Also, since $(F + G)' = F' + G' = f + g$, then $F + G$ is an antiderivative of $f + g$ on (a, b). Now

$$\int_a^b [f(x) + g(x)]\,dx = \left[F(x) + G(x) \right]_a^b = [F(b) + G(b)] - [F(a) + G(a)]$$

$$= [F(b) - F(a)] + [G(b) - G(a)]$$

$$= \int_a^b f(x)\,dx + \int_a^b g(x)\,dx \quad\blacksquare$$

EXAMPLE 1   Using Property (1) of the Definite Integral

$$\int_0^1 (x^2 + e^x)\,dx = \int_0^1 x^2\,dx + \int_0^1 e^x\,dx = \left[\frac{x^3}{3}\right]_0^1 + \left[e^x\right]_0^1$$

$$= \left[\frac{1^3}{3} - 0\right] + \left[e^1 - e^0\right] = \frac{1}{3} + e - 1 = e - \frac{2}{3} \quad\blacksquare$$

NOW WORK Problem 13.

THEOREM The Integral of a Constant Times a Function

Suppose a function f is continuous on the closed interval $[a, b]$. If k is a constant, then

$$\int_a^b kf(x)\,dx = k \int_a^b f(x)\,dx \tag{2}$$

IN WORDS A constant factor can be factored out of an integral.

You are asked to prove this theorem in Problem 93.

EXAMPLE 2 Using Property (2) of the Definite Integral

$$\int_1^e \frac{3}{x}\,dx = 3\int_1^e \frac{1}{x}\,dx = 3\left[\ln|x|\right]_1^e = 3\,(\ln e - \ln 1) = 3(1-0) = 3 \qquad \blacksquare$$

NOW WORK Problem 15.

The two properties above can be extended as follows:

THEOREM

Suppose each of the functions $f_1, f_2, \ldots, f_n$ is continuous on the closed interval $[a, b]$. If $k_1, k_2, \ldots, k_n$ are constants, then

$$\boxed{\begin{aligned} &\int_a^b \left[k_1 f_1(x) + k_2 f_2(x) + \cdots + k_n f_n(x)\right]\,dx \\ &= k_1 \int_a^b f_1(x)\,dx + k_2 \int_a^b f_2(x)\,dx + \cdots + k_n \int_a^b f_n(x)\,dx \end{aligned}} \qquad (3)$$

You are asked to prove this theorem in Problem 94.

EXAMPLE 3 Using Property (3) of the Definite Integral

Find $\displaystyle\int_1^2 \frac{3x^3 - 6x^2 - 5x + 4}{2x}\,dx$.

Solution The function $f(x) = \dfrac{3x^3 - 6x^2 - 5x + 4}{2x}$ is continuous on the closed interval $[1, 2]$. Using algebra and properties of the definite integral, we get

$$\int_1^2 \frac{3x^3 - 6x^2 - 5x + 4}{2x}\,dx = \int_1^2 \left[\frac{3}{2}x^2 - 3x - \frac{5}{2} + \frac{2}{x}\right]\,dx$$

$$= \int_1^2 \frac{3}{2}x^2\,dx - \int_1^2 3x\,dx - \int_1^2 \frac{5}{2}\,dx + \int_1^2 \frac{2}{x}\,dx$$

$$= \frac{3}{2}\int_1^2 x^2\,dx - 3\int_1^2 x\,dx - \frac{5}{2}\int_1^2 dx + 2\int_1^2 \frac{1}{x}\,dx$$

$$= \frac{3}{2}\left[\frac{x^3}{3}\right]_1^2 - 3\left[\frac{x^2}{2}\right]_1^2 - \frac{5}{2}\left[x\right]_1^2 + 2\left[\ln|x|\right]_1^2$$

$$= \frac{1}{2}(8-1) - \frac{3}{2}(4-1) - \frac{5}{2}(2-1) + 2(\ln 2 - \ln 1)$$

$$= -\frac{7}{2} + 2\ln 2 \qquad \blacksquare$$

NOW WORK Problem 27.

The next property states that a definite integral of a function f from a to b can be evaluated in pieces.

THEOREM

If a function f is continuous on an interval containing the numbers a, b, and c, then

$$\boxed{\int_a^b f(x)\,dx = \int_a^c f(x)\,dx + \int_c^b f(x)\,dx} \qquad (4)$$

A proof of this theorem is given in Appendix B.

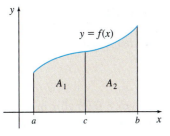

A = Area under the graph of f from a to b
= area A_1 + area A_2

$$\int_a^b f(x)\,dx = \int_a^c f(x)\,dx + \int_c^b f(x)\,dx$$

Figure 24

In particular, if f is continuous and nonnegative on a closed interval $[a, b]$ and if c is a number between a and b, then this property has a simple geometric interpretation, as seen in Figure 24.

EXAMPLE 4 Using Property (4) of the Definite Integral

(a) If f is continuous on the closed interval $[2, 7]$, then

$$\int_2^7 f(x)\,dx = \int_2^4 f(x)\,dx + \int_4^7 f(x)\,dx$$

(b) If g is continuous on the closed interval $[3, 25]$, then

$$\int_3^{10} g(x)\,dx = \int_3^{25} g(x)\,dx + \int_{25}^{10} g(x)\,dx$$ ∎

Example 4(b) illustrates that the number c need not lie between a and b.

NOW WORK Problem 43.

Property (4) is useful when integrating piecewise-defined functions.

EXAMPLE 5 Using Property (4) of the Definite Integral

Find the area A under the graph of

$$f(x) = \begin{cases} x^2 & \text{if } 0 \le x < 10 \\ 100 & \text{if } 10 \le x \le 15 \end{cases}$$

from 0 to 15.

Solution See Figure 25. Since f is nonnegative on the closed interval $[0, 15]$, then $\int_0^{15} f(x)\,dx$ equals the area A under the graph of f from 0 to 15. Since f is continuous on $[0, 15]$,

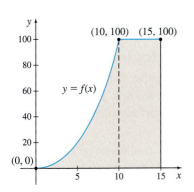

Figure 25 $A = \int_0^{15} f(x)\,dx$

$$\int_0^{15} f(x)\,dx = \int_0^{10} f(x)\,dx + \int_{10}^{15} f(x)\,dx = \int_0^{10} x^2\,dx + \int_{10}^{15} 100\,dx$$

$$= \left[\frac{x^3}{3}\right]_0^{10} + \left[100x\right]_{10}^{15} = \frac{1000}{3} + 500 = \frac{2500}{3}$$

The area under the graph of f is approximately 833.333 square units. ∎

NOW WORK Problem 35.

The next property establishes bounds on a definite integral.

THEOREM Bounds on an Integral

If a function f is continuous on a closed interval $[a, b]$ and if m and M denote the absolute minimum value and the absolute maximum value, respectively, of f on $[a, b]$, then

$$\boxed{m(b - a) \le \int_a^b f(x)\,dx \le M(b - a)} \tag{5}$$

A proof of this theorem is given in Appendix B.

If f is nonnegative on $[a, b]$, then the inequalities in (5) have a geometric interpretation. In Figure 26, the area of the shaded region is $\int_a^b f(x)\,dx$. The smaller rectangle has width $b - a$, height m, and area equal $m(b - a)$. The larger rectangle has width $b - a$, height M, and area $M(b - a)$. These three areas are numerically related by the inequalities in the theorem.

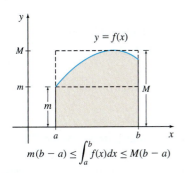

$$m(b - a) \le \int_a^b f(x)\,dx \le M(b - a)$$

Figure 26

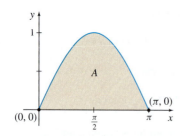

Figure 27 $f(x) = \sin x, 0 \leq x \leq \pi$.

NEED TO REVIEW? The Extreme Value Theorem is discussed in Section 4.2, pp. 265–266.

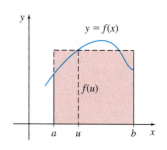

Figure 28 $\displaystyle\int_a^b f(x)\,dx = f(u)(b-a)$.

EXAMPLE 6   Using Property (5) of Definite Integrals

(a) Find an upper estimate and a lower estimate for the area A under the graph of $f(x) = \sin x$ from 0 to π.

(b) Find the actual area under the graph.

Solution The graph of f is shown in Figure 27. Since $f(x) \geq 0$ for all x in the closed interval $[0, \pi]$, the area A under its graph is given by the definite integral, $\int_0^\pi \sin x\,dx$.

(a) From the Extreme Value Theorem, f has an absolute minimum value and an absolute maximum value on the interval $[0, \pi]$. The absolute maximum of f occurs at $x = \dfrac{\pi}{2}$, and its value is $f\left(\dfrac{\pi}{2}\right) = \sin\dfrac{\pi}{2} = 1$. The absolute minimum occurs at $x = 0$ and at $x = \pi$; the absolute minimum value is $f(0) = \sin 0 = 0 = f(\pi)$. Using the inequalities in (5), the area under the graph of f is bounded as follows:

$$0 \leq \int_0^\pi \sin x\,dx \leq \pi$$

(b) The actual area under the graph is

$$A = \int_0^\pi \sin x\,dx = \left[-\cos x\right]_0^\pi = -\cos\pi + \cos 0 = 1 + 1 = 2 \text{ square units} \quad \blacksquare$$

NOW WORK **Problem 53.**

2 Work with the Mean Value Theorem for Integrals

Suppose f is a function that is continuous and nonnegative on a closed interval $[a, b]$. Figure 28 suggests that the area under the graph of f from a to b, $\int_a^b f(x)\,dx$, is equal to the area of some rectangle of width $b - a$ and height $f(u)$ for some choice (or choices) of u in the interval $[a, b]$.

In more general terms, for every function f that is continuous on a closed interval $[a, b]$, there is some number u (not necessarily unique) in the interval $[a, b]$ for which

$$\int_a^b f(x)\,dx = f(u)(b-a)$$

This result is known as the *Mean Value Theorem for Integrals*.

THEOREM Mean Value Theorem for Integrals

If a function f is continuous on a closed interval $[a, b]$, there is a real number u, $a \leq u \leq b$, for which

$$\boxed{\int_a^b f(x)\,dx = f(u)(b-a)} \tag{6}$$

Proof Let f be a function that is continuous on a closed interval $[a, b]$.

If f is a constant function, say, $f(x) = k$, on $[a, b]$, then

$$\int_a^b f(x)\,dx = \int_a^b k\,dx = k(b-a) = f(u)(b-a)$$

for any choice of u in $[a, b]$.

If f is not a constant function on $[a, b]$, then by the Extreme Value Theorem, f has an absolute maximum and an absolute minimum on $[a, b]$. Suppose f assumes its absolute minimum at the number c so that $f(c) = m$; and suppose f assumes its absolute maximum at the number C so that $f(C) = M$. Then by the Bounds on an

Integral Theorem (5), we have

$$m(b-a) \le \int_a^b f(x)\,dx \le M(b-a) \qquad \text{for all } x \text{ in } [a, b]$$

We divide each part by $(b-a)$ and replace m by $f(c)$ and M by $f(C)$. Then

$$f(c) \le \frac{1}{b-a}\int_a^b f(x)\,dx \le f(C)$$

NEED TO REVIEW? The Intermediate Value Theorem is discussed in Section 1.3, pp. 100–102.

Since $\dfrac{1}{b-a}\displaystyle\int_a^b f(x)\,dx$ is a real number between $f(c)$ and $f(C)$, it follows from the Intermediate Value Theorem that there is a real number u between c and C, for which

$$f(u) = \frac{1}{b-a}\int_a^b f(x)\,dx$$

That is, there is a real number u, $a \le u \le b$, for which

$$\int_a^b f(x)\,dx = f(u)\,(b-a) \qquad \blacksquare$$

EXAMPLE 7   Using the Mean Value Theorem for Integrals

Find the number(s) u guaranteed by the Mean Value Theorem for Integrals for $\int_2^6 x^2\,dx$.

Solution The Mean Value Theorem for Integrals states there is a number u, $2 \le u \le 6$, for which

$$\int_2^6 x^2\,dx = f(u)(6-2) = 4u^2 \qquad \textcolor{blue}{f(u) = u^2}$$

We integrate to obtain

$$\int_2^6 x^2\,dx = \left[\frac{x^3}{3}\right]_2^6 = \frac{1}{3}(216-8) = \frac{208}{3}$$

Then

$$\frac{208}{3} = 4u^2$$

$$u^2 = \frac{52}{3} \qquad \textcolor{blue}{2 \le u \le 6}$$

$$u = \sqrt{\frac{52}{3}} \approx 4.163 \qquad \textcolor{blue}{\text{Disregard the negative solution since } u > 0.} \qquad \blacksquare$$

NOW WORK Problem 55.

---

3 Find the Average Value of a Function

We know from the Mean Value Theorem for Integrals that if a function f is continuous on a closed interval $[a, b]$, there is a real number u, $a \le u \le b$, for which

$$\int_a^b f(x)\,dx = f(u)(b-a)$$

This means that if the function f is also nonnegative on the closed interval $[a, b]$, the area enclosed by a rectangle of height $f(u)$ and width $b-a$ equals the area under the graph of f from a to b. See Figure 29.

So if we replace $f(x)$ on $[a, b]$ by $f(u)$, we get a region with the same area. Consequently, $f(u)$ can be thought of as an *average value*, or *mean value*, *of f over* $[a, b]$.

We can obtain the average value of f over $[a, b]$ for any function f that is continuous on the closed interval $[a, b]$ by partitioning $[a, b]$ into n subintervals

$$[a, x_1], \quad [x_1, x_2], \quad \ldots, \quad [x_{i-1}, x_i], \quad \ldots, \quad [x_{n-1}, b]$$

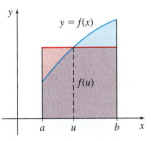

Figure 29 $\displaystyle\int_a^b f(x)\,dx = f(u)(b-a)$

each of length $\Delta x = \dfrac{b-a}{n}$, and choosing a number u_i in each of the n subintervals. Then an approximation of the average value of f over the interval $[a, b]$ is

$$\frac{f(u_1) + f(u_2) + \cdots + f(u_n)}{n} \tag{7}$$

Now we multiply (7) by $\dfrac{b-a}{b-a}$ to obtain

$$\frac{f(u_1) + f(u_2) + \cdots + f(u_n)}{n} = \frac{1}{b-a}\left[f(u_1)\frac{b-a}{n} + f(u_2)\frac{b-a}{n} + \cdots + f(u_n)\frac{b-a}{n}\right]$$

$$= \frac{1}{b-a}[f(u_1)\,\Delta x + f(u_2)\,\Delta x + \cdots + f(u_n)\,\Delta x]$$

$$= \frac{1}{b-a}\sum_{i=1}^{n} f(u_i)\,\Delta x$$

This sum approximates the average value of f. As the length of each subinterval gets smaller, the sums become better approximations to the average value of f on $[a, b]$. Furthermore, $\sum_{i=1}^{n} f(u_i)\,\Delta x$ are Riemann sums, so $\lim\limits_{n \to \infty} \sum_{i=1}^{n} f(u_i)\,\Delta x$ is a definite integral. This suggests the following definition:

IN WORDS The average value $\bar{y} = \dfrac{1}{b-a}\displaystyle\int_a^b f(x)\,dx$ of a function f equals the value $f(u)$ in the Mean Value Theorem for Integrals.

DEFINITION Average Value of a Function over an Interval

Let f be a function that is continuous on the closed interval $[a, b]$. The **average value** $\bar{y}$ of f over $[a, b]$ is

$$\boxed{\bar{y} = \frac{1}{b-a}\int_a^b f(x)\,dx} \tag{8}$$

EXAMPLE 8   Finding the Average Value of a Function

Find the average value of $f(x) = 3x - 8$ on the closed interval $[0, 2]$.

Solution   The average value of $f(x) = 3x - 8$ on the closed interval $[0, 2]$ is given by

$$\bar{y} = \frac{1}{b-a}\int_a^b f(x)\,dx = \frac{1}{2-0}\int_0^2 (3x - 8)\,dx$$

$$= \frac{1}{2}\left[\frac{3x^2}{2} - 8x\right]_0^2 = \frac{1}{2}(6 - 16) = -5$$

The average value of f on $[0, 2]$ is $\bar{y} = -5$. ∎

The function f and its average value are graphed in Figure 30.

Figure 30 $f(x) = 3x - 8,\, 0 \le x \le 2$

NOW WORK   Problem 61.

5.4 Assess Your Understanding

Concepts and Vocabulary

1. *True or False* $\int_2^3 (x^2 + x)\,dx = \int_2^3 x^2 dx + \int_2^3 x\,dx$

2. *True or False* $\int_0^3 5e^{x^2} dx = \int_0^3 5\,dx \cdot \int_0^3 e^{x^2} dx$

3. *True or False* $\int_0^5 (x^3 + 1)dx = \int_0^{-3}(x^3 + 1)dx + \int_{-3}^5 (x^3 + 1)dx$

4. If f is continuous on an interval containing the numbers a, b, and c, and if $\int_a^c f(x)\,dx = 3$ and $\int_c^b f(x)\,dx = -5$, then $\int_a^b f(x)\,dx = $ _____.

5. If a function f is continuous on the closed interval $[a, b]$, then $\bar{y} = \dfrac{1}{b-a}\displaystyle\int_a^b f(x)\,dx$ is the _____ _____ of f over $[a, b]$.

---

1. = NOW WORK problem $\boxed{\triangle}$ = Graphing technology recommended $\boxed{\text{CAS}}$ = Computer Algebra System recommended

6. *True or False* If a function f is continuous on a closed interval $[a, b]$ and if m and M denote the absolute minimum value and the absolute maximum value, respectively, of f on $[a, b]$, then

$$m \leq \int_a^b f(x)\, dx \leq M.$$

Skill Building

In Problems 7–12, find each definite integral given that
$\int_1^3 f(x)\, dx = 5,\ \int_1^3 g(x)\, dx = -2,\ \int_3^5 f(x)\, dx = 2,\ \int_3^5 g(x)\, dx = 1.$

7. $\displaystyle\int_1^3 [f(x) - g(x)]\, dx$

8. $\displaystyle\int_1^3 [f(x) + g(x)]\, dx$

9. $\displaystyle\int_1^3 [5f(x) - 3g(x)]\, dx$

10. $\displaystyle\int_1^3 [3f(x) + 4g(x)]\, dx$

11. $\displaystyle\int_1^5 [2f(x) - 3g(x)]\, dx$

12. $\displaystyle\int_1^5 [f(x) - g(x)]\, dx$

In Problems 13–32, find each definite integral using the Fundamental Theorem of Calculus and properties of the definite integral.

13. $\displaystyle\int_0^1 (t^2 - t^{3/2})\, dt$

14. $\displaystyle\int_{-2}^0 (x + x^2)\, dx$

15. $\displaystyle\int_{\pi/2}^{\pi} 4 \sin x\, dx$

16. $\displaystyle\int_0^1 3x^2 dx$

17. $\displaystyle\int_1^e -\frac{3}{x}\, dx$

18. $\displaystyle\int_e^8 \frac{1}{2x}\, dx$

19. $\displaystyle\int_{-\pi/4}^{\pi/4} (1 + 2\sec x\, \tan x)\, dx$

20. $\displaystyle\int_0^{\pi/4} (1 + \sec^2 x)\, dx$

21. $\displaystyle\int_1^4 (\sqrt{x} - 4x)\, dx$

22. $\displaystyle\int_0^1 (\sqrt[5]{t^2} + 1)\, dt$

23. $\displaystyle\int_{-2}^3 [(x - 1)(x + 3)]\, dx$

24. $\displaystyle\int_0^1 (z^2 + 1)^2 dz$

25. $\displaystyle\int_1^2 \frac{x^2 - 12}{x^4}\, dx$

26. $\displaystyle\int_1^e \frac{5s^2 + s}{s^2}\, ds$

27. $\displaystyle\int_1^4 \frac{x + 1}{\sqrt{x}}\, dx$

28. $\displaystyle\int_1^9 \frac{\sqrt{x} + 1}{x^2}\, dx$

29. $\displaystyle\int_1^2 \frac{2x^4 + 1}{x^4}\, dx$

30. $\displaystyle\int_1^3 \frac{2 - x^2}{x^4}\, dx$

31. $\displaystyle\int_0^{1/2} \left(5 + \frac{1}{\sqrt{1 - x^2}}\right) dx$

32. $\displaystyle\int_0^1 \left(1 + \frac{5}{1 + x^2}\right) dx$

In Problems 33–38, use properties of integrals and the Fundamental Theorem of Calculus to find each integral.

33. $\displaystyle\int_{-2}^1 f(x)\, dx$, where $f(x) = \begin{cases} 1 & \text{if } x < 0 \\ x^2 + 1 & \text{if } x \geq 0 \end{cases}$

34. $\displaystyle\int_{-1}^2 f(x)\, dx$, where $f(x) = \begin{cases} x + 1 & \text{if } x < 0 \\ x^2 + 1 & \text{if } x \geq 0 \end{cases}$

35. $\displaystyle\int_{-2}^2 f(x)\, dx$, where $f(x) = \begin{cases} 3x & \text{if } -2 \leq x < 0 \\ 2x^2 & \text{if } 0 \leq x \leq 2 \end{cases}$

36. $\displaystyle\int_0^4 h(x)\, dx$, where $h(x) = \begin{cases} x - 2 & \text{if } 0 \leq x \leq 2 \\ 2 - x & \text{if } 2 < x \leq 4 \end{cases}$

37. $\displaystyle\int_{-2}^1 H(x)\, dx$, where $H(x) = \begin{cases} 1 + x^2 & \text{if } -2 \leq x < 0 \\ 1 + 3x & \text{if } 0 \leq x \leq 1 \end{cases}$

38. $\displaystyle\int_{-\pi/2}^{\pi/2} f(x)\, dx$, where $f(x) = \begin{cases} x^2 + x & \text{if } -\dfrac{\pi}{2} \leq x \leq 0 \\ \sin x & \text{if } 0 < x < \dfrac{\pi}{4} \\ \dfrac{\sqrt{2}}{2} & \text{if } \dfrac{\pi}{4} \leq x \leq \dfrac{\pi}{2} \end{cases}$

In Problems 39–42, the domain of f is a closed interval $[a, b]$.
Find $\int_a^b f(x)\, dx$.

39.

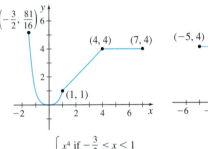

$f(x) = \begin{cases} x^4 & \text{if } -\dfrac{3}{2} \leq x < 1 \\ x & \text{if } 1 \leq x < 4 \\ 4 & \text{if } 4 \leq x \leq 7 \end{cases}$

40.

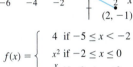

$f(x) = \begin{cases} 4 & \text{if } -5 \leq x < -2 \\ x^2 & \text{if } -2 \leq x \leq 0 \\ -\dfrac{x}{2} & \text{if } 0 < x \leq 2 \end{cases}$

41.

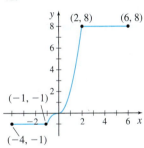

$f(x) = \begin{cases} -1 & \text{if } -4 \leq x \leq -1 \\ x^3 & \text{if } -1 < x < 2 \\ 8 & \text{if } 2 \leq x \leq 6 \end{cases}$

42.

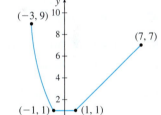

$f(x) = \begin{cases} x^2 & \text{if } -3 \leq x \leq -1 \\ 1 & \text{if } -1 < x \leq 1 \\ x & \text{if } 1 < x \leq 7 \end{cases}$

In Problems 43–46, use properties of definite integrals to verify each statement. Assume that all integrals involved exist.

43. $\displaystyle\int_3^{11} f(x)\, dx - \int_7^{11} f(x)\, dx = \int_3^7 f(x)\, dx$

44. $\displaystyle\int_{-2}^6 f(x)\, dx - \int_3^6 f(x)\, dx = \int_{-2}^3 f(x)\, dx$

45. $\displaystyle\int_0^4 f(x)\, dx - \int_6^4 f(x)\, dx = \int_0^6 f(x)\, dx$

46. $\displaystyle\int_{-1}^3 f(x)\, dx - \int_5^3 f(x)\, dx = \int_{-1}^5 f(x)\, dx$

In Problems 47–54, use the Bounds on an Integral Theorem to obtain a lower estimate and an upper estimate for each integral.

47. $\displaystyle\int_1^3 (5x + 1)\, dx$

48. $\displaystyle\int_0^1 (1 - x)\, dx$

49. $\displaystyle\int_{\pi/4}^{\pi/2} \sin x\, dx$

50. $\displaystyle\int_{\pi/6}^{\pi/3} \cos x\, dx$

51. $\displaystyle\int_0^1 \sqrt{1 + x^2}\, dx$

52. $\displaystyle\int_{-1}^1 \sqrt{1 + x^4}\, dx$

53. $\displaystyle\int_0^1 e^x\, dx$

54. $\displaystyle\int_1^{10} \frac{1}{x}\, dx$

In Problems 55–60, for each integral find the number(s) u guaranteed by the Mean Value Theorem for Integrals.

55. $\displaystyle\int_0^3 (2x^2 + 1)\, dx$

56. $\displaystyle\int_0^2 (2 - x^3)\, dx$

57. $\displaystyle\int_0^4 x^2\, dx$

58. $\displaystyle\int_0^4 (-x)\, dx$

59. $\displaystyle\int_0^{2\pi} \cos x\, dx$

60. $\displaystyle\int_{-\pi/4}^{\pi/4} \sec x \tan x\, dx$

In Problems 61–70, find the average value of each function f over the given interval.

61. $f(x) = e^x$ over $[0, 1]$

62. $f(x) = \dfrac{1}{x}$ over $[1, e]$

63. $f(x) = x^{2/3}$ over $[-1, 1]$

64. $f(x) = \sqrt{x}$ over $[0, 4]$

65. $f(x) = \sin x$ over $\left[0, \dfrac{\pi}{2}\right]$

66. $f(x) = \cos x$ over $\left[0, \dfrac{\pi}{2}\right]$

67. $f(x) = 1 - x^2$ over $[-1, 1]$

68. $f(x) = 16 - x^2$ over $[-4, 4]$

69. $f(x) = e^x - \sin x$ over $\left[0, \dfrac{\pi}{2}\right]$

70. $f(x) = x + \cos x$ over $\left[0, \dfrac{\pi}{2}\right]$

In Problems 71–74, find:
(a) The area under the graph of the function over the indicated interval.
(b) The average value of each function over the indicated interval.
(c) Interpret the results geometrically.

71. $[-1, 2]$

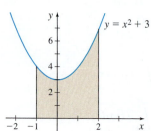

$y = x^2 + 3$

72. $[-2, 1]$

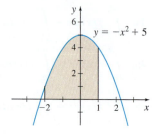

$y = -x^2 + 5$

73. $[-1, 2]$

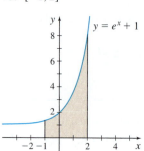

$y = e^x + 1$

74. $\left[0, \dfrac{3\pi}{4}\right]$

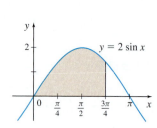

$y = 2 \sin x$

Applications and Extensions

In Problems 75–78, find each definite integral using the Fundamental Theorem of Calculus and properties of definite integrals.

75. $\displaystyle\int_{-2}^3 (x + |x|)\, dx$

76. $\displaystyle\int_0^3 |x - 1|\, dx$

77. $\displaystyle\int_0^2 |3x - 1|\, dx$

78. $\displaystyle\int_0^2 |2 - x|\, dx$

79. Average Temperature A rod 3 meters long is heated to $25x\,°C$, where x is the distance in meters from one end of the rod. Find the average temperature of the rod.

80. Average Daily Rainfall The rainfall per day, x days after the beginning of the year, is modeled by the function $r(x) = 0.00002(6511 + 366x - x^2)$, measured in centimeters. Find the average daily rainfall for the first 180 days of the year.

81. Structural Engineering A structural engineer designing a member of a structure must consider the forces that will act on that member. Most often, natural forces like snow, wind, or rain distribute force over the entire member. For practical purposes, however, an engineer determines the distributed force as a single resultant force acting at one point on the member. If the distributed force is given by the function $W = W(x)$, in newtons per meter (N/m), then the magnitude F_R of the resultant force is

$$F_R = \int_a^b W(x)\, dx$$

The position $\bar{x}$ of the resultant force measured in meters from the origin is given by

$$\bar{x} = \frac{\int_a^b x W(x)\, dx}{\int_a^b W(x)\, dx}$$

If the distributed force is $W(x) = 0.75x^3$, $0 \le x \le 5$, find:

(a) The magnitude of the resultant force.

(b) The position from the origin of the resultant force.

Source: Problem contributed by the students at Trine University, Avalon, IN.

82. Chemistry: Enthalpy In chemistry, **enthalpy** is a measure of the total energy of a system. For a nonreactive process with no phase change, the change in enthalpy ΔH is given by $\Delta H = \int_{T_1}^{T_2} C_p\, dT$, where C_p is the specific heat of the system in question. The specific heat per mol of the chemical benzene is

$$C_p = 0.126 + (2.34 \times 10^{-6})T,$$

where C_p is in kJ/(mol °C), and T is in degrees Celsius.

(a) What are the units of the change in enthalpy ΔH?

(b) What is the change in enthalpy ΔH associated with increasing the temperature of 1.0 mol of benzene from 20 °C to 40 °C?

(c) What is the change in enthalpy ΔH associated with increasing the temperature of 1.0 mol of benzene from 20 °C to 60 °C?

(d) Does the enthalpy of benzene increase, decrease, or remain constant as the temperature increases?

Source: Problem contributed by the students at Trine University, Avalon, IN.

83. **Average Mass Density** The mass density of a metal bar of length 3 meters is given by $\rho(x) = 1000 + x - \sqrt{x}$ kilograms per cubic meter, where x is the distance in meters from one end of the bar. What is the average mass density over the length of the entire bar?

84. **Average Velocity** The acceleration at time t of an object in rectilinear motion is given by $a(t) = 4\pi \cos t$. If the object's velocity is 0 at $t = 0$, what is the average velocity of the object over the interval $0 \le t \le \pi$?

85. **Average Area** What is the average area of all circles whose radii are between 1 and 3 m?

86. **Area**

(a) Use properties of integrals and the Fundamental Theorem of Calculus to find the area under the graph of $y = 3 - |x|$ from -3 to 3.

(b) Check your answer by using elementary geometry.

87. **Area**

(a) Use properties of integrals and the Fundamental Theorem of Calculus to find the area under the graph of $y = 1 - \left|\frac{1}{2}x\right|$ from -2 to 2.

(b) Check your answer by using elementary geometry. See the figure.

88. **Area** Let A be the area in the first quadrant that is enclosed by the graphs of $y = 3x^2$, $y = \frac{3}{x}$, the x-axis, and the line $x = k$, where $k > 1$, as shown in the figure.

(a) Find the area A as a function of k.

(b) When the area is 7, what is k?

(c) If the area A is increasing at the constant rate of 5 square units per second, at what rate is k increasing when $k = 15$?

89. **Rectilinear Motion** A car starting from rest accelerates at the rate of 3 m/s². Find its average speed over the first 8 seconds.

90. **Rectilinear Motion** A car moving at a constant velocity of 80 miles per hour begins to decelerate at the rate of 10 mi/h². Find its average speed over the next 10 minutes.

91. **Average Slope**

(a) Use the definition of average value of a function to find the *average slope* of the graph of $y = f(x)$, where $a \le x \le b$. (Assume that f' is continuous.)

(b) Give a geometric interpretation.

92. What theorem guarantees that the average slope found in Problem 91 is equal to $f'(u)$ for some u in $[a, b]$? What *different* theorem guarantees the same thing? (Do you see the connection between these theorems?)

93. Prove that if a function f is continuous on a closed interval $[a, b]$ and if k is a constant, then $\int_a^b kf(x)\, dx = k \int_a^b f(x)\, dx$.

94. Prove that if the functions $f_1, f_2, \ldots, f_n$ are continuous on a closed interval $[a, b]$ and if $k_1, k_2, \ldots, k_n$ are constants, then

$$\int_a^b [k_1 f_1(x) + k_2 f_2(x) + \cdots + k_n f_n(x)]\, dx$$

$$= k_1 \int_a^b f_1(x)\, dx + k_2 \int_a^b f_2(x)\, dx + \cdots + k_n \int_a^b f_n(x)\, dx$$

95. **Area** The area under the graph of $y = \cos x$ from $-\frac{\pi}{2}$ to $\frac{\pi}{2}$ is separated into two parts by the line $x = k$, $\frac{-\pi}{2} < k < \frac{\pi}{2}$, as shown in the figure. If the area under the graph of y from $-\frac{\pi}{2}$ to k is three times the area under the graph of y from k to $\frac{\pi}{2}$, find k.

96. **Displacement of a Damped Spring** The displacement x in meters of a damped spring from its equilibrium position at time t seconds is given by

$$x(t) = \frac{\sqrt{15}}{10} e^{-t} \sin\left(\sqrt{15}t\right) + \frac{3}{2} e^{-t} \cos\left(\sqrt{15}t\right)$$

(a) What is the displacement of the spring at $t = 0$?

(b) Graph the displacement for the first 2 seconds of the springs' motion.

(c) Find the average displacement of the spring for the first 2 seconds of its motion.

97. **Area** Let

$$f(x) = |x^4 + 3.44x^3 - 0.5041x^2 - 5.0882x + 1.1523|$$

be defined on the interval $[-3, 1]$. Find the area under the graph of f.

98. If f is continuous on $[a, b]$, show that the functions defined by

$$F(x) = \int_c^x f(t)\, dt \qquad G(x) = \int_d^x f(t)\, dt$$

for any choice of c and d in (a, b) always differ by a constant. Also show that

$$F(x) - G(x) = \int_c^d f(t)\, dt$$

99. Put It Together   Suppose $a < c < b$ and the function f is continuous on $[a, b]$ and differentiable on (a, b). Which of the following is *not* necessarily true?

(a) $\displaystyle\int_a^b f(x)\, dx = \int_a^c f(x)\, dx + \int_c^b f(x)\, dx.$

(b) There is a number d in (a, b) for which

$$f'(d) = \frac{f(b) - f(a)}{b - a}$$

(c) $\displaystyle\int_a^b f(x)\, dx \geq 0.$

(d) $\displaystyle\lim_{x \to c} f(x) = f(c).$

(e) If k is a real number, then $\int_a^b kf(x)\, dx = k \int_a^b f(x)\, dx.$

100. Minimizing Area   Find $b > 0$ so that the area enclosed in the first quadrant by the graph of $y = 1 + b - bx^2$ and the coordinate axes is a minimum.

101. Area   Find the area enclosed by the graph of $\sqrt{x} + \sqrt{y} = 1$ and the coordinate axes.

Challenge Problems

102. Average Speed   For a freely falling object starting from rest, $v_0 = 0$, find:

(a) The average speed $\bar{v}_t$ with respect to the time t in seconds over the closed interval $[0, 5]$.

(b) The average speed $\bar{v}_s$ with respect to the distance s of the object from its position at $t = 0$ over the closed interval $[0, s_1]$, where s_1 is the distance the object falls from $t = 0$ to $t = 5$ seconds.

(*Hint:* The derivation of the formulas for freely falling objects is given in Section 4.8, pp. 334–337.)

103. Average Speed   If an object falls from rest for 3 seconds, find:

(a) Its average speed with respect to time.

(b) Its average speed with respect to the distance it travels in 3 seconds.

104. Free Fall   For a freely falling object starting from rest, $v_0 = 0$, find:

(a) The average velocity $\bar{v}_t$ with respect to the time t over the closed interval $[0, t_1]$.

(b) The average velocity $\bar{v}_s$ with respect to the distance s of the object from its position at $t = 0$ over the closed interval $[0, s_1]$, where s_1 is the distance the object falls in time t_1. Assume $s(0) = 0$.

105. Put It Together

(a) What is the domain of $f(x) = 2|x - 1|x^2$?

(b) What is the range of f?

(c) For what values of x is f continuous?

(d) For what values of x is the derivative of f continuous?

(e) Find $\int_0^1 f(x)\, dx.$

106. Probability   A function f that is continuous on the closed interval $[a, b]$, and for which (i) $f(x) \geq 0$ for numbers x in $[a, b]$ and 0 elsewhere and (ii) $\int_a^b f(x)\, dx = 1$, is called a **probability density function**. If $a \leq c < d \leq b$, the probability of obtaining a value between c and d is defined as $\int_c^d f(x)\, dx.$

(a) Find a constant k so that $f(x) = kx$ is a probability density function on $[0, 2]$.

(b) Find the probability of obtaining a value between 1 and 1.5.

107. Cumulative Probability Distribution   Refer to Problem 106. If f is a probability density function, the **cumulative distribution function** F for f is defined as

$$F(x) = \int_a^x f(t)\, dt \qquad a \leq x \leq b$$

Find the cumulative distribution function F for the probability density function $f(x) = kx$ of Problem 106(a).

108. For the cumulative distribution function $F(x) = x - 1$, on the interval $[1, 2]$:

(a) Find the probability density function f corresponding to F.

(b) Find the probability of obtaining a value between 1.5 and 1.7.

109. Let $f(x) = x^3 - 6x^2 + 11x - 6$. Find $\int_1^3 |f(x)|\, dx.$

110. Show that for $x > 1$, $\ln x < 2(\sqrt{x} - 1)$.

(*Hint:* Use the result given in Problem 114.)

111. Prove that the average value of a line segment $y = m(x - x_1) + y_1$ on the interval $[x_1, x_2]$ equals the y-coordinate of the midpoint of the line segment from x_1 to x_2.

112. Prove that if a function f is continuous on a closed interval $[a, b]$ and if $f(x) \geq 0$ on $[a, b]$, then $\int_a^b f(x)\, dx \geq 0$.

113. (a) Prove that if f is continuous on a closed interval $[a, b]$ and $\int_a^b f(x)\, dx = 0$, there is at least one number c in $[a, b]$ for which $f(c) = 0$.

(b) Give a counterexample to the statement above if f is not required to be continuous.

114. Prove that if functions f and g are continuous on a closed interval $[a, b]$ and if $f(x) \geq g(x)$ on $[a, b]$, then

$$\int_a^b f(x)\, dx \geq \int_a^b g(x)\, dx.$$

115. Prove that if f is continuous on $[a, b]$, then

$$\left| \int_a^b f(x)\, dx \right| \leq \int_a^b |f(x)|\, dx.$$

Give a geometric interpretation of the inequality.

5.5 The Indefinite Integral; Growth and Decay Models

OBJECTIVES *When you finish this section, you should be able to:*

1 Find indefinite integrals (p. 379)
2 Use properties of indefinite integrals (p. 380)
3 Solve differential equations involving growth and decay (p. 382)

The Fundamental Theorem of Calculus establishes an important relationship between definite integrals and antiderivatives: the definite integral $\int_a^b f(x)\,dx$ can be found easily if an antiderivative of f can be found. Because of this, it is customary to use the integral symbol $\int$ as an instruction to find all antiderivatives of a function.

DEFINITION Indefinite Integral

The expression $\int f(x)\,dx$, called the **indefinite integral of** f, is defined as,

$$\int f(x)\,dx = F(x) + C$$

where F is any function for which $\dfrac{d}{dx}F(x) = f(x)$ and C is a number, called the **constant of integration**.

CAUTION In writing an indefinite integral $\int f(x)\,dx$, remember to include the "dx."

For example,

$$\int (x^2 + 1)\,dx = \frac{x^3}{3} + x + C \qquad \frac{d}{dx}\left(\frac{x^3}{3} + x + C\right) = \frac{3x^2}{3} + 1 + 0 = x^2 + 1$$

The process of finding either the indefinite integral $\int f(x)\,dx$ or the definite integral $\int_a^b f(x)\,dx$ is called **integration**, and in both cases the function f is called the **integrand**.

IN WORDS The definite integral $\int_a^b f(x)\,dx$ is a number; the indefinite integral $\int f(x)\,dx$ is a family of functions.

It is important to distinguish between the definite integral $\int_a^b f(x)\,dx$ and the indefinite integral $\int f(x)\,dx$. The definite integral is a *number* that depends on the limits of integration a and b. In contrast, the indefinite integral of f is a *family* of functions $F(x) + C$, C a constant, for which $F'(x) = f(x)$. For example,

$$\int_0^2 x^2\,dx = \left[\frac{x^3}{3}\right]_0^2 = \frac{8}{3} \qquad \int x^2\,dx = \frac{x^3}{3} + C$$

We summarize the antiderivatives of some important functions in Table 1 on page 380. Each entry is a result of a differentiation formula.

1 Find Indefinite Integrals

EXAMPLE 1   Finding Indefinite Integrals

Find:

(a) $\displaystyle\int x^4\,dx$ **(b)** $\displaystyle\int \sqrt{x}\,dx$ **(c)** $\displaystyle\int \frac{\sin x}{\cos^2 x}\,dx$

Solution **(a)** All the antiderivatives of $f(x) = x^4$ are $F(x) = \dfrac{x^5}{5} + C$, so

$$\int x^4\,dx = \frac{x^5}{5} + C$$

TABLE 1

Table of Integrals

$$\int dx = x + C$$

$$\int x^a \, dx = \frac{x^{a+1}}{a+1} + C; \quad a \neq -1$$

$$\int x^{-1} \, dx = \int \frac{1}{x} \, dx = \ln |x| + C$$

$$\int e^x \, dx = e^x + C$$

$$\int a^x \, dx = \frac{a^x}{\ln a} + C; \quad a > 0, a \neq 1$$

$$\int \sin x \, dx = -\cos x + C$$

$$\int \cos x \, dx = \sin x + C$$

$$\int \sec^2 x \, dx = \tan x + C$$

$$\int \sec x \tan x \, dx = \sec x + C$$

$$\int \csc x \cot x \, dx = -\csc x + C$$

$$\int \csc^2 x \, dx = -\cot x + C$$

$$\int \frac{1}{\sqrt{1-x^2}} \, dx = \sin^{-1} x + C, \quad |x| < 1$$

$$\int \frac{1}{1+x^2} \, dx = \tan^{-1} x + C$$

$$\int \frac{1}{x\sqrt{x^2-1}} \, dx = \sec^{-1} x + C, \quad |x| > 1$$

$$\int \sinh x \, dx = \cosh x + C$$

$$\int \cosh x \, dx = \sinh x + C$$

(b) All the antiderivatives of $f(x) = \sqrt{x} = x^{1/2}$ are $F(x) = \dfrac{x^{3/2}}{\dfrac{3}{2}} + C = \dfrac{2x^{3/2}}{3} + C$.

$$\int \sqrt{x} \, dx = \frac{2x^{3/2}}{3} + C$$

NEED TO REVIEW? Trigonometric identities are discussed in Appendix A.4, pp. A-32 to A-35.

(c) No integral in Table 1 corresponds to $f(x) = \dfrac{\sin x}{\cos^2 x}$, so we begin by using trigonometric identities to rewrite $\dfrac{\sin x}{\cos^2 x}$ in a form whose antiderivative is recognizable.

$$\frac{\sin x}{\cos^2 x} = \frac{\sin x}{\cos x \cdot \cos x} = \frac{1}{\cos x} \cdot \frac{\sin x}{\cos x} = \sec x \tan x$$

Then

$$\int \frac{\sin x}{\cos^2 x} \, dx = \int \sec x \tan x \, dx = \sec x + C \qquad \blacksquare$$

NOW WORK Problems 5 and 7.

2 Use Properties of Indefinite Integrals

Since the definite integral and the indefinite integral are closely related, properties of indefinite integrals are very similar to those of definite integrals:

• **Derivative of an Integral**:

$$\frac{d}{dx}\left[\int f(x) \, dx \right] = f(x) \qquad (1)$$

Property (1) is a consequence of the definition of $\int f(x)\,dx$. For example,

$$\frac{d}{dx}\int \sqrt{x^2+1}\,dx = \sqrt{x^2+1} \qquad \frac{d}{dt}\int e^t \cos t\,dt = e^t \cos t$$

- **Integral of the Sum of Two Functions:**

$$\boxed{\int [f(x)+g(x)]\,dx = \int f(x)\,dx + \int g(x)\,dx} \tag{2}$$

IN WORDS The indefinite integral of a sum of two functions equals the sum of the indefinite integrals.

The proof of property (2) follows directly from properties of derivatives, and is left as an exercise. See Problem 68.

- **Integral of a Constant Times a Function:** If k is a constant,

$$\boxed{\int kf(x)\,dx = k\int f(x)\,dx} \tag{3}$$

IN WORDS To find the indefinite integral of a constant k times a function f, find the indefinite integral of f and then multiply by k.

To prove property (3), differentiate the right side of (3).

$$\frac{d}{dx}\left[k\int f(x)\,dx\right] = k\underbrace{\left[\frac{d}{dx}\int f(x)\,dx\right]}_{\text{Constant Multiple Rule}} = \underbrace{kf(x)}_{\text{Property (1)}}$$

Constant Multiple Rule   Property (1)

EXAMPLE 2   **Using Properties of the Indefinite Integral**

$$\int (2x^{1/3} + 5x^{-1})\,dx = \int 2x^{1/3}\,dx + \int \frac{5}{x}\,dx = 2\int x^{1/3}\,dx + 5\int \frac{1}{x}\,dx$$

$$= 2\cdot\frac{x^{4/3}}{\frac{4}{3}} + 5\ln|x| + C = \frac{3x^{4/3}}{2} + 5\ln|x| + C \qquad \blacksquare$$

NOW WORK Problem 13.

Sometimes an appropriate algebraic manipulation is required before integrating.

EXAMPLE 3   **Using Properties of the Indefinite Integral**

(a) $$\int \left(\frac{12}{x^5} + \frac{1}{\sqrt{x}}\right) dx = 12\int \frac{1}{x^5}\,dx + \int \frac{1}{\sqrt{x}}\,dx = 12\int x^{-5}\,dx + \int x^{-1/2}\,dx$$

$$= 12\left(\frac{x^{-4}}{-4}\right) + \frac{x^{1/2}}{\frac{1}{2}} + C = -\frac{3}{x^4} + 2\sqrt{x} + C$$

(b) $$\int \frac{x^2+6}{x^2+1}\,dx = \int \frac{(x^2+1)+5}{x^2+1}\,dx = \int \left[\frac{x^2+1}{x^2+1} + \frac{5}{x^2+1}\right]dx$$

$$= \int \left[1 + \frac{5}{x^2+1}\right]dx \underset{\uparrow}{=} \int dx + \int \frac{5}{x^2+1}\,dx$$

Sum Property

$$= \int dx + 5\int \frac{1}{x^2+1}\,dx = x + 5\tan^{-1}x + C \qquad \blacksquare$$

NOW WORK Problem 33.

3 Solve Differential Equations Involving Growth and Decay

There are situations in science and nature, such as radioactive decay, population growth, and interest paid on an investment, in which a quantity A varies with time t in such a way that the rate of change of A with respect to t is proportional to A itself. These situations can be modeled by the differential equation

$$\boxed{\frac{dA}{dt} = kA} \tag{4}$$

where $k \neq 0$ is a real number.

- If $k > 0$, then $\dfrac{dA}{dt} = kA$, the rate of change of A with respect to t is positive, and the amount A is increasing.

- If $k < 0$, then $\dfrac{dA}{dt} = kA$, the rate of change of A with respect to t is negative, and the amount A is decreasing.

Suppose that the initial amount A_0 of the substance is known, giving us the boundary condition, or initial condition, $A = A(0) = A_0$ when $t = 0$.

We solve differential equations of the form $\dfrac{dA}{dt} = kA$ by writing $\dfrac{dA}{dt} = kA$ as $\dfrac{dA}{A} = k\,dt$.* Then we integrate both sides of the equation, on the left with respect to A and on the right with respect to t.

$$\int \frac{1}{A}\,dA = \int k\,dt$$

$$\ln|A| = kt + C$$

$$\ln A = kt + C \qquad {\color{blue} A > 0}$$

The initial condition requires that $A = A_0$ when $t = 0$. Then $\ln A_0 = C$, so

$$\ln A = kt + \ln A_0$$

$$\ln A - \ln A_0 = kt$$

$$\ln \frac{A}{A_0} = kt$$

$$\frac{A}{A_0} = e^{kt}$$

$$A = A_0 e^{kt}$$

The solution to the differential equation $\dfrac{dA}{dt} = kA$ is

$$\boxed{A = A_0 e^{kt}} \tag{5}$$

where A_0 is the initial amount.

Functions $A = A(t)$ whose rates of change are $\dfrac{dA}{dt} = kA$ are said to follow the **exponential law**, or the **law of uninhibited growth or decay**—or in a business context, **the law of continuously compounded interest.** Figure 31, on page 383, shows the graphs of the function $A(t) = A_0 e^{kt}$ for both $k > 0$ and $k < 0$.

---

*This technique for solving a differential equation, called **separating the variables**, is discussed in more detail in Chapter 16.

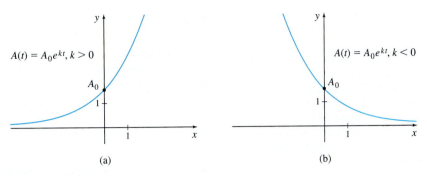

Figure 31

NOTE Example 4 is a model of uninhibited growth; it accurately reflects growth in early stages. After a time, growth no longer continues at a rate proportional to the number present. Factors, such as disease, lack of space, and dwindling food supply, begin to affect the rate of growth.

EXAMPLE 4 **Solving a Differential Equation for Growth**

Assume that a colony of bacteria grows at a rate proportional to the number of bacteria present. If the number of bacteria doubles in 5 hours (h), how long will it take for the number of bacteria to triple?

Solution Let $N(t)$ be the number of bacteria present at time t. Then the assumption that this colony of bacteria grows at a rate proportional to the number present can be modeled by

$$\frac{dN}{dt} = kN$$

where k is a positive constant of proportionality. To find k, we write the differential equation as $\frac{dN}{N} = k\,dt$ and integrate both sides. This differential equation is of form (4), and its solution is given by (5). So, we have

$$N(t) = N_0 e^{kt}$$

where N_0 is the initial number of bacteria in the colony. Since the number of bacteria doubles to $2N_0$ in $5\,\text{h}$,

$$N(5) = N_0\, e^{5k} = 2N_0$$

$$e^{5k} = 2$$

$$k = \frac{1}{5}\ln 2$$

The time t required for this colony to triple obeys the equation

$$N(t) = 3N_0$$

$$N_0 e^{kt} = 3N_0$$

$$e^{kt} = 3$$

$$t = \frac{1}{k}\ln 3 = 5\frac{\ln 3}{\ln 2} \approx 8$$
$$\underset{\underset{k = \frac{1}{5}\ln 2}{\uparrow}}{}$$

The number of bacteria will triple in about $8\,\text{h}$. ∎

NOW WORK **Problems 39 and 51.**

For a radioactive substance, the **rate of decay** is proportional to the amount of substance present at a given time t. That is, if $A = A(t)$ represents the amount of a radioactive substance at time t, then

$$\boxed{\dfrac{dA}{dt} = kA}$$

ORIGINS Willard F. Libby (1908–1980) grew up in California and went to college and graduate school at UC Berkeley. Libby was a physical chemist who taught at the University of Chicago and later at UCLA. While at Chicago, he developed the methods for using natural carbon-14 to date archaeological artifacts. Libby won the Nobel Prize in Chemistry in 1960 for this work.

where $k < 0$ and depends on the radioactive substance. The **half-life** of a radioactive substance is the time required for half of the substance to decay.

Carbon dating, a method for determining the age of an artifact, uses the fact that all living organisms contain two kinds of carbon: carbon-12 (a stable carbon) and a small proportion of carbon-14 (a radioactive isotope). When an organism dies, the amount of carbon-12 present remains unchanged, while the amount of carbon-14 begins to decrease. This change in the amount of carbon-14 present relative to the amount of carbon-12 present makes it possible to calculate how long ago the organism died.

EXAMPLE 5 Solving a Differential Equation for Decay

The skull of an animal found in an archaeological dig contains about 20% of the original amount of carbon-14. If the half-life of carbon-14 is 5730 years, how long ago did the animal die?

Solution Let $A = A(t)$ be the amount of carbon-14 present in the skull at time t. Then A satisfies the differential equation $\dfrac{dA}{dt} = kA$, whose solution is

$$A = A_0 e^{kt}$$

where A_0 is the amount of carbon-14 present at time $t = 0$. To determine the constant k, we use the fact that when $t = 5730$, half of the original amount A_0 remains.

$$\frac{1}{2}A_0 = A_0 e^{5730k}$$

$$\frac{1}{2} = e^{5730k}$$

$$5730k = \ln\frac{1}{2} = -\ln 2$$

$$k = -\frac{\ln 2}{5730}$$

The relationship between the amount A of carbon-14 present and the time t is

$$A(t) = A_0 e^{(-\ln 2/5730)t}$$

In this skull, 20% of the original amount of carbon-14 remains, so $A(t) = 0.20A_0$.

$$0.20A_0 = A_0 e^{(-\ln 2/5730)t}$$

$$0.20 = e^{(-\ln 2/5730)t}$$

Now, we take the natural logarithm of both sides.

$$\ln 0.20 = -\frac{\ln 2}{5730} \cdot t$$

$$t = -5730 \cdot \frac{\ln 0.20}{\ln 2} \approx 13,300$$

The animal died approximately 13,300 years ago. ∎

NOW WORK Problem 59.

5.5 Assess Your Understanding

Concepts and Vocabulary

1. $\dfrac{d}{dx}\left[\displaystyle\int f(x)\,dx\right] = $ _____

2. *True or False* If k is a constant, then

$$\int kf(x)\,dx = \left[\int k\,dx\right]\left[\int f(x)\,dx\right]$$

1. = NOW WORK problem   📈 = Graphing technology recommended   CAS = Computer Algebra System recommended

3. If a is a number, then $\int x^a \, dx =$ _____, provided $a \neq -1$.

4. *True or False* When integrating a function f, a constant of integration C is added to the result because $\int f(x) \, dx$ denotes all the antiderivatives of f.

Skill Building

In Problems 5–38, find each indefinite integral.

5. $\int x^{2/3} \, dx$

6. $\int t^{-4} \, dt$

7. $\int \dfrac{1}{\sqrt{1-x^2}} \, dx$

8. $\int \dfrac{1}{1+x^2} \, dx$

9. $\int \dfrac{5x^2 + 2x - 1}{x} \, dx$

10. $\int \dfrac{x+1}{x} \, dx$

11. $\int \dfrac{4}{3t} \, dt$

12. $\int 2e^u \, du$

13. $\int (4x^3 - 3x^2 + 5x - 2) \, dx$

14. $\int (3x^5 - 2x^4 - x^2 - 1) \, dx$

15. $\int \left(\dfrac{1}{x^3} + 1 \right) dx$

16. $\int \left(x - \dfrac{1}{x^2} \right) dx$

17. $\int (3\sqrt{z} + z) \, dz$

18. $\int (4\sqrt{x} + 1) \, dx$

19. $\int (4t^{3/2} + t^{1/2}) \, dt$

20. $\int \left(3x^{2/3} - \dfrac{1}{\sqrt{x}} \right) dx$

21. $\int u(u-1) \, du$

22. $\int t^2(t+1) \, dt$

23. $\int \dfrac{3x^5 + 1}{x^2} \, dx$

24. $\int \dfrac{x^2 + 2x + 1}{x^4} \, dx$

25. $\int \dfrac{t^2 - 4}{t - 2} \, dt$

26. $\int \dfrac{z^2 - 16}{z + 4} \, dz$

27. $\int (2x+1)^2 \, dx$

28. $\int 3(x^2+1)^2 \, dx$

29. $\int (x + e^x) \, dx$

30. $\int (2e^x - x^3) \, dx$

31. $\int 8(1 + x^2)^{-1} \, dx$

32. $\int \dfrac{-7}{1+x^2} \, dx$

33. $\int \dfrac{x^2 - 1}{2x^3} \, dx$

34. $\int \dfrac{x^2 + 4x - 1}{x^2} \, dx$

35. $\int \dfrac{\tan x}{\cos x} \, dx$

36. $\int \dfrac{1}{\sin^2 x} \, dx$

37. $\int \dfrac{2}{5\sqrt{1-x^2}} \, dx$

38. $\int -\dfrac{4}{x\sqrt{x^2-1}} \, dx$

Applications and Extensions

In Problems 39–50, solve each differential equation using the given boundary condition. (Hint: Rewrite each differential equation in the form $g(y)dy = f(x)dx$ and integrate both sides.)

39. $\dfrac{dy}{dx} = e^x$, $y = 4$ when $x = 0$

40. $\dfrac{dy}{dx} = \dfrac{1}{x}$, $y = 0$ when $x = 1$

41. $\dfrac{dy}{dx} = \dfrac{x^2 + x + 1}{x}$, $y = 0$ when $x = 1$

42. $\dfrac{dy}{dx} = x + e^x$, $y = 4$ when $x = 0$

43. $\dfrac{dy}{dx} = xy^{1/2}$, $y = 1$ when $x = 2$

44. $\dfrac{dy}{dx} = x^{1/2}y$, $y = 1$ when $x = 0$

45. $\dfrac{dy}{dx} = \dfrac{y-1}{x-1}$, $y = 2$ when $x = 2$

46. $\dfrac{dy}{dx} = \dfrac{y}{x}$, $y = 2$ when $x = 1$

47. $\dfrac{dy}{dx} = \dfrac{x}{\cos y}$, $y = \pi$ when $x = 2$

48. $\dfrac{dy}{dx} = y \sin x$, $y = e$ when $x = 0$

49. $\dfrac{dy}{dx} = \dfrac{4e^x}{y}$, $y = 2$ when $x = 0$

50. $\dfrac{dy}{dx} = 5ye^x$, $y = 1$ when $x = 0$

51. **Uninhibited Growth** The population of a colony of mosquitoes obeys the uninhibited growth equation $\dfrac{dN}{dt} = kN$. If there are 1500 mosquitoes initially, and there are 2500 mosquitoes after 24 h, what is the mosquito population after 3 days?

52. **Radioactive Decay** A radioactive substance follows the decay equation $\dfrac{dA}{dt} = kA$. If 25% of the substance disappears in 10 years, what is its half-life?

53. **Population Growth** The population of a suburb grows at a rate proportional to the population. Suppose the population doubles in size from 4000 to 8000 in an 18-month period and continues at the current rate of growth.

  (a) Write a differential equation that models the population P at time t in months.

  (b) Find the general solution to the differential equation.

  (c) Find the particular solution to the differential equation with the initial condition $P(0) = 4000$.

  (d) What will the population be in 4 years [$t = 48$]?

54. **Uninhibited Growth** At any time t in hours, the rate of increase in the area, in millimeters squared (mm^2), of a culture of bacteria is twice the area A of the culture.

  (a) Write a differential equation that models the area of the culture at time t.

  (b) Find the general solution to the differential equation.

  (c) Find the particular solution to the differential equation if $A = 10 \, \text{mm}^2$, when $t = 0$.

55. Radioactive Decay The amount A of the radioactive element radium in a sample decays at a rate proportional to the amount of radium present. The half-life of radium is 1690 years.

(a) Write a differential equation that models the amount A of radium present at time t.

(b) Find the general solution to the differential equation.

(c) Find the particular solution to the differential equation with the initial condition $A(0) = 8$ g.

(d) How much radium will be present in the sample in 100 years?

56. Radioactive Decay Carbon-14 is a radioactive element present in living organisms. After an organism dies, the amount A of carbon-14 present begins to decline at a rate proportional to the amount present at the time of death. The half-life of carbon-14 is 5730 years.

(a) Write a differential equation that models the rate of decay of carbon-14.

(b) Find the general solution to the differential equation.

(c) A piece of fossilized charcoal is found that contains 30% of the carbon-14 that was present when the tree it came from died. How long ago did the tree die?

57. World Population Growth Barring disasters (human-made or natural), the population P of humans grows at a rate proportional to its current size. According to the U.N. World Population studies, from 2005 to 2010 the population of the more developed regions of the world (Europe, North America, Australia, New Zealand, and Japan) grew at an annual rate of 0.409% per year.

(a) Write a differential equation that models the growth rate of the population.

(b) Find the general solution to the differential equation.

(c) Find the particular solution to the differential equation if in 2010 ($t = 0$), the population of the more developed regions of the world was 1.2359×10^9.

(d) If the rate of growth continues to follow this model, what is the projected population of the more developed regions in 2020?

Source: U.N. World Population Prospects, 2010 update.

58. Population Growth in Ecuador Barring disasters (human-made or natural), the population P of humans grows at a rate proportional to its current size. According to the U.N. World Population studies, from 2005 to 2010 the population of Ecuador grew at an annual rate of 1.490% per year. Assuming this growth rate continues:

(a) Write a differential equation that models the growth rate of the population.

(b) Find the general solution to the differential equation.

(c) Find the particular solution to the differential equation if in 2010 ($t = 0$), the population of Ecuador was 1.4465×10^7.

(d) If the rate of growth continues to follow this model, when will the projected population of Ecuador reach 20 million persons?

Source: U.N. World Population Prospects, 2010 update.

59. Oetzi the Iceman was found in 1991 by a German couple who were hiking in the Alps near the border of Austria and Italy. Carbon-14 testing determined that Oetzi died 5300 years ago. Assuming the half-life of carbon-14 is 5730 years, what percent of carbon-14 was left in his body? (An interesting note: In September 2010 the complete genome mapping of Oetzi was completed.)

60. Uninhibited Decay Radioactive beryllium is sometimes used to date fossils found in deep-sea sediments. (Carbon-14 dating cannot be used for fossils that lived underwater.) The decay of beryllium satisfies the equation $\dfrac{dA}{dt} = -\alpha A$, where $\alpha = 1.5 \times 10^{-7}$ and t is measured in years. What is the half-life of beryllium?

61. Decomposition of Sucrose Reacting with water in an acidic solution at $35\,°C$, sucrose ($C_{12}H_{22}O_{11}$) decomposes into glucose ($C_6H_{12}O_6$) and fructose ($C_6H_{12}O_6$) according to the law of uninhibited decay. An initial amount of 0.4 mol of sucrose decomposes to 0.36 mol in 30 min. How much sucrose will remain after 2 h? How long will it take until 0.10 mol of sucrose remains?

62. Chemical Dissociation Salt (NaCl) dissociates in water into sodium (Na^+) and chloride (Cl^-) ions at a rate proportional to its mass. The initial amount of salt is 25 kg, and after 10 h, 15 kg are left.

(a) How much salt will be left after 1 day?

(b) After how many hours will there be less than $\dfrac{1}{2}$ kg of salt left?

63. Voltage Drop The voltage of a certain condenser decreases at a rate proportional to the voltage. If the initial voltage is 20, and 2 seconds later it is 10, what is the voltage at time t? When will the voltage be 5?

64. Uninhibited Growth The rate of change in the number of bacteria in a culture is proportional to the number present. In a certain laboratory experiment, a culture has 10,000 bacteria initially, 20,000 bacteria at time t_1 minutes, and 100,000 bacteria at ($t_1 + 10$) minutes.

(a) In terms of t only, find the number of bacteria in the culture at any time t minutes ($t \geq 0$).

(b) How many bacteria are there after 20 min?

(c) At what time are 20,000 bacteria observed? That is, find the value of t_1.

65. Verify that $\int x\sqrt{x}\,dx \neq \left(\int x\,dx\right)\left(\int \sqrt{x}\,dx\right)$.

66. Verify that $\int x(x^2 + 1)\,dx \neq x \int (x^2 + 1)\,dx$.

67. Verify that $\displaystyle\int \frac{x^2 - 1}{x - 1}\,dx \neq \frac{\int (x^2 - 1)\,dx}{\int (x - 1)\,dx}$.

68. Prove that $\int [f(x) + g(x)]\,dx = \int f(x)\,dx + \int g(x)\,dx$.

69. Derive the integration formula $\displaystyle\int a^x\,dx = \frac{a^x}{\ln a} + C, a > 0,$ $a \neq 1$. (*Hint*: Begin with the derivative of $y = a^x$.)

70. Use the formula from Problem 69 to find:

(a) $\displaystyle\int 2^x\,dx$ **(b)** $\displaystyle\int 3^x\,dx$

Challenge Problems

71. (a) Find y' if $y = \ln\left|\tan\left(\dfrac{x}{2} + \dfrac{\pi}{4}\right)\right|$.

(b) Use the result to show that

$$\int \sec x\,dx = \ln\left|\tan\left(\frac{x}{2} + \frac{\pi}{4}\right)\right| + C$$

(c) Show that $\ln\left|\tan\left(\dfrac{x}{2} + \dfrac{\pi}{4}\right)\right| = \ln|\sec x + \tan x|$.

72. (a) Find y' if $y = x\sin^{-1}x + \sqrt{1-x^2}$.

(b) Use the result to show that

$$\int \sin^{-1}x\,dx = x\sin^{-1}x + \sqrt{1-x^2} + C$$

73. (a) Find y' if $y = \dfrac{1}{2}x\sqrt{a^2-x^2} + \dfrac{1}{2}a^2\sin^{-1}\left(\dfrac{x}{a}\right)$.

(b) Use the result to show that

$$\int \sqrt{a^2-x^2}\,dx = \frac{1}{2}x\sqrt{a^2-x^2} + \frac{1}{2}a^2\sin^{-1}\left(\frac{x}{a}\right) + C$$

74. (a) Find y' if $y = \ln|\csc x - \cot x|$.

(b) Use the result to show that

$$\int \csc x\,dx = \ln|\csc x - \cot x| + C$$

75. Gudermannian Function

(a) Graph $y = \text{gd}(x) = \tan^{-1}(\sinh x)$. This function is called the **gudermannian of** x (named after Christoph Gudermann).

(b) If $y = \text{gd}(x)$, show that $\cos y = \text{sech}\,x$ and $\sin y = \tanh x$.

(c) Show that if $y = \text{gd}(x)$, then y satisfies the differential equation $y' = \cos y$.

(d) Use the differential equation of (c) to obtain the formula

$$\int \sec y\,dy = \text{gd}^{-1}(y) + C$$

Compare this to $\int \sec x\,dx = \ln|\sec x + \tan x| + C$.

76. The formula $\dfrac{d}{dx}\displaystyle\int f(x)\,dx = f(x)$ says that if a function is integrated and the result is differentiated, the original function is returned. What about the other way around? Is the formula $\int f'(x)\,dx = f(x)$ correct? Be sure to justify your answer.

5.6 Method of Substitution; Newton's Law of Cooling

OBJECTIVES *When you finish this section, you should be able to:*

1 Find an indefinite integral using substitution (p. 387)
2 Find a definite integral using substitution (p. 391)
3 Integrate even and odd functions (p. 393)
4 Solve differential equations: Newton's Law of Cooling (p. 394)

1 Find an Indefinite Integral Using Substitution

Indefinite integrals that cannot be found using the formulas in Table 1 on page 380 sometimes can be found using the *method of substitution*. In the method of substitution, we use a change of variables to transform the integrand so one of the formulas in the table applies.

NEED TO REVIEW? Differentials are discussed in Section 3.4, pp. 230–232.

For example, to find $\int (x^2+5)^3 2x\,dx$, we use the substitution $u = x^2 + 5$. The differential of $u = x^2 + 5$ is $du = 2x\,dx$. Now we write $(x^2+5)^3 2x\,dx$ in terms of u and du, and integrate the simpler integral.

NOTE When using the method of substitution, once the substitution u is chosen, we write the original integral in terms of u and du.

$$\int \underbrace{(x^2+5)^3}_{u}\underbrace{2x\,dx}_{du} = \int u^3\,du = \frac{u^4}{4} + C = \frac{(x^2+5)^4}{4} + C$$
$$\uparrow$$
$$u = x^2 + 5$$

We can verify the answer by differentiating using the Power Rule for Functions.

$$\frac{d}{dx}\left[\frac{(x^2+5)^4}{4}+C\right] = \frac{1}{4}\left[4(x^2+5)^3(2x)\right] = (x^2+5)^3 2x$$

NEED TO REVIEW? The Chain Rule is discussed in Section 3.1, pp. 198–200.

The method of substitution is based on the Chain Rule, which states that if f and g are differentiable functions, then for the composite function $f \circ g$,

$$\frac{d}{dx}(f \circ g) = \frac{d}{dx}f(g(x)) = f'(g(x))\,g'(x)$$

The Chain Rule provides a template for finding integrals of the form

$$\int f'(g(x))g'(x)\,dx$$

If, in the integral, we let $u = g(x)$, then the differential $du = g'(x)\,dx$, and we have

$$\int f'(\underbrace{g(x)}_{u})\,\underbrace{g'(x)\,dx}_{du} = \int f'(u)\,du = f(u) + C = f(g(x)) + C$$

Replacing $g(x)$ by u and $g'(x)dx$ by du is called **substitution**. Substitution is a strategy for finding antiderivatives when the integrand is a composite function.

EXAMPLE 1   Finding Indefinite Integrals Using Substitution

Find:

(a) $\displaystyle\int \sin(3x+2)\,dx$ (b) $\displaystyle\int x\sqrt{x^2+1}\,dx$ (c) $\displaystyle\int \frac{e^{\sqrt{x}}}{\sqrt{x}}dx$

Solution (a) Since we know $\int \sin x\,dx$, we let $u = 3x+2$. Then $du = 3\,dx$ so $dx = \dfrac{du}{3}$.

$$\int \sin(\underbrace{3x+2}_{u})\,\underbrace{dx}_{\frac{du}{3}} = \int \sin u \frac{du}{3} = \frac{1}{3}\int \sin u\,du$$

$$= \frac{1}{3}(-\cos u) + C = \underset{\underset{u=3x+2}{\uparrow}}{=} -\frac{1}{3}\cos(3x+2) + C$$

(b) We let $u = x^2 + 1$. Then $du = 2x\,dx$, so $x\,dx = \dfrac{du}{2}$.

$$\int x\sqrt{x^2+1}\,dx = \int \sqrt{x^2+1}\,x\,dx = \int \sqrt{u}\,\frac{du}{2} = \frac{1}{2}\int u^{1/2}du = \frac{1}{2}\left(\frac{u^{3/2}}{\frac{3}{2}}\right) + C$$

$$= \frac{(x^2+1)^{3/2}}{3} + C$$

(c) We let $u = \sqrt{x} = x^{1/2}$. Then $du = \dfrac{1}{2}x^{-1/2}dx = \dfrac{dx}{2\sqrt{x}}$, so $\dfrac{dx}{\sqrt{x}} = 2\,du$.

$$\int \frac{e^{\sqrt{x}}}{\sqrt{x}}dx = \int e^{\sqrt{x}} \cdot \frac{dx}{\sqrt{x}} = \int e^u \cdot 2\,du = 2e^u + C = 2e^{\sqrt{x}} + C \qquad \blacksquare$$

NOW WORK   Problems 5 and 11.

When an integrand equals the product of an expression involving a function and its derivative (or a multiple of its derivative), then substitution is often a good strategy.

For example, for $\int \dfrac{e^{\sqrt{x}}}{\sqrt{x}}dx$, we used the substitution $u = \sqrt{x}$,

since $\dfrac{du}{dx} = \dfrac{d}{dx}\sqrt{x} = \dfrac{1}{2\sqrt{x}}$ is a multiple of $\dfrac{1}{\sqrt{x}}$.

Similarly, in (b) the factor x in the integrand makes the substitution $u = x^2 + 1$ work. On the other hand, if we try to use this same substitution to integrate $\int \sqrt{x^2 + 1}\,dx$, then

$$\int \sqrt{x^2 + 1}\,dx = \int \sqrt{u}\,\frac{du}{2x} \underset{\underset{x=\sqrt{u-1}}{\uparrow}}{=} \int \frac{\sqrt{u}}{2\sqrt{u-1}}\,du$$

and the resulting integral is *more* complicated than the original integral.

The idea behind substitution is to obtain an integral $\int h(u)\,du$ that is simpler than the original integral $\int f(x)\,dx$. When a substitution does not simplify the integral, try other substitutions. If none of these work, other integration methods should be tried. Some of these methods are explored in Chapter 7.

EXAMPLE 2   Finding Indefinite Integrals Using Substitution

Find:

(a) $\displaystyle\int \frac{5x^2 dx}{4x^3 - 1}$ (b) $\displaystyle\int \frac{e^x}{e^x + 4}\,dx$

Solution (a) Notice that the numerator equals the derivative of the denominator, except for a constant factor. So, we try substitution. Let $u = 4x^3 - 1$. Then $du = 12x^2 dx$

so $5x^2 dx = \dfrac{5}{12}\,du$.

$$\int \frac{5x^2 dx}{4x^3 - 1} = \int \frac{\frac{5}{12}\,du}{u} = \frac{5}{12}\int \frac{du}{u} = \frac{5}{12}\ln|u| + C = \frac{5}{12}\ln\left|4x^3 - 1\right| + C$$

(b) Here, the numerator equals the derivative of the denominator. So, we use the substitution $u = e^x + 4$. Then $du = e^x dx$.

$$\int \frac{e^x}{e^x + 4}\,dx = \int \frac{1}{e^x + 4}\cdot e^x dx = \int \frac{1}{u}\,du = \ln|u| + C = \underset{\underset{u=e^x+4>0}{\uparrow}}{\ln(e^x + 4) + C} \quad\blacksquare$$

NOW WORK   Problem 17.

EXAMPLE 3   Using Substitution To Establish an Integration Formula

Show that:

(a)
$$\boxed{\int \tan x\,dx = -\ln|\cos x| + C = \ln|\sec x| + C}$$

(b)
$$\boxed{\int \sec x\,dx = \ln|\sec x + \tan x| + C}$$

Solution (a) Since $\tan x = \dfrac{\sin x}{\cos x}$, we let $u = \cos x$. Then $du = -\sin x\,dx$ and

$$\int \tan x\,dx = \int \frac{\sin x}{\cos x}\,dx = \int -\frac{du}{u} = -\ln|u| + C = -\ln|\cos x| + C$$

$$= \underset{\underset{r\ln x = \ln x^r}{\uparrow}}{\ln|\cos x|^{-1}} + C = \ln\left|\frac{1}{\cos x}\right| + C = \ln|\sec x| + C$$

EXAMPLE 6   Finding a Definite Integral Using Substitution

Find $\displaystyle\int_0^2 x\sqrt{4-x^2}\,dx$.

Solution

Method 1: Use the related indefinite integral and then use the Fundamental Theorem of Calculus. The related indefinite integral $\displaystyle\int x\sqrt{4-x^2}\,dx$ can be found using the substitution $u=4-x^2$. Then $du=-2x\,dx$, so $x\,dx=-\dfrac{du}{2}$.

$$\int x\sqrt{4-x^2}\,dx = \int \sqrt{u}\left(-\frac{du}{2}\right) = -\frac{1}{2}\int u^{1/2}\,du = -\frac{1}{2}\cdot\frac{u^{3/2}}{\dfrac{3}{2}}+C$$

$$= -\frac{1}{3}(4-x^2)^{3/2}+C$$

Then by the Fundamental Theorem of Calculus,

$$\int_0^2 x\sqrt{4-x^2}\,dx = -\frac{1}{3}\left[(4-x^2)^{3/2}\right]_0^2 = -\frac{1}{3}\left[0-4^{3/2}\right]=\frac{8}{3}$$

Method 2: Find the definite integral directly by making a substitution in the integrand and changing the limits of integration. We let $u=4-x^2$; then $du=-2x\,dx$. Now use the function $u=4-x^2$ to change the limits of integration.

- The lower limit of integration is $x=0$ so, in terms of u, it becomes $u=4-0^2=4$.

- The upper limit of integration is $x=2$ so the upper limit becomes $u=4-2^2=0$.

Then

$$\int_0^2 x\sqrt{4-x^2}\,dx = \underset{\substack{\uparrow \\ u=4-x^2 \\ x\,dx=-\frac{1}{2}du}}{\int_4^0} \sqrt{u}\left(-\frac{du}{2}\right) = -\frac{1}{2}\int_4^0 \sqrt{u}\,du = -\frac{1}{2}\cdot\left[\frac{u^{3/2}}{\dfrac{3}{2}}\right]_4^0$$

$$= -\frac{1}{3}(0-8)=\frac{8}{3}$$

CAUTION When using substitution to find a definite integral directly, remember to change the limits of integration.

NOW WORK Problem 45.

EXAMPLE 7   Finding a Definite Integral Using Substitution

Find $\displaystyle\int_0^{\pi/2}\frac{1-\cos(2\theta)}{2}\,d\theta$.

Solution We use properties of integrals to simplify before integrating.

$$\int_0^{\pi/2}\frac{1-\cos(2\theta)}{2}\,d\theta = \frac{1}{2}\int_0^{\pi/2}[1-\cos(2\theta)]\,d\theta$$

$$= \frac{1}{2}\left[\int_0^{\pi/2}d\theta - \int_0^{\pi/2}\cos(2\theta)\,d\theta\right]$$

$$= \frac{1}{2}\int_0^{\pi/2}d\theta - \frac{1}{2}\int_0^{\pi/2}\cos(2\theta)\,d\theta$$

$$= \frac{1}{2}\Big[\theta\Big]_0^{\pi/2} - \frac{1}{2}\int_0^{\pi/2}\cos(2\theta)\,d\theta$$

$$= \frac{\pi}{4} - \frac{1}{2}\int_0^{\pi/2}\cos(2\theta)\,d\theta$$

In the integral on the right, we use the substitution $u = 2\theta$. Then $du = 2\,d\theta$ so $d\theta = \dfrac{du}{2}$. Now we change the limits of integration:

- when $\theta = 0$ then $u = 2(0) = 0$
- when $\theta = \dfrac{\pi}{2}$ then $u = 2\left(\dfrac{\pi}{2}\right) = \pi$

Now

$$\int_0^{\pi/2} \cos(2\theta)\,d\theta = \int_0^\pi \cos u\,\frac{du}{2} = \frac{1}{2}\Big[\sin u\Big]_0^\pi = \frac{1}{2}(\sin\pi - \sin 0) = 0$$

Then,

$$\int_0^{\pi/2}\frac{1-\cos(2\theta)}{2}\,d\theta = \frac{\pi}{4} - \frac{1}{2}\int_0^{\pi/2}\cos(2\theta)\,d\theta = \frac{\pi}{4} \qquad \blacksquare$$

NOW WORK **Problem 53.**

3 Integrate Even and Odd Functions

NEED TO REVIEW? Even and odd functions are discussed in Section P.1, pp. 9–10.

Integrals of even and odd functions can be simplified due to symmetry. Figure 32 illustrates the conclusions of the theorem that follows.

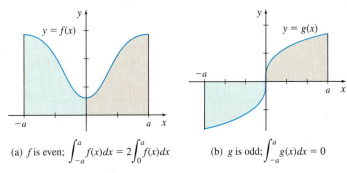

(a) f is even; $\displaystyle\int_{-a}^a f(x)\,dx = 2\int_0^a f(x)\,dx$ (b) g is odd; $\displaystyle\int_{-a}^a g(x)\,dx = 0$

Figure 32

THEOREM The Integrals of Even and Odd Functions

Let a function f be continuous on a closed interval $[-a, a]$, $a > 0$.

- If f is an even function, then

$$\boxed{\int_{-a}^a f(x)\,dx = 2\int_0^a f(x)\,dx}$$

- If f is an odd function, then

$$\boxed{\int_{-a}^a f(x)\,dx = 0}$$

The property for even functions is proved here; the proof for odd functions is left as an exercise. See Problem 123.

Proof f is an even function: Since f is continuous on the closed interval $[-a, a]$, $a > 0$, and 0 is in the interval $[-a, a]$, we have

$$\int_{-a}^a f(x)\,dx = \int_{-a}^0 f(x)\,dx + \int_0^a f(x)\,dx = -\int_0^{-a} f(x)\,dx + \int_0^a f(x)\,dx \qquad (2)$$

In $-\int_0^{-a} f(x)\,dx$, we use the substitution $u = -x$. Then $du = -dx$. Also, if $x = 0$, then $u = 0$, and if $x = -a$, then $u = a$. Therefore,

$$-\int_0^{-a} f(x)\,dx = \int_0^a f(-u)\,du = \int_0^a f(u)\,du = \int_0^a f(x)\,dx \qquad (3)$$

$$\underset{\substack{\uparrow \\ f \text{ is even} \\ f(-u) = f(u)}}{}$$

Combining (2) and (3), we obtain

$$\int_{-a}^{a} f(x)\,dx = \int_{0}^{a} f(x)\,dx + \int_{0}^{a} f(x)\,dx = 2\int_{0}^{a} f(x)\,dx \qquad\blacksquare$$

To use the theorem involving even and odd functions, three conditions must be met:

- The function f must be even or odd.
- The function f must be continuous on the closed interval $[-a, a]$, $a > 0$.
- The limits of integration must be $-a$ and a, $a > 0$.

EXAMPLE 8   Integrating an Even or Odd Function

Find:

(a) $\displaystyle\int_{-3}^{3} (x^7 - 4x^3 + x)\,dx$ **(b)** $\displaystyle\int_{-2}^{2} (x^4 - x^2 + 3)\,dx$

Solution **(a)** If $f(x) = x^7 - 4x^3 + x$, then $f(-x) = (-x)^7 - 4(-x)^3 + (-x) = -(x^7 - 4x^3 + x) = -f(x)$. Since f is an odd function,

$$\int_{-3}^{3} (x^7 - 4x^3 + x)\,dx = 0$$

(b) If $g(x) = x^4 - x^2 + 3$, then $g(-x) = (-x)^4 - (-x)^2 + 3 = x^4 - x^2 + 3 = g(x)$. Since g is an even function,

$$\int_{-2}^{2} (x^4 - x^2 + 3)\,dx = 2\int_{0}^{2} (x^4 - x^2 + 3)\,dx = 2\left[\frac{x^5}{5} - \frac{x^3}{3} + 3x\right]_{0}^{2}$$

$$= 2\left[\frac{32}{5} - \frac{8}{3} + 6\right] = \frac{292}{15} \qquad\blacksquare$$

NOW WORK Problems 63 and 67.

EXAMPLE 9   Using Properties of Integrals

If f is an even function and $\int_{0}^{2} f(x)\,dx = -6$ and $\int_{-5}^{0} f(x)\,dx = 8$, find $\int_{2}^{5} f(x)\,dx$.

Solution $\int_{2}^{5} f(x)\,dx = \int_{2}^{0} f(x)\,dx + \int_{0}^{5} f(x)\,dx$

Now $\int_{2}^{0} f(x)\,dx = -\int_{0}^{2} f(x)\,dx = 6$.

Since f is even, $\int_{0}^{5} f(x)\,dx = \int_{-5}^{0} f(x)\,dx = 8$. Then

$$\int_{2}^{5} f(x)\,dx = \int_{2}^{0} f(x)\,dx + \int_{0}^{5} f(x)\,dx = 6 + 8 = 14 \qquad\blacksquare$$

4 Solve Differential Equations: Newton's Law of Cooling

Suppose an object is heated to a temperature u_0. Then at time $t = 0$, the object is put into a medium with a constant lower temperature causing the object to cool. Newton's Law of Cooling states that the rate of change of the temperature of the object with respect to time is continuous and proportional to the difference between the temperature of the object and the ambient temperature (the temperature of the surrounding medium). That is, if $u = u(t)$ is the temperature of the object at time t and if T is the (constant) ambient temperature, then Newton's Law of Cooling is modeled by the differential equation

$$\boxed{\frac{du}{dt} = k[u(t) - T]} \qquad (4)$$

where k is a constant that depends on the object. Since the ambient temperature T is lower than $u(0) = u_0$, the object cools and its temperature decreases so that $\dfrac{du}{dt} < 0$. Then, since $u(t) > T$, k is a negative constant.

We find u as a function of t by solving the differential equation $\dfrac{du}{dt} = k(u - T)$. We rewrite the differential equation as $\dfrac{du}{u - T} = k\,dt$ and integrate both sides.

$$\int \frac{du}{u - T} = \int k\,dt$$

$$\ln|u - T| = kt + C$$

To find C, we use the boundary condition that at time $t = 0$, the initial temperature of the object is $u(0) = u_0$, Then

$$\ln|u_0 - T| = k \cdot 0 + C$$

$$C = \ln|u_0 - T|$$

Using this expression for C, we obtain

$$\ln|u - T| = kt + \ln|u_0 - T|$$

$$\ln|u - T| - \ln|u_0 - T| = kt$$

$$\ln\left|\frac{u - T}{u_0 - T}\right| = kt$$

$$\frac{u - T}{u_0 - T} = e^{kt}$$

$$u - T = (u_0 - T)e^{kt}$$

$$\boxed{u = (u_0 - T)\,e^{kt} + T} \tag{5}$$

EXAMPLE 10   Using Newton's Law of Cooling

An object is heated to $90\,°C$ and allowed to cool in a room with a constant ambient temperature of $20\,°C$. If after 10 min the temperature of the object is $60\,°C$, what will its temperature be after 20 min?

Solution When $t = 0$, $u(0) = u_0 = 90\,°C$, and when $t = 10$ min, $u(10) = 60\,°C$. Given that the ambient temperature T is $20\,°C$, we substitute these values into equation (5).

$$u(t) = (u_0 - T)e^{kt} + T$$

$$60 = (90 - 20)e^{10k} + 20 \qquad \text{\small $u = 60$ when $t = 10$; $T = 20$; $u_0 = 90$}$$

$$\frac{40}{70} = e^{10k}$$

$$k = \frac{1}{10}\ln\frac{4}{7} = 0.10\ln\frac{4}{7}$$

The temperature u is

$$u(t) = 70e^{[0.1\ln(4/7)]t} + 20$$

Then when $t = 20$, the temperature u of the object is

$$u(20) = 70e^{[0.1\ln(4/7)](20)} + 20 = 70e^{2\ln(4/7)} + 20 \approx 42.86\,°C \qquad \blacksquare$$

NOW WORK Problem 109.

110. Newton's Law of Cooling   A thermometer reading $70\,°F$ is taken outside where the ambient temperature is $22\,°F$. Four minutes later the reading is $32\,°F$.

   (a) Write the differential equation that models the temperature $u = u(t)$ of the thermometer at time t.

   (b) Find the general solution of the differential equation.

   (c) Find the particular solution to the differential equation, using the initial condition that when $t = 0$ min, then $u = 70\,°F$.

   (d) Find the thermometer reading 7 min after the thermometer was brought outside.

   (e) Find the time it takes for the reading to change from $70\,°F$ to within $\dfrac{1}{2}\,°F$ of the air temperature.

111. Forensic Science   At 4 p.m., a body was found floating in water whose temperature is $12\,°C$. When the woman was alive, her body temperature was $37\,°C$ and now it is $20\,°C$. Suppose the rate of change of the temperature $u = u(t)$ of the body with respect to the time t in hours (h) is proportional to $u(t) - T$, where T is the water temperature and the constant of proportionality is -0.159.

   (a) Write a differential equation that models the temperature $u = u(t)$ of the body at time t.

   (b) Find the general solution of the differential equation.

   (c) Find the particular solution to the differential equation, using the initial condition that at the time of death, when $t = 0$ h, her body temperature was $u = 37\,°C$.

   (d) At what time did the woman drown?

   (e) How long does it take for the woman's body to cool to $15°C$?

112. Newton's Law of Cooling   A pie is removed from a $350\,°F$ oven to cool in a room whose temperature is $72\,°F$.

   (a) Write the differential equation that models the temperature $u = u(t)$ of the pie at time t.

   (b) Find the general solution of the differential equation.

   (c) Find the particular solution to the differential equation, using the initial condition that when $t = 0$ min, then $u = 350\,°F$.

   (d) If $u(5) = 200\,°F$, what is the temperature of the pie after 15 min?

   (e) How long will it take for the pie to be $100\,°F$ and ready to eat?

113. Electric Potential   The electric field strength a distance z from the axis of a ring of radius R carrying a charge Q is given by the formula

$$E(z) = \frac{Qz}{(R^2 + z^2)^{3/2}}$$

If the electric potential V is related to E by $E = -\dfrac{dV}{dz}$, what is $V(z)$?

114. Impulse During a Rocket Launch   The impulse J due to a force F is the product of the force times the amount of time t for which the force acts. When the force varies over time,

$$J = \int_{t_1}^{t_2} F(t)\,dt.$$

We can model the force acting on a rocket during launch by an exponential function $F(t) = Ae^{bt}$, where A and b are constants that depend on the characteristics of the engine. At the instant lift-off occurs ($t = 0$), the force must equal the weight of the rocket.

   (a) Suppose the rocket weighs 25,000 N (a mass of about 2500 kg or a weight of 5500 lb), and 30 seconds after lift-off the force acting on the rocket equals twice the weight of the rocket. Find A and b.

   (b) Find the impulse delivered to the rocket during the first 30 seconds after the launch.

115. Air Resistance on a Falling Object   If an object of mass m is dropped, the air resistance on it when it has speed v can be modeled as $F_{air} = -kv$, where the constant k depends on the shape of the object and the condition of the air. The minus sign is necessary because the direction of the force is opposite to the velocity. Using Newton's Second Law of Motion, this force leads to a downward acceleration $a(t) = ge^{-kt/m}$. (You are asked to prove this in Problem 135.) Using the equation for $a(t)$, find:

   (a) $v(t)$, if the object starts from rest $v_0 = v(0) = 0$, with the positive direction downward.

   (b) $s(t)$, if the object starts from the position $s_0 = s(0) = 0$, with the positive direction downward.

   (c) What limits do $a(t)$, $v(t)$, and $s(t)$ approach if the object falls for a very long time ($t \to \infty$)? Interpret each result and explain if it is physically reasonable.

   (d) Graph $a = a(t)$, $v = v(t)$, $s = s(t)$. Do the graphs support the conclusions obtained in part (c)? Use $g = 9.8$ m/s^2, $k = 5$, and $m = 10$ kg.

116. Area   Let $f(x) = k \sin(kx)$, where k is a positive constant.

   (a) Find the area of the region under one arch of the graph of f.

   (b) Find the area of the triangle formed by the x-axis and the tangent lines to one arch of f at the points where the graph of f crosses the x-axis.

117. Use an appropriate substitution to show that

$$\int_0^1 x^m (1 - x)^n\,dx = \int_0^1 x^n (1 - x)^m\,dx,$$

where m, n are positive integers.

118. Properties of Integrals Find $\int_{-1}^{1} f(x)\, dx$ for the function given below:

$$f(x) = \begin{cases} x+1 & \text{if } x < 0 \\ \cos(\pi x) & \text{if } x \geq 0 \end{cases}$$

119. If f is continuous on $[a, b]$, show that

$$\int_{a}^{b} f(x)\, dx = \int_{a}^{b} f(a+b-x)\, dx$$

120. If $\int_{0}^{1} f(x)\, dx = 2$, find:

(a) $\int_{0}^{0.5} f(2x)\, dx$ (b) $\int_{0}^{3} f\left(\frac{1}{3}x\right) dx$

(c) $\int_{0}^{1/5} f(5x)\, dx$

(d) Find the upper and lower limits of integration so that

$$\int_{a}^{b} f\left(\frac{x}{4}\right) dx = 8.$$

(e) Generalize (d) so that $\int_{a}^{b} f(kx)\, dx = \frac{1}{k} \cdot 2$ for $k > 0$.

121. If $\int_{0}^{2} f(s)\, ds = 5$, find:

(a) $\int_{-1}^{1} f(s+1)\, ds$ (b) $\int_{-3}^{-1} f(s+3)\, ds$

(c) $\int_{4}^{6} f(s-4)\, ds$

(d) Find the upper and lower limits of integration so that

$$\int_{a}^{b} f(s-2)\, ds = 5.$$

(e) Generalize (d) so that $\int_{a}^{b} f(s-k)\, ds = 5$ for $k > 0$.

122. Find $\int_{0}^{b} |2x|\, dx$ for any real number b.

123. If f is an odd function, show that $\int_{-a}^{a} f(x)\, dx = 0$.

124. Find the constant k, where $0 \leq k \leq 3$, for which

$$\int_{0}^{3} \frac{x}{\sqrt{x^2 + 16}}\, dx = \frac{3k}{\sqrt{k^2 + 16}}$$

125. If n is a positive integer, for what number $c > 0$ is

$$\int_{0}^{c} x^{n-1}\, dx = \frac{1}{n}$$

126. If f is a continuous function defined on the interval $[0, 1]$, show that

$$\int_{0}^{\pi} x\, f(\sin x)\, dx = \frac{\pi}{2} \int_{0}^{\pi} f(\sin x)\, dx$$

127. Prove that $\int \csc x\, dx = \ln|\csc x - \cot x| + C$.
[*Hint:* Multiply and divide the integrand by $(\csc x - \cot x)$.]

128. Describe a method for finding $\int_{a}^{b} |f(x)|\, dx$ in terms of $F(x) = \int f(x)\, dx$ when $f(x)$ has finitely many zeros.

129. Find $\int \sqrt[n]{a+bx}\, dx$, where a and b are real numbers, $b \neq 0$, and $n \geq 2$ is an integer.

130. If f is continuous for all x, which of the following integrals have the same value?

(a) $\int_{a}^{b} f(x)\, dx$ (b) $\int_{0}^{b-a} f(x+a)\, dx$

(c) $\int_{a+c}^{b+c} f(x+c)\, dx$

Challenge Problems

131. Find $\displaystyle\int \frac{x^6 + 3x^4 + 3x^2 + x + 1}{(x^2 + 1)^2}\, dx$.

132. Find $\displaystyle\int \frac{\sqrt[4]{x}}{\sqrt{x} + \sqrt[3]{x}}\, dx$.

133. Find $\displaystyle\int \frac{3x+2}{x\sqrt{x+1}}\, dx$.

134. Find $\displaystyle\int \frac{dx}{(x \ln x)[\ln(\ln x)]}$.

135. Air Resistance on a Falling Object (Refer to Problem 115.) If an object of mass m is dropped, the air resistance on it when it has speed v can be modeled as

$$F_{\text{air}} = -kv,$$

where the constant k depends on the shape of the object and the condition of the air. The minus sign is necessary because the direction of the force is opposite to the velocity. Using Newton's Second Law of Motion, show that the downward acceleration of the object is

$$a(t) = ge^{-kt/m},$$

where g is the acceleration due to gravity.
(*Hint:* The velocity of the object obeys the differential equation

$$m\frac{dv}{dt} = mg - kv$$

Solve the differential equation for v and use the fact that

$$ma = mg - kv.)$$

136. A **separable differential equation** can be written in the form

$$\frac{dy}{dx} = \frac{f(x)}{g(y)}, \text{ where } f \text{ and } g \text{ are continuous. Then}$$

$$\int g(y)\, dy = \int f(x)\, dx$$

and integrating (if possible) will give a solution to the differential equation. Use this technique to solve parts (a)–(c) below. (You may need to leave your answer in implicit form.)

(a) $\dfrac{y^2}{x}\dfrac{dy}{dx} = 1 + x^2$ (b) $\dfrac{dy}{dx} = y\dfrac{x^2 - 2x + 1}{y + 3}$

(c) $y\dfrac{dy}{dx} = \dfrac{x^2}{y+4}$; if $y = 2$ when $x = 8$

Chapter Review

THINGS TO KNOW

5.1 Area

Definitions:

- Partition of an interval $[a, b]$ (p. 345)
- Area A under the graph of a function f from a to b (p. 348)

5.2 The Definite Integral

Definitions:

- Riemann sums (pp. 353–354)
- The definite integral (p. 355)
- $\int_a^a f(x)\,dx = 0$ (p. 356)
- $\int_a^b f(x)\,dx = -\int_b^a f(x)\,dx$ (p. 356)

Theorems:

- If a function f is continuous on a closed interval $[a, b]$, then the definite integral $\int_a^b f(x)\,dx$ exists. (p. 356)
- $\int_a^b h\,dx = h(b - a)$, h a constant (p. 357)

5.3 The Fundamental Theorem of Calculus

Fundamental Theorem of Calculus: Let f be a function that is continuous on a closed interval $[a, b]$.

- Part 1: The function I defined by $I(x) = \int_a^x f(t)\,dt$ has the properties that it is continuous on $[a, b]$ and differentiable on (a, b). Moreover, $I'(x) = \dfrac{d}{dx}\left[\int_a^x f(t)\,dt\right] = f(x)$, for all x in (a, b). (p. 362)
- Part 2: If F is any antiderivative of f on $[a, b]$, then
$$\int_a^b f(x)\,dx = F(b) - F(a). \text{ (p. 364)}$$

5.4 Properties of the Definite Integral

Properties of definite integrals:

If the functions f and g are continuous on the closed interval $[a, b]$ and k is a constant, then

- Integral of a sum:
$$\int_a^b [f(x) + g(x)]\,dx = \int_a^b f(x)\,dx + \int_a^b g(x)\,dx \quad \text{(p. 369)}$$

- Integral of a constant times a function:
$$\int_a^b kf(x)\,dx = k\int_a^b f(x)\,dx \quad \text{(p. 369)}$$

- $\int_a^b [k_1 f_1(x) + k_2 f_2(x) + \cdots + k_n f_n(x)]\,dx$
$$= k_1 \int_a^b f_1(x)\,dx + k_2 \int_a^b f_2(x)\,dx + \cdots + k_n \int_a^b f_n(x)\,dx.$$
(p. 370)

- If f is continuous on an interval containing the numbers a, b, and c, then
$$\int_a^b f(x)\,dx = \int_a^c f(x)\,dx + \int_c^b f(x)\,dx \quad \text{(p. 370)}$$

- **Bounds on an Integral:** If a function f is continuous on a closed interval $[a, b]$ and if m and M denote the absolute minimum and absolute maximum values, respectively, of f on $[a, b]$, then
$$m(b - a) \le \int_a^b f(x)\,dx \le M(b - a) \quad \text{(p. 371)}$$

- **Mean Value Theorem for Integrals:** If a function f is continuous on a closed interval $[a, b]$, then there is a real number u, where $a \le u \le b$, for which
$$\int_a^b f(x)\,dx = f(u)(b - a) \quad \text{(p. 372)}$$

Definition: The average value of a function over an interval $[a, b]$ is
$$\bar{y} = \frac{1}{b - a} \int_a^b f(x)\,dx \quad \text{(p. 374)}$$

5.5 The Indefinite Integral; Growth and Decay Models

The indefinite integral of f: $\displaystyle\int f(x)\,dx = F(x) + C$

if and only if $\dfrac{d}{dx}[F(x) + C] = f(x)$,

where C is the constant of integration. (p. 379)

Basic integration formulas: See Table 1. (p. 380)

Properties of indefinite integrals:

- Derivative of an integral:
$$\frac{d}{dx}\left[\int f(x)\,dx\right] = f(x) \quad \text{(p. 380)}$$

- Integral of a sum:
$$\int [f(x) + g(x)]\,dx = \int f(x)\,dx + \int g(x)\,dx \quad \text{(p. 381)}$$

- Integral of a constant k times a function:
$$\int kf(x)\,dx = k\int f(x)\,dx \quad \text{(p. 381)}$$

5.6 Method of Substitution; Newton's Law of Cooling

Method of substitution: (p. 388)

Method of substitution (definite integrals):

- Find the related indefinite integral using substitution. Then use the Fundamental Theorem of Calculus. (p. 391)
- Find the definite integral directly by making a substitution in the integrand and using the substitution to change the limits of integration. (p. 391)

Basic integration formulas:

- $\displaystyle\int \frac{g'(x)}{g(x)}\,dx = \ln|g(x)| + C$ (p. 390)

- $\displaystyle\int \tan x\,dx = \ln|\sec x| + C$ (p. 389)

- $\displaystyle\int \sec x\,dx = \ln|\sec x + \tan x| + C$ (p. 389)

- If f is an even function, then $\displaystyle\int_{-a}^a f(x)\,dx = 2\int_0^a f(x)\,dx$ (p. 393)

- If f is an odd function, then $\displaystyle\int_{-a}^a f(x)\,dx = 0.$ (p. 393)

OBJECTIVES

REVIEW EXERCISES

1. **Area**   Approximate the area under the graph of $f(x) = 2x + 1$ from 0 to 4 by finding s_n and S_n for $n = 4$ and $n = 8$.

2. **Area**   Approximate the area under the graph of $f(x) = x^2$ from 0 to 8 by finding s_n and S_n for $n = 4$ and $n = 8$ subintervals.

3. **Area**   Find the area A under the graph of $y = f(x) = 9 - x^2$ from 0 to 3 by using lower sums s_n (rectangles that lie below the graph of f).

4. **Area**   Find the area A under the graph of $y = f(x) = 8 - 2x$ from 0 to 4 using upper sums S_n (rectangles that lie above the graph of f).

5. **Riemann Sums**
   (a) Find the Riemann sum of $f(x) = x^2 - 3x + 3$ on the closed interval $[-1, 3]$ using a regular partition with four subintervals and the numbers $u_1 = -1$, $u_2 = 0$, $u_3 = 2$, and $u_4 = 3$.

   (b) Find the Riemann sums of f by partitioning $[-1, 3]$ into n subintervals of equal length and choosing u_i as the right endpoint of the ith subinterval $[x_{i-1}, x_i]$. Write the limit of the Riemann sums as a definite integral. Do not evaluate.

   (c) Find the limit as n approaches ∞ of the Riemann sums found in (b).

   (d) Find the definite integral from (b) using the Fundamental Theorem of Calculus. Compare the answer to the limit found in (c).

6. **Units of an Integral**   In the definite integral $\int_a^b a(t)\, dt$, where a represents acceleration measured in meters per second squared and t is measured in seconds, what are the units of $\int_a^b a(t)\, dt$?

In Problems 7–10, find each derivative using the Fundamental Theorem of Calculus.

7. $\dfrac{d}{dx} \displaystyle\int_0^x t^{2/3} \sin t\, dt$

8. $\dfrac{d}{dx} \displaystyle\int_e^x \ln t\, dt$

9. $\dfrac{d}{dx} \displaystyle\int_{x^2}^1 \tan t\, dt$

10. $\dfrac{d}{dx} \displaystyle\int_a^{2\sqrt{x}} \dfrac{t}{t^2 + 1}\, dt$

In Problems 11–20, find each integral.

11. $\displaystyle\int_1^{\sqrt{2}} x^{-2}\, dx$

12. $\displaystyle\int_1^{e^2} \dfrac{1}{x}\, dx$

13. $\displaystyle\int_0^1 \dfrac{1}{1 + x^2}\, dx$

14. $\displaystyle\int \dfrac{1}{x\sqrt{x^2 - 1}}\, dx$

15. $\displaystyle\int_0^{\ln 2} 4e^x\, dx$

16. $\displaystyle\int_0^2 (x^2 - 3x + 2)\, dx$

17. $\displaystyle\int_1^4 2^x\, dx$

18. $\displaystyle\int_0^{\pi/4} \sec x \tan x\, dx$

19. $\displaystyle\int \left(\dfrac{1 + 2xe^x}{x} \right) dx$

20. $\displaystyle\int \dfrac{1}{2} \sin x\, dx$

21. **Interpreting an Integral**   The function $v = v(t)$ is the speed v, in kilometers per hour, of a train at a time t, in hours. Interpret the integral $\int_0^{16} v(t)\, dt = 460$.

22. **Interpreting an Integral**   The function f is the rate of change of the volume V of oil, in liters per hour, draining from a storage tank at time t (in hours). Interpret the integral $\int_0^2 f(t)\, dt = 100$.

In Problems 23–26, find each integral.

23. $\int_{-2}^{2} f(x)\,dx$,   where $f(x) = \begin{cases} 3x + 2 & \text{if } -2 \le x < 0 \\ 2x^2 + 2 & \text{if } 0 \le x \le 2 \end{cases}$

24. $\int_{-1}^{4} |x|\,dx$

25. $\int_{-\pi/2}^{\pi/2} \sin x\,dx$

26. $\int_{-3}^{3} \dfrac{x^2}{x^2 + 9}\,dx$

Bounds on an Integral   In Problems 27 and 28, find lower and upper bounds for each integral.

27. $\int_{0}^{2} e^{x^2}\,dx$

28. $\int_{0}^{1} \dfrac{1}{1 + x^2}\,dx$

In Problems 29 and 30, for each integral find the number(s) u guaranteed by the Mean Value Theorem for Integrals.

29. $\int_{0}^{\pi} \sin x\,dx$

30. $\int_{-3}^{3} (x^3 + 2x)\,dx$

In Problems 31–34, find the average value of each function over the given interval.

31. $f(x) = \sin x$ over $\left[-\dfrac{\pi}{2}, \dfrac{\pi}{2}\right]$

32. $f(x) = x^3$ over $[1, 4]$

33. $f(x) = e^x$ over $[-1, 1]$

34. $f(x) = 6x^{2/3}$ over $[0, 8]$

35. Find $\dfrac{d}{dx} \int \sqrt{\dfrac{1}{1 + 4x^2}}\,dx$

36. Find $\dfrac{d}{dx} \int \ln x\,dx$.

In Problems 37 and 38, solve each differential equation using the given boundary condition.

37. $\dfrac{dy}{dx} = 3xy$; $y = 4$ when $x = 0$

38. $\cos y \dfrac{dy}{dx} = \dfrac{\sin y}{x}$; $y = \dfrac{\pi}{3}$ when $x = -1$

In Problems 39–51, find each integral.

39. $\int \dfrac{y\,dy}{(y - 2)^3}$

40. $\int \dfrac{x}{(2 - 3x)^3}\,dx$

41. $\int \sqrt{\dfrac{1 + x}{x^5}}\,dx, \; x > 0$

42. $\int_{\pi^2/4}^{4\pi^2} \dfrac{1}{\sqrt{x}} \sin \sqrt{x}\,dx$

43. $\int_{1}^{2} \dfrac{1}{t^4} \left(1 - \dfrac{1}{t^3}\right)^3 dt$

44. $\int \dfrac{e^x + 1}{e^x - 1}\,dx$

45. $\int \dfrac{dx}{\sqrt{x}\,(1 - 2\sqrt{x})}$

46. $\int_{1/5}^{3} \dfrac{\ln(5x)}{x}\,dx$

47. $\int_{-1}^{1} \dfrac{5^{-x}}{2^x}\,dx$

48. $\int e^{x+e^x}\,dx$

49. $\int_{0}^{1} \dfrac{x\,dx}{\sqrt{2 - x^4}}$

50. $\int_{4}^{5} \dfrac{dx}{x\sqrt{x^2 - 9}}$

51. $\int \sqrt[3]{x^3 + 3\cos x}\,(x^2 - \sin x)\,dx$

52. Find $f''(x)$ if $f(x) = \int_{0}^{x} \sqrt{1 - t^2}\,dt$.

53. Suppose that $F(x) = \int_{0}^{x} \sqrt{t}\,dt$ and $G(x) = \int_{1}^{x} \sqrt{t}\,dt$. Explain why $F(x) - G(x)$ is constant. Find the constant.

54. If $\int_{0}^{2} f(x + 2)\,dx = 3$, find $\int_{2}^{4} f(x)\,dx$.

55. If $\int_{1}^{2} f(x - c)\,dx = 5$, where c is a constant, find $\int_{1-c}^{2-c} f(x)\,dx$.

56. Area   Find the area under the graph of $y = \cosh x$ from $x = 0$ to $x = 2$.

57. Water Supply   A sluice gate of a dam is opened and water is released from the reservoir at a rate of $r(t) = 100 + \sqrt{t}$ gallons per minute, where t measures the time in minutes since the gate has been opened. If the gate is opened at 7 a.m. and is left open until 9:24 a.m., how much water is released?

58. Forensic Science   A body was found in a meat locker whose ambient temperature is $10\,°C$. When the person was alive, his body temperature was $37\,°C$ and now it is $25\,°C$. Suppose the rate of change of the temperature $u = u(t)$ of the body with respect time t in hour (h) is proportional to $u(t) - T$, where T is the ambient temperature and the constant of proportionality is -0.294.

(a) Write a differential equation that models the temperature $u = u(t)$ of the body at time t.

(b) Find the general solution of the differential equation.

(c) Find the particular solution of the differential equation, using the initial condition that at the time of death, $u(0) = 37\,°C$.

(d) If the body was found at 1 a.m., when did the person die?

(e) How long will it take for the body to cool to $12\,°C$?

59. Radioactive Decay   The amount A of the radioactive element radium in a sample decays at a rate proportional to the amount of radium present. Given the half-life of radium is 1690 years:

(a) Write a differential equation that models the amount A of radium present at time t.

(b) Find the general solution of the differential equation.

(c) Find the particular solution of the differential equation with the initial condition $A(0) = 10\,g$.

(d) How much radium will be present in the sample at $t = 300$ years?

60. Population Growth in China   Barring disasters (human-made or natural), the population P of humans grows at a rate proportional to its current size. According to the U.N. World Population studies, from 2005 to 2010 the population of China grew at an annual rate of 0.510% per year.

(a) Write a differential equation that models the growth rate of the population.

(b) Find the general solution of the differential equation.

(c) Find the particular solution of the differential equation if in 2010 ($t = 0$), the population of China was 1.341335×10^9.

(d) If the rate of growth continues to follow this model, when will the projected population of China reach 2 billion persons?

Source: U.N. World Population Prospects, 2010 update.

CHAPTER 5 PROJECT Managing the Klamath River

There is a gauge on the Klamath River, just downstream from the dam at Keno, Oregon. The U.S. Geological Survey has posted flow rates for this gauge every month since 1930. The averages of these monthly measurements since 1930 are given in Table 2. Notice that the data in Table 2 measure the rate of change of the volume V in cubic feet of water each second over one year; that is, the table

gives $\dfrac{dV}{dt} = V'(t)$ in cubic feet per second, where t is in months.

TABLE 2

Month (t)	Flow Rate (ft³/s)
January (1)	1911.79
February (2)	2045.40
March (3)	2431.73
April (4)	2154.14
May (5)	1592.73
June (6)	945.17
July (7)	669.46
August (8)	851.97
September (9)	1107.30
October (10)	1325.12
November (11)	1551.70
December (12)	1766.33

Source: USGS Surface-Water Monthly Statistics, available at http:// waterdata.usgs.gov/or/nwis/monthly.

1. Find the factor that will convert the data in Table 2 from seconds to days. [*Hint:* 1 day $= 1\,\mathrm{d} = 24$ hours $= 24\,\mathrm{h}(60\,\mathrm{min/h}) = (24)(60\ \mathrm{min})\,(60\,\mathrm{s/min}).$]

   If we assume February has 28.25 days, to account for a leap year, then 1 year $= 365.25$ days. If $V'(t)$ is the rate of flow of water, in cubic feet per day, the total flow of water over 1 year is given by

   $$V = V(t) = \int_0^{365.25} V'(t)\,dt.$$

2. Approximate the total annual flow using a Riemann sum. (*Hint:* Use $\Delta t_1 = 31$, $\Delta t_2 = 28.25$, etc.)

3. The solution to Problem 2 finds the sum of 12 rectangles whose widths are Δt_i, $1 \le i \le 12$, and whose heights are the flow rate for the ith month. Using the horizontal axis for time and the vertical axis for flow rate in ft³/day, plot the points of Table 2 as follows: (January 1, flow rate for January), (February 1, flow rate for February), ... ,(December 1, flow rate for December) and add the point (December 31, flow rate for January). Beginning with the point at January 1, connect each consecutive pair of points with a line segment, creating 12 trapezoids whose

bases are Δt_i, $1 \le i \le 12$. Approximate the total annual flow $V = V(t) = \int_0^{365.25} V'(t)\,dt$ by summing the areas of these trapezoids.

4. Using the horizontal axis for time and the vertical axis for flow rate in ft³/day, plot the points of Table 2 as follows: (January 31, flow rate for January), (February 28, flow rate for February), ... , (December 31, flow rate for December). Then add the point (January 1, flow rate for December) to the left of (January 31, flow rate for January). Connect consecutive points with a line segment, creating 12 trapezoids whose bases are Δt_i, $1 \le i \le 12$. Approximate the total annual flow $V = V(t) = \int_0^{365.25} V'(t)\,dt$ by summing the areas of these trapezoids.

5. Why did we add the extra point in Problems 3 and 4? How do you justify the choice?

6. Compare the three approximations. Discuss which might be the most accurate.

7. Consult Chapter 7 and read about Simpson's Rule (p. 514). Can you see a way to use it to approximate the total annual flow?

8. Another way to approximate $V = V(t) = \int_0^{365.25} V'(t)\,dt$ is to fit a polynomial function to the data. We could find a polynomial of degree 11 that passes through every point of the data, but a polynomial of degree 6 is sufficient to capture the essence of the behavior. The polynomial function f of degree 6 is

   $$f(t) = 2.2434817 \times 10^{-10}t^6 - 2.5288956 \times 10^{-7}t^5$$
   $$+\ 0.00010598313t^4 - 0.019872628t^3 + 1.557403t^2$$
   $$-\ 39.387734t + 2216.2455$$

   Find the total annual flow using $f(t) = V'(t)$.

9. Use technology to graph the polynomial function f over the closed interval $[0, 12]$. How well does the graph fit the data?

10. A manager could approximate the rate of flow of the river for every minute of every day using the function

    $$g(t) = 1529.403 + 510.330 \sin\frac{2\pi t}{365.25} + 489.377 \cos\frac{2\pi t}{365.25}$$
    $$-\ 47.049 \sin\frac{4\pi t}{365.25} - 249.059 \cos\frac{4\pi t}{365.25}$$

    where t represents the day of the year in the interval $[0, 365.25]$. The function g represents the best fit to the data that has the form of a sum of trigonometric functions with the period 1 year. We could fit the data perfectly using more terms, but the improvement in results would not be worth the extra work in handling that approximation. Use g to approximate the total annual flow.

11. Use technology to graph the function g over the closed interval $[0, 12]$. How well does the graph fit the data?

12. Compare the five approximations to the annual flow of the river. Discuss the advantages and disadvantages of using one over another. What method would you recommend to measure the annual flow of the Klamath River?

EXAMPLE 1 Finding the Area Between the Graphs of Two Functions

Find the area of the region enclosed by the graphs of $f(x) = e^x$ and $g(x) = \sqrt{x}$ and the lines $x = 0$ and $x = 1$.

Solution We begin by graphing the two functions and identifying the area A to be found. See Figure 3.

From the graph, we see that $f(x) \geq g(x)$ on the interval $[0, 1]$. Then, using the definition of area, we have

$$A = \int_a^b [f(x) - g(x)]\, dx = \int_0^1 (e^x - \sqrt{x})\, dx$$

$$= \int_0^1 e^x\, dx - \int_0^1 x^{1/2}\, dx = \left[e^x\right]_0^1 - \left[\frac{x^{3/2}}{\frac{3}{2}}\right]_0^1$$

$$= (e^1 - e^0) - \frac{2}{3}(1 - 0) = e - 1 - \frac{2}{3} = e - \frac{5}{3} \text{ square units} \quad \blacksquare$$

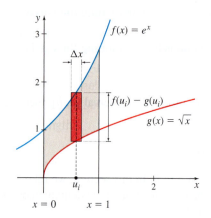

Figure 3

NOW WORK   Problem 3.

The definition of area (1) holds whether the graphs of f and g lie above the x-axis, below the x-axis, or partially above and partially below the x-axis as long as $f(x) \geq g(x)$ on $[a, b]$. It is critical to graph f and g on $[a, b]$ to determine the relationship between the graphs of f and g before setting up the integral. The key is to always subtract the smaller value from the larger value. This ensures that the height of each rectangle is positive.

EXAMPLE 2 Finding the Area Between the Graphs of Two Functions

Find the area of the region enclosed by the graphs of $f(x) = 10x - x^2$ and $g(x) = 3x - 8$.

Solution First we graph the two functions. See Figure 4.

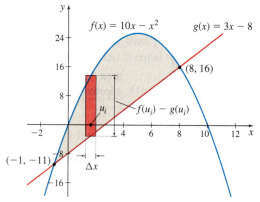

Figure 4 The graphs intersect at $(-1, -11)$ and $(8, 16)$. So, the limits of integration are $a = -1$ and $b = 8$.

The area we seek lies between the points of intersection of the graphs. Finding the x-values of the points of intersection will identify the limits of integration. We solve the equation $f(x) = g(x)$ to find these values.

$$10x - x^2 = 3x - 8 \qquad f(x) = g(x)$$

$$x^2 - 7x - 8 = 0$$

$$(x + 1)(x - 8) = 0$$

$$x = -1 \quad \text{or} \quad x = 8$$

The limits of integration are $a = -1$ and $b = 8$. Since $f(x) \geq g(x)$ on $[-1, 8]$, the area A is given by

$$A = \int_a^b [f(x) - g(x)] \, dx = \int_{-1}^8 [(10x - x^2) - (3x - 8)] \, dx$$

$$= \int_{-1}^8 (-x^2 + 7x + 8) \, dx = \left[\frac{-x^3}{3} + \frac{7x^2}{2} + 8x \right]_{-1}^8$$

$$= \left(-\frac{512}{3} + 224 + 64 \right) - \left(\frac{1}{3} + \frac{7}{2} - 8 \right) = \frac{243}{2} = 121.5 \text{ square units} \quad \blacksquare$$

NOW WORK Problem 7.

EXAMPLE 3 **Finding the Area Between the Graphs of Two Functions**

Find the area of the region enclosed by the graphs of $f(x) = \sin x$ and $g(x) = \cos x$ from the y-axis to their first point of intersection in the first quadrant.

Solution First we graph the two functions. See Figure 5.

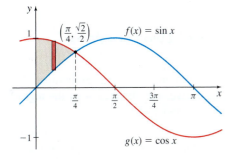

Figure 5

The points of intersection of the two graphs satisfy the equation $f(x) = g(x)$.

$$\sin x = \cos x \qquad f(x) = g(x)$$

$$\tan x = 1$$

The first point of intersection in the first quadrant occurs at $x = \tan^{-1} 1 = \dfrac{\pi}{4}$. The graphs intersect at the point $\left(\dfrac{\pi}{4}, \dfrac{\sqrt{2}}{2} \right)$, so the area A we seek lies between $x = 0$ and $x = \dfrac{\pi}{4}$. Since $\cos x \geq \sin x$ on $\left[0, \dfrac{\pi}{4} \right]$, the area A is given by

$$A = \int_0^{\pi/4} (\cos x - \sin x) \, dx = \left[\sin x + \cos x \right]_0^{\pi/4} = \left(\frac{\sqrt{2}}{2} + \frac{\sqrt{2}}{2} \right) - (0 + 1)$$

$$= \sqrt{2} - 1 \text{ square units} \quad \blacksquare$$

Suppose $y = g(x) \leq 0$ for all numbers x in the interval $[a, b]$, as illustrated in Figure 6(a). Then by the definition of area (1), the area A of the region enclosed by the graph of $f(x) = 0$ (the x-axis), the graph of g, and the lines $x = a$ and $x = b$, is given by

$$A = \int_a^b [f(x) - g(x)] \, dx = \int_a^b [0 - g(x)] \, dx = -\int_a^b g(x) \, dx$$

By symmetry, this area A is equal to the area under the graph of $y = -g(x) \geq 0$ from a to b. See Figure 6(b).

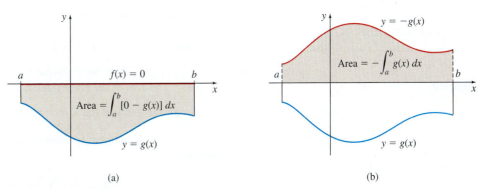

(a)   (b)

Figure 6

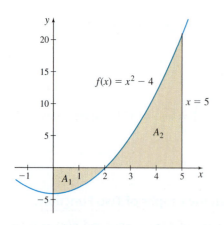

Figure 7 The area A enclosed by the graph of f, the x-axis, and the lines $x = 0$ and $x = 5$ is the sum of the areas A_1 and A_2.

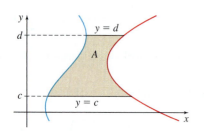

Figure 8

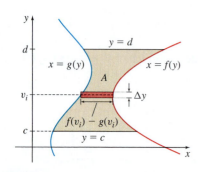

Figure 9

EXAMPLE 4   Finding the Area Between a Graph and the x-Axis

Find the area of the region enclosed by the graph of $f(x) = x^2 - 4$, the x-axis, and the lines $x = 0$ and $x = 5$.

Solution First we graph the function f. On the interval $[0, 5]$, the graph of f intersects the x-axis at $x = 2$. As shown in Figure 7, $f(x) \le 0$ on the interval $[0, 2]$, and $f(x) \ge 0$ on the interval $[2, 5]$. So, the area A is the sum of the areas A_1 and A_2, where

$$A_1 = \int_0^2 [0 - f(x)]\, dx = \int_0^2 -(x^2 - 4)\, dx = \left[-\frac{x^3}{3} + 4x \right]_0^2 = -\frac{8}{3} + 8 = \frac{16}{3}$$

$$A_2 = \int_2^5 f(x)\, dx = \int_2^5 (x^2 - 4)\, dx = \left[\frac{x^3}{3} - 4x \right]_2^5 = \left(\frac{125}{3} - 20 \right) - \left(\frac{8}{3} - 8 \right)$$

$$= \frac{81}{3} = 27$$

The area A we seek is $A = A_1 + A_2 = \dfrac{16}{3} + \dfrac{81}{3} = \dfrac{97}{3}$ square units. ∎

NOW WORK Problem **11**.

2 Find the Area Between the Graphs of Two Functions by Partitioning the y-Axis

In the previous examples, we found the area by partitioning the x-axis. Sometimes it is necessary to partition the y-axis. Look at Figure 8. We seek the area A of the region enclosed by the graphs and the horizontal lines $y = c$ and $y = d$.

The graphs that form the left and right borders of the region are not the graphs of functions (they fail the Vertical-line Test). Approximating A by partitioning the x-axis is not practical. However, both graphs in Figure 8 can be represented as functions of y since each satisfies the Horizontal-line Test.

> **THEOREM Horizontal-line Test**
>
> A set of points in the xy-plane is the graph of a function of the form $x = f(y)$ if and only if every horizontal line intersects the graph in at most one point.

The graphs in Figure 9 satisfy the Horizontal-line Test, so each is a function of y. In Figure 9, the graphs are labeled $x = g(y)$ and $x = f(y)$, where $x = g(y)$ is the horizontal distance from the y-axis to $g(y)$, and $x = f(y)$ is the horizontal distance from the y-axis to $f(y)$. Since the graph of f lies to the right of the graph of g, we know $f(y) \ge g(y)$.

Now to find the area A of the region enclosed by the two graphs and the horizontal lines $y = c$ and $y = d$, we partition the interval $[c, d]$ on the y-axis into n subintervals:

$$[y_0, y_1], [y_1, y_2], \ldots, [y_{i-1}, y_i], \ldots, [y_{n-1}, y_n] \qquad y_0 = c \quad y_n = d$$

each of width $\Delta y = \dfrac{d - c}{n}$. For each i, $i = 1, 2, \ldots n$, we select a number v_i in the subinterval $[y_{i-1}, y_i]$. Then we construct n rectangles, each of height Δy and width $f(v_i) - g(v_i)$. The area of the ith rectangle is $[f(v_i) - g(v_i)]\Delta y$.

The sum of the areas of the n rectangles, $\sum\limits_{i=1}^{n} [f(v_i) - g(v_i)]\Delta y$, approximates the area A we seek. As the number of subintervals increases, the approximation to the area improves, and

$$A = \lim_{n \to \infty} \sum_{i=1}^{n} [f(v_i) - g(v_i)]\, \Delta y$$

The approximating sums are Riemann sums, so if f and g are continuous on the interval $[c, d]$, then the limit is a definite integral and

$$A = \int_c^d [f(y) - g(y)]\, dy$$

Area

The area A of the region enclosed by the graphs of $x = f(y)$ and $x = g(y)$, and the horizontal lines $y = c$ and $y = d$, where f and g are continuous on the interval $[c, d]$ and $f(y) \geq g(y)$ for all numbers y in $[c, d]$, is

$$\boxed{A = \int_c^d [f(y) - g(y)]\, dy}$$

EXAMPLE 5 Finding Area by Partitioning the y-Axis

Find the area A of the region enclosed by the graphs of $x = f(y) = y + 2$ and $x = g(y) = y^2$.

Solution First we graph the two functions and identify the region whose area A we seek. See Figure 10.

The graphs intersect when $y + 2 = y^2$. Then $y^2 - y - 2 = (y - 2)(y + 1) = 0$. So, $y = 2$ or $y = -1$. When $y = 2$, $x = 4$; when $y = -1$, $x = 1$. The graphs intersect at the points $(4, 2)$ and $(1, -1)$.

Notice in Figure 10 that the graph of f is to the right of the graph of g; that is, $f(y) \geq g(y)$ for $-1 \leq y \leq 2$. This indicates that we can partition the y-axis and form rectangles from the left graph $(x = y^2)$ to the right graph $(x = y + 2)$ as y varies from -1 to 2. The area A of the region between the graphs is

$$A = \int_{-1}^2 [f(y) - g(y)]dy = \int_{-1}^2 [(y + 2) - y^2]dy = \left[\frac{y^2}{2} + 2y - \frac{y^3}{3}\right]_{-1}^2$$

$$= \left(2 + 4 - \frac{8}{3}\right) - \left(\frac{1}{2} - 2 + \frac{1}{3}\right) = 4.5 \text{ square units} \qquad\blacksquare$$

NOW WORK Problem 15.

There are times when either the x-axis or the y-axis can be partitioned.

EXAMPLE 6 Finding the Area Between the Graphs of Two Functions

Find the area A of the region enclosed by the graphs of $y = \sqrt{4 - 4x}$, $y = \sqrt{4 - x}$, and the x-axis:

(a) by partitioning the x-axis.

(b) by partitioning the y-axis.

Solution **(a)** We begin by graphing the two equations and identifying the region whose area we seek. See Figure 11.

If we partition the x-axis, the area A of the region we seek must be expressed as the sum of the two areas A_1 and A_2 marked in the figure. [Do you see why? The bottom graph changes at $x = 1$ from $y = \sqrt{4 - 4x}$ to $y = 0$ (the x-axis)].

Area A_1 is the region enclosed by the graphs of $y = \sqrt{4 - x}$, and $y = \sqrt{4 - 4x}$, and the line $x = 1$. Area A_2 is the region enclosed by the graph $y = \sqrt{4 - x}$, the x-axis,

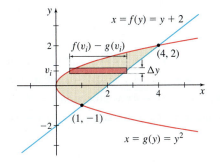

Figure 10

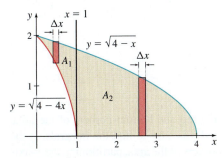

Figure 11 The area A is the sum of the areas A_1 and A_2.

and the line $x = 1$. Then

$$A = A_1 + A_2 = \int_0^1 (\sqrt{4-x} - \sqrt{4-4x})\, dx + \int_1^4 \sqrt{4-x}\, dx$$

$$= \int_0^1 \sqrt{4-x}\, dx - \int_0^1 \sqrt{4-4x}\, dx + \int_1^4 \sqrt{4-x}\, dx$$

$$= \int_0^4 \sqrt{4-x}\, dx - \int_0^1 \sqrt{4-4x}\, dx \qquad \int_0^1 \sqrt{4-x}\, dx + \int_1^4 \sqrt{4-x}\, dx = \int_0^4 \sqrt{4-x}\, dx$$

To find the first integral, we use the substitution $u = 4 - x$. Then $du = -dx$, and

$$\int_0^4 \sqrt{4-x}\, dx = -\int_4^0 u^{1/2}\, du = \int_0^4 u^{1/2}\, du = \left[\frac{2}{3}u^{3/2}\right]_0^4 = \frac{2}{3}(8-0) = \frac{16}{3}$$

For the other integral, we use the substitution $u = 4 - 4x$. Then $du = -4\, dx$, or equivalently, $dx = -\dfrac{du}{4}$, and

$$\int_0^1 \sqrt{4-4x}\, dx = -\frac{1}{4}\int_4^0 u^{1/2}\, du = \frac{1}{4}\int_0^4 u^{1/2}\, du = \frac{1}{4}\left[\frac{2}{3}u^{3/2}\right]_0^4 = \frac{1}{6}(8-0) = \frac{4}{3}$$

The area $A = \dfrac{16}{3} - \dfrac{4}{3} = 4$ square units.

(b) Figure 12 shows the graphs of $y = \sqrt{4-4x}$ and $y = \sqrt{4-x}$ and the region whose area A we seek. Since the graphs of $y = \sqrt{4-4x}$ and $y = \sqrt{4-x}$ satisfy the Horizontal-line Test for $0 \le y \le 2$, we can express $y = \sqrt{4-4x}$ as a function $x = f(y)$ and $y = \sqrt{4-x}$ as a function $x = g(y)$. To find $x = f(y)$, we solve $y = \sqrt{4-4x}$ for x, where $x \ge 0$:

$$y = \sqrt{4-4x}$$

$$y^2 = 4 - 4x$$

$$x = \frac{4-y^2}{4}$$

$$x = f(y) = 1 - \frac{y^2}{4}$$

To express $y = \sqrt{4-x}$ as a function $x = g(y)$, we solve for x, where $x \ge 0$:

$$y = \sqrt{4-x}$$

$$y^2 = 4 - x$$

$$x = g(y) = 4 - y^2$$

The graph of $x = g(y) = 4 - y^2$ is to the right of the graph of $x = f(y) = 1 - \dfrac{y^2}{4}$, $0 \le y \le 2$. So, $g(y) \ge f(y)$. Then

$$A = \int_0^2 [g(y) - f(y)]\, dy = \int_0^2 \left[(4-y^2) - \left(1 - \frac{y^2}{4}\right)\right] dy = \int_0^2 \left(3 - \frac{3y^2}{4}\right) dy$$

$$= \left[3y - \frac{y^3}{4}\right]_0^2 = 6 - \frac{8}{4} = 4 \text{ square units} \qquad \blacksquare$$

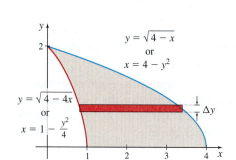

Figure 12

NOW WORK Problem **31**.

When the graphs of functions of x form the top and bottom borders of the area, partitioning the x-axis is usually easier, provided it is easy to find the integrals. If the graphs of functions of y form the left and right borders of the area, partitioning the y-axis is usually easier, provided the integrals with respect to y are easily found. Example 7 shows that sometimes the first choice of a partition does not work.

EXAMPLE 7 **Finding the Area Under a Graph**

Find the area A of the region enclosed by the graph of $y = \ln x$, the x-axis, and the line $x = e$.

Solution We begin by graphing $y = \ln x$ and $x = e$ and identifying the region whose area A we seek. See Figure 13.

Since the graph of $y = \ln x$ forms the top border of the area A, it appears that partitioning the x-axis from 1 to e is easier. This leads to the integral

$$A = \int_1^e \ln x \, dx$$

But at this place in the text, we do not have the tools to find $\int \ln x \, dx$.

So instead, we partition the y-axis from 0 to 1. The left graph is $y = \ln x$ or, equivalently, $x = e^y$, and the right graph is $x = e$. Then

$$A = \int_0^1 (e - e^y) \, dy = \left[ey - e^y \right]_0^1 = (e - e) - (0 - 1) = 1 \text{ square unit} \quad \blacksquare$$

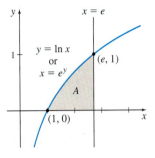

Figure 13

NOW WORK Problem 39.

6.1 Assess Your Understanding

Concepts and Vocabulary

1. Express the area between the graphs of $y = x^2$ and $y = \sqrt{x}$ as an integral using a partition of the x-axis. Do not find the integral.

2. Express the area between the graph of $x = y^2$ and the line $x = 1$ as an integral using a partition of the y-axis. Do not find the integral.

Skill Building

In Problems 3–12, find the area of the region enclosed by the graphs of the given equations by partitioning the x-axis.

3. $y = x$, $y = 2x$, $x = 1$

4. $y = x$, $y = 3x$, $x = 3$

5. $y = x^2$, $y = x$

6. $y = x^2$, $y = 4x$

7. $y = e^x$, $y = e^{-x}$, $x = \ln 2$

8. $y = e^x$, $y = -x + 1$, $x = 1$

9. $y = x^2$, $y = x^4$

10. $y = x$, $y = x^3$

11. $y = \cos x$, $y = \dfrac{1}{2}$, $0 \le x \le \dfrac{\pi}{3}$

12. $y = \sin x$, $y = \dfrac{1}{2}$, $\dfrac{\pi}{6} \le x \le \dfrac{5\pi}{6}$

In Problems 13–20, find the area of the region enclosed by the graphs of the given equations by partitioning the y-axis.

13. $x = y^2$, $x = 2 - y$

14. $x = y^2$, $x = y + 2$

15. $x = 9 - y^2$, $x = 5$

16. $x = 16 - y^2$, $x = 7$

17. $x = y^2 + 4$, $y = x - 6$

18. $x = y^2 + 6$, $y = 8 - x$

19. $y = \ln x$, $x = 1$, $y = 2$

20. $y = \ln x$, $x = e$, $y = 0$

In Problems 21–24, find the area of the shaded region in the graph.

21.

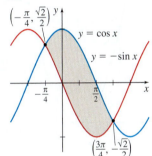

22.

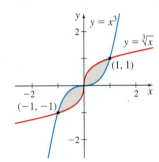

23.

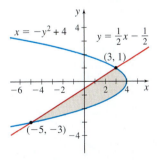

24.

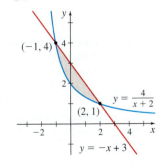

In Problems 25–32, find the area A of the region enclosed by the graphs of the given equations:
(a) by partitioning the x-axis.
(b) by partitioning the y-axis.

25. $y = \sqrt{x}$, $y = x^3$

26. $y = \sqrt{x}$, $y = x^2$

27. $y = x^2 + 1$, $y = x + 1$

28. $y = x^2 + 1$, $y = 4x + 1$

---

1. = NOW WORK problem   = Graphing technology recommended   **CAS** = Computer Algebra System recommended

29. $y = \sqrt{9 - x}$, $y = \sqrt{9 - 3x}$, x-axis

30. $y = \sqrt{16 - 2x}$, $y = \sqrt{16 - 4x}$, x-axis

31. $y = \sqrt{2x - 6}$, $y = \sqrt{x - 2}$, x-axis

32. $y = \sqrt{2x - 5}$, $y = \sqrt{4x - 17}$, x-axis

In Problems 33–46, find the area of the region enclosed by the graphs of the given equations.

33. $y = 4 - x^2$, $y = x^2$

34. $y = 9 - x^2$, $y = x^2$

35. $x = y^2 - 4$, $x = 4 - y^2$

36. $x = y^2$, $x = 16 - y^2$

37. $y = \ln x^2$, the x-axis, and the line $x = e$

38. $y = \ln x$, $y = 1 - x$, and the line $y = 1$

39. $y = \cos x$, $y = 1 - \dfrac{3}{\pi}x$, $x = \dfrac{\pi}{3}$

40. $y = \sin x$, $y = 1$, $0 \le x \le \dfrac{\pi}{2}$

41. $y = e^{2x}$ and the lines $x = 1$ and $y = 1$

42. $y = e^x$, $y = e^{3x}$, $x = 2$

43. $y^2 = 4x$, $4x - 3y - 4 = 0$

44. $y^2 = 4x + 1$, $x = y + 1$

45. $y = \sin x$, $y = \dfrac{2x}{\pi}$, $x \ge 0$

46. $y = \cos x$, $x \ge 0$, $y = \dfrac{3x}{\pi}$

Applications and Extensions

47. **An Archimedean Result** Show that the area of the shaded region in the figure is two-thirds of the area of the parallelogram ABCD. (This illustrates a result due to Archimedes concerning sectors of parabolas.)

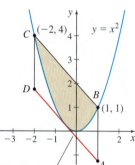

This line is parallel to the line joining $(-2, 4)$ and $(1, 1)$ and is tangent to $y = x^2$.

48. **Equal Areas** Find h so that the area of the region enclosed by the graphs of $y = x$, $y = 8x$, and $y = \dfrac{1}{x^2}$ is equal to that of an isosceles triangle of base 1 and height h. See the figure.

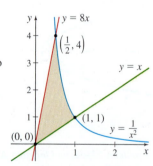

49. **Cost of Health Care** The cost of health care varies from one country to another. Between 2000 and 2010, the average cost of health insurance for a family of four in the United States was modeled by

$$A(x) = 8020.6596(1.0855^x) \qquad 0 \le x \le 10$$

where $x = 0$ corresponds to the year 2000 and $A(x)$ is measured in U.S. dollars. During the same years, the average cost of health care in Canada was given by

$$C(x) = 4944.6424(1.0711^x) \qquad 0 \le x \le 10$$

where $x = 0$ corresponds to the year 2000 and $C(x)$ is measured in U.S. dollars.

(a) Find the area between the graphs from $x = 0$ to $x = 10$. Round the answer to the nearest dollar.

(b) Interpret the answer to (a).

50. **Area** Find the area in the first quadrant enclosed by the graphs of $y = \sin(2x)$ and $y = \cos(2x)$, $0 \le x \le \dfrac{\pi}{8}$.

51. **Area** Find the area of the region enclosed by the graphs of $y = \sin^{-1} x$, $y = x$, $0 \le x \le \dfrac{1}{2}$. Which axis did you choose to partition? Explain your choice.

CAS 52. **Area**

(a) Graph $y = x^2$ and $y = \sin x$.

(b) Find the points of intersection of the graphs.

(c) Find the area enclosed by the graphs of $y = x^2$ and $y = \sin x$ using a partition of the x-axis.

(d) Find the area enclosed by the graphs of $y = x^2$ and $y = \sin x$ using a partition of the y-axis.

CAS 53. (a) Graph $y = \sin^{-1} x$, $x + y = 1$, and $y = 0$.

(b) Find the points of intersection of the graphs.

(c) Find the area enclosed by the graphs.

CAS 54. (a) Graph $y = \cos^{-1} x$, $y = x^3 + 1$, and $y = 0$.

(b) Find the points of intersection of the graphs.

(c) Find the area enclosed by the graphs.

CAS 55. (a) Graph $2x + y = 3$, $y = \dfrac{1}{1 + x^2}$, and $y = 1$.

(b) Find the points of intersection of the graphs.

(c) Find the area enclosed by the graphs.

CAS 56. (a) Graph $y = \dfrac{5}{1 + x^2}$, $x = \sqrt{y + 2}$, and $x = 0$.

(b) Find the points of intersection of the graphs.

(c) Find the area enclosed by the graphs.

CAS 57. (a) Graph $y = \sin^{-1} x$, $y = 1 - x^2$, and $y = 0$.

(b) Find the points of intersection of the graphs.

(c) Find the area enclosed by the graphs.

Challenge Problems

58. Find the area enclosed by the graph of $y^2 = x^2 - x^4$.

59. (a) Express the area A of the region in the first quadrant enclosed by the y-axis and the graphs of $y = \tan x$ and $y = k$ for $k > 0$ as a function of k.

(b) What is the value of A when $k = 1$?

(c) If the line $y = k$ is moving upward at the rate of $\dfrac{1}{10}$ unit per second, at what rate is A changing when $k = 1$?

60. The area A of the shaded region in the figure is $A = \dfrac{t}{2}$. Prove this as follows:

(a) Let A^* be twice the area outlined in the figure, and (x, y) be the coordinates of P. Explain why

$$A^* = x\sqrt{x^2 - 1} - 2\int_1^x \sqrt{u^2 - 1}\, du$$

(b) Differentiate A^* to show that $\dfrac{dA^*}{dx} = \dfrac{1}{\sqrt{x^2 - 1}}$.

(c) Show that $A^* = \cosh^{-1} x + C$, and explain why $C = 0$.

(d) Why does it follow that $A^* = t$?

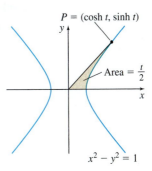

$P = (\cosh t, \sinh t)$

Area $= \dfrac{t}{2}$

$x^2 - y^2 = 1$

6.2 Volume of a Solid of Revolution: Disks and Washers

OBJECTIVES *When you finish this section, you should be able to:*

1 Use the disk method to find the volume of a solid formed by revolving a region about the x-axis (p. 415)

2 Use the disk method to find the volume of a solid formed by revolving a region about the y-axis (p. 416)

3 Use the washer method to find the volume of a solid formed by revolving a region about the x-axis (p. 418)

4 Use the washer method to find the volume of a solid formed by revolving a region about the y-axis (p. 420)

5 Find the volume of a solid formed by revolving a region about a line parallel to a coordinate axis (p. 421)

An example of a solid of revolution is the *right circular cylinder*, which is generated by revolving the region bounded by a horizontal line $y = A$, $A > 0$, the x-axis, and the lines $x = a$ and $x = b$ about the x-axis. See Figure 14(a). Another familiar example of a solid of revolution, pictured in Figure 14(b), is a *right circular cone* that is generated by revolving the region in the first quadrant bounded by a line $y = mx$, $m > 0$, the x-axis, and the line $x = h$ about the x-axis.

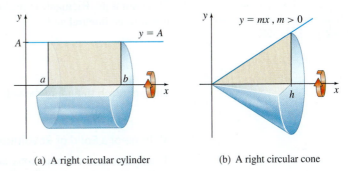

(a) A right circular cylinder

(b) A right circular cone

Figure 14

Although the volume of these simple solids of revolution can be found using geometry formulas, to find the volume of most solids of revolution requires the use of calculus.

Consider a region that is bounded by the graph of a function $y = f(x)$ that is continuous and nonnegative on the interval $[a, b]$, the x-axis, and the lines $x = a$ and $x = b$. See Figure 15(a). Suppose this region is revolved about the x-axis. The result is a **solid of revolution**. See Figure 15(b).

To find the volume of a solid of revolution, we begin by partitioning the interval $[a, b]$ into n subintervals:

$$[a, x_1], [x_1, x_2], \ldots, [x_{i-1}, x_i], \ldots, [x_{n-1}, b]$$

each of width $\Delta x = \dfrac{b - a}{n}$ and selecting a number u_i in each subinterval. Refer to Figure 15(c). The circle obtained by slicing the solid at u_i has radius $r_i = f(u_i)$ and area $A_i = \pi r_i^2 = \pi [f(u_i)]^2$. The volume V_i of the disk obtained from the ith subinterval is $V_i = \pi [f(u_i)]^2 \Delta x$, and an approximation to the volume V of the solid of revolution is

$$V \approx \pi \sum_{i=1}^{n} [f(u_i)]^2 \, \Delta x$$

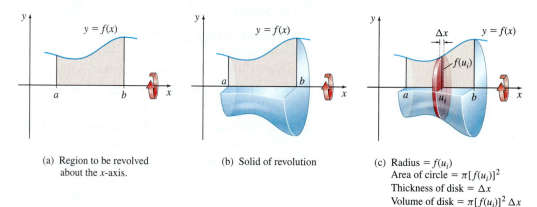

(a) Region to be revolved about the x-axis.

(b) Solid of revolution

(c) Radius $= f(u_i)$
Area of circle $= \pi [f(u_i)]^2$
Thickness of disk $= \Delta x$
Volume of disk $= \pi [f(u_i)]^2 \, \Delta x$

Figure 15

IN WORDS The volume V of a disk is

$V = $ (Area of the circular cross section)

$\quad \times$ (Thickness)

$\quad = \pi (\text{Radius})^2 (\text{Thickness})$

NEED TO REVIEW? Riemann sums and the definite integral are discussed in Sections 5.2, pp. 353–359.

As the number n of subintervals increases, the sums $\pi \sum_{i=1}^{n} [f(u_i)]^2 \Delta x$ become better approximations to the volume V of the solid, and

$$V = \pi \lim_{n \to \infty} \sum_{i=1}^{n} [f(u_i)]^2 \, \Delta x$$

These sums are Riemann sums. So if f is continuous on the interval $[a, b]$, then the limit is a definite integral and

$$V = \pi \int_a^b [f(x)]^2 \, dx$$

This approach to finding the volume of a solid of revolution is called the **disk method**.

Volume of a Solid of Revolution Using the Disk Method

If a function f is continuous and nonnegative on a closed interval $[a, b]$, then the volume V of the solid of revolution obtained by revolving the region bounded by the graph of f, the x-axis, and the lines $x = a$ and $x = b$ about the x-axis is

$$V = \pi \int_a^b [f(x)]^2 \, dx$$

1 Use the Disk Method to Find the Volume of a Solid Formed by Revolving a Region About the x-Axis

EXAMPLE 1   Using the Disk Method: Revolving About the x-Axis

Find the volume of the solid of revolution generated by revolving the region bounded by the graph of $y = \sqrt{x}$, the x-axis, and the line $x = 5$ about the x-axis.

Solution We begin by graphing the region to be revolved. See Figure 16(a). Figure 16(b) shows a typical disk and Figure 16(c) shows the solid of revolution. Using the disk method, the volume V of the solid of revolution is

$$V = \pi \int_0^5 (\sqrt{x})^2 \, dx = \pi \int_0^5 x \, dx = \pi \left[\frac{x^2}{2} \right]_0^5 = \frac{25}{2} \pi \text{ cubic units}$$

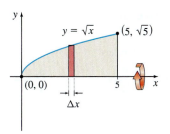

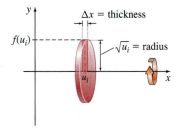

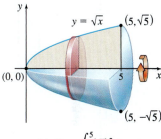

(a) The region to be revolved about the x-axis.

(b) Radius $= \sqrt{u_i}$
Area of circle $= \pi(\sqrt{u_i})^2$
Thickness of disk $= \Delta x$
Volume of disk $= \pi(\sqrt{u_i})^2 \Delta x$

(c) $V = \pi \int_0^5 (\sqrt{x})^2 \, dx$

DF Figure 16

■

NOW WORK Problem 5.

The disk method does not require the function f to be nonnegative. As Figure 17 illustrates, the volume V of the solid of revolution obtained by revolving about the x-axis the region bounded by the graph of a function f that is continuous on the interval $[a, b]$, the x-axis, and the lines $x = a$, and $x = b$ equals the volume of the solid of revolution obtained by revolving about the x-axis the region bounded by the graph of $y = |f(x)|$, the x-axis, and the lines $x = a$ and $x = b$. Since $|f(x)|^2 = [f(x)]^2$, the formula for V does not change.

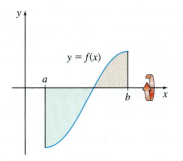

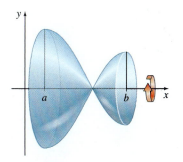

(a) The region to be revolved about the x-axis.

(b) The absolute value of the region to be revolved about the x-axis.

(c) The volume V is the same in both cases:

$$V = \pi \int_a^b [f(x)]^2 \, dx$$

Figure 17

EXAMPLE 2   Using the Disk Method: Revolving About the x-Axis

Find the volume of the solid of revolution generated by revolving the region bounded by the graph of $y = x^3$, the x-axis, and the lines $x = -1$ and $x = 2$ about the x-axis.

Solution   Figure 18(a) shows the graph of the region to be revolved about the x-axis. Figure 18(b) illustrates a typical disk and Figure 18(c) shows the solid of revolution.

Using the disk method, the volume V of the solid of revolution is

$$V = \pi \int_{-1}^{2} x^6 \, dx = \pi \left[\frac{x^7}{7} \right]_{-1}^{2} = \frac{\pi}{7}(128 + 1) = \frac{129}{7}\pi \text{ cubic units}$$

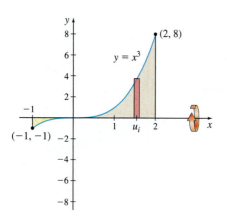

(a) The region to be revolved about the x-axis.

(b) Radius $= u_i^3$
Area of circle $= \pi (u_i^3)^2 = \pi u_i^6$
Thickness of disk $= \Delta x$
Volume of disk $= \pi u_i^6 \, \Delta x$

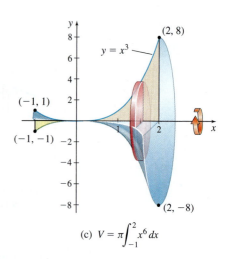

(c) $V = \pi \displaystyle\int_{-1}^{2} x^6 \, dx$

Figure 18

NOW WORK   Problem 27.

2 Use the Disk Method to Find the Volume of a Solid Formed by Revolving a Region About the y-Axis

A solid of revolution can be generated by revolving a region about any line. In particular, the volume V of the solid of revolution generated by revolving the region bounded by the graph of $x = g(y)$, where g is continuous and nonnegative on the closed interval $[c, d]$, the y-axis, and the horizontal lines $y = c$ and $y = d$ about the y-axis can be obtained by partitioning the y-axis and taking slices perpendicular to the y-axis of thickness Δy. See Figure 19(a) on page 417.

Again, the cross sections are circles, as shown in Figure 19(b). The radius of a typical cross section is $r_i = g(v_i)$, and its area A_i is $A_i = \pi r_i^2 = \pi [g(v_i)]^2$. The volume of the disk of radius r_i and thickness $\Delta y = \dfrac{d - c}{n}$ is $V_i = \pi r_i^2 \Delta y = \pi [g(v_i)]^2 \, \Delta y$. By summing the volumes of all the disks and taking the limit, the volume V of the solid of revolution is given by

$$V = \pi \int_{c}^{d} [g(y)]^2 \, dy$$

See Figure 19(c).

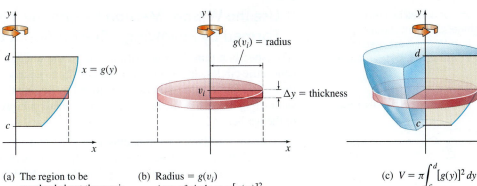

(a) The region to be revolved about the y-axis.

(b) Radius $= g(v_i)$
Area of circle $= \pi[g(v_i)]^2$
Thickness of disk $= \Delta y$
Volume of disk $= \pi[g(v_i)]^2 \Delta y$

(c) $V = \pi \displaystyle\int_c^d [g(y)]^2 \, dy$

DF Figure 19

EXAMPLE 3   **Using the Disk Method: Revolving About the y-Axis**

Use the disk method to find the volume of the solid of revolution generated by revolving the region bounded by the graph of $y = x^3$, the y-axis, and the lines $y = 1$ and $y = 8$ about the y-axis.

Solution   Figure 20(a) shows the region to be revolved. Since the solid is formed by revolving the region about the y-axis, we write $y = x^3$ as $x = \sqrt[3]{y} = y^{1/3}$. Figure 20(b) illustrates a typical disk, and Figure 20(c) shows the solid of revolution. Using the disk method, the volume V of the solid of revolution is

$$V = \pi \int_1^8 [y^{1/3}]^2 \, dy = \pi \int_1^8 y^{2/3} \, dy = \pi \left[\frac{y^{5/3}}{\frac{5}{3}} \right]_1^8$$

$$= \frac{3\pi}{5}(32 - 1) = \frac{93}{5}\pi \text{ cubic units}$$

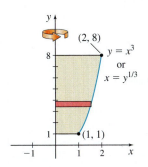

(a) The region to be revolved about the y-axis.

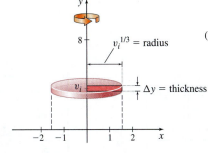
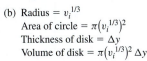

(b) Radius $= v_i^{1/3}$
Area of circle $= \pi\left(v_i^{1/3}\right)^2$
Thickness of disk $= \Delta y$
Volume of disk $= \pi\left(v_i^{1/3}\right)^2 \Delta y$

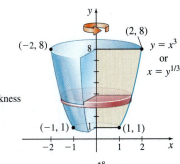

(c) $V = \pi \displaystyle\int_1^8 \left(y^{1/3}\right)^2 dy$

Figure 20

NOW WORK Problem 15.

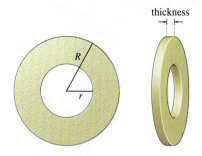

Figure 21 A washer.

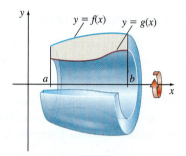

Figure 22 Solid formed by revolving the region bounded by the graphs of f and g about the x-axis. Notice that the interior of the solid is hollow.

IN WORDS The volume V of a washer is
$$V = \pi[(\text{Outer radius})^2 - (\text{Inner radius})^2]$$
$$\times (\text{Thickness})$$

3 Use the Washer Method to Find the Volume of a Solid Formed by Revolving a Region About the x-Axis

A **washer** is a thin flat ring with a hole in the middle. It can be represented by two concentric circles, an outer circle of radius R and an inner circle of radius r, as shown in Figure 21. The volume of a washer is found by finding the area of the outer circle, πR^2, subtracting the area of the inner circle, πr^2, then multiplying the result by the thickness of the washer.

Suppose we have two functions $y = f(x)$ and $y = g(x)$, $f(x) \geq g(x) \geq 0$, that are continuous on the closed interval $[a, b]$. If the region bounded by the graphs of f and g and the lines $x = a$ and $x = b$ is revolved about the x-axis, a solid of revolution with a hollow interior is generated, as illustrated in Figure 22. We seek a formula for finding its volume V.

Figure 23(a) shows the region to be revolved about the x-axis. To find the volume V of the resulting solid of revolution using the washer method, we begin by partitioning the interval $[a, b]$ into n subintervals:

$$[a, x_1], [x_1, x_2], \ldots, [x_{i-1}, x_i], \ldots, [x_{n-1}, b]$$

each of width $\Delta x = \dfrac{b - a}{n}$. In each subinterval $[x_{i-1}, x_i]$, $i = 1, 2, \ldots, n$, we select a number u_i. We then slice the solid at $x = u_i$. The slice is a washer that has outer radius $R_i = f(u_i)$, inner radius $r_i = g(u_i)$, and thickness Δx. A typical washer is shown in Figure 23(b). The area A_i between the concentric circles that form the washer is the difference between the area of the outer circle and the area of the inner circle.

$$A_i = \pi R_i^2 - \pi r_i^2 = \pi[f(u_i)]^2 - \pi[g(u_i)]^2 = \pi\{[f(u_i)]^2 - [g(u_i)]^2\}$$

The volume V_i of the washer is

$$V_i = A_i \Delta x = \pi\{[f(u_i)]^2 - [g(u_i)]^2\}\Delta x$$

and the volume V of the solid can be approximated by

$$V \approx \pi \sum_{i=1}^{n}\{[f(u_i)]^2 - [g(u_i)]^2\}\Delta x$$

As the number n of subintervals increases, the sums $\sum_{i=1}^{n} V_i$ become better approximations to the volume V of the solid of revolution. Since the sums are Riemann sums and since f and g are continuous, the limit as $n \to \infty$ is a definite integral. See Figure 23(c).

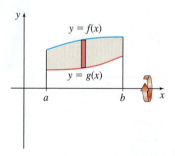

(a) The region to be revolved about the x-axis.

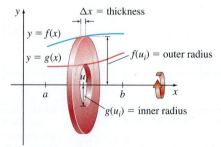

(b) Outer radius $= f(u_i)$
Inner radius $= g(u_i)$
Thickness of washer $= \Delta x$
Volume $= \pi\{[f(u_i)]^2 - [g(u_i)]^2\}\,\Delta x$

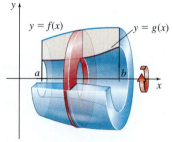

(c) $V = \pi \displaystyle\int_a^b \{[f(x)]^2 - [g(x)]^2\}\,dx$

Figure 23

Volume of a Solid of Revolution Using the Washer Method

If the functions $y = f(x)$ and $y = g(x)$ are continuous on the closed interval $[a, b]$ and if $f(x) \geq g(x) \geq 0$ on $[a, b]$, then the volume V of the solid of revolution obtained by revolving the region bounded by the graphs of f and g and the lines $x = a$ and $x = b$ about the x-axis is

$$V = \pi \int_a^b \left\{ [f(x)]^2 - [g(x)]^2 \right\} dx$$

EXAMPLE 4   Using the Washer Method: Revolving About the x-Axis

Find the volume V of the solid of revolution generated by revolving the region bounded by the graphs of $y = \dfrac{2}{x}$ and $y = 3 - x$ about the x-axis.

Solution We begin by graphing the two functions. See Figure 24(a). The x-coordinates of the points of intersection of the graphs satisfy the equation

$$\frac{2}{x} = 3 - x$$

$$x^2 - 3x + 2 = 0$$

$$(x - 1)(x - 2) = 0$$

So, the region to be revolved lies between $x = 1$ and $x = 2$. Notice that the graph of $y = 3 - x$ lies above the graph of $y = \dfrac{2}{x}$ on the interval $[1, 2]$.

As illustrated in Figure 24(b), if we partition the x-axis, the volume V_i of a typical washer is

$$V_i = \pi \left[(\text{Outer radius})^2 - (\text{Inner radius})^2 \right] \Delta x = \pi \left[(3 - u_i)^2 - \left(\frac{2}{u_i} \right)^2 \right] \Delta x$$

The volume V of the solid of revolution is

$$V = \pi \int_1^2 \left[(3 - x)^2 - \left(\frac{2}{x} \right)^2 \right] dx = \pi \int_1^2 \left(9 - 6x + x^2 - \frac{4}{x^2} \right) dx$$

$$= \pi \left[9x - 3x^2 + \frac{x^3}{3} + \frac{4}{x} \right]_1^2 = \frac{\pi}{3} \text{ cubic units}$$

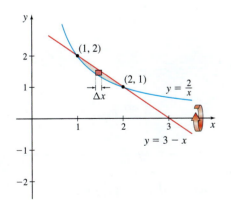

(a) The region to be revolved about the x-axis.

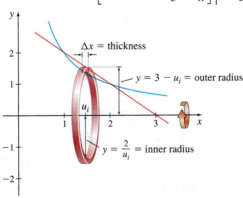

(b) Outer radius $= 3 - u_i$
   Inner radius $= \dfrac{2}{u_i}$
   Thickness of washer $= \Delta x$
   Volume $= \pi \left[(3 - u_i)^2 - \left(\frac{2}{u_i} \right)^2 \right] \Delta x$

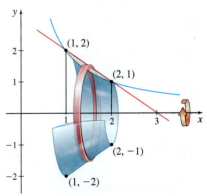

(c) $V = \pi \int_1^2 \left[(3 - x)^2 - \left(\frac{2}{x} \right)^2 \right] dx$

DF Figure 24

NOW WORK Problem 33.

4 Use the Washer Method to Find the Volume of a Solid Formed by Revolving a Region About the *y*-Axis

If we use the washer method and the region bounded by the graphs of two functions is revolved about the *y*-axis, we partition the *y*-axis, and the thickness of a typical washer will be Δy.

EXAMPLE 5 Using the Washer Method: Revolving About the *y*-Axis

Find the volume V of the solid of revolution generated by revolving the region enclosed by the graphs of $y = 2x$ and $y = x^2$ about the *y*-axis.

Solution We begin by graphing the two functions. See Figure 25(a). The *x*-coordinates of the points of intersection of the graphs satisfy the equation

$$2x = x^2$$
$$x^2 - 2x = 0$$
$$x(x - 2) = 0$$
$$x = 0 \quad \text{or} \quad x = 2$$

The points of intersection are $(0, 0)$ and $(2, 4)$. The limits of integration are from $y = 0$ to $y = 4$.

Since the solid is formed by revolving the region about the *y*-axis from $y = 0$ to $y = 4$, we write $y = 2x$ as $x = \dfrac{y}{2}$ and $y = x^2$ as $x = \sqrt{y}$. The outer radius is $\sqrt{y}$ and the inner radius is $\dfrac{y}{2}$.

As Figure 25(b) illustrates, if we partition the *y*-axis, the volume V_i of a typical washer is

$$V_i = \pi \left[(\text{Outer radius})^2 - (\text{Inner radius})^2 \right] \Delta y = \pi \left[(\sqrt{v_i})^2 - \left(\frac{v_i}{2} \right)^2 \right] \Delta y$$

The volume V of the solid of revolution shown in Figure 25(c) is

$$V = \pi \int_0^4 \left[(\sqrt{y})^2 - \left(\frac{y}{2} \right)^2 \right] dy = \pi \int_0^4 \left(y - \frac{y^2}{4} \right) dy$$

$$= \pi \left[\frac{y^2}{2} - \frac{y^3}{12} \right]_0^4 = \frac{8\pi}{3} \text{ cubic units}$$

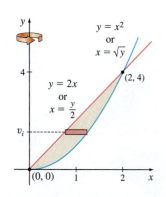

(a) The region to be revolved about the *y*-axis.

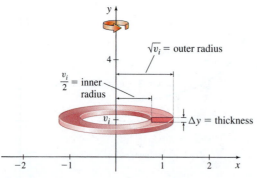

(b) Outer radius = $\sqrt{v_i}$

Inner radius = $\dfrac{v_i}{2}$

Thickness of washer = Δy

Volume = $\pi \left[\sqrt{v_i}^2 - \left(\frac{v_i}{2} \right)^2 \right] \Delta y$

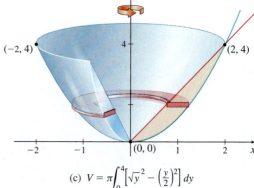

(c) $V = \pi \int_0^4 \left[\sqrt{y}^2 - \left(\frac{y}{2} \right)^2 \right] dy$

DF **Figure 25**

NOW WORK Problem 21.

5 Find the Volume of a Solid Formed by Revolving a Region About a Line Parallel to a Coordinate Axis

Earlier we stated that a solid of revolution can be generated by revolving a region about any line. Now we investigate how to find the volume of a solid formed by revolving a region about a line parallel to either the x-axis or the y-axis.

EXAMPLE 6 **Using the Washer Method: Revolving About a Horizontal Line**

Find the volume V of the solid of revolution generated by revolving the region bounded by the graphs of $y = 2x$ and $y = x^2$ about the line $y = -5$.

Solution The region to be revolved is the same as in Example 5, but it is now being revolved about the line $y = -5$. Figure 26 illustrates the region, a typical washer, and the solid of revolution.

Look at Figure 26(b). The outer radius of the washer is $2u_i + 5$ and the inner radius is $u_i^2 + 5$. Partitioning the x-axis, the volume V_i of a typical washer is

$$V_i = \pi \left[(2u_i + 5)^2 - \left(u_i^2 + 5 \right)^2 \right] \Delta x$$

The volume V of the solid of revolution as shown in Figure 26(c) is

$$V = \pi \int_0^2 \left[(2x + 5)^2 - (x^2 + 5)^2 \right] dx = \pi \int_0^2 (-x^4 - 6x^2 + 20x) \, dx$$

$$= \pi \left[-\frac{x^5}{5} - 2x^3 + 10x^2 \right]_0^2 = \frac{88}{5} \pi \text{ cubic units}$$

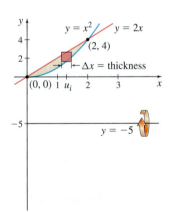

(a) The region to be revolved about the line $y = -5$.

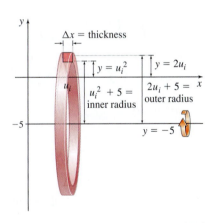

(b) Outer radius $= 2u_i + 5$
Inner radius $= u_i^2 + 5$
Thickness of washer $= \Delta x$
Volume $= \pi[(2u_i + 5)^2 - (u_i^2 + 5)^2] \, \Delta x$

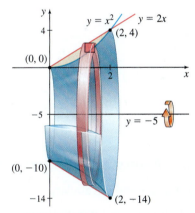

(c) $V = \pi \int_0^2 [(2x + 5)^2 - (x^2 + 5)^2] \, dx$

Figure 26

NOW WORK Problem 39.

EXAMPLE 7 **Using the Washer Method: Revolving About a Vertical Line**

Find the volume of the solid of revolution generated by revolving the region bounded by the graphs of $y = 2x$ and $y = x^2$ about the line $x = 2$.

Solution This example is similar to Example 5 except that the region is revolved about the line $x = 2$. Figure 27 shows the graph of the region, a typical washer, and the solid of revolution.

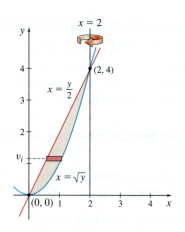

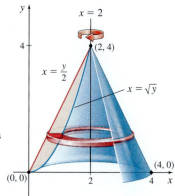

(a) The region to be revolved about the line $x = 2$.

(b) Outer radius $= 2 - \dfrac{v_i}{2}$
Inner radius $= 2 - \sqrt{v_i}$
Thickness of washer $= \Delta y$
Volume $= \pi\left[\left(2 - \dfrac{v_i}{2}\right)^2 - \left(2 - \sqrt{v_i}\right)^2\right]\Delta y$

(c) $V = \pi\displaystyle\int_0^4\left[\left(2 - \dfrac{y}{2}\right)^2 - \left(2 - \sqrt{y}\right)^2\right]dy$

Figure 27

Since the region is revolved about the vertical line $x = 2$, we express $y = 2x$ and $y = x^2$ as $x = \dfrac{y}{2}$ and $x = \sqrt{y}$. The outer radius is $2 - \dfrac{y}{2}$ and the inner radius is $2 - \sqrt{y}$. The volume V of the solid of revolution is

$$V = \pi\int_0^4\left[\left(2 - \frac{y}{2}\right)^2 - (2 - \sqrt{y})^2\right]dy$$

$$= \pi\int_0^4\left[\left(4 - 2y + \frac{y^2}{4}\right) - (4 - 4\sqrt{y} + y)\right]dy$$

$$= \pi\int_0^4\left(\frac{y^2}{4} - 3y + 4\sqrt{y}\right)dy = \pi\left[\frac{y^3}{12} - \frac{3y^2}{2} + \frac{8y^{3/2}}{3}\right]_0^4 = \frac{8}{3}\pi \text{ cubic units} \quad\blacksquare$$

NOW WORK   **Problem 41.**

6.2 Assess Your Understanding

Concepts and Vocabulary

1. If a function f is continuous on a closed interval $[a, b]$, then the volume V of the solid of revolution obtained by revolving the region bounded by the graph of f, the x-axis, and the lines $x = a$ and $x = b$ about the x-axis, is found using the formula $V = \underline{\hspace{1.5cm}}$.

2. *True or False*   When the region bounded by the graphs of the functions f and g and the lines $x = a$ and $x = b$ is revolved about the x-axis, the cross section exposed by making a slice at u_i perpendicular to the x-axis is two concentric circles, and the area A_i between the circles is $A_i = \pi[f(u_i) - g(u_i)]^2$.

3. *True or False*   If the functions f and g are continuous on the closed interval $[a, b]$ and if $f(x) \geq g(x) \geq 0$ on the interval, then the volume V of the solid of revolution obtained by revolving the region bounded by the graphs of f and g and the lines $x = a$ and $x = b$ about the x-axis is $V = \pi\int_a^b[f(x) - g(x)]^2\,dx$.

4. *True or False*   If the region bounded by the graphs of $y = x^2$ and $y = 2x$ is revolved about the line $y = 6$, the volume V of the solid

of revolution generated is found by finding the integral

$$V = \pi\int_6^7(4x^2 - x^4)\,dx.$$

Skill Building

In Problems 5–10, find the volume of the solid of revolution generated by revolving the region shown below about the indicated axis.

5. $y = 2\sqrt{x}$ about the x-axis

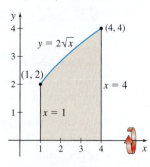

6. $y = x^4$ about the y-axis

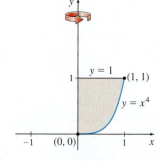

7. $y = \dfrac{1}{x}$ about the y-axis

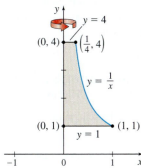

8. $y = x^{2/3}$ about the y-axis

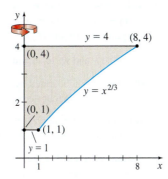

9. $y = \sec x$ about the x-axis

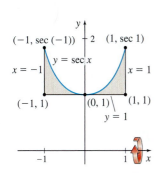

10. $y = x^2$ about the y-axis

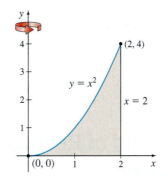

In Problems 11–16, use the disk method to find the volume of the solid of revolution generated by revolving the region bounded by the graphs of the given equations about the indicated axis.

11. $y = 2x^2$, the x-axis, $x = 1$; about the x-axis

12. $y = \sqrt{x}$, the x-axis, $x = 4$, $x = 9$; about the x-axis

13. $y = e^{-x}$, the x-axis, $x = 0$, $x = 2$; about the x-axis

14. $y = e^x$, the x-axis, $x = -1$, $x = 1$; about the x-axis

15. $y = x^2$, $x \geq 0$, $y = 1$, $y = 4$; about the y-axis

16. $y = 2\sqrt{x}$, the y-axis, $y = 4$; about the y-axis

In Problems 17–22, use the washer method to find the volume of the solid of revolution generated by revolving the region bounded by the graphs of the given equations about the indicated axis.

17. $y = x^2$, $x \geq 0$, the y-axis, $y = 4$; about the x-axis

18. $y = 2x^2$, $x \geq 0$, the y-axis, $y = 2$; about the x-axis

19. $y = 2\sqrt{x}$, the y-axis, $y = 4$; about the x-axis

20. $y = x^{2/3}$, the x-axis, $x = 8$; about the y-axis

21. $y = x^3$, the x-axis, $x = 2$; about the y-axis

22. $y = 2x^4$, the x-axis, $x = 1$; about the y-axis

In Problems 23–38, find the volume of the solid of revolution generated by revolving the region bounded by the graphs of the given equations about the indicated axis.

23. $y = \dfrac{1}{x}$, the x-axis, $x = 1$, $x = 2$; about the x-axis

24. $y = \dfrac{1}{x}$, the x-axis, $x = 1$, $x = 2$; about the y-axis

25. $y = \sqrt{x}$, the y-axis, $y = 9$; about the y-axis

26. $y = \sqrt{x}$, the y-axis, $y = 9$; about the x-axis

27. $y = (x - 2)^3$, the x-axis, $x = 0$, $x = 3$; about the x-axis

28. $y = (x - 2)^3$, the x-axis, $x = 0$, $x = 3$; about the y-axis

29. $y = (x + 1)^2$, $x \geq 0$, $y = 16$; about the y-axis

30. $y = (x + 1)^2$, $x \leq 0$, $y = 16$; about the x-axis

31. $x = y^4 - 1$, the y-axis; about the y-axis

32. $y = x^4 - 1$, the x-axis; about the x-axis

33. $y = 4x$, $y = x^3$, $x \geq 0$; about the x-axis

34. $y = 2x + 1$, $y = x$, $x = 0$, $x = 3$; about the x-axis

35. $y = 1 - x$, $y = e^x$, $x = 1$; about the x-axis

36. $y = \cos x$, $y = \sin x$, $x = 0$, $x = \dfrac{\pi}{4}$; about the x-axis

37. $y = \csc x$, $y = 0$, $x = \dfrac{\pi}{2}$, $x = \dfrac{3\pi}{4}$; about the x-axis

38. $y = \sec x$, $y = 0$, $x = 0$, $x = \dfrac{\pi}{3}$; about the x-axis

In Problems 39–46, find the volume of the solid of revolution generated by revolving the region bounded by the graphs of the given equations about the indicated line.

39. $y = e^x$, $y = 0$, $x = 0$, $x = 2$; about $y = -1$

40. $y = \dfrac{1}{x}$, $y = 0$, $x = 1$, $x = 4$; about $y = 4$

41. $y = x^2$, the x-axis, $x = 1$; about $x = 1$

42. $y = x^3$, $x = 0$, $y = 1$; about $x = -1$

43. $y = \sqrt{x}$, the x-axis, $x = 4$; about $x = -4$

44. $y = \dfrac{1}{\sqrt{x}}$, the x-axis, $x = 1$, $x = 4$; about $x = 4$

45. $y = \dfrac{1}{x^2}$, $y = 0$, $x = 1$, $x = 4$; about $y = 4$

46. $y = \sqrt{x}$, $y = 0$, $0 \leq x \leq 4$; about $y = -4$

Applications and Extensions

47. Volume of a Solid of Revolution A region in the first quadrant is bounded by the x-axis and the graph of $y = kx - x^2$, where $k > 0$.

   (a) In terms of k, find the volume generated when the region is revolved around the x-axis.

   (b) In terms of k, find the volume generated when the region is revolved around the y-axis.

   (c) Find the number k for which the volumes found in parts (a) and (b) are equal.

48. Volume of a Solid of Revolution

   (a) Find all numbers b for which the graphs of $y = 2x + b$ and $y^2 = 4x$ intersect in two distinct points.

   (b) If $b = -4$, find the area enclosed by the graphs of $y = 2x - 4$ and $y^2 = 4x$.

(c) If $b = 0$, find the volume of the solid generated by revolving about the x-axis the region bounded by the graphs of $y = 2x$ and $y^2 = 4x$.

49. **Volume of a Solid of Revolution**   Find the volume of the solid of revolution generated by revolving the region bounded by the graphs of $y = \cos x$ and $y = 0$ from $x = 0$ to $x = \dfrac{\pi}{2}$ about the line $y = 1$. $\left[\textit{Hint: } \cos^2 x = \dfrac{1 + \cos(2x)}{2}.\right]$

50. **Volume of a Solid of Revolution**   Find the volume of the solid of revolution generated by revolving the region bounded by the graphs of $y = \cos x$ and $y = 0$ from $x = 0$ to $x = \dfrac{\pi}{2}$ about the line $y = -1$. (See the hint in Problem 49.)

Challenge Problems

51. **Volume of a Solid of Revolution**   The graph of the function $P(x) = kx^2$ is symmetric with respect to the y-axis and contains the points $(0, 0)$ and (b, e^{-b^2}), where $b > 0$.

   (a) Find k and write an equation for $P = P(x)$.

   (b) The region bounded by P, the y-axis, and the line $y = e^{-b^2}$ is revolved about the y-axis to form a solid. Find its volume.

   (c) For what number b is the volume of the solid in (b) a maximum? Justify your answer.

52. **Volume of a Solid of Revolution**   Find the volume of the solid generated by revolving the region bounded by the catenary $y = a \cosh\left(\dfrac{x}{a}\right) + b - a$, the x-axis, $x = 0$, and $x = 1$ about the x-axis. $\left[\textit{Hint: } \cosh^2 x = \dfrac{\cosh(2x) + 1}{2}.\right]$

6.3 Volume of a Solid of Revolution: Cylindrical Shells

OBJECTIVES *When you finish this section, you should be able to:*

1 Use the shell method to find the volume of a solid formed by revolving a region about the y-axis (p. 425)

2 Use the shell method to find the volume of a solid formed by revolving a region about the x-axis (p. 428)

3 Use the shell method to find the volume of a solid formed by revolving a region about a line parallel to a coordinate axis (p. 430)

NOTE Sewer lines and the water pipes in a house are cylindrical shells.

There are solids of revolution for which the volume is difficult to find using the disk or washer method. In these situations, the volume can often be found using *cylindrical shells*.

A **cylindrical shell** is the solid between two concentric cylinders, as shown in Figure 28. If the inner radius of the cylinder is r and the outer radius is R, the volume V of a cylindrical shell of height h is

$$\boxed{V = \pi R^2 h - \pi r^2 h}$$

That is, the volume of a cylindrical shell equals the volume of the larger cylinder, which has radius R, minus the volume of the smaller cylinder, which has radius r. It is convenient to write this formula as

$$V = \pi(R^2 - r^2)h = \pi(R + r)(R - r)h = 2\pi\left(\frac{R + r}{2}\right)h(R - r)$$

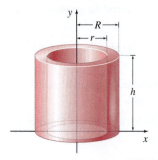

Figure 28 $V = \pi R^2 h - \pi r^2 h$.

$$\boxed{V = 2\pi\left(\frac{R + r}{2}\right)h(R - r)}$$

$$V = 2\pi \text{ (Average radius) (Height) (Thickness)}$$

1 Use the Shell Method to Find the Volume of a Solid Formed by Revolving a Region About the y-Axis

Suppose a function $y = f(x)$ is nonnegative and continuous on the closed interval $[a, b]$, where $a \geq 0$. We seek the volume V of the solid generated by revolving the region bounded by the graph of f, the x-axis, and the lines $x = a$ and $x = b$ about the y-axis. See Figure 29.

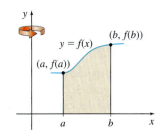

Figure 29

We begin by partitioning the interval $[a, b]$ into n subintervals:

$$[a, x_1], [x_1, x_2], \ldots, [x_{i-1}, x_i], \ldots, [x_{n-1}, b]$$

each of width $\Delta x = \dfrac{b - a}{n}$. We concentrate on the rectangle whose base is the subinterval $[x_{i-1}, x_i]$ and whose height is $f(u_i)$, where $u_i = \dfrac{x_{i-1} + x_i}{2}$ is the midpoint of the subinterval. See Figure 30(a). When this rectangle is revolved about the y-axis, it generates a cylindrical shell of average radius u_i, height $f(u_i)$, and thickness Δx, as shown in Figure 30(b). The volume V_i of this cylindrical shell is

$$V_i = 2\pi (\text{Average radius})(\text{Height})(\text{Thickness}) = 2\pi u_i f(u_i)\Delta x$$

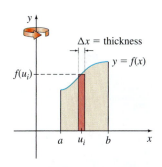

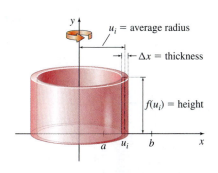

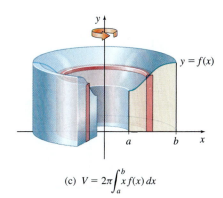

(a) The region to be revolved about the y-axis.

(b) Average radius $= u_i$
     Height $= f(u_i)$
     Thickness $= \Delta x$
     Volume $= 2\pi u_i\, f(u_i)\, \Delta x$

(c) $V = 2\pi \displaystyle\int_a^b x f(x)\, dx$

Figure 30

The sum of the volumes of the n cylindrical shells approximates the volume V of the solid generated by revolving the region bounded by the graph of $y = f(x)$, the x-axis, and the lines $x = a$ and $x = b$ about the y-axis. That is,

$$V \approx \sum_{i=1}^{n}[2\pi u_i f(u_i)\Delta x] = 2\pi \sum_{i=1}^{n}[u_i f(u_i)\Delta x]$$

As the number n of subintervals increases, the sums $2\pi \displaystyle\sum_{i=1}^{n}[u_i f(u_i)\Delta x]$ become better approximations to the volume V of the solid. These sums are Riemann sums, and since f is continuous on $[a, b]$, the limit is a definite integral. See Figure 30(c). When cylindrical shells are used to find the volume of a solid of revolution, we refer to the process as the **shell method**.

Volume of a Solid of Revolution About the y-Axis: the Shell Method

If $y = f(x)$ is a function that is continuous and nonnegative on the closed interval $[a, b]$, where $a \geq 0$, then the volume V of the solid generated by revolving the region bounded by the graph of f, the x-axis, and the lines $x = a$ and $x = b$ about the y-axis is

$$V = 2\pi \int_a^b x f(x)\, dx$$

It can be shown that the shell method and the washer method of Section 6.2 are equivalent; that is, they both give the same answer.* The advantage of having two equivalent, yet different, formulas is flexibility. There are times when one of the two methods is easier to use, as Example 1 illustrates.

EXAMPLE 1   Finding the Volume of a Solid: Revolving About the y-Axis

Find the volume V of the solid generated by revolving the region bounded by the graphs of $f(x) = x^2 + 2x$, the x-axis, and the line $x = 1$ about the y-axis.

Solution   Using the shell method: In the shell method, when a region is revolved about the y-axis, we partition the x-axis and use vertical shells. Figure 31(a) illustrates the region to be revolved and a typical rectangle of height $f(u_i)$ and thickness Δx that will become a shell with average radius u_i when it is revolved about the y-axis. The volume of a typical shell is $V_i = 2\pi(\text{Average radius})(\text{Height})(\text{Thickness}) = 2\pi u_i f(u_i)\Delta x$, as shown in Figure 31(b). Figure 31(c) illustrates the solid of revolution. The volume V of the solid of revolution is

$$V = 2\pi \int_0^1 x f(x)\,dx = 2\pi \int_0^1 [x(x^2 + 2x)]\,dx = 2\pi \int_0^1 (x^3 + 2x^2)\,dx$$

$$= 2\pi \left[\frac{x^4}{4} + \frac{2x^3}{3}\right]_0^1 = \frac{11\pi}{6} \text{ cubic units}$$

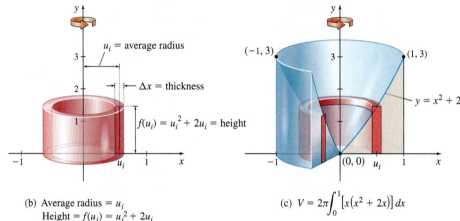

(a) The region to be revolved about the y-axis.

(b) Average radius $= u_i$
Height $= f(u_i) = u_i^2 + 2u_i$
Thickness $= \Delta x$
Volume $= 2\pi u_i (u_i^2 + 2u_i)\Delta x$

(c) $V = 2\pi \int_0^1 [x(x^2 + 2x)]\,dx$

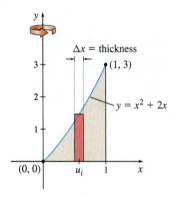

DF Figure 31 The shell method.

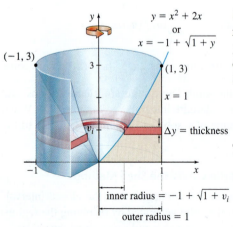

Outer radius $= 1$
Inner radius $= -1 + \sqrt{1 + v_i}$
Thickness of washer $= \Delta y$
Volume $= \pi[1^2 - (-1 + \sqrt{1 + v_i})^2]\Delta y$

Figure 32 The washer method.

Using the washer method: Using the washer method, a revolution about the y-axis requires integration with respect to y. This means we need to find the inverse function of $y = f(x)$. We treat $x^2 + 2x - y = 0$ as a quadratic equation in the variable x and use the quadratic formula with $a = 1$, $b = 2$, and $c = -y$ to obtain $x = g(y) = -1 \pm \sqrt{1 + y}$. Since $x \geq 0$, we use the $+$ sign.

See Figure 32. The volume of a typical washer is

$$V_i = \pi[(\text{Outer radius})^2 - (\text{Inner radius})^2] \times (\text{Thickness}).$$

The volume V of the solid of revolution is

$$V = \pi \int_0^3 [1^2 - (-1 + \sqrt{1+y})^2]\,dy = \pi \int_0^3 [1 - (1 - 2\sqrt{1+y} + 1 + y)]\,dy$$

$$= \pi \int_0^3 [2\sqrt{1+y} - 1 - y]\,dy = \pi \int_0^3 (2\sqrt{1+y})\,dy - \pi \int_0^3 (1 + y)\,dy$$

---

*This topic is discussed in detail in an article by Charles A. Cable (February 1984), "The Disk and Shell Method," *American Mathematical Monthly*, 91(2), 139.

The two integrals are found as follows:

- $\pi \displaystyle\int_0^3 (2\sqrt{1+y})\,dy = 2\pi \int_1^4 u^{1/2}\,du = 2\pi \left[\dfrac{u^{3/2}}{\dfrac{3}{2}} \right]_1^4 = \dfrac{28\pi}{3}$

  $\uparrow$
  Let $u = 1 + y$;
  then $du = dy$

- $\pi \displaystyle\int_0^3 (1+y)\,dy = \pi \left[y + \dfrac{y^2}{2} \right]_0^3 = \dfrac{15\pi}{2}$

The volume V is

$$V = \dfrac{28\pi}{3} - \dfrac{15\pi}{2} = \dfrac{11\pi}{6} \text{ cubic units}$$ ■

NOW WORK Problem 5.

Example 1 gives a clue to when the shell method is preferable to the washer method. When it is difficult, or impossible, to solve $y = f(x)$ for x, we use the shell method. For example, if the function in Example 1 had been $y = f(x) = x^5 + x^2 + 1$, we would not have been able to solve for x, so the practical choice is the shell method.

The next example illustrates the importance of sketching a graph before using a formula. Notice the limits of integration and how we determined them when we use the washer method.

EXAMPLE 2   Finding the Volume of a Solid: Revolving About the y-Axis

Find the volume V of the solid generated by revolving the region bounded by the graphs of $f(x) = x^2$ and $g(x) = 12 - x$ to the right of $x = 1$ about the y-axis.

Solution   *Using the shell method:* Figure 33(a) shows the graph of the region to be revolved and a typical rectangle.

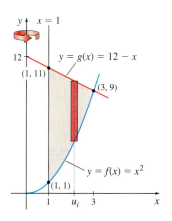

(a) Region to be revolved about the y-axis.

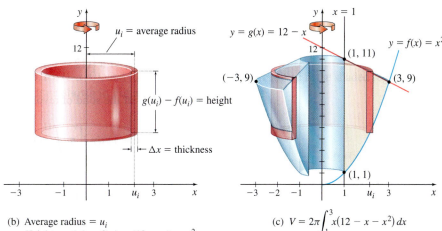

(b) Average radius $= u_i$
Height $= g(u_i) - f(u_i) = (12 - u_i) - u_i^2$
Thickness of shell $= \Delta x$
Volume $= 2\pi u_i (12 - u_i - u_i^2)\,\Delta x$

(c) $V = 2\pi \displaystyle\int_1^3 x(12 - x - x^2)\,dx$

Figure 33 The shell method.

As shown in Figure 33(b), in the shell method, we partition the x-axis and use vertical shells. A typical shell has height $h_i = g(u_i) - f(u_i) = (12 - u_i) - u_i^2 = 12 - u_i - u_i^2$, and volume $V_i = 2\pi u_i (12 - u_i - u_i^2)\Delta x$. Figure 33(c) shows the solid of revolution. Notice that the integration takes place from $x = 1$ to $x = 3$. The volume V of the solid of revolution is

$$V = 2\pi \int_1^3 x(12 - x - x^2)\,dx = 2\pi \int_1^3 (12x - x^2 - x^3)\,dx = 2\pi \left[6x^2 - \dfrac{x^3}{3} - \dfrac{x^4}{4} \right]_1^3$$

$$= 2\pi \left[\left(54 - 9 - \dfrac{81}{4} \right) - \left(6 - \dfrac{1}{3} - \dfrac{1}{4} \right) \right] = \dfrac{116\pi}{3} \text{ cubic units}$$

respect to y, so we express the equation of the ellipse as

$$x = g(y) = \frac{a}{b}\sqrt{b^2 - y^2}$$

The volume of a typical shell is

$$V_i = 2\pi \text{ (Average radius)(Height)(Thickness)} = 2\pi v_i g(v_i) \Delta y$$

See Figure 36(b).

Figure 36(c) shows the solid of revolution. The volume V of the solid of revolution is

$$V = 2\pi \int_0^b y\, g(y)\, dy = 2\pi \int_0^b y\left(\frac{a}{b}\sqrt{b^2 - y^2}\right) dy = 2\pi \frac{a}{b}\left(-\frac{1}{2}\right)\int_{b^2}^0 \sqrt{u}\, du$$

$$\uparrow$$
Let $u = b^2 - y^2$;
then $du = -2y\, dy$

$$= \frac{\pi a}{b}\int_0^{b^2} u^{1/2}\, du = \frac{\pi a}{b}\left[\frac{u^{3/2}}{\frac{3}{2}}\right]_0^{b^2} = \frac{2\pi a}{3b}(b^3) = \frac{2\pi ab^2}{3} \text{ cubic units} \qquad \blacksquare$$

NOW WORK Problem 33.

Using symmetry, the volume V of the ellipsoid generated by revolving $y = \frac{b}{a}\sqrt{a^2 - x^2}$, $-a \le x \le a$, about the x-axis is twice the volume found in Example 3, namely $V = \frac{4}{3}\pi ab^2$.

If, in Example 3, $a = b$, then the solid generated is a hemisphere whose volume is $\frac{2\pi a^3}{3}$ cubic units. By symmetry, the volume of a sphere of radius R, $(a = b = R)$, is $\frac{4\pi R^3}{3}$ cubic units.

3 Use the Shell Method to Find the Volume of a Solid Formed by Revolving a Region About a Line Parallel to a Coordinate Axis

EXAMPLE 4   Using the Shell Method: Revolving About the Line $x = 2$

Find the volume V of the solid of revolution generated by revolving the region bounded by the graph of $y = 2x - 2x^2$ and the x-axis about the line $x = 2$.

Solution The region bounded by the graph of $y = 2x - 2x^2$ and the x-axis is illustrated in Figure 37(a). A typical shell formed by revolving the region about the line $x = 2$, as shown in Figure 37(b), has an average radius of $2 - u_i$, height $f(u_i) = 2u_i - 2u_i^2$, and thickness Δx. The solid of revolution is depicted in Figure 37(c).

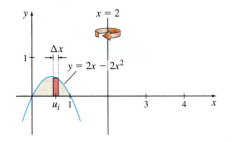

(a) Region to be revolved about the line $x = 2$.

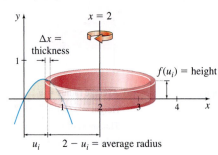

(b) Average radius $= 2 - u_i$
   Height $= f(u_i) = 2u_i - 2u_i^2$
   Thickness of shell $= \Delta x$
   Volume $= 2\pi(2 - u_i)f(u_i)\,\Delta x$

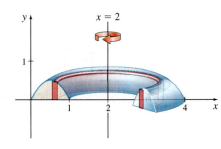

(c) $V = 2\pi \int_0^1 (2 - x)(2x - 2x^2)\, dx$

DF Figure 37

The volume V of the solid is

$$V = 2\pi \int_0^1 (2-x)(2x-2x^2)\,dx = 4\pi \int_0^1 (x^3 - 3x^2 + 2x)\,dx$$

$$= 4\pi \left[\frac{x^4}{4} - x^3 + x^2\right]_0^1 = \pi \text{ cubic units}$$ ■

NOW WORK Problem 17.

Summary The Volume V of a Solid of Revolution

	Washer Method	Shell Method
• **Revolution about the x-axis**	Partition the x-axis; use vertical washers.	Partition the y-axis; use horizontal shells.
• **Revolution about the y-axis**	Partition the y-axis; use horizontal washers.	Partition the x-axis; use vertical shells.

6.3 Assess Your Understanding

Concepts and Vocabulary

1. *True or False* In using the shell method to find the volume of a solid revolved about the x-axis, the integration occurs with respect to x.

2. *True or False* The volume of a cylindrical shell of outer radius R, inner radius r, and height h is given by $V = \pi r^2 h$.

3. *True or False* The volume of a solid of revolution can be found using either washers or cylindrical shells only if the region is revolved about the y-axis.

4. *True or False* If $y = f(x)$ is a function that is continuous and nonnegative on the closed interval $[a, b]$, $a \geq 0$, then the volume V of the solid generated by revolving the region bounded by the graph of f and the x-axis from $x = a$ to $x = b$ about the y-axis is $V = 2\pi \int_a^b x f(x)\,dx$.

Skill Building

In Problems 5–16, use the shell method to find the volume of the solid of revolution generated by revolving the region bounded by the graphs of the given equations about the indicated axis.

5. $y = x^2 + 1$, the x-axis, $0 \leq x \leq 1$; about the y-axis

6. $y = x^3$, $y = x^2$; about the y-axis

7. $y = \sqrt{x}$, $y = x^2$; about the y-axis

8. $y = \dfrac{1}{x}$, the x-axis, $x = 1$, $x = 4$; about the y-axis

9. $y = x^3$, the y-axis, $y = 8$; about the x-axis

10. $y = \sqrt{x}$, the y-axis, $y = 2$; about the x-axis

11. $x = \sqrt{y}$, the y-axis, $y = 1$; about the x-axis

12. $x = 4\sqrt{y}$, the y-axis, $y = 4$; about the x-axis

13. $y = x$, $y = x^2$; about the x-axis

14. $y = x$, $y = x^3$; in the first quadrant; about the x-axis

15. $y = e^{-x^2}$, and the x-axis, from $x = 0$ to $x = 2$; about the y-axis

16. $y = x^3 + x$ and the x-axis, $0 \leq x \leq 1$; about the y-axis

In Problems 17–22, use the shell method to find the volume of the solid of revolution generated by revolving the region bounded by the graphs of the given equations about the indicated line.

17. $y = x^2$, $y = 4x - x^2$; about $x = 4$

18. $y = x^2$, $y = 4x - x^2$; about $x = 3$

19. $y = x^2$, $y = 0$, $x = 1$, $x = 2$; about $x = 1$

20. $y = x^2$, $y = 0$, $x = 1$, $x = 2$; about $x = -2$

21. $x = y - y^2$, the y-axis; about $y = -1$

22. $x = y - y^2$, the y-axis; about $y = 1$

In Problems 23–26, use either the shell method or the disk/washer method to find the volume of the solid of revolution generated by revolving the shaded region in each graph:

(a) about the x-axis.
(b) about the y-axis.
(c) Explain why you chose the method you used.

23.

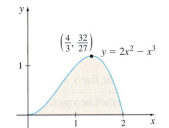

24.

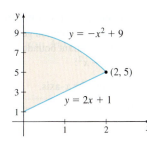

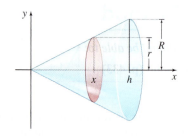

Figure 39

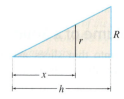

Figure 40

NEED TO REVIEW? Similar triangles are discussed in Appendix A.2, pp. A-13 to A-14.

NEED TO REVIEW? Similar triangles are discussed in Appendix A.2, pp. A-13 to A-14.

NOTE A right circular cone is a solid of revolution, so the formula for its volume also can be verified using the disk method.

EXAMPLE 1 Using the Slicing Method to Find the Volume of a Cone

Use the slicing method to verify that the volume of a right circular cone having radius R and height h is $V = \dfrac{1}{3}\pi R^2 h$.

Solution We position the cone with its vertex at the origin and its axis on the x-axis, as shown in Figure 39.

The cone extends from $x = 0$ to $x = h$. The cross section at any number x is a circle. To obtain its area A, we need the radius r of the circle. Embedded in Figure 39 are two similar triangles, one with sides x and r; the other with sides h and R, as shown in Figure 40. Because these triangles are similar (AAA), corresponding sides are in proportion. That is,

$$\frac{r}{x} = \frac{R}{h}$$

$$r = \frac{R}{h}x$$

So, r is a function of x and the area A of the circular cross section is

$$A = A(x) = \pi [r(x)]^2 = \pi \left(\frac{R}{h}x\right)^2 = \frac{\pi R^2}{h^2}x^2$$

Since A is a continuous function of x (where the slice was made), we can apply the slicing method. The volume V of the right circular cone is

$$V = \int_a^b A(x)\,dx = \int_0^h \frac{\pi R^2}{h^2}x^2\,dx = \frac{\pi R^2}{h^2}\int_0^h x^2\,dx$$

$$= \frac{\pi R^2}{h^2}\left[\frac{x^3}{3}\right]_0^h = \frac{\pi R^2 h}{3} \quad \text{cubic units} \qquad \blacksquare$$

NOW WORK Problem 13.

In Example 1, the cone was positioned so that its axis coincided with the x-axis and its vertex was at the origin. The way in which a solid is positioned relative to the x-axis is important if the area A of the slice is to be easily found, expressed as a function of x, and integrated.

EXAMPLE 2 Using the Slicing Method to Find the Volume of a Solid

A solid has a circular base of radius 3 units. Find the volume V of the solid if every plane cross section that is perpendicular to a fixed diameter is an equilateral triangle.

Solution Position the circular base so that its center is at the origin, and the fixed diameter is along the x-axis. See Figure 41(a). Then the equation of the circular base is $x^2 + y^2 = 9$. Each cross section of the solid is an equilateral triangle with sides $= 2y$, height h, and area $A = \sqrt{3}y^2$. See Figure 41(b). Since $y^2 = 9 - x^2$, the volume V_i of a typical slice is

$$V_i = (\text{Area of the cross section})(\text{Thickness}) = A(x_i)\,\Delta x = \sqrt{3}\left(9 - x_i^2\right)\Delta x$$

as shown in Figure 41(c).

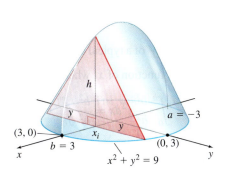
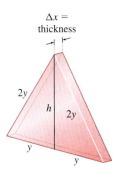

(a) Solid placed with the center of its circular base at the origin and its fixed diameter on the x-axis.

(b) Cross section:
$$h^2 + y^2 = 4y^2$$
$$h^2 = 3y^2$$
$$h = \sqrt{3}y$$
Area $= \frac{1}{2}$(Base)(Height) $= \sqrt{3}y^2$

(c) Slice:
$$V = (\text{Area})(\text{Thickness})$$
$$= \sqrt{3}y^2\,\Delta x = \sqrt{3}\left(9 - x^2\right)\Delta x$$

DF **Figure 41**

The volume V of the solid is

$$V = \int_a^b A(x)\,dx = \int_{-3}^{3} \sqrt{3}(9 - x^2)\,dx = 2\sqrt{3}\int_0^3 (9 - x^2)\,dx = 2\sqrt{3}\left[9x - \frac{x^3}{3}\right]_0^3$$

↑ The integrand is an even function.

NEED TO REVIEW? Even and odd functions are discussed in Section P.1, pp. 9–10, and the integrals of even and odd functions are discussed in Section 5.6, pp. 393–394.

$$= 36\sqrt{3} \text{ cubic units} \qquad\blacksquare$$

NOW WORK Problem 3.

Using the Slicing Method to Find the Volume
EXAMPLE 3 of a Pyramid

Find the volume V of a pyramid of height h with a square base, each side of length b.

Solution We position the pyramid with its vertex at the origin and its axis along the positive x-axis. Then the area A of a typical cross section at x is a square. Let s denote the length of the side of the square at x. See Figure 42(a). We form two triangles: one with height x and side s, the other with height h and side b. These triangles are similar (*AAA*), as shown in Figure 42(b). Then we have

$$\frac{x}{s} = \frac{h}{b} \qquad \text{or} \qquad s = \frac{b}{h}x$$

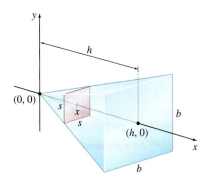
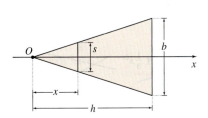

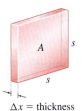

(a) Pyramid placed symmetric to the x-axis and with its vertex at the origin.

(b) Similar triangles

(c) Cross section at x:
Area $= A = s^2 = \frac{b^2}{h^2}x^2$

(d) Volume $= V = A\Delta x = \frac{b^2}{h^2}x^2\Delta x$

Figure 42

along a vertical line through the center of the cylinder, find the volume of the wedge removed. (*Hint:* Vertical cross sections of the wedge are right triangles.)

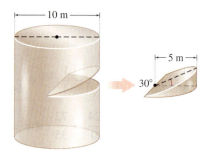

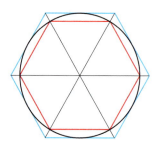

⎰⎱ **17. Volume of a Solid** Find the volume of the cylindrical solid with a bulge in the middle if slices taken perpendicular to the x-axis are circles whose diameters extend from the graph of $y = e^{-x^2}$ to the graph of $y = -e^{-x^2}$, $-1 \leq x \leq 1$.

18. In your own words, explain why the disk method is a special case of the slicing method.

Challenge Problems

19. Volume of a Bore A hole of radius 2 centimeters is bored completely through a solid metal sphere of radius 5 cm. If the axis of the hole passes through the center of the sphere, find the volume of the metal removed by the drilling. See the figure.

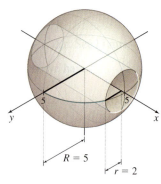

20. Volume The axes of two pipes of equal radii r intersect at right angles. Find their common volume.

21. Volume of a Cone Find the volume of a cone with height h and an elliptical base whose major axis has length $2a$ and minor axis has length $2b$. (*Hint:* The area of this ellipse is πab.)

22. Volume of a Solid Find the volume of a parallelepiped with edge lengths a, b, and c, where the edges having lengths a and b make an acute angle θ with each other, and the edge of length c makes an acute angle of ϕ with the diagonal of the parallelogram formed by a and b.

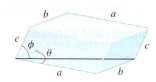

6.5 Arc Length

OBJECTIVES *When you finish this section, you should be able to:*

1 Find the arc length of the graph of a function $y = f(x)$ (p. 439)

2 Find the arc length of the graph of a function using a partition of the y-axis (p. 441)

We have already seen that a definite integral can be used to find the area of a plane region and the volume of certain solids. In this section, we will see that a definite integral can be used to find the length of a graph.

The idea of finding the length of a graph had its beginning with Archimedes. Ancient people knew that the ratio of the circumference C of any circle to its diameter d equaled the constant that we call π. That is, $\dfrac{C}{d} = \pi$. But Archimedes is credited as being the first person to analytically investigate the numerical value of π.* He drew a circle of diameter 1 unit, then inscribed (and circumscribed) the circle with regular polygons, and computed the perimeters of the polygons. He began with two hexagons (6 sides) and then two dodecagons (12 sides). See Figure 45. He continued until he had two polygons, each with 96 sides. He knew that the circumference C of the circle was larger than the perimeter of the inscribed polygon and smaller than the circumscribed polygon. In this way, he proved that $3\dfrac{10}{71} < \pi < 3\dfrac{1}{7}$. In essence, Archimedes approximated the length of the circle by representing it by smaller and smaller line segments and summing the lengths of the line segments.

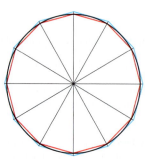

(a) Inscribed and circumscribed hexagons.

(b) Inscribed and circumscribed dodecagons.

Figure 45

---

*Backmann, Peter, **The History of** π. St Martin's Press, N.Y. 1971 p. 62.

ORIGINS Archimedes (287–212 BC) was a Greek astronomer, physicist, engineer, and mathematician. His contributions include the Archimedes screw, which is still used to lift water or grain, and the Archimedes principle, which states that the weight of an object can be determined by measuring the volume of water it displaces. Archimedes used the principle to measure the amount of gold in the king's crown; modern-day cooks can use it to measure the amount of butter to add to a recipe. Archimedes is also credited with using the Method of Exhaustion to find the area under a parabola and to approximate the value of π.

Our approach to finding the length of the graph of a function is similar to that used by Archimedes and follows the ideas we used to find area and volume.

1 Find the Arc Length of the Graph of a Function $y = f(x)$

Suppose we want to find the length of the graph of a function $y = f(x)$ from $x = a$ to $x = b$. We assume the derivative f' of f is continuous on some interval containing a and b.* See Figure 46.

We begin by partitioning the closed interval $[a, b]$ into n subintervals:

$$[a, x_1], [x_1, x_2], \ldots, [x_{i-1}, x_i], \ldots, [x_{n-1}, b] \qquad x_0 = a \quad x_n = b$$

each of width $\Delta x = \dfrac{b - a}{n}$. Corresponding to each number $a, x_1, x_2, \ldots, x_{i-1}, x_i, \ldots,$ x_{n-1}, b in the partition, there are points $P_0, P_1, P_2, \ldots, P_{i-1}, P_i, \ldots, P_{n-1}, P_n$ on the graph of f, as illustrated in Figure 47.

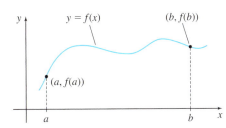

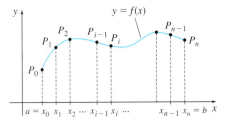

Figure 46 f' is continuous on an interval containing a and b.

Figure 47

When each point P_{i-1} is connected to the next point P_i by a line segment, the sum L of the lengths of the line segments approximates the length of the graph of $y = f(x)$ from $x = a$ to $x = b$. The sum L is given by

$$L = d(P_0, P_1) + d(P_1, P_2) + \cdots + d(P_{n-1}, P_n) = \sum_{i=1}^{n} d(P_{i-1}, P_i)$$

where $d(P_{i-1}, P_i)$ denotes the length of the line segment joining point P_{i-1} to point P_i. See Figure 48.

NEED TO REVIEW? The distance formula is discussed in Appendix A.3, p. A-16.

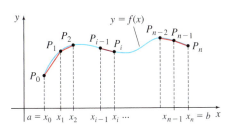

Figure 48 After the interval $[a, b]$ is partitioned, connect consecutive points with line segments.

Using the distance formula, the length of the ith line segment is

$$d(P_{i-1}, P_i) = \sqrt{(x_i - x_{i-1})^2 + (y_i - y_{i-1})^2} = \sqrt{(\Delta x)^2 + (\Delta y_i)^2}$$

$$= \sqrt{\frac{(\Delta x)^2 + (\Delta y_i)^2}{(\Delta x)^2}} \cdot \underset{\underset{\Delta x > 0}{\uparrow}}{\sqrt{(\Delta x)^2}} = \sqrt{1 + \left(\frac{\Delta y_i}{\Delta x}\right)^2}\, \Delta x$$

where $\Delta x = \dfrac{b - a}{n}$ and $\Delta y_i = y_i - y_{i-1} = f(x_i) - f(x_{i-1})$. Then the sum of the lengths of the line segments is

$$L = \sum_{i=1}^{n} \left[\sqrt{1 + \left(\frac{\Delta y_i}{\Delta x}\right)^2}\, \Delta x \right]$$

This sum is not a Riemann sum. To put this sum in the form of a Riemann sum, we need to express $\sqrt{1 + \left(\dfrac{\Delta y_i}{\Delta x}\right)^2}$ in terms of u_i, $x_{i-1} \le u_i \le x_i$. To do this, we first observe that the function f has a derivative f' that is continuous on an interval

---

*We will discuss shortly the reason for this restriction.

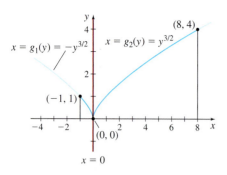

Figure 52 $f(x) = x^{2/3}, -1 \leq x \leq 8$.

From Figure 52, we observe that the length of the graph of f from $x = -1$ to $x = 8$ is the same as the length of the graph of $x = g_1(y) = -y^{3/2}$ from $y = 0$ to $y = 1$ plus the length of the graph of $x = g_2(y) = y^{3/2}$ from $y = 0$ to $y = 4$.

We investigate $x = g_1(y) = -y^{3/2}$ first. Since $g_1'(y) = -\dfrac{3}{2}y^{1/2}$ is continuous for all $y \geq 0$, we can use arc length formula (2) to find the arc length s_1 of g_1 from $y = 0$ to $y = 1$.

$$s_1 = \int_0^1 \sqrt{1 + [g_1'(y)]^2}\,dy \underset{\underset{g_1'(y) = -\frac{3}{2}y^{1/2}}{\uparrow}}{=} \int_0^1 \sqrt{1 + \left(-\frac{3}{2}y^{1/2}\right)^2}\,dy$$

$$= \int_0^1 \sqrt{1 + \frac{9}{4}y}\,dy = \frac{1}{2}\int_0^1 \sqrt{4 + 9y}\,dy$$

We use the substitution $u = 4 + 9y$. Then $du = 9\,dy$ or, equivalently, $dy = \dfrac{du}{9}$. The limits of integration are $u = 4$ when $y = 0$, and $u = 13$ when $y = 1$.

$$s_1 = \frac{1}{2}\int_4^{13}\sqrt{u}\,\frac{du}{9} = \frac{1}{18}\left[\frac{u^{3/2}}{\frac{3}{2}}\right]_4^{13} = \frac{1}{27}\left(13\sqrt{13} - 8\right)$$

Now we investigate $x = g_2(y)$. Since $g_2'(y) = \dfrac{3}{2}y^{1/2}$ is continuous for all $y \geq 0$, we can use arc length formula (2) to find the arc length s_2 of g_2 from $y = 0$ to $y = 4$.

$$s_2 = \int_0^4 \sqrt{1 + [g_2'(y)]^2}\,dy = \int_0^4 \sqrt{1 + \left(\frac{3}{2}y^{1/2}\right)^2}\,dy$$

$$= \int_0^4 \sqrt{1 + \frac{9}{4}y}\,dy = \frac{1}{2}\int_0^4 \sqrt{4 + 9y}\,dy \qquad \begin{array}{l}\text{Let } u = 4 + 9y \\ du = 9\,dy\end{array}$$

$$= \frac{1}{2}\int_4^{40}\sqrt{u}\,\frac{du}{9} = \frac{1}{18}\left[\frac{u^{3/2}}{\frac{3}{2}}\right]_4^{40} = \frac{1}{27}\left(80\sqrt{10} - 8\right)$$

The arc length s of $y = f(x) = x^{2/3}$ from $x = -1$ to $x = 8$ is the sum

$$s = s_1 + s_2 = \frac{1}{27}\left(13\sqrt{13} - 8\right) + \frac{1}{27}\left(80\sqrt{10} - 8\right)$$

$$= \frac{1}{27}\left(80\sqrt{10} + 13\sqrt{13} - 16\right)$$ ∎

NOW WORK Problem 23.

6.5 Assess Your Understanding

Concepts and Vocabulary

1. *True or False* If a function f has a derivative that is continuous on an interval containing a and b, the length s of the graph of $y = f(x)$ from $x = a$ to $x = b$ is given by the formula $s = \int_a^b \sqrt{1 + [f'(x)]^2}\,dx$.

2. *True or False* If the derivative of a function $y = f(x)$ is not continuous at some number in the interval $[a, b]$, its arc length from $x = a$ to $x = b$ can sometimes be found by partitioning the y-axis.

Skill Building

In Problems 3–6, use the arc length formula to find the length of each line between the points indicated. Verify your answer by using the distance formula.

3. $y = 3x - 1$, from $(1, 2)$ to $(3, 8)$

4. $y = -4x + 1$, from $(-1, 5)$ to $(1, -3)$

5. $2x - 3y + 4 = 0$, from $(1, 2)$ to $(4, 4)$

6. $3x + 4y - 12 = 0$, from $(0, 3)$ to $(4, 0)$

1. = NOW WORK problem   = Graphing technology recommended   CAS = Computer Algebra System recommended

In Problems 7–22, find the arc length of each graph by partitioning the x-axis.

7. $y = x^{2/3} + 1$, from $x = 1$ to $x = 8$

8. $y = x^{2/3} + 6$, from $x = 1$ to $x = 8$

9. $y = x^{3/2}$, from $x = 0$ to $x = 4$

10. $y = x^{3/2} + 4$, from $x = 1$ to $x = 4$

11. $9y^2 = 4x^3$, from $x = 0$ to $x = 1$; $y \geq 0$

12. $y = \dfrac{x^3}{6} + \dfrac{1}{2x}$, from $x = 1$ to $x = 3$

13. $y = \dfrac{2}{3}(x^2 + 1)^{3/2}$, from $x = 1$ to $x = 4$

14. $y = \dfrac{1}{3}(x^2 + 2)^{3/2}$, from $x = 2$ to $x = 4$

15. $y = \dfrac{2}{9}\sqrt{3}(3x^2 + 1)^{3/2}$, from $x = -1$ to $x = 2$

16. $y = (1 - x^{2/3})^{3/2}$, from $x = \dfrac{1}{8}$ to $x = 1$

17. $8y = x^4 + \dfrac{2}{x^2}$, from $x = 1$ to $x = 2$

18. $9y^2 = 4(1 + x^2)^3$, from $x = 0$ to $x = 2\sqrt{2}$; $y \geq 0$

19. $y = \ln(\sin x)$, from $x = \dfrac{\pi}{6}$ to $x = \dfrac{\pi}{3}$

20. $y = \ln(\cos x)$, from $x = \dfrac{\pi}{6}$ to $x = \dfrac{\pi}{3}$

21. $(x + 1)^3 = 4y^2$, from $x = -1$ to $x = 16$, $y \geq 0$

22. $y = x^{3/2} + 8$, from $x = 0$ to $x = 4$

In Problems 23–26, find the arc length of each graph by partitioning the y-axis.

23. $y = x^{2/3}$, from $x = 0$ to $x = 1$

24. $y = x^{2/3}$, from $x = -1$ to $x = 0$

25. $(x + 1)^2 = 4y^3$, from $y = 0$ to $y = 1$; $x \geq -1$

26. $x = \dfrac{2}{3}(y - 5)^{3/2}$, from $y = 5$ to $y = 6$

In Problems 27–32, (a) use the arc length formula (1) to set up the integral for arc length.
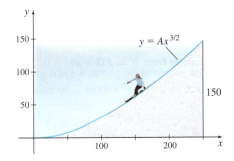*(b) If you have access to a graphing utility or a CAS, find the length. Do not attempt to integrate by hand.*

27. $y = x^2$, from $x = 0$ to $x = 2$

28. $x = y^2$, from $y = 1$ to $y = 3$

29. $y = \sqrt{25 - x^2}$, from $x = 0$ to $x = 4$

30. $x = \sqrt{4 - y^2}$, from $y = 0$ to $y = 1$

31. $y = \sin x$, from $x = 0$ to $x = \dfrac{\pi}{2}$

32. $x = y + \ln y$, from $y = 1$ to $y = 4$

Applications and Extensions

33. Length of a Hypocycloid   Find the total length of the hypocycloid $x^{2/3} + y^{2/3} = a^{2/3}$, $a > 0$.

34. Distance Along a Curved Path   Find the distance between $(1, 1)$ and $(3, 3\sqrt{3})$ along $y^2 = x^3$.

35. Perimeter   Find the perimeter of the region bounded by the graphs of $y^3 = x^2$ and $y = x$.

36. Perimeter   Find the perimeter of the region bounded by the graphs of $y = 3(x - 1)^{3/2}$ and $y = 3(x - 1)$.

37. Length of a Graph   Find the length of $6xy = y^4 + 3$ from $y = 1$ to $y = 2$.

38. Length of a Graph   Find the length of the hyperbolic function $y = \cosh x$ from $(0, 1)$ to $(2, \cosh 2)$.

39. Length of an Graph   Find the length of $y = \ln(\csc x)$ from $x = \dfrac{\pi}{4}$ to $x = \dfrac{\pi}{2}$.

40. Length of an Elliptical Arc   Set up the integral for the arc length of the ellipse $\dfrac{x^2}{a^2} + \dfrac{y^2}{b^2} = 1$ from $x = 0$ to $x = \dfrac{a}{2}$ in quadrant I. This integral, which is approximated by numerical techniques (see Chapter 7), is called an **elliptical integral of the second kind**.

41. Length of a Circular Arc   In each case below, $P_1 = (x_1, y_1)$ and $P_2 = (x_2, y_2)$ are points on the circle $x^2 + y^2 = 1$ that do not lie on a coordinate axis. Express the length of the counterclockwise arc $P_1 P_2$ in terms of integrals of the form

$$\int_u^v \frac{1}{\sqrt{1 - t^2}}\, dt, \qquad -1 < u < v < 1$$

(a) when P_1 is in quadrant I and P_2 is in quadrant II.

(b) when P_1 and P_2 are both in quadrant III and $y_1 < y_2$.

(c) when P_1 is in quadrant II and P_2 is in quadrant IV.

42. Modeling a Ski Slope   A ski slope is built on a mountainside and curves upward from ground level to a height h. The shape of the ski slope is modeled by the equation $y = Ax^{3/2}$, where x is the horizontal distance from the bottom of the ski slope measured along the base of the mountain and y is the vertical height of the ski slope at the distance x.

(a) Find an expression, in terms of A and h, for the length of the ski slope.

(b) Find A if the ski slope is 150 m high and has a horizontal distance of 250 m along the base.

(c) If a skier skis directly downhill from the top of the ski slope to the bottom, how far does she travel?

(d) Describe a simple way to check if the distance obtained in part (c) is reasonable.

Consider a freely hanging cable or chain that is being lengthened or shortened. The weight of the cable is a function of its length. As the cable is let out, its weight increases proportionally to its length, and when it is pulled in, it decreases proportionally to its length.

EXAMPLE 1 Finding the Work Done in Pulling in a Rope

A 60-ft rope weighing 8 lb per linear foot is used for mooring a cruise ship. See Figure 54. As the ship prepares to leave port, the rope is released, and it hangs freely over the side of the ship. How much work is done by the deckhand who winds in the rope?

Solution We position an x-axis parallel to the side of the ship with the origin O of the axis even with the bottom of the rope and $x = 60$ even with the ship's deck. The work done by the deckhand depends on the weight of the rope and the length of rope hanging over the edge.

Partition the interval $[0, 60]$ into n subintervals, each of length $\Delta x = \dfrac{60}{n}$, and choose a number u_i in each subinterval. Now think of the rope as n short segments, each of length Δx. Then

$$\text{Weight of the } i\text{th segment} = 8\Delta x \text{ lb}$$
$$\text{Distance the } i\text{th segment is lifted} = (60 - u_i) \text{ ft}$$
$$\text{Work done in lifting the } i\text{th segment} = 8(60 - u_i)\Delta x \text{ ft lb} \qquad W = Fx$$

The work W required to lift the 60 ft of rope is

$$W = \int_0^{60} 8(60 - x)\, dx = \left[480x - 4x^2\right]_0^{60} = 14{,}400 \text{ ft lb}$$

Figure 54

NOW WORK Problem 11.

2 Find the Work Done by a Spring Force

A common example of the work done by a variable force is found in stretching or compressing a spring that is fixed at one end. A spring is said to be in **equilibrium** when it is neither extended nor compressed. The **spring force** F needed to extend or compress a spring depends on the *stiffness* of the spring, which is measured by its **spring constant** k, a positive real number. Since a spring always attempts to return to equilibrium, a spring force F is often called a **restoring force**. A spring force F varies with the distance x that the free end of the spring is moved from its equilibrium length and obeys **Hooke's Law**:

$$\boxed{F(x) = -kx}$$

where k is the spring constant measured in newtons per meter in SI units or pounds per foot in U.S. units. The minus sign in Hooke's Law indicates that the direction of a spring force is opposite from the direction of the displacement.

As Figure 55 illustrates, the distance x in Hooke's Law is measured from the equilibrium, or unstretched, position of the spring. This distance x is positive if the spring is stretched from its equilibrium position and is negative if the spring is compressed from its equilibrium position. As a result, a spring force F is negative if the spring is stretched and F is positive if the spring is compressed.

In applied problems involving springs, the value of k, the spring constant, is often unknown. When we know information about how the spring behaves, the value of k can be found. The next example illustrates this.

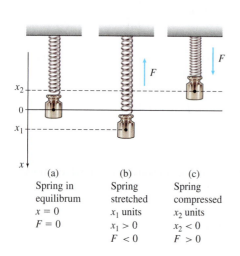

Figure 55

EXAMPLE 2   **Analyzing a Spring Force**

Suppose a spring in equilibrium is 0.8 m long and a spring force of 2 N stretches the spring to a length of 1.2 m.

(a) Find the spring constant k and the spring force F.

(b) What spring force is required to stretch the spring to a length of 3 m?

(c) How much work is done by the spring force in stretching it from equilibrium to 3 m?

Solution We position an axis parallel to the spring and place the origin at the free end of the spring in equilibrium, as in Figure 56.

(a) When the spring is stretched to a length of 1.2 m, then $x = 0.4$. Using Hooke's Law, we get

$$-2 = -k(0.4) \qquad \text{Hooke's law: } F(x) = -kx; \; F = -2; \; x = 0.4$$

$$k = \frac{2}{0.4} = 5 \, \text{N/m}$$

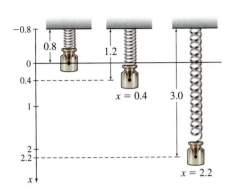

Figure 56

The spring constant is $k = 5$. The spring force F is $F = -kx = -5x$.

(b) The spring force F required to stretch the spring to a length of 3 m, that is, a distance $x = 3 - 0.8 = 2.2$ m from equilibrium, is

$$F = -5x = (-5)(2.2) = -11 \, \text{N}$$

(c) The work W done by the spring force F when stretching the spring from equilibrium $(x = 0)$ to $x = 2.2$ is

$$W = \int_0^{2.2} F(x)\, dx = -5 \int_0^{2.2} x \, dx = -5 \left[\frac{x^2}{2} \right]_0^{2.2} = -\frac{5}{2}(4.84) = -12.1 \, \text{J}$$

$$\underset{F(x) = -5x}{\uparrow}$$

■

EXAMPLE 3   **Finding the Work Done by a Spring Force**

Suppose an 0.8 m-long spring has a spring constant of $k = 5 \, \text{N/m}$.

(a) What spring force is required to compress the spring from its equilibrium position to a length of 0.5 m?

(b) How much work is done by the spring force when compressing the spring from equilibrium to a length of 0.5 m?

(c) How much work is done by the spring force when compressing the spring from 1.2 to 0.5 m?

(d) How much work is done by the spring force when compressing the spring from 1 to 0.6 m?

Solution We begin by positioning an axis parallel to the spring and placing the origin at the free end of the spring in equilibrium. See Figure 57.

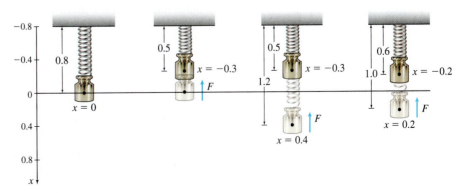

Figure 57

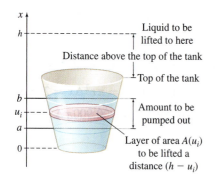

Figure 60

In general, to find the work required to pump liquid from a container, think of the liquid as thin layers of thickness Δx and area $A(x)$. If the liquid is to be lifted a height h above the bottom of the tank, the work required is

$$W = \int_a^b \rho g A(x)(h - x)\, dx$$

where ρ is the density of the liquid, g is the acceleration due to gravity, and the liquid to be lifted fills the container from $x = a$ to $x = b$. See Figure 60.

NOW WORK   Problem 17.

Application to Gravitational Force

In the mid-1600s, Isaac Newton proposed his theory of gravity. **Newton's Law of Universal Gravitation** states that every body in the universe attracts every other body; he called this force **gravitation**. Newton theorized that the gravitational force F attracting two bodies is proportional to the masses of both bodies and inversely proportional to the square of the distance x between them. That is,

$$F = F(x) = G\frac{m_1 m_2}{x^2}$$

where G is the **gravitational constant**. The widely accepted value of G,

$$G = 6.67 \times 10^{-11}\ \frac{\text{Nm}^2}{\text{kg}^2}$$

is a result of the work done by Henry Cavendish in 1798.

ORIGINS Henry Cavendish (1731–1810) was an English chemist and physicist. Known for his precision and accuracy, Cavendish calculated the density of Earth by measuring the force of the attraction between pairs of lead balls. An immediate result of his experiments was the first calculation of the gravitational constant G.
In a paper appearing in *Science* (2007, January 5), 315(5808), 74–77, the measurement was reevaluated using atom interferometry to be

$$G = 6.693 \times 10^{-11}\ \frac{\text{N m}^2}{\text{kg}^2}.$$

One of the conclusions of Newton's Law of Universal Gravitation is that the force required to move an object, say, a rocket of mass m kilograms that is at a point x meters above the center of Earth, is $F(x) = G\dfrac{Mm}{x^2}$, where M kilograms is the mass of Earth. Then the work required to move an object of mass m from the surface of Earth to a distance r meters from the center of Earth (where R meters is the radius of Earth) is

$$W = \int_R^r \left[G\frac{Mm}{x^2}\right] dx = -G\left[\frac{Mm}{x}\right]_R^r = G\frac{Mm}{R} - G\frac{Mm}{r}$$
$$= GMm\left(\frac{1}{R} - \frac{1}{r}\right)\ \text{J}$$

Physicists know that $GM = gR^2$, where $g \approx 9.8\ \text{m/s}^2$ is the acceleration due to gravity of Earth and $R \approx 6.37 \times 10^6$ m. If d is the distance the object is to be moved above the surface of Earth, then $r = R + d$. So, the work required to move an object of mass m to a distance d meters above the surface of Earth is

$$W = gRm\left(1 - \frac{R}{R + d}\right)$$

Observe that although the distance d may be extremely large, the work W required to move an object d meters will never be greater than $gRm \approx (6.24 \times 10^7)m$ J.

6.6 Assess Your Understanding

Concepts and Vocabulary

1. *True or False*   Work is the energy transferred to or from an object by a force acting on the object.

2. The work W done by a constant force F in moving an object a distance x along a straight line in the direction of F is _____.

3. A unit of work is called a _____ in SI units and a _____-_____ in the customary U.S. system of units.

4. The work W done by a continuously varying force $F = F(x)$ acting on an object, which moves the object along a straight line in the direction of F from $x = a$ to $x = b$, is given by the definite integral _____.

---

1. = NOW WORK problem   = Graphing technology recommended   CAS = Computer Algebra System recommended

5. A spring is said to be in _____ when it is neither extended nor compressed.

6. *True or False* The force F required to extend or compress a spring when it is either extended or compressed x units is $F = -kx$, where k is the spring constant.

7. *True or False* The mass density ρ of a fluid is defined as mass per unit volume (kg/m^3) and is a constant that depends on the type of fluid.

8. *True or False* Newton's Law of Universal Gravitation affirms that every body in the universe attracts every other body, and that the force F attracting two bodies is proportional to the product of the masses of both bodies and inversely proportional to the square of the distance between them.

Skill Building

9. How much work is done by a variable force $F(x) = (40 - x)$ N that moves an object along a straight line in the direction of F from $x = 5$ m to $x = 20$ m?

10. How much work is done by a variable force $F(x) = \dfrac{1}{x}$ N that moves an object along a straight line in the direction of F from $x = 1$ m to $x = 2$ m?

11. A 40 m chain weighing 3 kg/m hangs over the side of a bridge. How much work is done by a winch that winds the entire chain in?

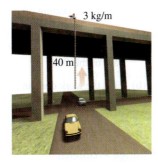

12. A 120 ft chain weighing 240 lb is dangling from the roof of an apartment building. How much work is done in pulling the entire chain up to the roof?

13. A force of 3 N is required to keep a spring extended $\dfrac{1}{4}$ m beyond its equilibrium length. What is its spring constant?

14. A force of 6 lb is required to keep a spring compressed to $\dfrac{1}{2}$ ft shorter than its equilibrium length. What is its spring constant?

15. A spring with spring constant $k = 5$ has an equilibrium length of 0.8 m. How much work is required to stretch the spring to 1.4 m?

16. A spring with spring constant $k = 0.3$ has an equilibrium length of 1.2 m. How much work is required to compress the spring to 1 m?

17. Pumping Water out of a Pool A swimming pool in the shape of a right circular cylinder, with height 4 ft and radius 12 ft, is full of water. See the figure below. How much work is required to pump all the water over the top of the pool? (The weight of water is 62.42 lb/ft³.)

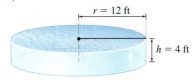

18. Pumping Gasoline out of a Tank A gasoline storage tank in the shape of a right circular cylinder, with height 10 m and radius 8 m, is full of gasoline. How much work is required to pump all the gasoline over the top of the tank? (The density of gasoline is $\rho = 720 \text{ kg/m}^3$.)

19. Pumping Corn Slurry A container in the shape of an inverted pyramid with a square base of 2 m by 2 m and height of 5 m is filled to a depth of 4 m with corn slurry. How much work is required to pump all the slurry over the top of the container? (The density of corn slurry is $\rho = 17.9 \text{ kg/m}^3$.)

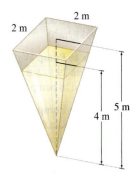

20. Pumping Olive Oil from a Vat A vat in the shape of an inverted pyramid with a rectangular base that measures 2 m by 0.5 m and height that measures 4 m is filled to a depth of 2 m with olive oil. How much work is required to pump all the olive oil over the top of the vat? (The density of olive oil is $\rho = 0.9 \text{ g/cm}^3$.)

Applications and Extensions

21. Work to Lift an Elevator How much work is required if six cables, each weighing 0.36 lb/in., lift a 10,000-lb elevator 400 ft? Assume the cables work together and equally share the weight of the elevator.

22. Work by Gravity A cable with a uniform linear mass density of 9 kg/m is being unwound from a cylindrical drum. If 15 m are already unwound, what is the work done by gravity in unwinding another 60 m of the cable?

23. Work to Lift a Bucket and Chain A uniform chain 10 m long and with mass 20 kg is hanging from the top of a building 10 m high. If a bucket filled with cement of mass 75 kg is attached to the end of the chain, how much work is required to pull the bucket and chain to the top of the building?

24. Work to Lift a Bucket and Chain In Problem 23, if a uniform chain 10 m long and with mass 15 kg is used instead, how much work is required to pull the bucket and chain to the top of the building?

25. Work of a Spring A spring, whose equilibrium length is 1 m, extends to a length of 3 m when a force of 3 N is applied. Find the work needed to extend the spring to a length of 2 m from its equilibrium length.

26. Work of a Spring A spring, whose equilibrium length is 2 m, is compressed to a length of $\dfrac{1}{2}$ m when a force of 10 N is applied. Find the work required to compress the spring to a length of 1 m.

27. Work of a Spring A spring, whose equilibrium length is 4 ft, extends to a length of 8 ft when a force of 2 lb is applied. If 9 ft-lb of work is required to extend this spring from its equilibrium position, what is its total length?

28. Work of a Spring If 8 ft-lb of work is used on the spring in Problem 27, how far is it extended?

The weight density of water is about $62.5 \, \text{lb/ft}^3$, so if the plate were suspended horizontally at a depth of 4 ft, the weight of the water on the plate is

$$F = \rho g h A = (62.5)(4)A = 250A \, \text{lb}$$

and the pressure of the water on the plate is

$$P = \frac{F}{A} = 250 \, \text{lb/ft}^2$$

If the plate has an area of $2 \, \text{ft}^2$, the force F of the water exerted on one side of the plate is $F = AP = (2)(250) = 500 \, \text{lb}$. In fact, as the horizontal plate at a depth h changes in size and or shape, the hydrostatic force F varies, but the hydrostatic pressure P remains constant. Nevertheless, the deeper the plate is in the fluid, the greater the pressure is on the plate.

If a plate is suspended vertically in a fluid, the pressure at the bottom of the plate is greater than the pressure at the top. To find the force of the fluid on one side of the plate, suppose the plate is suspended vertically in a fluid of mass density ρ. Suppose further that the plate is placed in a rectangular coordinate system and is enclosed by $y = c$, $y = d$, $x = g(y)$, and $x = f(y)$, where f and g are functions that are continuous on the closed interval $[c, d]$ and $f(y) \geq g(y)$ for all numbers y in $[c, d]$. See Figure 61.

The surface of the fluid is the line $y = H$, where $H \geq d$, so that the top of the plate is at a depth of $(H - d)$ and the bottom of the plate is at a depth of $(H - c)$.

Partition the interval $[c, d]$ into n subintervals:

$$[c, y_1], [y_1, y_2], \ldots, [y_{i-1}, y_i], \ldots, [y_{n-1}, d] \qquad c = y_0 \quad d = y_n$$

each of length $\Delta y = \dfrac{d - c}{n}$, and let v_i be a number in the ith subinterval $[y_{i-1}, y_i]$, $i = 1, 2, 3, \ldots, n$. If the length Δy is small, then all points in the horizontal ith slice of the plate are roughly the same distance $(H - v_i)$ from the surface, and the pressure P_i of the fluid on this portion of the plate is approximately $P_i = \rho g(H - v_i)$. The force F_i due to hydrostatic pressure on the ith subinterval of the plate is approximately

$$F_i = \rho g h A_i \approx \rho g(H - v_i)[f(v_i) - g(v_i)]\Delta y \qquad A_i = [f(v_i) - g(v_i)]\Delta y$$

An approximation to the total force F on the plate can be found by summing the forces from each subinterval. That is,

$$F \approx \sum_{i=1}^{n} \underbrace{\rho g}_{\text{Weight density of slice}} \underbrace{(H - v_i)}_{\text{Depth of slice}} \underbrace{[f(v_i) - g(v_i)] \, \Delta y}_{\text{Area of the slice}}$$

As the number of subintervals increases, that is, as $n \to \infty$, the sums F become better approximations of the force due to fluid pressure on the plate. Since these are Riemann sums and the functions f and g are continuous on $[c, d]$, the limit is a definite integral. That is,

$$F = \lim_{n \to \infty} \sum_{i=1}^{n} \rho g(H - v_i)[f(v_i) - g(v_i)]\Delta y = \int_{c}^{d} \rho g(H - y)[f(y) - g(y)] \, dy$$

Hydrostatic Force

The hydrostatic force F due to the pressure exerted by a fluid of mass density ρ on a plate of the type illustrated in Figure 61, where the functions f and g are continuous on the closed interval $[c, d]$, is

$$\boxed{F = \int_{c}^{d} \rho g(H - y)[f(y) - g(y)] \, dy}$$

IN WORDS The pressure P due to a static fluid on a horizontal plate depends on the depth of the plate in the fluid.

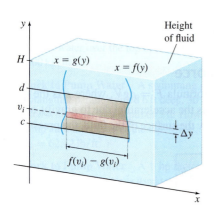

Figure 61 A plate suspended vertically in fluid.

EXAMPLE 1 Finding Hydrostatic Force

A trough, whose cross section is a trapezoid, measures 2 m across at the bottom and 4 m across at the top, and is 2 m deep. If the trough is filled with a liquid of mass density ρ, what is the force due to hydrostatic pressure on one end of the trough?

Solution We position the trough in a rectangular coordinate system, as shown in Figure 62.

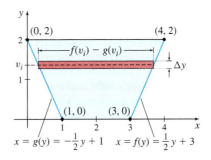

Figure 62

The sides of the end of the trough are lines that pass through the points $(0, 2)$, $(1, 0)$ and the points $(3, 0)$, $(4, 2)$, respectively. The equations of these lines are

$$y - 2 = \frac{0 - 2}{1 - 0}(x - 0) = -2x \qquad \text{or equivalently} \qquad x = g(y) = -\frac{1}{2}y + 1$$

and

$$y - 0 = \frac{2 - 0}{4 - 3}(x - 3) = 2x - 6 \qquad \text{or equivalently} \qquad x = f(y) = \frac{1}{2}y + 3$$

Then the hydrostatic force on the ith interval of the trough is

$$F_i = \underbrace{\rho \cdot g}_{\text{Weight}} \underbrace{(H - v_i)}_{\text{Depth}} \underbrace{[f(v_i) - g(v_i)]\Delta y}_{\text{Area}} \underset{\underset{H = 2}{\uparrow}}{=} \rho g (2 - v_i)[f(v_i) - g(v_i)]\Delta y$$

The liquid fills the trough from $y = 0$ to $y = 2$, so the hydrostatic force F due to the pressure of the liquid on an end of the trough is

$$F = \int_0^2 \rho g (2 - y)[f(y) - g(y)]\, dy$$

$$= \int_0^2 \rho g (2 - y)\left[\left(\frac{1}{2}y + 3\right) - \left(-\frac{1}{2}y + 1\right)\right] dy = \rho g \int_0^2 (2 - y)(y + 2)\, dy$$

$$= \rho g \int_0^2 (-y^2 + 4)\, dy = \rho g \left[-\frac{y^3}{3} + 4y\right]_0^2 = \rho g \left(-\frac{8}{3} + 8\right) = \frac{16}{3}\rho g \text{ N} \quad \blacksquare$$

NOW WORK Problem 13.

So far we have positioned the coordinate system so that the submerged plate is located in the first quadrant. As the next example illustrates, the coordinates may be placed in any convenient position. Keep in mind that the essential idea behind the formula for force due to hydrostatic pressure is

$$\text{Hydrostatic force} = \rho g \times \text{Depth} \times \text{Area}$$

6.8 Center of Mass; Centroid; the Pappus Theorem

OBJECTIVES *When you finish this section, you should be able to:*

1. Find the center of mass of a finite system of objects (p. 458)
2. Find the centroid of a homogeneous lamina (p. 460)
3. Find the volume of a solid of revolution using the Pappus Theorem (p. 464)

Anytime you have been able to balance an object on your fingertip, you have located its *center of mass*. See Figure 64. The **center of mass** of an object or a system of objects is the point that acts as if all the mass is concentrated at that point.

Figure 64 The center of mass of an object is the point where the object is balanced.

1 Find the Center of Mass of a Finite System of Objects

We begin with a system of two objects with masses m_1 and m_2 that are placed on the ends of a nearly weightless rod of length d that is hung from a wire. When the wire is placed at the center of mass of the objects, the rod will be horizontal. Mathematically, the rod is balanced, or in equilibrium, when

$$m_1 d_1 = m_2 d_2$$

where d_1 and d_2 are the distances of the objects from the vertical wire, as shown in Figure 65. The quantities $m_1 d_1$ and $m_2 d_2$, called **moments**, represent the tendency of the objects to rotate about the balancing point. When $m_1 d_1 = m_2 d_2$, the tendency of the objects to rotate is equal, so no rotation occurs and equilibrium is attained.

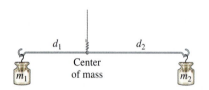

Figure 65 The rod is balanced when $m_1 d_1 = m_2 d_2$.

If the rod were placed on a positive x-axis, as in Figure 66, with mass m_1 at x_1, mass m_2 at x_2, and the center of mass at $\bar{x}$, then the rod is balanced when

$$m_1(\bar{x} - x_1) = m_2(x_2 - \bar{x})$$

$$m_1\bar{x} - m_1 x_1 = m_2 x_2 - m_2\bar{x}$$

$$(m_1 + m_2)\bar{x} = m_1 x_1 + m_2 x_2$$

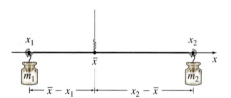

Figure 66 The center of mass is located at $\bar{x}$.

The *center of mass* $\bar{x}$ of the two objects satisfies the equation

$$\boxed{\bar{x} = \frac{m_1 x_1 + m_2 x_2}{m_1 + m_2}}$$

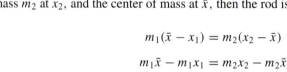

EXAMPLE 1 Finding the Center of Mass of a System of Objects on a Line

Find the center of mass of the system when a mass of 90 kg is placed at 6 and a mass of 40 kg is placed at 2.

Solution The system is shown in Figure 67, where the two masses are placed on a weightless seesaw. The center of mass $\bar{x}$ will be at some number where a fulcrum balances the two masses. Then for equilibrium,

$$\bar{x} = \frac{m_1 x_1 + m_2 x_2}{m_1 + m_2} = \frac{40(2) + 90(6)}{40 + 90} = \frac{620}{130} \approx 4.769$$

The center of mass is at $\bar{x} \approx 4.769$. ∎

DF Figure 67 At the center of mass $\bar{x}$ the system is in equilibrium.

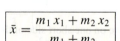
NOW WORK Problem 7.

The formula for the center of mass of two objects can be extended to any number n of objects.

Center of Mass of a System of n Objects on a Line

If n objects with masses $m_1, m_2, \ldots, m_n$ are placed on a line at $x_1, x_2, \ldots, x_n$, respectively, then for equilibrium

$$(m_1 + m_2 + \cdots + m_n)\bar{x} = m_1 x_1 + m_2 x_2 + \cdots + m_n x_n$$

and the **center of mass** is $\bar{x}$, where

$$\bar{x} = \frac{m_1 x_1 + m_2 x_2 + \cdots + m_n x_n}{m_1 + m_2 + \cdots + m_n} = \frac{\sum_{i=1}^{n}(m_i x_i)}{\sum_{i=1}^{n} m_i} = \frac{\sum_{i=1}^{n}(m_i x_i)}{M}$$

where $M = \sum_{i=1}^{n} m_i$ is the mass of the system.

The numbers $m_1 x_1, m_2 x_2, \ldots, m_n x_n$ are called the **moments about the origin** of the masses $m_1, m_2, \ldots, m_n$. So, the center of mass $\bar{x}$ is found by adding the moments about the origin and dividing by the total mass M of all the objects.

Now suppose the objects are not in a line, but are scattered in a plane.

Center of Mass of a System of n Objects in a Plane

If n objects $m_1, m_2, \ldots, m_n$ are located at the points $(x_1, y_1), (x_2, y_2), \ldots, (x_n, y_n)$ in a plane, then the **center of mass of the system** is located at the point $(\bar{x}, \bar{y})$, where

$$\bar{x} = \frac{\sum_{i=1}^{n}(m_i x_i)}{M} \qquad \bar{y} = \frac{\sum_{i=1}^{n}(m_i y_i)}{M}$$

and $M = \sum_{i=1}^{n} m_i$ is the mass of the system.

The sum $M_y = \sum_{i=1}^{n}(m_i x_i)$ is called the **moment of the system about the y-axis** and the sum $M_x = \sum_{i=1}^{n}(m_i y_i)$ is called the **moment of the system about the x-axis**. The center of mass formulas then can be written as

$$\bar{x} = \frac{M_y}{M} \qquad \bar{y} = \frac{M_x}{M} \tag{1}$$

Physically, M_y measures the tendency of the system to rotate about the y-axis; M_x measures the tendency of the system to rotate about the x-axis.

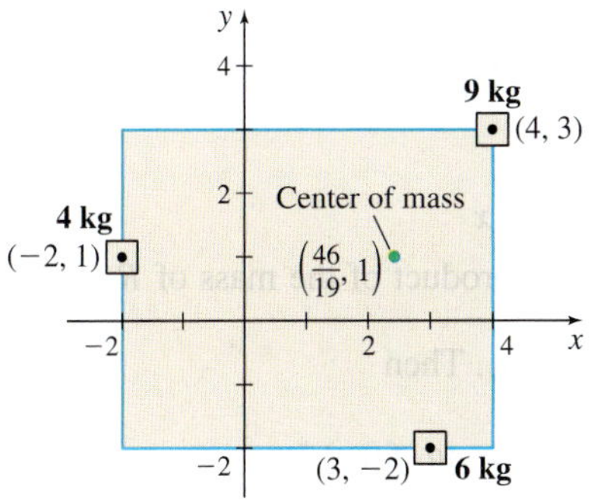

Figure 68 The center of mass of the system is $\left(\dfrac{46}{19}, 1\right)$.

Finding the Center of Mass of a System
EXAMPLE 2 of Objects in a Plane

Find the center of mass of the system of objects having masses 4, 6, and 9 kg, located at the points $(-2, 1)$, $(3, -2)$, and $(4, 3)$, respectively.

Solution See Figure 68. Where is a good estimate for the center of mass? Certainly, it will lie within the rectangle $-2 \leq x \leq 4$; $-2 \leq y \leq 3$.

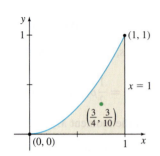

Figure 71 $f(x) = x^2, 0 \le x \le 1$.

EXAMPLE 3 **Finding the Centroid of a Homogeneous Lamina**

Find the centroid of the homogeneous lamina bounded by the graph of $f(x) = x^2$, the x-axis, and the line $x = 1$.

Solution The lamina is shown in Figure 71.

The area A of the lamina is

$$A = \int_0^1 x^2 dx = \left[\frac{x^3}{3}\right]_0^1 = \frac{1}{3}$$

Using formulas (2), the centroid of the lamina is

$$\bar{x} = \frac{1}{A}\int_0^1 xf(x)\,dx = \frac{1}{\frac{1}{3}}\int_0^1 x \cdot x^2\,dx = 3\int_0^1 x^3\,dx = 3\left[\frac{x^4}{4}\right]_0^1 = \frac{3}{4}$$

$$\bar{y} = \frac{1}{2A}\int_0^1 [f(x)]^2\,dx = \frac{1}{2 \cdot \frac{1}{3}}\int_0^1 (x^2)^2\,dx = \frac{3}{2}\int_0^1 x^4\,dx = \frac{3}{2}\left[\frac{x^5}{5}\right]_0^1 = \frac{3}{10}$$

The centroid of the lamina is $\left(\dfrac{3}{4}, \dfrac{3}{10}\right)$. ∎

EXAMPLE 4 **Finding the Centroid of a Homogeneous Lamina**

Find the centroid of one-quarter of a circular plate of radius R.

Solution We place the quarter-circle in the first quadrant, as shown in Figure 72(a). The equation of the quarter circle can be expressed as $f(x) = \sqrt{R^2 - x^2}$, where $0 \le x \le R$.

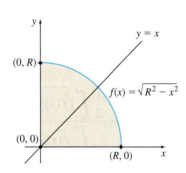

(a) The region is symmetric with respect to the line $y = x$.

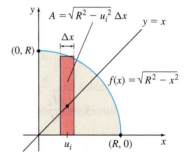

(b) The centroid lies on the line $y = x$.

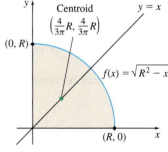

(c) The centroid is $\left(\frac{4}{3\pi}R, \frac{4}{3\pi}R\right)$.

Figure 72

If you guessed that because the quarter of the circular plate is symmetric with respect to the line $y = x$, the centroid will lie on this line, you are correct. See Figure 72(b). So, $\bar{x} = \bar{y}$. The area A of the quarter circular region is $A = \dfrac{\pi R^2}{4}$.

$$\bar{x} = \frac{1}{A}\int_a^b [xf(x)]\,dx = \frac{1}{A}\int_0^R x\sqrt{R^2 - x^2}\,dx$$

$$\bar{y} = \frac{1}{2A}\int_a^b [f(x)]^2\,dx = \frac{1}{2A}\int_0^R (R^2 - x^2)\,dx$$

Since $\bar{x} = \bar{y}$, and $\bar{y}$ is easier to find, we evaluate $\bar{y}$.

$$\bar{x} = \bar{y} = \frac{1}{2A} \int_0^R (R^2 - x^2)\, dx = \frac{1}{2\left(\dfrac{\pi R^2}{4}\right)} \left[R^2 x - \frac{x^3}{3}\right]_0^R = \frac{2}{\pi R^2}\left(\frac{2}{3}R^3\right) = \frac{4}{3\pi}R$$

The centroid of the lamina, as shown in Figure 72(c), is $(\bar{x}, \bar{y}) = \left(\dfrac{4}{3\pi}R, \dfrac{4}{3\pi}R\right)$. ∎

Example 4 illustrates the **symmetry principle**: If a homogeneous lamina is symmetric about a line L or a point P, then the centroid of the lamina lies on L or at the point P.

EXAMPLE 5 Finding the Centroid of a Homogeneous Lamina

Find the centroid of the lamina bounded by the graph of $y = f(x) = x^2 + 1$, the x-axis, and the lines $x = -2$ and $x = 2$.

Solution Figure 73(a) shows the graph of the lamina.

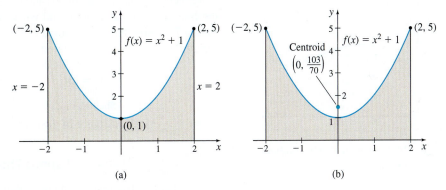

(a)   (b)

Figure 73

We notice two properties of f.

- The graph of f is symmetric about the y-axis, so by the symmetry principle, $\bar{x} = 0$.
- f is an even function, so $\displaystyle\int_{-2}^{2} f(x)\, dx = 2\int_0^2 f(x)\, dx.$

The area A of the region is

$$A = 2\int_0^2 (x^2 + 1)\, dx = 2\left[\frac{x^3}{3} + x\right]_0^2 = 2\left(\frac{8}{3} + 2\right) = \frac{28}{3}$$

Using (2), we get

$$\bar{y} = \frac{1}{2A}\int_a^b [f(x)]^2\, dx = \frac{1}{2A}\int_{-2}^{2}[f(x)]^2\, dx = \frac{1}{2A}\cdot 2\int_0^2 [f(x)]^2\, dx$$

$$= \frac{3}{28}\int_0^2 (x^2 + 1)^2\, dx = \frac{3}{28}\int_0^2 (x^4 + 2x^2 + 1)\, dx$$

$$= \frac{3}{28}\left[\frac{x^5}{5} + \frac{2x^3}{3} + x\right]_0^2 = \frac{3}{28}\left(\frac{32}{5} + \frac{16}{3} + 2\right) = \frac{103}{70} \approx 1.471$$

The centroid of the lamina, as shown in Figure 73(b), is $(\bar{x}, \bar{y}) = \left(0, \dfrac{103}{70}\right)$. ∎

Notice that the centroid in Example 5 does not lie within the lamina.

NOW WORK Problem 19.

In general, laminas are not homogeneous. The Challenge Problems at the end of the section investigate the center of mass of a lamina for which the density of the material varies with respect to x. The more general case, where the density of a lamina varies with both x and y, requires double integration and is treated in Chapter 14.

3 Find the Volume of a Solid of Revolution Using the Pappus Theorem

IN WORDS The volume of a solid formed by revolving a plane region around an axis equals the product of area of the region and the distance its centroid travels around the axis.

THEOREM The Pappus Theorem for Volume

Let R be a plane region of area A and let V be the volume of the solid of revolution obtained by revolving R about a line that does not intersect R. Then the volume V of the solid of revolution is

$$V = 2\pi A d$$

where d is the distance from the centroid of R to the line.

Proof We give a proof for the special case where the region R is bounded by the graph of a function f that is continuous and nonnegative on the interval $[a, b]$, the x-axis, and the lines $x = a$ and $x = b$, and where R is revolved about the y-axis, as shown in Figure 74. Using the shell method to find the volume of the solid of revolution and the centroid formula (2) for $\bar{x}$, we find

$$V = 2\pi \underset{\substack{\uparrow \\ \text{Shell Method}}}{\int_a^b} x f(x)\, dx = 2\pi (A\bar{x}) \underset{\substack{\uparrow \\ (2)}}{=} 2\pi A d$$

where $\bar{x} = d$ is the distance of the centroid $(\bar{x}, \bar{y})$ of R from the y-axis. ∎

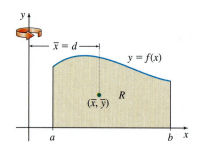

Figure 74 The region R to be revolved about the y-axis.

EXAMPLE 6   Using the Pappus Theorem for Volume

Use the Pappus Theorem to find the volume of the solid formed by revolving the region enclosed by the circle $(x - 3)^2 + y^2 = 1$ about the y-axis.

Solution By symmetry, the centroid of a circular region is the center of the circle. Here, the centroid is the point $(3, 0)$. See Figure 75(a).

ORIGINS The Greek mathematician Pappus of Alexandria (c. 300 AD) produced a mathematical collection containing a record of much of classical Greek mathematics. In it he shows a relationship between volume and centroids.

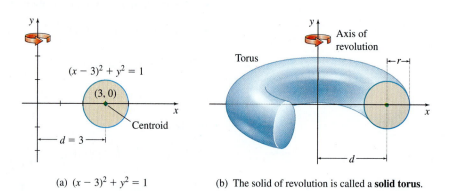

(a) $(x - 3)^2 + y^2 = 1$ (b) The solid of revolution is called a **solid torus**.

DF Figure 75

NOTE A **torus** is a doughnut-shaped surface.

The distance d from the centroid to the axis of revolution, which is the y-axis, is $d = 3$. The area of the circle is $A = \pi R^2 = \pi \cdot 1^2 = \pi$. It follows from the Pappus Theorem that the volume V of the solid of revolution [Figure 75(b)] is

$$V = 2\pi \cdot (\text{Area of the circle}) \cdot d = 2\pi \cdot \pi \cdot 3 = 6\pi^2 \approx 59.218 \text{ cubic units.} \qquad \blacksquare$$

IN WORDS A disk moving in a circle generates a solid torus. The volume of the torus equals the product of the circumference of the circle traveled by the centroid of the generating disk and the area of the disk.

NOW WORK Problem **23**.

6.8 Assess Your Understanding

Concepts and Vocabulary

1. *Multiple Choice* The [(a) midpoint of mass, (b) mass point, (c) center of mass] of a system is the point that acts as if all the mass is concentrated at that point.

2. *Multiple Choice* If an object of mass m is located on the x-axis a distance d from the origin, then the moment of the mass about the origin is [(a) 0, (b) d, (c) md, (d) m].

3. *True or False* A homogeneous lamina is made of material that has a constant density.

4. *Multiple Choice* If a homogeneous lamina is symmetric about a line L, then its [(a) centroid, (b) moment of mass, (c) axis] lies on L.

5. *True or False* The centroid of a lamina R always lies within R.

6. *Multiple Choice* The center of mass of a homogeneous lamina R is located at the [(a) centroid, (b) midpoint, (c) equilibrium point] of R, the geometric center of the lamina.

Skill Building

In Problems 7–10, find the center of mass of each system of masses.

7. $m_1 = 20$, $m_2 = 50$ located, respectively, at 4 and 10

8. $m_1 = 10$, $m_2 = 3$ located, respectively, at -2 and 3

9. $m_1 = 4$, $m_2 = 3$, $m_3 = 3$, $m_4 = 5$ located, respectively, at $-1, 2, 4, 3$

10. $m_1 = 7$, $m_2 = 3$, $m_3 = 2$, $m_4 = 4$ located, respectively, at $6, -2, -4, -1$

In Problems 11–14, find the moments M_x and M_y and the center of mass of each system of masses.

11. $m_1 = 4$, $m_2 = 8$, $m_3 = 1$ located, respectively, at the points $(0, 2), (2, 1), (4, 8)$

12. $m_1 = 6$, $m_2 = 2$, $m_3 = 10$ located, respectively, at the points $(-1, -1), (12, 6), (-1, -2)$

13. $m_1 = 4$, $m_2 = 3$, $m_3 = 3$, $m_4 = 5$ located, respectively, at the points $(-1, 2), (2, 3), (4, 5), (3, 6)$

14. $m_1 = 8$, $m_2 = 6$, $m_3 = 3$, $m_4 = 5$ located, respectively, at the points $(-4, 4), (0, 5), (6, 4), (-3, -5)$

In Problems 15–22, find the centroid of each homogeneous lamina bounded by the graphs of the given equations.

15. $y = 2x + 3$, $y = 0$, $x = -1$, $x = 2$

16. $y = \dfrac{3 - x}{2}$, $y = 0$, $x = -1$, $x = 3$

17. $y = x^2$, $y = 0$, $x = 3$

18. $y = x^3$, $x = 2$, $y = 0$

19. $y = 4x - x^2$ and the x-axis

20. $y = x^2 + x + 1$, $x = 0$, $x = 4$, and the x-axis

21. $y = \sqrt{x}$, $x = 4$, $y = 0$

22. $y = \sqrt[3]{x}$, $x = 8$, $y = 0$

In Problems 23 and 24, use the Pappus Theorem to find the volume of the solid of revolution.

23. The solid torus formed by revolving the circle $(x - 4)^2 + y^2 = 9$ about the y-axis

24. The solid torus formed by revolving the circle $x^2 + (y - 2)^2 = 1$ about the x-axis

Applications and Extensions

Centroid *In Problems 25–30, find the centroid of each homogeneous lamina. (Hint: The moments of the union of two or more nonoverlapping regions equals the sum of the moments of the individual regions.)*

25.

26.
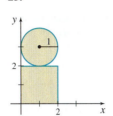

27.
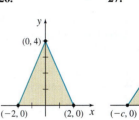

28.
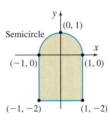

29.
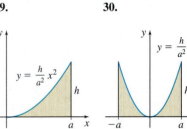

30.
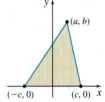

31. **Center of Mass of a Baseball Bat** Inspection of a baseball bat shows that it gets progressively thicker, that is, more massive, beginning at the handle and ending at the top. If the bat is aligned so that the x-axis runs through the center of the bat, the density λ, in kilograms per meter, can be modeled by $\lambda = kx$, where k is the constant of proportionality. The mass M of a bat of length L meters is $\int_0^L \lambda \, dx$. The center of mass $\bar{x}$ of the bat is $\dfrac{\int_0^L x\lambda \, dx}{M}$.

   (a) Find the mass of the bat.

   (b) Use the result of (a) to find the constant of proportionality k.

   (c) Find the center of mass of the bat.

   (d) Where is the "sweet spot" of the bat (the best place to make contact with the ball)? Explain why?

   (e) Give an explanation for the representation of the mass M by $\int_0^L \lambda \, dx$.

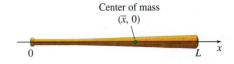

Center of mass
$(\bar{x}, 0)$

1. = NOW WORK problem   = Graphing technology recommended   CAS = Computer Algebra System recommended

32. The Pappus Theorem for Volume Find the volume of the solid formed by revolving the region bounded by the graphs of $y = \sqrt{x-1}$, $y = 0$, and $x = 2$ about the y-axis.

33. The Pappus Theorem for Volume Find the volume of the solid formed by revolving the triangular region bounded by the lines $x = 3$, $y = 2$, and $x + 2y = 9$ about the y-axis.

34. The Pappus Theorem for Volume Find the volume of a right circular cone of radius R and height H.

35. The Pappus Theorem for Volume Find the volume of the solid formed by revolving the triangular region whose vertices are at the points $(1, 1)$, $(5, 3)$, and $(3, 3)$ about the x-axis.

36. Centroid of a Right Triangle Find the centroid of a triangular region with vertices at $(0, 0)$, $(0, 4)$, and $(3, 0)$.

Challenge Problems

37. Centroid of a Triangle

(a) Show that the centroid of a triangular region is located at the intersection of the medians. [*Hint*: Place the vertices of the triangle at $(a, 0)$, $(b, 0)$, and $(0, c)$, $a < 0, b > 0, c > 0$.

(b) Show that the centroid of a triangular region divides each median into two segments whose lengths are in the ratio 2:1.

38. (a) Derive formulas for finding the centroid of a lamina bounded by the graphs of $y = f(x)$ and $y = g(x)$ and the lines $x = a$ and $x = b$, as shown in the figure below.

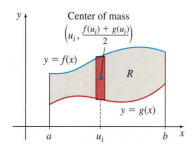

In Problems 39–44, use the result of Problem 38 to find the centroid of each homogeneous lamina enclosed by the graphs of the given equations.

39. $y = x^{2/3}$ and $y = x^2$ from $x = -1$ to $x = 1$

40. $y = \dfrac{1}{x}$, and the lines $y = \dfrac{1}{4}$, and $x = 1$

41. $y = -x^2 + 2$ and $y = |x|$

42. $y = \sqrt{x}$ and $y = x^2$

43. $y = 4 - x^2$ and $y = x + 2$.

44. $y = 9 - x^2$ and $y = |x|$.

45. Laminas with Variable Density Suppose a lamina is determined by a region R enclosed by the graph of $y = f(x)$, the x-axis, and the lines $x = a$ and $x = b$, where f is continuous and nonnegative on the closed interval $[a, b]$. Suppose further that the density of the material at (x, y) is $\rho = \rho(x)$, where ρ is continuous on $[a, b]$. Show that the center of mass $(\bar{x}, \bar{y})$ of R is given by

$$\bar{x} = \frac{M_y}{M} = \frac{\displaystyle\int_a^b \rho(x)xf(x)\,dx}{\displaystyle\int_a^b \rho(x)f(x)\,dx}$$

$$\bar{y} = \frac{M_x}{M} = \frac{\dfrac{1}{2}\displaystyle\int_a^b \rho(x)[f(x)]^2\,dx}{\displaystyle\int_a^b \rho(x)f(x)\,dx}$$

(*Hint*: Partition the interval $[a, b]$ into n subintervals, and use an argument similar to the one used for the centroid of a homogeneous lamina. Assume that if Δx is small, then R_i is homogeneous.)

46. Lamina with Variable Density Use the results of Problem 45 to find the center of mass of a lamina enclosed by $f(x) = \dfrac{1}{2}x^2$, the x-axis, and the lines $x = 1$ and $x = e$. Suppose the density at any point (x, y) of the lamina is $\rho(x) = \dfrac{1}{x^3}$.

47. Lamina with Variable Density Use the results of Problem 45 to find the center of mass of the lamina enclosed by $y = \sqrt{x+1}$, and the lines $x = 0$ and $y = 3$, where the density of the lamina is $\rho(x) = x$.

Laminas with Variable Density *In Problems 48–51, use the results of Problem 45 to find the center of mass of each lamina enclosed by the graph of f on the given interval and having density $\rho = \rho(x)$.*

48. $f(x) = x$; $[0, 3]$; $\rho(x) = x$

49. $f(x) = 3x - 1$; $[1, 4]$; $\rho(x) = x$

50. $f(x) = 2x$; $[0, 1]$; $\rho(x) = x + 1$

51. $f(x) = x$; $[1, 4]$; $\rho(x) = x + 1$

Chapter Review

THINGS TO KNOW

6.1 Area Between Graphs

Area A between two graphs:

- $A = \int_a^b [f(x) - g(x)]\, dx$, where $f(x) \geq g(x)$ for all x in the interval $[a, b]$ (p. 405)
- $A = \int_c^d [f(y) - g(y)]\, dy$, where $f(y) \geq g(y)$ for all y in the interval $[c, d]$ (p. 409)

6.2 Volume of a Solid of Revolution: Disks and Washers

Volume V of a solid of revolution: The disk method:

- $V = \pi \int_a^b [f(x)]^2\, dx$, where the region is revolved about the x-axis (p. 414)
- $V = \pi \int_c^d [g(y)]^2\, dy$, where the region is revolved about the y-axis (p. 416)

Volume V of a solid of revolution: The washer method:

- $V = \pi \int_a^b \{[f(x)]^2 - [g(x)]^2\}\, dx$, where the region is revolved about the x-axis (p. 419)
- $V = \pi \int_c^d \{[f(y)]^2 - [g(y)]^2\}\, dy$, where the region is revolved about the y-axis (p. 420)

6.3 Volume of a Solid of Revolution: Cylindrical Shells

- Cylindrical shell: The solid region between two concentric cylinders (p. 424)

Volume V of a solid of revolution: The shell method:

- $V = 2\pi \int_a^b x f(x)\, dx$, where the region is revolved about the y-axis (p. 425)
- $V = 2\pi \int_c^d y g(y)\, dy$, where the region is revolved about the x-axis (p. 429)

6.4 Volume of a Solid: Slicing Method

Volume V of a solid: The slicing method:

- $V = \int_a^b A(x)\, dx$, where the area $A = A(x)$ of the cross section of a solid is known and is continuous on $[a, b]$ (p. 433)

6.5 Arc Length

Arc length formulas:

- $s = \int_a^b \sqrt{1 + [f'(x)]^2}\, dx$ (p. 440)
- $s = \int_c^d \sqrt{1 + [g'(y)]^2}\, dy$ (p. 441)

6.6 Work

- **Work** is the energy transferred to or from an object by a force acting on the object. (p. 444)

Work formulas:

- F is a continuously varying force that moves an object from a to b: $W = \int_a^b F(x)\, dx$ (p. 445)
- F is a spring force; then $F(x) = -kx$ (Hooke's Law) (p. 446)
- F is the force required to pump a liquid; $F(x) = \rho g V(x)$ (p. 448)
- F is the attraction between two bodies; $F(x) = G\dfrac{m_1 m_2}{x^2}$ (p. 450)

6.7 Hydrostatic Pressure and Force

- Pressure P is the force F exerted per unit area A: $P = \dfrac{F}{A}$ (p. 453)

Hydrostatic force:

- $F = \displaystyle\int_c^d \rho g (H - y)[f(y) - g(y)]\, dy$ (p. 454)

6.8 Center of Mass; Centroid; The Pappus Theorem

- The center of mass of an object or system of objects is the point that acts as if all the mass is concentrated at that point. (p. 458)
- The moments of a system represent the tendency of the system to rotate about the center of mass. (p. 458)
- A lamina is a thin, flat sheet of material. (p. 460)
- If a lamina has constant density, it is homogeneous and its center of mass is located at the centroid. (p. 460)

The centroid of a homogeneous lamina:

- The lamina of area A enclosed by the graph of f, the x-axis, and the lines $x = a$ and $x = b$:

$$\bar{x} = \frac{1}{A} \int_a^b [x f(x)]\, dx \quad \text{and} \quad \bar{y} = \frac{1}{2A} \int_a^b [f(x)]^2\, dx,$$

where $A = \displaystyle\int_a^b f(x)\, dx$ (p. 461)

Properties of laminas:

- Symmetry principle: If a homogeneous lamina is symmetric about a line L or a point P, then the centroid of the lamina lies on L or at a point P. (p. 463)

The Pappus Theorem for Volume

- Let R be a plane region and let V be the volume of the solid of revolution obtained by revolving the area A of R about a line that does not intersect R. Then the volume V of the solid of revolution is $V = 2\pi A d$, where d is the distance from the centroid of R to the line. (p. 464)

OBJECTIVES

REVIEW EXERCISES

Area *In Problems 1–5, find the area A enclosed by the graphs of the given equations.*

1. $y = e^x$, $x = 0$, $y = 4$
2. $y = x^2$, $y = 18 - x^2$
3. $x = 2y^2$, $x = 2$
4. $y = \dfrac{1}{x}$, $x + y = 4$
5. $y = 4 - x^2$, $y = 3x$

Volume of a Solid of Revolution *In Problems 6–15, find the volume of the solid of revolution generated by revolving the region bounded by the graphs of the given equations about the given line.*

6. $y = x^2$, $y = 4x - x^2$; about the x-axis
7. $y = x^2 - 5x + 6$, $y = 0$; about the y-axis
8. $x = y^2 - 4$, $x = 0$; about the y-axis
9. $xy = 1$, $x = 1$, $x = 2$, $y = 0$; about the x-axis
10. $y = x^2 - 4$, $y = 0$; about the line $y = -4$
11. $y = 4x - x^2$ and the x-axis; about the x-axis
12. $y^2 = 8x$, $y \geq 0$, and $x = 2$; about the line $x = 2$

13. $y = \dfrac{x^3}{2}$, $y = 0$, $x = 2$; about the y-axis
14. $y = e^x$, $y = 1$, $x = 1$; about the x-axis
15. $y^2 = x^3$, $y = 8$, $x = 0$; about the line $x = 4$

Arc Length *In Problems 16–18, find the arc length of each graph.*

16. $y = x^{3/2} + 4$ from $x = 2$ to $x = 5$
17. $y = \dfrac{x^3}{6} + \dfrac{1}{2x}$ from $x = 2$ to $x = 6$
18. $2y^3 = x^2$ from $y = 0$ to $y = 2$

19. Center of Mass Find the center of mass of the system of masses: $m_1 = 1, m_2 = 3, m_3 = 8, m_4 = 1$ located, respectively, at $-1, 2, 14,$ and 0.

20. Moments and Center of Mass Find the moments M_x and M_y and the center of mass of the system of masses: $m_1 = 2, m_2 = 2, m_3 = 3, m_4 = 2$ located, respectively, at the points $(-4, 4), (2, 3), (4, 4), (-3, -5)$.

21. Area of a Triangle Use integration to find the area of the triangle formed by the lines $x - y + 1 = 0$, $3x + y - 13 = 0$, and $x + 3y - 7 = 0$.

22. Volume A solid has a circular base of radius 4 units. Slices taken perpendicular to a fixed diameter are equilateral triangles. Find its volume.

23. Volume Find the volume of the solid generated by revolving the region bounded by the graph of $4x^2 + 9y^2 = 36$ in the first quadrant about the x-axis.

24. Volume of a Cone Find the volume of an elliptical cone with base $\dfrac{x^2}{4} + y^2 = 1$ and height 5. (*Hint:* The area A of an ellipse with semi-axes a and b is $A = \pi ab$.)

25. Volume The base of a solid is enclosed by $4x + 5y = 20$, $x = 0$, $y = 0$. Every cross section perpendicular to the base along the x-axis is a semicircle. Find the volume of the solid.

26. Arc Length Find the point P on $y = \dfrac{2}{3}x^{3/2}$ to the right of the y-axis so that the length of the curve from $(0, 0)$ to P is $\dfrac{52}{3}$.

27. Work Find the work done in raising an 800-lb anchor 150 ft with a chain weighing 20 lb/ft.

28. Work Find the work done in raising a container of 1000 kg of silver ore from a mine 1200 m deep with a cable weighing 3 kg/m.

29. Work Pumping Water A hemispherical water tank has a diameter of 12 m. It is filled to a depth of 4 m. How much work is done in pumping all the water over the edge? (Use $\rho = 1000 \, \text{kg/m}^3$.)

30. Work of a Spring A spring with an unstretched length of 0.6 m requires a force of 4 N to stretch it to 0.8 m. How much work is done in stretching it to 1.4 m?

31. Work of a Spring Find the unstretched length of a spring if the work required to stretch the spring from 1.0 to 1.4 m is half the work required to stretch it from 1.2 to 1.8 m.

32. Hydrostatic Force A trough of trapezoidal cross section is 2 ft wide at the bottom, 4 ft wide at the top, and 3 ft deep. What is the force due to liquid pressure on the end, if it is full of water?

33. Hydrostatic Force A cylindrical tank is on its side. It has a diameter of 10 m and is full to a depth of 5 m with gasoline that has a density of 737 kg/m³. What is the force due to liquid pressure on the end?

34. Hydrostatic Force A dam is built in the shape of a trapezoid 1000 ft long at the top, 700 ft long at the bottom, and 80 ft deep. Determine the force of water on the dam if:

(a) the reservoir behind the dam is full.

(b) the reservoir behind the dam has a depth of 60 ft.

35. Centroid Find the centroid of the lamina bounded by the graphs of $y = \sqrt{x}$, $y = 0$, and $x = 9$.

36. Pappus Theorem for Volume Find the volume of the torus with an outer diameter of 5 cm and an inner diameter of 2 cm.

37. Volume of a Solid of Revolution Find the volume generated when the triangular region bounded by the lines $x = 3$, $y = 0$, and $2x + y - 12 = 0$ is revolved about the y-axis.

38. Volume of a Solid of Revolution Find the volume generated when the region bounded by the graphs of $y = 3\sqrt{x}$ and $y = -x^2 + 6x - 2$ is revolved about each of the following lines:

(a) $y = 0$ **(b)** $x = 0$ **(c)** $y = -2$

(d) $y = 8$ **(e)** $x = -3$ **(f)** $x = 5$

CHAPTER 6 PROJECT   **Determining the Amount of Concrete Needed for a Cooling Tower**

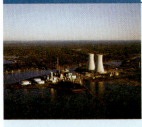

A common design for cooling towers is modeled by a branch of a hyperbola rotated about an axis. The design is used because of its strength and efficiency. Not only is the shape stronger than either a cone or cylinder, it takes less material to build. This shape also maximizes the natural upward draft of hot air without the need for fans.*

How much concrete is needed to build a cooling tower that has a base 460 ft wide that is 442 ft below the vertex if the top of the tower is 310 ft wide and 123 ft above the vertex? Assume the walls are a constant 5 in = 0.42 ft thick. See Figure 76.

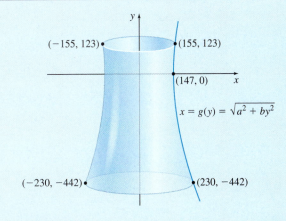

Figure 76

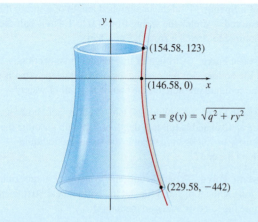

Figure 77

1. Show that the equation of the right branch of the hyperbola in Figure 76 is given by

$$x = g_2(y) = \sqrt{a^2 + by^2}$$

where $a = 147$ and $b = 0.16$.

2. Find the volume of the solid of revolution obtained by revolving the area enclosed by $x = g_2(y)$ from $y = -442$ to $y = 123$ about the y-axis.

3. Using the fact that the walls are 0.42 ft thick, show that the equation of the interior hyperbola is given by

$$x = g_1(y) = \sqrt{q^2 + ry^2}$$

where $q = 146.58$ and $r = 0.16$. See Figure 77.

4. Find the volume of the solid of revolution obtained by revolving the area enclosed by $x = g_1(y)$ from $y = -442$ to $y = 123$ about the y-axis.

5. Now determine the volume (in cubic feet) of concrete required for the cooling tower.

6. Do an Internet search to find the cost of concrete. How much will the concrete used in the cooling tower cost?

7. Write a report for an audience that is not familiar with cooling towers summarizing the findings in Problems 1–6.

For pictures of the cooling towers and power plant, as well as a time-lapse video of the towers being built, log onto https://www.dom.com/about/stations/fossil/brayton-point-power-station.jsp

---

*The design of an unsupported, reinforced concrete hyperbolic cooling tower was patented in 1918 (UK patent 198,863) by Frederic von Herson and Gerard Kupeers of the Netherlands.

Source: Excerpts from *The Providence Journal*, Frieda Squires, May 2010.

Techniques of Integration

The Birds of Rügen Island

Where else can a variety of shore and grassland birds, such as Redshanks, Lapwings, and Pied Avocet, find an undisturbed habitat? A department of the Western Pomeranian National Park administrates the Western Pomerania Lagoon Area and Jasmund National Park, in which Rügen Island lies. This area on the Baltic Sea off the coast of Germany includes some of the most important breeding grounds for water and mud flat birds of the Baltic Sea. How does a scientist model the population of birds that live in a protected area like Rügen Island?

CHAPTER 7 PROJECT See the Chapter Project on page 535 for a discussion of how we can use calculus to model bird populations.

We have developed a sizable collection of basic integration formulas that are listed below. Some of the integrals are the result of simple antidifferentiation. Others, such as $\int \tan x \, dx$, were found using the method of substitution, a *technique of integration*. In this chapter, we further expand the list of basic integrals by exploring more techniques of integration.

Integration, unlike differentiation, has no hard and fast rules. It is often difficult, sometimes even impossible, to integrate a function that appears to be simple.

As you study the integration techniques presented in this chapter, it is important to recognize the form of the integrand for each technique. Then, with practice, integration becomes easier.

$$\int x^a \, dx = \frac{x^{a+1}}{a+1} + C, a \neq -1$$

$$\int \frac{1}{x} dx = \ln |x| + C$$

$$\int e^x \, dx = e^x + C$$

$$\int a^x \, dx = \frac{a^x}{\ln a} + C, a > 0, a \neq 1$$

$$\int \sin x \, dx = -\cos x + C$$

$$\int \cos x \, dx = \sin x + C$$

$$\int \sec^2 x \, dx = \tan x + C$$

$$\int \csc^2 x \, dx = -\cot x + C$$

$$\int \sec x \tan x \, dx = \sec x + C$$

$$\int \csc x \cot x \, dx = -\csc x + C$$

$$\int \tan x \, dx = \ln |\sec x| + C$$

$$\int \cot x \, dx = \ln |\sin x| + C$$

$$\int \sec x \, dx = \ln |\sec x + \tan x| + C$$

$$\int \csc x \, dx = \ln |\csc x - \cot x| + C$$

$$\int \frac{dx}{\sqrt{a^2 - x^2}} = \sin^{-1} \frac{x}{a} + C, \quad a > 0$$

$$\int \frac{dx}{x\sqrt{x^2 - a^2}} = \frac{1}{a} \sec^{-1} \frac{x}{a} + C, a > 0$$

$$\int \frac{dx}{a^2 + x^2} = \frac{1}{a} \tan^{-1} \frac{x}{a} + C, a > 0$$

$$\int \sinh x \, dx = \cosh x + C$$

$$\int \cosh x \, dx = \sinh x + C$$

$$\int \text{sech}^2 x \, dx = \tanh x + C$$

$$\int \text{csch}^2 x \, dx = -\coth x + C$$

$$\int \text{sech} \, x \tanh x \, dx = -\text{sech} \, x + C$$

$$\int \text{csch} \, x \coth x \, dx = -\text{csch} \, x + C$$

7.1 Integration by Parts

OBJECTIVES *When you finish this section, you should be able to:*

1 Integrate by parts (p. 472)

2 Derive a formula using integration by parts (p. 476)

Integration by parts is a technique of integration based on the Product Rule for derivatives: If $u = f(x)$ and $v = g(x)$ are functions that are differentiable on an open interval (a, b), then

$$\frac{d}{dx}[f(x) \cdot g(x)] = f(x) \, g'(x) + f'(x) \, g(x)$$

Integrating both sides gives

$$\int \frac{d}{dx}[f(x) \cdot g(x)] \, dx = \int [f(x) \, g'(x) + f'(x) \, g(x)] \, dx$$

$$f(x) \cdot g(x) = \int f(x) \, g'(x) \, dx + \int f'(x) g(x) \, dx$$

Solving this equation for $\int f(x) g'(x) \, dx$ yields

$$\boxed{\int f(x) \, g'(x) \, dx = f(x) \cdot g(x) - \int f'(x) \, g(x) \, dx}$$

which is known as the **integration by parts formula**.

Let $u = f(x)$, $v = g(x)$. Then we can use their differentials, $du = f'(x) \, dx$ and $dv = g'(x) \, dx$ to obtain the formula in the form we usually use:

$$\boxed{\int u \, dv = uv - \int v \, du}$$

IN WORDS The integral of the product $u \, dv$ equals u times v minus the integral of the product $v \, du$.

1 Integrate by Parts

The goal of integration by parts is to choose u and dv so that $\int v \, du$ is easier to integrate than $\int u \, dv$.

EXAMPLE 1   **Integrating by Parts**

Find $\int x \, e^x \, dx$.

Solution Choose u and dv so that

$$\int u \, dv = \int x \, e^x \, dx$$

Suppose we choose

$$u = x \quad \text{and} \quad dv = e^x \, dx$$

Then

$$du = dx \quad \text{and} \quad v = \int dv = \int e^x \, dx = e^x$$

Notice that we did not add a constant. Only a particular antiderivative of dv is required at this stage; we will add the constant of integration at the end. Using the integration by parts formula, we have

$$\int \underbrace{x}_{u} \, \underbrace{e^x \, dx}_{dv} = \underbrace{x \, e^x}_{uv} - \int \underbrace{e^x}_{v} \, \underbrace{dx}_{du} = x \, e^x - e^x + C = e^x(x - 1) + C \quad \blacksquare$$

We intentionally chose $u = x$ and $dv = e^x dx$ so that $\int v\,du$ in the formula is easy to integrate. Suppose, instead, we chose

$$u = e^x \qquad \text{and} \qquad dv = x\,dx$$

Then

$$du = e^x dx \quad \text{and} \quad v = \int x\,dx = \frac{x^2}{2}$$

The integration by parts formula yields

$$\int x\,e^x\,dx = \int \underbrace{e^x}_{u}\ \underbrace{x\,dx}_{dv} = \underbrace{e^x\,\frac{x^2}{2}}_{uv} - \int \underbrace{\frac{x^2}{2}}_{v}\ \underbrace{e^x\,dx}_{du}$$

IN WORDS We choose dv so that it can be easily integrated and choose u so that $\int v\,du$ is simpler than $\int u\,dv$.

For this choice of u and v, the integral on the right is more complicated than the original integral, indicating an unwise choice of u and dv.

NOW WORK **Problem 3.**

EXAMPLE 2 **Integrating by Parts**

Find $\displaystyle\int x \sin x\,dx$.

Solution We use the integration by parts formula with

$$u = x \qquad \text{and} \qquad dv = \sin x\,dx \qquad \int u\,dv = \int x \sin x\,dx$$

Then

$$du = dx \quad \text{and} \quad v = \int \sin x\,dx = -\cos x$$

and

$$\int x \sin x\,dx = -x \cos x + \int \cos x\,dx = -x \cos x + \sin x + C$$
$$\underset{\displaystyle \int u\,dv = uv - \int v\,du}{\uparrow}$$

∎

NOW WORK **Problem 5.**

Unfortunately, there are no exact rules for choosing u and dv. But the following guidelines are helpful:

Integration by Parts: Guidelines for Choosing u and dv

- dx is always part of dv.
- dv should be easy to integrate.
- u and dv are chosen so that $\int v\,du$ is no more difficult to integrate than the original integral $\int u\,dv$.
- If the new integral is more complicated, try different choices for u and dv.

Choosing u and dv often involves trial and error. If a selection does not appear to work, try a different choice. If no choice works, it may be that some other technique of integration should be used.

EXAMPLE 3 Integrating by Parts to Find $\int \ln x \, dx$

Derive the formula

$$\int \ln x \, dx = x \ln x - x + C$$

Solution We use the integration by parts formula with

$$u = \ln x \qquad \text{and} \qquad dv = dx$$

Then

$$du = \frac{1}{x} \, dx \qquad \text{and} \qquad v = \int dx = x$$

Now

$$\int \ln x \, dx = x \cdot \ln x - \int x \cdot \frac{1}{x} \, dx = x \ln x - \int dx = x \ln x - x + C \qquad \blacksquare$$

> **NOTE** The integral $\int \ln x \, dx$ can be found in the list of integrals on the inserts at the front and back of the book and is a basic integral.

EXAMPLE 4 Integrating by Parts to Find $\int \tan^{-1} x \, dx$

Derive the formula

$$\int \tan^{-1} x \, dx = x \tan^{-1} x - \frac{1}{2} \ln(1 + x^2) + C$$

Solution We use the integration by parts formula with

$$u = \tan^{-1} x \qquad \text{and} \qquad dv = dx$$

Then

$$du = \frac{1}{1 + x^2} \, dx \qquad \text{and} \qquad v = \int dx = x$$

and

$$\int \tan^{-1} x \, dx = x \cdot \tan^{-1} x - \int \frac{x}{1 + x^2} \, dx$$

> **NEED TO REVIEW?** The method of substitution is discussed in Section 5.6, pp. 387–393.

To find the integral $\int \frac{x}{1 + x^2} \, dx$, we use the substitution $t = 1 + x^2$. Then $dt = 2x \, dx$, or equivalently, $x \, dx = \frac{dt}{2}$.

$$\int \frac{x}{1 + x^2} \, dx = \frac{1}{2} \int \frac{dt}{t} = \frac{1}{2} \ln |t| = \frac{1}{2} \ln(1 + x^2)$$

As a result, $\int \tan^{-1} x \, dx = x \tan^{-1} x - \frac{1}{2} \ln(1 + x^2) + C$. $\qquad \blacksquare$

NOW WORK Problem 9.

The next example shows that sometimes it is necessary to integrate by parts more than once.

EXAMPLE 5 Integrating by Parts

Find $\int x^2 e^x \, dx$.

Solution We use the integration by parts formula with

$$u = x^2 \qquad \text{and} \qquad dv = e^x \, dx$$

Then

$$du = 2x \, dx \qquad \text{and} \qquad v = \int e^x \, dx = e^x$$

and

$$\int x^2 e^x \, dx = x^2 e^x - 2 \int x e^x \, dx$$

The integral on the right is simpler than the original integral. To find it, we use integration by parts a second time. (Refer to Example 1.)

$$\int x e^x \, dx = x e^x - e^x$$

Then

$$\int x^2 e^x \, dx = x^2 e^x - 2(x e^x - e^x) + C = e^x (x^2 - 2x + 2) + C \qquad \blacksquare$$

NOW WORK Problem 13.

Table 1 provides guidelines to help choose u and dv for several types of integrals that are found using integration by parts. In the table, n is always a positive integer.

TABLE 1 Guidelines for Choosing u and dv

Integral; n is a positive integer	u	dv
$\int x^n e^{ax} \, dx$ $\int x^n \cos(ax) \, dx$ $\int x^n \sin(ax) \, dx$	$u = x^n$	$dv =$ what remains
$\int x^n \sin^{-1} x \, dx$	$u = \sin^{-1} x$	
$\int x^n \cos^{-1} x \, dx$	$u = \cos^{-1} x$	$dv = x^n \, dx$
$\int x^n \tan^{-1} x \, dx$	$u = \tan^{-1} x$	
$\int x^m (\ln x)^n \, dx$; m is a real number, $m \neq -1$	$u = (\ln x)^n$	$dv = x^m \, dx$

Integration by parts is also used to find certain definite integrals.

EXAMPLE 6 **Finding the Area Under the Graph of $f(x) = x \ln x$**

Find the area under the graph of $f(x) = x \ln x$ from 1 to 2.

Solution See Figure 1 for the graph of $f(x) = x \ln x$. The area A under the graph of f from 1 to 2 is $A = \int_1^2 x \ln x \, dx$. We use the integration by parts formula with

$$u = \ln x \qquad \text{and} \qquad dv = x \, dx$$

Then

$$du = \frac{1}{x} \, dx \qquad \text{and} \qquad v = \int x \, dx = \frac{x^2}{2}$$

and

$$A = \int_1^2 x \ln x \, dx = \left[\frac{x^2}{2} \ln x \right]_1^2 - \int_1^2 \frac{x^2}{2} \left(\frac{1}{x} dx \right) = 2 \ln 2 - \frac{1}{2} \int_1^2 x \, dx$$

$$= 2 \ln 2 - \frac{1}{2} \left[\frac{x^2}{2} \right]_1^2 = 2 \ln 2 - \frac{3}{4} \qquad \blacksquare$$

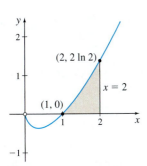

Figure 1 $f(x) = x \ln x$

NOW WORK Problem 37.

2 Derive a Formula Using Integration by Parts

Integration by parts is also used to derive general formulas involving integrals.

EXAMPLE 7 **Deriving a Formula**

Derive the formula

$$\int e^{ax} \cos(bx)\, dx = \frac{e^{ax}[b \sin(bx) + a \cos(bx)]}{a^2 + b^2} + C \qquad b \neq 0 \qquad (1)$$

Solution We use the integration by parts formula with

$$u = e^{ax} \qquad \text{and} \qquad dv = \cos(bx)\, dx$$

Then

$$du = ae^{ax}\, dx \qquad \text{and} \qquad v = \int \cos(bx)\, dx = \frac{1}{b} \sin(bx)$$

and

$$\int e^{ax} \cos(bx)\, dx = e^{ax} \frac{\sin(bx)}{b} - \frac{a}{b} \int e^{ax} \sin(bx)\, dx \qquad (2)$$

The new integral on the right, $\int e^{ax} \sin(bx)\, dx$, is different from the original integral, but it is essentially of the same form. We use integration by parts again with this integral by choosing

$$u = e^{ax} \qquad \text{and} \qquad dv = \sin(bx)\, dx$$

Then

$$du = ae^{ax}\, dx \qquad \text{and} \qquad v = \int \sin(bx)\, dx = -\frac{1}{b} \cos(bx)$$

and

$$\int e^{ax} \sin(bx)\, dx = -\frac{1}{b} e^{ax} \cos(bx) + \frac{a}{b} \int e^{ax} \cos(bx)\, dx \qquad (3)$$

Substituting the result from (3) into (2), we obtain

$$\int e^{ax} \cos(bx)\, dx = \frac{1}{b} e^{ax} \sin(bx) - \frac{a}{b} \left[-\frac{1}{b} e^{ax} \cos(bx) + \frac{a}{b} \int e^{ax} \cos(bx)\, dx \right]$$

$$\int e^{ax} \cos(bx)\, dx = \frac{1}{b} e^{ax} \sin(bx) + \frac{a}{b^2} e^{ax} \cos(bx) - \frac{a^2}{b^2} \int e^{ax} \cos(bx)\, dx$$

Now we solve for $\int e^{ax} \cos(bx)\, dx$ and simplify.

$$\int e^{ax} \cos(bx)\, dx + \frac{a^2}{b^2} \int e^{ax} \cos(bx)\, dx = \frac{1}{b} e^{ax} \sin(bx) + \frac{a}{b^2} e^{ax} \cos(bx)$$

$$\left(1 + \frac{a^2}{b^2}\right) \int e^{ax} \cos(bx)\, dx = \frac{1}{b^2} e^{ax}[b \sin(bx) + a \cos(bx)]$$

$$\int e^{ax} \cos(bx)\, dx = \frac{e^{ax}[b \sin(bx) + a \cos(bx)]}{a^2 + b^2} + C \qquad ■$$

For example, to find $\int e^{4x} \cos(5x)\, dx$, we use (1) with $a = 4$ and $b = 5$.

$$\int e^{4x} \cos(5x)\, dx = \frac{e^{4x}[5 \sin (5x) + 4 \cos (5x)]}{41} + C$$

NOW WORK Problem 41.

EXAMPLE 8   **Deriving a Formula**

Derive the formula

$$\int \sec^{n} x\, dx = \frac{\sec^{n-2} x \tan x}{n-1} + \frac{n-2}{n-1} \int \sec^{n-2} x\, dx \qquad n \geq 3 \qquad (4)$$

Solution We begin by writing $\sec^{n} x = \sec^{n-2} x \sec^{2} x$, and choose

$$u = \sec^{n-2} x \qquad \text{and} \qquad dv = \sec^{2} x\, dx$$

This choice makes $\int dv$ easy to integrate. Then

$$du = [(n-2) \sec^{n-3} x \cdot \sec x \tan x]\, dx = [(n-2) \sec^{n-2} x \tan x]\, dx$$

$$v = \int \sec^{2} x\, dx = \tan x$$

Using integration by parts, we get

$$\int \sec^{n} x\, dx = \sec^{n-2} x \tan x - (n-2) \int \sec^{n-2} x \tan^{2} x\, dx$$

To express the integrand on the right in terms of $\sec x$, we use the trigonometric identity, $\tan^{2} x + 1 = \sec^{2} x$, and replace $\tan^{2} x$ by $\sec^{2} x - 1$, obtaining

$$\int \sec^{n} x\, dx = \sec^{n-2} x \tan x - (n-2) \int \sec^{n-2} x (\sec^{2} x - 1)\, dx$$

$$\int \sec^{n} x\, dx = \sec^{n-2} x \tan x - (n-2) \int \sec^{n} x\, dx + (n-2) \int \sec^{n-2} x\, dx$$

Moving the middle term on the right to the left, we obtain

$$(n-1) \int \sec^{n} x\, dx = \sec^{n-2} x \tan x + (n-2) \int \sec^{n-2} x\, dx$$

Finally, divide both sides by $n - 1$:

$$\int \sec^{n} x\, dx = \frac{\sec^{n-2} x \tan x}{n-1} + \frac{n-2}{n-1} \int \sec^{n-2} x\, dx \qquad \blacksquare$$

Formula (4) is called a **reduction formula** because repeated applications of the formula eventually lead to an elementary integral. For this reduction formula, when n is even, repeated applications lead eventually to

$$\int \sec^{2} x\, dx = \tan x + C$$

When n is odd, repeated applications eventually lead to the integral

$$\int \sec x\, dx = \ln | \sec x + \tan x | + C$$

For example, if $n = 3$,

$$\int \sec^3 x \, dx = \frac{\sec x \tan x}{2} + \frac{1}{2} \int \sec x \, dx = \frac{\sec x \tan x}{2} + \frac{1}{2} \ln |\sec x + \tan x| + C$$

NOW WORK Problem 59.

7.1 Assess Your Understanding

Concepts and Vocabulary

1. *True or False*   Integration by parts is based on the Product Rule for derivatives.

2. The integration by parts formula states that $\int u \, dv = $ _____.

Skill Building

In Problems 3–30, use integration by parts to find each integral.

3. $\int x e^{2x} \, dx$

4. $\int x e^{-3x} \, dx$

5. $\int x \cos x \, dx$

6. $\int x \sin(3x) \, dx$

7. $\int \sqrt{x} \ln x \, dx$

8. $\int x^{-2} \ln x \, dx$

9. $\int \cot^{-1} x \, dx$

10. $\int \sin^{-1} x \, dx$

11. $\int (\ln x)^2 \, dx$

12. $\int x (\ln x)^2 \, dx$

13. $\int x^2 \sin x \, dx$

14. $\int x^2 \cos x \, dx$

15. $\int x \cos^2 x \, dx$

16. $\int x \sin^2 x \, dx$

17. $\int x \sinh x \, dx$

18. $\int x \cosh x \, dx$

19. $\int \cosh^{-1} x \, dx$

20. $\int \sinh^{-1} x \, dx$

21. $\int \sin(\ln x) \, dx$

22. $\int \cos(\ln x) \, dx$

23. $\int (\ln x)^3 \, dx$

24. $\int (\ln x)^4 \, dx$

25. $\int x^2 (\ln x)^2 \, dx$

26. $\int x^3 (\ln x)^2 \, dx$

27. $\int x^2 \tan^{-1} x \, dx$

28. $\int x \tan^{-1} x \, dx$

29. $\int 7^x x \, dx$

30. $\int 2^{-x} x \, dx$

In Problems 31–38, use integration by parts to find each definite integral.

31. $\int_0^\pi e^x \cos x \, dx$

32. $\int_0^1 x^2 e^{-x} \, dx$

33. $\int_0^2 x^2 e^{-3x} \, dx$

34. $\int_0^{\pi/4} x \tan^2 x \, dx$

35. $\int_1^9 \ln \sqrt{x} \, dx$

36. $\int_{\pi/4}^{3\pi/4} x \csc^2 x \, dx$

37. $\int_1^e (\ln x)^2 \, dx$

38. $\int_0^{\pi/4} x \sec^2 x \, dx$

Applications and Extensions

Area Between Two Graphs   *In Problems 39 and 40, find the area of the region enclosed by the graphs of f and g.*

39. $f(x) = 3 \ln x$ and $g(x) = x \ln x$, $x \ge 1$

40. $f(x) = 4x \ln x$ and $g(x) = x^2 \ln x$, $x \ge 1$

41. **Area Under a Graph**   Find the area under the graph of $y = e^x \sin x$ from 0 to π.

42. **Volume of a Solid of Revolution**   Find the volume of the solid of revolution generated by revolving the region bounded by the graph of $y = \cos x$ and the x-axis from $x = 0$ to $x = \dfrac{\pi}{2}$ about the y-axis. See the figure below.

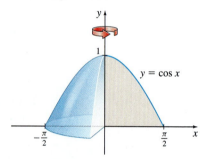

43. **Volume of a Solid of Revolution**   Find the volume of the solid of revolution generated by revolving the region bounded by the graph of $y = \sin x$ and the x-axis from $x = 0$ to $x = \dfrac{\pi}{2}$ about the y-axis.

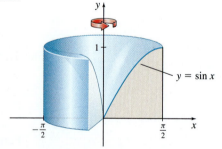

44. **Volume of a Solid of Revolution**   Find the volume of the solid of revolution generated by revolving the region bounded by the graph of $y = x\sqrt{\sin x}$ and the x-axis from $x = 0$ to $x = \dfrac{\pi}{2}$ about the x-axis.

45. Volume of a Solid of Revolution Find the volume of the solid of revolution generated by revolving the region bounded by the graph of $y = \ln x$ and the x-axis from $x = 1$ to $x = e$ about the x-axis.

46. Area

(a) Graph the functions $f(x) = x^3 e^{-3x}$ and $g(x) = x^2 e^{-3x}$ on the same set of coordinate axes.

(b) Find the area enclosed by the graphs of f and g.

(CAS) 47. Damped Spring The displacement x of a damped spring at time t, $0 \le t \le 5$, is given by
$x = x(t) = 3e^{-t} \cos(2t) + 2e^{-t} \sin(2t)$.

(a) Graph $x = x(t)$.

(b) Find the least positive number t that satisfies $x(t) = 0$.

(c) Find the area under the graph of $x = x(t)$ from $t = 0$ to the value of t found in (b).

48. A function $y = f(x)$ is continuous and differentiable on the interval $(2, 6)$. If $\int_3^5 f(x)\,dx = 18$ and $f(3) = 8$ and $f(5) = 11$, then find $\int_3^5 xf'(x)\,dx$.

In Problems 49–54, find each integral by first making a substitution and then integrating by parts.

49. $\displaystyle\int \sin\sqrt{x}\,dx$

50. $\displaystyle\int e^{\sqrt{x}}\,dx$

51. $\displaystyle\int \cos x \ln(\sin x)\,dx$

52. $\displaystyle\int e^x \ln(2 + e^x)\,dx$

53. $\displaystyle\int e^{4x} \cos e^{2x}\,dx$

54. $\displaystyle\int \cos x \tan^{-1}(\sin x)\,dx$

55. Find $\displaystyle\int x^3 e^{x^2}\,dx$. (*Hint:* Let $u = x^2$, $dv = xe^{x^2}\,dx$.)

56. Find $\displaystyle\int x^n \ln x\,dx$; $n \ne -1$, n real.

57. Find $\displaystyle\int xe^x \cos x\,dx$.

58. Find $\displaystyle\int xe^x \sin x\,dx$.

In Problems 59–62, derive each reduction formula where $n > 1$ is an integer.

59. $\displaystyle\int x^n \sin^{-1} x\,dx = \frac{x^{n+1}}{n+1}\sin^{-1}x - \frac{1}{n+1}\int \frac{x^{n+1}}{\sqrt{1-x^2}}\,dx$

60. $\displaystyle\int \frac{dx}{(x^2+1)^{n+1}} = \left(1 - \frac{1}{2n}\right)\int \frac{dx}{(x^2+1)^n} + \frac{x}{2n(x^2+1)^n}$

61. $\displaystyle\int \sin^n x\,dx = -\frac{\sin^{n-1} x \cos x}{n} + \frac{n-1}{n}\int \sin^{n-2} x\,dx$

62. $\displaystyle\int \sin^n x \cos^m x\,dx = -\frac{\sin^{n-1} x \cos^{m+1} x}{n+m}$
$+ \frac{n-1}{n+m}\int \sin^{n-2} x \cos^m x\,dx$

where $m \ne -n$, $m \ne -1$

63. (a) Find $\int x^2 e^{5x}\,dx$.

(b) Using integration by parts, derive a reduction formula for $\int x^n e^{kx}\,dx$, where $k \ne 0$ and $n \ge 2$ is an integer, in which the resulting integrand involves x^{n-1}.

64. (a) Assuming there is a function p for which $\int x^3 e^x\,dx = p(x)e^x$, show that $p(x) + p'(x) = x^3$.

(b) Use integration by parts to find a polynomial p of degree 3 for which $\int x^3 e^x\,dx = p(x)e^x + C$.

65. (a) Use integration by parts with $u = \sin x$ and $dv = \cos x\,dx$ to find a function f for which $\int \sin x \cos x\,dx = f(x) + C_1$.

(b) Use integration by parts with $u = \cos x$ and $dv = \sin x\,dx$ to find a function g for which $\int \sin x \cos x\,dx = g(x) + C_2$.

(c) Use the trigonometric identity $\sin(2x) = 2\sin x \cos x$ and substitution to find a function h for which
$$\int \sin x \cos x\,dx = h(x) + C_3.$$

(d) Compare the functions f and g. Find a relationship between C_1 and C_2.

(e) Compare the functions f and h. Find a relationship between C_1 and C_3.

66. Derive the formula
$$\int \ln(x + \sqrt{x^2+a^2})\,dx = x\ln(x + \sqrt{x^2+a^2}) - \sqrt{x^2+a^2} + C$$

67. Derive the formula
$$\int e^{ax}\sin(bx)\,dx = \frac{e^{ax}[a\sin(bx) - b\cos(bx)]}{a^2+b^2} + C, a > 0, b > 0$$

68. Suppose $F(x) = \int_0^x t\,g'(t)\,dt$ for all $x \ge 0$.
Show that $F(x) = xg(x) - \int_0^x g(t)\,dt$.

69. Use **Wallis' formulas**, given below, to find each definite integral.

• $\displaystyle\int_0^{\pi/2} \sin^n x\,dx = \int_0^{\pi/2} \cos^n x\,dx$ $n > 1$ an integer

$= \begin{cases} \dfrac{(n-1)(n-3)\cdots(4)(2)}{n(n-2)\cdots(5)(3)(1)} & n > 1 \text{ is odd} \\[2ex] \dfrac{(n-1)(n-3)\cdots(5)(3)(1)}{n(n-2)\cdots(4)(2)}\left(\dfrac{\pi}{2}\right) & n > 1 \text{ is even} \end{cases}$

(a) $\displaystyle\int_0^{\pi/2} \sin^6 x\,dx$

(b) $\displaystyle\int_0^{\pi/2} \sin^5 x\,dx$

(c) $\displaystyle\int_0^{\pi/2} \cos^8 x\,dx$

(d) $\displaystyle\int_0^{\pi/2} \cos^6 x\,dx$

Challenge Problems —————————

70. Derive Wallis' formulas given in Problem 69. (*Hint:* Use the result of Problem 61.)

71. (a) If n is a positive integer, use integration by parts to show that there is a polynomial p of degree n for which
$$\int x^n e^x\,dx = p(x)e^x + C$$

(b) Show that $p(x) + p'(x) = x^n$.

(c) Show that p can be written in the form

$$p(x) = \sum_{k=0}^{n}(-1)^k \frac{n!}{(n-k)!}x^{n-k}$$

72. Show that for any positive integer n,

$$\int_0^1 e^{x^2}\,dx$$

$$= e \cdot \left[1 - \frac{2}{3} + \frac{4}{15} - \frac{8}{105} + \cdots + \frac{(-1)^n 2^n}{(2n+1)(2n-1)\cdots 3 \cdot 1}\right]$$

$$+ (-1)^{n+1} \cdot \frac{2^{n+1}}{(2n+1)(2n-1)\cdots 3 \cdot 1}\int_0^1 x^{2n+2}e^{x^2}\,dx$$

73. Use integration by parts to show that if f is a polynomial of degree $n \geq 1$, then $\int f(x)e^x dx = g(x)e^x + C$ for some polynomial $g(x)$ of degree n.

74. Start with the identity $f(b) - f(a) = \int_a^b f'(t)\,dt$ and derive the following generalizations of the Mean Value Theorem for Integrals:

(a) $f(b) - f(a) = f'(a)(b-a) - \int_a^b f''(t)(t-b)\,dt$

(b) $f(b) - f(a) = f'(a)(b-a) + \dfrac{f''(a)}{2}(b-a)^2$
$$+ \int_a^b \frac{f'''(t)}{2}(t-b)^2\,dt$$

75. If $y = f(x)$ has the inverse function given by $x = f^{-1}(y)$, show that

$$\int_a^b f(x)\,dx + \int_{f(a)}^{f(b)} f^{-1}(y)\,dy = bf(b) - af(a)$$

76. (a) When integration by parts is used to find $\int e^x \cosh x\,dx$, what happens? Explain.

(b) Find $\int e^x \cosh x\,dx$ without using integration by parts.

7.2 Integrals Containing Trigonometric Functions

OBJECTIVES *When you finish this section, you should be able to:*

1 Find integrals of the form $\int \sin^n x\,dx$ or $\int \cos^n x\,dx$, $n \geq 2$ an integer (p. 480)
2 Find integrals of the form $\int \sin^m x \cos^n x\,dx$ (p. 483)
3 Find integrals of the form $\int \tan^m x \sec^n x\,dx$ or $\int \cot^m x \csc^n x\,dx$ (p. 483)
4 Find integrals of the form $\int \sin(ax)\sin(bx)\,dx$, $\int \sin(ax)\cos(bx)\,dx$, or $\int \cos(ax)\cos(bx)\,dx$ (p. 485)

In this section, we develop techniques to find certain trigonometric integrals. When studying these techniques, concentrate on the strategies used in the examples rather than trying to memorize the results.

1 Find Integrals of the form $\int \sin^n x\,dx$ or $\int \cos^n x\,dx$, $n \geq 2$ an Integer

Although we could use integration by parts to obtain reduction formulas for integrals of the form $\int \sin^n x\,dx$ or $\int \cos^n x\,dx$, $n \geq 2$ an integer, these integrals also can be found using other, often easier, techniques. We consider two cases:

- $n \geq 3$ an odd integer
- $n \geq 2$ an even integer.

Suppose we want to find $\int \sin^n x\,dx$ when $n \geq 3$ is an odd integer. We begin by writing the integral in the form

$$\int \sin^n x\,dx = \int \sin^{n-1} x \sin x\,dx$$

Since n is odd, $(n-1)$ is even and we can use the identity $\sin^2 x = 1 - \cos^2 x$.

Then the substitution $u = \cos x$, $du = -\sin x\,dx$, leads to an integral involving integer powers of u.

EXAMPLE 1 Finding the Integral $\int \sin^5 x\, dx$

Find $\int \sin^5 x\, dx$.

Solution Since the exponent 5 is odd, we write $\int \sin^5 x\, dx = \int \sin^4 x \sin x\, dx$, and use the identity $\sin^2 x = 1 - \cos^2 x$.

$$\int \sin^5 x\, dx = \int \sin^4 x \sin x\, dx = \int (\sin^2 x)^2 \sin x\, dx = \int (1 - \cos^2 x)^2 \sin x\, dx$$

$$= \int (1 - 2\cos^2 x + \cos^4 x) \sin x\, dx$$

Now we use the substitution $u = \cos x$. Then $du = -\sin x\, dx$, and

$$\int \sin^5 x\, dx = -\int (1 - 2u^2 + u^4)\, du = -u + \frac{2}{3}u^3 - \frac{1}{5}u^5 + C$$

$$= -\cos x + \frac{2}{3}\cos^3 x - \frac{1}{5}\cos^5 x + C \qquad \blacksquare$$

A similar technique is used to find $\int \cos^n x\, dx$, when $n \geq 3$ is an odd integer. In this case, we write

$$\boxed{\int \cos^n x\, dx = \int \cos^{n-1} x \cos x\, dx}$$

and use the trigonometric identity $\cos^2 x = 1 - \sin^2 x$. Then we use the substitution $u = \sin x$. For example,

$$\int \cos^3 x\, dx = \int \cos^2 x \cos x\, dx = \int (1 - \sin^2 x) \cos x\, dx$$

$$= \int (1 - u^2)\, du = u - \frac{u^3}{3} + C = \sin x - \frac{\sin^3 x}{3} + C$$

$$\underset{\substack{u = \sin x \\ du = \cos x\, dx}}{\uparrow}$$

NOW WORK Problem 3.

To find $\int \sin^n x\, dx$ or $\int \cos^n x\, dx$ when $n \geq 2$ is an even integer, the preceding strategy does not work. (Try it for yourself.) Instead, we use one of the identities below:

$$\boxed{\sin^2 x = \frac{1 - \cos(2x)}{2} \qquad \cos^2 x = \frac{1 + \cos(2x)}{2}}$$

to obtain a simpler integrand.

EXAMPLE 2 Finding the Integral $\int \sin^2 x\, dx$

Find $\int \sin^2 x\, dx$.

Solution Since the exponent of $\sin x$ is an even integer, we use the identity

$$\sin^2 x = \frac{1 - \cos(2x)}{2}$$

Then

$$\int \sin^2 x\, dx = \frac{1}{2}\int [1 - \cos(2x)]\, dx = \frac{1}{2}\int dx - \frac{1}{2}\int \cos(2x)\, dx$$

$$= \frac{1}{2}x + C_1 - \frac{1}{2}\int \cos u\, \frac{du}{2} \qquad u = 2x,\, du = 2\, dx.$$

$$= \frac{1}{2}x + C_1 - \frac{1}{4}\sin(2x) + C_2$$

NEED TO REVIEW? The method of substitution is discussed in Section 5.6, pp. 387–393.

NEED TO REVIEW? Trigonometric identities are discussed in Appendix A.4, pp. A-32 to A-35.

NOTE Usually we will just add the constant of integration at the end of the integration to avoid letting $C = C_1 + C_2$.

Since C_1 and C_2 are constants, we write the solution as

$$\int \sin^2 x \, dx = \frac{1}{2}x - \frac{1}{4}\sin(2x) + C$$

where $C = C_1 + C_2$. ∎

NOW WORK Problem 5.

EXAMPLE 3 Finding the Average Value of a Function

Find the average value $\bar{y}$ of the function $f(x) = \cos^4 x$ over the closed interval $[0, \pi]$.

NEED TO REVIEW? The average value of a function is discussed in Section 5.4, pp. 373–374.

Solution The average value $\bar{y}$ of a function f over $[a, b]$ is $\bar{y} = \dfrac{1}{b-a}\displaystyle\int_a^b f(x)\,dx$. For $f(x) = \cos^4 x$ on $[0, \pi]$, we have

$$\bar{y} = \frac{1}{\pi - 0}\int_0^\pi \cos^4 x \, dx = \underset{\underset{\cos^4 x = (\cos^2 x)^2}{\uparrow}}{\frac{1}{\pi}\int_0^\pi (\cos^2 x)^2 \, dx} = \underset{\underset{\cos^2 x = \frac{1+\cos(2x)}{2}}{\uparrow}}{\frac{1}{4\pi}\int_0^\pi [1 + \cos(2x)]^2 \, dx}$$

$$= \frac{1}{4\pi}\int_0^\pi \left[1 + 2\cos(2x) + \cos^2(2x)\right] dx$$

$$= \frac{1}{4\pi}\left[\int_0^\pi dx + 2\int_0^\pi \cos(2x)\,dx + \int_0^\pi \cos^2(2x)\,dx\right] \qquad (1)$$

Now

$$\int_0^\pi dx = \pi - 0 = \pi \quad \text{and} \quad \int_0^\pi \cos(2x)\,dx = \underset{\underset{du = 2\,dx}{\underset{u = 2x}{\uparrow}}}{\int_0^{2\pi} \cos u \, \frac{du}{2}} = \left[\frac{1}{2}\sin u\right]_0^{2\pi} = 0$$

To find $\int_0^\pi \cos^2(2x)\,dx$, we use the identity $\cos^2 \theta = \dfrac{1 + \cos(2\theta)}{2}$ again to write $\cos^2(2x) = \dfrac{1 + \cos(4x)}{2}$. Then

$$\int_0^\pi \cos^2(2x)\,dx = \int_0^\pi \frac{1 + \cos(4x)}{2}\,dx = \frac{1}{2}\left[\int_0^\pi dx + \int_0^\pi \cos(4x)\,dx\right]$$

$$= \frac{1}{2}\left[\pi + \int_0^{4\pi} \cos u \, \frac{du}{4}\right] \qquad u = 4x; du = 4\,dx$$

So, from (1),

$$\bar{y} = \frac{1}{4\pi}\left[\pi + 0 + \frac{\pi}{2}\right] = \frac{3}{8}$$ ∎

If f is nonnegative on an interval $[a, b]$, the average value of f over the interval $[a, b]$ represents the height of a rectangle with width $b - a$ whose area equals the area under the graph of f from a to b. Figure 2 shows the graph of f from Example 3 and the rectangle of height $\bar{y} = \dfrac{3}{8}$ and base π whose area is equal to the area under the graph of f.

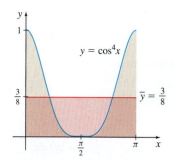

Figure 2

NOW WORK Problem 57.

2 Find Integrals of the Form $\int \sin^m x \cos^n x \, dx$

Integrals of the form $\int \sin^m x \cos^n x \, dx$ are found using variations of previous techniques. We discuss two cases:

- At least one of the exponents m or n is a positive odd integer.
- Both exponents are positive even integers.

EXAMPLE 4   **Finding the Integral $\int \sin^5 x \sqrt{\cos x} \, dx$**

Find $\displaystyle\int \sin^5 x \sqrt{\cos x}\, dx = \int \sin^5 x \, \cos^{1/2} x \, dx.$

Solution The exponent of $\sin x$ is 5, a positive, odd integer. We factor $\sin x$ from $\sin^5 x$ and write

$$\int \sin^5 x \cos^{1/2} x \, dx = \int \sin^4 x \, \cos^{1/2} x \, \sin x \, dx = \int (\sin^2 x)^2 \, \cos^{1/2} x \, \sin x \, dx$$

$$= \int (1 - \cos^2 x)^2 \cos^{1/2} x \, \sin x \, dx$$

Now we use the substitution $u = \cos x$.

$$\int \sin^5 x \, \cos^{1/2} x \, dx = \int \underbrace{(1 - \cos^2 x)^2 \, \cos^{1/2} x}_{\substack{u = \cos x \\ du = -\sin x \, dx}} \sin x \, dx = \int (1 - u^2)^2 u^{1/2} (-du)$$

$$= -\int (u^{1/2} - 2u^{5/2} + u^{9/2})\, du = -\frac{2}{3} u^{3/2} + \frac{4}{7} u^{7/2} - \frac{2}{11} u^{11/2} + C$$

$$= u^{3/2}\left[-\frac{2}{3} + \frac{4}{7} u^2 - \frac{2}{11} u^4 \right] + C$$

$$= \underbrace{(\cos x)^{3/2}}_{u = \cos x}\left[-\frac{2}{3} + \frac{4}{7} \cos^2 x - \frac{2}{11} \cos^4 x \right] + C \qquad\blacksquare$$

NOW WORK   Problem 11.

If m and n are both positive even integers in $\int \sin^m x \cos^n x \, dx$, we use the trigonometric identity $\sin^2 x + \cos^2 x = 1$ to obtain a sum of integrals, each integral involving only even powers of either $\sin x$ or $\cos x$. For example,

$$\int \sin^2 x \cos^4 x \, dx = \int (1 - \cos^2 x) \cos^4 x \, dx = \int \cos^4 x \, dx - \int \cos^6 x \, dx$$

The two integrals on the right are now of the form $\int \cos^n x \, dx$, n a positive even integer, and we can integrate them using the techniques discussed in Examples 2 and 3.

3 Find Integrals of the Form $\int \tan^m x \, \sec^n x \, dx$ or $\int \cot^m x \, \csc^n x \, dx$

We consider three cases involving integrals of the form $\int \tan^m x \, \sec^n x \, dx$:

- The exponent m on the tangent function is a positive odd integer.
- The exponent n on the secant function is a positive even integer.
- The tangent function is raised to a positive even integer m and the secant function is raised to a positive odd integer n.

The idea is to express the integrand so that we can use either the substitution $u = \tan x$ and $du = \sec^2 x \, dx$ or the substitution $u = \sec x$ and $du = \sec x \tan x \, dx$, while leaving an even power of one of the functions. Then we use a Pythagorean identity to express the integrand in terms of only one trigonometric function.

EXAMPLE 5 Finding the Integral $\int \tan^3 x \, \sec^4 x \, dx$

Find $\int \tan^3 x \, \sec^4 x \, dx$.

Solution Here, $\tan x$ is raised to the odd power 3. We factor $\tan x$ from $\tan^3 x$ and use the identity $\tan^2 x = \sec^2 x - 1$.

$$\int \tan^3 x \, \sec^4 x \, dx = \int \tan^2 x \, \tan x \, \sec^4 x \, dx \qquad \text{Factor } \tan x \text{ from } \tan^3 x.$$

$$= \int (\sec^2 x - 1) \tan x \, \sec^4 x \, dx \qquad \tan^2 x = \sec^2 x - 1$$

$$= \int (\sec^2 x - 1) \sec^3 x \, \sec x \tan x \, dx \qquad \text{Factor } \sec x \text{ from } \sec^4 x.$$

$$= \int (u^2 - 1) u^3 \, du \qquad \begin{array}{l} \text{Substitute } u = \sec x; \\ du = \sec x \tan x \, dx. \end{array}$$

$$= \int (u^5 - u^3) \, du = \frac{u^6}{6} - \frac{u^4}{4} + C = \underset{\substack{\uparrow \\ u = \sec x}}{\frac{\sec^6 x}{6}} - \frac{\sec^4 x}{4} + C \qquad \blacksquare$$

NOW WORK Problem 19.

EXAMPLE 6 Finding the Integral $\int \tan^2 x \, \sec^4 x \, dx$

Find $\int \tan^2 x \, \sec^4 x \, dx$.

Solution Here, $\sec x$ is raised to a positive even power. We factor $\sec^2 x$ from $\sec^4 x$ and use the identity $\sec^2 x = 1 + \tan^2 x$. Then

$$\int \tan^2 x \, \sec^4 x \, dx = \int \tan^2 x \, \sec^2 x \cdot \sec^2 x \, dx \qquad \text{Factor } \sec^2 x \text{ from } \sec^4 x.$$

$$= \int \tan^2 x (1 + \tan^2 x) \sec^2 x \, dx \qquad \sec^2 x = 1 + \tan^2 x$$

$$= \int u^2 (1 + u^2) \, du \qquad \begin{array}{l} \text{Substitute } u = \tan x; \\ du = \sec^2 x \, dx. \end{array}$$

$$= \int (u^2 + u^4) \, du = \frac{u^3}{3} + \frac{u^5}{5} + C = \underset{\substack{\uparrow \\ u = \tan x}}{\frac{\tan^3 x}{3}} + \frac{\tan^5 x}{5} + C \qquad \blacksquare$$

NOW WORK Problem 21.

When the tangent function is raised to a positive even integer m and the secant function is raised to a positive odd integer n, the approach is slightly different. Rather than factoring, we begin by using the identity $\tan^2 x = \sec^2 x - 1$.

EXAMPLE 7   **Finding the Integral $\int \tan^2 x \, \sec x \, dx$**

Find $\int \tan^2 x \, \sec x \, dx$.

Solution Here, $\tan x$ is raised to an even power and $\sec x$ to an odd power. We use the identity $\tan^2 x = \sec^2 x - 1$ to write

$$\int \tan^2 x \, \sec x \, dx = \int (\sec^2 x - 1) \sec x \, dx = \int (\sec^3 x - \sec x) dx$$

$$= \int \sec^3 x \, dx - \int \sec x \, dx \qquad (2)$$

Next we integrate $\int \sec^3 x \, dx$ by parts. Choose

$$u = \sec x \qquad \text{and} \qquad dv = \sec^2 x \, dx$$
$$du = \sec x \tan x \, dx \qquad \qquad v = \int \sec^2 x \, dx = \tan x$$

Then

$$\int \sec^3 x \, dx = \sec x \, \tan x - \int \tan^2 x \, \sec x \, dx \qquad \textcolor{blue}{\int u \, dv = uv - \int v \, du}$$

$$= \sec x \, \tan x - \int (\sec^2 x - 1) \, \sec x \, dx \qquad \textcolor{blue}{\tan^2 x = \sec^2 x - 1}$$

$$= \sec x \, \tan x - \int \sec^3 x \, dx + \int \sec x \, dx \qquad \textcolor{blue}{\text{Write the integral as the sum of two integrals.}}$$

$$2 \int \sec^3 x \, dx = \sec x \, \tan x + \int \sec x \, dx \qquad \textcolor{blue}{\text{Add } \int \sec^3 x \, dx \text{ to both sides.}}$$

$$\int \sec^3 x \, dx = \frac{1}{2}[\sec x \, \tan x + \ln | \sec x + \tan x |] \qquad \textcolor{blue}{\text{Solve for } \int \sec^3 x \, dx;}$$
$$\textcolor{blue}{\int \sec x \, dx = \ln |\sec x + \tan x|.}$$

NOTE Substituting $n = 3$ into the reduction formula for $\int \sec^n x \, dx$ (derived in Example 8 of Section 7.1) could also have been used to find $\int \sec^3 x \, dx$.

Now we substitute this result in (2).

$$\int \tan^2 x \, \sec x \, dx = \int \sec^3 x \, dx - \int \sec x \, dx$$

$$= \frac{1}{2} [\sec x \, \tan x + \ln | \sec x + \tan x |] - \ln | \sec x + \tan x | + C$$

$$= \frac{1}{2} [\sec x \, \tan x - \ln | \sec x + \tan x |] + C \qquad \blacksquare$$

NOW WORK Problem 23.

To find integrals of the form $\int \cot^m x \, \csc^n x \, dx$, we use the same strategies, but with the identity $\csc^2 x = 1 + \cot^2 x$.

4 Find Integrals of the Form $\int \sin(ax) \, \sin(bx) \, dx$, $\int \sin(ax) \, \cos(bx) \, dx$, or $\int \cos(ax) \, \cos(bx) \, dx$

Trigonometric integrals of the form

$$\int \sin(ax) \, \sin(bx) \, dx \qquad \int \sin(ax) \, \cos(bx) \, dx \qquad \int \cos(ax) \, \cos(bx) \, dx$$

are integrated using the product-to-sum identities:

- $2 \sin A \sin B = \cos(A - B) - \cos(A + B)$
- $2 \sin A \cos B = \sin(A + B) + \sin(A - B)$
- $2 \cos A \cos B = \cos(A - B) + \cos(A + B)$

These identities transform the integrand into a sum of sines and/or cosines.

EXAMPLE 8 **Finding the Integral** $\int \sin(3x)\sin(2x)\,dx$

Find $\int \sin(3x)\sin(2x)\,dx$.

Solution We use the product-to-sum identity $2\sin A \sin B = \cos(A-B) - \cos(A+B)$. Then

$$2\sin(3x)\sin(2x) = \cos(3x-2x) - \cos(3x+2x)$$

$$\sin(3x)\sin(2x) = \frac{1}{2}[\cos x - \cos(5x)]$$

Then

$$\int \sin(3x)\sin(2x)\,dx = \frac{1}{2}\int [\cos x - \cos(5x)]\,dx = \frac{1}{2}\int \cos x\,dx - \frac{1}{2}\int \cos(5x)\,dx$$

$$= \frac{1}{2}\sin x - \frac{1}{2}\int \cos u\,\frac{du}{5} = \frac{1}{2}\sin x - \frac{1}{10}\sin(5x) + C$$

$$\uparrow$$
$$u = 5x$$
$$du = 5\,dx$$

∎

NOW WORK Problem 27.

7.2 Assess Your Understanding

Concepts and Vocabulary

1. *True or False* To find $\int \cos^5 x\,dx$, factor out $\cos x$ and use the identity $\cos^2 x = 1 - \sin^2 x$.

2. *True or False* To find $\int \sin(2x)\cos(3x)\,dx$, use a product-to-sum identity.

Skill Building

In Problems 3–10, find each integral.

3. $\int \cos^5 x\,dx$

4. $\int \sin^3 x\,dx$

5. $\int \sin^6 x\,dx$

6. $\int \cos^4 x\,dx$

7. $\int \sin^2(\pi x)\,dx$

8. $\int \cos^4(2x)\,dx$

9. $\int_0^\pi \cos^5 x\,dx$

10. $\int_{-\pi/3}^{\pi/3} \sin^3 x\,dx$

In Problems 11–18, find each integral.

11. $\int \sin^3 x \cos^2 x\,dx$

12. $\int \sin^4 x \cos^3 x\,dx$

13. $\int \sin^2 x \cos^2 x\,dx$

14. $\int \sin^4 x \cos^2 x\,dx$

15. $\int \sin x \cos^{1/3} x\,dx$

16. $\int \cos^3 x \sin^{1/2} x\,dx$

17. $\int \sin^2\left(\frac{x}{2}\right)\cos^3\left(\frac{x}{2}\right)\,dx$

18. $\int \sin^3(4x)\cos^3(4x)\,dx$

In Problems 19–26, find each integral.

19. $\int \tan^3 x \sec^2 x\,dx$

20. $\int \tan x \sec^5 x\,dx$

21. $\int \tan^2 x \sec^2 x\,dx$

22. $\int \tan^5 x \sec^2 x\,dx$

23. $\int \tan^2 x \sec^3 x\,dx$

24. $\int \tan^4 x \sec x\,dx$

25. $\int \cot^3 x \csc x\,dx$

26. $\int \cot^3 x \csc^2 x\,dx$

In Problems 27–34, find each integral.

27. $\int \sin(3x)\cos x\,dx$

28. $\int \sin x \cos(3x)\,dx$

29. $\int \cos x \cos(3x)\,dx$

30. $\int \cos(2x)\cos x\,dx$

31. $\int \sin(2x)\sin(4x)\,dx$

32. $\int \sin(3x)\sin x\,dx$

33. $\int_0^{\pi/2} \sin(2x)\sin x\,dx$

34. $\int_0^\pi \cos x \cos(4x)\,dx$

In Problems 35–56, find each integral.

35. $\int \sin^2 x \cos x\,dx$

36. $\int \sin^3 x \cos x\,dx$

37. $\int \frac{\sin x\,dx}{\cos^2 x}$

38. $\int \frac{\cos x\,dx}{\sin^4 x}$

39. $\int \cos^3(3x)\,dx$

40. $\int \sin^5(3x)\,dx$

41. $\int_0^\pi \sin^3 x \cos^5 x\,dx$

42. $\int_0^{\pi/2} \sin^3 x \cos^3 x\,dx$

43. $\int \tan^3 x\,dx$

44. $\int \cot^5 x\,dx$

45. $\int \frac{\sec^6 x}{\tan^3 x}\,dx$

46. $\int \tan^{1/2} x \sec^2 x\,dx$

47. $\int \csc^2 x \cot^5 x\,dx$

48. $\int \cot x \csc^2 x\,dx$

---

1. = NOW WORK problem   = Graphing technology recommended   CAS = Computer Algebra System recommended

49. $\displaystyle\int \cot(2x)\csc^4(2x)\,dx$ **50.** $\displaystyle\int \cot^2(2x)\csc^3(2x)\,dx$

51. $\displaystyle\int_0^{\pi/4} \tan^4 x \sec^3 x\,dx$ **52.** $\displaystyle\int_0^{\pi/4} \tan^2 x \sec x\,dx$

53. $\displaystyle\int \sin\left(\frac{x}{2}\right)\cos\left(\frac{3x}{2}\right)dx$ **54.** $\displaystyle\int \cos(-x)\sin(4x)\,dx$

55. $\displaystyle\int \sin\left(\frac{x}{2}\right)\sin\left(\frac{3x}{2}\right)dx$ **56.** $\displaystyle\int \cos(\pi x)\cos(3\pi x)\,dx$

Applications and Extensions

57. Volume of a Solid of Revolution Find the volume of the solid of revolution generated by revolving the region bounded by the graph of $y = \sin x$ and the x-axis from $x = 0$ to $x = \pi$ about the x-axis. See the figure below.

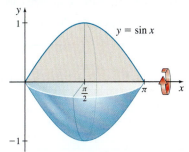

58. Volume of a Solid of Revolution
Find the volume of the solid of revolution generated by revolving the region bounded by the graphs of $y = \cos x$, $y = \sin x$, and $x = 0$ from $x = 0$ to $x = \dfrac{\pi}{4}$ about the x-axis.

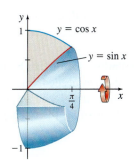

59. Average Value

  (a) Find the average value of $f(x) = \sin x \cos^4 x$ over the interval $[0, \pi]$.

  (b) Give a geometric interpretation to the average value.

  (c) Use graphing technology to graph f and the average value on the same screen.

60. Rectilinear Motion The acceleration a of an object at time t is given by $a(t) = \cos^2 t \sin t$ m/s^2. At $t = 0$, the object is at the origin and its speed is 5 m/s. Find its distance from the origin at any time t.

61. Area and Volume Let A be the area of the region in the first quadrant bounded by the graphs of $y = \sec x$, $y = 2\sin x$, and the y-axis.

  (a) Find A.

  (b) Find the volume of the solid of revolution generated by revolving the region about the x-axis.

62. (a) Use technology to graph the function $f(x) = \sin^n x$, $0 \le x \le \pi$, for $n = 5$, $n = 10$, $n = 20$, and $n = 50$.

(b) Find $\int_0^{\pi} \sin^n x\,dx$ correct to three decimal places for $n = 5$, $n = 10$, $n = 20$, and $n = 50$.

(c) What does (a) suggest about the shape of the graph of $f(x) = \sin^n x$, $0 \le x \le \pi$, as $n \to \infty$.

CAS **(d)** Find $\displaystyle\lim_{n\to\infty} \int_0^{\pi} \sin^n x\,dx$.

63. Find $\displaystyle\int \sin^4 x\,dx$.

  (a) Using the methods of this section.

  (b) Using the reduction formula given in Problem 61 in Section 7.1.

  (c) Verify that both results are equivalent.

CAS **(d)** Use a CAS to find $\displaystyle\int \sin^4 x\,dx$.

64. (a) Use the substitution $u = \sin x$ to find a function f for which

$$\int \sin x \cos x\,dx = f(x) + C_1.$$

(b) Use the substitution $u = \cos x$ to find a function g for which

$$\int \sin x \cos x\,dx = g(x) + C_2.$$

(c) Use the trigonometric identity $\sin(2x) = 2\sin x \cos x$ to find a function h for which

$$\int \sin x \cos x\,dx = h(x) + C_3.$$

(d) Compare the functions f and g. Find a relationship between C_1 and C_2.

(e) Compare the functions f and h. Find a relationship between C_1 and C_3.

65. Derive a formula for $\displaystyle\int \sin(mx)\sin(nx)\,dx$, $m \ne n$.

66. Derive a formula for $\displaystyle\int \sin(mx)\cos(nx)\,dx$, $m \ne n$.

67. Derive a formula for $\displaystyle\int \cos(mx)\cos(nx)\,dx$, $m \ne n$.

Challenge Problems

68. Use the substitution $\sqrt{x} = \sin y$ to find $\displaystyle\int_0^{1/2} \frac{\sqrt{x}}{\sqrt{1-x}}\,dx$.

$\left(\textit{Hint: } \sin^2 y = \dfrac{1 - \cos(2y)}{2}.\right)$

69. Use an appropriate substitution to show that

$$\int_0^{\pi/2} \sin^n \theta\,d\theta = \int_0^{\pi/2} \cos^n \theta\,d\theta.$$

70. (a) What is wrong with the following?

$$\int_0^{\pi} \cos^4 x\,dx = \int_0^{\pi} (\cos x)^3 \cos x\,dx = \int_0^{\pi} (\cos^2 x)^{3/2} \cos x\,dx$$

$$= \int_0^{\pi} (1 - \sin^2 x)^{3/2}\cos x\,dx = \int_0^0 (1 - u^2)^{3/2}\,du = 0$$

$$\begin{array}{ll} u = \sin x & x = 0 \Rightarrow u = 0 \\ du = \cos x\,dx & x = \pi \Rightarrow u = 0 \end{array}$$

(b) Find $\displaystyle\int_0^{\pi} \cos^4 x\,dx$.

7.3 Integration Using Trigonometric Substitution: Integrands Containing $\sqrt{a^2 - x^2}$, $\sqrt{x^2 + a^2}$, or $\sqrt{x^2 - a^2}$, $a > 0$

OBJECTIVES *When you finish this section, you should be able to:*

1 Find integrals containing $\sqrt{a^2 - x^2}$ (p. 488)
2 Find integrals containing $\sqrt{x^2 + a^2}$ (p. 489)
3 Find integrals containing $\sqrt{x^2 - a^2}$ (p. 491)
4 Use trigonometric substitution to find definite integrals (p. 492)

When an integrand contains a square root of the form $\sqrt{a^2 - x^2}$, $\sqrt{x^2 + a^2}$, or $\sqrt{x^2 - a^2}$, $a > 0$, an appropriate trigonometric substitution will eliminate the radical and sometimes transform the integral into a trigonometric integral like those studied earlier.

The substitutions to use for each of the three types of radicals are given in Table 2.

TABLE 2

Integrand Contains	Substitution	Based on the Identity
$\sqrt{a^2 - x^2}$	$x = a \sin\theta,\ -\dfrac{\pi}{2} \le \theta \le \dfrac{\pi}{2}$	$1 - \sin^2\theta = \cos^2\theta$
$\sqrt{x^2 + a^2}$	$x = a \tan\theta,\ -\dfrac{\pi}{2} < \theta < \dfrac{\pi}{2}$	$\tan^2\theta + 1 = \sec^2\theta$
$\sqrt{x^2 - a^2}$	$x = a \sec\theta,\ 0 \le \theta < \dfrac{\pi}{2},\ \pi \le \theta < \dfrac{3\pi}{2}$	$\sec^2\theta - 1 = \tan^2\theta$

CAUTION Be careful to use the restrictions on each substitution. They guarantee the substitution is a one-to-one function, which is a requirement for using substitution.

NEED TO REVIEW? Right triangle trigonometry is discussed in Appendix A.4, pp. A-27 to A-31.

Although the substitutions to use can be memorized, it is often easier to draw a right triangle and derive them as needed. Each substitution is based on the Pythagorean Theorem. By placing the sides a and x on a right triangle appropriately, the third side of the triangle will represent one of the three types of radicals, as shown in Figure 3.

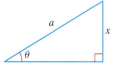

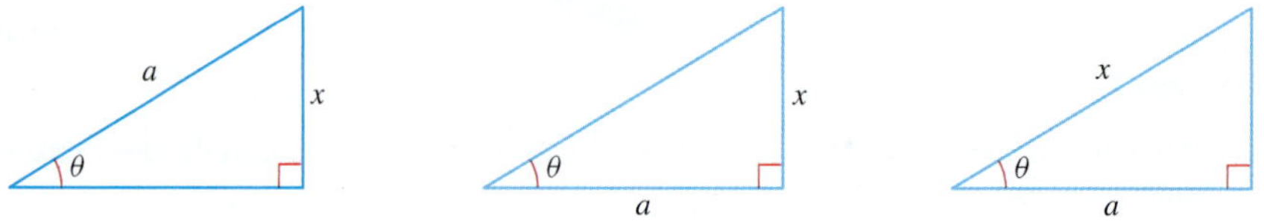

(a) Missing side is $\sqrt{a^2 - x^2}$
$\sin\theta = \dfrac{x}{a}$
$x = a \sin\theta$

(b) Missing side is $\sqrt{x^2 + a^2}$
$\tan\theta = \dfrac{x}{a}$
$x = a \tan\theta$

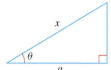

(c) Missing side is $\sqrt{x^2 - a^2}$
$\sec\theta = \dfrac{x}{a}$
$x = a \sec\theta$

Figure 3

1 Find Integrals Containing $\sqrt{a^2 - x^2}$

When an integrand contains a radical of the form $\sqrt{a^2 - x^2}$, $a > 0$, we use the substitution $x = a \sin\theta$, $-\dfrac{\pi}{2} \le \theta \le \dfrac{\pi}{2}$. Substituting $x = a \sin\theta$, $-\dfrac{\pi}{2} \le \theta \le \dfrac{\pi}{2}$, in the expression $\sqrt{a^2 - x^2}$ eliminates the radical as follows:

$$\sqrt{a^2 - x^2} = \sqrt{a^2 - a^2 \sin^2\theta} \qquad \text{Let } x = a \sin\theta;\ -\tfrac{\pi}{2} \le \theta \le \tfrac{\pi}{2}.$$

$$= a\sqrt{1 - \sin^2\theta} \qquad \text{Factor out } a^2;\ a > 0.$$

$$= a\sqrt{\cos^2\theta} \qquad 1 - \sin^2\theta = \cos^2\theta$$

$$= a \cos\theta \qquad \cos\theta \ge 0 \text{ since } -\tfrac{\pi}{2} \le \theta \le \tfrac{\pi}{2}$$

Finding an Integral Containing $\sqrt{4-x^2}$

Find $\displaystyle\int \frac{dx}{x^2\sqrt{4-x^2}}$.

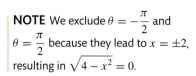

NOTE We exclude $\theta = -\dfrac{\pi}{2}$ and $\theta = \dfrac{\pi}{2}$ because they lead to $x = \pm 2$, resulting in $\sqrt{4-x^2} = 0$.

Solution The integrand contains the square root $\sqrt{4-x^2}$ that is of the form $\sqrt{a^2-x^2}$, where $a = 2$. We use the substitution $x = 2\sin\theta$, $-\dfrac{\pi}{2} < \theta < \dfrac{\pi}{2}$. Then $dx = 2\cos\theta\, d\theta$. Since

$$\underset{\underset{x=2\sin\theta}{\uparrow}}{\sqrt{4-x^2}} = \sqrt{4-4\sin^2\theta} = 2\sqrt{1-\sin^2\theta} = 2\underset{\underset{\cos\theta>0\ \text{since}\ -\frac{\pi}{2}<\theta<\frac{\pi}{2}}{\uparrow}}{\sqrt{\cos^2\theta}} = 2\cos\theta$$

we have

$$\int \frac{dx}{x^2\sqrt{4-x^2}} = \int \frac{2\cos\theta\, d\theta}{(4\sin^2\theta)(2\cos\theta)} = \int \frac{d\theta}{4\sin^2\theta} = \frac{1}{4}\int \csc^2\theta\, d\theta = -\frac{1}{4}\cot\theta + C$$

The original integral is a function of x, but the solution above is a function of θ. To express $\cot\theta$ in terms of x, refer to the right triangles drawn in Figure 4.

Using the Pythagorean Theorem, the third side of each triangle is $\sqrt{2^2-x^2} = \sqrt{4-x^2}$. So,

$$\cot\theta = \frac{\sqrt{4-x^2}}{x} \qquad -\frac{\pi}{2} < \theta < \frac{\pi}{2}$$

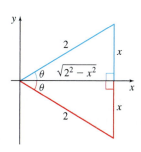

Figure 4 $\sin\theta = \dfrac{x}{2},\ -\dfrac{\pi}{2} < \theta < \dfrac{\pi}{2}$

Then

$$\int \frac{dx}{x^2\sqrt{4-x^2}} = -\frac{1}{4}\cot\theta + C = -\frac{1}{4}\frac{\sqrt{4-x^2}}{x} + C = -\frac{\sqrt{4-x^2}}{4x} + C \qquad \blacksquare$$

Alternatively, trigonometric identities can be used to express $\cot\theta$ in terms of x. Using identities, we get

$$\cot\theta = \frac{\cos\theta}{\sin\theta} = \underset{\substack{\uparrow \\ \cos^2\theta = 1-\sin^2\theta \\ \cos\theta > 0}}{\frac{\sqrt{1-\sin^2\theta}}{\sin\theta}} = \underset{\substack{\uparrow \\ x = 2\sin\theta \\ \sin\theta = \frac{x}{2}}}{\frac{\sqrt{1-\left(\frac{x}{2}\right)^2}}{\frac{x}{2}}} = \frac{2\sqrt{1-\frac{x^2}{4}}}{x} = \frac{\sqrt{4-x^2}}{x}$$

Problem 7.

2 Find Integrals Containing $\sqrt{x^2+a^2}$

When an integrand contains a radical of the form $\sqrt{x^2+a^2}$, $a > 0$, we use the substitution $x = a\tan\theta$, $-\dfrac{\pi}{2} < \theta < \dfrac{\pi}{2}$. Substituting $x = a\tan\theta$, $-\dfrac{\pi}{2} < \theta < \dfrac{\pi}{2}$, in the expression $\sqrt{x^2+a^2}$ eliminates the radical as follows:

$$\underset{\underset{x=a\tan\theta}{\uparrow}}{\sqrt{x^2+a^2}} = \sqrt{a^2\tan^2\theta + a^2} = \underset{\substack{\uparrow \\ \text{Factor out } a, \\ a > 0}}{a\sqrt{\tan^2\theta + 1}} = a\sqrt{\sec^2\theta} = \underset{\substack{\uparrow \\ \sec\theta > 0 \\ \text{since } -\frac{\pi}{2} < \theta < \frac{\pi}{2}}}{a\sec\theta}$$

EXAMPLE 2   Finding an Integral Containing $\sqrt{x^2 + 9}$

Find $\displaystyle\int \frac{dx}{(x^2 + 9)^{3/2}}$.

Solution The integral contains a square root $(x^2 + 9)^{3/2} = \left(\sqrt{x^2 + 9}\right)^3$ that is of the form $\sqrt{x^2 + a^2}$, where $a = 3$. We use the substitution $x = 3\tan\theta$, $-\dfrac{\pi}{2} < \theta < \dfrac{\pi}{2}$. Then $dx = 3\sec^2\theta\, d\theta$. Since

$$(x^2 + 9)^{3/2} = \underset{\underset{x = 3\tan\theta}{\uparrow}}{(9\tan^2\theta + 9)^{3/2}} = 9^{3/2}(\tan^2\theta + 1)^{3/2} = \underset{\underset{\tan^2\theta + 1 = \sec^2\theta}{\uparrow}}{27(\sec^2\theta)^{3/2}} = \underset{\underset{\sec\theta > 0}{\uparrow}}{27\sec^3\theta}$$

we have

$$\int \frac{dx}{(x^2 + 9)^{3/2}} = \int \frac{3\sec^2\theta\, d\theta}{27\sec^3\theta} = \frac{1}{9}\int \frac{d\theta}{\sec\theta} = \frac{1}{9}\int \cos\theta\, d\theta = \frac{1}{9}\sin\theta + C$$

To express the solution in terms of x, use either the right triangles in Figure 5 or identities.

From the right triangles, the hypotenuse is $\sqrt{x^2 + 3^2} = \sqrt{x^2 + 9}$. So, $\sin\theta = \dfrac{x}{\sqrt{x^2 + 9}}$. Then

$$\int \frac{dx}{(x^2 + 9)^{3/2}} = \frac{1}{9}\sin\theta + C = \frac{x}{9\sqrt{x^2 + 9}} + C \qquad \blacksquare$$

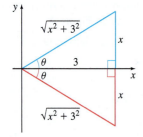

Figure 5 $\tan\theta = \dfrac{x}{3}$, $-\dfrac{\pi}{2} < \theta < \dfrac{\pi}{2}$

NOW WORK Problem 15.

NOTE An integral containing $\sqrt{x^2 + a^2}$, $a > 0$, can also be found using the substitution $x = a\sinh\theta$, since

$$\sqrt{x^2 + a^2} = \sqrt{a^2\sinh^2\theta + a^2}$$
$$= a\sqrt{\sinh^2\theta + 1}$$
$$= a\sqrt{\cosh^2\theta} = a\cosh\theta$$

Try it!

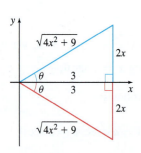

Figure 6 $\tan\theta = \dfrac{2x}{3}$, $-\dfrac{\pi}{2} < \theta < \dfrac{\pi}{2}$

EXAMPLE 3   Finding the Integral $\int (4x^2 + 9)^{1/2}dx$

Find $\displaystyle\int (4x^2 + 9)^{1/2}dx$.

Solution $\displaystyle\int (4x^2 + 9)^{1/2}dx = \int \sqrt{(2x)^2 + 3^2}\, dx$

We use the substitution $2x = 3\tan\theta$, $-\dfrac{\pi}{2} < \theta < \dfrac{\pi}{2}$. Then $dx = \dfrac{3}{2}\sec^2\theta\, d\theta$ and

$$\int (4x^2 + 9)^{1/2}dx = \frac{3}{2}\int \sqrt{9\tan^2\theta + 9}\,\sec^2\theta\, d\theta = \frac{9}{2}\int \sqrt{\tan^2\theta + 1}\,\sec^2\theta\, d\theta$$

$$= \frac{9}{2}\int \sec^3\theta\, d\theta$$

$$= \frac{9}{2}\left[\frac{1}{2}\sec\theta\tan\theta + \frac{1}{2}\ln|\sec\theta + \tan\theta|\right] + C$$

To express the solution in terms of x, refer to the right triangles drawn in Figure 6.

Using the Pythagorean Theorem, the hypotenuse of each triangle is $\sqrt{(2x)^2 + 9} = \sqrt{4x^2 + 9}$. So,

$$\sec\theta = \frac{\sqrt{4x^2 + 9}}{3} \quad \text{and} \quad \tan\theta = \frac{2x}{3} \qquad -\frac{\pi}{2} < \theta < \frac{\pi}{2}$$

Then

$$\int (4x^2 + 9)^{1/2} dx = \frac{9}{4} \left[\sec\theta \tan\theta + \ln|\sec\theta + \tan\theta| \right] + C$$

$$= \frac{9}{4} \left[\frac{\sqrt{4x^2 + 9}}{3} \cdot \frac{2x}{3} + \ln\left| \frac{\sqrt{4x^2 + 9}}{3} + \frac{2x}{3} \right| \right] + C$$

$$= \frac{9}{4} \left[\frac{2x\sqrt{4x^2 + 9}}{9} + \ln\frac{2x + \sqrt{4x^2 + 9}}{3} \right] + C \qquad \blacksquare$$

In general, if the integral contains $\sqrt{b^2x^2 + a^2}$, we use the substitution $bx = a\tan\theta$ $\left(x = \frac{a}{b}\tan\theta \right)$, $-\frac{\pi}{2} < \theta < \frac{\pi}{2}$.

NOW WORK Problem 45.

3 Find Integrals Containing $\sqrt{x^2 - a^2}$

When an integrand contains $\sqrt{x^2 - a^2}$, $a > 0$, we use the substitution $x = a\sec\theta$, $0 \le \theta < \frac{\pi}{2}, \pi \le \theta < \frac{3\pi}{2}$. Then

$$\underset{\underset{x = a\sec\theta}{\uparrow}}{\sqrt{x^2 - a^2}} = \underset{\underset{a > 0}{\uparrow}}{\sqrt{a^2\sec^2\theta - a^2}} = a\sqrt{\sec^2\theta - 1} = \underset{\underset{\tan\theta \ge 0, \text{ since } 0 \le \theta < \frac{\pi}{2}, \pi \le \theta < \frac{3\pi}{2}}{\uparrow}}{a\sqrt{\tan^2\theta}} = a\tan\theta$$

NOTE The substitution $x = a\cosh\theta$ can also be used for integrands containing $\sqrt{x^2 - a^2}$.

EXAMPLE 4 Finding an Integral Containing $\sqrt{x^2 - 4}$

Find $\displaystyle\int \frac{\sqrt{x^2 - 4}}{x}\, dx$.

Solution The integrand contains the square root $\sqrt{x^2 - 4}$ that is of the form $\sqrt{x^2 - a^2}$, where $a = 2$. We use the substitution $x = 2\sec\theta$, $0 \le \theta < \frac{\pi}{2}, \pi \le \theta < \frac{3\pi}{2}$. Then $dx = 2\sec\theta\tan\theta\, d\theta$. Since

$$\underset{\underset{x = 2\sec\theta}{\uparrow}}{\sqrt{x^2 - 4}} = \sqrt{4\sec^2\theta - 4} = 2\sqrt{\sec^2\theta - 1} = 2\sqrt{\tan^2\theta} = \underset{\substack{\tan\theta \ge 0 \\ \text{since } 0 \le \theta < \frac{\pi}{2}, \pi \le \theta < \frac{3\pi}{2}}}{2\tan\theta}$$

we have

$$\int \frac{\sqrt{x^2 - 4}}{x}\, dx = \int \frac{(2\tan\theta)(2\sec\theta\tan\theta\, d\theta)}{2\sec\theta} = 2\int \tan^2\theta\, d\theta = \underset{\underset{\tan^2\theta = \sec^2\theta - 1}{\uparrow}}{2\int (\sec^2\theta - 1)\, d\theta}$$

$$= 2\int \sec^2\theta\, d\theta - 2\int d\theta = 2\tan\theta - 2\theta + C$$

To express the solution in terms of x, we use either the right triangles in Figure 7 or trigonometric identities.

Using identities, we find,

$$\tan\theta = \underset{\underset{\substack{\tan^2\theta = \sec^2\theta - 1 \\ \tan\theta \ge 0}}{\uparrow}}{\sqrt{\sec^2\theta - 1}} = \sqrt{\frac{x^2}{4} - 1} = \frac{1}{2}\sqrt{x^2 - 4}$$

Also since $\sec\theta = \frac{x}{2}$, $0 \le \theta < \frac{\pi}{2}, \pi \le \theta < \frac{3\pi}{2}$, the inverse function $\theta = \sec^{-1}\frac{x}{2}$ is defined.

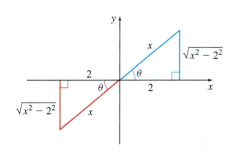

Figure 7 $\sec\theta = \dfrac{x}{2}$,

$0 < \theta < \dfrac{\pi}{2}, \pi < \theta < \dfrac{3\pi}{2}$.

Then

$$\int \frac{\sqrt{x^2 - 4}}{x}\,dx = 2\tan\theta - 2\theta + C = \sqrt{x^2 - 4} - 2\sec^{-1}\frac{x}{2} + C \qquad \blacksquare$$

NOW WORK Problem **29.**

4 Use Trigonometric Substitution to Find Definite Integrals

Trigonometric substitution is also useful when finding certain types of definite integrals.

EXAMPLE 5   **Finding the Area Enclosed by an Ellipse**

Find the area A enclosed by the ellipse $\dfrac{x^2}{4} + \dfrac{y^2}{9} = 1$.

Solution Figure 8 illustrates the ellipse. Since the ellipse is symmetric with respect to both the x-axis and the y-axis, the total area A of the ellipse is four times the shaded area in the first quadrant, where $0 \le x \le 2$ and $0 \le y \le 3$.

We begin by expressing y as a function of x.

$$\frac{x^2}{4} + \frac{y^2}{9} = 1$$

$$\frac{y^2}{9} = 1 - \frac{x^2}{4} = \frac{4 - x^2}{4}$$

$$y^2 = \frac{9}{4}(4 - x^2)$$

$$y = \frac{3}{2}\sqrt{4 - x^2} \qquad y \ge 0$$

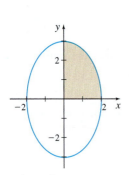

Figure 8 $\dfrac{x^2}{4} + \dfrac{y^2}{9} = 1$

So, the area A of the ellipse is four times the area under the graph of $y = \dfrac{3}{2}\sqrt{4 - x^2}$, $0 \le x \le 2$. That is,

$$A = 4\int_0^2 \frac{3}{2}\sqrt{4 - x^2}\,dx = 6\int_0^2 \sqrt{4 - x^2}\,dx$$

Since the integrand contains a square root of the form $\sqrt{a^2 - x^2}$ with $a = 2$, we use the substitution $x = 2\sin\theta$, $-\dfrac{\pi}{2} \le \theta \le \dfrac{\pi}{2}$. Then $dx = 2\cos\theta\,d\theta$. The new limits of integration are:

- When $x = 0$, $2\sin\theta = 0$, so $\theta = 0$.
- When $x = 2$, $2\sin\theta = 2$, so $\sin\theta = 1$ and $\theta = \dfrac{\pi}{2}$.

Then

$$A = 6\int_0^2 \sqrt{4 - x^2}\,dx = 6\int_0^{\pi/2} \sqrt{4 - 4\sin^2\theta}\cdot 2\cos\theta\,d\theta$$

$$= 6\int_0^{\pi/2} 2\sqrt{1 - \sin^2\theta}\cdot 2\cos\theta\,d\theta = 24\int_0^{\pi/2} \sqrt{\cos^2\theta}\cdot\cos\theta\,d\theta$$

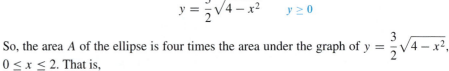

$$= 24\int_0^{\pi/2} \cos^2\theta\,d\theta = \frac{24}{2}\int_0^{\pi/2}[1 + \cos(2\theta)]\,d\theta$$

$$\underset{\cos\theta \ge 0}{\uparrow} \qquad \underset{\cos^2\theta = \frac{1 + \cos(2\theta)}{2}}{\uparrow}$$

$$= 12\left[\theta + \frac{1}{2}\sin(2\theta)\right]_0^{\pi/2} = 12\left(\frac{\pi}{2} + 0\right) = 6\pi$$

The area of the ellipse is 6π square units. ■

NOW WORK Problem **63.**

NEED TO REVIEW? The two approaches to finding a definite integral using the method of substitution are discussed in Section 5.6, pp. 391–393.

In Example 5, we changed the limits of integration to find the definite integral, so there was no need to change back to the variable x. But it is not always easy to obtain new limits of integration, as we see in the next example.

EXAMPLE 6   **Use Trigonometric Substitution to Find a Definite Integral**

Find the area under the graph of $y = \sqrt{x^2 - 1}$ (the upper half of the right branch of the hyperbola $y^2 = x^2 - 1$) from 1 to 3. See Figure 9.

Solution The area A we seek is $A = \int_1^3 \sqrt{x^2 - 1}\,dx$. The integral contains a square root of the form $\sqrt{x^2 - a^2}$, where $a = 1$, so we use the trigonometric substitution $x = \sec\theta$, $0 \le \theta < \dfrac{\pi}{2}$, $\pi \le \theta < \dfrac{3\pi}{2}$. Then $dx = \sec\theta\tan\theta\,d\theta$. Since the upper limit $x = 3$ does not result in a nice angle ($\theta = \sec^{-1} 3$), we find the indefinite integral first and then use the Fundamental Theorem of Calculus.

With $x = \sec\theta$ and $dx = \sec\theta\tan\theta\,d\theta$, we have

$$A = \int \sqrt{x^2 - 1}\,dx = \int \sqrt{\sec^2\theta - 1}\,\sec\theta\tan\theta\,d\theta = \int \tan\theta \cdot \sec\theta\tan\theta\,d\theta$$

$$= \int \tan^2\theta\,\sec\theta\,d\theta$$

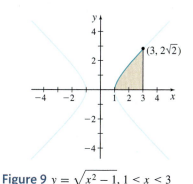

Figure 9 $y = \sqrt{x^2 - 1},\, 1 \le x \le 3$

Since $\tan\theta$ is raised to an even power and $\sec\theta$ to an odd power, we use the identity $\tan^2\theta = \sec^2\theta - 1$. Then

$$A = \int \sqrt{x^2 - 1}\,dx = \int \tan^2\theta\,\sec\theta\,d\theta = \int (\sec^2\theta - 1)\sec\theta\,d\theta$$

$$= \int \sec^3\theta\,d\theta - \int \sec\theta\,d\theta$$

$$= \frac{1}{2}[\sec\theta\tan\theta + \ln|\sec\theta + \tan\theta|] - \ln|\sec\theta + \tan\theta| + C$$

$$= \frac{1}{2}\sec\theta\tan\theta - \frac{1}{2}\ln|\sec\theta + \tan\theta| + C$$

RECALL $\int \sec^3\theta\,d\theta =$ $\dfrac{1}{2}[\sec\theta\tan\theta + \ln|\sec\theta + \tan\theta|]$. Either integrate by parts, or use the reduction formula.

Now we express $\tan\theta$ in terms of $x = \sec\theta$, and apply the Second Fundamental Theorem of Calculus.

$$A = \int_1^3 \sqrt{x^2 - 1}\,dx = \left[\frac{1}{2}x\sqrt{x^2 - 1} - \frac{1}{2}\ln\left|x + \sqrt{x^2 - 1}\right|\right]_1^3$$

$$\underset{\substack{\uparrow \\ \sec\theta = x \\ \tan\theta = \sqrt{x^2 - 1}}}{}$$

$$= \frac{3}{2}\sqrt{8} - \frac{1}{2}\ln(3 + \sqrt{8}) = 3\sqrt{2} - \frac{1}{2}\ln(3 + 2\sqrt{2}) \quad\blacksquare$$

NOW WORK   Problem **53**.

7.3 Assess Your Understanding

Concepts and Vocabulary

1. *True or False*   To find $\int \sqrt{a^2 - x^2}\,dx$, the substitution $x = a\sin\theta$, $-\dfrac{\pi}{2} \le \theta \le \dfrac{\pi}{2}$, can be used.

2. *Multiple Choice*   To find $\int \sqrt{x^2 + 16}\,dx$, use the substitution $x = [$**(a)** $4\sin\theta$, **(b)** $\tan\theta$, **(c)** $4\sec\theta$, **(d)** $4\tan\theta]$.

3. *Multiple Choice*   To find $\int \sqrt{x^2 - 9}\,dx$, use the substitution $x = [$**(a)** $\sec\theta$, **(b)** $3\sin\theta$, **(c)** $3\sec\theta$, **(d)** $3\tan\theta]$.

4. *Multiple Choice*   To find $\int \sqrt{25 - 4x^2}\,dx$, use the substitution $x = \left[\text{(a)} \dfrac{5}{2}\tan\theta, \text{(b)} \dfrac{5}{2}\sin\theta, \text{(c)} \dfrac{2}{5}\sin\theta, \text{(d)} \dfrac{2}{5}\sec\theta\right]$.

---

1. = NOW WORK problem   = Graphing technology recommended   CAS = Computer Algebra System recommended

Skill Building

In Problems 5–14, find each integral. Each of these integrals contains a term of the form $\sqrt{a^2 - x^2}$.

5. $\displaystyle\int \sqrt{4 - x^2}\, dx$

6. $\displaystyle\int \sqrt{16 - x^2}\, dx$

7. $\displaystyle\int \frac{x^2}{\sqrt{16 - x^2}}\, dx$

8. $\displaystyle\int \frac{x^2}{\sqrt{36 - x^2}}\, dx$

9. $\displaystyle\int \frac{\sqrt{4 - x^2}}{x^2}\, dx$

10. $\displaystyle\int \frac{\sqrt{9 - x^2}}{x^2}\, dx$

11. $\displaystyle\int x^2 \sqrt{4 - x^2}\, dx$

12. $\displaystyle\int x^2 \sqrt{1 - 16x^2}\, dx$

13. $\displaystyle\int \frac{dx}{(4 - x^2)^{3/2}}$

14. $\displaystyle\int \frac{dx}{(1 - x^2)^{3/2}}$

In Problems 15–26, find each integral. Each of these integrals contains a term of the form $\sqrt{x^2 + a^2}$.

15. $\displaystyle\int \sqrt{4 + x^2}\, dx$

16. $\displaystyle\int \sqrt{1 + x^2}\, dx$

17. $\displaystyle\int \frac{dx}{\sqrt{x^2 + 16}}$

18. $\displaystyle\int \frac{dx}{\sqrt{x^2 + 25}}$

19. $\displaystyle\int \sqrt{1 + 9x^2}\, dx$

20. $\displaystyle\int \sqrt{9 + 4x^2}\, dx$

21. $\displaystyle\int \frac{x^2}{\sqrt{4 + 9x^2}}\, dx$

22. $\displaystyle\int \frac{x^2}{\sqrt{x^2 + 16}}\, dx$

23. $\displaystyle\int \frac{dx}{x^2 \sqrt{x^2 + 4}}$

24. $\displaystyle\int \frac{dx}{x^2 \sqrt{4x^2 + 1}}$

25. $\displaystyle\int \frac{dx}{(x^2 + 4)^{3/2}}$

26. $\displaystyle\int \frac{dx}{(x^2 + 1)^{3/2}}$

In Problems 27–36, find each integral. Each of these integrals contains a term of the form $\sqrt{x^2 - a^2}$.

27. $\displaystyle\int \frac{x^2}{\sqrt{x^2 - 25}}\, dx$

28. $\displaystyle\int \frac{x^2}{\sqrt{x^2 - 16}}\, dx$

29. $\displaystyle\int \frac{\sqrt{x^2 - 1}}{x}\, dx$

30. $\displaystyle\int \frac{\sqrt{x^2 - 1}}{x^2}\, dx$

31. $\displaystyle\int \frac{dx}{x^2 \sqrt{x^2 - 36}}$

32. $\displaystyle\int \frac{dx}{x^2 \sqrt{x^2 - 9}}$

33. $\displaystyle\int \frac{dx}{\sqrt{4x^2 - 9}}$

34. $\displaystyle\int \frac{dx}{\sqrt{9x^2 - 4}}$

35. $\displaystyle\int \frac{dx}{(x^2 - 9)^{3/2}}$

36. $\displaystyle\int \frac{dx}{(25x^2 - 1)^{3/2}}$

In Problems 37–48, find each integral.

37. $\displaystyle\int \frac{x^2\, dx}{(x^2 - 9)^{3/2}}$

38. $\displaystyle\int \frac{x^2\, dx}{(x^2 - 4)^{3/2}}$

39. $\displaystyle\int \frac{x^2\, dx}{16 + x^2}$

40. $\displaystyle\int \frac{x^2\, dx}{1 + 16x^2}$

41. $\displaystyle\int \sqrt{4 - 25x^2}\, dx$

42. $\displaystyle\int \sqrt{9 - 16x^2}\, dx$

43. $\displaystyle\int \frac{dx}{(4 - 25x^2)^{3/2}}$

44. $\displaystyle\int \frac{dx}{(1 - 9x^2)^{3/2}}$

45. $\displaystyle\int \sqrt{4 + 25x^2}\, dx$

46. $\displaystyle\int \sqrt{9 + 16x^2}\, dx$

47. $\displaystyle\int \frac{dx}{x^3 \sqrt{x^2 - 16}}$

48. $\displaystyle\int \frac{dx}{x^3 \sqrt{x^2 - 1}}$

In Problems 49–58, find each definite integral.

49. $\displaystyle\int_0^1 \sqrt{1 - x^2}\, dx$

50. $\displaystyle\int_0^{1/2} \sqrt{1 - 4x^2}\, dx$

51. $\displaystyle\int_0^1 \sqrt{1 + x^2}\, dx$

52. $\displaystyle\int_0^2 \frac{x^2}{\sqrt{9 + x^2}}\, dx$

53. $\displaystyle\int_4^5 \frac{x^2}{\sqrt{x^2 - 9}}\, dx$

54. $\displaystyle\int_1^2 \frac{x^2}{\sqrt{4x^2 - 1}}\, dx$

55. $\displaystyle\int_0^2 \frac{x^2\, dx}{(16 - x^2)^{3/2}}$

56. $\displaystyle\int_0^1 \frac{x^2\, dx}{(25 - x^2)^{3/2}}$

57. $\displaystyle\int_0^3 \frac{x^2\, dx}{9 + x^2}$

58. $\displaystyle\int_0^1 \frac{x^2}{25 + x^2}\, dx$

Applications and Extensions

59. **Area of an Ellipse** Find $\int \sqrt{a^2 - x^2}\, dx$ and use it to find the area enclosed by the ellipse $\dfrac{x^2}{a^2} + \dfrac{y^2}{b^2} = 1$.

60. **Area of a Semicircle**

   (a) Find $\int_0^2 \sqrt{4 - x^2}\, dx$ by interpreting the integral as a certain area and using elementary geometry.

   (b) Find $\int_0^2 \sqrt{4 - x^2}\, dx$ using a trigonometric substitution.

   (c) Find the area of the semicircle $y = \sqrt{a^2 - x^2}$, $-a \leq x \leq a$, using integration.

61. **Average Value** Find the average value of the function
   $$f(x) = \frac{1}{\sqrt{9 - 4x^2}}$$ over the interval $[0, 1]$.

62. **Average Value** Find the average value of the function $f(x) = \sqrt{x^2 - 4}$ the interval $[2, 7]$.

63. **Area Under a Graph** Find the area under the graph of
   $$y = \frac{x^3}{\sqrt{9 - x^2}}$$ from $x = 0$ to $x = 2$.

64. **Area Under a Graph** Find the area under the graph of $y = x\sqrt{16 - x^2}$, $x \geq 0$.

65. Area Under a Graph Find the area under the graph of

$$y = \frac{x^2}{\sqrt{x^2 - 1}} \text{ from } x = 3 \text{ to } x = 5.$$

66. Hydrostatic Force A round window of radius 2 meters (m) is built into the side of a large, fresh-water aquarium tank. If the center of the window is 3 m below the water line, find the force due to hydrostatic pressure on the window. (*Hint*: The mass density of fresh water is $\rho = 1000 \, \text{kg}/\text{m}^3$.)

67. Area of a Lune A **lune** is a crescent-shaped area formed when two circles intersect.

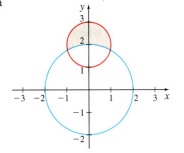

(a) Find the area of the smaller lune formed by the intersection of the two circles $x^2 + y^2 = 4$ and $x^2 + (y - 2)^2 = 1$, as shown in the figure.

(b) What is the area of the larger lune?

68. Area Find the area enclosed by the hyperbola $\dfrac{x^2}{9} - \dfrac{y^2}{16} = 1$ and the line $x = 6$.

69. Arc Length Find the length of the graph of the parabola $y = 5x - x^2$ that lies above the x-axis.

70. Arc Length Find the length of the graph of $y = \ln x$ from $x = \dfrac{\sqrt{3}}{3}$ to $x = \sqrt{3}$.

71. Volume of a Solid of Revolution Find the volume of the solid of revolution generated by revolving the region bounded by the graph of $y = \dfrac{1}{x^2 + 4}$ and the x-axis from $x = 0$ to $x = 1$ about the x-axis. See the figure.

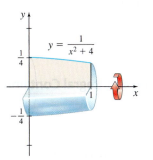

72. Volume of a Solid of Revolution Find the volume of the solid of revolution generated by revolving the region bounded by the graphs of $y = \dfrac{1}{\sqrt{9 - x^2}}$, $y = 0$, $x = 0$, and $x = 2$ about the x-axis.

In Problems 73–78, find each integral. (Hint: Begin with a substitution.)

73. $\displaystyle\int \frac{dx}{\sqrt{1 - (x - 2)^2}}$

74. $\displaystyle\int \sqrt{4 - (x + 2)^2} \, dx$

75. $\displaystyle\int \frac{dx}{\sqrt{(x - 1)^2 - 4}}$

76. $\displaystyle\int \frac{dx}{(x - 2)\sqrt{(x - 2)^2 + 9}}$

77. $\displaystyle\int e^x \sqrt{25 - e^{2x}} \, dx$

78. $\displaystyle\int e^x \sqrt{4 + e^{2x}} \, dx$

In Problems 79 and 80, use integration by parts and then the methods of this section to find each integral.

79. $\displaystyle\int x \sin^{-1} x \, dx$

80. $\displaystyle\int x \cos^{-1} x \, dx$

81. Find $\int \sqrt{x^2 + a^2} \, dx$

(a) By using a trigonometric substitution

(b) By using substitution with a hyperbolic function

In Problems 82–86, use a trigonometric substitution to derive each formula. Assume $a > 0$.

82. $\displaystyle\int \frac{dx}{\sqrt{a^2 - x^2}} = \sin^{-1} \frac{x}{a} + C$

83. $\displaystyle\int \frac{dx}{a^2 + x^2} = \frac{1}{a} \tan^{-1} \frac{x}{a} + C$

84. $\displaystyle\int \frac{dx}{x\sqrt{x^2 - a^2}} = \frac{1}{a} \sec^{-1} \frac{x}{a} + C$

85. $\displaystyle\int \frac{dx}{\sqrt{x^2 - a^2}} = \ln \left| \frac{x + \sqrt{x^2 - a^2}}{a} \right| + C$

86. $\displaystyle\int \frac{dx}{\sqrt{x^2 + a^2}} = \ln \left| x + \sqrt{x^2 + a^2} \right| + C$

Challenge Problems

87. Find $\displaystyle\int \frac{dx}{\sqrt{3x - x^2}}$

88. Derive the formula

$$\int \sqrt{x^2 - a^2} \, dx = \frac{1}{2}x\sqrt{x^2 - a^2} - \frac{1}{2}a^2 \ln \left| x + \sqrt{x^2 - a^2} \right| + C,$$

$a > 0$.

89. Find $\displaystyle\int \frac{dx}{\sqrt{x^2 + a^2}}$, $a > 0$, using the substitution $u = \sinh^{-1} \frac{x}{a}$.

Express your answer in logarithmic form.

90. Find $\displaystyle\int \frac{\sec^2 x}{\sqrt{\tan^2 x - 6\tan x + 8}} \, dx$.

EXAMPLE 3   **Finding an Integral Containing $2x - x^2$**

Find $\displaystyle\int \frac{dx}{\sqrt{2x - x^2}}$.

Solution   The integrand contains the quadratic expression $2x - x^2$, so we complete the square.

$$2x - x^2 = -x^2 + 2x = -(x^2 - 2x) = -(x^2 - 2x + 1) + 1$$
$$= -(x-1)^2 + 1 = 1 - (x-1)^2$$

RECALL $\displaystyle\int \frac{dx}{\sqrt{a^2 - x^2}} = \sin^{-1}\frac{x}{a} + C$

Then

$$\int \frac{dx}{\sqrt{2x - x^2}} = \int \frac{dx}{\sqrt{1 - (x-1)^2}} = \underset{\substack{\uparrow \\ u = x - 1 \\ du = dx}}{\int \frac{du}{\sqrt{1 - u^2}}} = \sin^{-1} u + C = \sin^{-1}(x - 1) + C$$

■

NOW WORK   **Problem 23.**

7.4 Assess Your Understanding

Skill Building

In Problems 1–32, find each integral.

1. $\displaystyle\int \frac{dx}{x^2 + 4x + 5}$

2. $\displaystyle\int \frac{dx}{x^2 + 2x + 5}$

3. $\displaystyle\int \frac{dx}{x^2 + 4x + 8}$

4. $\displaystyle\int \frac{dx}{x^2 - 6x + 10}$

5. $\displaystyle\int \frac{2\,dx}{3 + 2x + 2x^2}$

6. $\displaystyle\int \frac{3\,dx}{x^2 + 6x + 10}$

7. $\displaystyle\int \frac{x\,dx}{2x^2 + 2x + 3}$

8. $\displaystyle\int \frac{3x\,dx}{x^2 + 6x + 10}$

9. $\displaystyle\int \frac{dx}{\sqrt{8 + 2x - x^2}}$

10. $\displaystyle\int \frac{dx}{\sqrt{5 - 4x - 2x^2}}$

11. $\displaystyle\int \frac{dx}{\sqrt{4x - x^2}}$

12. $\displaystyle\int \frac{dx}{\sqrt{x^2 - 6x - 10}}$

13. $\displaystyle\int \frac{dx}{(x+1)\sqrt{x^2 + 2x + 2}}$

14. $\displaystyle\int \frac{dx}{(x-4)\sqrt{x^2 - 8x + 17}}$

15. $\displaystyle\int \frac{dx}{\sqrt{24 - 2x - x^2}}$

16. $\displaystyle\int \frac{dx}{\sqrt{9x^2 + 6x + 10}}$

17. $\displaystyle\int \frac{x - 5}{\sqrt{x^2 - 2x + 5}}dx$

18. $\displaystyle\int \frac{x + 1}{x^2 - 4x + 3}dx$

19. $\displaystyle\int_1^3 \frac{dx}{\sqrt{x^2 - 2x + 5}}$

20. $\displaystyle\int_{1/2}^1 \frac{x^2\,dx}{\sqrt{2x - x^2}}$

21. $\displaystyle\int \frac{e^x\,dx}{\sqrt{e^{2x} + e^x + 1}}$

22. $\displaystyle\int \frac{\cos x\,dx}{\sqrt{\sin^2 x + 4\sin x + 3}}$

23. $\displaystyle\int \frac{2x - 3}{\sqrt{4x - x^2 - 3}}dx$

24. $\displaystyle\int \frac{x + 3}{\sqrt{x^2 + 2x + 2}}dx$

25. $\displaystyle\int \frac{dx}{(x^2 - 2x + 10)^{3/2}}$

26. $\displaystyle\int \frac{dx}{\sqrt{x^2 - 2x + 10}}$

27. $\displaystyle\int \frac{dx}{\sqrt{x^2 + 2x - 3}}$

28. $\displaystyle\int x\sqrt{x^2 - 4x - 1}\,dx$

29. $\displaystyle\int \frac{\sqrt{5 + 4x - x^2}}{x - 2}dx$

30. $\displaystyle\int \sqrt{5 + 4x - x^2}\,dx$

31. $\displaystyle\int \frac{x\,dx}{\sqrt{x^2 + 2x - 3}}$

32. $\displaystyle\int \frac{x\,dx}{\sqrt{x^2 - 4x + 3}}$

Applications and Extensions

33. Show that if $k > 0$, then

$$\int \frac{dx}{\sqrt{(x+h)^2 + k}} = \ln\left[\sqrt{(x+h)^2 + k} + x + h\right] + C$$

34. Show that if $a > 0$ and $b^2 - 4ac > 0$, then

$$\int \frac{dx}{\sqrt{ax^2 + bx + c}} = \frac{1}{\sqrt{a}}\ln\left|\sqrt{ax^2 + bx + c} + \sqrt{a}x + \frac{b}{2\sqrt{a}}\right| + C$$

Challenge Problem

35. Find $\displaystyle\int \sqrt{\frac{a + x}{a - x}}\,dx$.

---

1. = NOW WORK problem   = Graphing technology recommended   CAS = Computer Algebra System recommended

7.5 Integration of Rational Functions Using Partial Fractions

OBJECTIVES *When you finish this section, you should be able to:*

1 Integrate a rational function whose denominator contains only distinct linear factors (p. 500)

2 Integrate a rational function whose denominator contains a repeated linear factor (p. 502)

3 Integrate a rational function whose denominator contains a distinct irreducible quadratic factor (p. 503)

4 Integrate a rational function whose denominator contains a repeated irreducible quadratic factor (p. 504)

NEED TO REVIEW? Rational functions are discussed in Section P.2, pp. 19–20.

In this section, we integrate rational functions using a technique called *partial fractions*. Although the integral of any rational function can be found using partial fractions, the process requires being able to factor the denominator of the rational function.

Recall that a rational function R in lowest terms is the ratio of two polynomial functions p and $q \neq 0$, where p and q have no common factors. The domain of R is the set of all real numbers for which $q \neq 0$. The rational function R is called **proper** when the degree of the polynomial p is less than the degree of the polynomial q; otherwise, R is an **improper** rational function.

Every improper rational function R can be reduced by long division to the sum of a polynomial function and a proper rational function. For example, the rational function

$R(x) = \dfrac{x^2}{x-1}$ is improper, but by long division,

$$R(x) = \frac{x^2}{x-1} = x + 1 + \frac{1}{x-1}$$

Then we can find $\int R(x)\,dx$ by using properties of an indefinite integral to obtain

$$\int \frac{x^2}{x-1}\,dx = \int \left(x + 1 + \frac{1}{x-1}\right) dx = \int x\,dx + \int dx + \int \frac{1}{x-1}\,dx$$

$$= \frac{x^2}{2} + x + \ln|x-1| + C$$

For these reasons, the discussion that follows deals only with proper rational functions in lowest terms. Such rational functions can be written as the sum of simpler functions, called **partial fractions.**

For example, since $\dfrac{2}{x-1} + \dfrac{3}{x+4} = \dfrac{5x+5}{(x-1)(x+4)}$, the rational function

$$R(x) = \frac{5x+5}{(x-1)(x+4)}$$

can be expressed as the sum $R(x) = \dfrac{2}{x-1} + \dfrac{3}{x+4}$. Then

$$R(x) = \int \frac{5x+5}{(x-1)(x+4)}\,dx = \int \left(\frac{2}{x-1} + \frac{3}{x+4}\right) dx$$

$$= 2\int \frac{1}{x-1}\,dx + 3\int \frac{1}{x+4}\,dx$$

$$= 2\ln|x-1| + 3\ln|x+4| + C$$

We use a technique called **partial fraction decomposition** to write a proper rational function R as the sum of simpler rational terms. The resulting partial fractions depend on the nature of the factors of the denominator q. It can be shown that any polynomial q whose coefficients are real numbers can be factored (over the real numbers) into products of linear and/or irreducible quadratic factors. This means *the integral of every rational function can be expressed in terms of algebraic, logarithmic, and/or inverse trigonometric functions.*

The rest of this section presents systematic methods of decomposing rational functions into sums of partial fractions, that is, into forms that we can integrate.

1 Integrate a Rational Function Whose Denominator Contains Only Distinct Linear Factors

Case 1: If the denominator q contains only distinct linear factors, say, $x - a_1$, $x - a_2, \ldots, x - a_n$, then $\dfrac{p}{q}$ can be written as

$$\frac{p(x)}{q(x)} = \frac{A_1}{x - a_1} + \frac{A_2}{x - a_2} + \cdots + \frac{A_n}{x - a_n} \tag{1}$$

where $A_1, A_2, \ldots, A_n$ are real numbers.

To find $\displaystyle\int \frac{p(x)}{q(x)} dx$, we integrate both sides of (1). Then

$$\int \frac{p(x)}{q(x)} dx = \int \frac{A_1}{x - a_1} dx + \int \frac{A_2}{x - a_2} dx + \cdots + \int \frac{A_n}{x - a_n} dx$$

$$= A_1 \ln|x - a_1| + A_2 \ln|x - a_2| + \cdots + A_n \ln|x - a_n| + C$$

All that remains is to find the numbers $A_1, \ldots, A_n$.

A procedure for finding the numbers $A_1, \ldots, A_n$ is illustrated in Example 1.

EXAMPLE 1   Integrating a Rational Function Whose Denominator Contains Only Distinct Linear Factors

Find $\displaystyle\int \frac{x \, dx}{x^2 - 5x + 6}$.

Solution   The integrand is a proper rational function in lowest terms. We begin by factoring the denominator: $x^2 - 5x + 6 = (x - 2)(x - 3)$. Since the factors are linear and distinct, we apply Case 1 and allow for the terms $\dfrac{A}{x - 2}$ and $\dfrac{B}{x - 3}$.

$$\frac{x}{(x - 2)(x - 3)} = \frac{A}{x - 2} + \frac{B}{x - 3}$$

Now we clear fractions by multiplying both sides of the equation by $(x - 2)(x - 3)$.

$$x = A(x - 3) + B(x - 2)$$
$$x = (A + B)x - (3A + 2B) \qquad \text{Group like terms.}$$

This is an identity in x, so the coefficients of like powers of x must be equal.

$$1 = A + B \qquad \text{The coefficient of } x \text{ equals 1.}$$
$$0 = -3A - 2B \qquad \text{The constant term on the left is 0.}$$

This is a system of two equations containing two variables. Solving the second equation for B, we get $B = -\dfrac{3}{2}A$. Substituting for B in the first equation produces the solution $A = -2$ from which $B = 3$. So,

$$\frac{x}{(x - 2)(x - 3)} = \frac{-2}{x - 2} + \frac{3}{x - 3}$$

Then

$$\int \frac{x}{(x-2)(x-3)} dx = \int \frac{-2}{x-2} dx + \int \frac{3}{x-3} dx$$

$$= -2 \ln |x-2| + 3 \ln |x-3| + C = \ln \left| \frac{(x-3)^3}{(x-2)^2} \right| + C \quad \blacksquare$$

NOW WORK **Problem 11.**

Alternatively, we can find the unknown numbers in the decomposition of $\frac{p}{q}$ by substituting convenient values of x into the identity obtained after clearing fractions.* In Example 1, after clearing fractions, the identity is

$$x = A(x-3) + B(x-2)$$

When $x = 3$, the term involving A drops out, leaving $3 = B \cdot 1$, so that $B = 3$. When $x = 2$, the term involving B drops out, leaving $2 = A \cdot (-1)$, so that $A = -2$.

EXAMPLE 2   Deriving Formulas Involving a Rational Function

Derive these formulas:

$$\text{(a)} \quad \int \frac{dx}{x^2 - a^2} = \frac{1}{2a} \ln \left| \frac{x-a}{x+a} \right| + C \quad a \neq 0$$

$$\text{(b)} \quad \int \frac{dx}{a^2 - x^2} = \frac{1}{2a} \ln \left| \frac{x+a}{x-a} \right| + C \quad a \neq 0$$

Solution **(a)** The factored denominator $x^2 - a^2 = (x-a)(x+a)$ contains only distinct linear factors. So, $\frac{1}{x^2 - a^2}$ can be decomposed into partial fractions of the form

$$\frac{1}{x^2 - a^2} = \frac{1}{(x-a)(x+a)} = \frac{A}{x-a} + \frac{B}{x+a}$$

$$1 = A(x+a) + B(x-a) \qquad \textcolor{blue}{\text{Multiply both sides by } (x-a)(x+a).}$$

This is an identity in x. When $x = a$, the term involving B drops out. Then $1 = A(2a)$, so $A = \frac{1}{2a}$. When $x = -a$, the term involving A drops out. Then $1 = B(-2a)$, so $B = -\frac{1}{2a}$. Then,

$$\frac{1}{x^2 - a^2} = \frac{A}{x-a} + \frac{B}{x+a} = \frac{1}{2a(x-a)} - \frac{1}{2a(x+a)}$$

$$\int \frac{dx}{x^2 - a^2} = \frac{1}{2a} \int \frac{dx}{x-a} - \frac{1}{2a} \int \frac{dx}{x+a} = \frac{1}{2a} \left(\int \frac{dx}{x-a} - \int \frac{dx}{x+a} \right)$$

$$= \frac{1}{2a} \left(\ln |x-a| - \ln |x+a| \right) + C$$

$$= \frac{1}{2a} \ln \left| \frac{x-a}{x+a} \right| + C$$

---

*This method is discussed in detail in H. J. Straight & R. Dowds (1984, June–July), *American Mathematical Monthly*, 91(6), 365.

(b) Using the result from (a), we get

$$\int \frac{dx}{a^2 - x^2} = -\int \frac{dx}{x^2 - a^2} = -\frac{1}{2a} \ln \left| \frac{x-a}{x+a} \right| + C = \frac{1}{2a} \ln \left| \frac{x-a}{x+a} \right|^{-1} + C$$

$$= \frac{1}{2a} \ln \left| \frac{x+a}{x-a} \right| + C \qquad ■$$

NOW WORK   Problem 37.

2 Integrate a Rational Function Whose Denominator Contains a Repeated Linear Factor

> *Case 2*: If the denominator q has a repeated linear factor $(x-a)^n$, $n \geq 2$ an integer, then the decomposition of $\dfrac{p}{q}$ includes the terms
>
> $$\boxed{\frac{A_1}{x-a}, \frac{A_2}{(x-a)^2}, \ldots, \frac{A_n}{(x-a)^n}}$$
>
> where $A_1, A_2, \ldots, A_n$ are real numbers.

EXAMPLE 3   Integrating a Rational Function Whose Denominator Contains a Repeated Linear Factor

Find $\displaystyle\int \frac{dx}{x(x-1)^2}$.

Solution Since x is a distinct linear factor of the denominator q, and $(x-1)^2$ is a repeated linear factor of the denominator, the decomposition of $\dfrac{1}{x(x-1)^2}$ into partial fractions has the three terms $\dfrac{A}{x}$, $\dfrac{B}{x-1}$, and $\dfrac{C}{(x-1)^2}$.

$$\frac{1}{x(x-1)^2} = \frac{A}{x} + \frac{B}{x-1} + \frac{C}{(x-1)^2} \qquad \text{\color{teal}Write the identity.}$$

$$1 = A(x-1)^2 + B \cdot x(x-1) + C \cdot x \qquad \text{\color{teal}Multiply both sides by } x(x-1)^2.$$

We find A, B, and C by choosing values of x that cause one or more terms to drop out. When $x = 1$, we have $1 = C \cdot 1$, so $C = 1$. When $x = 0$, we have $1 = A(0-1)^2$, so $A = 1$. Now using $A = 1$ and $C = 1$, we have

$$1 = (x-1)^2 + B \cdot x(x-1) + 1 \cdot x$$

Suppose we let $x = 2$. (Any choice other than 0 and 1 will also work.) Then

$$1 = 1 + 2B + 2$$

$$B = -1$$

Then

$$\frac{1}{x(x-1)^2} = \frac{1}{x} + \frac{-1}{(x-1)} + \frac{1}{(x-1)^2} \qquad \text{\color{teal}} A=1 \quad B=-1 \quad C=1$$

So,

$$\int \frac{dx}{x(x-1)^2} = \int \frac{dx}{x} - \int \frac{dx}{x-1} + \int \frac{dx}{(x-1)^2}$$

$$= \ln|x| - \ln|x-1| - \frac{1}{x-1} + C_1 \qquad \blacksquare$$

NOTE To avoid confusion, we use C_1 for the constant of integration whenever C appears in the partial fraction decomposition.

NOW WORK Problem 15.

3 Integrate a Rational Function Whose Denominator Contains a Distinct Irreducible Quadratic Factor

A quadratic polynomial $ax^2 + bx + c$ is called **irreducible** if it cannot be factored into real linear factors. This happens if the discriminant $b^2 - 4ac < 0$. For example, $x^2 + x + 1$ and $x^2 + 4$ are irreducible.

NEED TO REVIEW? The discriminant of a quadratic equation is discussed in Appendix A.1, pp. A-3 to A-4.

> ***Case 3:*** If the denominator q contains a nonrepeated irreducible quadratic factor $ax^2 + bx + c$, then the decomposition of $\frac{p}{q}$ includes the term
>
> $$\boxed{\frac{Ax + B}{ax^2 + bx + c}}$$
>
> where A and B are real numbers.

EXAMPLE 4 **Integrating a Rational Function Whose Denominator Contains a Distinct Irreducible Quadratic Factor**

Find $\int \frac{3x}{x^3 - 1}\, dx$.

Solution The denominator is $x^3 - 1 = (x-1)(x^2 + x + 1)$. Since $x - 1$ is a nonrepeated linear factor, by Case 1 the decomposition of $\frac{3x}{x^3 - 1} = \frac{3x}{(x-1)(x^2 + x + 1)}$ has the term $\frac{A}{x-1}$.

The discriminant of the quadratic equation $x^2 + x + 1 = 0$ is negative, so $x^2 + x + 1$ is an irreducible quadratic factor of q, and the decomposition of $\frac{p}{q}$ also contains the term $\frac{Bx + C}{x^2 + x + 1}$. Then

$$\frac{3x}{x^3 - 1} = \frac{A}{x-1} + \frac{Bx + C}{x^2 + x + 1}$$

Clearing the denominators, we have

$$3x = A(x^2 + x + 1) + (Bx + C)(x - 1)$$

This is an identity in x. When $x = 1$, we have $3 = 3A$, so $A = 1$. With $A = 1$, the identity becomes

$$3x = (x^2 + x + 1) + (Bx + C)(x - 1)$$

$$-x^2 + 2x - 1 = (Bx + C)(x - 1)$$

$$-(x - 1)^2 = (Bx + C)(x - 1)$$

$$-(x - 1) = Bx + C$$

$$B = -1, \quad C = 1$$

So,

$$\frac{3x}{x^3 - 1} = \frac{1}{x - 1} + \frac{-x + 1}{x^2 + x + 1}$$

$$\int \frac{3x}{x^3 - 1}\, dx = \int \left(\frac{1}{x - 1} + \frac{-x + 1}{x^2 + x + 1} \right) dx = \int \frac{1}{x - 1}\, dx + \int \frac{-x + 1}{x^2 + x + 1}\, dx$$

$$= \ln |x - 1| - \int \frac{x - 1}{x^2 + x + 1}\, dx \qquad (2)$$

To find the integral on the right, we complete the square in the denominator and use substitution.

$$\int \frac{x - 1}{x^2 + x + 1}\, dx = \int \frac{x - 1}{\left(x + \frac{1}{2}\right)^2 + \frac{3}{4}}\, dx \underset{\substack{\uparrow \\ u = x + \frac{1}{2}}}{=} \int \frac{u - \frac{3}{2}}{u^2 + \frac{3}{4}}\, du = \int \frac{u}{u^2 + \frac{3}{4}}\, du - \frac{3}{2} \int \frac{du}{u^2 + \frac{3}{4}}$$

$$= \frac{1}{2} \ln \left(u^2 + \frac{3}{4} \right) - \frac{3}{2} \left[\frac{2}{\sqrt{3}} \tan^{-1} \left(\frac{2}{\sqrt{3}} u \right) \right] \qquad \int \frac{du}{u^2 + a^2} = \frac{1}{a} \tan^{-1} \left(\frac{u}{a} \right)$$

$$= \frac{1}{2} \ln \left(u^2 + \frac{3}{4} \right) - \sqrt{3} \tan^{-1} \left(\frac{2}{\sqrt{3}} u \right)$$

$$= \frac{1}{2} \ln (x^2 + x + 1) - \sqrt{3} \tan^{-1} \left(\frac{2x + 1}{\sqrt{3}} \right)$$

Then from (2),

$$\int \frac{3x}{x^3 - 1}\, dx = \ln |x - 1| - \frac{1}{2} \ln (x^2 + x + 1) + \sqrt{3} \tan^{-1} \left(\frac{2x + 1}{\sqrt{3}} \right) + C_1 \qquad ■$$

NOW WORK **Problem 21.**

4 Integrate a Rational Function Whose Denominator Contains a Repeated Irreducible Quadratic Factor

Case 4: If the denominator q contains a repeated irreducible quadratic polynomial $(x^2 + bx + c)^n$, $n \geq 2$ an integer, then the decomposition of $\dfrac{p}{q}$ includes the terms

$$\frac{A_1 x + B_1}{x^2 + bx + c}, \frac{A_2 x + B_2}{(x^2 + bx + c)^2}, \ldots, \frac{A_n x + B_n}{(x^2 + bx + c)^n}$$

where $A_1, B_1, A_2, B_2, \ldots, A_n, B_n$ are real numbers.

EXAMPLE 5 Integrating a Rational Function Whose Denominator Contains a Repeated Irreducible Quadratic Factor

Find $\displaystyle\int \frac{x^3 + 1}{(x^2 + 4)^2}\, dx$.

Solution The denominator is a repeated, irreducible quadratic, so the decomposition of $\dfrac{x^3 + 1}{(x^2 + 4)^2}$ is

$$\frac{x^3 + 1}{(x^2 + 4)^2} = \frac{Ax + B}{x^2 + 4} + \frac{Cx + D}{(x^2 + 4)^2}$$

Clearing fractions and combining terms give

$$x^3 + 1 = (Ax + B)(x^2 + 4) + Cx + D$$

$$x^3 + 1 = Ax^3 + Bx^2 + (4A + C)x + 4B + D$$

Equating coefficients, we get

$$A = 1 \qquad B = 0 \qquad 4A + C = 0 \qquad 4B + D = 1$$
$$C = -4 \qquad\qquad D = 1$$

Then

$$\frac{x^3 + 1}{(x^2 + 4)^2} = \frac{x}{x^2 + 4} + \frac{-4x + 1}{(x^2 + 4)^2}$$

and

$$\int \frac{x^3 + 1}{(x^2 + 4)^2}dx = \int \frac{x}{x^2 + 4}dx + \int \frac{-4x + 1}{(x^2 + 4)^2}dx$$

$$= \int \frac{x}{x^2 + 4}dx - 4 \int \frac{x}{(x^2 + 4)^2}dx + \int \frac{dx}{(x^2 + 4)^2} \qquad (3)$$

In the first two integrals on the right in (3), we use the substitution $u = x^2 + 4$, $x \geq 0$. Then $du = 2x\,dx$, and

$$\bullet \qquad \int \frac{x}{x^2 + 4}dx = \frac{1}{2}\int \frac{du}{u} = \frac{1}{2}\ln |u| = \frac{1}{2}\ln(x^2 + 4) \qquad (4)$$

$$\bullet \quad -4 \int \frac{x}{(x^2 + 4)^2}dx = -2 \int \frac{du}{u^2} = \frac{2}{u} = \frac{2}{x^2 + 4} \qquad (5)$$

In the third integral on the right in (3), we use the trigonometric substitution $x = 2\tan\theta$, $-\dfrac{\pi}{2} < \theta < \dfrac{\pi}{2}$. Then $dx = 2\sec^2\theta\,d\theta$, and

$$\int \frac{dx}{(x^2 + 4)^2} = \int \frac{2\sec^2\theta\,d\theta}{\left(4\tan^2\theta + 4\right)^2} = \frac{2}{16}\int \frac{\sec^2\theta\,d\theta}{(\sec^2\theta)^2} = \frac{1}{8}\int \cos^2\theta\,d\theta$$

$$= \frac{1}{8}\int \frac{1 + \cos(2\theta)}{2}d\theta$$

$$= \frac{1}{16}\left[\theta + \frac{1}{2}\sin(2\theta)\right] = \frac{1}{16}(\theta + \sin\theta\cos\theta)$$

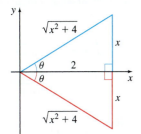

Figure 10 $\tan\theta = \dfrac{x}{2}, -\dfrac{\pi}{2} < \theta < \dfrac{\pi}{2}$

To express the solution in terms of x, either use the triangles in Figure 10 or use trigonometric identities as follows:

$$\sin\theta\cos\theta = \frac{\sin\theta}{\cos\theta}\cdot\cos^2\theta = \frac{\tan\theta}{\sec^2\theta} = \frac{\tan\theta}{\tan^2\theta + 1} = \frac{\dfrac{x}{2}}{\dfrac{x^2}{4} + 1} = \frac{2x}{x^2 + 4}$$

Then

$$\int \frac{dx}{(x^2 + 4)^2} = \frac{1}{16}(\theta + \sin\theta\cos\theta) = \frac{1}{16}\left(\tan^{-1}\frac{x}{2} + \frac{2x}{x^2 + 4}\right) \qquad (6)$$

$$\underset{x = 2\tan\theta;\ \theta = \tan^{-1}\dfrac{x}{2}}{\uparrow}$$

Now use the results of (3), (4), (5), and (6):

$$\int \frac{x^3 + 1}{(x^2 + 4)^2}dx = \frac{1}{2}\ln(x^2 + 4) + \frac{2}{x^2 + 4} + \frac{1}{16}\tan^{-1}\frac{x}{2} + \frac{x}{8(x^2 + 4)} + C_1 \qquad \blacksquare$$

NOW WORK Problem 23.

7.5 Assess Your Understanding

Concepts and Vocabulary

1. *Multiple Choice* A rational function $R(x) = \dfrac{p(x)}{q(x)}$ is proper when the degree of p is [(a) less than, (b) equal to, (c) greater than] the degree of q.

2. *True or False* Every improper rational function can be written as the sum of a polynomial and a proper rational function.

3. *True or False* Sometimes the integration of a proper rational function leads to a logarithm.

4. *True or False* The decomposition of $\dfrac{7x + 1}{(x + 1)^4}$ into partial fractions has three terms: $\dfrac{A}{x + 1} + \dfrac{B}{(x + 1)^2} + \dfrac{C}{(x + 1)^3}$, where A, B, and C are real numbers.

Skill Building

In Problems 5–8, find each integral by first writing the integrand as the sum of a polynomial and a proper rational function.

5. $\displaystyle\int \dfrac{x^2 + 1}{x + 1}\,dx$

6. $\displaystyle\int \dfrac{x^2 + 4}{x - 2}\,dx$

7. $\displaystyle\int \dfrac{x^3 + 3x - 4}{x - 2}\,dx$

8. $\displaystyle\int \dfrac{x^3 - 3x^2 + 4}{x + 3}\,dx$

In Problems 9–14, find each integral. (Hint: Each of the denominators contains only distinct linear factors.)

9. $\displaystyle\int \dfrac{dx}{(x - 2)(x + 1)}$

10. $\displaystyle\int \dfrac{dx}{(x + 4)(x - 1)}$

11. $\displaystyle\int \dfrac{x\,dx}{(x - 1)(x - 2)}$

12. $\displaystyle\int \dfrac{3x\,dx}{(x + 2)(x - 4)}$

13. $\displaystyle\int \dfrac{x\,dx}{(3x - 2)(2x + 1)}$

14. $\displaystyle\int \dfrac{dx}{(2x + 3)(4x - 1)}$

In Problems 15–18, find each integral. (Hint: Each of the denominators contains a repeated linear factor.)

15. $\displaystyle\int \dfrac{x - 3}{(x + 2)(x + 1)^2}\,dx$

16. $\displaystyle\int \dfrac{x + 1}{x^2(x - 2)}\,dx$

17. $\displaystyle\int \dfrac{x^2\,dx}{(x - 1)^2(x + 1)}$

18. $\displaystyle\int \dfrac{x^2 + x}{(x + 2)(x - 1)^2}\,dx$

In Problems 19–22, find each integral. (Hint: Each of the denominators contains an irreducible quadratic factor.)

19. $\displaystyle\int \dfrac{dx}{x(x^2 + 1)}$

20. $\displaystyle\int \dfrac{dx}{(x + 1)(x^2 + 4)}$

21. $\displaystyle\int \dfrac{x^2 + 2x + 3}{(x + 1)(x^2 + 2x + 4)}\,dx$

22. $\displaystyle\int \dfrac{x^2 - 11x - 18}{x(x^2 + 3x + 3)}\,dx$

In Problems 23–26, find each integral. (Hint: Each of the denominators contains a repeated irreducible quadratic factor.)

23. $\displaystyle\int \dfrac{2x + 1}{(x^2 + 16)^2}\,dx$

24. $\displaystyle\int \dfrac{x^2 + 2x + 3}{(x^2 + 4)^2}\,dx$

25. $\displaystyle\int \dfrac{x^3\,dx}{(x^2 + 16)^3}$

26. $\displaystyle\int \dfrac{x^2\,dx}{(x^2 + 4)^3}$

In Problems 27–36, find each integral.

27. $\displaystyle\int \dfrac{x\,dx}{x^2 + 2x - 3}$

28. $\displaystyle\int \dfrac{x^2 - x - 8}{(x + 1)(x^2 + 5x + 6)}\,dx$

29. $\displaystyle\int \dfrac{10x^2 + 2x}{(x - 1)^2(x^2 + 2)}\,dx$

30. $\displaystyle\int \dfrac{x + 4}{x^2(x^2 + 4)}\,dx$

31. $\displaystyle\int \dfrac{7x + 3}{x^3 - 2x^2 - 3x}\,dx$

32. $\displaystyle\int \dfrac{x^5 + 1}{x^6 - x^4}\,dx$

33. $\displaystyle\int \dfrac{x^2}{(x - 2)(x - 1)^2}\,dx$

34. $\displaystyle\int \dfrac{x^2 + 1}{(x + 3)(x - 1)^2}\,dx$

35. $\displaystyle\int \dfrac{2x + 1}{x^3 - 1}\,dx$

36. $\displaystyle\int \dfrac{dx}{x^3 - 8}$

In Problems 37–40, find each definite integral.

37. $\displaystyle\int_0^1 \dfrac{dx}{x^2 - 9}$

38. $\displaystyle\int_2^4 \dfrac{dx}{x^2 - 25}$

39. $\displaystyle\int_{-2}^3 \dfrac{dx}{16 - x^2}$

40. $\displaystyle\int_1^2 \dfrac{dx}{9 - x^2}$

Applications and Extensions

In Problems 41–54, find each integral. (Hint: Make a substitution before using partial fraction decomposition.)

41. $\displaystyle\int \dfrac{\cos\theta}{\sin^2\theta + \sin\theta - 6}\,d\theta$

42. $\displaystyle\int \dfrac{\sin x}{\cos^2 x - 2\cos x - 8}\,dx$

43. $\displaystyle\int \dfrac{\sin\theta}{\cos^3\theta + \cos\theta}\,d\theta$

44. $\displaystyle\int \dfrac{4\cos\theta}{\sin^3\theta + 2\sin\theta}\,d\theta$

45. $\displaystyle\int \dfrac{e^t}{e^{2t} + e^t - 2}\,dt$

46. $\displaystyle\int \dfrac{e^x}{e^{2x} + e^x - 6}\,dx$

47. $\displaystyle\int \dfrac{e^x}{e^{2x} - 1}\,dx$

48. $\displaystyle\int \dfrac{dx}{e^x - e^{-x}}$

49. $\displaystyle\int \dfrac{dt}{e^{2t} + 1}$

50. $\displaystyle\int \dfrac{dt}{e^{3t} + e^t}$

51. $\displaystyle\int \dfrac{\sin x \cos x}{(\sin x - 1)^2}\,dx$

52. $\displaystyle\int \dfrac{\cos x \sin x}{(\cos x - 2)^2}\,dx$

53. $\displaystyle\int \dfrac{\cos x}{(\sin^2 x + 9)^2}\,dx$

54. $\displaystyle\int \dfrac{\sin x}{(\cos^2 x + 4)^2}\,dx$

---

1. = NOW WORK problem   〰 = Graphing technology recommended   CAS = Computer Algebra System recommended

55. Area Find the area under the graph of $y = \dfrac{4}{x^2 - 4}$ from $x = 3$ to $x = 5$, as shown in the figure below.

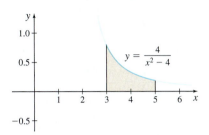

56. Area Find the area under the graph of $y = \dfrac{x - 4}{(x + 3)^2}$ from $x = 4$ to $x = 6$, as shown in the figure below.

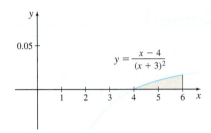

57. Area Find the area under the graph of $y = \dfrac{8}{x^3 + 1}$ from $x = 0$ to $x = 2$, as shown in the figure below.

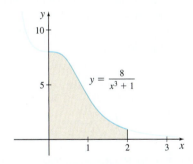

58. Volume of a Solid of Revolution Find the volume of the solid of revolution generated by revolving the region bounded by the graph of $y = \dfrac{x}{x^2 - 4}$ and the x-axis from $x = 3$ to $x = 5$ about the x-axis, as shown in the figure below.

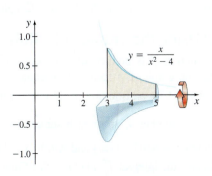

59. (a) Find the zeros of $q(x) = x^3 + 3x^2 - 10x - 24$.

(b) Factor q.

(c) Find the integral $\displaystyle\int \dfrac{3x - 7}{x^3 + 3x^2 - 10x - 24}\,dx$.

Challenge Problems

In Problems 60–71, simplify each integrand as follows: If the integrand involves fractional powers such as $x^{p/q}$ and $x^{r/s}$, make the substitution $x = u^n$, where n is the least common denominator of $\dfrac{p}{q}$ and $\dfrac{r}{s}$. Then find each integral.

60. $\displaystyle\int \dfrac{x\,dx}{3 + \sqrt{x}}$

61. $\displaystyle\int \dfrac{dx}{\sqrt{x} + 2}$

62. $\displaystyle\int \dfrac{dx}{x - \sqrt[3]{x}}$

63. $\displaystyle\int \dfrac{x\,dx}{\sqrt[3]{x} - 1}$

64. $\displaystyle\int \dfrac{dx}{\sqrt{x} + \sqrt[3]{x}}$

65. $\displaystyle\int \dfrac{dx}{3\sqrt{x} - \sqrt[3]{x}}$

66. $\displaystyle\int \dfrac{dx}{\sqrt[3]{2 + 3x}}$

67. $\displaystyle\int \dfrac{dx}{\sqrt[4]{1 + 2x}}$

68. $\displaystyle\int \dfrac{x\,dx}{(1 + x)^{3/4}}$

69. $\displaystyle\int \dfrac{dx}{(1 + x)^{2/3}}$

70. $\displaystyle\int \dfrac{\sqrt[3]{x} + 1}{\sqrt[3]{x} - 1}\,dx$

71. $\displaystyle\int \dfrac{dx}{\sqrt{x}(1 + \sqrt[3]{x})^2}$

Weierstrass Substitution *In Problems 72–87, use the following substitution, called a **Weierstrass substitution**. If an integrand is a rational expression of $\sin x$ or $\cos x$ or both, the substitution*

$$z = \tan \dfrac{x}{2} \qquad -\dfrac{\pi}{2} < \dfrac{x}{2} < \dfrac{\pi}{2}$$

or equivalently,

$$\sin x = \dfrac{2z}{1 + z^2} \qquad \cos x = \dfrac{1 - z^2}{1 + z^2} \qquad dx = \dfrac{2\,dz}{1 + z^2}$$

will transform the integrand into a rational function of z.

72. $\displaystyle\int \dfrac{dx}{1 - \sin x}$

73. $\displaystyle\int \dfrac{dx}{1 + \sin x}$

74. $\displaystyle\int \dfrac{dx}{1 - \cos x}$

75. $\displaystyle\int \dfrac{dx}{3 + 2\cos x}$

76. $\displaystyle\int \dfrac{2\,dx}{\sin x + \cos x}$

77. $\displaystyle\int \dfrac{dx}{1 - \sin x + \cos x}$

78. $\displaystyle\int \dfrac{\sin x}{3 + \cos x}\,dx$

79. $\displaystyle\int \dfrac{dx}{\tan x - 1}$

80. $\displaystyle\int \dfrac{dx}{\tan x - \sin x}$

81. $\displaystyle\int \dfrac{\sec x}{\tan x - 2}\,dx$

82. $\displaystyle\int \dfrac{\cot x}{1 + \sin x}\,dx$

83. $\displaystyle\int \dfrac{\sec x}{1 + \sin x}\,dx$

84. $\displaystyle\int_0^{\pi/2} \dfrac{dx}{\sin x + 1}$

85. $\displaystyle\int_{\pi/4}^{\pi/3} \dfrac{\csc x}{3 + 4\tan x}\,dx$

86. $\displaystyle\int_0^{\pi/2} \dfrac{\cos x}{2 - \cos x}\,dx$

87. $\displaystyle\int_0^{\pi/4} \dfrac{4\,dx}{\tan x + 1}$

88. Use a Weierstrass substitution to derive the formula

$$\int \csc x \, dx = \ln \sqrt{\frac{1 - \cos x}{1 + \cos x}} + C$$

89. Show that the result obtained in Problem 88 is equivalent to

$$\int \csc x \, dx = \ln |\csc x - \cot x| + C$$

90. Since $\dfrac{d}{dx} \tanh^{-1} x = \dfrac{1}{1 - x^2}$, we might expect that

$$\int_2^3 \frac{dx}{1 - x^2} = \tanh^{-1} 3 - \tanh^{-1} 2.$$

Why is this incorrect? What is the correct result?

91. Show that the two formulas below are equivalent.

$$\int \sec x \, dx = \ln |\sec x + \tan x| + C$$

$$\int \sec x \, dx = \ln \left| \frac{1 + \tan \dfrac{x}{2}}{1 - \tan \dfrac{x}{2}} \right| + C$$

$$\left[\textit{Hint:} \quad \tan \frac{x}{2} = \frac{\sin \dfrac{x}{2}}{\cos \dfrac{x}{2}} = \frac{\sin^2 \left(\dfrac{x}{2}\right)}{\sin \left(\dfrac{x}{2}\right) \cos \left(\dfrac{x}{2}\right)} = \frac{1 - \cos x}{\sin x}. \right]$$

92. Use the methods of this section to find $\displaystyle\int \frac{dx}{1 + x^4}$.

(*Hint:* Factor $1 + x^4$ into irreducible quadratics.)

7.6 Integration Using Numerical Techniques

OBJECTIVES *When you finish this section, you should be able to:*

1 Approximate an integral using the Trapezoidal Rule (p. 508)

2 Approximate an integral using Simpson's Rule (p. 514)

Most of the functions we have encountered so far belong to the class known as **elementary functions**. These functions include polynomial, exponential, logarithm, trigonometric, inverse trigonometric, hyperbolic, and inverse hyperbolic functions, as well as functions formed by combining one or more of these functions using addition, subtraction, multiplication, division, or composition.

We have found that the derivatives f', f'', and so on, of an elementary function f are also elementary functions. The integral of an elementary function $\int f(x) \, dx$, however, is not always an elementary function. Some examples of integrals of elementary functions that cannot be written in terms of elementary functions are

$$\int e^{x^2} dx \qquad \int e^{-x^2} dx \qquad \int \frac{\sin x}{x} dx$$

$$\int \frac{\cos x}{x} dx \qquad \int \frac{e^x \, dx}{x} \qquad \int \frac{dx}{\sqrt{1 - x^3}}$$

Recall that to find a definite integral using the Fundamental Theorem of Calculus requires an antiderivative of the integrand f. When it is not possible to express an antiderivative of f in terms of elementary functions, or when the integrand f is defined by a table of data or by a graph, the Fundamental Theorem is not useful. In such situations, there are numerical techniques we can use to approximate the definite integral.

In Chapter 5 we used Riemann sums to approximate definite integrals. Two other commonly used numerical techniques to approximate integrals are the *Trapezoidal Rule* and *Simpson's Rule*. A fourth method, based on *series,* is discussed in Chapter 8.

1 Approximate an Integral Using the Trapezoidal Rule

As we learned in Section 5.2, if a function f is nonnegative and continuous on an interval $[a, b]$, then the area A under the graph of f from a to b equals the definite integral $\int_a^b f(x) \, dx$. We can approximate A by partitioning $[a, b]$ into n subintervals, each of width $\Delta x = \dfrac{b - a}{n}$; choosing a number u_i from each subinterval $[x_{i-1}, x_i]$, $i = 1, 2, \ldots, n$; and forming n rectangles of height $f(u_i)$ and width Δx. The sum of the areas of these rectangles approximates the integral $\int_a^b f(x) \, dx$. See Figure 11.

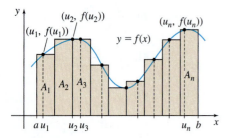

Figure 11 $A = \int_a^b f(x) \, dx$

$$\approx \sum_{i=1}^n A_i = \sum_{i=1}^n f(u_i) \Delta x$$

The Trapezoidal Rule approximates the definite integral $\int_a^b f(x)\,dx$ by replacing the rectangles with trapezoids. Then the area A under the graph of f from a to b, $\int_a^b f(x)\,dx$, is approximated by the sum of the areas of these trapezoids.

The Trapezoidal Rule

If a function f is continuous on the closed interval $[a, b]$, then

$$\int_a^b f(x)\,dx \approx \frac{b-a}{2n}\left[f(x_0) + 2f(x_1) + 2f(x_2) + \cdots + 2f(x_{n-1}) + f(x_n)\right]$$

where the closed interval $[a, b]$ is partitioned into n subintervals $[x_0, x_1], [x_1, x_2], \ldots,$ $[x_{n-1}, x_n]$, each of width $\Delta x = \dfrac{b-a}{n}$.

Although the Trapezoidal Rule does not require f to be nonnegative on the interval $[a, b]$, the proof given here assumes that $f(x) \geq 0$. Proofs of the Trapezoidal Rule with the nonnegative restriction relaxed can be found in numerical analysis books.

RECALL The area A of a trapezoid with width Δx and parallel heights l_1 and l_2 is $A = \dfrac{1}{2}(l_1 + l_2)\Delta x$.

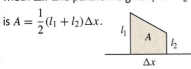

Proof Assume f is nonnegative and continuous on $[a, b]$. We seek the area A of the region bounded by the graph of f, the x-axis, and the lines $x = a$ and $x = b$. We partition the closed interval $[a, b]$ into n subintervals, $[a, x_1], [x_1, x_2], \ldots, [x_{i-1}, x_i], \ldots,$ $[x_{n-1}, b]$, each of width $\Delta x = \dfrac{b-a}{n}$. The y-coordinates corresponding to the numbers $x_0 = a, x_1, x_2, \ldots, x_{i-1}, x_i, \ldots, x_n = b$ are $f(x_0), f(x_1), f(x_2), \ldots, f(x_{i-1}), f(x_i),$ $\ldots, f(x_{n-1}), f(x_n)$. We connect each pair of consecutive points $(x_{i-1}, f(x_{i-1}))$ and $(x_i, f(x_i))$, $i = 1, 2, \ldots, n$, with a line segment to form n trapezoids. The trapezoid whose base is $\Delta x = x_i - x_{i-1} = \dfrac{b-a}{n}$ has parallel sides of lengths $f(x_{i-1})$ and $f(x_i)$. Based on Figure 12, its area A_i is

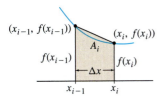

$$A_i = \frac{1}{2}\left[f(x_{i-1}) + f(x_i)\right]\Delta x$$

The sum of the areas of the n trapezoids approximates the area A under the graph of f. See Figure 13.

Then

$$A \approx \frac{1}{2}\left[f(x_0) + f(x_1)\right]\Delta x + \frac{1}{2}\left[f(x_1) + f(x_2)\right]\Delta x + \cdots + \frac{1}{2}\left[f(x_{n-1}) + f(x_n)\right]\Delta x$$

$$\approx \frac{1}{2}\left[f(x_0) + 2f(x_1) + 2f(x_2) + \cdots + 2f(x_{n-1}) + f(x_n)\right]\Delta x$$

Now substitute $\Delta x = \dfrac{b-a}{n}$ to obtain the Trapezoidal Rule. ∎

Figure 12

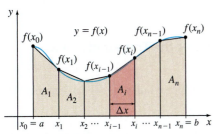

Figure 13 $\displaystyle\sum_{i=1}^{n} A_i \approx \int_a^b f(x)\,dx$

EXAMPLE 1 Approximating $\displaystyle\int_0^{\pi} \sin x\,dx$ Using the Trapezoidal Rule

(a) Use the Trapezoidal Rule with $n = 4$ and $n = 6$ to approximate $\int_0^{\pi} \sin x\,dx$, rounded to three decimal places.

(b) Compare each approximation to the exact value of $\int_0^{\pi} \sin x\,dx$.

Solution **(a)** We partition the interval $[0, \pi]$ into four subintervals, each of width

$$\Delta x = \frac{\pi - 0}{4} = \frac{\pi}{4}.$$

$$\left[0, \frac{\pi}{4}\right] \qquad \left[\frac{\pi}{4}, \frac{\pi}{2}\right] \qquad \left[\frac{\pi}{2}, \frac{3\pi}{4}\right] \qquad \left[\frac{3\pi}{4}, \pi\right]$$

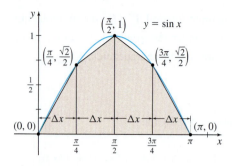

DF Figure 14 $n = 4$.

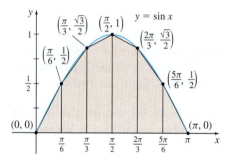

DF Figure 15 $n = 6$.

The values of $f(x) = \sin x$ corresponding to each endpoint are

$$f(0) = 0 \qquad f\left(\frac{\pi}{4}\right) = \frac{\sqrt{2}}{2} \qquad f\left(\frac{\pi}{2}\right) = 1 \qquad f\left(\frac{3\pi}{4}\right) = \frac{\sqrt{2}}{2} \qquad f(\pi) = 0$$

See Figure 14. Now we use the Trapezoidal Rule:

$$\int_0^\pi \sin x\, dx \approx \frac{\pi - 0}{2 \cdot 4}\left[\sin(0) + 2\sin\left(\frac{\pi}{4}\right) + 2\sin\left(\frac{\pi}{2}\right) + 2\sin\left(\frac{3\pi}{4}\right) + \sin(\pi)\right]$$

$$= \frac{\pi}{8}\left[0 + 2\left(\frac{\sqrt{2}}{2}\right) + 2\,(1) + 2\left(\frac{\sqrt{2}}{2}\right) + 0\right] = \frac{\pi}{8}\left(2 + 2\sqrt{2}\right)$$

$$= \frac{\pi}{4}\left(1 + \sqrt{2}\right) \approx 1.896$$

To approximate $\int_0^\pi \sin x\, dx$ using six subintervals, we partition $[0, \pi]$ into six subintervals, each of width $\Delta x = \dfrac{\pi - 0}{6} = \dfrac{\pi}{6}$, namely,

$$\left[0, \frac{\pi}{6}\right] \qquad \left[\frac{\pi}{6}, \frac{\pi}{3}\right] \qquad \left[\frac{\pi}{3}, \frac{\pi}{2}\right] \qquad \left[\frac{\pi}{2}, \frac{2\pi}{3}\right] \qquad \left[\frac{2\pi}{3}, \frac{5\pi}{6}\right] \qquad \left[\frac{5\pi}{6}, \pi\right]$$

See Figure 15. Then we use the Trapezoidal Rule.

$$\int_0^\pi \sin x\, dx \approx \frac{\pi}{2 \cdot 6}\left[0 + 2\cdot\frac{1}{2} + 2\cdot\frac{\sqrt{3}}{2} + 2\cdot 1 + 2\cdot\frac{\sqrt{3}}{2} + 2\cdot\frac{1}{2} + 0\right]$$

$$= \frac{\pi}{12}(4 + 2\sqrt{3}) = \frac{\pi}{6}(2 + \sqrt{3}) \approx 1.954$$

(b) The exact value of the integral is

$$\int_0^\pi \sin x\, dx = \Big[-\cos x\Big]_0^\pi = -\cos\pi + \cos 0 = 1 + 1 = 2$$

The approximation using the Trapezoidal Rule with four subintervals underestimates the integral by 0.104. The approximation using six subintervals underestimates the integral by 0.046. Notice that the approximation using six subintervals is more accurate than the approximation using four subintervals. ■

EXAMPLE 2 **Approximating $\displaystyle\int_1^2 \frac{e^x}{x}\, dx$ Using the Trapezoidal Rule**

Use the Trapezoidal Rule with $n = 4$ and $n = 6$ to approximate $\displaystyle\int_1^2 \frac{e^x}{x}\, dx$. Express the answer rounded to three decimal places.

Solution   We begin by partitioning the interval $[1, 2]$ into four subintervals, each of width $\Delta x = \dfrac{2 - 1}{4} = \dfrac{1}{4}$:

$$\left[1, \frac{5}{4}\right] \qquad \left[\frac{5}{4}, \frac{3}{2}\right] \qquad \left[\frac{3}{2}, \frac{7}{4}\right] \qquad \left[\frac{7}{4}, 2\right]$$

The values of $f(x) = \dfrac{e^x}{x}$ corresponding to each endpoint are

$$f(1) = e \quad f\left(\frac{5}{4}\right) = \frac{4e^{5/4}}{5} \quad f\left(\frac{3}{2}\right) = \frac{2e^{3/2}}{3} \quad f\left(\frac{7}{4}\right) = \frac{4e^{7/4}}{7} \quad f(2) = \frac{e^2}{2}$$

Then, using the Trapezoidal Rule, we get

$$\int_1^2 \frac{e^x}{x}\,dx \approx \frac{1}{2\cdot 4}\left[e + 2\cdot\frac{4e^{5/4}}{5} + 2\cdot\frac{2e^{3/2}}{3} + 2\cdot\frac{4e^{7/4}}{7} + \frac{e^2}{2}\right] \approx 3.069$$

To approximate $\int_1^2 \frac{e^x}{x}\,dx$ using six subintervals, we partition $[1, 2]$ into six subintervals, each of width $\Delta x = \dfrac{2-1}{6} = \dfrac{1}{6}$:

$$\left[1, \frac{7}{6}\right] \quad \left[\frac{7}{6}, \frac{4}{3}\right] \quad \left[\frac{4}{3}, \frac{3}{2}\right] \quad \left[\frac{3}{2}, \frac{5}{3}\right] \quad \left[\frac{5}{3}, \frac{11}{6}\right] \quad \left[\frac{11}{6}, 2\right]$$

Then, using the Trapezoidal Rule, we get

$$\int_1^2 \frac{e^x}{x}\,dx \approx \frac{1}{2\cdot 6}\left[f(1) + 2f\left(\frac{7}{6}\right) + 2f\left(\frac{4}{3}\right) + 2f\left(\frac{3}{2}\right) + 2f\left(\frac{5}{3}\right)\right.$$

$$\left. + 2f\left(\frac{11}{6}\right) + f(2)\right] \approx 3.063 \quad \blacksquare$$

Since $f(x) = \dfrac{e^x}{x} > 0$ on the interval $[1, 2]$, the integral $\int_1^2 \frac{e^x}{x}\,dx$ equals the area under the graph of f from 1 to 2. Figure 16 shows the graph of $y = \dfrac{e^x}{x}$, the area $A = \int_1^2 \frac{e^x}{x}\,dx$, and the approximation to the area using the Trapezoidal Rule with four subintervals.

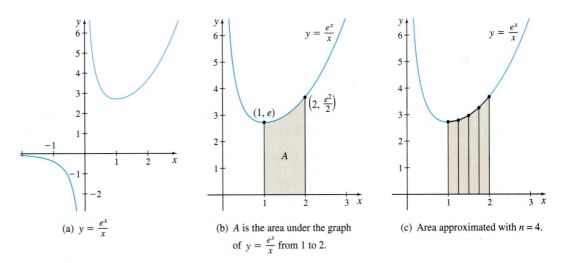

(a) $y = \dfrac{e^x}{x}$

(b) A is the area under the graph of $y = \dfrac{e^x}{x}$ from 1 to 2.

(c) Area approximated with $n = 4$.

Figure 16

NOW WORK **Problem** 9(a).

We cannot find the exact value of the integral $\int_1^2 \frac{e^x}{x}\,dx$. The significance of the Trapezoidal Rule is that not only can it be used to approximate such integrals, but it also provides an estimate of the *error*, the difference between the exact value of the integral and the approximate value.

THEOREM Error in the Trapezoidal Rule

Let f be a function for which f'' is continuous on some open interval containing a and b. The error in using the Trapezoidal Rule to approximate $\int_a^b f(x)\,dx$ can be estimated* using the formula

$$\text{Error} \leq \frac{(b-a)^3 M}{12n^2}$$

where M is the absolute maximum value of $|f''|$ on the closed interval $[a, b]$.

NOTE The error formula gives an upper bound to the error. A derivation of the error formula, which uses an extension of the Mean Value Theorem, may be found in numerical analysis books.

To find M in the error estimate usually involves a lot of computation, so we will use technology to obtain M.

EXAMPLE 3   Estimating the Error in Using the Trapezoidal Rule

Estimate the error that results from using the Trapezoidal Rule with $n = 4$ and with $n = 6$ to approximate $\int_1^2 \dfrac{e^x}{x}\,dx$. See Example 2.

NOTE A CAS can be used to find $f''(x)$.

Solution   To estimate the error resulting from approximating $\int_1^2 \dfrac{e^x}{x}\,dx$ using the Trapezoidal Rule, we need to find the absolute maximum value of $|f''|$ on the interval $[1, 2]$. We begin by finding f'':

$$f(x) = \frac{e^x}{x}$$

$$f'(x) = \frac{d}{dx}\frac{e^x}{x} = \frac{xe^x - e^x}{x^2} = \frac{e^x}{x} - \frac{e^x}{x^2}$$

$$f''(x) = \frac{d}{dx}\left(\frac{e^x}{x} - \frac{e^x}{x^2}\right) = \frac{d}{dx}\frac{e^x}{x} - \frac{d}{dx}\frac{e^x}{x^2} = \left(\frac{e^x}{x} - \frac{e^x}{x^2}\right) - \frac{e^x x^2 - 2e^x x}{x^4}$$

$$= \frac{e^x}{x} - \frac{e^x}{x^2} - \frac{e^x}{x^2} + \frac{2e^x}{x^3}$$

$$= e^x\left(\frac{1}{x} - \frac{2}{x^2} + \frac{2}{x^3}\right)$$

Since $|f''|$ is continuous on the interval $[1, 2]$, the Extreme Value Theorem guarantees that $|f''|$ has an absolute maximum on $[1, 2]$. We find the absolute maximum of $|f''|$ using graphing technology. As seen from Figure 17, the absolute maximum is at the left endpoint 1. The absolute maximum value of f'' is $\left|e^1\left(\dfrac{1}{1} - \dfrac{2}{1^2} + \dfrac{2}{1^3}\right)\right| = e$. So, $M = e$.

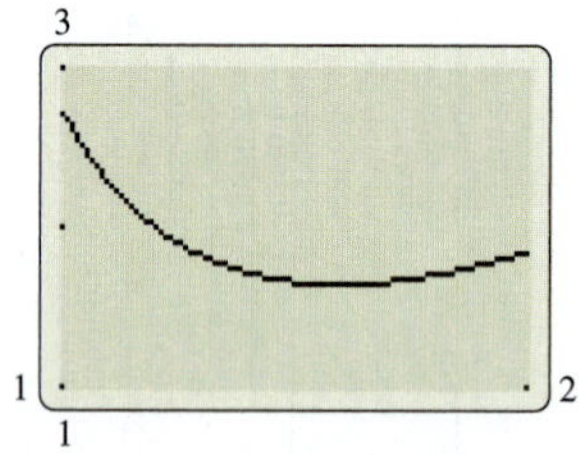

Figure 17 The absolute maximum of $|f''|$ on $[1, 2]$ is at $x = 1$.

When $n = 4$, the error in using the Trapezoidal Rule is

$$\text{Error} \leq \frac{(b-a)^3 M}{12n^2} = \frac{(2-1)^3(e)}{(12)(4^2)} = \frac{e}{192} \approx 0.014$$

That is,

$$3.069 - 0.014 \leq \int_1^2 \frac{e^x}{x}\,dx \leq 3.069 + 0.014$$

$$3.055 \leq \int_1^2 \frac{e^x}{x}\,dx \leq 3.083$$

---

*The usefulness of this result is that we can find an upper bound to the error without knowing the exact value of the integral.

When $n = 6$, the error in using the Trapezoidal Rule is

$$\text{Error} \leq \frac{(b-a)^3 M}{12n^2} = \frac{(2-1)^3 (e)}{(12)(6^2)} = \frac{e}{432} \approx 0.006$$

That is,

$$3.063 - 0.006 \leq \int_1^2 \frac{e^x}{x} dx \leq 3.063 + 0.006$$

$$3.057 \leq \int_1^2 \frac{e^x}{x} dx \leq 3.069 \qquad \blacksquare$$

NOW WORK   **Problem 9(b).**

EXAMPLE 4   Obtaining a Desired Accuracy Using the Trapezoidal Rule

Find the number of subintervals n needed to guarantee that the Trapezoidal Rule approximates $\int_1^2 \frac{e^x}{x} dx$ to within 0.0001.

Solution   To be sure that the approximation of $\int_1^2 \frac{e^x}{x} dx$ is correct to within 0.0001, we require the error be less than 0.0001. That is,

$$\text{Error} \leq \frac{(b-a)^3 M}{12n^2} = \frac{(2-1)^3 M}{12n^2} = \frac{e}{12n^2} < 0.0001$$
$$\underset{\uparrow}{}\\ M = e$$

$$n^2 > \frac{e}{(0.0001)(12)} = \frac{e}{0.0012}$$

$$n > \sqrt{\frac{e}{0.0012}} \approx 47.6$$

To ensure that the error is less than 0.0001, we round up to 48 subintervals. $\blacksquare$

NOW WORK   **Problem 9(c).**

The Trapezoidal Rule is very useful in applications when only experimental (empirical) data are available.

EXAMPLE 5   Using the Trapezoidal Rule with Empirical Data

A 140-foot (ft) tree trunk is cut into 20-ft logs. The diameter of each cross-sectional cut is measured and its area A recorded in the table below. (x is the distance in feet of the cut from the top of the trunk.)

x (ft)	0	20	40	60	80	100	120	140
A (ft^2)	120	124	128	130	132	136	144	158

Find the approximate volume of the tree trunk.

Solution   The volume of the tree trunk is $V = \int_0^{140} A(x)\, dx$, where $A(x)$ is the area of a slice at x. Since only eight data points are given, the function $A(x)$ is not explicitly known.

To approximate the volume, we partition the interval $[0, 140]$ into seven subintervals, each of width $\Delta x = \dfrac{140}{7} = 20$. This partition corresponds to the given data. Using the Trapezoidal Rule, the approximate volume of the tree trunk is

$$V = \int_0^{140} A(x)\,dx \approx \frac{1}{2}(20)\,[A(0) + 2\,A\,(20) + 2\,A(40) + 2\,A(60) + 2\,A(80)$$

$$+\, 2\,A(100) + 2\,A(120) + A\,(140)]$$

$$V \approx 10\,[120 + 2(124) + 2(128) + 2(130) + 2(132) + 2(136) + 2(144) + 158]$$

$$= 18{,}660$$

The volume of the tree trunk is approximately $18{,}660\,\text{ft}^3$. ∎

NOW WORK Problem 25.

2 Approximate an Integral Using Simpson's Rule

NOTE With the Trapezoidal Rule, we approximate a function f with line segments (first degree polynomials), whereas with Simpson's Rule, we approximate f with parabolic arcs (second degree polynomials).

Another way to approximate the definite integral of a function f that is continuous on a closed interval $[a, b]$, is called *Simpson's Rule*. Simpson's Rule, which approximates $\int_a^b f(x)\,dx$ using parabolic arcs, instead of line segments as in the Trapezoidal Rule, often gives a better approximation to the area A.

Before stating Simpson's Rule, we derive a formula for the area A under a parabolic arc. Let the graph in Figure 18 represent the parabola $y = ax^2 + bx + c$. Draw the lines $x = -h$ and $x = h$, so that the points $(-h, y_0)$, $(0, y_1)$, and (h, y_2) lie on the parabola.

The area A enclosed by the parabola, the x-axis, and the lines $x = -h$ and $x = h$ is given by

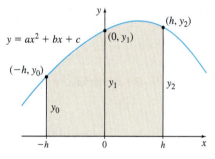

Figure 18

$$A = \int_{-h}^{h} y\,dx = \int_{-h}^{h} (ax^2 + bx + c)\,dx = \left[a\frac{x^3}{3} + b\frac{x^2}{2} + cx\right]_{-h}^{h}$$

$$= \frac{2}{3}ah^3 + 2ch = \frac{h}{3}(2ah^2 + 6c) \tag{1}$$

Since the parabola contains the points $(-h, y_0)$, $(0, y_1)$, and (h, y_2), each point must satisfy the equation $y = ax^2 + bx + c$, and we have the system of equations:

$$y_0 = ah^2 - bh + c \qquad x = -h$$

$$y_1 = c \qquad x = 0$$

$$y_2 = ah^2 + bh + c \qquad x = h$$

Adding the first and third equations gives

$$y_0 + y_2 = 2ah^2 + 2c$$

Add $4y_1 = 4c$ to each side.

$$y_0 + y_2 + 4y_1 = 2ah^2 + 6c$$

Then substitute $y_0 + y_2 + 4y_1$ for $2ah^2 + 6c$ in formula (1) for the area A to obtain

$$A = \frac{h}{3}(y_0 + 4y_1 + y_2) \tag{2}$$

This formula for A depends on only y_0, y_1, y_2, and the distance h.

Now assume that a function f is nonnegative and continuous on $[a, b]$. Then the area A under the graph of f, $A = \int_a^b f(x)\,dx$, can be approximated by partitioning the interval $[a, b]$ into an even number n of subintervals, each of width $\Delta x = \dfrac{b - a}{n}$. The

y-coordinates corresponding to the numbers $x_0 = a, x_1, x_2, \ldots, x_{i-1}, x_i, \ldots, x_n = b$ are $f(x_0), f(x_1), f(x_2), \ldots, f(x_{i-1}), f(x_i), \ldots, f(x_{n-1}), f(x_n)$. We connect each triple of points $(x_{i-1}, f(x_{i-1}))$ $(x_i, f(x_i))$, and $(x_{i+1}, f(x_{i+1}))$, $i = 1, 3, 5, \ldots, n - 1$ with parabolic arcs, and use (2) to find the area A_i enclosed by the arc centered at $(x_i, f(x_i))$ and the lines $x = x_{i-1}$ and $x = x_{i+1}$. With $h = \Delta x = \dfrac{b-a}{n}$, we have

$$A_i = \frac{b-a}{3n}[f(x_{i-1}) + 4f(x_i) + f(x_{i+1})]$$

The sum of these areas gives an approximation to $A = \int_a^b f(x)\,dx$.

ORIGINS *Simpson's Rule* is named after the British mathematician Thomas Simpson, who lived from 1710 to 1761.

Simpson's Rule

If a function f is continuous on the closed interval $[a, b]$, then

$$\int_a^b f(x)\,dx \approx \frac{b-a}{3n}[f(x_0) + 4f(x_1) + 2f(x_2) + 4f(x_3) + 2f(x_4)$$
$$+ \cdots + 2f(x_{n-2}) + 4f(x_{n-1}) + f(x_n)]$$

where the closed interval $[a, b]$ is partitioned into an even number n of subintervals $[x_0, x_1], [x_1, x_2], \ldots, [x_{n-1}, x_n]$, each of width $\dfrac{b-a}{n}$.

A proof of Simpson's Rule that does not assume f is nonnegative can be found in most numerical analysis books.

EXAMPLE 6   **Approximating an Integral Using Simpson's Rule**

Use Simpson's Rule with $n = 4$ to approximate $\displaystyle\int_\pi^{2\pi} \frac{\sin x}{x}\,dx$. Express the answer rounded to three decimal places.

Solution   We partition the interval $[\pi, 2\pi]$ into the four subintervals

$$\left[\pi, \frac{5\pi}{4}\right] \qquad \left[\frac{5\pi}{4}, \frac{3\pi}{2}\right] \qquad \left[\frac{3\pi}{2}, \frac{7\pi}{4}\right] \qquad \left[\frac{7\pi}{4}, 2\pi\right]$$

each of width $\dfrac{\pi}{4}$. The value of $f(x) = \dfrac{\sin x}{x}$ corresponding to each endpoint is

$$f(\pi) = 0 \quad f\left(\frac{5\pi}{4}\right) = -\frac{2\sqrt{2}}{5\pi} \quad f\left(\frac{3\pi}{2}\right) = -\frac{2}{3\pi} \quad f\left(\frac{7\pi}{4}\right) = -\frac{2\sqrt{2}}{7\pi} \quad f(2\pi) = 0$$

Then using Simpson's Rule, we get

$$\int_\pi^{2\pi} \frac{\sin x}{x}\,dx \approx \frac{2\pi - \pi}{3 \cdot 4}\left[f(\pi) + 4f\left(\frac{5\pi}{4}\right) + 2f\left(\frac{3\pi}{2}\right) + 4f\left(\frac{7\pi}{4}\right) + f(2\pi)\right]$$

$$= \left(\frac{\pi}{12}\right)\left[0 + 4\left(-\frac{2\sqrt{2}}{5\pi}\right) + 2\left(-\frac{2}{3\pi}\right) + 4\left(-\frac{2\sqrt{2}}{7\pi}\right) + 0\right] \approx -0.434 \quad ∎$$

The graph of $y = \dfrac{\sin x}{x}$ is shown in Figure 19 on page 516.

7.6 Assess Your Understanding

Concepts and Vocabulary

1. *True or False*   The Trapezoidal Rule approximates an integral $\int_a^b f(x)\,dx$ by replacing the graph of f with line segments.

2. *True or False*   Simpson's Rule approximates an integral by using parabolic arcs.

Skill Building

For Problems 3 and 5, use the graph below to approximate the area A. Round answers to three decimal places.

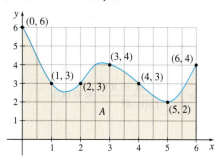

For Problems 4 and 6, use the graph below to approximate the area A. Round answers to three decimal places.

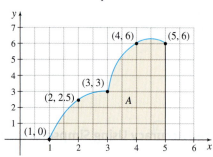

3. Use the Trapezoidal Rule with $n = 3$ and $n = 6$ to approximate the area under the graph.

4. Use the Trapezoidal Rule with $n = 2$ and $n = 4$ to approximate the area under the graph.

5. Use Simpson's Rule with $n = 2$ and $n = 6$ to approximate the area under the graph.

6. Use Simpson's Rule with $n = 2$ and $n = 4$ to approximate the area under the graph.

In Problems 7–12:
(a) *Use the Trapezoidal Rule to approximate each integral.*
(b) *Estimate the error in using the approximation. Express the answer rounded to three decimal places.*
(c) *Find the number n of subintervals needed to guarantee that an approximation is correct to within 0.0001.*

7. $\displaystyle\int_{\pi/2}^{\pi} \frac{\sin x}{x}\,dx;\quad n = 3$

8. $\displaystyle\int_{3\pi/2}^{2\pi} \frac{\cos x}{x}\,dx;\quad n = 3$

9. $\displaystyle\int_0^1 e^{-x^2}\,dx;\quad n = 4$

10. $\displaystyle\int_0^1 e^{x^2}\,dx;\quad n = 4$

11. $\displaystyle\int_{-1}^0 \frac{dx}{\sqrt{1 - x^3}};\quad n = 4$

12. $\displaystyle\int_0^1 \frac{dx}{\sqrt{1 + x^3}};\quad n = 3$

In Problems 13–18:
(a) *Use Simpson's Rule to approximate each integral.*
(b) *Estimate the error in using the approximation. Express the answer rounded to three decimal places.*
(c) *Find the number n of subintervals needed to guarantee that an approximation is correct to within 0.0001.*

13. $\displaystyle\int_1^2 \frac{e^x}{x}\,dx;\quad n = 4$

14. $\displaystyle\int_{3\pi/2}^{2\pi} \frac{\cos x}{x}\,dx;\quad n = 4$

15. $\displaystyle\int_0^1 e^{-x^2}\,dx;\quad n = 4$

16. $\displaystyle\int_0^1 e^{x^2}\,dx;\quad n = 4$

17. $\displaystyle\int_{-1}^0 \frac{dx}{\sqrt{1 - x^3}};\quad n = 4$

18. $\displaystyle\int_0^1 \frac{dx}{\sqrt{1 + x^2}};\quad n = 4$

Applications and Extensions

19. (a) Show that $\displaystyle\int_1^2 \frac{dx}{x} = \ln 2$.

(b) Use the Trapezoidal Rule with $n = 5$ to approximate $\displaystyle\int_1^2 \frac{dx}{x}$.

(c) Use Simpson's Rule with $n = 6$ to approximate $\displaystyle\int_1^2 \frac{dx}{x}$.

20. Area   Selected measurements of a function f are given in the table below. Use Simpson's Rule to approximate the area enclosed by the graph of f, the x-axis, and the lines $x = 2$ and $x = 4.4$.

x	2.0	2.4	2.8	3.2	3.6	4.0	4.4
y	3.03	4.61	5.80	6.59	7.76	8.46	9.19

21. Arc Length   Approximate the arc length of the graph of $y = \sin x$ from $x = 0$ to $x = \dfrac{\pi}{2}$

(a) using Simpson's Rule with $n = 4$.

(b) using the Trapezoidal Rule with $n = 3$.

22. Arc Length   Approximate the arc length of the graph of $y = e^x$ from $x = 0$ to $x = 4$

(a) using Simpson's Rule with $n = 4$.

(b) using the Trapezoidal Rule with $n = 8$.

23. Work   A gas expands from a volume of 1 cubic inch (in.³) to 2.5 in.³; values of the volume V and pressure p (in pounds per square inch) during the expansion are given in the table below. Find the total work W done in the expansion using Simpson's Rule. (*Hint:* $W = \int_a^b p\,dV$).

V	1	1.25	1.5	1.75	2	2.25	2.5
p	68.7	55.0	45.8	39.3	34.4	30.5	27.5

24. Work   In the table below, F is the force in pounds acting on an object in its direction of motion and x is the displacement of the object in feet. Use the Trapezoidal Rule to approximate the work done by the force in moving the object from $x = 0$ to $x = 50$.

x	0	5	10	15	20	25	30	35	40	45	50
F	100	80	66	56	50	45	40	36	33	30	28

---

1. = NOW WORK problem   = Graphing technology recommended   CAS = Computer Algebra System recommended

25. Volume In the table below, S is the area in square meters of the cross section of a railroad track cutting through a mountain, and x meters is the corresponding distance along the line. Use the Trapezoidal Rule to find the number of cubic meters of earth removed to make the cutting from $x = 0$ to $x = 150$. See the figure below.

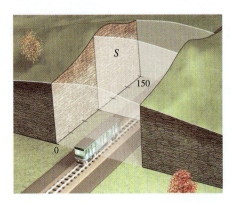

x	0	25	50	75	100	125	150
S	105	118	142	120	110	90	78

26. Area Use Simpson's Rule to approximate the surface area of the pond pictured in the figure.

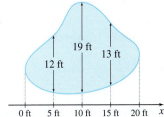

27. Volume The area of the horizontal section of a reservoir is A square meters at a height x meters from the bottom. Corresponding values of A and x are given in the table below. Approximate the volume of water in the reservoir using the Trapezoidal Rule and also using Simpson's Rule. See the figure.

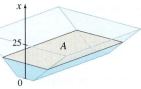

x	0	2.5	5	7.5	10	12.5	15	17.5	20	22.5	25
A	0	2510	3860	4870	5160	5590	5810	6210	6890	7680	8270

28. Area A series of soundings taken across a river channel is given in the table below, where x meters is the distance from one shore and y meters is the corresponding depth of the water. Find its area by the Trapezoidal Rule.

x	0	10	20	30	40	50	60	70	80
y	5	10	13.2	15	15.6	12	6	4	0

29. Volume of a Solid of Revolution Use the Trapezoidal Rule with $n = 3$ to approximate the volume of the solid of revolution formed by revolving the region shown in the figure below about the x-axis.

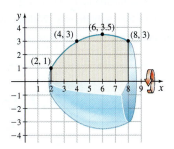

30. Distance Traveled The speed v, in meters per second, of an object at time t is given in the table below.

t	0	0.5	1	1.5	2	2.5	3
v	5.1	5.3	5.6	6.1	6.8	6.7	6.5

(a) Use the Trapezoidal Rule to approximate the distance s traveled by the object from $t = 0$ to $t = 3$.

(b) Use Simpson's Rule to approximate the distance s traveled by the object from $t = 0$ to $t = 3$.

31. Volume of a Solid of Revolution Approximate the volume of the solid of revolution in the figure below generated by revolving the region bounded by the graph of $y = \sin x$ and the y-axis from $x = 0$ to $x = \dfrac{\pi}{2}$ about the y-axis

(a) using Simpson's Rule with $n = 4$.

(b) using the Trapezoidal Rule with $n = 3$.

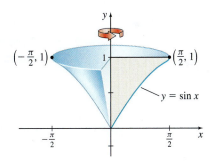

32. Arc Length Use the Trapezoidal Rule to find the arc length of the ellipse $9x^2 + 100y^2 = 900$ in the first quadrant from $x = 0$ to $x = 8$. Partition the interval into four equal subintervals, and round the answer to three decimal places.

33. Approximate $\displaystyle\int_0^\pi f(x)\,dx$ if $f(x) = \begin{cases} \dfrac{\sin x}{x} & \text{if } x \neq 0 \\ 1 & \text{if } x = 0 \end{cases}$

(a) Use the Trapezoidal Rule with $n = 6$.

(b) Use Simpson's Rule with $n = 6$.

[CAS] **34. Approximate** $\displaystyle\int_{-1}^1 5e^{-x^2}\,dx$.

(a) Use the Trapezoidal Rule with $n = 20$ subintervals.

(b) Use Simpson's Rule with $n = 20$ subintervals.

(c) Use a CAS to find the integral.

35. Let T_n be the approximation to $\int_a^b f(x)\,dx$ given by the Trapezoidal Rule with n subintervals. Without using the error formula given in the text, show that $\lim\limits_{n \to \infty} T_n = \int_a^b f(x)\,dx$.

36. Show that if $f(x) = Ax^3 + Bx^2 + Cx + D$, then Simpson's Rule gives the exact value of $\int_a^b f(x)\,dx$.

7.7 Integration Using Tables and Computer Algebra Systems

OBJECTIVES *When you finish this section, you should be able to:*

1 Use a Table of Integrals (p. 520)

2 Use a computer algebra system (p. 522)

While it is important to be able to use techniques of integration, to save time or to check one's work, a Table of Integrals or a computer algebra system (CAS) is often useful.

1 Use a Table of Integrals

The inserts in the back of the book contain a list of integration formulas called a **Table of Integrals.** Many of the integration formulas in the list were derived in this chapter. Although the table seems long, it is far from complete. A more comprehensive table of integrals may be found in Daniel Zwillinger (Ed.), *Standard Mathematical Tables and Formulae*, 32nd ed., Boca Raton, FL: CRC Press.

EXAMPLE 1   Using a Table of Integrals

Use a Table of Integrals to find $\displaystyle \int \frac{dx}{\sqrt{\left(4x - x^2\right)^3}}$.

Solution Look through the headings in the Table of Integrals until you locate *Integrals Containing* $\sqrt{2ax - x^2}$. Continue in the subsection until you find a form that closely resembles the integrand given. You should find Integral 82:

$$\int \frac{dx}{\left(2ax - x^2\right)^{3/2}} = \frac{x - a}{a^2\sqrt{2ax - x^2}} + C$$

This is the integral we seek with $a = 2$. So,

$$\int \frac{dx}{\sqrt{\left(4x - x^2\right)^3}} = \frac{x - 2}{4\sqrt{4x - x^2}} + C$$

∎

NOW WORK Problem 3.

Some integrals in the table are reduction formulas.

EXAMPLE 2   Using a Table of Integrals

Use a Table of Integrals to find $\displaystyle \int x^2 \tan^{-1} x\,dx$.

Solution   Find the subsection of the table titled *Integrals Containing Inverse Trigonometric Functions*. Then look for an integral whose form closely resembles the problem. You should find Integral 114:

$$\int x^n \tan^{-1} x\,dx = \frac{1}{n + 1}\left(x^{n+1}\tan^{-1}x - \int \frac{x^{n+1}\,dx}{1 + x^2}\right) \qquad n \neq -1$$

This is the integral we seek with $n = 2$.

$$\int x^2 \tan^{-1} x \, dx = \frac{1}{3}\left(x^3 \tan^{-1} x - \int \frac{x^3 \, dx}{1 + x^2}\right)$$

We find the integral on the right by using the substitution $u = 1 + x^2$. Then $du = 2x \, dx$ and

$$\int \frac{x^3}{1 + x^2} dx = \int \frac{x^2 x \, dx}{1 + x^2} = \int \frac{u - 1}{u} \frac{du}{2} = \frac{1}{2}\int \left(1 - \frac{1}{u}\right) du = \frac{1}{2}u - \frac{1}{2}\ln|u|$$

$$= \frac{1 + x^2}{2} - \frac{\ln(1 + x^2)}{2}$$

So,

$$\int x^2 \tan^{-1} x \, dx = \frac{1}{3}\left(x^3 \tan^{-1} x - \int \frac{x^3}{1 + x^2} dx\right)$$

$$= \frac{1}{3}\left[x^3 \tan^{-1} x - \frac{1 + x^2}{2} + \frac{1}{2}\ln(1 + x^2)\right] + C \qquad \blacksquare$$

NOW WORK Problem 11.

Sometimes the given integral is found in the tables after a substitution is made.

EXAMPLE 3   Using a Table of Integrals

Use a Table of Integrals to find $\displaystyle\int \frac{3x + 5}{\sqrt{3x + 6}} dx$.

Solution Find the subsection of the table titled *Integrals Containing* $\sqrt{ax + b}$ (the square root of a linear expression). Then look for an integral whose form closely resembles the problem. The closest one is an integral with x in the numerator and $\sqrt{ax + b}$ in the denominator,

Integral 42: $$\int \frac{x \, dx}{\sqrt{a + bx}} = \frac{2}{3b^2}(bx - 2a)\sqrt{a + bx} + C \qquad (1)$$

To express the given integral as one with a single variable in the numerator, use the substitution $u = 3x + 5$. Then $du = 3 \, dx$. Since

$$\sqrt{3x + 6} = \sqrt{(3x + 5) + 1} = \sqrt{u + 1}$$

we find

$$\int \frac{3x + 5}{\sqrt{3x + 6}} dx \underset{\underset{u = 3x + 5, \frac{1}{3}du = dx}{\uparrow}}{=} \frac{1}{3}\int \frac{u \, du}{\sqrt{u + 1}}$$

which is in the form of (1) with $a = 1$ and $b = 1$ So,

$$\int \frac{3x + 5}{\sqrt{3x + 6}} dx = \frac{1}{3}\int \frac{u \, du}{\sqrt{u + 1}} \underset{\underset{(1)}{\uparrow}}{=} \frac{1}{3}\cdot\frac{2}{3}(u - 2)\sqrt{1 + u} + C$$

$$\underset{\underset{u = 3x + 5}{\uparrow}}{=} \frac{2}{9}[(3x + 5) - 2]\sqrt{1 + (3x + 5)} + C = \frac{2}{3}(x + 1)\sqrt{3x + 6} + C \qquad \blacksquare$$

NOW WORK Problem 5.

NEED TO REVIEW? Computer algebra systems are discussed in Section P.8, pp. 64–67.

2 Use a Computer Algebra System

A computer algebra system (or CAS) is computer software that can perform symbolic manipulation of mathematical expressions. Some graphing utilities, such as a TI-89 or a TI Nspire, also have a CAS capability. CAS software packages such as Mathematica, Maple, Matlab, and muPAD offer more extensive symbolic manipulation capabilities, as well as tools for visualization and numerical approximation. These packages offer intuitive interfaces so that the user can obtain results without needing to write a program. A simple, online CAS, based on Mathematica, can be found at WolframAlpha.com.

A CAS is often used instead of a Table of Integrals. When using a CAS to find an integral, keep in mind:

- A CAS can find indefinite integrals or definite integrals.
- For indefinite integrals, the constant of integration is often omitted.
- For definite integrals, there is often an option to specify whether the solution should be expressed in exact form or as an approximate solution, and to what precision.
- Absolute value symbols are often omitted from logarithmic answers.
- The symbol "log" is often used to mean ln.
- An indefinite integral is sometimes expressed in a different, but equivalent, algebraic form from that found in a table or by hand. With simplification, integrals found by hand, with a Table of Integrals, or with a CAS will be the same.
- When the CAS fails to find a result, either because the integral is infeasible or beyond the capability of the CAS, most systems show this by repeating the integral.
- Many CAS products have the capability to handle improper integrals (Section 7.8).

EXAMPLE 4   Finding an Integral Using a CAS

Find $\int x^2 (2x^3 - 4)^5 dx$.

Solution Using WolframAlpha to find the integral, input

$$\text{integrate } x\hat{\ }2((2x\hat{\ }3) - 4)\hat{\ }5$$

The output is

$$\int x^2 (2x^3 - 4)^5 dx = 32 \left(\frac{x^{18}}{18} - \frac{2x^{15}}{3} + \frac{10x^{12}}{3} - \frac{80x^9}{9} + \frac{40x^6}{3} - \frac{32x^3}{3} \right) + C$$

Using Mathematica returns the same result without the constant C. ∎

To find $\int x^2 (2x^3 - 4)^5 dx$ by hand, we use substitution with $u = 2x^3 - 4$. Then $du = 6x^2 dx$. So,

$$\int x^2 (2x^3 - 4)^5 dx = \frac{1}{6} \int u^5 du = \frac{1}{6} \cdot \frac{u^6}{6} = \frac{(2x^3 - 4)^6}{36} + C$$

If this solution is expanded using the Binomial Theorem, it will differ from the CAS answer by a constant.

NOW WORK   Problems 19, 21, and 27 using a CAS and compare your answers to Problems 3, 5, and 11.

7.7 Assess Your Understanding

Skill Building

In Problems 1–16, find each integral using the Table of Integrals found at the back of the book.

1. $\displaystyle\int e^{2x}\cos x\,dx$

2. $\displaystyle\int e^{5x+1}\sin(2x+3)\,dx$

3. $\displaystyle\int x\sqrt{4x+3}\,dx$

4. $\displaystyle\int \frac{dx}{(x^2-1)^{3/2}}$

5. $\displaystyle\int (x+1)\sqrt{4x+5}\,dx$

6. $\displaystyle\int \frac{dx}{[(2x+3)^2-1]^{3/2}}$

7. $\displaystyle\int \frac{dx}{x\sqrt{4x+6}}$

8. $\displaystyle\int \frac{dx}{x\sqrt{8+x}}$

9. $\displaystyle\int \frac{\sqrt{4x+6}}{x}\,dx$

10. $\displaystyle\int \frac{\sqrt{8+x}}{x^2}\,dx$

11. $\displaystyle\int \frac{x^3}{(\ln x)^2}\,dx$

12. $\displaystyle\int x^3(\ln x)^2\,dx$

13. $\displaystyle\int \sin^{-1}(2x)\,dx$

14. $\displaystyle\int \tan^{-1}(-3x)\,dx$

15. $\displaystyle\int_1^2 \frac{x^3}{\sqrt{3x-x^2}}\,dx$

16. $\displaystyle\int_1^e \frac{1}{x^2\sqrt{x^2+2}}\,dx$

[CAS] *In Problems 17–32:*
  (a) *Redo Problems 1–16 using a CAS.*
  (b) *Compare the result to the answer obtained using a Table of Integrals.*
  (c) *If the results are different, verify that they are equivalent.*

[CAS] *In Problems 33–38, use a CAS to investigate whether each indefinite integral can be expressed using elementary functions.*

33. $\displaystyle\int \sqrt{1+x^3}\,dx$

34. $\displaystyle\int \sqrt{1+\sin x}\,dx$

35. $\displaystyle\int e^{-x^2}\,dx$

36. $\displaystyle\int \frac{\cos x}{x}\,dx$

37. $\displaystyle\int x\tan x\,dx$

38. $\displaystyle\int \sqrt{1+e^x}\,dx$

7.8 Improper Integrals

OBJECTIVES *When you finish this section, you should be able to:*
1 **Find integrals with an infinite limit of integration (p. 524)**
2 **Interpret an improper integral geometrically (p. 525)**
3 **Integrate functions over [a, b] that are not defined at an endpoint (p. 527)**
4 **Use the Comparison Test for Improper Integrals (p. 529)**

In Chapter 5, the definition of the definite integral $\int_a^b f(x)\,dx$ required that both a and b be real numbers. We also required that the function f be defined on the closed interval $[a, b]$. Here, we take up instances for which:

- Integrals have infinite limits of integration:

$$\int_1^\infty \frac{1}{\sqrt{x}}\,dx \qquad \int_{-\infty}^0 \frac{x-3}{x^3-8}\,dx \qquad \int_{-\infty}^\infty \frac{x}{(x^2+1)^2}\,dx$$

- The integrand is not defined at a number in the interval of integration:

$$\int_0^1 \frac{1}{\sqrt{x}}\,dx \qquad \int_0^{\pi/2} \tan x\,dx \qquad \int_{-1}^1 \frac{1}{x^3}\,dx$$

Integrals like these are called *improper integrals*.

---

1. = NOW WORK problem   **[∕∖]** = Graphing technology recommended   **[CAS]** = Computer Algebra System recommended

DEFINITION Improper Integral

If a function f is continuous on the interval $[a, \infty)$, then $\int_a^\infty f(x)\,dx$, called an **improper integral**, is defined as

$$\int_a^\infty f(x)\,dx = \lim_{b \to \infty} \int_a^b f(x)\,dx$$

provided the limit exists as a real number. If $\lim_{b \to \infty} \int_a^b f(x)\,dx$ exists as a real number, the improper integral $\int_a^\infty f(x)\,dx$ is said to **converge**. If the limit does not exist or if the limit is infinite, the improper integral $\int_a^\infty f(x)\,dx$ is said to **diverge**.

If a function f is continuous on the interval $(-\infty, b]$, then $\int_{-\infty}^b f(x)\,dx$, called an **improper integral**, is defined as

$$\int_{-\infty}^b f(x)\,dx = \lim_{a \to -\infty} \int_a^b f(x)\,dx$$

provided the limit exists as a real number. If $\lim_{a \to -\infty} \int_a^b f(x)\,dx$ exists as a real number, the improper integral $\int_{-\infty}^b f(x)\,dx$ **converges**. If the limit does not exist or if the limit is infinite, the improper integral $\int_{-\infty}^b f(x)\,dx$ **diverges**.

1 Find Integrals with an Infinite Limit of Integration

EXAMPLE 1   Integrating Functions over Infinite Intervals

Determine whether the following improper integrals converge or diverge:

(a) $\displaystyle\int_1^\infty \frac{1}{x}\,dx$ **(b)** $\displaystyle\int_{-\infty}^0 e^x\,dx$ **(c)** $\displaystyle\int_{\pi/2}^\infty \sin x\,dx$

NEED TO REVIEW? Limits at infinity are discussed in Section 1.5, pp. 120–122.

Solution (a) $\displaystyle\int_1^\infty \frac{1}{x}\,dx$: $\displaystyle\lim_{b \to \infty} \int_1^b \frac{1}{x}\,dx = \lim_{b \to \infty} \left[\ln |x|\right]_1^b = \lim_{b \to \infty} [\ln b - \ln 1] = \infty$.

The limit is infinite, so $\displaystyle\int_1^\infty \frac{1}{x}\,dx$ diverges.

(b) $\displaystyle\int_{-\infty}^0 e^x\,dx$: $\displaystyle\lim_{a \to -\infty} \int_a^0 e^x\,dx = \lim_{a \to -\infty} \left[e^x\right]_a^0 = \lim_{a \to -\infty} (1 - e^a) = 1 - 0 = 1$.

Since the limit exists, $\displaystyle\int_{-\infty}^0 e^x\,dx$ converges and equals 1.

(c) $\displaystyle\int_{\pi/2}^\infty \sin x\,dx$: $\displaystyle\lim_{b \to \infty} \int_{\pi/2}^b \sin x\,dx = \lim_{b \to \infty} \left[-\cos x\right]_{\pi/2}^b$

$$= \lim_{b \to \infty} [-\cos b + 0]$$

$$= -\lim_{b \to \infty} \cos b$$

This limit does not exist, since as $b \to \infty$, the value of $\cos b$ oscillates between -1 and 1. So, $\displaystyle\int_{\pi/2}^\infty \sin x\,dx$ diverges. ∎

NOW WORK   Problem 17.

If both limits of integration are infinite, then the following definition is used.

> If a function f is continuous for all x and if, for any number c, *both* improper integrals $\int_{-\infty}^{c} f(x)\,dx$ and $\int_{c}^{\infty} f(x)\,dx$ converge, then the improper integral $\int_{-\infty}^{\infty} f(x)\,dx$ **converges,** and
>
> $$\int_{-\infty}^{\infty} f(x)\,dx = \int_{-\infty}^{c} f(x)\,dx + \int_{c}^{\infty} f(x)\,dx$$
>
> If *either* or *both* of the integrals on the right diverge, then the improper integral $\int_{-\infty}^{\infty} f(x)\,dx$ **diverges**.

NOTE The decision to use 0 to break up the integral is arbitrary. Usually, the choice made simplifies finding the integral.

EXAMPLE 2   Integrating Functions over Infinite Intervals

Determine whether $\int_{-\infty}^{\infty} 4x^3\,dx$ converges or diverges.

Solution We begin by writing $\int_{-\infty}^{\infty} 4x^3\,dx = \int_{-\infty}^{0} 4x^3\,dx + \int_{0}^{\infty} 4x^3\,dx$ and evaluate each improper integral on the right.

$$\int_{-\infty}^{0} 4x^3\,dx: \quad \lim_{a \to -\infty}\left(4\int_{a}^{0} x^3\,dx\right) = \lim_{a \to -\infty}\left[x^4\right]_{a}^{0} = \lim_{a \to -\infty}(0 - a^4) = -\infty$$

There is no need to continue. $\int_{-\infty}^{\infty} 4x^3\,dx$ diverges. ∎

CAUTION The definition requires that two improper integrals converge in order for $\int_{-\infty}^{\infty} f(x)\,dx$ to converge. Do not set $\int_{-\infty}^{\infty} f(x)\,dx = \lim_{a \to \infty} \int_{-a}^{a} f(x)\,dx$. If this were done in Example 2, the result would have been $\int_{-a}^{a} 4x^3\,dx = \left[x^4\right]_{-a}^{a} = a^4 - a^4 = 0$, and we would have incorrectly concluded that $\int_{-\infty}^{\infty} f(x)\,dx$ converges and equals 0.

NOW WORK Problem 23.

2 Interpret an Improper Integral Geometrically

If a function f is continuous and nonnegative on the interval $[a, \infty)$, then $\int_{a}^{\infty} f(x)\,dx$ can be interpreted geometrically. For each number $b > a$, the definite integral $\int_{a}^{b} f(x)\,dx$ represents the area under the graph of $y = f(x)$ from a to b, as shown in Figure 21(a). As $b \to \infty$, this area approaches the area under the graph of $y = f(x)$ over the interval $[a, \infty)$, as shown in Figure 21(b). The area under the graph of $y = f(x)$ to the right of a is defined by $\int_{a}^{\infty} f(x)\,dx$, provided the improper integral converges. If it diverges, there is no area defined.

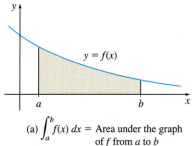

(a) $\int_{a}^{b} f(x)\,dx$ = Area under the graph of f from a to b

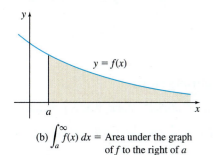

(b) $\int_{a}^{\infty} f(x)\,dx$ = Area under the graph of f to the right of a

Figure 21

EXAMPLE 3   Determining If the Area Under a Graph Is Defined

Determine if the area under the graph of $y = \dfrac{1}{x^2}$ to the right of $x = 1$ is defined.

Solution See Figure 22. To determine if the area is defined, we examine $\int_{1}^{\infty} \dfrac{1}{x^2}\,dx$.

$$\int_{1}^{\infty} \frac{1}{x^2}\,dx: \quad \lim_{b \to \infty} \int_{1}^{b} \frac{1}{x^2}\,dx = \lim_{b \to \infty}\left[-\frac{1}{x}\right]_{1}^{b} = \lim_{b \to \infty}\left(-\frac{1}{b} + 1\right) = 1$$

The area under the graph $f(x) = \dfrac{1}{x^2}$ to the right of 1 is defined and equals 1 square unit. ∎

Figure 22

NOW WORK Problem 71.

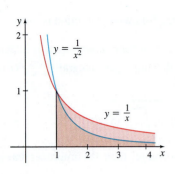

Figure 23

In Example 1(a), we found that $\displaystyle\int_1^\infty \frac{1}{x}\,dx$ diverges, but Example 3 shows $\displaystyle\int_1^\infty \frac{1}{x^2}\,dx$ converges. Yet, as Figure 23 illustrates, the graphs of $y = \dfrac{1}{x}$ and $y = \dfrac{1}{x^2}$ on the interval $[1, \infty]$ are very similar. The difference is that $y = \dfrac{1}{x^2}$ approaches 0 more rapidly than $y = \dfrac{1}{x}$ as $x \to \infty$. The next result generalizes these conclusions and is used in Chapter 8.

THEOREM

$\displaystyle\int_1^\infty \frac{dx}{x^p}$ converges if $p > 1$ and diverges if $p \le 1$.

Proof We consider two cases: $p = 1$ and $p \ne 1$.

$$p = 1; \qquad \int_1^\infty \frac{dx}{x^p} = \int_1^\infty \frac{dx}{x}: \quad \lim_{b \to \infty} \int_1^b \frac{dx}{x} = \lim_{b \to \infty} \ln b = \infty \qquad \text{From Example 1(a)}$$

$$p \ne 1; \qquad \int_1^\infty \frac{dx}{x^p}: \quad \lim_{b \to \infty} \int_1^b \frac{dx}{x^p} = \lim_{b \to \infty} \left[\frac{x^{-p+1}}{-p+1} \right]_1^b$$

$$= \lim_{b \to \infty} \left[\frac{1}{1-p}(b^{-p+1} - 1) \right]$$

$$= \frac{1}{1-p} \lim_{b \to \infty} \left(\frac{1}{b^{p-1}} - 1 \right)$$

If $p > 1$, then $p - 1 > 0$, and $\displaystyle\lim_{b \to \infty} \frac{1}{b^{p-1}} = 0$. So,

$$\int_1^\infty \frac{dx}{x^p} = \frac{1}{1-p}(0 - 1) = \frac{1}{p-1}$$

and the improper integral $\displaystyle\int_1^\infty \frac{dx}{x^p}$, $p > 1$, converges.

If $p < 1$, then $p - 1 < 0$, and $\displaystyle\lim_{b \to \infty} \frac{1}{b^{p-1}} = \lim_{b \to \infty} b^{1-p} = \infty$. So, the improper integral $\displaystyle\int_1^\infty \frac{dx}{x^p}$, $p < 1$, diverges. ∎

NOTE Interesting Fact: The area under the graph of $y = \dfrac{1}{x}$ to the right of 1 is not defined [see Example 1(a)], but the volume obtained by revolving the region with this same area about the x-axis is defined.

EXAMPLE 4   Finding the Volume of Gabriel's Horn

Find the volume of the solid of revolution, called **Gabriel's Horn,** that is generated by revolving the region bounded by the graph of $y = \dfrac{1}{x}$ and the x-axis to the right of 1 about the x-axis. Use the disk method.

Solution Figure 24 illustrates the region being revolved and the solid of revolution that it generates. Using the disk method, the volume V is

$$V = \pi \int_1^\infty \left(\frac{1}{x} \right)^2 dx: \quad \pi \lim_{b \to \infty} \int_1^b \frac{1}{x^2}\,dx = \pi \lim_{b \to \infty} \left[-\frac{1}{x} \right]_1^b$$

$$= \pi \lim_{b \to \infty} \left(-\frac{1}{b} + 1 \right) = \pi$$

The volume of the solid of revolution is π cubic units. ∎

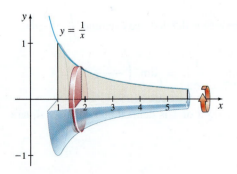

Figure 24 Gabriel's Horn

NOW WORK   Problem **73.**

3 Integrate Functions over $[a, b]$ That Are Not Defined at an Endpoint

DEFINITION

If a function f is continuous on $(a, b]$, but is not defined at a, then $\int_a^b f(x)\,dx$ is an **improper integral**, and

$$\int_a^b f(x)\,dx = \lim_{t \to a^+} \int_t^b f(x)\,dx$$

provided the limit exists as a real number.

If a function f is continuous on $[a, b)$, but is not defined at b, then $\int_a^b f(x)\,dx$ is an **improper integral**, and

$$\int_a^b f(x)\,dx = \lim_{t \to b^-} \int_a^t f(x)\,dx$$

provided the limit exists as a real number.

When the limit exists as a real number, the improper integral is said to **converge**; otherwise, it is said to **diverge**.

Be sure to use the correct one-sided limit. If f is not defined at the left endpoint a, then for t to be in the interval of integration, $t > a$ and t must approach a from the right. Similarly, if f is not defined at the right endpoint b, then for t to be in the interval of integration, $t < b$ and t must approach b from the left. Figure 25 may help you remember the correct one-sided limit to use.

If a function f is continuous and nonnegative on the interval $(a, b]$, then the improper integral $\int_a^b f(x)\,dx$ may be interpreted geometrically. For each number t, $a < t \le b$, the integral $\int_t^b f(x)\,dx$ represents the area under the graph of $y = f(x)$ from t to b, as shown in Figure 26(a). As $t \to a^+$, this area approaches the area under the graph of $y = f(x)$ from a to b, as shown in Figure 26(b). That is, if $\int_a^b f(x)\,dx$ converges, its value is defined to be the area A under the graph of f from a to b.

IN WORDS These improper integrals, sometimes called **improper integrals of the second kind**, have finite limits of integration, but the integrand is not defined at one (but not both) of them.

f is continuous on $(a, b]$
f is not defined at a

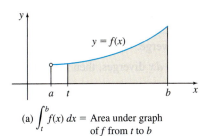

f is continuous on $[a, b)$
f is not defined at b

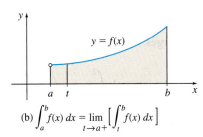

Figure 25

(a) $\int_t^b f(x)\,dx = $ Area under graph of f from t to b

(b) $\int_a^b f(x)\,dx = \lim_{t \to a^+} \left[\int_t^b f(x)\,dx \right]$

Figure 26

EXAMPLE 5 Determining If the Area Under a Graph Is Defined

Determine if the area under the graph of $y = \dfrac{1}{\sqrt{x}}$ from 0 to 4 is defined.

Solution Figure 27 shows the area under the graph of $y = \dfrac{1}{\sqrt{x}}$ from 0 to 4. The area is given by $\displaystyle\int_0^4 \frac{1}{\sqrt{x}}\,dx$. Since the integrand $f(x) = \dfrac{1}{\sqrt{x}}$ is continuous on $(0, 4]$ but is not defined at 0, $\displaystyle\int_0^4 \frac{1}{\sqrt{x}}\,dx$ is an improper integral.

$$\int_0^4 \frac{1}{\sqrt{x}}\,dx : \quad \lim_{t \to 0^+} \int_t^4 \frac{1}{\sqrt{x}}\,dx = \lim_{t \to 0^+} \int_t^4 x^{-1/2}\,dx = \lim_{t \to 0^+} \left[\frac{x^{1/2}}{\frac{1}{2}} \right]_t^4$$

$$= \lim_{t \to 0^+} (2 \cdot 2 - 2\sqrt{t}) = 4 - 2 \lim_{t \to 0^+} \sqrt{t} = 4$$

So, $\displaystyle\int_0^4 \frac{1}{\sqrt{x}}\,dx$ converges, and the area under the graph of $y = \dfrac{1}{\sqrt{x}}$ from 0 to 4 is defined and equals 4. ∎

Figure 27 $y = \dfrac{1}{\sqrt{x}}, 0 < x \le 4$.

NOW WORK Problem 27.

7.8 Assess Your Understanding

Concepts and Vocabulary

1. *Multiple Choice* If a function f is continuous on the interval $[a, \infty)$, then $\int_a^\infty f(x)\, dx$ is called [(a) a definite, (b) an infinite, (c) an improper, (d) a proper] integral.

2. *Multiple Choice* If the $\lim\limits_{b \to \infty} \int_a^b f(x)\, dx$ does not exist, the improper integral $\int_a^\infty f(x)\, dx$ [(a) converges, (b) diverges, (c) equals ab].

3. *True or False* If a function f is continuous for all x, then the improper integral $\int_{-\infty}^\infty f(x)\, dx$ always converges.

4. *True or False* If a function f is continuous and nonnegative on the interval $[a, \infty)$, and $\int_a^\infty f(x)\, dx$ converges, then $\int_a^\infty f(x)\, dx$ represents the area under the graph of $y = f(x)$ for $x \geq a$.

5. *True or False* If a function f is continuous for all x, then the improper integral $\int_{-\infty}^\infty f(x)\, dx = \lim\limits_{a \to \infty} \int_{-a}^a f(x)\, dx$.

6. To determine whether the improper integral $\int_a^b f(x)\, dx$ converges or diverges, where f is continuous on $[a, b)$, but is not defined at b, requires finding what limit?

Skill Building

In Problems 7–14, determine whether each integral is improper. For those that are improper, state the reason.

7. $\displaystyle\int_0^\infty x^2\, dx$

8. $\displaystyle\int_0^5 x^3\, dx$

9. $\displaystyle\int_2^3 \frac{dx}{x-1}$

10. $\displaystyle\int_1^2 \frac{dx}{x-1}$

11. $\displaystyle\int_0^1 \frac{1}{x}\, dx$

12. $\displaystyle\int_{-1}^1 \frac{x\, dx}{x^2+1}$

13. $\displaystyle\int_0^1 \frac{x}{x^2-1}\, dx$

14. $\displaystyle\int_0^\infty e^{-2x}\, dx$

In Problems 15–24, determine whether each improper integral converges or diverges. If it converges, find its value.

15. $\displaystyle\int_1^\infty \frac{dx}{x^3}$

16. $\displaystyle\int_{-\infty}^{-10} \frac{dx}{x^2}$

17. $\displaystyle\int_0^\infty e^{2x}\, dx$

18. $\displaystyle\int_0^\infty e^{-x}\, dx$

19. $\displaystyle\int_{-\infty}^{-1} \frac{4}{x}\, dx$

20. $\displaystyle\int_1^\infty \frac{4}{x}\, dx$

21. $\displaystyle\int_3^\infty \frac{dx}{(x-1)^4}$

22. $\displaystyle\int_{-\infty}^0 \frac{dx}{(x-1)^4}$

23. $\displaystyle\int_{-\infty}^\infty \frac{dx}{x^2+4}$

24. $\displaystyle\int_{-\infty}^\infty \frac{dx}{x^2+1}$

In Problems 25–32, determine whether each improper integral converges or diverges. If it converges, find its value.

25. $\displaystyle\int_0^1 \frac{dx}{x^2}$

26. $\displaystyle\int_0^1 \frac{dx}{x^3}$

27. $\displaystyle\int_0^1 \frac{dx}{x}$

28. $\displaystyle\int_4^6 \frac{dx}{x-4}$

29. $\displaystyle\int_0^4 \frac{dx}{\sqrt{4-x}}$

30. $\displaystyle\int_1^5 \frac{x\, dx}{\sqrt{5-x}}$

31. $\displaystyle\int_{-1}^1 \frac{dx}{\sqrt[3]{x}}$

32. $\displaystyle\int_0^3 \frac{dx}{(x-2)^2}$

In Problems 33–62, determine whether each improper integral converges or diverges. If it converges, find its value.

33. $\displaystyle\int_0^\infty \cos x\, dx$

34. $\displaystyle\int_0^\infty \sin(\pi x)\, dx$

35. $\displaystyle\int_{-\infty}^0 e^x\, dx$

36. $\displaystyle\int_{-\infty}^0 e^{-x}\, dx$

37. $\displaystyle\int_0^{\pi/2} \frac{x\, dx}{\sin x^2}$

38. $\displaystyle\int_0^1 \frac{\ln x\, dx}{x}$

39. $\displaystyle\int_0^1 \frac{dx}{1-x^2}$

40. $\displaystyle\int_1^2 \frac{dx}{\sqrt{x^2-1}}$

41. $\displaystyle\int_0^1 \frac{x\, dx}{(1-x^2)^2}$

42. $\displaystyle\int_0^{2a} \frac{dx}{(x-a)^2}$

43. $\displaystyle\int_0^{\pi/4} \tan(2x)\, dx$

44. $\displaystyle\int_0^{\pi/2} \csc x\, dx$

45. $\displaystyle\int_0^\infty \frac{x\, dx}{\sqrt{x+1}}$

46. $\displaystyle\int_2^\infty \frac{dx}{x\sqrt{x^2-1}}$

47. $\displaystyle\int_{-\infty}^\infty \frac{dx}{x^2+4x+5}$

48. $\displaystyle\int_{-\infty}^\infty \frac{dx}{e^x+e^{-x}}$

49. $\displaystyle\int_{-\infty}^2 \frac{dx}{\sqrt{4-x}}$

50. $\displaystyle\int_{-\infty}^1 \frac{x\, dx}{\sqrt{2-x}}$

51. $\displaystyle\int_2^4 \frac{2x\, dx}{\sqrt[3]{x^2-4}}$

52. $\displaystyle\int_0^\pi \frac{1}{1-\cos x}\, dx$

53. $\displaystyle\int_{-1}^1 \frac{1}{x^3}\, dx$

54. $\displaystyle\int_0^2 \frac{dx}{x-1}$

55. $\displaystyle\int_0^2 \frac{dx}{(x-1)^{1/3}}$

56. $\displaystyle\int_{-1}^1 \frac{dx}{x^{5/3}}$

57. $\displaystyle\int_1^2 \frac{dx}{(2-x)^{3/4}}$

58. $\displaystyle\int_0^4 \frac{dx}{\sqrt{8x-x^2}}$

59. $\displaystyle\int_a^{3a} \frac{2x\, dx}{(x^2-a^2)^{3/2}}, a > 0$

60. $\displaystyle\int_0^3 \frac{x\, dx}{(9-x^2)^{3/2}}$

61. $\displaystyle\int_0^\infty xe^{-x^2}\, dx$

62. $\displaystyle\int_0^\infty e^{-x}\sin x\, dx$

In Problems 63–70:

(a) Use the Comparison Test for Improper Integrals to determine whether each improper integral converges or diverges. (Hint: Use the fact that $\displaystyle\int_1^\infty \frac{dx}{x^p}$ converges if $p > 1$ and diverges if $p \leq 1$.)

(CAS) (b) If the integral converges, use a CAS to find its value.

63. $\displaystyle\int_1^\infty \frac{1}{\sqrt{x^2-1}}\, dx$

64. $\displaystyle\int_2^\infty \frac{2}{\sqrt{x^2-4}}\, dx$

65. $\displaystyle\int_1^\infty \frac{1+e^{-x}}{x}\, dx$

66. $\displaystyle\int_1^\infty \frac{3e^{-x}}{x}\, dx$

67. $\displaystyle\int_1^\infty \frac{\sin^2 x}{x^2}\, dx$

68. $\displaystyle\int_1^\infty \frac{\cos^2 x}{x^2}\, dx$

69. $\displaystyle\int_1^\infty \frac{dx}{(x+1)\sqrt{x}}$

70. $\displaystyle\int_1^\infty \frac{dx}{x\sqrt{1+x^2}}$

---

1. = NOW WORK problem   N = Graphing technology recommended   (CAS) = Computer Algebra System recommended

Applications and Extensions

71. Area Between Graphs Find the area, if it is defined, of the region enclosed by the graphs of $y = \dfrac{1}{x+1}$ and $y = \dfrac{1}{x+2}$ on the interval $[0, \infty)$. See the figure.

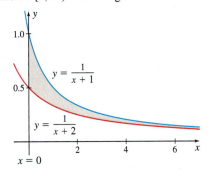

72. Area Between Graphs Find the area, if it is defined, under the graph of $y = \dfrac{1}{1+x^2}$ to the right of $x = 0$. See the figure below.

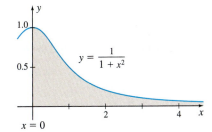

73. Volume of a Solid of Revolution Find the volume, if it is defined, of the solid of revolution generated by revolving the region bounded by the graph of $y = e^{-x}$ and the x-axis to the right of $x = 0$ about the x-axis. See the figure below.

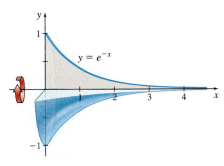

74. Volume of a Solid of Revolution Find the volume, if it is defined, of the solid of revolution generated by revolving the region bounded by the graph of $y = \dfrac{1}{\sqrt{x}}$ and the x-axis to the right of $x = 1$ about the x-axis. See the figure below.

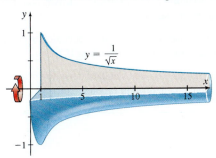

75. Area Between Graphs Find the area, if it is defined, between the graph of $y = \dfrac{8a^3}{x^2 + 4a^2}$, $a > 0$, and its horizontal asymptote.

76. Drug Reaction The rate of reaction r to a given dose of a drug at time t hours after administration is given by $r(t) = t\,e^{-t^2}$ (measured in appropriate units).

    (a) Why is it reasonable to define the total *reaction* as the area under the graph of $y = r(t)$ on $[0, \infty)$?

    (b) Find the total reaction to the given dose of the drug.

77. Present Value of Money The present value PV of a capital asset that provides a perpetual stream of revenue that flows continuously at a rate of $R(t)$ dollars per year is given by

$$PV = \int_0^\infty R(t)e^{-rt}\,dt$$

where r, expressed as a decimal, is the annual rate of interest compounded continuously.

    (a) Find the present value of an asset if it provides a constant return of $100 per year and $r = 8\%$.

    (b) Find the present value of an asset if it provides a return of $R(t) = 1000 + 80t$ dollars per year and $r = 7\%$.

78. Electrical Engineering In a problem in electrical theory, the integral $\int_0^\infty Ri^2\,dt$ occurs, where the current $i = I e^{-Rt/L}$, t is time, and R, I, and L are constants. Find the integral.

79. Magnetic Potential The magnetic potential u at a point on the axis of a circular coil is given by

$$u = \frac{2\pi N I r}{10} \int_x^\infty \frac{dy}{(r^2 + y^2)^{3/2}}$$

where N, I, r, and x are constants. Find the integral.

80. Electrical Engineering The field intensity F around a long ("infinite") straight wire carrying electric current is given by the integral

$$F = \frac{r I m}{10} \int_{-\infty}^\infty \frac{dy}{(r^2 + y^2)^{3/2}}$$

where r, I, and m are constants. Find the integral.

81. Work The force F of gravitational attraction between two point masses m and M that are r units apart is $F = \dfrac{GmM}{r^2}$, where G is the universe gravitational constant. Find the work done in moving the mass m along a straight-line path from $r = 1$ unit to $r = \infty$.

82. For what numbers a does $\int_0^1 x^a\,dx$ converge?

In Problems 83–86, use integration by parts and perhaps L'Hôpital's Rule to find each improper integral.

83. $\displaystyle\int_0^\infty x e^{-x}\,dx$ **84.** $\displaystyle\int_0^1 x \ln x\,dx$

85. $\displaystyle\int_0^\infty e^{-x}\cos x\,dx$ **86.** $\displaystyle\int_0^\infty \tan^{-1} x\,dx$

87. Show that $\int_0^\infty \sin x\,dx$ and $\int_{-\infty}^0 \sin x\,dx$ each diverge, yet $\lim\limits_{t \to \infty} \int_{-t}^t \sin x\,dx = 0$.

88. Find a function f for which $\int_0^\infty f(x)\,dx$ and $\int_{-\infty}^0 f(x)\,dx$ each diverge, yet $\lim_{t\to\infty}\int_{-t}^t f(x)\,dx = 1$.

89. Use the Comparison Test for Improper Integrals to show that $\int_0^\infty \dfrac{1}{\sqrt{2+\sin x}}\,dx$ diverges.

90. Use the Comparison Test for Improper Integrals to show that $\int_2^\infty \dfrac{\ln x}{\sqrt{x^2-1}}\,dx$ diverges.

91. If n is a positive integer, show that:

(a) $\displaystyle\int_0^\infty x^n e^{-x}\,dx = n\int_0^\infty x^{n-1} e^{-x}\,dx$

(b) $\displaystyle\int_0^\infty x^n e^{-x}\,dx = n!$

92. Show that $\displaystyle\int_e^\infty \dfrac{dx}{x(\ln x)^p}$ converges if $p > 1$ and diverges if $p \le 1$.

93. Show that $\displaystyle\int_a^b \dfrac{dx}{(x-a)^p}$ converges if $0 < p < 1$ and diverges if $p \ge 1$.

94. Show that $\displaystyle\int_a^b \dfrac{dx}{(b-x)^p}$ converges if $0 < p < 1$ and diverges if $p \ge 1$.

95. Refer to Problems 93 and 94. Discuss the convergence or divergence of the integrals if $p \le 0$. Support your explanation with an example.

96. Comparison Test for Improper Integrals   Show that if two functions f and g are nonnegative and continuous on the interval $[a, \infty)$, and if $f(x) \ge g(x)$ for all numbers $x > c$, where $c \ge a$, then

• If $\int_a^\infty f(x)\,dx$ converges, then $\int_a^\infty g(x)\,dx$ also converges.

• If $\int_a^\infty g(x)\,dx$ diverges, then $\int_a^\infty f(x)\,dx$ also diverges.

*Laplace transforms are useful in solving a special class of differential equations. The **Laplace transform** $L\{f(x)\}$ of a function f is defined as*

$$L\{f(x)\} = \int_0^\infty e^{-sx} f(x)\,dx, \; x \ge 0, \; s \text{ a complex number}$$

In Problems 97–102, find the Laplace transform of each function.

97. $f(x) = x$

98. $f(x) = \cos x$

99. $f(x) = \sin x$

100. $f(x) = e^x$

101. $f(x) = e^{ax}$

102. $f(x) = 1$

Challenge Problems

103. Find the arc length of $y = \sqrt{x - x^2} - \sin^{-1}\sqrt{x}$.

104. Find $\displaystyle\int_{-\infty}^a e^{(x-e^x)}\,dx$.

105. Find $\displaystyle\int_{-\infty}^\infty e^{(x-e^x)}\,dx$.

106. (a) Show that the area of the region in the first quadrant bounded by the graph of $y = e^{-x}$ and the x-axis is divided into two equal parts by the line $x = \ln 2$.

(b) If the two regions with equal areas described in (a) are rotated about the x-axis, are the resulting volumes equal? If they are unequal, which one is larger and by how much?

*In Problems 107 and 108, use the following definition: A **probability density function** is a function f, whose domain is the set of all real numbers, with the following properties:*

• $f(x) \ge 0$ for all x

• $\int_{-\infty}^\infty f(x)\,dx = 1$

107. Uniform Density Function   Show that the function f below is a probability density function.

$$f(x) = \begin{cases} 0 & \text{if } x < a \\ \dfrac{1}{b-a} & \text{if } a \le x \le b, a < b \\ 0 & \text{if } x > b \end{cases}$$

108. Exponential Density Function   Show that the function f is a probability density function for $a > 0$.

$$f(x) = \begin{cases} \dfrac{1}{a} e^{-x/a} & \text{if } x \ge 0 \\ 0 & \text{if } x < 0 \end{cases}$$

*In Problems 109 and 110, use the following definition: The **expected value** or **mean** μ associated with a probability density function f is defined by*

$$\mu = \int_{-\infty}^\infty x f(x)\,dx$$

The expected value can be thought of as a weighted average of its various probabilities.

109. Find the expected value μ of the uniform density function f given in Problem 107.

110. Find the expected value μ of the exponential probability density function f defined in Problem 108.

*In Problems 111 and 112, use the following definitions: The **variance** σ^2 of a probability density function f is defined as*

$$\sigma^2 = \int_{-\infty}^\infty (x-\mu)^2 f(x)\,dx$$

*The variance is the average of the squared deviation from the mean. The **standard deviation** σ of a probability density function f is the square root of its variance σ^2.*

111. Find the variance σ^2 and standard deviation σ of the uniform density function defined in Problem 107.

112. Find the variance σ^2 and standard deviation σ of the exponential density function defined in Problem 108.

Chapter Review

THINGS TO KNOW

7.1 Integration by Parts
- Integration by parts formula $\int u\,dv = uv - \int v\,du$ (p. 472)
- Guidelines for choosing u and dv (Table 1, p. 475)

7.2 Integrals Containing Trigonometric Functions
Procedures:
- Trigonometric integrals of the form $\int \sin^n x\,dx$ or $\int \cos^n x\,dx$, $n \geq 2$ an integer (pp. 480, 481)
- Trigonometric integrals of the form $\int \sin^m x \cos^n x\,dx$ (p. 483)
- Trigonometric integrals of the form $\int \tan^m x \sec^n x\,dx$ or $\int \cot^m x \csc^n x\,dx$ (pp. 483–485)
- Trigonometric integrals of the form $\int \sin(ax) \sin(bx)\,dx$, $\int \sin(ax) \cos(bx)\,dx$, or $\int \cos(ax) \cos(bx)\,dx$ (p. 485)

7.3 Integration Using Trigonometric Substitution: Integrands Containing $\sqrt{a^2 - x^2}$, $\sqrt{x^2 + a^2}$, or $\sqrt{x^2 - a^2}$
See Table 2 (p. 488):
- Integrals containing $\sqrt{a^2 - x^2}$: Use the substitution $x = a \sin\theta$, $-\frac{\pi}{2} \leq \theta \leq \frac{\pi}{2}$. (p. 488)
- Integrals containing $\sqrt{a^2 + x^2}$: Use the substitution $x = a \tan\theta$, $-\frac{\pi}{2} < \theta < \frac{\pi}{2}$. (p. 489)
- Integrals containing $\sqrt{x^2 - a^2}$: Use the substitution $x = a \sec\theta$, $0 \leq \theta < \frac{\pi}{2}$ or $\pi \leq \theta < \frac{3\pi}{2}$. (p. 491)

7.4 Substitution: Integrands Containing $ax^2 + bx + c$ (p. 496)

7.5 Integration of Rational Functions Using Partial Fractions
Definitions:
- Proper and improper rational functions (p. 499)
- Partial fractions (p. 499)
- Irreducible quadratic factor (p. 503)

Partial Fraction Decomposition of $R(x) = \dfrac{p(x)}{q(x)}$, $q(x) \neq 0$:
- Case 1: a proper rational function whose denominator contains only distinct linear factors (p. 500)
- Case 2: a proper rational function whose denominator contains a repeated linear factor $(x - a)^n$, $n \geq 2$, (p. 502)
- Case 3: a proper rational function whose denominator contains a distinct irreducible quadratic factor $x^2 + bx + c$. (p. 503)
- Case 4: a proper rational function whose denominator contains a repeated irreducible quadratic factor $(x^2 + bx + c)^n$, $n \geq 2$ an integer, (p. 504)

7.6 Integration Using Numerical Techniques
- Trapezoidal Rule: (p. 509)
- Error in the Trapezoidal Rule $\leq \dfrac{(b-a)^3 M}{12n^2}$, where M is the absolute maximum value of $|f''|$ on the closed interval $[a, b]$. (p. 512)
- Simpson's Rule: (p. 515)
- Error in Simpson's Rule $\leq \dfrac{(b-a)^5 M}{180n^4}$, where M is the absolute maximum value of $|f^{(4)}(x)|$ on the closed interval $[a, b]$. (p. 516)

7.7 Integration Using Tables and Computer Algebra Systems (p. xx)

7.8 Improper Integrals
- Improper integrals of the form $\int_a^\infty f(x)\,dx$; $\int_{-\infty}^b f(x)\,dx$; and $\int_{-\infty}^\infty f(x)\,dx$ (pp. 524, 525)
- Converge; diverge (p. 524)
- Improper integrals $\int_a^b f(x)\,dx$, where $f(a)$, $f(b)$, or $f(c)$ is not defined, $a < c < b$. (pp. 527 and 528)
- **Theorem** $\int_1^\infty \dfrac{dx}{x^p}$ converges if $p > 1$ and diverges if $p \leq 1$. (p. 526)

Comparison Test for Improper Integrals (p. 529)

OBJECTIVES

Section	You should be able to ...	Examples	Review Exercises
7.1	**1** Integrate by parts (p. 472)	1–6	8, 9, 16, 19, 21, 25, 47, 48
	2 Derive a formula using integration by parts (p. 476)	7, 8	37, 38
7.2	**1** Find integrals of the form $\int \sin^n x\, dx$ or $\int \cos^n x\, dx$, $n \geq 2$ an integer (p. 480)	1–3	5, 31
	2 Find integrals of the form $\int \sin^m x \cos^n x\, dx$ (p. 483)	4	10, 28
	3 Find integrals of the form $\int \tan^m x \sec^n x\, dx$ or $\int \cot^m x \csc^n x\, dx$ (p. 483)	5–7	3, 4
	4 Find integrals of the form $\int \sin(ax) \sin(bx)\, dx$, $\int \sin(ax) \cos(bx)\, dx$, or $\int \cos(ax) \cos(bx)\, dx$ (p. 485)	8	32, 33
7.3	**1** Find integrals containing $\sqrt{a^2 - x^2}$ (p. 488)	1	6, 11
	2 Find integrals containing $\sqrt{x^2 + a^2}$ (p. 489)	2, 3	18, 24, 26
	3 Find integrals containing $\sqrt{x^2 - a^2}$ (p. 491)	4	7, 30
	4 Use trigonometric substitution to find definite integrals (p. 492)	5, 6	34, 35
7.4	**1** Find an integral that contains a quadratic expression (p. 496)	1–3	1, 14, 20
7.5	**1** Integrate a rational function whose denominator contains only distinct linear factors (p. 500)	1, 2	12, 27, 29
	2 Integrate a rational function whose denominator contains a repeated linear factor (p. 502)	3	17, 23
	3 Integrate a rational function whose denominator contains a distinct irreducible quadratic factor (p. 503)	4	2, 15
	4 Integrate a rational function whose denominator contains a repeated irreducible quadratic factor (p. 504)	5	22
7.6	**1** Approximate an integral using the Trapezoidal Rule (p. 508)	1–5	49(a), 50
	2 Approximate an integral using Simpson's Rule (p. 514)	6–8	49(b)
7.7	**1** Use a Table of Integrals (p. 520)	1–3	36(a)
	2 Use a computer algebra system (p. 522)	4	36(b)
7.8	**1** Find integrals with an infinite limit of integration (p. 524)	1, 2	39, 42, 44
	2 Interpret an improper integral geometrically (p. 525)	3, 4	51, 52
	3 Integrate functions over $[a, b]$ that are not defined at an endpoint (p. 527)	5–7	40, 41, 43
	4 Use the Comparison Test for Improper Integrals (p. 529)	8	45, 46

REVIEW EXERCISES

In Problems 1–35, find each integral.

1. $\displaystyle\int \frac{dx}{x^2 + 4x + 20}$

2. $\displaystyle\int \frac{y+1}{y^2 + y + 1}\, dy$

3. $\displaystyle\int \sec^3 \phi \tan \phi\, d\phi$

4. $\displaystyle\int \cot^2 \theta\ \csc \theta\, d\theta$

5. $\displaystyle\int \sin^3 \phi\, d\phi$

6. $\displaystyle\int \frac{x^2}{\sqrt{4 - x^2}}\, dx$

7. $\displaystyle\int \frac{dx}{\sqrt{(x+2)^2 - 1}}$

8. $\displaystyle\int_0^{\pi/4} x \sin(2x)\, dx$

9. $\displaystyle\int v \csc^2 v\, dv$

10. $\displaystyle\int \sin^2 x \cos^3 x\, dx$

11. $\displaystyle\int (4 - x^2)^{3/2}\, dx$

12. $\displaystyle\int \frac{3x^2 + 1}{x^3 + 2x^2 - 3x}\, dx$

13. $\displaystyle\int \frac{e^{2t}\, dt}{e^t - 2}$

14. $\displaystyle\int \frac{dy}{5 + 4y + 4y^2}$

15. $\displaystyle\int \frac{x\, dx}{x^4 - 16}$

16. $\displaystyle\int x^3 e^{x^2}\, dx$

17. $\displaystyle\int \frac{y^2\, dy}{(y+1)^3}$

18. $\displaystyle\int \frac{dx}{x^2 \sqrt{x^2 + 25}}$

19. $\displaystyle\int x \sec^2 x\, dx$

20. $\displaystyle\int \frac{dx}{\sqrt{16 + 4x - 2x^2}}$

21. $\displaystyle\int \ln(1 - y)\, dy$

22. $\displaystyle\int \frac{x^3 - 2x - 1}{(x^2 + 1)^2}\, dx$

23. $\displaystyle\int \frac{3x^2 + 2}{x^3 - x^2}\, dx$

24. $\displaystyle\int \frac{dy}{\sqrt{2 + 3y^2}}$

25. $\displaystyle\int x^2 \sin^{-1} x\, dx$

26. $\displaystyle\int \sqrt{16 + 9x^2}\, dx$

27. $\displaystyle\int \frac{dx}{x^2 + 2x}$

28. $\displaystyle\int \sin^4 y \cos^4 y\, dy$

29. $\displaystyle\int \frac{w - 2}{1 - w^2}\, dw$

30. $\displaystyle\int \frac{x}{\sqrt{x^2 - 4}}\, dx$

31. $\displaystyle\int \frac{1}{\sqrt{x}} \cos^2 \sqrt{x}\, dx$

32. $\displaystyle\int \sin\left(\frac{\pi}{2} x\right) \sin(\pi x)\, dx$

33. $\displaystyle\int \sin x \cos(2x)\, dx$

34. $\displaystyle\int_0^1 \frac{x^2}{\sqrt{4 - x^2}}\, dx$

35. $\displaystyle\int_0^{\sqrt{3}} \frac{x\, dx}{\sqrt{1 + x^2}}$

36. (a) Find $\displaystyle\int \frac{\cos^2(2x)\,dx}{\sin^3(2x)}$ using a Table of Integrals.

(b) Find $\displaystyle\int \frac{\cos^2(2x)\,dx}{\sin^3(2x)}$ using a computer algebra system (CAS).

(c) Verify the results from (a) and (b) are equivalent.

In Problems 37 and 38, derive each formula where $n > 1$ is an integer.

37. $\displaystyle\int x^n \tan^{-1} x\,dx = \frac{x^{n+1}}{n+1}\tan^{-1}x - \frac{1}{n+1}\int \frac{x^{n+1}}{1+x^2}dx$

38. $\displaystyle\int x^n(ax+b)^{1/2}dx = \frac{2x^n(ax+b)^{3/2}}{(2n+3)a}$
$\displaystyle\qquad\qquad - \frac{2bn}{(2n+3)a}\int x^{n-1}(ax+b)^{1/2}dx$

In Problems 39–42, determine whether each improper integral converges or diverges. If it converges, find its value.

39. $\displaystyle\int_1^\infty \frac{e^{-\sqrt{x}}}{\sqrt{x}}dx$

40. $\displaystyle\int_0^1 \frac{\sin\sqrt{x}}{\sqrt{x}}dx$

41. $\displaystyle\int_0^1 \frac{x\,dx}{\sqrt{1-x^2}}$

42. $\displaystyle\int_{-\infty}^0 xe^x\,dx$

43. Show that $\displaystyle\int_0^{\pi/2} \frac{\sin x}{\cos x}dx$ diverges.

44. Show that $\displaystyle\int_1^\infty \frac{\sqrt{1+x^{1/8}}}{x^{3/4}}dx$ diverges.

In Problems 45 and 46, use the Comparison Test for Improper Integrals to determine whether each improper integral converges or diverges.

45. $\displaystyle\int_1^\infty \frac{1+e^{-x}}{x}dx$

46. $\displaystyle\int_0^\infty \frac{x}{(1+x)^3}dx$

47. If $\int x^2 \cos x\,dx = f(x) - \int 2x \sin x\,dx$, find f.

48. Area and Volume

(a) Find the area A of the region R bounded by the graphs of $y = \ln x$, the x-axis, and the line $x = e$.

(b) Find the volume of the solid of revolution generated by revolving R about the x-axis.

(c) Find the volume of the solid of revolution generated by revolving R about the y-axis.

49. Arc Length Approximate the arc length of $y = \cos x$ from $x = 0$ to $x = \dfrac{\pi}{2}$.

(a) using the Trapezoidal Rule with $n = 3$.

(b) using Simpson's Rule with $n = 4$.

50. Distance The velocity v (in meters per second) of a particle at time t is given in the table. Use the Trapezoidal Rule to approximate the distance traveled from $t = 1$ to $t = 4$.

t (s)	1	1.5	2	2.5	3	3.5	4
v (m/s)	3	4.3	4.6	5.1	5.8	6.2	6.6

51. Area Find the area, if it exists, of the region bounded by the graphs of $y = x^{-2/3}$, $y = 0$, $x = 0$, and $x = 1$.

52. Volume Find the volume, if it exists, of the solid of revolution generated when the region bounded by the graphs of $y = x^{-2/3}$, $y = 0$, $x = 0$, and $x = 1$ is revolved about the x-axis.

CHAPTER 7 PROJECT   The Birds of Rügen Island

Let $P = P(t)$ denote the population of rare birds on Rügen Island, where t is the time, in years. Suppose M equals the maximum sustainable number of birds and m equals the minimum population, below which the species becomes extinct. The population P can be modeled by the differential equation

$$\frac{dP}{dt} = k(M - P)(P - m)$$

where k is a positive constant.

1. Suppose the maximum population M is 1200 birds and the minimum population m is 100 birds. If $k = 0.001$, write the differential equation that models the population $P = P(t)$.

2. Solve the differential equation from Problem 1. (*Hint:*
$$\frac{dP}{(M-P)(P-m)} = k\,dt;$$ now integrate both sides and use partial fractions for the left integral.)

3. If the population at time $t = 0$ is 300 birds, find the particular solution of the differential equation.

4. How many birds will exist in 5 years?

5. Using graphing technology, graph the solution found in Problem 3.

6. The graph from Problem 5 seems to have an inflection point. Verify this and find it.

7. What conclusions can you draw about the rate of change of population based on your answer to Problem 6?

8. Write an essay about using the given differential equation to model the bird population. What assumptions are being made? What situations are being ignored?

• To watch a brief video, "Discover Germany: Rügen," go to http://www.youtube.com/watch?v=E1Ra6QHLIt4

8.1 Sequences

OBJECTIVES *When you finish this section, you should be able to:*

1 Write the terms of a sequence (p. 539)

2 Find the *n*th term of a sequence (p. 539)

3 Use properties of convergent sequences (p. 542)

4 Use a related function or the Squeeze Theorem to show a sequence converges (p. 543)

5 Determine whether a sequence converges or diverges (p. 545)

NEED TO REVIEW? Sequences are discussed in Appendix A.5, pp. A-38 to A-39.

The study of infinite sums of numbers has important applications in physics and engineering since it provides an alternate way of representing functions. In particular, *infinite series* may be used to approximate irrational numbers, such as e, π, and $\ln 2$. The theory of infinite series is developed through the use of a special kind of function called a *sequence*.

DEFINITION

A **sequence** is a function whose domain is the set of positive integers and whose range is a subset of the real numbers.

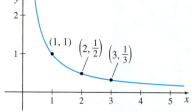

(a) $f(x) = \dfrac{1}{x}$, $x > 0$

(b) $f(n) = \dfrac{1}{n}$, n is a positive integer

Figure 1

To get an idea of what this means, consider the graph of the function $f(x) = \dfrac{1}{x}$ for $x > 0$, as shown in Figure 1(a). If all the points on the graph are removed *except* the points $(1, 1)$, $\left(2, \dfrac{1}{2}\right)$, $\left(3, \dfrac{1}{3}\right)$, and so on, as shown in Figure 1(b), then these points are the *graph of a sequence*. Notice the graph consists of points, one point for each positive integer.

A sequence is often represented by listing its values in order. For example, the sequence in Figure 1(b) can be written as

$$f(1), f(2), f(3), f(4), f(5), \ldots$$

or as the list

$$1, \frac{1}{2}, \frac{1}{3}, \frac{1}{4}, \frac{1}{5}, \ldots$$

The list never ends, as the three dots at the end (called **ellipsis**) indicate. The numbers in the list are called the **terms** of the sequence. Using subscripted letters to represent the terms of a sequence, this sequence can be written as

$$s_1 = f(1) = 1 \qquad s_2 = f(2) = \frac{1}{2} \qquad s_3 = f(3) = \frac{1}{3} \cdots \qquad s_n = f(n) = \frac{1}{n} \cdots$$

IN WORDS $\{s_n\}_{n=1}^{\infty} = \left\{\dfrac{1}{n}\right\}_{n=1}^{\infty}$ or $\{s_n\} = \left\{\dfrac{1}{n}\right\}$ is the name of the sequence; $s_n = \dfrac{1}{n}$ is the *n*th term of the sequence.

It is easy to obtain any term of this sequence because $s_n = f(n) = \dfrac{1}{n}$. In general, whenever a rule for the **nth term** of a sequence is known, then any term of the sequence can be found. We also use the nth term to identify the sequence. When the nth term is enclosed in braces, it represents the sequence. The notation $\{s_n\} = \left\{\dfrac{1}{n}\right\}$ or $\{s_n\}_{n=1}^{\infty} = \left\{\dfrac{1}{n}\right\}_{n=1}^{\infty}$ both represent the sequence $1, \dfrac{1}{2}, \dfrac{1}{3}, \dfrac{1}{4}, \dfrac{1}{5}, \ldots$.

1 Write the Terms of a Sequence

EXAMPLE 1   **Writing the Terms of a Sequence**

Write the first three terms of each sequence:

(a) $\{b_n\}_{n=1}^{\infty} = \left\{ \dfrac{1}{3n-2} \right\}_{n=1}^{\infty}$ **(b)** $\{c_n\} = \left\{ \dfrac{2n-1}{n^3} \right\}$

Solution   **(a)** The nth term of this sequence is $b_n = \dfrac{1}{3n-2}$. The first three terms are

$$b_1 = \frac{1}{3\cdot 1 - 2} = 1 \qquad b_2 = \frac{1}{3\cdot 2 - 2} = \frac{1}{4} \qquad b_3 = \frac{1}{3\cdot 3 - 2} = \frac{1}{7}$$

(b) The nth term of this sequence is $c_n = \dfrac{2n-1}{n^3}$. Then

$$c_1 = \frac{2(1)-1}{1^3} = 1 \qquad c_2 = \frac{2(2)-1}{2^3} = \frac{3}{8} \qquad c_3 = \frac{2(3)-1}{3^3} = \frac{5}{27} \qquad \blacksquare$$

NOW WORK   **Problem 15.**

For simplicity, from now on we use the notation $\{s_n\}$, rather than $\{s_n\}_{n=1}^{\infty}$, to represent a sequence.

EXAMPLE 2   **Writing the Terms of a Sequence**

Write the first five terms of the sequence

$$\{a_n\} = \left\{ (-1)^n \left(\frac{1}{2} \right)^n \right\}$$

Solution   The nth term of this sequence is $a_n = (-1)^n \left(\dfrac{1}{2} \right)^n$. So,

$$a_1 = (-1)^1 \left(\frac{1}{2} \right)^1 = -\frac{1}{2} \quad a_2 = (-1)^2 \left(\frac{1}{2} \right)^2 = \frac{1}{4} \quad a_3 = (-1)^3 \left(\frac{1}{2} \right)^3 = -\frac{1}{8}$$

$$a_4 = \frac{1}{16} \quad a_5 = -\frac{1}{32}$$

The first five terms of the sequence $\{a_n\}$ are $-\dfrac{1}{2}, \dfrac{1}{4}, -\dfrac{1}{8}, \dfrac{1}{16}, -\dfrac{1}{32}$. $\blacksquare$

Notice the terms of the sequence $\{a_n\}$ given in Example 2 *alternate* between positive and negative due to the factor $(-1)^n$, which equals -1 when n is odd and equals 1 when n is even. Sequences of the form $\{(-1)^n a_n\}$, $\{(-1)^{n-1} a_n\}$, or $\{(-1)^{n+1} a_n\}$, where $a_n > 0$ for all n, are called **alternating sequences**.

NOW WORK   **Problem 17.**

2 Find the nth Term of a Sequence

The rule defining a sequence $\{s_n\}$ is often expressed by an explicit formula for its nth term s_n. There are times, however, when a sequence is indicated using the first few terms, suggesting a natural choice for the nth term.

We state, without proof, some properties of convergent sequences.

THEOREM Properties of Convergent Sequences

If $\{s_n\}$ and $\{t_n\}$ are convergent sequences and if c is a number, then

- Constant multiple property:

$$\lim_{n \to \infty} (cs_n) = c \lim_{n \to \infty} s_n \qquad (1)$$

- Sum and difference properties:

$$\lim_{n \to \infty} (s_n \pm t_n) = \lim_{n \to \infty} s_n \pm \lim_{n \to \infty} t_n \qquad (2)$$

- Product property:

$$\lim_{n \to \infty} (s_n \cdot t_n) = \left(\lim_{n \to \infty} s_n \right) \left(\lim_{n \to \infty} t_n \right) \qquad (3)$$

- Quotient property:

$$\lim_{n \to \infty} \frac{s_n}{t_n} = \frac{\lim\limits_{n \to \infty} s_n}{\lim\limits_{n \to \infty} t_n} \qquad \text{provided} \quad \lim_{n \to \infty} t_n \neq 0 \qquad (4)$$

- Power property:

$$\lim_{n \to \infty} s_n^p = \left[\lim_{n \to \infty} s_n \right]^p \qquad p \geq 2 \text{ is an integer} \qquad (5)$$

- Root property:

$$\lim_{n \to \infty} \sqrt[p]{s_n} = \sqrt[p]{\lim_{n \to \infty} s_n} \qquad p \geq 2 \text{ and } s_n \geq 0 \text{ if } p \text{ is even} \qquad (6)$$

3 Use Properties of Convergent Sequences

EXAMPLE 5   Using Properties of Convergent Sequences

Use properties of convergent sequences to find $\lim\limits_{n \to \infty} s_n$.

(a) $\{s_n\} = \left\{ \dfrac{2}{n} + 3 \right\}$ **(b)** $\{s_n\} = \left\{ \dfrac{4}{n^2} \right\}$ **(c)** $\{s_n\} = \left\{ \sqrt[3]{\dfrac{16n^2 + 3n}{2n^2}} \right\}$

Solution (a) $\lim\limits_{n \to \infty} s_n = \lim\limits_{n \to \infty} \left(\dfrac{2}{n} + 3 \right) = \lim\limits_{n \to \infty} \dfrac{2}{n} + \lim\limits_{n \to \infty} 3$

$$= 2 \lim_{n \to \infty} \frac{1}{n} + \lim_{n \to \infty} 3 = 2 \cdot 0 + 3 = 3$$
$$\uparrow$$
$$\lim_{n \to \infty} \frac{1}{n} = 0; \ \lim_{n \to \infty} 3 = 3$$

(b) $\lim\limits_{n \to \infty} s_n = \lim\limits_{n \to \infty} \dfrac{4}{n^2} = 4 \lim\limits_{n \to \infty} \dfrac{1}{n^2} = 4 \lim\limits_{n \to \infty} \dfrac{1}{n} \cdot \lim\limits_{n \to \infty} \dfrac{1}{n} = 4 \cdot 0 \cdot 0 = 0$

$$\uparrow$$
$$\lim_{n \to \infty} \frac{1}{n} = 0$$

(c) $\displaystyle\lim_{n\to\infty} s_n = \lim_{n\to\infty} \sqrt[3]{\frac{16n^2 + 3n}{2n^2}} = \sqrt[3]{\lim_{n\to\infty}\left(\frac{16n^2 + 3n}{2n^2}\right)} = \sqrt[3]{\lim_{n\to\infty}\left(8 + \frac{3}{2n}\right)}$

$\displaystyle = \sqrt[3]{\lim_{n\to\infty} 8 + \lim_{n\to\infty}\frac{3}{2n}} = \sqrt[3]{8 + \frac{3}{2}\lim_{n\to\infty}\frac{1}{n}} = \sqrt[3]{8 + \frac{3}{2}\cdot 0} = \sqrt[3]{8} = 2$ ∎

$$\lim_{n\to\infty} 8 = 8 \qquad\qquad \lim_{n\to\infty}\frac{1}{n} = 0$$

NOW WORK Problem 37.

The next result is also useful for showing a sequence converges. You are asked to prove it in Problem 136.

THEOREM

Let $\{s_n\}$ be a sequence of real numbers. If $\displaystyle\lim_{n\to\infty} s_n = L$ and if f is a function that is continuous at L and is defined for all numbers s_n, then $\displaystyle\lim_{n\to\infty} f(s_n) = f(L)$.

EXAMPLE 6   Showing a Sequence Converges

Show $\left\{\ln\left(\dfrac{2}{n} + 3\right)\right\}$ converges and find its limit.

Solution Since $\displaystyle\lim_{n\to\infty}\left(\frac{2}{n} + 3\right) = 3$ [from Example 5(a)], the sequence $\{s_n\} = \left\{\dfrac{2}{n} + 3\right\}$ converges to 3. The function $f(x) = \ln x$ is continuous on its domain, so it is continuous at 3. Then

$$\lim_{n\to\infty} f(s_n) = \lim_{n\to\infty} f\left(\frac{2}{n} + 3\right) = \lim_{n\to\infty} \ln\left(\frac{2}{n} + 3\right) = \ln\left[\lim_{n\to\infty}\left(\frac{2}{n} + 3\right)\right] = \ln 3$$

So, the sequence $\left\{\ln\left(\dfrac{2}{n} + 3\right)\right\}$ converges to $\ln 3$.   ∎

NOW WORK Problem 45.

4 Use a Related Function or the Squeeze Theorem to Show a Sequence Converges

Sometimes a sequence $\{s_n\}$ can be associated with a *related function* f, which can be helpful in determining whether the sequence converges.

DEFINITION Related Function of a Sequence

A related function f of the sequence $\{s_n\}$ has the following two properties:

- f is defined on the open interval $(0, \infty)$; that is, the domain of f is the set of positive real numbers.
- $f(n) = s_n$ for all integers $n \geq 1$.

EXAMPLE 7   Identifying a Related Function of a Sequence

If $\{s_n\} = \left\{\dfrac{n}{e^n}\right\}$, then a related function is given by $f(x) = \dfrac{x}{e^x}$, where $x > 0$, as shown in Figure 6.   ∎

There is a connection between the convergence of certain sequences $\{s_n\}$ and the behavior at infinity of a related function f of the sequence $\{s_n\}$. The following result, which we state without proof, explains this connection.

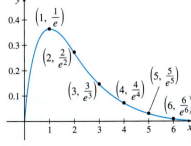

DF Figure 6 $f(x) = \dfrac{x}{e^x}$; $\{s_n\} = \left\{\dfrac{n}{e^n}\right\}$

THEOREM

Let $\{s_n\}$ be a sequence of real numbers and let f be a related function of $\{s_n\}$. Suppose L is a real number.

$$\text{If } \lim_{x \to \infty} f(x) = L \qquad \text{then} \qquad \lim_{n \to \infty} s_n = L \qquad (7)$$

EXAMPLE 8   Using a Related Function to Show a Sequence Converges

Show that $\left\{ \dfrac{3n^2 + 5n - 2}{6n^2 - 6n + 5} \right\}$ converges and find its limit.

NEED TO REVIEW? Limits at infinity are discussed in Section 1.5, pp. 120–125.

Solution The function

$$f(x) = \frac{3x^2 + 5x - 2}{6x^2 - 6x + 5} \quad x > 0$$

is a related function of the sequence $\left\{ \dfrac{3n^2 + 5n - 2}{6n^2 - 6n + 5} \right\}$. Since

$$\lim_{x \to \infty} f(x) = \lim_{x \to \infty} \frac{3x^2 + 5x - 2}{6x^2 - 6x + 5} = \lim_{x \to \infty} \frac{\dfrac{3x^2}{6x^2} + \dfrac{5x}{6x^2} - \dfrac{2}{6x^2}}{1 - \dfrac{6x}{6x^2} + \dfrac{5}{6x^2}}$$

$$= \lim_{x \to \infty} \frac{\dfrac{1}{2} + \dfrac{5}{6x} - \dfrac{1}{3x^2}}{1 - \dfrac{1}{x} + \dfrac{5}{6x^2}} = \frac{\dfrac{1}{2} + 0 - 0}{1 - 0 + 0} = \frac{1}{2}$$

the sequence $\left\{ \dfrac{3n^2 + 5n - 2}{6n^2 - 6n + 5} \right\}$ converges and $\displaystyle\lim_{n \to \infty} \frac{3n^2 + 5n - 2}{6n^2 - 6n + 5} = \frac{1}{2}$. ∎

NOW WORK   Problem 51.

NEED TO REVIEW? L'Hôpital's Rule is discussed in Section 4.5, pp. 299–302.

We can sometimes find the limit of a sequence $\{s_n\}$ by applying L'Hôpital's Rule to its related function f, provided f meets the necessary requirements.

EXAMPLE 9   Using L'Hôpital's Rule to Show a Sequence Converges

Show that $\left\{ \dfrac{n}{e^n} \right\}$ converges and find its limit.

Solution We begin with the related function $f(x) = \dfrac{x}{e^x}$, $x > 0$. To find $\displaystyle\lim_{x \to \infty} f(x)$, we use L'Hôpital's Rule.

$$\lim_{x \to \infty} f(x) = \lim_{x \to \infty} \frac{x}{e^x} \underset{\substack{\uparrow \\ \text{Use L'Hôpital's Rule}}}{=} \lim_{x \to \infty} \frac{1}{e^x} = 0$$

Since $\displaystyle\lim_{x \to \infty} f(x) = 0$, the sequence $\left\{ \dfrac{n}{e^n} \right\}$ converges and $\displaystyle\lim_{n \to \infty} \frac{n}{e^n} = 0$. ∎

NOW WORK   Problem 57.

Be careful! A related function f can be used to show a sequence $\{s_n\}$ converges only if $\displaystyle\lim_{x \to \infty} f(x) = L$, where L is a *real number*. If $\displaystyle\lim_{x \to \infty} f(x)$ is infinite, then $\{s_n\}$ diverges. If $\displaystyle\lim_{x \to \infty} f(x)$ does not exist, then we cannot use the theorem relating $\displaystyle\lim_{n \to \infty} s_n$ and $\displaystyle\lim_{x \to \infty} f(x)$.

NEED TO REVIEW? The Squeeze Theorem for functions is discussed in Section 1.4, pp. 106–107.

THEOREM The Squeeze Theorem for Sequences

Suppose $\{a_n\}$, $\{b_n\}$, and $\{s_n\}$ are sequences and N is a positive integer. If $a_n \leq s_n \leq b_n$ for every integer $n > N$, and if $\lim\limits_{n \to \infty} a_n = \lim\limits_{n \to \infty} b_n = L$, then $\lim\limits_{n \to \infty} s_n = L$.

Figure 7 illustrates the Squeeze Theorem for sequences.

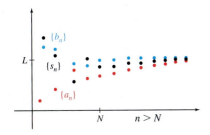

Figure 7 $a_n \leq s_n \leq b_n$ for $n > N$.

EXAMPLE 10   Using the Squeeze Theorem for Sequences

Show that $\{s_n\} = \left\{ (-1)^n \dfrac{1}{n} \right\}$ converges and find its limit.

Solution We seek two sequences that "squeeze" $\{s_n\} = \left\{ (-1)^n \dfrac{1}{n} \right\}$ as n becomes large. We begin with $|s_n|$:

$$|s_n| = \left| (-1)^n \frac{1}{n} \right| \leq \frac{1}{n}$$

$$-\frac{1}{n} \leq (-1)^n \frac{1}{n} \leq \frac{1}{n}$$

Notice that s_n is bounded by $\{a_n\} = \left\{ -\dfrac{1}{n} \right\}$ and $\{b_n\} = \left\{ \dfrac{1}{n} \right\}$. Since $a_n \leq s_n \leq b_n$ for all n and $\lim\limits_{n \to \infty} a_n = \lim\limits_{n \to \infty} \left(-\dfrac{1}{n} \right) = 0$, and $\lim\limits_{n \to \infty} b_n = \lim\limits_{n \to \infty} \dfrac{1}{n} = 0$, then by the Squeeze Theorem, the sequence $\{s_n\} = \left\{ (-1)^n \dfrac{1}{n} \right\}$ converges and $\lim\limits_{n \to \infty} s_n = 0$. ∎

NOW WORK   Problem 59.

5 Determine Whether a Sequence Converges or Diverges

A sequence $\{s_n\}$ diverges if $\lim\limits_{n \to \infty} s_n$ does not exist. This can happen if

- there is no single number L that the terms of the sequence approach as $n \to \infty$
- $\lim\limits_{n \to \infty} s_n = \infty$

DEFINITION Divergence of a Sequence to Infinity

The sequence $\{s_n\}$ **diverges to infinity**, that is,

$$\lim_{n \to \infty} s_n = \infty$$

if, given any positive number M, there is a positive integer N so that whenever $n > N$, then $s_n > M$.

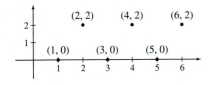

Figure 8 $\{1 + (-1)^n\}$.

EXAMPLE 11   Showing a Sequence Diverges

Show that the following sequences diverge:

(a) $\{1 + (-1)^n\}$ **(b)** $\{n\}$

Solution (a) The terms of the sequence are $0, 2, 0, 2, 0, 2, \ldots$. See Figure 8. Since the terms alternate between 0 and 2, the terms of the sequence $\{1 + (-1)^n\}$ do not approach a single number L. So, the sequence $\{1 + (-1)^n\}$ is divergent.

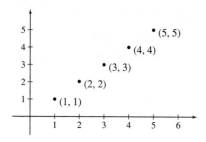

Figure 9 $\{s_n\} = \{n\}$.

(b) The terms of the sequence $\{s_n\} = \{n\}$ are 1, 2, 3, 4, …. Given any positive number M, we choose a positive integer $N > M$. Then whenever $n > N$, we have $s_n = n > N > M$. That is, the sequence $\{n\}$ diverges to infinity. See Figure 9. ∎

NOW WORK   **Problem 63.**

The next results are useful throughout the chapter. You are asked to prove them in Problems 132–135.

The sequence $\{s_n\} = \{r^n\}$, where r is a real number,
- converges to 0, if $-1 < r < 1$.
- converges to 1, if $r = 1$.
- diverges for all other numbers.

EXAMPLE 12   Determine Whether $\{r^n\}$ Converges or Diverges

Determine whether the following sequences converge or diverge:

(a) $\{s_n\} = \left\{\left(\dfrac{3}{4}\right)^n\right\}$ **(b)** $\{t_n\} = \left\{\left(\dfrac{4}{3}\right)^n\right\}$

Solution (a) The sequence $\{s_n\} = \left\{\left(\dfrac{3}{4}\right)^n\right\}$ converges to 0 because $-1 < \dfrac{3}{4} < 1$.

(b) The sequence $\{t_n\} = \left\{\left(\dfrac{4}{3}\right)^n\right\}$ diverges because $\dfrac{4}{3} > 1$. ∎

NOW WORK   **Problem 67.**

There are other ways to show that a sequence converges or diverges. To explore these, we need to define a *bounded sequence* and a *monotonic sequence*.

Bounded Sequences

A sequence $\{s_n\}$ is **bounded from above** if every term of the sequence is less than or equal to some number M. That is,

$$s_n \leq M \qquad \text{for all } n$$

Figure 10

See Figure 10.

Similarly, a sequence $\{s_n\}$ is **bounded from below** if every term of the sequence is greater than or equal to some number m. That is,

$$s_n \geq m \qquad \text{for all } n$$

See Figure 11.

For example, since $s_n = \cos n \leq 1$ for all n, the sequence $\{s_n\} = \{\cos n\}$ is bounded from above by 1, and since $s_n = \cos n \geq -1$ for all n, $\{s_n\}$ is bounded from below by -1.

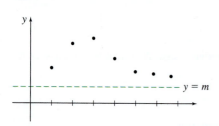

Figure 11

EXAMPLE 13   Determining Whether a Sequence Is Bounded from Above or Bounded from Below

(a) The sequence $\{s_n\} = \left\{\dfrac{3n}{n+2}\right\}$ is bounded both from above and below because

$$\frac{3n}{n+2} = \frac{3}{1+\dfrac{2}{n}} < 3 \qquad \text{and} \qquad \frac{3n}{n+2} > 0 \quad \text{for all } n \geq 1$$

See Figure 12(a).

(b) The sequence $\{a_n\} = \left\{\dfrac{4n}{3}\right\}$ is bounded from below because $\dfrac{4n}{3} > 1$ for all $n \geq 1$.

It is not bounded from above because $\displaystyle\lim_{n\to\infty} \dfrac{4n}{3} = \dfrac{4}{3} \lim_{n\to\infty} n = \infty$. See Figure 12(b).

(c) The sequence $\{b_n\} = \left\{(-1)^{n+1}n\right\}$ is neither bounded from above nor bounded from below. If n is odd, $\displaystyle\lim_{n\to\infty} b_n = \lim_{n\to\infty} n = \infty$, and if n is even, $\displaystyle\lim_{n\to\infty} b_n = \lim_{n\to\infty} (-n) = -\infty$. See Figure 12(c). ■

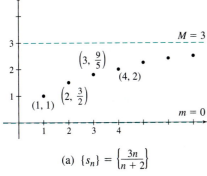
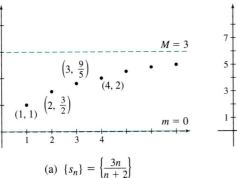
(a) $\{s_n\} = \left\{\dfrac{3n}{n+2}\right\}$

(b) $\{a_n\} = \left\{\dfrac{4n}{3}\right\}$

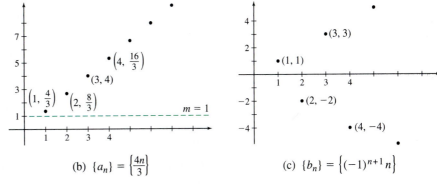
(c) $\{b_n\} = \left\{(-1)^{n+1}n\right\}$

DF Figure 12

NOW WORK Problem 73.

A sequence $\{s_n\}$ is **bounded** if it is bounded both from above and from below. For a bounded sequence $\{s_n\}$, there is a positive number K for which

$$|s_n| \leq K \qquad \text{for all } n \geq 1$$

For example, the sequence $\{s_n\} = \left\{\dfrac{3n}{n+2}\right\}$ [from Example 13(a)] is bounded,

since $\left|\dfrac{3n}{n+2}\right| \leq 3$ for all integers $n \geq 1$.

THEOREM Boundedness Theorem

A convergent sequence is bounded.

Proof If $\{s_n\}$ is a convergent sequence, there is a number L for which $\displaystyle\lim_{n\to\infty} s_n = L$. We use the definition of the limit of a sequence with $\varepsilon = 1$. Then there is a positive integer N so that

$$|s_n - L| < 1 \qquad \text{for all } n > N$$

Then, for all $n > N$,

$$|s_n| = |s_n - L + L| \leq |s_n - L| + |L| < 1 + |L|$$

$$\underset{\substack{\uparrow \\ \text{Triangle Inequality}}}{}$$

If we choose K to be the largest number in the finite collection

$$|s_1|, \quad |s_2|, \quad |s_3|, \quad \ldots, \quad |s_N|, \quad 1 + |L|$$

it follows that $|s_n| \leq K$ for *all* integers $n \geq 1$. That is, the sequence $\{s_n\}$ is bounded. ■

A restatement of the boundedness theorem provides a test for divergent sequences.

THEOREM Test for Divergence of a Sequence

If a sequence is not bounded from above or if it is not bounded from below, then it diverges.

For example, the sequences $\{-n\}$, $\{2^n\}$, and $\{\ln n\}$ are either not bounded from above or not bounded from below, so they are divergent.

NEED TO REVIEW? The Triangle Inequality is discussed in Appendix A.1, p. A-7

CAUTION The converse of the boundedness theorem is not true. Bounded sequences may converge or they may diverge. For example, the sequence $\{1 + (-1)^n\}$ in Example 11(a) is bounded, but it diverges.

Summary

To determine whether a sequence converges:

- Look at a few terms of the sequence to see if a trend is developing. For example, the first five terms of the sequence

$$\left\{1 + \frac{(-1)^n}{n^2}\right\} \text{ are } 1 - 1, \ 1 + \frac{1}{4}, \ 1 - \frac{1}{9}, \ 1 + \frac{1}{16},$$

and $1 - \dfrac{1}{25}$. The pattern suggests that the sequence converges to 1.

- Find the limit of the nth term using any available limit technique, including basic limits, limit properties, or a related function (possibly using L'Hôpital's Rule). For example, for the

sequence $\left\{\dfrac{\ln n}{n}\right\}$, we examine the limit of the related function $f(x) = \dfrac{\ln x}{x}$.

$$\lim_{x \to \infty} f(x) = \lim_{x \to \infty} \left(\frac{\ln x}{x}\right) = \lim_{x \to \infty} \frac{\frac{1}{x}}{1} = 0$$

and conclude the sequence $\left\{\dfrac{\ln n}{n}\right\}$ converges to 0.

- Show that the sequence is bounded and monotonic.

8.1 Assess Your Understanding

Concepts and Vocabulary

1. *True or False* A sequence is a function whose domain is the set of positive real numbers.

2. *True or False* If the sequence $\{s_n\}$ is convergent, then $\lim_{n \to \infty} s_n = 0$.

3. *True or False* If $f(x)$ is a related function of the sequence $\{s_n\}$ and there is a real number L for which $\lim_{x \to \infty} f(x) = L$, then $\{s_n\}$ converges.

4. *Multiple Choice* If there is a positive number K for which $|s_n| \le K$ for all integers $n \ge 1$, then $\{s_n\}$ is [(a) increasing, (b) bounded, (c) decreasing, (d) convergent.]

5. *True or False* A bounded sequence is convergent.

6. *True or False* An unbounded sequence is divergent.

7. *True or False* A sequence $\{s_n\}$ is decreasing if and only if $s_n \le s_{n+1}$ for all integers $n \ge 1$.

8. *True or False* A sequence must be monotonic to be convergent.

9. *True or False* To use an algebraic ratio to show that the sequence $\{s_n\}$ is increasing, show that $\dfrac{s_{n+1}}{s_n} \ge 0$ for all $n \ge 1$.

10. *Multiple Choice* If the derivative of a related function f of a sequence $\{s_n\}$ is negative, then the sequence $\{s_n\}$ is [(a) bounded, (b) decreasing, (c) increasing, (d) convergent.]

11. *True or False* When determining whether a sequence $\{s_n\}$ converges or diverges, the beginning terms of the sequence can be ignored.

12. *True or False* Sequences that are both bounded and monotonic diverge.

Skill Building

In Problems 13–22, the nth term of a sequence $\{s_n\}$ is given. Write the first four terms of each sequence.

13. $s_n = \dfrac{n+1}{n}$

14. $s_n = \dfrac{2}{n^2}$

15. $s_n = \ln n$

16. $s_n = \dfrac{n}{\ln(n+1)}$

17. $s_n = \dfrac{(-1)^{n+1}}{2n+1}$

18. $s_n = \dfrac{1-(-1)^n}{2}$

19. $s_n = \begin{cases} (-1)^{n+1} & \text{if } n \text{ is even} \\ 1 & \text{if } n \text{ is odd} \end{cases}$

20. $s_n = \begin{cases} n^2 + n & \text{if } n \text{ is even} \\ 4n + 1 & \text{if } n \text{ is odd} \end{cases}$

21. $s_n = \dfrac{n!}{2^n}$

22. $s_n = \dfrac{n!}{n^2}$

In Problems 23–32, the first few terms of a sequence are given. Find an expression for the nth term of each sequence, assuming the indicated pattern continues for all n.

23. $2, 4, 6, 8, 10, \ldots$

24. $1, 3, 5, 7, 9, \ldots$

25. $2, 4, 8, 16, 32, \ldots$

26. $1, 8, 27, 64, 125, \ldots$

27. $\dfrac{1}{2}, -\dfrac{1}{3}, \dfrac{1}{4}, -\dfrac{1}{5}, \dfrac{1}{6}, \ldots$

28. $1, -2, 3, -4, 5, \ldots$

29. $\dfrac{1}{2}, \dfrac{2}{3}, \dfrac{3}{4}, \dfrac{4}{5}, \ldots$

30. $\dfrac{1}{2}, \dfrac{4}{3}, \dfrac{9}{4}, \dfrac{16}{5}, \ldots$

31. $1, 1, 2, 6, 24, 120, 720, \ldots$

32. $1, 1, \dfrac{1}{2}, \dfrac{1}{6}, \dfrac{1}{24}, \dfrac{1}{120}, \ldots$

In Problems 33–44, use properties of convergent sequences to find the limit of each sequence.

33. $\left\{\dfrac{3}{n}\right\}$

34. $\left\{\dfrac{-2}{n}\right\}$

35. $\left\{1 - \dfrac{1}{n}\right\}$

36. $\left\{\dfrac{1}{n} + 4\right\}$

37. $\left\{\dfrac{4n+2}{n}\right\}$

38. $\left\{\dfrac{2n+1}{n}\right\}$

39. $\left\{\left(\dfrac{2-n}{n^2}\right)^4\right\}$

40. $\left\{\left(\dfrac{n^3 - 2n}{n^3}\right)^2\right\}$

41. $\left\{\sqrt{\dfrac{n+1}{n^2}}\right\}$

42. $\left\{\sqrt[3]{8 - \dfrac{1}{n}}\right\}$

43. $\left\{\left(1 - \dfrac{1}{n}\right)\left(1 - \dfrac{1}{n^2}\right)\right\}$

---

1. = NOW WORK problem   📉 = Graphing technology recommended   CAS = Computer Algebra System recommended

44. $\left\{\left(1 - \dfrac{1}{n}\right)\left(1 - \dfrac{1}{n^2}\right)\left(1 - \dfrac{1}{n^3}\right)\right\}$

In Problems 45–50, show that each sequence converges. Find its limit.

45. $\left\{\ln\dfrac{n+1}{3n}\right\}$ **46.** $\left\{\ln\dfrac{n^2+2}{2n^2+3}\right\}$ **47.** $\left\{e^{(4/n)-2}\right\}$

48. $\left\{e^{3+(6/n)}\right\}$ **49.** $\left\{\sin\dfrac{1}{n}\right\}$ **50.** $\left\{\cos\dfrac{1}{n}\right\}$

In Problems 51–62, use a related function or the Squeeze Theorem for sequences to show each sequence converges. Find its limit.

51. $\left\{\dfrac{n^2-4}{n^2+n-2}\right\}$ **52.** $\left\{\dfrac{n+2}{n^2+6n+8}\right\}$

53. $\left\{\dfrac{n^2}{2n+1} - \dfrac{n^2}{2n-1}\right\}$ **54.** $\left\{\dfrac{6n^4-5}{7n^4+3}\right\}$

55. $\left\{\dfrac{\sqrt{n}+2}{\sqrt{n}+5}\right\}$ **56.** $\left\{\dfrac{\sqrt{n}}{e^n}\right\}$ **57.** $\left\{\dfrac{n^2}{3^n}\right\}$

58. $\left\{\dfrac{(n-1)^2}{e^n}\right\}$ **59.** $\left\{\dfrac{(-1)^n}{3n^2}\right\}$ **60.** $\left\{\dfrac{(-1)^n}{\sqrt{n}}\right\}$

61. $\left\{\dfrac{\sin n}{n}\right\}$ **62.** $\left\{\dfrac{\cos n}{n}\right\}$

In Problems 63–72, determine whether each sequence converges or diverges.

63. $\{\cos(\pi n)\}$ **64.** $\left\{\cos\left(\dfrac{\pi}{2}n\right)\right\}$ **65.** $\{\sqrt{n}\}$

66. $\{n^2\}$ **67.** $\left\{\left(-\dfrac{1}{3}\right)^n\right\}$ **68.** $\left\{\left(\dfrac{1}{3}\right)^n\right\}$

69. $\left\{\left(\dfrac{5}{4}\right)^n\right\}$ **70.** $\left\{\left(\dfrac{\pi}{2}\right)^n\right\}$ **71.** $\left\{\dfrac{n+(-1)^n}{n}\right\}$

72. $\left\{\dfrac{1}{n}+(-1)^n\right\}$

In Problems 73–80, determine whether each sequence is bounded from above, bounded from below, both, or neither.

73. $\left\{\dfrac{\ln n}{n}\right\}$ **74.** $\left\{\dfrac{\sin n}{n}\right\}$ **75.** $\left\{n+\dfrac{1}{n}\right\}$

76. $\left\{\dfrac{3}{n+1}\right\}$ **77.** $\left\{\dfrac{n^2}{n+1}\right\}$ **78.** $\left\{\dfrac{2^n}{n^2}\right\}$

79. $\left\{\left(-\dfrac{1}{2}\right)^n\right\}$ **80.** $\{n^{1/2}\}$

In Problems 81–88, determine whether each sequence is monotonic. If the sequence is monotonic, is it increasing, nondecreasing, decreasing, or nonincreasing?

81. $\left\{\dfrac{3^n}{(n+1)^3}\right\}$ **82.** $\left\{\dfrac{2n+1}{n}\right\}$ **83.** $\left\{\dfrac{\ln n}{\sqrt{n}}\right\}$

84. $\left\{\dfrac{\sqrt{n}+1}{n}\right\}$ **85.** $\left\{\left(\dfrac{1}{3}\right)^n\right\}$ **86.** $\left\{\dfrac{n^2}{5^n}\right\}$

87. $\left\{\dfrac{n!}{3^n}\right\}$ **88.** $\left\{\dfrac{n!}{n^2}\right\}$

In Problems 89–94, show that each sequence converges by showing it is either increasing (nondecreasing) and bounded from above or decreasing (nonincreasing) and bounded from below.

89. $\{ne^{-n}\}$ **90.** $\{\tan^{-1}n\}$ **91.** $\left\{\dfrac{n}{n+1}\right\}$

92. $\left\{\dfrac{n}{n^2+1}\right\}$ **93.** $\left\{2-\dfrac{1}{n}\right\}$ **94.** $\left\{\dfrac{n}{2^n}\right\}$

In Problems 95–114, determine whether each sequence converges or diverges. If it converges, find its limit.

95. $\left\{\dfrac{3}{n}+6\right\}$ **96.** $\left\{2-\dfrac{4}{n}\right\}$

97. $\left\{\ln\left(\dfrac{n+1}{3n}\right)\right\}$ **98.** $\left\{\cos\left(n\pi+\dfrac{\pi}{2}\right)\right\}$

99. $\{(-1)^n\sqrt{n}\}$ **100.** $\left\{\dfrac{(-1)^n}{2n}\right\}$

101. $\left\{\dfrac{3^n+1}{4^n}\right\}$ **102.** $\left\{n+\sin\dfrac{1}{n}\right\}$

103. $\left\{\dfrac{\ln(n+1)}{n+1}\right\}$ **104.** $\left\{\dfrac{\ln(n+1)}{\sqrt{n}}\right\}$

105. $\{0.5^n\}$ **106.** $\{(-2)^n\}$

107. $\left\{\cos\dfrac{\pi}{n}\right\}$ **108.** $\left\{\sin\dfrac{\pi}{n}\right\}$

109. $\left\{\cos\left(\dfrac{n}{e^n}\right)\right\}$ **110.** $\left\{\sin\left(\dfrac{(n+1)^3}{e^n}\right)\right\}$

111. $\{e^{1/n}\}$ **112.** $\left\{\dfrac{1}{ne^{-n}}\right\}$

113. $\left\{1+\left(\dfrac{1}{2}\right)^n\right\}$ **114.** $\left\{1-\left(\dfrac{1}{2}\right)^n\right\}$

Applications and Extensions

In Problems 115–124, determine whether each sequence converges or diverges.

115. $\left\{\dfrac{n^2\tan^{-1}n}{n^2+1}\right\}$ **116.** $\left\{n\sin\dfrac{1}{n}\right\}$

117. $\left\{\dfrac{n+\sin n}{n+\cos(4n)}\right\}$ **118.** $\left\{\dfrac{n^2}{2n+1}\sin\dfrac{1}{n}\right\}$

119. $\{\ln n - \ln(n+1)\}$ **120.** $\left\{\ln n^2+\ln\dfrac{1}{n^2+1}\right\}$

121. $\left\{\dfrac{n^2}{\sqrt{n^2+1}}\right\}$ **122.** $\left\{\dfrac{5^n}{(n+1)^2}\right\}$

123. $\left\{\dfrac{2^n}{(2)(4)(6)\cdots(2n)}\right\}$ **124.** $\left\{\dfrac{3^{n+1}}{(3)(6)(9)\cdots(3n)}\right\}$

125. The nth term of a sequence is $s_n = \dfrac{1}{n^2+n\cos n+1}$. Does the sequence $\{s_n\}$ converge or diverge? (*Hint:* Show that the derivative of $\dfrac{1}{x^2+x\cos x+1}$ is negative for $x>1$.)

$\dfrac{1}{8}$, $\dfrac{1}{16}$, and so forth. Therefore,

$$1 = \frac{1}{2} + \frac{1}{4} + \frac{1}{8} + \frac{1}{16} + \cdots$$

Surprised?

Now look at this result from a different point of view by starting with the infinite sum

$$\frac{1}{2} + \frac{1}{4} + \frac{1}{8} + \frac{1}{16} + \cdots \qquad (1)$$

One way we might add the fractions is by using *partial sums* to see whether a trend develops. The first five partial sums are

$$\frac{1}{2} = 0.5$$

$$\frac{1}{2} + \frac{1}{4} = \frac{3}{4} = 0.75$$

$$\frac{1}{2} + \frac{1}{4} + \frac{1}{8} = \frac{3}{4} + \frac{1}{8} = \frac{7}{8} = 0.875$$

$$\frac{1}{2} + \frac{1}{4} + \frac{1}{8} + \frac{1}{16} = \frac{7}{8} + \frac{1}{16} = \frac{15}{16} = 0.9375$$

$$\frac{1}{2} + \frac{1}{4} + \frac{1}{8} + \frac{1}{16} + \frac{1}{32} = \frac{15}{16} + \frac{1}{32} = \frac{31}{32} = 0.96875$$

Each of these sums uses more terms from (1), and each sum seems to be getting closer to 1. The infinite sum in (1) is an example of an *infinite series.*

DEFINITION Infinite Series

If $a_1, a_2, \ldots, a_n, \ldots$ is an infinite collection of numbers, the expression

$$\sum_{k=1}^{\infty} a_k = a_1 + a_2 + \cdots + a_n + \cdots$$

is called an **infinite series** or, simply, a **series.**

NEED TO REVIEW? Sums and summation notation are discussed in Appendix A.5, pp. A-40 to A-42.

The numbers $a_1, a_2, \ldots, a_n, \ldots$ are called the **terms** of the series, and the number a_n is called the **nth term** or **general term** of the series. The symbol $\sum$ stands for summation; k is the **index of summation.** Although the index of summation can begin at any integer, in most of our work with series, it will begin at 1.

1 Determine Whether a Series Has a Sum

To define a sum of an infinite series $\displaystyle\sum_{k=1}^{\infty} a_k$, we make use of the sequence $\{S_n\}$ defined by

$$S_1 = a_1$$

$$S_2 = a_1 + a_2 = \sum_{k=1}^{2} a_k$$

$$\vdots$$

$$S_n = a_1 + a_2 + \cdots + a_n = \sum_{k=1}^{n} a_k$$

$$\vdots$$

This sequence $\{S_n\}$ is called the **sequence of partial sums** of the series $\displaystyle\sum_{k=1}^{\infty} a_k$.

For example, consider again the series

$$\sum_{k=1}^{\infty} \frac{1}{2^k} = \frac{1}{2} + \frac{1}{2^2} + \frac{1}{2^3} + \frac{1}{2^4} + \cdots = \frac{1}{2} + \frac{1}{4} + \frac{1}{8} + \frac{1}{16} + \cdots$$

As it turns out, the partial sums S_n can each be written as 1 minus a power of $\frac{1}{2}$, as follows:

$$S_1 = a_1 = \frac{1}{2} = 1 - \frac{1}{2}$$

$$S_2 = S_1 + a_2 = \left(1 - \frac{1}{2}\right) + \frac{1}{4} = 1 - \frac{1}{4} = 1 - \frac{1}{2^2}$$

$$S_3 = S_2 + a_3 = \left(1 - \frac{1}{4}\right) + \frac{1}{8} = 1 - \frac{1}{8} = 1 - \frac{1}{2^3}$$

$$S_4 = S_3 + a_4 = \left(1 - \frac{1}{8}\right) + \frac{1}{16} = 1 - \frac{1}{16} = 1 - \frac{1}{2^4}$$

$$\vdots$$

$$S_n = 1 - \frac{1}{2^n}$$

$$\vdots$$

The nth partial sum is $S_n = 1 - \frac{1}{2^n}$, and as n increases, the sequence $\{S_n\}$ of partial sums approaches a limit. That is,

$$\lim_{n \to \infty} S_n = \lim_{n \to \infty} \left(1 - \frac{1}{2^n}\right) = \lim_{n \to \infty} 1 - \lim_{n \to \infty} \frac{1}{2^n} = 1 - 0 = 1$$

We agree to call this limit the *sum of the series*, and we write

$$\boxed{\sum_{k=1}^{\infty} \frac{1}{2^k} = \frac{1}{2} + \frac{1}{4} + \frac{1}{8} + \frac{1}{16} + \cdots = 1}$$

DEFINITION Convergence, Divergence of an Infinite Series

If the sequence $\{S_n\}$ of partial sums of an infinite series $\sum\limits_{k=1}^{\infty} a_k$ has a limit S, then the series **converges** and is said to have the **sum** S. That is, if $\lim\limits_{n \to \infty} S_n = S$, then

$$\boxed{\sum_{k=1}^{\infty} a_k = a_1 + a_2 + \cdots + a_n + \cdots = S}$$

An infinite series **diverges** if the sequence of partial sums diverges.

EXAMPLE 1   Finding the Sum of a Series

Show that

$$\sum_{k=1}^{\infty} \frac{1}{k(k+1)} = \frac{1}{1 \cdot 2} + \frac{1}{2 \cdot 3} + \frac{1}{3 \cdot 4} + \cdots = \frac{1}{2} + \frac{1}{6} + \frac{1}{12} + \cdots = 1$$

Solution We begin with the sequence $\{S_n\}$ of partial sums,

$$S_1 = \frac{1}{1 \cdot 2}$$

$$S_2 = \frac{1}{1 \cdot 2} + \frac{1}{2 \cdot 3}$$

$$S_3 = \frac{1}{1 \cdot 2} + \frac{1}{2 \cdot 3} + \frac{1}{3 \cdot 4}$$

$$\vdots$$

$$S_n = \frac{1}{1 \cdot 2} + \frac{1}{2 \cdot 3} + \frac{1}{3 \cdot 4} + \cdots + \frac{1}{n(n+1)}$$

$$\vdots$$

Since

$$\frac{1}{n(n+1)} = \frac{1}{n} - \frac{1}{n+1} \qquad \text{Use partial fractions.}$$

S_n can be written as

$$S_n = \left(\frac{1}{1} - \frac{1}{2}\right) + \left(\frac{1}{2} - \frac{1}{3}\right) + \cdots + \left(\frac{1}{n-1} - \frac{1}{n}\right) + \left(\frac{1}{n} - \frac{1}{n+1}\right)$$

After removing parentheses notice that all the terms except the first and last cancel, so that

$$S_n = 1 - \frac{1}{n+1}$$

Then

$$\lim_{n \to \infty} S_n = \lim_{n \to \infty} \left(1 - \frac{1}{n+1}\right) = 1$$

NOTE Sums for which the middle terms cancel, as in Example 1, are called **telescoping sums**.

The series $\displaystyle\sum_{k=1}^{\infty} \frac{1}{k(k+1)}$ converges, and its sum is 1. ■

NOW WORK Problem 11.

EXAMPLE 2 Showing a Series Diverges

Show that the series $\displaystyle\sum_{k=1}^{\infty} (-1)^k = -1 + 1 - 1 + \cdots$ diverges.

Solution The sequence $\{S_n\}$ of partial sums for this series is

$$S_1 = -1$$
$$S_2 = -1 + 1 = 0$$
$$S_3 = -1 + 1 - 1 = -1$$
$$S_4 = -1 + 1 - 1 + 1 = 0$$
$$\vdots$$
$$S_n = \begin{cases} -1 & \text{if } n \text{ is odd} \\ 0 & \text{if } n \text{ is even} \end{cases}$$

Since $\displaystyle\lim_{n \to \infty} S_n$ does not exist, the sequence $\{S_n\}$ of partial sums diverges. Therefore, the series diverges. ■

NOW WORK Problem 49.

EXAMPLE 3 Determining Whether a Series Converges or Diverges

Determine whether the series $\displaystyle\sum_{k=1}^{\infty} k = 1 + 2 + 3 + \cdots$ converges or diverges.

Solution The sequence $\{S_n\}$ of partial sums is

$$S_1 = 1$$
$$S_2 = 1 + 2$$
$$S_3 = 1 + 2 + 3$$
$$\vdots$$
$$S_n = 1 + 2 + 3 + \cdots + n$$

To express S_n in a way that will make it easy to find $\lim\limits_{n \to \infty} S_n$, we use the formula for the sum of the first n integers:

$$S_n = \sum_{k=1}^{n} k = 1 + 2 + 3 + \cdots + n = \frac{n(n+1)}{2}$$

RECALL

$$\sum_{k=1}^{n} k = 1 + 2 + \cdots + n = \frac{n(n+1)}{2}.$$

(See Appendix A.5, p. A-41.)

Since $\lim\limits_{n \to \infty} S_n = \lim\limits_{n \to \infty} \dfrac{n(n+1)}{2} = \infty$, the sequence $\{S_n\}$ of partial sums diverges. So, the series $\sum\limits_{k=1}^{\infty} k$ diverges. ■

NOW WORK   Problem 43.

2 Analyze a Geometric Series

Geometric series occur in a large variety of applications including biology, finance, and probability. They are also useful in analyzing other infinite series.

DEFINITION Geometric Series

A series of the form

$$\sum_{k=0}^{\infty} ar^{k} = \sum_{k=1}^{\infty} ar^{k-1} = a + ar + ar^2 + \cdots + ar^{n-1} + \cdots$$

where $a \neq 0$ is called a **geometric series**.

In a geometric series, the ratio r of any two consecutive terms is a fixed real number.

To investigate the conditions for convergence of a geometric series, we examine the nth partial sum:

$$S_n = a + ar + ar^2 + \cdots + ar^{n-1} \tag{1}$$

If $r = 0$, the nth partial sum is $S_n = a$ and $\lim\limits_{n \to \infty} S_n = a$. The sequence of partial sums converges when $r = 0$.

If $r = 1$, the series becomes $\sum\limits_{k=1}^{\infty} a = a + a + \cdots + a + \cdots$, and the nth partial sum is

$$S_n = a + a + \cdots + a = na$$

Since $a \neq 0$, $\lim\limits_{n \to \infty} S_n = \infty$ or $-\infty$, so the sequence $\{S_n\}$ of partial sums diverges when $r = 1$.

If $r = -1$, the series is $\sum\limits_{k=1}^{\infty} a(-1)^{k-1} = a - a + a - a + \cdots$ and the nth partial sum is

$$S_n = \begin{cases} 0 & \text{if } n \text{ is even} \\ a & \text{if } n \text{ is odd} \end{cases}$$

Since $a \neq 0$, $\lim\limits_{n \to \infty} S_n$ does not exist. The sequence $\{S_n\}$ of partial sums diverges when $r = -1$.

Suppose $r \neq 0$, $r \neq 1$, and $r \neq -1$. Since $r \neq 0$, we multiply both sides of (1) by r to obtain

$$rS_n = ar + ar^2 + \cdots + ar^n$$

Now subtract rS_n from S_n.

$$S_n - rS_n = (a + ar + ar^2 + \cdots + ar^{n-1}) - (ar + ar^2 + \cdots + ar^{n-1} + ar^n)$$
$$= a - ar^n$$
$$S_n(1 - r) = a(1 - r^n)$$

Since $r \neq 1$, the nth partial sum of the geometric series can be expressed as

$$S_n = \frac{a(1-r^n)}{1-r} = \frac{a-ar^n}{1-r} = \frac{a}{1-r} - \frac{ar^n}{1-r}$$

Now,

$$\lim_{n\to\infty} S_n = \lim_{n\to\infty}\left[\frac{a}{1-r} - \frac{ar^n}{1-r}\right] = \lim_{n\to\infty}\frac{a}{1-r} - \lim_{n\to\infty}\frac{ar^n}{1-r} = \frac{a}{1-r} - \frac{a}{1-r}\lim_{n\to\infty} r^n$$

We now use the fact that if $|r| < 1$, then $\lim\limits_{n\to\infty} r^n = 0$ (refer to page 546 in Section 8.1). We conclude that if $|r| < 1$, then $\lim\limits_{n\to\infty} S_n = \frac{a}{1-r}$. So, a geometric series converges to $S = \frac{a}{1-r}$ if $-1 < r < 1$.

If $|r| > 1$, use the fact that $\lim\limits_{x\to\infty} r^n$ does not exist to conclude that a geometric series diverges if $r < -1$ or $r > 1$.

This proves the following theorem:

THEOREM Convergence of a Geometric Series

- If $|r| < 1$, the geometric series $\sum\limits_{k=1}^{\infty} ar^{k-1}$ converges, and its sum is

$$\sum_{k=1}^{\infty} ar^{k-1} = \frac{a}{1-r}$$

- If $|r| \geq 1$, the geometric series $\sum\limits_{k=1}^{\infty} ar^{k-1}$ diverges.

EXAMPLE 4 Determining Whether a Geometric Series Converges

Determine whether each geometric series converges or diverges. If it converges, find its sum.

(a) $\displaystyle\sum_{k=1}^{\infty} 8\left(\frac{2}{5}\right)^{k-1}$ (b) $\displaystyle\sum_{k=1}^{\infty}\left(-\frac{5}{9}\right)^{k-1}$ (c) $\displaystyle\sum_{k=1}^{\infty} 3\left(\frac{3}{2}\right)^{k-1}$

(d) $\displaystyle\sum_{k=1}^{\infty}\frac{1}{2^k}$ (e) $\displaystyle\sum_{k=0}^{\infty}\left(\frac{1}{3}\right)^{k-1}$

Solution We compare each series to $\sum\limits_{k=1}^{\infty} ar^{k-1}$.

(a) In this series $a = 8$ and $r = \frac{2}{5}$. Since $|r| = \frac{2}{5} < 1$, the series converges and

$$\sum_{k=1}^{\infty} 8\left(\frac{2}{5}\right)^{k-1} = \frac{8}{1-\dfrac{2}{5}} = 8\left(\frac{5}{3}\right) = \frac{40}{3}$$

(b) Here, $a = 1$ and $r = -\frac{5}{9}$. Since $|r| = \frac{5}{9} < 1$, the series converges and

$$\sum_{k=1}^{\infty}\left(-\frac{5}{9}\right)^{k-1} = \frac{1}{1-\left(-\dfrac{5}{9}\right)} = \frac{9}{14}$$

(c) Here, $a = 3$ and $r = \dfrac{3}{2}$. Since $|r| = \dfrac{3}{2} > 1$, the series $\displaystyle\sum_{k=1}^{\infty} 3\left(\dfrac{3}{2}\right)^{k-1}$ diverges.

(d) $\displaystyle\sum_{k=1}^{\infty} \dfrac{1}{2^k}$ is not in the form $\displaystyle\sum_{k=1}^{\infty} ar^{k-1}$. To place it in this form, we proceed as follows:

$$\sum_{k=1}^{\infty} \frac{1}{2^k} = \sum_{k=1}^{\infty} \left(\frac{1}{2}\right)^k \underset{\substack{\uparrow \\ \text{Write in the form } \sum\limits_{k=1}^{\infty} ar^{k-1}}}{=} \sum_{k=1}^{\infty} \left[\frac{1}{2} \cdot \left(\frac{1}{2}\right)^{k-1}\right]$$

So, $\displaystyle\sum_{k=1}^{\infty} \dfrac{1}{2^k}$ is a geometric series with $a = \dfrac{1}{2}$ and $r = \dfrac{1}{2}$. Since $|r| < 1$, the series converges, and its sum is

$$\sum_{k=1}^{\infty} \frac{1}{2^k} = \frac{\dfrac{1}{2}}{1 - \dfrac{1}{2}} = 1$$

which agrees with the sum we found earlier.

(e) $\displaystyle\sum_{k=0}^{\infty} \left(\dfrac{1}{3}\right)^{k-1}$ starts at 0, so it is not in the form, $\displaystyle\sum_{k=1}^{\infty} ar^{k-1}$. To place it in this form, change the index to l, where $l = k + 1$. Then when $k = 0$, $l = 1$ and

$$\sum_{k=0}^{\infty} \left(\frac{1}{3}\right)^{k-1} = \sum_{l=1}^{\infty} \left(\frac{1}{3}\right)^{l-2} = \sum_{l=1}^{\infty} \left(\frac{1}{3}\right)^{-1} \left(\frac{1}{3}\right)^{l-1} = \sum_{l=1}^{\infty} 3\left(\frac{1}{3}\right)^{l-1}$$

That is, $\displaystyle\sum_{k=0}^{\infty} \left(\dfrac{1}{3}\right)^{k-1} = \displaystyle\sum_{l=1}^{\infty} 3\left(\dfrac{1}{3}\right)^{l-1}$ is a geometric series with $a = 3$ and $r = \dfrac{1}{3}$. Since $|r| < 1$, the series converges, and its sum is

$$\sum_{k=0}^{\infty} \left(\frac{1}{3}\right)^{k-1} = \frac{3}{1 - \dfrac{1}{3}} = \frac{9}{2} \qquad \blacksquare$$

NOW WORK Problems 21 and 29.

EXAMPLE·5 **Writing a Repeating Decimal as a Fraction**

Express the repeating decimal $0.090909\ldots$ as a quotient of two integers.

Solution We write the infinite decimal $0.090909\ldots$ as an infinite series:

$$0.090909\ldots = 0.09 + 0.0009 + 0.000009 + 0.00000009 + \cdots$$

$$= \frac{9}{100} + \frac{9}{10000} + \frac{9}{1000000} + \cdots$$

$$= \frac{9}{100}\left(1 + \frac{1}{100} + \frac{1}{10000} + \frac{1}{1000000} + \cdots\right)$$

$$= \sum_{k=1}^{\infty} \frac{9}{100}\left(\frac{1}{100}\right)^{k-1}$$

This is a geometric series with $a = \dfrac{9}{100}$ and $r = \dfrac{1}{100}$. Since $|r| < 1$, the series converges and its sum is

$$\sum_{k=1}^{\infty} \frac{9}{100}\left(\frac{1}{100}\right)^{k-1} = \frac{\dfrac{9}{100}}{1 - \dfrac{1}{100}} = \frac{9}{99} = \frac{1}{11}$$

So, $0.090909\ldots = \dfrac{1}{11}$. ∎

NOW WORK Problem 59.

EXAMPLE 6   Using a Geometric Series with a Bouncing Ball

A ball is dropped from a height of 12 m. Each time it strikes the ground, it bounces back to a height three-fourths the distance from which it fell. Find the total distance traveled by the ball. See Figure 18.

Solution Let h_n denote the height of the ball on the nth bounce. Then

$$h_0 = 12$$

$$h_1 = \frac{3}{4}(12)$$

$$h_2 = \frac{3}{4}\left[\frac{3}{4}(12)\right] = \left(\frac{3}{4}\right)^2(12)$$

$$\vdots$$

$$h_n = \left(\frac{3}{4}\right)^n(12)$$

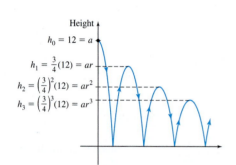

Height

$h_0 = 12 = a$

$h_1 = \frac{3}{4}(12) = ar$

$h_2 = \left(\frac{3}{4}\right)^2(12) = ar^2$

$h_3 = \left(\frac{3}{4}\right)^3(12) = ar^3$

DF Figure 18

After the first bounce, the ball travels up a distance $h_1 = \dfrac{3}{4}(12)$ and then the same distance back down. Between the first and the second bounce, the total distance traveled is therefore $h_1 + h_1 = 2h_1$. The *total* distance H traveled by the ball is

$$H = h_0 + 2h_1 + 2h_2 + 2h_3 + \cdots = h_0 + \sum_{k=1}^{\infty}(2h_k) = 12 + \sum_{k=1}^{\infty} 2\left[12\left(\frac{3}{4}\right)^k\right]$$

$$= 12 + \sum_{k=1}^{\infty} 24\left[\frac{3}{4}\left(\frac{3}{4}\right)^{k-1}\right]$$

$$= 12 + \sum_{k=1}^{\infty} 18\left(\frac{3}{4}\right)^{k-1}$$

The sum is a geometric series with $a = 18$ and $r = \dfrac{3}{4}$. The series converges and

$$H = 12 + \sum_{k=1}^{\infty} 18\left(\frac{3}{4}\right)^{k-1} = 12 + \frac{18}{1 - \dfrac{3}{4}} = 84$$

The ball travels a total distance of 84 m. ∎

NOW WORK Problem 63.

3 Analyze the Harmonic Series

Another useful series, even though it diverges, is the *harmonic series*.

DEFINITION Harmonic Series

The infinite series

$$\sum_{k=1}^{\infty} \frac{1}{k} = 1 + \frac{1}{2} + \frac{1}{3} + \cdots$$

is called the **harmonic series**.

THEOREM

The harmonic series $\sum_{k=1}^{\infty} \frac{1}{k}$ diverges.

RECALL An unbounded sequence diverges (p. 547).

Proof To show that the harmonic series diverges, we look at the partial sums whose indexes of summation are powers of 2. That is, we investigate the sequence S_1, S_2, S_4, S_8, and so on. We can show that this sequence is not bounded as follows:

$$S_1 = 1 > \frac{1}{2} = 1\left(\frac{1}{2}\right)$$

$$S_2 = 1 + \frac{1}{2} > \frac{1}{2} + \frac{1}{2} = 2\left(\frac{1}{2}\right)$$

$$S_4 = 1 + \frac{1}{2} + \left(\frac{1}{3} + \frac{1}{4}\right) > 2\left(\frac{1}{2}\right) + \left(\frac{1}{4} + \frac{1}{4}\right) = 3\left(\frac{1}{2}\right)$$

$$S_8 = 1 + \frac{1}{2} + \left(\frac{1}{3} + \frac{1}{4}\right) + \left(\frac{1}{5} + \frac{1}{6} + \frac{1}{7} + \frac{1}{8}\right)$$

$$> 3\left(\frac{1}{2}\right) + \left(\frac{1}{8} + \frac{1}{8} + \frac{1}{8} + \frac{1}{8}\right) = 4\left(\frac{1}{2}\right)$$

$$\vdots$$

$$S_{2^{n-1}} > n\left(\frac{1}{2}\right)$$

We conclude that the sequence $\{S_{2^{n-1}}\}$ is not bounded, so the sequence $\{S_n\}$ of partial sums is not bounded. It follows that the sequence $\{S_n\}$ of partial sums diverges. Therefore, the harmonic series $\sum_{k=1}^{\infty} \frac{1}{k}$ diverges. ∎

Summary

- A series $\sum_{k=1}^{\infty} a_k$ converges if and only if its sequence $\{S_n\}$ of partial sums converges.

- The geometric series $\sum_{k=1}^{\infty} ar^{k-1}$, $a \neq 0$, converges when $|r| < 1$ and diverges for $|r| \geq 1$. If the geometric series converges, its sum is $S = \dfrac{a}{1-r}$.

- The harmonic series $\sum_{k=1}^{\infty} \frac{1}{k}$ diverges.

Using a Geometric Series in a Biology Application[*]

This application deals with the rate of occurrence of *retinoblastoma*, a rare type of eye cancer in children. An *allele (allelomorph)* is a gene that gives rise to one of a pair of contrasting characteristics, such as smooth or rough, tall or short. Each person normally has two such genes for each characteristic. An individual may have two "tall" genes, two "short" genes, or one of each. In reproduction, each parent gives one of the two types to the child.

The tendency to develop retinoblastoma apparently depends on the mutation of both copies of a gene, called RB1. The mutation rate from a normal RB1 allele to the mutant RB1 in each generation is approximately $m = 0.00002 = 2 \times 10^{-5}$. In this example, we ignore the very unlikely possibility of mutation from an abnormal RB1 to a normal RB1 gene. At the beginning of the twentieth century, retinoblastoma was nearly always fatal, but by the early 1950s, approximately 70% of children affected with the disease survived, although they usually became blind in one or both eyes. The current (2009) survival rate is about 95% and the goals of treatment are to prevent the tumor cells from growing and spreading and to preserve vision.

Assume that survivors reproduce at about half the normal rate. (The assumption is based on scientific guesswork.) Then the productive proportion of persons affected with the retinoblastoma in 2009 was $r = (0.5)(0.95) = 0.475$. This rate is remarkable, considering that in 1900, $r \approx 0$, and in 1950 $r \approx 0.35$.

Starting with zero inherited cases in an early generation, for the nth consecutive generation, we obtain a rate of

m	due to mutation in the nth generation
mr	due to mutation in the $(n-1)$st generation
mr^2	due to mutation in the $(n-2)$nd generation
$\vdots$	
mr^n	due to mutation in the zero (original) generation

Then, the total rate of occurrence of the disease in the nth generation is

$$p_n = m + mr + \cdots + mr^n = \frac{m(1 - r^{n+1})}{1 - r}$$

from which

$$p = \lim_{n \to \infty} p_n = \frac{m}{1 - r} = \frac{2 \times 10^{-5}}{1 - 0.475} = 3.810 \times 10^{-5}$$

indicating that the total rate of persons affected with the disease will be almost twice the mutation rate.

Notice that if $r = 0$, as in 1900, then $p = m = 2 \times 10^{-5}$, and that if $r = 0.35$, as in 1950, then $p = 3.08 \times 10^{-5}$. We see that with better medical care, retinoblastoma has become more frequent. As medical care improves, the rate of occurrence of the disease can be expected to become even greater. As Neel and Schull pointed out, with improved medical care, the frequency of an abnormal gene RB1 increases rapidly at first, then more slowly, until an equilibrium point is reached.

---

[*]Adapted from J. L. Young & M. A. Smith (1999), Retinoblastoma. In L. A. G. Ries, M. A. Smith, J. G. Gurney, M. Linet, T. Tamra, J. L. Young, & G. R. Bunin (Eds.), *Cancer incidence and survival among children and adolescents: United States SEER Program, 1975–1995*, NIH Pub. No. 99-4649, Bethesda, MD: National Cancer Institute.

Children's Hospital of Philadelphia (2009), *Retinoblastoma*, http://www.CHOP.edu.
J. V. Neel & W. J. Schull (1958), *Human heredity*, 3rd ed. (pp. 333–334), Chicago: University of Chicago Press. Reprinted by permission.

8.2 Assess Your Understanding

Concepts and Vocabulary

1. *Multiple Choice*   If $a_1, a_2, \ldots, a_n, \ldots$ is an infinite collection of numbers, the expression $\sum\limits_{k=1}^{\infty} a_k = a_1 + a_2 + \cdots + a_n + \cdots$ is called [(**a**) an infinite sequence, (**b**) an infinite series, (**c**) a partial sum].

2. *Multiple Choice*   If $\sum\limits_{k=1}^{\infty} a_k$ is an infinite series, then the sequence $\{S_n\}$ where $S_n = \sum\limits_{k=1}^{n} a_k$, is called the sequence of [(**a**) fractional parts, (**b**) early terms, (**c**) completeness, (**d**) partial sums] of the infinite series.

3. *True or False*   A series converges if and only if its sequence of partial sums converges.

4. *True or False*   A geometric series $\sum\limits_{k=1}^{\infty} ar^{k-1}$, $a \neq 0$, converges if $|r| \leq 1$.

5. The sum of a convergent geometric series $\sum\limits_{k=1}^{\infty} ar^{k-1}$, $a \neq 0$, is $S = $ _____.

6. *True or False*   The harmonic series $\sum\limits_{k=1}^{\infty} \dfrac{1}{k}$ converges because $\lim\limits_{n \to \infty} \dfrac{1}{n} = 0$.

Skill Building

In Problems 7–10, find the fourth partial sum of each series.

7. $\sum\limits_{k=1}^{\infty} \left(\dfrac{3}{4}\right)^{k-1}$
8. $\sum\limits_{k=1}^{\infty} \dfrac{(-1)^{k+1}}{3^{k-1}}$
9. $\sum\limits_{k=1}^{\infty} k$
10. $\sum\limits_{k=1}^{\infty} \ln k$

In Problems 11–16, find the sum of each telescoping series.

11. $\sum\limits_{k=1}^{\infty} \left(\dfrac{1}{k+2} - \dfrac{1}{k+3}\right)$
12. $\sum\limits_{k=1}^{\infty} \left[\dfrac{1}{k^2} - \dfrac{1}{(k+1)^2}\right]$

13. $\sum\limits_{k=1}^{\infty} \left(\dfrac{1}{3^{k+1}} - \dfrac{1}{3^k}\right)$
14. $\sum\limits_{k=1}^{\infty} \left(\dfrac{1}{4^{k+1}} - \dfrac{1}{4^k}\right)$

15. $\sum\limits_{k=1}^{\infty} \dfrac{1}{4k^2 - 1}$ $\left[\text{Hint: } \dfrac{1}{4k^2-1} = \dfrac{1}{2}\left(\dfrac{1}{2k-1} - \dfrac{1}{2k+1}\right)\right]$

16. $\sum\limits_{k=1}^{\infty} \dfrac{1}{k(k+1)(k+2)}$

$\left[\text{Hint: } \dfrac{1}{k(k+1)(k+2)} = \dfrac{1}{2}\left(\dfrac{1}{k(k+1)} - \dfrac{1}{(k+1)(k+2)}\right)\right]$

In Problems 17–38, determine whether each geometric series converges or diverges. If it converges, find its sum.

17. $\sum\limits_{k=1}^{\infty} (\sqrt{2})^{k-1}$
18. $\sum\limits_{k=1}^{\infty} (0.33)^{k-1}$

19. $\sum\limits_{k=1}^{\infty} 5\left(\dfrac{1}{6}\right)^{k-1}$
20. $\sum\limits_{k=1}^{\infty} 4\,(1.1)^{k-1}$

21. $\sum\limits_{k=0}^{\infty} 7\left(\dfrac{1}{3}\right)^{k}$
22. $\sum\limits_{k=0}^{\infty} \left(\dfrac{7}{4}\right)^{k}$

23. $\sum\limits_{k=1}^{\infty} (-0.38)^{k-1}$
24. $\sum\limits_{k=1}^{\infty} (-0.38)^{k}$

25. $\sum\limits_{k=0}^{\infty} \dfrac{2^{k+1}}{3^k}$
26. $\sum\limits_{k=0}^{\infty} \dfrac{5^k}{6^{k+1}}$

27. $\sum\limits_{k=0}^{\infty} \dfrac{1}{4^{k+1}}$
28. $\sum\limits_{k=0}^{\infty} \dfrac{4^{k+1}}{3^k}$

29. $\sum\limits_{k=1}^{\infty} \sin^{k-1}\left(\dfrac{\pi}{2}\right)$
30. $\sum\limits_{k=1}^{\infty} \tan^{k-1}\left(\dfrac{\pi}{4}\right)$

31. $\sum\limits_{k=1}^{\infty} \left(-\dfrac{3}{2}\right)^{k-1}$
32. $\sum\limits_{k=1}^{\infty} \left(-\dfrac{2}{3}\right)^{k-1}$

33. $1 + \dfrac{1}{3} + \dfrac{1}{9} + \cdots + \left(\dfrac{1}{3}\right)^{n} + \cdots$

34. $1 + \dfrac{1}{4} + \dfrac{1}{16} + \cdots + \left(\dfrac{1}{4}\right)^{n} + \cdots$

35. $1 + 2 + 4 + \cdots + 2^n + \cdots$

36. $1 - \dfrac{1}{2} + \dfrac{1}{4} - \dfrac{1}{8} + \cdots + \dfrac{(-1)^{n-1}}{2^{n-1}} + \cdots$

37. $\left(\dfrac{1}{7}\right)^{2} + \left(\dfrac{1}{7}\right)^{3} + \cdots + \left(\dfrac{1}{7}\right)^{n} + \cdots$

38. $\left(\dfrac{3}{4}\right)^{5} + \left(\dfrac{3}{4}\right)^{6} + \cdots + \left(\dfrac{3}{4}\right)^{n} + \cdots$

In Problems 39–58, determine whether each series converges or diverges. If it converges, find its sum.

39. $\sum\limits_{k=0}^{\infty} \dfrac{1}{k+1}$
40. $\sum\limits_{k=4}^{\infty} k^{-1}$

41. $\sum\limits_{k=1}^{\infty} \dfrac{1}{100^k}$
42. $\sum\limits_{k=1}^{\infty} e^{-k}$

43. $\sum\limits_{k=1}^{\infty} (-10k)$
44. $\sum\limits_{k=1}^{\infty} \dfrac{3k}{5}$

45. $\sum\limits_{k=1}^{\infty} \cos^{k-1}\left(\dfrac{2\pi}{3}\right)$
46. $\sum\limits_{k=1}^{\infty} \sin^{k-1}\left(\dfrac{\pi}{6}\right)$

47. $\sum\limits_{k=1}^{\infty} \dfrac{\tan^k\left(\dfrac{\pi}{4}\right)}{k}$
48. $\sum\limits_{k=1}^{\infty} \dfrac{\sin^k\left(\dfrac{\pi}{2}\right)}{k}$

49. $\sum\limits_{k=1}^{\infty} \cos(\pi k)$
50. $\sum\limits_{k=1}^{\infty} \sin\left(\dfrac{\pi k}{2}\right)$

1. = NOW WORK problem   [graph icon] = Graphing technology recommended   CAS = Computer Algebra System recommended

51. $\displaystyle\sum_{k=1}^{\infty} 2^{-k}3^{k+1}$

52. $\displaystyle\sum_{k=1}^{\infty} 3^{1-k}2^{1+k}$

53. $\displaystyle\sum_{k=1}^{\infty} \left(-\frac{1}{3}\right)^k$

54. $\displaystyle\sum_{k=1}^{\infty} \frac{\pi}{3^k}$

55. $\displaystyle\sum_{k=1}^{\infty} \ln\frac{k}{k+1}$

56. $\displaystyle\sum_{k=1}^{\infty} \left[e^{2k-1} - e^{2(k+1)^{-1}}\right]$

57. $\displaystyle\sum_{k=1}^{\infty} \left(\sin\frac{1}{k} - \sin\frac{1}{k+1}\right)$

58. $\displaystyle\sum_{k=1}^{\infty} \left(\tan\frac{1}{k} - \tan\frac{1}{k+1}\right)$

In Problems 59–62, express each repeating decimal as a rational number by using a geometric series.

59. $0.5555\ldots$

60. $0.727272\ldots$

61. $4.28555\ldots$ *(Hint: $4.28555\ldots = 4.28 + 0.00555\ldots$.)*

62. $7.162162\ldots$

Applications and Extensions

63. Distance a Ball Travels   A ball is dropped from a height of 18 ft. Each time it strikes the ground, it bounces back to two-thirds of the previous height. Find the total distance traveled by the ball.

64. Diminishing Returns   A rich man promises to give you $1000 on January 1, 2015. Each day thereafter he will give you $\frac{9}{10}$ of what he gave you the previous day.

(a) What is the total amount you will receive?

(b) What is the first date on which the amount you receive is less than 1 cent?

65. Stocking a Lake   Mirror Lake is stocked periodically with rainbow trout. In year n of the stocking program, the population is given by $p_n = 3000r^n + h\displaystyle\sum_{k=1}^{n} r^{k-1}$, where h is the number of fish added by the program per year and r, $0 < r < 1$, is the percent of fish removed each year.

(a) What does a manager expect the steady rainbow trout population to be as $n \to \infty$?

(b) If $r = 0.5$, how many fish h should be added annually to obtain a steady population of 4000 rainbow trout?

66. Marginal Propensity to Consume   Suppose that individuals in the United States spend 90% of every additional dollar that they earn. Then according to economists, an individual's **marginal propensity to consume** is 0.90. For example, if Jane earns an additional dollar, she will spend $0.9(1) = \$0.90$ of it. The individual who earns Jane's $0.90 will spend 90% of it or $0.81. The process of spending continues and results in the series

$$\sum_{k=1}^{\infty} 0.90^{k-1} = 1 + 0.90 + 0.90^2 + 0.90^3 + \ldots$$

The sum of this series is called the **multiplier.** What is the multiplier if the marginal propensity to consume is 90%?

67. Stock Pricing   One method of pricing a stock is to discount the stream of future dividends of the stock. Suppose a stock currently pays $\$P$ annually in dividends, and historically, the dividend has

increased by $i\%$ annually. If an investor wants an annual rate of return of $r\%$, this method of pricing a stock states that the stock should be priced at the present value of an infinite stream of payments:

$$\text{Price} = P + P\frac{1+i}{1+r} + P\left(\frac{1+i}{1+r}\right)^2 + P\left(\frac{1+i}{1+r}\right)^3 + \ldots$$

(a) Find the price of a stock priced using this method.

(b) Suppose an investor desires a 9% return on a stock that currently pays an annual dividend of $4.00, and, historically, the dividend has been increased by 3% annually. What is the highest price the investor should pay for the stock?

68. Koch's Snowflake   The area inside the fractal known as the **Koch snowflake** can be described as the sum of the areas of infinitely many equilateral triangles. See the figure. For all but the center (largest) triangle, a triangle in the Koch snowflake is $\frac{1}{9}$ the area of the next largest triangle in the fractal. Suppose the largest (center) triangle has an area of 1 square unit. Then the area of the snowflake is given by the series

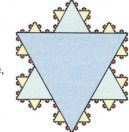

$$1 + 3\left(\frac{1}{9}\right) + 12\left(\frac{1}{9}\right)^2 + 48\left(\frac{1}{9}\right)^3 + 192\left(\frac{1}{9}\right)^4 + \ldots$$

Find the area of the Koch snowflake by finding the sum of the series.

69. Zeno's paradox   is about a race between Achilles and a tortoise. The tortoise is allowed a certain lead at the start of the race. Zeno claimed the tortoise must win such a race. He reasoned that for Achilles to overtake the tortoise, at some time he must cover $\frac{1}{2}$ of the distance that originally separated them. Then, when he covers another $\frac{1}{4}$ of the original distance separating them, he will still have $\frac{1}{4}$ of that distance remaining, and so on. Therefore by Zeno's reasoning, Achilles never catches the tortoise. Use a series argument to explain this paradox. Assume that the difference in speed between Achilles and the tortoise is a constant v meters per second.

70. Probability   A coin-flipping game involves two people who successively flip a coin. The first person to obtain a head is the winner. In probability, it turns out that the person who flips first has the probability of winning given by the series below. Find this probability.

$$\frac{1}{2} + \frac{1}{8} + \frac{1}{32} + \cdots + \frac{1}{2^{2n-1}} + \cdots$$

71. Controlling Salmonella   *Salmonella* is a common enteric bacterium infecting both humans and farm animals with salmonellosis. Barn surfaces contaminated with salmonella can be the major source of salmonellosis spread in a farm. While cleaning barn surfaces is used as a control measure on pig and cattle farms, the efficiency of cleaning has been a concern. Suppose, on average, there are p kilograms (kg) of feces produced each day and cleaning is performed with the constant efficiency e, $0 < e < 1$. At the end of each day, $(1 - e)$ kg of feces from the previous day is added to the amount of the present day. Let $T(n)$ be the total accumulated fecal material on day n. Farmers are concerned when $T(n)$ exceeds the threshold level L.

(a) Express $T(n)$ as a geometric series.

(b) Find $\lim_{n\to\infty} T(n)$.

(c) Determine the minimum cleaning efficiency e_{min} required to guarantee $T(n) \leq L$ for all n.

(d) Suppose that $p = 120$ kg and $L = 180$ kg. Using (b) and (c), find e_{min} and $T(365)$ for $e = \dfrac{4}{5}$.

> **Source**: R. Gautam, G. Lahodny, M. Bani-Yaghoub, & R. Ivanek Based on their paper "Understanding the role of cleaning in the control of Salmonella Typhimurium in a grower finisher pig herd: a modeling approach."

72. Show that $0.9999\ldots = 1$.

In Problems 73 and 74 use a geometric series to prove the given statement.

73. $\dfrac{x}{x - 1} = \displaystyle\sum_{k=1}^{\infty} \dfrac{1}{x^{k-1}}$ for $|x| > 1$

74. $\dfrac{1}{1 + x} = \displaystyle\sum_{k=0}^{\infty} (-1)^k x^k$ for $|x| < 1$

75. Find the smallest number n for which $\displaystyle\sum_{k=1}^{n} \dfrac{1}{k} \geq 3$.

76. Find the smallest number n for which $\displaystyle\sum_{k=1}^{n} \dfrac{1}{k} \geq 4$.

77. Show that the series $\displaystyle\sum_{k=1}^{\infty} \dfrac{\sqrt{k+1} - \sqrt{k}}{\sqrt{k(k+1)}}$ converges and has the sum 1.

78. Show that $\displaystyle\sum_{k=1}^{\infty} \dfrac{1}{k(k+2)} = \dfrac{3}{4}$.

79. Show that $\displaystyle\sum_{k=1}^{\infty} \dfrac{1}{k(k+1)(k+2)} = \dfrac{1}{4}$.

80. Show that $\displaystyle\sum_{k=1}^{\infty} \dfrac{1}{k(k+1)(k+2)(k+3)} = \dfrac{1}{18}$.

81. Show that $\displaystyle\sum_{k=1}^{\infty} \dfrac{1}{k(k+1)(k+2)\cdots(k+a)} = \dfrac{1}{a}\left(\dfrac{1}{a!}\right)$; $a \geq 1$ is an integer.

82. Solve for x: $\dfrac{x}{2 + 2x} = x + x^2 + x^3 + \cdots$, $|x| < 1$.

83. Show that the sum of any convergent geometric series whose first term and common ratio are rational is rational.

84. The sum S_n of the first n terms of a geometric series is given by the formula

$$S_n = a + ar + ar^2 + \cdots + ar^{n-1} = \frac{a(r^n - 1)}{r - 1} \qquad a > 0, \quad r \neq 1$$

Find $\displaystyle\lim_{r\to 1} \dfrac{a(r^n - 1)}{r - 1}$ and compare the result with a geometric series in which $r = 1$.

Challenge Problems

The following discussion relates to Problems 85 and 86.

An interesting relationship between the nth partial sum of the harmonic series and $\ln n$ was discovered by Euler. In particular, he showed that

$$\gamma = \lim_{n\to\infty}\left(1 + \frac{1}{2} + \frac{1}{3} + \cdots + \frac{1}{n} - \ln n\right)$$

exists and is approximately equal to 0.5772. **Euler's number**, as γ is called, appears in many interesting areas of mathematics. For example, it is involved in the evaluation of the exponential integral, $\displaystyle\int_x^\infty \dfrac{e^{-t}}{t}\, dt$, which is important in applied mathematics. It is also related to two special functions—the gamma function and Riemann's zeta function (see Challenge Problem 75, Section 8.3). Surprisingly, it is still unknown whether Euler's number is rational or irrational.

85. The harmonic series diverges quite slowly. For example, the partial sums S_{10}, S_{20}, S_{50}, and S_{100} have approximate values 2.92897, 3.59774, 4.49921, and 5.18738, respectively. In fact, the sum of the first million terms of the harmonic series is about 14.4. With this in mind, what would you conjecture about the rate of convergence of the limit defining γ? Test your conjecture by calculating approximate values for γ by using the partial sums given above.

86. Use the approximate value of Euler's number 0.5772 to approximate

$$1 + \frac{1}{2} + \frac{1}{3} + \cdots + \frac{1}{1{,}000{,}000{,}000}$$

87. Show that a real number has a repeating decimal if and only if it is rational.

88. (a) Suppose $\displaystyle\sum_{k=1}^{\infty} s_k$ is a series with the property that $s_n \geq 0$ for all integers $n \geq 1$. Show that $\displaystyle\sum_{k=1}^{\infty} s_k$ converges if and only if the sequence $\{S_n\}$ of partial sums is bounded.

(b) Use the result of (a) to show the harmonic series diverges.

8.3 Properties of Series; the Integral Test

OBJECTIVES *When you finish this section, you should be able to:*

1 Use the Test for Divergence (p. 567)
2 Work with properties of series (p. 567)
3 Use the Integral Test (p. 569)
4 Analyze a *p*-series (p. 570)

We have been determining whether a series $\sum_{k=1}^{\infty} a_k$ converges or diverges by finding a single compact expression for the sequence $\{S_n\}$ of partial sums as a function of n, and then examining $\lim_{n \to \infty} S_n$. For most series $\sum_{k=1}^{\infty} a_k$, however, this is not possible. As a result, we develop alternate methods for determining whether $\sum_{k=1}^{\infty} a_k$ is convergent or divergent. Most of these alternate methods only tell us whether a series converges or diverges, but provide no information about the sum of a convergent series. Fortunately, in many applications involving series, it is more important to know whether or not a series converges. Knowing the sum of a convergent series, although desirable, is not always necessary.

The next result gives a property of convergent series that is used often.

THEOREM

If the series $\sum_{k=1}^{\infty} a_k$ converges, then $\lim_{n \to \infty} a_n = 0$.

Proof The nth partial sum of $\sum_{k=1}^{\infty} a_k$ is $S_n = \sum_{k=1}^{n} a_k$. Since $S_{n-1} = \sum_{k=1}^{n-1} a_k$, it follows that

$$a_n = S_n - S_{n-1}$$

Since the series $\sum_{k=1}^{\infty} a_k$ converges, the sequence $\{S_n\}$ of partial sums has a limit S. Then $\lim_{n \to \infty} S_n = S$ and $\lim_{n \to \infty} S_{n-1} = S$, so

$$\lim_{n \to \infty} a_n = \lim_{n \to \infty} (S_n - S_{n-1}) = \lim_{n \to \infty} S_n - \lim_{n \to \infty} S_{n-1} = S - S = 0 \qquad \blacksquare$$

IN WORDS If $\sum_{k=1}^{\infty} a_k$ converges, then $\lim_{n \to \infty} a_n$ equals 0. If $\lim_{n \to \infty} a_n$ equals 0, then the series $\sum_{k=1}^{\infty} a_k$ may converge or diverge.

So if a series $\sum_{k=1}^{\infty} a_k$ converges, then $\lim_{n \to \infty} a_n = 0$. But there are many divergent series $\sum_{k=1}^{\infty} a_k$ for which $\lim_{n \to \infty} a_n = 0$. For example, the limit of the nth term of the harmonic series $\sum_{k=1}^{\infty} \frac{1}{k}$ is $\lim_{n \to \infty} \frac{1}{n} = 0$, and yet it diverges.

By restating the theorem, we obtain a useful test for divergence.

THEOREM Test for Divergence

The infinite series $\sum_{k=1}^{\infty} a_k$ diverges if $\lim_{n \to \infty} a_n \neq 0$.

Be careful! In testing $\sum_{k=1}^{\infty} a_k$ for convergence/divergence, if $\lim_{n \to \infty} a_n \neq 0$, the series diverges, but if $\lim_{n \to \infty} a_n = 0$, the series may converge or it may diverge.

1 Use the Test for Divergence

EXAMPLE 1   Using the Test for Divergence

(a) $\displaystyle\sum_{k=1}^{\infty} 87$ diverges, since $\displaystyle\lim_{n \to \infty} 87 = 87 \neq 0$.

(b) $\displaystyle\sum_{k=1}^{\infty} k$ diverges, since $\displaystyle\lim_{n \to \infty} n = \infty \neq 0$.

(c) $\displaystyle\sum_{k=1}^{\infty} (-1)^k$ diverges, since $\displaystyle\lim_{n \to \infty} (-1)^n$ does not exist.

(d) $\displaystyle\sum_{k=1}^{\infty} 2^k$ diverges, since $\displaystyle\lim_{n \to \infty} 2^n = \infty \neq 0$. ∎

NOW WORK   Problem 17.

2 Work with Properties of Series

Next we investigate some properties of convergent and divergent series. Knowing these properties can help to determine whether a series converges or diverges.

THEOREM

If two infinite series are identical after a certain term, then either both series converge or both series diverge. If both series converge, they do not necessarily have the same sum.

Proof   Consider the two series

$$\sum_{k=1}^{\infty} a_k = a_1 + a_2 + \cdots + a_p + a_{p+1} + \cdots + a_n + \cdots$$

$$\sum_{k=1}^{\infty} b_k = b_1 + b_2 + \cdots + b_p + a_{p+1} + \cdots + a_n + \cdots$$

Notice that after the first p terms, the remaining terms of both series are identical. The sequence $\{S_n\}$ of partial sums of $\displaystyle\sum_{k=1}^{\infty} a_k$ and the sequence $\{T_n\}$ of partial sums of $\displaystyle\sum_{k=1}^{\infty} b_k$ are given by

$$S_n = a_1 + a_2 + \cdots + a_p + a_{p+1} + \cdots + a_n$$
$$T_n = b_1 + b_2 + \cdots + b_p + a_{p+1} + \cdots + a_n$$

Now,

$$S_n - T_n = (a_1 + \cdots + a_p) - (b_1 + \cdots + b_p) \qquad \text{After the } p\text{th term, the terms are the same.}$$

$$S_n = T_n + (a_1 + \cdots + a_p) - (b_1 + \cdots + b_p) \qquad \text{For } n > p.$$

$$\lim_{n \to \infty} S_n = \lim_{n \to \infty} \left[T_n + (a_1 + \cdots + a_p) - (b_1 + \cdots + b_p) \right]$$

$$\lim_{n \to \infty} S_n = \lim_{n \to \infty} T_n + \lim_{n \to \infty} \left[(a_1 + \cdots + a_p) - (b_1 + \cdots + b_p) \right]$$

$$\lim_{n \to \infty} S_n = \lim_{n \to \infty} T_n + k \qquad \begin{aligned} & k = (a_1 + \cdots + a_p) \\ & \quad -(b_1 + \cdots + b_p) \\ & \text{is some number.} \end{aligned}$$

Consequently, either both limits exist (both series converge) or neither limit exists (both series diverge). No other possibility can occur. ∎

THEOREM Sum and Difference of Convergent Series

If $\displaystyle\sum_{k=1}^{\infty} a_k = S$ and $\displaystyle\sum_{k=1}^{\infty} b_k = T$ are two convergent series, then the series $\displaystyle\sum_{k=1}^{\infty}(a_k + b_k)$

and the series $\displaystyle\sum_{k=1}^{\infty}(a_k - b_k)$ also converge. Moreover,

$$\sum_{k=1}^{\infty}(a_k + b_k) = \sum_{k=1}^{\infty} a_k + \sum_{k=1}^{\infty} b_k = S + T$$

$$\sum_{k=1}^{\infty}(a_k - b_k) = \sum_{k=1}^{\infty} a_k - \sum_{k=1}^{\infty} b_k = S - T$$

The proof of the sum part of this theorem is left as an exercise. See Problem 65.

THEOREM Constant Multiple of a Series

Let c be a nonzero real number. If $\displaystyle\sum_{k=1}^{\infty} a_k = S$ is a convergent series, then the series

$\displaystyle\sum_{k=1}^{\infty}(ca_k)$ also converges. Moreover,

$$\sum_{k=1}^{\infty}(ca_k) = c\sum_{k=1}^{\infty} a_k = cS$$

If the series $\displaystyle\sum_{k=1}^{\infty} a_k$ diverges, then the series $\displaystyle\sum_{k=1}^{\infty}(ca_k)$ also diverges.

IN WORDS Multiplying each term of a series by a nonzero constant does not affect the convergence (or divergence) of the series.

The proof of the convergent part of this theorem is left as an exercise. See Problem 66.

EXAMPLE 2   **Using Properties of Series**

Determine whether each series converges or diverges. If it converges, find its sum.

(a) $\displaystyle\sum_{k=4}^{\infty} \frac{1}{k}$ **(b)** $\displaystyle\sum_{k=1}^{\infty} \frac{2}{k}$ **(c)** $\displaystyle\sum_{k=1}^{\infty}\left(\frac{1}{2^{k-1}} + \frac{1}{3^{k-1}}\right)$

Solution **(a)** Except for the first three terms, the series $\displaystyle\sum_{k=4}^{\infty} \frac{1}{k} = \frac{1}{4} + \frac{1}{5} + \frac{1}{6} + \cdots$

is identical to the harmonic series, which diverges. So, it follows that $\displaystyle\sum_{k=4}^{\infty} \frac{1}{k}$ also diverges.

(b) $\displaystyle\sum_{k=1}^{\infty} \frac{2}{k} = \sum_{k=1}^{\infty}\left(2 \cdot \frac{1}{k}\right)$. Since the harmonic series $\displaystyle\sum_{k=1}^{\infty} \frac{1}{k}$ diverges, the series

$\displaystyle\sum_{k=1}^{\infty}\left(2 \cdot \frac{1}{k}\right) = \sum_{k=1}^{\infty} \frac{2}{k}$ diverges.

(c) Since the series $\displaystyle\sum_{k=1}^{\infty} \frac{1}{2^{k-1}}$ and the series $\displaystyle\sum_{k=1}^{\infty} \frac{1}{3^{k-1}}$ are both convergent geometric

series, the series defined by $\displaystyle\sum_{k=1}^{\infty}\left(\frac{1}{2^{k-1}} + \frac{1}{3^{k-1}}\right)$ is also convergent. The sum is

$$\sum_{k=1}^{\infty}\left(\frac{1}{2^{k-1}} + \frac{1}{3^{k-1}}\right) = \sum_{k=1}^{\infty} \frac{1}{2^{k-1}} + \sum_{k=1}^{\infty} \frac{1}{3^{k-1}} = \frac{1}{1-\frac{1}{2}} + \frac{1}{1-\frac{1}{3}} = 2 + \frac{3}{2} = \frac{7}{2} \quad\blacksquare$$

NOW WORK   **Problem 39.**

For series that have only positive terms, it is possible to construct tests for convergence that use only the nth term of the series and do not require knowledge of the form of the sequence $\{S_n\}$ of partial sums.

For example, consider an infinite series

$$\sum_{k=1}^{\infty} a_k = a_1 + a_2 + \cdots + a_n + \cdots$$

where each term is positive. Suppose $\{S_n\}$ is the sequence of partial sums. Since each term $a_n > 0$, then $S_n = S_{n-1} + a_n > S_{n-1}$. That is, the sequence $\{S_n\}$ of partial sums is increasing. If $\{S_n\}$ is bounded from above, then the sequence $\{S_n\}$ of partial sums converges (p. 549, Section 8.1). The *General Convergence Test* follows from this conclusion.

THEOREM General Convergence Test

An infinite series of positive terms converges if and only if its sequence of partial sums is bounded. The sum of such an infinite series will not exceed an upper bound.

We use this theorem to develop other tests for convergence of positive series; the first one we discuss is the *Integral Test*.

THEOREM Integral Test

Let f be a function that is continuous, positive, and decreasing on the interval $[1, \infty)$. Let $a_k = f(k)$ for all positive integers k. Then the series $\sum_{k=1}^{\infty} a_k$ converges if and only if the improper integral $\int_1^{\infty} f(x)\, dx$ converges.

NEED TO REVIEW? Improper Integrals are discussed in Section 7.8, pp. 523–529.

A proof of the Integral Test is given at the end of the section.

3 Use the Integral Test

The Integral Test is used when the nth term of the series is related to a function f that not only is continuous, positive, and decreasing on the interval $[1, \infty)$, but also has an antiderivative that can be readily found.

EXAMPLE 3 Using the Integral Test

Determine whether the series $\sum_{k=1}^{\infty} a_k = \sum_{k=1}^{\infty} \dfrac{4}{k^2+1}$ converges or diverges.

Solution The function $f(x) = \dfrac{4}{x^2+1}$ is defined on the interval $[1, \infty)$ and is continuous, positive, and decreasing for all numbers $x \geq 1$. Also, $a_k = f(k)$ for all positive integers k. Using the Integral Test, we find

CAUTION In Example 3 the fact that $\int_1^{\infty} \dfrac{4}{x^2+1}\, dx = \pi$ does not mean that the sum S of the series is π. Do not confuse the value of the improper integral $\int_1^{\infty} f(x)\, dx$ with the sum S of the series. In general, they are *not* equal.

$$\int_1^{\infty} \frac{4}{x^2+1}\, dx : \lim_{b\to\infty}\int_1^b \frac{4}{x^2+1}\, dx = \lim_{b\to\infty}\left[4\int_1^b \frac{1}{x^2+1}\, dx\right] = 4\lim_{b\to\infty}\left[\tan^{-1}x\right]_1^b$$

$$= 4\lim_{b\to\infty}\left[\tan^{-1}b - \tan^{-1}1\right] = 4\left[\frac{\pi}{2}-\frac{\pi}{4}\right] = \pi$$

Since the improper integral converges, the series $\sum_{k=1}^{\infty}\dfrac{4}{k^2+1}$ converges. ∎

EXAMPLE 4   Using the Integral Test

Determine whether the series $\displaystyle\sum_{k=1}^{\infty} a_k = \sum_{k=1}^{\infty} \frac{2k}{k^2+1}$ converges or diverges.

Solution The function $f(x) = \dfrac{2x}{x^2+1}$ is continuous, positive, and decreasing since $f'(x) \leq 0$ for all numbers $x \geq 1$, and $a_k = f(k)$ for all positive integers k. Using the Integral Test, we find

$$\int_1^{\infty} \frac{2x}{x^2+1}\,dx : \lim_{b \to \infty} \int_1^b \frac{2x}{x^2+1}\,dx = \lim_{b \to \infty}\left[\ln(x^2+1)\right]_1^b$$

$$= \lim_{b \to \infty}\left[\ln(b^2+1) - \ln 2\right] = \infty$$

Since the improper integral diverges, the series $\displaystyle\sum_{k=1}^{\infty} \frac{2k}{k^2+1}$ also diverges. ■

NOW WORK Problem 21.

To use the Integral Test, the lower limit of integration does not need to be 1, as we see in the next example.

EXAMPLE 5   Using the Integral Test

Determine whether the series $\displaystyle\sum_{k=2}^{\infty} \frac{1}{k(\ln k)^2}$ converges or diverges.

Solution The function $f(x) = \dfrac{1}{x(\ln x)^2}$ is continuous, positive, and decreasing on the interval $[2, \infty)$. Also $\dfrac{1}{k(\ln k)^2} = f(k)$ for all integers greater than or equal to 2. Using the Integral Test, we investigate the improper integral $\displaystyle\int_2^{\infty} \frac{dx}{x(\ln x)^2}$. We find an antiderivative of $\dfrac{1}{x(\ln x)^2}$ by using the substitution $u = \ln x$, $du = \dfrac{1}{x}\,dx$. Then,

$$\int \frac{dx}{x(\ln x)^2} = \int \frac{du}{u^2} = -\frac{1}{u} + C = -\frac{1}{\ln x} + C$$

Now we find the improper integral

$$\int_2^{\infty} \frac{dx}{x(\ln x)^2} : \lim_{b \to \infty}\left[-\frac{1}{\ln x}\right]_2^b = \lim_{b \to \infty}\left(-\frac{1}{\ln b}\right) + \frac{1}{\ln 2} = \frac{1}{\ln 2}$$

Since the improper integral $\displaystyle\int_2^{\infty} \frac{dx}{x(\ln x)^2}$ converges, the series $\displaystyle\sum_{k=2}^{\infty} \frac{1}{k(\ln k)^2}$ converges. ■

NOW WORK Problem 27.

4 Analyze a p-Series

Another important series is the *p-series.*

DEFINITION p-series

A *p*-**series** is an infinite series of the form

$$\sum_{k=1}^{\infty} \frac{1}{k^p} = 1 + \frac{1}{2^p} + \frac{1}{3^p} + \cdots + \frac{1}{n^p} + \cdots$$

where p is a positive real number.

A p-series is sometimes referred to as a **hyperharmonic series**, since the harmonic series is a special case of a p-series when $p = 1$. Some examples of p-series are

$$p = 1: \quad \sum_{k=1}^{\infty} \frac{1}{k} = 1 + \frac{1}{2} + \frac{1}{3} + \cdots + \frac{1}{n} + \cdots \qquad \text{\color{blue}{The harmonic series}}$$

$$p = \frac{1}{2}: \quad \sum_{k=1}^{\infty} \frac{1}{k^{1/2}} = 1 + \frac{1}{2^{1/2}} + \frac{1}{3^{1/2}} + \cdots + \frac{1}{n^{1/2}} + \cdots$$

$$p = 3: \quad \sum_{k=1}^{\infty} \frac{1}{k^3} = 1 + \frac{1}{2^3} + \frac{1}{3^3} + \cdots + \frac{1}{n^3} + \cdots$$

The following theorem establishes the values of p for which a p-series converges and for which it diverges.

THEOREM Convergence/Divergence of a p-Series

The p-series

$$\sum_{k=1}^{\infty} \frac{1}{k^p} = 1 + \frac{1}{2^p} + \frac{1}{3^p} + \cdots + \frac{1}{n^p} + \cdots$$

converges if $p > 1$ and diverges if $0 < p \leq 1$.

Proof The function $f(x) = \dfrac{1}{x^p}$, $p > 0$, is continuous, positive, and decreasing for all numbers $x \geq 1$, and $f(k) = \dfrac{1}{k^p}$ for all positive integers k. So, the series $\displaystyle\sum_{k=1}^{\infty} \frac{1}{k^p}$ converges if and only if the improper integral $\displaystyle\int_{1}^{\infty} \frac{1}{x^p}\, dx$ converges.

In Section 7.8, page 526, we proved that $\displaystyle\int_{1}^{\infty} \frac{dx}{x^p}$ converges if $p > 1$ and diverges if $p \leq 1$. So, the p-series $\displaystyle\sum_{k=1}^{\infty} \frac{1}{k^p}$ converges if $p > 1$ and diverges if $0 < p \leq 1$. ∎

EXAMPLE 6 Analyzing a p-Series

(a) The series

$$\sum_{k=1}^{\infty} \frac{1}{k^3} = 1 + \frac{1}{2^3} + \frac{1}{3^3} + \cdots + \frac{1}{n^3} + \cdots$$

converges, since it is a p-series where $p = 3$.

(b) The series

$$\sum_{k=1}^{\infty} \frac{1}{\sqrt{k}} = 1 + \frac{1}{\sqrt{2}} + \frac{1}{\sqrt{3}} + \cdots + \frac{1}{\sqrt{n}} + \cdots$$

diverges, since it is a p-series where $p = \dfrac{1}{2}$. ∎

NOW WORK Problem 35.

The theorem describing the criterion for the convergence or divergence of a p-series has an interesting corollary that establishes a pair of bounds for the sum of a convergent p-series.

IN WORDS The corollary provides an interval of width 1 that contains the sum of a convergent p-series. But it gives no information about the actual sum.

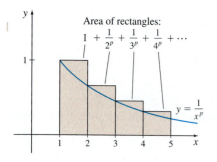

Figure 19

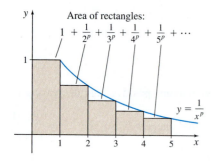

Figure 20

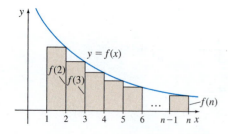

Figure 21

COROLLARY Bounds for the Sum of a Convergent p-Series

If $p > 1$, then

$$\frac{1}{p-1} < \sum_{k=1}^{\infty} \frac{1}{k^p} < 1 + \frac{1}{p-1}$$

Proof From Figure 19 we have

$$\left(\text{Area under the graph of } y = \frac{1}{x^p}\right) = \int_1^{\infty} \frac{dx}{x^p} < 1 + \frac{1}{2^p} + \frac{1}{3^p} + \cdots = \sum_{k=1}^{\infty} \frac{1}{k^p} \quad (1)$$

From Figure 20 we have

$$1 + \left(\text{Area under the graph of } y = \frac{1}{x^p}\right) = 1 + \int_1^{\infty} \frac{dx}{x^p} > 1 + \frac{1}{2^p} + \frac{1}{3^p} + \cdots = \sum_{k=1}^{\infty} \frac{1}{k^p} \quad (2)$$

Combining (1) and (2), we have

$$\int_1^{\infty} \frac{dx}{x^p} < \sum_{k=1}^{\infty} \frac{1}{k^p} < 1 + \int_1^{\infty} \frac{dx}{x^p} \quad (3)$$

Since $p > 1$,

$$\int_1^{\infty} \frac{dx}{x^p} = \lim_{b \to \infty} \int_1^b \frac{dx}{x^p} = \lim_{b \to \infty} \left[\frac{x^{-p+1}}{-p+1}\right]_1^b = \lim_{b \to \infty} \frac{b^{1-p} - 1}{1 - p} \underset{p > 1}{=} \frac{0 - 1}{1 - p} = \frac{1}{p-1}$$

Using this result in (3), we get

$$\frac{1}{p-1} < \sum_{k=1}^{\infty} \frac{1}{k^p} < 1 + \frac{1}{p-1} \qquad ■$$

For example, based on the corollary, if $p = 2$, then $1 < \sum_{k=1}^{\infty} \frac{1}{k^2} < 2$, and if $p = 3$, then $\frac{1}{2} < \sum_{k=1}^{\infty} \frac{1}{k^3} < \frac{3}{2}$.

Since $\sum_{k=1}^{\infty} \frac{1}{k^p}$ converges for $p > 1$, the series has a sum S. However, the exact sum has been found only for positive even integers p. In 1752 Euler was the first to show that $\sum_{k=1}^{\infty} \frac{1}{k^2} = \frac{\pi^2}{6}$. But the sum of other convergent p-series, such as $\sum_{k=1}^{\infty} \frac{1}{k^3}$, is not known.

Proof of the Integral Test Suppose $\int_1^{\infty} f(x)\, dx$ converges. Then $\lim_{n \to \infty} \int_1^n f(x)\, dx$ exists and equals some number L. As Figure 21 shows, $\int_1^n f(x)\, dx$ underestimates the area under the graph of f on $[1, \infty)$. As a result,

$$\int_1^n f(x)\, dx < L \qquad \text{for any number } n$$

But f is decreasing on the interval $[1, n]$. So, the sum of the areas of the $n - 1$ rectangles drawn in Figure 21 underestimates the area under the graph of $y = f(x)$ from 1 to n. That is,

$$f(2) + f(3) + \cdots + f(n) < \int_1^n f(x)\, dx < L$$

Now add $f(1) = a_1$ to each part, and use the fact that $a_1 = f(1)$, $a_2 = f(2)$, ..., $a_n = f(n)$. Then

$$a_1 + a_2 + a_3 + \cdots + a_n = f(1) + f(2) + \cdots + f(n) < f(1) + \int_1^n f(x)\, dx < f(1) + L$$

This means that the partial sums of $\sum_{k=1}^{\infty} a_k$ are bounded from above. Since the partial sums are also increasing, $\sum_{k=1}^{\infty} a_k$ converges.

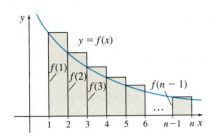

Figure 22

Now suppose $\int_1^\infty f(x)\,dx$ diverges. Since f is decreasing on $[1, n]$, the sum of the areas of the $n - 1$ rectangles shown in Figure 22 overestimates the area under the graph of $y = f(x)$ from 1 to n. That is,

$$f(1) + f(2) + \cdots + f(n-1) > \int_1^n f(x)\,dx$$

Since $\int_1^n f(x)\,dx \to \infty$ as $n \to \infty$, the nth partial sum S_n must also approach infinity, and it follows that the series $\sum_{k=1}^\infty a_k$ diverges. ∎

8.3 Assess Your Understanding

Concepts and Vocabulary

1. *Multiple Choice* If the series $\sum_{k=1}^\infty a_k$, $a_k > 0$, converges then $\lim_{n\to\infty} a_n = $ [(a) 0, (b) a_1, (c) a_n, (d) ∞].

2. *True or False* If $\lim_{n\to\infty} a_n = 0$, then the series $\sum_{k=1}^\infty a_k$ converges.

3. *True or False* The series $\sum_{k=1}^\infty k^3$ diverges.

4. *True or False* If the first 100 terms of two infinite series are different, but from the 101st term on they are identical, then either both series converge or both series diverge.

5. *True or False* If $\sum_{k=1}^\infty (a_k + b_k)$ converges, then $\sum_{k=1}^\infty a_k$ converges and $\sum_{k=1}^\infty b_k$ converges.

6. *True or False* If $\sum_{k=1}^\infty a_k = S$ is a convergent series and c is a nonzero real number, then $\sum_{k=1}^\infty (ca_k) = cS$.

7. *True or False* Let f be a function defined on the interval $[1, \infty)$ that is continuous, positive, and decreasing on its domain. Let $a_k = f(k)$ for all positive integers k. Then the series $\sum_{k=1}^\infty a_k$ converges if and only if the improper integral $\int_1^\infty f(x)\,dx$ converges.

8. *True or False* For an infinite series of positive terms, if its sequence of partial sums is not bounded, then you cannot tell if the series converges or diverges.

9. The p-series $\sum_{k=1}^\infty \dfrac{1}{k^p}$ converges if _____ and diverges if _____.

10. Does $\sum_{k=1}^\infty \dfrac{1}{k^{3/2}}$ converge or diverge?

11. Does $\sum_{k=1}^\infty \dfrac{1}{k^{-1/2}}$ converge or diverge?

12. *True or False* If $p > 1$, then the convergent p-series $\sum_{k=1}^\infty \dfrac{1}{k^p}$ is bounded by $\dfrac{1}{p-1} < \sum_{k=1}^\infty \dfrac{1}{k^p} < 1$.

Skill Building

In Problems 13–18, use the Test for Divergence to show each series diverges.

13. $\displaystyle\sum_{k=1}^\infty 16$

14. $\displaystyle\sum_{k=1}^\infty \dfrac{k+9}{k}$

15. $\displaystyle\sum_{k=1}^\infty \ln k$

16. $\displaystyle\sum_{k=1}^\infty e^k$

17. $\displaystyle\sum_{k=1}^\infty \dfrac{k^2}{k^2+4}$

18. $\displaystyle\sum_{k=1}^\infty \dfrac{k^2+3}{\sqrt{k}}$

In Problems 19–28, use the Integral Test to determine whether each series converges or diverges.

19. $\displaystyle\sum_{k=1}^\infty \dfrac{1}{k^{1.01}}$

20. $\displaystyle\sum_{k=1}^\infty \dfrac{1}{k^{0.9}}$

21. $\displaystyle\sum_{k=1}^\infty \dfrac{\ln k}{k}$

22. $\displaystyle\sum_{k=2}^\infty \dfrac{1}{k\sqrt{\ln k}}$

23. $\displaystyle\sum_{k=1}^\infty k e^{-k^2}$

24. $\displaystyle\sum_{k=1}^\infty k e^{-k}$

25. $\displaystyle\sum_{k=1}^\infty \dfrac{1}{k^2+1}$

26. $\displaystyle\sum_{k=2}^\infty \dfrac{1}{k\sqrt{k^2-1}}$

27. $\displaystyle\sum_{k=2}^\infty \dfrac{1}{k\ln k}$

28. $\displaystyle\sum_{k=2}^\infty \dfrac{1}{k(\ln k)^3}$

In Problems 29–38, determine whether each p-series converges or diverges.

29. $\displaystyle\sum_{k=1}^\infty \dfrac{1}{k^2}$

30. $\displaystyle\sum_{k=1}^\infty \dfrac{1}{k^4}$

31. $\displaystyle\sum_{k=1}^\infty \dfrac{1}{k^{1/3}}$

32. $\displaystyle\sum_{k=1}^\infty \dfrac{1}{k^{2/3}}$

33. $\displaystyle\sum_{k=1}^\infty \dfrac{1}{k^e}$

34. $\displaystyle\sum_{k=1}^\infty \dfrac{1}{k^\pi}$

35. $1 + \dfrac{1}{2\sqrt{2}} + \dfrac{1}{3\sqrt{3}} + \dfrac{1}{4\sqrt{4}} + \cdots$

1. = NOW WORK problem   📐 = Graphing technology recommended   CAS = Computer Algebra System recommended

36. $1 + \dfrac{1}{\sqrt[3]{2}} + \dfrac{1}{\sqrt[3]{3}} + \dfrac{1}{\sqrt[3]{4}} + \cdots$

37. $1 + \dfrac{1}{4\sqrt{2}} + \dfrac{1}{9\sqrt{3}} + \dfrac{1}{16\sqrt{4}} + \cdots$

38. $1 + \dfrac{1}{8} + \dfrac{1}{27} + \dfrac{1}{64} + \cdots$

In Problems 39–54, determine whether each series converges or diverges.

39. $\displaystyle\sum_{k=1}^{\infty} \dfrac{10}{k}$

40. $\displaystyle\sum_{k=1}^{\infty} \dfrac{2}{1+k}$

41. $\displaystyle\sum_{k=1}^{\infty} \dfrac{k^2+1}{4k+1}$

42. $\displaystyle\sum_{k=1}^{\infty} \dfrac{k^3}{k^3+3}$

43. $\displaystyle\sum_{k=1}^{\infty} \left(k + \dfrac{1}{k}\right)$

44. $\displaystyle\sum_{k=1}^{\infty} \left(\dfrac{1}{3^k} - \dfrac{1}{4^k}\right)$

45. $\displaystyle\sum_{k=1}^{\infty} \left(\dfrac{1}{3k} - \dfrac{1}{4k}\right)$

46. $\displaystyle\sum_{k=1}^{\infty} \left(k - \dfrac{10}{k}\right)$

47. $\displaystyle\sum_{k=1}^{\infty} \sin\left(\dfrac{\pi}{2}k\right)$

48. $\displaystyle\sum_{k=1}^{\infty} \sec(\pi k)$

49. $\displaystyle\sum_{k=3}^{\infty} \dfrac{k+1}{k-2}$

50. $\displaystyle\sum_{k=5}^{\infty} \dfrac{2k^5+3}{k^5-4k^4}$

51. $\displaystyle\sum_{k=2}^{\infty} \dfrac{1}{k(\ln k)^{1/2}}$

52. $\displaystyle\sum_{k=2}^{\infty} \dfrac{1}{k(\ln k)^2}$

53. $\displaystyle\sum_{k=3}^{\infty} \dfrac{2k}{k^2-4}$

54. $\displaystyle\sum_{k=1}^{\infty} \dfrac{1}{(2k-1)(2k)}$

Applications and Extensions

55. Integral Test Use the Integral Test to show that the series
$$\sum_{k=2}^{\infty} \dfrac{1}{k(\ln k)^p}$$ converges if and only if $p > 1$.

56. Integral Test Use the Integral Test to show that the series
$$\sum_{k=3}^{\infty} \dfrac{1}{k(\ln k)\,[\ln(\ln k)^p]}$$ converges if and only if $p > 1$.

57. Faulty Logic Let $S = 1 + 2 + 4 + 8 + \cdots$. Then
$$2S = 2 + 4 + 8 + 16 + \cdots = -1 + (1 + 2 + 4 + \cdots) = -1 + S$$

Therefore,
$$S = 1 + 2 + 4 + 8 + \cdots = -1$$

What went wrong here?

58. Find examples to show that the series $\displaystyle\sum_{k=1}^{\infty}(a_k + b_k)$ and
$\displaystyle\sum_{k=1}^{\infty}(a_k - b_k)$ may converge or diverge if $\displaystyle\sum_{k=1}^{\infty} a_k$ and $\displaystyle\sum_{k=1}^{\infty} b_k$ each diverge.

CAS **59. Approximating π^2** The p-series $\displaystyle\sum_{k=1}^{\infty} \dfrac{1}{k^2}$ converges.

  (a) Find the sum of the series in exact form.

  (b) Use the first hundred terms of the series to approximate π^2.

60. The p-series $\displaystyle\sum_{k=1}^{\infty} \dfrac{1}{k^3}$ converges.

  (a) Provide an interval of width 1 that contains the sum $\displaystyle\sum_{k=1}^{\infty} \dfrac{1}{k^3}$.

CAS   **(b)** Use the first hundred terms of the series to approximate the sum.

CAS *In Problems 61–63, use the Integral Test to determine whether each series converges or diverges.*

61. $\displaystyle\sum_{k=1}^{\infty}\left(k^6 e^{-k}\right)$ **62.** $\displaystyle\sum_{k=1}^{\infty} \dfrac{k+3}{k^2+6k+7}$ **63.** $\displaystyle\sum_{k=2}^{\infty} \dfrac{5k+6}{k^3-1}$

CAS **64.** The p-series $\displaystyle\sum_{k=1}^{\infty} \dfrac{1}{k^p}$ converges for $p > 1$ and diverges for
$0 < p \le 1$. The series $\displaystyle\sum_{k=1}^{\infty} a_k = \sum_{k=1}^{\infty} \dfrac{1}{k^{0.99}}$ diverges and the
series $\displaystyle\sum_{k=1}^{\infty} b_k = \sum_{k=1}^{\infty} \dfrac{1}{k^{1.01}}$ converges.

  (a) Find the partial sums S_{10}, S_{1000}, and $S_{100,000}$ for each series.

  (b) Explain the results found in (a).

65. Show that the sum of two convergent series is a convergent series.

66. Show that for a nonzero real number c, if $\displaystyle\sum_{k=1}^{\infty} a_k = S$ is a
convergent series, then the series $\displaystyle\sum_{k=1}^{\infty}(ca_k) = cS$.

67. Suppose $\displaystyle\sum_{k=N+1}^{\infty} a_k = S$ and $a_1 + a_2 + \cdots + a_N = K$.

Prove that $\displaystyle\sum_{k=1}^{\infty} a_k$ converges and its sum is $S + K$.

68. If $\displaystyle\sum_{k=1}^{\infty} a_k$ converges and $\displaystyle\sum_{k=1}^{\infty} b_k$ diverges, then prove that
$$\sum_{k=1}^{\infty}(a_k + b_k)\ \text{diverges.}$$

69. Suppose $\displaystyle\sum_{k=1}^{\infty} a_k$ converges, and $a_n > 0$ for all n.

Show that $\displaystyle\sum_{k=1}^{\infty} \dfrac{a_k}{1+a_k}$ converges.

Challenge Problems

70. Determine whether the series $\displaystyle\sum_{k=1}^{\infty} \dfrac{1}{k \ln\left(1 + \dfrac{1}{k}\right)}$ converges.

71. Integral Test Use the Integral Test to show that the series
$$\sum_{k=2}^{\infty} \dfrac{1}{(\ln k)^p}$$ diverges for all numbers p.

72. For what numbers p and q does the series $\displaystyle\sum_{k=2}^{\infty} \dfrac{(\ln k)^q}{k^p}$ converge?

73. Find all real numbers x for which $\displaystyle\sum_{k=1}^{\infty} k^x$ converges. Express your answer using interval notation.

74. Consider the finite sum $S_n = \sum\limits_{k=1}^{n} \dfrac{1}{1+k^2}$.

   (a) By comparing S_n with an appropriate integral, show that $S_n \le \tan^{-1} n$ for $n \ge 1$.

   (b) Use (a) to deduce that $\sum\limits_{k=1}^{\infty} \dfrac{1}{1+k^2}$ converges.

   (c) Prove that $\dfrac{\pi}{4} \le \sum\limits_{k=1}^{\infty} \dfrac{1}{1+k^2} \le \dfrac{\pi}{2}$.

75. Riemann's zeta function is defined as

$$\zeta(s) = 1 + \frac{1}{2^s} + \frac{1}{3^s} + \cdots \qquad \text{for } s > 1$$

As mentioned on page 572, Euler showed that $\zeta(2) = \dfrac{\pi^2}{6}$. He also found the value of the zeta function for many other even values of s. As of now, no one knows the value of the zeta function for odd values of s. However, it is not too difficult to approximate these values, as this problem demonstrates.

   (a) Find $\sum\limits_{k=1}^{10} \dfrac{1}{k^3}$.

   (b) Using integrals in a way analogous to their use in the proof of the Integral Test, find upper and lower bounds for $\sum\limits_{k=1}^{\infty} \dfrac{1}{k^3}$.

   (c) What can you conclude about $\zeta(3)$?

76. (a) By considering graphs like those shown on the right, show that if f is decreasing, positive, and continuous, then

$$f(n+1)+\cdots+f(m) \le \int_n^m f(x)\,dx \le f(n)+\cdots+f(m-1)$$

(b) Under the assumption of (a), prove that if $\sum\limits_{k=1}^{\infty} f(k)$ converges, then

$$\sum_{k=n+1}^{\infty} f(k) \le \int_n^{\infty} f(x)\,dx \le \sum_{k=n}^{\infty} f(k)$$

(c) Let $f(x) = \dfrac{1}{x^2}$. Use the inequality in (b) to determine *exactly* how many terms of the series $\sum\limits_{k=1}^{\infty} \dfrac{1}{k^2}$ one must take in order to have $|\,\text{Error}\,| < \left(\dfrac{1}{2}\right) 10^{-2}$. How many terms must one take to have $|\,\text{Error}\,| < \left(\dfrac{1}{2}\right) 10^{-10}$?

Source: Based on an article by R. P. Boas, Jr., *American Mathematical Monthly*, "Partial Sums of Infinite Series, and How They Grow" Vol. 84, No. 4 (April 1977), pp. 237–258.

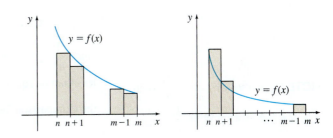

8.4 Comparison Tests

OBJECTIVES *When you finish this section, you should be able to:*

1 Use Comparison Tests for Convergence and Divergence (p. 576)

2 Use the Limit Comparison Test (p. 577)

We can determine whether a series converges or diverges by comparing it to a series whose behavior we already know. Suppose $\sum\limits_{k=1}^{\infty} b_k$ is a series of positive terms that is known to converge to B, and we want to determine if the series of positive terms $\sum\limits_{k=1}^{\infty} a_k$ converges. If term-by-term $a_k \le b_k$, that is, if

$$a_1 \le b_1, \quad a_2 \le b_2, \ldots, a_n \le b_n, \ldots$$

then it follows that

$$\left(n\text{th partial sum of } \sum_{k=1}^{\infty} a_k\right) \le \left(n\text{th partial sum of } \sum_{k=1}^{\infty} b_k\right) < B$$

This shows that the sequence of partial sums $\{S_n\}$ of $\sum\limits_{k=1}^{\infty} a_k$ is bounded, so by the General

RECALL The General Convergence Test: An infinite series of positive terms converges if and only if its sequence of partial sums is bounded.

Convergence Test, $\sum\limits_{k=1}^{\infty} a_k$ must also converge. This proves the *Comparison Test for Convergence.*

1 Use Comparison Tests for Convergence and Divergence

It is important to remember that the early terms in a series have no effect on the convergence or divergence of a series. In fact, the Comparison Test for Convergence is true if $0 < a_n \leq b_n$ for all $n \geq N$, where N is some suitably selected integer. We use this in the next example by ignoring the first term.

EXAMPLE 1   Using the Comparison Test for Convergence

Show that the series below converges:

$$\sum_{k=1}^{\infty} \frac{1}{k^k} = 1 + \frac{1}{2^2} + \frac{1}{3^3} + \cdots + \frac{1}{n^n} + \cdots$$

Solution We know that the geometric series $\sum_{k=1}^{\infty} \frac{1}{2^k}$ converges since $|r| = \frac{1}{2} < 1$. Now, since $\frac{1}{n^n} \leq \frac{1}{2^n}$ for all $n \geq 2$, except for the first term, each term of the series $\sum_{k=1}^{\infty} \frac{1}{k^k}$ is less than or equal to the corresponding term of the convergent geometric series. So, by the Comparison Test for Convergence, the series $\sum_{k=1}^{\infty} \frac{1}{k^k}$ converges. ■

NOTE Do you see why $\frac{1}{n^n} \leq \frac{1}{2^n}$ when $n \geq 2$? $n \geq 2 \Rightarrow n^n \geq 2^n \Rightarrow \frac{1}{n^n} \leq \frac{1}{2^n}$.

NOW WORK   Problem 5.

In Example 1, you may have asked, "How did you know that the given series should be compared to $\sum_{k=1}^{\infty} \frac{1}{2^k}$?" The easy answer is to try a series that you know converges, such as a geometric series with $|r| < 1$ or a p-series with $p > 1$. The honest answer is that only practice and experience will guide you.

There is also a comparison test for divergence. Suppose $\sum_{k=1}^{\infty} c_k$ is a series of positive terms we know diverges, and $\sum_{k=1}^{\infty} a_k$ is the series of positive terms to be tested. If term-by-term $a_k \geq c_k$, that is, if

$$a_1 \geq c_1, \ a_2 \geq c_2, \ \ldots, \ a_n \geq c_n, \ldots$$

then it follows that

$$\left(n\text{th partial sum of } \sum_{k=1}^{\infty} a_k \right) \geq \left(n\text{th partial sum of } \sum_{k=1}^{\infty} c_k \right)$$

But we know that the sequence of partial sums of $\sum_{k=1}^{\infty} c_k$ is unbounded. So, the sequence of partial sums of $\sum_{k=1}^{\infty} a_k$ is also unbounded, and $\sum_{k=1}^{\infty} a_k$ diverges.

EXAMPLE 2 **Using the Comparison Test for Divergence**

Show that the series $\sum_{k=1}^{\infty} \dfrac{k+3}{k(k+2)}$ diverges.

Solution Since $\dfrac{k+3}{k(k+2)}$ has a factor $\dfrac{1}{k}$, we choose to compare the given series to the

harmonic series $\sum_{k=1}^{\infty} \dfrac{1}{k}$, which diverges.

$$\frac{n+3}{n(n+2)} = \left(\frac{n+3}{n+2}\right)\left(\frac{1}{n}\right) > \frac{1}{n}$$

It follows from the Comparison Test for Divergence that $\sum_{k=1}^{\infty} \dfrac{k+3}{k(k+2)}$ diverges. ∎

NOW WORK Problem 9.

2 Use the Limit Comparison Test

The Comparison Tests for Convergence and Divergence are algebraic tests that require certain inequalities hold. Obtaining such inequalities can be challenging. The *Limit Comparison Test* requires certain *conditions* on the limit of a ratio. You may find this comparison test easier to use.

THEOREM Limit Comparison Test

Suppose $\sum_{k=1}^{\infty} a_k$ and $\sum_{k=1}^{\infty} b_k$ are both series of positive terms.

If $\lim_{n \to \infty} \dfrac{a_n}{b_n} = L, 0 < L < \infty$, then both series converge or both series diverge.

The Limit Comparison Test gives no guidance if $\lim_{n \to \infty} \dfrac{a_n}{b_n} = 0$ or $\lim_{n \to \infty} \dfrac{a_n}{b_n} = \infty$

or $\lim_{n \to \infty} \dfrac{a_n}{b_n}$ does not exist.

Proof If $\lim_{n \to \infty} \dfrac{a_n}{b_n} = L > 0$, then for any $\varepsilon > 0$, there is a positive real number N so that

$\left|\dfrac{a_n}{b_n} - L\right| < \varepsilon$, for all $n > N$. If we choose $\varepsilon = \dfrac{L}{2}$, then

$$\left|\frac{a_n}{b_n} - L\right| < \frac{L}{2} \qquad \text{for all } n > N$$

$$\frac{L}{2} < \frac{a_n}{b_n} < \frac{3L}{2} \qquad \text{for all } n > N$$

Since $b_n > 0$,

$$\frac{L}{2}b_n < a_n < \frac{3L}{2}b_n \qquad \text{for all } n > N$$

If $\sum_{k=1}^{\infty} a_k$ converges, by the Comparison Test for Convergence, the series $\sum_{k=1}^{\infty} \left(\dfrac{L}{2}b_k\right)$

$= \dfrac{L}{2}\sum_{k=1}^{\infty} b_k$ also converges. It follows that the series $\sum_{k=1}^{\infty} b_k$ is also convergent.

If $\sum_{k=1}^{\infty} b_k$ converges, then so does $\sum_{k=1}^{\infty} \left(\dfrac{3L}{2}b_k\right)$. Since $a_n < \dfrac{3L}{2}b_n$, for all $n > N$,

by the Comparison Test for Convergence, $\sum_{k=1}^{\infty} a_k$ also converges.

Therefore, $\sum_{k=1}^{\infty} a_k$ and $\sum_{k=1}^{\infty} b_k$ converge together. Since one cannot converge and the other diverge, they must diverge together. ∎

RECALL If c is a nonzero real number and if $\sum_{k=1}^{\infty} a_k = S$ is a convergent series, then the series $\sum_{k=1}^{\infty} (ca_k)$ also converges. Furthermore,

$$\sum_{k=1}^{\infty} (ca_k) = c\sum_{k=1}^{\infty} a_k = cS.$$

The Limit Comparison Test is quite versatile for comparing algebraically complex series to a p-series. To find the correct choice of the p-series to use in the comparison, examine the behavior of the terms of the series for large values of n. For example,

- To test $\displaystyle\sum_{k=1}^{\infty} \frac{1}{3k^2 + 5k + 2}$, use the p-series $\displaystyle\sum_{k=1}^{\infty} \frac{1}{k^2}$, because for large n,

$$\frac{1}{3n^2 + 5n + 2} = \frac{1}{n^2\left(3 + \dfrac{5}{n} + \dfrac{2}{n^2}\right)} \underset{\substack{\uparrow \\ \text{for large } n}}{\approx} \frac{1}{n^2}\left(\frac{1}{3}\right)$$

- To test $\displaystyle\sum_{k=1}^{\infty} \frac{2k^2 + 5}{3k^3 - 5k^2 + 2}$, use the p-series $\displaystyle\sum_{k=1}^{\infty} \frac{1}{k}$, because for large n,

$$\frac{2n^2 + 5}{3n^3 - 5n^2 + 2} = \frac{n^2\left(2 + \dfrac{5}{n^2}\right)}{n^3\left(3 - \dfrac{5}{n} + \dfrac{2}{n^3}\right)} = \frac{1}{n}\left(\frac{2 + \dfrac{5}{n^2}}{3 - \dfrac{5}{n} + \dfrac{2}{n^3}}\right) \underset{\substack{\uparrow \\ \text{for large } n}}{\approx} \frac{1}{n}\left(\frac{2}{3}\right)$$

- To test $\displaystyle\sum_{k=1}^{\infty} \frac{\sqrt{3k + 1}}{\sqrt{4k^2 - 2k + 1}}$, use the p-series $\displaystyle\sum_{k=1}^{\infty} \frac{1}{k^{1/2}}$, because for large n,

$$\frac{\sqrt{3n + 1}}{\sqrt{4n^2 - 2n + 1}} = \frac{\sqrt{n}\sqrt{3 + \dfrac{1}{n}}}{\sqrt{n^2}\sqrt{4 - \dfrac{2}{n} + \dfrac{1}{n^2}}} = \frac{1}{\sqrt{n}}\left(\frac{\sqrt{3 + \dfrac{1}{n}}}{\sqrt{4 - \dfrac{2}{n} + \dfrac{1}{n^2}}}\right) \underset{\substack{\uparrow \\ \text{for large } n}}{\approx} \frac{1}{n^{1/2}}\left(\frac{\sqrt{3}}{2}\right)$$

EXAMPLE 3 Using the Limit Comparison Test

Determine whether the series $\displaystyle\sum_{k=1}^{\infty} \frac{1}{2k^{3/2} + 5}$ converges or diverges.

Solution We choose an appropriate p-series to use for comparison by examining the behavior of the series for large values of n:

$$\frac{1}{2n^{3/2} + 5} = \frac{1}{n^{3/2}\left(2 + \dfrac{5}{n^{3/2}}\right)} = \frac{1}{n^{3/2}}\left(\frac{1}{2 + \dfrac{5}{n^{3/2}}}\right) \underset{\substack{\uparrow \\ \text{for large } n}}{\approx} \frac{1}{n^{3/2}}\left(\frac{1}{2}\right)$$

This leads us to choose the p-series $\displaystyle\sum_{k=1}^{\infty} \frac{1}{k^{3/2}}$, which converges, and use the Limit Comparison Test with

$$a_n = \frac{1}{2n^{3/2} + 5} \qquad \text{and} \qquad b_n = \frac{1}{n^{3/2}}$$

$$\lim_{n\to\infty} \frac{a_n}{b_n} = \lim_{n\to\infty} \frac{\dfrac{1}{2n^{3/2} + 5}}{\dfrac{1}{n^{3/2}}} = \lim_{n\to\infty} \frac{n^{3/2}}{2n^{3/2} + 5} = \lim_{n\to\infty} \frac{1}{2 + \dfrac{5}{n^{3/2}}} = \frac{1}{2}$$

Since the limit is a positive number and the p-series $\displaystyle\sum_{k=1}^{\infty} \frac{1}{k^{3/2}}$ converges, then by the Limit Comparison Test, $\displaystyle\sum_{k=1}^{\infty} \frac{1}{2k^{3/2} + 5}$ also converges. ■

NOW WORK Problem 15.

EXAMPLE 4   **Using the Limit Comparison Test**

Determine whether the series $\displaystyle\sum_{k=1}^{\infty} \frac{3\sqrt{k}+2}{\sqrt{k^3+3k^2+1}}$ converges or diverges.

Solution We choose a p-series for comparison by examining how the terms of the series behave for large values of n:

$$\frac{3\sqrt{n}+2}{\sqrt{n^3+3n^2+1}} = \frac{\sqrt{n}\left(3+\dfrac{2}{\sqrt{n}}\right)}{\sqrt{n^3}\left(\sqrt{1+\dfrac{3}{n}+\dfrac{1}{n^3}}\right)} = \frac{n^{1/2}}{n^{3/2}}\frac{3+\dfrac{2}{\sqrt{n}}}{\sqrt{1+\dfrac{3}{n}+\dfrac{1}{n^3}}} \approx \underset{\substack{\uparrow \\ \text{for large } n}}{\frac{1}{n}}(3)$$

So, we compare the series $\displaystyle\sum_{k=1}^{\infty} \frac{3\sqrt{k}+2}{\sqrt{k^3+3k^2+1}}$ to the harmonic series $\displaystyle\sum_{k=1}^{\infty} \frac{1}{k}$, which diverges, and use the Limit Comparison Test with

$$a_n = \frac{3\sqrt{n}+2}{\sqrt{n^3+3n^2+1}} \qquad \text{and} \qquad b_n = \frac{1}{n}$$

$$\lim_{n\to\infty} \frac{a_n}{b_n} = \lim_{n\to\infty} \frac{\dfrac{3\sqrt{n}+2}{\sqrt{n^3+3n^2+1}}}{\dfrac{1}{n}} = \lim_{n\to\infty} \frac{n\left(3\sqrt{n}+2\right)}{\sqrt{n^3+3n^2+1}} = \lim_{n\to\infty} \frac{3n^{3/2}+2n}{\sqrt{n^3+3n^2+1}}$$

$$= \lim_{n\to\infty} \frac{3n^{3/2}}{n^{3/2}} = 3$$

Since the limit is a positive real number and the p-series $\displaystyle\sum_{k=1}^{\infty} \frac{1}{k}$ diverges, then by the

Limit Comparison Test, $\displaystyle\sum_{k=1}^{\infty} \frac{3\sqrt{k}+2}{\sqrt{k^3+3k^2+1}}$ also diverges. ∎

NOW WORK   **Problem 17.**

Using comparison tests to determine whether a series converges or diverges requires that you know convergent and divergent series to use in the comparison. Table 3 lists some series we have already encountered that are useful for this purpose.

TABLE 3

Series	Convergent	Divergent
The geometric series $\displaystyle\sum_{k=1}^{\infty} ar^{k-1}$	$\lvert r \rvert < 1$	$\lvert r \rvert \geq 1$
The harmonic series $\displaystyle\sum_{k=1}^{\infty} \frac{1}{k}$		Divergent
The p-series $\displaystyle\sum_{k=1}^{\infty} \frac{1}{k^p}$	$p > 1$	$0 < p \leq 1$
The series $\displaystyle\sum_{k=1}^{\infty} \frac{1}{k^k}$	Convergent	

8.4 Assess Your Understanding

Concepts and Vocabulary

1. *Multiple Choice* If each term of a series $\sum\limits_{k=1}^{\infty} a_k$ of positive terms is greater than or equal to the corresponding term of a known divergent series $\sum\limits_{k=1}^{\infty} c_k$ of positive terms, then the series $\sum\limits_{k=1}^{\infty} a_k$ is [**(a)** convergent, **(b)** divergent].

2. *True or False* Suppose $\sum\limits_{k=1}^{\infty} a_k$ and $\sum\limits_{k=1}^{\infty} b_k$ are both series of positive terms. The series $\sum\limits_{k=1}^{\infty} a_k$ and the series $\sum\limits_{k=1}^{\infty} b_k$ both converge or both diverge if $\lim\limits_{n \to \infty} \dfrac{a_n}{b_n} = 0$.

3. *True or False* If $\sum\limits_{k=1}^{\infty} a_k$ and $\sum\limits_{k=1}^{\infty} b_k$ are both series of positive terms and if $\lim\limits_{n \to \infty} \dfrac{a_n}{b_n} = L$, where L is a positive real number, then the series to be tested converges.

4. *True or False* Since the p-series $\sum\limits_{k=1}^{\infty} \dfrac{1}{k^{3/2}}$ converges and $\lim\limits_{n \to \infty} \dfrac{\dfrac{1}{2n^{3/2} + 5}}{\dfrac{1}{n^{3/2}}} = \dfrac{1}{2}$, then by the Limit Comparison Test, the series $\sum\limits_{k=1}^{\infty} \dfrac{1}{2k^{3/2} + 5}$ converges to $\dfrac{1}{2}$.

Skill Building

In Problems 5–14, use the Comparison Tests for Convergence or Divergence to determine whether each series converges or diverges.

5. $\sum\limits_{k=1}^{\infty} \dfrac{1}{k(k+1)}$: by comparing it with $\sum\limits_{k=1}^{\infty} \dfrac{1}{k^2}$

6. $\sum\limits_{k=1}^{\infty} \dfrac{1}{(k+2)^2}$: by comparing it with $\sum\limits_{k=1}^{\infty} \dfrac{1}{k^2}$

7. $\sum\limits_{k=2}^{\infty} \dfrac{4^k}{7^k + 1}$: by comparing it with $\sum\limits_{k=2}^{\infty} \left(\dfrac{4}{7}\right)^k$

8. $\sum\limits_{k=1}^{\infty} \dfrac{1}{(2k-1)(2^k)}$: by comparing it with $\sum\limits_{k=1}^{\infty} \dfrac{1}{2^k}$

9. $\sum\limits_{k=2}^{\infty} \dfrac{1}{\sqrt{k(k-1)}}$: by comparing it with $\sum\limits_{k=2}^{\infty} \dfrac{1}{k}$

10. $\sum\limits_{k=2}^{\infty} \dfrac{\sqrt{k}}{k-1}$: by comparing it with $\sum\limits_{k=2}^{\infty} \dfrac{1}{\sqrt{k}}$

11. $\sum\limits_{k=1}^{\infty} \dfrac{1}{k(k+1)(k+2)}$

12. $\sum\limits_{k=1}^{\infty} \dfrac{6}{5k-2}$

13. $\sum\limits_{k=1}^{\infty} \dfrac{\sin^2 k}{k^\pi}$

14. $\sum\limits_{k=1}^{\infty} \dfrac{\cos^2 k}{k^2 + 1}$

In Problems 15–28, use the Limit Comparison Test to determine whether each series converges or diverges.

15. $\sum\limits_{k=1}^{\infty} \dfrac{1}{(k+1)(k+2)}$

16. $\sum\limits_{k=1}^{\infty} \dfrac{1}{k^2 + 1}$

17. $\sum\limits_{k=1}^{\infty} \dfrac{1}{\sqrt{k^2 + 1}}$

18. $\sum\limits_{k=1}^{\infty} \dfrac{\sqrt{k}}{k+4}$

19. $\sum\limits_{k=1}^{\infty} \dfrac{3\sqrt{k} + 2}{2k^2 + 5}$

20. $\sum\limits_{k=2}^{\infty} \dfrac{3\sqrt{k} + 2}{2k - 3}$

21. $\sum\limits_{k=2}^{\infty} \dfrac{1}{k\sqrt{k^2 - 1}}$

22. $\sum\limits_{k=1}^{\infty} \dfrac{k}{(2k-1)^2}$

23. $\sum\limits_{k=1}^{\infty} \dfrac{3k+4}{k2^k}$

24. $\sum\limits_{k=2}^{\infty} \dfrac{k-1}{k2^k}$

25. $\sum\limits_{k=1}^{\infty} \dfrac{1}{2^k + 1}$

26. $\sum\limits_{k=1}^{\infty} \dfrac{5}{3^k + 2}$

27. $\sum\limits_{k=1}^{\infty} \dfrac{k+5}{k^{k+1}}$

28. $\sum\limits_{k=1}^{\infty} \dfrac{5}{k^k + 1}$

In Problems 29–40, use any of the comparison tests to determine whether each series converges or diverges.

29. $\sum\limits_{k=1}^{\infty} \dfrac{6k}{5k^2 + 2}$

30. $\sum\limits_{k=2}^{\infty} \dfrac{6k+3}{2k^3 - 2}$

31. $\sum\limits_{k=1}^{\infty} \dfrac{7+k}{(1+k^2)^4}$

32. $\sum\limits_{k=1}^{\infty} \left(\dfrac{7+k}{1+k^2}\right)^4$

33. $\sum\limits_{k=1}^{\infty} \dfrac{e^{1/k}}{k}$

34. $\sum\limits_{k=1}^{\infty} \dfrac{1}{1+e^k}$

35. $\sum\limits_{k=1}^{\infty} \dfrac{\left(1+\dfrac{1}{k}\right)^2}{e^k}$

36. $\sum\limits_{k=1}^{\infty} \dfrac{1}{k\,2^k}$

37. $\sum\limits_{k=1}^{\infty} \dfrac{1+\sqrt{k}}{k}$

38. $\sum\limits_{k=1}^{\infty} \left(\dfrac{1+3\sqrt{k}}{k^2}\right)$

39. $\sum\limits_{k=1}^{\infty} \left(\dfrac{1}{2}\right)^k \sin^2 k$

40. $\sum\limits_{k=1}^{\infty} \dfrac{\tan^{-1} k}{k^3}$

Applications and Extensions

In Problems 41–48, determine whether each series converges or diverges.

41. $\sum\limits_{k=2}^{\infty} \dfrac{2}{k^3 \ln k}$

42. $\sum\limits_{k=2}^{\infty} \dfrac{1}{\sqrt{k}(\ln k)^4}$

43. $\sum\limits_{k=2}^{\infty} \dfrac{\ln k}{k+3}$

44. $\sum\limits_{k=2}^{\infty} \dfrac{(\ln k)^2}{k^{5/2}}$

45. $\sum\limits_{k=1}^{\infty} \sin \dfrac{1}{k}$

46. $\sum\limits_{k=1}^{\infty} \tan \dfrac{1}{k}$

47. $\sum\limits_{k=1}^{\infty} \dfrac{1}{k!}$

48. $\sum\limits_{k=1}^{\infty} \dfrac{k!}{k^k}$

---

1. = NOW WORK problem   〰 = Graphing technology recommended   CAS = Computer Algebra System recommended

49. It is known that $\displaystyle\sum_{k=1}^{\infty} \frac{1}{k^2}$ is a convergent p-series.

   (a) Use the Comparison Test for Convergence with $\displaystyle\sum_{k=1}^{\infty} \frac{1}{k^2}$

     to show that $\displaystyle\sum_{k=1}^{\infty} \frac{1}{k^2 + 1}$ converges.

   (b) Explain why the Comparison Test for Convergence and

     $\displaystyle\sum_{k=2}^{\infty} \frac{1}{k^2}$ cannot be used to show $\displaystyle\sum_{k=2}^{\infty} \frac{1}{k^2 - 1}$ converges.

   (c) Can the Limit Comparison Test be used to show $\displaystyle\sum_{k=2}^{\infty} \frac{1}{k^2 - 1}$

     converges? If it cannot, explain why.

50. Show that any series of the form $\displaystyle\sum_{k=1}^{\infty} \frac{d_k}{10^k}$, where the d_k are

digits $(0, 1, 2, \ldots, 9)$, converges.

51. Suppose the series $\displaystyle\sum_{k=1}^{\infty} a_k$ of positive terms is to be tested for

convergence or divergence, and the series $\displaystyle\sum_{k=1}^{\infty} d_k$ of positive terms

diverges. Show that if $\displaystyle\lim_{n \to \infty} \frac{a_n}{d_n} = \infty$, then $\displaystyle\sum_{k=1}^{\infty} a_k$ diverges.

52. Suppose the series $\displaystyle\sum_{k=1}^{\infty} a_k$ of positive terms is to be tested for

convergence or divergence, and the series $\displaystyle\sum_{k=1}^{\infty} d_k$ of positive terms

converges. Show that if $\displaystyle\lim_{n \to \infty} \frac{a_n}{d_n} = 0$, then $\displaystyle\sum_{k=1}^{\infty} a_k$ converges.

In Problems 53–56, use the results of Problems 51 and 52 to determine whether each of the following series converges or diverges.

53. $\displaystyle\sum_{k=2}^{\infty} \frac{1}{\ln k}$ **54.** $\displaystyle\sum_{k=2}^{\infty} \left(\frac{1}{\ln k} \right)^2$

55. $\displaystyle\sum_{k=1}^{\infty} \frac{\ln k}{k^2}$ **56.** $\displaystyle\sum_{k=2}^{\infty} \frac{1}{(k \ln k)^2}$

57. Explain why the Limit Comparison Test and the series $\displaystyle\sum_{k=1}^{\infty} \frac{1}{e^k}$,

which converges, cannot be used to determine if the series

$\displaystyle\sum_{k=2}^{\infty} \frac{1}{\ln k}$ converges or diverges.

58. (a) Show that $\displaystyle\sum_{k=1}^{\infty} \frac{\ln k}{k^p}$ converges for $p > 1$.

   (b) Show that $\displaystyle\sum_{k=1}^{\infty} \frac{(\ln k)^r}{k^p}$, where r is a positive number,

     converges for $p > 1$.

59. (a) Show that $\displaystyle\sum_{k=1}^{\infty} \frac{\ln k}{k^p}$ diverges for $0 < p \le 1$.

   (b) Show that $\displaystyle\sum_{k=1}^{\infty} \frac{(\ln k)^r}{k^p}$, where r is a positive real number

     and $0 < p \le 1$, diverges.

In Problems 60–63, use the results of Problems 58 and 59 to determine whether each of the following series converges or diverges.

60. $\displaystyle\sum_{k=1}^{\infty} \frac{\ln k}{k}$ **61.** $\displaystyle\sum_{k=1}^{\infty} \frac{\sqrt{\ln k}}{\sqrt{k}}$

62. $\displaystyle\sum_{k=2}^{\infty} \frac{\ln k}{k^3}$ **63.** $\displaystyle\sum_{k=2}^{\infty} \frac{1}{\sqrt{k^3 \ln k}}$

64. Use the Comparison Tests for Convergence or Divergence to show

that the p-series $\displaystyle\sum_{k=1}^{\infty} \frac{1}{k^p}$:

   (a) converges if $p > 1$.

   (b) diverges if $0 < p \le 1$.

65. If the series $\displaystyle\sum_{k=1}^{\infty} a_k$ of positive terms converges, show that the

series $\displaystyle\sum_{k=1}^{\infty} \frac{a_k}{k}$ also converges.

66. Show that $\displaystyle\sum_{k=1}^{\infty} \frac{1}{1 + 2^k}$ converges.

67. It is known that the harmonic series $\displaystyle\sum_{k=1}^{\infty} b_k = \sum_{k=1}^{\infty} \frac{1}{k}$ diverges.

Find two different series $\displaystyle\sum_{k=1}^{\infty} a_k$, one that converges and one that

diverges, so that $\displaystyle\lim_{n \to \infty} \frac{a_n}{b_n} = 0$. These two examples show that the

Limit Comparison Test gives no guidance if $\displaystyle\lim_{n \to \infty} \frac{a_n}{b_n} = 0$.

Challenge Problems

In Problems 68 and 69, determine whether each series converges or diverges.

68. $\displaystyle\sum_{k=2}^{\infty} \frac{\ln(2k + 1)}{\sqrt{k^2 - 2}\sqrt{k^3 - 2k - 3}}$

69. $\displaystyle\sum_{k=1}^{\infty} \frac{\sqrt{k}}{\sqrt{(k^3 - k + 1)\ln(2k + 1)}}$

70. Show that the series $\displaystyle\sum_{k=1}^{\infty} \frac{1 + \sin k}{4^k}$ converges.

71. Show that the series $\displaystyle\sum_{k=1}^{\infty} \frac{1}{k^{1+1/k}}$ diverges.

72. (a) Determine whether the series $\displaystyle\sum_{k=1}^{\infty} \frac{k^2 - 3k - 2}{k^2(k + 1)^2}$ converges or

     diverges.

   (b) If it converges, find its sum.

8.5 Alternating Series; Absolute Convergence

OBJECTIVES *When you finish this section, you should be able to:*

1 Determine whether an alternating series converges (p. 583)
2 Approximate the sum of a convergent alternating series (p. 584)
3 Determine whether a series converges (p. 586)

In Sections 8.3 and 8.4, the series have only positive terms. We now investigate series that have some positive and some negative terms. The properties of series with positive terms often help to explain the properties of series that include terms with mixed signs.

The most common series with both positive and negative terms are *alternating series,* in which the terms alternate between positive and negative.

DEFINITION Alternating Series

An **alternating series** is a series either of the form

$$\sum_{k=1}^{\infty} (-1)^{k+1} a_k = a_1 - a_2 + a_3 - a_4 + \cdots$$

or of the form

$$\sum_{k=1}^{\infty} (-1)^{k} a_k = -a_1 + a_2 - a_3 + a_4 - \cdots$$

where $a_k > 0$ for all integers $k \geq 1$.

Two examples of alternating series are

$$1 - \frac{1}{2} + \frac{1}{3} - \frac{1}{4} + \cdots = \sum_{k=1}^{\infty} \frac{(-1)^{k+1}}{k} \quad \text{and} \quad -1 + \frac{1}{3!} - \frac{1}{5!} + \frac{1}{7!} - \cdots = \sum_{k=1}^{\infty} \frac{(-1)^{k}}{(2k-1)!}$$

To determine whether an alternating series converges, we use the *Alternating Series Test.*

THEOREM Alternating Series Test

If the numbers a_k, where $a_k > 0$, of an alternating series

$$\sum_{k=1}^{\infty} (-1)^{k+1} a_k = a_1 - a_2 + a_3 - a_4 + \cdots$$

satisfy the two conditions:

- $\lim_{n \to \infty} a_n = 0$
- the a_k are nonincreasing; that is, $a_1 \geq a_2 \geq a_3 \geq \cdots \geq a_n \geq a_{n+1} \geq \cdots$

then the alternating series converges.

NOTE The Alternating Series Test is credited to Leibniz and is often called the **Leibniz Test** or **Leibniz Criterion.** The test also works if the a_k are *eventually* nonincreasing.

Proof First consider partial sums with an even number of terms, and group them into pairs:

$$S_{2n} = a_1 - a_2 + a_3 - a_4 + \cdots + a_{2n-1} - a_{2n} = (a_1 - a_2) + (a_3 - a_4) + \cdots + (a_{2n-1} - a_{2n})$$

Since $a_1 \geq a_2 \geq a_3 \geq a_4 \geq \cdots \geq a_{2n-1} \geq a_{2n} \geq a_{2n+1} \geq \cdots$, the difference within each pair of parentheses is either 0 or positive. So, the sequence $\{S_{2n}\}$ of partial sums is nondecreasing. That is,

$$S_2 \leq S_4 \leq S_6 \leq \cdots \leq S_{2n} \leq S_{2n+2} \leq \cdots \tag{1}$$

For partial sums with an odd number of terms, group the terms a little differently, and write

$$S_{2n+1} = a_1 - a_2 + a_3 - a_4 + a_5 - \cdots - a_{2n} + a_{2n+1}$$
$$= a_1 - (a_2 - a_3) - (a_4 - a_5) - \cdots - (a_{2n} - a_{2n+1})$$

Again, the difference within each pair of parentheses is either 0 or positive. But this time the difference is subtracted from the previous term, so the sequence $\{S_{2n+1}\}$ of partial sums is nonincreasing. That is,

$$S_1 \geq S_3 \geq S_5 \geq \cdots \geq S_{2n-1} \geq S_{2n+1} \geq \cdots \tag{2}$$

Now, since $S_{2n+1} - S_{2n} = a_{2n+1} > 0$, we have

$$S_{2n} < S_{2n+1} \tag{3}$$

Combine (1), (2), and (3) as follows:

$$\underbrace{S_2 \leq S_4 \leq S_6 \leq \cdots \leq S_{2n}}_{(1)} < \underbrace{S_{2n+1} \leq \cdots \leq S_5 \leq S_3 \leq S_1}_{(2)}$$
$$\underbrace{}_{(3)}$$

Figure 23 shows the sequences S_{2n} and S_{2n+1}.

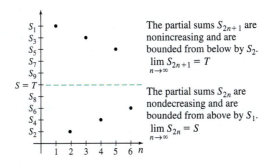

The partial sums S_{2n+1} are nonincreasing and are bounded from below by S_2.
$$\lim_{n \to \infty} S_{2n+1} = T$$

The partial sums S_{2n} are nondecreasing and are bounded from above by S_1.
$$\lim_{n \to \infty} S_{2n} = S$$

Figure 23

RECALL A bounded, monotonic sequence converges.

The sequence S_2, S_4, S_6, ... is nondecreasing and bounded above by S_1. So, this sequence has a limit; call it T. Similarly, the sequence S_1, S_3, S_5, ... is nonincreasing and is bounded from below by S_2, so it, too, has a limit; call it S. Since $S_{2n+1} - S_{2n} = a_{n+1}$, we have

$$S - T = \lim_{n \to \infty} S_{2n+1} - \lim_{n \to \infty} S_{2n} = \lim_{n \to \infty} (S_{2n+1} - S_{2n}) = \lim_{n \to \infty} a_{2n+1}$$

Since $\lim_{n \to \infty} a_{2n+1} = 0$, we find $S - T = 0$. This means that $S = T$, and the sequences $\{S_{2n}\}$ and $\{S_{2n+1}\}$ both converge to S. It follows that

$$\lim_{n \to \infty} S_n = S$$

and so the alternating series converges. ■

1 Determine Whether an Alternating Series Converges

EXAMPLE 1   Showing an Alternating Series Converges

Show that the **alternating harmonic series** $\displaystyle\sum_{k=1}^{\infty} \frac{(-1)^{k+1}}{k} = 1 - \frac{1}{2} + \frac{1}{3} - \frac{1}{4} + \cdots$ converges.

Solution   We check the two conditions of the Alternating Series Test. Since $\displaystyle\lim_{n \to \infty} a_n = \lim_{n \to \infty} \frac{1}{n} = 0$, the first condition is met. Since $a_{n+1} = \dfrac{1}{n+1} < \dfrac{1}{n} = a_n$, the a_n are nonincreasing, so the second condition is met. By the Alternating Series Test, the series converges. ■

NOTE In Section 8.9 we show that the sum of the alternating harmonic series is ln 2.

NOW WORK   Problem 7.

When using the Alternating Series Test, we check the limit of the nth term first. If $\lim\limits_{n \to \infty} a_n \neq 0$, the Test for Divergence tells us immediately that the series diverges. For example, in testing the alternating series

$$\sum_{k=2}^{\infty} (-1)^k \frac{k}{k-1} = 2 - \frac{3}{2} + \frac{4}{3} - \frac{5}{4} + \cdots$$

we find $\lim\limits_{n \to \infty} a_n = \lim\limits_{n \to \infty} \frac{n}{n-1} = 1$. There is no need to look further. By the Test for Divergence, the series diverges.

If $\lim\limits_{n \to \infty} a_n = 0$, then check whether the terms a_k are nonincreasing. Recall there are three ways to verify if the a_k are nonincreasing:

- Use the Algebraic Difference test and show that $a_{n+1} - a_n \leq 0$ for all $n \geq 1$.
- Use the Algebraic Ratio test and show that $\dfrac{a_{n+1}}{a_n} \leq 1$ for all $n \geq 1$.
- Use the derivative of the related function f, defined for $x > 0$ and for which $a_n = f(n)$ for all n, and show that $f'(x) \leq 0$ for all $x > 0$.

EXAMPLE 2 **Showing an Alternating Series Converges**

Show that the alternating series $\sum\limits_{k=0}^{\infty} \dfrac{(-1)^k}{(2k)!} = 1 - \dfrac{1}{2} + \dfrac{1}{24} - \dfrac{1}{720} + \cdots$ converges.

Solution We begin by confirming that $\lim\limits_{n \to \infty} a_n = \lim\limits_{n \to \infty} \dfrac{1}{(2n)!} = 0$. Next using the Algebraic Ratio test, we verify that the terms $a_k = \dfrac{1}{(2k)!}$ are nonincreasing. Since

$$\frac{a_{n+1}}{a_n} = \frac{\dfrac{1}{[2(n+1)]!}}{\dfrac{1}{(2n)!}} = \frac{(2n)!}{(2n+2)!} = \frac{(2n)!}{(2n+2)(2n+1)(2n)!}$$

$$= \frac{1}{(2n+2)(2n+1)} < 1 \qquad \text{for all } n \geq 1$$

the terms a_k are nonincreasing. By the Alternating Series Test, the series converges. ∎

NOW WORK Problem 13.

2 Approximate the Sum of a Convergent Alternating Series

An important property of certain convergent alternating series is that we can estimate the error made by using a partial sum S_n to approximate the sum S of the series.

THEOREM **Error Estimate for a Convergent Alternating Series**

For a convergent alternating series $\sum\limits_{k=1}^{\infty} (-1)^{k+1} a_k$ that satisfies the two conditions of the Alternating Series Test, the error E_n in using the sum S_n of the first n terms as an approximation to the sum S of the series is numerically less than or equal to the $(n+1)$st term of the series. That is,

$$\boxed{|E_n| \leq a_{n+1}}$$

Proof We follow the ideas used in the proof of the Alternating Series Test.

If n is even, $S_n < S$, and

$$0 < S - S_n = a_{n+1} - (a_{n+2} - a_{n+3}) - \cdots \leq a_{n+1}$$

If n is odd, $S_n > S,$ and
$$0 < S_n - S = a_{n+1} - (a_{n+2} - a_{n+3}) - \cdots \leq a_{n+1}$$
In either case,
$$|E_n| = |S - S_n| \leq a_{n+1} \qquad \blacksquare$$

In Example 1, we found that the alternating harmonic series converges. Table 4 shows how slowly the series converges. For example, using an additional 900 terms from 99 to 999 reduces the estimated error by only $|E_{999} - E_{99}| = 0.009$.

TABLE 4

Series	Number of Terms	Estimate of the Sum	Maximum Error
$\displaystyle\sum_{k=1}^{\infty} \frac{(-1)^{k+1}}{k}$	$n = 3$	$S_3 = 1 - \dfrac{1}{2} + \dfrac{1}{3} \approx 0.833$	$|E_3| \leq \dfrac{1}{4} = 0.25$
	$n = 9$	$S_9 = \displaystyle\sum_{k=1}^{9} \frac{(-1)^{k+1}}{k} \approx 0.746$	$|E_9| \leq \dfrac{1}{10} = 0.1$
	$n = 99$	$S_{99} = \displaystyle\sum_{k=1}^{99} \frac{(-1)^{k+1}}{k} \approx 0.7$	$|E_{99}| \leq \dfrac{1}{100} = 0.01$
	$n = 999$	$S_{999} = \displaystyle\sum_{k=1}^{999} \frac{(-1)^{k+1}}{k} \approx 0.694$	$|E_{999}| \leq \dfrac{1}{1000} = 0.001$

EXAMPLE 3   **Approximating the Sum of a Convergent Alternating Series**

Approximate the sum S of the alternating series correct to within 0.0001.
$$\sum_{k=0}^{\infty} \frac{(-1)^k}{(2k)!} = 1 - \frac{1}{2!} + \frac{1}{4!} - \frac{1}{6!} + \frac{1}{8!} - \cdots$$

Solution   We demonstrated in Example 2 that this series converges by showing that it satisfies the conditions of the Alternating Series Test. The fifth term of the series, $\dfrac{1}{8!} = \dfrac{1}{40,320} \approx 0.000025$, is the first term less than or equal to 0.0001. This term represents an upper estimate to the error when the sum S of the series is approximated by adding the first four terms. So, the sum
$$S \approx \sum_{k=0}^{3} \frac{(-1)^k}{(2k)!} = 1 - \frac{1}{2!} + \frac{1}{4!} - \frac{1}{6!} = 1 - \frac{1}{2} + \frac{1}{24} - \frac{1}{720} = \frac{389}{720} \approx 0.5403$$

is correct to within 0.0001. $\blacksquare$

NOW WORK   **Problem 19.**

Absolute and Conditional Convergence

The ideas of *absolute convergence* and *conditional convergence* are used to describe the convergence or divergence of a series $\displaystyle\sum_{k=1}^{\infty} a_k$ in which the terms a_k are sometimes positive and sometimes negative (not necessarily alternating).

> **DEFINITION** Absolute Convergence
>
> A series $\displaystyle\sum_{k=1}^{\infty} a_k$ is **absolutely convergent** if the series
> $$\sum_{k=1}^{\infty} |a_k| = |a_1| + |a_2| + \cdots + |a_n| + \cdots$$
> is convergent.

THEOREM Absolute Convergence Test

If a series $\sum\limits_{k=1}^{\infty} a_k$ is absolutely convergent, then it is convergent.

Proof For any n,

$$-|a_n| \le a_n \le |a_n|$$

Adding $|a_n|$ yields

$$0 \le a_n + |a_n| \le 2|a_n|$$

Since $\sum\limits_{k=1}^{\infty} |a_k|$ converges, then $\sum\limits_{k=1}^{\infty} (2|a_k|) = 2\sum\limits_{k=1}^{\infty} |a_k|$ also converges. By the Comparison Test for Convergence, $\sum\limits_{k=1}^{\infty} (a_k + |a_k|)$ converges. But $a_n = (a_n + |a_n|) - |a_n|$.

Since $\sum\limits_{k=1}^{\infty} a_k$ is the difference of two convergent series, it also converges. ∎

3 Determine Whether a Series Converges

EXAMPLE 4   Determining Whether a Series Converges

Determine whether the series $1 - \dfrac{1}{2} - \dfrac{1}{4} + \dfrac{1}{8} - \dfrac{1}{16} - \dfrac{1}{32} + \dfrac{1}{64} - \cdots$ converges.

Solution The series $1 - \dfrac{1}{2} - \dfrac{1}{4} + \dfrac{1}{8} - \dfrac{1}{16} - \dfrac{1}{32} + \dfrac{1}{64} - \cdots$ converges absolutely, since

$1 + \dfrac{1}{2} + \dfrac{1}{4} + \cdots + \dfrac{1}{2^{n-1}} + \cdots = \sum\limits_{k=1}^{\infty} \left(\dfrac{1}{2}\right)^{k-1}$, a geometric series with $r = \dfrac{1}{2}$, converges.

So by the Absolute Convergence Test, the series $1 - \dfrac{1}{2} - \dfrac{1}{4} + \dfrac{1}{8} - \dfrac{1}{16} - \dfrac{1}{32} + \dfrac{1}{64} - \cdots$ converges. ∎

NOW WORK Problem 31.

EXAMPLE 5   Determining Whether a Series Converges

Determine whether the series $\sum\limits_{k=1}^{\infty} \dfrac{\sin k}{k^2} = \dfrac{\sin 1}{1^2} + \dfrac{\sin 2}{2^2} + \dfrac{\sin 3}{3^2} + \cdots$ converges.

Solution This series has both positive and negative terms, but it is not an alternating series. Use the Absolute Convergence Test to investigate the series $\sum\limits_{k=1}^{\infty} \left|\dfrac{\sin k}{k^2}\right|$. Since

$$\left|\dfrac{\sin n}{n^2}\right| \le \dfrac{1}{n^2}$$

for all n, and since $\sum\limits_{k=1}^{\infty} \dfrac{1}{k^2}$ is a convergent p-series, then by the Comparison Test for Convergence, the series $\sum\limits_{k=1}^{\infty} \left|\dfrac{\sin k}{k^2}\right|$ converges. Since $\sum\limits_{k=1}^{\infty} \dfrac{\sin k}{k^2}$ is absolutely convergent, it follows that $\sum\limits_{k=1}^{\infty} \dfrac{\sin k}{k^2}$ is convergent. ∎

NOW WORK Problem 29.

The converse of the Absolute Convergence Test is not true. That is, if a series is convergent, the series may or may not be absolutely convergent. There are convergent series that are not absolutely convergent. For instance, we have shown that the alternating harmonic series

$$\sum_{k=1}^{\infty} \frac{(-1)^{k+1}}{k} = 1 - \frac{1}{2} + \frac{1}{3} - \frac{1}{4} + \frac{1}{5} - \cdots$$

IN WORDS If a series is absolutely convergent, then it is convergent. But just because a series converges does not mean it is absolutely convergent.

converges. The series of absolute values is the harmonic series $\sum_{k=1}^{\infty} \frac{1}{k}$, which is divergent.

DEFINITION Conditional Convergence

A series that is convergent without being absolutely convergent is called **conditionally convergent**.

| EXAMPLE 6 | Determining Whether a Series Is Absolutely or Conditionally Convergent or Divergent |

Determine the numbers p for which the series $\sum_{k=1}^{\infty} \frac{(-1)^{k+1}}{k^p}$ is absolutely convergent, conditionally convergent, or divergent.

Solution We begin by testing the series for absolute convergence. The series of absolute values is $\sum_{k=1}^{\infty} \left| \frac{(-1)^{k+1}}{k^p} \right| = \sum_{k=1}^{\infty} \frac{1}{k^p}$. This is a p-series, which converges if $p > 1$ and diverges if $p \leq 1$. So, $\sum_{k=1}^{\infty} \frac{(-1)^{k+1}}{k^p}$ is absolutely convergent if $p > 1$.

It remains to determine what happens when $p \leq 1$. We use the Alternating Series Test, and begin by investigating $\lim_{n \to \infty} a_n$:

- $\lim_{n \to \infty} \frac{1}{n^p} = 0 \quad 0 < p \leq 1$
- $\lim_{n \to \infty} \frac{1}{n^p} = 1 \quad p = 0$
- $\lim_{n \to \infty} \frac{1}{n^p} = \infty \quad p < 0$

Consequently, $\sum_{k=1}^{\infty} \frac{(-1)^{k+1}}{k^p}$ diverges if $p \leq 0$.

Continuing, we check the second condition of the Alternating Series Test when $0 < p \leq 1$. Using the related function $f(x) = \frac{1}{x^p}$, for $x > 0$, we have $f'(x) = -\frac{p}{x^{p+1}}$. Since $f'(x) < 0$ for $0 < p \leq 1$, the second condition of the Alternating Series Test is satisfied. We conclude that $\sum_{k=1}^{\infty} \frac{(-1)^{k+1}}{k^p}$ is conditionally convergent if $0 < p \leq 1$.

To summarize, the alternating p-series, $\sum_{k=1}^{\infty} \frac{(-1)^{k+1}}{k^p}$ is absolutely convergent if $p > 1$, conditionally convergent if $0 < p \leq 1$, and divergent if $p \leq 0$. ■

NOW WORK Problem 33.

As illustrated in Figure 24 on p. 588, a series $\sum_{k=1}^{\infty} a_k$ is either convergent or divergent. If it is convergent, it is either absolutely convergent or conditionally convergent. The flowchart in Figure 25 can be used to determine whether a series is absolutely convergent, conditionally convergent, or divergent.

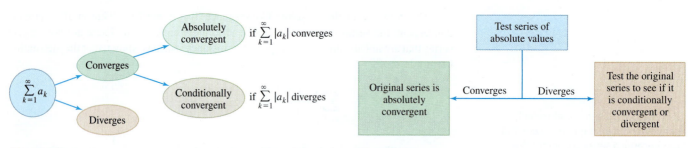

Figure 24

Figure 25

Absolute convergence of a series is stronger than conditional convergence, and absolutely convergent series and conditionally convergent series have different properties. We state, but do not prove, some of these properties now.

> **Properties of Absolutely Convergent and Conditionally Convergent Series**
>
> - If a series $\sum\limits_{k=1}^{\infty} a_k$ is absolutely convergent, any rearrangement of its terms results in a series that is also absolutely convergent. Both series will have the same sum. See Problem 66.
>
> - If two series $\sum\limits_{k=1}^{\infty} a_k$ and $\sum\limits_{k=1}^{\infty} b_k$ are absolutely convergent, the series $\sum\limits_{k=1}^{\infty} (a_k + b_k)$ is also absolutely convergent. Moreover,
>
> $$\sum_{k=1}^{\infty} (a_k + b_k) = \sum_{k=1}^{\infty} a_k + \sum_{k=1}^{\infty} b_k$$
>
> See Problem 62.
>
> - If two series $\sum\limits_{k=1}^{\infty} a_k$ and $\sum\limits_{k=1}^{\infty} b_k$ are absolutely convergent, then the series
>
> $$\sum_{k=1}^{\infty} c_k = \sum_{k=1}^{\infty} (a_1 b_k + a_2 b_{k-1} + \cdots + a_k b_1)$$
>
> is also absolutely convergent.
>
> The sum of the series $\sum\limits_{k=1}^{\infty} c_k$ equals the product of the sums of the two original series.
>
> - If a series converges absolutely, then the series consisting of just the positive terms converges, as does the series consisting of just the negative terms. See Problem 63.
>
> - If a series converges conditionally, then the series consisting of just the positive terms diverges, as does the series consisting of just the negative terms. See Problems 51, 52, and 64.
>
> - If a series $\sum\limits_{k=1}^{\infty} a_k$ is conditionally convergent, its terms may be rearranged to form a new series that converges to *any* sum we like. In fact, they may be rearranged to form a *divergent series*. See Problems 53–55.

8.5 Assess Your Understanding

Concepts and Vocabulary

1. *True or False*   The series $\sum\limits_{k=1}^{\infty} (-1)^k \cos(k\pi)$ is an alternating series.

2. *True or False* $\sum\limits_{k=1}^{\infty} [1 + (-1)^k]$ is an alternating series.

3. *True or False*   In an alternating series, $\sum\limits_{k=1}^{\infty} (-1)^k a_k$, if $\lim\limits_{n \to \infty} a_n = 0$, then the series is convergent.

---

1. = NOW WORK problem   = Graphing technology recommended   CAS = Computer Algebra System recommended

4. *True or False*   If the alternating series $\sum_{k=1}^{\infty} (-1)^{k+1} a_k$ satisfies the two conditions of the Alternating Series Test, then the error E_n in using the sum S_n of the first n terms as an approximation of the sum S of the series is $|E_n| \leq a_n$.

5. *True or False*   A series that is not absolutely convergent is divergent.

6. *True or False*   If a series is absolutely convergent, then the series converges.

Skill Building

In Problems 7–18, use the Alternating Series Test to determine whether each alternating series converges or diverges.

7. $\sum_{k=1}^{\infty} (-1)^{k+1} \dfrac{1}{k^2}$

8. $\sum_{k=1}^{\infty} (-1)^{k+1} \dfrac{1}{2\sqrt{k}}$

9. $\sum_{k=1}^{\infty} (-1)^{k+1} \dfrac{k}{2k+1}$

10. $\sum_{k=1}^{\infty} (-1)^{k+1} \dfrac{k+1}{k}$

11. $\sum_{k=1}^{\infty} (-1)^{k+1} \dfrac{k^2}{5k^2+2}$

12. $\sum_{k=1}^{\infty} (-1)^{k+1} \dfrac{k+1}{k^2}$

13. $\sum_{k=1}^{\infty} \dfrac{(-1)^{k+1}}{(k+1)2^k}$

14. $\sum_{k=2}^{\infty} (-1)^k \dfrac{1}{k \ln k}$

15. $\sum_{k=2}^{\infty} (-1)^k \dfrac{1}{1+2^{-k}}$

16. $\sum_{k=0}^{\infty} (-1)^k \dfrac{1}{k!}$

17. $\sum_{k=1}^{\infty} (-1)^{k+1} \left(\dfrac{k}{k+1} \right)^k$

18. $\sum_{k=1}^{\infty} (-1)^{k+1} \dfrac{k^2}{(k+1)^3}$

In Problems 19–26, approximate the sum of each series using the first three terms and find an upper estimate to the error in using this approximation.

19. $\sum_{k=1}^{\infty} (-1)^{k+1} \dfrac{1}{k^2}$

20. $\sum_{k=0}^{\infty} (-1)^k \dfrac{1}{k!}$

21. $\sum_{k=1}^{\infty} (-1)^{k+1} \dfrac{1}{k^4}$

22. $\sum_{k=1}^{\infty} (-1)^{k+1} \left(\dfrac{1}{\sqrt{k}} \right)^k$

23. $\sum_{k=0}^{\infty} (-1)^k \dfrac{1}{k!} \left(\dfrac{1}{3} \right)^k$

24. $\sum_{k=0}^{\infty} (-1)^k \dfrac{1}{k!} \left(\dfrac{1}{2} \right)^k$

25. $\sum_{k=0}^{\infty} (-1)^k \dfrac{1}{2k+1} \left(\dfrac{1}{3} \right)^{2k+1}$

26. $\sum_{k=1}^{\infty} (-1)^{k+1} \dfrac{1}{k^k}$

In Problems 27–42, determine whether each series is absolutely convergent, conditionally convergent, or divergent.

27. $\sum_{k=1}^{\infty} \dfrac{(-1)^{k+1}}{2k}$

28. $\sum_{k=1}^{\infty} \dfrac{(-1)^{k+1}}{3k-4}$

29. $\sum_{k=1}^{\infty} (-1)^{k+1} \dfrac{\sin k}{k^2+1}$

30. $\sum_{k=1}^{\infty} (-1)^{k+1} \dfrac{\cos k}{k^2}$

31. $\sum_{k=1}^{\infty} (-1)^{k+1} \left(\dfrac{1}{5} \right)^k$

32. $\sum_{k=1}^{\infty} (-1)^{k+1} \dfrac{5^k}{k!}$

33. $\sum_{k=1}^{\infty} (-1)^{k+1} \dfrac{e^k}{k}$

34. $\sum_{k=1}^{\infty} \dfrac{(-1)^{k+1} 2^k}{k^2}$

35. $\sum_{k=1}^{\infty} \dfrac{(-1)^{k+1}}{k(k+1)}$

36. $\sum_{k=1}^{\infty} \dfrac{(-1)^{k+1}}{k\sqrt{k+3}}$

37. $\sum_{k=1}^{\infty} \dfrac{(-1)^{k+1} \sqrt{k}}{k^2+1}$

38. $\sum_{k=1}^{\infty} \dfrac{(-1)^{k+1} \sqrt{k}}{k+1}$

39. $\sum_{k=1}^{\infty} (-1)^{k+1} \dfrac{\ln k}{k}$

40. $\sum_{k=1}^{\infty} \dfrac{(-1)^{k+1} \ln k}{k^3}$

41. $\sum_{k=1}^{\infty} (-1)^{k+1} \dfrac{1}{k \, e^k}$

42. $\sum_{k=1}^{\infty} \dfrac{(-1)^{k+1}}{e^k}$

Applications and Extensions

In Problems 43–50, determine whether each series is absolutely convergent, conditionally convergent, or divergent.

43. $\sum_{k=2}^{\infty} \dfrac{(-1)^k}{k \ln k}$

44. $\sum_{k=2}^{\infty} \dfrac{(-1)^k}{k (\ln k)^2}$

45. $\sum_{k=1}^{\infty} \dfrac{\sin k}{k^2}$

46. $\sum_{k=1}^{\infty} \dfrac{(-1)^{k+1} \tan^{-1} k}{k}$

47. $\sum_{k=1}^{\infty} \dfrac{(-1)^{k+1}}{k^{1/k}}$

48. $\sum_{k=1}^{\infty} (-1)^{k+1} \left(\dfrac{k}{k+1} \right)^k$

49. $1 - \dfrac{1}{2!} + \dfrac{1}{3!} - \dfrac{1}{4!} + \dfrac{1}{5!} - \cdots$

50. $1 - \dfrac{1}{3^2} + \dfrac{1}{5^2} - \dfrac{1}{7^2} + \cdots$

51. Show that the positive terms of $\sum_{k=1}^{\infty} \dfrac{(-1)^{k+1}}{k}$ diverge.

52. Show that the negative terms of $\sum_{k=1}^{\infty} \dfrac{(-1)^{k+1}}{k}$ diverge.

53. Show that the terms of the series $\sum_{k=1}^{\infty} \dfrac{(-1)^{k+1}}{k}$ can be rearranged so the resulting series converges to 0.

54. Show that the terms of the series $\sum_{k=1}^{\infty} \dfrac{(-1)^{k+1}}{k}$ can be rearranged so the resulting series converges to 2.

55. Show that the terms of the series $\sum_{k=1}^{\infty} \dfrac{(-1)^{k+1}}{k}$ can be rearranged so the resulting series diverges.

56. Show that the series

$$e^{-x} \cos x + e^{-2x} \cos(2x) + e^{-3x} \cos(3x) + \cdots$$

is absolutely convergent for all positive values of x.
(*Hint:* Use the fact that $|\cos \theta| \leq 1$.)

57. Determine whether the series below converges (absolutely or conditionally) or diverges.

$$1 + r \cos \theta + r^2 \cos(2\theta) + r^3 \cos(3\theta) + \cdots$$

58. What is wrong with the following argument?

$$A = 1 - \dfrac{1}{2} + \dfrac{1}{3} - \dfrac{1}{4} + \dfrac{1}{5} - \dfrac{1}{6} + \dfrac{1}{7} - \dfrac{1}{8} + \cdots$$

$$\left(\dfrac{1}{2} \right) A = \dfrac{1}{2} - \dfrac{1}{4} + \dfrac{1}{6} - \dfrac{1}{8} + \cdots$$

So,

$$A + \left(\dfrac{1}{2} \right) A = 1 + \dfrac{1}{3} - \dfrac{1}{2} + \dfrac{1}{5} + \dfrac{1}{7} - \dfrac{1}{4} + \cdots$$

The series on the right is a rearrangement of the terms of the series A. So its sum is A, meaning

$$A + \left(\frac{1}{2}\right) A = A$$
$$A = 0$$

But,

$$A = \left(1 - \frac{1}{2}\right) + \left(\frac{1}{3} - \frac{1}{4}\right) + \left(\frac{1}{5} - \frac{1}{6}\right) + \cdots > 0$$

Bernoulli's Error   *In Problems 59–61, consider an incorrect argument given by Jakob Bernoulli to prove that*

$$\sum_{k=1}^{\infty} \frac{1}{k(k+1)} = \frac{1}{2} + \frac{1}{6} + \frac{1}{12} + \cdots = 1$$

Bernoulli's argument went as follows:

Let $N = 1 + \dfrac{1}{2} + \dfrac{1}{3} + \dfrac{1}{4} + \cdots$. *Then*

$$N - 1 = \frac{1}{2} + \frac{1}{3} + \frac{1}{4} + \frac{1}{5} + \cdots.$$

Now, subtract term-by-term to get

$$N - (N - 1) = \left(1 - \frac{1}{2}\right) + \left(\frac{1}{2} - \frac{1}{3}\right) + \left(\frac{1}{3} - \frac{1}{4}\right) + \cdots \quad (4)$$

$$1 = \frac{1}{2} + \frac{1}{6} + \frac{1}{12} + \cdots$$

59. What is wrong with Bernoulli's argument?

60. In general, what can be said about the convergence or divergence of a series formed by taking the term-by-term difference (or sum) of two divergent series? Support your answer with examples.

61. Although the method is wrong. Bernoulli's conclusion is correct; that is, it is true that $\displaystyle\sum_{k=1}^{\infty} \frac{1}{k(k+1)} = 1$. Prove it! [*Hint:* Look at the partial sums using the form of the series in (4).]

62. Show that if two series $\displaystyle\sum_{k=1}^{\infty} a_k$ and $\displaystyle\sum_{k=1}^{\infty} b_k$ are absolutely convergent, the series $\displaystyle\sum_{k=1}^{\infty} (a_k + b_k)$ is also absolutely convergent.

Moreover, $\displaystyle\sum_{k=1}^{\infty} (a_k + b_k) = \sum_{k=1}^{\infty} a_k + \sum_{k=1}^{\infty} b_k$.

63. Show that if a series converges absolutely, then the series consisting of just the positive terms converges, as does the series consisting of just the negative terms.

64. Show that if a series converges conditionally, then the series consisting of just the positive terms diverges, as does the series consisting of just the negative terms.

65. Determine whether the series $\displaystyle\sum_{k=1}^{\infty} c_k$, where

$$c_k = \begin{cases} \dfrac{1}{a^k} & \text{if } k \text{ is even} \\[2mm] -\dfrac{1}{b^k} & \text{if } k \text{ is odd} \end{cases} \quad a > 1, \ b > 1,$$

converges absolutely, converges conditionally, or diverges.

66. Prove that if a series $\displaystyle\sum_{k=1}^{\infty} a_k$ is absolutely convergent, any rearrangement of its terms results in a series that is also absolutely convergent.
(*Hint:* Use the triangle inequality: $|a + b| \le |a| + |b|$.)

67. Suppose that the terms of the series $\displaystyle\sum_{k=1}^{\infty} a_k$ are alternating, $|a_{k+1}| \le |a_k|$ for all integers k, and $\displaystyle\lim_{n\to\infty} a_n = 0$. Show that the series converges.

Challenge Problems

In Problems 68–71, determine whether each series is absolutely convergent, conditionally convergent, or divergent.

68. $\displaystyle\sum_{k=1}^{\infty} \sin \frac{(-1)^k}{k}$

69. $\displaystyle\sum_{k=2}^{\infty} \frac{(-1)^k}{\sqrt[p]{k^3 + 1}}, \ p > 2$

70. $\displaystyle\sum_{k=2}^{\infty} (-1)^k k^{(1-k)/k}$

71. $\displaystyle\sum_{k=2}^{\infty} \frac{(-1)^k}{(\ln k)^{\ln k}}$

In Problems 72–74, use the result of Problem 67 to determine whether each series converges or diverges.

72. $\displaystyle\sum_{k=1}^{\infty} (-1)^{k+1} \frac{\sqrt{k}}{(k+1)}$

73. $\displaystyle\sum_{k=1}^{\infty} (-1)^{k+1} \frac{k}{(k+1)^2}$

74. $\displaystyle\sum_{k=2}^{\infty} (-1)^k \ln \frac{k+1}{k}$

75. Let $\{a_n\}$ be a sequence that is decreasing and is bounded from below by 0. Define

$$R_n = \sum_{k=n+1}^{\infty} (-1)^{k+1} a_k \quad \text{and} \quad \Delta a_k = a_k - a_{k+1}$$

Suppose that the sequence $\{\Delta a_k\}$ decreases.

(a) Show that the series $\displaystyle\sum_{k=1}^{\infty} (-1)^{k+1} \Delta a_k$ is a convergent alternating series.

(b) Show that

$$|R_n| = \frac{a_n}{2} + \frac{1}{2} \sum_{k=1}^{\infty} (-1)^k \Delta a_{k+n-1}$$

and $\displaystyle\sum_{k=1}^{\infty} (-1)^k \Delta a_{k+n-1} < 0$. Deduce $|R_n| < \frac{1}{2} a_n$.

(c) Show that

$$|R_n| = \frac{a_{n+1}}{2} + \frac{1}{2} \sum_{k=1}^{\infty} (-1)^{k+1} \Delta a_{k+n}$$

(d) Conclude that $\dfrac{a_{n+1}}{2} < |R_n|$.

Source: Based on R. Johnsonbaugh, Summing an alternating series, *American Mathematical Monthly*, Vol. 86, No. 8 (Oct. 1979), pp. 637–648.

8.6 Ratio Test; Root Test

OBJECTIVES *When you finish this section, you should be able to:*

1 Use the Ratio Test (p. 591)

2 Use the Root Test (p. 593)

One of the most practical tests for convergence of a series makes use of the ratio of two consecutive terms.

THEOREM Ratio Test

Let $\displaystyle\sum_{k=1}^{\infty} a_k$ be a series of nonzero terms.

1. Let $\displaystyle\lim_{n\to\infty}\left|\frac{a_{n+1}}{a_n}\right| = L$, a number.

- If $L < 1$, then the series $\displaystyle\sum_{k=1}^{\infty} a_k$ converges absolutely and so $\displaystyle\sum_{k=1}^{\infty} a_k$ is convergent.
- If $L = 1$, the test fails to indicate whether the series converges or diverges.
- If $L > 1$, then the series $\displaystyle\sum_{k=1}^{\infty} a_k$ diverges.

2. If $\displaystyle\lim_{n\to\infty}\left|\frac{a_{n+1}}{a_n}\right| = \infty$, then the series $\displaystyle\sum_{k=1}^{\infty} a_k$ diverges.

A proof of the Ratio Test is given at the end of the section.

1 Use the Ratio Test

EXAMPLE 1   Using the Ratio Test

Use the Ratio Test to determine whether each series converges or diverges.

(a) $\displaystyle\sum_{k=1}^{\infty}\frac{k}{4^k}$ **(b)** $\displaystyle\sum_{k=1}^{\infty}\frac{2^k}{k}$ **(c)** $\displaystyle\sum_{k=1}^{\infty}\frac{3k+1}{k^2}$ **(d)** $\displaystyle\sum_{k=1}^{\infty}\frac{1}{k!}$

Solution   (a) $\displaystyle\sum_{k=1}^{\infty}\frac{k}{4^k}$ is a series of nonzero terms; $a_{n+1} = \dfrac{n+1}{4^{n+1}}$ and $a_n = \dfrac{n}{4^n}$. The absolute value of their ratio is

$$\left|\frac{a_{n+1}}{a_n}\right| = \frac{\dfrac{n+1}{4^{n+1}}}{\dfrac{n}{4^n}} = \frac{n+1}{4^{n+1}} \cdot \frac{4^n}{n} = \frac{n+1}{4n}$$

Then

$$\lim_{n\to\infty}\left|\frac{a_{n+1}}{a_n}\right| = \lim_{n\to\infty}\frac{n+1}{4n} = \frac{1}{4} < 1$$

Since the limit is less than 1, the series converges.

(b) $\displaystyle\sum_{k=1}^{\infty}\frac{2^k}{k}$ is a series of nonzero terms; $a_{n+1} = \dfrac{2^{n+1}}{n+1}$ and $a_n = \dfrac{2^n}{n}$. The absolute value of their ratio is

$$\left|\frac{a_{n+1}}{a_n}\right| = \frac{2^{n+1}}{n+1} \cdot \frac{n}{2^n} = \frac{2n}{n+1}$$

Then

$$\lim_{n\to\infty}\left|\frac{a_{n+1}}{a_n}\right| = \lim_{n\to\infty}\frac{2n}{n+1} = 2 > 1$$

Since the limit is greater than 1, the series diverges.

(c) $\displaystyle\sum_{k=1}^{\infty} \frac{3k+1}{k^2}$ is a series of nonzero terms; $a_{n+1} = \dfrac{3(n+1)+1}{(n+1)^2} = \dfrac{3n+4}{(n+1)^2}$ and

$a_n = \dfrac{3n+1}{n^2}$. The absolute value of their ratio is

$$\left|\frac{a_{n+1}}{a_n}\right| = \left[\frac{3n+4}{(n+1)^2}\right]\left(\frac{n^2}{3n+1}\right) = \left(\frac{3n+4}{3n+1}\right)\left(\frac{n^2}{n^2+2n+1}\right) = \frac{3n^3+4n^2}{3n^3+7n^2+5n+1}$$

Then

$$\lim_{n\to\infty}\left|\frac{a_{n+1}}{a_n}\right| = \lim_{n\to\infty}\frac{3n^3+4n^2}{3n^3+7n^2+5n+1} = 1$$

The Ratio Test gives no information about this series. Another test must be used.

$\left(\text{You can show that the series diverges by comparing it to the harmonic series }\displaystyle\sum_{k=1}^{\infty}\frac{1}{k}.\right)$

(d) $\displaystyle\sum_{k=1}^{\infty}\frac{1}{k!}$ is a series of nonzero terms; $a_{n+1} = \dfrac{1}{(n+1)!}$ and $a_n = \dfrac{1}{n!}$. The absolute

value of their ratio is

$$\left|\frac{a_{n+1}}{a_n}\right| = \frac{n!}{(n+1)!} = \frac{1}{n+1}$$

Then

$$\lim_{n\to\infty}\left|\frac{a_{n+1}}{a_n}\right| = \lim_{n\to\infty}\frac{1}{n+1} = 0$$

NOTE In Section 8.9, we show that $\displaystyle\sum_{k=1}^{\infty}\frac{1}{k!} = e$.

Since the limit is less than 1, the series $\displaystyle\sum_{k=1}^{\infty}\frac{1}{k!}$ converges. ∎

NOW WORK Problem 7.

As Example 1 illustrates, the Ratio Test is useful in determining whether a series containing factorials and/or powers converges or diverges.

EXAMPLE 2 **Using the Ratio Test**

Use the Ratio Test to determine whether the series $\displaystyle\sum_{k=1}^{\infty}\frac{k!}{k^k}$ converges or diverges.

Solution $\displaystyle\sum_{k=1}^{\infty}\frac{k!}{k^k}$ is a series of nonzero terms. Since $a_{n+1} = \dfrac{(n+1)!}{(n+1)^{n+1}}$ and $a_n = \dfrac{n!}{n^n}$,

the absolute value of their ratio is

$$\left|\frac{a_{n+1}}{a_n}\right| = \frac{\dfrac{(n+1)!}{(n+1)^{n+1}}}{\dfrac{n!}{n^n}} = \frac{(n+1)!}{(n+1)^{n+1}}\cdot\frac{n^n}{n!} = \frac{n^n}{(n+1)^n}$$

$$= \left(\frac{n}{n+1}\right)^n = \left(\frac{1}{1+\dfrac{1}{n}}\right)^n = \frac{1}{\left(1+\dfrac{1}{n}\right)^n}$$

So,

$$\lim_{n\to\infty}\left|\frac{a_{n+1}}{a_n}\right| = \lim_{n\to\infty}\frac{1}{\left(1+\dfrac{1}{n}\right)^n} = \frac{\displaystyle\lim_{n\to\infty} 1}{\displaystyle\lim_{n\to\infty}\left(1+\dfrac{1}{n}\right)^n} = \frac{1}{e}$$

NEED TO REVIEW? The number e expressed as a limit is discussed in Section 3.3, pp. 227–228.

Since $\dfrac{1}{e} < 1$, the series converges. ∎

NOW WORK Problem 15.

We conclude the discussion of the Ratio Test with these observations:

- To test $\displaystyle\sum_{k=1}^{\infty} a_k$ for convergence, it is important to check whether the *limit of the ratio* $\left| \dfrac{a_{n+1}}{a_n} \right|$, not the ratio itself, is less than 1.

  For example, for the harmonic series $\displaystyle\sum_{k=1}^{\infty} \dfrac{1}{k}$, which diverges,

  the ratio $\left| \dfrac{a_{n+1}}{a_n} \right| = \left| \dfrac{n}{n+1} \right| < 1$, but $\displaystyle\lim_{n \to \infty} \left| \dfrac{a_{n+1}}{a_n} \right| = \lim_{n \to \infty} \dfrac{n}{n+1} = 1.$

- For divergence, it is sufficient to show that the ratio $\left| \dfrac{a_{n+1}}{a_n} \right| > 1$ for all n.

- The Ratio Test is not conclusive when $\displaystyle\lim_{n \to \infty} \left| \dfrac{a_{n+1}}{a_n} \right| = 1$. It is also not conclusive when $\displaystyle\lim_{n \to \infty} \left| \dfrac{a_{n+1}}{a_n} \right| \neq \infty$ does not exist.

- If the general term a_n of an infinite series involves n, either exponentially or factorially, the Ratio Test often answers the question of convergence or divergence.

The *Root Test* works well for series of nonzero terms whose nth term involves an nth power.

THEOREM Root Test

Let $\displaystyle\sum_{k=1}^{\infty} a_k$ be a series of nonzero terms. Suppose $\displaystyle\lim_{n \to \infty} \sqrt[n]{|a_n|} = L$, a number.

- If $L < 1$, then $\displaystyle\sum_{k=1}^{\infty} a_k$ is absolutely convergent, so the series $\displaystyle\sum_{k=1}^{\infty} a_k$ converges.

- If $L > 1$, then $\displaystyle\sum_{k=1}^{\infty} a_k$ diverges.

- If $L = 1$, the test is inconclusive.

The proof of the Root Test is similar to the proof of the Ratio Test. It is left as an exercise (Problem 55).

2 Use the Root Test

EXAMPLE 3 Using the Root Test

Use the Root Test to determine whether the series $\displaystyle\sum_{k=1}^{\infty} \dfrac{e^k}{k^k}$ converges or diverges.

Solution $\displaystyle\sum_{k=1}^{\infty} \dfrac{e^k}{k^k}$ is a series of nonzero terms. The nth term is $a_n = \dfrac{e^n}{n^n} = \left(\dfrac{e}{n} \right)^n$. Since a_n involves an nth power, we use the Root Test.

$$\lim_{n \to \infty} \sqrt[n]{|a_n|} = \lim_{n \to \infty} \sqrt[n]{\left(\dfrac{e}{n} \right)^n} = \lim_{n \to \infty} \dfrac{e}{n} = 0 < 1$$

The series $\displaystyle\sum_{k=1}^{\infty} \dfrac{e^k}{k^k}$ converges. ∎

NOW WORK Problem 27.

EXAMPLE 4 **Using the Root Test**

Use the Root Test to determine whether the series $\displaystyle\sum_{k=1}^{\infty}\left(\frac{8k+3}{5k-2}\right)^{k}$ converges or diverges.

Solution $\displaystyle\sum_{k=1}^{\infty}\left(\frac{8k+3}{5k-2}\right)^{k}$ is a series of nonzero terms. The nth term is $a_n = \left(\frac{8n+3}{5n-2}\right)^{n}$.

Since a_n involves an nth power, we use the Root Test.

$$\lim_{n\to\infty}\sqrt[n]{|a_n|} = \lim_{n\to\infty}\sqrt[n]{\left(\frac{8n+3}{5n-2}\right)^{n}} = \lim_{n\to\infty}\frac{8n+3}{5n-2} = \frac{8}{5} > 1$$

The series diverges. ∎

NOW WORK **Problem 23.**

Proof of the Ratio Test Let $\displaystyle\sum_{k=1}^{\infty} a_k$ be a series of nonzero terms.

Case 1: $\displaystyle\lim_{n\to\infty}\left|\frac{a_{n+1}}{a_n}\right| = L$, a number.

- $0 \le L < 1$ Let r be any number for which $L < r < 1$. Since $\displaystyle\lim_{n\to\infty}\left|\frac{a_{n+1}}{a_n}\right| = L$

  and $L < r$, then by the definition of the limit of a sequence, we can find a

  number N so that for any number $n > N$, the ratio $\left|\frac{a_{n+1}}{a_n}\right|$ can be made as close as

  we please to L and be less than r. Then

$$\left|\frac{a_{N+1}}{a_N}\right| < r \qquad \text{or} \qquad |a_{N+1}| < r \cdot |a_N|$$

$$\left|\frac{a_{N+2}}{a_{N+1}}\right| < r \qquad \text{or} \qquad |a_{N+2}| < r \cdot |a_{N+1}| < r^2 \cdot |a_N|$$

$$\left|\frac{a_{N+3}}{a_{N+2}}\right| < r \qquad \text{or} \qquad |a_{N+3}| < r \cdot |a_{N+2}| < r^3 \cdot |a_N|$$

  Each term of the series $|a_{N+1}| + |a_{N+2}| + \cdots$ is less than the corresponding term
  of the geometric series $|a_N|r + |a_N|r^2 + |a_N|r^3 + \cdots$. Since $|r| < 1$, the
  geometric series converges. By the Comparison Test for Convergence, the series
  $|a_{N+1}| + |a_{N+2}| + \cdots$ also converges. So, the series $\displaystyle\sum_{k=1}^{\infty}|a_k|$ converges. By the
  Absolute Convergence Test, the series $\displaystyle\sum_{k=1}^{\infty} a_k$ converges.

- $L = 1$ To show that the test fails for $L = 1$, we exhibit two series, one that
  diverges and another that converges, to show that no conclusion can be drawn.
  Consider $\displaystyle\sum_{k=1}^{\infty}\frac{1}{k}$ and $\displaystyle\sum_{k=1}^{\infty}\frac{1}{k^2}$. The first is the harmonic series, which diverges. The
  second is a p-series with $p > 1$, which converges. It is left to you to show that
  $\displaystyle\lim_{n\to\infty}\left|\frac{a_{n+1}}{a_n}\right| = 1$ in each case. (See Problems 45 and 46.)

- $L > 1$ Let r be any number for which $1 < r < L$. Since $\displaystyle\lim_{n\to\infty}\left|\frac{a_{n+1}}{a_n}\right| = L$,

  there is a number N so that for any number $n > N$, the ratio $\left|\frac{a_{n+1}}{a_n}\right|$ can be made as close as we please to L and will be greater than r. That is, for all numbers $n > N$, the ratio $\left| \dfrac{a_{n+1}}{a_n} \right| > r > 1$ so that $|a_{n+1}| > |a_n|$. After the Nth term, the absolute value of the terms are positive and increasing. So, $\lim_{n \to \infty} |a_n| \neq 0$ and, therefore, $\lim_{n \to \infty} a_n \neq 0$. By the Test for Divergence, the series diverges.

Case 2: $\lim_{n \to \infty} \left| \dfrac{a_{n+1}}{a_n} \right| = \infty$ The proof that this series diverges is left as an exercise (Problem 54). ∎

8.6 Assess Your Understanding

Concepts and Vocabulary

1. *True or False* The Ratio Test can be used to show that the series $\sum_{k=1}^{\infty} \cos(k\pi)$ diverges.

2. *True or False* In using the Ratio Test, if $\lim_{n \to \infty} \left| \dfrac{a_{n+1}}{a_n} \right| = L$, then the sum of the series $\sum_{k=1}^{\infty} a_k$ equals L.

3. *True or False* In using the Ratio Test, if $\lim_{n \to \infty} \left| \dfrac{a_{n+1}}{a_n} \right| = 1$, then the Ratio Test indicates that the series $\sum_{k=1}^{\infty} a_k$ converges.

4. *True or False* The Root Test works well if the nth term of a series of nonzero terms involves an nth root.

Skill Building

In Problems 5–22, use the Ratio Test to determine whether each series converges or diverges.

5. $\sum_{k=1}^{\infty} \dfrac{4k^2 - 1}{2^k}$

6. $\sum_{k=1}^{\infty} \dfrac{1}{(2k+1)2^k}$

7. $\sum_{k=1}^{\infty} k \left(\dfrac{2}{3} \right)^k$

8. $\sum_{k=1}^{\infty} \dfrac{5^k}{k^2}$

9. $\sum_{k=1}^{\infty} \dfrac{10^k}{(2k)!}$

10. $\sum_{k=1}^{\infty} \dfrac{(2k)!}{5^k 3^{k-1}}$

11. $\sum_{k=1}^{\infty} \dfrac{k}{(2k-2)!}$

12. $\sum_{k=1}^{\infty} \dfrac{(k+1)!}{3^k}$

13. $\sum_{k=1}^{\infty} \dfrac{2^k}{k(k+1)}$

14. $\sum_{k=1}^{\infty} \dfrac{k!}{k^2(k+1)^2}$

15. $\sum_{k=1}^{\infty} \dfrac{k^3}{k!}$

16. $\sum_{k=1}^{\infty} \dfrac{k!}{k^{k+1}}$

17. $\sum_{k=1}^{\infty} \dfrac{3^{k-1}}{k \cdot 2^k}$

18. $\sum_{k=1}^{\infty} \dfrac{k(k+2)}{3^k}$

19. $\sum_{k=1}^{\infty} \dfrac{k}{e^k}$

20. $\sum_{k=1}^{\infty} \dfrac{e^k}{k^3}$

21. $\sum_{k=1}^{\infty} k \cdot 2^k$

22. $\sum_{k=1}^{\infty} \dfrac{4^k}{k}$

In Problems 23–34, use the Root Test to determine whether each series converges or diverges.

23. $\sum_{k=1}^{\infty} \left(\dfrac{2k+1}{5k+1} \right)^k$

24. $\sum_{k=1}^{\infty} \left(\dfrac{3k-1}{2k+1} \right)^k$

25. $\sum_{k=1}^{\infty} \left(\dfrac{k}{5} \right)^k$

26. $\sum_{k=1}^{\infty} \dfrac{\pi^{2k}}{k^k}$

27. $\sum_{k=2}^{\infty} \left(\dfrac{\ln k}{k} \right)^k$

28. $\sum_{k=2}^{\infty} \left(\dfrac{1}{\ln k} \right)^k$

29. $\sum_{k=1}^{\infty} \left(\dfrac{\sqrt{k^2+1}}{3k} \right)^k$

30. $\sum_{k=1}^{\infty} \left(\dfrac{\sqrt{4k^2+1}}{k} \right)^k$

31. $\sum_{k=1}^{\infty} \dfrac{k^2}{2^k}$

32. $\sum_{k=1}^{\infty} \dfrac{k^3}{3^k}$

33. $\sum_{k=1}^{\infty} \dfrac{k^4}{5^k}$

34. $\sum_{k=1}^{\infty} \dfrac{k}{3^k}$

In Problems 35–44, determine whether each series converges or diverges.

35. $\sum_{k=1}^{\infty} \dfrac{10}{(3k+1)^k}$

36. $\sum_{k=1}^{\infty} \left(1 + \dfrac{1}{k} \right)^{k^2}$

37. $\sum_{k=1}^{\infty} \dfrac{(k+1)(k+2)}{k!}$

38. $\sum_{k=1}^{\infty} \dfrac{k!}{(3k+1)!}$

39. $\sum_{k=1}^{\infty} \dfrac{k \ln k}{2^k}$

40. $\sum_{k=1}^{\infty} \left[\ln \left(e^3 + \dfrac{1}{k} \right) \right]^k$

41. $\sum_{k=1}^{\infty} \sin^k \left(\dfrac{1}{k} \right)$

42. $\sum_{k=1}^{\infty} \dfrac{k^k}{2^{k^2}}$

43. $\sum_{k=1}^{\infty} \dfrac{\left(1 + \dfrac{1}{k} \right)^{2k}}{e^k}$

44. $\sum_{k=2}^{\infty} \dfrac{2^k(k+1)}{k^2(k+2)}$

Applications and Extensions

45. For the divergent series $\sum_{k=1}^{\infty} \dfrac{1}{k}$, show that $\lim_{n \to \infty} \left| \dfrac{a_{n+1}}{a_n} \right| = 1$.

1. = NOW WORK problem   📈 = Graphing technology recommended   CAS = Computer Algebra System recommended

46. For the convergent series $\sum_{k=1}^{\infty} \frac{1}{k^2}$, show that $\lim\limits_{n \to \infty} \left| \frac{a_{n+1}}{a_n} \right| = 1$.

47. Give an example of a convergent series $\sum_{k=1}^{\infty} a_k$ for which

$$\lim_{n \to \infty} \left| \frac{a_{n+1}}{a_n} \right| \neq \infty \text{ does not exist.}$$

48. Give an example of a divergent series $\sum_{k=1}^{\infty} a_k$ for which

$$\lim_{n \to \infty} \left| \frac{a_{n+1}}{a_n} \right| \neq \infty \text{ does not exist.}$$

49. (a) Show that the series $\sum_{k=1}^{\infty} \frac{(-1)^k 3^k}{k!}$ converges.

CAS (b) Use technology to find the sum of the series.

50. Determine whether the following series is convergent or divergent:

$$\frac{1}{3} - \frac{2^3}{3^2} + \frac{3^3}{3^3} - \frac{4^3}{3^4} + \cdots + \frac{(-1)^{n-1} n^3}{3^n} + \cdots$$

51. Show that $\lim\limits_{n \to \infty} \frac{n!}{n^n} = 0$, where n denotes a positive integer.

52. Show that the Root Test is inconclusive for $\sum_{k=1}^{\infty} \frac{1}{k}$ and $\sum_{k=1}^{\infty} \frac{1}{k^2}$.

53. Use the Ratio Test to find the real numbers x for which the series $\sum_{k=1}^{\infty} \frac{x^k}{k^2}$ converges or diverges.

54. Prove that $\sum_{k=1}^{\infty} a_k$ diverges if $\lim\limits_{n \to \infty} \left| \frac{a_{n+1}}{a_n} \right| = \infty$.

55. Prove the Root Test.

56. The terms of the series

$$\frac{1}{4} + \frac{1}{2} + \frac{1}{8} + \frac{1}{4} + \frac{1}{16} + \frac{1}{8} + \frac{1}{32} + \cdots$$

are $a_{2k} = \frac{1}{2^k}$ and $a_{2k-1} = \frac{1}{2^{k+1}}$.

(a) Show that using the Ratio Test to determine whether the series converges is inconclusive.

(b) Show that using the Root Test to determine whether the series converges is conclusive.

(c) Does the series converge?

Challenge Problems ─────────

57. Show that the following series converges:

$$1 + \frac{2}{2^2} + \frac{3}{3^3} + \frac{1}{4^4} + \frac{2}{5^5} + \frac{3}{6^6} + \cdots$$

58. Show that $\sum_{k=1}^{\infty} \frac{(k+1)^2}{(k+2)!}$ converges. $\left(\textit{Hint: Use the Limit}\right.$

Comparison Test and the convergent series $\sum_{k=1}^{\infty} \frac{1}{k!} . \bigg)$

59. Suppose $0 < a < b < 1$. Use the Root Test to show that the series

$$a + b + a^2 + b^2 + a^3 + b^3 + \cdots$$

converges.

60. Show that if the Ratio Test indicates a series converges, then so will the Root Test. The converse is not true. Refer to Problem 56.

8.7 Summary of Tests

OBJECTIVES *When you finish this section, you should be able to:*

1 Choose an appropriate test to determine whether a series converges (p. 596)

In the previous sections, we have discussed a variety of tests that can be used to determine if a series converges or diverges. In the exercises following each section, the instructions indicate which test to use. Unfortunately, in practice, series do not come with instructions. This section summarizes the tests that we have discussed and gives some clues as to what test has the best chance of answering the fundamental question, 'Does the series converge or diverge?'

1 Choose an Appropriate Test to Determine Whether a Series Converges

The following outline is a guide to help you choose a test to use when determining the convergence or divergence of a series. Table 5 lists the tests we have discussed. Table 6 describes important series we have analyzed.

Guide to Choosing a Test to Determine Whether a Series Is Convergent

1. Check to see if the series is a geometric series or a p-series. If yes, then use the conclusion given for these series in Table 6.

2. Find $\lim_{n \to \infty} a_n$ of the series $\sum_{k=1}^{\infty} a_k$. If $\lim_{n \to \infty} a_n \neq 0$, then by the Test for Divergence, the series diverges.

3. If the series $\sum_{k=1}^{\infty} a_k$ has only positive terms and meets the conditions of the Integral Test, find the related function f. Use the integral test if $\int_1^{\infty} f(x)\, dx$ is easy to find.

4. If the series $\sum_{k=1}^{\infty} a_k$ has only positive terms and the nth term is a quotient of sums or differences of powers of n, the Limit Comparison Test with an appropriate p-series will usually work.

5. If the series $\sum_{k=1}^{\infty} a_k$ has only positive terms and the preceding attempts fail, then try the Comparison Test for Convergence or the Comparison Test for Divergence.

6. *Series with some negative terms.*

   - For an alternating series, use the Alternating Series Test. It is sometimes better to use the Absolute Convergence Test first.

   - For other series containing negative terms, always use the Absolute Convergence Test first.

7. If the series $\sum_{k=1}^{\infty} a_k$ has nonzero terms that involve products, factorials, or powers, the Ratio Test is a good choice.

8. If the series $\sum_{k=1}^{\infty} a_k$ has nonzero terms and the nth term involves an nth power, try the Root Test.

TABLE 5 Tests for Convergence and Divergence of Series

Test Name	Description	Comment
Test for Divergence for all series (p. 566)	$\sum_{k=1}^{\infty} a_k$ diverges if $\lim_{n \to \infty} a_n \neq 0$.	No information is obtained about convergence if $\lim_{n \to \infty} a_n = 0$.
Integral Test (p. 569) for series of positive terms	$\sum_{k=1}^{\infty} a_k$ converges (diverges) if $\int_1^{\infty} f(x)\, dx$ converges (diverges), where f is continuous, positive, and decreasing for $x \geq 1$; and $f(k) = a_k$ for all k.	Good to use if f is easy to integrate.
Comparison Test for Convergence for series of positive terms (p. 576)	$\sum_{k=1}^{\infty} a_k$ converges if $0 < a_k \leq b_k$ and the series $\sum_{k=1}^{\infty} b_k$ converges.	$\sum_{k=1}^{\infty} b_k$ must have positive terms and be convergent.
Comparison Test for Divergence for series of positive terms (p. 576)	$\sum_{k=1}^{\infty} a_k$ diverges if $a_k \geq c_k > 0$ and the series $\sum_{k=1}^{\infty} c_k$ diverges.	$\sum_{k=1}^{\infty} c_k$ must have positive terms and be divergent.

Continued

TABLE 5 (*Continued*)

Test Name	Description	Comment
Limit Comparison Test (p. 577) for series of positive terms	$\sum_{k=1}^{\infty} a_k$ converges (diverges) if $\sum_{k=1}^{\infty} b_k$ converges (diverges), and $\lim_{n \to \infty} \dfrac{a_n}{b_n} = L$, a positive real number.	$\sum_{k=1}^{\infty} b_k$ must have positive terms, whose convergence (divergence) can be determined.
Alternating Series Test (p. 582)	$\sum_{k=1}^{\infty} (-1)^{k+1} a_k, \; a_k > 0$, converges if • $\lim_{n \to \infty} a_n = 0$ and • the a_k are nonincreasing.	The error made by using the nth partial sum as an approximation to the sum S of the series is less than the $(n+1)$st term of the series.
Absolute Convergence Test (p. 586)	If $\sum_{k=1}^{\infty} |a_k|$ converges, then $\sum_{k=1}^{\infty} a_k$ converges.	The converse is not true. That is, if $\sum_{k=1}^{\infty} |a_k|$ diverges, $\sum_{k=1}^{\infty} a_k$ may converge.
Ratio Test (p. 591) for series with nonzero terms	$\sum_{k=1}^{\infty} a_k$ converges if $\lim_{n \to \infty} \left| \dfrac{a_{n+1}}{a_n} \right| < 1$. $\sum_{k=1}^{\infty} a_k$ diverges if $\lim_{n \to \infty} \left| \dfrac{a_{n+1}}{a_n} \right| > 1$ or if $\lim_{n \to \infty} \left| \dfrac{a_{n+1}}{a_n} \right| = \infty$.	Good to use if a_n includes factorials or powers. It provides no information if $\lim_{n \to \infty} \left| \dfrac{a_{n+1}}{a_n} \right| = 1$.
Root Test (p. 593) for series with nonzero terms	$\sum_{k=1}^{\infty} a_k$ converges if $\lim_{n \to \infty} \sqrt[n]{|a_n|} < 1$. $\sum_{k=1}^{\infty} a_k$ diverges if $\lim_{n \to \infty} \sqrt[n]{|a_n|} > 1$ or if $\lim_{n \to \infty} \sqrt[n]{|a_n|} = \infty$.	Good to use if a_n involves nth powers. It provides no information if $\lim_{n \to \infty} \sqrt[n]{|a_n|} = 1$.

TABLE 6 Important Series

Series Name	Series Description	Comments
Geometric series (p. 557)	$\sum_{k=1}^{\infty} ar^{k-1} = a + ar + ar^2 + \cdots, \; a \neq 0$	Converges to $\dfrac{a}{1-r}$ if $|r| < 1$; diverges if $|r| \geq 1$.
Harmonic series (p. 561)	$\sum_{k=1}^{\infty} \dfrac{1}{k} = 1 + \dfrac{1}{2} + \dfrac{1}{3} + \cdots$	Diverges.
p-series (p. 570)	$\sum_{k=1}^{\infty} \dfrac{1}{k^p} = 1 + \dfrac{1}{2^p} + \dfrac{1}{3^p} + \cdots$	Converges if $p > 1$; diverges if $0 < p \leq 1$.
k-to-the-k series (p. 576)	$\sum_{k=1}^{\infty} \dfrac{1}{k^k} = 1 + \dfrac{1}{2^2} + \dfrac{1}{3^3} + \dfrac{1}{4^4} + \cdots$	Converges.
Factorial series (p. 591)	$\sum_{k=0}^{\infty} \dfrac{1}{k!} = 1 + 1 + \dfrac{1}{2} + \dfrac{1}{6} + \dfrac{1}{24} + \cdots$	Converges.
Alternating harmonic series (p. 583)	$\sum_{k=1}^{\infty} \dfrac{(-1)^{k+1}}{k} = 1 - \dfrac{1}{2} + \dfrac{1}{3} - \dfrac{1}{4} + \cdots$	Converges.

8.7 Assess Your Understanding

Concepts and Vocabulary

1. *True or False*   The series $\sum\limits_{k=1}^{\infty} \dfrac{1}{k^p}$ converges if $p \geq 1$.

2. *True or False*   According to the Test for Divergence, an infinite series $\sum\limits_{k=1}^{\infty} a_k$ converges if $\lim\limits_{n \to \infty} a_n = 0$.

3. *True or False*   If a series is absolutely convergent, then it is convergent.

4. *True or False*   If a series is not absolutely convergent, then it is divergent.

5. *True or False*   According to the Ratio Test, a series $\sum\limits_{k=1}^{\infty} a_k$ of nonzero terms converges if $\left| \dfrac{a_{n+1}}{a_n} \right| < 1$.

6. To use the Comparison Test for Convergence to show that a series $\sum\limits_{k=1}^{\infty} a_k$ of positive terms converges, find a series $\sum\limits_{k=1}^{\infty} b_k$ that is known to converge and show that $0 < \underline{\quad} \leq \underline{\quad}$.

Skill Building

In Problems 7–39, determine whether each series converges (absolutely or conditionally) or diverges. Use any applicable test.

7. $\sum\limits_{k=1}^{\infty} \dfrac{9k^3 + 5k^2}{k^{5/2} + 4}$

8. $\sum\limits_{k=1}^{\infty} \dfrac{(-1)^{k+1}}{\sqrt{2k+1}}$

9. $6 + 2 + \dfrac{2}{3} + \dfrac{2}{9} + \dfrac{2}{27} + \cdots$

10. $\sum\limits_{k=1}^{\infty} \dfrac{1}{k^2} \sin \dfrac{\pi}{k}$

11. $\sum\limits_{k=1}^{\infty} \dfrac{3k+2}{k^3+1}$

12. $1 + \dfrac{2^2+1}{2^3+1} + \dfrac{3^2+1}{3^3+1} + \dfrac{4^2+1}{4^3+1} + \cdots$

13. $\sum\limits_{k=1}^{\infty} \dfrac{k+4}{k\sqrt{3k-2}}$

14. $\sum\limits_{k=1}^{\infty} \dfrac{\sin k}{k^3}$

15. $\sum\limits_{k=1}^{\infty} \dfrac{3^{2k-1}}{k^2+2k}$

16. $\sum\limits_{k=1}^{\infty} \dfrac{5^k}{k!}$

17. $\sum\limits_{k=1}^{\infty} \left(1 + \dfrac{2}{k}\right)^k$

18. $\sum\limits_{k=1}^{\infty} \dfrac{k^2+4}{e^k}$

19. $\dfrac{2}{3} - \dfrac{3}{4} \cdot \dfrac{1}{2} + \dfrac{4}{5} \cdot \dfrac{1}{3} - \dfrac{5}{6} \cdot \dfrac{1}{4} + \cdots$

20. $2 + \dfrac{3}{2} \cdot \dfrac{1}{4} + \dfrac{4}{3} \cdot \dfrac{1}{4^2} + \dfrac{5}{4} \cdot \dfrac{1}{4^3} + \cdots$

21. $1 + \dfrac{1 \cdot 3}{2!} + \dfrac{1 \cdot 3 \cdot 5}{3!} + \dfrac{1 \cdot 3 \cdot 5 \cdot 7}{4!} + \cdots$

22. $\dfrac{1}{\sqrt{1 \cdot 2 \cdot 3}} + \dfrac{1}{\sqrt{2 \cdot 3 \cdot 4}} + \dfrac{1}{\sqrt{3 \cdot 4 \cdot 5}} + \cdots$

23. $\sum\limits_{k=1}^{\infty} \dfrac{k!}{(2k)!}$

24. $\sum\limits_{k=1}^{\infty} k^3 e^{-k^4}$

25. $\sum\limits_{k=1}^{\infty} \dfrac{1}{\sqrt{k}+100}$

26. $\sum\limits_{k=1}^{\infty} \dfrac{k^2+5k}{3+5k^2}$

27. $\sum\limits_{k=1}^{\infty} \dfrac{1}{\sqrt[3]{k^4+4}}$

28. $\sum\limits_{k=1}^{\infty} \dfrac{1}{11} \left(\dfrac{-3}{2}\right)^k$

29. $\dfrac{1}{3} - \dfrac{2}{4} + \dfrac{3}{5} - \dfrac{4}{6} + \cdots$

30. $\sum\limits_{k=1}^{\infty} \dfrac{k(-4)^{3k}}{5^k}$

31. $\sum\limits_{k=1}^{\infty} \left(-\dfrac{1}{k}\right)^k$

32. $\sum\limits_{k=1}^{\infty} \dfrac{5}{2^k+1}$

33. $\sum\limits_{k=1}^{\infty} e^{-k^2}$

34. $\dfrac{\sin \sqrt{1}}{1^{3/2}} + \dfrac{\sin \sqrt{2}}{2^{3/2}} + \dfrac{\sin \sqrt{3}}{3^{3/2}} + \cdots$

35. $\sum\limits_{k=2}^{\infty} \dfrac{(-1)^{k-1}}{k(\ln k)^3}$

36. $\sum\limits_{k=1}^{\infty} \dfrac{1}{(2k)^k}$

37. $\sum\limits_{k=2}^{\infty} \left(\dfrac{\ln k}{1000}\right)^k$

38. $\sum\limits_{k=1}^{\infty} \dfrac{1}{\cosh^2 k}$

39. $\sum\limits_{k=1}^{\infty} \dfrac{\tan^{-1} k}{k^2}$

In Problems 40–42, determine whether each series converges or diverges. If it converges, find its sum.

40. $\sum\limits_{k=1}^{\infty} \left(\sqrt{k+1} - \sqrt{k}\right)$

41. $\sum\limits_{k=4}^{\infty} \left(\dfrac{1}{k-3} - \dfrac{1}{k}\right)$

42. $\sum\limits_{k=2}^{\infty} \ln \dfrac{k}{k+1}$

43. Determine whether $1 + \dfrac{1 \cdot 2}{1 \cdot 3} + \dfrac{1 \cdot 2 \cdot 3}{1 \cdot 3 \cdot 5} + \dfrac{1 \cdot 2 \cdot 3 \cdot 4}{1 \cdot 3 \cdot 5 \cdot 7} + \cdots$ converges or diverges.

44. (a) Show that the series $\sum\limits_{k=1}^{\infty} \left[\left(\dfrac{2}{3}\right)^k - \dfrac{2}{k^2+2k}\right]$ converges.

    (b) Find the sum of the series.

45. (a) Show that the series $\sum\limits_{k=1}^{\infty} \left[\left(-\dfrac{1}{4}\right)^k + \dfrac{3}{k(k+1)}\right]$ converges.

    (b) Find the sum of the series.

Challenge Problems

46. (a) Determine whether the series
$$1 - 1 - \dfrac{1}{2} + \dfrac{1}{3} + \dfrac{1}{3} - \dfrac{1}{9} - \dfrac{1}{4} + \dfrac{1}{27} + \dfrac{1}{5} - \dfrac{1}{81} - \cdots$$
converges or diverges.

    (b) Find the sum of the series if it converges.

In Problems 47 and 48, determine whether each series converges or diverges.

47. $\sum\limits_{k=1}^{\infty} \dfrac{\ln k}{2k^3 - 1}$

48. $\sum\limits_{k=1}^{\infty} \sin^3 \left(\dfrac{1}{k}\right)$

---

1. = NOW WORK problem   = Graphing technology recommended   CAS = Computer Algebra System recommended

THEOREM Convergence/Divergence of a Power Series

(a) If the power series $\displaystyle\sum_{k=0}^{\infty} a_k x^k$ converges for a number $x_0 \neq 0$, then it converges absolutely for all numbers x for which $|x| < |x_0|$.

(b) If the power series $\displaystyle\sum_{k=0}^{\infty} a_k x^k$ diverges for a number x_1, then it diverges for all numbers x for which $|x| > |x_1|$.

> **RECALL** If a series $\displaystyle\sum_{k=0}^{\infty} a_k$ converges, then $\displaystyle\lim_{n \to \infty} a_n = 0$.

Proof Part (a) Assume that $\displaystyle\sum_{k=0}^{\infty} a_k x_0^k$ converges. Then

$$\lim_{n \to \infty} \left(a_n x_0^n \right) = 0$$

Using the definition of the limit of a sequence and choosing $\varepsilon = 1$, there is a positive integer N for which $|a_n x_0^n| < 1$ for all $n > N$. Now for any number x for which $|x| < |x_0|$, we have

$$|a_n x^n| = \left| \frac{a_n x^n x_0^n}{x_0^n} \right| = |a_n x_0^n| \left| \frac{x}{x_0} \right|^n < \left| \frac{x}{x_0} \right|^n \qquad \text{for } n > N$$

$$\underset{\uparrow}{\qquad\qquad\qquad\qquad\quad} |a_n x_0^n| < 1$$

The series $\displaystyle\sum_{k=0}^{\infty} \left| \frac{x}{x_0} \right|^k$ is a convergent geometric series since $r = \left| \frac{x}{x_0} \right| < 1$ ($|x| < |x_0|$).

Therefore, by the Comparison Test for Convergence, the series $\displaystyle\sum_{k=0}^{\infty} \left| a_k x^k \right|$ converges, and so the power series $\displaystyle\sum_{k=0}^{\infty} a_k x^k$ converges absolutely for all numbers x such that $|x| < |x_0|$.

Part (b) Suppose the series converges for some number x, $|x| > |x_1|$. Then it must converge for x_1 [by Part (a)], which contradicts the hypothesis of the theorem. Therefore, the series diverges for all x such that $|x| > |x_1|$. ∎

The next result is a consequence of the previous theorem. It states that every power series belongs to one of three categories.

THEOREM

For a power series $\displaystyle\sum_{k=0}^{\infty} a_k (x - c)^k$, exactly one of the following is true:

- The series converges only if $x = c$.
- The series converges absolutely for all x.
- There is a positive number R for which the series converges absolutely for all x, $|x - c| < R$, and diverges for all x, $|x - c| > R$. The behavior of the series at $|x - c| = R$ must be determined separately.

In the theorem, the number R is called the **radius of convergence.** If the series converges only for $x = c$, then $R = 0$; if the series converges absolutely for all x, then $R = \infty$. If the series converges absolutely for $|x - c| < R$, $0 < R < \infty$, we call the set of all numbers x for which the power series converges the **interval of convergence** of the power series. Once the radius R of convergence is determined, we test the endpoints $x = c - R$ and $x = c + R$ to find the interval of convergence.

As Example 1 illustrates, the Ratio Test is a useful method for determining the radius of convergence of a power series. However, the test gives no information about convergence or divergence at the endpoints of the interval of convergence. At an endpoint, a power series may be absolutely convergent, conditionally convergent, or divergent.

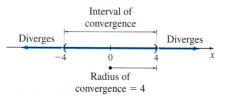

Interval of convergence

Diverges ← −4 ⟨ 0 4 ⟩ → Diverges x

Radius of convergence $= 4$

Figure 26

For example, the series $\displaystyle\sum_{k=0}^{\infty} \frac{k x^k}{4^k}$ in Example 1(b) converges absolutely for $|x| < 4$ and diverges for $|x| \geq 4$. So, the radius of convergence is $R = 4$, and the interval of convergence is the open interval $(-4, 4)$, as shown in Figure 26.

2 Find the Interval of Convergence of a Power Series

EXAMPLE 2   **Finding the Interval of Convergence of a Power Series**

Find the radius of convergence and the interval of convergence of the power series

$$\sum_{k=1}^{\infty} \frac{x^{2k}}{k}$$

Solution We use the Ratio Test with $a_n = \dfrac{x^{2n}}{n}$ and $a_{n+1} = \dfrac{x^{2(n+1)}}{n+1}$. Then

$$\lim_{n\to\infty}\left|\frac{a_{n+1}}{a_n}\right| = \lim_{n\to\infty}\left|\frac{\dfrac{x^{2(n+1)}}{n+1}}{\dfrac{x^{2n}}{n}}\right| = \lim_{n\to\infty}\left|\frac{x^{2n+2}}{n+1}\cdot\frac{n}{x^{2n}}\right| = x^2 \lim_{n\to\infty}\frac{n}{n+1} = x^2$$

The series converges absolutely if $x^2 < 1$, or equivalently, if $-1 < x < 1$.

To determine the behavior at the endpoints, we investigate $x = -1$ and $x = 1$. When $x = 1$ or $x = -1$, $\dfrac{x^{2k}}{k} = \dfrac{(x^2)^k}{k} = \dfrac{1^k}{k} = \dfrac{1}{k}$, so the series reduces to the harmonic series $\sum_{k=1}^{\infty}\dfrac{1}{k}$, which diverges. Consequently, the radius of convergence is $R = 1$, and the interval of convergence is $-1 < x < 1$, as shown in Figure 27. ∎

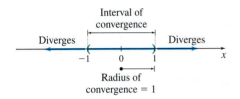

Interval of convergence

Diverges ——— Diverges

−1   0   1

x

Radius of convergence = 1

Figure 27

NOW WORK **Problem 19(a).**

EXAMPLE 3   **Finding the Interval of Convergence of a Power Series**

Find the radius of convergence R and the interval of convergence of the power series

$$\sum_{k=0}^{\infty}\frac{x^k}{(k+2)^{2k}}$$

Solution We use the Root Test. Then

$$\lim_{n\to\infty}\sqrt[n]{\left|\frac{x^n}{(n+2)^{2n}}\right|} = \lim_{n\to\infty}\frac{|x|}{(n+2)^2} = |x| \lim_{n\to\infty}\frac{1}{(n+2)^2} = 0$$

The series converges absolutely for all x. The radius of convergence is $R = \infty$, and the interval of convergence is $(-\infty, \infty)$. ∎

NOW WORK **Problems 19(b) and 21.**

EXAMPLE 4   **Finding the Interval of Convergence of a Power Series**

Find the radius of convergence R and the interval of convergence of the power series

$$\sum_{k=0}^{\infty}(-1)^k\frac{(x-2)^k}{k+1}$$

Solution $\sum_{k=0}^{\infty}(-1)^k\dfrac{(x-2)^k}{k+1}$ is a power series centered at 2. We use the Ratio Test with

$$a_n = (-1)^n\frac{(x-2)^n}{n+1} \text{ and } a_{n+1} = (-1)^{n+1}\frac{(x-2)^{n+1}}{n+2}. \text{ Then}$$

$$\lim_{n\to\infty}\left|\frac{a_{n+1}}{a_n}\right| = \lim_{n\to\infty}\left|\frac{\dfrac{(-1)^{n+1}(x-2)^{n+1}}{n+2}}{\dfrac{(-1)^n(x-2)^n}{n+1}}\right| = \lim_{n\to\infty}\left|\frac{(n+1)(x-2)}{n+2}\right|$$

$$= |x-2|\lim_{n\to\infty}\frac{n+1}{n+2} = |x-2|$$

The series converges absolutely if $|x - 2| < 1$, or equivalently if $1 < x < 3$. The radius of convergence is $R = 1$. We check the endpoints $x = 1$ and $x = 3$ separately.

If $x = 1$,

$$\sum_{k=0}^{\infty}(-1)^k \frac{(x-2)^k}{k+1} = \sum_{k=0}^{\infty}(-1)^k \frac{(-1)^k}{k+1} = \sum_{k=0}^{\infty}\frac{(-1)^{2k}}{k+1} = \sum_{k=0}^{\infty}\frac{1}{k+1}$$

$$= 1 + \frac{1}{2} + \frac{1}{3} + \cdots + \frac{1}{n+1} + \cdots$$

which is the divergent harmonic series.

If $x = 3$,

$$\sum_{k=0}^{\infty}(-1)^k \frac{(x-2)^k}{k+1} = \sum_{k=0}^{\infty}(-1)^k \frac{1}{k+1} = 1 - \frac{1}{2} + \frac{1}{3} - \frac{1}{4} + \cdots + \frac{(-1)^n}{n+1} + \cdots$$

which is the convergent alternating harmonic series.

The series $\sum_{k=0}^{\infty}(-1)^k \frac{(x-2)^k}{k+1}$ converges for $1 < x \le 3$, as shown in Figure 28. ∎

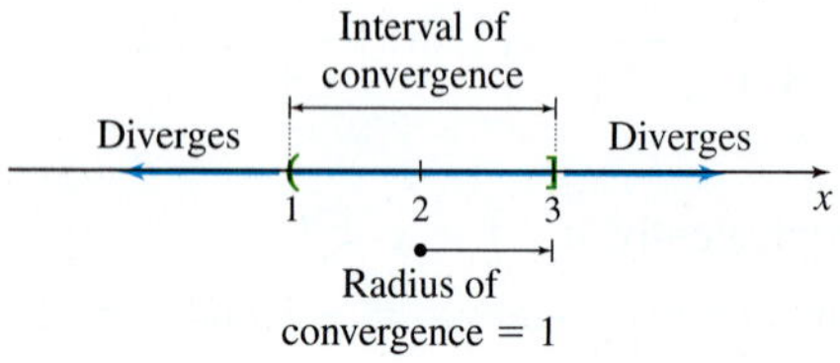

Interval of convergence

Diverges   Diverges

Radius of convergence = 1

Figure 28

NOW WORK   Problem 31.

3 Define a Function Using a Power Series

A power series $\sum_{k=0}^{\infty} a_k x^k$ defines a function f whose domain is the interval of convergence of the power series. If I is the interval of convergence of $\sum_{k=0}^{\infty} a_k x^k$, the function f defined by the power series $\sum_{k=0}^{\infty} a_k x^k$ is

$$\boxed{f(x) = a_0 + a_1 x + a_2 x^2 + \cdots + a_n x^n + \cdots}$$

The domain of f is the interval of convergence I.

If f is defined by the power series $\sum_{k=0}^{\infty} a_k x^k$, whose interval of convergence is I, and if x_0 is a number in I, then f can be evaluated at x_0 by finding the sum of the series

$$f(x_0) = \sum_{k=0}^{\infty} a_k x_0^k = a_0 + a_1 x_0 + a_2 x_0^2 + \cdots + a_n x_0^n + \cdots$$

EXAMPLE 5   Analyzing a Function Defined by a Power Series

A function f is defined by the power series $f(x) = \sum_{k=0}^{\infty} x^k$.

(a) Find the domain of f.

(b) Evaluate $f\left(\frac{1}{2}\right)$ and $f\left(-\frac{1}{3}\right)$.

(c) Find f.

Solution   (a) $\sum_{k=0}^{\infty} x^k$ is a power series centered at 0 with $a_k = 1$. Then

$$f(x) = 1 + x + x^2 + x^3 + x^4 + \cdots$$

The domain of f equals the interval of convergence of the power series. Since the series $\sum_{k=0}^{\infty} x^k$ is a geometric series, it converges for $|x| < 1$. The radius of convergence is 1, and the interval of convergence is $(-1, 1)$. The domain of f is the open interval $(-1, 1)$.

(b) The numbers $\dfrac{1}{2}$ and $-\dfrac{1}{3}$ are in the interval $(-1, 1)$, so they are in the domain of f.

Then $f\left(\dfrac{1}{2}\right)$ is a geometric series with $r = \dfrac{1}{2}$, $a = 1$, and

$$f\left(\frac{1}{2}\right) = 1 + \frac{1}{2} + \left(\frac{1}{2}\right)^2 + \left(\frac{1}{2}\right)^3 + \cdots = \frac{a}{1-r} = \frac{1}{1 - \dfrac{1}{2}} = 2$$

Similarly,

$$f\left(-\frac{1}{3}\right) = 1 - \frac{1}{3} + \left(-\frac{1}{3}\right)^2 + \left(-\frac{1}{3}\right)^3 + \cdots = \frac{1}{1 + \dfrac{1}{3}} = \frac{3}{4}$$

(c) Since f is defined by a geometric series, we can find f by summing the series.

$$f(x) = \sum_{k=0}^{\infty} x^k = 1 + x + x^2 + \cdots + x^n + \cdots = \underset{\substack{\uparrow \\ a=1;\, r=x}}{\frac{1}{1-x}} \quad -1 < x < 1 \qquad \blacksquare$$

In Example 5, the function f defined by the power series $\displaystyle\sum_{k=0}^{\infty} x^k$ was found to be

$$f(x) = \frac{1}{1-x} = \sum_{k=0}^{\infty} x^k, \quad -1 < x < 1.$$ In this case, we say that the function f is **represented** by the power series.

NOW WORK **Problem 45.**

EXAMPLE 6   **Representing a Function by a Power Series Centered at 0**

Represent each of the following functions by a power series centered at 0:

(a) $h(x) = \dfrac{1}{1 - 2x^2}$ **(b)** $g(x) = \dfrac{1}{3 + x}$ **(c)** $F(x) = \dfrac{x^2}{1 - x}$

Solution We use the function $f(x) = \dfrac{1}{1-x}, -1 < x < 1$, represented by the geometric series $\displaystyle\sum_{k=0}^{\infty} x^k$.

(a) In the geometric series for $f(x) = \dfrac{1}{1-x}$, we replace x by $2x^2$. This series

converges if $|2x^2| < 1$, or equivalently if $-\dfrac{\sqrt{2}}{2} < x < \dfrac{\sqrt{2}}{2}$. Then on the open interval

$\left(-\dfrac{\sqrt{2}}{2}, \dfrac{\sqrt{2}}{2}\right)$, the function $h(x) = \dfrac{1}{1 - 2x^2}$ is represented by the power series

$$h(x) = \frac{1}{1 - 2x^2} = 1 + (2x^2) + (2x^2)^2 + (2x^2)^3 + \cdots$$

$$= 1 + 2x^2 + 4x^4 + 8x^6 + \cdots + 2^n x^{2n} + \cdots = \sum_{k=0}^{\infty} (2x^2)^k = \sum_{k=0}^{\infty} 2^k x^{2k}$$

(b) We begin by writing

$$g(x) = \frac{1}{3+x} = \frac{1}{3}\left(\frac{1}{1 + \dfrac{x}{3}}\right) = \frac{1}{3}\left[\frac{1}{1 - \left(-\dfrac{x}{3}\right)}\right]$$

Now in the geometric series for $f(x) = \dfrac{1}{1-x}$, replace x by $-\dfrac{x}{3}$. This series converges if $\left| -\dfrac{x}{3} \right| < 1$, or equivalently if $-3 < x < 3$. Then in the open interval $(-3, 3)$,

$g(x) = \dfrac{1}{3+x}$ is represented by the power series

$$g(x) = \frac{1}{3+x} = \frac{1}{3}\left[1 + \left(-\frac{x}{3}\right) + \left(-\frac{x}{3}\right)^2 + \left(-\frac{x}{3}\right)^3 + \cdots \right] = \frac{1}{3}\sum_{k=0}^{\infty}(-1)^k \left(\frac{x}{3}\right)^k$$

$$= \sum_{k=0}^{\infty} \frac{(-1)^k x^k}{3^{k+1}}$$

(c) $F(x) = \dfrac{x^2}{1-x} = x^2\left(\dfrac{1}{1-x}\right)$. Now for all numbers in the interval $(-1, 1)$,

$$\frac{1}{1-x} = 1 + x + x^2 + \cdots + x^n + \cdots$$

So for any number x in the interval of convergence, $-1 < x < 1$, we have

$$F(x) = x^2\left(1 + x + x^2 + \cdots + x^n + \cdots\right) = x^2 + x^3 + x^4 + \cdots + x^{n+2} + \cdots = \sum_{k=2}^{\infty} x^k \;\blacksquare$$

NOTE Notice that the power series for F begins at $k = 2$.

NOW WORK **Problem 55.**

4 Use Properties of Power Series

The function f represented by a power series has properties similar to those of a polynomial. We state three of these properties without proof.

THEOREM Properties of Power Series

Let $\displaystyle\sum_{k=0}^{\infty} a_k x^k$ be a power series in x having a nonzero radius of convergence R. Define the function f as

$$f(x) = \sum_{k=0}^{\infty} a_k x^k = a_0 + a_1 x + a_2 x^2 + \cdots + a_n x^n + \cdots \quad -R < x < R$$

- *Continuity property:*

$$\lim_{x \to x_0}\left(\sum_{k=0}^{\infty} a_k x^k\right) = \sum_{k=0}^{\infty}\left(\lim_{x \to x_0} a_k x^k\right) = \sum_{k=0}^{\infty} a_k x_0^k \quad -R < x_0 < R$$

- *Differentiation property:*

$$\frac{d}{dx}\left(\sum_{k=0}^{\infty} a_k x^k\right) = \sum_{k=0}^{\infty}\left(\frac{d}{dx} a_k x^k\right) = \sum_{k=1}^{\infty} k a_k x^{k-1}$$

- *Integration property:*

$$\int_0^x \left(\sum_{k=0}^{\infty} a_k t^k\right) dt = \sum_{k=0}^{\infty}\left(\int_0^x a_k t^k \, dt\right) = \sum_{k=0}^{\infty} \frac{a_k x^{k+1}}{k+1}$$

The differentiation and integration properties of power series state that a power series can be differentiated and integrated term-by-term and that the resulting series represent the derivative and integral, respectively, of the function represented by the original power series. Moreover, it can be shown that the power series obtained by differentiating (or

CAUTION The theorem states that the *radii of convergence* of the three series $\sum\limits_{k=0}^{\infty} a_k x^k$, $\sum\limits_{k=1}^{\infty} ka_k x^{k-1}$, and $\sum\limits_{k=0}^{\infty} \dfrac{a_k x^{k+1}}{k+1}$ are the same. This does not imply that the *intervals of convergence* are the same; the endpoints of the interval must be checked separately for each series. For example, the interval of convergence of $\sum\limits_{k=0}^{\infty} \dfrac{x^k}{k}$ is $[-1, 1)$, but the interval of convergence of its derivative $\sum\limits_{k=1}^{\infty} x^{k-1}$ is $(-1, 1)$.

integrating) a power series whose radius of convergence is R, converges and has the same radius of convergence R as the original power series. (See Problem 86.)

The differentiation and integration properties can be used to obtain new functions defined by a power series.

EXAMPLE 7   **Using the Differentiation Property of Power Series**

Use the differentiation property of power series to find the derivative of

$$f(x) = \frac{1}{1-x} = \sum_{k=0}^{\infty} x^k$$

Solution   The function $f(x) = \dfrac{1}{1-x}$, defined on the open interval $(-1, 1)$, is represented by the power series

$$f(x) = \frac{1}{1-x} = 1 + x + x^2 + \cdots + x^n + \cdots = \sum_{k=0}^{\infty} x^k$$

Using the differentiation property, we find that

$$f'(x) = \frac{1}{(1-x)^2} = 1 + 2x + 3x^2 + \cdots + nx^{n-1} + \cdots = \sum_{k=1}^{\infty} kx^{k-1}$$

whose radius of convergence is 1. ∎

NOW WORK   Problem 59(a).

EXAMPLE 8   **Finding the Power Series Representation for $\ln \dfrac{1}{1-x}$**

(a) Find the power series representation for $\ln \dfrac{1}{1-x}$.

(b) Find $\ln 2$.

Solution   (a) If $y = \ln \dfrac{1}{1-x}$, then $y' = \dfrac{1}{1-x}$. That is, y' is represented by the geometric series $\sum\limits_{k=0}^{\infty} x^k$, which converges on the interval $(-1, 1)$. So, if we use the integration property of power series for $y' = \dfrac{1}{1-x}$, we obtain a series for $y = \ln \dfrac{1}{1-x}$.

$$y' = \frac{1}{1-x} = 1 + x + x^2 + \cdots + x^n + \cdots = \sum_{k=0}^{\infty} x^k$$

$$\int_0^x \frac{1}{1-t}\, dt = \int_0^x (1 + t + t^2 + \cdots + t^n + \cdots)\, dt$$

$$\ln \frac{1}{1-x} = x + \frac{x^2}{2} + \frac{x^3}{3} + \cdots + \frac{x^{n+1}}{n+1} + \cdots = \sum_{k=0}^{\infty} \frac{x^{k+1}}{k+1}$$

The radius of convergence of this series is 1.

To find the interval of convergence, we investigate the endpoints. When $x = 1$,

$$x + \frac{x^2}{2} + \frac{x^3}{3} + \cdots + \frac{x^{n+1}}{n+1} + \cdots = 1 + \frac{1}{2} + \frac{1}{3} + \cdots$$

the harmonic series, which diverges. When $x = -1$,

$$x + \frac{x^2}{2} + \frac{x^3}{3} + \cdots + \frac{x^{n+1}}{n+1} + \cdots = -1 + \frac{1}{2} - \frac{1}{3} + \cdots + (-1)^{n+1}\frac{1}{n+1} + \cdots$$

an alternating harmonic series, which converges. The interval of convergence of the power series $\displaystyle\sum_{k=1}^{\infty}\frac{x^k}{k}$ is $[-1, 1)$. So,

$$\ln\frac{1}{1-x} = x + \frac{x^2}{2} + \frac{x^3}{3} + \cdots = \sum_{k=0}^{\infty}\frac{x^{k+1}}{k+1} \qquad -1 \le x < 1 \qquad (1)$$

(b) To find $\ln 2$, notice that when $x = -1$, we have $\ln\dfrac{1}{1-x} = \ln\dfrac{1}{2} = -\ln 2$. In (1) let $x = -1$. Then

$$-\ln 2 = -1 + \frac{1}{2} - \frac{1}{3} + \cdots$$

$$\ln 2 = 1 - \frac{1}{2} + \frac{1}{3} \cdots \qquad \blacksquare$$

The sum of the alternating harmonic series is $\ln 2$.

NOW WORK   Problems 59(b) and 65.

EXAMPLE 9   Finding the Power Series Representation for $\tan^{-1}x$

Show that the power series representation for $\tan^{-1}x$ is

$$\tan^{-1}x = x - \frac{x^3}{3} + \frac{x^5}{5} - \frac{x^7}{7} + \cdots + (-1)^n\frac{x^{2n+1}}{2n+1} + \cdots = \sum_{k=0}^{\infty}\frac{(-1)^k x^{2k+1}}{2k+1}$$

Find the radius of convergence and the interval of convergence.

ORIGINS The power series representation for $\tan^{-1}x$ is called **Gregory's series** (or Gregory–Leibniz series or Madhava–Gregory series). James Gregory (1638–1675) was a Scottish mathematician. His mother, Janet Anderson, was his teacher and taught him geometry. Gregory, like many other mathematicians of his time, was searching for a good way to approximate π, a result of which was Gregory's series. Gregory was also a major contributor to the theory of optics, and he is credited with inventing the reflective telescope.

Solution If $y = \tan^{-1}x$, then $y' = \dfrac{1}{1+x^2}$, which is the sum of the geometric series $\displaystyle\sum_{k=0}^{\infty}(-1)^k x^{2k}$. That is,

$$\frac{1}{1+x^2} = \sum_{k=0}^{\infty}(-1)^k x^{2k} = 1 - x^2 + x^4 - x^6 + \cdots$$

This series converges when $|x^2| < 1$, or equivalently for $-1 < x < 1$. We use the integration property of power series to integrate $y' = \dfrac{1}{1+x^2}$ and obtain a series for $y = \tan^{-1}x$.

$$\int_0^x \frac{dt}{1+t^2} = \int_0^x (1 - t^2 + t^4 - \cdots)\,dt$$

$$\tan^{-1}x = x - \frac{x^3}{3} + \frac{x^5}{5} - \frac{x^7}{7} + \cdots + (-1)^n\frac{x^{2n+1}}{2n+1} + \cdots = \sum_{k=0}^{\infty}(-1)^k\frac{x^{2k+1}}{2k+1}$$

The radius of convergence is 1. To find the interval of convergence, we check $x = -1$ and $x = 1$. For $x = -1$,

$$-1 + \frac{1}{3} - \frac{1}{5} + \frac{1}{7} - \cdots$$

For $x = 1$, we get

$$1 - \frac{1}{3} + \frac{1}{5} - \frac{1}{7} + \cdots$$

Both of these series satisfy the two conditions of the Alternating Series Test, and so each one converges. The interval of convergence is the closed interval $[-1, 1]$. $\blacksquare$

Since $\tan^{-1} 1 = \dfrac{\pi}{4}$, we can use Gregory's series to approximate π. Then

$$\frac{\pi}{4} = \sum_{k=0}^{\infty} (-1)^k \frac{1^{2k+1}}{2k+1} = \sum_{k=0}^{\infty} \frac{(-1)^k}{2k+1} = 1 - \frac{1}{3} + \frac{1}{5} - \cdots$$

While we now have an approximation for π, unfortunately the series converges very slowly, requiring many terms to get close to π. See Problem 82.

NOW WORK Problem 69.

8.8 Assess Your Understanding

Concepts and Vocabulary

1. *True or False* Every power series $\displaystyle\sum_{k=0}^{\infty} a_k(x-c)^k$ converges for at least one number.

2. *True or False* Let b_n denote the nth term of the power series $\displaystyle\sum_{k=0}^{\infty} a_k x^k$. If $\displaystyle\lim_{n \to \infty} \left| \frac{b_{n+1}}{b_n} \right| < 1$ for every number x, then $\displaystyle\sum_{k=0}^{\infty} a_k x^k$ is absolutely convergent on the interval $(-\infty, \infty)$.

3. *True or False* If the radius of convergence of a power series $\displaystyle\sum_{k=0}^{\infty} a_k x^k$ is 0, then the power series converges only for $x = 0$.

4. *True or False* If a power series converges at one endpoint of its interval of convergence, then it must converge at its other endpoint.

5. *True or False* The power series $\displaystyle\sum_{k=0}^{\infty} a_k x^k$ and $\displaystyle\sum_{k=0}^{\infty} a_k(x-3)^k$ have the same radius of convergence.

6. *True or False* The power series $\displaystyle\sum_{k=0}^{\infty} a_k x^k$ and $\displaystyle\sum_{k=0}^{\infty} a_k(x-3)^k$ have the same interval of convergence.

7. *True or False* If the power series $\displaystyle\sum_{k=0}^{\infty} a_k x^k$ converges for $x = 8$, then it converges for $x = -8$.

8. *True or False* If the power series $\displaystyle\sum_{k=0}^{\infty} a_k x^k$ converges for $x = 3$, then it converges for $x = 1$.

9. *True or False* If the power series $\displaystyle\sum_{k=0}^{\infty} a_k x^k$ converges for $x = -4$, then it converges for $x = 3$.

10. *True or False* If the power series $\displaystyle\sum_{k=0}^{\infty} a_k x^k$ converges for $x = 3$, then it diverges for $x = 5$.

11. *True or False* A possible interval of convergence for the power series $\displaystyle\sum_{k=0}^{\infty} a_k x^k$ is $[-2, 4]$.

12. *True or False* If the power series $\displaystyle\sum_{k=0}^{\infty} a_k x^k$ diverges for a number x_1, then it converges for all numbers x for which $|x| < |x_1|$.

Skill Building

In Problems 13–16, find all numbers x for which each power series converges.

13. $\displaystyle\sum_{k=0}^{\infty} k x^k$

14. $\displaystyle\sum_{k=0}^{\infty} \frac{k x^k}{3^k}$

15. $\displaystyle\sum_{k=0}^{\infty} \frac{(x+1)^k}{3^k}$

16. $\displaystyle\sum_{k=1}^{\infty} \frac{(x-2)^k}{k^2}$

In Problems 17–26:

(a) Use the Ratio Test to find the radius of convergence and the interval of convergence of each power series.

(b) Use the Root Test to find the radius of convergence and the interval of convergence of each power series.

(c) Which test, the Ratio Test or the Root Test, did you find easier to use? Give the reasons why.

17. $\displaystyle\sum_{k=0}^{\infty} \frac{x^k}{2^k(k+1)}$

18. $\displaystyle\sum_{k=0}^{\infty} (-1)^k \frac{x^k}{2^k(k+1)}$

19. $\displaystyle\sum_{k=0}^{\infty} \frac{x^k}{k+5}$

20. $\displaystyle\sum_{k=0}^{\infty} \frac{x^k}{1+k^2}$

21. $\displaystyle\sum_{k=0}^{\infty} \frac{k^2 x^k}{3^k}$

22. $\displaystyle\sum_{k=0}^{\infty} \frac{2^k x^k}{3^k}$

23. $\displaystyle\sum_{k=0}^{\infty} \frac{k x^k}{2k+1}$

24. $\displaystyle\sum_{k=0}^{\infty} (6x)^k$

25. $\displaystyle\sum_{k=0}^{\infty} (x-3)^k$

26. $\displaystyle\sum_{k=0}^{\infty} \frac{k(2x)^k}{3^k}$

In Problems 27–44, find the radius of convergence and the interval of convergence of each power series.

27. $\displaystyle\sum_{k=1}^{\infty} \frac{x^k}{k^3}$

28. $\displaystyle\sum_{k=2}^{\infty} \frac{x^k}{\ln k}$

29. $\displaystyle\sum_{k=1}^{\infty} \frac{(x-2)^k}{k^3}$

30. $\displaystyle\sum_{k=0}^{\infty} \frac{k(x-2)^k}{3^k}$

31. $\displaystyle\sum_{k=0}^{\infty} \frac{(-1)^k}{(2k+1)!} x^{2k+1}$

32. $\displaystyle\sum_{k=1}^{\infty} (kx)^k$

33. $\displaystyle\sum_{k=1}^{\infty} \frac{k x^k}{\ln(k+1)}$

34. $\displaystyle\sum_{k=1}^{\infty} \frac{x^k}{\ln(k+1)}$

35. $\displaystyle\sum_{k=0}^{\infty} \frac{k(k+1)x^k}{4^k}$

---

1. = NOW WORK problem   = Graphing technology recommended   **CAS** = Computer Algebra System recommended

36. $\displaystyle\sum_{k=1}^{\infty} \frac{(-1)^k (x-5)^k}{k(k+1)}$

37. $\displaystyle\sum_{k=0}^{\infty} (-1)^k \frac{(x-3)^{2k}}{9^k}$

38. $\displaystyle\sum_{k=0}^{\infty} \frac{x^k}{e^k}$

39. $\displaystyle\sum_{k=0}^{\infty} (-1)^k \frac{(2x)^k}{k!}$

40. $\displaystyle\sum_{k=0}^{\infty} \frac{(x+1)^k}{k!}$

41. $\displaystyle\sum_{k=0}^{\infty} (-1)^k \frac{(x-1)^{4k}}{k!}$

42. $\displaystyle\sum_{k=1}^{\infty} \frac{(x+1)^k}{k(k+1)(k+2)}$

43. $\displaystyle\sum_{k=1}^{\infty} \frac{k^k x^k}{k!}$

44. $\displaystyle\sum_{k=0}^{\infty} \frac{3^k (x-2)^k}{k!}$

45. A function f is defined by the power series

$$f(x) = \sum_{k=0}^{\infty} \frac{x^k}{3^k}.$$

   (a) Find the domain of f.

   (b) Evaluate $f(2)$ and $f(-1)$.

   (c) Find f.

46. A function f is defined by the power series

$$f(x) = \sum_{k=0}^{\infty} (-1)^k \left(\frac{x}{2}\right)^k.$$

   (a) Find the domain of f.

   (b) Evaluate $f(0)$ and $f(1)$.

   (c) Find f.

47. A function f is defined by the power series

$$f(x) = \sum_{k=0}^{\infty} \frac{(x-2)^k}{2^k}.$$

   (a) Find the domain of f.

   (b) Evaluate $f(1)$ and $f(2)$.

   (c) Find f.

48. A function f is defined by the power series

$$f(x) = \sum_{k=0}^{\infty} (-1)^k (x+3)^k.$$

   (a) Find the domain of f.

   (b) Evaluate $f(-3)$ and $f(-2.5)$.

   (c) Find f.

49. If $\displaystyle\sum_{k=0}^{\infty} a_k x^k$ converges for $x = 3$, what, if anything, can be said about the convergence at $x = 2$? Can anything be said about the convergence at $x = 5$?

50. If $\displaystyle\sum_{k=0}^{\infty} a_k (x-2)^k$ converges for $x = 6$, at what other numbers x must the series necessarily converge?

51. If the series $\displaystyle\sum_{k=0}^{\infty} a_k x^k$ converges for $x = 6$ and diverges for $x = -8$, what, if anything, can be said about the truth of the

following statements?

   (a) The series converges for $x = 2$.

   (b) The series diverges for $x = 7$.

   (c) The series is absolutely convergent for $x = 6$.

   (d) The series converges for $x = -6$.

   (e) The series diverges for $x = 10$.

   (f) The series is absolutely convergent for $x = 4$.

52. If the radius of convergence of the power series $\displaystyle\sum_{k=0}^{\infty} a_k (x-3)^k$ is $R = 5$, what, if anything, can be said about the truth of the following statements?

   (a) The series converges for $x = 2$.

   (b) The series diverges for $x = 7$.

   (c) The series diverges for $x = 8$.

   (d) The series converges for $x = -6$.

   (e) The series converges for $x = -2$.

In Problems 53–58:

(a) Use a geometric series to represent each function as a power series centered at 0.

(b) Determine the radius of convergence and the interval of convergence of each series.

53. $f(x) = \dfrac{1}{1+x^3}$

54. $f(x) = \dfrac{1}{1-x^2}$

55. $f(x) = \dfrac{1}{6-2x}$

56. $f(x) = \dfrac{4}{x+2}$

57. $f(x) = \dfrac{x}{1+x^3}$

58. $f(x) = \dfrac{4x^2}{x+2}$

In Problems 59–62:

(a) Use the differentiation property of power series to find $f'(x)$ for each series.

(b) Use the integration property of power series to find the indefinite integral of each series.

59. $f(x) = \displaystyle\sum_{k=0}^{\infty} \frac{(-1)^k x^{2k+1}}{(2k+1)!}$

60. $f(x) = \displaystyle\sum_{k=0}^{\infty} \frac{(-1)^k x^{2k}}{(2k)!}$

61. $f(x) = \displaystyle\sum_{k=0}^{\infty} \frac{x^k}{k!}$

62. $f(x) = \displaystyle\sum_{k=0}^{\infty} \frac{(-1)^k x^k}{k!}$

In Problems 63–70, find a power series representation of f. Use a geometric series and properties of a power series.

63. $f(x) = \dfrac{1}{(1+x)^2}$

64. $f(x) = \dfrac{1}{(1-x)^3}$

65. $f(x) = \dfrac{2}{3(1-x)^2}$

66. $f(x) = \dfrac{1}{(1-x)^4}$

67. $f(x) = \ln\left(\dfrac{1}{1+x}\right)$

68. $f(x) = \ln(1-2x)$

69. $f(x) = \ln(1-x^2)$

70. $f(x) = \ln(1+x^2)$

Applications and Extensions

In Problems 71–78, find all x for which each power series converges.

71. $\displaystyle\sum_{k=1}^{\infty} \frac{x^k}{k}$

72. $\displaystyle\sum_{k=1}^{\infty} \frac{(x-4)^k}{k}$

73. $\displaystyle\sum_{k=1}^{\infty} \frac{x^k}{2k+1}$

74. $\displaystyle\sum_{k=1}^{\infty} \frac{x^k}{k^2}$

75. $\displaystyle\sum_{k=0}^{\infty} x^{k^2}$

76. $\displaystyle\sum_{k=1}^{\infty} \frac{k^a}{a^k}(x-a)^k, \quad a \neq 0$

77. $\displaystyle\sum_{k=0}^{\infty} \frac{(k!)^2}{(2k)!}(x-1)^k$

78. $\displaystyle\sum_{k=0}^{\infty} \frac{\sqrt{k!}}{(2k)!}x^k$

79. (a) In the geometric series $\dfrac{1}{1-x} = \displaystyle\sum_{k=0}^{\infty} x^k$, $-1 < x < 1$, replace x by x^2 to obtain the power series representation for $\dfrac{1}{1-x^2}$.

(b) What is its interval of convergence?

80. (a) Integrate the power series found in Problem 79 for $\dfrac{1}{1-x^2}$

to obtain the power series $\dfrac{1}{2}\ln\dfrac{1+x}{1-x}$.

(b) What is its interval of convergence?

81. Use the power series found in Problem 80 to get an approximation for ln 2 correct to three decimal places.

(CAS) 82. Use the first 1000 terms of Gregory's series to approximate $\dfrac{\pi}{4}$.

What is the approximation for π?

83. If $R > 0$ is the radius of convergence of $\displaystyle\sum_{k=1}^{\infty} a_k x^k$, show

that $\displaystyle\lim_{n \to \infty}\left|\frac{a_{n+1}}{a_n}\right| = \frac{1}{R}$, provided this limit exists.

84. If R is the radius of convergence of $\displaystyle\sum_{k=1}^{\infty} a_k x^k$, show that the

radius of convergence of $\displaystyle\sum_{k=1}^{\infty} a_k x^{2k}$ is $\sqrt{R}$.

85. Prove that if a power series is absolutely convergent at one endpoint of its interval of convergence, then the power series is absolutely convergent at the other endpoint.

86. Suppose $\displaystyle\sum_{k=0}^{\infty} a_k x^k$ converges for $|x| < R$ and that $\displaystyle\lim_{n\to\infty}\left|\frac{a_{n+1}}{a_n}\right|$

exists. Show that $\displaystyle\sum_{k=1}^{\infty} k a_k x^{k-1}$ and $\displaystyle\sum_{k=0}^{\infty} \frac{a_k}{k+1}x^{k+1}$ also converge

for $|x| < R$.

Challenge Problems

87. Consider the differential equation

$$(1+x^2)\,y'' - 4xy' + 6y = 0$$

Assuming there is a solution $y(x) = \displaystyle\sum_{k=0}^{\infty} a_k x^k$, substitute and

obtain a formula for a_k. Your answer should have the form

$$y(x) = a_0(1 - 3x^2) + a_1\left(x - \frac{1}{3}x^3\right) \qquad a_0, a_1 \text{ real numbers}$$

88. If the series $\displaystyle\sum_{k=0}^{\infty} a_k 3^k$ converges, show that the series $\displaystyle\sum_{k=1}^{\infty} k a_k 2^k$

also converges.

89. Find the interval of convergence of the series $\displaystyle\sum_{k=1}^{\infty} \frac{(x-2)^k}{k(3^k)}$.

90. Let a power series $S(x)$ be convergent for $|x| < R$. Assume

that $S(x) = \displaystyle\sum_{k=0}^{\infty} a_k x^k$ with partial sums $S_n(x) = \displaystyle\sum_{k=0}^{n} a_k x^k$.

Suppose for any number $\varepsilon > 0$, there is a number N so that

when $n > N$, $|S(x) - S_n(x)| < \dfrac{\varepsilon}{3}$ for all $|x| < R$.

Show that $S(x)$ is continuous for all $|x| < R$.

91. Find the power series in x, denoted by $f(x)$, for which $f''(x) + f(x) = 0$ and $f(0) = 0$, $f'(0) = 1$. What is the radius of convergence of the series?

92. The **Bessel function of order m** of the first kind, where m is a nonnegative integer, is defined as

$$J_m(x) = \sum_{k=0}^{\infty}(-1)^k \frac{1}{(k+m)!\,k!}\left(\frac{x}{2}\right)^{2k+m}$$

Show that:

(a) $J_0(x) = x^{-1}\dfrac{d}{dx}(x J_1(x))$

(b) $J_1(x) = x^{-2}\dfrac{d}{dx}(x^2 J_2(x))$

8.9 Taylor Series; Maclaurin Series

OBJECTIVES *When you finish this section, you should be able to:*

1 Express a function as a Taylor series or a Maclaurin series (p. 613)

2 Determine the convergence of a Taylor/Maclaurin series (p. 614)

3 Find Taylor/Maclaurin expansions (p. 616)

4 Work with a binomial series (p. 619)

We saw in Section 8.8 that it is often possible to obtain a power series representation for a function by starting with a known series and differentiating, integrating, or substituting.

But what if you have no initial series? In other words, so far we have seen that functions can be represented by power series, and if we know the sum of the power series, then we know the function. In this section, we investigate what the power series representation of a function must look like *if it has a power series representation.*

Consider the power series in $(x - c)$:

$$\sum_{k=0}^{\infty} a_k(x - c)^k = a_0 + a_1(x - c) + a_2(x - c)^2 + \cdots + a_n(x - c)^n + \cdots$$

and suppose its interval of convergence is the open interval $(c - R, c + R)$, $R > 0$. We define the function f as the series

$$f(x) = \sum_{k=0}^{\infty} a_k(x - c)^k = a_0 + a_1(x - c) + a_2(x - c)^2 + \cdots + a_n(x - c)^n + \cdots \quad (1)$$

The coefficients a_0, a_1, ... can be expressed in terms of f and its derivatives in the following way. Repeatedly differentiate the function using the differentiation property of a power series,

$$f'(x) = \sum_{k=1}^{\infty} k\, a_k(x - c)^{k-1} = a_1 + 2a_2(x - c) + 3a_3(x - c)^2 + 4a_4(x - c)^3 + \cdots$$

$$f''(x) = \sum_{k=2}^{\infty} [k(k - 1)]\, a_k(x - c)^{k-2}$$
$$= (2 \cdot 1)\, a_2 + (3 \cdot 2)\, a_3(x - c) + (4 \cdot 3)\, a_4(x - c)^2 + \cdots$$

$$f'''(x) = \sum_{k=3}^{\infty} [k(k - 1)(k - 2)]a_k(x - c)^{k-3}$$
$$= (3 \cdot 2 \cdot 1)\, a_3 + (4 \cdot 3 \cdot 2)\, a_4(x - c) + \cdots$$

and for any positive integer n,

$$f^{(n)}(x) = \sum_{k=n}^{\infty} k(k - 1)(k - 2) \cdots (k - n + 1)a_k(x - c)^{k-n}$$
$$= [n \cdot (n - 1) \cdot \ldots \cdot 1]\, a_n + [(n + 1) \cdot n \cdot (n - 1) \cdot \ldots \cdot 2]\, a_{n+1}(x - c) + \cdots$$
$$= n!\, a_n + (n + 1)!\, a_{n+1}(x - c) + \cdots$$

Now we let $x = c$ in each derivative and solve for a_k.

$$f(c) = a_0 \qquad a_0 = f(c)$$
$$f'(c) = a_1 \qquad a_1 = f'(c)$$
$$f''(c) = 2a_2 \qquad a_2 = \frac{f''(c)}{2!}$$
$$f'''(c) = 3!\, a_3 \qquad a_3 = \frac{f'''(c)}{3!}$$
$$\vdots \qquad \vdots$$
$$f^{(n)}(c) = n!a_n \qquad a_n = \frac{f^{(n)}(c)}{n!}$$

If we substitute for a_k in (1), we obtain

$$f(x) = f(c) + f'(c)(x - c) + \frac{f''(c)}{2!}(x - c)^2 + \cdots + \frac{f^{(n)}(c)}{n!}(x - c)^n + \cdots$$

and have proved the following result.

THEOREM Taylor Series; Maclaurin Series

Suppose f is a function that has derivatives of all orders on the open interval $(c - R, c + R)$, $R > 0$. If f can be represented by the power series $\sum_{k=0}^{\infty} a_k (x - c)^k$, whose radius of convergence is R, then

$$f(x) = f(c) + f'(c)(x - c) + \frac{f''(c)}{2!}(x - c)^2 + \cdots + \frac{f^{(n)}(c)}{n!}(x - c)^n + \cdots$$

$$= \sum_{k=0}^{\infty} \frac{f^{(k)}(c)}{k!}(x - c)^k \qquad (2)$$

for all numbers x in the open interval $(c - R, \ c + R)$. A power series that has the form of equation (2) is called a **Taylor series** of the function f.

When $c = 0$, the Taylor series representation of a function f

$$f(x) = f(0) + f'(0)\,x + \frac{f''(0)}{2!}x^2 + \cdots + \frac{f^{(n)}(0)}{n!}x^n + \cdots = \sum_{k=0}^{\infty} \frac{f^{(k)}(0)}{k!}x^k$$

is called a **Maclaurin series**.

ORIGINS Colin Maclaurin (1698–1746) was a Scottish mathematician. An orphan, he and his brother were raised by an uncle. Maclaurin was 11 years old when he entered the University of Glasgow, and within a year he had taught himself Euclid's elements (geometry). Maclaurin was 19 when he was appointed professor of Mathematics at the University of Aberdeen, Scotland. Maclaurin was an avid supporter of Newton's mathematics. In response to a claim that Newton's calculus lacked rigor, Maclaurin wrote the first systematic treatise on Newton's methods. In it he used Taylor series centered about 0, which are now known as Maclaurin series.

MACLAURIN,

The Taylor series in $(x - c)$ of a function f is referred to as the **Taylor expansion of f about c**; the Maclaurin series of a function f is called the **Maclaurin expansion of f about 0**.

In a Taylor series, all the derivatives are evaluated at c, and the interval of convergence has its center at c. In a Maclaurin series, all the derivatives are evaluated at 0, and the interval of convergence has its center at 0.

1 Express a Function as a Taylor Series or a Maclaurin Series

The next example shows what a Maclaurin expansion of $f(x) = e^x$ must look like (if there is one).

EXAMPLE 1   Expressing a Function as a Maclaurin Series

Assuming that $f(x) = e^x$ can be represented by a power series in x, find its Maclaurin series.

Solution To express a function f as a Maclaurin series, we begin by evaluating f and its derivatives at 0.

$$f(x) = e^x \qquad f(0) = 1$$
$$f'(x) = e^x \qquad f'(0) = 1$$
$$f''(x) = e^x \qquad f''(0) = 1$$
$$\vdots \qquad\qquad \vdots$$

Then we use the definition of a Maclaurin series.

$$f(x) = \sum_{k=0}^{\infty} \frac{f^{(k)}(0)}{k!}x^k = 1 + x + \frac{x^2}{2!} + \frac{x^3}{3!} + \cdots + \frac{x^n}{n!} + \cdots = \sum_{k=0}^{\infty} \frac{x^k}{k!} \qquad (3)$$

$\uparrow$
$f^{(k)}(0)=1$

∎

NOW WORK Problem **3**.

But how can we be sure $f(x) = e^x$ can be represented by a power series? We know [Example 1(a), p. 600] that the power series $\sum_{k=0}^{\infty} \dfrac{x^k}{k!}$ converges absolutely for all numbers x. But does the series $\sum_{k=0}^{\infty} \dfrac{x^k}{k!}$ converge to e^x? To answer these questions, we need to investigate the convergence of a Taylor series.

2 Determine the Convergence of a Taylor/Maclaurin Series

The conditions on the function f that guarantee that its power series representation actually converges to f are based on the Taylor polynomial $P_n(x)$ of f discussed in Section 3.5 (pp. 240–241). There, we found that the Taylor polynomial $P_n(x)$ of a function f whose first n derivatives are continuous on an open interval containing the number c is given by

$$P_n(x) = f(c) + f'(c)(x - c) + \frac{f''(c)}{2!}(x - c)^2 + \cdots + \frac{f^{(n)}(c)}{n!}(x - c)^n$$

To use $P_n(x)$ to approximate the function f for x close to c, we investigate the difference between $f(x)$ and $P_n(x)$, called the **remainder R_n**.

THEOREM Taylor's Formula with Remainder

Let f be a function whose first $n + 1$ derivatives are continuous on an open interval I containing the number c. Then for every x in I, there is a number u between x and c for which

$$f(x) = f(c) + f'(c)(x - c) + \frac{f''(c)}{2!}(x - c)^2 + \cdots + \frac{f^{(n)}(c)}{n!}(x - c)^n + R_n(x)$$

where

$$R_n(x) = \frac{f^{(n+1)}(u)}{(n + 1)!}(x - c)^{n+1}$$

is the remainder after n terms.

The proof of this result appears in Appendix B.

The Taylor series in $x - c$ of a function f is $\sum_{k=0}^{\infty} \dfrac{f^{(k)}(c)}{k!}(x - c)^k$. Notice that the nth partial sum of the Taylor series in $x - c$ of f is precisely the Taylor polynomial $P_n(x)$ of f at c. If the Taylor series $\sum_{k=0}^{\infty} \dfrac{f^{(k)}(c)}{k!}(x - c)^k$ converges to $f(x)$, it follows that

$$f(x) = \lim_{n \to \infty} P_n(x)$$

But Taylor's Formula with Remainder states that

$$f(x) = P_n(x) + R_n(x)$$

So, if the Taylor series converges, we must have

$$f(x) = \lim_{n \to \infty} [P_n(x) + R_n(x)] = \lim_{n \to \infty} P_n(x) + \lim_{n \to \infty} R_n(x) = f(x) + \lim_{n \to \infty} R_n(x)$$

That is, $\lim_{n \to \infty} R_n(x) = 0$.

THEOREM Convergence of a Taylor Series

If a function f has derivatives of all orders in an open interval $I = (c - R, c + R)$, $R > 0$, centered at c, and if

$$\lim_{n \to \infty} R_n(x) = 0$$

for all numbers x in I, then

$$f(x) = \sum_{k=0}^{\infty} \frac{f^{(k)}(c)}{k!} (x - c)^k = f(c) + f'(c)(x - c) + \frac{f''(c)}{2!}(x - c)^2$$
$$+ \cdots + \frac{f^{(n)}(c)}{n!}(x - c)^n + \cdots$$

for all numbers x in I.

At first glance, the convergence theorem appears simple, but in practice it is not always easy to show that $\lim_{n \to \infty} R_n(x) = 0$. One reason is that the term $f^{(n+1)}(u)$, which appears in $R_n(x)$, depends on n, making the limit difficult to find.

EXAMPLE 2   Determining the Convergence of a Maclaurin Series

Show that $1 + x + \dfrac{x^2}{2!} + \dfrac{x^3}{3!} + \cdots + \dfrac{x^n}{n!} + \cdots$ converges to e^x for every number x. That is, prove that

$$e^x = 1 + \frac{x}{1!} + \frac{x^2}{2!} + \frac{x^3}{3!} + \cdots + \frac{x^n}{n!} + \cdots = \sum_{k=0}^{\infty} \frac{x^k}{k!}$$

for all real numbers.

Solution To prove that $1 + \dfrac{x}{1!} + \dfrac{x^2}{2!} + \dfrac{x^3}{3!} + \cdots = e^x$ for every number x, we need to show that $\lim_{n \to \infty} R_n(x) = 0$. Since $f^{(n+1)}(x) = e^x$, we have

$$R_n(x) = \frac{f^{(n+1)}(u)x^{n+1}}{(n+1)!} = \frac{e^u x^{n+1}}{(n+1)!}$$

where u is between 0 and x. To show that $\lim_{n \to \infty} R_n(x) = 0$, we consider two cases: $x > 0$ and $x < 0$.

Case 1: When $x > 0$, then $0 < u < x$, so that $1 < e^u < e^x$ and, for every positive integer n,

$$0 < R_n(x) = \frac{e^u x^{n+1}}{(n+1)!} < \frac{e^x x^{n+1}}{(n+1)!}$$

By the Ratio Test, the series $\sum_{k=0}^{\infty} \dfrac{x^{k+1}}{(k+1)!}$ converges for all x. It follows that $\lim_{n \to \infty} \dfrac{x^{n+1}}{(n+1)!} = 0$, and, therefore,

$$\lim_{n \to \infty} \frac{e^x x^{n+1}}{(n+1)!} = e^x \lim_{n \to \infty} \frac{x^{n+1}}{(n+1)!} = 0$$

By the Squeeze Theorem, $\lim_{n \to \infty} R_n(x) = 0$.

Case 2: When $x < 0$, then $x < u < 0$ and $e^x < e^u < 1$, so that

$$0 \le |R_n(x)| = \frac{e^u \cdot |x|^{n+1}}{(n+1)!} < \frac{|x|^{n+1}}{(n+1)!}$$

NEED TO REVIEW? The Squeeze Theorem is discussed in Section 1.4, pp. 106–107.

EXAMPLE 6   Finding the Maclaurin Expansion for $f(x) = e^x \cos x$

Find the first five terms of the Maclaurin expansion for $f(x) = e^x \cos x$.

Solution The Maclaurin expansion for $f(x) = e^x \cos x$ is obtained by multiplying the Maclaurin expansion for e^x by the Maclaurin expansion for $\cos x$. That is,

$$e^x \cos x = \left(1 + x + \frac{x^2}{2!} + \frac{x^3}{3!} + \frac{x^4}{4!} + \frac{x^5}{5!} + \cdots\right)\left(1 - \frac{x^2}{2!} + \frac{x^4}{4!} - \cdots\right)$$

Then

$$e^x \cos x = 1\left(1 - \frac{x^2}{2!} + \frac{x^4}{4!} - \cdots\right) + x\left(1 - \frac{x^2}{2!} + \frac{x^4}{4!} - \cdots\right)$$

$$+ \frac{x^2}{2!}\left(1 - \frac{x^2}{2!} + \frac{x^4}{4!} - \cdots\right) + \frac{x^3}{3!}\left(1 - \frac{x^2}{2!} + \frac{x^4}{4!} - \cdots\right)$$

$$+ \frac{x^4}{4!}\left(1 - \frac{x^2}{2!} + \frac{x^4}{4!} - \cdots\right) + \frac{x^5}{5!}\left(1 - \frac{x^2}{2!} + \frac{x^4}{4!} - \cdots\right) + \cdots$$

$$= \left(1 - \frac{x^2}{2} + \frac{x^4}{24}\right) + \left(x - \frac{x^3}{2} + \frac{x^5}{24}\right) + \left(\frac{x^2}{2} - \frac{x^4}{4}\right) + \left(\frac{x^3}{6} - \frac{x^5}{12}\right)$$

$$+ \frac{x^4}{24} + \frac{x^5}{120} + \cdots$$

$$= 1 + x + \left(-\frac{1}{2} + \frac{1}{2}\right)x^2 + \left(-\frac{1}{2} + \frac{1}{6}\right)x^3 + \left(\frac{1}{24} - \frac{1}{4} + \frac{1}{24}\right)x^4$$

$$+ \left(\frac{1}{24} - \frac{1}{12} + \frac{1}{120}\right)x^5 + \cdots$$

$$= 1 + x - \frac{1}{3}x^3 - \frac{1}{6}x^4 - \frac{1}{30}x^5 + \cdots$$ ■

NOW WORK Problem 29.

EXAMPLE 7   Finding the Taylor Expansion for $f(x) = \cos x$ about $\dfrac{\pi}{2}$

Find the Taylor expansion for $f(x) = \cos x$ about $\dfrac{\pi}{2}$.

Solution To express $f(x) = \cos x$ as a Taylor expansion about $\dfrac{\pi}{2}$, we evaluate f and its derivatives at $\dfrac{\pi}{2}$.

$$f(x) = \cos x \qquad f\left(\frac{\pi}{2}\right) = 0$$

$$f'(x) = -\sin x \qquad f'\left(\frac{\pi}{2}\right) = -1$$

$$f''(x) = -\cos x \qquad f''\left(\frac{\pi}{2}\right) = 0$$

$$f'''(x) = \sin x \qquad f'''\left(\frac{\pi}{2}\right) = 1$$

For derivatives of odd order, $f^{(2n+1)}\left(\dfrac{\pi}{2}\right) = (-1)^{n+1}$. For derivatives of even order, $f^{(2n)}\left(\dfrac{\pi}{2}\right) = 0$. The Taylor expansion for $f(x) = \cos x$ about $\dfrac{\pi}{2}$ is

$$f(x) = \cos x = f\left(\frac{\pi}{2}\right) + f'\left(\frac{\pi}{2}\right)\left(x - \frac{\pi}{2}\right) + \frac{f''\left(\frac{\pi}{2}\right)}{2!}\left(x - \frac{\pi}{2}\right)^2$$

$$+ \frac{f'''\left(\frac{\pi}{2}\right)}{3!}\left(x - \frac{\pi}{2}\right)^3 + \cdots$$

$$= -\left(x - \frac{\pi}{2}\right) + \frac{1}{3!}\left(x - \frac{\pi}{2}\right)^3 - \frac{1}{5!}\left(x - \frac{\pi}{2}\right)^5 + \cdots$$

$$= \sum_{k=0}^{\infty} \frac{(-1)^{k+1}}{(2k+1)!}\left(x - \frac{\pi}{2}\right)^{2k+1}$$

The radius of convergence is ∞; the interval of convergence is $(-\infty, \infty)$. ∎

NOW WORK Problem 15.

4 Work with a Binomial Series

In algebra, the Binomial Theorem states that if m is a positive integer and a and b are real numbers, then

$$(a + b)^m = a^m + \binom{m}{1} a^{m-1}b + \binom{m}{2} a^{m-2}b^2 + \cdots + \binom{m}{m-2} a^2 b^{m-2}$$

$$+ \binom{m}{m-1} ab^{m-1} + b^m = \sum_{k=0}^{m} \binom{m}{k} a^{m-k}b^k$$

where

$$\binom{m}{k} = \frac{m!}{k!\,(m-k)!} = \frac{m(m-1)\cdots(m-k+1)}{k!}$$

NEED TO REVIEW? The Binomial Theorem is discussed in Appendix A.5, p. A-43.

In the Binomial Theorem, m is a positive integer. To generalize the result, we find a Maclaurin series for the function $f(x) = (1 + x)^m$, where m is *any real number*.

EXAMPLE 8   Finding the Maclaurin Expansion for $f(x) = (1 + x)^m$

Find the Maclaurin series for $f(x) = (1 + x)^m$, where m is any real number.

Solution We begin by finding the derivatives of f at 0:

$$f(x) = (1 + x)^m \qquad\qquad f(0) = 1$$
$$f'(x) = m(1 + x)^{m-1} \qquad\qquad f'(0) = m$$
$$f''(x) = m(m-1)(1 + x)^{m-2} \qquad\qquad f''(0) = m(m-1)$$
$$\vdots \qquad\qquad\qquad\qquad \vdots$$
$$f^{(n)}(x) = m(m-1)(m-2)\cdots(m-n+1)(1+x)^{m-n} \qquad f^{(n)}(0) = m(m-1)(m-2)\cdots(m-n+1)$$

The Maclaurin expansion for f is

$$(1 + x)^m = 1 + m x + \frac{m(m-1)}{2!}x^2 + \cdots$$

$$+ \frac{m(m-1)(m-2)\cdots(m-n+1)}{n!}x^n + \cdots$$

$$= \binom{m}{0} + \binom{m}{1}x + \binom{m}{2}x^2 + \cdots + \binom{m}{n}x^n + \cdots = \sum_{k=0}^{\infty} \binom{m}{k} x^k$$

where

$$\binom{m}{0} = 1 \quad \text{and} \quad \binom{m}{k} = \frac{m(m-1)(m-2)\cdots(m-k+1)}{k!}$$ ■

The series

$$(1+x)^m = \sum_{k=0}^{\infty} \binom{m}{k} x^k$$

$$= 1 + mx + \frac{m(m-1)}{2!} x^2 + \frac{m(m-1)(m-2)}{3!} x^3 + \cdots + \binom{m}{n} x^n + \cdots$$

is called a **binomial series** because of its similarity in form to the Binomial Theorem, and $\binom{m}{k}$ is called the **binomial coefficient of x^k**.

The following result, which we state without proof, gives the conditions under which the binomial series $(1+x)^m$ converges.

THEOREM Convergence of a Binomial Series

The binomial series

$$(1+x)^m = \sum_{k=0}^{\infty} \binom{m}{k} x^k$$

converges

- for all x if m is a nonnegative integer. (In this case, there are only $m+1$ nonzero terms.)
- on the open interval $(-1, 1)$ if $m \leq -1$.
- on the half-open interval $(-1, 1]$ if $-1 < m < 0$.
- on the closed interval $[-1, 1]$ if $m > 0$, but m is not an integer.

EXAMPLE 9 Using a Binomial Series

Represent the function $f(x) = \sqrt{x+1}$ as a Maclaurin series, and find its interval of convergence.

Solution Write $\sqrt{x+1} = (1+x)^{1/2}$ and use the binomial series with $m = \dfrac{1}{2}$. The result is

$$(1+x)^{1/2} = 1 + \frac{1}{2}x + \frac{\left(\dfrac{1}{2}\right)\left(-\dfrac{1}{2}\right)}{2!} x^2 + \frac{\dfrac{1}{2}\left(-\dfrac{1}{2}\right)\left(-\dfrac{3}{2}\right)}{3!} x^3 + \cdots$$

$$= 1 + \frac{1}{2}x - \frac{1}{8}x^2 + \frac{1}{16}x^3 - \cdots$$

Since $m = \dfrac{1}{2} > 0$ and m is not an integer, the series converges on the closed interval $[-1, 1]$. ■

NOW WORK Problem 37.

EXAMPLE 10 **Using a Binomial Series**

Represent the function $f(x) = \sin^{-1} x$ by a Maclaurin series.

Solution Recall that

$$\sin^{-1} x = \int_0^x \frac{dt}{\sqrt{1 - t^2}} = \int_0^x (1 - t^2)^{-1/2} dt$$

We write the integrand as a binomial series, with $x = -t^2$ and $m = -\dfrac{1}{2}$.

$$(1 - t^2)^{-1/2} = 1 + \left(-\frac{1}{2}\right)(-t^2) + \frac{\left(-\frac{1}{2}\right)\left(-\frac{3}{2}\right)}{2!}(-t^2)^2$$

$$+ \frac{\left(-\frac{1}{2}\right)\left(-\frac{3}{2}\right)\left(-\frac{5}{2}\right)}{3!}(-t^2)^3 + \cdots$$

$$= 1 + \frac{1}{2}t^2 + \frac{3}{8}t^4 + \frac{5}{16}t^6 + \cdots$$

Now we use the integration property of a power series to obtain

$$\int_0^x \frac{dt}{\sqrt{1 - t^2}} = \int_0^x \left(1 + \frac{1}{2}t^2 + \frac{3}{8}t^4 + \frac{5}{16}t^6 + \cdots\right) dt$$

$$\sin^{-1} x = x + \left(\frac{1}{2}\right)\left(\frac{x^3}{3}\right) + \left(\frac{3}{8}\right)\left(\frac{x^5}{5}\right) + \left(\frac{5}{16}\right)\left(\frac{x^7}{7}\right) + \cdots$$

$$= x + \frac{x^3}{6} + \frac{3}{40}x^5 + \frac{5}{112}x^7 + \cdots$$ ∎

In Problem 57, you are asked to show that the interval of convergence is $[-1, 1]$.

Summary

TABLE 7

Series	Comment
$\dfrac{1}{1 - x} = \displaystyle\sum_{k=0}^{\infty} x^k = 1 + x + x^2 + x^3 + \cdots$	Converges for $\|x\| < 1$.
$\tan^{-1} x = \displaystyle\sum_{k=0}^{\infty} (-1)^k \frac{x^{2k+1}}{2k + 1} = x - \frac{x^3}{3} + \frac{x^5}{5} - \cdots$	Converges on the interval $[-1, 1]$.
$e^x = \displaystyle\sum_{k=0}^{\infty} \frac{x^k}{k!} = 1 + \frac{x}{1!} + \frac{x^2}{2!} + \frac{x^3}{3!} + \cdots$	Converges for all real numbers x.
$\sin x = \displaystyle\sum_{k=0}^{\infty} \frac{(-1)^k x^{2k+1}}{(2k + 1)!} = x - \frac{x^3}{3!} + \frac{x^5}{5!} - \frac{x^7}{7!} + \cdots$	Converges for all real numbers x.
$\cos x = \displaystyle\sum_{k=0}^{\infty} \frac{(-1)^k x^{2k}}{(2k)!} = 1 - \frac{x^2}{2!} + \frac{x^4}{4!} - \frac{x^6}{6!} + \cdots$	Converges for all real numbers x.
$\ln(1 + x) = \displaystyle\sum_{k=0}^{\infty} \frac{(-1)^k x^{k+1}}{k + 1}$	Converges on the interval $(-1, 1]$.
$(1 + x)^m = \displaystyle\sum_{k=0}^{\infty} \binom{m}{k} x^k$	For convergence, see the theorem on page 620.

8.9 Assess Your Understanding

Concepts and Vocabulary

1. The series representation of a function f given by the power series $f(x) = f(c) + f'(c)(x-c) + \dfrac{f''(c)(x-c)^2}{2!} + \cdots + \dfrac{f^{(n)}(c)\,(x-c)^n}{n!} + \cdots$ is called a(n) _____ _____ about c.

2. If $c = 0$ in the Taylor expansion of a function f, then the expansion is called a(n) _____ expansion.

Skill Building

In Problems 3–14, assuming each function can be represented by a power series, find the Maclaurin expansion of each function.

3. $f(x) = \ln(1-x)$

4. $f(x) = \ln(1+x)$

5. $f(x) = \dfrac{1}{1-x}$

6. $f(x) = \dfrac{1}{1-3x}$

7. $f(x) = \dfrac{1}{(1+x)^2}$

8. $f(x) = (1+x)^{-3}$

9. $f(x) = \dfrac{1}{1+x^2}$

10. $f(x) = \dfrac{1}{1+2x^3}$

11. $f(x) = e^{3x}$

12. $f(x) = e^{x/2}$

13. $f(x) = \sin(\pi x)$

14. $f(x) = \cos(2x)$

In Problems 15–22, assuming each function can be represented by a power series, find the Taylor expansion of each function about the given number c.

15. $f(x) = e^x$; $c = 1$

16. $f(x) = e^{2x}$; $c = -1$

17. $f(x) = \ln x$; $c = 1$

18. $f(x) = \sqrt{x}$; $c = 1$

19. $f(x) = \dfrac{1}{x}$; $c = 1$

20. $f(x) = \dfrac{1}{\sqrt{x}}$; $c = 4$

21. $f(x) = \sin x$; $c = \dfrac{\pi}{6}$

22. $f(x) = \cos x$; $c = -\dfrac{\pi}{2}$

In Problems 23–26, assuming each function can be represented by a power series, find the Taylor expansion of each function about the given number c. Comment on the result.

23. $f(x) = 3x^3 + 2x^2 + 5x - 6$; $c = 0$

24. $f(x) = 4x^4 - 2x^3 - x$; $c = 0$

25. $f(x) = 3x^3 + 2x^2 + 5x - 6$; $c = 1$

26. $f(x) = 4x^4 - 2x^3 + x$; $c = 1$

In Problems 27 and 28, find the Maclaurin expansion for each function.

27. $f(x) = \sinh x$

28. $f(x) = e^{-x^2}$

In Problems 29–32, use properties of power series to find the first five nonzero terms of the Maclaurin expansion.

29. $f(x) = xe^x$

30. $f(x) = xe^{-x}$

31. $f(x) = e^{-x}\sin x$

32. $f(x) = e^{-x}\cos x$

In Problems 33 and 34, use a Maclaurin series for f to obtain the first four nonzero terms of the Maclaurin expansion for g.

33. $f(x) = \dfrac{1}{\sqrt{1-x^2}}$; $g(x) = \sin^{-1} x$

34. $f(x) = \tan x$; $g(x) = \ln(\cos x)$

In Problems 35–42, use a binomial series to represent each function, and find the interval of convergence.

35. $f(x) = \sqrt{1+x^2}$

36. $f(x) = \dfrac{1}{\sqrt{1-x}}$

37. $f(x) = (1+x)^{1/5}$

38. $f(x) = (1-x)^{5/3}$

39. $f(x) = \dfrac{1}{(1+x^2)^{1/2}}$

40. $f(x) = \dfrac{1}{(1+x)^{3/4}}$

41. $f(x) = \dfrac{2x}{\sqrt{1-x}}$

42. $f(x) = \dfrac{x}{1+x^3}$

Applications and Extensions

43. Find the Maclaurin expansion for $f(x) = \sin^2 x$.

44. Find the Maclaurin expansion for $f(x) = \cos^2 x$.

45. Obtain the Maclaurin expansion of $\cos x$ by integrating the Maclaurin series for $\sin x$.

46. Find the Maclaurin expansion for $f(x) = \ln\dfrac{1}{1-x}$. Compare the result to the power series representation of $f(x) = \ln\dfrac{1}{1-x}$ found in Section 8.8, Example 8, page 607.

47. Find the first five nonzero terms of the Maclaurin expansion for $f(x) = \sec x$.

48. **Probability** The **standard normal distribution** $p(x) = \dfrac{1}{\sqrt{2\pi}}e^{-x^2/2}$ is important in probability and statistics. If a random variable Z has a standard normal distribution, then the probability that an observation of Z is between $Z = a$ and $Z = b$ is given by
$$P(a \le Z \le b) = \dfrac{1}{\sqrt{2\pi}} \int_a^b e^{-x^2/2}\, dx$$

(a) Find the Maclaurin expansion for $p(x) = \dfrac{1}{\sqrt{2\pi}}e^{-x^2/2}$.

(b) Use properties of power series to find a power series representation for P.

(c) Use the first four terms of the series representation for P to approximate $P(-0.5 \le Z \le 0.3)$.

(d) Use technology to approximate $P(-0.5 \le Z \le 0.3)$.

In Problems 49–52, use a Maclaurin expansion to find each integral.

49. $\displaystyle\int \dfrac{1}{1+x^2}\,dx$

50. $\displaystyle\int \sec x\, dx$

51. $\displaystyle\int e^{x^{1/3}}\,dx$

52. $\displaystyle\int \ln(1+x)\, dx$

1. = NOW WORK problem   = Graphing technology recommended   CAS = Computer Algebra System recommended

53. Even Functions Show that if f is an even function, then the Maclaurin expansion for f has only even powers of x.

54. Odd Functions Show that if f is an odd function, then the Maclaurin expansion for f has only odd powers of x.

55. Show that $(1 + x)^m = \sum_{k=0}^{\infty} \binom{m}{k} x^k$, when m is a nonnegative integer, by showing that $R_n(x) \to 0$ as $n \to \infty$.

56. Show that the series $\sum_{k=0}^{\infty} \binom{m}{k} x^k$ converges absolutely for $|x| < 1$ and diverges for $|x| > 1$ if $m < 0$. (*Hint:* Use the Ratio Test.)

57. Show that the interval of convergence of the Maclaurin expansion for $f(x) = \sin^{-1} x$ is $[-1, 1]$.

58. Euler's Error Euler believed $\dfrac{1}{2} = 1 - 1 + 1 - 1 + 1 - 1 + \cdots$. He based his argument to support this equation on his belief in the identification of a series and the values of the function from which it was derived.

  (a) Write the Maclaurin expansion for $\dfrac{1}{1 + x}$. Do this without calculating any derivatives.

  (b) Evaluate both sides of the equation you derived in (a) at $x = 1$ to arrive at the formula above.

  (c) Criticize the procedure used in (b).

Challenge Problems

59. Find the exact sum of the infinite series:

$$\frac{x^3}{1(3)} - \frac{x^5}{3(5)} + \frac{x^7}{5(7)} - \frac{x^9}{7(9)} + \cdots \quad \text{for } x = 1$$

60. Find an elementary expression for $\sum_{k=1}^{\infty} \dfrac{x^{k+1}}{k(k+1)}$.

  Hint: Integrate the series for $\ln \dfrac{1}{1 - x}$.

61. Show that $\sum_{k=1}^{\infty} \dfrac{k}{(k + 1)!} = 1$

62. Let $s_n = \dfrac{1}{1!} + \dfrac{1}{2!} + \cdots + \dfrac{1}{n!}, \quad n = 1, 2, 3, \ldots$.

  (a) Show that $n! \geq 2^{n-1}$.

  (b) Show that $0 < s_n \leq 1 + \dfrac{1}{2} + \left(\dfrac{1}{2}\right)^2 + \cdots + \left(\dfrac{1}{2}\right)^{n-1}$.

  (c) Show that $0 < s_n < s_{n+1} < 2$. Then, conclude that $S = \lim_{n \to \infty} s_n$ and $S \leq 2$.

  (d) Let $t_n = \left[1 + \dfrac{1}{n}\right]^n$. Show that

$$t_n = 1 + 1 + \frac{1}{2!}\left[1 - \frac{1}{n}\right] + \frac{1}{3!}\left[1 - \frac{1}{n}\right]\left[1 - \frac{2}{n}\right] + \cdots$$

$$+ \frac{1}{n!}\left[1 - \frac{1}{n}\right]\left[1 - \frac{2}{n}\right]\cdots\left[1 - \frac{n-1}{n}\right] < s_n + 1$$

  (e) Show that $0 < t_n < t_{n+1} < 3$. Then, conclude that $e = \lim_{n \to \infty} t_n \leq 3$.

63. Show that $\left[1 + \dfrac{1}{n}\right]^n < e$ for all $n > 0$.

64. From the fact that $\sin t \leq t$ for all $t \geq 0$, use integration repeatedly to prove

$$1 - \frac{x^2}{2!} \leq \cos x \leq 1 - \frac{x^2}{2!} + \frac{x^4}{4!} \quad \text{for all } x \geq 0$$

65. Find the first four nonzero terms of the Maclaurin expansion for $f(x) = (1 + x)^x$.

66. Show that $f(x) = \begin{cases} e^{-1/x^2} & x \neq 0 \\ 0 & x = 0 \end{cases}$ has a Maclaurin expansion at $x = 0$. Then show that the Maclaurin series does not converge to f.

8.10 Approximations Using Taylor/Maclaurin Expansions

OBJECTIVES *When you finish this section, you should be able to:*

1 Approximate functions and their graphs (p. 623)

2 Approximate the number e; approximate logarithms (p. 625)

3 Approximate definite integrals (p. 627)

1 Approximate Functions and Their Graphs

If we know the power series representation of a function or if we know the function to which a Taylor or Maclaurin series converges, we can use the first several terms of the series to approximate both the function and its graph.

EXAMPLE 1   Approximating $y = \sin x$

(a) Approximate $y = \sin x$ by using the first four nonzero terms of its Maclaurin expansion.
(b) Graph $y = \sin x$ along with the approximation found in (a).
(c) Use (a) to approximate $\sin 0.1$.
(d) What is the error in using this approximation?

Solution (a) The Maclaurin expansion for $y = \sin x$ was found in Example 3 (p. 616) of Section 8.9.

$$y = \sin x = x - \frac{x^3}{3!} + \frac{x^5}{5!} - \frac{x^7}{7!} + \cdots = \sum_{k=0}^{\infty} (-1)^k \frac{x^{2k+1}}{(2k+1)!}$$

Using the first four nonzero terms of the Maclaurin expansion, we can approximate $\sin x$ as

$$\sin x \approx x - \frac{x^3}{3!} + \frac{x^5}{5!} - \frac{x^7}{7!} \tag{1}$$

(b) The graphs of $y = \sin x$ and the approximation in (1) are given in Figure 29.
(c) Using (1), we get

$$\sin 0.1 \approx 0.1 - \frac{0.1^3}{3!} + \frac{0.1^5}{5!} - \frac{0.1^7}{7!} \approx 0.0998$$

(d) Since the Maclaurin expansion for $y = \sin x$ at $x = 0$ is an alternating series that satisfies the conditions of the Alternating Series Test, the error E in using the first four terms as an approximation is less than or equal to the absolute value of the 5th term at $x = 0.1$. That is,

$$E \le \left| \frac{0.1^9}{9!} \right| = 2.756 \times 10^{-15} \qquad \blacksquare$$

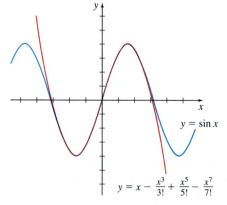

$y = \sin x$

$y = x - \frac{x^3}{3!} + \frac{x^5}{5!} - \frac{x^7}{7!}$

DF Figure 29

NOW WORK   Problem 1.

EXAMPLE 2   Approximating $y = \sqrt{1+x}$

(a) Write the Maclaurin expansion for $y = \sqrt{1+x}$ using a binomial series.
(b) Approximate $\sqrt{1+x}$ using the first five terms of the Maclaurin series.
(c) Graph $y = \sqrt{1+x}$ together with the first five terms of the approximation found in (b).
(d) Comment on the graphs in (c).
(e) Use the first five terms of the approximation for $\sqrt{1+x}$ to approximate $\sqrt{1.2}$. What is the error in using this approximation.

Solution (a) For $-1 \le x \le 1$, we have

$$\sqrt{1+x} = (1+x)^{1/2} = \sum_{k=0}^{\infty} \binom{\frac{1}{2}}{k} x^k$$

$$= 1 + \frac{1}{2}x + \frac{\frac{1}{2}\left(-\frac{1}{2}\right)}{2!} x^2 + \frac{\frac{1}{2}\left(-\frac{1}{2}\right)\left(-\frac{3}{2}\right)}{3!} x^3 + \cdots + \binom{\frac{1}{2}}{n} x^n + \cdots$$

(b) From part (a), we have

$$\sqrt{1+x} \approx 1 + \frac{1}{2}x - \frac{1}{8}x^2 + \frac{1}{16}x^3 - \frac{5}{128}x^4 \tag{2}$$

(c) The graphs of $y = \sqrt{1+x}$ and the first five terms of the approximation in (2) are given in Figure 30.

(d) On the interval of convergence $[-1, 1]$, the graphs are almost identical. Outside this interval, the graphs diverge.

(e) Using (2) and $x = 0.2$, we have

$$\sqrt{1.2} \approx 1 + \frac{1}{2}(0.2) - \frac{1}{8}(0.2)^2 + \frac{1}{16}(0.2)^3 - \frac{5}{128}(0.2)^4 = 1.0954375$$

Since the binomial series for $y = \sqrt{1+x}$ at $x = 0.2$ is an alternating series that satisfies the conditions of the Alternating Series Test, the error E in using the first five terms of the series as an approximation of $\sqrt{1.2}$ is less than or equal to the absolute value of the 6th term at $x = 0.2$. That is,

$$E \le \left| \frac{\frac{1}{2}\left(-\frac{1}{2}\right)\left(-\frac{3}{2}\right)\left(-\frac{5}{2}\right)\left(-\frac{7}{2}\right)}{5!} x^5 \right|_{x = 0.2} = \frac{7}{256}(0.2)^5 = 8.75 \times 10^{-6} \quad \blacksquare$$

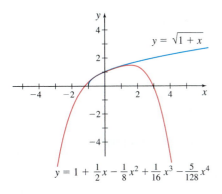

Figure 30

$y = \sqrt{1+x}$

$y = 1 + \frac{1}{2}x - \frac{1}{8}x^2 + \frac{1}{16}x^3 - \frac{5}{128}x^4$

NOW WORK Problem 3.

2 Approximate the Number e; Approximate Logarithms

The important number e, the base of the natural logarithm, is an irrational number, so its decimal expansion neither terminates nor repeats. Here, we explore one way that e can be approximated. Another important irrational number is π. We explore several ways to approximate π in Problems 18, 19, and 23.

EXAMPLE 3 **Approximating e**

Approximate e correct to within 0.000001.

Solution In Example 2 (p. 615) of Section 8.9, we showed that the Maclaurin series $\sum_{k=0}^{\infty} \frac{x^k}{k!}$ converges to e^x for every number x. If $x = 1$, we have

$$e = e^1 = 1 + 1 + \frac{1}{2!} + \frac{1}{3!} + \cdots + \frac{1}{n!} + \cdots = \sum_{k=0}^{\infty} \frac{1}{k!}$$

But this series converges very slowly.

For faster convergence, we can use $x = -1$. Then

$$e^{-1} = 1 - 1 + \frac{1}{2!} - \frac{1}{3!} + \cdots + \frac{(-1)^n}{n!} + \cdots = \sum_{k=0}^{\infty} \frac{(-1)^k}{k!}$$

Since this series is an alternating series that satisfies the conditions of the Alternating Series Test, we can use the error estimate for an alternating series. Since we want e correct to within 0.000001, we need the error $E < 0.000001$. That is, we need

$\frac{1}{n!} < 0.000001$, or equivalently, $n! > 10^6$. Since $10! = 3.6288 \times 10^6 > 10^6$, but $9! = 362,880 < 10^6$, using 10 terms of the Maclaurin series will approximate e^{-1} correct to within 0.000001.

$$\sum_{k=0}^{9} \frac{(-1)^k}{k!} = 1 - 1 + \frac{1}{2} - \frac{1}{3!} + \frac{1}{4!} - \frac{1}{5!} + \frac{1}{6!} - \frac{1}{7!} + \frac{1}{8!} - \frac{1}{9!} = \frac{16,687}{45,360}$$

Then

$$e = \frac{1}{e^{-1}} = \left(\frac{16,687}{45,360} \right)^{-1} \approx 2.718\,28$$

correct to within 0.000001. ∎

NOW WORK Problem 19.

Approximate Logarithms

In Example 8 (p. 607) of Section 8.8, we integrated the geometric series $\dfrac{1}{1-x} = \displaystyle\sum_{k=0}^{\infty} x^k$, and found that

$$\ln \frac{1}{1-x} = x + \frac{x^2}{2} + \frac{x^3}{3} + \cdots + \frac{x^n}{n} + \cdots \tag{3}$$

for $-1 \le x < 1$. It might appear that this expansion can be used to compute the logarithms of numbers, but the series converges only for $-1 \le x < 1$, and unless x is close to 0, the series converges so slowly that too many terms would be required for practical use.

A more useful formula for computing logarithms is obtained by multiplying (3) by (-1) and using the fact that $-\ln \dfrac{1}{A} = \ln \left(\dfrac{1}{A} \right)^{-1} = \ln A$. Then

$$\ln(1-x) = -x - \frac{x^2}{2} - \frac{x^3}{3} - \cdots \tag{4}$$

Now we replace x by $-x$. The result is

$$\ln(1+x) = x - \frac{x^2}{2} + \frac{x^3}{3} - \cdots \tag{5}$$

Subtracting (4) from (5), we obtain

$$\ln \frac{1+x}{1-x} = 2 \left(x + \frac{x^3}{3} + \frac{x^5}{5} + \cdots \right) \tag{6}$$

This series converges for $-1 < x < 1$.

If N is a positive integer, let $x = \dfrac{1}{2N+1}$. Then $0 < x < 1$, and

$$\frac{1+x}{1-x} = \frac{1 + \dfrac{1}{2N+1}}{1 - \dfrac{1}{2N+1}} = \frac{2N+1+1}{2N+1-1} = \frac{N+1}{N}$$

Substituting into (6), we obtain

$$\ln \frac{N+1}{N} = 2 \left(x + \frac{x^3}{3} + \frac{x^5}{5} + \cdots \right)$$

Since $x = \dfrac{1}{2N+1}$,

$$\ln(N+1) = \ln N + 2 \left[\frac{1}{2N+1} + \frac{1}{3} \left(\frac{1}{2N+1} \right)^3 + \frac{1}{5} \left(\frac{1}{2N+1} \right)^5 + \cdots \right] \tag{7}$$

This series converges for all positive integers N.

EXAMPLE 4 **Approximating a Logarithm**

(a) Approximate $\ln 2$ using the first three terms of (7).
(b) Approximate $\ln 3$ using the first three terms of (7).

Solution **(a)** In (7), we let $N = 1$ and use the first three terms of the series; then

$$\ln 2 \approx 2 \left[\frac{1}{3} + \frac{1}{3} \left(\frac{1}{3} \right)^3 + \frac{1}{5} \left(\frac{1}{3} \right)^5 \right] \approx 0.693004$$

(The first three terms of this series approximates $\ln 2$ correct to within 0.001.)
(b) To find $\ln 3$, let $N = 2$ and use $\ln 2 = 0.693004$ in (7).

$$\ln 3 \approx \ln 2 + 2 \left[\frac{1}{5} + \frac{1}{3} \left(\frac{1}{5} \right)^3 + \frac{1}{5} \left(\frac{1}{5} \right)^5 \right]$$

$$\approx 0.693004 + 2 \left[\frac{1}{5} + \frac{1}{3} \left(\frac{1}{5} \right)^3 + \frac{1}{5} \left(\frac{1}{5} \right)^5 \right] \approx 1.098465 \qquad ■$$

Continuing in this way, a table of natural logarithms can be formed.

NOW WORK **Problem 15.**

3 Approximate Definite Integrals

We have already discussed the approximation of definite integrals using Riemann sums, the trapezoidal rule, and Simpson's rule. Taylor/Maclaurin series can also be used to approximate definite integrals.

EXAMPLE 5 **Using a Maclaurin Expansion to Approximate an Integral**

Use a Maclaurin expansion to approximate $\int_0^{1/2} e^{-x^2} dx$ correct to within 0.001.

Solution Replace x by $-x^2$ in the Maclaurin expansion for e^x to obtain the Maclaurin expansion for e^{-x^2}.

$$e^x = 1 + x + \frac{x^2}{2!} + \frac{x^3}{3!} + \cdots$$

$$e^{-x^2} = 1 - x^2 + \frac{x^4}{2!} - \frac{x^6}{3!} + \frac{x^8}{4!} - \cdots$$

Now use the integration property of power series.

$$\int_0^{1/2} e^{-x^2} dx = \int_0^{1/2} \left(1 - x^2 + \frac{x^4}{2!} - \frac{x^6}{3!} + \cdots \right) dx$$

$$= \left[x - \frac{x^3}{3} + \frac{x^5}{2!5} - \frac{x^7}{3!7} + \cdots \right]_0^{1/2}$$

$$= \frac{1}{2} - \frac{1}{3(2)^3} + \frac{1}{2!5(2)^5} - \frac{1}{3!7(2)^7} + \cdots$$

$$\approx 0.5 - 0.041666 + 0.003125 - 0.000186 + \cdots$$

Since this series satisfies the two conditions of the Alternating Series Test, the error due to using the first three terms as an approximation is less than the 4th term,

$$\frac{1}{3!7(2)^7} = 1.860119048 \times 10^{-4}. \text{ So,}$$

$$\int_0^{1/2} e^{-x^2}dx \approx \frac{1}{2} - \frac{1}{3(2)^3} + \frac{1}{2!5(2)^5} \approx 0.461458$$

correct to within 0.001. ∎

NOW WORK Problem 9.

Notice that only three terms were needed to obtain the desired accuracy. To obtain this same accuracy using Simpson's rule or the trapezoidal rule would have required a very fine partition of $\left[0, \frac{1}{2}\right]$. In practice, efficiency of technique is an important consideration for solving numerical problems.

8.10 Assess Your Understanding

Skill Building

1. **(a)** Approximate $y = \cos x$ by using the first four nonzero terms of its Maclaurin expansion.

   (b) Graph $y = \cos x$ along with the approximation found in (a).

   (c) Use the approximation in (a) to approximate $\cos \frac{\pi}{90}$, $(2°)$.

   (d) What is the error in using this approximation?

   (e) If the approximation in (a) is used, for what values of x is the error less than 0.0001?

2. **(a)** Approximate $y = e^x$ by using the first four nonzero terms of its Maclaurin expansion.

   (b) Graph $y = e^x$ along with the approximation found in (a).

   (c) Use the approximation in (a) to approximate $e^{1/2}$.

   (d) What is the error in using this approximation?

   (e) If the approximation in (a) is used, for what values of x is the error less than 0.0001?

3. **(a)** Represent the function $f(x) = \sqrt[3]{1+x}$ as a Maclaurin series.

   (b) What is the interval of convergence?

   (c) Approximate $y = \sqrt[3]{1+x}$ using the first five terms of its Maclaurin series.

   (d) Graph $y = \sqrt[3]{1+x}$ along with the approximation found in (c).

   (e) Comment on the graphs in (d) and the result of the approximation.

   (f) Use the result from (c) to approximate $\sqrt[3]{0.9}$. What is the error in using this approximation?

4. **(a)** Represent $y = \frac{1}{\sqrt{4+x}}$ as a Maclaurin series.

   (b) What is the interval of convergence?

   (c) Approximate $y = \frac{1}{\sqrt{4+x}}$ using the first four terms of its Maclaurin series.

   (d) Graph $y = \frac{1}{\sqrt{4+x}}$ along with the approximation found in (c).

   (e) Comment on the graphs in (d) and the result of the approximation.

   (f) Use the result from (c) to approximate $\frac{1}{\sqrt{4.2}}$. What is the error in using this approximation?

5. **(a)** Represent $y = \tan^{-1} x$ as a Maclaurin series.

   (b) What is the interval of convergence?

   (c) Approximate $y = \tan^{-1} x$ using the first five nonzero terms of its Maclaurin series.

   (d) Graph $y = \tan^{-1} x$ along with the approximation found in (c).

   (e) Comment on the graphs in (d) and the result of the approximation.

6. **(a)** Represent $y = \frac{1}{1-x}$ as a Maclaurin series.

   (b) What is the interval of convergence?

   (c) Approximate $y = \frac{1}{1-x}$ using the first four nonzero terms of its Maclaurin series.

   (d) Graph $y = \frac{1}{1-x}$ along with the approximation found in (c).

   (e) Comment on the graphs in (d) and the result of the approximation.

In Problems 7–14, use properties of power series to approximate each integral using the first four terms of a Maclaurin series.

7. $\displaystyle\int_0^1 \sin x^2 dx$

8. $\displaystyle\int_0^1 \cos x^2\, dx$

9. $\displaystyle\int_0^1 \frac{x^2}{1+\cos x}dx$

10. $\displaystyle\int_0^{0.1} \frac{x}{\ln(2+x)}dx$

1. = NOW WORK problem   ◪ = Graphing technology recommended   CAS = Computer Algebra System recommended

11. $\displaystyle\int_0^{0.2} \sqrt[3]{1+x^4}\,dx$

12. $\displaystyle\int_0^{1/2} \sqrt[3]{1+x}\,dx$

13. $\displaystyle\int_0^{1/2} \frac{1}{\sqrt[3]{1+x^2}}\,dx$

14. $\displaystyle\int_0^{0.2} \frac{1}{\sqrt{1+x^3}}\,dx$

Applications and Extensions

15. Use the recursive formula (7) for $\ln(N+1)$ to show $\ln 4 \approx 1.38629$.

16. **(a)** Write the first three nonzero terms and the general term of the Maclaurin series for $f(x) = 3\sin\left(\dfrac{x}{2}\right)$.

   (b) What is the interval of convergence for the series found in (a)?

   (c) What is the minimum number of terms of the series in (a) that are necessary to approximate f on the interval $(-2,\ 2)$ with an error less than or equal to 0.1?

17. **Calculators Are Perfect, Right?**   They always get the right answer with what looks like little or no effort at all. This is very misleading because calculators use a lot of the ideas learned in this chapter to obtain these answers. One advantage we have over a calculator is that we can approximate answers up to any accuracy we choose. So let's prove you are smarter than any calculator by approximating each of the following correct to within 0.001 using a Maclaurin series.

   (a) $F(x) = \sin x, \quad x = 4°$

   (b) $F(x) = \cos x, \quad x = 15°$

   (c) $F(x) = \tan^{-1} x, \quad x = 0.05$

   Source: Contributed by the students at Lander University, Greenwood, SC.

18. **Leibniz Formula for π**   Leibniz derived the following formula for $\dfrac{\pi}{4}$: $\dfrac{\pi}{4} = 1 - \dfrac{1}{3} + \dfrac{1}{5} - \dfrac{1}{7} + \dfrac{1}{9} - \cdots$.

   (a) Find $\displaystyle\int_0^1 \frac{1}{1+x^2}\,dx$.

   (b) Expand the integrand in (a) into a power series and integrate it term-by-term to get Leibniz's formula.

   (c) Find the sum of the first 10 terms in the above series. Does it appear that Leibniz's formula is useful for approximating π?

   (d) How many terms are required to approximate π correct to 10 decimal places?

19. **Approximating π**   The series approximation of π using Gregory's series converges very slowly. (See Problem 82, Section 8.8. p. 611.) A more rapidly convergent series is obtained by using the identity

$$\tan^{-1} 1 = \tan^{-1}\left(\frac{1}{2}\right) + \tan^{-1}\left(\frac{1}{3}\right)$$

Use $x = \dfrac{1}{2}$ and $x = \dfrac{1}{3}$ in **Gregory's series**, together with this identity, to approximate π using the first four terms.

20. **Faster than light?**   At low speeds, the kinetic energy K, that is, the energy due to the motion of an object of mass m and speed v, is given by the formula $K = K(v) = \dfrac{1}{2}mv^2$. But this formula is only an approximation to the general formula, and works only for speeds much less than the speed of light, c. The general formula, which holds for all speeds, is

$$K_{\text{gen}}(v) = mc^2\left(\frac{1}{\sqrt{1 - \dfrac{v^2}{c^2}}} - 1\right)$$

The formula for K was used very successfully for many years before Einstein arrived at the general formula, so K must be essentially correct for low speeds. Use a binomial expansion to show that $\dfrac{1}{2}mv^2$ is a first approximation to K_{gen} for v close to 0.

Challenge Problems

21. Let $a_k = (-1)^{k+1} \displaystyle\int_0^{\pi/k} \sin(kx)\,dx$.

   (a) Find a_k.

   (b) Show that the infinite series $\displaystyle\sum_{k=1}^{\infty} a_k$ converges.

   (c) Show that $1 \le \displaystyle\sum_{k=1}^{\infty} a_k \le \frac{3}{2}$.

22. Find the Maclaurin expansion for $f(x) = xe^{x^3}$.

23. **(a)** Use the Maclaurin expansion for $f(x) = \sin^{-1} x$ to find numerical series for $\dfrac{\pi}{2}$ and for $\dfrac{\pi}{6}$.

   (b) Use the result of (a) to approximate $\dfrac{\pi}{2}$ correct to within 0.001.

   (c) Use the result of (a) to approximate $\dfrac{\pi}{6}$ correct to within 0.001.

Chapter Review

THINGS TO KNOW

8.1 Sequences

Definitions:

- A sequence is a function whose domain is the set of positive integers and whose range is a subset of real numbers. (p. 538)
- nth term of a sequence (p. 538)
- Limit of a sequence (p. 541)
- Convergence; divergence of a sequence (p. 541)
- Related function of a sequence (p. 543)
- Divergence of a sequence to infinity (p. 545)
- Bounded sequence (p. 546)
- Monotonic sequence (p. 548)

Properties of a Convergent Sequence: (p. 542)

If $\{s_n\}$ and $\{t_n\}$ are convergent sequences and if c is a number, then

- Constant multiple property: $\lim\limits_{n \to \infty} (c s_n) = c \lim\limits_{n \to \infty} s_n$
- Sum and difference properties:

  $\lim\limits_{n \to \infty} (s_n \pm t_n) = \lim\limits_{n \to \infty} s_n \pm \lim\limits_{n \to \infty} t_n$
- Product property: $\lim\limits_{n \to \infty} (s_n \cdot t_n) = \left(\lim\limits_{n \to \infty} s_n \right) \left(\lim\limits_{n \to \infty} t_n \right)$
- Quotient property: $\lim\limits_{n \to \infty} \dfrac{s_n}{t_n} = \dfrac{\lim\limits_{n \to \infty} s_n}{\lim\limits_{n \to \infty} t_n}$

  provided $\lim\limits_{n \to \infty} t_n \neq 0$
- Power property: $\lim\limits_{n \to \infty} s_n^p = \left[\lim\limits_{n \to \infty} s_n \right]^p$

  where $p \geq 2$ is an integer
- Root property: $\lim\limits_{n \to \infty} \sqrt[p]{s_n} = \sqrt[p]{\lim\limits_{n \to \infty} s_n}$,

  where $p \geq 2$ and $s_n \geq 0$ if p is even

Theorems:

- Let $\{s_n\}$ be a sequence of real numbers. If $\lim\limits_{n \to \infty} s_n = L$ and if f is a function that is continuous at L and is defined for all numbers s_n, then $\lim\limits_{n \to \infty} f(s_n) = f(L)$. (p. 543)
- The Squeeze Theorem for sequences (p. 545)
- A convergent sequence is bounded (p. 547)
- If a sequence is not bounded from above or if it is not bounded from below, then it diverges. (p. 547)
- An increasing (or nondecreasing) sequence $\{s_n\}$ that is bounded from above converges. (p. 549)
- A decreasing (or nonincreasing) sequence $\{s_n\}$ that is bounded from below converges. (p. 549)
- Let $\{s_n\}$ be a sequence and let f be a related function of $\{s_n\}$. Suppose L is a real number. If $\lim\limits_{x \to \infty} f(x) = L$, then $\lim\limits_{n \to \infty} s_n = L$. (p. 544)
- The sequence $\{r^n\}$, where r is a real number,
  - converges to 0, for $-1 < r < 1$.
  - converges to 1, for $r = 1$.
  - diverges for all other numbers (p. 546).

Procedure: Ways to show a sequence is monotonic (p. 548)

Summary: How to determine if a sequence converges (p. 550)

8.2 Infinite Series

- If $a_1, a_2, \ldots, a_n, \ldots$ is an infinite collection of numbers, the expression $\sum\limits_{k=1}^{\infty} a_k = a_1 + a_2 + \cdots + a_n + \cdots$ is called an infinite series or, simply, a series. (p. 554)
- nth term or general term of a series (p. 554)
- Partial sum $S_n = \sum\limits_{k=1}^{n} a_k$, where S_n is the sum of the first n terms of the series $\sum\limits_{k=1}^{\infty} a_k$ (p. 554)
- Convergence, divergence of a series (p. 555)
- Geometric series $\sum\limits_{k=1}^{\infty} ar^{k-1} = a + ar + ar^2 + \cdots, a \neq 0$ (p. 557)

  $\sum\limits_{k=1}^{\infty} ar^{k-1}$ converges if $|r| < 1$, and its sum is $\dfrac{a}{1-r}$

  $\sum\limits_{k=1}^{\infty} ar^{k-1}$ diverges if $|r| \geq 1$. (p. 558)
- Harmonic series $\sum\limits_{k=1}^{\infty} \dfrac{1}{k} = 1 + \dfrac{1}{2} + \dfrac{1}{3} + \cdots$ (p. 561)

  The harmonic series diverges. (p. 561)

Summary: Series and convergence of series (p. 561)

8.3 Properties of Series; the Integral Test

- If the series $\sum\limits_{k=1}^{\infty} a_k$ converges, then $\lim\limits_{n \to \infty} a_n = 0$. (p. 566)
- The Test for Divergence:

  The infinite series $\sum\limits_{k=1}^{\infty} a_k$ diverges if $\lim\limits_{n \to \infty} a_n \neq 0$. (p. 566)
- If two infinite series are identical after a certain term, then either both series converge or both series diverge. If both series converge, they do not necessarily have the same sum. (p. 567)
- The General Convergence Test (p. 569)
- The Integral Test (p. 569)
- A p-series $\sum\limits_{k=1}^{\infty} \dfrac{1}{k^p} = 1 + \dfrac{1}{2^p} + \dfrac{1}{3^p} + \cdots + \dfrac{1}{n^p} + \cdots$, where p is a positive real number. (p. 570) The p-series $\sum\limits_{k=1}^{\infty} \dfrac{1}{k^p}$ converges if $p > 1$ and diverges if $0 < p \leq 1$. (p. 571)
- Bounds on the sum of a p-series:

  If $p > 1$, then $\dfrac{1}{p-1} < \sum\limits_{k=1}^{\infty} \dfrac{1}{k^p} < 1 + \dfrac{1}{p-1}$. (p. 572)

Properties of Convergent Series: If $\sum\limits_{k=1}^{\infty} a_k$ and $\sum\limits_{k=1}^{\infty} b_k$ are two convergent series and if $c \neq 0$ is a number, then

- Sum and difference properties:

$$\sum_{k=1}^{\infty} (a_k \pm b_k) = \sum_{k=1}^{\infty} a_k \pm \sum_{k=1}^{\infty} b_k \quad \text{(p. 568)}$$

- Constant multiple property:

$$\sum_{k=1}^{\infty}(ca_k) = c\sum_{k=1}^{\infty} a_k \quad (\text{p. 568})$$

- If $\sum_{k=1}^{\infty} a_k$ diverges, then $\sum_{k=1}^{\infty}(ca_k)$ also diverges. (p. 568)

8.4 Comparison Tests
Theorems:

- Comparison Test for Convergence: If $0 < a_k \le b_k$, for all k, and $\sum_{k=1}^{\infty} b_k$ converges, then $\sum_{k=1}^{\infty} a_k$ converges. (p. 576)
- Comparison Test for Divergence: If $0 < c_k \le a_k$, for all k, and $\sum_{k=1}^{\infty} c_k$ diverges, then $\sum_{k=1}^{\infty} a_k$ diverges. (p. 576)
- Limit Comparison Test: Suppose $\sum_{k=1}^{\infty} a_k$ and $\sum_{k=1}^{\infty} b_k$ are both series of positive terms. If $\lim_{n\to\infty}\dfrac{a_n}{b_n} = L, 0 < L < \infty$, then both series converge or both diverge. (p. 577)

Summary: Table 3: Series often used for comparisons (p. 579)

8.5 Alternating Series; Absolute Convergence
Definitions:

- Alternating series (p. 582)
- A series $\sum_{k=1}^{\infty} a_k$ is absolutely convergent if the series $\sum_{k=1}^{\infty}|a_k|$ is convergent. (p. 585)
- A series that is convergent without being absolutely convergent is conditionally convergent. (p. 587)

Theorems:

- Alternating Series Test: (p. 582)
- Error estimate (p. 584)
- Absolute Convergence Test: If a series $\sum_{k=1}^{\infty} a_k$ is absolutely convergent, then it is convergent. (p. 586)

Properties of Absolutely Convergent and Conditionally Convergent series: (p. 588)

- The alternating harmonic series $\sum_{k=1}^{\infty}\dfrac{(-1)^{k+1}}{k}$ converges. (p. 583).

8.6 Ratio Test, Root Test
- Ratio Test (p. 591)
- Root Test (p. 593)

8.7 Summary of Tests
- Guide for choosing a test (p. 597)
- Tests for convergence and divergence (Table 5; pp. 597–598)

8.8 Power Series
Definitions:

- Power series: $\sum_{k=0}^{\infty} a_k x^k$ or $\sum_{k=0}^{\infty} a_k(x-c)^k$, where c is a constant. (p. 600)
- Radius of convergence (p. 602)
- Interval of convergence (p. 602)

Theorems:

- If a power series centered at 0 converges for a number $x_0 \ne 0$, then it converges absolutely for all numbers x for which $|x| < |x_0|$. (p. 602)
- If a power series centered at 0 diverges for a number x_1, then it diverges for all numbers x for which $|x| > |x_1|$. (p. 602)
- For a power series centered at c, exactly one of the following is true (p. 602):
  - The series converges for only $x = c$.
  - The series converges absolutely for all x.
  - There is a positive number R for which the series converges absolutely for all x, $|x-c| < R$, and diverges for all x, $|x-c| > R$.

Properties of Power Series: (p. 606)

Let $f(x) = \sum_{k=0}^{\infty} a_k x^k$ be a power series in x having a nonzero radius of convergence R.

- *Continuity property:*
$$\lim_{x\to x_0}\left(\sum_{k=0}^{\infty} a_k x^k\right) = \sum_{k=0}^{\infty}\left(\lim_{x\to x_0} a_k x^k\right) = \sum_{k=0}^{\infty} a_k x_0^k$$

- *Differentiation property:*
$$\frac{d}{dx}\left(\sum_{k=0}^{\infty} a_k x^k\right) = \sum_{k=0}^{\infty}\left(\frac{d}{dx} a_k x^k\right) = \sum_{k=1}^{\infty} k a_k x^{k-1}$$

- *Integration property:*
$$\int_0^x\left(\sum_{k=0}^{\infty} a_k t^k\right)dt = \sum_{k=0}^{\infty}\left(\int_0^x a_k t^k\, dt\right) = \sum_{k=0}^{\infty}\frac{a_k x^{k+1}}{k+1}$$

8.9 Taylor Series; Maclaurin Series
Theorems:

- Taylor series: (p. 613)
$$f(x) = f(c) + f'(c)(x-c) + \frac{f''(c)}{2!}(x-c)^2$$
$$+\cdots + \frac{f^{(n)}(c)}{n!}(x-c)^n + \cdots = \sum_{k=0}^{\infty}\frac{f^{(k)}(c)}{k!}(x-c)^k$$

- Maclaurin series (p. 613)
$$f(x) = f(0) + f'(0)x + \frac{f''(0)x^2}{2!} + \cdots + \frac{f^{(n)}(0)x^n}{n!} + \cdots$$
$$= \sum_{k=0}^{\infty}\frac{f^{(k)}(0)}{k!}x^k$$

- Taylor's formula with remainder (p. 614)
- Convergence of a Taylor series (p. 615)
- Binomial series (p. 620)
- Convergence of a binomial series (p. 620)

8.10 Approximations Using Taylor/Maclaurin Expansions
(pp. 623–628)

OBJECTIVES

Section	You should be able to …	Example	Review Exercises
8.1	1 Write the terms of a sequence (p. 539)	1, 2	1, 2
	2 Find the nth term of a sequence (p. 539)	3, 4	3
	3 Use properties of convergent sequences (p. 542)	5, 6	4, 5
	4 Use a related function or the Squeeze Theorem to show a sequence converges (p. 543)	7–10	6, 7
	5 Determine whether a sequence converges or diverges (p. 545)	11–15	8–13
8.2	1 Determine whether a series has a sum (p. 554)	1–3	14, 15
	2 Analyze a geometric series (p. 557)	4–6	17–20
	3 Analyze the harmonic series (p. 561)		16
8.3	1 Use the Test for Divergence (p. 567)	1	21
	2 Work with properties of series (p. 567)	2	25–27
	3 Use the Integral Test (p. 569)	3–5	22, 23
	4 Analyze a p-series (p. 570)	6	24
8.4	1 Use Comparison Tests for Convergence and Divergence (p. 576)	1, 2	28
	2 Use the Limit Comparison Test (p. 577)	3, 4	28–30
8.5	1 Determine whether an alternating series converges (p. 583)	1, 2	31–33
	2 Approximate the sum of a convergent alternating series (p. 584)	3	31–33
	3 Determine whether a series converges (p. 586)	4–6	34–37
8.6	1 Use the Ratio Test (p. 591)	1, 2	38, 39
	2 Use the Root Test (p. 593)	3, 4	40, 41
8.7	1 Choose an appropriate test to determine whether a series converges (p. 596)		42–52
8.8	1 Determine whether a power series converges (p. 600)	1	53(a)–58(a)
	2 Find the interval of convergence of a power series (p. 603)	2–4	53(b)–58(b)
	3 Define a function using a power series (p. 604)	5, 6	59, 60
	4 Use properties of power series (p. 606)	7–9	61
8.9	1 Express a function as a Taylor series or a Maclaurin series (p. 613)	1	64
	2 Determine the convergence of a Taylor/Maclaurin series (p. 614)	2	
	3 Find Taylor/Maclaurin expansions (p. 616)	3–7	62, 63, 65, 66
	4 Work with a binomial series (p. 619)	8–10	67–69
8.10	1 Approximate functions and their graphs (p. 623)	1, 2	70
	2 Approximate the number e; approximate logarithms (p. 625)	3, 4	71
	3 Approximate definite integrals (p. 627)	5	72, 73

REVIEW EXERCISES

In Problems 1 and 2, the nth term of a sequence $\{s_n\}$ is given. Write the first five terms of each sequence.

1. $s_n = \dfrac{(-1)^{n+1}}{n^4}$ **2.** $s_n = \dfrac{2^n}{3^n}$

3. Find an expression for the nth term of the sequence, $2, -\dfrac{3}{2}, \dfrac{9}{8},$

$-\dfrac{27}{32}, \dfrac{81}{128}, \ldots$, assuming the indicated pattern continues for all n.

In Problems 4 and 5, use properties of convergent sequences to find the limit of each sequence.

4. $\left\{1 + \dfrac{n}{n^2+1}\right\}$ **5.** $\left\{\ln\dfrac{n+2}{n}\right\}$

In Problems 6 and 7, use a related function or the Squeeze Theorem for sequences to show each sequence converges. Find its limit.

6. $\left\{\tan^{-1} n\right\}$ **7.** $\left\{\dfrac{(-1)^n}{(n+1)^2}\right\}$

8. Determine if the sequence $\left\{\dfrac{e^n}{(n+2)^2}\right\}$ is monotonic. If it is monotonic, is it increasing, nondecreasing, decreasing, or nonincreasing? Is it bounded from above and/or from below? Does it converge?

In Problems 9–12, determine whether each sequence converges or diverges. If it converges, find its limit.

9. $\{n!\}$ **10.** $\left\{\left(\dfrac{5}{8}\right)^n\right\}$

11. $\left\{\left(-\dfrac{1}{2}\right)^n\right\}$ **12.** $\left\{(-1)^n + e^{-n}\right\}$

13. Show that sequence $\left\{1 + \dfrac{2}{n}\right\}$ converges by showing it is either bounded from above and increasing or is bounded from below and decreasing.

14. Find the fifth partial sum of $\sum_{k=1}^{\infty} \frac{(-1)^k}{4^{k-1}}$.

15. Find the sum of the telescoping series $\sum_{k=1}^{\infty} \left(\frac{4}{k+4} - \frac{4}{k+5} \right)$.

In Problems 16–19, determine whether each series converges or diverges. If it converges, find its sum.

16. $\sum_{k=1}^{\infty} \frac{\cos^2(k\pi)}{k}$

17. $\sum_{k=1}^{\infty} -(\ln 2)^k$

18. $\sum_{k=0}^{\infty} \frac{e}{3^k}$

19. $\sum_{k=1}^{\infty} (4^{1/3})^k$

20. Express $0.123123123\ldots$ as a rational number using a geometric series.

21. Show that the series $\sum_{k=1}^{\infty} \frac{3k-2}{k}$ diverges.

In Problems 22 and 23, use the Integral Test to determine whether each series converges or diverges.

22. $\sum_{k=1}^{\infty} \frac{\ln k}{k^2}$

23. $\sum_{k=1}^{\infty} \frac{1}{4k^2+9}$

24. Determine whether the *p*-series $\sum_{k=1}^{\infty} \frac{1}{k^{5/2}}$ converges or diverges. If it converges, find bounds for the sum.

In Problems 25–27, determine whether each series converges or diverges.

25. $\sum_{k=5}^{\infty} \left[\frac{1}{k^5} \cdot \frac{1}{2^k} \right]$

26. $\sum_{k=1}^{\infty} \left[\frac{3}{5^k} - \left(\frac{2}{3} \right)^{k-1} \right]$

27. $\sum_{k=1}^{\infty} \frac{3}{k^5}$

In Problems 28–30, use a Comparison Test to determine whether each series converges or diverges.

28. $\sum_{k=1}^{\infty} \frac{1}{\sqrt{k}+1}$

29. $\sum_{k=1}^{\infty} \frac{k+1}{k^{k+1}}$

30. $\sum_{k=1}^{\infty} \frac{4}{k \, 3^k}$

In Problems 31–33, determine whether each alternating series converges or diverges. If the series converges, approximate the sum of each series correct to within 0.001.

31. $\sum_{k=1}^{\infty} (-1)^{k+1} \frac{k+2}{k(k+1)}$

32. $\sum_{k=1}^{\infty} (-1)^{k+1} \frac{k^2}{e^k}$

33. $\sum_{k=1}^{\infty} (-1)^k \frac{3}{\sqrt[3]{k}}$

In Problems 34–37, determine whether each series converges (absolutely or conditionally) or diverges.

34. $\sum_{k=1}^{\infty} \sin\left(\frac{\pi}{2}k \right)$

35. $\sum_{k=1}^{\infty} \frac{(-1)^{k+1}}{\sqrt{k}}$

36. $\sum_{k=1}^{\infty} \frac{\cos k}{k^3}$

37. $\frac{1}{2} - \frac{4}{2^3+1} + \frac{9}{3^3+1} - \frac{16}{4^3+1} + \cdots$

In Problems 38 and 39, use the Ratio Test to determine whether each series converges or diverges.

38. $\sum_{k=1}^{\infty} \frac{2^k}{k!}$

39. $\sum_{k=1}^{\infty} \frac{k!}{e^{k^2}}$

In Problems 40 and 41, use the Root Test to determine whether each series converges or diverges.

40. $\sum_{k=1}^{\infty} \frac{2^k}{(k+3)^{k+1}}$

41. $\sum_{k=1}^{\infty} (-1)^k (e^{-k} - 1)^k$

In Problems 42–52, determine whether each series converges or diverges.

42. $\sum_{k=1}^{\infty} (-1)^{k+1} \frac{2^{k+1}}{3^k}$

43. $\sum_{k=1}^{\infty} \ln\left(1 + \frac{1}{k} \right)$

44. $\sum_{k=5}^{\infty} \frac{3}{k\sqrt{k-4}}$

45. $\sum_{k=1}^{\infty} \frac{1}{\left(1 + \frac{k^2+1}{k^2} \right)^k}$

46. $\sum_{k=1}^{\infty} \frac{2 \cdot 4 \cdot 6 \cdots (2k)}{1 \cdot 3 \cdot 5 \cdots (2k-1)}$

47. $\sum_{k=1}^{\infty} \frac{k^2}{(1+k^3) \ln \sqrt[3]{1+k^3}}$

48. $\sum_{k=1}^{\infty} \frac{k^{10}}{2^k}$

49. $\sum_{k=1}^{\infty} \frac{\left(1 + \frac{1}{k^2} \right)^{k^2}}{2^k}$

50. $\sum_{k=1}^{\infty} \left(\frac{k^2+1}{k} \right)^k$

51. $\sum_{k=1}^{\infty} \frac{k!}{3k^k}$

52. $\sum_{k=1}^{\infty} (-1)^{k+1} \frac{k+2}{3k-2}$

In Problems 53–58,
(a) Find the radius of convergence of each power series.
(b) Find the interval of convergence of each power series.

53. $\sum_{k=1}^{\infty} \frac{(x-3)^{3k-1}}{k^2}$

54. $\sum_{k=1}^{\infty} \frac{x^k}{\sqrt[3]{k}}$

55. $\sum_{k=0}^{\infty} (-1)^k \frac{1}{k!(k+1)} \left(\frac{x}{2} \right)^{2k+1}$

56. $\sum_{k=1}^{\infty} \frac{k^k}{(k!)^2} x^k$

57. $\sum_{k=1}^{\infty} \frac{(x-1)^k}{k}$

58. $\sum_{k=0}^{\infty} \frac{3^k x^k}{5^k}$

In Problems 59 and 60, express each function as a power series centered at 0.

59. $f(x) = \frac{2}{x+3}$

60. $f(x) = \frac{1}{1-3x}$

61. (a) Use properties of a power series to find the power series representation for $\int \frac{1}{1-3x^2} dx$.

(b) Use (a) to approximate $\int_0^{1/2} \frac{1}{1-3x^2} dx$ correct to within 0.001.

62. Find the Taylor expansion of $f(x) = \dfrac{1}{1 - 2x}$ about $c = 1$.

63. Find the Taylor expansion of $f(x) = e^{x/2}$ about $c = 1$.

64. Find the Maclaurin expansion of $f(x) = 2x^3 - 3x^2 + x + 5$. Comment on the result.

65. Find the Taylor expansion of $f(x) = \tan x$ about $c = \dfrac{\pi}{4}$.

66. Find the first five terms of the Maclaurin expansion for $f(x) = e^{-x} \sin x$.

In Problems 67–69, use a binomial series to represent each function. Then determine its interval of convergence.

67. $f(x) = \dfrac{1}{(x + 1)^4}$

68. $f(x) = \sqrt[3]{x^2 - 1}$

69. $f(x) = \dfrac{1}{\sqrt{1 - x}}$

70. (a) Approximate $y = \cos x$ by using the first four nonzero terms of the Taylor expansion for $y = \cos x$ about $\dfrac{\pi}{2}$.

(b) Graph $y = \cos x$ along with the Taylor expansion found in (a).

(c) Use (a) to approximate $\cos 88°$.

(d) What is the error in using this approximation?

(e) If the approximation in (a) is used, for what values of x is the error less than 0.0001?

71. Use the Maclaurin expansion for $f(x) = e^x$ to approximate $e^{0.3}$ correct to three decimal points

In Problems 72 and 73, use properties of power series to approximate each integral using the first four terms of a Maclaurin series.

72. $\displaystyle\int_0^{1/2} \dfrac{dx}{\sqrt{1 - x^3}}$

73. $\displaystyle\int_0^{1/2} e^{x^2}\, dx$

CHAPTER 8 PROJECT   How Calculators Calculate

The sine function is used in many scientific applications, so a calculator/computer must be able to evaluate it with lightning-fast speed.

While we know how to find the exact value of the sine function for many numbers, such as 0, $\dfrac{\pi}{6}$, $\dfrac{\pi}{2}$, and so on, we have no methodology for finding the exact value of $\sin 3$ (which should be close to $\sin \pi$) or $\sin 1.5$ (which should be close to $\sin \dfrac{\pi}{2}$). Since the sine function can be evaluated at any real number, we first use some of its properties to restrict its domain to something more manageable.

1. Explain why we can evaluate $\sin x$ for any x using only the interval $\left[-\dfrac{\pi}{2}, \dfrac{\pi}{2}\right]$. (We could restrict that domain further, but this will work for now. See Problem 6 below.)

2. Use the Maclaurin expansion for $\sin x$ to find an approximation for $\sin \dfrac{1}{2}$ correct to within 10^{-5}. Compare your approximation to the one your calculator/computer provides. How many terms of the series do you need to obtain this accuracy?

3. Find an approximation for $\sin \dfrac{3}{2}$ correct to within 10^{-5}. How many terms of the series do you need to obtain this accuracy?

4. Explain why the approximation in Problem 3 requires more terms than that of Problem 2.

5. Represent $\sin x$ as a Taylor expansion about $\dfrac{\pi}{4}$.

6. Explain why we can evaluate $\sin x$ for any x using only the interval $\left[0, \dfrac{\pi}{2}\right]$.

7. Use the result of Problem 5 to find an approximation for $\sin \dfrac{1}{2}$ correct to within 10^{-5}. Compare the result with the values your calculator/computer supplies for $\sin \dfrac{1}{2}$, as well as with the result from the Maclaurin approximation obtained in Problem 2. How many terms of the series do you need for the approximation?

8. Use the Taylor expansion to find an approximation for $\sin \dfrac{3}{2}$ correct to within 10^{-5}. Compare the result with the value your calculator/computer supplies for $\sin \dfrac{3}{2}$, as well as with the result from the Maclaurin approximation obtained in Problem 3. How many terms of the series do you need for the approximation?

The answers to Problems 3 and 8 reveal why Maclaurin series or Taylor series are not used to approximate the value of most functions. But often the methods used are similar. For example, a Chebyshev polynomial approximation to the sine function on the interval $\left[-\dfrac{\pi}{2}, \dfrac{\pi}{2}\right]$ still has the form of a Maclaurin series, but it was designed to converge more uniformly than the Maclaurin series, so that it can be expected to give answers near $\dfrac{\pi}{2}$ that are roughly as accurate as those near zero.

Chebyshev polynomials are commonly found in mathematical libraries for calculators/computers. For example, the widely used

Gnu Compiler Collection* uses Chebyshev polynomials to evaluate trigonometric functions. The Chebyshev polynomial approximation of degree 7 for the sine function is

$$S_7(x) = 0.9999966013x - 0.1666482357x^3$$
$$+ 0.008306286146x^5 - 0.1836274858 \times 10^{-3}x^7 \quad (1)$$

The Chebyshev polynomials are designed to remain close to a function across an entire closed interval. They seek to keep the approximation within a specified distance of the function being approximated at every point of that interval. If S_n is a Chebyshev approximation of degree n to the sine function on $\left[-\dfrac{\pi}{2}, \dfrac{\pi}{2}\right]$, then the error estimate in using $S_n(x)$ is given by

$$\max_{-\pi/2 \leq x \leq \pi/2} |\sin x - S_n(x)| \leq \frac{\left(\dfrac{\pi}{2}\right)^{n+1}}{2^n(n+1)!} \quad (2)$$

Like most error estimates of this type, it gives an upper bound to the error.

9. Use the Chebyshev polynomial approximation in (1) for $x = \dfrac{1}{2}$ and $x = \dfrac{3}{2}$. Compare the results with the values your calculator/computer supplies for $\sin \dfrac{1}{2}$ and $\sin \dfrac{3}{2}$, as well as with the results from the Maclaurin approximations and the Taylor series approximations obtained in Problems 2, 3, 7, and 8.

10. Define $E_7(x) = |\sin x - S_7(x)|$. Use graphing technology to graph E_7 on $\left[-\dfrac{\pi}{2}, \dfrac{\pi}{2}\right]$.

11. Find the local maximum and local minimum values of E_7 on $\left[-\dfrac{\pi}{2}, \dfrac{\pi}{2}\right]$. Compare these numbers with the error estimate in equation (2), and discuss the characteristics of the error.

12. Which approximation for the sine function would be preferable: the Maclaurin approximation, the Taylor approximation, or the Chebyshev approximation? Why?

---

*For more information on the Gnu Compiler Collection (GCC), go to https://www.gnu.org/software/gcc/

2 Find a Rectangular Equation for a Curve Represented Parametrically

The plane curve in Figure 2 should look familiar. To identify it, we find the corresponding rectangular equation by eliminating the parameter t from the parametric equations

$$x(t) = 3t^2 \qquad y(t) = 2t \qquad -2 \le t \le 2$$

We begin by solving for t in $y = 2t$, obtaining $t = \dfrac{y}{2}$. Then we substitute $t = \dfrac{y}{2}$ in the other equation.

$$x = 3t^2 = 3\left(\frac{y}{2}\right)^2 = \frac{3y^2}{4}$$

The equation $x = \dfrac{3y^2}{4}$ is a parabola with its vertex at the origin and its axis of symmetry along the x-axis. We refer to this equation as the **rectangular equation** of the curve to distinguish it from the parametric equations.

Notice that the plane curve represented by the parametric equations $x(t) = 3t^2$, $y(t) = 2t$, $-2 \le t \le 2$, is only part of the parabola $x = \dfrac{3y^2}{4}$. In general, the graph of the rectangular equation obtained by eliminating the parameter will contain more points than the plane curve defined using parametric equations. Therefore, when graphing a plane curve represented by a rectangular equation, we may have to restrict the graph to match the parametric equations.

EXAMPLE 2   Finding a Rectangular Equation for a Plane Curve Represented Parametrically

Find a rectangular equation of the curve whose parametric equations are

$$x(t) = R \cos t \qquad y(t) = R \sin t$$

where $R > 0$ is a constant. Graph the plane curve and indicate its orientation.

Solution   The presence of the sine and cosine functions in the parametric equations suggests using the Pythagorean Identity $\cos^2 t + \sin^2 t = 1$. Then

$$\left(\frac{x}{R}\right)^2 + \left(\frac{y}{R}\right)^2 = 1 \qquad \cos t = \frac{x}{R} \quad \sin t = \frac{y}{R}$$

$$x^2 + y^2 = R^2$$

The graph of the rectangular equation is a circle with center at the origin and radius R. In the parametric equations, as the parameter t increases, the points (x, y) on the circle are traced out in the counterclockwise direction, as shown in Figure 3. ■

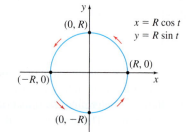

Figure 3 The orientation is counterclockwise.

$x = R \cos t$
$y = R \sin t$

NOW WORK   Problems 9 and 15.

When analyzing the parametric equations in Example 2, notice there are no restrictions on t, so t varies from $-\infty$ to ∞. As a result, the circle is repeated each time t increases by 2π.

If we wanted to describe a curve consisting of exactly one revolution in the counterclockwise direction, the domain must be restricted to an interval of length 2π. For example, we could use

$$x(t) = R \cos t \qquad y(t) = R \sin t \quad 0 \le t \le 2\pi$$

Now the plane curve starts when $t = 0$ at $(R, 0)$ and ends when $t = 2\pi$ at $(R, 0)$.

If we wanted the curve to consist of exactly three revolutions in the counterclockwise direction, the domain must be restricted to an interval of length 6π. For example, we could use $x(t) = R \cos t$, $y(t) = R \sin t$, with $-2\pi \le t \le 4\pi$, or $0 \le t \le 6\pi$, or $2\pi \le t \le 8\pi$, or any arbitrary interval of length 6π.

EXAMPLE 3 **Finding a Rectangular Equation for a Plane Curve Represented Parametrically**

Find rectangular equations for the plane curves represented by each of the parametric equations. Graph each curve and indicate its orientation.

(a) $x(t) = R \cos t$ $y(t) = R \sin t$ $0 \le t \le \pi$ and $R > 0$
(b) $x(t) = R \sin t$ $y(t) = R \cos t$ $0 \le t \le \pi$ and $R > 0$

Solution **(a)** We eliminate the parameter t using a Pythagorean Identity.

$$\cos^2 t + \sin^2 t = 1$$

$$\left(\frac{x}{R}\right)^2 + \left(\frac{y}{R}\right)^2 = 1 \qquad \cos t = \frac{x}{R}, \quad \sin t = \frac{y}{R}$$

$$x^2 + y^2 = R^2 \tag{1}$$

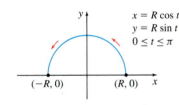

The rectangular equation represents a circle with radius R and center at the origin. In the parametric equations, $0 \le t \le \pi$, so the curve begins when $t = 0$ at the point $(R, 0)$, passes through the point $(0, R)$ when $t = \frac{\pi}{2}$, and ends when $t = \pi$ at the point $(-R, 0)$. The curve is an upper semicircle of radius R with counterclockwise orientation, as shown in Figure 4. If we solve equation (1) for y, we obtain the rectangular equation of the semicircle

$$y = \sqrt{R^2 - x^2} \qquad \text{where} \ -R \le x \le R$$

Figure 4 $y = \sqrt{R^2 - x^2}, -R \le x \le R.$

(b) We eliminate the parameter t as we did in (a), and again we obtain

$$x^2 + y^2 = R^2 \tag{2}$$

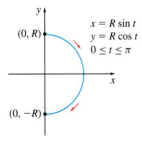

The rectangular equation represents a circle with radius R and center at $(0, 0)$. But in the parametric equations, $0 \le t \le \pi$, so now the curve begins when $t = 0$ at the point $(0, R)$, passes through the point $(R, 0)$ when $t = \frac{\pi}{2}$, and ends at the point $(0, -R)$ when $t = \pi$. The curve is a right semicircle of radius R with a *clockwise* orientation, as shown in Figure 5. If we solve equation (2) for x, we obtain the rectangular equation of the semicircle

$$x = \sqrt{R^2 - y^2} \qquad \text{where} \ -R \le y \le R \qquad ■$$

Figure 5 $x = \sqrt{R^2 - y^2}, -R \le y \le R.$

NOTE In Example 3 (a) we expressed the rectangular equation as a function $y = f(x)$, and in (b) we expressed the rectangular equation as a function $x = g(y)$.

Parametric equations are not unique; that is, different parametric equations can represent the same graph. Two examples of other parametric equations that represent the graph in Figure 5 are

- $x(t) = R \sin t$ $y(t) = -R \cos(\pi - t)$ $0 \le t \le \pi$

- $x(t) = R \sin(3t)$ $y(t) = R \cos(3t)$ $0 \le t \le \frac{\pi}{3}$

There are other examples.

NOW WORK Problem **17.**

For the motion to be counterclockwise, as t increases from 0, the value of x must decrease and the value of y must increase. This requires $\omega > 0$. [Do you see why? If $\omega > 0$, then $x = 2\cos(\omega t)$ is decreasing when $t > 0$ is near zero, and $y = \sin(\omega t)$ is increasing when $t > 0$ is near zero.]

NEED TO REVIEW? The period of a sinusoidal graph is discussed in Section P. 6, p. 50.

Finally, since 1 revolution takes 1 second, the period is $\dfrac{2\pi}{\omega} = 1$, so $\omega = 2\pi$. Parametric equations that satisfy the conditions given in (a) are

$$x(t) = 2\cos(2\pi t) \qquad y(t) = \sin(2\pi t) \quad 0 \le t \le 1$$

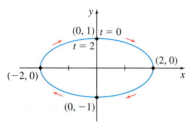

Figure 9 $\dfrac{x^2}{4} + y^2 = 1$

(b) Figure 9 shows the graph of the ellipse $\dfrac{x^2}{4} + y^2 = 1$.

Since the motion begins at the point $(0, 1)$, we want $x = 0$ and $y = 1$ when $t = 0$. We let

$$x(t) = 2\sin(\omega t) \qquad \text{and} \qquad y(t) = \cos(\omega t)$$

for some constant ω. This choice for $x = x(t)$ and $y = y(t)$ satisfies the equation $\dfrac{x^2}{4} + y^2 = 1$ and also satisfies the requirement that when $t = 0$, then $x = 2\sin 0 = 0$ and $y = \cos 0 = 1$.

For the motion to be clockwise, as t increases from 0, the value of x must increase and the value of y must decrease. This requires that $\omega > 0$.

Finally, since 1 revolution takes 2 seconds, the period is $\dfrac{2\pi}{\omega} = 2$, or $\omega = \pi$. Parametric equations that satisfy the conditions given in (b) are

$$x(t) = 2\sin(\pi t) \qquad y(t) = \cos(\pi t) \quad 0 \le t \le 2$$ ∎

NOW WORK **Problem 55.**

The Cycloid

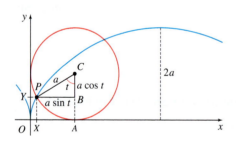

Figure 10

RECALL For a circle of radius r, a central angle of θ radians subtends an arc whose length s is $s = r\theta$.

Suppose that a circle rolls along a horizontal line without slipping. As the circle rolls along the line, a point P on the circle traces out a curve called a **cycloid**, as shown in Figure 10. Deriving the equation of the cycloid in rectangular coordinates is complicated, but the derivation in terms of parametric equations is relatively easy.

We begin with a circle of radius a. Suppose the fixed line on which the circle rolls is the x-axis. Let the origin be one of the points at which the point P comes into contact with the x-axis. Figure 10 shows the position of this point P after the circle has rolled a bit. The angle t (in radians) measures the angle through which the circle has rolled. Since the circle does not slip, it follows that

$$\text{Arc } AP = d(O, A)$$

The length of the arc AP is

$$at = d(O, A) \qquad s = r\theta, \quad r = a, \quad \theta = t$$

The x-coordinate of the point $P = (x, y)$ is

$$x = d(O, X) = d(O, A) - d(X, A) = at - a\sin t = a(t - \sin t)$$

The y-coordinate of the point P is

$$y = d(O, Y) = d(A, C) - d(B, C) = a - a\cos t = a(1 - \cos t)$$

THEOREM Parametric Equations of a Cycloid

Parametric equations of a cycloid are

$$\boxed{x(t) = a(t - \sin t) \qquad y(t) = a(1 - \cos t) \qquad -\infty < t < \infty}$$

Applications to Mechanics

NOTE In Greek, *brachistochrone* means "the shortest time," and *tautochrone* "equal time."

If $a < 0$ in the parametric equations of the cycloid, the result is an inverted cycloid, as shown in Figure 11(a). The inverted cycloid arises as a result of some remarkable applications in the field of mechanics. We discuss two of them: the *brachistochrone* and *tautochrone*.

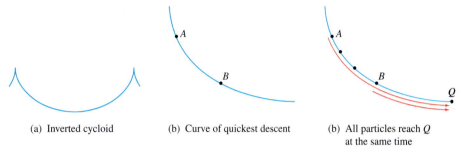

(a) Inverted cycloid   (b) Curve of quickest descent   (b) All particles reach Q at the same time

Figure 11

The **brachistochrone** is the curve of quickest descent. If an object is constrained to follow some path from a point A to a lower point B (not on the same vertical line) and is acted on only by gravity, the time needed to make the descent is minimized if the path is an inverted cycloid. See Figure 11(b). For example, in sliding packages from a loading dock onto a ship, a ramp in the shape of an inverted cycloid might be used so the packages get to the ship in the least amount of time. This discovery, which is attributed to many famous mathematicians (including Johann Bernoulli and Blaise Pascal), was a significant step in creating the branch of mathematics known as the *calculus of variations*.

Suppose Q is the lowest point on an inverted cycloid. If several objects placed at various positions on an inverted cycloid simultaneously begin to slide down the cycloid, they will reach the point Q at the same time, as indicated in Figure 11(c). This is referred to as the **tautochrone property** of the cycloid. It was used by the Dutch mathematician, physicist, and astronomer Christian Huygens (1629–1695) to construct a pendulum clock with a bob that swings along an inverted cycloid, as shown in Figure 12. In Huygens' clock, the bob was made to swing along an inverted cycloid by suspending the bob on a thin wire constrained by two plates shaped like cycloids. In a clock of this design, the period of the pendulum is independent of its amplitude.

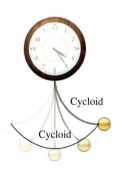

Cycloid

Cycloid

Figure 12

9.1 Assess Your Understanding

Concepts and Vocabulary

1. Let $x = x(t)$ and $y = y(t)$ be two functions whose common domain is some interval I. The collection of points defined by $(x, y) = (x(t), y(t))$ is called a plane _____. The variable t is called a(n) _____.

2. *Multiple Choice* The parametric equations $x(t) = 2 \sin t$, $y(t) = 3 \cos t$ define a(n) [**(a)** line, **(b)** hyperbola, **(c)** ellipse, **(d)** parabola].

3. *Multiple Choice* The parametric equations $x(t) = a \sin t$, $y(t) = a \cos t$ define a [**(a)** line, **(b)** hyperbola, **(c)** parabola, **(d)** circle].

4. If a circle rolls along a horizontal line without slipping, a point P on the circle traces out a curve called a(n) _____.

5. *True or False* The parametric equations defining a curve are unique.

6. *True or False* Plane curves represented using parametric equations have an orientation.

Skill Building

In Problems 7–20:

(a) *Find the rectangular equation of each plane curve with the given parametric equations.*

(b) *Graph the plane curve represented by the parametric equations and indicate its orientation.*

7. $x(t) = 2t + 1$, $y(t) = t + 2$; $-\infty < t < \infty$

8. $x(t) = t - 2$, $y(t) = 3t + 1$; $-\infty < t < \infty$

9. $x(t) = 2t + 1$, $y(t) = t + 2$; $0 \le t \le 2$

10. $x(t) = t - 2$, $y(t) = 3t + 1$; $0 \le t \le 2$

11. $x(t) = e^t$, $y(t) = t$; $-\infty < t < \infty$

12. $x(t) = t$, $y(t) = \dfrac{1}{t}$; $-\infty < t < \infty, t \ne 0$

13. $x(t) = \sin t$, $y(t) = \cos t$; $0 \le t \le 2\pi$

14. $x(t) = \cos t$, $y(t) = \sin t$; $0 \le t \le \pi$

1. = NOW WORK problem   = Graphing technology recommended   CAS = Computer Algebra System recommended

(a) $\left[0, \dfrac{5\pi}{6}\right]$ **(b)** $[\pi, 2\pi]$ **(c)** $\left[0, \dfrac{\pi}{2}\right]$ **(d)** $\left[\dfrac{\pi}{6}, \dfrac{5\pi}{6}\right]$

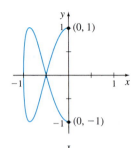

I

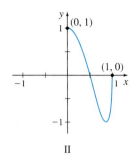

II

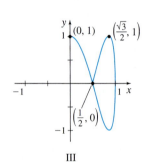

III

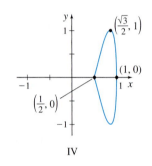

IV

📶 **65. (a)** Graph the plane curve represented by the parametric
equations $x(t) = 7(t - \sin t)$, $y(t) = 7(1 - \cos t)$,
$0 \le t \le 2\pi$.

(b) Find the coordinates of the point (x, y) on the curve when
$t = 2.1$.

📶 **66. (a)** Graph the plane curve represented by the parametric
equations $x(t) = t - e^t$, $y(t) = 2e^{t/2}$, $-8 \le t \le 2$.

(b) Find the coordinates of the point (x, y) on the curve when
$t = 1.5$.

67. Find parametric equations for the ellipse $\dfrac{x^2}{a^2} + \dfrac{y^2}{b^2} = 1$.

68. Find parametric equations for the hyperbola $\dfrac{x^2}{a^2} - \dfrac{y^2}{b^2} = 1$.

*Problems 69–71 involve projectile motion. When an object is propelled
upward at an inclination θ to the horizontal with initial speed v_0, the
resulting motion is called* **projectile motion.** *See the figure. Parametric
equations that model the path of the projectile, ignoring air resistance,
are given by*

$$x(t) = (v_0 \cos \theta)t \qquad y(t) = -\dfrac{1}{2}gt^2 + (v_0 \sin \theta)t + h$$

*where t is time, g is the constant acceleration due to gravity
(approximately 32 ft/s² or 9.8 m/s²), and h is the height from which
the projectile is released.*

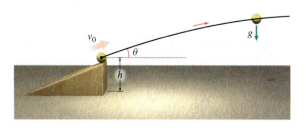

69. Trajectory of a Baseball   A baseball is hit with an initial speed
of 125 ft/s at an angle of 40° to the horizontal. The ball is hit at a
height of 3 ft above the ground.

(a) Find parametric equations that model the position of the ball
as a function of the time t in seconds.

(b) What is the height of the baseball after 2 seconds?

(c) What horizontal distance x has the ball traveled after
2 seconds?

(d) How long does it take the baseball to travel $x = 300$ ft?

(e) What is the height of the baseball at the time found in (d)?

(f) How long is the ball in the air before it hits the ground?

(g) How far has the baseball traveled horizontally when it hits the
ground?

70. Trajectory of a Baseball   A pitcher throws a baseball with an
initial speed of 145 ft/s at an angle of 20° to the horizontal. The
ball leaves his hand at a height of 5 ft.

(a) Find parametric equations that model the position of the ball
as a function of the time t in seconds.

(b) What is the height of the baseball after $\dfrac{1}{2}$ second?

(c) What horizontal distance x has the ball traveled after $\dfrac{1}{2}$
second?

(d) How long does it take the baseball to travel $x = 60$ ft?

(e) What is the height of the baseball at the time found in (d)?

71. Trajectory of a Football   A quarterback throws a football with
an initial speed of 80 ft/s at an angle of 35° to the horizontal. The
ball leaves the quarterback's hand at a height of 6 ft.

(a) Find parametric equations that model the position of the ball
as a function of time t in seconds.

(b) What is the height of the football after 1 second?

(c) What horizontal distance x has the ball traveled after
1 second?

(d) How long does it take the football to travel $x = 120$ ft?

(e) What is the height of the football at the time found in (d)?

72. The plane curve represented by the parametric equations $x(t) = t$,
$y(t) = t^2$, and the plane curve represented by the parametric
equations $x(t) = t^2$, $y(t) = t^4$ (where, in each case, time t is the
parameter) appear to be identical, but they differ in an important
aspect. Identify the difference, and explain its meaning.

73. The circle $x^2 + y^2 = 4$ can be represented by the parametric
equations $x(\theta) = 2 \cos \theta$, $y(\theta) = 2 \sin \theta$, $0 \le \theta \le 2\pi$, or by the
parametric equations $x(\theta) = 2 \sin \theta$, $y(\theta) = 2 \cos \theta$, $0 \le \theta \le 2\pi$.
But the plane curves represented by each pair of parametric
equations are different. Identify and explain the difference.

74. Uniform Motion   A train leaves a station at 7:15 a.m. and
accelerates at the rate of 3 mi/h². Mary, who can run 6 mi/h,
arrives at the station 2 seconds after the train has left. Find
parametric equations that model the motion of the train and of
Mary as a function of time. (*Hint:* The position s at time t of an
object having acceleration a is $s = \dfrac{1}{2}at^2$.)

Challenge Problems

75. Find parametric equations for the circle $x^2 + y^2 = R^2$, using as the parameter the slope m of the line through the point $(-R, 0)$ and a general point $P = (x, y)$ on the circle.

76. Find parametric equations for the parabola $y = x^2$, using as the parameter the slope m of the line joining the point $(1, 1)$ to a general point $P = (x, y)$ on the parabola.

77. Hypocycloid   Let a circle of radius b roll, without slipping, inside a fixed circle with radius a, where $a > b$. A fixed point P on the circle of radius b traces out a curve, called a **hypocycloid**, as shown in the figure. If $A = (a, 0)$ is the initial position of the point P and if t denotes the angle from the

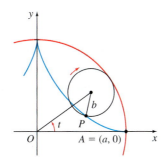

positive x-axis to the line segment from the origin to the center of the circle, show that the parametric equations of the hypocycloid are

$$x(t) = (a - b)\cos t + b\cos\left(\frac{a - b}{b}t\right)$$

$$y(t) = (a - b)\sin t - b\sin\left(\frac{a - b}{b}t\right) \quad 0 \le t \le 2\pi$$

78. Hypocycloid   Show that the rectangular equation of a hypocycloid with $a = 4b$ is $x^{2/3} + y^{2/3} = a^{2/3}$.

79. Epicycloid   Suppose a circle of radius b rolls on the outside of a second circle, as shown in the figure. Find the parametric equations of the curve, called an **epicycloid**, traced out by the point P.

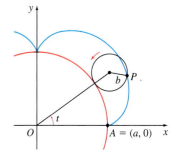

9.2 Tangent Lines; Arc Length

OBJECTIVES *When you finish this section, you should be able to:*

1 **Find an equation of the tangent line at a point on a plane curve (p. 648)**

2 **Find the arc length of a plane curve (p. 651)**

Recall that if a function f is differentiable, the derivative $f'(x) = \dfrac{dy}{dx}$ is the slope of the tangent line to the graph of f at a point (x, y). To obtain a formula for the slope of the tangent line to a curve when it is represented by parametric equations, $x = x(t)$, $y = y(t)$, $a \le t \le b$, we require the curve to be *smooth*.

DEFINITION Smooth Curve

Let C denote a plane curve represented by the parametric equations

$$x = x(t) \qquad y = y(t) \quad a \le t \le b$$

Suppose each function $x(t)$ and $y(t)$ is continuous on the closed interval $[a, b]$ and differentiable on the open interval (a, b). If both $\dfrac{dx}{dt}$ and $\dfrac{dy}{dt}$ are continuous and are never simultaneously 0 on (a, b), then C is called a **smooth curve**.

A smooth curve $x = x(t)$, $y = y(t)$, for which $\dfrac{dx}{dt}$ is never 0, can be represented by the rectangular equation $y = F(x)$, where F is differentiable. (You are asked to prove this in Problem 64.) Suppose $(x(t), y(t))$ is a point on the curve. Then

$$y = F(x)$$

$$y(t) = F(x(t))$$

Now, we use the Chain Rule to obtain

$$\frac{dy}{dt} = \frac{dy}{dx} \cdot \frac{dx}{dt}$$

lines is given by

$$\frac{dy}{dx} = \frac{\dfrac{dy}{dt}}{\dfrac{dx}{dt}} = \frac{\dfrac{d}{dt}(t^2)}{\dfrac{d}{dt}(t^3 - 4t)} = \frac{2t}{3t^2 - 4}$$

When $t = -2$, the slope of the tangent line is

$$\frac{dy}{dx} = \frac{2(-2)}{3(-2)^2 - 4} = \frac{-4}{8} = -\frac{1}{2}$$

and an equation of the tangent line at the point $(0, 4)$ is

$$y - 4 = -\frac{1}{2}(x - 0)$$

$$y = -\frac{1}{2}x + 4$$

When $t = 2$, the slope of the tangent line at the point $(0, 4)$ is

$$\frac{dy}{dx} = \frac{2(2)}{3(2)^2 - 4} = \frac{4}{8} = \frac{1}{2}$$

and an equation of the tangent line at $(0, 4)$ is

$$y - 4 = \frac{1}{2}(x - 0)$$

$$y = \frac{1}{2}x + 4$$ ∎

NOW WORK Problem 45.

EXAMPLE 4 **Projectile Motion**

A projectile is fired at an angle θ, $0 < \theta < \dfrac{\pi}{2}$, to the horizontal with an initial speed of v_0 m/s. Assuming no air resistance, the position of the projectile after t seconds is given by the parametric equations $x(t) = (v_0 \cos\theta)t$, $y(t) = (v_0 \sin\theta)t - \dfrac{1}{2}gt^2$, $t \geq 0$, where g is the acceleration due to gravity.

(a) Find the slope of the tangent line to the motion of the projectile as a function of t.
(b) At what time is the projectile at its maximum height?

Solution **(a)** The slope of the tangent line is given by $\dfrac{dy}{dx}$.

$$\frac{dy}{dx} = \frac{\dfrac{dy}{dt}}{\dfrac{dx}{dt}} = \frac{\dfrac{d}{dt}\left[(v_0 \sin\theta)t - \dfrac{1}{2}gt^2\right]}{\dfrac{d}{dt}[(v_0 \cos\theta)t]} = \frac{v_0 \sin\theta - gt}{v_0 \cos\theta} = \tan\theta - \frac{gt}{v_0 \cos\theta}$$

(b) The projectile is at its maximum height when the slope of the tangent line equals 0. That is, when

$$\frac{dy}{dx} = \frac{v_0 \sin\theta - gt}{v_0 \cos\theta} = 0$$

$$v_0 \sin\theta - gt = 0$$

$$t = \frac{v_0 \sin\theta}{g}$$ ∎

2 Find the Arc Length of a Plane Curve

NEED TO REVIEW? Arc length is discussed in Section 6.5, pp. 438–442.

For a function $y = f(x)$ that has a derivative that is continuous on some interval containing a and b, the arc length s of f from a to b is given by $s = \int_a^b \sqrt{1 + [f'(x)]^2}\, dx$. For a smooth curve represented by the parametric equations $x = x(t)$, $y = y(t)$, we have the following formula for arc length.

> **THEOREM Arc Length Formula for Parametric Equations**
>
> For a smooth curve C represented by the parametric equations
> $$x = x(t) \qquad y = y(t) \quad a \le t \le b$$
> the arc length s of C from $t = a$ to $t = b$ is given by the formula
> $$s = \int_a^b \sqrt{\left(\frac{dx}{dt}\right)^2 + \left(\frac{dy}{dt}\right)^2}\, dt$$

A partial proof of this theorem is given at the end of the section.

EXAMPLE 5   Finding Arc Length for Parametric Equations

Find the length s of the curve represented by the parametric equations
$$x(t) = t^3 + 2 \qquad\qquad y(t) = 2t^{9/2}$$
from the point where $t = 1$ to the point where $t = 3$. Figure 16 shows the graph of the curve.

Solution We begin by finding the derivatives $\dfrac{dx}{dt}$ and $\dfrac{dy}{dt}$.

$$\frac{dx}{dt} = 3t^2 \qquad \text{and} \qquad \frac{dy}{dt} = 9t^{7/2}$$

The curve is smooth for $1 \le t \le 3$. Now using the arc length formula for parametric equations, we have

$$s = \int_a^b \sqrt{\left(\frac{dx}{dt}\right)^2 + \left(\frac{dy}{dt}\right)^2}\, dt = \int_1^3 \sqrt{(3t^2)^2 + (9t^{7/2})^2}\, dt$$

$$= \int_1^3 \sqrt{9t^4 + 81\,t^7}\, dt = \int_1^3 3t^2\sqrt{1 + 9t^3}\, dt$$

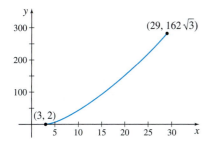

Figure 16 $x(t) = t^3 + 2$, $y(t) = 2t^{9/2}$, $1 \le t \le 3$

NEED TO REVIEW? The substitution method for definite integrals is discussed in Section 5.6, pp. 391–393.

We use the substitution $u = 1 + 9t^3$. Then $du = 27t^2 dt$, or equivalently, $3t^2 dt = \dfrac{du}{9}$. Changing the limits of integration, we find that when $t = 1$, then $u = 10$; and when $t = 3$, then $u = 1 + 9 \cdot 3^3 = 244$. The arc length s is

$$s = \int_1^3 3t^2\sqrt{1 + 9t^3}\, dt = \int_{10}^{244} \sqrt{u}\left(\frac{du}{9}\right) = \frac{1}{9}\int_{10}^{244} u^{1/2}du = \frac{1}{9}\left[\frac{u^{3/2}}{\frac{3}{2}}\right]_{10}^{244}$$

$$= \frac{2}{27}[244^{3/2} - 10^{3/2}] = \frac{4}{27}\left[244\sqrt{61} - 5\sqrt{10}\right] \qquad \blacksquare$$

NOW WORK Problem 31.

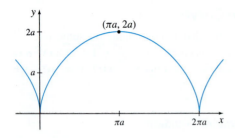

Figure 17 $x(t) = a(t - \sin t)$, $y(t) = a(1 - \cos t)$, $a > 0$.

EXAMPLE 6   **Finding the Arc Length of a Cycloid**

Find the length s of one arch of the cycloid:

$$x(t) = a(t - \sin t) \qquad y(t) = a(1 - \cos t) \quad a > 0$$

Figure 17 shows the graph of the cycloid.

Solution   One arch of the cycloid is obtained when t varies from 0 to 2π. Since

$$\frac{dx}{dt} = a - a\cos t = a(1 - \cos t) \qquad \text{and} \qquad \frac{dy}{dt} = a\sin t$$

are both continuous and are never simultaneously 0 on $(0, 2\pi)$, the cycloid is smooth on $[0, 2\pi]$. The arc length s is

$$s = \int_a^b \sqrt{\left(\frac{dx}{dt}\right)^2 + \left(\frac{dy}{dt}\right)^2}\, dt = \int_0^{2\pi} \sqrt{a^2(1 - \cos t)^2 + a^2 \sin^2 t}$$

$$= a \int_0^{2\pi} \sqrt{(1 - 2\cos t + \cos^2 t) + \sin^2 t}\, dt = a \int_0^{2\pi} \sqrt{1 - 2\cos t + 1}\, dt$$

$$= \sqrt{2}a \int_0^{2\pi} \sqrt{1 - \cos t}\, dt$$

To integrate $\sqrt{1 - \cos t}$ we use a half-angle identity. Since $\sin\dfrac{t}{2} \geq 0$ if $0 \leq t \leq 2\pi$, we have $\sin\dfrac{t}{2} = \sqrt{\dfrac{1 - \cos t}{2}}$. Then

$$\sqrt{1 - \cos t} = \sqrt{2} \sin\frac{t}{2}$$

Now the arc length s from $t = 0$ to $t = 2\pi$ is

$$s = \sqrt{2}a \int_0^{2\pi} \sqrt{1 - \cos t}\, dt = \sqrt{2}a \int_0^{2\pi} \sqrt{2} \sin\frac{t}{2}\, dt = 2a \left[-2\cos\frac{t}{2}\right]_0^{2\pi} = 8a \quad ■$$

NOW WORK   Problem 37.

Partial Proof of the Arc Length Formula   The proof for parametric equations is similar to the one we used in Chapter 6 to prove the arc length formula for a function $y = f(x)$, $a \leq x \leq b$.

First, we partition the closed interval $[a, b]$ into n subintervals:

$$[a, t_1], [t_1, t_2], \dots, [t_{i-1}, t_i], \dots, [t_{n-1}, b]$$

each of length $\Delta t = \dfrac{b - a}{n}$. Corresponding to each number $a, t_1, t_2, \dots, t_{n-1}, b$, there is a point $P_0, P_1, P_2, \dots, P_n$, on the curve, as shown in Figure 18.

We join each point P_{i-1} to the next point P_i with a line segment. The sum of the lengths of the line segments is an approximation to the length of the curve from $t = a$ to $t = b$. This sum can be written as

$$d(P_0, P_1) + d(P_1, P_2) + \cdots + d(P_{n-1}, P_n) = \sum_{i=1}^{n} d(P_{i-1}, P_i)$$

where $d(P_{i-1}, P_i)$ is the length of the line segment joining the points P_{i-1} and P_i. Using the distance formula, the length of each line segment is $d(P_{i-1}, P_i) = \sqrt{[x(t_i) - x(t_{i-1})]^2 + [y(t_i) - y(t_{i-1})]^2}$. Then the sum of the lengths of the line segments is

$$\sum_{i=1}^{n} d(P_{i-1}, P_i) = \sum_{i=1}^{n} \sqrt{[x(t_i) - x(t_{i-1})]^2 + [y(t_i) - y(t_{i-1})]^2}$$

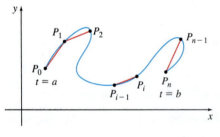

Figure 18 $x = x(t)$, $y = y(t)$, $a \leq t \leq b$

NEED TO REVIEW? The Mean Value Theorem is discussed in Section 4.3, pp. 277–278.

Since the curve is smooth, the functions $x = x(t)$ and $y = y(t)$ satisfy the conditions of the Mean Value Theorem on $[a, b]$, and so satisfy the same conditions on each subinterval $[t_{i-1}, t_i]$. This means that there are numbers u_i and v_i in each open interval (t_{i-1}, t_i) for which

$$x(t_i) - x(t_{i-1}) = \left[\frac{dx}{dt}(u_i)\right]\Delta t \qquad y(t_i) - y(t_{i-1}) = \left[\frac{dy}{dt}(v_i)\right]\Delta t$$

The sum of the lengths of the segments can be written as

$$\sum_{i=1}^{n} d(P_{i-1}, P_i) = \sum_{i=1}^{n} \sqrt{\left[\frac{dx}{dt}(u_i)\right]^2 + \left[\frac{dy}{dt}(v_i)\right]^2}\,\Delta t$$

These sums are not Riemann sums, since the numbers u_i and v_i are not necessarily equal. However, there is a result (usually given in advanced calculus) that states the limit of the sums $\sum_{i=1}^{n} \sqrt{\left[\frac{dx}{dt}(u_i)\right]^2 + \left[\frac{dy}{dt}(v_i)\right]^2}\,\Delta t$, as $\Delta t \to 0$, is a definite integral.* That is, the length s of the curve from $t = a$ to $t = b$ is

$$s = \int_a^b \sqrt{\left(\frac{dx}{dt}\right)^2 + \left(\frac{dy}{dt}\right)^2}\,dt \qquad \blacksquare$$

---

*This is where the continuity of the derivatives is used, to guarantee the existence of the definite integral.

9.2 Assess Your Understanding

Concepts and Vocabulary

1. *Multiple Choice* Let C denote a curve represented by the parametric equations $x = x(t)$, $y = y(t)$, $a \le t \le b$, where each function $x(t)$ and $y(t)$ is continuous on the closed interval $[a, b]$ and differentiable on the open interval (a, b). If both $\frac{dx}{dt}$ and $\frac{dy}{dt}$ are continuous and never simultaneously 0 on (a, b), then C is called a [(a) smooth, (b) differentiable, (c) parametric] curve.

2. *True or False* If C is smooth curve, represented by the parametric equations $x = x(t)$ and $y = y(t)$, $a \le t \le b$, then the slope of the tangent line to C at the point (x, y) is given by the formula $\dfrac{dy}{dx} = \dfrac{\frac{dy}{dt}}{\frac{dx}{dt}}$, provided $\frac{dx}{dt} \ne 0$.

3. If in the formula for the slope of a tangent line, $\frac{dy}{dt} = 0$ $\left(\text{but } \frac{dx}{dt} \ne 0\right)$, then the curve has a(n) _____ tangent line at the point $(x(t), y(t))$. If $\frac{dx}{dt} = 0$ $\left(\text{but } \frac{dy}{dt} \ne 0\right)$, then the curve has a(n) _____ tangent line at the point $(x(t), y(t))$.

4. *True or False* For a smooth curve C represented by the parametric equations $x = x(t)$, $y = y(t)$, $a \le t \le b$, the length s

of C from $t = a$ to $t = b$ is given by the formula

$$s = \int_a^b \sqrt{\frac{d^2x}{dt^2} + \frac{d^2y}{dt^2}}\,dt.$$

Skill Building

In Problems 5–12, find $\dfrac{dy}{dx}$. Assume $\dfrac{dx}{dt} \ne 0$.

5. $x(t) = e^t \cos t, \quad y(t) = e^t \sin t$

6. $x(t) = 1 + e^{-t}, \quad y(t) = e^{3t}$

7. $x(t) = t + \dfrac{1}{t}, \quad y(t) = 4 + t$

8. $x(t) = t + \dfrac{1}{t}, \quad y(t) = t - \dfrac{1}{t}$

9. $x(t) = \cos t + t \sin t, \quad y(t) = \sin t - t \cos t$

10. $x(t) = \cos^3 t, \quad y(t) = \sin^3 t$

11. $x(t) = \cot^2 t, \quad y(t) = \cot t$

12. $x(t) = \sin t, \quad y(t) = \sec^2 t$

In Problems 13–26, for each pair of parametric equations:
(a) Find an equation of the tangent line to the curve at the given number.
(b) Graph the curve and the tangent line.

---

1. = NOW WORK problem   = Graphing technology recommended   **CAS** = Computer Algebra System recommended

13. $x(t) = 2t^2$, $y(t) = t$ at $t = 2$

14. $x(t) = t$, $y(t) = 3t^2$ at $t = -2$

15. $x(t) = 3t$, $y(t) = 2t^2 - 1$ at $t = 1$

16. $x(t) = 2t$, $y(t) = t^2 - 2$ at $t = 2$

17. $x(t) = \sqrt{t}$, $y(t) = \dfrac{1}{t}$ at $t = 4$

18. $x(t) = \dfrac{2}{t^2}$, $y(t) = \dfrac{1}{t}$ at $t = 1$

19. $x(t) = \dfrac{t}{t+2}$, $y(t) = \dfrac{4}{t+2}$ at $t = 0$

20. $x(t) = \dfrac{t^2}{1+t}$, $y(t) = \dfrac{1}{1+t}$ at $t = 0$

21. $x(t) = e^t$, $y(t) = e^{-t}$ at $t = 0$

22. $x(t) = e^{2t}$, $y(t) = e^t$ at $t = 0$

23. $x(t) = \sin t$, $y(t) = \cos t$ at $t = \dfrac{\pi}{4}$

24. $x(t) = \sin^2 t$, $y(t) = \cos t$ at $t = \dfrac{\pi}{4}$

25. $x(t) = 4\sin t$, $y(t) = 3\cos t$ at $t = \dfrac{\pi}{3}$

26. $x(t) = 2\sin t - 1$, $y(t) = \cos t + 2$ at $t = \dfrac{\pi}{6}$

In Problems 27–30, for each smooth curve, find any points where the tangent line is either horizontal or vertical.

27. $x(t) = t^2$, $y(t) = t^3 - 4t$

28. $x(t) = t^3 - 9t$, $y(t) = t^2$

29. $x(t) = 1 - \cos t$, $y(t) = 1 - \sin t$, $0 \le t \le 2\pi$

30. $x(t) = -3\cos t + \cos(3t)$, $y(t) = \sin t$, $0 \le t \le 2\pi$

In Problems 31–38, find the length of each curve over the given interval.

31. $x(t) = t^3$, $y(t) = t^2$; $0 \le t \le 2$

32. $x(t) = 3t^2 + 1$, $y(t) = t^3 - 1$; $0 \le t \le 2$

33. $x(t) = t - 1$, $y(t) = \dfrac{1}{2}t^2$; $0 \le t \le 2$

34. $x(t) = t^2$, $y(t) = 2t$; $1 \le t \le 3$

35. $x(t) = 4\sin t$, $y(t) = 4\cos t$; $-\dfrac{\pi}{2} \le t \le \dfrac{\pi}{2}$

36. $x(t) = 6\sin t$, $y(t) = 6\cos t$; $-\dfrac{\pi}{2} \le t \le \dfrac{\pi}{2}$

37. $x(t) = 2\sin t - 1$, $y(t) = 2\cos t + 1$; $0 \le t \le 2\pi$

38. $x(t) = e^t \sin t$, $y(t) = e^t \cos t$; $0 \le t \le \pi$

In Problems 39–44:
(a) Use the arc length formula for parametric equations to set up the integral for finding the length of each curve over the given interval.
CAS *(b) Find the length s of each curve over the given interval.*
(c) Graph each curve over the given interval.

39. $x(t) = 2\cos(2t)$, $y(t) = t^2$; $0 \le t \le 2\pi$

40. $x(t) = t^2$, $y(t) = \sin t$; $0 \le t \le 2\pi$

41. $x(t) = t^2$, $y(t) = \sqrt{t + 2}$; $-2 \le t \le 2$

42. $x(t) = t - \cos t$, $y(t) = 1 - \sin t$; $0 \le t \le \pi$

43. $x(t) = 3\cos t + \cos(3t)$, $y(t) = 3\sin t - \sin(3t)$; $0 \le t \le 2\pi$
    (a hypocycloid)

44. $x(t) = 5\cos t - \cos(5t)$, $y(t) = 5\sin t - \sin(5t)$; $0 \le t \le 2\pi$
    (an epicycloid)

Applications and Extensions

45. Tangent Lines

   (a) Find all the points on the plane curve C represented by $x(t) = t^2 + 2$, $y(t) = t^3 - 4t$, where the tangent line is horizontal or where it is vertical.

   (b) Show that C has two tangent lines at the point $(6, 0)$.

   (c) Find equations of these tangent lines.

(d) Graph C.

46. Tangent Lines

   (a) Find all the points on the plane curve C represented by $x(t) = 2t^2$, $y(t) = 8t - t^3$, where the tangent line is horizontal or where it is vertical.

   (b) Show that C has two tangent lines at the point $(16, 0)$.

   (c) Find equations of these tangent lines.

(d) Graph C.

47. Arc Length

   (a) Find the arc length of one arch of the four-cusped hypocycloid
$$x(t) = b\sin^3 t, \; y(t) = b\cos^3 t, \, 0 \le t \le \dfrac{\pi}{2}.$$

(b) Graph the portion of the curve for $0 \le t \le \dfrac{\pi}{2}$.

48. Arc Length

   (a) Find the arc length of the spiral
$$x(t) = t\cos t, \; y(t) = t\sin t, \, 0 \le t \le \pi.$$

(b) Graph the portion of the curve for $0 \le t \le \pi$.

Distance Traveled   *In Problems 49–54, find the distance a particle travels along the given path over the indicated time interval.*

49. $x(t) = 3t$, $y(t) = t^2 - 3$; $0 \le t \le 2$

50. $x(t) = t^2$, $y(t) = 3t$; $0 \le t \le 2$

51. $x(t) = \dfrac{t^2}{2} + 1$, $y(t) = \dfrac{1}{3}(2t + 3)^{3/2}$; $0 \le t \le 2$

52. $x(t) = a\cos t$, $y(t) = a\sin t$; $0 \le t \le \pi$

53. $x(t) = \cos(2t)$, $y(t) = \sin^2 t$; $0 \le t \le \dfrac{\pi}{2}$

54. $x(t) = \dfrac{1}{t}$, $y(t) = \ln t$; $1 \le t \le 2$

In Problems 55–57, find the speed at time t of an object moving along each curve.

55. $x(t) = 20t, \quad y(t) = -16t^2$

56. $x(t) = t + \cos t, \quad y(t) = 2t - \sin t$

57. $x(t) = 20\sin(2t), \quad y(t) = 6\cos t$

58. Arc Length along a Circle Use the arc length formula for parametric equations to show that for a circle of radius r, the length s of the arc subtended by a central angle of θ radians is $s = r\theta$.

59. If the smooth curve C, represented by the parametric equations $x = x(t), y = y(t), a \le t \le b$, is the graph of a function $y = f(x)$, then x can be used in place of the parameter t, and the parametric equations for C are $x = t, y = f(t), a \le t \le b$. Show that the arc length formula for these parametric equations takes the form

$$s = \int_a^b \sqrt{\left(\frac{dx}{dt}\right)^2 + \left(\frac{dy}{dt}\right)^2}\, dt = \int_a^b \sqrt{1 + [f'(x)]^2}\, dx$$

Using Differentials to Approximate Arc Length *Problems 60–63 use the following discussion:*

For a smooth curve C represented by the parametric equations $x = x(t), y = y(t), a \le t \le b$, the arc length s satisfies the equation $\dfrac{ds}{dt} = \sqrt{\left(\dfrac{dx}{dt}\right)^2 + \left(\dfrac{dy}{dt}\right)^2}$, so that

$$\left(\frac{ds}{dt}\right)^2 = \left(\frac{dx}{dt}\right)^2 + \left(\frac{dy}{dt}\right)^2.$$ In terms of differentials, this can be written as

$$(ds)^2 = (dx)^2 + (dy)^2$$

$$ds = \sqrt{(dx)^2 + (dy)^2}$$

Geometrically, the differential $ds = \sqrt{(dx)^2 + (dy)^2}$ is the length of the hypotenuse of a right triangle with sides of lengths dx and dy, as shown in the figure.

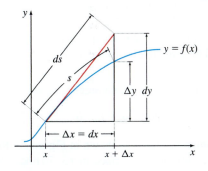

Because $\dfrac{dy}{dx}$ is the slope of the tangent line, the hypotenuse lies on the tangent line to the curve at x. Then the differential ds can be used to approximate the arc length s between two nearby points. Using a differential to approximate arc length is particularly useful when it is difficult, or impossible, to find the arc length using

$$s = \int_a^b \sqrt{\left(\frac{dx}{dt}\right)^2 + \left(\frac{dy}{dt}\right)^2}\, dt.$$

Use the differential ds to approximate each arc length.

60. $x(t) = t^{1/3}, y(t) = t^2$ from $t = 1$ to $t = 1.1$

61. $x(t) = \sqrt{t}, y(t) = t^3$ from $t = 1$ to $t = 1.2$

62. $x(t) = a\sin t, y(t) = b\cos t$ from $t = 0$ to $t = 0.1$

63. $x(t) = e^{at}, y(t) = e^{bt}$ from $t = 0$ to $t = 0.2$

Challenge Problems _____

64. Show that a smooth curve $x = f(t), y = g(t)$, for which $\dfrac{dx}{dt}$ is never 0, can be represented by a rectangular equation $y = F(x)$, where F is differentiable. [*Hint:* Use the fact that $t = f^{-1}(x)$ exists and is differentiable.]

65. Find the point on the curve $x = \dfrac{4}{3}t^3 + 3t^2, \ y = t^3 - 4t^2$, for which the length of the curve from $(0, 0)$ to (x, y) is $\dfrac{80\sqrt{2} - 40}{3}$.

66. Higher-Order Derivatives Find an expression for $\dfrac{d^2y}{dx^2}$ if $x = f(t), y = g(t)$, where f and g have second-order derivatives.

67. Find $\dfrac{d^2y}{dx^2}$ if $x(\theta) = a\cos^3\theta, y(\theta) = a\sin^3\theta$.

68. Consider the cycloid defined by $x(t) = a(t - \sin t), y(t) = a(1 - \cos t), a > 0$. Discuss the behavior of the tangent line to the cycloid at $t = 0$.

9.3 Surface Area of a Solid of Revolution

In Chapter 6, we used a definite integral to find the *volume* of a solid of revolution. To obtain the solid of revolution, we revolved a region around an axis. Here, we are interested in finding the *surface area* of the solid of revolution. To obtain the surface, we revolve a smooth curve about an axis.

Consider a line segment of length L that lies above the x-axis. See Figure 19(a). If the region bounded by the line segment and the x-axis from a to b is revolved about an axis, the resulting solid of revolution is a **frustum**. The frustum of a right circular cone has slant height L and base radii r_1 and r_2. See Figure 19(b).

NOTE A frustum is a portion of a solid of revolution that lies between two parallel planes.

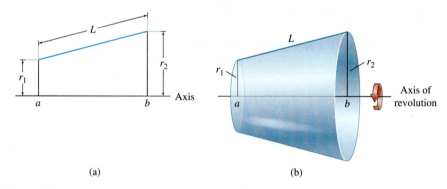

(a)   (b)

Figure 19

The surface area S of the frustum is

$$S = 2\pi \left(\frac{r_1 + r_2}{2} \right) L$$

That is, the surface area S equals the product of 2π, the average radius of the frustum, and the slant height of the frustum. This equation forms the basis for the formula for the surface area of a solid of revolution.

Suppose C is a smooth curve represented by the parametric equations

$$x = x(t) \qquad y = y(t) \quad a \le t \le b$$

where $y = y(t) \ge 0$ on the closed interval $[a, b]$, as shown in Figure 20. Revolving C about the x-axis generates a solid of revolution with surface area S. To find S, we partition the interval $[a, b]$ into n subintervals:

$$[a, t_1], [t_1, t_2], \ldots, [t_{i-1}, t_i], \ldots, [t_{n-1}, b]$$

each of length $\Delta t = \dfrac{b - a}{n}$. Corresponding to each number $a, t_1, t_2, \ldots, t_{i-1}, t_i, \ldots,$ t_{n-1}, b, there is a point $P_0, P_1, \ldots, P_{i-1}, P_i, \ldots, P_{n-1}, P_n$ on the curve C. We join each point P_{i-1} to the next point P_i with a line segment and focus on the line segment joining the points P_{i-1} and P_i. See Figure 21(a). When this line segment of length $d(P_{i-1}, P_i)$ is revolved about the x-axis, it generates a frustum of a right circular cone whose surface

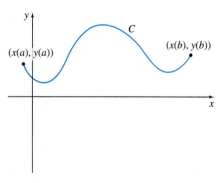

Figure 20

area S_i is

$$S_i = 2\pi \left[\frac{y(t_{i-1}) + y(t_i)}{2} \right] [d(P_{i-1}, P_i)] \tag{1}$$

See Figure 21(b).

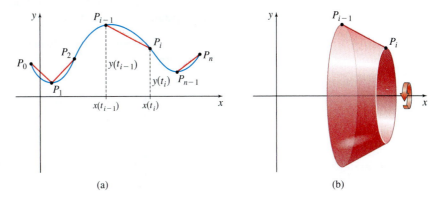

Figure 21

We follow the same reasoning used for finding the arc length of a smooth curve, (Section 9.2). Using the distance formula, the length of the ith line segment is

$$d(P_{i-1}, P_i) = \sqrt{[x(t_i) - x(t_{i-1})]^2 + [y(t_i) - y(t_{i-1})]^2}$$

Now we apply the Mean Value Theorem to $x(t)$ and $y(t)$. There are numbers u_i and v_i in each open interval (t_{i-1}, t_i) for which

$$x(t_i) - x(t_{i-1}) = \left[\frac{dx}{dt}(u_i) \right] \Delta t \qquad \text{and} \qquad y(t_i) - y(t_{i-1}) = \left[\frac{dy}{dt}(v_i) \right] \Delta t$$

So,

$$d(P_{i-1}, P_i) = \sqrt{\left\{ \left[\frac{dx}{dt}(u_i) \right] \Delta t \right\}^2 + \left\{ \left[\frac{dy}{dt}(v_i) \right] \Delta t \right\}^2}$$

$$= \sqrt{\left[\frac{dx}{dt}(u_i) \right]^2 + \left[\frac{dy}{dt}(v_i) \right]^2} \, \Delta t$$

where u_i and v_i are numbers in the ith subinterval.

Now we replace $d(P_{i-1}, P_i)$ in equation (1) with $\sqrt{\left[\frac{dx}{dt}(u_i) \right]^2 + \left[\frac{dy}{dt}(v_i) \right]^2} \, \Delta t$.

Then the surface area generated by the sum of the line segments is

$$\sum_{i=1}^{n} S_i = \sum_{i=1}^{n} 2\pi \left[\frac{y(t_{i-1}) + y(t_i)}{2} \right] \sqrt{\left[\frac{dx}{dt}(u_i) \right]^2 + \left[\frac{dy}{dt}(v_i) \right]^2} \, \Delta t$$

These sums approximate the surface area generated by revolving C about the x-axis and lead to the following result.

THEOREM Surface Area of a Solid of Revolution

The surface area S of a solid of revolution generated by revolving the smooth curve C represented by $x = x(t)$, $y = y(t)$, $a \le t \le b$, where $y(t) \ge 0$, about the x-axis is

$$S = 2\pi \int_a^b y(t) \sqrt{\left(\frac{dx}{dt} \right)^2 + \left(\frac{dy}{dt} \right)^2} \, dt \tag{2}$$

1 Find the Surface Area of a Solid of Revolution Obtained from Parametric Equations

EXAMPLE 1 Finding the Surface Area of a Solid of Revolution Obtained from Parametric Equations

Find the surface area of the solid generated by revolving the smooth curve C represented by the parametric equations $x(t) = 2t^3$, $y(t) = 3t^2$, $0 \le t \le 1$, about the x-axis.

Solution We begin by graphing the smooth curve C and revolving it about the x-axis. See Figure 22.

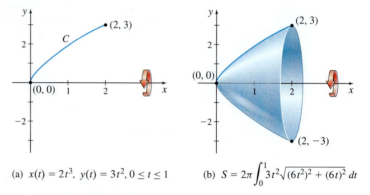

(a) $x(t) = 2t^3$, $y(t) = 3t^2, 0 \le t \le 1$ (b) $S = 2\pi \int_0^1 3t^2 \sqrt{(6t^2)^2 + (6t)^2} \, dt$

Figure 22

We use formula (2) with $\dfrac{dx}{dt} = \dfrac{d}{dt}\left(2t^3\right) = 6t^2$ and $\dfrac{dy}{dt} = \dfrac{d}{dt}\left(3t^2\right) = 6t$. Then

$$S = 2\pi \int_a^b y(t) \sqrt{\left(\frac{dx}{dt}\right)^2 + \left(\frac{dy}{dt}\right)^2} \, dt = 2\pi \int_0^1 3t^2 \sqrt{(6t^2)^2 + (6t)^2} \, dt$$

$$= 2\pi \int_0^1 3t^2 \sqrt{36t^4 + 36t^2} \, dt = 36\pi \int_0^1 t^3 \sqrt{t^2 + 1} \, dt = \frac{36\pi}{2} \int_1^2 (u-1)\sqrt{u} \, du$$

$$\underset{\substack{\uparrow \\ u = t^2 + 1; du = 2t \, dt \\ \text{when } t = 0, u = 1; \text{ when } t = 1, u = 2}}{}$$

$$= 18\pi \left[\frac{2}{5} u^{5/2} - \frac{2}{3} u^{3/2} \right]_1^2 = \frac{24\pi}{5} (\sqrt{2} + 1)$$

The surface area of the solid of revolution is $\dfrac{24\pi}{5}(\sqrt{2} + 1) \approx 36.405$ square units. ∎

NOW WORK Problem 5.

To help remember the formula for the surface area of a solid of revolution, think of the integrand as the product of the slant height $\sqrt{\left(\dfrac{dx}{dt}\right)^2 + \left(\dfrac{dy}{dt}\right)^2}$ and the circumference $2\pi y$ of the circle traced by a point (x, y) on the corresponding sub-arc. Also keep in mind that the limits of integration are parameter values, not x values.

If the curve is revolved about the y-axis, we have a similar formula for the surface area of the solid of revolution.

THEOREM Surface Area of a Solid of Revolution

The surface area S of a solid of revolution generated by revolving the smooth curve C represented by $x = x(t)$, $y = y(t)$, $a \le t \le b$, where $x = x(t) \ge 0$, about the y-axis is

$$S = 2\pi \int_a^b x(t) \sqrt{\left(\frac{dx}{dt}\right)^2 + \left(\frac{dy}{dt}\right)^2} \, dt$$

2 Find the Surface Area of a Solid of Revolution Obtained from a Rectangular Equation

Let C be a smooth curve represented by a rectangular equation $y = f(x)$, $a \leq x \leq b$, where $f(x) \geq 0$ on $[a, b]$. A set of parametric equations for this curve is $x(t) = t$ and $y(t) = f(t)$. Then $\dfrac{dx}{dt} = 1$, $dx = dt$, and $\dfrac{dy}{dt} = \dfrac{dy}{dx}\dfrac{dx}{dt} = \dfrac{dy}{dx} \cdot 1 = f'(x)$. The surface area S of the solid of revolution obtained by revolving C, $a \leq x \leq b$, about the x-axis

is $S = 2\pi \displaystyle\int_a^b y(t)\sqrt{\left(\dfrac{dx}{dt}\right)^2 + \left(\dfrac{dy}{dt}\right)^2}\, dt$. So in terms of $y = f(x)$, $a \leq x \leq b$, we have

$$S = 2\pi \int_a^b f(x)\sqrt{1 + [f'(x)]^2}\, dx \tag{3}$$

EXAMPLE 2   Finding the Surface Area of a Solid of Revolution Obtained from a Rectangular Equation

Find the surface area of the solid generated by revolving the curve represented by $y = \sqrt{x}$, from $x = 0$ to $x = 1$, about the x-axis.

Solution We begin with the graph of $y = \sqrt{x}$, $0 \leq x \leq 1$, and revolve it about the x-axis, as shown in Figure 23.

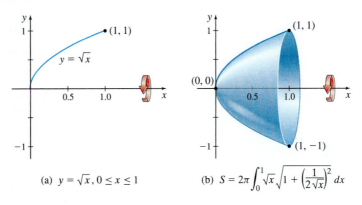

(a) $y = \sqrt{x}, 0 \leq x \leq 1$ (b) $S = 2\pi \displaystyle\int_0^1 \sqrt{x}\sqrt{1 + \left(\dfrac{1}{2\sqrt{x}}\right)^2}\, dx$

Figure 23

We use formula (3) with $f(x) = \sqrt{x}$ and $f'(x) = \dfrac{1}{2\sqrt{x}}$. The surface area S we seek is

$$S = 2\pi \int_a^b f(x)\sqrt{1 + [f'(x)]^2}\, dx = 2\pi \int_0^1 \sqrt{x}\sqrt{1 + \left(\dfrac{1}{2\sqrt{x}}\right)^2}\, dx$$

$$= 2\pi \int_0^1 \sqrt{x\left(1 + \dfrac{1}{4x}\right)}\, dx = 2\pi \int_0^1 \dfrac{1}{2}\sqrt{4x + 1}\, dx$$

$$= \pi \int_1^5 \sqrt{u}\left(\dfrac{du}{4}\right) = \dfrac{\pi}{4}\left[\dfrac{u^{3/2}}{\dfrac{3}{2}}\right]_1^5 = \dfrac{\pi}{6}(5\sqrt{5} - 1)$$

$u = 4x + 1;\ \dfrac{du}{4} = dx$

when $x = 0$, $u = 1$; when $x = 1$, $u = 5$

The surface area of the solid of revolution is $\dfrac{\pi}{6}(5\sqrt{5} - 1) \approx 5.330$ square units. ∎

NOW WORK Problem 11.

9.3 Assess Your Understanding

Concepts and Vocabulary

1. *True or False* When a smooth curve C represented by the parametric equations $x = x(t)$, $y = y(t)$, $y \geq 0$, $a \leq t \leq b$, is revolved about the x-axis, the surface area S of the solid of revolution is given by $S = 2\pi \int_a^b x(t) \sqrt{\left(\dfrac{dx}{dt}\right)^2 + \left(\dfrac{dy}{dt}\right)^2}\, dt$.

2. The surface area S of a solid of revolution generated by revolving the smooth curve C represented by $x = x(t)$, $y = y(t)$, $a \leq t \leq b$, where $x(t) \geq 0$, about the y-axis is $S =$ _____.

Skill Building

In Problems 3–14, find the surface area of the solid generated by revolving each curve about the x-axis.

3. $x(t) = 3t^2$, $y(t) = 6t$; $0 \leq t \leq 1$

4. $x(t) = t^2$, $y(t) = 2t$; $0 \leq t \leq 3$

5. $x(\theta) = \cos^3 \theta$, $y(\theta) = \sin^3 \theta$; $0 \leq \theta \leq \dfrac{\pi}{2}$

6. $x(t) = t - \sin t$, $y(t) = 1 - \cos t$; $0 \leq t \leq \pi$

7. $y = x^3$, $0 \leq x \leq 1$

8. $y = 4x^3$, $0 \leq x \leq 2$

9. $y = \dfrac{x^4}{8} + \dfrac{1}{4x^2}$, $1 \leq x \leq 2$

10. $y = \sqrt{x} + a$; $a \geq 0$, $1 \leq x \leq 9$

11. $y = e^x$, $0 \leq x \leq 1$

12. $y = e^{-x}$, $0 \leq x \leq 1$

13. $y = \sqrt{a^2 - x^2}$, $-a \leq x \leq a$

14. $y = \dfrac{a}{2}(e^{x/a} + e^{-x/a})$, $0 \leq x \leq a$

In Problems 15–20, find the surface area of the solid generated by revolving each curve about the y-axis.

15. $x(t) = 3t^2$, $y(t) = 2t^3$; $0 \leq t \leq 1$

16. $x(t) = 2t + 1$, $y(t) = t^2 + 3$; $0 \leq t \leq 3$

17. $x(t) = 2\sin t$, $y(t) = 2\cos t$; $0 \leq t \leq \dfrac{\pi}{2}$

18. $x(t) = 3\cos t$, $y(t) = 2\sin t$; $0 \leq t \leq \dfrac{\pi}{2}$

19. $x = \dfrac{1}{4}y^2$, $0 \leq y \leq 2$

20. $x^{2/3} + y^{2/3} = a^{2/3}$; $x \geq 0$, $0 \leq y \leq a$

[CAS] 21. Find the surface area of the solid generated by revolving one arch of the cycloid $x(t) = 6(t - \sin t)$, $y(t) = 6(1 - \cos t)$ about the x-axis.

[CAS] 22. Find the surface area of the solid generated by revolving the graph of $y = \ln x$, $1 \leq x \leq 10$, about the x-axis.

Applications and Extensions

23. **Gabriel's Horn** The surface formed by revolving the region between the graph of $y = \dfrac{1}{x}$, $x \geq 1$, and the x-axis about the x-axis is called **Gabriel's horn.** See the figure.

    (a) Find the surface area of Gabriel's horn.

    (b) Find the volume of Gabriel's horn.

    Interesting Note: The volume of Gabriel's horn is finite, but the surface area of Gabriel's horn is infinite.

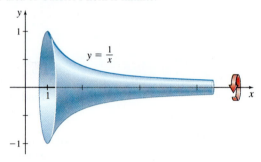

24. **Surface Area** Find the surface area of the solid of revolution obtained by revolving the graph of $y = e^{-x}$, $x \geq 0$, about the x-axis.

25. **Surface Area of a Catenoid** When an arc of a catenary $y = \cosh x$, $a \leq x \leq b$, is revolved about the x-axis, it generates a surface called a **catenoid**, which has the least surface area of all surfaces generated by rotating curves having the same endpoints. Find its surface area. See the figure.

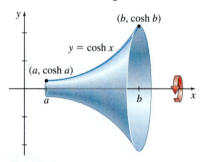

26. **Surface Area of a Sphere** Find a formula for the surface area of a sphere of radius R.

27. **Surface Area** Show that the surface area S of a right circular cone of altitude h and base b is $S = \pi b \sqrt{h^2 + b^2}$.

Challenge Problems

28. **Searchlight** The reflector of a searchlight is formed by revolving an arc of a parabola about its axis. Find the surface area of the reflector if it measures 1 m across its widest point and is $\dfrac{1}{4}$ m deep.

---

29. Surface Area of a Bead   A sphere of radius R has a hole of radius $a < R$ drilled through its center. The axis of the hole coincides with a diameter of the sphere. Find the surface area of the part of the sphere that remains.

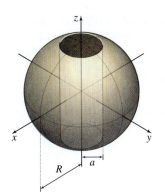

30. Surface Area of a Plug   A plug is made to repair the hole in the sphere in Problem 29. What is the surface area of the plug?

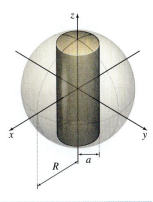

9.4 Polar Coordinates

OBJECTIVES *When you finish this section, you should be able to:*

1 Plot points using polar coordinates (p. 661)
2 Convert between rectangular coordinates and polar coordinates (p. 664)
3 Identify and graph polar equations (p. 666)

In a rectangular coordinate system, there are two perpendicular axes, one horizontal (the x-axis) and one vertical (the y-axis). The point of intersection of the axes is the origin and is labeled O. A point is represented by a pair of numbers (x, y), where x and y equal the signed distance of the point from the y-axis and the x-axis, respectively. A polar coordinate system is constructed by selecting a point O, called the **pole,** and a ray with its vertex at the pole, called the **polar axis.** It is customary to have the pole coincide with the origin and the polar axis coincide with the positive x-axis of the rectangular coordinate system, as shown in Figure 24.

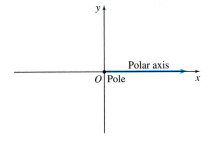

Figure 24

1 Plot Points Using Polar Coordinates

A point P in the polar coordinate system is represented by an ordered pair of numbers (r, θ), called the **polar coordinates** of P. If $r > 0$, then r is the distance of the point from the **pole** (the origin), and θ is an angle (measured in radians or degrees) whose initial side is the **polar axis** (the positive x-axis) and whose terminal side is a ray from the pole through the point P. See Figure 25.

As an example, suppose the polar coordinates of a point P are $\left(3, \dfrac{2\pi}{3}\right)$. We locate P by drawing an angle of $\dfrac{2\pi}{3}$ radians with its vertex at the pole and its initial side along the polar axis. Then P is the point on the terminal side of the angle that is 3 units from the pole. See Figure 26.

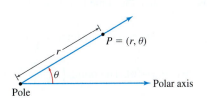

Figure 25 Polar Coordinates.

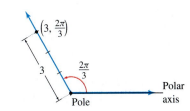

Figure 26

If $r = 0$, then the point $P = (0, \theta)$ is at the pole for any θ. If $r < 0$, the location of the point P is *not* on the terminal side of θ. Instead, P is on the extension through the pole of the ray forming the terminal side of θ and at a distance $|r|$ from the pole. See Figure 27. So, the point $\left(-3, \dfrac{2\pi}{3}\right)$ is located 3 units from the pole on the extension through the pole of the ray that forms the angle $\dfrac{2\pi}{3}$ with the polar axis. See Figure 28.

NOTE The points (r, θ) and $(-r, \theta)$ are reflections about the pole.

Figure 27   **Figure 28**

EXAMPLE 1 **Plotting Points Using Polar Coordinates**

Plot the points with the following polar coordinates:

(a) $\left(3, \dfrac{5\pi}{3}\right)$ (b) $\left(2, -\dfrac{\pi}{4}\right)$ (c) $(3, 0)$ (d) $\left(-2, \dfrac{\pi}{4}\right)$

Solution Figure 29 shows the points.

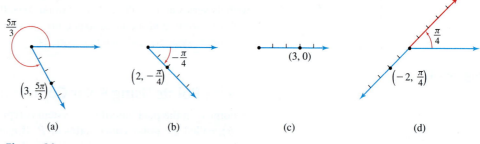

(a)   (b)   (c)   (d)

Figure 29

◼

NOW WORK Problems 17 and 23.

The polar coordinates $\left(3, \dfrac{2\pi}{3}\right)$ of the point P shown in Figure 30(a) on page 663 are one of many possible polar coordinates of P. For example, since the angles $\dfrac{8\pi}{3}$ and $-\dfrac{4\pi}{3}$ have the same terminal side as the angle $\dfrac{2\pi}{3}$, the point P can be represented by any of the polar coordinates $\left(3, \dfrac{2\pi}{3}\right)$, $\left(3, \dfrac{8\pi}{3}\right)$, and $\left(3, -\dfrac{4\pi}{3}\right)$, as shown in Figures 30(a)–(c). The point $\left(3, \dfrac{2\pi}{3}\right)$ can also be represented by the polar coordinates $\left(-3, -\dfrac{\pi}{3}\right)$, as illustrated in Figure 30(d).

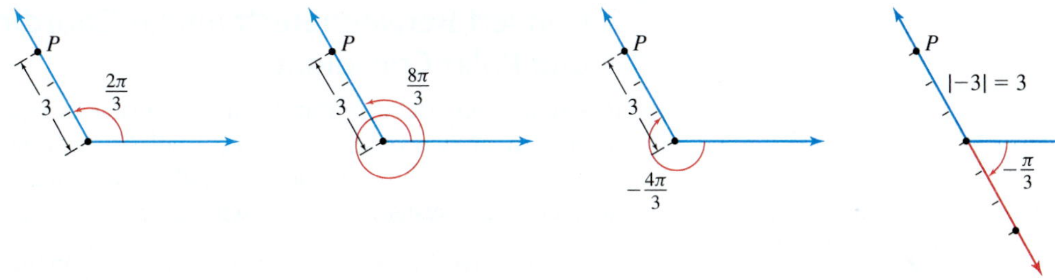

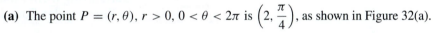

(a) $P = \left(3, \frac{2\pi}{3}\right)$ (b) $P = \left(3, \frac{8\pi}{3}\right)$ (c) $P = \left(3, -\frac{4\pi}{3}\right)$ (d) $P = \left(-3, -\frac{\pi}{3}\right)$

Figure 30

So, there is a major difference between rectangular and polar coordinate systems. In a rectangular system, each point in the plane corresponds to exactly one pair of rectangular coordinates; in a polar coordinate system, every point in the plane can be represented by infinitely many polar coordinates.

EXAMPLE 2 **Plotting a Point Using Polar Coordinates**

Plot the point P whose polar coordinates are $\left(-2, -\frac{3\pi}{4}\right)$. Then find three other polar coordinates of the same point with the properties:

(a) $r > 0$ and $0 < \theta < 2\pi$ (b) $r > 0$ and $-2\pi < \theta < 0$
(c) $r < 0$ and $0 < \theta < 2\pi$

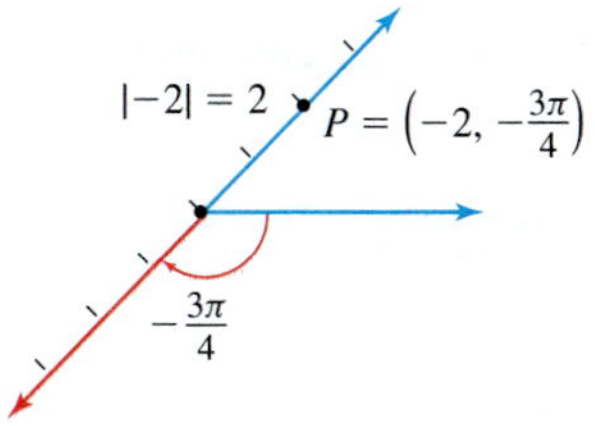

Solution The point $\left(-2, -\frac{3\pi}{4}\right)$ is located by first drawing the angle $-\frac{3\pi}{4}$. Then P is on the extension of the terminal side of θ through the pole at a distance 2 units from the pole, as shown in Figure 31.

(a) The point $P = (r, \theta)$, $r > 0$, $0 < \theta < 2\pi$ is $\left(2, \frac{\pi}{4}\right)$, as shown in Figure 32(a).

(b) The point $P = (r, \theta)$, $r > 0$, $-2\pi < \theta < 0$ is $\left(2, -\frac{7\pi}{4}\right)$, as shown in Figure 32(b).

(c) The point $P = (r, \theta)$, $r < 0$, $0 < \theta < 2\pi$ is $\left(-2, \frac{5\pi}{4}\right)$, as shown in Figure 32(c).

Figure 31

$|-2| = 2$ $P = \left(-2, -\frac{3\pi}{4}\right)$

$-\frac{3\pi}{4}$

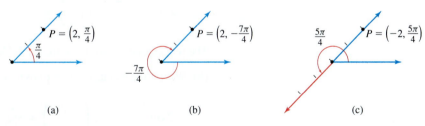

(a)   (b)   (c)

Figure 32

NOW WORK **Problem 25.**

Summary
- A point P with polar coordinates (r, θ) also can be represented by

$$(r, \theta + 2n\pi) \quad \text{or} \quad (-r, \theta + (2n + 1)\pi) \quad n \text{ is an integer}$$

- The polar coordinates of the pole are $(0, \theta)$, where θ is any angle.

2 Convert Between Rectangular Coordinates and Polar Coordinates

It is sometimes useful to transform coordinates or equations in rectangular form into polar form, or vice versa. To do this, recall that the origin in the rectangular coordinate system coincides with the pole in the polar coordinate system and the positive x-axis in the rectangular system coincides with the polar axis in the polar system. The positive y-axis in the rectangular system is the ray $\theta = \dfrac{\pi}{2}$ in the polar system.

Suppose a point P has the polar coordinates (r, θ) and the rectangular coordinates (x, y), as shown in Figure 33. If $r > 0$, then P is on the terminal side of θ, and

$$\cos \theta = \frac{x}{r} \quad \text{and} \quad \sin \theta = \frac{y}{r}$$

If $r < 0$, then $P = (r, \theta)$ can be represented as $(-r, \pi + \theta)$, where $-r > 0$. Since

$$\cos(\theta + \pi) = -\cos \theta = \frac{x}{-r} \quad \text{and} \quad \sin(\theta + \pi) = -\sin \theta = \frac{y}{-r}$$

then, whether $r > 0$ or $r < 0$,

$$x = r \cos \theta \quad \text{and} \quad y = r \sin \theta$$

If $r = 0$, then P is the pole and the same relationships hold.

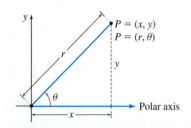

Figure 33

> **NEED TO REVIEW?** The definitions of the trigonometric functions and their properties are discussed in Appendix A.4, pp. A27–A32.

THEOREM Conversion from Polar Coordinates to Rectangular Coordinates

If P is a point with polar coordinates (r, θ), then the rectangular coordinates (x, y) of P are given by

$$\boxed{x = r \cos \theta \quad \text{and} \quad y = r \sin \theta}$$

EXAMPLE 3 Converting from Polar Coordinates to Rectangular Coordinates

Find the rectangular coordinates of each point whose polar coordinates are:

(a) $\left(4, \dfrac{\pi}{3}\right)$ **(b)** $\left(-2, \dfrac{3\pi}{4}\right)$ **(c)** $\left(-3, -\dfrac{5\pi}{6}\right)$

Solution **(a)** We use the equations $x = r \cos \theta$ and $y = r \sin \theta$ with $r = 4$ and $\theta = \dfrac{\pi}{3}$.

$$x = 4 \cos \frac{\pi}{3} = 4 \left(\frac{1}{2}\right) = 2 \quad \text{and} \quad y = 4 \sin \frac{\pi}{3} = 4 \left(\frac{\sqrt{3}}{2}\right) = 2\sqrt{3}$$

The rectangular coordinates are $(2, 2\sqrt{3})$.

(b) We use the equations $x = r \cos \theta$ and $y = r \sin \theta$ with $r = -2$ and $\theta = \dfrac{3\pi}{4}$.

$$x = -2 \cos \frac{3\pi}{4} = -2 \left(-\frac{\sqrt{2}}{2}\right) = \sqrt{2} \quad \text{and} \quad y = -2 \sin \frac{3\pi}{4} = -2 \left(\frac{\sqrt{2}}{2}\right) = -\sqrt{2}$$

The rectangular coordinates are $(\sqrt{2}, -\sqrt{2})$.

(c) We use the equations $x = r \cos \theta$ and $y = r \sin \theta$ with $r = -3$ and $\theta = -\dfrac{5\pi}{6}$.

$$x = -3 \cos \left(-\frac{5\pi}{6}\right) = -3 \left(-\frac{\sqrt{3}}{2}\right) = \frac{3\sqrt{3}}{2}$$

$$y = -3 \sin \left(-\frac{5\pi}{6}\right) = -3 \left(-\frac{1}{2}\right) = \frac{3}{2}$$

The rectangular coordinates are $\left(\dfrac{3\sqrt{3}}{2}, \dfrac{3}{2}\right)$. ∎

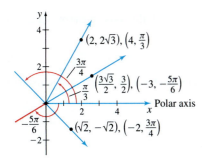

Figure 34

The points $\left(4, \frac{\pi}{3}\right)$, $\left(-2, \frac{3\pi}{4}\right)$, and $\left(-3, -\frac{5\pi}{6}\right)$ and the points $(2, 2\sqrt{3})$, $(\sqrt{2}, -\sqrt{2})$, and $\left(\frac{3\sqrt{3}}{2}, \frac{3}{2}\right)$ are graphed in their respective coordinate systems in Figure 34.

NOW WORK   Problems 33 and 39.

Now suppose P has the rectangular coordinates (x, y). To represent P in polar coordinates, refer again to Figure 33. The point P lies on a circle with radius r and center at $(0, 0)$, so $x^2 + y^2 = r^2$. Since $\tan\theta = \frac{y}{x}$, if $x \neq 0$, we have

$$r^2 = x^2 + y^2 \qquad \text{and} \qquad \tan\theta = \frac{y}{x} \quad x \neq 0$$

If $x = 0$, the point $P = (x, y)$ is on the y-axis. So, $r = y$ and $\theta = \frac{\pi}{2}$.

THEOREM Conversion from Rectangular Coordinates to Polar Coordinates

If P is any point in the plane with rectangular coordinates (x, y), the polar coordinates (r, θ) of P are given by

$$r = \sqrt{x^2 + y^2} \qquad \text{and} \qquad \tan\theta = \frac{y}{x} \quad \text{if } x \neq 0$$

$$r = y \qquad \text{and} \qquad \theta = \frac{\pi}{2} \quad \text{if } x = 0$$

Be careful when applying this theorem. If $x \neq 0$, $\tan\theta = \frac{y}{x}$ and $\theta = \tan^{-1}\frac{y}{x}$, $-\frac{\pi}{2} < \theta < \frac{\pi}{2}$, placing θ in quadrants I or IV. If the point lies in quadrant II or III, we must find $\tan^{-1}\left(\frac{y}{x}\right)$ in quadrant I or IV, respectively, and then add π to the result. It is advisable to plot the point (x, y) at the start to identify the quadrant that contains the point.

NEED TO REVIEW? The inverse tangent function is discussed in Section P.7, pp. 58–60.

EXAMPLE 4 Converting from Rectangular Coordinates to Polar Coordinates

Find polar coordinates of each point whose rectangular coordinates are:

(a) $(4, -4)$ **(b)** $(-1, -\sqrt{3})$ **(c)** $(4, -1)$

Solution **(a)** The point $(4, -4)$, plotted in Figure 35, is in quadrant IV. The distance from the pole to the point $(4, -4)$ is

$$r = \sqrt{x^2 + y^2} = \sqrt{4^2 + (-4)^2} = \sqrt{32} = 4\sqrt{2}$$

Since the point $(4, -4)$ is in quadrant IV, $-\frac{\pi}{2} < \theta < 0$. So,

$$\theta = \tan^{-1}\left(\frac{y}{x}\right) = \tan^{-1}\left(\frac{-4}{4}\right) = \tan^{-1}(-1) = -\frac{\pi}{4}$$

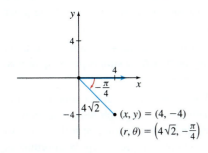

Figure 35

A pair of polar coordinates for this point is $\left(4\sqrt{2}, -\dfrac{\pi}{4}\right)$. Other possible representations include $\left(-4\sqrt{2}, \dfrac{3\pi}{4}\right)$ and $\left(4\sqrt{2}, \dfrac{7\pi}{4}\right)$.

(b) The point $(-1, -\sqrt{3})$, plotted in Figure 36, is in quadrant III. The distance from the pole to the point $(-1, -\sqrt{3})$ is

$$r = \sqrt{(-1)^2 + (-\sqrt{3})^2} = \sqrt{1+3} = 2$$

Since the point $(-1, -\sqrt{3})$ lies in quadrant III and the inverse tangent function gives an angle in quadrant I, we add π to $\tan^{-1}\left(\dfrac{y}{x}\right)$ to obtain an angle in quadrant III.

$$\theta = \tan^{-1}\left(\dfrac{-\sqrt{3}}{-1}\right) + \pi = \tan^{-1}(\sqrt{3}) + \pi = \dfrac{\pi}{3} + \pi = \dfrac{4\pi}{3}$$

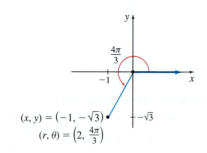

$(x, y) = (-1, -\sqrt{3})$
$(r, \theta) = \left(2, \dfrac{4\pi}{3}\right)$

Figure 36

A pair of polar coordinates for the point is $\left(2, \dfrac{4\pi}{3}\right)$. Other possible representations include $\left(-2, \dfrac{\pi}{3}\right)$ and $\left(2, -\dfrac{2\pi}{3}\right)$.

(c) The point $(4, -1)$, plotted in Figure 37, lies in quadrant IV. The distance from the pole to the point $(4, -1)$ is

$$r = \sqrt{x^2 + y^2} = \sqrt{4^2 + (-1)^2} = \sqrt{17}$$

Since the point $(4, -1)$ is in quadrant IV, $-\dfrac{\pi}{2} < \theta < 0$. So,

$$\theta = \tan^{-1}\left(\dfrac{y}{x}\right) = \tan^{-1}\left(\dfrac{-1}{4}\right) \approx -0.245 \text{ radians}$$

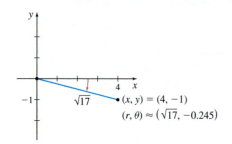

$(x, y) = (4, -1)$
$(r, \theta) \approx (\sqrt{17}, -0.245)$

Figure 37

A pair of polar coordinates for this point is $(\sqrt{17}, -0.245)$. Other possible representations for the point include $\left(\sqrt{17}, \tan^{-1}\left(-\dfrac{1}{4}\right) + 2\pi\right) \approx (\sqrt{17}, 6.038)$ and $\left(-\sqrt{17}, \tan^{-1}\left(-\dfrac{1}{4}\right) + \pi\right) \approx (-\sqrt{17}, 2.897)$. ■

NOW WORK Problems **41** and **47.**

3 Identify and Graph Polar Equations

Just as a rectangular grid is used to plot points given by rectangular coordinates, a *polar grid* is used to plot points given by polar coordinates. A **polar grid** consists of concentric circles centered at the pole and of rays with vertices at the pole, as shown in Figure 38. An equation whose variables are polar coordinates is called a **polar equation**, and the **graph of a polar equation** is the set of all points for which at least one of the polar coordinate representations satisfies the equation.

Polar equations of circles with their center at the pole, lines containing the pole, horizontal and vertical lines, and circles containing the pole have simple polar equations. In Section 9.5, we graph other important polar equations.

There are occasions when geometry is all that is needed to graph a polar equation. But usually other methods are required. One method used to graph polar equations is to convert the equation to rectangular coordinates. In the discussion that follows, (x, y) represents the rectangular coordinates of a point P, and (r, θ) represents polar coordinates of the point P.

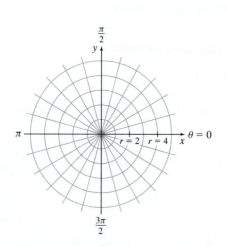

Figure 38 Polar grid

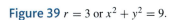

EXAMPLE 5 Identifying and Graphing a Polar Equation

Identify and graph each equation:

(a) $r = 3$ **(b)** $\theta = \dfrac{\pi}{4}$

Solution **(a)** If r is fixed at 3 and θ is allowed to vary, the graph is a circle with its center at the pole and radius 3, as shown in Figure 39. To confirm this, we convert the polar equation $r = 3$ to a rectangular equation.

$$r = 3$$
$$r^2 = 9 \qquad \text{Square both sides.}$$
$$x^2 + y^2 = 9 \qquad r^2 = x^2 + y^2$$

Figure 39 $r = 3$ or $x^2 + y^2 = 9$.

(b) If θ is fixed at $\dfrac{\pi}{4}$ and r is allowed to vary, the result is a line containing the pole, making an angle of $\dfrac{\pi}{4}$ with the polar axis. That is, the graph of $\theta = \dfrac{\pi}{4}$ is a line containing the pole with slope $\tan \theta = \tan \dfrac{\pi}{4} = 1$, as shown in Figure 40. To confirm this, we convert the polar equation to a rectangular equation.

$$\theta = \frac{\pi}{4}$$
$$\tan \theta = \tan \frac{\pi}{4}$$
$$\frac{y}{x} = 1$$
$$y = x$$

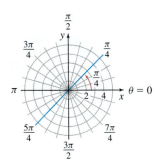

Figure 40 $\theta = \dfrac{\pi}{4}$ or $y = x$.

NOW WORK Problems 51 and 61.

EXAMPLE 6 Identifying and Graphing Polar Equations

Identify and graph the equations:

(a) $r \sin \theta = 2$ **(b)** $r = 4 \sin \theta$

Solution **(a)** Here, both r and θ are allowed to vary, so the graph of the equation is not as obvious. If we use the fact that $y = r \sin \theta$, the equation $r \sin \theta = 2$ becomes $y = 2$. So, the graph of $r \sin \theta = 2$ is the horizontal line $y = 2$ that lies 2 units above the pole, as shown in Figure 41.

(b) To convert the equation $r = 4 \sin \theta$ to rectangular coordinates, we multiply the equation by r to obtain

$$r^2 = 4r \sin \theta$$

Now we use the formulas $r^2 = x^2 + y^2$ and $y = r \sin \theta$. Then

$$r^2 = 4r \sin \theta$$
$$x^2 + y^2 = 4y \qquad\qquad r^2 = x^2 + y^2, \ \ r \sin \theta = y$$
$$x^2 + (y^2 - 4y) = 0$$
$$x^2 + \left(y^2 - 4y + 4\right) = 4 \qquad\qquad \text{Complete the square in } y.$$
$$x^2 + (y - 2)^2 = 4 \qquad\qquad \text{Factor.}$$

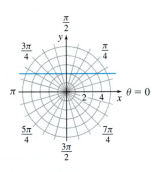

Figure 41 $r \sin \theta = 2$ or $y = 2$.

This is the standard form of the equation of a circle with its center at the point $(0, 2)$ and radius 2 in rectangular coordinates. See Figure 42. Notice that the circle passes through the pole. ■

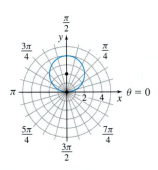

Figure 42 $r = 4 \sin \theta$ or $x^2 + (y - 2)^2 = 4$.

NOW WORK Problems 65 and 67.

Table 1 summarizes and extends the results of Examples 5 and 6.

TABLE 1

Description	Line passing through the pole making an angle α with the polar axis	Vertical line	Horizontal line
Rectangular equation	$y = (\tan \alpha)x$	$x = a$	$y = b$
Polar equation	$\theta = \alpha$	$r \cos \theta = a$	$r \sin \theta = b$
Typical graph	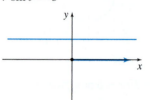	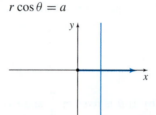	

Description	Circle, center at the pole, radius a	Circle, passing through the pole, tangent to the line $\theta = \dfrac{\pi}{2}$, center on the polar axis, radius a	Circle, passing through the pole, tangent to the polar axis, center on the line $\theta = \dfrac{\pi}{2}$, radius a
Rectangular equation	$x^2 + y^2 = a^2, a > 0$	$x^2 + y^2 = \pm 2ax, \quad a > 0$	$x^2 + y^2 = \pm 2ay, \quad a > 0$
Polar equation	$r = a, a > 0$	$r = \pm 2a \cos \theta, \quad a > 0$	$r = \pm 2a \sin \theta, \quad a > 0$
Typical graph	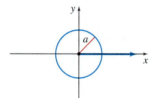	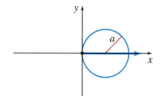	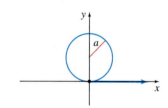

9.4 Assess Your Understanding

Concepts and Vocabulary

1. In a polar coordinate system, the origin is called the _____, and the _____ coincides with the positive x-axis of the rectangular coordinate system.

2. *True or False* Another representation in polar coordinates for the point $\left(2, \dfrac{\pi}{3}\right)$ is $\left(2, \dfrac{4\pi}{3}\right)$.

3. *True or False* In a polar coordinate system, each point in the plane has exactly one pair of polar coordinates.

4. *True or False* In a rectangular coordinate system, each point in the plane has exactly one pair of rectangular coordinates.

5. *True or False* In polar coordinates (r, θ), the number r can be negative.

6. *True or False* If (r, θ) are the polar coordinates of the point P, then $|r|$ is the distance of the point P from the pole.

7. To convert the point (r, θ) in polar coordinates to a point (x, y) in rectangular coordinates, use the formulas $x =$ _____ and $y =$ _____.

8. An equation whose variables are polar coordinates is called a(n) _____ _____.

Skill Building

In Problems 9–16, match each point in polar coordinates with A, B, C, or D on the graph.

9. $\left(2, -\dfrac{11\pi}{6}\right)$ 10. $\left(-2, -\dfrac{\pi}{6}\right)$

11. $\left(-2, \dfrac{\pi}{6}\right)$ 12. $\left(2, \dfrac{7\pi}{6}\right)$

13. $\left(2, \dfrac{5\pi}{6}\right)$ 14. $\left(-2, \dfrac{5\pi}{6}\right)$

15. $\left(-2, \dfrac{7\pi}{6}\right)$ 16. $\left(2, \dfrac{11\pi}{6}\right)$

In Problems 17–24, polar coordinates of a point are given. Plot each point in a polar coordinate system.

17. $\left(4, \dfrac{\pi}{3}\right)$ 18. $\left(-4, \dfrac{\pi}{3}\right)$ 19. $\left(-4, -\dfrac{\pi}{3}\right)$

20. $\left(4, -\dfrac{\pi}{3}\right)$ 21. $\left(\sqrt{2}, \dfrac{\pi}{4}\right)$ 22. $\left(7, -\dfrac{7\pi}{4}\right)$

23. $\left(-6, \dfrac{4\pi}{3}\right)$ 24. $\left(5, \dfrac{\pi}{2}\right)$

---

1. = NOW WORK problem   **⚙** = Graphing technology recommended   **CAS** = Computer Algebra System recommended

In Problems 25–32, polar coordinates of a point are given. Find other polar coordinates (r, θ) of the point for which:

(a) $r > 0, -2\pi \le \theta < 0$
(b) $r < 0, 0 \le \theta < 2\pi$
(c) $r > 0, 2\pi \le \theta < 4\pi$

25. $\left(5, \dfrac{2\pi}{3}\right)$ **26.** $\left(4, \dfrac{3\pi}{4}\right)$ **27.** $(-2, 3\pi)$

28. $(-3, 4\pi)$ **29.** $\left(1, \dfrac{\pi}{2}\right)$ **30.** $(2, \pi)$

31. $\left(-3, -\dfrac{\pi}{4}\right)$ **32.** $\left(-2, -\dfrac{2\pi}{3}\right)$

In Problems 33–40, polar coordinates of a point are given. Find the rectangular coordinates of each point.

33. $\left(6, \dfrac{\pi}{6}\right)$ **34.** $\left(-6, \dfrac{\pi}{6}\right)$ **35.** $\left(-6, -\dfrac{\pi}{6}\right)$

36. $\left(6, -\dfrac{\pi}{6}\right)$ **37.** $\left(5, \dfrac{\pi}{2}\right)$ **38.** $\left(8, \dfrac{\pi}{4}\right)$

39. $\left(2\sqrt{2}, -\dfrac{\pi}{4}\right)$ **40.** $\left(-5, -\dfrac{\pi}{3}\right)$

In Problems 41–50, rectangular coordinates of a point are given. Plot the point. Find polar coordinates (r, θ) of each point for which $r > 0$ and $0 \le \theta < 2\pi$.

41. $(5, 0)$ **42.** $(2, -2)$ **43.** $(-2, 2)$ **44.** $(-2, -2\sqrt{3})$
45. $(\sqrt{3}, 1)$ **46.** $(0, -3)$ **47.** $(-\sqrt{3}, 1)$ **48.** $(3\sqrt{2}, -3\sqrt{2})$
49. $(3, 2)$ **50.** $(-6.5, 1.2)$

In Problems 51–58, match each of the graphs (A) through (H) to one of the following polar equations.

51. $r = 2$ **52.** $\theta = \dfrac{\pi}{4}$ **53.** $r = 2 \cos \theta$

54. $r \cos \theta = 2$ **55.** $r = -2 \cos \theta$ **56.** $r = 2 \sin \theta$

57. $\theta = \dfrac{3\pi}{4}$ **58.** $r \sin \theta = 2$

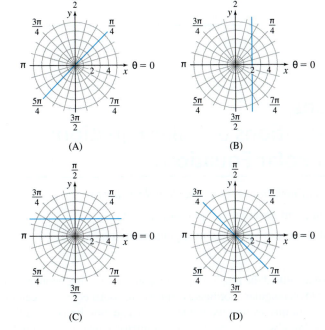

(A)   (B)

(C)   (D)

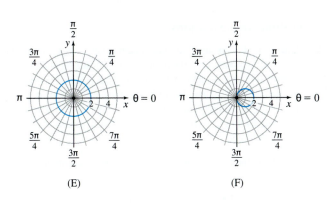

(E)   (F)

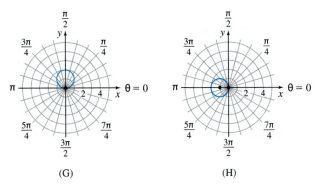

(G)   (H)

In Problems 59–74, identify and graph each polar equation. Convert to a rectangular equation if necessary.

59. $r = 4$ **60.** $r = 2$ **61.** $\theta = \dfrac{\pi}{3}$

62. $\theta = -\dfrac{\pi}{4}$ **63.** $r \sin \theta = 4$ **64.** $r \cos \theta = 4$

65. $r \cos \theta = -2$ **66.** $r \sin \theta = -2$ **67.** $r = 2 \cos \theta$

68. $r = 2 \sin \theta$ **69.** $r = -4 \sin \theta$ **70.** $r = -4 \cos \theta$

71. $r \sec \theta = 4$ **72.** $r \csc \theta = 8$ **73.** $r \csc \theta = -2$

74. $r \sec \theta = -4$

In Problems 75–82, the letters x and y represent rectangular coordinates. Write each equation in polar coordinates r and θ.

75. $\dfrac{x^2}{4} + \dfrac{y^2}{9} = 1$ **76.** $x - 4y + 4 = 0$

77. $x^2 + y^2 - 4x = 0$ **78.** $y = -6$

79. $x^2 = 1 - 4y$ **80.** $y^2 = 1 - 4x$

81. $xy = 1$ **82.** $x^2 + y^2 - 2x + 4y = 0$

In Problems 83–94, the letters r and θ represent polar coordinates. Write each equation in rectangular coordinates x and y.

83. $r = \cos \theta$ **84.** $r = 2 + \cos \theta$ **85.** $r^2 = \sin \theta$

86. $r^2 = 1 - \sin \theta$ **87.** $r = \dfrac{4}{1 - \cos \theta}$ **88.** $r = \dfrac{3}{3 - \cos \theta}$

89. $r^2 = \theta$ **90.** $\theta = -\dfrac{\pi}{4}$ **91.** $r = 2$

92. $r = -5$ **93.** $\tan \theta = 4$ **94.** $\cot \theta = 3$

Applications and Extensions

95. Chicago In Chicago, the road system is based on a rectangular coordinate system, with the intersection of Madison and State Streets at the origin, and east as the positive x-axis. Intersections are indicated by the number of blocks they are from the origin. For example, Wrigley Field is located at 1060 West Addison, which is 10 blocks west of State Street and 36 blocks north of Madison Street.

    (a) Write the location of Wrigley Field using rectangular coordinates.

    (b) Write the location of Wrigley Field using polar coordinates. Use east as the polar axis.

    (c) U.S. Cellular Field is located at 35th Street and Princeton, which is 3 blocks west of State Street and 35 blocks south of Madison Street. Write the location of U.S. Cellular Field using rectangular coordinates.

    (d) Write the location of U.S. Cellular Field using polar coordinates.

96. Show that the formula for the distance d between two points $P_1 = (r_1, \theta_1)$ and $P_2 = (r_2, \theta_2)$ is

$$d = \sqrt{r_1^2 + r_2^2 - 2r_1r_2\cos(\theta_2 - \theta_1)}$$

97. Horizontal Line Show that the graph of the equation $r\sin\theta = a$ is a horizontal line a units above the pole if $a > 0$, and $|a|$ units below the pole if $a < 0$.

98. Vertical Line Show that the graph of the equation $r\cos\theta = a$ is a vertical line a units to the right of the pole if $a > 0$, and $|a|$ units to the left of the pole if $a < 0$.

99. Circle Show that the graph of the equation $r = 2a\sin\theta$, $a > 0$, is a circle of radius a with its center at the rectangular coordinates $(0, a)$.

100. Circle Show that the graph of the equation $r = -2a\sin\theta$, $a > 0$, is a circle of radius a with its center at the rectangular coordinates $(0, -a)$.

101. Circle Show that the graph of the equation $r = 2a\cos\theta$, $a > 0$, is a circle of radius a with its center at the rectangular coordinates $(a, 0)$.

102. Circle Show that the graph of the equation $r = -2a\cos\theta$, $a > 0$, is a circle of radius a with its center at the rectangular coordinates $(-a, 0)$.

103. Exploring Using Graphing Technology

    (a) Use a square screen to graph $r_1 = \sin\theta$, $r_2 = 2\sin\theta$, and $r_3 = 3\sin\theta$.

    (b) Describe how varying the constant a, $a > 0$, alters the graph of $r = a\sin\theta$.

    (c) Graph $r_1 = -\sin\theta$, $r_2 = -2\sin\theta$, and $r_3 = -3\sin\theta$.

    (d) Describe how varying the constant a, $a < 0$, alters the graph of $r = a\sin\theta$.

104. Exploring Using Graphing Technology

    (a) Use a square screen to graph $r_1 = \cos\theta$, $r_2 = 2\cos\theta$, and $r_3 = 3\cos\theta$.

    (b) Describe how varying the constant a, $a > 0$, alters the graph of $r = a\cos\theta$.

    (c) Graph $r_1 = -\cos\theta$, $r_2 = -2\cos\theta$, and $r_3 = -3\cos\theta$.

    (d) Describe how varying the constant a, $a < 0$, alters the graph of $r = a\cos\theta$.

Challenge Problems

105. Show that $r = a\sin\theta + b\cos\theta$, a, b not both zero, is the equation of a circle. Find the center and radius of the circle.

106. Express $r^2 = \cos(2\theta)$ in rectangular coordinates free of radicals.

107. Prove that the area of the triangle with vertices $(0, 0)$, (r_1, θ_1), (r_2, θ_2) is

$$A = \frac{1}{2}r_1r_2\,\sin(\theta_2 - \theta_1)\qquad 0 \leq \theta_1 < \theta_2 \leq \pi$$

9.5 Polar Equations; Parametric Equations of Polar Equations; Arc Length of Polar Equations

OBJECTIVES *When you finish this section, you should be able to:*

1 Graph a polar equation; find parametric equations (p. 671)

2 Find the arc length of a curve represented by a polar equation (p. 675)

In the previous section, we identified the graphs of polar equations by using geometry or by converting the equation to rectangular coordinates. Since many polar equations cannot be identified in these ways, in this section we graph a polar equation by constructing a table and plotting points. We also show how to find parametric equations for a polar equation. We end the section by finding the length of a curve represented by polar coordinates.

1 Graph a Polar Equation; Find Parametric Equations

EXAMPLE 1   Graphing a Polar Equation (Cardioid); Finding Parametric Equations

(a) Graph the polar equation $r = 1 - \sin\theta, 0 \leq \theta \leq 2\pi$.

(b) Find parametric equations for $r = 1 - \sin\theta$.

Solution (a) The polar equation $r = 1 - \sin\theta$ contains $\sin\theta$, which has the period 2π. We construct Table 2 using common values of θ that range from 0 to 2π, plot the points (r, θ), and trace out the graph, beginning at the point $(1, 0)$ and ending at the point $(1, 2\pi)$, as shown in Figure 43(a). Figure 43(b) shows the graph using technology.

TABLE 2

θ	$r = 1 - \sin\theta$	(r, θ)
0	$1 - 0 = 1$	$(1, 0)$
$\dfrac{\pi}{6}$	$1 - \dfrac{1}{2} = \dfrac{1}{2}$	$\left(\dfrac{1}{2}, \dfrac{\pi}{6}\right)$
$\dfrac{\pi}{2}$	$1 - 1 = 0$	$\left(0, \dfrac{\pi}{2}\right)$
$\dfrac{5\pi}{6}$	$1 - \dfrac{1}{2} = \dfrac{1}{2}$	$\left(\dfrac{1}{2}, \dfrac{5\pi}{6}\right)$
π	$1 - 0 = 1$	$(1, \pi)$
$\dfrac{7\pi}{6}$	$1 - \left(-\dfrac{1}{2}\right) = \dfrac{3}{2}$	$\left(\dfrac{3}{2}, \dfrac{7\pi}{6}\right)$
$\dfrac{3\pi}{2}$	$1 - (-1) = 2$	$\left(2, \dfrac{3\pi}{2}\right)$
$\dfrac{11\pi}{6}$	$1 - \left(-\dfrac{1}{2}\right) = \dfrac{3}{2}$	$\left(\dfrac{3}{2}, \dfrac{11\pi}{6}\right)$
2π	$1 - 0 = 1$	$(1, 2\pi)$

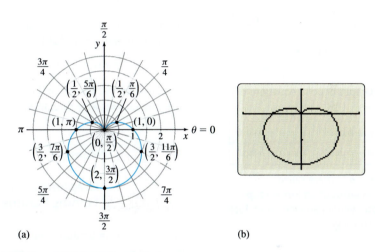

(a)   (b)

Figure 43 The cardioid $r = 1 - \sin\theta$.

NOTE Graphs of polar equations of the form

$r = a(1 + \cos\theta)$	$r = a(1 + \sin\theta)$
$r = a(1 - \cos\theta)$	$r = a(1 - \sin\theta)$

where $a > 0$, are called **cardioids**. A cardioid contains the pole and is heart-shaped (giving the curve its name).

(b) We obtain parametric equations for $r = 1 - \sin\theta$ by using the conversion formulas $x = r\cos\theta$ and $y = r\sin\theta$:

$$x = r\cos\theta = (1 - \sin\theta)\cos\theta \qquad y = r\sin\theta = (1 - \sin\theta)\sin\theta$$

Here, θ is the parameter, and if $0 \leq \theta \leq 2\pi$, then the graph is traced out exactly once in the counterclockwise direction. ∎

NOW WORK Problem 5.

NOTE Limaçon (pronounced "leema sown") is a French word for "snail."

EXAMPLE 2   Graphing a Polar Equation (Limaçon Without an Inner Loop); Finding Parametric Equations

(a) Graph the polar equation $r = 3 + 2\cos\theta, 0 \leq \theta \leq 2\pi$.

(b) Find parametric equations for $r = 3 + 2\cos\theta$.

Solution (a) The polar equation $r = 3 + 2\cos\theta$ contains $\cos\theta$, which has the period 2π. We construct Table 3 using common values of θ that range from 0 to 2π, plot the points (r, θ), and trace out the graph, beginning at the point $(5, 0)$ and ending at the point $(5, 2\pi)$, as shown in Figure 44(a). Figure 44(b) shows the graph using technology.

TABLE 3

θ	$r = 3 + 2\cos\theta$	(r, θ)
0	$3 + 2(1) = 5$	$(5, 0)$
$\dfrac{\pi}{3}$	$3 + 2\left(\dfrac{1}{2}\right) = 4$	$\left(4, \dfrac{\pi}{3}\right)$
$\dfrac{\pi}{2}$	$3 + 2(0) = 3$	$\left(3, \dfrac{\pi}{2}\right)$
$\dfrac{2\pi}{3}$	$3 + 2\left(-\dfrac{1}{2}\right) = 2$	$\left(2, \dfrac{2\pi}{3}\right)$
π	$3 + 2(-1) = 1$	$(1, \pi)$
$\dfrac{4\pi}{3}$	$3 + 2\left(-\dfrac{1}{2}\right) = 2$	$\left(2, \dfrac{4\pi}{3}\right)$
$\dfrac{3\pi}{2}$	$3 + 2(0) = 3$	$\left(3, \dfrac{3\pi}{2}\right)$
$\dfrac{5\pi}{3}$	$3 + 2\left(\dfrac{1}{2}\right) = 4$	$\left(4, \dfrac{5\pi}{3}\right)$
2π	$3 + 2(1) = 5$	$(5, 2\pi)$

NOTE Graphs of polar equations of the form

$r = a + b\cos\theta$	$r = a + b\sin\theta$
$r = a - b\cos\theta$	$r = a - b\sin\theta$

where $a > b > 0$, are called **limaçons without an inner loop**. A limaçon without an inner loop does not pass through the pole.

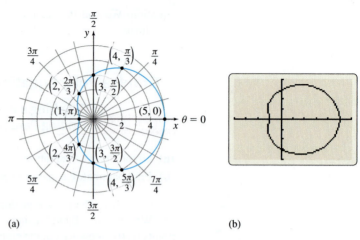

(a)   (b)

Figure 44 The limaçon $r = 3 + 2\cos\theta$.

(b) We obtain parametric equations for $r = 3 + 2\cos\theta$ by using the conversion formulas $x = r\cos\theta$ and $y = r\sin\theta$:

$$x = r\cos\theta = (3 + 2\cos\theta)\cos\theta \qquad y = r\sin\theta = (3 + 2\cos\theta)\sin\theta$$

Here, θ is the parameter, and if $0 \le \theta \le 2\pi$, then the graph is traced out exactly once in the counterclockwise direction. ∎

NOW WORK Problem 7.

EXAMPLE 3   Graphing a Polar Equation (Limaçon with an Inner Loop); Finding Parametric Equations

(a) Graph the polar equation $r = 1 + 2\cos\theta$, $0 \le \theta \le 2\pi$.

(b) Find parametric equations for $r = 1 + 2\cos\theta$.

Solution (a) The polar equation $r = 1 + 2\cos\theta$ contains $\cos\theta$, which has the period 2π. We construct Table 4 using common values of θ that range from 0 to 2π, plot the points (r, θ), and trace out the graph, beginning at the point $(3, 0)$ and ending at the point $(3, 2\pi)$, as shown in Figure 45(a). Figure 45(b) shows the graph using technology.

TABLE 4

θ	$r = 1 + 2\cos\theta$	(r, θ)
0	$1 + 2(1) = 3$	$(3, 0)$
$\dfrac{\pi}{3}$	$1 + 2\left(\dfrac{1}{2}\right) = 2$	$\left(2, \dfrac{\pi}{3}\right)$
$\dfrac{\pi}{2}$	$1 + 2(0) = 1$	$\left(1, \dfrac{\pi}{2}\right)$
$\dfrac{2\pi}{3}$	$1 + 2\left(-\dfrac{1}{2}\right) = 0$	$\left(0, \dfrac{2\pi}{3}\right)$
π	$1 + 2(-1) = -1$	$(-1, \pi)$
$\dfrac{4\pi}{3}$	$1 + 2\left(-\dfrac{1}{2}\right) = 0$	$\left(0, \dfrac{4\pi}{3}\right)$
$\dfrac{3\pi}{2}$	$1 + 2(0) = 1$	$\left(1, \dfrac{3\pi}{2}\right)$
$\dfrac{5\pi}{3}$	$1 + 2\left(\dfrac{1}{2}\right) = 2$	$\left(2, \dfrac{5\pi}{3}\right)$
2π	$1 + 2(1) = 3$	$(3, 2\pi)$

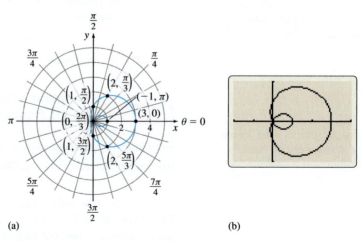

(a)   (b)

Figure 45 The limaçon $r = 1 + 2\cos\theta$.

NOTE Graphs of polar equations of the form

$r = a + b\cos\theta$	$r = a + b\sin\theta$
$r = a - b\cos\theta$	$r = a - b\sin\theta$

where $0 < a < b$, are called **limaçons with an inner loop**. A limaçon with an inner loop passes through the pole twice.

(b) We obtain parametric equations for $r = 1 + 2\cos\theta$ by using the conversion formulas $x = r\cos\theta$ and $y = r\sin\theta$:

$$x = r\cos\theta = (1 + 2\cos\theta)\cos\theta \qquad y = r\sin\theta = (1 + 2\cos\theta)\sin\theta$$

Here, θ is the parameter, and if $0 \le \theta \le 2\pi$, then the graph is traced out exactly once in the counterclockwise direction. ∎

NOW WORK Problem 9.

EXAMPLE 4 Graphing a Polar Equation (Rose); Finding Parametric Equations

(a) Graph the polar equation $r = 2\cos(2\theta)$, $0 \le \theta \le 2\pi$.

(b) Find parametric equations for $r = 2\cos(2\theta)$.

Solution **(a)** The polar equation $r = 2\cos(2\theta)$ contains $\cos(2\theta)$, which has the period π. So, we construct Table 5 using common values of θ that range from 0 to 2π, noting that the values for $\pi \le \theta \le 2\pi$ repeat the values for $0 \le \theta \le \pi$. Then we plot the points (r, θ) and trace out the graph. Figure 46(a) on page 674 shows the graph from the point $(2, 0)$ to the point $(2, \pi)$. Figure 46(b) completes the graph from the point $(2, \pi)$ to the point $(2, 2\pi)$.

TABLE 5

θ	$r = 2\cos(2\theta)$	(r, θ)	θ	$r = 2\cos(2\theta)$	(r, θ)
0	$2(1) = 2$	$(2, 0)$			
$\dfrac{\pi}{6}$	$2\left(\dfrac{1}{2}\right) = 1$	$\left(1, \dfrac{\pi}{6}\right)$	$\dfrac{7\pi}{6}$	$2\left(\dfrac{1}{2}\right) = 1$	$\left(1, \dfrac{7\pi}{6}\right)$
$\dfrac{\pi}{4}$	$2(0) = 0$	$\left(0, \dfrac{\pi}{4}\right)$	$\dfrac{5\pi}{4}$	$2(0) = 0$	$\left(0, \dfrac{5\pi}{4}\right)$
$\dfrac{\pi}{3}$	$2\left(-\dfrac{1}{2}\right) = -1$	$\left(-1, \dfrac{\pi}{3}\right)$	$\dfrac{4\pi}{3}$	$2\left(-\dfrac{1}{2}\right) = -1$	$\left(-1, \dfrac{4\pi}{3}\right)$
$\dfrac{\pi}{2}$	$2(-1) = -2$	$\left(-2, \dfrac{\pi}{2}\right)$	$\dfrac{3\pi}{2}$	$2(-1) = -2$	$\left(-2, \dfrac{3\pi}{2}\right)$
$\dfrac{2\pi}{3}$	$2\left(-\dfrac{1}{2}\right) = -1$	$\left(-1, \dfrac{2\pi}{3}\right)$	$\dfrac{5\pi}{3}$	$2\left(-\dfrac{1}{2}\right) = -1$	$\left(-1, \dfrac{5\pi}{3}\right)$
$\dfrac{3\pi}{4}$	$2(0) = 0$	$\left(0, \dfrac{3\pi}{4}\right)$	$\dfrac{7\pi}{4}$	$2(0) = 0$	$\left(0, \dfrac{7\pi}{4}\right)$
$\dfrac{5\pi}{6}$	$2\left(\dfrac{1}{2}\right) = 1$	$\left(1, \dfrac{5\pi}{6}\right)$	$\dfrac{11\pi}{6}$	$2\left(\dfrac{1}{2}\right) = 1$	$\left(1, \dfrac{11\pi}{6}\right)$
π	$2(1) = 2$	$(2, \pi)$	2π	$2(1) = 2$	$(2, 2\pi)$

NOTE Graphs of polar equations of the form $r = a\cos(n\theta)$ or $r = a\sin(n\theta)$, $a > 0$, n an integer, are called **roses**. If n is an even integer, the rose has $2n$ petals and passes through the pole $4n$ times. If n is an odd integer, the rose has n petals and passes through the pole $2n$ times.

(b) Parametric equations for $r = 2\cos(2\theta)$:

$$x = r\cos\theta = 2\cos(2\theta)\cos\theta \qquad y = r\sin\theta = 2\cos(2\theta)\sin\theta$$

where θ is the parameter, and if $0 \le \theta \le 2\pi$, then the graph is traced out exactly once in the counterclockwise direction. ∎

NOW WORK Problem 11.

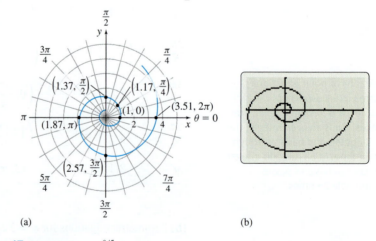

(a) $r = 2 \cos(2\theta), 0 \leq \theta \leq \pi$ (b) $r = 2 \cos(2\theta), 0 \leq \theta \leq 2\pi$

Figure 46 A rose with four petals.

EXAMPLE 5 Graphing a Polar Equation (Spiral); Finding Parametric Equations

(a) Graph the equation $r = e^{\theta/5}$.

(b) Find parametric equations for $r = e^{\theta/5}$.

Solution The polar equation $r = e^{\theta/5}$ lacks the symmetry you may have observed in the previous examples. Since there is no number θ for which $r = 0$, the graph does not contain the pole. Also observe that:

- r is positive for all θ.
- r increases as θ increases.
- $r \to 0$ as $\theta \to -\infty$.
- $r \to \infty$ as $\theta \to \infty$.

NOTE Graphs of polar equations of the form $r = e^{\theta/a}$, $a > 0$, are called **logarithmic spirals**, since the equation can be written as $\theta = a \ln r$. A logarithmic spiral spirals infinitely both toward the pole and away from it.

We use a calculator to obtain Table 6. Figure 47(a) shows part of the graph $r = e^{\theta/5}$. Figure 47(b) shows the graph for $\theta = -\dfrac{3\pi}{2}$ to $\theta = 2\pi$ using technology.

TABLE 6

θ	$r = e^{\theta/5}$	(r, θ)
$-\dfrac{3\pi}{2}$	0.39	$\left(0.39, -\dfrac{3\pi}{2}\right)$
$-\pi$	0.53	$(0.53, -\pi)$
$-\dfrac{\pi}{2}$	0.73	$\left(0.73, -\dfrac{\pi}{2}\right)$
$-\dfrac{\pi}{4}$	0.85	$\left(0.85, -\dfrac{\pi}{4}\right)$
0	1	$(1, 0)$
$\dfrac{\pi}{4}$	1.17	$\left(1.17, \dfrac{\pi}{4}\right)$
$\dfrac{\pi}{2}$	1.37	$\left(1.37, \dfrac{\pi}{2}\right)$
π	1.87	$(1.87, \pi)$
$\dfrac{3\pi}{2}$	2.57	$\left(2.57, \dfrac{3\pi}{2}\right)$
2π	3.51	$(3.51, 2\pi)$

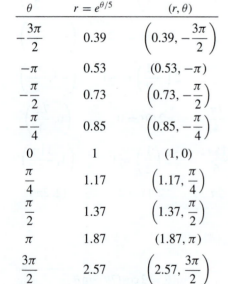

(a)   (b)

Figure 47 The spiral $r = e^{\theta/5}$.

(b) We obtain parametric equations for $r = e^{\theta/5}$ by using the conversion formulas $x = r \cos \theta$ and $y = r \sin \theta$:

$$x = r \cos \theta = e^{\theta/5} \cos \theta \qquad y = r \sin \theta = e^{\theta/5} \sin \theta$$

where θ is the parameter and θ is any real number. ∎

TABLE 7 Library of Polar Equations

Name	Cardioid	Limaçon without inner loop	Limaçon with inner loop
Polar equations	$r = a \pm a \cos\theta, \quad a > 0$ $r = a \pm a \sin\theta, \quad a > 0$	$r = a \pm b \cos\theta, \quad 0 < b < a$ $r = a \pm b \sin\theta, \quad 0 < b < a$	$r = a \pm b \cos\theta, \quad 0 < a < b$ $r = a \pm b \sin\theta, \quad 0 < a < b$
Typical graph			

Name	Lemniscate (See p. 677)	Rose with three petals	Rose with four petals
Polar equations	$r^2 = a^2 \cos(2\theta), \quad a > 0$ $r^2 = a^2 \sin(2\theta), \quad a > 0$	$r = a \sin(3\theta), \quad a > 0$ $r = a \cos(3\theta), \quad a > 0$	$r = a \sin(2\theta), \quad a > 0$ $r = a \cos(2\theta), \quad a > 0$
Typical graph			

2 Find the Arc Length of a Curve Represented by a Polar Equation

Suppose a curve C is represented by the polar equation $r = f(\theta)$, $\alpha \leq \theta \leq \beta$, where both f and its derivative $f'(\theta) = \dfrac{dr}{d\theta}$ are continuous on an interval containing α and β. Using θ as the parameter, parametric equations for the curve C are

$$x(\theta) = r \cos\theta = f(\theta)\cos\theta \qquad y(\theta) = r \sin\theta = f(\theta)\sin\theta$$

Then

$$\frac{dx}{d\theta} = -f(\theta)\sin\theta + f'(\theta)\cos\theta \qquad \frac{dy}{d\theta} = f(\theta)\cos\theta + f'(\theta)\sin\theta$$

After simplification,

$$\left(\frac{dx}{d\theta}\right)^2 + \left(\frac{dy}{d\theta}\right)^2 = [f(\theta)]^2 + [f'(\theta)]^2 = r^2 + \left(\frac{dr}{d\theta}\right)^2 \qquad r = f(\theta)$$

Since we are using parametric equations, the length s of C from $\theta = \alpha$ to $\theta = \beta$ is

$$s = \int_\alpha^\beta \sqrt{\left(\frac{dx}{d\theta}\right)^2 + \left(\frac{dy}{d\theta}\right)^2}\, d\theta = \int_\alpha^\beta \sqrt{r^2 + \left(\frac{dr}{d\theta}\right)^2}\, d\theta$$

THEOREM Arc Length of the Graph of a Polar Equation

If a curve C is represented by the polar equation $r = f(\theta)$, $\alpha \leq \theta \leq \beta$, and if $f'(\theta) = \dfrac{dr}{d\theta}$ is continuous on an interval containing α and β, then the arc length s of C from $\theta = \alpha$ to $\theta = \beta$ is

$$s = \int_\alpha^\beta \sqrt{r^2 + \left(\frac{dr}{d\theta}\right)^2}\, d\theta$$

EXAMPLE 6   **Finding the Arc Length of a Logarithmic Spiral**

Find the arc length s of the logarithmic spiral represented by $r = f(\theta) = e^{3\theta}$ from $\theta = 0$ to $\theta = 2$.

Solution   We use the arc length formula $s = \displaystyle\int_{\alpha}^{\beta} \sqrt{r^2 + \left(\dfrac{dr}{d\theta}\right)^2}\, d\theta$ with $r = e^{3\theta}$.

Then $\dfrac{dr}{d\theta} = 3e^{3\theta}$ and

$$s = \int_{0}^{2} \sqrt{(e^{3\theta})^2 + (3e^{3\theta})^2}\, d\theta = \int_{0}^{2} \sqrt{10e^{6\theta}}\, d\theta = \sqrt{10} \int_{0}^{2} e^{3\theta}\, d\theta$$

$$= \sqrt{10} \left[\frac{e^{3\theta}}{3}\right]_{0}^{2} = \frac{\sqrt{10}}{3}(e^{6} - 1)$$ ∎

NOW WORK   **Problem 19.**

9.5 Assess Your Understanding

Concepts and Vocabulary

1. *True or False*   A cardioid passes through the pole.

2. *Multiple Choice*   The equations for cardioids and limaçons are very similar. They all have the form

$r = a \pm b\cos\theta$ or $r = a \pm b\sin\theta$, $a > 0, b > 0$

The equations represent a limaçon with an inner loop if [(a) $a < b$, (b) $a > b$, (c) $a = b$]; a cardioid if [(a) $a < b$, (b) $a > b$, (c) $a = b$]; and a limaçon without an inner loop if [(a) $a < b$, (b) $a > b$, (c) $a = b$].

3. *True or False*   The graph of $r = \sin(4\theta)$ is a rose.

4. The rose $r = \cos(3\theta)$ has _____ petals.

Skill Building

In Problems 5–12, for each polar equation:
(a) Graph the equation.
(b) Find parametric equations that represent the equation.

5. $r = 2 + 2\cos\theta$ **6.** $r = 3 - 3\sin\theta$

7. $r = 4 - 2\cos\theta$ **8.** $r = 2 + \sin\theta$

9. $r = 1 + 2\sin\theta$ **10.** $r = 2 - 3\cos\theta$

11. $r = \sin(3\theta)$ **12.** $r = 4\cos(4\theta)$

In Problems 13–18, graph each pair of polar equations on the same polar grid. Find polar coordinates of the point(s) of intersection and label the point(s) on the graph.

13. $r = 8\cos\theta$, $r = 2\sec\theta$ **14.** $r = 8\sin\theta$, $r = 4\csc\theta$

15. $r = \sin\theta$, $r = 1 + \cos\theta$ **16.** $r = 3$, $r = 2 + 2\cos\theta$

17. $r = 1 + \sin\theta$, $r = 1 + \cos\theta$

18. $r = 1 + \cos\theta$, $r = 3\cos\theta$

In Problems 19–22, find the arc length of each curve.

19. $r = f(\theta) = e^{\theta/2}$ from $\theta = 0$ to $\theta = 2$

20. $r = f(\theta) = e^{2\theta}$ from $\theta = 0$ to $\theta = 2$

21. $r = f(\theta) = \cos^2 \dfrac{\theta}{2}$ from $\theta = 0$ to $\theta = \pi$

22. $r = f(\theta) = \sin^2 \dfrac{\theta}{2}$ from $\theta = 0$ to $\theta = \pi$

Applications and Extensions

In Problems 23–26, the polar equation for each graph is either $r = a + b\cos\theta$ or $r = a + b\sin\theta$, $a > 0$, $b > 0$. Select the correct equation and find the values of a and b.

23.
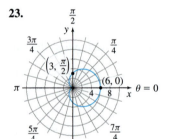

24.
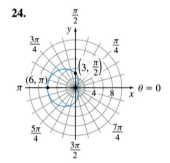

25.
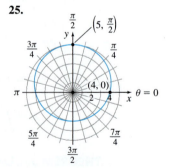

26.
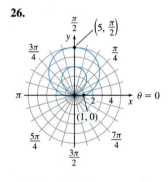

In Problems 27–32, find an equation of the tangent line to each curve at the given number. (Hint: Find parametric equations that represent each polar equation. See Section 9.2.)

27. $r = 2\cos(3\theta)$ at $\theta = \dfrac{\pi}{6}$ **28.** $r = 3\sin(3\theta)$ at $\theta = \dfrac{\pi}{3}$

29. $r = 2 + \cos\theta$ at $\theta = \dfrac{\pi}{4}$ **30.** $r = 3 - \sin\theta$ at $\theta = \dfrac{\pi}{6}$

---

1. = NOW WORK problem   〔N〕 = Graphing technology recommended   〔CAS〕 = Computer Algebra System recommended

31. $r = 4 + 5\sin\theta$ at $\theta = \dfrac{\pi}{4}$ **32.** $r = 1 - 2\cos\theta$ at $\theta = \dfrac{\pi}{4}$

Lemniscates Graphs of polar equations of the form $r^2 = a^2\cos(2\theta)$ or $r^2 = a^2\sin(2\theta)$, where $a \neq 0$, are called **lemniscates**. A lemniscate passes through the pole twice and is shaped like the infinity symbol ∞.

In Problems 33–36, for each equation:
(a) *Graph the lemniscate.*
(b) *Find parametric equations that represent the equation.*

33. $r^2 = 4\sin(2\theta)$ **34.** $r^2 = 9\cos(2\theta)$

35. $r^2 = \cos(2\theta)$ **36.** $r^2 = 16\sin(2\theta)$

In Problems 37–48:
(a) *Graph each polar equation.*
(b) *Find parametric equations that represent each equation.*

37. $r = \dfrac{2}{1 - \cos\theta}$ (parabola)   **38.** $r = \dfrac{1}{1 - \cos\theta}$ (parabola)

39. $r = \dfrac{1}{3 - 2\cos\theta}$ (ellipse)   **40.** $r = \dfrac{2}{1 - 2\cos\theta}$ (hyperbola)

41. $r = \theta; \theta \geq 0$ (spiral of Archimedes)

42. $r = \dfrac{3}{\theta}; \theta > 0$ (reciprocal spiral)

43. $r = \csc\theta - 2; \ 0 < \theta < \pi$ (conchoid)

44. $r = 3 - \dfrac{1}{2}\csc\theta$ (conchoid)

45. $r = \sin\theta\tan\theta$ (cissoid)   **46.** $r = \cos\dfrac{\theta}{2}$

47. $r = \tan\theta$ (kappa curve)   **48.** $r = \cot\theta$ (kappa curve)

49. Show that $r = 4(\cos\theta + 1)$ and $r = 4(\cos\theta - 1)$ have the same graph.

50. Show that $r = 5(\sin\theta + 1)$ and $r = 5(\sin\theta - 1)$ have the same graph.

51. Arc Length Find the arc length of the spiral $r = \theta$ from $\theta = 0$ to $\theta = 2\pi$.

52. Arc Length Find the arc length of the spiral $r = 3\theta$ from $\theta = 0$ to $\theta = 2\pi$.

53. Perimeter Find the perimeter of the cardioid $r = f(\theta) = 1 - \cos\theta, \ -\pi \leq \theta \leq \pi$.

54. Exploring Using Graphing Technology

(a) Graph $r_1 = 2\cos(4\theta)$. Clear the screen and graph $r_2 = 2\cos(6\theta)$. How many petals does each of the graphs have?

(b) Clear the screen and graph, in order, each on a clear screen, $r_1 = 2\cos(3\theta)$, $r_2 = 2\cos(5\theta)$, and $r_3 = 2\cos(7\theta)$. What do you notice about the number of petals? Do the results support the definition of a rose?

55. Exploring Using Graphing Technology Graph $r_1 = 3 - 2\cos\theta$. Clear the screen and graph $r_2 = 3 + 2\cos\theta$. Clear the screen and graph $r_3 = 3 + 2\sin\theta$. Clear the screen and graph $r_4 = 3 - 2\sin\theta$. Describe the pattern.

56. Horizontal and Vertical Tangent Lines Find the horizontal and vertical tangent lines of the cardioid $r = 1 - \sin\theta$ discussed in Example 1.

57. Horizontal and Vertical Tangent Lines Find the horizontal and vertical tangent lines of the cardioid $r = 3 + 3\cos\theta$.

CAS 58. Horizontal and Vertical Tangent Lines Find the horizontal and vertical tangent lines of the limaçon with an inner loop $r = 1 + 2\cos\theta$ discussed in Example 3.

CAS 59. Horizontal and Vertical Tangent Lines Find the horizontal and vertical tangent lines of the rose with four petals $r = 2\cos(2\theta), 0 \leq \theta \leq 2\pi$, discussed in Example 4.

CAS 60. Horizontal and Vertical Tangent Lines Find the horizontal and vertical tangent lines of the spiral $r = e^{\theta/5}$ discussed in Example 5.

CAS 61. Horizontal and Vertical Tangent Lines Find the horizontal and vertical tangent lines of the lemniscate $r^2 = 4\sin(2\theta)$.

62. Test for Symmetry Symmetry with respect to the polar axis can be tested by replacing θ with $-\theta$. If an equivalent equation results, the graph is symmetric with respect to the polar axis.

(a) Explain why this test is valid.

(b) Use the test to show that $r = 3 + 2\cos\theta$ is symmetric with respect to the polar axis.

63. Test for Symmetry Symmetry with respect to the pole can be tested by replacing r by $-r$ or by replacing θ by $\theta + \pi$. If either substitution produces an equivalent equation, the graph is symmetric with respect to the pole.

(a) Explain why these tests are valid.

(b) Show that $r^2 = 4\sin(2\theta)$ is symmetric with respect to the pole.

64. Test for Symmetry Symmetry with respect to the line $\theta = \dfrac{\pi}{2}$ can be tested by replacing θ by $\pi - \theta$. If an equivalent equation results, the graph is symmetric with respect to the line $\theta = \dfrac{\pi}{2}$.

(a) Explain why this test is valid.

(b) Use the test to show that $r = 2\cos(2\theta)$ is symmetric with respect to the line $\theta = \dfrac{\pi}{2}$.

Challenge Problems

Tests for Symmetry The three tests for symmetry described in Problems 62–64 are *sufficient* conditions for symmetry, but they are not *necessary* conditions. That is, an equation may fail these tests and still have a graph that is symmetric with respect to the polar axis, the line $\theta = \dfrac{\pi}{2}$, or the pole.

65. Testing for Symmetry The graph of $r = \sin(2\theta)$ (a rose with four petals) is symmetric with respect to the polar axis, the line $\theta = \dfrac{\pi}{2}$, and the pole. Show that the test for symmetry with respect to the pole (see Problem 63) works, but the test for symmetry with respect to the polar axis fails (see Problem 62).

66. Arc Length Find the entire arc length of the curve

$$r = a \sin^3 \frac{\theta}{3}, \ a > 0. \ (\textit{Hint: Use parametric equations.})$$

CAS **67. Arc Length of a Rose Petal**

(a) Graph the rose $r = 4\sin(5\theta)$.

(b) The petal in the first quadrant begins when $\theta = 0$ and ends when $\theta = \alpha$. Find α.

(c) Find the length s of the petal described in (b).

9.6 Area in Polar Coordinates

OBJECTIVES *When you finish this section, you should be able to:*

1 **Find the area of a region enclosed by the graph of a polar equation (p. 678)**

2 **Find the area of a region enclosed by the graphs of two polar equations (p. 681)**

3 **Find the surface area of a solid of revolution obtained from the graph of a polar equation (p. 682)**

In this section, we find the area of a region enclosed by the graph of a polar equation and two rays that have the pole as a common vertex. The technique used is similar to that used in Chapter 5, except, instead of approximating the area using rectangles, we approximate the area using sectors of a circle. Figure 48 illustrates the area of a sector of a circle.

RECALL The area A of the sector of a circle of radius r formed by a central angle of θ radians is $A = \dfrac{1}{2}r^2\theta$.

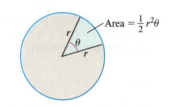

Figure 48

1 Find the Area of a Region Enclosed by the Graph of a Polar Equation

In Figure 49, $r = f(\theta)$ is a function that is nonnegative and continuous on the interval $\alpha \le \theta \le \beta$. Let A denote the area of the region enclosed by the graph of $r = f(\theta)$ and the rays $\theta = \alpha$ and $\theta = \beta$, where $0 \le \alpha < \beta \le 2\pi$. It is helpful to think of the region R as being "swept out" by rays, beginning with the ray $\theta = \alpha$ and continuing to the ray $\theta = \beta$.

We partition the closed interval $[\alpha, \beta]$ into n subintervals:

$$[\alpha, \theta_1], [\theta_1, \theta_2], \ldots, [\theta_{i-1}, \theta_i], \ldots, [\theta_{n-1}, \beta]$$

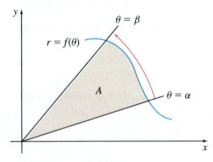

Figure 49

each of length $\Delta\theta = \dfrac{\beta - \alpha}{n}$. As shown in Figure 50, we select an angle $\theta_i{}^*$ in each subinterval $[\theta_{i-1}, \theta_i]$. The quantity $\dfrac{1}{2}[f(\theta_i{}^*)]^2\Delta\theta$ is the area of the circular sector with radius $r = f(\theta_i{}^*)$ and central angle $\Delta\theta$. The sums of the areas of these sectors

$$\sum_{i=1}^{n} \frac{1}{2}[f(\theta_i{}^*)]^2\Delta\theta$$

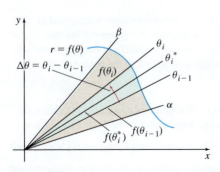

Figure 50

are an approximation of the area A we seek. As the number n of subintervals increases, the sums $\displaystyle\sum_{i=1}^{n} \frac{1}{2}\,[f(\theta_i{}^*)]^2\,\Delta\theta$ become better approximations to the total area A. The

sums $\displaystyle\sum_{i=1}^{n} \frac{1}{2}[f(\theta_i{}^*)]^2\Delta\theta$ are Riemann sums, and since $r = f(\theta)$ is continuous, the limit exists and equals a definite integral.

$$\lim_{n\to\infty} \sum_{i=1}^{n} \frac{1}{2}[f(\theta_i{}^*)]^2\Delta\theta = \int_{\alpha}^{\beta} \frac{1}{2}[f(\theta)]^2 d\theta = \int_{\alpha}^{\beta} \frac{1}{2}r^2 d\theta$$

NEED TO REVIEW? The definite integral is discussed in Section 5.2, pp. 353–359.

THEOREM Area in Polar Coordinates

If $r = f(\theta)$ is nonnegative and continuous on the closed interval $[\alpha, \beta]$, where $\alpha < \beta$ and $\beta - \alpha \leq 2\pi$, then the area A of the region enclosed by the graph of $r = f(\theta)$ and the rays $\theta = \alpha$ and $\theta = \beta$ is given by

$$A = \int_{\alpha}^{\beta} \frac{1}{2} r^2 \, d\theta.$$

Be sure to graph the equation $r = f(\theta)$, $\alpha \leq \theta \leq \beta$, before using the formula. In drawing the graph, include the rays $\theta = \alpha$, indicating the start, and $\theta = \beta$, indicating the end, of the region whose area is to be found. These rays determine the limits of integration in the area formula.

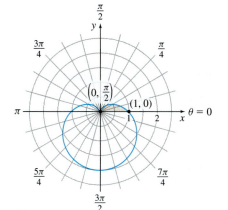

(a)

(b)

Figure 51 $r = 1 - \sin\theta, 0 \leq \theta \leq \dfrac{\pi}{2}$

EXAMPLE 1   Finding the Area Enclosed by a Part of a Cardioid

Find the area of the region enclosed by the cardioid $r = 1 - \sin\theta, 0 \leq \theta \leq \dfrac{\pi}{2}$.

Solution The cardioid represented by $r = 1 - \sin\theta$ is shown in Figure 51(a) and the area enclosed by the cardioid from $\theta = 0$ to $\theta = \dfrac{\pi}{2}$ is shaded.

The region is swept out beginning with the ray $\theta = 0$ and ending with the ray $\theta = \dfrac{\pi}{2}$, as shown in Figure 51(b). The limits of integration are 0 and $\dfrac{\pi}{2}$ and the area A is

$$A = \int_{\alpha}^{\beta} \frac{1}{2} r^2 \, d\theta = \int_{0}^{\pi/2} \frac{1}{2}(1 - \sin\theta)^2 \, d\theta = \frac{1}{2}\int_{0}^{\pi/2}(1 - 2\sin\theta + \sin^2\theta)\, d\theta$$

$$= \frac{1}{2}\int_{0}^{\pi/2}\left\{1 - 2\sin\theta + \frac{1}{2}[1 - \cos(2\theta)]\right\} d\theta \quad \sin^2\theta = \frac{1 - \cos(2\theta)}{2}$$

$$= \frac{1}{2}\int_{0}^{\pi/2}\left[\frac{3}{2} - 2\sin\theta - \frac{1}{2}\cos(2\theta)\right] d\theta$$

$$= \frac{1}{2}\left[\frac{3}{2}\theta + 2\cos\theta - \frac{1}{4}\sin(2\theta)\right]_{0}^{\pi/2} = \frac{3\pi - 8}{8} \qquad \blacksquare$$

NOW WORK Problem 9.

EXAMPLE 2   Finding the Area Enclosed by a Cardioid

Find the area enclosed by the cardioid $r = 1 - \sin\theta$.

Solution Look again at the cardioid in Figure 51(a). The region enclosed by the cardioid is swept out beginning with the ray $\theta = 0$ and ending with the ray $\theta = 2\pi$. So, the limits of integration are 0 and 2π, and the area A is

$$A = \int_{0}^{2\pi} \frac{1}{2}(1 - \sin\theta)^2 \, d\theta = \frac{1}{2}\left[\frac{3}{2}\theta + 2\cos\theta - \frac{1}{4}\sin(2\theta)\right]_{0}^{2\pi} = \frac{3\pi}{2} \qquad \blacksquare$$

NOW WORK Problem 13.

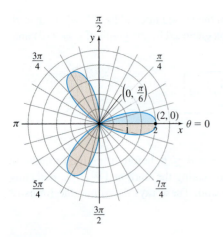

Figure 52 $r = 2\cos(3\theta)$

NEED TO REVIEW? Solving trigonometric equations is discussed in Section P.7, pp. 61–63.

EXAMPLE 3 Finding the Area Enclosed by a Rose

Find the area enclosed by the graph of $r = 2\cos(3\theta)$, a rose with three petals.

Solution Figure 52 shows the rose. The area of the blue shaded region in quadrant I equals one-sixth of the area A enclosed by the graph.* The shaded region in quadrant I is swept out beginning with the ray $\theta = 0$. It ends at the point $(0, \theta)$, $0 < \theta < \dfrac{\pi}{2}$, where θ is the solution of the equation.

$$2\cos(3\theta) = 0 \qquad 0 < \theta < \frac{\pi}{2}$$

$$\cos(3\theta) = 0$$

$$3\theta = \frac{\pi}{2} + 2k\pi$$

$$\theta = \frac{\pi}{6} + \frac{2k\pi}{3}$$

Since $0 < \theta < \dfrac{\pi}{2}$, we have $\theta = \dfrac{\pi}{6}$. The area of the shaded region in quadrant I swept out by the rays $\theta = 0$ and $\theta = \dfrac{\pi}{6}$ is given by $\displaystyle\int_0^{\pi/6} \frac{1}{2}r^2 d\theta$, and the area A of the region we seek is 6 times this area.

$$A = 6\int_0^{\pi/6} \frac{1}{2}r^2\, d\theta = 3\int_0^{\pi/6} 4\cos^2(3\theta)\, d\theta = 12\int_0^{\pi/6} \cos^2(3\theta)\, d\theta$$

$$= 12\int_0^{\pi/6} \frac{1 + \cos(6\theta)}{2}\, d\theta \qquad \cos^2(3\theta) = \frac{1 + \cos(6\theta)}{2}$$

$$= 6\left[\theta + \frac{1}{6}\sin(6\theta)\right]_0^{\pi/6} = 6\left(\frac{\pi}{6}\right) = \pi \qquad ■$$

NOW WORK Problem 17.

EXAMPLE 4 Finding the Area Enclosed by a Limaçon

Find the area of the region enclosed by the limaçon $r = 2 + \cos\theta$.

Solution Figure 53 shows the graph of $r = 2 + \cos\theta$, a limaçon without an inner loop.

We see that the region above the polar axis equals the region below it, so the area A of the region enclosed by the limaçon equals twice the area of the region enclosed by $r = 2 + \cos\theta$ and swept out by the rays $\theta = 0$ and $\theta = \pi$.

$$A = 2\int_0^{\pi} \frac{1}{2}r^2 d\theta = \int_0^{\pi} (2 + \cos\theta)^2\, d\theta = \int_0^{\pi} (4 + 4\cos\theta + \cos^2\theta)\, d\theta$$

$$= \int_0^{\pi} \left[4 + 4\cos\theta + \frac{1 + \cos(2\theta)}{2}\right] d\theta \qquad \cos^2\theta = \frac{1 + \cos(2\theta)}{2}$$

$$= \left[4\theta + 4\sin\theta + \frac{\theta}{2} + \frac{1}{4}\sin(2\theta)\right]_0^{\pi} = \frac{9\pi}{2} \qquad ■$$

NOW WORK Problem 21.

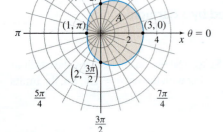

Figure 53 $r = 2 + \cos\theta$

---

*We need to exploit symmetry here since there are intervals on which $r < 0$, and the area formula requires that $r > 0$.

2 Find the Area of a Region Enclosed by the Graphs of Two Polar Equations

To find the area A of the region enclosed by the graphs of two polar equations, we begin by graphing the equations and finding their points of intersection, if any.

EXAMPLE 5 **Finding the Area of the Region Enclosed by the Graphs of Two Polar Equations**

Find the area of the region that lies outside the cardioid $r = 1 + \cos\theta$ and inside the circle $r = 3\cos\theta$.

Solution We begin by graphing each equation. See Figure 54(a). Then we find the points of intersection of the two graphs by solving the equation,

$$3\cos\theta = 1 + \cos\theta$$

$$2\cos\theta = 1$$

$$\cos\theta = \frac{1}{2}$$

$$\theta = -\frac{\pi}{3} \quad \text{or} \quad \theta = \frac{\pi}{3}$$

The graphs intersect at the points $\left(\frac{3}{2}, -\frac{\pi}{3}\right)$ and $\left(\frac{3}{2}, \frac{\pi}{3}\right)$.

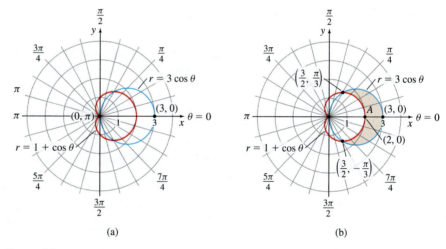

(a)   (b)

Figure 54

The area A of the region that lies outside the cardioid and inside the circle is shown as the shaded portion in Figure 54(b). Notice that the area A is the difference between the area of the region enclosed by the circle $r = 3\cos\theta$ swept out by the rays $\theta = -\frac{\pi}{3}$ and $\theta = \frac{\pi}{3}$, and the area of the region enclosed by the cardioid $r = 1 + \cos\theta$ swept out by the same rays. So,

$$A = \int_{-\pi/3}^{\pi/3} \frac{1}{2}(3\cos\theta)^2 d\theta - \int_{-\pi/3}^{\pi/3} \frac{1}{2}(1 + \cos\theta)^2 d\theta = \frac{1}{2}\int_{-\pi/3}^{\pi/3}[9\cos^2\theta - (1 + 2\cos\theta + \cos^2\theta)]\,d\theta$$

$$= \frac{1}{2}\int_{-\pi/3}^{\pi/3}(8\cos^2\theta - 1 - 2\cos\theta)\,d\theta = \frac{1}{2}\int_{-\pi/3}^{\pi/3}\left[8\left(\frac{1 + \cos(2\theta)}{2}\right) - 1 - 2\cos\theta\right]d\theta$$

$$= \frac{1}{2}\int_{-\pi/3}^{\pi/3}[3 + 4\cos(2\theta) - 2\cos\theta]\,d\theta = \frac{1}{2}\Big[3\theta + 2\sin(2\theta) - 2\sin\theta\Big]_{-\pi/3}^{\pi/3} = \pi \qquad ■$$

NOW WORK Problem **23**.

CAUTION The circle and the cardioid shown in Figure 54(a) actually intersect in a third point, the pole. The pole is not identified when we solve the equation to find the points of intersection because the pole has coordinates $(0, \pi)$ on $r = 1 + \cos\theta$, but it has coordinates $\left(0, \dfrac{\pi}{2}\right)$ and $\left(0, \dfrac{3\pi}{2}\right)$ on $r = 3\cos\theta$. This demonstrates the importance of graphing polar equations when looking for their points of intersection. Since the pole presents particular difficulties, let $r = 0$ in each equation to determine whether the graph passes through the pole.

3 Find the Surface Area of a Solid of Revolution Obtained from a Polar Equation

Suppose a smooth curve C is given by the polar equation $r = f(\theta)$, $\alpha \leq \theta \leq \beta$. Parametric equations for this curve are

$$x(\theta) = r\cos\theta = f(\theta)\cos\theta \qquad y(\theta) = r\sin\theta = f(\theta)\sin\theta$$

Then

$$\frac{dx}{d\theta} = f'(\theta)\cos\theta - f(\theta)\sin\theta \qquad \frac{dy}{d\theta} = f'(\theta)\sin\theta + f(\theta)\cos\theta$$

$$\left(\frac{dx}{d\theta}\right)^2 + \left(\frac{dy}{d\theta}\right)^2 = [f(\theta)]^2 + [f'(\theta)]^2 = r^2 + \left(\frac{dr}{d\theta}\right)^2$$

Since the surface area S of the solid of revolution obtained by revolving C about the polar axis (x-axis) when using parametric equations is

$$S = 2\pi \int_\alpha^\beta y(\theta)\sqrt{\left(\frac{dx}{d\theta}\right)^2 + \left(\frac{dy}{d\theta}\right)^2}\, d\theta$$

the surface area S using a polar equation is

$$S = 2\pi \int_\alpha^\beta r\sin\theta\sqrt{r^2 + \left(\frac{dr}{d\theta}\right)^2}\, d\theta = 2\pi \int_\alpha^\beta f(\theta)\sin\theta\sqrt{[f(\theta)]^2 + [f'(\theta)]^2}\, d\theta \qquad (1)$$

EXAMPLE 6   Finding the Surface Area of a Solid of Revolution

Find the surface area of the solid of revolution generated by revolving the arc of the circle $r = a$, $a > 0$, $0 \leq \theta \leq \dfrac{\pi}{4}$, about the polar axis.

Solution See Figure 55 on page 683. We find the surface area S using formula (1). Since $r = f(\theta) = a$, $f'(\theta) = 0$. Then

$$S = 2\pi \int_\alpha^\beta f(\theta)\sin\theta\sqrt{[f(\theta)]^2 + [f'(\theta)]^2}\, d\theta$$

$$= 2\pi \int_0^{\pi/4} a\sin\theta\sqrt{a^2}\, d\theta = 2\pi a^2 \int_0^{\pi/4} \sin\theta\, d\theta$$

$$= 2\pi a^2 \left[-\cos\theta\right]_0^{\pi/4} = 2\pi a^2 \left(-\frac{\sqrt{2}}{2} + 1\right) = \pi a^2 (2 - \sqrt{2}) \qquad \blacksquare$$

NOW WORK Problem 27.

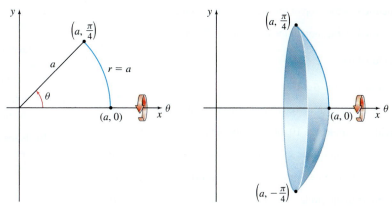

Figure 55

9.6 Assess Your Understanding

Concepts and Vocabulary

1. The area A of the sector of a circle of radius r and central angle θ is $A =$ _____.

2. *True or False*   The area enclosed by the graph of a polar equation and two rays that have the pole as a common vertex is found by approximating the area using sectors of a circle.

3. *True or False*   The area A enclosed by the graph of the equation $r = f(\theta)$, $r \geq 0$, and the rays $\theta = \alpha$ and $\theta = \beta$, is given by $A = \int_{\alpha}^{\beta} f(\theta)\, d\theta$.

4. *True or False*   If $x(\theta) = r \cos \theta$, $y(\theta) = r \sin \theta$ are parametric equations of the polar equation $r = f(\theta)$, then
$$\left(\frac{dx}{d\theta}\right)^2 + \left(\frac{dy}{d\theta}\right)^2 = r^2 + \left(\frac{dr}{d\theta}\right)^2.$$

Skill Building

In Problems 5–8, find the area of the shaded region.

5. $r = \cos(2\theta)$

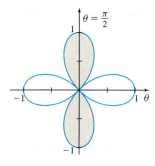

6. $r = 2 \sin(3\theta)$

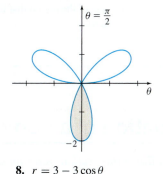

7. $r = 2 + 2 \sin \theta$

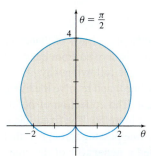

8. $r = 3 - 3 \cos \theta$

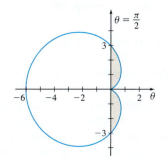

In Problems 9–12, find the area of the region enclosed by the graph of each polar equation swept out by the given rays.

9. $r = 3 \cos \theta$; $\theta = 0$ to $\theta = \dfrac{\pi}{3}$

10. $r = 3 \sin \theta$; $\theta = 0$ to $\theta = \dfrac{\pi}{4}$

11. $r = a\theta$; $\theta = 0$ to $\theta = 2\pi$ 12. $r = e^{a\theta}$; $\theta = 0$ to $\theta = \dfrac{\pi}{2}$

In Problems 13–18, find the area of the region enclosed by the graph of each polar equation.

13. $r = 1 + \cos \theta$ 14. $r = 2 - 2 \sin \theta$ 15. $r = 3 + \sin \theta$

16. $r = 3(2 - \sin \theta)$ 17. $r = 8 \sin(3\theta)$ 18. $r = \cos(4\theta)$

In Problems 19–22, find the area of the region enclosed by one loop of the graph of each polar equation.

19. $r = 4 \sin(2\theta)$ 20. $r = 5 \cos(3\theta)$

21. $r^2 = 4 \cos(2\theta)$ 22. $r = a^2 \cos(2\theta)$

In Problems 23–26, find the area of each region described.

23. Inside $r = 2 \sin \theta$; outside $r = 1$

24. Inside $r = 4 \cos \theta$; outside $r = 2$

25. Inside $r = \sin \theta$; outside $r = 1 - \cos \theta$

26. Inside $r^2 = 4 \cos(2\theta)$; outside $r = \sqrt{2}$

In Problems 27–30, find the surface area of the solid of revolution generated by revolving each curve about the polar axis.

27. $r = \sin \theta$, $0 \leq \theta \leq \dfrac{\pi}{2}$ 28. $r = 1 + \cos \theta$, $0 \leq \theta \leq \pi$

29. $r = e^{\theta}$, $0 \leq \theta \leq \pi$ 30. $r = 2a \cos \theta$, $0 \leq \theta \leq \dfrac{\pi}{2}$

Applications and Extensions

In Problems 31–48, find the area of the region:

31. enclosed by the small loop of the limaçon $r = 1 + 2 \cos \theta$.

32. enclosed by the small loop of the limaçon $r = 1 + 2 \sin \theta$.

---

1. = NOW WORK problem   = Graphing technology recommended   [CAS] = Computer Algebra System recommended

33. enclosed by the loop of the graph of $r = 2 - \sec\theta$.

34. enclosed by the loop of the graph of $r = 5 + \sec\theta$.

35. enclosed by $r = 2\sin^2\dfrac{\theta}{2}$.

36. enclosed by $r = 6\cos^2\theta$.

37. inside the circle $r = 8\cos\theta$ and to the right of the line $r = 2\sec\theta$.

38. inside the circle $r = 10\sin\theta$ and above the line $r = 2\csc\theta$.

39. outside the circle $r = 3$ and inside the cardioid $r = 2 + 2\cos\theta$.

40. inside the circle $r = \sin\theta$ and outside the cardioid $r = 1 + \cos\theta$.

41. common to the circle $r = \cos\theta$ and the cardioid $r = 1 - \cos\theta$.

42. common to the circles $r = \cos\theta$ and $r = \sin\theta$.

43. common to the inside of the cardioid $r = 1 + \sin\theta$ and the outside of the cardioid $r = 1 + \cos\theta$.

44. common to the inside of the lemniscate $r^2 = 8\cos(2\theta)$ and the outside of the circle $r = 2$.

45. enclosed by the rays $\theta = 0$ and $\theta = 1$ and $r = e^{-\theta}$, $0 \le \theta \le 1$.

46. enclosed by the rays $\theta = 0$ and $\theta = 1$ and $r = e^{\theta}$, $0 \le \theta \le 1$.

47. enclosed by the rays $\theta = 1$ and $\theta = \pi$ and $r = \dfrac{1}{\theta}$, $1 \le \theta \le \pi$.

48. inside the outer loop but outside the inner loop of $r = 1 + 2\sin\theta$.

49. Area   Find the area of the loop of the graph of $r = \sec\theta + 2$.

50. Surface Area of a Sphere   Develop a formula for the surface area of a sphere of radius R.

51. Surface Area of a Bead   A sphere of radius R has a hole of radius $a < R$ drilled through it. See the figure. The axis of the hole coincides with a diameter of the sphere.

   (a) Find the surface area of that part of the sphere that remains.

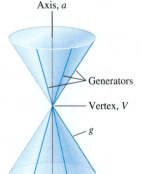

(b) Is the area found in Problem 29, Section 9.3, the same as the area found here? If not, justify the difference.

52. Surface Area of a Plug   A plug is made to repair the hole in the sphere in Problem 51.

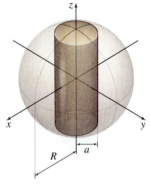

   (a) What is the surface area of the plug?

   (b) Is the area found in Problem 30, Section 9.3, the same as the area found here? If not, justify the difference.

53. Area   Find the area enclosed by the loop of the **strophoid**
$$r = \sec\theta - 2\cos\theta,$$
$$-\frac{\pi}{2} < \theta < \frac{\pi}{2}$$
as shown in the figure.

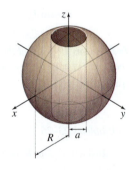

[CAS] **54. Area**

   (a) Graph the limaçon $r = 2 - 3\cos\theta$.

   (b) Find the area enclosed by the inner loop. Round the answer to three decimal places.

Challenge Problems

55. Show that the area enclosed by the graph of $r\theta = a$ and the rays $\theta = \theta_1$ and $\theta = \theta_2$ is proportional to the difference of the radii, $r_1 - r_2$, where $r_1 = \dfrac{a}{\theta_1}$ and $r_2 = \dfrac{a}{\theta_2}$.

56. Find the area of the region that lies outside the circle $r = 1$ and inside the rose $r = 3\sin(3\theta)$.

57. Find the area of the region that lies inside the circle $r = 2$ and outside the rose $r = 3\sin(2\theta)$.

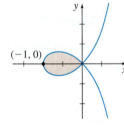

Figure 56

9.7 The Polar Equation of a Conic

OBJECTIVE   *When you finish this section, you should be able to:*

1 Express a conic as a polar equation (p. 686)

The polar equation of a conic is used to explain and to derive Kepler's Laws of Planetary Motion, which are discussed in Chapter 11. The word "conic" is derived from the word "cone," which is a geometric figure that can be constructed in the following way: Let a and g be two distinct lines that intersect at a point V. Fix the line a and revolve the line g about a while keeping the angle between the lines constant. The collection of points swept out by the line g is called a **right circular cone**. See Figure 56. The fixed line a is called the **axis** of the cone; the point V is its **vertex**. Any line passing through V that makes the same angle with a as the original line g is called a **generator** of the cone.

Each generator lies entirely on the cone. The cone consists of two parts, called **nappes**, that intersect at the vertex.

Conics, an abbreviation for **conic sections**, are curves that result when a right circular cone and a plane intersect. The conics discussed here arise when the plane does not contain the vertex of the cone, as shown in Figure 57. If the plane contains the vertex, the intersection of the plane and the cone is a point, a line, or a pair of lines. These are called **degenerate cases**.

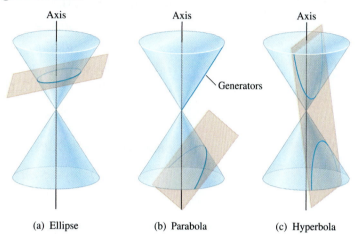

(a) Ellipse   (b) Parabola   (c) Hyperbola

Figure 57

A conic is:

- an **ellipse**, when the plane intersects the axis a at an angle greater than the angle between a and g. An ellipse lies on only one nappe of the cone. See Figure 57(a).
- a **parabola**, when the plane intersects the axis a at the same angle as the line g; that is, the plane is parallel to exactly one generator. A parabola lies on only one nappe of the cone. See Figure 57(b).
- a **hyperbola**, when the plane intersects the axis a at an angle smaller than the angle between a and g. A hyperbola lies on both nappes of the cone. See Figure 57(c).

In Appendix A.3, pp. A-22 to A-25, we discuss the rectangular equations of a parabola, an ellipse, and a hyperbola. To obtain the polar equations, we use a unified definition that simultaneously defines all three conics.

DEFINITION

Let D denote a fixed line called the **directrix**; let F denote a fixed point called the **focus**, which is not on D; and let e be a fixed positive number called the **eccentricity**. A **conic** is the set of points P in the plane for which the ratio of the distance from F to P to the distance from D to P equals e. That is, a **conic** is the collection of points P for which

$$\boxed{\frac{d(F, P)}{d(D, P)} = e} \tag{1}$$

- If $e = 1$, the conic is a parabola.
- If $e < 1$, the conic is an ellipse.
- If $e > 1$, the conic is a hyperbola.

Figure 58 illustrates the definition.

- In a parabola, the **axis** is the line through the focus perpendicular to the directrix.
- In an ellipse, the **major axis** is the line through the focus perpendicular to the directrix.
- In a hyperbola, the **transverse axis** is the line through the focus perpendicular to the directrix.

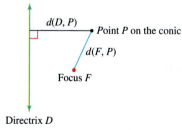

Figure 58 $\dfrac{d(F, P)}{d(D, P)} = e.$

1 Express a Conic as a Polar Equation

The equations for the conics in polar coordinates are derived by positioning the focus F at the pole (the origin) and the directrix D either parallel to or perpendicular to the polar axis.

Suppose the directrix D is perpendicular to the polar axis at a distance p units to the left of the pole (the focus F), as shown in Figure 59. If $P = (r, \theta)$ is any point on the conic, then by (1),

$$\frac{d(F, P)}{d(D, P)} = e \quad \text{or} \quad d(F, P) = e \cdot d(D, P)$$

Now drop the perpendicular from the point P to the polar axis. Using the point of intersection Q, we have $d(O, Q) = r \cos \theta$. Then

$$d(D, P) = p + d(O, Q) = p + r \cos \theta$$

Since $d(F, P) = d(O, P) = r$,

$$d(F, P) = e \cdot d(D, P)$$
$$r = e(p + r \cos \theta) \qquad \color{blue}{d(F, P) = r; d(D, P) = p + r \cos \theta}$$
$$r - er \cos \theta = ep$$
$$r = \frac{ep}{1 - e \cos \theta}$$

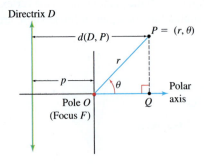

Directrix D

Figure 59

THEOREM The Polar Equation of a Conic

The polar equation of a conic with focus at the pole and directrix perpendicular to the polar axis at a distance p to the left of the pole is

$$r = \frac{ep}{1 - e \cos \theta} \tag{2}$$

where e is the eccentricity of the conic.

EXAMPLE 1 Identifying and Graphing the Polar Equation of a Conic

(a) Identify and graph the equation $r = \dfrac{4}{2 - \cos \theta}$.

(b) Convert the polar equation to a rectangular equation.

(c) Find parametric equations for the polar equation.

Solution **(a)** We divide the numerator and the denominator by 2 to express the equation in the form $r = \dfrac{ep}{1 - e \cos \theta}$.

$$r = \frac{4}{2 - \cos \theta} = \frac{2}{1 - \frac{1}{2} \cos \theta} \qquad \color{blue}{r = \frac{ep}{1 - e \cos \theta}}$$

Now we see that $e = \dfrac{1}{2}$. Since $ep = 2$, we find $p = \dfrac{2}{e} = \dfrac{2}{\frac{1}{2}} = 4$. This is an ellipse,

because $e = \dfrac{1}{2} < 1$. One focus is at the pole, and the directrix is perpendicular to the polar axis a distance of $p = 4$ units to the left of the pole. The major axis is along the polar axis.

Letting $\theta = 0$ and $\theta = \pi$, we can find the vertices of the ellipse.

$$r = \frac{4}{2 - \cos 0} = \frac{4}{2 - 1} = 4 \quad \text{and} \quad r = \frac{4}{2 - \cos \pi} = \frac{4}{2 - (-1)} = \frac{4}{3}$$

So, the vertices of the ellipse are the points whose polar coordinates are $(4, 0)$ and $\left(\dfrac{4}{3}, \pi\right)$. The y-intercepts of the ellipse are at $\theta = \dfrac{\pi}{2}$ and $\theta = \dfrac{3\pi}{2}$, which give rise to the points $\left(2, \dfrac{\pi}{2}\right)$ and $\left(2, \dfrac{3\pi}{2}\right)$. The graph of the ellipse is shown in Figure 60.

(b) To obtain a rectangular equation of the ellipse, we eliminate the fraction and then square the resulting polar equation.

$$r = \frac{4}{2 - \cos\theta}$$

$$r(2 - \cos\theta) = 4$$

$$2r - r\cos\theta = 4$$

$$2r = 4 + r\cos\theta$$

$$4r^2 = (4 + r\cos\theta)^2 \qquad \text{Square the equation.}$$

$$4(x^2 + y^2) = (4 + x)^2 \qquad r^2 = x^2 + y^2, \quad x = r\cos\theta$$

$$4x^2 + 4y^2 = 16 + 8x + x^2$$

$$3\left(x^2 - \frac{8}{3}x\right) + 4y^2 = 16$$

$$3\left(x - \frac{4}{3}\right)^2 + 4y^2 = 16 + 3\left(\frac{16}{9}\right) = \frac{64}{3} \qquad \text{Complete the square in } x.$$

This is the equation of an ellipse in rectangular coordinates, with its center at the point $\left(\dfrac{4}{3}, 0\right)$ in rectangular coordinates.

(c) Parametric equations of the ellipse $r = \dfrac{4}{2 - \cos\theta}$ are

$$x(\theta) = r\cos\theta = \frac{4\cos\theta}{2 - \cos\theta} \qquad y(\theta) = r\sin\theta = \frac{4\sin\theta}{2 - \cos\theta} \quad 0 \le \theta \le 2\pi$$

where θ is the parameter. ∎

There are four possibilities for the position of the directrix relative to the polar axis. These are summarized in Table 8.

TABLE 8 Polar Equations of Conics (Focus at the Pole, Eccentricity e)

Equation	Description
$r = \dfrac{ep}{1 - e\cos\theta}$	Directrix is perpendicular to the polar axis a distance p units to the left of the pole.
$r = \dfrac{ep}{1 + e\cos\theta}$	Directrix is perpendicular to the polar axis a distance p units to the right of the pole.
$r = \dfrac{ep}{1 + e\sin\theta}$	Directrix is parallel to the polar axis a distance p units above the pole.
$r = \dfrac{ep}{1 - e\sin\theta}$	Directrix is parallel to the polar axis a distance p units below the pole.

Eccentricity

If $e = 1$, the conic is a parabola; the axis of symmetry is perpendicular to the directrix.

If $e < 1$, the conic is an ellipse; the major axis perpendicular to the directrix.

If $e > 1$, the conic is a hyperbola; the transverse axis is perpendicular to the directrix.

NOW WORK Problems **5** and **13**.

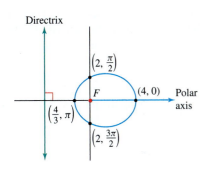

Directrix

$\left(2, \frac{\pi}{2}\right)$

F $(4, 0)$ Polar axis

$\left(\frac{4}{3}, \pi\right)$

$\left(2, \frac{3\pi}{2}\right)$

Figure 60

EXAMPLE 2 **Analyzing the Orbit of an Exoplanet**

Many hundreds of planets beyond our solar system have been discovered. They are known as **exoplanets**. One of these exoplanets, HD 190360b, orbits the star HD 190360 (most stars do not have names) in an elliptical orbit given by the polar equation

$$r = \frac{3.575}{1 - 0.316 \cos\theta}$$

where the star HD 190360 is at the pole, the major axis is along the polar axis, and r is measured in astronomical units (AU). (One AU $= 1.5 \times 10^{11}$ m, which is the average distance from Earth to the Sun.)

(a) What is the eccentricity of the exoplanet HD 190360b's orbit?

(b) Find the distance from the exoplanet HD 190360b to the star HD 190360 at periapsis (shortest distance).

(c) Find the distance from the exoplanet HD 190360b to the star HD 190360 at apoapsis (greatest distance).

Solution **(a)** The equation of HD 190360b's orbit is in the form $r = \dfrac{ep}{1 - e \cos\theta}$. So, the eccentricity of the orbit is $e = 0.316$. See Figure 61.

(b) Periapsis occurs when r is a minimum. Since r is minimum when $\cos\theta = -1$, periapsis occurs for

$$r = \frac{3.575}{1 - 0.316 \cos\theta} = \frac{3.575}{1 - 0.316(-1)} = \frac{3.575}{1.316} \approx 2.717$$

The periapsis is at the point $(2.717, \pi)$. The exoplanet HD 190360b is approximately 2.717 AU from the star HD 190360 at periapsis.

(c) The apoapsis occurs when r is a maximum. Since r is maximum when $\cos\theta = 1$, the apoapsis occurs for

$$r = \frac{3.575}{1 - 0.316 \cos\theta} = \frac{3.575}{1 - 0.316(1)} = \frac{3.575}{0.684} \approx 5.227$$

The apoapsis is at the point $(5.227, 0)$. The exoplanet HD 190360b is approximately 5.227 AU from the star HD 190360 at apoapsis. ■

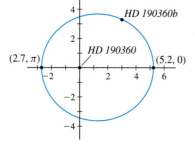

Figure 61 $r = \dfrac{3.575}{1 - 0.316 \cos\theta}$. The orbit of exoplanet HD 190360b about star HD 190360.

NOW WORK **Problem 33.**

9.7 Assess Your Understanding

Concepts and Vocabulary

1. *Multiple Choice* A(n) [**(a)** parabola, **(b)** ellipse, **(c)** hyperbola] is the set of points P in the plane for which the distance from a fixed point called the focus to P equals the distance from a fixed line called the directrix to P.

2. In your own words, explain what is meant by the eccentricity e of a conic.

3. Identify the graphs of each of these polar equations:

$r = \dfrac{2}{1 + \sin\theta}$ and $r = \dfrac{2}{1 + \cos\theta}$. How are they the same? How are they different?

4. *True or False* The polar equation of a conic with focus at the pole and directrix perpendicular to the polar axis at a distance p to the left of the pole is $r = \dfrac{ep}{1 - p \cos\theta}$, where e is the eccentricity of the conic.

Skill Building

In Problems 5–12, identify each conic. Find its eccentricity e and the position of its directrix.

5. $r = \dfrac{1}{1 + \cos\theta}$

6. $r = \dfrac{3}{1 - \sin\theta}$

7. $r = \dfrac{4}{2 - 3\sin\theta}$

8. $r = \dfrac{2}{1 + 2\cos\theta}$

9. $r = \dfrac{3}{4 - 2\cos\theta}$

10. $r = \dfrac{6}{8 + 2\cos\theta}$

11. $r = \dfrac{4}{3 + 3\sin\theta}$

12. $r = \dfrac{1}{6 + 2\sin\theta}$

In Problems 13–20, for each polar equation:

(a) Identify and graph the equation.

(b) Convert the polar equation to a rectangular equation.

(c) Find parametric equations for the polar equation.

---

1. = NOW WORK problem   📈 = Graphing technology recommended   CAS = Computer Algebra System recommended

13. $r = \dfrac{8}{4 + 3\sin\theta}$ **14.** $r = \dfrac{10}{5 + 4\cos\theta}$

15. $r = \dfrac{9}{3 - 6\cos\theta}$ **16.** $r = \dfrac{12}{4 + 8\sin\theta}$

17. $r(3 - 2\sin\theta) = 6$ **18.** $r(2 - \cos\theta) = 2$

19. $r = \dfrac{6\sec\theta}{2\sec\theta - 1}$ **20.** $r = \dfrac{3\csc\theta}{\csc\theta - 1}$

Applications and Extensions

In Problems 21–26, find the slope of the tangent line to the graph of each conic at θ.

21. $r = \dfrac{9}{4 - \cos\theta}$, $\theta = 0$ **22.** $r = \dfrac{3}{1 - \sin\theta}$, $\theta = 0$

23. $r = \dfrac{8}{4 + \sin\theta}$, $\theta = \dfrac{\pi}{2}$ **24.** $r = \dfrac{10}{5 + 4\sin\theta}$, $\theta = \pi$

25. $r(2 + \cos\theta) = 4$, $\theta = \pi$ **26.** $r(3 - 2\sin\theta) = 6$, $\theta = \dfrac{\pi}{2}$

In Problems 27–32, find a polar equation for each conic. For each equation, a focus is at the pole.

27. $e = \dfrac{4}{5}$; directrix is perpendicular to the polar axis 3 units to the left of the pole.

28. $e = \dfrac{2}{3}$; directrix is parallel to the polar axis 3 units above the pole.

29. $e = 1$; directrix is parallel to the polar axis 1 unit above the pole.

30. $e = 1$; directrix is parallel to the polar axis 2 units below the pole.

31. $e = 6$; directrix is parallel to the polar axis 2 units below the pole.

32. $e = 5$; directrix is perpendicular to the polar axis 3 units to the right of the pole.

33. Halley's Comet   As with most comets, Halley's comet has a highly elliptical orbit about the Sun, given by the polar equation

$$r = \dfrac{1.155}{1 - 0.967\cos\theta}$$

where the Sun is at the pole, the semimajor axis is along the polar axis, and r is measured in AU (astronomical unit). One AU = 1.5×10^{11} m, which is the average distance from Earth to the Sun.

  (a) What is the eccentricity of the comet's orbit?

  (b) Find the distance from Halley's comet to the Sun at perihelion (shortest distance from the Sun).

  (c) Find the distance from Halley's comet to the Sun at aphelion (greatest distance from the Sun).

  (d) Graph the orbit of Halley's comet.

34. Orbit of Mercury   The planet Mercury travels around the Sun in an elliptical orbit given approximately by

$$r = \dfrac{(3.442)\,10^7}{1 - 0.206\cos\theta}$$

where r is measured in miles and the Sun is at the pole.

  (a) What is the eccentricity of Mercury's orbit?

  (b) Find the distance from Mercury to the Sun at perihelion (shortest distance from the Sun).

  (c) Find the distance from Mercury to the Sun at aphelion (greatest distance from the Sun).

  (d) Graph the orbit of Mercury.

35. The Effect of Eccentricity

  (a) Graph the conic $r = \dfrac{2e}{1 - e\cos\theta}$ for the following values of e: (i) $e = 0.2$, (ii) $e = 0.6$, (iii) $e = 0.9$, (iv) $e = 1$, (v) $e = 2$, (vi) $e = 4$.

  (b) Describe how the shape of the conic changes as $e > 1$ gets larger.

  (c) Describe how the shape of the conic changes as $e < 1$ gets closer to 0.

36. Show that the polar equation for a conic with its focus at the pole and whose directrix is perpendicular to the polar axis at a distance p units to the right of the pole is given by

$$r = \dfrac{ep}{1 + e\cos\theta}$$

37. Show that the polar equation for a conic with its focus at the pole and whose directrix is parallel to the polar axis at a distance p units above the pole is given by $r = \dfrac{ep}{1 + e\sin\theta}$.

38. Show that the polar equation for a conic with its focus at the pole and whose directrix is parallel to the polar axis at a distance p units below the pole is given by $r = \dfrac{ep}{1 - e\sin\theta}$.

Challenge Problems

39. Show that the surface area of the solid generated by revolving the first-quadrant arc of the ellipse $\dfrac{x^2}{a^2} + \dfrac{y^2}{b^2} = 1$, $x \ge 0$, $y \ge 0$, about the x-axis is

$$S = \pi b^2 + \dfrac{\pi ab}{e}\,\sin^{-1} e$$

where e is the eccentricity of the ellipse.

40. In this section, one focus of each conic has been at the pole. Write the general equation for a conic in polar coordinates if there is no focus at the pole. That is, suppose the focus F has polar coordinates (r_1, θ_1), and the directrix D is given by $r\cos(\theta + \theta_1) = -d$, where $d > 0$. Let the eccentricity be e.

Chapter Review

THINGS TO KNOW

9.1 Parametric Equations

- Parametric equations $x = x(t)$, $y = y(t)$, where t is the parameter (p. 637)
- Plane curve C: the collection of points $(x, y) = (x(t), y(t))$ (p. 637)
- Orientation: the direction a point on the curve defined for $a \le t \le b$ moves as t moves from a to b (p. 637)
- Convert between parametric equations and rectangular equations (p. 638–640)
- Cycloid: the curve represented by $x(t) = a(t - \sin t)$, $y(t) = a(1 - \cos t)$ (p. 642)

9.2 Tangent Lines; Arc Length

- Smooth curve (p. 647)
- Slope of a tangent line to a smooth curve C

$$\frac{dy}{dx} = \frac{\dfrac{dy}{dt}}{\dfrac{dx}{dt}}, \quad \frac{dx}{dt} \ne 0 \text{ (p. 648)}$$

- Vertical tangent line to a smooth curve C:

$$\frac{dx}{dt} = 0, \text{ but } \frac{dy}{dt} \ne 0 \text{ (p. 648)}$$

- Horizontal tangent line to a smooth curve C:

$$\frac{dy}{dt} = 0, \text{ but } \frac{dx}{dt} \ne 0 \text{ (p. 648)}$$

- Arc length formula for parametric equations:

$$s = \int_a^b \sqrt{\left(\frac{dx}{dt}\right)^2 + \left(\frac{dy}{dt}\right)^2} \, dt \text{ (p. 651)}$$

9.3 Surface Area of a Solid of Revolution

The surface area S of the solid of revolution generated by revolving a smooth curve C represented by the parametric equations $x = x(t)$, $y = y(t)$, $a \le t \le b$

- about the x-axis: $S = 2\pi \int_a^b y(t) \sqrt{\left(\dfrac{dx}{dt}\right)^2 + \left(\dfrac{dy}{dt}\right)^2} \, dt$ (p. 657)
- about the y-axis: $S = 2\pi \int_a^b x(t) \sqrt{\left(\dfrac{dx}{dt}\right)^2 + \left(\dfrac{dy}{dt}\right)^2} \, dt$ (p. 658)
- about the x-axis: if the smooth curve C is represented by a rectangular equation $y = f(x)$:
$S = 2\pi \int_a^b f(x) \sqrt{1 + [f'(x)]^2} \, dx$ (p. 659)

9.4 Polar Coordinates

- Polar coordinates (r, θ); pole O; polar axis (p. 661)
- A point P with polar coordinates (r, θ) also can be represented by $(r, \theta + 2n\pi)$ or $(-r, \theta + (2n + 1)\pi)$, n an integer. (p. 663)
- The polar coordinates of the pole are $(0, \theta)$, where θ is any angle. (p. 663)
- For polar coordinates (r, θ) and rectangular coordinates (x, y):

  - $x = r \cos \theta$, $y = r \sin \theta$ (p. 664)
  - $r^2 = x^2 + y^2$ and $\tan \theta = \dfrac{y}{x}$, $x \ne 0$ (p. 665)
  - $r = y$ and $\theta = \dfrac{\pi}{2}$ if $x = 0$ (p. 665)

- Table 1 gives polar equations for some lines and circles. (p. 668)

9.5 Polar Equations; Parametric Equations of Polar Equations; Arc Length of Polar Equations

- Library of Polar Equations (Table 7) (p. 675)
- Arc length s of a curve represented by a polar equation from

$$\theta = \alpha \text{ to } \theta = \beta, \alpha \le \theta \le \beta: s = \int_\alpha^\beta \sqrt{r^2 + \left(\frac{dr}{d\theta}\right)^2} \, d\theta$$

(p. 675)

9.6 Area in Polar Coordinates

- The area A enclosed by the graph of $r = f(\theta)$ and the rays

$$\theta = \alpha \text{ and } \theta = \beta: A = \int_\alpha^\beta \frac{1}{2} r^2 \, d\theta \text{ (p. 679)}$$

- The surface area S of the solid of revolution obtained by revolving $r = f(\theta)$, $\alpha \le \theta \le \beta$, $0 \le \alpha < \beta \le 2\pi$, about the polar axis:

$$S = 2\pi \int_\alpha^\beta f(\theta) \sin \theta \sqrt{[f(\theta)]^2 + [f'(\theta)]^2} \, d\theta$$

$$= 2\pi \int_\alpha^\beta r \sin \theta \sqrt{r^2 + \left(\frac{dr}{d\theta}\right)^2} \, d\theta$$

(p. 682)

9.7 The Polar Equation of a Conic

- The polar equation of a conic: Table 8 (p. 687)
- Eccentricity: Parabola: $e = 1$; ellipse: $e < 1$; hyperbola: $e > 1$ (p. 687)

OBJECTIVES

Section	You should be able to ...	Examples	Review Exercises
9.1	**1** Graph parametric equations (p. 637)	1	1(b)–6(b)
	2 Find a rectangular equation for a curve represented parametrically (p. 638)	2–4	1(a)–6(a), 1(c)–6(c)
	3 Use time as the parameter in parametric equations (p. 640)	5	13, 48, 49
	4 Convert a rectangular equation to parametric equations (p. 641)	6, 7	11, 12
9.2	**1** Find an equation of the tangent line at a point on a plane curve (p. 648)	1–4	7–10, 50, 51
	2 Find arc length of a plane curve (p. 651)	5, 6	52–55
9.3	**1** Find the surface area of a solid of revolution obtained from parametric equations (p. 658)	1	63, 64
	2 Find the surface area of a solid of revolution obtained from a rectangular equation (p. 659)	2	65, 66
9.4	**1** Plot points using polar coordinates (p. 661)	1, 2	14–17
	2 Convert between rectangular coordinates and polar coordinates (p. 664)	3, 4	14–31
	3 Identify and graph polar equations (p. 666)	5, 6	32–35
9.5	**1** Graph a polar equation; find parametric equations (p. 671)	1–5	36–45
	2 Find the arc length of a curve represented by a polar equation (p. 675)	6	56–59
9.6	**1** Find the area of a region enclosed by the graph of a polar equation (p. 678)	1–4	60
	2 Find the area of a region enclosed by the graphs of two polar equations (p. 681)	5	61, 62
	3 Find the surface area of a solid of revolution obtained from the graph of a polar equation (p. 682)	6	67
9.7	**1** Express a conic as a polar equation (p. 686)	1, 2	46, 47

REVIEW EXERCISES

In Problems 1–6:

(a) *Find the rectangular equation of each curve.*

(b) *Graph each plane curve whose parametric equations are given and show its orientation.*

(c) *Determine the restrictions on x and y that make the rectangular equation identical to the plane curve.*

1. $x(t) = 4t - 2$, $y(t) = 1 - t$; $-\infty < t < \infty$

2. $x(t) = 2t^2 + 6$, $y(t) = 5 - t$; $-\infty < t < \infty$

3. $x(t) = e^t$, $y(t) = e^{-t}$ $-\infty < t < \infty$

4. $x(t) = \ln t$, $y(t) = t^3$; $t > 0$

5. $x(t) = \sec^2 t$, $y(t) = \tan^2 t$; $0 \le t \le \dfrac{\pi}{4}$

6. $x(t) = t^{3/2}$, $y(t) = 2t + 4$; $t \ge 0$

In Problems 7–10, for the parametric equations below:

(a) *Find an equation of the tangent line to the curve at t.*

(b) *Graph the curve and the tangent line.*

7. $x(t) = t^2 - 4$, $y(t) = t$ at $t = 1$

8. $x(t) = 3 \sin t$, $y(t) = 4 \cos t + 2$ at $t = \dfrac{\pi}{4}$

9. $x(t) = \dfrac{1}{t^2}$, $y(t) = \sqrt{t^2 + 1}$ at $t = 3$

10. $x(t) = \dfrac{t^2}{1+t}$, $y(t) = \dfrac{t}{1+t}$ at $t = 0$

In Problems 11 and 12, find two different pairs of parametric equations for each rectangular equation.

11. $y = -2x + 4$ **12.** $y = 2x$

13. Describe the motion of an object that moves so that at time t (in seconds) it has coordinates $x(t) = 2 \cos t$, $y(t) = \sin t$, $0 \le t \le 2\pi$.

In Problems 14–17, the polar coordinates of a point are given. Plot each point in a polar coordinate system, and find its rectangular coordinates.

14. $\left(3, \dfrac{\pi}{6}\right)$ **15.** $\left(-2, \dfrac{4\pi}{3}\right)$

16. $\left(3, -\dfrac{\pi}{2}\right)$ **17.** $\left(-4, -\dfrac{\pi}{4}\right)$

In Problems 18–21, the rectangular coordinates of a point are given. Find two pairs of polar coordinates (r, θ) for each point, one with $r > 0$ and the other with $r < 0$, $0 \le \theta < 2\pi$.

18. $(2, 0)$ **19.** $(3, 4)$ **20.** $(-5, 12)$ **21.** $(-3, 3)$

In Problems 22–27, the letters r, θ represent polar coordinates. Write each equation in terms of the rectangular coordinates x, y.

22. $r = 4 \sin(2\theta)$ **23.** $r = e^{\theta/2}$ **24.** $r = \dfrac{1}{1 + 2\cos\theta}$

25. $r = a - \sin\theta$ **26.** $r^2 = 4\cos(2\theta)$ **27.** $r = \theta$

In Problems 28–31, the letters x, y represent rectangular coordinates. Write each equation in terms of the polar coordinates r, θ.

28. $x^2 + y^2 = x$ **29.** $(x^2 + y^2)^2 = x^2 - y^2$

30. $y^2 = (x^2 + y^2) \cos^2[(x^2 + y^2)^{1/2}]$

31. $\dfrac{x^2}{2^2} + \dfrac{y^2}{3^2} = 1$

In Problems 32–35, identify and graph each polar equation. Convert it to a rectangular equation if necessary.

32. $r \sin\theta = 1$ **33.** $r \sec\theta = 2$

34. $r = \sin\theta$ **35.** $r = -5 \cos\theta$

In Problems 36–45, for each equation:

(a) *Graph the equation.*
(b) *Find parametric equations that represent the equation.*

36. $r = 1 - \sin\theta$ **37.** $r = 4\cos(2\theta)$

38. $r = \dfrac{1}{2} - \sin\theta$ **39.** $r = \dfrac{4}{1 - 2\cos\theta}$

40. $r = 4\sin(3\theta)$ **41.** $r = 2 - 2\cos\theta$

42. $r = 2 - \sin\theta$ **43.** $r = e^{0.5\theta}$

44. $r^2 = 1 - \sin^2\theta$ **45.** $r^2 = 1 + \sin^2\theta$

In Problems 46 and 47, for each polar equation:

(a) *Identify and graph the equation.*
(b) *Convert the polar equation to a rectangular equation.*
(c) *Find parametric equations for the polar equation.*

46. $r = \dfrac{2}{1 - \cos\theta}$ **47.** $r = \dfrac{1}{1 - \dfrac{1}{6}\cos\theta}$

In Problems 48 and 49, find parametric equations for an object that moves along the ellipse $\dfrac{x^2}{16} + \dfrac{y^2}{9} = 1$ with the motion described.

48. The motion begins at $(4, 0)$, is counterclockwise, and requires 4 seconds for a complete revolution.

49. The motion begins at $(0, 3)$, is clockwise, and requires 5 seconds for a complete revolution.

In Problems 50 and 51, find the points (if any) on the curve at which the tangent line is vertical or horizontal.

50. $x(t) = t^3 - 1, \quad y(t) = 2t^2 + 1$

51. $x(t) = 1 - \sin t, \quad y(t) = 2 + 3\cos t, \quad 0 \le t \le 2\pi$

In Problems 52–55, find the arc length of each plane curve.

52. $x(t) = \sinh^{-1} t, \; y(t) = \sqrt{t^2 + 1}$ from $t = 0$ to $t = 1$

53. $x(t) = \tan t, \; y(t) = \dfrac{1}{3}(\sec^2 t + 1)$ from $t = 0$ to $t = \dfrac{\pi}{4}$

54. $x(t) = e^t, \; y(t) = \dfrac{1}{2}e^{2t} - \dfrac{1}{4}t$ from $t = 0$ to $t = 2$

55. $x = \dfrac{1}{2}y^2 - \dfrac{1}{4}\ln y$ from $y = 1$ to $y = 2$

In Problems 56–59, find the arc length of each curve represented by a polar equation.

56. $r = 2\sin\theta$ from $\theta = 0$ to $\theta = \pi$

57. $r = e^{-\theta}$ from $\theta = 0$ to $\theta = 2\pi$

58. $r = 3\theta$ from $\theta = 0$ to $\theta = 2\pi$

59. $r = 2\sin^2\dfrac{\theta}{2}$ from $\theta = -\dfrac{\pi}{2}$ to $\theta = \dfrac{\pi}{2}$

60. Area Find the area of the region inside the circle $r = 4\sin\theta$ and above the line $r = 3\csc\theta$.

61. Area Find the area of the region that lies inside the rose $r = 4\cos(2\theta)$ and outside the circle $r = \sqrt{2}$.

62. Area Find the area of the region common to the graphs of $r = \cos\theta$ and $r = 1 - \cos\theta$.

63. Surface Area Find the surface area of the solid generated by revolving the smooth curve represented by the parametric equations $x(t) = \sinh^{-1} t$, $y(t) = \sqrt{t^2 + 1}$ from $0 \le t \le 1$ about the x-axis.

64. Surface Area Find the surface area of the solid generated by revolving the smooth curve represented by $x(t) = e^t - t$, $y(t) = 4e^{t/2}$ from $t = 0$ to $t = 1$ about the x-axis.

65. Surface Area Find the surface area of the solid of revolution generated by revolving the curve represented by $x = y^2 + 1$, $0 \le y \le 2$, about the y-axis.

66. Surface Area Find the surface area of the solid generated by revolving the curve represented by $y = \cos\left(\dfrac{x}{2}\right), 0 \le x \le \pi$ about the x-axis.

67. Surface Area Find the surface area of the solid of revolution generated by revolving the arc of the circle $r = 4$, $0 \le \theta \le \dfrac{\pi}{3}$, about the x-axis.

CHAPTER 9 PROJECT   **Polar Graphs and Microphones**

Microphones can be configured to have different sensitivities which are best modeled using various polar patterns. The oldest example is an **omnidirectional microphone** which records sound equally from all directions. Omnidirectional microphones can be a good option when recording jungle sounds or general ambient noise, or when recording for your cell phone. The recording pattern for this microphone is virtually circular, and a good polar model is

$$r = k$$

where $k > 0$ is the sensitivity parameter. Notice that by adjusting k (say using 1, 4, or 8) we obtain concentric circles. Rotating any one of

these circles around a diameter results in a sphere. Notice that the microphone receiver is at the center of the sphere and that sound from all directions is received equally.

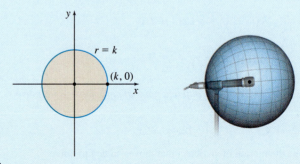

For this model, say when $k = 1$, we can convert the polar equation to parametric equations obtaining

$$x = \cos t \quad y = \sin t \quad 0 \leq t \leq 2\pi$$

Sound technicians have several things to consider when using microphones, including finding the best location to place the microphone and the best location to place any monitors. Since an omnidirectional microphone collects sound equally from all directions, the microphone should be placed directly in the center of the sounds being recorded with the most important sounds positioned closest to the microphone. Unfortunately, monitors cannot be used with an omnidirectional microphone since there is no place to put them where the microphone does not pick up their sound. To help remedy this and other possible issues, different types of microphone models are used.

A **cardioid microphone** is used when we want to pick up sounds mostly from the front of the microphone. Cardioid microphones are considered "unidirectional". A polar model for a cardioid microphone is

$$r = k(1 + \cos \theta) \quad 0 \leq \theta \leq 2\pi \quad k > 0$$

Notice that the microphone is placed with the transducer at the pole and pointing toward the polar axis. A cardioid microphone is formed by rotating the cardioid about the polar axis. See the figure.

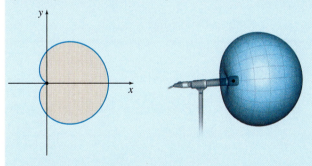

1. Sketch several cardioids using $k = 1, 4, 8$.
2. Explain why any noise created by the person's hand or the microphone stand is unlikely to be picked up by a cardioid microphone.
3. Using $k = 1$, convert the cardioid polar equation to a pair of parametric equations.
4. To determine the percentage of the sound that comes from the "front" of the microphone, we consider the sound coming at the

microphone with angles in the polar wedge $-\dfrac{\pi}{4} \leq \theta \leq \dfrac{\pi}{4}$. With $k = 1$, determine the proportion of the surface area coming in from the front relative to that coming in from all directions.

5. Determine the proportion of the sound that is picked up from behind the microphone in the polar wedge $\dfrac{3\pi}{4} \leq \theta \leq \dfrac{5\pi}{4}$.

Since less than 1% of all the sound received comes from the backside of a cardioid microphone, monitors can be placed directly behind the microphone.

A **hypercardioid microphone** is used to pick up sounds mostly from in front of the microphone, but it also receives some sound from the rear. A polar model for such a hypercardioid microphone is

$$r = \frac{k\,|1 + 2\cos\theta|}{3} \quad 0 \leq \theta \leq 2\pi \quad k > 0$$

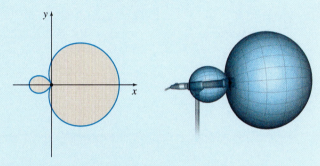

6. Sketch several hypercardioid graphs using $k = 1, 4,$ and 8.
7. Express the hypercardioid polar equation with $k = 1$, as a pair of parametric equations.
8. Determine the proportion of the sound that is picked up from the front of a hypercardioid microphone $r = \dfrac{|1 + 2\cos\theta|}{3}$ by finding percentage of sound coming at the microphone from the polar wedge $-\dfrac{\pi}{4} \leq \theta \leq \dfrac{\pi}{4}$.
9. Determine the proportion of the sound that is picked up from behind the hypercardioid microphone in the polar wedge $\dfrac{3\pi}{4} \leq \theta \leq \dfrac{5\pi}{4}$.
10. Compare the results of Problems 4, 5, 8, and 9, and describe how the microphones differ.

Appendix A Precalculus Used in Calculus

The topics reviewed here are not exhaustive of the precalculus used in calculus. However, they do represent a large body of the material you will see in calculus. If you encounter difficulty with any of this material, consult a textbook in precalculus for more detail and explanation.

A.1 Algebra Used in Calculus

OBJECTIVES *When you finish this section, you should be able to:*

1 Factor and simplify algebraic expressions (p. A-1)

2 Complete the square (p. A-2)

3 Solve equations (p. A-3)

4 Solve inequalities (p. A-5)

5 Work with exponents (p. A-8)

6 Work with logarithms (p. A-10)

1 Factor and Simplify Algebraic Expressions

EXAMPLE 1 **Factoring Algebraic Expressions**

Factor each expression completely:

(a) $2(x+3)(x-2)^3 + (x+3)^2(3)(x-2)^2$

(b) $\dfrac{4}{3}x^{1/3}(2x+1) + 2x^{4/3}$

Solution (a) In expression (a), $(x+3)$ and $(x-2)^2$ are **common factors**, factors found in each term. Factor them out.

$$2(x+3)(x-2)^3 + (x+3)^2(3)(x-2)^2$$
$$= (x+3)(x-2)^2[2(x-2) + 3(x+3)] \qquad \text{Factor out } (x+3)(x-2)^2.$$
$$= (x+3)(x-2)^2(5x+5) \qquad \text{Simplify.}$$
$$= 5(x+3)(x-2)^2(x+1) \qquad \text{Factor out 5.}$$

(b) We begin by writing the term $2x^{4/3}$ as a fraction with a denominator of 3.

$$\frac{4}{3}x^{1/3}(2x+1) + 2x^{4/3} = \frac{4x^{1/3}(2x+1)}{3} + \frac{6x^{4/3}}{3}$$
$$= \frac{4x^{1/3}(2x+1) + 6x^{4/3}}{3} \qquad \text{Add the two fractions.}$$
$$= \frac{2x^{1/3}[2(2x+1) + 3x]}{3} \qquad \text{2 and } x^{1/3} \text{ are common factors.}$$
$$= \frac{2x^{1/3}(7x+2)}{3} \qquad \text{Simplify.} \qquad \blacksquare$$

NOTE The set of real numbers that a variable can assume is called the **domain of the variable.**

Solution **(a)** First, notice that the domain of the variable is $\{x \mid x \neq 1,\ x \neq 2\}$. Now clear the equation of rational expressions by multiplying both sides by $(x - 1)(x - 2)$.

$$\frac{3}{x - 2} = \frac{1}{x - 1} + \frac{7}{(x - 1)(x - 2)}$$

$$(x - 1)(x - 2)\frac{3}{x - 2} = (x - 1)(x - 2)\left[\frac{1}{x - 1} + \frac{7}{(x - 1)(x - 2)}\right] \qquad \text{Multiply both sides by } (x - 1)(x - 2).$$

$$3x - 3 = (x - 1)(x - 2)\frac{1}{x - 1} + (x - 1)(x - 2)\frac{7}{(x - 1)(x - 2)} \qquad \text{Distribute on both sides.}$$

$$3x - 3 = (x - 2) + 7 \qquad \text{Simplify.}$$

$$3x - 3 = x + 5$$

$$2x = 8$$

$$x = 4$$

Since 4 is in the domain of the variable, the solution is 4.

(b) We group the terms of $x^3 - x^2 - 4x + 4 = 0$, and factor by grouping.

$$x^3 - x^2 - 4x + 4 = 0$$

$$(x^3 - x^2) - (4x - 4) = 0 \qquad \text{Group the terms.}$$

$$x^2(x - 1) - 4(x - 1) = 0 \qquad \text{Factor out the common factor from each group.}$$

$$(x^2 - 4)(x - 1) = 0 \qquad \text{Factor out the common factor } (x - 1).$$

$$(x - 2)(x + 2)(x - 1) = 0 \qquad x^2 - 4 = (x - 2)(x + 2)$$

$$x - 2 = 0 \text{ or } x + 2 = 0 \text{ or } x - 1 = 0 \qquad \text{Set each factor equal to 0.}$$

$$x = 2 \qquad\qquad x = -2 \qquad\quad x = 1 \qquad \text{Solve.}$$

The solutions are -2, 1, and 2.

CAUTION Squaring both sides of an equation may lead to extraneous solutions. Check all apparent solutions.

(c) We square both sides of the equation since the index of a square root is 2.

$$\sqrt{x - 1} = x - 7$$

$$(\sqrt{x - 1})^2 = (x - 7)^2 \qquad \text{Square both sides.}$$

$$x - 1 = x^2 - 14x + 49$$

$$x^2 - 15x + 50 = 0 \qquad \text{Put in standard form.}$$

$$(x - 10)(x - 5) = 0 \qquad \text{Factor.}$$

$$x = 10 \text{ or } x = 5 \qquad \text{Set each factor equal to 0 and solve.}$$

Check: $x = 10$: $\sqrt{x - 1} = \sqrt{10 - 1} = \sqrt{9} = 3$ and $x - 7 = 10 - 7 = 3$

$x = 5$: $\sqrt{x - 1} = \sqrt{5 - 1} = \sqrt{4} = 2$ and $x - 7 = 5 - 7 = -2$

The apparent solution 5 is extraneous; the only solution of the equation is 10.

RECALL $|a| = a$ if $a \geq 0$; $|a| = -a$ if $a < 0$. If $|x| = b$, $b \geq 0$, then $x = b$ or $x = -b$.

(d) $|1 - x| = 2$

$$1 - x = 2 \quad \text{or} \quad 1 - x = -2 \qquad \text{The expression inside the absolute value bars equals 2 or } -2.$$

$$-x = 1 \qquad\qquad -x = -3 \qquad \text{Simplify.}$$

$$x = -1 \qquad\qquad x = 3 \qquad \text{Simplify.}$$

The solutions are -1 and 3. ■

4 Solve Inequalities

In expressing the solution to an inequality, *interval notation* is often used.

NOTE Every interval of real numbers contains both rational numbers, numbers that can be expressed as the quotient of two integers, and irrational numbers, numbers that are not rational.

DEFINITION

Let a and b represent two real numbers with $a < b$.

A **closed interval**, denoted by $[a, b]$, consists of all real numbers x for which $a \leq x \leq b$.

An **open interval**, denoted by (a, b), consists of all real numbers x for which $a < x < b$.

The **half-open**, or **half-closed**, **intervals** are:

- $(a, b]$, consisting of all real numbers x for which $a < x \leq b$, and
- $[a, b)$, consisting of all real numbers x for which $a \leq x < b$.

In each of these definitions, a is called the **left endpoint** and b the **right endpoint** of the interval.

The symbol ∞ (read "infinity") is *not* a real number, but a notational device used to indicate unboundedness in the positive direction. The symbol $-\infty$ (read "negative infinity") also is not a real number, but a notational device used to indicate unboundedness in the negative direction. Using the symbols ∞ and $-\infty$, we define five other kinds of intervals:

- $[a, \infty)$, consisting of all real numbers x for which $x \geq a$
- (a, ∞), consisting of all real numbers x for which $x > a$
- $(-\infty, a]$, consisting of all real numbers x for which $x \leq a$
- $(-\infty, a)$, consisting of all real numbers x for which $x < a$
- $(-\infty, \infty)$, consisting of all real numbers x

Notice that ∞ and $-\infty$ are never included as endpoints, since neither is a real number.

Table 1 summarizes interval notation, corresponding inequality notation, and their graphs.

TABLE 1

Interval	Inequality	Graph
The open interval (a, b)	$a < x < b$	
The closed interval $[a, b]$	$a \leq x \leq b$	
The half-open interval $[a, b)$	$a \leq x < b$	
The half-open interval $(a, b]$	$a < x \leq b$	
The interval $[a, \infty)$	$a \leq x < \infty$	
The interval (a, ∞)	$a < x < \infty$	
The interval $(-\infty, a]$	$-\infty < x \leq a$	
The interval $(-\infty, a)$	$-\infty < x < a$	

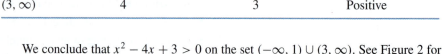

EXAMPLE 6   Solving Inequalities

Solve each inequality and graph the solution:

(a) $4x + 7 \geq 2x - 3$ (b) $x^2 - 4x + 3 > 0$ (c) $x^2 + x + 1 < 0$ (d) $\dfrac{1+x}{1-x} > 0$

Solution (a)

$$4x + 7 \geq 2x - 3$$
$$4x \geq 2x - 10 \qquad \text{Subtract 7 from both sides.}$$
$$2x \geq -10 \qquad \text{Subtract } 2x \text{ from both sides.}$$
$$x \geq -5 \qquad \text{Divide both sides by 2.}$$
$$\qquad\qquad\qquad \text{(The direction of the inequality symbol is unchanged.)}$$

The solution using interval notation is $[-5, \infty)$. See Figure 1 for the graph of the solution.

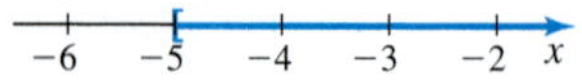

Figure 1 $x \geq 5$

(b) This is a quadratic inequality. The related quadratic equation $x^2 - 4x + 3 = (x - 1)(x - 3) = 0$ has two solutions, 1 and 3. We use these numbers to partition the number line into three intervals. Now select a test number in each interval, and determine the value of $x^2 - 4x + 3$ at the test number. See Table 2.

NOTE The test number can be any real number in the interval, but it cannot be an endpoint.

TABLE 2

Interval	Test Number	Value of $x^2 - 4x + 3$	Sign of $x^2 - 4x + 3$
$(-\infty, 1)$	0	3	Positive
$(1, 3)$	2	-1	Negative
$(3, \infty)$	4	3	Positive

We conclude that $x^2 - 4x + 3 > 0$ on the set $(-\infty, 1) \cup (3, \infty)$. See Figure 2 for the graph of the solution.

Figure 2 $x < 1$ or $x > 3$

(c) The quadratic equation $x^2 + x + 1 = 0$ has no real solution, since its discriminant is negative. [See Example 4(b)]. When this happens, the quadratic inequality is either positive for all real numbers or negative for all real numbers. To see which is true, evaluate $x^2 + x + 1$ at some number, say, 0. At 0, $x^2 + x + 1 = 1$, which is positive. So, $x^2 + x + 1 > 0$ for all real numbers x. The inequality $x^2 + x + 1 < 0$ has no solution.

(d) The only solution of the rational equation $\dfrac{1+x}{1-x} = 0$ is $x = -1$; also, the expression $\dfrac{1+x}{1-x}$ is not defined for $x = 1$. We use the solution -1 and the value 1, at which the expression is undefined, to partition the real number line into three intervals. Now select a test number in each interval, and evaluate the rational expression $\dfrac{1+x}{1-x}$ at each test number. See Table 3.

TABLE 3

Interval	Test Number	Value of $\dfrac{1+x}{1-x}$	Sign of $\dfrac{1+x}{1-x}$
$(-\infty, -1)$	-2	$-\dfrac{1}{3}$	Negative
$(-1, 1)$	0	1	Positive
$(1, \infty)$	2	-3	Negative

We conclude that $\dfrac{1+x}{1-x} > 0$ on the interval $(-1, 1)$. See Figure 3 for the graph of the solution. ∎

Figure 3 $-1 < x < 1$

NOTE If $x > 0$ and $y > 0$, the Triangle Inequality is an algebraic statement for the fact that the length of any side of a triangle is less than the sum of the lengths of the other two sides.

Below we state an important relationship involving the absolute value of a sum.

THEOREM Triangle Inequality

If x and y are real numbers, then

$$|x + y| \leq |x| + |y|$$

Use the following theorem as a guide to solving inequalities involving absolute values.

THEOREM

If a is a positive number and if u is an algebraic expression, then

$$|u| < a \text{ is equivalent to } -a < u < a \tag{1}$$
$$|u| > a \text{ is equivalent to } u < -a \quad \text{or} \quad u > a \tag{2}$$

Similar relationships hold for the nonstrict inequalities $|u| \leq a$ and $|u| \geq a$.

EXAMPLE 7   Solving Inequalities Involving Absolute Value

Solve each inequality and graph the solution:

(a) $|3 - 4x| < 11$ **(b)** $|2x + 4| - 1 \leq 9$ **(c)** $\left| \dfrac{4x + 1}{2} - \dfrac{3}{5} \right| > 1$

Solution (a) The absolute value is less than the number 11, so statement (1) applies.

$$
\begin{aligned}
|3 - 4x| &< 11 \\
-11 < 3 - 4x &< 11 &&\text{Apply statement (1).} \\
-14 < -4x &< 8 &&\text{Subtract 3 from each part.} \\
\frac{-14}{-4} > x &> \frac{8}{-4} &&\text{Divide each part by } -4, \text{ which reverses the inequality signs.} \\
-2 < x &< \frac{7}{2} &&\text{Simplify and rearrange the ordering.}
\end{aligned}
$$

RECALL Multiplying (or dividing) an inequality by a negative quantity reverses the direction of the inequality sign.

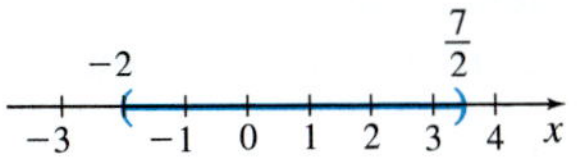

Figure 4 $-2 < x < \dfrac{7}{2}$

The solutions are all the numbers in the open interval $\left(-2, \dfrac{7}{2} \right)$. See Figure 4 for the graph of the solution.

(b) We begin by putting $|2x + 4| - 1 \leq 9$ into the form $|u| \leq a$.

$$
\begin{aligned}
|2x + 4| - 1 &\leq 9 \\
|2x + 4| &\leq 10 &&\text{Add 1 to each side.} \\
-10 \leq 2x + 4 &\leq 10 &&\text{Apply statement (1), but use } \leq. \\
-14 \leq 2x &\leq 6 \\
-7 \leq x &\leq 3
\end{aligned}
$$

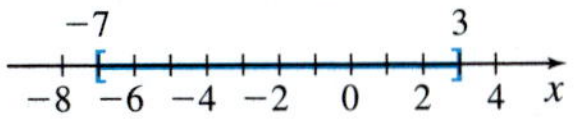

Figure 5 $-7 \leq x \leq 3$

The solutions are all the numbers in the closed interval $[-7, 3]$. See Figure 5 for the graph of the solution.

(c) $\left| \dfrac{4x + 1}{2} - \dfrac{3}{5} \right| > 1$ is in the form of statement (2). We begin by simplifying the expression inside the absolute value.

$$\left| \frac{4x + 1}{2} - \frac{3}{5} \right| = \left| \frac{5(4x + 1)}{10} - \frac{2(3)}{10} \right| = \left| \frac{20x + 5 - 6}{10} \right| = \left| \frac{20x - 1}{10} \right|$$

The original inequality is equivalent to the inequality below.

$$\left|\frac{20x - 1}{10}\right| > 1$$

$$\frac{20x - 1}{10} < -1 \qquad \text{or} \qquad \frac{20x - 1}{10} > 1 \qquad \text{\color{blue}Apply statement (2).}$$

$$20x - 1 < -10 \qquad \text{or} \qquad 20x - 1 > 10$$

$$20x < -9 \qquad \text{or} \qquad 20x > 11$$

$$x < -\frac{9}{20} \qquad \text{or} \qquad x > \frac{11}{20}$$

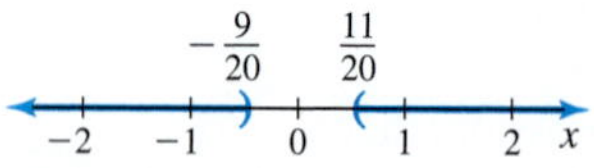

Figure 6 $x < -\dfrac{9}{20}$ or $x > \dfrac{11}{20}$

The solutions are all the numbers in the set $\left(-\infty, -\dfrac{9}{20}\right) \cup \left(\dfrac{11}{20}, \infty\right)$. See Figure 6 for the graph of the solution. ■

5 Work with Exponents

Integer exponents provide a shorthand notation for repeated multiplication. For example,

$$2^3 = 2 \cdot 2 \cdot 2 = 8 \quad \text{or} \quad \left(\frac{1}{3}\right)^4 = \frac{1}{3} \cdot \frac{1}{3} \cdot \frac{1}{3} \cdot \frac{1}{3} = \frac{1}{81}$$

DEFINITION

If a is a real number and n is a positive integer, then the symbol a^n represents the product of n factors of a. That is,

$$a^n = \underbrace{a \cdot a \cdot \ldots \cdot a}_{n \text{ factors}}$$

Here, it is understood that $a^1 = a$.

Then $a^2 = a \cdot a$ and $a^3 = a \cdot a \cdot a$, and so on. In the expression a^n, the number a is called the **base** and the number n is called the **exponent** or **power**. We read a^n as "a raised to the power n" or as "a to the nth power." We usually read a^2 as "a squared" and a^3 as "a cubed."

DEFINITION

If $a \neq 0$, then

$$a^0 = 1$$

If n is a positive integer, then

$$a^{-n} = \frac{1}{a^n} \qquad a \neq 0$$

The following properties, called the *Laws of Exponents*, can be proved using the preceding definitions.

THEOREM Laws of Exponents

In each of these properties a and b are real numbers, and u and v are integers.

$$a^u a^v = a^{u+v} \qquad\qquad (a^u)^v = a^{uv} \qquad\qquad (ab)^u = a^u b^u$$

$$\frac{a^u}{a^v} = a^{u-v} = \frac{1}{a^{v-u}} \quad \text{if } a \neq 0 \qquad\qquad \left(\frac{a}{b}\right)^u = \frac{a^u}{b^u} \quad \text{if } b \neq 0$$

DEFINITION

The **principal nth root of a real number** a, where $n \geq 2$ is an integer, symbolized by $\sqrt[n]{a}$, is defined as the solution of the equation $b^n = a$.

$$\sqrt[n]{a} = b \text{ is equivalent to } a = b^n$$

If n is even, then both $a \geq 0$ and $b \geq 0$, and if n is odd, then a and b are any real numbers and both have the same sign.

IN WORDS The symbol $\sqrt[n]{a}$ means "the number that, when raised to the nth power, equals a."

If a is negative and n is even, then $\sqrt[n]{a}$ is not defined. When $\sqrt[n]{a}$ is defined, the principal nth root of a number is unique.

The symbol $\sqrt[n]{a}$ for the principal nth root of a is called a **radical**; the integer n is called the **index**, and a is called the **radicand**. If the index of a radical is 2, we call $\sqrt[2]{a}$ the **square root** of a and omit the index 2 by writing $\sqrt{a}$. If the index is 3, we call $\sqrt[3]{a}$ the **cube root** of a.

Radicals are used to define rational exponents.

DEFINITION

If a is a real number and $n \geq 2$ is an integer, then

$$a^{1/n} = \sqrt[n]{a}$$

provided that $a^{1/n} = \sqrt[n]{a}$ exists.

DEFINITION

If a is a real number and m and n are integers with $n \geq 2$, then

$$a^{m/n} = (a^{1/n})^m$$

provided that $\sqrt[n]{a}$ exists.

From the two definitions, we have

$$a^{m/n} = \sqrt[n]{a^m} = (\sqrt[n]{a})^m$$

In simplifying the rational expression $a^{m/n}$, either $\sqrt[n]{a^m}$ or $(\sqrt[n]{a})^m$ can be used. The choice depends on which is easier to simplify. Generally, taking the root first, as in $(\sqrt[n]{a})^m$, is easier.

But does a^x have meaning, where the base a is a positive real number and the exponent x is an irrational number? The answer is yes, and although a rigorous definition requires methods discussed in calculus, the basis for the definition is easy to follow: Select a rational number r that is formed by truncating (removing) all but a finite number of digits from the irrational number x. Then it is reasonable to expect that

$$a^x \approx a^r$$

For example, take the irrational number $\pi = 3.14159\ldots$. Then an approximation to a^π is

$$a^\pi \approx a^{3.14}$$

where the digits after the hundredths position have been truncated from the value for π. A better approximation would be

$$a^\pi \approx a^{3.14159}$$

where the digits after the hundred-thousandths position have been truncated. Continuing in this way, we can obtain approximations to a^π to any desired degree of accuracy.

It can be shown that the Laws of Exponents hold for real number exponents u and v.

6 Work with Logarithms

The definition of a *logarithm* is based on an exponential relationship.

DEFINITION

Suppose $y = a^x$, $a > 0$, $a \neq 1$, and x is a real number. The **logarithm with base a of y**, symbolized by $\log_a y$, is the exponent to which a must be raised to obtain y. That is,

$$\boxed{\log_a y = x \text{ is equivalent to } y = a^x}$$

As this definition states, a logarithm is a name for a certain exponent.

EXAMPLE 8   Working with Logarithms

(a) If $x = \log_3 y$, then $y = 3^x$. For example, $4 = \log_3 81$ is equivalent to $81 = 3^4$.

(b) If $x = \log_5 y$, then $y = 5^x$. For example,

$$-1 = \log_5\left(\frac{1}{5}\right) \text{ is equivalent to } \frac{1}{5} = 5^{-1}$$
∎

THEOREM Properties of Logarithms

In the properties given next, u and a are positive real numbers, $a \neq 1$, and r is any real number. The number $\log_a u$ is the exponent to which a must be raised to obtain u. That is,

$$\boxed{a^{\log_a u} = u}$$

The logarithm with base a of a raised to a power equals that power. That is,

$$\boxed{\log_a a^r = r}$$

EXAMPLE 9   Using Properties of Logarithms

(a) $2^{\log_2 \pi} = \pi$ (b) $\log_{0.2} 0.2^{(-\sqrt{2})} = -\sqrt{2}$ (c) $\log_{1/5}\left(\frac{1}{5}\right)^{kt} = kt$ ∎

THEOREM Properties of Logarithms

In the following properties, u, v, and a are positive real numbers, $a \neq 1$, and r is any real number:

• **The Log of a Product Equals the Sum of the Logs**

$$\boxed{\log_a(uv) = \log_a u + \log_a v} \tag{3}$$

• **The Log of a Quotient Equals the Difference of the Logs**

$$\boxed{\log_a\left(\frac{u}{v}\right) = \log_a u - \log_a v} \tag{4}$$

• **The Log of a Power Equals the Product of the Power and the Log**

$$\boxed{\log_a u^r = r \log_a u} \tag{5}$$

EXAMPLE 10   Using Properties (3), (4), and (5) of Logarithms

(a) $\log_a\left(x\sqrt{x^2+1}\right) = \log_a x + \log_a \sqrt{x^2+1}$ $\log_a(uv) = \log_a u + \log_a v$

$$= \log_a x + \log_a(x^2+1)^{1/2}$$

$$= \log_a x + \frac{1}{2}\log_a(x^2+1) \qquad \log_a u^r = r\log_a u$$

(b) $\log_a \dfrac{x^2}{(x-1)^3} \underset{\underset{\log_a\left(\frac{u}{v}\right)\,=\,\log_a u\,-\,\log_a v}{\uparrow}}{=} \log_a x^2 - \log_a (x-1)^3 \underset{\underset{\log_a u^r\,=\,r\log_a u}{\uparrow}}{=} 2\log_a x - 3\log_a (x-1)$

(c) $\log_a x + \log_a 9 + \log_a(x^2+1) - \log_a 5 = \log_a(9x) + \log_a(x^2+1) - \log_a 5$

$$= \log_a[9x(x^2+1)] - \log_a 5$$

$$= \log_a\left[\frac{9x(x^2+1)}{5}\right] \qquad \blacksquare$$

CAUTION In using properties (3) through (5), be careful about the values that the variable may assume. For example, the domain of the variable for $\log_a x$ is $x > 0$, and for $\log_a(x-1)$ it is $x > 1$. That is, the equality $\log_a x + \log_a(x-1) = \log_a[x(x-1)]$ is true only for $x > 1$.

CAUTION Common errors made by some students include:

- Expressing the logarithm of a sum as the sum of the logarithms.

$$\log_a(u+v) \quad \text{is } not \text{ equal to} \quad \log_a u + \log_a v$$

  Correct statement: $\log_a(uv) = \log_a u + \log_a v$ Property (3)

- Expressing the difference of logarithms as the quotient of logarithms.

$$\log_a u - \log_a v \quad \text{is } not \text{ equal to} \quad \frac{\log_a u}{\log_a v}$$

  Correct statement: $\log_a u - \log_a v = \log_a\left(\dfrac{u}{v}\right)$ Property (4)

- Expressing a logarithm raised to a power as the product of the power and the logarithm.

$$(\log_a u)^r \quad \text{is } not \text{ equal to} \quad r\log_a u$$

  Correct statement: $\log_a u^r = r\log_a u$ Property (5)

Since most calculators can calculate only logarithms with base 10, called **common logarithms** and abbreviated log, and logarithms with base $e \approx 2.718$, called **natural logarithms** and abbreviated ln, it is often useful to be able to change the bases of logarithms.

THEOREM Change-of-Base Formula

If $a \neq 1$, $b \neq 1$, and u are positive real numbers, then

$$\boxed{\log_a u = \frac{\log_b u}{\log_b a}}$$

For example, to approximate $\log_2 15$, we use the change-of-base formula. Then we use a calculator.

$$\log_2 15 \underset{\underset{\text{Change-of-Base Formula}}{\uparrow}}{=} \frac{\log 15}{\log 2} \approx 3.907$$

A.2 Geometry Used in Calculus

OBJECTIVES *When you finish this section, you should be able to:*

1 Use properties of triangles and the Pythagorean Theorem (p. A-11)

2 Work with congruent triangles and similar triangles (p. A-12)

3 Use geometry formulas (p. A-15)

1 Use Properties of Triangles and the Pythagorean Theorem

A **triangle** is a three-sided polygon. The lengths of the sides are labeled with lowercase letters, such as a, b, and c, and the angles are labeled with uppercase letters, such as A, B, and C, with angle A opposite side a, angle B opposite side b, and angle C opposite side c, as shown in Figure 7.

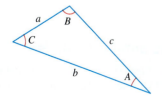

Figure 7

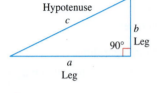

Hypotenuse
c

b
90° Leg

a
Leg

Figure 8

Refer to the triangle in Figure 7. The sum of the lengths of any two sides of a triangle is always greater than the length of the remaining side. The sum of the measures of the angles of a triangle, when measured in degrees, equals 180°. That is,

$$a + b > c \qquad a + c > b \qquad b + c > a \qquad A + B + C = 180°$$

An **isosceles triangle** is a triangle with two equal sides. In an isosceles triangle, the angles opposite the two equal sides are equal.

An **equilateral triangle** is a triangle with three equal sides. In an equilateral triangle, each angle measures 60°.

The *Pythagorean Theorem* is a statement about *right triangles*. A **right triangle** contains a **right angle**, that is, an angle measuring 90°. The side of the triangle opposite the 90° angle is called the **hypotenuse**; the remaining two sides are called **legs**. In Figure 8, c represents the length of the hypotenuse, and a and b represent the lengths of the legs.

THEOREM Pythagorean Theorem

In a right triangle, the square of the length of the hypotenuse is equal to the sum of the squares of the lengths of the legs. That is, in the right triangle shown in Figure 8,

$$\boxed{c^2 = a^2 + b^2}$$

EXAMPLE 1 Finding the Hypotenuse of a Right Triangle

In a right triangle, one leg has length 4 and the other has length 3. What is the length of the hypotenuse?

Solution Since the triangle is a right triangle, we use the Pythagorean Theorem, with $a = 4$ and $b = 3$ to find the length c of the hypotenuse.

$$c^2 = a^2 + b^2$$
$$c^2 = 4^2 + 3^2 = 16 + 9 = 25$$
$$c = \sqrt{25} = 5$$

∎

The converse of the Pythagorean Theorem is also true.

THEOREM Converse of the Pythagorean Theorem

In a triangle, if the square of the length of one side equals the sum of the squares of the lengths of the other two sides, the triangle is a right triangle. The 90° angle is opposite the longest side.

EXAMPLE 2 Using the Converse of the Pythagorean Theorem

Show that a triangle whose sides have lengths 5, 12, and 13 is a right triangle. Identify the hypotenuse.

Solution We square the lengths of the sides.

$$5^2 = 25 \qquad 12^2 = 144 \qquad 13^2 = 169$$

Notice that the sum of the first two squares (25 and 144) equals the third square (169). So, the triangle is a right triangle. The longest side, 13, is the hypotenuse. ∎

See Figure 9.

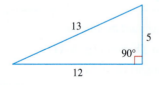

13

5
90°
12

Figure 9

2 Work with Congruent Triangles and Similar Triangles

The word *congruent* means "coinciding when superimposed." For example, two angles are congruent if they have the same measure, and two line segments are congruent if they have the same length.

IN WORDS Two triangles are congruent if they are the same size and shape.

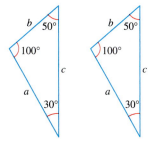

Figure 10 Congruent triangles.

DEFINITION Congruent Triangles

Two triangles are **congruent** if in each triangle, the corresponding angles have the same measure and the corresponding sides have the same length.

In Figure 10, corresponding angles are equal and the lengths of the corresponding sides are equal. So, these triangles are congruent.

It is not necessary to verify that all three angles and all three sides have the same measure to determine whether two triangles are congruent.

Determining Congruent Triangles

- **Angle-Side-Angle Case (ASA)** Two triangles are congruent if the measures of two angles from each triangle are equal and the lengths of the corresponding sides between the two equal angles are equal.

  For example, in Figure 11, the two triangles are congruent because both triangles have an angle measuring $40°$, an angle measuring $80°$, and the sides between these angles are both 10 units in length.

- **Side-Side-Side Case (SSS)** Two triangles are congruent if the lengths of the sides of one triangle are equal to the lengths of the sides of the other triangle.

  For example, in Figure 12, the two triangles are congruent because both triangles have sides with lengths 8, 15, and 20 units.

- **Side-Angle-Side Case (SAS)** Two triangles are congruent if the lengths of two corresponding sides of the triangles are equal and the angles between the two sides have the same measure.

  For example, in Figure 13, the two triangles are congruent because both triangles have sides of length 7 units and 8 units, and the angle between the two congruent sides in each triangle measures $40°$.

- **Angle-Angle-Side Case (AAS)** Two triangles are congruent if the measures of two angles and the length of a nonincluded side of one triangle are equal to the corresponding parts of the other triangle.

CAUTION Knowing that two triangles have equal angles is not sufficient to conclude that the triangles are congruent. Similarly, knowing that the lengths of two corresponding sides and the measure of a nonincluded angle of the triangles are equal is not sufficient to conclude that the triangles are congruent.

  For example, in Figure 14, the two triangles are congruent because both triangles have angles measuring $35°$ and $96°$, and the nonincluded side adjacent to the $35°$ angle has length 10 units.

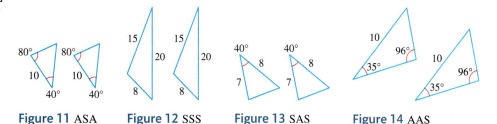

Figure 11 ASA   **Figure 12** SSS   **Figure 13** SAS   **Figure 14** AAS

We contrast congruent triangles with *similar* triangles.

DEFINITION Similar Triangles

Two triangles are **similar** if in each triangle corresponding angles have the same measure and corresponding sides are proportional in length, that is, the ratio of the lengths of the corresponding sides of each triangle equals the same constant.

IN WORDS Two triangles are similar if they have the same shape, but (possibly) different sizes.

A.3 Analytic Geometry Used in Calculus

OBJECTIVES *When you finish this section, you should be able to:*

1 Use the distance formula (p. A-16)

2 Graph equations, find intercepts, and test for symmetry (p. A-16)

3 Work with equations of a line (p. A-18)

4 Work with the equation of a circle (p. A-21)

5 Graph parabolas, ellipses, and hyperbolas (p. A-22)

Figure 20

NOTE If (x, y) are the coordinates of a point P, then x is called the x-**coordinate** of P and y is called the y-**coordinate** of P.

1 Use the Distance Formula

Any point P in the xy-plane can be located using an **ordered pair** (x, y) of real numbers. The ordered pair (x, y), called the **coordinates** of P, gives enough information to locate the point P in the plane. For example, to locate the point with coordinates $(-3, 1)$, move 3 units along the x-axis to the left of O, and then move straight up 1 unit. Then **plot** the point by placing a dot at this location. See Figure 20 in which the points with coordinates $(-3, 1)$, $(-2, -3)$, $(3, -2)$, and $(3, 2)$ are plotted.

THEOREM Distance Formula

The distance between two points $P_1 = (x_1, y_1)$ and $P_2 = (x_2, y_2)$, denoted by $d(P_1, P_1)$, is

$$d(P_1, P_2) = \sqrt{(x_2 - x_1)^2 + (y_2 - y_1)^2}$$

EXAMPLE 1   Using the Distance Formula

Find the distance d between the points $(-3, 5)$ and $(3, 2)$.

Solution We use the distance formula with $P_1 = (x_1, y_1) = (-3, 5)$ and $P_2 = (x_2, y_2) = (3, 2)$.

$$d = \sqrt{[3 - (-3)]^2 + (2 - 5)^2} = \sqrt{6^2 + (-3)^2} = \sqrt{36 + 9} = \sqrt{45} = 3\sqrt{5} \approx 6.708$$

■

2 Graph Equations, Find Intercepts, and Test for Symmetry

An **equation in two variables**, say, x and y, is a statement in which two expressions involving x and y are equal. The expressions are called the **sides** of the equation. Since an equation is a statement, it may be true or false, depending on the value of the variables. Any values of x and y that result in a true statement are said to **satisfy** the equation.

For example, the following are equations in two variables x and y:

$$x^2 + y^2 = 5 \qquad 2x - y = 6 \qquad y = 2x + 5 \qquad x^2 = y$$

The equation, $x^2 + y^2 = 5$ is satisfied for $x = 1$ and $y = 2$, since $1^2 + 2^2 = 1 + 4 = 5$. Other choices of x and y, such as $x = -1$ and $y = -2$, also satisfy this equation. It is not satisfied for $x = 2$ and $y = 3$, since $2^2 + 3^2 = 4 + 9 = 13 \neq 5$.

The **graph of an equation in two variables** x and y consists of the set of points in the xy-plane whose coordinates (x, y) satisfy the equation.

EXAMPLE 2   Graphing an Equation by Plotting Points

Graph the equation $y = x^2$.

Solution Table 4 lists several points on the graph. In Figure 21(a) the points are plotted and then they are connected with a smooth curve to obtain the graph (a *parabola*). Figure 21(b) shows the graph using graphing technology.

TABLE 4

x	$y = x^2$	(x, y)
-4	16	$(-4, 16)$
-3	9	$(-3, 9)$
-2	4	$(-2, 4)$
-1	1	$(-1, 1)$
0	0	$(0, 0)$
1	1	$(1, 1)$
2	4	$(2, 4)$
3	9	$(3, 9)$
4	16	$(4, 16)$

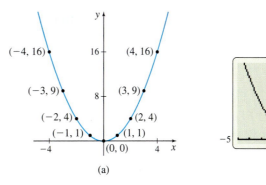

(a)   (b)

Figure 21 $y = x^2$

The graphs in Figure 21 do not show all the points whose coordinates (x, y) satisfy the equation $y = x^2$. For example, the point $(6, 36)$ satisfies the equation $y = x^2$, but it is not shown. It is important when graphing to present enough of the graph so that any viewer will "see" the rest of the graph as an obvious continuation of what is actually there.

The points, if any, at which a graph crosses or touches a coordinate axis are called the **intercepts** of the graph. See Figure 22. The x-coordinate of a point where a graph crosses or touches the x-axis is an **x-intercept**. At an x-intercept, the y-coordinate equals 0. The y-coordinate of a point where a graph crosses or touches the y-axis is a **y-intercept**. At a y-intercept, the x-coordinate equals 0. When graphing an equation, all its intercepts should be displayed or be easily inferred from the part of the graph that is displayed.

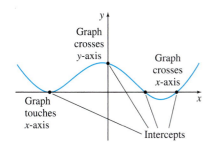

Figure 22

EXAMPLE 3 **Finding the Intercepts of a Graph**

Find the x-intercept(s) and the y-intercept(s) of the graph of $y = x^2 - 4$.

Solution To find the x-intercept(s), we let $y = 0$ and solve the equation

$$x^2 - 4 = 0$$
$$(x + 2)(x - 2) = 0 \qquad \text{Factor.}$$
$$x = -2 \quad \text{or} \quad x = 2 \qquad \text{Set each factor equal to 0 and solve.}$$

The equation has two solutions, -2 and 2. The x-intercepts are -2 and 2.

To find the y-intercept(s), we let $x = 0$ in the equation.

$$y = 0^2 - 4 = -4$$

The y-intercept is -4. ∎

DEFINITION Symmetry

- A graph is **symmetric with respect to the x-axis** if, for every point (x, y) on the graph, the point $(x, -y)$ is also on the graph. See Figure 23.
- A graph is **symmetric with respect to the y-axis** if, for every point (x, y) on the graph, the point $(-x, y)$ is also on the graph. See Figure 24.
- A graph is **symmetric with respect to the origin** if, for every point (x, y) on the graph, the point $(-x, -y)$ is also on the graph. See Figure 25.

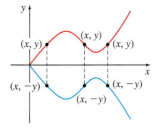

Figure 23 Symmetry with respect to the x-axis.

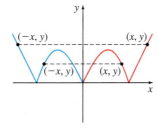

Figure 24 Symmetry with respect to the y-axis.

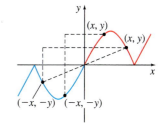

Figure 25 Symmetry with respect to the origin.

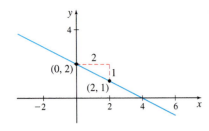

Figure 29 $2x + 4y = 8$

The coefficient of x is the slope. The slope is $-\dfrac{1}{2}$. The constant 2 is the y-intercept, so the point $(0, 2)$ is on the graph. Now use the slope $-\dfrac{1}{2}$. Starting at the point $(0, 2)$, we move 2 units to the right and then 1 unit down to the point $(2, 1)$. We plot this point and draw a line through the two points. See Figure 29. ■

When two lines in the plane do not intersect, they are **parallel**.

THEOREM Criterion for Parallel Lines

Two nonvertical lines are parallel if and only if their slopes are equal and they have different y-intercepts.

EXAMPLE 7 Finding an Equation of a Line that Is Parallel to a Given Line

Find an equation of the line that contains the point $(1, 2)$ and is parallel to the line $y = 5x$.

Solution Since the two lines are parallel, the slope of the line we seek equals the slope of the line $y = 5x$, which is 5. Since the line we seek also contains the point $(1, 2)$, we use the point-slope form to obtain the equation.

$$y - y_1 = m(x - x_1)$$
$$y - 2 = 5(x - 1)$$
$$y - 2 = 5x - 5$$
$$y = 5x - 3$$

The line $y = 5x - 3$ contains the point $(1, 2)$ and is parallel to the line $y = 5x$. See Figure 30. ■

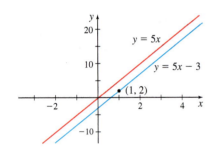

Figure 30 Parallel lines

When two lines intersect at a right angle $(90°)$, they are **perpendicular**. See Figure 31.

THEOREM Criterion for Perpendicular Lines

Two nonvertical lines are perpendicular if and only if the product of their slopes is -1.

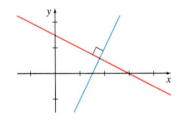

Figure 31 Perpendicular lines

EXAMPLE 8 Finding an Equation of a Line that Is Perpendicular to a Given Line

Find an equation of the line that contains the point $(-1, 3)$ and is perpendicular to the line $4x + y = -1$.

Solution We begin by writing the equation of the given line in slope-intercept form to find its slope.

$$4x + y = -1$$
$$y = -4x - 1$$

This line has a slope of -4. Any line perpendicular to this line will have slope $\dfrac{1}{4}$. Because the point $(-1, 3)$ is on this line, we use the point-slope form of a line.

$$y - y_1 = m(x - x_1)$$
$$y - 3 = \frac{1}{4}[x - (-1)]$$
$$y - 3 = \frac{1}{4}x + \frac{1}{4}$$
$$y = \frac{1}{4}x + \frac{13}{4}$$

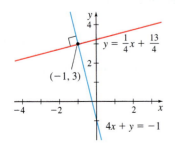

Figure 32

Figure 32 shows the graphs of $4x + y = -1$ and $y = \dfrac{1}{4}x + \dfrac{13}{4}$. ■

4 Work with the Equation of a Circle

An advantage of a coordinate system is that it enables a geometric statement to be translated into an algebraic statement, and vice versa. Consider, for example, the following geometric statement that defines a circle.

DEFINITION Circle

A **circle** is the set of points in the xy-plane that is a fixed distance r from a fixed point (h, k). The fixed distance r is called the **radius**, and the fixed point (h, k) is called the **center** of the circle.

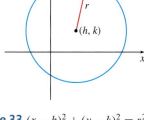

Figure 33 shows the graph of a circle. To find an equation that describes a circle, let (x, y) represent the coordinates of any point on the circle with radius r and center (h, k). Then the distance between the point (x, y) and (h, k) equals r. That is, by the distance formula,

$$\sqrt{(x - h)^2 + (y - k)^2} = r$$

Now we square both sides to obtain

$$(x - h)^2 + (y - k)^2 = r^2$$

Figure 33 $(x - h)^2 + (y - k)^2 = r^2$

This form of the equation of a circle is called the **standard form**.

THEOREM

The standard form of a circle with radius r and center at the origin $(0, 0)$ is

$$x^2 + y^2 = r^2$$

DEFINITION Unit Circle

The circle with radius $r = 1$ and center at the origin is called the **unit circle** and has the equation

$$x^2 + y^2 = 1$$

See Figure 34.

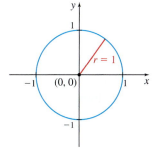

Figure 34 The unit circle: $x^2 + y^2 = 1$

EXAMPLE 9 **Graphing a Circle**

Graph the equation $(x + 3)^2 + (y - 2)^2 = 16$.

Solution This is the standard form of an equation of a circle. To graph the circle, we first identify the center and the radius of the circle.

$$(x + 3)^2 + (y - 2)^2 = 16$$
$$(x - (-3))^2 + (y - 2)^2 = 4^2$$

$$hkr^2$$

The circle has its center at the point $(-3, 2)$; its radius is 4 units. To graph the circle, we plot the center $(-3, 2)$. Then we locate the four points on the circle that are 4 units to the left, to the right, above, and below the center. These four points are used as guides to obtain the graph. See Figure 35. ∎

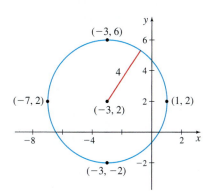

Figure 35 $(x + 3)^2 + (y - 2)^2 = 16$

DEFINITION General Form of the Equation of a Circle

The equation

$$x^2 + y^2 + ax + by + c = 0$$

is called the **general form of the equation of a circle**.

NEED TO REVIEW? Completing the square is discussed in Section A.1, p. A-2.

If the equation of a circle is given in the general form, we use the method of completing the square to put the equation in standard form in order to identify the center and radius.

EXAMPLE 10   Graphing a Circle Whose Equation Is in General Form

Graph the equation

$$x^2 + y^2 + 4x - 6y + 12 = 0$$

Solution We complete the square in both x and y to put the equation in standard form. First, we group the terms involving x, group the terms involving y, and put the constant on the right side of the equation. The result is

$$(x^2 + 4x) + (y^2 - 6y) = -12$$

Next, we complete the square of each expression in parentheses. Remember that any number added to the left side of an equation must be added to the right side.

$$(x^2 + 4x + 4) + (y^2 - 6y + 9) = -12 + 4 + 9 = 1$$

$$\text{Add } \left(\frac{4}{2}\right)^2 = 4 \quad \text{Add } \left(\frac{-6}{2}\right)^2 = 9$$

$$(x + 2)^2 + (y - 3)^2 = 1 \qquad \text{Factor.}$$

This is the standard form of the equation of a circle with radius 1 and center at the point $(-2, 3)$. To graph the circle, we plot the center $(-2, 3)$ and points 1 unit to the right, to the left, above and below the point $(-2, 3)$, as shown in Figure 36. ■

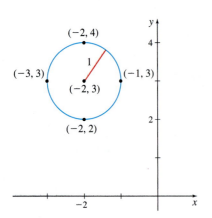

Figure 36 $x^2 + y^2 + 4x - 6y + 12 = 0$

5 Graph Parabolas, Ellipses, and Hyperbolas

The graph of the equation $y = x^2$ is an example of a *parabola*. See Figure 21 on page A-26. In fact, every parabola has a shape like the graph of $y = x^2$.

A **parabola** is the graph of an equation in one of the four forms:

$$\boxed{y^2 = 4ax \qquad y^2 = -4ax \qquad x^2 = 4ay \qquad x^2 = -4ay}$$

where $a > 0$.

In each of the parabolas in Figure 37, the **vertex** V is at the origin $(0, 0)$. When the equation is of the form $y^2 = 4ax$, the point $(a, 0)$ is called the **focus**, and the parabola is symmetric with respect to the x-axis. When the equation is of the form $x^2 = 4ay$, the point $(0, a)$ is the focus, and the parabola is symmetric with respect to the y-axis.

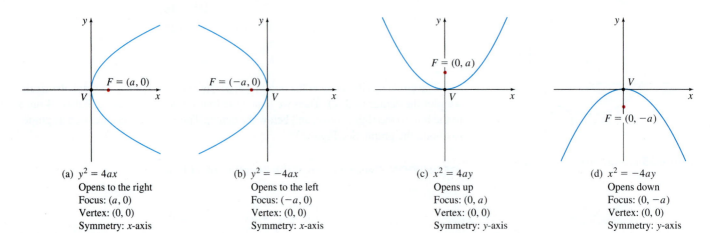

(a) $y^2 = 4ax$
Opens to the right
Focus: $(a, 0)$
Vertex: $(0, 0)$
Symmetry: x-axis

(b) $y^2 = -4ax$
Opens to the left
Focus: $(-a, 0)$
Vertex: $(0, 0)$
Symmetry: x-axis

(c) $x^2 = 4ay$
Opens up
Focus: $(0, a)$
Vertex: $(0, 0)$
Symmetry: y-axis

(d) $x^2 = -4ay$
Opens down
Focus: $(0, -a)$
Vertex: $(0, 0)$
Symmetry: y-axis

Figure 37

EXAMPLE 11 **Graphing a Parabola with Vertex at the Origin**

Graph the equation $x^2 = 16y$.

Solution The graph of $x^2 = 16y$ is a parabola of the form

$$x^2 = 4ay$$

where $a = 4$. Look at Figure 37(c). The graph will open up, with focus at $(0, 4)$. The vertex is at $(0, 0)$. To graph the parabola, we let $y = a$; that is, let $y = 4$. (Any other positive number will also work.)

$$x^2 = 16y = 16(4) = 64$$
$$\uparrow$$
$$y = a = 4$$
$$x = -8 \quad \text{or} \quad x = 8$$

The points $(-8, 4)$ and $(8, 4)$ are on the parabola and establish its opening. See Figure 38. ∎

Figure 38 $x^2 = 16y$

An **ellipse** is the graph of an equation in one of the two forms:

$$\frac{x^2}{a^2} + \frac{y^2}{b^2} = 1 \qquad \frac{x^2}{b^2} + \frac{y^2}{a^2} = 1$$

where $a \geq b > 0$.

NOTE If $a = b$, then $\dfrac{x^2}{a^2} + \dfrac{y^2}{a^2} = 1$ is the equation of a circle of radius a.

An ellipse is oval-shaped. The line segment dividing the ellipse in half the long way is the **major axis**; its length is $2a$. The two points of intersection of the ellipse and the major axis are the **vertices** of the ellipse. The midpoint of the vertices is the **center** of the ellipse. The line segment through the center of the ellipse and perpendicular to the major axis is the **minor axis**. Along the major axis c units from the center of the ellipse, where $c^2 = a^2 - b^2$, $c > 0$, are two points, called **foci**, that determine the shape of the ellipse. See Figure 39.

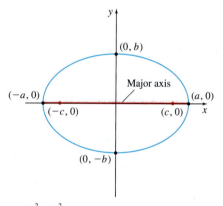

(a) $\dfrac{x^2}{a^2} + \dfrac{y^2}{b^2} = 1, a \geq b > 0$
Major axis: along the x-axis
Center: $(0, 0)$
Vertices: $(-a, 0), (a, 0)$
Foci: $(-c, 0), (c, 0)$, where $c^2 = a^2 - b^2, c > 0$
Symmetry: x-axis, y-axis, origin

(b) $\dfrac{x^2}{b^2} + \dfrac{y^2}{a^2} = 1, a \geq b > 0$
Major axis: along the y-axis
Center: $(0, 0)$
Vertices: $(0, -a), (0, a)$
Foci: $(0, -c), (0, c)$, where $c^2 = a^2 - b^2, c > 0$
Symmetry: x-axis, y-axis, origin

Figure 39

EXAMPLE 12 **Graphing an Ellipse with Center at the Origin**

Graph the equation: $9x^2 + 4y^2 = 36$

Solution To put the equation in standard form, we divide each side by 36.

$$\frac{x^2}{4} + \frac{y^2}{9} = 1$$

The graph of this equation is an ellipse. Since the larger number is under y^2, the major axis is along the y-axis. The center is at the origin $(0, 0)$. It is easiest to graph an ellipse

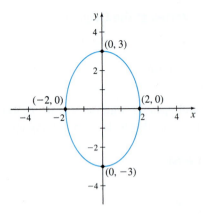

Figure 40 $9x^2 + 4y^2 = 36$

by finding its intercepts:

x-intercepts: Let $y = 0$ y-intercepts: Let $x = 0$

$$\frac{x^2}{4} + \frac{0^2}{9} = 1 \qquad\qquad\qquad 0^2 + \frac{y^2}{9} = 1$$

$$x^2 = 4 \qquad\qquad\qquad\qquad\qquad y^2 = 9$$

$$x = -2 \text{ and } x = 2 \qquad\qquad y = -3 \text{ and } y = 3$$

The points $(-2, 0)$, $(2, 0)$, $(0, -3)$, $(0, 3)$ are the intercepts of the ellipse. See Figure 40. ∎

A **hyperbola** is the graph of an equation in one of the two forms:

$$\frac{x^2}{a^2} - \frac{y^2}{b^2} = 1 \qquad \frac{y^2}{b^2} - \frac{x^2}{a^2} = 1$$

where $a > 0$ and $b > 0$.

Two points, called **foci**, determine the shape of the hyperbola. The midpoint of the line segment containing the foci is called the **center** of the hyperbola. The foci are located c units from the center, where $c^2 = a^2 + b^2$, $c > 0$. The line containing the foci is called the **transverse axis**. The **vertices** of a hyperbola are its intercepts.

See Figure 41. The graph of a hyperbola consists of two branches. For the hyperbola $\frac{x^2}{a^2} - \frac{y^2}{b^2} = 1$, the branches of the graph open left and right; for the hyperbola $\frac{y^2}{b^2} - \frac{x^2}{a^2} = 1$, the branches of the graph open up and down.

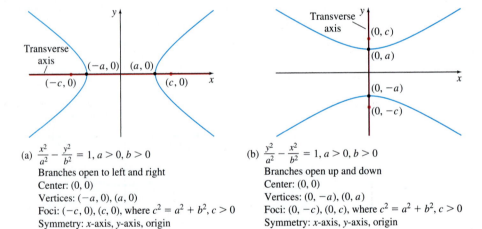

(a) $\frac{x^2}{a^2} - \frac{y^2}{b^2} = 1$, $a > 0$, $b > 0$
Branches open to left and right
Center: $(0, 0)$
Vertices: $(-a, 0)$, $(a, 0)$
Foci: $(-c, 0)$, $(c, 0)$, where $c^2 = a^2 + b^2$, $c > 0$
Symmetry: x-axis, y-axis, origin

(b) $\frac{y^2}{a^2} - \frac{x^2}{b^2} = 1$, $a > 0$, $b > 0$
Branches open up and down
Center: $(0, 0)$
Vertices: $(0, -a)$, $(0, a)$
Foci: $(0, -c)$, $(0, c)$, where $c^2 = a^2 + b^2$, $c > 0$
Symmetry: x-axis, y-axis, origin

Figure 41

EXAMPLE 13   Graphing a Hyperbola With Center at the Origin

Graph the equation $\frac{y^2}{4} - \frac{x^2}{5} = 1$.

Solution The graph of $\frac{y^2}{4} - \frac{x^2}{5} = 1$ is a hyperbola. The hyperbola consists of two branches, one opening up, the other opening down, like the graph in Figure 41(b). The hyperbola has no x-intercepts. To find the y-intercepts, we let $x = 0$ and solve for y.

$$\frac{y^2}{4} = 1$$

$$y^2 = 4$$

$$y = -2 \qquad \text{or} \qquad y = 2$$

The y-intercepts are -2 and 2, so the vertices are $(0, -2)$ and $(0, 2)$. The transverse axis is the vertical line $x = 0$. To graph the hyperbola, let $y = \pm 3$ (or any numbers ≥ 2 or ≤ -2).

Then

$$\frac{y^2}{4} - \frac{x^2}{5} = 1$$

$$\frac{9}{4} - \frac{x^2}{5} = 1 \qquad \textcolor{blue}{y = \pm 3}$$

$$\frac{x^2}{5} = \frac{5}{4}$$

$$x^2 = \frac{25}{4}$$

$$x = -\frac{5}{2} \text{ or } x = \frac{5}{2}$$

The points $\left(-\frac{5}{2}, 3\right)$, $\left(-\frac{5}{2}, -3\right)$, $\left(\frac{5}{2}, 3\right)$, and $\left(\frac{5}{2}, -3\right)$ are on the hyperbola. See Figure 42 for the graph. ∎

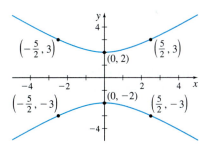

Figure 42 $\dfrac{y^2}{4} - \dfrac{x^2}{5} = 1$

A.4 Trigonometry Used in Calculus

OBJECTIVES *When you finish this section, you should be able to:*

1 Work with angles, arc length of a circle, and circular motion (p. A-25)

2 Define and evaluate trigonometric functions (p. A-27)

3 Determine the domain and the range of the trigonometric functions (p. A-31)

4 Use basic trigonometry identities (p. A-32)

5 Use sum and difference, double-angle and half-angle, and sum-to-product and product-to-sum formulas (p. A-34)

6 Solve triangles using the Law of Sines and the Law of Cosines (p. A-35)

1 Work with Angles, Arc Length of a Circle, and Circular Motion

A **ray**, or **half-line**, is the portion of a line that starts at a point V on the line and extends indefinitely in one direction. The point V of a ray is called its **vertex**. See Figure 43.

Figure 43

If two rays are drawn with a common vertex, they form an **angle**. We call one ray of an angle the **initial side** and the other ray the **terminal side**. The angle formed is identified by showing the direction and amount of rotation from the initial side to the terminal side. If the rotation is in the counterclockwise direction, the angle is **positive**; if the rotation is clockwise, the angle is **negative**. See Figure 44.

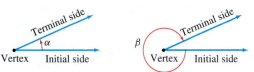

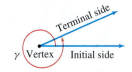

(a) Counterclockwise rotation
   Positive angle

(b) Clockwise rotation
   Negative angle

(c) Counterclockwise rotation
   Positive angle

Figure 44

An angle θ is in **standard position** if its vertex is at the origin of a rectangular coordinate system and its initial side is on the positive x-axis. See Figure 45 on page A-26.

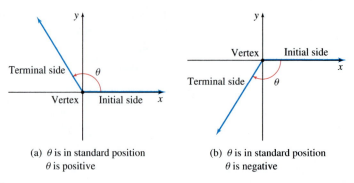

(a) θ is in standard position
θ is positive

(b) θ is in standard position
θ is negative

Figure 45

Suppose an angle θ is in standard position. If its terminal side coincides with a coordinate axis, we say that θ is a **quadrantal angle**. If its terminal side does not coincide with a coordinate axis, we say that θ **lies in a quadrant**. See Figure 46.

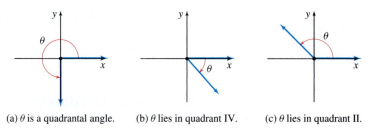

(a) θ is a quadrantal angle.

(b) θ lies in quadrant IV.

(c) θ lies in quadrant II.

Figure 46

Angles are measured by determining the amount of rotation needed for the initial side to coincide with the terminal side. The two commonly used measures for angles are *degrees* and *radians*.

The angle formed by rotating the initial side exactly once in the counterclockwise direction until it coincides with itself (one revolution) measures 360 degrees, abbreviated 360°. See Figure 47.

A **central angle** is a positive angle whose vertex is at the center of a circle. The rays of a central angle subtend (intersect) an arc on the circle. If the radius of the circle is r and the arc subtended by the central angle is also of length r, then the measure of the angle is **1 radian**. See Figure 48.

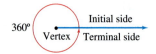

Figure 47 1 revolution counterclockwise is 360°.

THEOREM Arc Length

For a circle of radius r, a central angle of θ radians subtends an arc whose length s is

$$s = r\theta$$

See Figure 49.

Consider a circle of radius r. A central angle of one revolution will subtend an arc equal to the circumference of the circle, as shown in Figure 50. Because the circumference of a circle equals $2\pi r$, we use $s = 2\pi r$ in the formula for arc length to find the radian measure of an angle of one revolution.

$$s = r\theta$$
$$2\pi r = r\theta \qquad \text{For one revolution: } s = 2\pi r$$
$$\theta = 2\pi \text{ radians} \qquad \text{Solve for } \theta.$$

Since one revolution is equivalent to 360°, we have 360° = 2π radians so that

$$180° = \pi \text{ radians}$$

In calculus, radians are generally used to measure angles, unless degrees are specifically mentioned. Table 5 lists the radian and degree measures for some common angles.

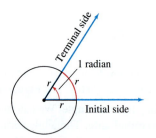

Figure 48 1 Radian

Figure 49 $s = r\theta$

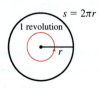

Figure 50 1 revolution $= 2\pi$ radians

NOTE If the measure of an angle is given as a number, it is understood to mean radians. If the measure of an angle is given in degrees, then it will be marked either with the symbol ° or the word "degrees."

TABLE 5

Radians	0	$\dfrac{\pi}{6}$	$\dfrac{\pi}{4}$	$\dfrac{\pi}{3}$	$\dfrac{\pi}{2}$	$\dfrac{2\pi}{3}$	$\dfrac{3\pi}{4}$	$\dfrac{5\pi}{6}$	π	$\dfrac{3\pi}{2}$	2π
Degrees	0°	30°	45°	60°	90°	120°	135°	150°	180°	270°	360°

THEOREM Area of a Sector

The area A of the sector of a circle of radius r formed by a central angle of θ radians is

$$A = \frac{1}{2}r^2\theta$$

See Figure 51.

Figure 51 $A = \dfrac{1}{2}r^2\theta$

DEFINITION Linear Speed, Angular Speed

Suppose an object moves around a circle of radius r at a constant speed. If s is the distance around the circle traveled in time t, then the **linear speed** v of the object is

$$v = \frac{s}{t}$$

The **angular speed** ω (the Greek letter omega) of an object moving at a constant speed around a circle of radius r is

$$\omega = \frac{\theta}{t}$$

where θ is the angle (measured in radians) swept out in time t.

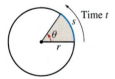

Figure 52 $v = \dfrac{s}{t}; \; \omega = \dfrac{\theta}{t}$

See Figure 52.

Angular speed is used to describe the turning rate of an engine. For example, an engine idling at 900 rpm (revolutions per minute) rotates at an angular speed of

$$900\,\frac{\text{revolutions}}{\text{minute}} = 900\,\frac{\text{revolutions}}{\text{minute}} \cdot 2\pi\,\frac{\text{radians}}{\text{revolution}} = 1800\pi\,\frac{\text{radians}}{\text{minute}}$$

There is an important relationship between linear speed and angular speed:

$$\text{linear speed} = v = \frac{s}{t} \underset{s=r\theta}{=} \frac{r\theta}{t} = r\left(\frac{\theta}{t}\right) \underset{\omega=\frac{\theta}{t}}{=} r\omega$$

RECALL In the formula $s = r\theta$ for the arc length of a circle, the angle θ must be measured in radians.

So,

$$v = r\omega$$

where ω is measured in radians per unit time.

When using the equation $v = r\omega$, remember that $v = \dfrac{s}{t}$ has the dimensions of length per unit of time (such as meters per second or miles per hour), r has the same length dimension as s, and ω has the dimensions of radians per unit of time. If the angular speed is given in terms of *revolutions* per unit of time (as is often the case), be sure to convert it to *radians* per unit of time, using the fact that one revolution = 2π radians, before using the formula $v = r\omega$.

2 Define and Evaluate Trigonometric Functions

There are two common approaches to trigonometry: One uses right triangles and the other uses the unit circle. We suggest you first review the approach you are most familiar with and then read the other approach. The two approaches are given side-by-side on pages A-28 and A-29.

Right Triangle Approach

Suppose θ is an **acute angle**; that is, $0 < \theta < \dfrac{\pi}{2}$, as shown in Figure 53(a). Using the angle θ, we form a right triangle, like the one illustrated in Figure 53(b), with hypotenuse of length c and legs of lengths a and b. The three sides of the right triangle can be used to form exactly six ratios:

$$\frac{b}{c}, \quad \frac{a}{c}, \quad \frac{b}{a}, \quad \frac{c}{b}, \quad \frac{c}{a}, \quad \frac{a}{b}$$

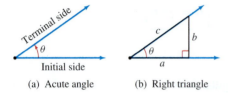

(a) Acute angle (b) Right triangle

Figure 53

Because the ratios depend only on the angle θ and not on the triangle itself, each ratio is given a name that involves θ. See Figure 54.

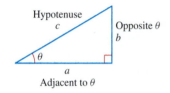

Figure 54

DEFINITION Trigonometric Functions

The six ratios of a right triangle are called the **trigonometric functions of an acute angle** θ and are defined as follows:

Function name	Abbreviation	Value	Relation using words
sine of θ	$\sin \theta$	$\dfrac{b}{c}$	$\dfrac{\text{opposite}}{\text{hypotenuse}}$
cosine of θ	$\cos \theta$	$\dfrac{a}{c}$	$\dfrac{\text{adjacent}}{\text{hypotenuse}}$
tangent of θ	$\tan \theta$	$\dfrac{b}{a}$	$\dfrac{\text{opposite}}{\text{adjacent}}$
cosecant of θ	$\csc \theta$	$\dfrac{c}{b}$	$\dfrac{\text{hypotenuse}}{\text{opposite}}$
secant of θ	$\sec \theta$	$\dfrac{c}{a}$	$\dfrac{\text{hypotenuse}}{\text{adjacent}}$
cotangent of θ	$\cot \theta$	$\dfrac{a}{b}$	$\dfrac{\text{adjacent}}{\text{opposite}}$

To extend the definition of the trigonometric functions to include angles that are not acute, the angle is placed in standard position in a rectangular coordinate system, as shown in Figure 55.

Unit Circle Approach

Let t be any real number. We position a t-axis perpendicular to the x-axis so that $t = 0$ coincides with the point $(1, 0)$ of the xy-plane, and positive values of t are above the x-axis. See Figure 56(a).

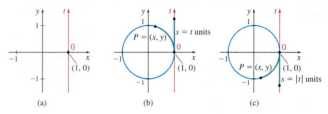

(a)   (b)   (c)

Figure 56

Now look at the unit circle in Figure 56(b). If $t > 0$, we begin at the point $(1, 0)$ on the unit circle and travel $s = t$ units in the counterclockwise direction along the circle to arrive at the point $P = (x, y)$. In this sense, the length $s = t$ units on the t-axis is being **wrapped** around the unit circle.

If $t < 0$, we begin at the point $(1, 0)$ on the unit circle and travel $s = |t|$ units in the clockwise direction along the circle to the point $P = (x, y)$, as shown in Figure 56(c).

If $t > 2\pi$ or if $t < -2\pi$, it will be necessary to travel around the circle more than once before arriving at the point P.

This discussion tells us that, for any real number t, there is a unique point $P = (x, y)$ on the unit circle, called **the point on the unit circle that corresponds to t**. The coordinates of the point $P = (x, y)$ are used to define the *six trigonometric functions of t*.

DEFINITION Trigonometric Functions of a Number t

Let t be a real number and let $P = (x, y)$ be the point on the unit circle that corresponds to t.

- The **sine function** is defined as
$$\sin t = y$$

- The **cosine function** is defined as
$$\cos t = x$$

- If $x \neq 0$, the **tangent function** is defined as
$$\tan t = \frac{y}{x} \qquad x \neq 0$$

- If $y \neq 0$, the **cosecant function** is defined as
$$\csc t = \frac{1}{y} \qquad y \neq 0$$

- If $x \neq 0$, the **secant function** is defined as
$$\sec t = \frac{1}{x} \qquad x \neq 0$$

- If $y \neq 0$, the **cotangent function** is defined as
$$\cot t = \frac{x}{y} \qquad y \neq 0$$

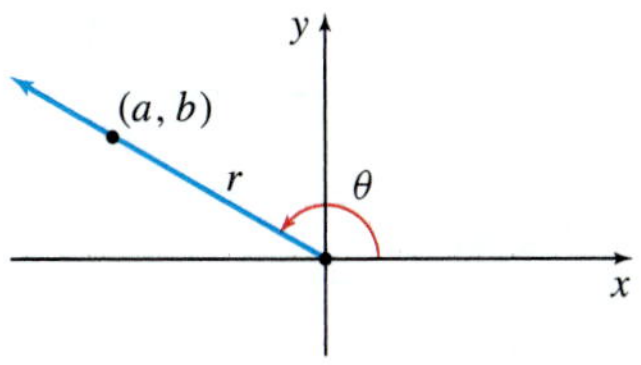

Figure 55

DEFINITION **Trigonometric Functions of θ**

Let θ be any angle in standard position and let $P = (a, b)$ be any point, except the origin, on the terminal side of θ. If $r = \sqrt{a^2 + b^2}$ denotes the distance of the point P from the origin, then the **six trigonometric functions of θ** are defined as the ratios

$$\sin\theta = \frac{b}{r} \qquad \cos\theta = \frac{a}{r} \qquad \tan\theta = \frac{b}{a}$$

$$\csc\theta = \frac{r}{b} \qquad \sec\theta = \frac{r}{a} \qquad \cot\theta = \frac{a}{b}$$

provided no denominator equals 0. If a denominator equals 0, that trigonometric function of the angle θ is not defined.

Notice in these definitions that, if $x = 0$, that is, if the point P is on the y-axis, then the tangent function and the secant function are not defined. Also, if $y = 0$, that is, if the point P is on the x-axis, then the cosecant function and the cotangent function are not defined.

Because the unit circle is used in these definitions, the trigonometric functions are sometimes called **circular functions**.

Suppose θ is an angle in standard position, whose terminal side is the ray from the origin through point P. Since the unit circle has radius $r = 1$ unit, then, using the formula for arc length, we find that

$$s = r\theta = \theta$$

So, if $s = |t|$ units, then $\theta = t$ radians and the trigonometric functions of the angle θ are defined as

$$\sin\theta = \sin t \quad \cos\theta = \cos t \quad \tan\theta = \tan t$$

$$\csc\theta = \csc t \quad \sec\theta = \sec t \quad \cot\theta = \cot t$$

Evaluating Trigonometric Functions

Using properties of right triangles, we can find the values of the six trigonometric functions of $\frac{\pi}{6} = 30°$, $\frac{\pi}{4} = 45°$, and $\frac{\pi}{3} = 60°$ given in Table 6.

TABLE 6

θ (Radians)	θ (Degrees)	$\sin\theta$	$\cos\theta$	$\tan\theta$	$\csc\theta$	$\sec\theta$	$\cot\theta$
$\dfrac{\pi}{6}$	$30°$	$\dfrac{1}{2}$	$\dfrac{\sqrt{3}}{2}$	$\dfrac{\sqrt{3}}{3}$	2	$\dfrac{2\sqrt{3}}{3}$	$\sqrt{3}$
$\dfrac{\pi}{4}$	$45°$	$\dfrac{\sqrt{2}}{2}$	$\dfrac{\sqrt{2}}{2}$	1	$\sqrt{2}$	$\sqrt{2}$	1
$\dfrac{\pi}{3}$	$60°$	$\dfrac{\sqrt{3}}{2}$	$\dfrac{1}{2}$	$\sqrt{3}$	$\dfrac{2\sqrt{3}}{3}$	2	$\dfrac{\sqrt{3}}{3}$

The values of the trigonometric functions at the quadrantal angles, $\left(0 = 0°,\right.$ $\frac{\pi}{2} = 90°$, $\frac{3\pi}{2} = 270°$, and $2\pi = 360°\left.\right)$ are given in Table 7.

TABLE 7

Radians	Degrees	$\sin\theta$	$\cos\theta$	$\tan\theta$	$\csc\theta$	$\sec\theta$	$\cot\theta$
0	$0°$	0	1	0	Not defined	1	Not defined
$\dfrac{\pi}{2}$	$90°$	1	0	Not defined	1	Not defined	0
π	$180°$	0	-1	0	Not defined	-1	Not defined
$\dfrac{3\pi}{2}$	$270°$	-1	0	Not defined	-1	Not defined	0
2π	$360°$	0	1	0	Not defined	1	Not defined

Table 8 lists the signs of the six trigonometric functions for each quadrant.

TABLE 8

Quadrant	$\sin \theta$, $\csc \theta$	$\cos \theta$, $\sec \theta$	$\tan \theta$, $\cot \theta$	Conclusion
I	Positive	Positive	Positive	All trigonometric functions are positive.
II	Positive	Negative	Negative	Only sine and its reciprocal, cosecant, are positive.
III	Negative	Negative	Positive	Only tangent and its reciprocal, cotangent, are positive.
IV	Negative	Positive	Negative	Only cosine and its reciprocal, secant, are positive.

For nonacute angles θ that lie in a quadrant, the acute angle formed by the terminal side of θ and the x-axis, called a **reference angle**, is often used.

Figure 57 illustrates the reference angle for some general angles θ. Note that a reference angle is always an acute angle.

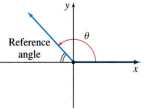 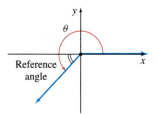 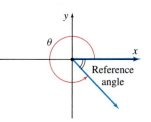 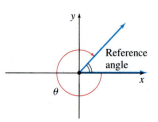

Figure 57

Although there are formulas for calculating reference angles, usually it is easier to find the reference angle for a given angle by making a quick **sketch** of the angle. The advantage of using reference angles is that, except for the correct sign, the values of the trigonometric functions of a general angle θ equal the values of the trigonometric functions of its reference angle.

THEOREM Reference Angles

If θ is an angle that lies in a quadrant and if A is its reference angle, then

$$\sin \theta = \pm \sin A \qquad \cos \theta = \pm \cos A \qquad \tan \theta = \pm \tan A$$
$$\csc \theta = \pm \csc A \qquad \sec \theta = \pm \sec A \qquad \cot \theta = \pm \cot A$$

where the $+$ or $-$ sign depends on the quadrant in which θ lies.

EXAMPLE 1   Using Reference Angles

Find the exact value of: **(a)** $\cos \dfrac{17\pi}{6}$ **(b)** $\tan \left(-\dfrac{\pi}{3} \right)$

Solution (a) Refer to Figure 58(a) on page A-31. The angle $\dfrac{17\pi}{6}$ is in quadrant II, where the cosine function is negative. The reference angle for $\dfrac{17\pi}{6}$ is $\dfrac{\pi}{6}$. Since $\cos \dfrac{\pi}{6} = \dfrac{\sqrt{3}}{2}$,

$$\cos \frac{17\pi}{6} = -\cos \frac{\pi}{6} = -\frac{\sqrt{3}}{2}$$

(b) Refer to Figure 58(b). The angle $-\dfrac{\pi}{3}$ is in quadrant IV, where the tangent function is negative. The reference angle for $-\dfrac{\pi}{3}$ is $\dfrac{\pi}{3}$. Since $\tan \dfrac{\pi}{3} = \sqrt{3}$,

$$\tan \left(-\frac{\pi}{3} \right) = -\tan \frac{\pi}{3} = -\sqrt{3}$$

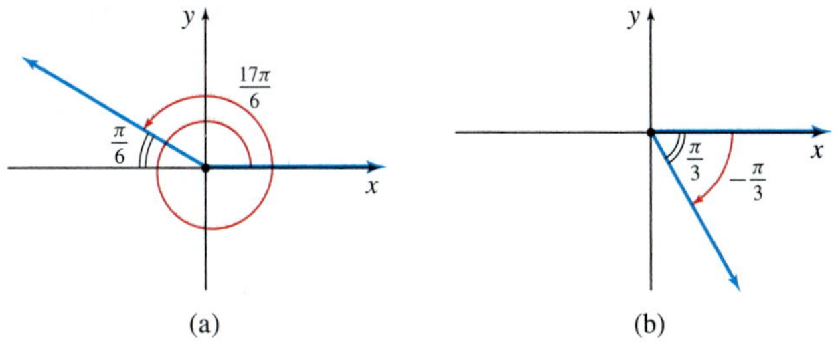

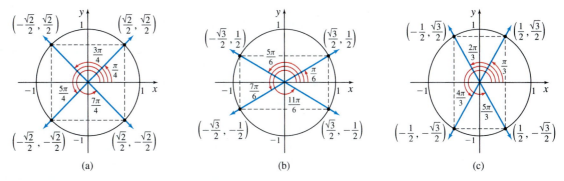

(a)   (b)

Figure 58

The use of symmetry provides information about integer multiples of the angles $\dfrac{\pi}{4} = 45°$, $\dfrac{\pi}{6} = 30°$, and $\dfrac{\pi}{3} = 60°$. See the unit circles in Figure 59(a)–(c).

(a)   (b)   (c)

Figure 59

EXAMPLE 2   Finding the Exact Values of a Trigonometry Function

Use reference angles or the symmetry shown in Figure 59 to obtain:

(a) $\sin \dfrac{7\pi}{4} = -\dfrac{\sqrt{2}}{2}$ (b) $\cos \dfrac{7\pi}{6} = -\dfrac{\sqrt{3}}{2}$ (c) $\tan \dfrac{2\pi}{3} = \dfrac{\dfrac{\sqrt{3}}{2}}{-\dfrac{1}{2}} = -\sqrt{3}$

(d) $\cos \left(-\dfrac{3\pi}{4}\right) = -\dfrac{\sqrt{2}}{2}$ (e) $\sin \left(-\dfrac{\pi}{6}\right) = -\dfrac{1}{2}$ (f) $\sin \dfrac{7\pi}{3} = \dfrac{\sqrt{3}}{2}$ ■

3 Determine the Domain and the Range of the Trigonometric Functions

Suppose θ is an angle in standard position and $P = (x, y)$ is the point on the unit circle corresponding to θ. Then the six trigonometric functions are

$$\sin \theta = y \qquad \cos \theta = x \qquad \tan \theta = \dfrac{y}{x}$$
$$\csc \theta = \dfrac{1}{y} \qquad \sec \theta = \dfrac{1}{x} \qquad \cot \theta = \dfrac{x}{y}$$

For $\sin \theta$ and $\cos \theta$, there is no concern about dividing by 0, so θ can be any angle. It follows that the domain of the sine and cosine functions is all real numbers.

- The domain of the sine function is all real numbers.
- The domain of the cosine function is all real numbers.

For the tangent and secant functions, x cannot be 0 since this results in division by 0. On the unit circle, there are two such points, $(0, 1)$ and $(0, -1)$. These two points correspond to $\dfrac{\pi}{2} = 90°$ and $\dfrac{3\pi}{2} = 270°$ or, more generally, to any angle that is an odd integer multiple of $\dfrac{\pi}{2}$, such as $\pm \dfrac{\pi}{2} = \pm 90°$, $\pm \dfrac{3\pi}{2} = \pm 270°$, $\pm \dfrac{5\pi}{2} = \pm 450°$, and so on. These angles must be excluded from the domain of the tangent function and the domain of the secant function.

- The domain of the tangent function is all real numbers, except odd integer multiples of $\dfrac{\pi}{2} = 90°$.
- The domain of the secant function is all real numbers, except odd integer multiples of $\dfrac{\pi}{2} = 90°$.

For the cotangent and cosecant functions, y cannot be 0 since this results in division by 0. On the unit circle, there are two such points, $(1, 0)$ and $(-1, 0)$. These two points correspond to $0 = 0°$ and $\pi = 180°$ or, more generally, to any angle that is an integer multiple of π, such as $0 = 0°$, $\pm\pi = \pm180°$, $\pm2\pi = \pm360°$, $\pm3\pi = \pm540°$, and so on. These angles must be excluded from the domain of the cotangent function and the domain of the cosecant function.

- The domain of the cotangent function is all real numbers, except integer multiples of $\pi = 180°$.
- The domain of the cosecant function is all real numbers, except integer multiples of $\pi = 180°$.

Now we investigate the range of the six trigonometric functions. If $P = (x, y)$ is the point on the unit circle corresponding to the angle θ, then it follows that $-1 \le x \le 1$ and $-1 \le y \le 1$. Since $\sin\theta = y$ and $\cos\theta = x$, we have

$$-1 \le \sin\theta \le 1 \qquad \text{and} \qquad -1 \le \cos\theta \le 1$$

The range of both the sine function and the cosine function is the closed interval $[-1, 1]$. Using absolute value notation, we have $|\sin\theta| \le 1$ and $|\cos\theta| \le 1$.

If θ is not an integer multiple of $\pi = 180°$, then $\csc\theta = \dfrac{1}{y}$. Since $y = \sin\theta$ and $|y| = |\sin\theta| \le 1$, it follows that $|\csc\theta| = \dfrac{1}{|\sin\theta|} = \dfrac{1}{|y|} \ge 1$. That is, $\dfrac{1}{y} \le -1$ or $\dfrac{1}{y} \ge 1$. Since $\csc\theta = \dfrac{1}{y}$, the range of the cosecant functions consists of all real numbers in the set $(-\infty, -1] \cup [1, \infty)$. That is,

$$\csc\theta \le -1 \qquad \text{or} \qquad \csc\theta \ge 1$$

If θ is not an odd integer multiple of $\dfrac{\pi}{2} = 90°$, then $\sec\theta = \dfrac{1}{x}$. Since $x = \cos\theta$ and $|x| = |\cos\theta| \le 1$, it follows that $|\sec\theta| = \dfrac{1}{|\cos\theta|} = \dfrac{1}{|x|} \ge 1$. That is, $\dfrac{1}{x} \le -1$ or $\dfrac{1}{x} \ge 1$. Since $\sec\theta = \dfrac{1}{x}$, the range of the secant function consists of all real numbers in the set $(-\infty, -1] \cup [1, \infty)$. That is,

$$\sec\theta \le -1 \qquad \text{or} \qquad \sec\theta \ge 1$$

The range of both the tangent function and the cotangent function is all real numbers:

$$-\infty < \tan\theta < \infty \qquad \text{and} \qquad -\infty < \cot\theta < \infty$$

4 Use Basic Trigonometry Identities

DEFINITION Identity

Two expressions a and b involving a variable x are **identically equal** if

$$a = b$$

for every value of x for which both expressions are defined. Such an equation is referred to as an **identity**. An equation that is not an identity is called a **conditional equation**.

For example, the equation $2x + 3 = x$ is a conditional equation because it is true only for $x = -3$. However, the equation $x^2 + 2x = (x + 1)^2 - 1$ is true for any value of x, so it is an identity.

The basic trigonometry identities listed below are consequences of the definition of the six trigonometric functions.

Basic Trigonometry Identities

- **Quotient Identities**

$$\tan \theta = \frac{\sin \theta}{\cos \theta} \qquad \cot \theta = \frac{\cos \theta}{\sin \theta}$$

- **Reciprocal Identities**

$$\csc \theta = \frac{1}{\sin \theta} \qquad \sec \theta = \frac{1}{\cos \theta} \qquad \cot \theta = \frac{1}{\tan \theta}$$

- **Pythagorean Identities**

$$\sin^2 \theta + \cos^2 \theta = 1 \qquad \tan^2 \theta + 1 = \sec^2 \theta \qquad \cot^2 \theta + 1 = \csc^2 \theta$$

- **Even/Odd Identities**

$$\cos(-\theta) = \cos \theta \qquad \sin(-\theta) = -\sin \theta \qquad \tan(-\theta) = -\tan \theta$$

EXAMPLE 3   Using Basic Trigonometry Identities

(a) $\tan \dfrac{\pi}{9} - \dfrac{\sin \dfrac{\pi}{9}}{\cos \dfrac{\pi}{9}} = \tan \dfrac{\pi}{9} - \tan \dfrac{\pi}{9} = 0$

$\uparrow$ $\dfrac{\sin \theta}{\cos \theta} = \tan \theta$

(b) $\sin^2 \dfrac{\pi}{12} + \dfrac{1}{\sec^2 \dfrac{\pi}{12}} = \sin^2 \dfrac{\pi}{12} + \cos^2 \dfrac{\pi}{12} = 1$

$\uparrow$ $\cos \theta = \dfrac{1}{\sec \theta}$ $\uparrow$ $\sin^2 \theta + \cos^2 \theta = 1$

EXAMPLE 4   Using Basic Trigonometry Identities

Given that $\sin \theta = \dfrac{1}{3}$ and $\cos \theta < 0$, find the exact value of each of the remaining five trigonometric functions.

Solution We begin by solving the identity $\sin^2 \theta + \cos^2 \theta = 1$ for $\cos \theta$.

$$\sin^2 \theta + \cos^2 \theta = 1$$
$$\cos^2 \theta = 1 - \sin^2 \theta$$
$$\cos \theta = \pm\sqrt{1 - \sin^2 \theta}$$

Because $\cos \theta < 0$, we choose the negative value of the radical, and use the fact that $\sin \theta = \dfrac{1}{3}$.

$$\cos \theta = -\sqrt{1 - \sin^2 \theta} = -\sqrt{1 - \frac{1}{9}} = -\sqrt{\frac{8}{9}} = -\frac{2\sqrt{2}}{3}$$

$\uparrow$ $\sin \theta = \dfrac{1}{3}$

The values of $\sin\theta$ and $\cos\theta$ are now known, so we use the quotient and reciprocal identities to obtain

$$\tan\theta = \frac{\sin\theta}{\cos\theta} = \frac{\frac{1}{3}}{-\frac{2\sqrt{2}}{3}} = -\frac{1}{2\sqrt{2}} = -\frac{\sqrt{2}}{4} \qquad \cot\theta = \frac{1}{\tan\theta} = -2\sqrt{2}$$

$$\sec\theta = \frac{1}{\cos\theta} = \frac{1}{-\frac{2\sqrt{2}}{3}} = -\frac{3}{2\sqrt{2}} = -\frac{3\sqrt{2}}{4} \qquad \csc\theta = \frac{1}{\sin\theta} = \frac{1}{\frac{1}{3}} = 3 \qquad \blacksquare$$

5 Use Sum and Difference, Double-Angle and Half-Angle, and Sum-to-Product and Product-to-Sum Formulas

THEOREM Sum and Difference Formulas for Sine, Cosine, and Tangent

- $\sin(A + B) = \sin A \cos B + \cos A \sin B$
- $\sin(A - B) = \sin A \cos B - \cos A \sin B$
- $\cos(A + B) = \cos A \cos B - \sin A \sin B$
- $\cos(A - B) = \cos A \cos B + \sin A \sin B$
- $\tan(A + B) = \dfrac{\tan A + \tan B}{1 - \tan A \tan B}$
- $\tan(A - B) = \dfrac{\tan A - \tan B}{1 + \tan A \tan B}$

EXAMPLE 5   Using Trigonometry Identities

If $\sin A = \dfrac{4}{5}$, $\dfrac{\pi}{2} < A < \pi$, and $\sin B = -\dfrac{2}{\sqrt{5}} = -\dfrac{2\sqrt{5}}{5}$, $\pi < B < \dfrac{3\pi}{2}$, find the exact value of:

(a) $\cos A$ **(b)** $\cos B$ **(c)** $\cos(A + B)$ **(d)** $\sin(A + B)$

Solution (a) We use a Pythagorean identity to find $\cos A$.

$$\cos A = -\sqrt{1 - \sin^2 A} = -\sqrt{1 - \left(\frac{4}{5}\right)^2} = -\sqrt{1 - \frac{16}{25}} = -\sqrt{\frac{9}{25}} = -\frac{3}{5}$$

$\uparrow$
A is in quadrant II
$\cos A < 0$

(b) We also find $\cos B$ using a Pythagorean identity.

$$\cos B = -\sqrt{1 - \sin^2 B} = -\sqrt{1 - \frac{4}{5}} = -\sqrt{\frac{1}{5}} = -\frac{\sqrt{5}}{5}$$

(c) We use the results from (a) and (b) and a sum formula to find $\cos(A + B)$.

$$\cos(A + B) = \cos A \cos B - \sin A \sin B$$

$$= \left(-\frac{3}{5}\right)\left(-\frac{\sqrt{5}}{5}\right) - \left(\frac{4}{5}\right)\left(-\frac{2\sqrt{5}}{5}\right) = \frac{11\sqrt{5}}{25}$$

(d) We use a sum formula to find $\sin(A + B)$.

$$\sin(A + B) = \sin A \cos B + \cos A \sin B = \left(\frac{4}{5}\right)\left(-\frac{\sqrt{5}}{5}\right) + \left(-\frac{3}{5}\right)\left(-\frac{2\sqrt{5}}{5}\right) = \frac{2\sqrt{5}}{25}$$

$\blacksquare$

THEOREM Double-Angle Formulas

• $\sin(2\theta) = 2\sin\theta\cos\theta$	• $\cos(2\theta) = \cos^2\theta - \sin^2\theta$	• $\tan(2\theta) = \dfrac{2\tan\theta}{1 - \tan^2\theta}$
• $\cos(2\theta) = 1 - 2\sin^2\theta$	• $\cos(2\theta) = 2\cos^2\theta - 1$	

The double-angle formulas in the first row are derived directly from the sum formulas by letting $\theta = A = B$. A Pythagorean identity is used to obtain those in the second row.

By rearranging the double-angle formulas, other identities can be obtained that are used in calculus.

$$\bullet\ \sin^2\theta = \frac{1 - \cos(2\theta)}{2} \qquad \bullet\ \cos^2\theta = \frac{1 + \cos(2\theta)}{2}$$

If, in the double-angle formulas, we replace θ with $\dfrac{1}{2}\theta$, we obtain these identities.

$$\bullet\ \sin^2\frac{\theta}{2} = \frac{1 - \cos\theta}{2} \qquad \bullet\ \cos^2\frac{\theta}{2} = \frac{1 + \cos\theta}{2} \qquad \bullet\ \tan^2\frac{\theta}{2} = \frac{1 - \cos\theta}{1 + \cos\theta}$$

Taking the square root produces the **half-angle formulas**.

$$\bullet\ \sin\frac{\theta}{2} = \pm\sqrt{\frac{1 - \cos\theta}{2}} \qquad \bullet\ \cos\frac{\theta}{2} = \pm\sqrt{\frac{1 + \cos\theta}{2}} \qquad \bullet\ \tan\frac{\theta}{2} = \pm\sqrt{\frac{1 - \cos\theta}{1 + \cos\theta}}$$

where the $+$ or $-$ sign is determined by the quadrant of the angle $\dfrac{\theta}{2}$.

The **product-to-sum and sum-to-product identities** are also used in calculus. They are a result of combining the sum and difference formulas.

$$\bullet\ \sin A\, \cos B = \frac{1}{2}[\sin(A + B) + \sin(A - B)]$$

$$\bullet\ \cos A\, \cos B = \frac{1}{2}[\cos(A + B) + \cos(A - B)]$$

$$\bullet\ \sin B\, \sin B = \frac{1}{2}[\cos(A - B) - \cos(A + B)]$$

6 Solve Triangles Using the Law of Sines and the Law of Cosines

If no angle of a triangle is a right angle, the triangle is called **oblique**. An oblique triangle either has three acute angles or two acute angles and one obtuse angle (an angle between $\dfrac{\pi}{2}$ and π). See Figure 60.

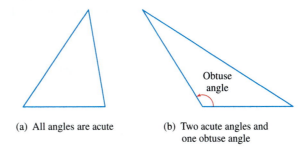

(a) All angles are acute   (b) Two acute angles and one obtuse angle

Figure 60 Two oblique triangles.

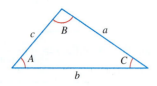

Figure 61

NOTE To solve an oblique triangle, you must know the length of at least one side. Knowing only angles results in a family of similar triangles.

In the discussion that follows, we label an oblique triangle so that side a is opposite angle A, side b is opposite angle B, and side c is opposite angle C, as shown in Figure 61.

To **solve an oblique triangle** means to find the lengths of its sides and the measures of its angles. To do this, we need to know the length of one side along with (i) two angles; (ii) one angle and one other side; or (iii) the other two sides. There are four possibilities to consider:

Case 1: One side and two angles are known (ASA or SAA).
Case 2: Two sides and the angle opposite one of them are known (SSA).
Case 3: Two sides and the included angle are known (SAS).
Case 4: Three sides are known (SSS).

Figure 62 illustrates the four cases.

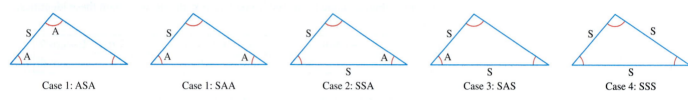

Case 1: ASA   Case 1: SAA   Case 2: SSA   Case 3: SAS   Case 4: SSS

Figure 62

The *Law of Sines* is used to solve triangles for which Case 1 or 2 holds.

THEOREM Law of Sines

For a triangle with sides a, b, c and opposite angles A, B, C, respectively,

$$\frac{\sin A}{a} = \frac{\sin B}{b} = \frac{\sin C}{c} \tag{1}$$

The Law of Sines actually consists of three equalities:

$$\frac{\sin A}{a} = \frac{\sin B}{b} \qquad \frac{\sin A}{a} = \frac{\sin C}{c} \qquad \frac{\sin B}{b} = \frac{\sin C}{c}$$

Formula (1) is a compact way to write these three equations.

For Case 1 (ASA or SAA), two angles are given. The third angle can be found using the fact that the sum of the angles in a triangle measures $180°$. Then use the Law of Sines (twice) to find the two missing sides.

Case 2 (SSA), which applies to triangles for which two sides and the angle opposite one of them are known, is referred to as the **ambiguous case**, because the information can result in one triangle, two triangles, or no triangle at all.

EXAMPLE 6   **Using the Law of the Sines to Solve a Triangle**

Solve the triangle with side $a = 6$, side $b = 8$, and angle $A = 35°$, which is opposite side a.

Solution Because two sides and an opposite angle are known (SSA), we use the Law of Sines to find angle B.

$$\frac{\sin A}{a} = \frac{\sin B}{b}$$

Since $a = 6$, $b = 8$, and $A = 35°$, we have

$$\frac{\sin 35°}{6} = \frac{\sin B}{8}$$

$$\sin B = \frac{8 \sin 35°}{6} \approx 0.765$$

$$B_1 \approx 49.9° \quad \text{or} \quad B_2 \approx 180° - 49.9° = 130.1°$$

For both choices of B, we have $A + B < 180°$. So, there are two triangles, one containing the angle $B_1 \approx 49.9°$ and the other containing the angle $B_2 \approx 130.1°$. The third angle C is either

$$C_1 = 180° - A - B_1 \approx 95.1° \quad \text{or} \quad C_2 = 180° - A - B \approx 14.9°$$

$$\uparrow \qquad\qquad\qquad\qquad\qquad\qquad \uparrow$$

$$A = 35° \qquad\qquad\qquad\qquad\qquad A = 35°$$
$$B_1 = 49.9° \qquad\qquad\qquad\qquad B_2 = 130.1°$$

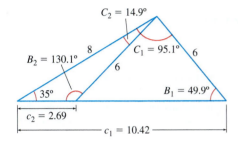

Figure 63

The third side c satisfies the Law of Sines, so

$$\frac{\sin A}{a} = \frac{\sin C_1}{c_1} \qquad\qquad \frac{\sin A}{a} = \frac{\sin C_2}{c_2}$$

$$\frac{\sin 35°}{6} = \frac{\sin 95.1°}{c_1} \qquad\qquad \frac{\sin 35°}{6} = \frac{\sin 14.9°}{c_2}$$

$$c_1 = \frac{6 \sin 95.1°}{\sin 35°} \approx 10.42 \qquad c_2 = \frac{6 \sin 14.9°}{\sin 35°} \approx 2.69$$

The two solved triangles are illustrated in Figure 63. ∎

Cases 3 and 4 are solved using the *Law of Cosines*.

THEOREM Law of Cosines

For a triangle with sides a, b, c and opposite angles A, B, C, respectively,

> • $c^2 = a^2 + b^2 - 2ab \cos C$
> • $b^2 = a^2 + c^2 - 2ac \cos B$
> • $a^2 = b^2 + c^2 - 2bc \cos A$

NOTE If the triangle is a right triangle, the Law of Cosines reduces to the Pythagorean Theorem.

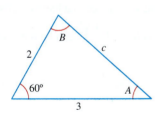

Figure 64

EXAMPLE 7   Using the Law of Cosines to Solve a Triangle

Solve the triangle: $a = 2$, $b = 3$, $C = 60°$ shown in Figure 64.

Solution Because two sides a and b and the included angle $C = 60°$ (SAS) are known, we use the Law of Cosines to find the third side c.

$$c^2 = a^2 + b^2 - 2ab \cos C = 2^2 + 3^2 - 2 \cdot 2 \cdot 3 \cdot \cos 60° = 13 - \left(12 \cdot \frac{1}{2}\right) = 7$$

$$c = \sqrt{7}$$

Side c is of length $\sqrt{7}$. To find either the angle A or B, use the Law of Cosines. For A,

$$a^2 = b^2 + c^2 - 2bc \cos A$$

$$2bc \cos A = b^2 + c^2 - a^2$$

$$\cos A = \frac{b^2 + c^2 - a^2}{2bc} = \frac{9 + 7 - 4}{2 \cdot 3\sqrt{7}} = \frac{12}{6\sqrt{7}} = \frac{2\sqrt{7}}{7}$$

$$A \approx 40.9°$$

Then to find the third angle, use the fact that the sum of the angles of a triangle, when measured in degrees, equals $180°$. That is,

$$40.9° + B + 60° = 180°$$

$$B = 79.1° \qquad\qquad\qquad ∎$$

A.5 Sequences; Summation Notation; the Binomial Theorem

OBJECTIVES *When you finish this section you should be able to:*

1 Write the first several terms of a sequence (p. A-38)
2 Write the terms of a recursively defined sequence (p. A-39)
3 Use summation notation (p. A-40)
4 Find the sum of the first n terms of a sequence (p. A-41)
5 Use the Binomial Theorem (p. A-42)

1 Write the First Several Terms of a Sequence

DEFINITION Sequence

A **sequence** is a rule that assigns a real number to each positive integer.

A sequence is often represented by listing its values in order. For example, the sequence whose rule is to assign to each positive integer its reciprocal may be represented as

$$s_1 = 1, \quad s_2 = \frac{1}{2}, \quad s_3 = \frac{1}{3}, \quad s_4 = \frac{1}{4}, \dots, \quad s_n = \frac{1}{n}, \dots$$

or merely as the list

$$1, \quad \frac{1}{2}, \quad \frac{1}{3}, \quad \frac{1}{4}, \quad \dots, \quad \frac{1}{n}, \quad \dots$$

The list never ends, as the dots (**ellipsis**) indicate. The real numbers in this ordered list are called the **terms** of the sequence.

Notice that when we use subscripted letters to represent the terms of a sequence, s_1 is the 1st term, s_2 is the 2nd term, $\dots$, s_n is the nth term, and so on. If we have a rule for the nth term, as in the sequence above, it is used to identify the sequence. We represent such a sequence by enclosing the nth term in braces. The notation $\{s_n\} = \left\{ \dfrac{1}{n} \right\}$

or $\{s_n\}_{n=1}^{\infty} = \left\{ \dfrac{1}{n} \right\}_{n=1}^{\infty}$ both represent the sequence $1, \dfrac{1}{2}, \dfrac{1}{3}, \dfrac{1}{4}, \dfrac{1}{5}, \dots$.

For simplicity, we use the notation $\{s_n\}$ to represent a sequence that has a rule for the nth term. For example, the sequence whose nth term is $b_n = \left(\dfrac{1}{2} \right)^n$ can be represented either as

$$b_1 = \frac{1}{2}, \quad b_2 = \frac{1}{4}, \quad b_3 = \frac{1}{8}, \dots, \quad b_n = \frac{1}{2^n}, \dots,$$

or as

$$\{b_n\} = \left\{ \left(\frac{1}{2} \right)^n \right\}$$

EXAMPLE 1   Writing the Terms of a Sequence

Write the first six terms of the sequence: $\{b_n\} = \left\{ (-1)^{n-1} \dfrac{2}{n} \right\}$

Solution $b_1 = (-1)^0 \dfrac{2}{1} = 2 \qquad b_2 = (-1)^1 \dfrac{2}{2} = -1 \qquad b_3 = (-1)^2 \dfrac{2}{3} = \dfrac{2}{3}$

$$b_4 = -\frac{1}{2} \qquad\qquad b_5 = \frac{2}{5} \qquad\qquad b_6 = -\frac{1}{3}$$

The first six terms of the sequence are $2, -1, \dfrac{2}{3}, -\dfrac{1}{2}, \dfrac{2}{5}, -\dfrac{1}{3}$. ∎

EXAMPLE 2   **Determining a Sequence from a Pattern**

The pattern on the left suggests the nth term of the sequence on the right.

Pattern	nth term	Sequence
$e, \dfrac{e^2}{2}, \dfrac{e^3}{3}, \dfrac{e^4}{4}, \ldots$	$a_n = \dfrac{e^n}{n}$	$\{a_n\} = \left\{\dfrac{e^n}{n}\right\}$
$1, \dfrac{1}{3}, \dfrac{1}{9}, \dfrac{1}{27}, \ldots$	$b_n = \dfrac{1}{3^{n-1}}$	$\{b_n\} = \left\{\dfrac{1}{3^{n-1}}\right\}$
$1, 3, 5, 7, \ldots$	$c_n = 2n - 1$	$\{c_n\} = \{2n - 1\}$
$1, 4, 9, 16, 25, \ldots$	$d_n = n^2$	$\{d_n\} = \{n^2\}$
$1, -\dfrac{1}{2}, \dfrac{1}{3}, -\dfrac{1}{4}, \dfrac{1}{5}, \ldots$	$e_n = (-1)^{n+1}\left(\dfrac{1}{n}\right)$	$\{e_n\} = \left\{(-1)^{n+1}\left(\dfrac{1}{n}\right)\right\}$

2 Write the Terms of a Recursively Defined Sequence

A second way of defining a sequence is to assign a value to the first (or the first few) term(s) and specify the nth term by a formula or an equation that involves one or more of the terms preceding it. Such a sequence is defined **recursively**, and the rule or formula is called a **recursive formula**.

EXAMPLE 3   **Writing the Terms of a Recursively Defined Sequence**

Write the first five terms of the recursively defined sequence.

$$s_1 = 1 \qquad s_n = 4s_{n-1} \qquad n \geq 2$$

Solution The first term is given as $s_1 = 1$. To get the second term, we use $n = 2$ in the formula to obtain $s_2 = 4s_1 = 4 \cdot 1 = 4$. To obtain the third term, we use $n = 3$ in the formula, getting $s_3 = 4s_2 = 4 \cdot 4 = 16$. Each new term requires that we know the value of the preceding term. The first five terms are

$$s_1 = 1$$
$$s_2 = 4s_{2-1} = 4s_1 = 4 \cdot 1 = 4$$
$$s_3 = 4s_{3-1} = 4s_2 = 4 \cdot 4 = 16$$
$$s_4 = 4s_{4-1} = 4s_3 = 4 \cdot 16 = 64$$
$$s_5 = 4s_{5-1} = 4s_4 = 4 \cdot 64 = 256$$

EXAMPLE 4   **Writing the Terms of a Recursively Defined Sequence**

Write the first five terms of the recursively defined sequence.

$$u_1 = 1 \qquad u_2 = 1 \qquad u_n = u_{n-2} + u_{n-1} \qquad n \geq 3$$

Solution We are given the first two terms. To obtain the third term requires that we know each of the previous two terms.

$$u_1 = 1$$
$$u_2 = 1$$
$$u_3 = u_1 + u_2 = 1 + 1 = 2$$
$$u_4 = u_2 + u_3 = 1 + 2 = 3$$
$$u_5 = u_3 + u_4 = 2 + 3 = 5$$

The sequence defined in Example 4 is called a **Fibonacci sequence**; its terms are called **Fibonacci numbers**. These numbers appear in a wide variety of applications.

3 Use Summation Notation

It is often important to be able to find the sum of the first n terms of a sequence $\{a_n\}$, namely,

$$a_1 + a_2 + a_3 + \cdots + a_n$$

Rather than write down all these terms, we introduce **summation notation**. Using summation notation, the sum is written

$$a_1 + a_2 + a_3 + \cdots + a_n = \sum_{k=1}^{n} a_k$$

The symbol $\sum$ (the uppercase Greek letter sigma, which is an S in our alphabet) is simply an instruction to sum, or add up, the terms. The integer k is called the **index of summation**; it tells us where to start the sum ($k = 1$) and where to end it ($k = n$). The expression $\sum_{k=1}^{n} a_k$ is read, "The sum of the terms a_k from $k = 1$ to $k = n$."

EXAMPLE 5 Expanding Summation Notation

Write out each sum:

(a) $\displaystyle\sum_{k=1}^{n} \frac{1}{k}$ (b) $\displaystyle\sum_{k=1}^{n} k!$

Solution (a) $\displaystyle\sum_{k=1}^{n} \frac{1}{k} = 1 + \frac{1}{2} + \frac{1}{3} + \cdots + \frac{1}{n}$ (b) $\displaystyle\sum_{k=1}^{n} k! = 1! + 2! + \cdots + n!$ ∎

EXAMPLE 6 Writing a Sum in Summation Notation

Express each sum using summation notation:

(a) $1^2 + 2^2 + 3^2 + \cdots + 9^2$ (b) $1 + \frac{1}{2} + \frac{1}{4} + \frac{1}{8} + \cdots + \frac{1}{2^{n-1}}$

Solution (a) The sum $1^2 + 2^2 + 3^2 + \cdots + 9^2$ has 9 terms, each of the form k^2; it starts at $k = 1$ and ends at $k = 9$:

$$1^2 + 2^2 + 3^2 + \cdots + 9^2 = \sum_{k=1}^{9} k^2$$

(b) The sum

$$1 + \frac{1}{2} + \frac{1}{4} + \frac{1}{8} + \cdots + \frac{1}{2^{n-1}}$$

has n terms, each of the form $\frac{1}{2^{k-1}}$. It starts at $k = 1$ and ends at $k = n$.

$$1 + \frac{1}{2} + \frac{1}{4} + \frac{1}{8} + \cdots + \frac{1}{2^{n-1}} = \sum_{k=1}^{n} \frac{1}{2^{k-1}}$$ ∎

The index of summation does not need to begin at 1 or end at n. For example, the sum in Example 6(b) could be expressed as

$$\sum_{k=0}^{n-1} \frac{1}{2^k} = 1 + \frac{1}{2} + \frac{1}{4} + \frac{1}{8} + \cdots + \frac{1}{2^{n-1}}$$

Letters other than k can be used as the index. For example,

$$\sum_{j=1}^{n} j! \quad \text{and} \quad \sum_{i=1}^{n} i!$$

each represent the same sum as Example 5(b).

4 Find the Sum of the First n Terms of a Sequence

When working with summation notation, the following properties are useful for finding the sum of the first n terms of a sequence.

THEOREM Properties Involving Summation Notation

If $\{a_n\}$ and $\{b_n\}$ are two sequences and c is a real number, then:

$$\bullet \quad \sum_{k=1}^{n}(ca_k) = c\sum_{k=1}^{n}a_k \tag{1}$$

$$\bullet \quad \sum_{k=1}^{n}(a_k + b_k) = \sum_{k=1}^{n}a_k + \sum_{k=1}^{n}b_k \tag{2}$$

$$\bullet \quad \sum_{k=1}^{n}(a_k - b_k) = \sum_{k=1}^{n}a_k - \sum_{k=1}^{n}b_k \tag{3}$$

$$\bullet \quad \sum_{k=j+1}^{n}a_k = \sum_{k=1}^{n}a_k - \sum_{k=1}^{j}a_k, \quad \text{where } 0 < j < n \tag{4}$$

The proof of property (1) follows from the distributive property of real numbers. The proofs of properties (2) and (3) are based on the commutative and associative properties of real numbers. Property (4) states that the sum from $j + 1$ to n equals the sum from 1 to n minus the sum from 1 to j. It can be helpful to use this property when the index of summation begins at a number larger than 1.

THEOREM Formulas for Sums of the First n Terms of a Sequence

$$\bullet \quad \sum_{k=1}^{n}1 = \underbrace{1 + 1 + 1 + \cdots + 1}_{n \text{ terms}} = n \tag{5}$$

$$\bullet \quad \sum_{k=1}^{n}k = 1 + 2 + 3 + \cdots + n = \frac{n(n+1)}{2} \tag{6}$$

$$\bullet \quad \sum_{k=1}^{n}k^2 = 1^2 + 2^2 + 3^2 + \cdots + n^2 = \frac{n(n+1)(2n+1)}{6} \tag{7}$$

$$\bullet \quad \sum_{k=1}^{n}k^3 = 1^3 + 2^3 + 3^3 + \cdots + n^3 = \left[\frac{n(n+1)}{2}\right]^2 \tag{8}$$

Notice the difference between formulas (5) and (6). In formula (5), the constant 1 is being summed from 1 to n, while in (6) the index of summation k is being summed from 1 to n.

EXAMPLE 7   **Finding Sums**

Find each sum:

(a) $\displaystyle\sum_{k=1}^{100}\frac{1}{2}$ **(b)** $\displaystyle\sum_{k=1}^{5}(3k)$ **(c)** $\displaystyle\sum_{k=1}^{10}(k^3 + 1)$ **(d)** $\displaystyle\sum_{k=6}^{20}(4k^2)$

Solution **(a)** $\displaystyle\sum_{k=1}^{100}\frac{1}{2} \underset{\underset{(1)}{\uparrow}}{=} \frac{1}{2}\sum_{k=1}^{100}1 \underset{\underset{(5)}{\uparrow}}{=} \frac{1}{2}\cdot 100 = 50$

(b) $\displaystyle\sum_{k=1}^{5}(3k) = 3\sum_{k=1}^{5}k = 3\left(\dfrac{5(5+1)}{2}\right) = 3(15) = 45$

         ↑ (1)   ↑ (6)

(c) $\displaystyle\sum_{k=1}^{10}(k^3+1) = \sum_{k=1}^{10}k^3 + \sum_{k=1}^{10}1 = \left[\dfrac{10(10+1)}{2}\right]^2 + 10 = 3025 + 10 = 3035$

        ↑ (2)   ↑ (8) and (5)

(d) Notice that the index of summation starts at 6.

$$\sum_{k=6}^{20}(4k^2) = 4\sum_{k=6}^{20}k^2 = 4\left[\sum_{k=1}^{20}k^2 - \sum_{k=1}^{5}k^2\right]$$

        ↑ (1)   ↑ (4)

$$= 4\left[\dfrac{20(21)(41)}{6} - \dfrac{5(6)(11)}{6}\right] = 4[2870 - 55] = 11,260 \qquad\blacksquare$$

   ↑ (7)

5 Use the Binomial Theorem

You already know formulas for expanding $(x+a)^n$ for $n=2$ and $n=3$. The *Binomial Theorem* is a formula for the expansion of $(x+a)^n$ for any positive integer n. If $n=1$, 2, 3, and 4, the expansion of $(x+a)^n$ is straightforward.

* $(x+a)^1 = x + a$ Two terms, beginning with x^1 and ending with a^1

* $(x+a)^2 = x^2 + 2ax + a^2$ Three terms, beginning with x^2 and ending with a^2

* $(x+a)^3 = x^3 + 3ax^2 + 3a^2x + a^3$ Four terms, beginning with x^3 and ending with a^3

* $(x+a)^4 = x^4 + 4ax^3 + 6a^2x^2 + 4a^3x + a^4$ Five terms, beginning with x^4 and ending with a^4

Notice that each expansion of $(x+a)^n$ begins with x^n, ends with a^n, and has $n+1$ terms. As you look at the terms from left to right, the powers of x are decreasing by one, while the powers of a are increasing by one. As a result, we might conjecture that the expansion of $(x+a)^n$ would look like this:

$$(x+a)^n = x^n + \underline{\quad} ax^{n-1} + \underline{\quad} a^2x^{n-2} + \cdots + \underline{\quad} a^{n-1}x + a^n$$

where the blanks are numbers to be found. This, in fact, is the case.

Before filling in the blanks, we introduce the symbol $\dbinom{n}{j}$.

DEFINITION

If j and n are integers with $0 \le j \le n$, the symbol $\dbinom{n}{j}$ is defined as

$$\binom{n}{j} = \dfrac{n!}{j!(n-j)!}$$

EXAMPLE 8   Evaluating $\binom{n}{j}$

(a) $\dbinom{3}{1} = \dfrac{3!}{1!(3-1)!} = \dfrac{3!}{1!2!} = \dfrac{3 \cdot 2 \cdot 1}{1(2 \cdot 1)} = \dfrac{6}{2} = 3$

(b) $\dbinom{4}{2} = \dfrac{4!}{2!(4-2)!} = \dfrac{4!}{2!2!} = \dfrac{4 \cdot 3 \cdot 2 \cdot 1}{(2 \cdot 1)(2 \cdot 1)} = \dfrac{24}{4} = 6$

(c) $\dbinom{8}{7} = \dfrac{8!}{7!(8-7)!} = \dfrac{8!}{7!1!} = \underset{\underset{8! \,=\, 8 \cdot 7!}{\uparrow}}{\dfrac{8 \cdot 7!}{7! \cdot 1!}} = \dfrac{8}{1} = 8$

RECALL $0! = 1$

(d) $\dbinom{n}{0} = \dfrac{n!}{0!(n-0)!} = \dfrac{n!}{0!n!} = \dfrac{n!}{1 \cdot n!} = 1$

(e) $\dbinom{n}{n} = \dfrac{n!}{n!(n-n)!} = \dfrac{n!}{n!0!} = \dfrac{n!}{n! \cdot 1} = 1$ ∎

THEOREM   Binomial Theorem

Let x and a be real numbers. For any positive integer n,

$$(x+a)^n = \binom{n}{0}x^n + \binom{n}{1}ax^{n-1} + \binom{n}{2}a^2x^{n-2} + \cdots + \binom{n}{j}a^jx^{n-j}$$

$$+ \cdots + \binom{n}{n}a^n = \sum_{j=0}^{n}\binom{n}{j}a^jx^{n-j}$$

Because of its appearance in the Binomial Theorem, the symbol $\binom{n}{j}$ is called a **binomial coefficient**.

EXAMPLE 9   Expanding a Binomial

Use the Binomial Theorem to expand $(x+2)^5$.

Solution We use the Binomial Theorem with $a = 2$ and $n = 5$. Then

$$(x+2)^5 = \binom{5}{0}x^5 + \binom{5}{1}2x^4 + \binom{5}{2}2^2x^3 + \binom{5}{3}2^3x^2 + \binom{5}{4}2^4x + \binom{5}{5}2^5$$

$$= 1 \cdot x^5 + 5 \cdot 2x^4 + 10 \cdot 4x^3 + 10 \cdot 8x^2 + 5 \cdot 16x + 1 \cdot 32$$

$$= x^5 + 10x^4 + 40x^3 + 80x^2 + 80x + 32$$ ∎

B.1 Limit Theorems and Proofs

Uniqueness of a Limit

The limit of a function f, if it exists, is unique; that is, a function can only have one limit.

> **THEOREM** A Limit Is Unique
>
> If a function f is defined on an open interval containing the number c, except possibly at c itself, and if $\lim_{x \to c} f(x) = L_1$ and $\lim_{x \to c} f(x) = L_2$, then $L_1 = L_2$.

RECALL An indirect proof (a proof by contradiction) begins by assuming the conclusion is false. Then we show that this assumption leads to a contradiction.

RECALL The Triangle Inequality: If x and y are real numbers, then $|x + y| \le |x| + |y|$. See Appendix A.1, p. A-7.

Proof Assume that $L_1 \ne L_2$. We will show that this assumption leads to a contradiction. By the definition of the limit of a function f, $\lim_{x \to c} f(x) = L_1$ if, for any given number $\varepsilon > 0$, there is a number $\delta_1 > 0$, so that

$$\text{whenever } 0 < |x - c| < \delta_1 \quad \text{then } |f(x) - L_1| < \varepsilon \tag{1}$$

Similarly, $\lim_{x \to c} f(x) = L_2$ if, for any given number $\varepsilon > 0$, there is a number $\delta_2 > 0$, so that

$$\text{whenever } 0 < |x - c| < \delta_2 \quad \text{then } |f(x) - L_2| < \varepsilon \tag{2}$$

Now

$$L_1 - L_2 = L_1 - f(x) + f(x) - L_2$$

so, by applying the Triangle Inequality, we get

$$|L_1 - L_2| = |L_1 - f(x) + f(x) - L_2| \le |L_1 - f(x)| + |f(x) - L_2| \tag{3}$$

For any given number $\varepsilon > 0$, let δ be the smaller of δ_1 and δ_2. Then from (1)–(3), we can conclude that whenever $0 < |x - c| < \delta \le \delta_1$ and $0 < |x - c| < \delta \le \delta_2$, we have

$$|L_1 - L_2| < \varepsilon + \varepsilon = 2\varepsilon \tag{4}$$

In particular, (4) is true for $\varepsilon = \dfrac{1}{2}|L_1 - L_2| > 0$. (Remember $L_1 \ne L_2$.) Then from (4),

$$|L_1 - L_2| < 2\varepsilon = |L_1 - L_2|$$

which is a contradiction. Therefore, $L_1 = L_2$, and the limit, if it exists, is unique. ∎

Algebra of Limits

> **THEOREM** Limit of a Sum
>
> If f and g are functions for which $\lim_{x \to c} f(x)$ and $\lim_{x \to c} g(x)$ both exist, then $\lim_{x \to c} [f(x) + g(x)]$ exists and
>
> $$\lim_{x \to c}[f(x) + g(x)] = \lim_{x \to c} f(x) + \lim_{x \to c} g(x)$$

Proof Suppose $\lim\limits_{x \to c} f(x) = L$ and $\lim\limits_{x \to c} g(x) = M$. We need to show that for any number $\varepsilon > 0$, there is a number $\delta > 0$, so that

$$\text{whenever } 0 < |x - c| < \delta \quad \text{then } |[f(x) + g(x)] - [L + M]| < \varepsilon$$

Since $\lim\limits_{x \to c} f(x) = L$, by the definition of a limit, given the number $\dfrac{\varepsilon}{2} > 0$, there is a number $\delta_1 > 0$, so that

$$\text{whenever } 0 < |x - c| < \delta_1 \quad \text{then } |f(x) - L| < \frac{\varepsilon}{2}$$

Since $\lim\limits_{x \to c} g(x) = M$, for this same number $\dfrac{\varepsilon}{2}$, there is a number $\delta_2 > 0$, so that

$$\text{whenever } 0 < |x - c| < \delta_2 \quad \text{then } |g(x) - M| < \frac{\varepsilon}{2}$$

Let δ be the smaller of δ_1 and δ_2. Then $\delta \leq \delta_1$ and $\delta \leq \delta_2$. Using this δ,

$$\text{whenever } 0 < |x - c| < \delta \quad \text{then } |f(x) - L| < \frac{\varepsilon}{2}$$

$$\text{whenever } 0 < |x - c| < \delta \quad \text{then } |g(x) - M| < \frac{\varepsilon}{2}$$

That is, whenever $0 < |x - c| < \delta$,

$$|[f(x) + g(x)] - [L + M]| = |[f(x) - L] + [g(x) - M]|$$

$$\leq |f(x) - L| + |g(x) - M| \quad \text{Use the Triangle Inequality.}$$

$$< \frac{\varepsilon}{2} + \frac{\varepsilon}{2} = \varepsilon$$

So, $\lim\limits_{x \to c} [f(x) + g(x)] = L + M$. ∎

THEOREM Limit of a Product

If f and g are functions for which $\lim\limits_{x \to c} f(x)$ and $\lim\limits_{x \to c} g(x)$ both exist, then $\lim\limits_{x \to c} [f(x) \cdot g(x)]$ exists and

$$\lim_{x \to c} [f(x) \cdot g(x)] = \lim_{x \to c} f(x) \cdot \lim_{x \to c} g(x)$$

Proof Suppose $\lim\limits_{x \to c} f(x) = L$ and $\lim\limits_{x \to c} g(x) = M$. We need to show that for any number $\varepsilon > 0$, there is a number $\delta > 0$, so that

$$\text{whenever } 0 < |x - c| < \delta \quad \text{then } |f(x) \cdot g(x) - L \cdot M| < \varepsilon$$

Subtracting and adding $f(x) \cdot M$ in the expression $f(x) \cdot g(x) - L \cdot M$ result in terms involving $g(x) - M$ and $f(x) - L$:

$$|f(x) \cdot g(x) - L \cdot M| = |f(x) \cdot g(x) - f(x) \cdot M + f(x) \cdot M - L \cdot M|$$

$$= |f(x) \cdot [g(x) - M] + [f(x) - L] \cdot M|$$

$$\leq |f(x)| \cdot |g(x) - M| + |f(x) - L| \cdot |M| \tag{5}$$

$$\uparrow$$
Use the Triangle Inequality.

Since $\lim\limits_{x \to c} f(x) = L$ and $\lim\limits_{x \to c} g(x) = M$, then there is a number $\delta_1 > 0$, so that whenever $0 < |x - c| < \delta_1$, then

$$|f(x) - L| < 1, \quad \text{from which} \quad |f(x)| < 1 + |L| \tag{6}$$

Also given a number $\varepsilon > 0$, there is a number δ_2, so that whenever $0 < |x - c| < \delta_2$, then

$$|g(x) - M| < \frac{\varepsilon}{1 + |L| + |M|} \tag{7}$$

Given a number $\varepsilon > 0$, there is a number δ_3, so that whenever $0 < |x - c| < \delta_3$, then

$$|f(x) - L| < \frac{\varepsilon}{1 + |L| + |M|} \tag{8}$$

Choose δ to be the minimum of δ_1, δ_2, and δ_3 and combine (5)–(8). Then for any given $\varepsilon > 0$, there is a $\delta > 0$, so that whenever $0 < |x - c| < \delta$, we have

$$|f(x) \cdot g(x) - L \cdot M| < [1 + |L|]\frac{\varepsilon}{1 + |L| + |M|} + |M|\frac{\varepsilon}{1 + |L| + |M|}$$

$$< [1 + |L| + |M|]\frac{\varepsilon}{1 + |L| + |M|} = \varepsilon$$

That is, $\lim_{x \to c} [f(x) \cdot g(x)] = \lim_{x \to c} f(x) \cdot \lim_{x \to c} g(x)$. ∎

THEOREM Squeeze Theorem

If the functions f, g, and h have the property that for all x in an open interval containing c, except possibly at c itself,

$$f(x) \leq g(x) \leq h(x)$$

and if

$$\lim_{x \to c} f(x) = \lim_{x \to c} h(x) = L$$

then

$$\lim_{x \to c} g(x) = L$$

Proof Since $\lim_{x \to c} f(x) = \lim_{x \to c} h(x) = L$, then for any number $\varepsilon > 0$, there are positive numbers δ_1 and δ_2, so that

$$\text{whenever } 0 < |x - c| < \delta_1 \quad \text{then } |f(x) - L| < \varepsilon$$
$$\text{whenever } 0 < |x - c| < \delta_2 \quad \text{then } |h(x) - L| < \varepsilon$$

Choose δ to be the smaller of the numbers δ_1 and δ_2. Then $0 < |x - c| < \delta$ implies that both $|f(x) - L| < \varepsilon$ and $|h(x) - L| < \varepsilon$. In other words, $0 < |x - c| < \delta$ implies that both

$$L - \varepsilon < f(x) < L + \varepsilon \quad \text{and} \quad L - \varepsilon < h(x) < L + \varepsilon$$

Since $f(x) \leq g(x) \leq h(x)$ for all $x \neq c$ in the open interval, it follows that whenever $0 < |x - c| < \delta$ and x is in the open interval, we have

$$L - \varepsilon < f(x) \leq g(x) \leq h(x) < L + \varepsilon$$

Then for any given number $\varepsilon > 0$, there is a positive number δ, so that whenever $0 < |x - c| < \delta$, then $L - \varepsilon < g(x) < L + \varepsilon$, or equivalently, $|g(x) - L| < \varepsilon$. That is, $\lim_{x \to c} g(x) = L$. ∎

B.2 Theorems and Proofs Involving Inverse Functions

THEOREM Continuity of the Inverse Function

If f is a one-to-one function that is continuous on its domain, then its inverse function f^{-1} is also continuous on its domain.

Proof Let (a, b) be the largest open interval included in the domain of f. If f is continuous on its domain, then it is continuous on (a, b). Since f is one-to-one, then f is either increasing on (a, b) or decreasing on (a, b).

Suppose f is increasing on (a, b). Then f^{-1} is also increasing. Let $y_0 = f(x_0)$. We need to show that f^{-1} is continuous at y_0, given that f is continuous at x_0. The number $f^{-1}(y_0) = x_0$ is in the open interval (a, b). Choose $\varepsilon > 0$ sufficiently small, so that $f^{-1}(y_0) - \varepsilon$ and $f^{-1}(y_0) + \varepsilon$ are also in (a, b). Then choose δ, so that

$$f[f^{-1}(y_0) - \varepsilon] < y_0 - \delta \qquad \text{and} \qquad y_0 + \delta < f[f^{-1}(y_0) + \varepsilon]$$

Then whenever $y_0 - \delta < y < y_0 + \delta$, we have

$$f[f^{-1}(y_0) - \varepsilon] < y < f[f^{-1}(y_0) + \varepsilon]$$

This means whenever $0 < |y - y_0| < \delta$, then

$$f^{-1}(y_0) - \varepsilon < f^{-1}(y) < f^{-1}(y_0) + \varepsilon \qquad \text{or equivalently,} \qquad |f^{-1}(y) - f^{-1}(y_0)| < \varepsilon$$

That is, $\lim_{y \to y_0} f^{-1}(y) = f^{-1}(y_0)$, so f^{-1} is continuous at y_0.

The case where f is decreasing on (a, b) is proved in a similar way. ∎

THEOREM Derivative of the Inverse Function

Let $y = f(x)$ and $x = g(y)$ be inverse functions. If f is differentiable on an open interval containing x_0 and if $f'(x_0) \neq 0$, then g is differentiable at $y_0 = f(x_0)$ and

$$\frac{d}{dy} g(y_0) = \frac{1}{f'(x_0)}$$

where the notation $f'(x_0)$ means the value of $f'(x)$ at x_0 and the notation $\dfrac{d}{dy} g(y_0)$ means the value of $\dfrac{d}{dy} g(y)$ at y_0.

Proof Since f and g are inverses of one another, then $f(x) = y$ if and only if $x = g(y)$. So, we have the following identity, where $g(y_0) = x_0$:

$$\frac{g(y) - g(y_0)}{y - y_0} = \frac{x - x_0}{f(x) - f(x_0)} = \frac{1}{\dfrac{f(x) - f(x_0)}{x - x_0}}$$

By the continuity of an inverse function, the continuity of f at x_0 implies the continuity of g at y_0, and $y \to y_0$ as $x \to x_0$.

Now take the limits of both sides of the above identity. Since $f'(x_0) \neq 0$, we have

$$g'(y_0) = \lim_{y \to y_0} \frac{g(y) - g(y_0)}{y - y_0} = \frac{1}{\displaystyle\lim_{x \to x_0} \frac{f(x) - f(x_0)}{x - x_0}} = \frac{1}{f'(x_0)} \qquad ∎$$

B.3 Derivative Theorems and Proofs

A proof of the Chain Rule when Δu is never 0 appears in Chapter 3. Here, we consider the case when Δu may be 0.

THEOREM Chain Rule

If a function g is differentiable at x_0, and a function f is differentiable at $g(x_0)$, then the composite function $f \circ g$ is differentiable at x_0 and

$$(f \circ g)'(x_0) = f'(g(x_0)) \cdot g'(x_0).$$

Using Leibniz notation, if $y = f(u)$ and $u = g(x)$, then

$$\frac{dy}{dx} = \frac{dy}{du} \cdot \frac{du}{dx},$$

where $\dfrac{dy}{du}$ is evaluated at $u_0 = g(x_0)$ and $\dfrac{du}{dx}$ is evaluated at x_0.

Proof For the fixed number x_0, let $\Delta u = g(x_0 + \Delta x) - g(x_0)$. Since the function $u = g(x)$ is differentiable at x_0, it is also continuous at x_0, and therefore $\Delta u \to 0$ as $\Delta x \to 0$.

For the fixed number $u_0 = g(x_0)$, let $\Delta y = f(u_0 + \Delta u) - f(u_0)$. Since the function $y = f(u)$ is differentiable at the number u_0, we can write $\lim_{\Delta u \to 0} \dfrac{\Delta y}{\Delta u} = \dfrac{dy}{du}(u_0)$. This implies that for any $\Delta u \neq 0$:

$$\frac{\Delta y}{\Delta u} = \frac{dy}{du}(u_0) + \alpha, \tag{1}$$

where $\alpha = \alpha(\Delta u)$ is a function of Δu such that $\alpha \to 0$ as $\Delta u \to 0$. Now we define $\alpha = 0$ when $\Delta u = 0$, so that $\alpha = \alpha(\Delta u)$ is continuous at 0. Multiplying (1) by $\Delta u \neq 0$, we obtain

$$\Delta y = \frac{dy}{du}(u_0) \cdot \Delta u + \alpha(\Delta u) \cdot \Delta u. \tag{2}$$

Notice that the equation (2) is true for all Δu:

- If Δu is not equal to 0, then (2) is a consequence of (1).
- If Δu equals 0, then the left hand side of (2) is

$$\Delta y = f(u_0 + \Delta u) - f(u_0) = f(u_0) - f(u_0) = 0$$

and the right-hand side of 2 is also 0.

Now divide (2) by $\Delta x \neq 0$:

$$\frac{\Delta y}{\Delta x} = \frac{dy}{du}(u_0) \cdot \frac{\Delta u}{\Delta x} + \alpha(\Delta u) \cdot \frac{\Delta u}{\Delta x}. \tag{3}$$

Since the function $u = g(x)$ is differentiable at x_0, $\lim_{\Delta x \to 0} \dfrac{\Delta u}{\Delta x} = \dfrac{du}{dx}(x_0)$. Also, since $\Delta u \to 0$ when $\Delta x \to 0$ and $\alpha(\Delta u)$ is continuous at $\Delta u = 0$, we conclude that $\alpha(\Delta u) \to 0$ as $\Delta x \to 0$. So we can take the limit of (3) as $\Delta x \to 0$, which proves that the derivative $\dfrac{dy}{dx}(x_0)$ exists and is equal to

$$\frac{dy}{dx}(x_0) = \lim_{\Delta x \to 0} \frac{\Delta y}{\Delta x} = \lim_{\Delta x \to 0} \left[\frac{dy}{du}(u_0) \cdot \frac{\Delta u}{\Delta x} + \alpha(\Delta u) \cdot \frac{\Delta u}{\Delta x} \right]$$

$$= \frac{dy}{du}(u_0) \cdot \left[\lim_{\Delta x \to 0} \frac{\Delta u}{\Delta x} \right] + \left[\lim_{\Delta x \to 0} \alpha(\Delta u) \right] \cdot \left[\lim_{\Delta x \to 0} \frac{\Delta u}{\Delta x} \right]$$

$$= \frac{dy}{du}(u_0) \cdot \frac{du}{dx}(x_0) + 0 \cdot \frac{du}{dx}(x_0) = \frac{dy}{du}(u_0) \cdot \frac{du}{dx}(x_0). \qquad \blacksquare$$

Partial Proof of L'Hôpital's Rule

To prove L'Hôpital's Rule requires an extension of the Mean Value Theorem, called *Cauchy's Mean Value Theorem*.

THEOREM Cauchy's Mean Value Theorem

If the functions f and g are continuous on the closed interval $[a, b]$ and differentiable on the open interval (a, b), and if $g'(x) \neq 0$ on (a, b), then there is a number c in (a, b) for which

$$\frac{f'(c)}{g'(c)} = \frac{f(b) - f(a)}{g(b) - g(a)}$$

ORIGINS The theorem was named after the French mathematician Augustin Cauchy (1789–1857).

Notice that under the conditions of Cauchy's Mean Value Theorem, $g(b) \neq g(a)$, because otherwise, by Rolle's Theorem, $g'(c) = 0$ for some c in the interval (a, b).

Proof Define the function h as

$$h(x) = [g(b) - g(a)][f(x) - f(a)] - [g(x) - g(a)][f(b) - f(a)] \qquad a \le x \le b$$

Then h is continuous on $[a, b]$ and differentiable on (a, b) and $h(a) = h(b) = 0$. So by Rolle's Theorem, there is a number c in the interval (a, b) for which $h'(c) = 0$. That is,

$$h'(c) = [g(b) - g(a)]f'(c) - g'(c)[f(b) - f(a)] = 0$$

$$[g(b) - g(a)]f'(c) = g'(c)[f(b) - f(a)]$$

$$\frac{f'(c)}{g'(c)} = \frac{f(b) - f(a)}{g(b) - g(a)}$$

∎

> **NOTE** A special case of Cauchy's Mean Value Theorem is the Mean Value Theorem. To get the Mean Value Theorem, let $g(x) = x$. Then $g'(x) = 1$, $g(b) = b$, and $g(a) = a$, giving the Mean Value Theorem.

THEOREM L'Hôpital's Rule

Suppose the functions f and g are differentiable on an open interval I containing the number c, except possibly at c, and $g'(x) \ne 0$ for all $x \ne c$ in I. Let L denote either a real number or $\pm\infty$, and suppose $\dfrac{f(x)}{g(x)}$ is an indeterminate form at c of the type $\dfrac{0}{0}$ or $\dfrac{\infty}{\infty}$. If $\displaystyle\lim_{x \to c} \frac{f'(x)}{g'(x)} = L$, then $\displaystyle\lim_{x \to c} \frac{f(x)}{g(x)} = L$.

Partial Proof Suppose $\dfrac{f(x)}{g(x)}$ is an indeterminate form at c of the type $\dfrac{0}{0}$, and suppose $\displaystyle\lim_{x \to c} \frac{f'(x)}{g'(x)} = L$, where L is a real number. We need to prove $\displaystyle\lim_{x \to c} \frac{f(x)}{g(x)} = L$. First define the functions F and G as follows:

$$F(x) = \begin{cases} f(x) & \text{if } x \ne c \\ 0 & \text{if } x = c \end{cases}, \qquad G(x) = \begin{cases} g(x) & \text{if } x \ne c \\ 0 & \text{if } x = c \end{cases}$$

Both F and G are continuous at c, since $\displaystyle\lim_{x \to c} F(x) = \lim_{x \to c} f(x) = 0 = F(c)$ and $\displaystyle\lim_{x \to c} G(x) = \lim_{x \to c} g(x) = 0 = G(c)$. Also,

$$F'(x) = f'(x) \qquad \text{and} \qquad G'(x) = g'(x)$$

for all x in the interval I, except possibly at c. Since the conditions for Cauchy's Mean Value Theorem are met by F and G in either $[x, c]$ or $[c, x]$, there is a number u between c and x for which

$$\frac{F(x) - F(c)}{G(x) - G(c)} = \frac{F'(u)}{G'(u)} = \frac{f'(u)}{g'(u)}$$

Since $F(c) = 0$ and $G(c) = 0$, this simplifies to $\dfrac{f(x)}{g(x)} = \dfrac{f'(u)}{g'(u)}$.

Since u is between c and x, it follows that

$$\lim_{x \to c} \frac{f(x)}{g(x)} = \lim_{u \to c} \frac{f'(u)}{g'(u)} = L$$

A similar argument is used if L is infinite. The proof when $\dfrac{f(x)}{g(x)}$ is an indeterminate form at ∞ of the type $\dfrac{\infty}{\infty}$ is omitted here, but it may be found in books on advanced calculus. ∎

The use of L'Hôpital's Rule when $c = \infty$ for an indeterminate form of the type $\dfrac{0}{0}$ is justified by the following argument. In $\displaystyle\lim_{x \to \infty} \frac{f(x)}{g(x)}$, let $x = \dfrac{1}{u}$. Then as $x \to \infty$, $u \to 0^+$, and

$$\lim_{x \to \infty} \frac{f(x)}{g(x)} = \lim_{u \to 0^+} \frac{f\left(\frac{1}{u}\right)}{g\left(\frac{1}{u}\right)} = \lim_{u \to 0^+} \frac{\frac{d}{du} f\left(\frac{1}{u}\right)}{\frac{d}{du} g\left(\frac{1}{u}\right)} = \lim_{u \to 0^+} \frac{-\frac{1}{u^2} f'\left(\frac{1}{u}\right)}{-\frac{1}{u^2} g'\left(\frac{1}{u}\right)} = \lim_{x \to \infty} \frac{f'(x)}{g'(x)} = L$$

Chain Rule

$x = \dfrac{1}{u}$

Proof That Continuous Partial Derivatives Are Sufficient for Differentiability

THEOREM

Let $z = f(x, y)$ be a function of two variables whose domain is D. Let (x_0, y_0) be an interior point of D. If the partial derivatives f_x and f_y exist at each point of some disk centered at (x_0, y_0), and if f_x and f_y are each continuous at (x_0, y_0), then f is differentiable at (x_0, y_0).

Proof The proof depends on the Mean Value Theorem for derivatives. Let Δx and Δy be changes, not both 0, in x and in y, respectively, so that the point $(x_0 + \Delta x, y_0 + \Delta y)$ lies in some disk centered at (x_0, y_0). The change in z is

$$\Delta z = f(x_0 + \Delta x, y_0 + \Delta y) - f(x_0, y_0)$$

Adding and subtracting $f(x_0, y_0 + \Delta y)$ on the right-hand side, we obtain

$$\Delta z = f(x_0 + \Delta x, y_0 + \Delta y) - f(x_0, y_0 + \Delta y) + f(x_0, y_0 + \Delta y) - f(x_0, y_0) \quad (4)$$

The expression $f(x, y_0 + \Delta y)$ is a function of x alone, and its partial derivative $f_x(x, y_0 + \Delta y)$ exists in the disk centered at (x_0, y_0). Then by the Mean Value Theorem, there is a real number u between x_0 and $x_0 + \Delta x$ for which

$$f(x_0 + \Delta x, y_0 + \Delta y) - f(x_0, y_0 + \Delta y) = f_x(u, y_0 + \Delta y)\Delta x \quad (5)$$

Similarly, the expression $f(x_0, y)$ is a function of y alone, and the partial derivative $f_y(x_0, y)$ exists in the disk centered at (x_0, y_0). Again, by the Mean Value Theorem, there is a real number v between y_0 and $y_0 + \Delta y$ for which

$$f(x_0, y_0 + \Delta y) - f(x_0, y_0) = f_y(x_0, v)\Delta y \quad (6)$$

Substitute (5) and (6) back into equation (4) for Δz to obtain

$$\Delta z = f_x(u, y_0 + \Delta y)\Delta x + f_y(x_0, v)\Delta y$$

Now introduce the functions η_1 and η_2 defined by

$$\eta_1 = f_x(u, y_0 + \Delta y) - f_x(x_0, y_0) \qquad \text{and} \qquad \eta_2 = f_y(x_0, v) - f_y(x_0, y_0)$$

As $(\Delta x, \Delta y) \to (0, 0)$, then $u \to x_0$ and $v \to y_0$. Since f_x and f_y are continuous at (x_0, y_0), η_1 and η_2 have the desired property that

$$\lim_{(\Delta x, \Delta y) \to (0,0)} \eta_1 = \lim_{(\Delta x, \Delta y) \to (0,0)} [f_x(u, y_0 + \Delta y) - f_x(x_0, y_0)]$$

$$= f_x(x_0, y_0) - f_x(x_0, y_0) = 0$$

and

$$\lim_{(\Delta x, \Delta y) \to (0,0)} \eta_2 = \lim_{(\Delta x, \Delta y) \to (0,0)} \left[f_y(x_0, v) - f_y(x_0, y_0) \right]$$

$$= f_y(x_0, y_0) - f_y(x_0, y_0) = 0$$

As a result, Δz can be written as

$$\Delta z = f_x(u, y_0 + \Delta y)\Delta x + f_y(x_0, v)\Delta y$$
$$= [\eta_1 + f_x(x_0, y_0)]\Delta x + [\eta_2 + f_y(x_0, y_0)]\Delta y$$
$$= f_x(x_0, y_0)\Delta x + f_y(x_0, y_0)\Delta y + \eta_1\Delta x + \eta_2\Delta y$$

proving that f is differentiable at (x_0, y_0). ■

B.4 Integral Theorems and Proofs

THEOREM

If a function f is continuous on an interval containing the numbers a, b, and c, then

$$\int_a^b f(x)\,dx = \int_a^c f(x)\,dx + \int_c^b f(x)\,dx$$

Proof Since f is continuous on an interval containing a, b, and c, the three integrals above exist.

Part 1 Assume $a < b < c$. Since f is continuous on $[a, b]$ and on $[b, c]$, given any $\varepsilon > 0$, there is a number $\delta_1 > 0$, so that

$$\left| \sum_{i=1}^k f(u_i)\Delta x_i - \int_a^b f(x)\,dx \right| < \frac{\varepsilon}{2} \tag{1}$$

for every Riemann sum $\sum_{i=1}^k f(u_i)\Delta x_i$ for f on $[a, b]$, where $x_{i-1} \le u_i \le x_i$, $i = 1, 2, \ldots, k$, and whose partition P_1 of $[a, b]$ has norm $\|P_1\| < \delta_1$. There is also a number $\delta_2 > 0$ for which

$$\left| \sum_{i=k+1}^n f(u_i)\Delta x_i - \int_b^c f(x)\,dx \right| < \frac{\varepsilon}{2} \tag{2}$$

for every Riemann sum $\sum_{i=k+1}^n f(u_i)\Delta x_i$ for f on $[b, c]$, where $x_{i-1} \le u_i \le x_i$, $i = k+1, k+2, \ldots, n$, and whose partition P_2 of $[b, c]$ has norm $\|P_2\| < \delta_2$.

NOTE The partition P_1 is $x_0 = a, \ldots, x_k = b$. The partition P_2 is $x_k = b, \ldots, x_n = c$.

Let δ be the smaller of δ_1 and δ_2. Then (1) and (2) hold, with δ replacing δ_1 and δ_2. If (1) and (2) are added and if $\|P_1\| < \delta$ and $\|P_2\| < \delta$, then

$$\left| \sum_{i=1}^k f(u_i)\Delta x_i - \int_a^b f(x)\,dx \right| + \left| \sum_{i=k+1}^n f(u_i)\Delta x_i - \int_b^c f(x)\,dx \right| < \frac{\varepsilon}{2} + \frac{\varepsilon}{2} = \varepsilon$$

Using the Triangle Inequality, this result implies that for $\|P_1\| < \delta$ and $\|P_2\| < \delta$,

$$\left| \sum_{i=1}^k f(u_i)\Delta x_i - \int_a^b f(x)\,dx + \sum_{i=k+1}^n f(u_i)\Delta x_i - \int_b^c f(x)\,dx \right| < \varepsilon \tag{3}$$

Denote $P_1 \cup P_2$ by P^*. Then P^* is a partition of $[a, c]$ having the number $b = x_k$ as an endpoint of the kth subinterval. So,

$$\sum_{i=1}^k f(u_i)\Delta x_i + \sum_{i=k+1}^n f(u_i)\Delta x_i = \sum_{i=1}^n f(u_i)\Delta x_i$$

are Riemann sums for f on P^*. Since $\|P^*\| < \delta$ implies that $\|P_1\| < \delta$ and $\|P_2\| < \delta$, it follows from (3) that

$$\left| \sum_{i=1}^n f(u_i)\Delta x_i - \left[\int_a^b f(x)\,dx + \int_b^c f(x)\,dx \right] \right| < \varepsilon$$

for every Riemann sum $\sum_{i=1}^{n} f(u_i)\Delta x_i$ for f on $[a, c]$ whose partition P^* of $[a, c]$ has b as an endpoint of a subinterval of the partition and has norm $\|P^*\| < \delta$. Therefore,

$$\int_a^c f(x)\, dx = \int_a^b f(x)\, dx + \int_b^c f(x)\, dx$$

Part 2 There are six possible orderings (permutations) of the numbers a, b, and c:

$$a < b < c \qquad a < c < b \qquad b < a < c \qquad b < c < a \qquad c < a < b \qquad c < b < a$$

In Part 1, we showed that the theorem is true for the order $a < b < c$. Now consider any other order, say, $b < c < a$. From Part 1,

$$\int_b^c f(x)\, dx + \int_c^a f(x)\, dx = \int_b^a f(x)\, dx \qquad (4)$$

But,

$$\int_c^a f(x)\, dx = -\int_a^c f(x)\, dx \quad \text{and} \quad \int_b^a f(x)\, dx = -\int_a^b f(x)\, dx$$

Now we substitute this into (4).

$$\int_b^c f(x)\, dx - \int_a^c f(x)\, dx = -\int_a^b f(x)\, dx$$

$$\int_a^b f(x)\, dx + \int_b^c f(x)\, dx = \int_a^c f(x)\, dx$$

proving the theorem for $b < c < a$.

The proofs for the remaining four permutations of a, b, and c are similar. ■

THEOREM Fundamental Theorem of Calculus, Part 1

Let f be a function that is continuous on a closed interval $[a, b]$. The function I defined by

$$I(x) = \int_a^x f(t)\, dt$$

has the property that it is continuous on $[a, b]$ and differentiable on (a, b). Moreover,

$$I'(x) = \frac{d}{dx}\left[\int_a^x f(t)\, dt\right] = f(x)$$

for all x in (a, b).

Proof Let x and $x + h$, $h \neq 0$, be in the interval (a, b). Then

$$I(x) = \int_a^x f(t)\, dt \qquad I(x + h) = \int_a^{x+h} f(t)\, dt$$

and

$$I(x + h) - I(x) = \int_a^{x+h} f(t)\, dt + \int_x^a f(t)\, dt \qquad \int_x^a f(t)\, dt = -\int_a^x f(t)\, dt$$

$$= \int_x^a f(t)\, dt + \int_a^{x+h} f(t)\, dt = \int_x^{x+h} f(t)\, dt$$

Dividing both sides by $h \neq 0$, we get

$$\frac{I(x + h) - I(x)}{h} = \frac{1}{h}\int_x^{x+h} f(t)\, dt \qquad (5)$$

NEED TO REVIEW? The Mean Value Theorem for Integrals is discussed in Section 5.4, p. 372.

Now we use the Mean Value Theorem for Integrals in the integral on the right. There are two possibilities: either $h > 0$ or $h < 0$.

If $h > 0$, there is a number u, where $x \leq u \leq x + h$, for which

$$\int_x^{x+h} f(t)\,dt = f(u)h$$

$$\frac{1}{h} \int_x^{x+h} f(t)\,dt = f(u)$$

$$\frac{I(x+h) - I(x)}{h} = f(u) \qquad \text{From (5)}$$

Since $x \leq u \leq x + h$, as $h \to 0^+$, u approaches x^+, so

$$\lim_{h \to 0^+} \frac{I(x+h) - I(x)}{h} = \lim_{h \to 0^+} f(u) = \lim_{u \to x^+} f(u) = f(x)$$

f is continuous

Using a similar argument for $h < 0$, we obtain

$$\lim_{h \to 0^-} \frac{I(x+h) - I(x)}{h} = f(x)$$

Since the two one-sided limits are equal,

$$\lim_{h \to 0} \frac{I(x+h) - I(x)}{h} = f(x)$$

The limit is the derivative of the function I, meaning $I'(x) = f(x)$ for all x in (a, b). ∎

THEOREM Bounds on an Integral

If a function f is continuous on a closed interval $[a, b]$ and if m and M denote the absolute minimum and absolute maximum values of f on $[a, b]$, respectively, then

$$m(b - a) \leq \int_a^b f(x)\,dx \leq M(b - a)$$

The bounds on an integral theorem is proved by contradiction.

Proof **Part 1** $m(b - a) \leq \int_a^b f(x)\,dx$

Assume

$$m(b - a) > \int_a^b f(x)\,dx \tag{6}$$

Since f is continuous on $[a, b]$,

$$\lim_{\|P\| \to 0} \sum_{i=1}^n f(u_i)\Delta x_i = \int_a^b f(x)\,dx$$

By (6), $m(b - a) - \int_a^b f(x)\,dx > 0$. We choose ε, so that

$$\varepsilon = m(b - a) - \int_a^b f(x)\,dx > 0 \tag{7}$$

Then there is a number $\delta > 0$, so that for all partitions P of $[a, b]$ with norm $\|P\| < \delta$, we have

$$\left| \sum_{i=1}^n f(u_i)\Delta x_i - \int_a^b f(x)\,dx \right| < \varepsilon$$

which is equivalent to

$$\int_a^b f(x)\,dx - \varepsilon < \sum_{i=1}^n f(u_i)\Delta x_i < \int_a^b f(x)\,dx + \varepsilon$$

By (7), the right inequality can be expressed as

$$\sum_{i=1}^n f(u_i)\Delta x_i < \int_a^b f(x)\,dx + \varepsilon = \int_a^b f(x)\,dx + \left[m(b-a) - \int_a^b f(x)\,dx\right]$$

$$= m(b-a)$$

Consequently,

$$\sum_{i=1}^n f(u_i)\Delta x_i < m(b-a) = \sum_{i=1}^n m\,\Delta x_i$$

implying that for every partition P of $[a, b]$ with $\|P\| < \delta$,

$$f(u_i) < m$$

for some u_i in $[a, b]$. But this is impossible because m is the absolute minimum of f on $[a, b]$. Therefore, the assumption $m(b-a) > \int_a^b f(x)\,dx$ is false. That is,

$$m(b-a) \le \int_a^b f(x)\,dx$$

Part 2 To prove $\int_a^b f(x)\,dx \le M(b-a)$, use a similar argument. ∎

B.5 A Bounded Monotonic Sequence Converges

THEOREM

An increasing (or nondecreasing) sequence $\{s_n\}$ that is bounded from above converges. A decreasing (or nonincreasing) sequence $\{s_n\}$ that is bounded from below converges.

To prove this theorem, we need the following property of real numbers. The set of real numbers is defined by a collection of axioms. One of these axioms is the Completeness Axiom.

Completeness Axiom of Real Numbers

If S is a nonempty set of real numbers that has an upper bound, then it has a least upper bound. Similarly, if S has a lower bound, then it has a greatest lower bound.

As an example, consider the set S: $\{x \mid x^2 < 2,\ x > 0\}$. The set of upper bounds to S is the set $\{x \mid x^2 \ge 2,\ x > 0\}$.

(a) If our universe is the set of rational numbers, the set of upper bounds has no minimum (since $\sqrt{2}$ is not rational).

(b) If our universe is the set of real numbers, then by the Completeness Axiom, the set of upper bounds has a minimum ($\sqrt{2}$). That is, this axiom completes the set of real numbers by incorporating the set of irrational numbers with the set of rational numbers to form the set of real numbers.

We prove the theorem for a nondecreasing sequence $\{s_n\}$. The proofs for the other three cases are similar.

Proof Suppose $\{s_n\}$ is a nondecreasing sequence that is bounded from above. Since $\{s_n\}$ is bounded from above, there is a positive number K (an upper bound), so that $s_n \leq K$ for every n. From the Completeness Axiom, the set $\{s_n\}$ has a least upper bound L. That is, $s_n \leq L$ for every n.

Then for any $\varepsilon > 0$, $L - \varepsilon$ is not an upper bound of $\{s_n\}$. That is, $L - \varepsilon < s_N$ for some integer N. Since $\{s_n\}$ is nondecreasing, $s_N \leq s_n$ for all $n > N$. Then for all $n > N$,

$$L - \varepsilon < s_n \leq L < L + \varepsilon$$

That is, $|s_n - L| < \varepsilon$ for all $n > N$, so the sequence $\{s_n\}$ converges to L. ∎

B.6 Taylor's Formula with Remainder

THEOREM Taylor's Formula with Remainder

Let f be a function whose first $n + 1$ derivatives are continuous on an interval I containing the number c. Then for every x in the interval, there is a number u between x and c for which

$$f(x) = f(c) + f'(c)(x - c) + \frac{f''(c)}{2!}(x - c)^2 + \cdots + \frac{f^{(n)}(c)}{n!}(x - c)^n + R_n(x)$$

where

$$R_n(x) = \frac{f^{(n+1)}(u)}{(n + 1)!}(x - c)^{n+1}$$

Proof For a fixed number $x \neq c$ in the interval I, there is a number L (depending on x) for which

$$f(x) = f(c) + \frac{f'(c)}{1!}(x - c) + \frac{f''(c)}{2!}(x - c)^2 + \cdots + \frac{f^{(n)}(c)}{n!}(x - c)^n + \frac{L}{(n + 1)!}(x - c)^{n+1} \quad (1)$$

Define the function F to be

$$F(t) = f(x) - f(t) - \frac{f'(t)}{1!}(x - t) - \frac{f''(t)}{2!}(x - t)^2 - \cdots - \frac{f^{(n)}(t)}{n!}(x - t)^n - \frac{L}{(n + 1)!}(x - t)^{n+1} \quad (2)$$

The domain of F is $c \leq t \leq x$ if $x > c$ and $x \leq t \leq c$ if $x < c$. Since $f(t)$, $f'(t)$, $f''(t), \ldots, f^{(n)}(t)$ are each continuous, then F is continuous on its domain. Furthermore, F is differentiable and

$$\frac{dF}{dt} = F'(t) = -f'(t) + \left[f'(t) - \frac{f''(t)}{1!}(x - t) \right] + \left[\frac{f''(t)}{1!}(x - t) - \frac{f'''(t)}{2!}(x - t)^2 \right]$$

$$+ \cdots + \left[\frac{f^{(n)}(t)}{(n - 1)!}(x - t)^{n-1} - \frac{f^{(n+1)}(t)}{n!}(x - t)^n \right] + \frac{L}{n!}(x - t)^n$$

$$= -\frac{f^{(n+1)}(t)}{n!}(x - t)^n + \frac{L}{n!}(x - t)^n$$

for all t between x and c. From (1) and (2), we have

$$F(c) = f(x) - f(c) - \frac{f'(c)}{1!}(x - c) - \frac{f''(c)}{2!}(x - c)^2 - \cdots - \frac{f^{(n)}(c)}{n!}(x - c)^n - \frac{L}{(n + 1)!}(x - c)^{n+1} = 0$$

Then

$$F(x) = f(x) - f(x) - \frac{f'(x)}{1!}(x - x) - \cdots - \frac{f^{(n)}(x)}{n!}(x - x)^n - \frac{L}{(n + 1)!}(x - x)^{n+1} = 0$$

Now apply Rolle's Theorem to F. Then there is a number u between c and x for which

$$F'(u) = -\frac{f^{(n+1)}(u)}{n!}(x - u)^n + \frac{L}{n!}(x - u)^n = 0$$

Solving for L, we find $L = f^{(n+1)}(u)$. Now let $t = c$ and $L = f^{(n+1)}(u)$ in (2) and solve for $f(x)$. Then

$$f(x) = f(c) + f'(c)(x - c) + \frac{f''(c)}{2!}(x - c)^2 + \cdots + \frac{f^{(n)}(c)}{n!}(x - c)^n + R_n(x)$$

where

$$R_n(x) = \frac{f^{(n+1)}(u)}{(n + 1)!}(x - c)^{n+1}$$

■

Answers

Chapter P

Section P.1

1. Independent, dependent **2.** True **3.** False **4.** False **5.** False **6.** Vertical **7.** 5, −3 **8.** −2 **9.** (a) **10.** (a), (b) **11.** False **12.** 8
13. (a) −4 **(b)** $3x^2 - 2x - 4$ **(c)** $-3x^2 - 2x + 4$ **(d)** $3x^2 + 8x + 1$ **(e)** $3x^2 + 6xh + 3h^2 + 2x + 2h - 4$ **15. (a)** 4 **(b)** $|x| + 4$ **(c)** $-|x| - 4$
(d) $|x + 1| + 4$ **(e)** $|x + h| + 4$ **17.** $(-\infty, \infty)$ **19.** $(-\infty, -3] \cup [3, \infty)$ **21.** $\{x \mid x \neq -2, 0, 2\}$ **23.** −3 **25.** $\dfrac{1}{\sqrt{x + h + 7} + \sqrt{x + 7}}$

27. $2x + h + 2$ **29.** It is not a function. **31. (a)** Domain: $[-\pi, \pi]$, range: $[-1, 1]$ **(b)** Intercepts: $\left(-\dfrac{\pi}{2}, 0\right), \left(\dfrac{\pi}{2}, 0\right), (0, 1)$

(c) Symmetric with respect to the y-axis, but not with respect to the x-axis or the origin.

33. (a) $f(-1) = 2$, **(b)**   **(c)** Domain: $[-2, \infty)$, **35. (a)** $f(-1) = 0$, **(b)** 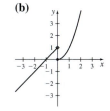 **(c)** Domain:
$f(0) = 3$, $f(1) = 5$,   range: $(-\infty, 4) \cup \{5\}$, $f(0) = 0$, $f(1) = 1$, $(-\infty, \infty)$, range:
$f(8) = -6$ intercepts: $(0, 3), (2, 0)$ $f(8) = 64$ $(-\infty, \infty)$, intercepts:
                                                                                                                                         $(-1, 0), (0, 0)$

37. $f(0) = 3$, $f(-6) = -3$ **39.** Negative **41.** $(-3, 6) \cup (10, 11]$ **43.** $[-3, 3]$ **45.** 3 **47.** Once **49.** −5, 8 **51.** $(4, 8)$ **53.** $(0, 8)$
55. $\{x \mid x \neq 6\}$ **57.** −3, $(4, -3)$ **59.** −2 **61.** Odd; symmetric with respect to the origin, but not with respect to the x-axis or the y-axis.
63. Neither; not symmetric to the x-axis, y-axis, or the origin. **65. (a)** −6 **(b)** −8 **(c)** −10 **(d)** $-2(x + 1)$

67. $f(x) = \begin{cases} -x, & -1 \leq x < 0 \\ 1 & \\ \frac{1}{2}x, & 0 \leq x \leq 2 \end{cases}$, domain: $[-1, 2]$, range: $[0, 1]$ **69.** $f(x) = \begin{cases} -1, & x < 1 \\ 0, & x = 1 \\ 2 - x, & 1 < x \leq 2 \end{cases}$, domain: $(-\infty, 2]$, range: $\{-1\} \cup [0, 1)$

71. $f(x) = \begin{cases} -x^3, & -2 < x < 1 \\ 0, & x = 1 \\ x^2, & 1 < x \leq 3 \end{cases}$, domain: $(-2, 3]$, range: $(-1, 9]$ **73. (a)** \$250.60 **(b)** \$2655.40 **(c)** Answers will vary.

Section P.2

1. (a) **2.** True **3.** True **4.** False **5.** False **6.** Zero **7. (b)** **8.** True **9.** True **10.** False **11.** D **13.** F **15.** C **17.** G
19. (a) 2 **(b)** 3 **(c)** −4 **21. (a)** 7 with multiplicity 1; −4 with multiplicity 3 **(b)** x-intercepts: 7, −4, y-intercept: −1344

(c) Crosses at 7 and at −4 **23.** (c), (e), (f) **25.** Domain: $\{x \mid x \neq -3\}$, intercept: $(0, 0)$ **27.** Domain: $(-\infty, \infty)$, intercepts: $(0, 0)$, $\left(\dfrac{1}{3}, 0\right)$

29. (a) $A(x) = 16x - x^3$ **(b)** $[0, 4]$
31. (a)   **(b)** 283.8 feet
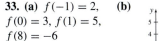   **(c)** 15.63 feet

Section P.3

1. $\{x \mid 0 \leq x \leq 5\}$ **2.** False **3.** True **4.** False **5.** True **6.** False **7.** True **8.** Horizontal, right **9.** y **10.** −5, −2, 2 **11. (a)** $(f + g)(x) =$

$5x + 1$, domain: $(-\infty, \infty)$ **(b)** $(f - g)(x) = x + 7$, domain: $(-\infty, \infty)$ **(c)** $(f \cdot g)(x) = 6x^2 - x - 12$, domain: $(-\infty, \infty)$ **(d)** $\left(\dfrac{f}{g}\right)(x) =$

$\dfrac{3x + 4}{2x - 3}$, domain: $\left\{x \mid x \neq \dfrac{3}{2}\right\}$ **13. (a)** $(f + g)(x) = \sqrt{x + 1} + \dfrac{2}{x}$, domain: $\{x \mid -1 \leq x < 0\} \cup \{x \mid x > 0\}$ **(b)** $(f - g)(x) = \sqrt{x + 1} - \dfrac{2}{x}$,

domain: $\{x \mid -1 \leq x < 0\} \cup \{x \mid x > 0\}$ **(c)** $(f \cdot g)(x) = \dfrac{2\sqrt{x + 1}}{x}$, domain: $\{x \mid -1 \leq x < 0\} \cup \{x \mid x > 0\}$ **(d)** $\left(\dfrac{f}{g}\right)(x) = \dfrac{1}{2}x\sqrt{x + 1}$,

domain: $\{x \mid x \geq -1\}$ **15. (a)** $(f \circ g)(4) = 98$ **(b)** $(g \circ f)(2) = 49$ **(c)** $(f \circ f)(1) = 4$ **(d)** $(g \circ g)(0) = 4$ **17. (a)** $(f \circ g)(1) = -1$
(b) $(f \circ g)(-1) = -1$ **(c)** $(g \circ f)(-1) = 8$ **(d)** $(g \circ f)(1) = 8$ **(e)** $(g \circ g)(-2) = 8$ **(f)** $(f \circ f)(-1) = -7$ **19. (a)** $(g \circ f)(-1) = 4$
(b) $(g \circ f)(6) = 2$ **(c)** $(f \circ g)(6) = 1$ **(d)** $(f \circ g)(4) = -2$

21. (a) $(f \circ g)(x) = 24x + 1$, domain: $(-\infty, \infty)$　**(b)** $(g \circ f)(x) = 24x + 8$, domain: $(-\infty, \infty)$　**(c)** $(f \circ f)(x) = 9x + 4$, domain: $(-\infty, \infty)$　**(d)** $(g \circ g)(x) = 64x$, domain: $(-\infty, \infty)$　**23. (a)** $(f \circ g)(x) = x$, domain: $\{x \mid x \geq 1\}$　**(b)** $(g \circ f)(x) = |x|$, domain: $(-\infty, \infty)$

(c) $(f \circ f)(x) = x^4 + 2x^2 + 2$, domain: $(-\infty, \infty)$　**(d)** $(g \circ g)(x) = \sqrt{\sqrt{x-1} - 1}$, domain: $\{x \mid x \geq 2\}$　**25. (a)** $(f \circ g)(x) = \dfrac{2}{2-x}$, domain:

$\{x \mid x \neq 0, 2\}$　**(b)** $(g \circ f)(x) = \dfrac{2(x-1)}{x}$, domain: $\{x \mid x \neq 0, 1\}$　**(c)** $(f \circ f)(x) = x$, domain: $\{x \mid x \neq 1\}$　**(d)** $(g \circ g)(x) = x$, domain:

$\{x \mid x \neq 0\}$　**27.** $f(x) = x^4, g(x) = 2x + 3$　**29.** $f(x) = \sqrt{x}, g(x) = x^2 + 1$　**31.** $f(x) = |x|, g(x) = 2x + 1$

33.

35.

37.

39.

41.

43.

45.

47. (a)

(b)

(c)

(d)

(e)

(f)

(g)

49. (a)

(b)

(c) Answers will vary.　**(d)**

(e) Answers will vary.

Section P.4

1. False　**2.** $[4, \infty)$　**3.** False　**4.** True　**5.** False　**6.** False　**7.** Answers will vary.　**8.** Answers will vary.　**9.** One-to-one
11. Not one-to-one　**13.** One-to-one　**15.** See Student Solutions Manual.　**17.** See Student Solutions Manual.
19. (a) One-to-one　**(b)** $\{(5, -3), (9, -2), (2, -1), (11, 0), (-5, 1)\}$,　**(c)** Domain of f: $\{-3, -2, -1, 0, 1\}$, range of f: $\{-5, 2, 5, 9, 11\}$,
domain of f^{-1}: $\{-5, 2, 5, 9, 11\}$, range of f^{-1}: $\{-3, -2, -1, 0, 1\}$
21. (a) Not one-to-one　**(b)** Does not apply　**(c)** Domain of f: $\{-10, -3, -2, 1, 2\}$, range of f: $\{0, 1, 2, 9\}$

23.

25.

27.

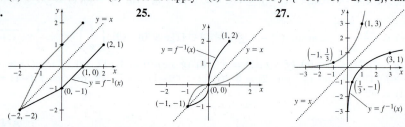

29. (a) $f^{-1}(x) = \dfrac{x-2}{4}$ **(b)** Both the domain and range of f are $(-\infty, \infty)$. Both the domain and range of f^{-1} are $(-\infty, \infty)$.

31. (a) $f^{-1}(x) = x^3 - 10$ **(b)** Both the domain and range of f are $(-\infty, \infty)$. Both the domain and range of f^{-1} are $(-\infty, \infty)$.

33. (a) $f^{-1}(x) = 2 + \dfrac{1}{x}$ **(b)** Domain of f: $\{x \mid x \neq 2\}$, range of f: $\{y \mid y \neq 0\}$, domain of f^{-1}: $\{x \mid x \neq 0\}$, range of f^{-1}: $\{y \mid y \neq 2\}$

35. (a) $f^{-1}(x) = \dfrac{3 - 2x}{x - 2}$ **(b)** Domain of f: $\{x \mid x \neq -2\}$, range of f: $\{y \mid y \neq 2\}$, domain of f^{-1}: $\{x \mid x \neq 2\}$, range of f^{-1}: $\{y \mid y \neq -2\}$

37. (a) $f^{-1}(x) = \sqrt{x - 4}$ **(b)** Domain of f: $\{x \mid x \geq 0\}$, range of f: $\{y \mid y \geq 4\}$, domain of f^{-1}: $\{x \mid x \geq 4\}$, range of f^{-1}: $\{y \mid y \geq 0\}$

Section P.5

1. $\left(-1, \dfrac{1}{a}\right), (0, 1), (1, a)$ **2.** False   **3.** 4   **4.** 3   **5.** False   **6.** False   **7.** 1   **8.** $\{x \mid x > 0\}$ **9.** $\left(\dfrac{1}{a}, -1\right), (1, 0), (a, 1)$ **10. (a)**   **11.** False

12. True   **13.** True   **14.** 1   **15.** Answers will vary.   **16.** 0   **17. (a)** $g(-1) = \dfrac{9}{4}, \left(-1, \dfrac{9}{4}\right)$ **(b)** $x = 3, (3, 66)$ **19. (a)**   **21. (c)**   **23. (b)**

25. Domain: $(-\infty, \infty)$, range: $(0, \infty)$ **27.** Domain: $(-\infty, \infty)$, range: $(0, \infty)$ **29.** Domain: $(-\infty, \infty)$, range: $(0, \infty)$

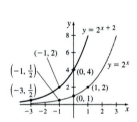   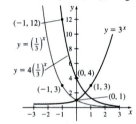   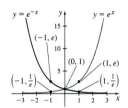

31. $\{x \mid x \neq 0\}$ **33.** $\{x \mid x > 1\}$ **35. (b)**   **37. (f)**   **39. (c)**

41. (a) $\{x \mid x > -4\}$ **(b and f)**

(c) $(-\infty, \infty)$

(d) $f^{-1}(x) = e^x - 4$

(e) $(-\infty, \infty)$

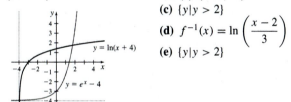

43. (a) $(-\infty, \infty)$ **(b and f)**

(c) $\{y \mid y > 2\}$

(d) $f^{-1}(x) = \ln\left(\dfrac{x-2}{3}\right)$

(e) $\{y \mid y > 2\}$

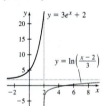

45. It shifts the x-intercept c units to the left.   **47.** $x = 0, x = 2$ **49.** $x = \dfrac{1}{2}$ **51.** $x = \dfrac{1 - \ln 4}{2}$ **53.** $x = \dfrac{\ln\left(\dfrac{9}{5}\right)}{\ln 8}$ **55.** $x = \dfrac{\ln 3}{\ln 36}$

57. $x = \dfrac{7}{2}$ **59.** $x = \dfrac{1}{2}$ **61.** $x = 3$ **63.** $x \approx 2.787$ **65.** $x \approx 1.315$

67. (a)   **(b)** $\left(\dfrac{3}{2}, \ln\dfrac{11}{2}\right)$ **(c)** $\left\{x \mid -\dfrac{1}{3} < x < \dfrac{3}{2}\right\}$

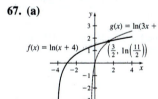

Section P.6

1. $2\pi, \pi$ **2.** All real numbers except odd multiples of $\dfrac{\pi}{2}$ **3.** $\{y \mid -1 \leq y \leq 1\}$ **4.** The period of $\tan x$ is π.   **5.** False   **6.** $3, \dfrac{\pi}{3}$ **7.** True

8. False   **9.** True   **10.** Origin   **11.** y-axis   **12.** Answers will vary.   **13.** -1 **15.** $-\dfrac{2\sqrt{3}}{3}$ **17.** 0   **19.** -1 **21. (f)**   **23. (a)**   **25. (d)**

27.   **29.**   **31.**

33. $\dfrac{1}{2}, 2$ **35.** $3, 2\pi$ **37.** $f(x) = 2\sin(2x)$ **39.** $f(x) = \dfrac{1}{2}\cos(2x)$ **41.** $f(x) = -\sin\left(\dfrac{3}{2}x\right)$ **43.** $f(x) = 1 - \cos\left(\dfrac{4\pi}{3}x\right)$

45. $f(x) = \cot x$ **47.** $f(x) = \tan\left(x - \dfrac{\pi}{2}\right)$

Section P.7

1. $\sin y$ **2.** False   **3.** False   **4.** True   **5.** False   **6.** True   **7.** True   **8.** True   **9.** $\dfrac{\pi}{4}$ **11.** $\dfrac{\pi}{3}$ **13.** $-\dfrac{\pi}{4}$ **15.** $\dfrac{\pi}{4}$ **17.** $\dfrac{3}{5}$ **19.** $\dfrac{\pi}{5}$

21. $\sqrt{1-u^2}$ **23.** $\sqrt{1+u^2}$ **25.** See Student Solutions Manual.   **27.** $\left\{\dfrac{5\pi}{6}, \dfrac{11\pi}{6}\right\}$ **29.** $\left\{\dfrac{7\pi}{6}, \dfrac{11\pi}{6}\right\}$ **31.** $\left\{\dfrac{\pi}{2}, \dfrac{7\pi}{6}, \dfrac{11\pi}{6}\right\}$

33. $\left\{\dfrac{\pi}{3}, \dfrac{2\pi}{3}, \dfrac{4\pi}{3}, \dfrac{5\pi}{3}\right\}$ **35.** $\left\{\dfrac{\pi}{2}\right\}$ **37.** $\left\{\dfrac{\pi}{6}, \dfrac{5\pi}{6}, \dfrac{3\pi}{2}\right\}$ **39.** $\left\{\dfrac{3\pi}{4}, \dfrac{7\pi}{4}\right\}$ **41.** $\left\{\dfrac{\pi}{3}, \dfrac{2\pi}{3}, \dfrac{4\pi}{3}, \dfrac{5\pi}{3}\right\}$

43. $\left\{0, \dfrac{\pi}{3}, \dfrac{\pi}{2}, \dfrac{2\pi}{3}, \pi, \dfrac{4\pi}{3}, \dfrac{3\pi}{2}, \dfrac{5\pi}{3}\right\}$ **45.** $\{1.373, 4.515\}$ **47.** $\{3.872, 5.553\}$

49. (a and c)   **(b)** $\left(\dfrac{\pi}{12}, \dfrac{7}{2}\right), \left(\dfrac{5\pi}{12}, \dfrac{7}{2}\right)$ **(c)** See **(a)**.   **(d)** $\left\{x \left|\ \dfrac{\pi}{12} < x < \dfrac{5\pi}{12}\right.\right\}$

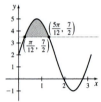

Chapter 1
Section 1.1

1. (c)   **2.** True   **3.** False   **4.** False   **5.** slope   **6.** False

7. $\displaystyle\lim_{x\to 1} 2x = 2$

	x approaches 1 from the left				x approaches 1 from the right		
x	0.9	0.99	0.999	$\to 1 \leftarrow$	1.001	1.01	1.1
$f(x) = 2x$	1.8	1.98	1.998	f approaches 2	2.002	2.02	2.2

9. $\displaystyle\lim_{x\to 0} (x^2 + 2) = 2$

	x approaches 0 from the left				x approaches 0 from the right		
x	−0.1	−0.01	−0.001	$\to 0 \leftarrow$	0.001	0.01	0.1
$f(x) = x^2 + 2$	2.01	2.0001	2.000001	f approaches 2	2.000001	2.0001	2.01

11. $\displaystyle\lim_{x\to -3} \dfrac{x^2 - 9}{x + 3} = -6$

	x approaches −3 from the left				x approaches −3 from the right		
x	−3.5	−3.1	−3.01	$\to -3 \leftarrow$	−2.99	−2.9	−2.5
$f(x) = \dfrac{x^2-9}{x+3}$	−6.5	−6.1	−6.01	f approaches −6	−5.99	−5.9	−5.5

13. $\displaystyle\lim_{x\to 0} \dfrac{2 - 2e^x}{x} = -2$

	x approaches 0 from the left				x approaches 0 from the right		
x	−0.2	−0.1	−0.01	$\to 0 \leftarrow$	0.01	0.1	0.2
$f(x) = \dfrac{2-2e^x}{x}$	−1.8127	−1.9033	−1.9900	f approaches −2	−2.0100	−2.1034	−2.2140

15. $\displaystyle\lim_{x\to 0} \dfrac{1 - \cos x}{x} = 0$

	x approaches 0 from the left				x approaches 0 from the right		
x	−0.2	−0.1	−0.01	$\to 0 \leftarrow$	0.01	0.1	0.2
$f(x) = \dfrac{1-\cos x}{x}$	−0.09967	−0.04996	−0.00500	f approaches 0	0.00500	0.04996	0.09967

17. (a) 2   **(b)** 2   **(c)** 2   **19. (a)** 3   **(b)** 6   **(c)** The limit does not exist.   **21.** 1   **23.** 1   **25.** The limit does not exist.
27. The limit does not exist.   **29.** 9   **31.** The limit does not exist.   **33.** 2   **35.** The limit does not exist.   **37.** Answers will vary.
39. Answers will vary.   **41.** 1   **43.** 0   **45.** 1   **47.** 0   **49.** 0

51. (a) $m_{\text{sec}} = 15$ **(b)** $m_{\text{sec}} = 3(x + 2)$ **(c)** $\displaystyle\lim_{x\to 2} m_{\text{sec}} = 12$ **(d)**

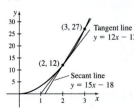

53. (a) $m_{sec} = 2 + \dfrac{1}{2}h$ for $h \neq 0$

(c) $\lim\limits_{h \to 0} m_{sec} = 2$ **(d)** $m_{tan} = 2$

(b)

h	-0.5	-0.1	-0.001	0.001	0.1	0.5
m_{sec}	1.75	1.95	1.9995	2.0005	2.05	2.25

(e)

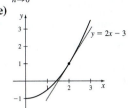

55. (a) These values suggest that $\lim\limits_{x \to 0} \cos \dfrac{\pi}{x}$ may be 1. **(b)** These values suggest that $\lim\limits_{x \to 0} \cos \dfrac{\pi}{x}$ may be -1 **(c)** $\lim\limits_{x \to 0} \cos \dfrac{\pi}{x}$ does not exit. See student solution manual. **(d)** See student solution manual.

57. (a) $\lim\limits_{x \to 2} \dfrac{x-8}{2} = -3$ **(b)** $1.8 \leq x \leq 2.2$ **(c)** $1.98 \leq x \leq 2.02$

x	1.9	1.99	1.999	$\to 2 \leftarrow$	2.001	2.01	2.1
	\multicolumn{3}{c}{x approaches 2 from the left}		\multicolumn{3}{c}{x approaches 2 from the right}	
$f(x) = \dfrac{x-8}{2}$	-3.05	-3.005	-3.0005	f approaches -3	-2.9995	-2.995	-2.95

59. (a) $C(w) = \begin{cases} 0.46, & 0 < w \leq 1 \\ 0.66, & 1 < w \leq 2 \\ 0.86, & 2 < w \leq 3 \\ 1.06, & 3 < w \leq 3.5 \end{cases}$ **(b)** $\{w | 0 < w \leq 3.5\}$ **(c)** See the graph below.

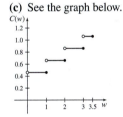

(d) $\lim\limits_{w \to 2^-} C(w) = 0.66$, $\lim\limits_{w \to 2^+} C(w) = 0.86$, $\lim\limits_{w \to 2} C(w)$ does not exist. **(e)** $\lim\limits_{w \to 0^+} C(w) = 0.46$ **(f)** $\lim\limits_{w \to 3.5^-} C(w) = 1.06$

61. (a) $\lim\limits_{t \to 7} S(t) = 100$ **(b)** Given any $\varepsilon > 0$, there exists a $\delta > 0$ such that $|S(t) - 100| < \varepsilon$ whenever $0 < |t - 7| < \delta$. **63.** No, the value of the function at $x = 2$ has no bearing on $\lim\limits_{x \to 2} f(x)$. **65. (a)** The graph of $f(x)$ is the horizontal line $y = -1$ excluding the point $(3, -1)$. **(b)** $\lim\limits_{x \to 3^-} f(x) = \lim\limits_{x \to 3^+} f(x) = -1$. **(c)** The graph suggests $\lim\limits_{x \to 3} f(x) = -1$. **67.** 0 **69.** 0

Section 1.2

1. (a) -3 **(b)** π **2.** 243 **3.** 4 **4. (a)** -1 **(b)** e **5. (a)** -2 **(b)** $\dfrac{7}{2}$ **6. (a)** -6 **(b)** 0 **7.** True **8.** 2 **9.** False **10.** True **11.** 14

13. 0 **15.** 1 **17.** 6 **19.** $\sqrt{11}$ **21.** $2\sqrt{78}$ **23.** $\sqrt{7 + \sqrt{3}}$ **25.** 0 **27.** 5 **29.** $\dfrac{-31}{8}$ **31.** 10 **33.** $\dfrac{13}{4}$ **35.** 4 **37.** 2 **39.** 2 **41.** $\dfrac{\sqrt{2}}{4}$

43. $\dfrac{1}{30}$ **45.** 5 **47.** 6 **49.** 9 **51.** -1 **53.** 8 **55.** 0 **57.** $\dfrac{1}{4}$ **59. (a)** 6 **(b)** -16 **(c)** -16 **(d)** 0 **(e)** $-\dfrac{1}{4}$. **(f)** 0 **61.** 6 **63.** -4

65. $\dfrac{1}{2}$ **67.** 4 **69.** $6x + 4$ **71.** $-\dfrac{2}{x^2}$ **73.** $\lim\limits_{x \to 1^-} f(x) = -1$, $\lim\limits_{x \to 1^+} f(x) = 2$, $\lim\limits_{x \to 1} f(x)$ does not exist **75.** $\lim\limits_{x \to 1^-} f(x) = 2$, $\lim\limits_{x \to 1^+} f(x) = 2$, $\lim\limits_{x \to 1} f(x) = 2$ **77.** $\lim\limits_{x \to 1^-} f(x) = 0$, $\lim\limits_{x \to 1^+} f(x) = 0$, $\lim\limits_{x \to 1} f(x) = 0$ **79.** $\lim\limits_{x \to 3^-} f(x) = 6$ $\lim\limits_{x \to 3^+} f(x) = 6$, $\lim\limits_{x \to 3} f(x) = 6$ **81.** The limit does not exist. **83.** $2x$ **85.** $-\dfrac{1}{x^2}$ **87.** $-\dfrac{1}{16}$ **89.** 6 **91.** 0

93. (a) $C(x) = \begin{cases} 9.00 & 0 \leq x \leq 10 \\ 9.00 + 0.95(x - 10) & 10 < x \leq 30 \\ 28.00 + 1.65(x - 30) & 30 < x \leq 100 \\ 143.50 + 2.20(x - 100) & x > 100 \end{cases}$

(e)

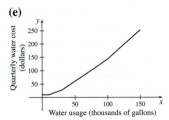

(b) $\{x | x \geq 0\}$ **(c)** $\lim\limits_{x \to 5} C(x) = 9.00$, $\lim\limits_{x \to 10} C(x) = 9.00$, $\lim\limits_{x \to 30} C(x) = 28.00$, $\lim\limits_{x \to 100} C(x) = 143.50$ **(d)** 9.00

95. (a) 0 **(b)** As the temperature of a gas approaches zero, the molecules in the gas stop moving.

97. $\lim\limits_{x \to 0} |x| = 0$ since $\lim\limits_{x \to 0^-} |x| = 0$ and $\lim\limits_{x \to 0^+} |x| = 0$. **99.** Answers will vary. **101.** Answers will vary. **103.** See student solution manual.

105. na^{n-1} **107.** $\dfrac{m}{n}$ **109.** $\dfrac{a+b}{2}$ **111.** 0.

Section 1.3

1. True **2.** False **3.** $f(c)$ is defined, $\lim_{x \to c} f(x)$ exists, $\lim_{x \to c} f(x) = f(c)$ **4.** True **5.** False. **6.** False **7.** False **8.** True **9.** The function is discontinuous. **10.** The function is continuous. **11.** True **12.** False **13. (a)** The function is not continuous at $c = -3$. **(b)** $\lim_{x \to -3} f(x) \neq f(-3)$ **(c)** The discontinuity is removable. **(d)** $f(-3) = -2$ **15. (a)** The function is not continuous at $c = 2$. **(b)** $\lim_{x \to 2^-} f(x) \neq f(2)$. **(c)** The discontinuity is removable. **17. (a)** The function is continuous at $c = 4$. **19.** The function is continuous at $c = -1$. **21.** The function is continuous at $c = -2$. **23.** The function is continuous at $c = 2$. **25.** The function is not continuous at $c = 1$. **27.** The function is not continuous at $c = 1$. **29.** The function is continuous at $c = 0$. **31.** The function is not continuous at $c = 0$. **33.** $f(2) = 4$ **35.** $f(1) = 2$ **37.** The function is continuous on the given interval. **39.** The function is not continuous on the given interval. The function is continuous on the set $\{x \mid x < 3\} \cup \{x \mid x > 3\}$. **41.** The function is continuous on $\{x \mid x \neq 0\}$. **43.** The function is continuous on the set of all real numbers. **45.** The function is continuous of $\{x \mid x \geq 0, x \neq 9\}$. **47.** The function is continuous on the set $\{x \mid x < 2\}$. **49.** The function is continuous on the set of all real numbers. **51.** f is continuous at 0 because $\lim_{x \to 0} f(x) = f(0)$. **53.** f is not continuous at 3 because $\lim_{x \to 3} f(x)$ does not exist. **55.** f is continuous at 1 because $\lim_{x \to 1} f(x) = f(1)$.

57. (a)

(b) Based on the graph, it appears that the function is continuous for all real numbers. **(c)** f is actually continuous at all real numbers except $x = 2$. **(d)** You should not use graphing technology alone to determine continuity.

59. Yes. A zero will exist on the given interval. **61.** The IVT gives no information. **63.** The IVT gives no information. **65.** 1.154 **67.** 0.211 **69.** 3.134 **71.** 1.157 **73. (a)** Since f is continuous on $[0, 1]$, $f(0) < 0$ and $f(1) > 0$, the IVT guarantees that f must have a zero on the interval $(0, 1)$. **(b)** 0.828 **75.** The function is not continuous at $c = 1$.
77. The function is continuous on the set of all real numbers.

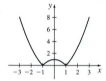

79. (a) $C(w) = \begin{cases} 0.46, & 0 < w \leq 1 \\ 0.66, & 1 < w \leq 2 \\ 0.86, & 2 < w \leq 3 \\ 1.06, & 3 < w \leq 3.5 \end{cases}$ **(b)** $\{w \mid 0 < w \leq 3.5\}$ **(c)** The function is continuous on the intervals $(0, 1]$, $(1, 2]$, $(2, 3]$, and $(3, 3.5]$.

(d) At $w = 1$, $w = 2$, and $w = 3$, the function has a jump discontinuity. **(e)** See student solution manual.

81. (a) $C(x) = \begin{cases} 22.13 + 0.62273x, & 0 < x \leq 50 \\ 29.207 + 0.48116x, & x > 50 \end{cases}$ **(b)** $\{x \mid x > 0\}$ **(c)** $C(x)$ is continuous on its domain. **(d)** No such points.

(e) There is no jump in cost for usage over 50 therms. **83. (a)** $\dfrac{Gm}{R^2}$ **(b)** 1.3 m/s² **(c)** The gravity on Europa is less than the gravity on Earth.

85. $A = 5$, $B = 7$

87. (a) $\{x \mid x < 2\} \cup \{x \mid x > 2, x \neq 8\}$ **(b)** f is discontinuous at $x = 8$. **(c)** The discontinuity at $x = 8$ is removable.
89. (a) Since f is continuous on $[0, 2]$, $f(0) < 0$ and $f(2) > 0$, the IVT guarantees that f must have a zero on the interval $(0, 2)$. **(b)** $x = 0.443$
91. This does not contradict the IVT. $f(-1)$ and $f(2)$ are both positive, so there is no guarantee that the function has a zero in the interval $(-1, 2)$.
93. This does not contradict the IVT. The IVT guarantees that f must have at least one zero on the interval $(-5, 0)$.

95. (a) This does not contradict the IVT. $f(-2)$ and $f(2)$ are both positive, so there is no guarantee that the function has a zero in the interval $(-2, 2)$. **(b)** The graph below indicates that f does not have a zero in the interval $(-2, 2)$.

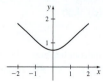

97. See student solution manual. **99. (a)** $x = -2, x = 2, x = 3$ **(b)** $p = 5$ **(c)** $h(x)$ is an even function. **101.** Answers will vary.
103. $(1.125, 1.25)$ **105.** $(0.125, 0.25)$ **107.** $(3.125, 3.25)$ **109.** $(1.125, 1.25)$ **111.** Since f is continuous on $[0, 1]$, $f(0) < 0$ and $f(1) > 0$, the IVT guarantees that f must have a zero on the interval $(0, 1)$. To the nearest tenth, the zero is $x = 0.8$ **113.** See student solution manual.
115. See student solution manual. **117.** See student solution manual. **119.** See student solution manual.

Section 1.4

1. 0 **2.** False **3.** L **4.** False **5.** 1 **7.** 1 **9.** 0 **11.** $\dfrac{\sqrt{3}}{2} + \dfrac{1}{2}$ **13.** 1 **15.** $\dfrac{3}{2}$ **17.** 0 **19.** 1 **21.** $\dfrac{1}{2}$ **23.** 7 **25.** 2 **27.** $\dfrac{1}{2}$ **29.** 5
31. 0 **33.** 1 **35.** f is continuous at $c = 0$. **37.** f is continuous at $c = \dfrac{\pi}{4}$. **39.** f is continuous on $\{x | x \neq 4\}$. **41.** f is continuous

on $\left\{ x | x \neq \dfrac{3\pi}{2} + 2k\pi \right\}$. **43.** f is continuous on $\{x | x > 0, x \neq 3\}$. **45.** f is continuous on the set of all real numbers. **47.** 0 **49.** 0

51. See student solution manual. **53.** See student solution manual. **55. (a)** $\displaystyle\lim_{\theta \to \pi/4^+} x(\theta) = 5\sqrt{2}t$, $\displaystyle\lim_{\theta \to \pi/4^+} y(\theta) = -16t^2 + 5\sqrt{2}t$ **(b)** See

student solution manual. **(c)** $\displaystyle\lim_{\theta \to \pi/2^-} x(\theta) = 0$, $\displaystyle\lim_{\theta \to \pi/2^+} y(\theta) = -16t^2 + 10t$ **(d)** See student solution manual. **57.** See student solution

manual. **59.** $f(0) = \pi$ **61.** Yes **63.** See student solution manual. **65.** See student solution manual. **67.** See student solution manual. **69.** 0

71. $\left(\dfrac{n}{n + 1}, 0 \right)$

Section 1.5

1. False **2. (a)** $-\infty$ **(b)** ∞ **(c)** $-\infty$ **3.** False **4.** vertical **5. (a)** 0 **(b)** 0 **(c)** ∞ **6.** False **7. (a)** 0 **(b)** ∞ **(c)** 0 **8.** True
9. 2 **11.** ∞ **13.** ∞ **15.** $x = -1, x = 3$ **17.** -3 **19.** ∞ **21.** 0 **23.** ∞ **25.** $x = -3, x = 0, x = 4$ **27.** $-\infty$ **29.** ∞ **31.** ∞

33. ∞ **35.** $-\infty$ **37.** $-\infty$ **39.** $-\infty$ **41.** $-\infty$ **43.** 0 **45.** $\dfrac{2}{5}$ **47.** 1 **49.** 0 **51.** $\dfrac{5}{4}$ **53.** 0 **55.** 0 **57.** $\sqrt{3}$ **59.** $-\infty$

61. $x = 0$ is a vertical asymptote. $y = 3$ is a horizontal asymptote. **63.** $x = -1$ and $x = 1$ are vertical asymptotes. $y = 1$ is a horizontal asymptote.

65. $x = \dfrac{3}{2}$ is a vertical asymptote. $y = \dfrac{\sqrt{2}}{2}$ and $y = -\dfrac{\sqrt{2}}{2}$ are horizontal asymptotes. **67. (a)** $\{x | x \neq -2, x \neq 0\}$ **(b)** $y = 0$ is a horizontal

asymptote. **(c)** $x = 0$ and $x = -2$ are vertical asymptotes. **(d)** See student solution manual. **69. (a)** $\left\{ x | x \neq \dfrac{3}{2}, x \neq 2 \right\}$

(b) $y = \dfrac{1}{2}$ is a horizontal asymptote. **(c)** $x = \dfrac{3}{2}$ is a vertical asymptote. **(d)** See student solution manual. **71. (a)** $\{x | x \neq 0, x \neq 1\}$

(b) There are no horizontal asymptotes. **(c)** $x = 0$ is a vertical asymptote. **(d)** See student solution manual. **73.** Answers will vary.
75. (a) T **(b)** u_0 **77.** ∞

79. (a) 500 **(b)**

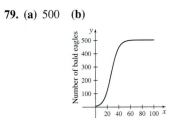

Time (years)

(c) See student solution manual.

81. (a)

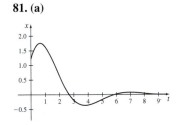

The graph suggests that $\displaystyle\lim_{t \to \infty} x(t) = 0$.

(b) 0 **(c)** Yes

83. (a) 0.26 moles **(b)** 395 min **(c)** 0 moles **(d)** See student solution manual.
85. See student solution manual. **87.** See student solution manual. **89.** See student solution manual.

91. (a)

x	100	10,000	1,000,000	1,000,000,000	$\to \infty$
$f(x) = \left(1 + \dfrac{1}{x}\right)^x$	2.70814	2.718146	2.718280	2.718282	2.718282

(b) $e \approx 2.718281828$ **(c)** See student solution manual.

93. (a) ∞ **(b)** See student solution manual.

Section 1.6

1. False **2.** True **3.** True **4.** False **5.** True **6.** False **7.** $\delta = 0.005$ **9.** $\delta = \dfrac{1}{12}$ **11.** $\delta = 0.02$ **13. (a)** $\delta = 0.025$ **(b)** $\delta = 0.0025$
(c) $\delta = 0.00025$ **(d)** $\delta = \dfrac{\varepsilon}{4}$ **15. (a)** $\delta = 0.1$ **(b)** $\delta = 0.01$ **(c)** $\delta = \varepsilon$ **17.** Given any $\varepsilon > 0$, let $\delta = \dfrac{\varepsilon}{3}$. See student solution manual for the
complete proof. **19.** Given any $\varepsilon > 0$, let $\delta = \dfrac{\varepsilon}{2}$. See student solution manual for the complete proof. **21.** Given any $\varepsilon > 0$, let $\delta = \dfrac{\varepsilon}{5}$. See student
solution manual for the complete proof. **23.** Given any $\varepsilon > 0$, let $\delta = \min\left\{ 1, \dfrac{\varepsilon}{3} \right\}$. See student solution manual for the complete proof. **25.** Given
any $\varepsilon > 0$, let $\delta = \min\left\{ 1, \dfrac{2\varepsilon}{7} \right\}$. See student solution manual for the complete proof. **27.** Given any $\varepsilon > 0$, let $\delta = \varepsilon^3$. See student solution

manual for the complete proof. **29.** Given any $\varepsilon > 0$, let $\delta = \min\left\{1, \dfrac{\varepsilon}{3}\right\}$. See student solution manual for the complete proof. **31.** Given any $\varepsilon > 0$, let $\delta = \min\{1, 6\varepsilon\}$. See student solution manual for the complete proof. **33.** See student solution manual. **35.** Given any $\varepsilon > 0$, let $\delta = \min\left\{1, \dfrac{234}{7}\varepsilon\right\}$. See student solution manual for the complete proof. **37.** Given any $\varepsilon > 0$, let $\delta = \min\{1, 26\varepsilon\}$. See student solution manual for the complete proof. **39.** Given any $\varepsilon > 0$, let $\delta = \dfrac{\varepsilon}{1 + |m|}$. See student solution manual for the complete proof. **41.** x must be within 0.05 of 3. **43.** See student solution manual. **45.** Given any $\varepsilon > 0$, let $M = \dfrac{1}{\varepsilon^2}$. See student solution manual for the complete proof. **47.** See student solution manual. **49.** See student solution manual. **51.** See student solution manual. **53.** Given any $\varepsilon > 0$, let $\delta = \min\left\{1, \dfrac{\varepsilon}{47}\right\}$. See student solution manual for the complete proof. **55.** $M = 101$ **57.** See student solution manual.

Review Exercises

1. $\displaystyle\lim_{x\to 0}\dfrac{1-\cos x}{1+\cos x} = 0.$

		x approaches 0 from the left				x approaches 0 from the right		
x		-0.1	-0.01	-0.001	$\to 0 \leftarrow$	0.001	0.01	0.1
$f(x) = \dfrac{1-\cos x}{1+\cos x}$		0.002504	0.000025	0.00000025	f approaches 0	0.00000025	0.000025	0.002504

3. $\displaystyle\lim_{x\to 2} f(x)$ does not exist. **5.** $-\dfrac{3}{x^2}$ **7.** 1 **9.** $-\pi$ **11.** 0 **13.** 27 **15.** 4 **17.** $\dfrac{2}{3}$ **19.** $-\dfrac{1}{6}$ **21.** 6 **23.** 5 **25.** -1 **27.** 8 **29.** $\displaystyle\lim_{x\to 2^-} f(x) = 7$, $\displaystyle\lim_{x\to 2^+} f(x) = 7$, $\displaystyle\lim_{x\to 2} f(x) = 7$ **31.** f is not continuous at $c = 1$. **33.** f is continuous at $c = 0$.

35. f is not continuous at $c = \frac{1}{2}$. **37.** (a) $2x - 3$ (b) -1 **39.** f is continuous on the set $\{x \mid x \neq 3\}$. **41.** f is continuous on the set $\{x \mid x \neq -2, x \neq 0\}$. **43.** f is continuous on the set of all real numbers. **45.** 0.215

47. $\displaystyle\lim_{x\to 0^+}\dfrac{|x|}{x}(1-x) = 1$, $\displaystyle\lim_{x\to 0^-}\dfrac{|x|}{x}(1-x) = -1$, $\displaystyle\lim_{x\to 0}\dfrac{|x|}{x}(1-x)$ does not exist. **49.** $\dfrac{1}{2\sqrt{x}}$ **51.** 1 **53.** $\dfrac{3}{4}$ **55.** 0 **57.** $-\infty$ **59.** 3

61. $x = -3$ is a vertical asymptote. $y = 4$ is a horizontal asymptote. **63.** Yes **65.** $f(0) = -\pi$

67. (a)

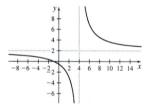

(b) $f(x) = \dfrac{2x+2}{x-4}$

69. See student solution manual.

Chapter 2
Section 2.1

1. True **2.** True **3.** Prime, slope, $(c, f(c))$ **4.** True **5.** 6 **6.** derivative

7.

Time interval	Δt	$\dfrac{\Delta s}{\Delta t}$
$[3, 3.1]$	0.1	61
$[3, 3.01]$	0.01	60.1
$[3, 3.001]$	0.001	60.01

The velocity appears to approach 60 m/s.

9. $v(t) = 4$ m/s at $t = 0$, $v(t) = 16$ m/s at $t = 2$, $v(t) = 6t + 4$ m/s at any t

11. $v(1) = 7$ cm/s, $v(4) = \dfrac{385}{16}$ cm/s

13. $y = -12x - 12$

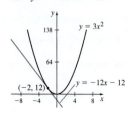

15. $y = 12x + 16$

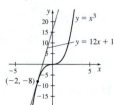

17. $y = -x + 2$

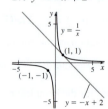

19. $y = -\dfrac{1}{36}x + \dfrac{7}{36}$

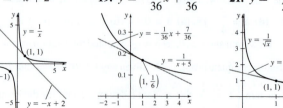

21. $y = -\dfrac{1}{2}x + \dfrac{3}{2}$

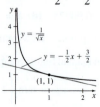

23. (a) 5 (b) 5 (c) 5 **25.** (a) 0 (b) $\dfrac{7}{16}$ (c) $\dfrac{c(c+6)}{(c+3)^2}$ **27.** $f'(1)=2$ **29.** $f'(0)=0$ **31.** $f'(-1)=-5$ **33.** $f'(4)=\dfrac{1}{4}$

35. $f'(0)=-7$ **37.** (a) $7<t<10$ and $11<t<13$ (b) $0<t<4$ (c) $4\le t\le 7$ and $10\le t\le 11$ (d) Average velocity ≈ 0.675 mi/min
(e) 0 mi/min **39.** -3 **41.** No **43.** $R'(50)=0.59$ **45.** (a) 250 magazines per day (b) 200 magazines per day
47. (a) 32 ft/s (b) $t\approx 7.914$ s (c) Average velocity ≈ 126.618 ft/s (d) $v(7.914)\approx 253.235$ ft/s
49. (a) 9.8 m/s (b) $t\approx 4.243$ s (c) Average velocity ≈ 20.789 m/s (d) $v(4.243)\approx 41.578$ m/s

51. (a) $\dfrac{\Delta V}{\Delta x}\approx 12.060\,\text{cm}^3/\text{cm}$ (b) $V'(2)=12\,\text{cm}^3/\text{cm}$ **53.** $f'(x)=2ax+b$ **55.** (a) $d'(t)$ is the rate of change of diameter (in centimeters)
with respect to time (in days). (b) $d'(1)>d'(20)$ (c) $d'(1)$ is the instantaneous rate of change of the peach's diameter on day 1 and $d'(20)$ is
the instantaneous rate of change of the peach's diameter on day 20.

Section 2.2

1. False **2.** False **3.** (b) Vertical **4.** derivative **5.** 0 **7.** 2 **9.** $-2c$ **11.** $f'(x)=0$, all real numbers **13.** $f'(x)=6x+1$, all real numbers
15. $f'(x)=\dfrac{5}{2\sqrt{x-1}},\, x>1$

17. $f'(x)=\dfrac{1}{3}$ **19.** $f'(x)=4x-5$ **21.** $f'(x)=3x^2-8$

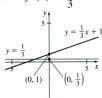
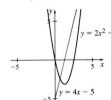
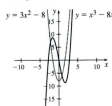

23. Not a graph of f and f' **25.** Graph of f and f'. The blue curve is the graph of f; the green curve is the graph of f'.
27.   **29.**

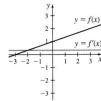
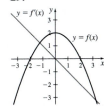

31. (B) **33.** (A) **35.** $f'(-8)=-\dfrac{1}{3}$ **37.** $f'(2)$ does not exist. **39.** $f'(1)=2$ **41.** $f'\left(\dfrac{1}{2}\right)$ does not exist. **43.** $f'(-1)$ does not exist.

45. (a) -2 and 4 (b) $0,2,6$ **47.** (a) $u_1'(1)$ does not exist. (b) $u_1(t)$ models a switch that is off when $t<1$ and on when $t\ge 1$. **49.** $f'(x)=m$
51. $f'(x)=-2/x^2$ **53.** $f(x)=x^2, c=2$ **55.** $f(x)=x^2, c=1$ **57.** $f(x)=\sin x, c=\dfrac{\pi}{6}$ **59.** $f(x)=2x^2-x-6, c=2$
61. (a) Continuous at 0 (b) $f'(0)=0$ (c)

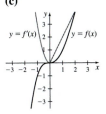

63. (a) $s'(4.99)=74.700$ ft/s; $s'(5.01)=0$ ft/s (b) Not continuous (c) The vehicle comes to a sudden stop. **65.** $\dfrac{dp}{dx}=-kp$

67. $(\sqrt{2},4)$ and $(-\sqrt{2},4)$ **69.** (a) $\pi\,\Delta r(2r+\Delta r)$ (b) $2\pi\,\Delta r$ (c) $2\pi r+\pi\,\Delta r$ (d) 2π (e) 2π **71.** See student solutions manual.
73. See student solutions manual. **75.** (a) Parallel (b) Perpendicular **77.** (a) $k=4$ (b) $f'(3)=12$ (c) $f'(x)=4x$

Section 2.3

1. $0; 3x^2$ **2.** nx^{n-1} **3.** True **4.** $k\left[\dfrac{d}{dx}f(x)\right]$ **5.** e^x **6.** True **7.** $f'(x)=3$ **9.** $f'(x)=2x+3$ **11.** $f'(u)=40u^4-5$

13. $f'(s)=3as^2+3s$ **15.** $f'(t)=\dfrac{5}{3}t^4$ **17.** $f'(t)=\dfrac{3}{5}t^2$ **19.** $f'(x)=\dfrac{3x^2+2}{7}$ **21.** $f'(x)=2ax+b$ **23.** $f'(x)=4e^x$

25. $f'(u)=10u-2e^u$ **27.** $\sqrt{3}$ **29.** $2\pi R$ **31.** $4\pi r^2$

33. (a) 3   **(b)** $y = 3x - 1$ **(c)**

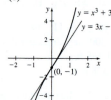

35. (a) 6   **(b)** $y = 6x + 1$ **(c)**

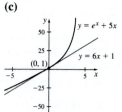

37. (a) $(2, -8)$ **(e)**

(b) $y = -8$

(c) $x > 2$

(d) $x < 2$

(f) f is increasing when $x > 2$ and decreasing when $x < 2$.

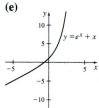

39. (a) None   **(e)**

(b) None

(c) All real numbers

(d) None

(f) f is increasing for all x.

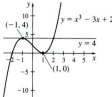

41. (a) $(1, 0), (-1, 4)$ **(e)**

(b) $y = 0, y = 4$

(c) $x < -1$ or $x > 1$

(d) $-1 < x < 1$

(f) f is increasing when $x < -1$ or $x > 1$ and decreasing when $-1 < x < 1$.

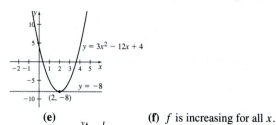

43. $v(0) = -1$ m/s and $v(5) = 74$ m/s   **45. (a)** $v(t) = 2t - 5$ **(b)** $t = \dfrac{5}{2}$ **47. (a)** $\dfrac{4}{5}$ **(b)** $-\dfrac{7}{20}$ **(c)** $-\dfrac{9}{5}$ **(d)** $\dfrac{23}{20}$ **(e)** $-\dfrac{21}{20}$ **(f)** $-\dfrac{23}{10}$

49. (a) $f'(x) = 24x^2 - 24x + 6$ **(b)** $f'(x) = 6(2x - 1)^2$ **51.** $\dfrac{5}{16}$ **53.** $20480\sqrt{3} \approx 35472.401$ **55.** $3ax^2$ **57.** $y = 5x - 3$

59. $y = x + 1$ **61.** $y = 3x + \dfrac{5}{3}, y = 3x - 9$

63. (a) $y = 45x - 65$ **(d)**

(b) $\left(\dfrac{1}{\sqrt{3}}, -\dfrac{5\sqrt{3}}{9} - 1 \right)$,

$\left(-\dfrac{1}{\sqrt{3}}, \dfrac{5\sqrt{3}}{9} - 1 \right)$ **(c)** At

$x = \dfrac{1}{\sqrt{3}}, y = x - \dfrac{8\sqrt{3}}{9} - 1$ and at

$x = -\dfrac{1}{\sqrt{3}}, y = x + \dfrac{8\sqrt{3}}{9} - 1$

65. See student solutions manual.   **67.** See student solutions manual.   **69.** $a = 3, b = 2, c = 0$ **71.** $(2, 4)$ **73. (a)** 0   **(b)** $-kR$ **(c)** $-2kR$

75. (a) $\dfrac{dL}{dT} = 4\sigma AT^3$ **(b)** $2.694 \times 10^{23} \dfrac{\text{W}}{\text{K}}$ **(c)** 2.694×10^{23} W   **77.** $F'(x) = mx$ **79.** See student solutions manual.

81. $\dfrac{1}{2}$

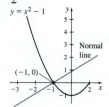

83. $-\dfrac{1}{5}$

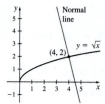

85. -4

87. $Q = \left(\dfrac{9}{4}, \dfrac{81}{16}\right)$ **89.** $y = -\dfrac{7}{64}x^4 + \dfrac{7}{16}x^3 + \dfrac{21}{16}x^2 - 4x$ **91. (a)** $c = 1$ **(b)** $y = 12x - 16$ and $y = 12x + 1$

Section 2.4

1. False   **2.** $f(x)g'(x) + f'(x)g(x)$ **3.** False   **4.** $\dfrac{\left[\dfrac{d}{dx}f(x)\right]g(x) - f(x)\left[\dfrac{d}{dx}g(x)\right]}{[g(x)]^2}$ **5.** True   **6.** $-\dfrac{\dfrac{d}{dx}g(x)}{[g(x)]^2}$ **7.** 0   **8.** $\dfrac{d^2 s}{dt^2}$

9. $f'(x) = e^x(x+1)$ **11.** $f'(x) = 5x^4 - 2x$ **13.** $f'(x) = 18x^2 + 6x - 10$ **15.** $s'(t) = 16t^7 - 24t^5 + 10t^4 - 4t^3 + 4t - 1$

17. $f'(x) = x^3 e^x + 3x^2 e^x + e^x + 3x^2$ **19.** $g'(s) = \dfrac{2}{(s+1)^2}$ **21.** $G'(u) = -\dfrac{4}{(1+2u)^2}$ **23.** $f'(x) = \dfrac{2(6x^2 + 16x + 3)}{(3x+4)^2}$

25. $f'(w) = -\dfrac{3w^2}{(w^3 - 1)^2}$ **27.** $s'(t) = -\dfrac{3}{t^4}$ **29.** $f'(x) = \dfrac{4}{e^x}$ **31.** $f'(x) = -\dfrac{40}{x^5} - \dfrac{6}{x^3}$ **33.** $f'(x) = 9x^2 + \dfrac{2}{3x^3}$ **35.** $s'(t) = -\dfrac{1}{t^2} + \dfrac{2}{t^3} - \dfrac{3}{t^4}$

37. $f'(x) = e^x\left(\dfrac{1}{x^2} - \dfrac{2}{x^3}\right)$ **39.** $f'(x) = e^{-x}\left(1 - x - \dfrac{1}{x} - \dfrac{1}{x^2}\right)$ **41.** $f'(x) = 6x + 1, f''(x) = 6$ **43.** $f'(x) = f''(x) = e^x$

45. $f'(x) = e^x(x+6), f''(x) = e^x(x+7)$ **47.** $f'(x) = 8x^3 + 3x^2 + 10, f''(x) = 24x^2 + 6x$ **49.** $f'(x) = 1 - \dfrac{1}{x^2}, f''(x) = \dfrac{2}{x^3}$ **51.** $f'(t) = $
$1 + \dfrac{1}{t^2}, f''(t) = -\dfrac{2}{t^3}$ **53.** $f'(x) = e^x(x^{-1} - x^{-2}), f''(x) = e^x(2x^{-3} - 2x^{-2} + x^{-1})$ **55. (a)** $y' = -\dfrac{1}{x^2}, y'' = \dfrac{2}{x^3}$ **(b)** $y' = \dfrac{5}{x^2}, y'' = -\dfrac{10}{x^3}$

57. $v(t) = 32t + 20, a(t) = 32$ **59.** $v(t) = 9.8t + 4, a(t) = 9.8$ **61.** $f^{(4)}(x) = 0$ **63.** 5040   **65.** e^u **67.** $-e^x$

69. (a) $\dfrac{3}{4}$ **(b)** $y = \dfrac{3}{4}x + \dfrac{1}{4}$ **(c)** $(0, 0), (2, 4)$ **(d)**

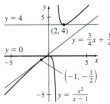

71. (a) $\dfrac{5}{4}$ **(d)**

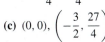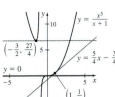

(b) $y = \dfrac{5}{4}x - \dfrac{3}{4}$

(c) $(0, 0), \left(-\dfrac{3}{2}, \dfrac{27}{4}\right)$

73. (a) $(-2, 5), (2, -27)$ **(e)**

(b) $y = 5, y = -27$

(c) $x < -2$ or $x > 2$

(d) $-2 < x < 2$

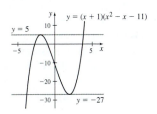

75. (a) $(0, 0), (-2, -4)$ **(e)**

(b) $y = 0, y = -4$

(c) $x < -2$ or $x > 0$

(d) $-2 < x < -1$ or $-1 < x < 0$

77. (a) $(-1, -e^{-1})$ **(e)**

(b) $y = -\dfrac{1}{e}$

(c) $x > -1$

(d) $x < -1$

79. (a) $(-1, -2e),$ **(e)**

$(3, 6e^{-3})$

(b) $y = -2e,$

$y = 6e^{-3}$

(c) $-1 < x < 3$

(d) $x < -1$ or $x > 3$

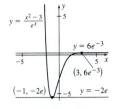

81. (a) $\dfrac{8}{5}$ **(b)** $-\dfrac{29}{10}$ **(c)** $\dfrac{1}{9}$ **(d)** $-\dfrac{19}{160}$ **(e)** $\dfrac{1}{9}$ **(f)** $\dfrac{12}{25}$ **83. (a)** $v(t) = -9.8t + 39.2$ **(b)** $t = 4$ seconds   **(c)** 78.4 m   **(d)** -9.8 m/s^2

(e) $t = 8$ s   **(f)** $v = -39.2$ m/s   **(g)** 156.8 m

85. (a) $0 \le x \le 100$ **(b)**

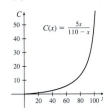

(c) \$13,333.33   **(d)** $C'(x) = \dfrac{550}{(110 - x)^2}$

(e) $C'(40) = 0.112 = \$112/\%, C'(60) = 0.220 = \$220/\%, C'(80) = 0.611 = \$611/\%,$
$C'(90) = 1.375 = \$1,375/\%$

(f) Cleanup costs rise dramatically as the percentage of pollutant removed is increased.

87. (a) $f'(t) = \dfrac{0.4 - 0.8t^2}{(2t^2 + 1)^2}$ mg/L/h

(b and c) $f'\left(\dfrac{1}{6}\right) \approx 0.339$ mg/L/h, $f'\left(\dfrac{1}{2}\right) \approx 0.089$ mg/L/h,

$f'(1) \approx -0.044$ mg/L/h

(d)

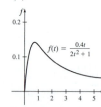

(e) Concentration is highest at approximately 45 min and is about 0.14 mg/L.

89. (a) $D'(p) = -200,000\dfrac{p+5}{(p^2 + 10p + 50)^2}$ **(b)** $D'(5) = -128$, $D'(10) = -48$, $D'(15) = -22.145$

91. (a) $\dfrac{dp}{dr} = -\dfrac{9nRT}{4\pi r^4}$ **(b)** As the radius increases, pressure within the container decreases. **(c)** $p'(1/4) \approx -415945.358$ Pa/m

93. (a) $v(t) = 3t^2 - 1$, $a(t) = 6t$, $J(t) = 6$, $S(t) = 0$ **(b)** $t = \pm\sqrt{3}/3$ **(c)** $a(2) = 12$ m/s^2, $a(5) = 30$ m/s^2 **(d)** No

(e) Snap is identically zero. **95. (a)** $a(t) = 1.6 + 1.998t$ m/s^2 **(b)** $J(t) = 1.998$ m/s^3

97. (a) $\dfrac{dJ}{dr} = -\dfrac{2I}{\pi r^3}$ **(b)** As radius increases, current density decreases. **(c)** -12.732 A/mm^3 **99.** See student solutions manual.

101. $y' = 6x^5 - 5x^4 + 20x^3 - 18x^2 + 2x - 5$ **103.** $y' = 9x^2(x^3 + 1)^2$ **105.** $y' = (x^5 + 2x^4 - 4x^2 - 5x + 6)/x^7$

107. See student solutions manual. **109.** $f'(x) = 2x(x^8 + 2x^4 - 2x^2 + 1)/(x^4 + 1)^2$

111. (a) $f'(x) = \dfrac{2}{x+1} - \dfrac{2x}{(x+1)^2}$ **(b)** $f'(x) = \dfrac{2}{(x+1)^2}$ **(c)** $f^{(5)}(x) = \dfrac{240}{(x+1)^6}$

113. (a) $[f_1'(x)\cdot f_2(x)\cdots f_n(x)] + [f_1(x)\cdot f_2'(x)\cdot f_3(x)\cdots f_n(x)] + \cdots + [f_1(x)\cdots f_{n-1}(x)\cdot f_n'(x)]$ **(b)** $-\dfrac{1}{f_1(x)\cdots f_n(x)}\left[\dfrac{f_1'(x)}{f_1(x)} + \cdots + \dfrac{f_n'(x)}{f_n(x)}\right]$

115. (a) $f_1(x) = \dfrac{x}{x-1}$, $f_2(x) = \dfrac{2x-1}{x}$, $f_3(x) = \dfrac{3x-1}{2x-1}$, $f_4(x) = \dfrac{5x-2}{3x-1}$, $f_5(x) = \dfrac{8x-3}{5x-2}$

(b) $a_0 = 1$, $a_1 = 1$, $a_2 = 2$, $a_3 = 3$, $a_4 = 5$, $a_5 = 8$ **(c)** $a_n = a_{n-1} + a_{n-2}$

(d) $f_0'(x) = 1$, $f_1'(x) = -\dfrac{1}{(x-1)^2}$, $f_2'(x) = \dfrac{1}{x^2}$, $f_3'(x) = -\dfrac{1}{(2x-1)^2}$, $f_4'(x) = \dfrac{1}{(3x-1)^2}$, $f_5'(x) = -\dfrac{1}{(5x-2)^2}$

Section 2.5

1. False **2.** False **3.** True **4.** False **5.** $f'(\pi) = 2$ **7.** $f'(\pi/3) = -4 + 2\sqrt{3} \approx -0.536$ **9.** $y' = 3\cos\theta + 2\sin\theta$
11. $y' = \cos^2 x - \sin^2 x = \cos 2x$ **13.** $y' = \cos t - t\sin t$ **15.** $y' = e^x(\sec^2 x + \tan x)$ **17.** $y' = \pi\sec^3 u + \pi\tan^2 u\sec u$
19. $y' = -(x\csc^2 x + \cot x)/x^2$ **21.** $y' = x(x\cos x + 2\sin x)$ **23.** $y' = t\sec^2 t + \tan t - \sqrt{3}\sec t\tan t$ **25.** $y' = -1/(1 - \cos\theta)$
27. $y' = (\cos t + t\cos t - \sin t)/(1+t)^2$ **29.** $y' = (\cos x - \sin x)/e^x$ **31.** $y' = -2/(\sin\theta - \cos\theta)^2$
33. $y' = (\sec t\tan t + t\tan^2 t - t - \tan t)/(1 + t\sin t)^2$ **35.** $y' = -\csc\theta(\csc^2\theta + \cot^2\theta)$ **37.** $y' = 2\sec^2 x/(1 - \tan x)^2$
39. $y'' = -\sin x$ **41.** $y'' = 2\tan\theta\sec^2\theta$ **43.** $y'' = 2\cos t - t\sin t$ **45.** $y'' = 2e^x\cos x$ **47.** $y'' = 3\cos u - 2\sin u$
49. $y'' = -(a\sin x + b\cos x)$

51. (a) $y = x$ **(b)** **53. (a)** $y = x$ **(b)** **55. (a)** $y = \sqrt{2}$ **(b)**

57. (a) $\left\{\left(\tan^{-1}(2) + 2n\pi, \sqrt{5}\right)\right\}$, $\left\{\left(\tan^{-1}(2) + (2n+1)\pi, -\sqrt{5}\right)\right\}$, where n is an integer **(b)**

59. (a) $\{(2n\pi, 1)\}$, $\{((2n+1)\pi, -1)\}$, where n is an integer **(b)**

61. $f^{(n)}(x) = \begin{cases} (-1)^{\frac{n}{2}}\sin(x) & n \text{ even} \\ (-1)^{\frac{n-1}{2}}\cos(x) & n \text{ odd} \end{cases}$

63. -1 **65. (a)** $v(t) = -\dfrac{1}{8}\sin t$ m/s **(b)** $\dfrac{\pi}{2} + n\pi$, where n is an integer **(c)** $a(t) = -\dfrac{1}{8}\cos t$ m/s^2 **(e)**

(d) $\dfrac{\pi}{2} + n\pi$, where n is an integer

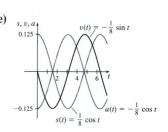

67. (a) $\dfrac{dy}{d\theta} = -8\sin\theta$ **(b)** $\dfrac{dy}{d\theta}\left(\dfrac{\pi}{12}\right) \approx -2.071$ ft/radian

69. (a)

(b) Max = 4, Min = $\dfrac{4}{3}$

(c) $x = \pi, 2\pi, 3\pi$

(d) $w'(\pi - 0.1) \approx 0.395$, $w'(2\pi - 0.1) \approx -0.045$

(e) A symmetric wave would give slopes having equal magnitudes but opposite signs.

71. (a) $10,000, $25,093, $15,411, $37,570   **(d)**

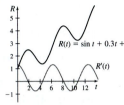

(b) $R'(t) = \cos(t) + 0.3$

(c) 1.054

(e) Answers will vary.

73. $n = 4$ **75.** See student solutions manual.   **77.** See student solutions manual.   **79.** See student solutions manual.

81. See student solutions manual.   **83. (a)** $f^*(x) = 4x + 2$ **(b)** $f^*(x) = -2\sin x$ **(c)** $f^* = 2f'$

Review Exercises

1. (a) $\dfrac{1}{2}$ **(b)** $\dfrac{1}{4}$ **(c)** $\dfrac{1}{2\sqrt{c}}$ **3.** 2   **5.** 5   **7.** 2

9. $f'(x) = 1$ **11.** $f'(x) = -\dfrac{3}{2x^4}$ **13.** Does not have a derivative at $x = 1$.   **15.** Not the graph of a function and its derivative.

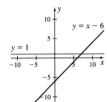
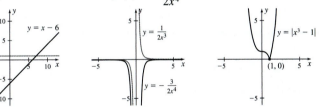

17.

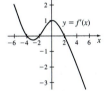

19. $f'(x) = 5x^4$ **21.** $f'(x) = x^3$ **23.** $f'(x) = 4x - 3$ **25.** $F'(x) = 14x$ **27.** $f'(x) = 5(3x^2 - 18x + 18)$ **29.** $f'(x) = 2 + 3x^{-2}$

31. $f'(x) = -\dfrac{35}{(x-5)^2}$ **33.** $f'(x) = 4x + 10x^{-3}$ **35.** $f'(x) = -ax^{-2} + 3bx^{-4}$ **37.** $f'(x) = -\dfrac{6(2x - 3)}{(x^2 - 3x)^3}$ **39.** $s'(t) = \dfrac{2t^2(t - 3)}{(t - 2)^2}$

41. $F'(z) = -\dfrac{2z}{(z^2 + 1)^2}$ **43.** $g'(z) = -\dfrac{2z - 1}{(1 - z + z^2)^2}$ **45.** $s'(t) = -e^t$ **47.** $f'(x) = -\dfrac{x}{e^x}$ **49.** $f'(x) = x\cos x + \sin x$

51. $G'(u) = \sec u(\sec u + \tan u)$ **53.** $f'(x) = e^x(\cos x + \sin x)$ **55.** $f'(x) = 2(\cos^2 x - \sin^2 x) = 2\cos 2x$ **57.** $f'(x) = 2\cos x\sin x = \sin 2x$

59. $f'(\theta) = -\dfrac{\sin\theta + \cos\theta}{2e^\theta}$ **61.** $f'(x) = 50x + 30$, $f''(x) = 50$ **63.** $g'(u) = \dfrac{1}{(2u + 1)^2}$, $g''(u) = -\dfrac{4}{(2u + 1)^3}$

65. $f'(u) = -\dfrac{\sin u + \cos u}{e^u}$, $f''(u) = \dfrac{2\sin u}{e^u}$

67. (a) $y = -7x + 5$ **(b)**

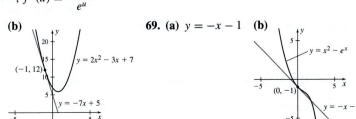

69. (a) $y = -x - 1$ **(b)**

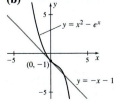

71. (a) $v(0) = -6$ m/s, $v(5) = 4$ m/s, $v(t) = 2t - 6$ m/s (b) $a(t) = 2$ m/s^2 **73.** (a) $\dfrac{dp}{dx} = -\dfrac{50{,}000}{(5x + 100)^2}$ (b) $R(x) = \dfrac{10{,}000x}{5x + 100} - 5x$

(c) $R'(10) = \dfrac{355}{9} \approx \$39.44/\text{lb}$, $R'(40) = \dfrac{55}{9} \approx \$6.11/\text{lb}$ **75.** $f'(3)$ does not exist. **77.** $8x - 16$, $f(x) = (4 - 2x)^2$

Chapter 3
Section 3.1

1. Chain **2.** True **3.** False **4.** $\tan u$; $1 + \cos x$ **5.** $100(x^3 + 4x + 1)^{99}(3x^2 + 4)$ **6.** $6xe^{3x^2+5}$ **7.** True **8.** $2x\cos(x^2)$

9. $\dfrac{dy}{dx} = 15x^2(x^3 + 1)^4$ **11.** $\dfrac{dy}{dx} = \dfrac{2x}{(x^2 + 2)^2}$ **13.** $\dfrac{dy}{dx} = 2\left(\dfrac{1}{x} + 1\right)(-x^{-2})$ **15.** $f'(x) = 6(3x + 5)$ **17.** $f'(x) = -18(6x - 5)^{-4}$

19. $g'(x) = 8x(x^2 + 5)^3$ **21.** $f'(u) = 3(u - 1/u)^2(1 + u^{-2})$ **23.** $g'(x) = 3(4x + e^x)^2(4 + e^x)$ **25.** $f'(x) = 2\tan x \sec^2 x$

27. $f'(z) = 2(\tan z + \cos z)(\sec^2 z - \sin z)$ **29.** $y' = 4x(x^2 + 4)(2x^3 - 1)^3 + 18x^2(x^2 + 4)^2(2x^3 - 1)^2$ **31.** $y' = \dfrac{2\sin x(x\cos x - \sin x)}{x^3}$

33. $y' = 4\cos(4x)$ **35.** $y' = 2\cos(x^2 + 2x - 1) \cdot (2x + 2)$ **37.** $y' = \cos(1/x) \cdot (-x^{-2})$ **39.** $y' = 4\sec(4x)\tan(4x)$ **41.** $y' = e^{x^{-1}}(-x^{-2})$
43. $y' = -(x^4 - 2x + 1)^{-2}(4x^3 - 2)$ **45.** $y' = 100(-(1 + 99e^{-x})^{-2}(-99e^{-x}))$ **47.** $y' = 2^{\sin x}\cos x \ln 2$ **49.** $y' = 6^{\sec x}\sec x \tan x \ln 6$

51. $y' = 5e^{3x} + 15xe^{3x}$ **53.** $y' = 2x\sin(4x) + 4x^2\cos(4x)$ **55.** $y' = -ae^{-ax}\sin(bx) + be^{-ax}\cos(bx)$ **57.** $y' = \dfrac{2ae^{ax}}{(e^{ax} + 1)^2}$

59. $\dfrac{dy}{dx} = -576x^{-5}(48x^{-4} + 1)^2$ **61.** $\dfrac{dy}{dx} = -64x^{-5}$ **63.** $y' = -2e^{-2x}\cos(3x) - 3e^{-2x}\sin(3x)$ **65.** $y' = -2xe^{x^2}\sin(e^{x^2})$

67. $y' = e^{\cos(4x)}(-4\sin(4x))$ **69.** $y' = 24\sin(3x)\cos(3x)$ **71.** $y' = 6x^2(x^3 + 1)$

73. (a) $y = 0$ **75.** (a) $y = \dfrac{16}{27} - \dfrac{7x}{27}$ **77.** (a) $y = 2x + 1$

(b) (b) (b)

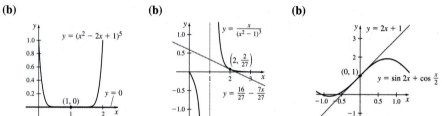

79. $\dfrac{d^2y}{dx^2} = -25x^8\cos(x^5) - 20x^3\sin(x^5)$ **81.** $h'(1) = -12$ **83.** $h'(0) = 0$ **85.** $\dfrac{dy}{dx}\Big|_{x=0} = 78$ **87.** $\dfrac{df}{dx} = f'(x^2 + 1)(2x)$

89. $\dfrac{df}{dx} = f'\left(\dfrac{x + 1}{x - 1}\right)\left(\dfrac{-2}{(x - 1)^2}\right)$ **91.** $\dfrac{df}{dx} = f'(\sin x)\cos x$ **93.** $\dfrac{d^2f}{dx^2} = f''(\cos x)\sin^2 x - f'(\cos x)\cos x$

95. (a) $v(t) = -A\omega\sin(\omega t + \varphi)$ (b) $t = \dfrac{k\pi - \varphi}{\omega}$, $k = 0, \pm1, \pm2, \ldots$ (c) $a(t) = -A\omega^2\cos(\omega t + \varphi)$

(d) $t = \dfrac{(2k - 1)\dfrac{\pi}{2} - \varphi}{\omega}$, $k = 0, \pm1, \pm2, \ldots$ **97.** $a(t) = \dfrac{-20\pi}{9}\sin(\pi t/6)$ **99.** $\dfrac{dR}{dT}\Big|_{T=320K} = -0.000004$

101. (a) 102 (c) $\dfrac{dA}{dt} = 18.9e^{-0.21t}$ (e)

(b) (d) $A'(5) = 6.614$; $A'(10) = 2.314$

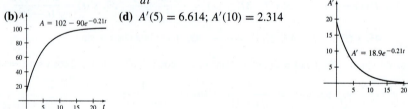

103. (a) $a(t) = ge^{-\frac{kt}{m}}$ (b) mg/k (c) 0 **105.** See student solutions manual. **107.** (a) $\omega(t) = -\dfrac{\pi\sqrt{k}\sin\left(\dfrac{\sqrt{k}t}{\sqrt{10}}\right)}{3\sqrt{10}}$

(b) $\omega(3) = -\dfrac{\pi\sqrt{k}\sin\left(\dfrac{3\sqrt{k}}{\sqrt{10}}\right)}{3\sqrt{10}}$ **109.** $F'(1) = 2$ **111.** See student solutions manual. **113.** See student solutions manual. **115.** See student

solutions manual. **117.** See student solutions manual. **119.** See student solutions manual. **121.** (a) $y^{(n)}(x) = a^n e^{ax}$
(b) $y^{(n)}(x) = (-1)^n a^n e^{-ax}$

123. (a) $\dfrac{d^{11}\cos(ax)}{dx^{11}} = a^{11}\sin(ax)$ **(b)** $\dfrac{d^{12}\cos(ax)}{dx^{12}} = a^{12}\cos(ax)$ **(c)** $f^{(n)}(x) = -a^n\sin(ax),\ n = 1 + 4k,\ k = 0, 1, 2, 3, \ldots;$

$f^{(n)}(x) = -a^n\cos(ax),\ n = 2 + 4k,\ k = 0, 1, 2, 3, \ldots;\ f^{(n)}(x) = a^n\sin(ax),\ n = 3 + 4k,\ k = 0, 1, 2, 3, \ldots;\ f^{(n)}(x) = a^n\cos(ax),\ n = 4 + 4k,$

$k = 0, 1, 2, 3, \ldots$ **125.** See student solutions manual. **127.** See student solutions manual. **129.** See student solutions manual.

131. See student solutions manual. **133. (a)** $-72000\pi^2(\text{m/sec}^2)$ **(b)** $72000\pi^2(\text{N})$

Section 3.2

1. True **2.** True **3.** True **4.** $\dfrac{1}{\sqrt{1-x^2}}$ **5.** False **6.** $x^{-2/3}$ **7.** $y' = -x/y$ **9.** $y' = (\cos x)/e^y$ **11.** $y' = \dfrac{-e^{x+y}}{e^{x+y}-1}$ **13.** $y' = -2y/x$

15. $y' = \dfrac{2x-y}{2y+x}$ **17.** $y' = -y^2/x^2$ **19.** $y' = \dfrac{-2x^{-2}y - 2x}{2y - 2x^{-1}}$ **21.** $y' = \dfrac{e^y\sin x - e^x\sin y}{e^x\cos y + e^y\cos x}$ **23.** $y' = \dfrac{-6x(x^2+y)^2}{3(x^2+y)^2 - 1}$

25. $y' = \dfrac{\sec^2(x-y)}{1+\sec^2(x-y)}$ **27.** $y' = \dfrac{\sin y}{1 - x\cos y}$ **29.** $y' = \dfrac{ye^{xy} - 2xy}{x^2 - xe^{xy}}$ **31.** $y' = \dfrac{2}{3}x^{-1/3}$ **33.** $y' = \dfrac{2}{3}x^{-1/3}$ **35.** $y' = \dfrac{1}{3}x^{-2/3} + \dfrac{1}{3}x^{-4/3}$

37. $y' = \dfrac{3}{2}x^2(x^3 - 1)^{-1/2}$ **39.** $y' = (x^2 - 1)^{1/2} + x^2(x^2 - 1)^{-1/2}$ **41.** $y' = e^{(x^2-9)^{1/2}} \cdot x(x^2 - 9)^{-1/2}$

43. $y' = \dfrac{3}{2}(x^2\cos x)^{1/2}(2x\cos x - x^2\sin x)$ **45.** $y' = 3x(x^2 - 3)^{1/2}(6x + 1)^{5/3} + 10(x^2 - 3)^{3/2}(6x + 1)^{2/3}$ **47.** $y' = \dfrac{4}{\sqrt{1 - 16x^2}}$

49. $y' = \dfrac{1}{x\sqrt{9x^2 - 1}}$ **51.** $y' = \dfrac{e^x}{\sqrt{1 - e^{2x}}}$ **53.** $y' = \sin^{-1}x + \dfrac{x}{\sqrt{1 - x^2}}$ **55.** $y' = \dfrac{\cos x}{1 + \sin^2 x}$ **57.** $y' = -xy^{-1},\ y'' = -y^{-1} - x^2y^{-3}$

59. $y' = -\dfrac{y^2 + 2xy}{x^2 + 2xy},\ y'' = \dfrac{6xy(x+y)(x^2 + xy + y^2)}{(x^2 + 2xy)^3}$ **61.** $y' = \dfrac{x}{(x^2 + 1)^{1/2}},\ y'' = \dfrac{1}{(x^2 + 1)^{3/2}}$

63. (a) $-1/2$ **(b)** $y = -\dfrac{x}{2} + \dfrac{5}{2}$ **65. (a)** 3 **(b)** $y = 3x - 8$ **67. (a)** $1/2$ **(b)** $y = \dfrac{x}{2} + 2$

(c)

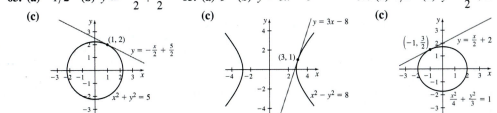

69. $y' = -4/9;\ y'' = -100/243$ **71.** $g'(4) = -1/2$ **73.** $f'(-2) = 2$ **75.** $g'(0) = 1/2,\ g'(3) = 1/5$ **77.** $y' = 3x/|x|$

79. $y' = \dfrac{2(2x - 1)}{|2x - 1|}$ **81.** $y' = \dfrac{-\sin x\cos x}{|\cos x|}$ **83.** $y' = \cos|x| \cdot \dfrac{|x|}{x}$ **85.** $y = -2x + 10$

87. (a) $(-3/2, \pm 3\sqrt{3}/2)$ **(b)** $(-4, 0),\ (1/2, \pm\sqrt{3}/2)$ **(c)** $(0, 0)$

89. (a) $y' = \dfrac{-y - 1}{x + 4y}$ **(b)** $y = -\dfrac{1}{3}x + \dfrac{5}{3}$ **(c)** $(6, -3),\ (2, 1)$ **91. (a)** $y = \dfrac{1}{2}x$

(d)

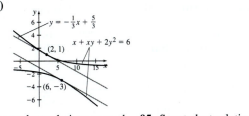

(b)

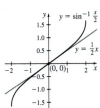

93. See student solutions manual. **95.** See student solutions manual.

97. (a) $y' = \dfrac{3x^2 - 2y}{2x - 3y^2}$ **(b)** $y = -x + 2$ **(c)** $\left(\dfrac{2^{4/3}}{3}, \dfrac{2^{5/3}}{3}\right)$ **(d)**

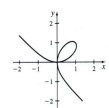

99. $\left(\dfrac{-5}{\sqrt{2}}, \dfrac{5}{\sqrt{2}}\right)$ **101.** See student solutions manual. **103.** See student solutions manual. **105.** See student solutions manual.

107. See student solutions manual. **109.** See student solutions manual.

111. (a) $\dfrac{dy}{dx} = -y/x$ for the first and $\dfrac{dy}{dx} = x/y$ for the second.   **(b)**

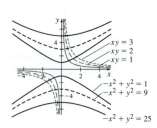

$xy = 3$
$xy = 2$
$xy = 1$
$-x^2 + y^2 = 1$
$-x^2 + y^2 = 9$
$-x^2 + y^2 = 25$

113. See student solutions manual.   **115.** See student solutions manual.

Section 3.3

1. $\dfrac{1}{x}$ **2.** True   **3.** True   **4.** False   **5.** $\dfrac{1}{x}$ **6.** e **7.** $y' = 5/x$ **9.** $y' = \dfrac{1}{u\ln 2}$ **11.** $y' = \dfrac{1}{x}\cos x - \sin x \ln x$ **13.** $y' = 1/x$

15. $y' = \dfrac{e^t + e^{-t}}{e^t - e^{-t}}$ **17.** $y' = \ln(x^2 + 4) + \dfrac{2x^2}{x^2 + 4}$ **19.** $y' = \ln\left(\sqrt{v^2 + 1}\right) + \dfrac{v^2}{v^2 + 1}$ **21.** $y' = \dfrac{1}{1 - x^2}$ **23.** $y' = \dfrac{1}{x\ln x}$

25. $y' = \dfrac{1}{x(x^2 + 1)}$ **27.** $y' = \dfrac{2x^4 - 5x^2 + 1}{x^5 - x}$ **29.** $y' = \cot\theta$ **31.** $y' = \dfrac{1}{\sqrt{x^2 + 4}}$ **33.** $y' = \dfrac{2x}{(1 + x^2)\ln 2}$ **35.** $y' = \dfrac{1}{x(1 + (\ln(x))^2)}$

37. $y' = \dfrac{1}{\tan^{-1} t} \cdot \dfrac{1}{1 + t^2}$ **39.** $y' = \dfrac{1}{2}(\ln x)^{-1/2} \cdot \dfrac{1}{x}$ **41.** $y' = \cos(\ln\theta) \cdot \dfrac{1}{\theta}$ **43.** $y' = \ln\sqrt{\cos(2x)} - x\tan(2x)$ **45.** $\dfrac{dy}{dx} = -\ln y \cdot x$

47. $\dfrac{dy}{dx} = \dfrac{x^2 + y^2 - 2x}{2y - x^2 - y^2}$ **49.** $\dfrac{dy}{dx} = \dfrac{y}{x(1 - y)}$ **51.** $y' = (x^2 + 1)^2(2x^3 - 1)^4\left(\dfrac{4x}{x^2 + 1} + \dfrac{24x^2}{2x^3 - 1}\right)$

53. $y' = \dfrac{x^2(x^3 + 1)}{\sqrt{x^2 + 1}}\left(\dfrac{2}{x} + \dfrac{3x^2}{x^3 + 1} - \dfrac{x}{x^2 + 1}\right)$ **55.** $y' = \dfrac{x\cos x}{(x^2 + 1)^3\sin x}\left(\dfrac{1}{x} - \tan x - \dfrac{6x}{x^2 + 1} - \cot x\right)$ **57.** $y' = (3x)^x(\ln(3x) + 1)$

59. $y' = 2x^{\ln x}\dfrac{\ln x}{x}$ **61.** $y' = x^{x^2}(2x\ln x + x)$ **63.** $y' = xe^x\left(e^x\ln x + \dfrac{e^x}{x}\right)$ **65.** $y' = x^{\sin x}\left(\cos x \ln x + \dfrac{\sin x}{x}\right)$

67. $y' = (\sin x)^x(\ln(\sin x) + x\cot x)$ **69.** $y' = (\sin x)^{\cos x}(-\sin x \ln(\sin x) + \cos x \cot x)$ **71.** $y' = \dfrac{-y}{x\ln x}$ **73.** $y = 5x - 1$

75. $y = -5x + 18$ **77.** e^2 **79.** $e^{1/3}$ **81.** $\dfrac{d^{10}y}{dx^{10}} = \dfrac{362880}{x}$ **83.** $\dfrac{dy}{dx} = 2$ **85.** $y' = x^x(\ln x + 1)$ **87.** $y' = \tan^{-1}(x/a)$

89. (a) See student solutions manual.   **(b)** \$6107.01   **(c)** 28.881 years   **(d)** See student solutions manual.   **91. (a)** $-\dfrac{10}{\ln 2}$
(b) Answers will vary.   **93.** See student solutions manual.   **95.** See student solutions manual.   **97.** See student solutions manual.

Section 3.4

1. (d)   **2.** $f(x_0) + f'(x_0)(x - x_0)$ **3.** True   **4.** $\left|\dfrac{\Delta Q}{Q(x_0)}\right|$ **5.** True   **6.** True   **7.** $dy = (3x^2 - 2)\,dx$ **9.** $dy = 12x(x^2 + 1)^{1/2}\,dx$

11. $dy = (6\cos(3x) + 1)\,dx$ **13.** $dy = -e^{-x}\,dx$ **15.** $dy = (e^x + xe^x)\,dx$ **17. (a)** $dy = e^x dx$ **(b) (i)** $dy = 1.3591$, $\Delta y = 1.7634$
(ii) $dy = 0.2718$, $\Delta y = 0.2859$ **(iii)** $dy = 0.0272$, $\Delta y = 0.0273$;   **(c) (i)** 0.4043   **(ii)** 0.0141   **(iii)** 0.000136   **19. (a)** $dy = \dfrac{2}{3\sqrt[3]{x}}\,dx$
(b) (i) $dy = 0.2646$, $\Delta y = 0.2546$ **(ii)** $dy = 0.0529$, $\Delta y = 0.0525$ **(iii)** $dy = 0.0053$, $\Delta y = 0.0053$;   **(c) (i)** 0.009952,   **(ii)** 0.000431,
(iii) 4×10^{-6} **21. (a)** $dy = -\sin x\,dx$ **(b) (i)** $dy = 0$, $\Delta y = 0.1224$ **(ii)** $dy = 0$, $\Delta y = 0.005$ **(iii)** $dy = 0$, $\Delta y = 0.00005$;
(c) (i) 0.122,   **(ii)** 0.005   **(iii)** 0.00005

23. (a) $L(x) = 405x - 567$ **25. (a)** $L(x) = \dfrac{x}{4} + 1$ **27. (a)** $L(x) = x - 1$ **29. (a)** $L(x) = -\dfrac{\sqrt{3}x}{2} + \dfrac{\pi}{2\sqrt{3}} + \dfrac{1}{2}$
(b)

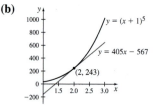

(b)

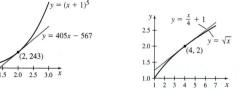

(b)

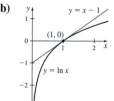

(b)

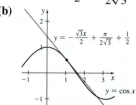

31. (a) 0.006   **(b)** −0.00006   **33. (a)** See student solutions manual.   **(b)** $c_3 = 1.155$ **35. (a)** See student solutions manual.   **(b)** $c_3 = 0.214$
37. (a) See student solutions manual.   **(b)** $c_3 = 3.136$ **39. (a)** See student solutions manual.   **(b)** $c_3 = 1.157$ **41. (a)** See student solutions
manual.   **(b)** $c_5 = -0.567$ **43. (a)** See student solutions manual.   **(b)** $c_5 = -0.800$ **45. (a)** See student solutions manual.   **(b)** $c_5 = 4.796$
47. 2π **49.** 3.6π **51. (a)** 1.6π **(b)** Underestimate   **(c)** See student solutions manual.   **53.** 30; 0.9%   **55.** 72 min   **57.** 685.914 N
59. 1/3 %   **61.** 6%   **63.** 1.310   **65.** 0.703   **67.** 2.718   **69.** Answers will vary.   **71.** See student solutions manual.   **73.** See student solutions
manual.   **75.** See student solutions manual.   **77.** 8.08

Section 3.5

1. True **2.** True **3.** $P_5(x) = 3(x-1)^3 + 11(x-1)^2 + 7(x-1) + 4$ **5.** $P_5(x) = 2(x+1)^4 - 14(x+1)^3 + 30(x+1)^2 - 25(x+1) + 7$

7. $P_5(x) = (x-2)^5 + 10(x-2)^4 + 40(x-2)^3 + 80(x-2)^2 + 80(x-2) + 32$ **9.** $P_5(x) = \dfrac{1}{5}(x-1)^5 - \dfrac{1}{4}(x-1)^4 + \dfrac{1}{3}(x-1)^3 - \dfrac{1}{2}(x-1)^2 + (x-1)$

11. $P_5(x) = -(x-1)^5 + (x-1)^4 - (x-1)^3 + (x-1)^2 - (x-1) + 1$ **13.** $P_5(x) = \dfrac{x^4}{24} - \dfrac{x^2}{2} + 1$ **15.** $P_5(x) = \dfrac{4x^5}{15} + \dfrac{2x^4}{3} + \dfrac{4x^3}{3} + 2x^2 + 2x + 1$

17. $P_5(x) = x^5 + x^4 + x^3 + x^2 + x + 1$ **19.** $P_5(x) = -6x^5 + 5x^4 - 4x^3 + 3x^2 - 2x + 1$

21. $P_5(x) = -\dfrac{1}{20}(x-1)^5 + \dfrac{1}{12}(x-1)^4 - \dfrac{1}{6}(x-1)^3 + \dfrac{1}{2}(x-1)^2 + (x-1)$

23. $P_5(x) = \dfrac{15}{4096}(x-1)^5 + \dfrac{3}{1024}(x-1)^4 - \dfrac{3}{64}(x-1)^3 + \dfrac{3}{16}(x-1)^2 + \dfrac{1}{2}(x-1) + 2$

25. $P_5(x) = \dfrac{64}{15}\left(x - \dfrac{\pi}{4}\right)^5 + \dfrac{10}{3}\left(x - \dfrac{\pi}{4}\right)^4 + \dfrac{8}{3}\left(x - \dfrac{\pi}{4}\right)^3 + 2\left(x - \dfrac{\pi}{4}\right)^2 + 2\left(x - \dfrac{\pi}{4}\right) + 1$

27. $P_5(x) = \dfrac{x^5}{5} - \dfrac{x^3}{3} + x$ **29.** $f(x) = 3(x-1)^2 + 1$ **31.** $f(x) = (x-1)^3 + 4(x-1)^2 + 5(x-1) - 6$

33. (a) $P_4(x) = \dfrac{28\left(x - \dfrac{1}{2}\right)^4}{27\sqrt{3}} + \dfrac{8\left(x - \dfrac{1}{2}\right)^3}{9\sqrt{3}} + \dfrac{2\left(x - \dfrac{1}{2}\right)^2}{3\sqrt{3}} + \dfrac{2\left(x - \dfrac{1}{2}\right)}{\sqrt{3}} + \dfrac{\pi}{6}$ **(b)**

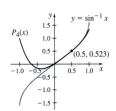

35. (a) $P_5(x) = \dfrac{-171}{16384}(x-1)^5 + \dfrac{75}{4096}(x-1)^4 + \dfrac{3}{256}(x-1)^3 - \dfrac{9}{64}(x-1)^2 + \dfrac{3}{8}(x-1) + \dfrac{1}{2}$ **(b)**

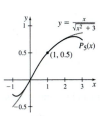

37. (a) $P_4(x) = 0.34 - 4.208 \times 10^{-5}x + 2.605 \times 10^{-9}x^2 - 1.075 \times 10^{-13}x^3 + 3.325 \times 10^{-18}x^4$ **(b)**

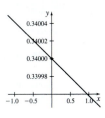

39. (a) $P_3(x) = 3.205 + 0.620x + 0.058x^2 + 0.0034x^3$ **(b)**

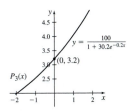

41. See student solutions manual. **43.** See student solutions manual.

Section 3.6

1. False **2.** $\sinh x / \cosh x$ **3.** False **4.** (a) **5.** False **6.** False **7.** True **9.** $\dfrac{3}{4}$ **11.** 1 **13.** 0 **15.** See student solutions manual.

17. See student solutions manual. **19.** See student solutions manual. **21.** See student solutions manual. **23.** See student solutions manual.

25. $y' = 3\cosh(3x)$ **27.** $y' = 2x\sinh(x^2 + 1)$ **29.** $y' = -\dfrac{\operatorname{csch}^2\left(\dfrac{1}{x}\right)}{x^2}$ **31.** $y' = 4\sinh(x)\sinh(4x) + \cosh(x)\cosh(4x)$

33. $y' = 2\sinh(x)\cosh(x)$ **35.** $y' = e^x \sinh(x) + e^x \cosh(x)$ **37.** $y' = 2x\,\text{sech}(x) - x^2 \tanh(x)\text{sech}(x)$ **39.** $y' = \dfrac{4}{\sqrt{16x^2 - 1}}$

41. $y' = \dfrac{2x}{1 - \left(1 - x^2\right)^2}$ **43.** $y' = \dfrac{x}{\sqrt{x^2 + 1}} + \sinh^{-1}(x)$ **45.** $y' = \dfrac{\sec^2(x)}{1 - \tan^2(x)}$ **47.** $y' = \dfrac{x}{\sqrt{x^2 - 1}\sqrt{x^2 - 2}}$

49. $P_6(x) = \dfrac{x^6}{720} + \dfrac{x^4}{24} + \dfrac{x^2}{2} + 1$ **51.** $\theta = \mp\sinh\left(\dfrac{25}{6}\right)$

53. (a) 625.092   **(b)** 580.744   **(c)** ± 0.384 **(d)** ± 2.766 **(e)** $\pm 82.026°$ **(f)**

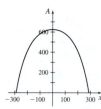

55. (a) See student solutions manual.   **(b)** $|x| > 1$ **57.** See student solutions manual.   **59.** See student solutions manual.
61. See student solutions manual.   **63.** See student solutions manual.   **65.** 0   **67.** See student solutions manual.

Review Exercises

1. $an(ax + b)^{n-1}$ **3.** $(1 - x)^{1/2} - \dfrac{x}{2}(1 - x)^{-1/2}$ **5.** $3x(x^2 + 4)^{1/2}$ **7.** $-\dfrac{a}{x\sqrt{2ax - x^2}}$ **9.** $(e^x - x)^{5x}\left(5\ln(e^x - x) + \dfrac{5x(e^x - 1)}{e^x - x}\right)$

11. $-\dfrac{2x}{(x - 1)^3}$ **13.** $\sec(2x) + 2x\sec(2x)\tan(2x)$ **15.** $\dfrac{1}{2}\left(a^2\sin(x/a)\right)^{-1/2} a\cos(x/a)$ **17.** $\cos v - \sin^2 v \cos v = \cos^3 v$ **19.** $1.05^x \ln 1.05$

21. $-\dfrac{1}{\sqrt{u^2 + 25}}$ **23.** $2\cot(2x)$ **25.** $\dfrac{2x - 2}{x^2 - 2x}$ **27.** $e^{-x}\left(-\ln x + \dfrac{1}{x}\right)$ **29.** $\dfrac{12}{x(144 - x^2)}$ **31.** $\dfrac{e^x(x^3 - x^2 - 12)}{(x - 2)^2}$ **33.** $\dfrac{\sqrt{x}}{1 + x}$

35. $\dfrac{1}{\sqrt{x - x^2}}$ **37.** $\tan^{-1}(x)$ **39.** $\dfrac{1}{2}\text{sech}^2(x/2) + \dfrac{8 - 2x^2}{(4 + x^2)^2}$ **41.** $\dfrac{1}{2}(\sinh(x))^{-1/2}\cosh(x)$ **43.** $\dfrac{1}{5y^4 + 1}$ **45.** $\dfrac{y(x\cos y - 1)}{x(1 + xy\sin y)}$

47. $\dfrac{1 + y\cos(xy)}{1 - x\cos(xy)}$ **49.** $y = \dfrac{10 - y}{6y + x}$; $y'' = \dfrac{2(y - 10)(3y + x + 30)}{(6y + x)^3}$ **51.** $y' = \dfrac{8x - e^y}{xe^y}$; $y'' = \dfrac{-64x^2 + 8xe^y + e^{2y}}{x^2 e^{2y}}$ **53.** $1/2$ **55.** $e^{2/5}$

57. 0   **59.** See student solutions manual.   **61.** $x \neq (2k - 1)\dfrac{\pi}{2}$, $k = 0, \pm 1, \pm 2, \ldots$ **63. (a)** Domain: $\left\{x : x \geq -\dfrac{1}{6}\right\}$; Range: $\{y : y \geq 0\}$

(b) $\dfrac{3}{5}$ **(c)** $\left(0, \dfrac{13}{5}\right)$ **(d)** $\left(\dfrac{4}{3}, 3\right)$ **65.** $dy = \dfrac{2xy - 3x^2}{4y - x^2}\,dx$ **67.** $L(x) = 2x - 1$

69. (a) $dy = dx$; $\Delta y = \tan(\Delta x)$ **(b) (i)** $dy = 0.5$, $\Delta y \approx 0.546$;   **(ii)** $dy = 0.1$, $\Delta y \approx 0.100$;   **(iii)** $dy = 0.01$, $\Delta y \approx 0.010$

71. $-\dfrac{1}{2}(1 + \ln 8)$ **73.** -2 **75.** $P_4(x) = \dfrac{x^3}{3} + x$ **77.** $P_6(x) = -\dfrac{1}{384}(x - 2)^6 + \dfrac{1}{160}(x - 2)^5 - \dfrac{1}{64}(x - 2)^4 + \dfrac{1}{24}(x - 2)^3 - \dfrac{1}{8}(x - 2)^2 + \dfrac{x - 2}{2} + \ln 2$

79. (a) See student solutions manual.   **(b)** 2.568   **81.** $y = 6x - 3$

Chapter 4
Section 4.1

1. $10 \text{ m}^3/\text{min}$ **2.** 0.5 m/min **3.** $-\dfrac{8}{3}$ **4.** -1 **5.** $5\pi^2$ **7.** 40   **9.** $900 \text{ cm}^3/\text{s}$ **11.** -0.000625 cm/h **13.** $-\dfrac{8}{\sqrt{2009}} \approx 0.178 \text{ cm/min}$

15. $\dfrac{2\sqrt{3}\pi}{45} \approx 0.242 \text{ cm}^2/\text{min}$ **17.** $-0.75 \text{ m}^2/\text{min}$ **19.** $-\dfrac{1}{6} \text{ rad/s}$ **21.** 0.2 m/min **23.** $\dfrac{4}{\pi} \approx 1.273 \text{ m/min}$

25. (a) $\dfrac{16}{3\pi} \approx 1.698 \text{ m/min}$ **(b)** $\dfrac{96 - 9\pi}{32} \approx 2.116 \text{ m}^3/\text{min}$ **27.** $\dfrac{18}{\sqrt{17}} \approx 4.366 \text{ units/s}$ **29.** $\dfrac{100}{\sqrt{2}} \approx 70.711 \text{ ft/s}$ **31.** $1.75 \text{ kg/cm}^2/\text{min}$

33. $100.8\pi \approx 316.673 \text{ ft}^2/\text{min}$ **35. (a)** $\dfrac{1.5}{\sqrt{55}} \approx 0.202 \text{ m/s}$ **(b)** $\dfrac{0.5}{\sqrt{3}} \approx 0.289 \text{ m/s}$ **(c)** $\dfrac{1.5}{\sqrt{7}} \approx 0.567 \text{ m/s}$ **37.** $\dfrac{24\pi}{5} \approx 15.08 \text{ km/s}$

39. 4 m/min **41.** He is rising at $\dfrac{25\sqrt{3}\pi}{2} \approx 68.018 \text{ ft/min}$. He is moving horizontally at $\dfrac{25\pi}{2} \approx 32.267 \text{ ft/min}$.   **43.** $\approx -1.671 \text{ m/s}$

45. (a) 150   **(b)** 200   **(c)** 50   **47.** $\approx -4.864 \text{ lb/s}$ **49.** $-100\pi \approx -314.159 \text{ ft/s}$ **51.** -316.8 rad/h
53. (a) The volume of the box decreases at a rate x^2 times the rate of decrease in the height.   **(b)** Answers will vary.   **55.** $\approx 32.071 \text{ ft/s}$
57. $\approx 0.165 \text{ in}^2/\text{min}$

Section 4.2

1. False **2.** (b) **3.** False **4.** False **5.** False **6.** False **7.** x_1: neither, x_2: local maximum, x_3: local minimum and absolute minimum, x_4: neither, x_5: local maximum, x_6: neither, x_7: local minimum, x_8: absolute maximum

9.   **11.**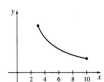

13. 4 **15.** 0, 2 **17.** $-1, 0, 1$ **19.** 0 **21.** 0 **23.** π **25.** $-\dfrac{\sqrt{2}}{2}, -1, 1, \dfrac{\sqrt{2}}{2}$ **27.** 0, 2 **29.** $-3, 0, 1$ **31.** 3, 4 **33.** $-3\sqrt{3}, -3, 3, 3\sqrt{3}$

35. 1 **37.** Absolute maximum 20 at $x = 10$, absolute minimum -16 at $x = 4$

39. Absolute maximum 16 at $x = 4$, absolute minimum -4 at $x = 2$ **41.** Absolute maximum 9 at $x = 2$, absolute minimum 0 at $x = 1$

43. Absolute maximum 1 at $x = -1, 1$, absolute minimum 0 at $x = 0$ **45.** Absolute maximum 4 at $x = 4$, absolute minimum 2 at $x = 1$

47. Absolute maximum π at $x = \pi$, absolute minimum 0 at $x = 0$ **49.** Absolute maximum $\dfrac{1}{2}$ at $x = \dfrac{\sqrt{2}}{2}$, absolute minimum $-\dfrac{1}{2}$ at $x = -\dfrac{\sqrt{2}}{2}$

51. Absolute maximum 0 at $x = 0$, absolute minimum $-\dfrac{1}{2}$ at $x = -1, \dfrac{1}{2}$

53. Absolute maximum $64\sqrt[3]{16}$ at $x = 5$, absolute minimum 0 at $x = -3, 1$

55. Absolute maximum $\dfrac{1}{3}$ at $x = 4$, absolute minimum -1 at $x = 2$ **57.** Absolute maximum $\dfrac{\sqrt[3]{18}}{3\sqrt{3}}$ at $x = 3\sqrt{3}$, absolute minimum 0 at $x = 3$

59. Absolute maximum 1 at $x = 0$, absolute minimum $e - 3$ at $x = 1$ **61.** Absolute maximum 9 at $x = 3$, absolute minimum 1 at $x = 0$

63. Absolute maximum 8 at $x = 2$, absolute minimum 0 at $x = 0$

65. (a) $f'(x) = 12x^3 - 6x^2 - 42x + 36$ (b) $-2, 1, \dfrac{3}{2}$ (c) 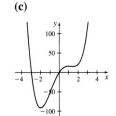 Absolute minimum at $x = -2$, local maximum at $x = 1$, local minimum at $x = \dfrac{3}{2}$

67. (a) $f'(x) = \dfrac{3x^4 + 5x^3 + 8x^2 - 10x - 96}{2\sqrt{x+5}(x^2+2)^{\frac{3}{2}}}$ (b) $\approx -2.364, \approx 1.977$ (c) 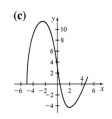 Local maximum at $x \approx -2.364$, local minimum at $x \approx 1.977$

69. (a) $f'(x) = 4x^3 - 37.2x^2 + 98.48x - 68.64$ (b) Absolute maximum 0 at $x = 0$, absolute minimum -37.2 at $x = 5$ (c)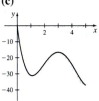

71. (a) 50 mi/h (b) 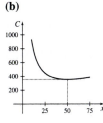 **73.** (a) $67.5°$, 18.7452 ft (b) 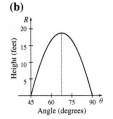 (c) Yes.

75. (a) See Student Solutions Manual. (b) 846.858 m **77.** $\dfrac{3\sqrt{6}}{2} \approx 3.674$, $\dfrac{9\sqrt{3}}{2} \approx 7.794$ **79.** 5 m

69. (a) Local maximum 0 at $x = 0$, local minimum $-6\sqrt[3]{2}$ at $x = -\sqrt{2}, \sqrt{2}$ **(b)** Concave up on $(-\infty, 0) \cup (0, \infty)$ **(c)** No inflection point

71. (a) Local minimum $\dfrac{1}{2} + \dfrac{1}{2}\ln 2$ at $x = \dfrac{\sqrt{2}}{2}$ **(b)** Concave up on $(0, \infty)$ **(c)** No inflection point

73. (a) Local maximum $\dfrac{16\sqrt{5}}{125}$ at $x = \dfrac{1}{2}$, local minimum $-\dfrac{16\sqrt{5}}{125}$ at $x = -\dfrac{1}{2}$ **(b)** Concave up on $\left(-\dfrac{\sqrt{3}}{2}, 0\right) \cup \left(\dfrac{\sqrt{3}}{2}, \infty\right)$, concave down on

$\left(-\infty, -\dfrac{\sqrt{3}}{2}\right) \cup \left(0, \dfrac{\sqrt{3}}{2}\right)$ **(c)** $\left(-\dfrac{\sqrt{3}}{2}, -\dfrac{16\sqrt{3}}{7^{5/2}}\right), \left(\dfrac{\sqrt{3}}{2}, \dfrac{16\sqrt{3}}{7^{5/2}}\right), (0, 0)$

75. (a) Local maximum $\dfrac{2\sqrt{3}}{9}$ at $x = \pm\dfrac{\sqrt{6}}{3}$, local minimum 0 at $x = 0$ **(b)** Concave up on $\left(-\dfrac{1}{6}\sqrt{27 - 3\sqrt{33}}, \dfrac{1}{6}\sqrt{27 - 3\sqrt{33}}\right)$, concave

down on $\left(-1, -\dfrac{1}{6}\sqrt{27 - 3\sqrt{33}}\right) \cup \left(\dfrac{1}{6}\sqrt{27 - 3\sqrt{33}}, 1\right)$ **(c)** $\left(\pm\dfrac{1}{6}\sqrt{27 - 3\sqrt{33}}, \dfrac{9 - \sqrt{33}}{12}\sqrt{\dfrac{3 + \sqrt{33}}{12}}\right)$

77. (a) Local maximum 1 at $x = k\pi + \dfrac{\pi}{2}$, local minimum 0 at $x = k\pi, k \in \mathbb{N}$ **(b)** Concave up on $\left(k\pi - \dfrac{\pi}{4}, k\pi + \dfrac{\pi}{4}\right)$, concave down on

$\left(k\pi + \dfrac{\pi}{4}, k\pi + \dfrac{3\pi}{4}\right)$ **(c)** $\left(\dfrac{(2k + 1)\pi}{4}, \dfrac{1}{2}\right)$

79. (a) Local maximum $\dfrac{5\pi}{3} + \sqrt{3}$ at $x = \dfrac{5\pi}{3}$, local minimum $\dfrac{\pi}{3} - \sqrt{3}$ at $x = \dfrac{\pi}{3}$ **(b)** Concave up on $(0, \pi)$, concave down on $(\pi, 2\pi)$ **(c)** (π, π)

81. (a) Local maximum -20 at $x = 3$, local minimum -21 at $x = 2$ **(c)** Answers will vary
83. (a) Local maximum $4e$ at $x = 1$, local minimum 0 at $x = 3$ **(c)** Answers will vary.

85. Answers will vary.   **87.** Answers will vary.   **89.** Answers will vary.   **91.** Answers will vary.   **93.** Answers will vary.
One possible answer below.   One possible answer below.   One possible answer below.   One possible answer below.   One possible answer below.

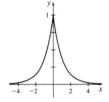 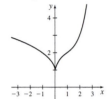 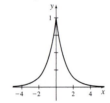 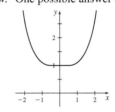 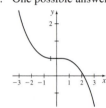

95. Answers will vary.
One possible answer below.

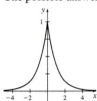

97. (a) Concave up on $\left(-\infty, 2 - \dfrac{\sqrt{2}}{2}\right) \cup \left(2 + \dfrac{\sqrt{2}}{2}, \infty\right)$, concave down on $\left(2 - \dfrac{\sqrt{2}}{2}, 2 + \dfrac{\sqrt{2}}{2}\right)$

(b) $\left(2 - \dfrac{\sqrt{2}}{2}, \dfrac{1}{\sqrt{e}}\right), \left(2 + \dfrac{\sqrt{2}}{2}, \dfrac{1}{\sqrt{e}}\right)$ **(c)** The figure displays the graph of f with the points of inflection marked with black squares. At the point on the left, the concavity changes from up to down, while at the point on the right, the concavity changes from down to up.

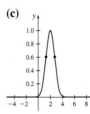

99. (a) Concave up on $(-\infty, 0.135) \cup (0.721, 5.144)$, concave down on $(0.135, 0.721) \cup (5.144, \infty)$
(b) $(0.135, 2.433), (0.721, 2.139), (5.144, 0.072)$
(c) The figure displays the graph of f with the points of inflection marked with black squares. From left to right, the concavity changes from up to down, down to up, and up to down at the inflection points.

101. See Student Solutions Manual. **103.** $a = -3, b = 9$ **105. (a)** $N'(t) = \dfrac{99,990,000e^{-t}}{(1 + 9999e^{-t})^2}$ **(b)** $N'(t)$ is increasing on the interval $(0, \ln 9999)$ and decreasing on the interval $(\ln 9999, \infty)$. **(c)** $\ln 9999$ **(d)** $(\ln 9999, 5000)$ **(e)** Answers will vary.

107. (a)

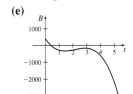

(b) Answers will vary.   **(c)** $P'(t) = \dfrac{84741024e^{-0.0166t}}{(1 + 7.518e^{-0.0166t})^2}$

(d) $P'(t)$ is increasing on the interval $(0, 121.524)$ and decreasing on the interval $(121.524, \infty)$.

(f) $(121.524, 2817937.751)$ **(g)** Answers will vary.

109. (a) Local minimum $B(2.80) \approx -328.53$, local maximum $B(5.71) \approx -170.80$.   **(b)** Both represent a budget deficit.   **(c)** Concave up on $(0, 4.26)$, concave down on $(4.26, 9)$, $(4.26, -249.27)$ is a point of inflection.   **(d)** To the left of the inflection point, the budget is increasing at an increasing rate. To the immediate right of the inflection point, the budget is increasing at a decreasing rate.

(e)   It's not an accurate predictor for the budget for the years 2010 and beyond.

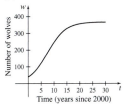

111. $a = -\dfrac{7}{8}, b = \dfrac{21}{4}, c = 0$, and $d = 5$ **113.** Local maximum 2 at $x = \dfrac{\pi}{3}$, local minimum -2 at $x = \dfrac{4\pi}{3}$. Inflection points $\left(\dfrac{5\pi}{6}, 0\right)$, $\left(\dfrac{11\pi}{6}, 0\right)$ **115.** $f''(x)$ does not exist at the critical number $x = 0$.   **117.** (e)   **119.** See Student Solutions Manual.   **121.** See Student Solutions Manual.   **123.** See Student Solutions Manual.   **125.** See Student Solutions Manual.   **127.** See Student Solutions Manual.   **129.** See Student Solutions Manual.   **131.** See Student Solutions Manual.   **133.** See Student Solutions Manual.   **135.** See Student Solutions Manual.

Section 4.5

1. False   **2.** False   **3.** False   **4.** False   **5.** Answers will vary.   **6.** Answers will vary.   **7.** Yes, $\dfrac{0}{0}$ **9.** No, $\dfrac{1}{0}$ **11.** Yes, $\dfrac{\infty}{\infty}$ **13.** No, $\dfrac{1}{0}$ **15.** Yes, $\dfrac{0}{0}$ **17.** Yes, $\dfrac{0}{0}$ **19.** Yes, $0 \cdot \infty$ **21.** Yes, $\infty - \infty$ **23.** Yes, ∞^0 **25.** No, $(-1)^0$ **27.** $\dfrac{0}{0}, 5$ **29.** $\dfrac{0}{0}, \dfrac{1}{2}$ **31.** $\dfrac{0}{0}, 2$ **33.** $\dfrac{0}{0}, -\pi$ **35.** $\dfrac{\infty}{\infty}, 0$ **37.** $\dfrac{\infty}{\infty}, 0$ **39.** $\dfrac{0}{0}, 1$ **41.** $\dfrac{0}{0}, -\dfrac{1}{6}$ **43.** $0 \cdot \infty, 0$ **45.** $0 \cdot \infty, 1$ **47.** $\infty - \infty, 0$ **49.** $\infty - \infty, -1$ **51.** $0^0, 1$ **53.** $\infty^0, 1$ **55.** $\infty^0, 1$ **57.** $1^\infty, 1$ **59.** 2   **61.** 0   **63.** 0   **65.** 2   **67.** 1   **69.** $\dfrac{1}{4}$ **71.** 0   **73.** 0   **75.** 1   **77.** $\dfrac{4a^2}{\pi}$ **79.** $\dfrac{1}{2}$ **81.** -1 **83.** 1   **85.** 1   **87.** 1   **89.** 1

91. (a) 366   **(b)** Answers will vary.   **(c)**

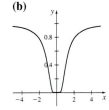

93. (a) $\dfrac{E}{R}, \dfrac{Et}{L}$ **(b)** Answers will vary.   **95.** See Student Solutions Manual.   **97.** See Student Solutions Manual.   **99.** 0

101. See Student Solutions Manual.   **103.** See Student Solutions Manual.

105. (a) See Student Solutions Manual.   **(b)**

107. See Student Solutions Manual.

109. (a) 0   **(b)** a **(c)** ∞ **(d)** Does not exist.   **(e)** 0   **(f)** a **(g)** 1   **(h)** a **(i)** Does not exist.   **(j)** e^{-1}

111. (a) $f''(x)$ **(b)** $f'''(x)$ **(c)** $\lim\limits_{h \to 0} \dfrac{\sum_{k=0}^{n}(-1)^k \dfrac{n!}{k!(n-k)!} f(x + (n-k)h)}{h^n} = f^{(n)}(x)$

113. See Student Solutions Manual.

Review Exercises

1. $-\dfrac{4}{5}$cm²/min   3. 318.953 mph   5. $(-8, -9)$: local minimum and absolute minimum, $(-5, 0)$: neither, $(-2, 9)$: local maximum and absolute

maximum, $(1, 0)$: local minimum, $(3, 4)$: local maximum, $(5, 0)$: local minimum   7. $0, \dfrac{\pi}{2}, \pi$ 9. Absolute maximum 21 at $x = -2$ and absolute

minimum -11 at $x = 2$ 11. $\left(2, \dfrac{3}{2}\right)$ 13. (a) Local maximum $\dfrac{203}{27}$ at $x = -\dfrac{4}{3}$ and local minimum -11 at $x = 2$

15. (a) Local maximum $16e^{-4}$ at $x = 2$ and local minimum 0 at $x = 0$

17.   19.   21.

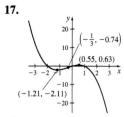 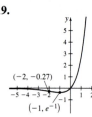 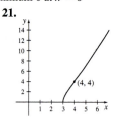

23. (a) Increasing on $(-1.207, \infty)$ and decreasing on $(-\infty, -1.207)$
(b) Concave up on $(-\infty, \infty)$
(c) No point of inflection

25. (B)   27.

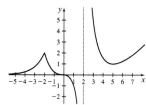

29. 4.708 in.   31. $F(x) = C$ 33. $F(x) = \sin x + C$ 35. $F(x) = 2\ln|x| + C$ 37. $F(x) = x^4 - 3x^3 + 5x^2 - 3x + C$ 39. $\dfrac{31}{5}$ cm/s

41. Yes, $\dfrac{0}{0}$ 43. Yes, $\infty - \infty$ 45. 9   47. 2   49. 1   51. $\dfrac{1}{6}$ 53. 1   55. $\dfrac{8}{9}$ 57. $y = e^x + 1$ 59. $y = 2\ln|x| + 4$ 61. 1075 items   63. $\dfrac{1}{e}$

Chapter 5
Section 5.1

1. Answers will vary.   2. False   3. $\dfrac{1}{2}$ 4. True

5. (a) Area is $\dfrac{35}{4}$.   (b) Area is $\dfrac{37}{4}$.   (c) $\dfrac{35}{4} < 9 < \dfrac{37}{4}$

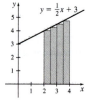 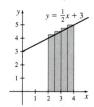

7. (a) 3   (b) 5   9. [1, 2], [2, 3], and [3, 4]   11. $\left[-1, -\dfrac{1}{2}\right], \left[-\dfrac{1}{2}, 0\right], \left[0, \dfrac{1}{2}\right], \left[\dfrac{1}{2}, 1\right], \left[1, \dfrac{3}{2}\right], \left[\dfrac{3}{2}, 2\right], \left[2, \dfrac{5}{2}\right], \left[\dfrac{5}{2}, 3\right], \left[3, \dfrac{7}{2}\right]$, and $\left[\dfrac{7}{2}, 4\right]$

13. (a) 18   (b) 48

15. (a)   (b) $\left[0, \dfrac{3}{n}\right], \left[\dfrac{3}{n}, 2 \cdot \dfrac{3}{n}\right], \ldots, \left[(n-1) \cdot \dfrac{3}{n}, 3\right]$ (c) See student solutions manual.   (d) See student solutions manual.

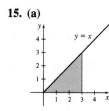

(e) See student solutions manual.

17. (a) $s_4 = 40, s_8 = 44$ (b) $S_4 = 56, S_8 = 52$ 19. (a) $s_4 = 34, s_8 = \dfrac{77}{2}$ (b) $S_4 = 50, S_8 = \dfrac{93}{2}$

21. (a) $s_4 = \dfrac{\sqrt{2}}{4}\pi \approx 1.111, s_8 \approx 1.582$ (b) $S_4 = \dfrac{\sqrt{2}+2}{4}\pi \approx 2.682, S_8 \approx 2.367$

23. $s_n = \displaystyle\sum_{i=1}^{n}\left(3(i-1)\dfrac{10}{n}\right)\dfrac{10}{n} = 150 - \dfrac{150}{n}; \lim_{n \to \infty} s_n = 150$

25. (a) $A = \lim_{n \to \infty} s_n = \lim_{n \to \infty} \left(20 - \dfrac{16}{n} \right) = 20$ **(b)** $A = \lim_{n \to \infty} S_n = \lim_{n \to \infty} \left(20 + \dfrac{16}{n} \right) = 20$ **(c)** Answers will vary.

27. (a) $A = \lim_{n \to \infty} s_n = \lim_{n \to \infty} \left(24 - \dfrac{24}{n} \right) = 24$ **(b)** $A = \lim_{n \to \infty} S_n = \lim_{n \to \infty} \left(24 + \dfrac{24}{n} \right) = 24$ **(c)** Answers will vary.

29. (a) $A = \lim_{n \to \infty} s_n = \lim_{n \to \infty} \left(\dfrac{32}{3} - \dfrac{16}{n} + \dfrac{16}{3n^2} \right) = \dfrac{32}{3}$ **(b)** $A = \lim_{n \to \infty} S_n = \lim_{n \to \infty} \left(\dfrac{32}{3} + \dfrac{16}{n} + \dfrac{16}{3n^2} \right) = \dfrac{32}{3}$ **(c)** Answers will vary.

31. (a) $A = \lim_{n \to \infty} s_n = \lim_{n \to \infty} \left(\dfrac{16}{3} - \dfrac{4}{n} - \dfrac{4}{3n^2} \right) = \dfrac{16}{3}$ **(b)** $A = \lim_{n \to \infty} S_n = \lim_{n \to \infty} \left(\dfrac{16}{3} + \dfrac{4}{n} - \dfrac{4}{3n^2} \right) = \dfrac{16}{3}$ **(c)** Answers will vary.

33. 10 **35.** 18 **37.** $\dfrac{58}{3}$ **39.** $A \approx 25990$ **41.** $A \approx 8.463$

43. (a)

(b) $\left[1, 1 + \dfrac{3}{n} \right], \left[1 + \dfrac{3}{n}, 1 + 2 \cdot \dfrac{3}{n} \right], \ldots, \left[1 + (n-1) \cdot \dfrac{3}{n}, 4 \right]$. **(c)** See student solutions manual.

(d) See student solutions manual.

(e)

n	5	10	50	100
s_n	4.754	5.123	5.456	5.500
S_n	6.554	6.023	5.636	5.590

(f) $5.500 \le A \le 5.590$

45. $s_n = \dfrac{b^3 - a^3}{3} - \dfrac{(b-a)^2}{2n}(a+b) + \dfrac{(b-a)^3}{6n^2}$, $A = \dfrac{b^3 - a^3}{3}$, $S_n = \dfrac{b^3 - a^3}{3} + \dfrac{(b-a)^2}{2n}(a+b) + \dfrac{(b-a)^3}{6n^2}$ **47.** $\dfrac{B_1 + B_2}{2} H$

Section 5.2

1. Partition **2.** (c) 2 **3.** True **4.** Lower limit; upper limit; integral sign; integrand **5.** 0 **6.** False **7.** True **8.** (b) 10 **9.** $\dfrac{13}{8}$ **11.** $\dfrac{11}{4}$

13. (a) $[-4, -1], [-1, 0], [0, 1], [1, 3], [3, 5], [5, 6]$. **(b)** 0 **(c)** 13 **15.** $\int_0^2 (e^x + 2)\, dx$ **17.** $\int_0^{2\pi} \cos x\, dx$ **19.** $\int_1^4 \dfrac{2}{x^2}\, dx$

21. $\int_1^e x \ln x\, dx$ **23.** $7e$ **25.** 3π **27.** 0 **29.** $\int_2^6 \left(2 + \sqrt{4 - (x-4)^2} \right) dx$ **31.** $\int_{-2}^4 (3 + \sin(1.5x))\, dx$ **33.** Yes; answers will vary.

35. No; answers will vary. **37.** No; answers will vary. **39. (a)** $\sum_{i=1}^{n} \left(\dfrac{8i^2}{n^3} - \dfrac{2}{n} \right)$ **(b)** $\int_0^2 (x^2 - 1)\, dx$ **(c)** $\dfrac{2}{3}$

41. (a) $\sum_{i=1}^{n} \dfrac{3}{n} \sqrt{\dfrac{3i}{n} + 1}$ **(b)** $\int_0^3 \sqrt{x+1}\, dx$ **(c)** $\dfrac{14}{3}$ **43. (a)** $\sum_{i=1}^{n} \dfrac{2}{n} e^{\frac{2i}{n}}$ **(b)** $\int_0^2 e^x\, dx$ **(c)** $e^2 - 1 \approx 6.389$ **45.** $-\dfrac{7}{2}$

47. (a)

n	10	50	100
Left	14.536	14.737	14.762
Right	15.030	14.836	14.812
Mid	14.789	14.787	14.787

(b) ≈ 14.787

49. (a)

n	10	50	100
Left	4.702	4.712	4.712
Right	4.702	4.712	4.712
Mid	4.717	4.713	4.712

(b) $\dfrac{3\pi}{2} \approx 4.712$

51. $\dfrac{4448}{6435} \approx 0.691$ **53.** joules **55.** meters

57. (a)

(b) $18 - \dfrac{9\pi}{2} \approx 3.863$ **(c)** See student solutions manual.

59. (a) $\dfrac{95}{2}$ **(b)** $\dfrac{71}{2}$ **(c)** $\dfrac{83}{2} = 41.5$ **(d)** $\int_1^5 x^2\, dx = 41.\overline{3}$

61. See student solutions manual. **63.** See student solutions manual. **65.** See student solutions manual.

Section 5.3

1. $f(x)$ **2.** $b - a$ **3.** False **4.** False **5.** $\sqrt{x^2 + 1}$ **7.** $(3 + t^2)^{3/2}$ **9.** $\ln x$ **11.** $6x^2 \sqrt{4x^6 + 1}$ **13.** $5x^4 \sec(x^5)$ **15.** $-\sin(x^2)$

17. $-10x(30x^2)^{2/3}$ **19.** 5 **21.** $\dfrac{15}{4}$ **23.** $\dfrac{2}{3}$ **25.** $\sqrt{3}$ **27.** $\sqrt{2} - 1$ **29.** $\dfrac{e-1}{e}$ **31.** 1 **33.** $\dfrac{\pi}{4}$ **35.** $\dfrac{99}{5}$ **37.** $-\dfrac{\sqrt{2}}{2}$ **39.** $\dfrac{15}{4}$

41. $e^2 - 1$ **43.** 160 **45.** $\dfrac{\pi}{6}$ **47.** $\dfrac{\pi}{3}$ **49.** $2\sqrt{r} - 2$, which goes to infinity as $r \to \infty$ **51.** Answers will vary.

53. (a) $\int_a^b H(t)\,dt$ **(b)** cm^3 **(c)** $100\,\mathrm{cm}^3$ of helium leaked out in the first 5 hours. **55. (a)** $\int_0^{5.2} 9.8t\,dt$ **(b)** $132.5\,\mathrm{m}$ **57.** $P(x) = 3x^2 + 6x + 6$

59. (a) $\dfrac{64}{3}$ and $\dfrac{128}{3}$ **(b)** Even **(c)** Answers will vary. **61. (a)** 1 and 2 **(b)** Even **(c)** Answers will vary. **63.** $\dfrac{1}{\sqrt[3]{2}}$ **65.** $F'(c)$ **67.** $e^4 - 1$

69. 4 **71.** See student solutions manual. **73. (a)** See student solutions manual. **(b)** See student solutions manual. **(c)** Answers will vary.

75. $a = 4$ **77. (a)** $\dfrac{\cos x}{\sqrt{1 - \sin^2(x)}} = \dfrac{\cos x}{|\cos x|}$ **(b)** Yes **(c)** Yes

Section 5.4

1. True **2.** False **3.** True **4.** -2 **5.** Average value **6.** False **7.** 7 **9.** 31 **11.** 17 **13.** $-\dfrac{1}{15}$ **15.** 4 **17.** -3 **19.** $\dfrac{\pi}{2}$ **21.** $-\dfrac{76}{3}$

23. $\dfrac{5}{3}$ **25.** -3 **27.** $\dfrac{20}{3}$ **29.** $\dfrac{55}{24}$ **31.** $\dfrac{(\pi + 15)}{6}$ **33.** $\dfrac{7}{3}$ **35.** $-\dfrac{2}{3}$ **37.** $\dfrac{43}{6}$ **39.** $\dfrac{679}{32}$ **41.** $\dfrac{131}{4}$ **43.** See student solutions

manual. **45.** See student solutions manual. **47.** 12 to 32 **49.** $\dfrac{\sqrt{2\pi}}{8}$ to $\dfrac{\pi}{4}$ **51.** 1 to $\sqrt{2}$ **53.** 1 to e **55.** $u = \sqrt{3}$ **57.** $u = \dfrac{4\sqrt{3}}{3}$

59. $u = \dfrac{\pi}{2}, \dfrac{3\pi}{2}$ **61.** $e - 1$ **63.** $\dfrac{3}{5}$ **65.** $\dfrac{2}{\pi}$ **67.** $\dfrac{2}{3}$ **69.** $\dfrac{2}{\pi}\left(\sqrt{e^\pi} - 2\right)$ **71. (a)** 12 **(b)** 4 **(c)** Answers will vary.

73. (a) $e^2 + 3 - \dfrac{1}{e}$ **(b)** $\dfrac{e^3 + 3e - 1}{3e} \approx 3.340$ **(c)** Answers will vary. **75.** 9 **77.** $\dfrac{13}{3}$ **79.** $37.5\,^\circ\mathrm{C}$ **81. (a)** $\dfrac{1875}{16} = 117.1875\,\mathrm{N}$ **(b)** 4 m

83. $\approx 1000.345\,\mathrm{kg/m^3}$ **85.** $\dfrac{13\pi}{3} \approx 13.614\,\mathrm{m}^2$ **87. (a)** 2 **(b)** $\dfrac{1}{2}bh = 2$ **89.** 12 m/s **91. (a)** $\dfrac{f(b) - f(a)}{b - a}$ **(b)** Answers will vary.

93. See student solutions manual. **95.** $k = \dfrac{\pi}{6}$ **97.** ≈ 8.296 **99.** Only (c) need not be true. **101.** $\dfrac{1}{6}$

103. (a) -14.7 m/s **(b)** See student solutions manual. **105. (a)** $(-\infty, \infty)$ **(b)** $[0, \infty)$ **(c)** $(-\infty, \infty)$ **(d)** All $x \neq 1$. **(e)** $\dfrac{1}{6}$

107. $F(x) = \dfrac{x^2}{4}$ **109.** $\dfrac{1}{2}$ **111.** See student solutions manual. **113. (a)** See student solutions manual. **(b)** Answers will vary.

115. See student solutions manual.

Section 5.5

1. $f(x)$ **2.** False **3.** $\dfrac{x^{\alpha+1}}{\alpha + 1} + C$ **4.** True **5.** $\dfrac{3}{5}x^{5/3} + C$ **7.** $\arcsin x + C$ **9.** $\dfrac{5}{2}x^2 - \ln|x| + 2x + C$ **11.** $\dfrac{4}{3}\ln|t| + C$

13. $x^4 - x^3 + \dfrac{5}{2}x^2 - 2x + C$ **15.** $x - \dfrac{1}{2x^2} + C$ **17.** $2\sqrt{z^3} + \dfrac{1}{2}z^2 + C$ **19.** $\dfrac{8}{5}\sqrt{t^5} + \dfrac{2}{3}\sqrt{t^3} + C$ **21.** $\dfrac{2u^3 - 3u^2}{6} + C$

23. $\dfrac{3}{4}x^4 - \dfrac{1}{x} + C$ **25.** $\dfrac{1}{2}t^2 + 2t + C$ **27.** $\dfrac{4}{3}x^2 + 2x^2 + x + C$ **29.** $\dfrac{1}{2}x^2 + e^x + C$ **31.** $8\arctan x + C$ **33.** $\dfrac{1}{2}\ln|x| + \dfrac{1}{4x^2} + C$

35. $\sec x + C$ **37.** $\dfrac{2}{5}\arcsin x + C$ **39.** $y = e^x + 3$ **41.** $y = \dfrac{1}{2}x^2 + x + \ln|x| - \dfrac{3}{2}$ **43.** $y = \dfrac{x^4}{16}$ **45.** $y = x$ **47.** $y = \arcsin\left(\dfrac{x^2 - 4}{2}\right) + \pi$

49. $y = -\sqrt{8e^x - 4}$ **51.** ≈ 6944 mosquitoes **53. (a)** $\dfrac{dP}{dt} = kP$ **(b)** $P = P_0 e^{kt}$ **(c)** $P = 4000e^{\frac{\ln 2}{18}t}$ **(d)** $\approx 25{,}398$ people

55. (a) $\dfrac{dA}{dt} = kA$ **(b)** $A = A_0 e^{kt}$ **(c)** $A = 8e^{-\frac{\ln 2}{1690}t}$ **(d)** ≈ 7.679 g

57. (a) $\dfrac{dP}{dt} = 0.00409P$ **(b)** $P = P_0 e^{0.00409t}$ **(c)** $1.2359 \times 10^9 e^{0.00409t}$ **(d)** $\approx 1.287 \times 10^9$ people **59.** $\approx 52.670\%$ remains.

61. ≈ 0.262 mol and ≈ 394.729 min **63.** $V = 20e^{\frac{-\ln 2}{2}t}$ and $t = 4$ s **65** See student solutions manual. **67.** See student solutions manual.
69. See student solutions manual. **71. (a)** $\sec x$ **(b)** See student solutions manual. **(c)** See student solutions manual.

73. (a) $\sqrt{a^2 - x^2}$ **(b)** See student solutions manual.
75. (a) **(b)** See student solutions manual. **(c)** See student solutions manual. **(d)** See student solutions manual.

Section 5.6

1. $\dfrac{1}{2}\sin u$ **2.** False **3. (c)** 0 **4.** True **5.** $\dfrac{1}{3}e^{3x+1} + C$ **7.** $\dfrac{1}{14}(t^2 - 1)^7 + C$ **9.** $\dfrac{1}{3}\sin^{-1}(x^3) + C$ **11.** $-\dfrac{1}{3}\cos(3x) + C$ **13.** $-\dfrac{1}{3}\cos^3 x + C$

15. $-e^{1/x} + C$ **17.** $\dfrac{1}{2}\ln|x^2 - 1| + C$ **19.** $2\sqrt{1 + e^x} + C$ **21.** $\dfrac{-2}{3\left(\sqrt{x} + 1\right)^3} + C$ **23.** $4(e^x - 1)^{3/4} + C$ **25.** $\dfrac{1}{2}\ln|2\sin x - 1| + C$

27. $\frac{1}{5}\ln|\sec 5x + \tan 5x| + C$ **29.** $\frac{2}{3}(\tan x)^{3/2} + C$ **31.** $\sec x + C$ **33.** $-e^{\cos x} + C$ **35.** $\frac{2}{5}(x-2)(x+3)^{3/2} + C$ **37.** $\frac{1}{3}\sin(3x) - \cos x + C$

39. $\frac{1}{5}\tan^{-1}\frac{x}{5} + C$ **41.** $\sin^{-1}\frac{x}{3} + C$ **43.** $\frac{1}{2}\cosh^2 x + C$ **45. (a-b)** $-\frac{2}{21}$ **(c)** Answers will vary.

47. (a-b) $\frac{1}{3}(e^2 - e)$ **(c)** Answers will vary.   **49. (a-b)** $\frac{232}{5}$ **(c)** Answers will vary.   **51. (a-b)** $\frac{1640}{\ln 3}$ **(c)** Answers will vary.   **53.** $\left(\frac{2}{3}\right)^{5/2}$

55. $\ln\sqrt{\frac{e^4+1}{2}}$ **57.** $\ln(\log_2 3)$ **59.** $\sin e^{\pi} - \sin 1$ **61.** $\frac{\pi}{8}$ **63.** $-\frac{32}{3}$ **65.** 0   **67.** $\frac{3\pi}{2}$ **69.** 50   **71.** $\tan^{-1}x + \ln\sqrt{x^2+1} + C$

73. $\frac{1}{14}\left(2\sqrt{x^2+3} - \frac{4}{x} + 9\right)^7 + C$ **75.** $\frac{2}{21}x^{3/2}(12x^2+7) + C$ **77.** $\frac{4}{9}(t^{3/2}+4)^{3/2} + C$ **79.** $\frac{3^{2x+1}}{2\ln 3} + C$ **81.** $-\sin^{-1}\frac{\cos x}{2} + C$ **83.** $\frac{1}{6}\ln\frac{187}{27}$

85. $(\ln 10)\ln|\ln x| + C$ **87.** $\ln 2$ **89.** $(\ln 2)\ln 3$ **91.** $b = \sqrt[3]{2}$ **93.** $\frac{1}{3}(x+1)^3 + C$ **95.** 0   **97.** $\frac{124}{15}$ **99.** $\frac{\pi}{3\sqrt{3}}$ **101.** $a^2\sinh 1 + a(b-a)$

103. $\frac{2\ln 2}{\pi}$ **105.** 8   **107.** 4   **109. (a)** $\frac{du}{dt} = k(u(t) - T)$ **(b)** $u = (u_0 - T)e^{kt} + T$ **(c)** $\approx 16.507\,°C$

111. (a) $\frac{du}{dt} = k(u(t) - T)$ **(b)** $u = (u_0 - T)e^{kt} + T$ **(c)** $u = 25e^{-0.159t} + 12$ **(d)** $\approx 8{:}50{:}01$ a.m.   **(e)** See student solutions manual.

113. $\frac{Q}{\sqrt{R^2 + z^2}} + C$

115. (a) $v(t) = \frac{mg}{k}(1 - e^{-kt/m})$ **(b)** $s(t) = \frac{mg}{k}\left(\frac{m}{k}(e^{-kt/m} - 1) + t\right)$ **(c)** $a \to 0,\ v \to \frac{mg}{k},\ s \to \infty$ **(d)** See student solutions

manual.   **117.** See student solutions manual.   **119.** See student solutions manual.

121. (a) 5   **(b)** 5   **(c)** 5   **(d)** $a = 2, b = 4$ **(e)** $a = k, b = k + 2$

123. Hint: Split integral; see student solutions manual. at 0   **125.** $c = 1$ **127.** See student solutions manual.   **129.** $\frac{n}{b(n+1)}(a+bx)^{\frac{n+1}{n}} + C$

131. $\frac{x^3}{3} + x - \frac{1}{2(x^2+1)} + C$ **133.** $2\ln\left|\sqrt{x+1} - 1\right| - 2\ln\left|\sqrt{x+1} + 1\right| + 6\sqrt{x+1} + C$ **135.** See student solutions manual.

Review Exercises

1. $s_4 = 16, S_4 = 24, s_8 = 18, S_8 = 22$ **3.** $s_n = \sum_{i=1}^{n}\frac{27}{n}\left(1 - \left(\frac{i}{n}\right)^2\right)$

5. (a) 14   **(b)** $\lim_{n\to\infty}\sum_{i=1}^{n}\frac{4}{n}\left(\left(-1 + \frac{4i}{n}\right)^2 - 3\left(-1 + \frac{4i}{n}\right) + 3\right) = \int_{-1}^{3}(x^2 - 3x + 3)\,dx$ **(c)** $\frac{28}{3}$ **(d)** $\frac{28}{3}$ **7.** $x^{2/3}\sin x$ **9.** $-2x\tan x^2$

11. $\frac{\sqrt{2}-1}{\sqrt{2}}$ **13.** $\frac{\pi}{4}$ **15.** 4   **17.** $\frac{14}{\ln 2}$ **19.** $2e^x + \ln|x| + C$ **21.** Position increased by 460 km in the first 16 hours.   **23.** $\frac{22}{3}$ **25.** 0

27. $2 \le \int_0^2 e^{x^2}\,dx \le 2e^4$ **29.** $\sin^{-1}\frac{2}{\pi} \approx 0.690$ and $\pi - \sin^{-1}\frac{2}{\pi} \approx 2.451$.   **31.** 0   **33.** $\frac{e^2-1}{2e}$ **35.** $\sqrt{\frac{1}{1+4x^2}}$ **37.** $y = 4e^{3x^2/2}$

39. $\frac{1-y}{(y-2)^2} + C$ **41.** $-\frac{2}{3}\left(\frac{1+x}{x}\right)^{3/2} + C$ **43.** $\frac{1}{12}$ **45.** $-\ln|1 - 2\sqrt{x}| + C$ **47.** $\frac{99}{10\ln 10}$ **49.** $\frac{\pi}{8}$ **51.** $\frac{1}{4}(x^3 + 3\cos x)^{4/3} + C$ **53.** $\frac{2}{3}$

55. 5   **57.** 15,552 gallons   **59. (a)** $\frac{dA}{dt} = A_0 e^{kt}$ **(b)** $A = A_0 e^{-t\frac{\ln 2}{1690}}$ **(c)** $A = 10e^{-t\frac{\ln 2}{1690}}$ **(d)** ≈ 8.842 g

Chapter 6
Section 6.1

1. $\int_0^1 (\sqrt{x} - x^2)\,dx$ **2.** $\int_{-1}^1 (1 - y^2)\,dy$ **3.** $\frac{1}{2}$ **5.** $\frac{1}{6}$ **7.** $\frac{1}{2}$ **9.** $\frac{4}{15}$ **11.** $\frac{\sqrt{3}}{2} - \frac{\pi}{6}$ **13.** $\frac{9}{2}$ **15.** $\frac{32}{3}$ **17.** $\frac{9}{2}$ **19.** $e^2 - 3$

21. $2\sqrt{2}$ **23.** $\frac{32}{3}$ **25. (a)** $\int_0^1 (\sqrt{x} - x^3)\,dx = \frac{5}{12}$ **(b)** $\int_0^1 (\sqrt[3]{y} - y^2)\,dy = \frac{5}{12}$

27. (a) $\int_0^1 ((x+1) - (x^2+1))\,dx = \frac{1}{6}$ **(b)** $\int_1^2 (\sqrt{y-1} - (y-1))\,dy = \frac{1}{6}$ **29. (a)** $\int_0^3 (\sqrt{9-x} - \sqrt{9-3x})\,dx + \int_3^9 \sqrt{9-x}\,dx = 12$

(b) $\int_0^3 \left((9 - y^2) - \left(3 - \frac{y^2}{3}\right)\right)dy = 12$ **31. (a)** $\int_2^3 \sqrt{x-2}\,dx + \int_3^4 (\sqrt{x-2} - \sqrt{2x-6})\,dx = \frac{2\sqrt{2}}{3}$

(b) $\int_0^{\sqrt{2}}\left(\left(\frac{y^2}{2} + 3\right) - (y^2 + 2)\right)dy = \frac{2\sqrt{2}}{3}$ **33.** $\frac{16\sqrt{2}}{3}$ **35.** $\frac{64}{3}$ **37.** 2   **39.** $\frac{\sqrt{3}}{2} - \frac{\pi}{6}$ **41.** $e^2 - \frac{3}{2}$ **43.** $\frac{125}{24}$ **45.** $1 - \frac{\pi}{4}$

47. See Student Solutions Manual. **49. (a)** $53,213 **(b)** Answers will vary. **51.** $\dfrac{\sqrt{3}}{2} + \dfrac{\pi}{12} - \dfrac{9}{8}$

53. (a)

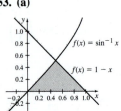

(b) The point of intersection is approximately (0.4890266, 0.5109734). **(c)** $A \approx 0.2526954$

55. (a)

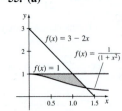

(b) The point of intersection is approximately (1.3171826, 0.3656347). **(c)** $A \approx 0.2951422$

57. (a)

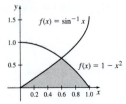

(b) The point of intersection is approximately (0.5985697, 0.6417144). **(c)** $A \approx 0.3247648$

59. (a) $A(k) = \int_0^{\arctan(k)} (k - \tan x)\, dx$ **(b)** $A(1) = \dfrac{\pi}{4} - \dfrac{\ln 2}{2}$ **(c)** $\dfrac{\pi}{40}$

Section 6.2

1. $\pi \int_a^b [f(x)]^2\, dx$ **2.** False **3.** False **4.** False **5.** $V = 30\pi$ **7.** $\dfrac{3\pi}{4}$ **9.** $2\pi(\tan 1 - 1)$ **11.** $\dfrac{4\pi}{5}$ **13.** $\left(\dfrac{1 - e^{-4}}{2}\right)\pi$ **15.** $\dfrac{15\pi}{2}$

17. $\dfrac{128\pi}{5}$ **19.** 32π **21.** $\dfrac{64\pi}{5}$ **23.** $\dfrac{\pi}{2}$ **25.** $\dfrac{59{,}049\pi}{5}$ **27.** $\dfrac{129\pi}{7}$ **29.** $\dfrac{117\pi}{2}$ **31.** $\dfrac{64\pi}{45}$ **33.** $\dfrac{512\pi}{21}$ **35.** $\left(\dfrac{e^2}{2} - \dfrac{5}{6}\right)\pi$ **37.** π

39. $\left(\dfrac{e^4}{2} + 2e^2 - \dfrac{5}{2}\right)\pi$ **41.** $\dfrac{\pi}{6}$ **43.** $\dfrac{1024\pi}{15}$ **45.** $\dfrac{363\pi}{64}$ **47. (a)** $\dfrac{k^5\pi}{30}$ **(b)** $\dfrac{k^4\pi}{6}$ **(c)** $k = 5$. **49.** $\left(2 - \dfrac{\pi}{4}\right)\pi$

51. (a) $k = \dfrac{e^{-b^2}}{b^2}$, $P(x) = \dfrac{e^{-b^2}}{b^2}x^2$ **(b)** $V = \dfrac{b^2 e^{-b^2}\pi}{2}$ **(c)** $b = 1$; answers will vary.

Section 6.3

1. False **2.** False **3.** False **4.** True **5.** $\dfrac{3\pi}{2}$ **7.** $\dfrac{3\pi}{10}$ **9.** $\dfrac{768\pi}{7}$ **11.** $\dfrac{4\pi}{5}$ **13.** $\dfrac{2\pi}{15}$ **15.** $\left(\dfrac{e^4 - 1}{e^4}\right)\pi$ **17.** 16π **19.** $\dfrac{17\pi}{6}$ **21.** $\dfrac{\pi}{2}$

23. (a) $\dfrac{128\pi}{105}$ **(b)** $\dfrac{16\pi}{5}$ **(c)** See Student Solutions Manual. **25. (a)** $\dfrac{206\pi}{15}$ **(b)** 12π **(c)** See Student Solutions Manual. **27.** $\dfrac{64\pi}{5}$

29. $\dfrac{4\pi}{15}$ **31.** $\dfrac{308\pi}{3}$ **33.** $\dfrac{80\pi}{3}$ **35.** $\dfrac{\pi}{6}$ **37.** $\dfrac{1177\pi}{10}$ **39.** $\dfrac{\pi}{2}$ **41. (a)** $\dfrac{16\pi}{5}$ **(b)** $\dfrac{8\pi}{3}$ **(c)** $\dfrac{16\pi}{3}$ **(d)** $\dfrac{14\pi}{15}$ **43.** $f(x) = \dfrac{x^2 + 2x}{\sqrt{\pi}}$

45. See Student Solutions Manual. **47.** See Student Solutions Manual.

Section 6.4

1. Answers will vary. **2.** False **3.** $\dfrac{128}{3}$ **5.** $\dfrac{256{,}000}{3}$ **7.** $\dfrac{9\pi}{35}$ **9.** $\dfrac{37\pi}{1320}$ **11.** $\dfrac{\pi}{40}$ **13.** See Student Solutions Manual. **15.** $\dfrac{1}{3}\pi r^2 h$

17. ≈ 3.7582496 **19.** $\left(\dfrac{500}{3} - 28\sqrt{21}\right)\pi$ **21.** $\dfrac{\pi}{3}bah$

Section 6.5

1. False **2.** True **3.** $2\sqrt{10}$ **5.** $\sqrt{13}$. **7.** $\dfrac{80\sqrt{10} - 13\sqrt{13}}{27}$ **9.** $\dfrac{80\sqrt{10} - 8}{27}$ **11.** $\dfrac{4\sqrt{2} - 2}{3}$ **13.** 45 **15.** 21 **17.** $\dfrac{33}{16}$

19. $\ln\left(2 + \sqrt{3}\right) - \ln\left(\sqrt{3}\right)$ **21.** $\dfrac{79}{2}$ **23.** $\dfrac{13\sqrt{13} - 8}{27}$ **25.** $\dfrac{20\sqrt{10} - 2}{27}$ **27. (a)** $\int_0^2 \sqrt{1 + (2x)^2}\, dx$ **(b)** $\sqrt{17} - \dfrac{\ln\left(\sqrt{17} - 4\right)}{4}$

29. (a) $\int_0^4 \sqrt{1 + \left(\dfrac{-x}{\sqrt{25 - x^2}}\right)^2}\, dx$ **(b)** $5\arcsin\left(\dfrac{4}{5}\right)$ **31. (a)** $\int_0^{\frac{\pi}{2}} \sqrt{1 + (\cos(x))^2}\, dx$ **(b)** ≈ 1.910098894 **33.** $6a$

35. ≈ 2.853923436 **37.** $\dfrac{17}{12}$ **39.** $\ln\left(2 + \sqrt{2}\right) - \ln\left(\sqrt{2}\right)$

41. (a) $\int_{x_1}^{x_2} \dfrac{1}{\sqrt{1-t^2}}\, dt$ **(b)** $\int_{x_2}^{x_1} \dfrac{1}{\sqrt{1-t^2}}\, dt$ or $\int_{y_1}^{y_2} \dfrac{1}{\sqrt{1-t^2}}\, dt$

(c) $\int_{y_2}^{y_1} \dfrac{1}{\sqrt{1-t^2}}\, dt + \int_{x_1}^{x_2} \dfrac{1}{\sqrt{1-t^2}}\, dt$ **43. (a)** $b + \dfrac{5}{2\sinh^{-1}\left(\dfrac{3}{4}\right)}$ **(b)** $\dfrac{15}{\sinh^{-1}\left(\dfrac{3}{4}\right)}$

45. (a) $h(x) = -\dfrac{46}{5625}(x)(x-150)$ **(b)** $\sqrt{14{,}089} + \dfrac{5625}{92}\ln(3) + \dfrac{5625}{46}\ln(5) - \dfrac{5625}{92}\ln\left(\sqrt{14{,}089}-92\right)$

47. (a) See Student Solutions Manual. **(b)** See Student Solutions Manual. **(c)** Answers will vary. **49.** $f(x) = \cosh x$

Section 6.6

1. True **2.** $W = Fx$ **3.** A unit of work is called a *joule* in SI units and a *foot-pound* in the customary U.S. system of units. **4.** $W = \int_a^b F(x)\, dx$

5. Equilibrium **6.** True **7.** True **8.** True **9.** $\dfrac{825}{2}$ J **11.** 2400 J **13.** $k = 12$ N/m **15.** 0.9 J **17.** $W \approx 225{,}905$ J **19.** $W \approx 122.197$ J

21. 6,073,600 ft lb **23.** 850 J **25.** $\dfrac{3}{4}$ J **27.** 10 ft **29.** $\dfrac{64{,}000\pi}{3}$ J **31.** $352{,}800\pi$ J **33.** $176{,}400\pi$ J **35.** 289,890 ft-lb **37.** 6.965×10^9 J

39. 475 ft-lb **41.** $\dfrac{80\pi}{3}$ **43. (a)** $W = 820 \int_0^2 12 \cdot (4-x) \cdot \sqrt{4x-x^2}\, dx$ **(b)** $W \approx 88{,}067$ J **45.** 2.092×10^5 in pounds

Section 6.7

1. Force **2.** True **3.** (b) **4.** True **5.** 367,500 N **7.** $16{,}333.\overline{3}$ N **9.** 352,800 N **11.** 22,500 lb **13.** $8166.\overline{6}$ N **15.** 26,353 N
17. 2.94×10^9 N **19.** 157,786.5 N **21.** 70.4375 N

Section 6.8

1. (c) **2.** (c) **3.** True **4.** (a) **5.** False **6.** (a) **7.** $\bar{x} = \dfrac{58}{7}$ **9.** $\bar{x} = \dfrac{29}{15}$ **11.** $M_y = 20$, $M_x = 24$, $(\bar{x}, \bar{y}) = \left(\dfrac{20}{13}, \dfrac{24}{13}\right)$

13. $M_y = 29$, $M_x = 62$, $(\bar{x}, \bar{y}) = \left(\dfrac{29}{15}, \dfrac{62}{15}\right)$ **15.** $\left(\dfrac{7}{8}, \dfrac{19}{8}\right)$ **17.** $\left(\dfrac{9}{4}, \dfrac{27}{10}\right)$ **19.** $\left(2, \dfrac{8}{5}\right)$ **21.** $\left(\dfrac{12}{5}, \dfrac{3}{4}\right)$ **23.** $72\pi^2$ **25.** $\left(1, \dfrac{4+3\pi}{4+\pi}\right)$

27. $\left(\dfrac{a}{3}, \dfrac{b}{3}\right)$ **29.** $\left(\dfrac{3a}{4}, \dfrac{3h}{10}\right)$ **31. (a)** $M = \dfrac{kL^2}{2}$ **(b)** $k = \dfrac{2M}{L^2}$ **(c)** $\bar{x} = \dfrac{2}{3}L$ **(d)** Answers will vary. **(e)** Answers will vary. **33.** $\dfrac{22\pi}{3}$

35. $\dfrac{28\pi}{3}$ **37.** See Student Solutions Manual. **39.** $\left(0, \dfrac{3}{7}\right)$ **41.** $\left(0, \dfrac{38}{35}\right)$ **43.** $\left(-\dfrac{1}{2}, \dfrac{12}{5}\right)$ **45.** See Student Solutions Manual.

47. $\left(\dfrac{836}{217}, \dfrac{80}{31}\right)$ **49.** $\left(\dfrac{227}{74}, \dfrac{607}{148}\right)$ **51.** $\left(\dfrac{113}{38}, \dfrac{113}{76}\right)$

Review Exercises

1. $8\ln 2 - 3$ **3.** $\dfrac{8}{3}$ **5.** $\dfrac{125}{6}$ **7.** $\dfrac{5\pi}{6}$ **9.** $\dfrac{\pi}{2}$ **11.** $\dfrac{512\pi}{15}$ **13.** $\dfrac{32\pi}{5}$ **15.** $\dfrac{3456\pi}{35}$ **17.** $\dfrac{209}{6}$ **19.** 9 **21.** 4 **23.** $\dfrac{256-28\sqrt{7}}{3}\pi$ **25.** $\dfrac{10\pi}{3}$

27. 345,000 ft-lb **29.** $2{,}508{,}800\pi$ J 0.3 m **31.** 0.3 m **33.** 601,883 N **35.** $\left(\dfrac{27}{5}, \dfrac{9}{8}\right)$ **37.** 72π

Chapter 7
Section 7.1

1. True **2.** $uv - \int v\, du$ **3.** $\dfrac{x}{2}e^{2x} - \dfrac{1}{4}e^{2x} + C$ **5.** $x\sin x + \cos x + C$ **7.** $\dfrac{2}{3}x^{1.5}\ln x - \dfrac{4}{9}x^{1.5} + C$ **9.** $x\cot^{-1}x + \dfrac{1}{2}\ln(x^2+1) + C$

11. $x(\ln x)^2 - 2x\ln x + 2x + C$ **13.** $2x\sin x - (x^2-2)\cos x + C$ **15.** $(2x^2 - \sin^2 x + \cos^2 x + 4x\sin x\cos x)/8 + C$

17. $x\cosh x - \sinh x + C$ **19.** $x\cosh^{-1}x - \sqrt{x^2-1} + C$ **21.** $\dfrac{x}{2}(\sin(\ln x) - \cos(\ln x)) + C$ **23.** $x(\ln x)^3 - 3x(\ln x)^2 + 6x\ln x - 6x + C$

25. $x^3(9(\ln x)^2 - 6\ln x + 2)/27 + C$ **27.** $(2x^3\tan^{-1}x - x^2 + \ln(x^2+1))/6 + C$ **29.** $7^x(x\ln 7 - 1)/(\ln 7)^2 + C$ **31.** $-(1+e^\pi)/2 \approx -12.070$

33. $(2 - 50e^{-6})/27 \approx 0.069$ **35.** $9\ln 3 - 4 \approx 5.888$ **37.** $(e-2)/27 \approx 0.71828$ **39.** $4.5\ln 3 - 4 \approx 0.944$ **41.** $(1+e^\pi)/2 \approx 12.070$

43. $2\pi \approx 6.283$ **45.** $e\pi - 2\pi \approx 2.257$

47. (a) **(b)** $\tan^{-1}\dfrac{2+\sqrt{13}}{3} \approx 1.079$ **(c)** $7 + 2\sqrt{13}e^{-\tan^{-1}\frac{2+\sqrt{13}}{3}} \approx 1.890$

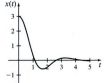

49. $2\sin\sqrt{x} - 2\sqrt{x}\cos\sqrt{x} + C$ **51.** $(\sin x)\ln(\sin x) - \sin x + C$ **53.** $(e^{2x}\sin e^{2x} + \cos e^{2x})/2 + C$ **55.** $e^{x^2}(x^2 - 1)/2 + C$
57. $e^x(x\sin x + x\cos x - \sin x)/2 + C$ **59.** See the Student Solutions Manual. **61.** See the Student Solutions Manual.
63. (a) $e^{5x}(x^2/5 - 2x/5^2 + 2/5^3) + C$ **(b)** $\int x^n e^{kx}\,dx = (x^n e^k x - n\int x^{n-1}e^{kx})/k$

65. (a) $0.5\sin^2 x + C_1$ **(b)** $-0.5\cos^2 x + C_2$ **(c)** $-\dfrac{\cos(2x)}{4} + C_3$ **(d)** $C_2 = 0.5 + C_1$ **(e)** $C_3 = 0.25 + C_1$
67. See the Student Solutions Manual. **69. (a)** $\dfrac{5\pi}{32} \approx 0.491$ **(b)** $\dfrac{8}{15} \approx 0.533$ **(c)** $\dfrac{35\pi}{256} \approx 0.430$ **(d)** $\dfrac{5\pi}{32} \approx 0.491$
71. See the Student Solutions Manual. **73.** See the Student Solutions Manual. **75.** See the Student Solutions Manual.

Section 7.2

1. True **2.** True **3.** $\dfrac{\sin^5 x}{5} - \dfrac{2\sin^3 x}{3} + \sin x + C$ **5.** $(-\sin 6x + 9\sin 4x - 45\sin 2x + 60x)/192 + C$ **7.** $\dfrac{x}{2} - \dfrac{\sin 2\pi x}{4\pi} + C$ **9.** 0

11. $\dfrac{\cos^5 x}{5} - \dfrac{\cos^3 x}{3} + C$ **13.** $\dfrac{x}{8} - \dfrac{\sin 4x}{32} + C$ **15.** $-\dfrac{3}{4}\cos^{4/3}x + C$ **17.** $-\dfrac{2}{5}\sin^5\left(\dfrac{x}{2}\right) + \dfrac{2}{3}\sin^3\left(\dfrac{x}{2}\right) + C$ **19.** $\dfrac{1}{4}\tan^4(x) + C$ **21.** $\dfrac{\tan^3 x}{3} + C$

23. $\left(-\sin 3x\sec^4 x + 7\tan x\sec^3 x + 4\ln\left|\cos\left(\dfrac{x}{2}\right) - \sin\left(\dfrac{x}{2}\right)\right| - 4\ln\left(\sin\left(\dfrac{x}{2}\right) + \cos\left(\dfrac{x}{2}\right)\right)\right)/32 + C$ **25.** $\csc(x) - \dfrac{\csc^3 x}{3} + C$

27. $-\dfrac{1}{4}\cos 2x - \dfrac{1}{8}\cos 4x + C$ **29.** $\dfrac{1}{4}\sin 2x + \dfrac{1}{8}\sin 4x + C$ **31.** $\dfrac{1}{4}\sin 2x - \dfrac{1}{12}\sin 6x + C$ **33.** $\dfrac{2}{3}$ **35.** $\dfrac{\sin^3 x}{3} + C$ **37.** $\dfrac{1}{\cos x} + C$

39. $\dfrac{1}{4}\sin 3x + \dfrac{1}{36}\sin 9x + C$ **41.** 0 **43.** $\dfrac{1}{2}\tan^2 x + \ln\cos x + C$ **45.** $(-\csc^2 x + \sec^2 x + 4\ln\sin x - 4\ln\cos x)/2 + C$ **47.** $-\dfrac{1}{6}\cot^6 x + C$

49. $-\dfrac{1}{8}\csc^4(2x) + C$ **51.** $\dfrac{7\sqrt{2}}{48} + \dfrac{\ln\left(1+\sqrt{2}\right)}{16} \approx 0.261$ **53.** $-\dfrac{1}{2}\cos x - \dfrac{1}{4}\cos 2x + C$ **55.** $\dfrac{1}{2}\sin x - \dfrac{1}{4}\sin 2x + C$ **57.** $\dfrac{\pi^2}{2} \approx 4.935$

59. (a) $\dfrac{2}{5\pi} \approx 0.127$. **(b)** Answers will vary. **(c)**

61. (a) $-2 + \sqrt{2} + \dfrac{1}{2}\ln\left(3 + 2\sqrt{2}\right) \approx 0.296$ **(b)** $\dfrac{4\pi - \pi^2}{2} \approx 1.348$

63. (a) $\dfrac{3x}{8} - \dfrac{1}{4}\sin 2x + \dfrac{1}{32}\sin 4x + C$ **(b)** $-\dfrac{1}{4}\sin^3 x\cos x - \dfrac{3}{8}\sin x\cos x + \dfrac{3}{8}x + C$ **(c)** See the Student Solutions Manual.

65. $\dfrac{1}{2(m-n)}\sin((m-n)x) - \dfrac{1}{2(m+n)}\sin((m+n)x) + C$ **67.** $\dfrac{1}{2(m-n)}\sin((m-n)x) + \dfrac{1}{2(m+n)}\sin((m+n)x) + C$
69. See the Student Solutions Manual.

Section 7.3

1. True **2.** (d) **3.** (c) **4.** (b) **5.** $\dfrac{x}{2}\sqrt{4-x^2} + 2\sin^{-1}\dfrac{x}{2} + C$ **7.** $-\dfrac{x}{2}\sqrt{16-x^2} + 8\sin^{-1}\dfrac{x}{4} + C$ **9.** $-\dfrac{1}{x}\sqrt{4-x^2} - \sin^{-1}\dfrac{x}{2} + C$

11. $\dfrac{x^3-2x}{4}\sqrt{4-x^2} + 2\sin^{-1}\dfrac{x}{2} + C$ **13.** $\dfrac{1}{4}\dfrac{x}{\sqrt{4-x^2}} + C$ **15.** $2\left[\dfrac{x\sqrt{x^2+4}}{4} + \ln\dfrac{x^2+\sqrt{x^2+4}}{2}\right] + C$ **17.** $\ln\dfrac{x+\sqrt{x^2+16}}{4} + C$

19. $\dfrac{x}{2}\sqrt{1+9x^2} + \dfrac{1}{6}\ln\left|\sqrt{1+9x^2}+3x\right| + C$ **21.** $\dfrac{1}{27}\left[\dfrac{3x\sqrt{4+9x^2}}{4} - \ln\dfrac{3x+\sqrt{4+9x^2}}{2}\right] + C$ **23.** $-\dfrac{\sqrt{x^2+4}}{4x} + C$ **25.** $\dfrac{x}{4\sqrt{x^2+4}} + C$

27. $\dfrac{x}{2}\sqrt{x^2-25} + \dfrac{25}{2}\ln\left(x+\sqrt{x^2-25}\right) + C$ **29.** $\sqrt{x^2-1} + \tan^{-1}\dfrac{1}{\sqrt{x^2-1}} + C$ **31.** $\dfrac{\sqrt{x^2-36}}{36x} + C$ **33.** $\dfrac{1}{2}\ln\left(2x+\sqrt{4x^2-9}\right) + C$

35. $\dfrac{-x}{9\sqrt{x^2-9}} + C$ **37.** $-\dfrac{x}{\sqrt{x^2-9}} + \ln\dfrac{x+\sqrt{x^2-9}}{3} + C$ **39.** $x - 4\tan^{-1}\dfrac{x}{4} + C$ **41.** $\dfrac{x}{2}\sqrt{4-25x^2} + \dfrac{2}{5}\sin^{-1}\dfrac{5x}{2} + C$

43. $\dfrac{x}{4\sqrt{4-25x^2}} + C$ **45.** $\dfrac{2}{5}\left[\dfrac{5x\sqrt{4+25x^2}}{4} + \ln\dfrac{5x+\sqrt{4+25x^2}}{2}\right] + C$ **47.** $\dfrac{\sqrt{x^2-16}}{32x^2} - \dfrac{1}{128}\tan^{-1}\dfrac{4}{\sqrt{x^2-16}} + C$ **49.** $\dfrac{\pi}{4} \approx 0.785$

51. $\dfrac{\sqrt{2}+\ln\left(1+\sqrt{2}\right)}{2} \approx 1.148$ **53.** $10 - 2\sqrt{7} + 9\ln 3 - \dfrac{9}{2}\ln\left(4+\sqrt{7}\right) \approx 6.073$ **55.** $\dfrac{1}{\sqrt{3}} - \dfrac{\pi}{6} \approx 0.0538$ **57.** $3 - \dfrac{3\pi}{4} \approx 0.644$

59. πab **61.** $\dfrac{1}{2}\sin^{-1}\dfrac{2}{3} \approx 0.365$ **63.** $18 - \dfrac{22\sqrt{5}}{3} \approx 1.602$ **65.** $-3\sqrt{(2)} - 0.5\ln\left(3+2\sqrt{(2)}\right) + 5\sqrt{(6)} + 0.5\ln\left(5+2\sqrt{(6)}\right) \approx 8.2697$

67. (a) $4\sin^{-1}\left(\dfrac{\sqrt{15}}{8}\right)+\sin^{-1}\left(\dfrac{\sqrt{15}}{4}\right)-\dfrac{\sqrt{15}}{2}\approx 1.403$ **(b)** $\pi-\left[4\sin^{-1}\left(\dfrac{\sqrt{15}}{8}\right)+\sin^{-1}\left(\dfrac{\sqrt{15}}{4}\right)-\dfrac{\sqrt{15}}{2}\right]\approx 1.739$

69. $\dfrac{5\sqrt{26}}{2}+\dfrac{1}{2}\ln\left(5+\sqrt{26}\right)\approx 13.904$ **71.** $\dfrac{\pi}{40}+\dfrac{\pi}{16}\tan^{-1}\dfrac{1}{2}\approx 0.170$ **73.** $-\sin^{-1}(2-x)+C$ **75.** $\ln\left(x-1+\sqrt{(x-1)^2-4}\right)+C$

77. $\dfrac{e^x}{2}\sqrt{25-e^{2x}}+\dfrac{25}{2}\sin^{-1}\dfrac{e^x}{5}+C$ **79.** $\dfrac{x}{4}\sqrt{1-x^2}+\dfrac{2x^2-1}{4}\sin^{-1}x+C$

81. (a) $\dfrac{x}{2}\sqrt{x^2+a^2}+\dfrac{a^2}{2}\ln\left|\dfrac{x+\sqrt{x^2+a^2}}{a}\right|+C$ **(b)** $\dfrac{x}{2}\sqrt{x^2+a^2}+\dfrac{a^2}{2}\sinh^{-1}\dfrac{x}{a}+C$ **83.** See the Student Solutions Manual.

85. See the Student Solutions Manual. **87.** $\sin^{-1}\left(\dfrac{2x}{3}-1\right)+C$ **89.** $\ln\left(\dfrac{x+\sqrt{x^2+a^2}}{a}\right)+C$

Section 7.4

1. $\tan^{-1}(x+2)+C$ **3.** $\dfrac{1}{2}\tan^{-1}\dfrac{x+2}{2}+C$ **5.** $\dfrac{2}{\sqrt{5}}\tan^{-1}\dfrac{2x+1}{\sqrt{5}}+C$ **7.** $\dfrac{1}{4}\ln(2x^2+2x+3)-\dfrac{\sqrt{5}}{10}\tan^{-1}\dfrac{2x+1}{\sqrt{5}}+C$ **9.** $-\sin^{-1}\dfrac{1-x}{3}+C$

11. $\sin^{-1}\left(\dfrac{x}{2}-1\right)+C$ **13.** $\ln(x+1)-\ln\left(\sqrt{x^2+2x+2}+1\right)+C$ **15.** $\sin^{-1}\dfrac{x+1}{5}+C$ **17.** $\sqrt{x^2-2x+5}-4\sinh^{-1}\dfrac{x-1}{2}+C$

19. $\sinh^{-1}1\approx 0.881$ **21.** $\sinh^{-1}\dfrac{2e^x+1}{\sqrt{3}}+C$ **23.** $-2\sqrt{4x-x^2-3}-\sin^{-1}(2-x)+C$ **25.** $\dfrac{x-1}{9\sqrt{x^2-2x+10}}+C$

27. $\ln\left(x+1+\sqrt{x^2+2x-3}\right)+C$ **29.** $\sqrt{5+4x-x^2}-3\ln\left(3+\sqrt{5+4x-x^2}\right)+3\ln(x-2)+C$

31. $\sqrt{x^2+2x-3}-\ln\left(x+1+\sqrt{x^2+2x-3}\right)+C$ **33.** See the Student Solutions Manual. **35.** $(x-a)\sqrt{\dfrac{a+x}{a-x}}+a\tan^{-1}\dfrac{x}{\sqrt{a^2-x^2}}+C$

Section 7.5

1. (a) **2.** True **3.** True **4.** False **5.** $\dfrac{x^2}{2}-x+2\ln(x+1)+C$ **7.** $\dfrac{x^3}{3}+x^2+7x+10\ln(x-2)+C$ **9.** $\dfrac{1}{3}\ln\dfrac{x-2}{x+1}+C$ **11.** $\ln\dfrac{(x-2)^2}{1-x}+C$

13. $\dfrac{2}{21}\ln(2-3x)+\dfrac{1}{14}\ln(2x+1)+C$ **15.** $\dfrac{4}{x+1}+5\ln\dfrac{x+1}{x+2}+C$ **17.** $\dfrac{1}{2-2x}+\dfrac{3}{4}\ln(x-1)+\dfrac{1}{4}\ln(x+1)+C$ **19.** $\ln x-\ln\sqrt{x^2+1}+C$

21. $\dfrac{1}{6}\ln(x^2+2x+4)+\dfrac{2}{3}\ln(x+1)+C$ **23.** $\dfrac{x-32}{32(x^2+16)}+\dfrac{1}{128}\tan^{-1}\dfrac{x}{4}+C$ **25.** $-\dfrac{x^2+8}{2(x^2+16)^2}+C$ **27.** $\dfrac{1}{4}\ln(1-x)+\dfrac{3}{4}\ln(x+3)+C$

29. $-\dfrac{7}{3}\ln(x^2+2)-\dfrac{4}{x-1}+\dfrac{14}{3}\ln(1-x)+\dfrac{2\sqrt{2}}{3}\tan^{-1}\dfrac{x}{\sqrt{2}}+C$ **31.** $\ln\dfrac{(x-3)^2}{x(x+1)}+C$ **33.** $\dfrac{1}{x-1}+4\ln(2-x)-3\ln(x-1)+C$

35. $-\dfrac{1}{2}\ln(x^2+x+1)+\ln(1-x)+\dfrac{1}{\sqrt{3}}\tan^{-1}\dfrac{2x+1}{\sqrt{3}}+C$ **37.** $-\dfrac{1}{3}\tanh^{-1}\dfrac{1}{3}\approx -0.116$ **39.** $\dfrac{1}{8}\ln 21\approx 0.381$ **41.** $\dfrac{1}{5}\ln\dfrac{2-\sin\theta}{3+\sin\theta}+C$

43. $\dfrac{1}{2}\ln(\cos(2\theta)+3)-\ln(\cos\theta)+C$ **45.** $\ln\dfrac{\sqrt[3]{1-e^t}}{2+e^t}+C$ **47.** $\ln\dfrac{\sqrt{1-e^x}}{\sqrt{1+e^x}}+C$ **49.** $t-\dfrac{1}{2}\ln(e^{2t}+1)+C$ **51.** $\ln(\sin x-1)-\dfrac{1}{\sin x-1}+C$

53. $\dfrac{\sin x}{18(\sin^2 x+9)}+\dfrac{1}{54}\tan^{-1}\dfrac{\sin x}{3}+C$ **55.** $\ln\dfrac{15}{7}\approx 0.762$ **57.** $\dfrac{4\pi\sqrt{3}}{3}+\dfrac{4}{3}\ln 3\approx 8.720$ **59. (a)** $-4,-2,3$

(b) $(x+4)(x+2)(x-3)$ **(c)** $\dfrac{2}{35}\ln(3-x)+\dfrac{13}{10}\ln(x+2)-\dfrac{19}{14}\ln(x+4)+C$ **61.** $2\sqrt{x}-4\ln\left(\sqrt{x}+2\right)+C$

63. $\dfrac{3}{5}x^{5/3}+\dfrac{3}{4}x^{4/3}+\dfrac{3}{2}x^{2/3}+x+3\sqrt[3]{x}+3\ln\left(1-\sqrt[3]{x}\right)+C$ **65.** $\dfrac{2}{3}\sqrt{x}+\dfrac{1}{3}\sqrt[3]{x}+\dfrac{2}{9}\sqrt[6]{x}+\dfrac{2}{27}\ln\left(1-3\sqrt[6]{x}\right)+C$ **67.** $\dfrac{2}{3}(2x+1)^{3/4}+C$

69. $3\sqrt[3]{x+1}+C$ **71.** $-\dfrac{3\sqrt[6]{x}}{\sqrt[3]{x}+1}+3\tan^{-1}\sqrt[6]{x}+C$ **73.** $\dfrac{-2}{1+\tan\dfrac{x}{2}}+C$ **75.** $\dfrac{2}{\sqrt{5}}\tan^{-1}\dfrac{\tan\dfrac{x}{2}}{\sqrt{5}}+C$ **77.** $-\ln\left(1-\tan\dfrac{x}{2}\right)+C$

79. $\dfrac{1}{2}\ln(\cos x-\sin x)-\dfrac{x}{2}+C$ **81.** $-\dfrac{2}{\sqrt{5}}\tanh^{-1}\dfrac{2\left(\tan\dfrac{x}{2}\right)+1}{\sqrt{5}}+C$ **83.** $\dfrac{\tan\dfrac{x}{2}}{\left(1+\tan\dfrac{x}{2}\right)^2}-\dfrac{1}{2}\ln\left(\left(\tan\dfrac{x}{2}\right)-1\right)+\dfrac{1}{2}\ln\left(1+\tan\dfrac{x}{2}\right)+C$

85. $\dfrac{8}{15}\tanh^{-1}\dfrac{8-5\sqrt{3}}{11}-\dfrac{1}{6}\ln 3+\dfrac{1}{15}\tanh^{-1}\dfrac{55{,}681}{39{,}409\sqrt{2}}\approx 0.041$ **87.** $\dfrac{\pi}{2}+\ln 2\approx 2.264$ **89.** See the Student Solutions Manual.

91. See the Student Solutions Manual.

Section 7.6

1. True **2.** True **3.** 22 and 20 **5.** 26 and $\dfrac{58}{3} \approx 19.333$ **7. (a)** 0.483 **(b)** 0.007 **(c)** 26 **9. (a)** 0.743 **(b)** 0.010 **(c)** 41

11. (a) 0.907 **(b)** 0.005 **(c)** 29 **13. (a)** 3.059 **(b)** 0.00053 **(c)** 6 **15. (a)** 0.747 **(b)** 0.00026 **(c)** 5

17. (a) 0.910 **(b)** 0.00033 **(c)** 6 **19. (a)** See the Student Solutions Manual. **(b)** 0.696 **(c)** 0.693 **21. (a)** 1.910 **(b)** 1.910

23. ≈ 62.983 **25.** 16787.5 m^3 **27.** T: $131,787.5 \text{ m}^3$, S: $132,625 \text{ m}^3$ **29.** $\dfrac{105}{2}\pi \approx 164.934$ **31. (a)** 1.518 **(b)** 1.971

33. (a) 1.583 **(b)** 1.765 **35.** See the Student Solutions Manual.

Section 7.7

1. $\dfrac{e^{2x}}{5}(\sin x + 2\cos x) + C$ **3.** $\dfrac{2x-1}{20}(4x+3)^{3/2} + C$ **5.** $\dfrac{6x+5}{60}(4x+5)^{3/2} + C$ **7.** $\dfrac{1}{\sqrt{6}}\ln\left|\dfrac{\sqrt{4x+6}-\sqrt{6}}{\sqrt{4x+6}+\sqrt{6}}\right| + C$

9. $2\sqrt{4x+6} + \dfrac{6}{\sqrt{6}}\ln\left|\dfrac{\sqrt{4x+6}-\sqrt{6}}{\sqrt{4x+6}+\sqrt{6}}\right| + C$ **11.** $4Ei(4\ln x) - \dfrac{x^4}{\ln x}$ **13.** $\dfrac{1}{2}\sqrt{1-4x^2} + x\sin^{-1}(2x) + C$

15. $-\dfrac{9\sqrt{2}}{4} + \dfrac{135}{8}\cos^{-1}\dfrac{2\sqrt{2}}{3} \approx 2.553$ **17. (a)** $\dfrac{e^{2x}}{5}(\sin x + 2\cos x) + C$ **(b)** They agree.

19. (a) $\dfrac{2x-1}{20}(4x+3)^{3/2} + C$ **(b)** They agree. **21. (a)** $\dfrac{6x+5}{60}(4x+5)^{3/2} + C$ **(b)** They agree. **23. (a)** $-\sqrt{\dfrac{2}{3}}\tanh^{-1}\sqrt{\dfrac{2x}{3}+1} + C$

(b) They disagree. **(c)** See Student Solutions Manual. **25. (a)** $2\sqrt{4x+6} - 2\sqrt{6}\tanh^{-1}\sqrt{\dfrac{2x}{3}+1} + C$ **(b)** They disagree.

(c) See Student Solutions Manual. **27. (a)** $\dfrac{1}{2}Ei(x) - \dfrac{x+1}{2x^2}e^x + C$ **(b)** They agree. **29. (a)** $\dfrac{1}{2}\sqrt{1-4x^2} + x\sin^{-1}(2x) + C$ **(b)** They agree.

31. (a) $-\dfrac{9\sqrt{2}}{4} + \dfrac{135}{8}\cos^{-1}\dfrac{2\sqrt{2}}{3} \approx 2.553$ **(b)** They agree. **33.** Yes **35.** No **37.** No

Section 7.8

1. (c) **2. (b)** **3.** False **4.** True **5.** False **6.** $\lim_{t\to b^-}\int_a^t f(x)\,dx$ **7.** Yes, ∞ as endpoint **9.** No **11.** Yes, function undefined at 0

13. Yes, function undefined at 1 **15.** Converges to $\dfrac{1}{2}$ **17.** Diverges **19.** Diverges **21.** Converges to $\dfrac{1}{24} \approx 0.0417$

23. Converges to $\dfrac{\pi}{2} \approx 1.571$ **25.** Diverges **27.** Diverges **29.** Converges to 4 **31.** Converges to 0 **33.** Diverges **35.** Converges to 1

37. Diverges **39.** Diverges **41.** Diverges **43.** Diverges **45.** Diverges **47.** Converges to π **49.** Diverges **51.** Converges to $\sqrt[3]{486} \approx 7.862$

53. Diverges **55.** Converges to 0 **57.** Converges to 4 **59.** Diverges **61.** Converges to $\dfrac{1}{2}$ **63.** Diverges **65.** Diverges **67. (a)** Converges

(b) ≈ 0.673 **69. (a)** Converges **(b)** $\dfrac{\pi}{2} \approx 1.571$ **71.** $\ln 2 \approx 0.693$ **73.** $\dfrac{\pi}{2} \approx 1.571$ **75.** $2\pi a^2$ **77. (a)** 1250 **(b)** $30,612.20

79. $\dfrac{2\pi NI}{10r}\left(1 - \dfrac{x}{\sqrt{r^2+x^2}}\right)$ **81.** GmM **83.** 1 **85.** $\dfrac{1}{2}$ **87.** See the Student Solutions Manual. **89.** See the Student Solutions Manual.

91. See the Student Solutions Manual. **93.** See the Student Solutions Manual. **95.** See the Student Solutions Manual. **97.** $\dfrac{1}{s^2}$ **99.** $\dfrac{s}{s^2+1}$

101. $\dfrac{1}{s-a}$ **103.** 2 **105.** 1 **107.** See the Student Solutions Manual. **109.** $\dfrac{a+b}{2}$ **111.** $\sigma^2 = \dfrac{(b-a)^2}{12}, \sigma = \dfrac{b-a}{2\sqrt{3}}$

Review Exercises

1. $\dfrac{1}{4}\tan^{-1}\dfrac{x+2}{4} + C$ **3.** $\dfrac{1}{3}\sec^3\phi + C$ **5.** $-\cos\phi + \dfrac{1}{3}\cos^3\phi + C$ **7.** $\ln\left(x+2+\sqrt{(x+2)^2-1}\right) + C$ **9.** $\ln(\sin x) - x\cot x + C$

11. $\dfrac{10x-x^3}{4}\sqrt{4-x^2} + 6\sin^{-1}\dfrac{x}{2} + C$ **13.** $e^t + 2\ln(2-e^t) + C$ **15.** $\dfrac{1}{16}\ln\dfrac{4-x^2}{4+x^2} + C$ **17.** $\dfrac{4y+3}{2(y+1)^2} + \ln(y+1) + C$

19. $x\tan x + \ln|\cos x| + C$ **21.** $(y-1)\ln(1-y) - y + C$ **23.** $\dfrac{2}{x} + 5\ln(1-x) - 2\ln x + C$ **25.** $\dfrac{1}{3}x^3\sin^{-1}x + \dfrac{x^2+2}{9}\sqrt{1-x^2} + C$

27. $\dfrac{1}{2}\ln\dfrac{x}{x+2} + C$ **29.** $\dfrac{1}{2}\ln(1-w) - \dfrac{3}{2}\ln(w+1) + C$ **31.** $\sqrt{x} + \dfrac{1}{2}\sin\left(2\sqrt{x}\right) + C$ **33.** $-\dfrac{1}{6}\cos^3 x + \dfrac{1}{2}\cos x + \dfrac{1}{2}\sin^2 x\cos x + C$ **35.** 1

37. See the Student Solutions Manual. **39.** Converges to $2e^{-1} \approx 0.736$ **41.** Converges to 1 **43.** See the Student Solutions Manual.

45. Diverges **47.** $f(x) = x^2\sin x$ **49. (a)** 1.910 **(b)** 1.910 **51.** 3

Chapter 8
Section 8.1

1. False **2.** False **3.** True **4.** (b) **5.** False **6.** True **7.** False **8.** False **9.** False **10.** (b) **11.** True **12.** False **13.** $2, \dfrac{3}{2}, \dfrac{4}{3}, \dfrac{5}{4}$

15. $0, \ln 2, \ln 3, \ln 4$ **17.** $\dfrac{1}{3}, -\dfrac{1}{5}, \dfrac{1}{7}, -\dfrac{1}{9}$ **19.** $1, -1, 1, -1$ **21.** $\dfrac{1}{2}, \dfrac{1}{2}, \dfrac{3}{4}, \dfrac{3}{2}$ **23.** $a_n = 2n$ **25.** $a_n = 2^n$ **27.** $a_n = \dfrac{(-1)^{n+1}}{n+1}$

29. $a_n = \dfrac{n}{n+1}$ **31.** $a_n = (n-1)!$ **33.** 0 **35.** 1 **37.** 4 **39.** 0 **41.** 0 **43.** 1 **45.** $\ln \dfrac{1}{3}$ **47.** e^{-2} **49.** 0 **51.** 1 **53.** $-\dfrac{1}{2}$ **55.** 1

57. 0 **59.** 0 **61.** 0 **63.** Diverges **65.** Diverges **67.** Converges **69.** Diverges **71.** Converges **73.** Bounded from above and from below
75. Bounded from below **77.** Bounded from below **79.** Bounded from above and from below **81.** Nonmonotonic **83.** Nonmonotonic
85. Monotonic decreasing **87.** Nonmonotonic **89.** Decreasing, bounded from below **91.** Increasing, bounded from above
93. Increasing, bounded from above **95.** Converges; 6 **97.** Converges; $-\ln 3$ **99.** Diverges **101.** Converges; 0 **103.** Converges; 0
105. Converges; 0 **107.** Converges; 1 **109.** Converges; 1 **111.** Converges; 1 **113.** Converges; 1 **115.** Converges **117.** Converges

119. Converges **121.** Diverges **123.** Converges **125.** Converges **127.** $p_n = 3000r^n + h\dfrac{r^n-1}{r-1}$ Converges $\lim\limits_{n\to\infty} p_n = \dfrac{h}{1-r}$
129. (a) $I_n = 0.95^n I_0$ (b) 77
131. (a) See Student Solutions Manual. (b) $\lim\limits_{n\to\infty} e^{n/(n+2)} = e$ (c)

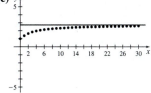

133. See Student Solutions Manual. **135.** See Student Solutions Manual. **137.** See Student Solutions Manual. **139.** Answers will vary.
141. See Student Solutions Manual. **143.** Diverges **145.** See Student Solutions Manual. **147.** See Student Solutions Manual.
149. See Student Solutions Manual. **151.** See Student Solutions Manual. **153.** See Student Solutions Manual.

Section 8.2

1. (b) **2.** (d) **3.** True **4.** True **5.** $\dfrac{a}{1-r}$ **6.** False **7.** $\dfrac{175}{64}$ **9.** 10 **11.** $\dfrac{1}{3}$ **13.** $-\dfrac{1}{3}$ **15.** $\dfrac{1}{2}$ **17.** Diverges **19.** Converges; 6

21. Converges; $\dfrac{21}{2}$ **23.** Converges; $\dfrac{50}{69}$ **25.** Converges; 6 **27.** Converges; $\dfrac{1}{3}$ **29.** Diverges **31.** Diverges **33.** Converges; $\dfrac{3}{2}$

35. Diverges **37.** Converges; $\dfrac{1}{42}$ **39.** Diverges **41.** Converges; $\dfrac{1}{99}$ **43.** Diverges **45.** Converges; $\dfrac{2}{3}$ **47.** Diverges **49.** Diverges

51. Diverges **53.** Converges; $-\dfrac{1}{4}$ **55.** Diverges **57.** Converges; $\sin(1)$ **59.** $\dfrac{5}{9}$ **61.** $\dfrac{3857}{900}$ **63.** 90 ft **65.** (a) $\dfrac{h}{1-r}$ (b) $h = 2000$

67. (a) $P\dfrac{1+r}{r-i}$ (b) \$6.67 **69.** See Student Solutions Manual. **71.** (a) $T(n) = p\sum\limits_{k=1}^{n}(1-e)^{k-1}$ (b) $\lim\limits_{n\to\infty} T(n) = \dfrac{p}{e}$ (c) $e_{\min} = \dfrac{p}{L}$

(d) $e_{\min} = \dfrac{2}{3}$; $T(365) = 150$ **73.** See Student Solutions Manual. **75.** $n = 11$ **77.** See Student Solutions Manual.

79. See Student Solutions Manual. **81.** See Student Solutions Manual. **83.** See Student Solutions Manual. **85.** See Student Solutions Manual.
87. See Student Solutions Manual.

Section 8.3

1. (a) 0 **2.** False **3.** True **4.** True **5.** False **6.** True **7.** True **8.** False **9.** $p > 1; 0 < p \le 1$ **10.** Converges **11.** Diverges
12. False **13.** Diverges **15.** Diverges **17.** Diverges **19.** Converges **21.** Diverges **23.** Converges **25.** Converges **27.** Diverges
29. Converges **31.** Diverges **33.** Converges **35.** Converges **37.** Converges **39.** Diverges **41.** Diverges **43.** Diverges **45.** Diverges
47. Diverges **49.** Diverges **51.** Diverges **53.** Diverges **55.** See Student Solutions Manual. **57.** See Student Solutions Manual.

59. (a) $\dfrac{\pi^2}{6}$ (b) See Student Solutions Manual. **61.** Converges **63.** Converges **65.** See Student Solutions Manual.

67. See Student Solutions Manual. **69.** See Student Solutions Manual. **71.** See Student Solutions Manual. **73.** $(-\infty, -1)$
75. (a) 1.2 (b) Upper is 1.5, lower is 0.5. (c) See Student Solutions Manual.

Section 8.4

1. (b) **2.** False **3.** True **4.** False **5.** Converges **7.** Converges **9.** Diverges **11.** Converges **13.** Converges **15.** Converges
17. Diverges **19.** Converges **21.** Converges **23.** Converges **25.** Converges **27.** Converges **29.** Diverges **31.** Converges
33. Diverges **35.** Converges **37.** Diverges **39.** Converges **41.** Converges **43.** Diverges **45.** Diverges **47.** Converges
49. See Student Solutions Manual. **51.** See Student Solutions Manual. **53.** Diverges **55.** Converges **57.** See Student Solutions Manual.

59. (a) See Student Solutions Manual.   **(b)** See Student Solutions Manual.   **61.** Diverges   **63.** Converges   **65.** See Student Solutions Manual.

67. Answers will vary; for example: $\sum \dfrac{1}{(k+1)\ln k}$ diverges; $\sum \dfrac{1}{k^2}$ converges   **69.** Diverges   **71.** See Student Solutions Manual.

Section 8.5

1. False   **2.** True   **3.** False   **4.** False   **5.** False   **6.** True   **7.** Converges   **9.** Diverges   **11.** Diverges   **13.** Converges   **15.** Diverges
17. Diverges   **19.** 0.8611; upper estimate to the error is 0.0625   **21.** 0.9498; upper estimate to the error is 0.0039
23. 0.7222; upper estimate to the error is 0.00617   **25.** 0.3218; upper estimate to the error is 0.000653   **27.** Conditionally convergent
29. Absolutely convergent   **31.** Absolutely convergent   **33.** Diverges   **35.** Absolutely convergent   **37.** Absolutely convergent
39. Conditionally convergent   **41.** Absolutely convergent   **43.** Conditionally convergent   **45.** Absolutely convergent   **47.** Diverges
49. Absolutely convergent   **51.** See Student Solutions Manual.   **53.** See Student Solutions Manual.   **55.** See Student Solutions Manual.
57. Absolutely convergent for $|r| < 1$ **59.** See Student Solutions Manual.   **61.** See Student Solutions Manual.   **63.** See Student Solutions Manual.   **65.** Absolutely convergent   **67.** See Student Solutions Manual.   **69.** Absolutely convergent if $2 < p < 3$; conditionally convergent if $p \geq 3$ **71.** Absolutely convergent   **73.** Converges   **75.** See Student Solutions Manual.

Section 8.6

1. False   **2.** False   **3.** False   **4.** False   **5.** Converges   **7.** Converges   **9.** Converges   **11.** Converges   **13.** Diverges   **15.** Converges
17. Diverges   **19.** Converges   **21.** Diverges   **23.** Converges   **25.** Diverges   **27.** Converges   **29.** Converges   **31.** Converges
33. Converges   **35.** Converges by the root test   **37.** Converges by the ratio test   **39.** Converges by the ratio test   **41.** Converges by the root test
43. Converges by the root test   **45.** See Student Solutions Manual.   **47.** Answers will vary.   **49. (a)** Converges by the ratio test   **(b)** ≈ 0.049787
51. See Student Solutions Manual.   **53.** Converges if $|x| < 1$; diverges if $|x| > 1$; ratio test is inconclusive if $|x| = 1$.
55. See Student Solutions Manual.   **57.** See Student Solutions Manual.   **59.** See Student Solutions Manual.

Section 8.7

1. False   **2.** False   **3.** True   **4.** False   **5.** False   **6.** a_k, b_k **7.** Diverges by the Test for Divergence   **9.** Absolutely convergent by the Geometric Series Test   **11.** Absolutely convergent by the Limit Comparison Test   **13.** Diverges by the Comparison Test for Divergence   **15.** Diverges by the Ratio Test   **17.** Diverges by the Test for Divergence   **19.** Conditionally convergent by the Alternating Series Test   **21.** Diverges by the Ratio Test   **23.** Absolutely convergent by the Ratio Test   **25.** Diverges by the Limit Comparison Test   **27.** Absolutely convergent by the Comparison Test for Convergence   **29.** Diverges by the Test for Divergence   **31.** Absolutely convergent by the Root Test   **33.** Absolutely convergent by the Root Test   **35.** Absoutely convergent by the Integral Test   **37.** Diverges by the Root Test   **39.** Absolutely convergent by the Comparison Test for Convergence   **41.** Converges; $\dfrac{11}{6}$ **43.** Diverges   **45. (a)** Use the Geometric Series Test and Limit Comparison Test.   **(b)** $\dfrac{14}{5}$ **47.** Converges

Section 8.8

1. True   **2.** True   **3.** True   **4.** False   **5.** True   **6.** False   **7.** False   **8.** True   **9.** True   **10.** False   **11.** False   **12.** False   **13.** $-1 < x < 1$
15. $-4 < x < 2$ **17. (a, b)** $R = 2; -2 \leq x < 2$ **(c)** Answers will vary.   **19. (a, b)** $R = 1; -1 \leq x < 1$ **(c)** Answers will vary.
21. (a, b) $R = 3; -3 < x < 3$ **(c)** Answers will vary.   **23. (a, b)** $R = 1; -1 < x < 1$ **(c)** Answers will vary.   **25. (a, b)** $R = 1; 2 < x < 4$
(c) Answers will vary.   **27.** $R = 1; -1 \leq x \leq 1$ **29.** $R = 1; 1 \leq x \leq 3$ **31.** $R = \infty; -\infty < x < \infty$ **33.** $R = 1; -1 < x < 1$

35. $R = 4; -4 < x < 4$ **37.** $R = 3; 0 < x < 6$ **39.** $R = \infty; -\infty < x < \infty$ **41.** $R = \infty; -\infty < x < \infty$ **43.** $R = \dfrac{1}{e}; -\dfrac{1}{e} < x < \dfrac{1}{e}$

45. (a) $-3 < x < 3$ **(b)** $f(2) = 3, f(-1) = \dfrac{3}{4}$ **(c)** $f(x) = \dfrac{3}{3-x}$ **47. (a)** $0 < x < 4$ **(b)** $f(1) = \dfrac{2}{3}, f(2) = 1$ **(c)** $f(x) = \dfrac{2}{4-x}$

49. Converges at $x = 2$; no information about $x = 5$ **51. (a)** True   **(b)** False   **(c)** False   **(d)** False   **(e)** True   **(f)** True

53. (a) $f(x) = \displaystyle\sum_{n=0}^{\infty} (-1)^n x^{3n}$ **(b)** $R = 1; -1 < x < 1$ **55. (a)** $f(x) = \dfrac{1}{6}\displaystyle\sum_{n=0}^{\infty} \left(\dfrac{x}{3}\right)^n$ **(b)** $R = 3; -3 < x < 3$

57. (a) $f(x) = \displaystyle\sum_{n=0}^{\infty} (-1)^n x^{3n+1}$ **(b)** $R = 1; -1 < x < 1$ **59. (a)** $f'(x) = \displaystyle\sum_{k=0}^{\infty} \dfrac{(-1)^k x^{2k}}{(2k)!}$

(b) $\displaystyle\int f(x)\, dx = C + \sum_{k=0}^{\infty} \dfrac{(-1)^k x^{2k+2}}{(2k+2)!} = 1 - \cos x + C$ **61. (a)** $f'(x) = \displaystyle\sum_{k=1}^{\infty} \dfrac{x^{k-1}}{(k-1)!}$ **(b)** $\displaystyle\int f(x)\,dx = C + \sum_{k=0}^{\infty} \dfrac{x^{k+1}}{(k+1)!} = e^x - 1 + C$

63. $f(x) = \displaystyle\sum_{k=1}^{\infty} (-1)^{k-1} k x^{k-1}$ **65.** $f(x) = \dfrac{2}{3}\displaystyle\sum_{k=1}^{\infty} k x^{k-1}$ **67.** $f(x) = \displaystyle\sum_{k=0}^{\infty} \dfrac{(-1)^{k+1} x^{k+1}}{k+1}$ **69.** $f(x) = -\displaystyle\sum_{k=0}^{\infty} \dfrac{x^{2k+2}}{k+1}$ **71.** $-1 \leq x < 1$

73. $-1 \leq x < 1$ **75.** $-1 < x < 1$ **77.** $-3 < x < 5$ **79. (a)** $\displaystyle\sum_{k=0}^{\infty} x^{2k}$ **(b)** $-1 < x < 1$ **81.** 0.693   **83.** See Student Solutions Manual.

85. See Student Solutions Manual.   **87.** See Student Solutions Manual.   **89.** $-1 \leq x < 5$ **91.** $\displaystyle\sum_{k=0}^{\infty} \dfrac{(-1)^k x^{2k+1}}{(2k+1)!}; R = \infty$

Section 8.9

1. Taylor series **2.** Maclaurin **3.** $f(x) = -\sum_{k=1}^{\infty} \dfrac{x^k}{k}$ **5.** $f(x) = \sum_{k=0}^{\infty} x^k$ **7.** $f(x) = \sum_{k=0}^{\infty} (-1)^k (k+1) x^k$ **9.** $f(x) = \sum_{k=0}^{\infty} (-1)^k x^{2k}$

11. $f(x) = \sum_{k=0}^{\infty} \dfrac{x^k 3^k}{k!}$ **13.** $f(x) = \sum_{k=0}^{\infty} \dfrac{(-1)^{k+1} \pi^{2k+1} x^{2k+1}}{(2k+1)!}$ **15.** $f(x) = \sum_{k=0}^{\infty} \dfrac{e(x-1)^k}{k!}$ **17.** $f(x) = \sum_{k=1}^{\infty} \dfrac{(-1)^{k+1}(x-1)^k}{k}$

19. $f(x) = \sum_{k=0}^{\infty} (-1)^k (x-1)^k$ **21.** $f(x) = \sum_{k=0}^{\infty} \dfrac{\sin\left(\frac{1}{6}(\pi + 3k\pi)\right) \left(x - \frac{\pi}{6}\right)^k}{k!}$ **23.** $f(x) = -6 + 5x + 2x^2 + 3x^3$

25. $f(x) = 4 + 18(x-1) + 11(x-1)^2 + 3(x-1)^3$ **27.** $f(x) = \sum_{k=0}^{\infty} \dfrac{x^{2k+1}}{(2k+1)!}$ **29.** $x + x^2 + \dfrac{x^3}{2} + \dfrac{x^4}{6} + \dfrac{x^5}{24}$ **31.** $x - x^2 + \dfrac{x^3}{3} - \dfrac{x^5}{30} + \dfrac{x^6}{90}$

33. $x + \dfrac{x^3}{6} + \dfrac{3x^5}{40} + \dfrac{5x^7}{112}$ **35.** $\sum_{k=0}^{\infty} \binom{1/2}{k} x^{2k} = 1 + \dfrac{x^2}{2} - \dfrac{x^4}{8} + \dfrac{x^6}{16} + \ldots$; interval: $[-1, 1]$ **37.** $\sum_{k=0}^{\infty} \binom{1/5}{k} x^k = 1 + \dfrac{x}{5} - \dfrac{2x^2}{25} + \dfrac{6x^3}{125} + \ldots$; interval:

$[-1, 1]$ **39.** $\sum_{k=0}^{\infty} \binom{-1/2}{k} x^{2k} = 1 - \dfrac{x^2}{2} + \dfrac{3x^4}{8} - \dfrac{5x^6}{16} + \ldots$; interval: $[-1, 1]$ **41.** $\sum_{k=1}^{\infty} 2\binom{-1/2}{k} (-1)^k x^{k+1} = 2x + x^2 + \dfrac{3}{4}x^3 + \dfrac{5}{8}x^4 + \ldots$; interval:

$(-1, 1]$ **43.** $\sin^2 x = \sum_{k=1}^{\infty} (-1)^{k+1} \dfrac{2^{2k-1} x^{2k}}{(2k)!}$ **45.** $\cos x = \sum_{k=0}^{\infty} \dfrac{(-1)^k x^{2k}}{(2k)!}$ **47.** $\sec x = 1 + \dfrac{x^2}{2} + \dfrac{5x^4}{24} + \dfrac{61x^6}{720} + \dfrac{277x^8}{8064}$

49. $\arctan x + C_1 = C_2 + \sum_{k=0}^{\infty} \dfrac{(-1)^{k+1} x^{2k+1}}{2k+1}$ **51.** $3e^{x^{1/3}}(x^{2/3} - 2x^{1/3} + 2) + C_1 = C_2 + \sum_{k=0}^{\infty} \dfrac{3x^{\frac{k}{3}+1}}{(3+k)k!}$ **53.** See Student Solutions Manual.

55. See Student Solutions Manual. **57.** See Student Solutions Manual. **59.** -0.5 **61.** See Student Solutions Manual.

63. See Student Solutions Manual. **65.** $1 + x^2 - \dfrac{x^3}{2} + \dfrac{5x^4}{6}$

Section 8.10

1. (a) $1 - \dfrac{x^2}{2!} + \dfrac{x^4}{4!} - \dfrac{x^6}{6!}$ **(b)** **(c)** $\cos(\pi/90) \approx 0.999$ **(d)** $|E| \leq 5.47 \times 10^{-17}$ **(e)** $|x| \leq 1.190$

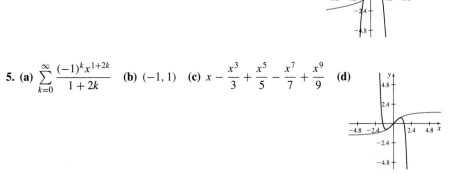

3. (a) $\sum_{k=0}^{\infty} \binom{1/3}{k} x^k$ **(b)** $[-1, 1]$ **(c)** $1 + \dfrac{x}{3} - \dfrac{x^2}{9} + \dfrac{5x^3}{81} - \dfrac{10x^4}{243}$ **(d)** **(e)** Answers will vary.
(f) ≈ 0.965; the error is $\leq 3.018 \times 10^{-7}$

5. (a) $\sum_{k=0}^{\infty} \dfrac{(-1)^k x^{1+2k}}{1+2k}$ **(b)** $(-1, 1)$ **(c)** $x - \dfrac{x^3}{3} + \dfrac{x^5}{5} - \dfrac{x^7}{7} + \dfrac{x^9}{9}$ **(d)** **(e)** Answers will vary.

7. 0.310 **9.** 0.195 **11.** 0.200 **13.** 0.487 **15.** See Student Solutions Manual. **17. (a)** 0.069 **(b)** 0.965 **(c)** $2.862°$

19. $1538665/489888 \approx 3.14085$ **21. (a)** $(-1)^{k+1} \dfrac{2}{k}$ **(b)** See Student Solutions Manual **(c)** See Student Solutions Manual

23. (a) See Student Solutions Manual. **(b)** 1.571 **(c)** 0.524

Review Exercises

1. $1, -\dfrac{1}{16}, \dfrac{1}{81}, -\dfrac{1}{256}, \dfrac{1}{625}$ **3.** $a_n = (-1)^{n+1}2^{3-2n}3^{n-1}$ **5.** 0   **7.** 0   **9.** Diverges   **11.** Converges; 0   **13.** See Student Solutions Manual.

15. $\dfrac{4}{5}$ **17.** Converges; $\dfrac{\ln 2}{\ln 2 - 1}$ **19.** Diverges   **21.** See Student Solutions Manual.   **23.** Converges   **25.** Converges   **27.** Converges

29. Converges   **31.** Converges; 1.079   **33.** Converges; -1.715 **35.** Conditionally convergent   **37.** Conditionally convergent   **39.** Converges

41. Diverges   **43.** Diverges   **45.** Converges   **47.** Diverges   **49.** Converges   **51.** Converges   **53.** (a) $R = 1$ (b) $2 \le x \le 4$

55. (a) $R = \infty$ (b) all x **57.** (a) $R = 1$ (b) $0 \le x < 2$ **59.** $f(x) = \displaystyle\sum_{k=0}^{\infty} (-1)^k \dfrac{2x^k}{3^{1+k}}$ **61.** (a) $\displaystyle\sum_{k=0}^{\infty} \dfrac{3^k}{2k+1} x^{2k+1}$ (b) 0.760

63. $\displaystyle\sum_{k=0}^{\infty} \dfrac{\sqrt{e}(x-1)^k}{2^k k!}$ **65.** $1 + 2\left(x - \dfrac{\pi}{4}\right) + 2\left(x - \dfrac{\pi}{4}\right)^2 + \dfrac{8}{3}\left(x - \dfrac{\pi}{4}\right)^3 + \dfrac{10}{3}\left(x - \dfrac{\pi}{4}\right)^4 + \dfrac{64}{15}\left(x - \dfrac{\pi}{4}\right)^5 + \ldots$

67. $\displaystyle\sum_{k=0}^{\infty} \binom{-4}{k} x^k; \; -1 < x < 1$ **69.** $\displaystyle\sum_{k=0}^{\infty} \binom{-1/2}{k}(-1)^k x^k; \; -1 \le x < 1$ **71.** 1.349   **73.** 0.545

Chapter 9
Section 9.1

1. Plane curve; parameter   **2.** (c)   **3.** (d)   **4.** Cycloid   **5.** False   **6.** True

7. (a) $x = 2y - 3$
(b)
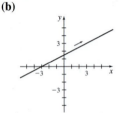

9. (a) $x = 2y - 3, x \in [1, 5]$
(b)
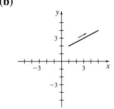

11. (a) $x = e^y$
(b)
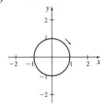

13. (a) $x^2 + y^2 = 1$
(b)
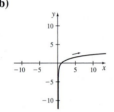

15. (a) $\dfrac{x^2}{4} + \dfrac{y^2}{9} = 1$
(b)
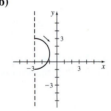

17. (a) $\dfrac{(x+3)^2}{4} + \dfrac{(y-1)^2}{4} = 1$
(b)
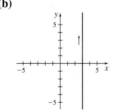

19. (a) $x = 3$
(b)
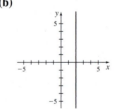

21. (a) $x = 2$
(b)
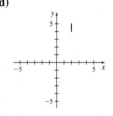

(c) $x = 2, y > 4$
(d)

23. (a) $x = y^2 + 5$
(b)
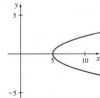

(c) $x \ge 5, y \ge 0$
(d)

25. (a) $y = (x-1)^3$
(b)

(c) $x \ge 2, y \ge 1$
(d)

27. (a) $x = \sec \arctan y$
(b)
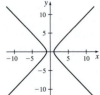

(c) $x \ge 1$
(d)

29. (a) $x = y^2$
(b)

(c) $x \ge 0, y \ge 0$
(d)

31. **(a)** $x = \left(\dfrac{y+1}{2}\right)^2$ **(c)** $x \geq 0$

(b) 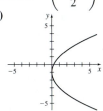   **(d)**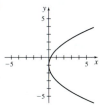

33. **(a)** $x = \left(\dfrac{y+1}{2}\right)^2$ **(c)** $x > 0$ **35.** **(a)** $\dfrac{x+2}{3} + \dfrac{y^2}{4} = 1$ **(c)** $-2 \leq x \leq 1, -2 \leq y \leq 2$

(b) 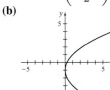   **(d)** 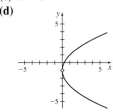   **(b)** 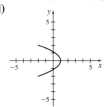   **(d)**

37. $y = \dfrac{2x}{1+x}, x \geq 0, 0 \leq y \leq 2$ **39.** $x = \dfrac{4}{\sqrt{4 - \left(\dfrac{y}{x}\right)^2}}, x \geq 2, y \geq 0$ **41.** $(x+2)^2 + \left(\left(\dfrac{4-y}{2}\right)^2\right) = 1, -3 \leq x \leq -1, 2 \leq y \leq 6$

43. $x = t, y = 4t - 2$; or $x = \dfrac{t+2}{4}, y = t$ **45.** $x = t, y = -2t^2 + 1$; or $x = t^3, y = -2x^6 + 1$

47. $x = \sqrt[3]{t}, y = 4t$; or $x = \dfrac{t}{\sqrt[3]{4}}, y = t^3$ **49.** $x = \dfrac{1}{3}\sqrt{t} - 3, y = t$; or $x = t^2 - 3, y = 9t^4$ **51.** $x = t + 2, y = t, 0 \leq t \leq 5$

53. $x = 3\cos t, y = 2\sin t, -\pi \leq t \leq \dfrac{\pi}{2}$ **55.** $x = 3\cos\dfrac{2\pi}{3}t, y = 2\sin\dfrac{2\pi}{3}t$ **57.** $x = 3\sin\pi t, y = 2\cos\pi t$

59. **(a)** 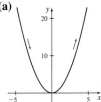   **(b)** 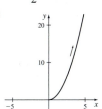   **(c)**    **(d)**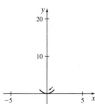

61. $I \rightarrow$ (d) counterclockwise $II \rightarrow$ (a) counterclockwise $III \rightarrow$ (b) counterclockwise $IV \rightarrow$ (c) counterclockwise

63. $I \rightarrow$ (c) from $(1, 0)$ to $(-1, 0)$ $II \rightarrow$ (b) from $(-1, 0)$ to $(1, 0)$ $III \rightarrow$ (a) clockwise $IV \rightarrow$ (d) from $\left(-\dfrac{\sqrt{2}}{2}, 1\right)$ to $(1, 0)$

65. **(a)**   **(b)** $x \approx 8.66, y \approx 10.5$

67. $x = a\sin t, y = b\cos t$ or $x = a\cos t, y = b\sin t$ **69.** **(a)** $x \approx 95.8t$ ft, $y \approx -16t^2 + 80.3t + 3$ ft   **(b)** $y \approx 99.6$ ft   **(c)** $x \approx 191.6$ ft
(d) $t \approx 3.13$ s   **(e)** $y \approx 97.6$ ft   **(f)** $t \approx 5.1$ s   **(g)** $x \approx 488.6$ ft   **71.** **(a)** $x \approx 65.5t$ ft, $y \approx -16t^2 + 45.9t + 6$ ft   **(b)** $y \approx 35.9$ ft
(c) $x \approx 65.5$ ft   **(d)** $t \approx 1.8$ s   **(e)** $y \approx 36.8$ ft   **73.** The first curve is counterclockwise from $(2, 0)$; the second is clockwise from $(0, 2)$.

75. $x = \dfrac{m^2}{2R} - R, y = \pm\dfrac{m}{2R}\sqrt{4R^2 - m^2}, 0 \leq m \leq 2R$ **77.** See Student Solutions Manual.

79. $x = (a+b)\cos t - b\cos\left(\dfrac{a+b}{b}t\right), y = (a+b)\sin t - b\sin\left(\dfrac{a+b}{b}t\right)$

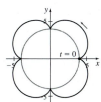

$a = 4, b = 1$

Section 9.2

1. (a)   **2.** True   **3.** Horizontal; vertical   **4.** False   **5.** $\dfrac{dy}{dx} = \dfrac{\sin t + \cos t}{\cos t - \sin t}$ **7.** $\dfrac{dy}{dx} = \dfrac{1}{1 - \dfrac{1}{t^2}}$ **9.** $\dfrac{dy}{dx} = \tan t$ **11.** $\dfrac{dy}{dx} = \dfrac{1}{2 \cot t}$

13. (a) $y = \dfrac{1}{8}x + 1$ **15.** (a) $y = \dfrac{4}{3}x - 3$ **17.** (a) $y = -\dfrac{1}{4}x + \dfrac{3}{4}$ **19.** (a) $y = -2x + 2$

(b)   (b)   (b)   (b)

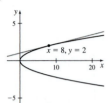   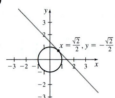

21. (a) $y = -x + 2$ **23.** (a) $y = -x + \sqrt{2}$ **25.** (a) $y = -\dfrac{3\sqrt{3}}{4}x + 6$ **27.** Horizontal at $\left(\dfrac{4}{3}, -\dfrac{8}{3}\sqrt{\dfrac{4}{3}}\right), \left(\dfrac{4}{3}, \dfrac{8}{3}\sqrt{\dfrac{4}{3}}\right)$;

(b)   (b)   (b)

vertical at $(0, 0)$

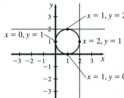   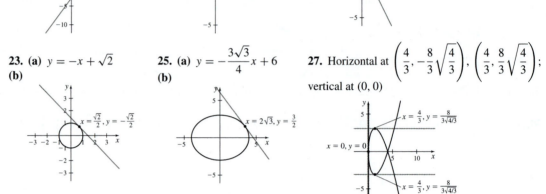

29. Horizontal at $(1, 0)$, $(1, 2)$; vertical at $(0, 1)$, $(2, 1)$ **31.** $\dfrac{8}{27}\left(10\sqrt{10} - 1\right)$ **33.** 2.96   **35.** 4π **37.** 4π

39. (a) $s = \int_0^{2\pi} \sqrt{(-4\sin 2t)^2 + (2t)^2}\, dt$ (b) $s \approx 44.5$ (c)

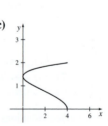

41. (a) $s = \int_{-2}^{2} \sqrt{(2t)^2 + \left(\dfrac{1}{2\sqrt{t+2}}\right)^2}\, dt$ (b) $s \approx 8.4$ (c)

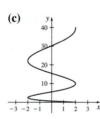

43. (a) $s = \int_0^{2\pi} \sqrt{(-3\sin t - 3\sin 3t)^2 + (3\cos t - 3\cos 3t)^2}\, dt$ (b) $s = 24$ (c)

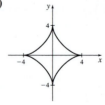

45. (a) Horizontal at $\left(\dfrac{4}{3}, -\dfrac{8}{3}\sqrt{\dfrac{4}{3}}\right)$, $\left(\dfrac{4}{3}, \dfrac{8}{3}\sqrt{\dfrac{4}{3}}\right)$; vertical at $(2, 0)$ **(d)**

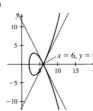

(b) $t = 2, t = -2$; see Student Solutions Manual
(c) $y = 2x - 12, y = -2x + 12$

47. (a) $s = \dfrac{3b}{2}$ **(b)** $b = 1$

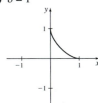

49. $s \approx 7.47$ **51.** $s \approx 4.95$ **53.** $s \approx 2.24$ **55.** $\sqrt{400 + 1024t}$ **57.** $\sqrt{1600\cos^2 2t + 36\sin^2 t}$ **59.** See Student Solutions Manual.

61. 0.73423 **63.** $\sqrt{e^{0.4a} - 2e^{0.2a} + e^{0.4b} - 2e^{0.2b} + 2}$ **65.** $t = 2\sqrt{\sqrt[3]{9 - 4\sqrt{2}} - 1}$ **67.** $\dfrac{1}{3a\cos^4\theta \sin\theta}$

Section 9.3

1. False   **2.** $S = 2\pi \int_a^b x(t)\sqrt{(dx/dt)^2 + (dy/dt)^2}\,dt$ **3.** 137.86   **5.** 3.77   **7.** 3.56   **9.** 14.469   **11.** 22.943   **13.** $4\pi a^2$ **15.** 36.405
17. 25.133   **19.** 5.28   **21.** 2412.743   **23. (a)** Infinite   **(b)** π **25.** $\pi(b + \sinh(b)\cosh(b) - a - \sinh(a)\cosh(a))$ **27.** See Student Solutions Manual.   **29.** See Student Solutions Manual.

Section 9.4

1. Pole, polar axis   **2.** False   **3.** False   **4.** True   **5.** True   **6.** True   **7.** $x = r\cos\theta, y = r\sin\theta$ **8.** Polar equation   **9.** A   **11.** C
13. B   **15.** A
17.   **19.** 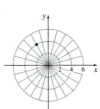   **21.** 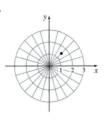   **23.**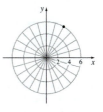

25. (a) $\left(5, -\dfrac{4\pi}{3}\right)$ **(b)** $\left(-5, \dfrac{5\pi}{3}\right)$ **(c)** $\left(5, \dfrac{8\pi}{3}\right)$ **27. (a)** $(2, -2\pi)$ **(b)** $(-2, \pi)$ **(c)** $(2, 2\pi)$ **29. (a)** $\left(1, -\dfrac{3\pi}{2}\right)$ **(b)** $\left(-1, \dfrac{3\pi}{2}\right)$
(c) $\left(1, \dfrac{5\pi}{2}\right)$ **31. (a)** $\left(3, -\dfrac{5\pi}{4}\right)$ **(b)** $\left(-3, \dfrac{7\pi}{4}\right)$ **(c)** $\left(3, \dfrac{11\pi}{4}\right)$ **33.** $(5.1962, 3)$ **35.** $(-5.1962, 3)$ **37.** $(0, 5)$ **39.** $(2, -2)$

41. 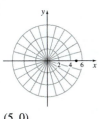   **43.** 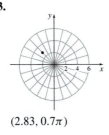   **45.** 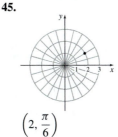   **47.** 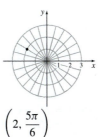   **49.**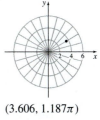

$(5, 0)$ $(2.83, 0.7\pi)$ $\left(2, \dfrac{\pi}{6}\right)$ $\left(2, \dfrac{5\pi}{6}\right)$ $(3.606, 1.187\pi)$

51. E   **53.** F   **55.** H   **57.** D

59. 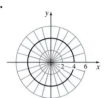 **61.** 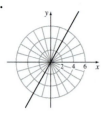 **63.** 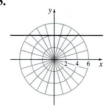 **65.** 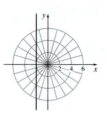 **67.**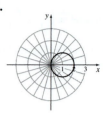

69. 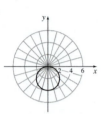 **71.** 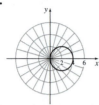 **73.**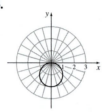

75. $r = 6 \dfrac{\sqrt{9\cos^2\theta + 4\sin^2\theta}}{9\cos^2\theta + 4\sin^2\theta}$ **77.** $r = 4\cos\theta$ **79.** $r^2\cos^2\theta + 4r\sin\theta - 1 = 0$ **81.** $r = \dfrac{\sqrt{\cos\theta\sin\theta}}{\cos\theta\sin\theta}$ **83.** $\left(x - \dfrac{1}{2}\right)^2 + y^2 = \dfrac{1}{4}$

85. $y = \sqrt{x^2 + y^2}(x^2 + y^2)$ **87.** $x = \dfrac{y^2}{8} - 2$ **89.** $x^2 + y^2 - \tan^{-1}\dfrac{y}{x} = 0$ **91.** $y = \sqrt{4 - x^2}$ **93.** $y = 4x$

95. (a) $x = -10, y = 36$ **(b)** $r = 37.4, \theta = 0.59\pi$ **(c)** $x = -3, y = -35$ **(d)** $r = 35, \theta = 1.47\pi$

97. If $a > 0$, If $a < 0$, **99.** **101.**

 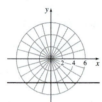 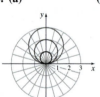 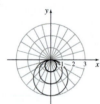

103. (a) **(c)**

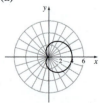 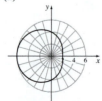

(b) Changes the radius **(d)** Changes the radius and the sign

105. See Student Solutions Manual. **107.** See Student Solutions Manual.

Section 9.5

1. True **2.** a, c, b **3.** True **4.** 3

5. (a) 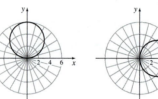 **7. (a)** 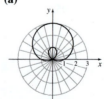 **9. (a)**   **11. (a)**

(b) $x = 2(1 + \cos\theta)\cos\theta$, $y = 2(1 + \cos\theta)\sin\theta$

(b) $x = 2(2 - \cos\theta)\cos\theta$, $y = 2(2 - \cos\theta)\sin\theta$

(b) $x = (1 + 2\sin\theta)\cos\theta$, $y = (1 + 2\sin\theta)\sin\theta$

(b) $x = \sin(3\theta)\cos\theta$, $y = \sin(3\theta)\sin\theta$

13.

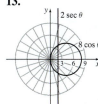

$r = -4, \theta = -\dfrac{2}{3}\pi;$

$r = -4, \theta = \dfrac{2}{3}\pi;$

$r = 4, \theta = -\dfrac{1}{3}\pi;$

$r = 4, \theta = \dfrac{1}{3}\pi$

15.

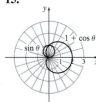

$r = 1, \theta = \dfrac{1}{2}\pi;$

$r = 0, \theta = \pi$

17.

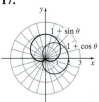

$r = \dfrac{1}{2}(2 + \sqrt{2}), \theta = \dfrac{1}{4}\pi;$

$r = \dfrac{1}{2}(2 - \sqrt{2}), \theta = -\dfrac{1}{4}\pi$

19. $\sqrt{5}(e-1) \approx 3.84219$ **21.** 2 **23.** $3 + 3\cos\theta$ **25.** $4 + \sin\theta$ **27.** $y = \dfrac{1}{\sqrt{3}}x$ **29.** $y = (\sqrt{2} - 2)x + \dfrac{5\sqrt{2}}{2} - \dfrac{1}{2}$

31. $y = -\left(1 + \dfrac{5\sqrt{2}}{4}\right)x + \dfrac{57\sqrt{2}}{8} + 10$

33. (a)

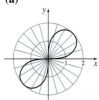

(b) $x = 2\sqrt{\sin(2\theta)}\cos\theta,$
$y = 2\sqrt{\sin(2\theta)}\sin\theta$

35. (a)

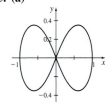

(b) $x = \sqrt{\cos(2\theta)}\cos\theta,$
$y = \sqrt{\cos(2\theta)}\sin\theta$

37. (a)

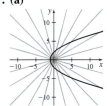

(b) $x = \dfrac{2\cos\theta}{1 - \cos\theta},$
$y = \dfrac{2\sin\theta}{1 - \cos\theta}$

39. (a)

(b) $x = \dfrac{\cos\theta}{3 - 2\cos\theta},$
$y = \dfrac{\sin\theta}{3 - 2\cos\theta}$

41. (a)

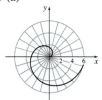

(b) $x = \theta\cos\theta,$
$y = \theta\sin\theta$

43. (a)

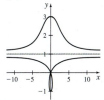

(b) $x = (\csc\theta - 2)\cos\theta,$
$y = (\csc\theta - 2)\sin\theta$

45. (a)

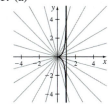

(b) $x = \sin^2\theta,$
$y = \sin^2\theta\tan\theta$

47. (a)

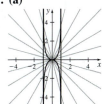

(b) $x = \sin\theta,$
$y = \sin\theta\tan\theta$

49. See Student Solutions Manual. **51.** $\pi\sqrt{1 + 4\pi^2} + \dfrac{1}{2}\text{arcsinh}(2\pi) \approx 21.2563$ **53.** 8 **55.** See Student Solutions Manual.

57. Horizontal: $y = \dfrac{9\sqrt{3}}{4}$; vertical: $x = -\dfrac{3}{4}, x = 6$ **59.** Horizontal: $y = \dfrac{4}{3\sqrt{6}}, y = -\dfrac{4}{3\sqrt{6}}, y = 2, y = -2$; vertical: $x = \dfrac{4}{3\sqrt{6}}, x = -\dfrac{4}{3\sqrt{6}},$

$x = 2, x = -2$ **61.** Horizontal: $y = \dfrac{3^{3/4}}{\sqrt{2}}, y = -\dfrac{3^{3/4}}{\sqrt{2}}$; vertical: $x = \dfrac{3^{3/4}}{\sqrt{2}}, x = -\dfrac{3^{3/4}}{\sqrt{2}}$

63. (a) See Student Solutions Manual. (b) $r^2 = (-r)^2$ and $\sin(2\theta) = \sin(2\theta + 2\pi)$ **65.** $\sin(2\theta) = \sin(2\theta + 2\pi)$ but $\sin(2\theta) = -\sin(-2\theta)$

67. (a) (b) $\alpha = \dfrac{\pi}{10}$ (c) 4.20201

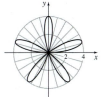

Section 9.6

1. $\dfrac{\theta r^2}{2}$ **2.** True   **3.** False   **4.** True   **5.** $\dfrac{\pi}{4} \approx 0.785398$ **7.** $2\left(4 + \dfrac{3\pi}{2}\right) \approx 17.42478$ **9.** $\dfrac{3}{16}\left(3\sqrt{3} + 4\pi\right) \approx 3.33047$ **11.** $\dfrac{4\pi^3 a^2}{3}$

13. $\dfrac{3\pi}{2} \approx 4.71239$ **15.** $\dfrac{19\pi}{2} \approx 29.8451$ **17.** $32\pi \approx 100.531$ **19.** $2\pi \approx 6.28318$ **21.** 2 **23.** $\dfrac{1}{6}\left(3\sqrt{3} + 4\pi\right) \approx 2.96042$

25. $1 - \dfrac{\pi}{4} \approx 0.214602$ **27.** $\dfrac{\pi^2}{2} \approx 4.9348$ **29.** $\dfrac{2}{5}\sqrt{2}(1 + e^{2\pi})\pi \approx 953.428$ **31.** $\pi - \dfrac{3\sqrt{3}}{2} \approx 0.543516$

33. $\sqrt{3} + \dfrac{4\pi}{3} - 4\log\left(2 + \sqrt{3}\right) \approx 0.653009$ **35.** $\dfrac{3\pi}{2} \approx 4.71239$ **37.** $4\sqrt{3} + \dfrac{32\pi}{3} \approx 40.4385$ **39.** $\dfrac{9\sqrt{3}}{2} - \pi \approx 4.65264$

41. $\dfrac{5\pi}{6} - \sqrt{3} \approx 0.88$ **43.** $2\sqrt{2} \approx 2.82843$ **45.** $\dfrac{1 - e^{-2}}{4} \approx 0.21617$ **47.** $\dfrac{\pi - 1}{2\pi} \approx 0.340845$ **49.** $\sqrt{3} + \dfrac{4\pi}{3} - 4\ln\left(2 + \sqrt{3}\right) \approx 0.653$

51. (a) $a\pi\sqrt{R^2 + 1}$ **(b)** Answers will vary.   **53.** $2 - \dfrac{\pi}{2} \approx 0.429204$ **55.** See Student Solutions Manual.   **57.** 0.276285

Section 9.7

1. (a)   **2.** Answers will vary.   **3.** They are different.   **4.** False   **5.** $e = 1$, directrix perpendicular to the polar axis $p = 1$ units to the right of the pole   **7.** $e = \dfrac{3}{2}$, directrix parallel to the polar axis $p = \dfrac{4}{3}$ units below the pole   **9.** $e = \dfrac{1}{2}$, directrix perpendicular to the polar axis $p = \dfrac{3}{2}$ units to the left of the pole   **11.** $e = 1$, directrix parallel to the polar axis $p = \dfrac{4}{3}$ units above the pole

13. (a) Ellipse

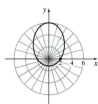

(b) $7\left(y + \dfrac{24}{7}\right)^2 + 16x^2 = \dfrac{1024}{7}$

(c) $x = \dfrac{8\cos\theta}{4 + 3\sin\theta}, \; y = \dfrac{8\sin\theta}{4 + 3\sin\theta}$

15. (a) Hyperbola

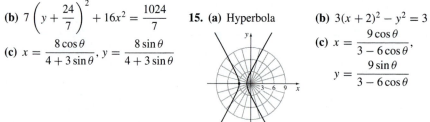

(b) $3(x + 2)^2 - y^2 = 3$

(c) $x = \dfrac{9\cos\theta}{3 - 6\cos\theta}, \; y = \dfrac{9\sin\theta}{3 - 6\cos\theta}$

17. (a) Ellipse

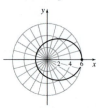

(b) $9x^2 + 5\left(y - \dfrac{12}{5}\right)^2 = \dfrac{324}{5}$ **(c)** $x = \dfrac{6\cos\theta}{3 - 2\sin\theta}, \; y = \dfrac{6\sin\theta}{3 - 2\sin\theta}$

19. (a) Ellipse

(b) $4y^2 + 3(x - 2)^2 = 48$ **(c)** $x = \dfrac{6\cos\theta}{2 - \cos\theta}, \; y = \dfrac{6\sin\theta}{2 - \cos\theta}$

21. $\dfrac{\pi}{2}$ **23.** 0 **25.** $\dfrac{\pi}{2}$ **27.** $r = \dfrac{12}{5 - 4\cos\theta}$ **29.** $r = \dfrac{1}{1 + \sin\theta}$ **31.** $r = \dfrac{12}{1 - 6\sin\theta}$

33. (a) 0.967 **(b)** 0.587 **(c)** 35 **(d)** See Student Solutions Manual.

35. See Student Solutions Manual.   **37.** See Student Solutions Manual.   **39.** See Student Solutions Manual.

Review Exercises

1. (a)

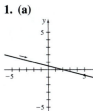

3. (a)

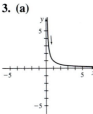

5. (a)

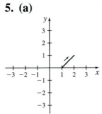

7. (a) $y = \dfrac{1}{2}x + \dfrac{5}{2}$
(b)

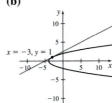

9. (a) $y = -\dfrac{81}{2\sqrt{10}}x + \dfrac{9}{2\sqrt{10}} + \sqrt{10}$
(b)

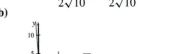

(b) $x = -4y + 2$
(c) No restrictions

(b) $y = \dfrac{1}{x}$
(c) $x \geq 0,\ y \geq 0$

(b) $x = \sec^2 \arctan \sqrt{y}$
(c) $1 \leq x \leq 2,\ 0 \leq y \leq 1$

11. $x = t,$
  $y = -2t + 4$ or
  $x = t^3,$
  $y = -2t^3 + 4$

13. $\dfrac{x^2}{4} + y = 1$

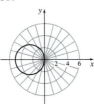

15. $\left(1, \sqrt{3}\right)$

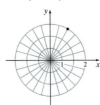

17. $\left(-2\sqrt{2}, 2\sqrt{2}\right)$

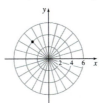

19. $(5, 0.295\pi),\ (-5, 1.295\pi)$ **21.** $\left(3\sqrt{2}, \dfrac{3\pi}{4}\right),\ \left(-3\sqrt{2}, \dfrac{7\pi}{4}\right)$ **23.** $x^2 + y^2 - e^{\tan^{-1}(y/x)} = 0$ **25.** $\dfrac{\left(y - x^2 - y^2\right)\sqrt{x^2 - y^2}}{x^2 - y^2} - a = 0$

27. $x^2 + y^2 - \left(\tan^{-1}\dfrac{y}{x}\right)^2 = 0$ **29.** $\cos^2\theta - \sin^2\theta = 0$ **31.** $r = \dfrac{6\sqrt{9\cos^2\theta + 4\sin^2\theta}}{9\cos^2\theta + 4\sin^2\theta}$

33.

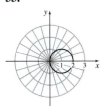

35.

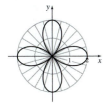

37. (a)

(b) $x = 4\cos\theta\cos(2\theta),\ y = 4\sin\theta\cos(2\theta)$

39. (a)

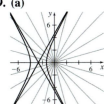

(b) $x = \dfrac{4\cos\theta}{1 - 2\cos\theta},\ y = \dfrac{4\sin\theta}{1 - 2\cos\theta}$

41. (a)

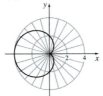

(b) $x = \cos\theta(2 - 2\cos\theta),$
  $y = \sin\theta(2 - 2\cos\theta)$

43. (a)

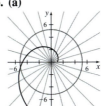

(b) $x = e^{0.5\theta}\cos\theta,$
  $y = e^{0.5\theta}\sin\theta$

45. (a)

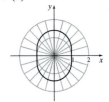

(b) $x = \cos\theta\sqrt{1 + \sin^2\theta},$
  $y = \sin\theta\sqrt{1 + \sin^2\theta}$

47. (a)

θ from 0 to 2π

(b) $35\left(x - \dfrac{6}{35}\right)^2 + 36y^2 = \dfrac{1296}{35}$

(c) $x = \dfrac{6\cos\theta}{6 - \cos\theta}$, $y = \dfrac{6\sin\theta}{6 - \cos\theta}$

49. $x = 4\sin\left(\dfrac{2\pi t}{5}\right)$, $y = 3\cos\left(\dfrac{2\pi t}{5}\right)$, $0 < t < 5$ **51.** Vertical: $(0, 2)$, $(2, 2)$; horizontal: $(1, 5)$, $(1, -1)$ **53.** 1.10202   **55.** 1.6732868

57. $\sqrt{2}(1 - e^{-2\pi}) \approx 1.41157$ **59.** $2(4 - 2\sqrt{2}) \approx 2.34315$ **61.** $2\sqrt{3} - \dfrac{2\pi}{3} \approx 1.36971$ **63.** $\dfrac{\pi^2}{2} \approx 4.9348$

65. $\dfrac{\pi}{32}\left(196\sqrt{17} + 15\arcsin 4\right) \approx 82.4226$ **67.** $16\pi \approx 50.2655$

Photo Credits

Index

Note: **Boldface** indicates a definition, *italics* indicates a figure, and 1*t* indicates a table.

TABLE OF DERIVATIVES

1. $\dfrac{d}{dx}c = 0,\quad c$ is a constant

2. $\dfrac{d}{dx}(u \pm v) = \dfrac{du}{dx} \pm \dfrac{dv}{dx}$

3. $\dfrac{d}{dx}(cu) = c\dfrac{du}{dx},\quad c$ a constant

4. $\dfrac{d}{dx}(uv) = u\dfrac{dv}{dx} + v\dfrac{du}{dx}$

5. $\dfrac{d}{dx}\left(\dfrac{u}{v}\right) = \dfrac{v\dfrac{du}{dx} - u\dfrac{dv}{dx}}{v^2}$

6. $\dfrac{dy}{dx} = \dfrac{dy}{du}\dfrac{du}{dx},\quad y = f(u),\quad u = g(x)$

7. $\dfrac{d}{dx}x^a = ax^{a-1}$

8. $\dfrac{d}{dx}\sin x = \cos x$

9. $\dfrac{d}{dx}\cos x = -\sin x$

10. $\dfrac{d}{dx}\tan x = \sec^2 x$

11. $\dfrac{d}{dx}\cot x = -\csc^2 x$

12. $\dfrac{d}{dx}\sec x = \sec x \tan x$

13. $\dfrac{d}{dx}\csc x = -\csc x \cot x$

14. $\dfrac{d}{dx}\ln x = \dfrac{1}{x}$

15. $\dfrac{d}{dx}e^x = e^x$

16. $\dfrac{d}{dx}a^x = a^x \ln a,\quad a > 0,\ a \neq 1$

17. $\dfrac{d}{dx}\log_a x = \dfrac{1}{x \ln a},\quad a > 0,\ a \neq 1$

18. $\dfrac{d}{dx}\sin^{-1} x = \dfrac{1}{\sqrt{1-x^2}}$

19. $\dfrac{d}{dx}\tan^{-1} x = \dfrac{1}{1+x^2}$

20. $\dfrac{d}{dx}\sec^{-1} x = \dfrac{1}{x\sqrt{x^2-1}}$

21. $\dfrac{d}{dx}\cos^{-1} x = -\dfrac{1}{\sqrt{1-x^2}}$

22. $\dfrac{d}{dx}\cot^{-1} x = -\dfrac{1}{1+x^2}$

23. $\dfrac{d}{dx}\csc^{-1} x = -\dfrac{1}{x\sqrt{x^2-1}}$

24. $\dfrac{d}{dx}\sinh x = \cosh x$

25. $\dfrac{d}{dx}\cosh x = \sinh x$

26. $\dfrac{d}{dx}\tanh x = \operatorname{sech}^2 x$

27. $\dfrac{d}{dx}\coth x = -\operatorname{csch}^2 x$

28. $\dfrac{d}{dx}\operatorname{sech} x = -\operatorname{sech} x \tanh x$

29. $\dfrac{d}{dx}\operatorname{csch} x = -\operatorname{csch} x \coth x$

30. $\dfrac{d}{dx}\sinh^{-1} x = \dfrac{1}{\sqrt{1+x^2}}$

31. $\dfrac{d}{dx}\cosh^{-1} x = \dfrac{1}{\sqrt{x^2-1}}$

32. $\dfrac{d}{dx}\tanh^{-1} x = \dfrac{1}{1-x^2}$

TABLE OF INTEGRALS
General Formulas

1. $\displaystyle \int [f(x) \pm g(x)]\,dx = \int f(x)\,dx \pm \int g(x)\,dx$

2. $\displaystyle \int kf(x)\,dx = k\int f(x)\,dx,\quad k$ a constant

3. $\displaystyle \int u\,dv = uv - \int v\,du$

Essential Integrals

1. $\displaystyle \int x^a\,dx = \dfrac{x^{a+1}}{a+1} + C,\quad a \neq -1$

2. $\displaystyle \int \dfrac{1}{x}\,dx = \ln|x| + C$

3. $\displaystyle \int e^x\,dx = e^x + C$

4. $\displaystyle \int \sin x\,dx = -\cos x + C$

5. $\displaystyle \int \cos x\,dx = \sin x + C$

6. $\displaystyle \int \tan x\,dx = \ln|\sec x| + C$

7. $\displaystyle \int \sec x\,dx = \ln|\sec x + \tan x| + C$

8. $\displaystyle \int \sec^2 x\,dx = \tan x + C$

9. $\displaystyle \int \sec x \tan x\,dx = \sec x + C$

10. $\displaystyle \int \dfrac{dx}{\sqrt{1-x^2}} = \sin^{-1} x + C$

11. $\displaystyle \int \dfrac{dx}{1+x^2} = \tan^{-1} x + C$

12. $\displaystyle \int \dfrac{dx}{x\sqrt{x^2-1}} = \sec^{-1} x + C$

13. $\displaystyle \int \cot x\,dx = \ln|\sin x| + C$

14. $\displaystyle \int \csc x\,dx = \ln|\csc x - \cot x| + C$

15. $\displaystyle \int \csc^2 x\,dx = -\cot x + C$

16. $\displaystyle \int \csc x \cot x\,dx = -\csc x + C$

17. $\displaystyle \int \dfrac{dx}{\sqrt{a^2-x^2}} = \sin^{-1}\dfrac{x}{a} + C,\quad a > 0$

18. $\displaystyle \int \dfrac{dx}{a^2+x^2} = \dfrac{1}{a}\tan^{-1}\dfrac{x}{a} + C,\quad a > 0$

19. $\displaystyle \int \dfrac{dx}{x\sqrt{x^2-a^2}} = \dfrac{1}{a}\sec^{-1}\dfrac{x}{a} + C,\quad a > 0$

20. $\displaystyle \int a^x\,dx = \dfrac{a^x}{\ln a} + C,\quad a > 0,\ a \neq 1$

Useful Integrals

21. $\displaystyle\int \sinh x \, dx = \cosh x + C$

22. $\displaystyle\int \cosh x \, dx = \sinh x + C$

23. $\displaystyle\int \text{sech}^2 x \, dx = \tanh x + C$

24. $\displaystyle\int \text{csch}^2 x \, dx = -\coth x + C$

25. $\displaystyle\int \text{sech}\, x \, \tanh x \, dx = -\text{sech}\, x + C$

26. $\displaystyle\int \text{csch}\, x \coth x \, dx = -\text{csch}\, x + C$

27. $\displaystyle\int \frac{dx}{a + bx} = \frac{1}{b}\ln|a + bx| + C$

28. $\displaystyle\int \frac{x\, dx}{a + bx} = \frac{1}{b^2}(a + bx - a\ln|a + bx|) + C$

29. $\displaystyle\int \frac{x\, dx}{(a + bx)^2} = \frac{a}{b^2(a + bx)} + \frac{1}{b^2}\ln|a + bx| + C$

30. $\displaystyle\int \frac{x^2\, dx}{(a + bx)^2} = \frac{1}{b^3}\left(a + bx - \frac{a^2}{a + bx} - 2a\ln|a + bx|\right) + C$

31. $\displaystyle\int \frac{dx}{x(a + bx)^2} = \frac{1}{a(a + bx)} - \frac{1}{a^2}\ln\left|\frac{a + bx}{x}\right| + C$

32. $\displaystyle\int \frac{dx}{x^2(a + bx)} = -\frac{1}{ax} + \frac{b}{a^2}\ln\left|\frac{a + bx}{x}\right| + C$

33. $\displaystyle\int x(a + bx)^n \, dx = \frac{(a + bx)^{n+1}}{b^2}\left(\frac{a + bx}{n + 2} - \frac{a}{n + 1}\right) + C, \quad n \neq -1, -2$

34. $\displaystyle\int \frac{x\, dx}{(a + bx)(c + dx)} = \frac{1}{bc - ad}\left(-\frac{a}{b}\ln|a + bx| + \frac{c}{d}\ln|c + dx|\right) + C, \quad bc - ad \neq 0$

35. $\displaystyle\int \frac{x\, dx}{(a + bx)^2(c + dx)} = \frac{1}{bc - ad}\left[\frac{a}{b(a + bx)} + \frac{c}{bc - ad}\ln\left|\frac{a + bx}{c + dx}\right|\right] + C, \quad bc - ad \neq 0$

36. $\displaystyle\int \frac{dx}{a^2 - x^2} = \frac{1}{2a}\ln\left|\frac{x + a}{x - a}\right| + C$

37. $\displaystyle\int \frac{dx}{x^2 - a^2} = \frac{1}{2a}\ln\left|\frac{x - a}{x + a}\right| + C$

38. $\displaystyle\int \frac{dx}{(a^2 \pm x^2)^n} = \frac{1}{2(n-1)a^2}\left[\frac{x}{(a^2 \pm x^2)^{n-1}} + (2n - 3)\int \frac{dx}{(a^2 \pm x^2)^{n-1}}\right], \quad n \neq 1$

39. $\displaystyle\int \frac{dx}{(x^2 - a^2)^n} = \frac{1}{2(n-1)a^2}\left[-\frac{x}{(x^2 - a^2)^{n-1}} - (2n - 3)\int \frac{dx}{(x^2 - a^2)^{n-1}}\right], \quad n \neq 1$

Integrals Containing $\sqrt{a+bx}$

40. $\displaystyle\int x\sqrt{a+bx}\,dx = \frac{2}{15b^2}(3bx - 2a)(a+bx)^{3/2} + C$

41. $\displaystyle\int x^n\sqrt{a+bx}\,dx = \frac{2}{b(2n+3)}[x^n(a+bx)^{3/2} - na\int x^{n-1}\sqrt{a+bx}\,dx]$

42. $\displaystyle\int \frac{x\,dx}{\sqrt{a+bx}} = \frac{2}{3b^2}(bx - 2a)\sqrt{a+bx} + C$

43. $\displaystyle\int \frac{x^2\,dx}{\sqrt{a+bx}} = \frac{2}{15b^2}(8a^2 - 4abx + 3b^2x^2)\sqrt{a+bx} + C$

44. $\displaystyle\int \frac{x^n\,dx}{\sqrt{a+bx}} = \frac{2x^n\sqrt{a+bx}}{b(2n+1)} - \frac{2na}{b(2n+1)}\int\frac{x^{n-1}\,dx}{\sqrt{a+bx}}$

45. $\displaystyle\int \frac{dx}{x\sqrt{a+bx}} = \begin{cases} \dfrac{1}{\sqrt{a}}\ln\left|\dfrac{\sqrt{a+bx}-\sqrt{a}}{\sqrt{a+bx}+\sqrt{a}}\right| + C, & a>0 \\[2ex] \dfrac{2}{\sqrt{-a}}\tan^{-1}\sqrt{\dfrac{a+bx}{-a}} + C, & a<0 \end{cases}$

46. $\displaystyle\int \frac{dx}{x^n\sqrt{a+bx}} = -\frac{\sqrt{a+bx}}{a(n-1)x^{n-1}} - \frac{b(2n-3)}{2a(n-1)}\int\frac{dx}{x^{n-1}\sqrt{a+bx}}$

47. $\displaystyle\int \frac{\sqrt{a+bx}}{x}\,dx = 2\sqrt{a+bx} + a\int\frac{dx}{x\sqrt{a+bx}}$

48. $\displaystyle\int \frac{\sqrt{a+bx}}{x^2}\,dx = -\frac{\sqrt{a+bx}}{x} + \frac{b}{2}\int\frac{dx}{x\sqrt{a+bx}}$

Integrals Containing $\sqrt{x^2 \pm a^2}$

49. $\displaystyle\int \sqrt{x^2 \pm a^2}\,dx = \frac{x}{2}\sqrt{x^2 \pm a^2} \pm \frac{a^2}{2}\ln\left|x + \sqrt{x^2 \pm a^2}\right| + C$

50. $\displaystyle\int x\sqrt{x^2 \pm a^2}\,dx = \frac{1}{3}(x^2 \pm a^2)^{3/2} + C$

51. $\displaystyle\int x^2\sqrt{x^2 \pm a^2}\,dx = \frac{x}{8}(2x^2 \pm a^2)\sqrt{x^2 \pm a^2} - \frac{a^4}{8}\ln\left|x + \sqrt{x^2 \pm a^2}\right| + C$

52. $\displaystyle\int \frac{\sqrt{x^2 + a^2}}{x}\,dx = \sqrt{x^2 + a^2} - a\ln\left|\frac{a + \sqrt{x^2 + a^2}}{x}\right| + C$

53. $\displaystyle\int \frac{\sqrt{x^2 - a^2}}{x}\,dx = \sqrt{x^2 - a^2} - a\sec^{-1}\frac{x}{a} + C$

54. $\displaystyle\int \frac{\sqrt{x^2 \pm a^2}}{x^2}\,dx = -\frac{\sqrt{x^2 \pm a^2}}{x} + \ln\left|x + \sqrt{x^2 \pm a^2}\right| + C$

55. $\displaystyle\int \frac{dx}{\sqrt{x^2 \pm a^2}} = \ln\left|x + \sqrt{x^2 \pm a^2}\right| + C$

56. $\displaystyle\int \frac{x^2\,dx}{\sqrt{x^2 \pm a^2}} = \frac{x}{2}\sqrt{x^2 \pm a^2} \pm \frac{a^2}{2}\ln\left|x + \sqrt{x^2 \pm a^2}\right| + C$

57. $\displaystyle\int \frac{dx}{x\sqrt{x^2 + a^2}} = -\frac{1}{a}\ln\left|\frac{a + \sqrt{x^2 + a^2}}{x}\right| + C$

58. $\displaystyle\int \frac{dx}{x\sqrt{x^2 - a^2}} = \frac{1}{a}\sec^{-1}\frac{x}{a} + C$

59. $\displaystyle\int \frac{dx}{x^2\sqrt{x^2 \pm a^2}} = \mp\frac{\sqrt{x^2 \pm a^2}}{a^2 x} + C$

60. $\displaystyle\int (x^2 \pm a^2)^{3/2}\,dx = \frac{x}{8}(2x^2 \pm 5a^2)\sqrt{x^2 \pm a^2} + \frac{3a^4}{8}\ln\left|x + \sqrt{x^2 \pm a^2}\right| + C$

61. $\displaystyle\int \frac{dx}{(x^2 \pm a^2)^{3/2}} = \pm\frac{x}{a^2\sqrt{x^2 \pm a^2}} + C$

Integrals Containing Inverse Trigonometric Functions

106. $\displaystyle\int \sin^{-1}x\ dx = x\sin^{-1}x + \sqrt{1-x^2} + C$

107. $\displaystyle\int \cos^{-1}x\ dx = x\cos^{-1}x - \sqrt{1-x^2} + C$

108. $\displaystyle\int \tan^{-1}x\ dx = x\tan^{-1}x - \frac{1}{2}\ln(1+x^2) + C$

109. $\displaystyle\int x\sin^{-1}x\ dx = \frac{2x^2-1}{4}\sin^{-1}x + \frac{x\sqrt{1-x^2}}{4} + C$

110. $\displaystyle\int x\cos^{-1}x\ dx = \frac{2x^2-1}{4}\cos^{-1}x - \frac{x\sqrt{1-x^2}}{4} + C$

111. $\displaystyle\int x\tan^{-1}x\ dx = \frac{x^2+1}{2}\tan^{-1}x - \frac{x}{2} + C$

112. $\displaystyle\int x^n \sin^{-1}x\ dx = \frac{1}{n+1}\left(x^{n+1}\sin^{-1}x - \int \frac{x^{n+1}dx}{\sqrt{1-x^2}} \right),\quad n\neq -1$

113. $\displaystyle\int x^n \cos^{-1}x\ dx = \frac{1}{n+1}\left(x^{n+1}\cos^{-1}x + \int \frac{x^{n+1}dx}{\sqrt{1-x^2}} \right),\quad n\neq -1$

114. $\displaystyle\int x^n \tan^{-1}x\ dx = \frac{1}{n+1}\left(x^{n+1}\tan^{-1}x - \int \frac{x^{n+1}dx}{1+x^2} \right),\quad n\neq -1$

Integrals Containing Exponential and Logarithmic Functions

115. $\displaystyle\int xe^{ax}\ dx = \frac{1}{a^2}(ax-1)e^{ax} + C$

116. $\displaystyle\int x^n e^{ax}\ dx = \frac{1}{a}x^n e^{ax} - \frac{n}{a}\int x^{n-1}e^{ax}\ dx$

117. $\displaystyle\int \frac{e^x}{x^n}\ dx = -\frac{e^x}{(n-1)x^{n-1}} + \frac{1}{n-1}\int \frac{e^x}{x^{n-1}}dx$

118. $\displaystyle\int \ln x\,dx = x\ln x - x + C$

119. $\displaystyle\int (\ln x)^n\ dx = x(\ln x)^n - n\int (\ln x)^{n-1}\ dx$

120. $\displaystyle\int x^n \ln x\,dx = \left(\frac{x^{n+1}}{n+1}\right)\left(\ln x - \frac{1}{n+1}\right) + C$

121. $\displaystyle\int \frac{(\ln x)^n}{x}\ dx = \frac{(\ln x)^{n+1}}{n+1} + C,\quad n\neq -1$

122. $\displaystyle\int \frac{dx}{x\ln x} = \ln|\ln x| + C$

123. $\displaystyle\int x^m (\ln x)^n\ dx = \frac{x^{m+1}(\ln x)^n}{m+1} - \frac{n}{m+1}\int x^m (\ln x)^{n-1}\ dx$

124. $\displaystyle\int \frac{x^m}{(\ln x)^n}\ dx = -\frac{x^{m+1}}{(n-1)(\ln x)^{n-1}} + \frac{m+1}{n-1}\int \frac{x^m}{(\ln x)^{n-1}}dx$

125. $\displaystyle\int e^{ax}\sin bx\ dx = \frac{e^{ax}}{a^2+b^2}(a\sin bx - b\cos bx) + C$

126. $\displaystyle\int e^{ax}\cos bx\ dx = \frac{e^{ax}}{a^2+b^2}(a\cos bx + b\sin bx) + C$

Integrals Containing Hyperbolic Functions

127. $\displaystyle\int \sinh x\ dx = \cosh x + C$

128. $\displaystyle\int \cosh x\ dx = \sinh x + C$

129. $\displaystyle\int \tanh x\ dx = \ln\cosh x + C$

130. $\displaystyle\int \coth x\ dx = \ln|\sinh x| + C$

131. $\displaystyle\int \text{sech}\ x\ dx = \tan^{-1}(\sinh x) + C$

132. $\displaystyle\int \text{csch}\ x\ dx = \ln\left|\tanh\frac{x}{2}\right| + C$

133. $\displaystyle\int \text{sech}^2 x\ dx = \tanh x + C$

134. $\displaystyle\int \text{csch}^2 x\ dx = -\coth x + C$

135. $\displaystyle\int \text{sech}\ x\tanh x\ dx = -\text{sech}\ x + C$

136. $\displaystyle\int \text{csch}\ x\coth x\ dx = -\text{csch}\ x + C$

137. $\displaystyle\int \sinh^2 x\ dx = \frac{\sinh 2x}{4} - \frac{x}{2} + C$

138. $\displaystyle\int \cosh^2 x\ dx = \frac{\sinh 2x}{4} + \frac{x}{2} + C$

139. $\displaystyle\int \tanh^2 x\ dx = x - \tanh x + C$

140. $\displaystyle\int \coth^2 x\ dx = x - \coth x + C$

141. $\displaystyle\int x\sinh x\ dx = x\cosh x - \sinh x + C$

142. $\displaystyle\int x\cosh x\ dx = x\sinh x - \cosh x + C$

143. $\displaystyle\int x^n \sinh x\ dx = x^n \cosh x - n\int x^{n-1}\cosh x\ dx$

144. $\displaystyle\int x^n \cosh x\ dx = x^n \sinh x - n\int x^{n-1}\sinh x\ dx$